ANNUAL REVIEW OF ASTRONOMY AND ASTROPHYSICS

ANNUAL REVIEW OF ASTRONOMY AND ASTROPHYSICS

VOLUME 38, 2000

GEOFFREY BURBIDGE, *Editor*
University of California at San Diego

ALLAN SANDAGE, *Associate Editor*
Observatories of the Carnegie Institution of Washington

FRANK H. SHU, *Associate Editor*
University of California at Berkeley

www.AnnualReviews.org   science@AnnualReviews.org   650-493-4400

ANNUAL REVIEWS
4139 El Camino Way • P.O. BOX 10139 • Palo Alto, California 94303-0139

ANNUAL REVIEWS
Palo Alto, California, USA

COPYRIGHT © 2000 BY ANNUAL REVIEWS, PALO ALTO, CALIFORNIA, USA.
ALL RIGHTS RESERVED. The appearance of the code at the bottom of the first page
of an article in this serial indicates the copyright owner's consent that copies of the article
may be made for personal or internal use, or for the personal or internal use of specific
clients. This consent is given on the condition that the copier pay the stated per-copy fee
of $14.00 per article through the Copyright Clearance Center, Inc. (222 Rosewood Drive,
Danvers, MA 01923) for copying beyond that permitted by Section 107 or 108 of the US
Copyright Law. The per-copy fee of $14.00 per article also applies to the copying, under
the stated conditions, of articles published in any *Annual Review* serial before January 1,
1978. Individual readers, and nonprofit libraries acting for them, are permitted to make
a single copy of an article without charge for use in research or teaching. This consent
does not extend to other kinds of copying such as copying for general distribution, for
advertising or promotional purposes, for creating new collective works, or for resale. For
such uses, written permission is required. Write to Permissions Dept., Annual Reviews,
4139 El Camino Way, P.O. Box 10139, Palo Alto, CA 94303-0139 USA.

International Standard Serial Number: 0066-4146
International Standard Book Number: 0-8243-0938-3
Library of Congress Catalog Card Number: 63-8846

Annual Review and publication titles are registered trademarks of Annual Reviews.
⊗ The paper used in this publication meets the minimum requirements of American National
Standards for Information Sciences—Permanence of Paper for Printed Library Materials.
ANSI Z39.48-1992.

Annual Reviews and the Editors of its publications assume no responsibility for the
statements expressed by the contributors to this *Annual Review*.

TYPESET BY TECHBOOKS, FAIRFAX, VA
PRINTED AND BOUND IN THE UNITED STATES OF AMERICA

PREFACE

This volume was planned at a meeting held on May 23, 1998, in San Francisco, California. Those who attended the meeting included Geoffrey Burbidge (Editor), Allan Sandage, and Frank Shu (Associate Editors), Ken Brecher, Anne Cowley, David Helfand, J. Miller, and Scott Tremaine (Editorial Committee members), Sandra Cooperman (Production Editor), and guests H. Spinard, and S. Woosley.

In the Preface to Volume 37, I pointed out that 24 articles were scheduled for this volume. In fact only 18 are included here. For the next volume, Volume 39 (2001), 22 articles are scheduled.

I wish to thank the authors, the Associate Editors, Sandra Cooperman, and the recent Production Editors, Larisa North and Nancy Donham, all of whom have contributed very positively to the production of this volume.

In previous years I have not commented on individual articles. However, in this volume there are two articles which for different reasons are worthy of note. In Volume 37, Allan Sandage wrote about some of the major scientific achievements which came from the first 50 years of operation of the Palomar Observatory. In this volume, George Wallerstein and J. B. Oke have written a second article covering much of the work which was not described in Dr. Sandage's article.

I also wish to comment on the background of the article entitled "*Gamma Ray Burst Afterglows.*" I orginally invited Dr. Jan van Paradijs, one of the greatest scientists in the field to write this article, and I was delighted when he accepted the invitation. To the great sorrow of all of us, he became mortally ill and died before the article was completed. His collaborators, Dr. Chryssa Kouveliotou, and Dr. Ralph Wijers were able to finish the article so that it could appear in this volume. Jan's death was a terrible loss for all of us.

Geoffrey Burbidge
Editor
February 2000

CONTENTS

RELATED ARTICLES

From the *Annual Review of Earth and Planetary Sciences*, Volume 28 (2000):
Mars 2000, Arden L. Albee,
Asteroid Fragmentation and Evolution of Asteroids, Eileen V. Ryan

From the *Annual Review of Nuclear and Particle Science*, Volume 49 (1999):
The Cosmic Microwave Background and Particle Physics, Marc Kamionkowski and Arthur Kosowsky

ANNUAL REVIEWS is a nonprofit scientific publisher established to promote the advancement of the sciences. Beginning in 1932 with the *Annual Review of Biochemistry*, the Company has pursued as its principal function the publication of high-quality, reasonably priced *Annual Review* volumes. The volumes are organized by Editors and Editorial Committees who invite qualified authors to contribute critical articles reviewing significant developments within each major discipline. The Editor-in-Chief invites those interested in serving as future Editorial Committee members to communicate directly with him. Annual Reviews is administered by a Board of Directors, whose members serve without compensation.

Donald E. Osterbrock

Annu. Rev. Astron. Astrophys. 2000. 38:1–33

A FORTUNATE LIFE IN ASTRONOMY

Donald E. Osterbrock

University of California Observatories/Lick Observatory, Department of Astronomy and Astrophysics, University of California, Santa Cruz, California 95064; e-mail: don@ucolick.org

Key Words   autobiography, gaseous nebulae, galaxies, active galactic nuclei, Seyfert galaxies

■ **Abstract**   I have had a very fortunate career in astronomy, benefiting greatly from numerous accidents of fate. I grew up in Cincinnati, Ohio, served in the US Army Air Force in World War II, and had all my further education at the University of Chicago, from PhB in the College to PhD in astronomy and astrophysics. There, as a postdoc at Princeton University, and as a young faculty member at Caltech and Mount Wilson and Palomar Observatories, I had excellent teachers and mentors. I have done research primarily on gaseous nebulae and active galactic nuclei, but also made a few early contributions on stellar interiors and the heating in the outer layers of the Sun. The major part of my scientific career was at the University of Wisconsin and Lick Observatory, but I also had three productive years at the Institute for Advanced Study.

CHILDHOOD AND EDUCATION

I have had a very fortunate career in astronomy, benefiting greatly from numerous accidents of fate that were completely unplanned on my part and that I often was not even aware of at the time. The first was my birth in 1924 in Cincinnati, Ohio, a pleasant old southern Midwestern city on the bank of a beautiful river. In my childhood and formative years the population of Cincinnati was still predominantly German-American, a culture in which hard work, education, science, poetry, music, and love of the outdoor life were all positive values. My father's parents and my mother's grandparents had all emigrated from Germany.

Both my parents had dropped out of day high school to take full-time jobs—my father as a stenographer and my mother as a laboratory assistant in one of the big Cincinnati soap companies. They completed their high school education at the public night school not far from their homes. My father then entered the University of Cincinnati as a student of electrical engineering in a "cooperative" program, in which he and all the other engineering students alternately attended classes during half of each trimester and worked full-time in "practical" engineering-type jobs in Cincinnati between these periods of formal education. He was very

0066-4146/00/0915-0001$14.00

1

good in physical sciences and especially in mathematics; after he graduated from the University of Cincinnati in 1918 he was hired as an instructor in electrical engineering. My parents were married later that same year; my mother quit her job, and my father rose steadily through the academic ranks, though the salaries at that city-funded university were never very good. In his later years my father taught nearly all the more mathematical electrical-engineering courses and ultimately became chairman of the department.

My parents were married six years before I was born; they had wanted a child before that, but I was their first—at least, the first to live. They pampered me tremendously, and at the same time they inculcated their values, including science and mathematics, in me. The grade school and high school I attended were good, as most of the Cincinnati public schools were at that time. I devoured numerous books on astronomy in the high school library, which drew me into the subject. My father took me to hear public lectures by famous visiting lecturers, such as Harlow Shapley and Otto Struve, and more mundane speakers, at the local astronomy club. As the first Harvard books on astronomy, written at a semi-technical level, came out, I devoured them voraciously. I had a small, crude, second-hand, amateur-made reflecting telescope, with which I "observed" the moon, planets, double stars, and bright nebulae. By then I really wanted to be an astronomer.

I was in my senior year in high school on December 7, 1941, when the Japanese attacked Pearl Harbor. I still remember hearing the news flash about the air raid which suddenly interrupted the live radio broadcast of a symphony concert to which our parents were listening, as my brother and I were reading the comic section of the Sunday newspaper. That same day my father predicted that it would be a long war against Germany, Italy, and Japan, but we would win it, and I would be in it. He was right on all three counts.

After graduation from high school and a few months at the University of Cincinnati, I joined the US Army in January 1943, and after basic training was sent to an Army Air Force pre-meteorology school at the University of Chicago. We were soldiers in an Air Force unit who fell out early each morning and marched to and from our separate classes. There we learned all the physics and mathematics required for a bachelor's degree, but without any electives, humanities, or social-science courses. The course lasted one year; we had excellent lecturers from the faculty and grad student TAs for our discussion sections. By the time we finished the school, the Air Force did not need any more weather forecasters, and I was sent to Victorville Army Air Field in California, where I learned on the job to be a weather observer. Then I went to Hickam Field, Hawaii, and after a few months there, to Okinawa. We Air Force weathermen were never in much danger and the war soon ended; six months later I was able to return to the States. In three years in the Army, I had met a much wider cross section of humanity than I would ever have done otherwise. I had matured, seen the world, and returned home with a sense of purpose, to an education financed by the taxpayers under the GI Bill of Rights.

In the fall of 1946 I went back to Chicago as a civilian student, took the necessary humanities courses for a bachelor's degree, and also took physics and astronomy graduate courses every quarter for three years. The Chicago physics department was outstanding; I learned quantum mechanics from Gregor Wentzel and nuclear physics from Enrico Fermi, two top-notch scientists and teachers. During my last year in Chicago several senior astronomy professors from Yerkes Observatory—Otto Struve, Subrahmanyan Chandrasekhar, Gerard P. Kuiper, and William W. Morgan—gave an excellent series of graduate courses on the campus, which I took. Thornton Page, a former Rhodes Scholar at Oxford and the only astronomer in residence on the campus, guided me in several reading courses, and I became especially interested in interstellar matter and gaseous nebulae, which were then hot research subjects. Page was almost the best teacher I ever had.

In 1949 I went on to Yerkes Observatory in Williams Bay, Wisconsin, about seventy-five miles from Chicago, where all the astronomy faculty members worked and taught the graduate courses. It was tremendously stimulating, and I believe at that time the outstanding graduate astronomy program in the country. Many visitors came from abroad to do research at Yerkes; we got to know them personally at the Van Biesbroecks' boarding house, where students and visitors lived and ate. There were unending discussions of research, astronomy, and astronomers at every meal. There was practically nothing to do in Williams Bay except learn astronomy, but it was an excellent place to do that.

In 1950 Otto Struve left for Berkeley, to my disappointment, but Bengt Strömgren came from Denmark to become director at the beginning of 1951. He was another excellent teacher of astrophysics who stimulated me especially by his stellar interiors course, in those heady days at the beginning of our quantitative understanding of stellar evolution. I learned to observe as a volunteer working with Morgan; he was then mapping the nearest spiral arms of our Galaxy by determining accurate spectroscopic absolute magnitudes of distant OB stars and extinction corrections from their measured color indices. He gave an important paper on the spiral arms at an American Astronomical Society meeting in December 1951; only the abstract was published, but in it he listed my friend and fellow grad student Stewart Sharpless and me as coauthors, although 90% of the work was Morgan's own (Morgan et al 1952, Federer 1952). It was an extremely generous action on his part, and I think that in the early years of my career that one abstract gave me more name recognition than all my own, more theoretical papers.

Chandra was my thesis adviser, who suggested the main subject for it—the gravitational interaction between interstellar clouds and stars, with a consequent gradual increase in the peculiar velocities of the stars. He was then working on his correlation methods, and this problem was a nice application of it. However, Lyman Spitzer and Martin Schwarzschild had begun working earlier on the same problem using standard mathematical techniques. They published their paper well before I completed my work, and my thesis merely confirmed their result (once Spitzer had straightened me out on an error in my final numerical result) (Osterbrock 1952).

In the spring of 1952, as I was finishing my thesis, I hoped that I would get a job in astronomy; in those days there was no AAS job register, no open recruitments, and no search committees. Everything worked by the "old boy" network. Chandra advised me not to go to one university which had expressed an interest in me, and soon after that I received a letter from Spitzer offering me a postdoctoral fellowship at Princeton. Undoubtedly Chandra had recommended me strongly to Spitzer and Schwarzschild, and Strömgren had supported him. Of course I took the fellowship.

POSTDOCTORAL FELLOWSHIP

When I arrived in Princeton in the summer of 1953, Schwarzschild had been publishing research papers on stellar evolution for five years. This was a new subject; it had been evident to him and to a few other theoretical astrophysicists since 1938, when Hans Bethe worked out the carbon cycle and proton-proton chain as the energy sources for stars, that their consequences would determine how stars evolved. With a succession of research visitors and PhD thesis students, Schwarzschild had found that under any physically reasonable assumptions, the central core of a star would become smaller and denser as it burned out its hydrogen, and the star itself would become a subgiant and then a giant. With Allan Sandage, then a visiting research fellow from Caltech, Schwarzschild had shown how the turn-off from the main sequence in a globular cluster like M 3 could be matched approximately, and the age of the cluster could be estimated (Sandage & Schwarzschild 1952).

All this was fascinating to me, for I had read George Gamow's *Birth and Death of the Sun* in 1946, and had studied atomic and nuclear physics with the world experts in Chicago. However, I did not wish to work on this project. At Yerkes, in Strömgren's course on stellar interiors, I had learned of the problem of understanding the intrinsically faint, low-mass, red-dwarf stars. The available models with convective cores which fitted the sun and more luminous main-sequence stars predicted central temperatures for the K and M dwarfs which were too high. From their luminosities it seemed evident that the relatively gentle proton-proton chain was their main energy source, but at the central temperatures given by the models, it would produce too much energy. In his final exam in the course, Strömgren gave us a problem whose solution showed that dwarf stars whose energy source was the proton-proton chain would not have central convective cores, but would be in radiative equilibrium at their centers. It was a new thought to us (although my notes show that Strömgren had mentioned it in a colloquium), for no one had published it up to then (although a result of TG Cowling hinted at it). Strömgren also told us that a few theorists, including himself and Ludwig Biermann, had conjectured that the red dwarfs in fact had deep outer convective zones, with less steep gradients than in the main-sequence models which fitted the sun. This would lead naturally to a lower central temperature, which might explain the apparent energy-overproduction problem.

This interested me greatly, and I felt I could work on it myself. Since Strömgren was devoting all his time to putting his ideas on quantitative spectral classification by narrow-band photoelectric photometry into practice, I asked him if I could work on the M-dwarf problem at Princeton. He very kindly said yes, and encouraged me to do so. Hence, when I arrived in Princeton I told Martin Schwarzschild that I would like to tackle it. He was well aware of the problem, but was studying evolution to and through the red-giant stage. He questioned me closely as to whether Bengt Strömgren had really said I could work on the red dwarfs, but after I assured him Bengt had, Martin was all for it and became my mentor and adviser. All his integration methods were mine to use, and he helped me greatly by discussing the problems that came up in the course of this work. In my long career in astronomy, I have met many outstanding scientists; all of them were interesting, but few of them were also outstanding human beings, genuinely helpful, self-effacing, and even self-sacrificing in their relations with other human beings lower on the totem pole. The two great exceptions I know to this "Golden Rule" were Bengt Strömgren and Martin Schwarzschild; they were my friends, supporters, and idols all my life, as they were to many other scientists of my age group.

Princeton had a very different atmosphere than Yerkes. The rickety old wooden observatory, where Henry Norris Russell (who was now retired but who came to his office frequently) had been a student, professor, and director, was on the same street as the massive, impressive stone and brick eating clubs, whose members were grooming themselves for the "service of the nation" as the bankers, corporate lawyers, and cabinet officers of the next generation. The director's house, where Spitzer lived with his family, immediately adjoined the observatory, so he could go directly from his front hall to his office without venturing outside. The astronomy department spring picnic, held in his yard, had a long tradition, and I still remember Fred Hoyle, a visitor that year, swinging exuberantly high in the air on an old automobile tire, tied with a long, sturdy rope to a branch far up in an ancient elm tree, as we all watched in awe and with some apprehension.

Schwarzschild's assistant, Richard Härm, showed me his numerical integration methods, some of which I had already learned from Paul Herget, my guru and mentor at the Cincinnati Observatory. Every step was done by hand with a Marchant or Fridén calculator (essentially glorified adding machines), and we wrote the intermediate results systematically in pencil on large sheets of quadrille paper. I had learned matching the solutions for the red-dwarf interiors and envelopes in the U, V plane in Chandra's course, but it was much more fun to be working on a real physical problem no one had solved before. After I had proved myself by doing several solutions, Spitzer and Schwarzschild assigned a graduate-student assistant—first Jack Rogerson, later Andy Skumanich—to do some of the integrations for me, so I got to teach them how to calculate stellar interiors. Peter Naur, a Danish student who had come to Yerkes with Strömgren (and then moved on to New York), and I wrote an analytic paper, essentially a generalization of the final exam question we had solved in his course earlier that year (Naur & Osterbrock 1953). Everything worked out wonderfully in the numerical integrations; they showed that the red-dwarf stars did have deep outer convective zones and radiative centers. The models I calculated on that basis did fit the masses, radii, and luminosities derived from observational data. I gave my first

AAS oral paper on the preliminary results at the meeting at Amherst in December 1952.

As I was finishing the written paper in the spring, I got a chance to use the Princeton electronic computer, then under development at the Institute for Advanced Study under John von Neumann. It consisted of a huge, air-cooled room full of racks of electronic equipment, including thousands of vacuum tubes; as a result it was usually out of operation, with one or more tubes burned out. Herman H. Goldstine was in direct charge of it, and even before I came to Princeton, Schwarzschild had had Malcolm Savedoff, then a graduate student, write the nonlinear differential equations of stellar structure in a form suitable for its machine language. Apparently the computer had never worked long enough for these programs to be used effectively—at least, no results from them had ever been published. But my problem was simpler; for the convective envelope, the three first-order differential equations reduce to two differential equations plus one algebraic equation (the adiabatic equation of state). The Institute computer could handle these by the spring of 1953 (in principle), so at Schwarzschild's suggestion I worked out and provided a network of starting boundary values to the Institute programmers. Weeks later the resulting solutions emerged; they agreed with our hand-calculated values but probably had higher precision, and there certainly were more of them, closer to one another in the adiabatic parameter I used for interpolating. These convective-envelope solutions went into the final paper, which I completed at Princeton, and were published several years later in a huge set of tables giving the numerical results from Schwarzschild's group (Härm & Schwarzschild 1955). My paper had an effect on the field; it was the first real solution of the red-dwarf problem incorporating convective envelopes, the essential feature of these stars (Osterbrock 1953). Later, my friend Nelson Limber—who had been a graduate student with me at Yerkes, then served in the Army during the Korean War, and then had gone on to Princeton to work as a postdoc with Schwarzschild—greatly improved on my treatment, including partial degeneracy, electron conductivity, and other relevant details. The general method has stood up well.

In my second semester at Princeton, Spitzer was working full-time on Project Matterhorn (the attempt to harness thermonuclear fission reactions for peaceful uses) and asked me to teach his graduate course on stellar atmospheres. I was glad to do so. At that time, Princeton accepted two entering astronomy graduate students each year and had a two-year cycle of courses, so there were four students in my class: Rogerson, Skumanich, George Field, and Leonard Searle. I had never taught before, nor been a TA (I was a grader for several physics courses at Chicago), but like other grad students of my time I had been exposed to a lot of teachers by then. Some of them had been very good, some of them had been very bad. I tried to emulate the good ones and to improve on them if I could. My aim was always to give the physical ideas as clearly as possible, without a lot of mathematical derivations, but enough so that the students could go back and fill in the details as they gained understanding of the material. Fermi had been the best teacher I ever had; this was the method he invariably used, along with many homework problems to help us grasp the ideas. I followed him in that too, and since all four of these students had long, highly successful careers in astrophysics, I felt I had not done them too much harm. Bev Oke was then an advanced grad student at Princeton, working on his thesis on O stars with Spitzer; after they returned to Princeton from Pasadena, where they had spent the first part of the academic year, I got to know Bev well. He invited my wife and me to our first dinner in academic gowns, at the Graduate College where he lived.

Toward the end of my year at Princeton, Hoyle arrived on the scene to work with Schwarzschild on red giants. The problem was that the calculated evolutionary tracks fitted the early stages well, near the main sequence, but then rapidly evolved to large, cool model stars with color indices much redder than any stars observed in clusters. Hoyle's office was next to mine, at the back of the building, and we often discussed our research with one another. Coming from Yerkes, I was used to hard workers, but Fred worked harder and longer than even the most extreme workaholics there (Chandra, Kuiper, and Struve). It was amazing to see Hoyle and Schwarzschild in action; Fred would come up with five or six new ideas every day, but Martin would carefully examine each one of them, and usually shoot them all down, or all but one, by the end of the day. He expressed fewer new ideas, but no doubt had examined all of his carefully before tossing them out for discussion. He and Fred made a good team. Hoyle very frequently asked me questions about the outer convective zones in my red-dwarf models; sometimes I became exasperated by his repeated queries on this one subject. But it turned out that he and Schwarzschild realized that convection was the missing link in the red-giant tracks. Like red dwarfs, these stars have deep outer convective zones which stop the evolving stars from expanding further and instead drive them up to higher luminosity at a roughly constant surface temperature (Hoyle & Schwarzschild 1955). So, although I had nothing to do with their breakthrough, my work on red dwarfs had brought this arcane branch of the theory of stellar structure to their attention!

That spring I received a letter from Jesse Greenstein offering me an instructor-ship at Caltech. Again, I had not expected it. He had been at Princeton for one day, a month or two earlier, and we had had a brief and inconclusive conversation. I had no idea Jesse was looking me over, and thinking back on it, I would not have hired myself on the basis of what I remembered saying to him. But no doubt Chandra, Schwarzschild, Strömgren, and Spitzer had all recommended me for the job, and who was I to say no!

MICHIGAN SUMMER SCHOOL

Driving west from Princeton to Pasadena, my wife and I stopped for a month at Ann Arbor, where I took part in a "summer school" (or workshop as we would call it today), organized by Leo Goldberg, then observatory director and chairman of the astronomy department at the University of Michigan. It was a wonderful opportunity for me and the thirty or so other postdocs and grad students who were there. The main lecturer was Walter Baade, whom I had heard once before at a symposium on the structure of our Galaxy in 1950, also at Ann Arbor. Baade was the outstanding observational astronomer in America at that time; his "discovery" (or recognition) of the two stellar populations, which turned out to be young stars and old, opened up whole new worlds of research, especially on stellar evolution and galactic structure. Morgan's work on mapping the spiral arms in our Galaxy near the sun, in which Sharpless and I had participated, was a direct outgrowth of the enthusiasm and inspiration he had gained from hearing Baade's invited lectures at the American Astronomical Society meetings in Ohio in 1947 and Pasadena in 1948. At the Michigan summer school in 1953, Baade gave a series of eleven lectures spread over four weeks, recounting all he knew then on the two populations and their role in galactic structure. A dynamic, engaging personality, he was always available for scientific discussions, and I had many talks with him, as of course did nearly everyone else who was there. The other lecturers were Gamow, speaking mostly on what we call "big-bang cosmology" today; Ed Salpeter, on thermonuclear reactions in stars; and George K. Batchelor, on turbulence. The first three topics were among the hottest in the astrophysics of the 1950s, and turbulence was especially interesting to me because it had been the subject of Chandra's theoretical seminars for most of the three years I was at Yerkes. Many of the best young astronomers and astrophysicists, especially from the Midwest, were at UM as participants. Among my contemporaries I met and got to know several who became outstanding research workers of the next forty years, including Allan Sandage, Vera Rubin, Gene Parker, and Bernard Pagel. In addition, Nancy Grace Roman, Margaret and Geoff Burbidge, Marshall Wrubel, Larry Helfer, Eugenio Mendoza, and Limber (all of whom I already knew from Yerkes) were there, as well as Oke; they also went on to outstanding careers—a few, alas, cut short by early deaths. Though only a month long, the Michigan Summer School was a formative experience from which all of us benefited enormously (Gingerich 1994).

CALTECH

The Caltech astrophysics group, or department (we were part of the Division of Physics, Mathematics, and Astronomy), that I joined was new and small. Greenstein had been brought in from Yerkes to head it when the 200-inch Hale telescope was dedicated at Palomar Observatory in 1948. He brought Guido Münch to Caltech in 1950, me in 1953, and Art Code in 1956; all three of us were former Chandra PhD thesis students who had observational experience with telescopes. Fritz Zwicky, originally a theoretical physicist who had switched to astrophysics in the 1930s, was in the department but did no teaching (except two graduate research courses in his own somewhat peculiar specialties, which were listed in the catalogue but which no student I knew took in the five years I was there).

He did not take part in the department meetings Greenstein had with the rest of us, nor did he participate in the PhD final oral examinations. The teaching load was light; generally I taught one course a quarter for five of six quarters in each two-year cycle. Greenstein was glad to hand the stellar interiors course over to me; I also taught advanced undergraduate astrophysics courses, which I liked, and sometimes (so far as I can remember) stellar dynamics, which I did not like.

Our Caltech group was part of Mount Wilson and Palomar Observatories, a joint operation at that time, with Ira S. ("Ike") Bowen as director. Caltech owned Palomar Observatory, and the Carnegie Institution of Washington owned Mount Wilson, but the two staffs used both observatories on an equal basis. The Mount Wilson astronomers all had faculty status at Caltech and would teach a one-quarter course in their research specialty for our grad students on an occasional basis. We also had funds to bring a distinguished visiting professor to teach a one-quarter graduate course each year, usually in the spring. Among them were such luminaries as Jan Oort, Ludwig Biermann, and Fred Hoyle.

Inspired by Baade's lectures at Ann Arbor and many discussions with him, I decided to go into research on color-magnitude diagrams of globular clusters, especially M 15, and a study of a local-group galaxy, NGC 6822, the dwarf irregular first identified as a star system by Edwin Hubble. Both programs were to be done by photographic photometry, using photoelectric standard stars. Greenstein, certainly the best boss I ever had (just as Bowen was the best director), was less than thrilled by this news but made no attempt to dissuade me. Both these fields were among the most popular with the Mount Wilson staff already, and there was little room left for an outsider, though I did not realize that at first. Baade taught me to take direct plates with the 100-inch, and later with the 200-inch. I became proficient with the telescopes but soon learned that the accurate photoelectric magnitudes of standard stars to the necessary faint levels, which in Baade's exciting vision would be provided for me by other staff members, were very hard and in fact impossible for me to get. I accumulated a lot of nice plates of M 15 and NGC 6822, but never got any results out of them. With the 60-inch I tried to find planetary nebulae in globular clusters, as Francis Pease had found one in M 15 years earlier, comparing an ultraviolet exposure (on which the planetary was bright) with a blue exposure (on which it was much fainter). But the planetary he had discovered proved to be the only one in the thirty-odd clusters for which I obtained such pairs of plates. This was actually consistent with the fact that the planetary nebulae are mostly old disk objects, while nearly all the globular clusters I could observe were extreme halo objects, but I did not publish this.

Baade realized that I was more interested in spectroscopy than in direct photography and suggested that I obtain the radial-velocity curve of NGC 3115, the nearly edge-on S0 galaxy that Oort had studied in detail. From the light distribution (interpreted as mass), various turning points and changes in slope were expected in the velocity curve. Rudolph Minkowski showed me how to use the fast "nebular" spectrographs of the 100-inch and 200-inch telescopes, but I soon saw that this program was hopeless. Without the digital data and sky subtraction we have today, isolating the absorption-line spectrum of a galaxy became rapidly more and more difficult as the distance from its nucleus increased. A mass of night-sky emission features plus scattered moonlight (which usually was present during parts of my nights) made it impossible to measure accurately the velocities of the galaxy H and K lines. On the other hand, I could see that these fast spectrographs were ideal for obtaining emission-line spectra of gaseous nebulae.

In 1954 I was excited to read a paper by Mike Seaton, a young British atomic theorist, who had calculated accurate new collision cross sections for electrons on O^- ions, which emit the strong nebular [O II] $\lambda 3727$ emission line (as it appears at low resolution, actually a close doublet, $\lambda\lambda 3726, 3729$). Because these lines are highly forbidden, meaning the transition probabilities for their emission are very slow, collisional deexcitation effects are important, and the ratio of intensities of the two lines depends strongly on the local electron density, as Seaton (1954) emphasized. Using the nebular spectrograph on the 100-inch telescope with a grating working in a high order, I could obtain good spectrograms that resolved the components of the doublet and could measure their strengths by the standard methods of photographic photometry. I quickly observed many points in the Orion nebula, clearly an object in which the electron density decreases outward from the Trapezium (photoionizing O stars). The measured line ratios matched qualitatively the expected variation with density decreasing outward, and allowed mean densities along each line of sight through the nebula to be calculated (Osterbrock 1955).

I wrote Seaton about these results, and soon we were collaborating by air mail, although we had never met in person. He improved the quantum-mechanical calculations, I obtained more measurements of the [O II] line ratios in H II regions and planetary nebulae, and we increased our empirical knowledge of both nebulae and O^+ (Seaton & Osterbrock 1957). Edith Flather, a dedicated assistant, and I further analyzed the [O II] measurements, and by comparing them with radio-frequency continuum measurements (analogous to surface brightness, but independent of extinction effects), we showed that the density distribution in the Orion nebula (for instance) is very clumpy, consisting basically of small, dense condensations in a much lower-density gas (Osterbrock & Flather 1959). We expressed this result in terms of the "filling factor," a concept Strömgren (1948) had introduced in one of his classic papers on H II regions. It seems to be a better approximation to reality in all nebulae I know than the "simple" picture of a homogeneous density distribution, which most theorists then favored. Today we realize that the Orion nebula is even more complicated, a thin ionized surface of a dense, dark molecular cloud. I went on measuring electron densities in as many kinds of nebulae as I could, including not only planetary nebulae and H II regions but also a Herbig-Haro object, the T Tauri nebulosity, a nova shell, and the "filaments" in supernova remnants, both young (the Crab nebula) and old (the Cygnus Loop and IC 443) (Osterbrock 1957, 1958a, 1958b).

I enjoyed this research because it represented an application of my favorite subject, atomic physics, to measuring quantitatively the physical conditions previously unknown in real objects in the universe; Greenstein liked it because it was new and interesting; and Bowen liked it too because it involved forbidden lines and fast spectrographs based on Schmidt cameras, both intimately connected to the great subjects of his career. However, I was ready to leave Caltech. Although the telescopic and instrumental opportunities were unsurpassed, the scientific atmosphere was extremely stimulating, and I liked all my colleagues, still my wife and I were both homesick for the Midwest and especially Wisconsin. Southern California seemed to us a concrete jungle of cars and smog (backyard burning was still the accepted trash-disposal method), and we felt isolated in a vast megalopolis, too far from the rivers and lakes, forests and woods, hills and fields we loved. Our Pasadena friends naturally thought we were crazy, but I quietly let Albert Whitford, then head of the small astronomy department of the University of Wisconsin (in Madison), know that if he ever had an opening, I would be interested in it. I thought there was little chance of a job there, but once again I was lucky. Within a year Whitford accepted the directorship of Lick Observatory, Code returned to Madison as his successor, and I accompanied Art to help build up a new department of astronomy there. My five years at Caltech were productive and pivotal for me; they had made me over from a budding theorist to an experienced astrophysical observer and confirmed my interest in gaseous nebulae and interstellar matter as my research fields.

I had my first two PhD thesis students at Caltech, George Abell and John Mathis. Abell had finished his graduate courses before I arrived in Pasadena. Married and with two small children, he had taken a full-time job as the main observer with the 48-inch Schmidt telescope on the Palomar Observatory Sky Survey, with the proviso that he could base his thesis on the plates he obtained on this program. He had begun his thesis on finding and cataloguing clusters of galaxies and discussing their properties and arrangement in space, with Baade and Minkowski as his unofficial advisers well before my arrival. They wanted me as the official adviser to be the buffer between Abell and Zwicky, who had pronounced views of his own on the subject and no love for the two Mount Wilson astronomers. I knew little about clusters of galaxies but discussed George's work with him and otherwise kept out of his way. He was a fantastically hard worker. I thought some of his statistical conclusions were overdrawn examples of after-the-fact discussions, yielding extremely small probabilities that any other picture could fit his data, and told him so, but he continued to believe that he was right. Certainly the main results—that there are many clusters, that there are "clusters of clusters" (higher-order clustering), and that intergalactic extinction is negligible out to the distances he was studying—were true. His catalogue of what are now called "Abell clusters" is still being widely used today. It was a very good thesis, to which I contributed very little (Abell 1958).

I had a lot more to do with Mathis's thesis, on the He I spectrum emitted in the recombination of ionized He^+ and subsequent downward radiative transitions in a photoionized nebula. Mathis did the best treatment of the He I statistical equilibrium up to that time, and he got the observational data himself with the 60-inch telescope on Mount Wilson to apply it to the Orion nebula. It was a good thesis that led to the best available value of the relative abundance of helium with respect to hydrogen in gaseous nebulae, and hence in newly formed ("young") stars (Mathis 1957a,b).

WISCONSIN

The University of Wisconsin had a strong research tradition, an outstanding one for a state university. It was best in the agricultural and life sciences, especially biochemistry, and in the School of Medicine. In astronomy George C. Comstock (first dean of its Graduate School), Joel Stebbins, and Whitford, successive directors of Wisconsin's Washburn Observatory, had been extremely productive research professors, but their department had been small and had never had a regular graduate program. Only four astronomy PhDs had been awarded in the history of the University of Wisconsin before Code and I arrived in 1958: to Morse Huffer, Olin Eggen, Ted Houck, and John Bahng. The main research instrument had been the ancient 15.6-inch refractor ("larger than the Great Harvard 15 1/2-inch refractor") on Observatory Hill on the campus, surrounded by the lights of downtown Madison and its West Side. Stebbins and Whitford were recognized world pioneers in photoelectric photometry, but during the later years of Stebbins's career and almost all of Whitford's, they did their most important observing at Lick, Mount Wilson, and Palomar Observatories.

However, after World War II, with the state treasury and the Wisconsin Alumni Research Foundation (WARF, Wisconsin's "little NSF") flush, the university administration, finally recognizing they had a real research asset on their hands, decided to put some money into astronomy. They provided the funds to build a new Pine Bluff Observatory, with a 36-inch Cassegrain reflector, in a relatively light-free area fifteen miles west of the campus. It was an adequate instrument, at least to start the kind of photoelectric observing programs that Whitford, who had begun planning the telescope, visualized carrying out. It was completed in the spring of 1958 and was dedicated at an AAS meeting that summer, a few days after my arrival to start as a new faculty member at the University of Wisconsin. At midnight on June 30, Baade, who gave his Henry Norris Russell lecture on stellar and galactic evolution at that meeting; Whitford, whose job at Wisconsin ended that night; and I, whose job began then, were exchanging toasts (in good Wisconsin beer) at the lounge in Baade's motel.

The new reflector was ideal for the nebular work I wanted to do. It had a photoelectric photometer mounted to it and a new one-channel scanning photoelectric spectrometer (or scanner). Whitford and Code had designed it, based on their experience with an earlier scanner they had designed, built, and used at Mount Wilson. Although for spectrophotometry of stars the aperture of the telescope is important, for nebulae whose angular size is comparable with or larger than the slit, the signal does not depend on aperture. Thus the 36-inch was fine for nebulae and allowed me to get all the light from fairly large planetary nebulae into the slit and thus to measure them.

Along with the new telescope, Wisconsin got a new graduate program in astronomy. Art was chair, and I came with him; Huffer and Houck, who had been there with Whitford, stayed; and Bob Bless, then a recent PhD from Michigan, came as a research associate to work with Art on absolute energy-calibration measurements of bright stars, comparing them with a standard lamp on a tower erected especially for that purpose. Bless joined the faculty a year later, and Mathis a year after that. Although the telescope and new department had been planned in the early 1950s, by 1958 the post-Sputnik era in American science was roaring along, feeding money into hard science and especially into space, astronomy, and physics. Once again I had been lucky; I found myself in a new growth industry as universities, astronomy, and our department continued to expand until the early 1970s.

We taught a full program of graduate courses in a two-year cycle, much as was done at Yerkes, Princeton, and Caltech. We tried to shift the various courses around among ourselves, but I frequently taught stellar interiors, gaseous nebulae and interstellar matter, and sometimes galactic structure, as well as undergraduate astronomy survey courses, which I enjoyed. In the late 1950s and the 1960s many Wisconsin students who took these courses could be interested in astronomy, always the most popular science to the general public, if handled right. My classes particularly liked the traditional lecture I gave on Halloween on the contributions of astrology to astronomy; I gave the lecture wearing a wizard's hat and a white lab coat, accompanied by clouds of white carbon-dioxide vapor released from dry ice at appropriate moments by the TAs. The students enjoyed this performance even more than I did. After Huffer retired (he then went on to teach seven more years in California), I even taught celestial mechanics once, but then we decided to drop the course.

As a new graduate department, we did not at first attract the top students who could get into Caltech, Princeton, Harvard, and Yerkes, but we did get some excellent prospects from less prestigious undergraduate schools. In our first few years we only admitted a limited number of students, and the department was small; I usually had only one or two students working with me as research assistants, and I developed close ties with them.

In my early years at Wisconsin, radio astronomers elsewhere were measuring the radio-frequency continuum ("free-free") radiation of gaseous nebulae, especially H II regions and a few of the brightest planetaries. My aim was to measure the corresponding optical line radiation ("bound-bound"), generally $H\beta$ (because $H\alpha$ was blended with [N II] ll6583, 6548 at our resolution), and by comparing it with the radio-frequency flux, to determine the extinction. The absolute-energy calibration, which Code's measurements provided, was essential for this program.

The H II regions were much larger than the slit of the scanner, so our instrument maker, Lloyd McElwain, constructed a small, 5-inch aperture refracting telescope for me, using an army surplus lens, a brass tube cannibalized from an old transit instrument, and an equatorial drive from some previous Wisconsin experimental apparatus. I helped McElwain dig the hole for the concrete pier he poured for it at Pine Bluff, and with a galvanized-metal shed with a rollaway roof I was in business.

We used a photoelectric photometer with interference filters at the nebular lines and in the continuum, and kept all the electronics wired up in a wheelbarrow, which could be rolled out to the 5-inch telescope when it was clear. Observing in the open in the Wisconsin late fall was cold, but we got results (Osterbrock & Stockhausen 1960, 1961).

Graduate students working with me measured smaller planetary nebulae with the scanner on the 36-inch, and larger ones with a photoelectric photometer on a 12-inch Cassegrain reflector at Pine Bluff, using the same interference filters (Capriotti & Daub 1960, O'Dell et al 1961). Bob O'Dell, in an excellent PhD thesis he began with Minkowski, a visiting professor at Wisconsin in 1960–61 (while I was on leave at the Institute for Advanced Study), and completed with me, discussed all this material in terms of the evolutionary picture of planetaries proposed by Iosef Shklovsky a few years earlier. The idea is simple; as the nebular shell expands it becomes larger and less dense, which predicts a relation between its mean electron density and surface brightness. O'Dell (1962) confirmed this relationship observationally and used the data to calibrate Shklovsky's method, based on these concepts, for determining the approximate distance to the planetary. It has been widely used ever since, improved of course over the years by more detailed theories and by more numerous and better calibrated measurements.

We also used the 36-inch with the scanner to measure the Balmer decrement, the ratio of emission-line strengths $H\alpha$: $H\beta$: $H\gamma$ in several bright planetaries, a subject of considerable debate at that time. A single line ratio depends on the extinction, but with two ratios this can be eliminated, assuming a standard reddening curve. Our measurements did not quite confirm the theoretical Balmer decrement, but were so close that I thought the unknown systematic errors in the calibration were probably the cause of the remaining discrepancy, rather than the theory. The latter depends only on what appear to be accurately calculated parameters of the simple one-electron system, atomic hydrogen (Osterbrock et al 1963). This is one case in which I agreed with Eddington's dictum that observational data should only be believed if it is confirmed by a good theory!

I was also interested in comets, as a result of an excellent series of lectures that Pol Swings had given at Yerkes while I was a student there. Later, at Palomar, I had photographed several of them with the 48-inch Schmidt telescope. We could use the same photoelectric photometer, but a different set of interference filters, to measure the fluxes in bands of C^2 and CN in comets, and thus get the densities of these molecules, their rate of production, and their rate of decay in the solar radiation field (Osterbrock & O'Dell 1962, Osterbrock & Stockhausen 1965). Although I did not continue this work myself, two of the Wisconsin graduate students I taught—O'Dell, who did his thesis with me, and Mike A'Hearn, who did not—took it up after they left Madison and became great experts in it.

My interest in galaxy research had been awakened by Baade at Mount Wilson and Palomar Observatories, where he had urged me to work on Seyfert galaxies, with their strong, broad emission lines, and on the relatively few elliptical galaxies with [OII] $\lambda3727$ in emission in their nuclei. The ellipticals were a puzzle to him;

in his dichotomy they were Population II objects, but nebular emission lines were a marker of OB stars and hence Population I. Minkowski and I tried to measure the electron density in M 87 (Virgo A), the giant elliptical with particularly strong $\lambda 3727$ in its spectrum. We immediately saw that the line profile is very broad, indicating high internal velocities, but we made an estimate of the electron density from the mean wavelength of the blended doublet (Minkowski & Osterbrock 1959). I obtained spectra of most other ellipticals then known to have strong $\lambda 3727$ emission lines. My relatively high-dispersion spectrograms showed the emission line broadened in all of them, which I suggested was probably due to rotation. I showed that the amount of gas required to produce the measured fluxes is actually quite small, and I discussed O subdwarfs, similar to the few known in globular clusters (especially M 13), as the possible photoionization source (Osterbrock 1960). Today we know that there are many varieties of old populations which Baade, in his pioneering research, lumped together, and it seems likely that these ellipticals are much more like active galactic nuclei than OB associations.

I had obtained all the spectral data for these papers with the Hale 200-inch reflector; the Wisconsin 36-inch was much too small for me to hope to continue this research there. However, in 1963–64 I went on leave for one year as a visiting professor at Yerkes Observatory, where I had access to the 82-inch McDonald Observatory reflector. It had fast photographic spectrographs at both the Cassegrain and prime foci, which I used to obtain spectrophotometric data on NGC 1068 and NGC 4151, two of the brightest Seyfert galaxies then known. Bob Parker, my first postdoc (he had been Münch's PhD thesis student at Caltech), and I reduced and discussed the NGC 1068 data, and showed quantitatively that the ionization of the gas covers a much wider range than in H II regions or most planetary nebulae. We knew that OB stars could not produce all these ions, and we tried to analyze the energy input to the gas in terms of fast particles (suggested by the line profiles), but this interpretation was relatively unconvincing (Osterbrock & Parker 1965). We did not realize it at the time, but our paper confirmed much of the earlier theoretical analysis of Seyfert galaxies by Lo Woltjer (1959), which we had missed when it was published. I reported our results at the first Seyfert-galaxy symposium, held at the University of Arizona in 1968 (Pacholczyk & Weymann 1968), where Oke and Wal Sargent also reported their data, taken with a photoelectric scanner, on NGC 4151. I hoped to use such an instrument, with a higher quantum efficiency than photographic plates, linear response, and digital output, on a larger telescope some day, as I had become quite interested in Seyfert galaxies.

Soon after Code and I had gone to Wisconsin, opportunities for ultraviolet astronomical observations from telescopes in rockets and artificial satellites above the Earth's atmosphere began to become available. Art shifted most of his research activities to this spectral region, along with Bless and Houck, while Mathis, Parker (who joined our faculty after one year as a postdoc), and I remained in ground-based research. However, at Art's request I drew up a list of the strongest expected ultraviolet nebular emission lines to use for planning purposes. My quantitative estimates showed that not only forbidden lines of the classical type like [Ne IV] λ2422, 2424 would be strong, but also permitted lines like Mg II λλ2796, 2803 and C IV λλ1548, 1551 as well as intercombination or "semi-forbidden" lines like C III] λ1909 (Osterbrock 1963). Greenstein had first mentioned the possible importance of the latter two types of transitions to me at Caltech, from his early experience with ultraviolet spectra of the sun, including the chromosphere, taken from rockets. In 1963 Maarten Schmidt, Greenstein, Oke, and Tom Matthews had recognized two quasars as objects with "nebular" emission-line spectra at large redshifts (Shields 1999). Many observers, especially at Palomar, began obtaining spectra of even larger-redshift quasars, and found my paper on expected strong ultraviolet emission lines useful as a finding list for identifications. Parker and I noted that the quasars had spectra "like" Seyfert galaxies or nebulae with a wide range of ionization, which implied that the abundances of the elements and the electron temperatures were similar in all these objects; on this basis we predicted the spectra of quasars in then still unobserved parts of the spectrum and to weaker lines (Osterbrock & Parker 1966).

Quasars and Seyfert galaxies obviously had very similar spectra, as many observers realized. This connection was perhaps made most explicit and convincing by Morgan (1971, see also Morgan & Dreiser 1983). He had become extremely interested in the classification of the forms of galaxies, which he discussed with me frequently during the year I was a visiting professor, and after it during my many visits to Williams Bay. His morphological classification scheme was based completely on "the thing itself" in his words—that is, the galaxy's appearance—without any theoretical interpretation at all; nevertheless he liked to try out his ideas on a person like me who understood some of the theory, but was sympathetic. Morgan's morphological types showed a better correlation with integrated spectral types or color indices than the types on the classical Hubble system. He also demonstrated convincingly that the so-called S0 galaxies were actually a mixture of at least two quite different types of systems, and he isolated and named the cD galaxies, a type now universally accepted and widely used (Morgan 1988). He and I published a joint paper on the forms and stellar contents of galaxies, based to a considerable extent on these discussions—most of the classification part was by Morgan and most of the emission-line part was by me. Among other things, it showed that the Seyfert galaxies tend to have "later" morphological types than starburst or star-forming galaxies, but "earlier" types than most field galaxies without emission lines in their spectra (Morgan & Osterbrock 1969).

Some of the graduate students who worked with me did theoretical theses, particularly in my later years in Madison, after we had finished most of what we could easily do with the 36-inch. The earliest, by Gene Capriotti (1964), was an excellent analysis of the effects of self-absorption of the Balmer emission line spectrum by excited H atoms in nebulae. It showed that this process was not at all effective in normal planetary nebulae and H II regions, contrary to the expectation of some theorists at that time. Later, Bob Williams (1967) calculated one of the first model nebulae using a power-law continuum, intended to mimic synchrotron radiation in the Crab nebula, as the input source. Some of his models were also intended to apply to Seyfert galaxies and quasars, and then a few years later Gordon

MacAlpine (1972) calculated similar but more advanced photoionization models with exponents for the power law chosen to match quasars themselves. Dipankar Mallik (1975) did his thesis with me on the temperature structure at the ionization front of a model H II region.

Others of my students obtained their observational data on telescopes elsewhere, beginning with Ralph Stockhausen, who did the radio-frequency observations mentioned earlier at NRAO in Green Bank. Joe Miller (1968) did an excellent thesis showing that the radial velocities of H II regions with known distances (from OB stars within them) do not give their distances (based on a galactic rotation model) to the accuracy many H I 21-cm line observers then believed. He obtained all the radial-velocity data from spectrograms he took at Kitt Peak National Observatory. Dan Weedman (1968) did his thesis on analyzing the expansion-velocity and ionization distributions within planetary nebulae using emission-line profiles he and I obtained at Mount Wilson with the coudé spectrograph of the 100-inch reflector. Holland Ford (1970, 1971) completed an excellent thesis with me, which he had begun with Code, on the kinematics of star clusters in both Magellanic Clouds, their spectral types, and their ages, using image-tube spectrograms he had taken at Cerro Tololo. My last PhD student at Wisconsin, Reggie Dufour, also used the CTIO telescopes for his thesis on the abundances of the elements in the H II regions of the Magellanic Clouds (Dufour 1975). He finished it with Mathis as his adviser, after I had left Madison. I was very lucky in having an excellent group of graduate students during my years at the University of Wisconsin.

By 1970, my own and my students' and postdocs' (including John Dyson, Louise Webster, and Tom Bohuski) nebular research had become widely known. I had given review talks at meetings, written and published review papers, and taught a graduate course on gaseous nebulae several times. Scientific book publishers were eager to find new authors, and I decided the time was ripe for me to write a monograph and graduate textbook on nebular research. In it, I tried to explain the physical ideas as clearly as I could and to work in new concepts gradually in terms of real nebulae. I was very lucky to have an editor, Bruce Armbruster, who fought to keep my own words in the text as I wanted them, rather than letting the copy editor rewrite them. The result, *Astrophysics of Gaseous Nebulae*, was widely accepted and used in many graduate programs in this country and abroad. I called it alternately "the little blue book" or "the good news" and liked to say that I required all my later students to understand it well enough to find at least one mistake or misprint in it in order to get their degrees. "Don't listen to what I say; listen to what I mean" was one of Richard Feynman's most famous sayings at Caltech, which I adopted and quoted frequently.

INSTITUTE FOR ADVANCED STUDY

Back in 1960–61, after only two years at Wisconsin, I got the chance to spend a whole year doing research at the Institute for Advanced Study in Princeton. Strömgren had left Yerkes to become the Institute's first professor of astrophysics, "the man who got Einstein's office" in Fuld Hall. He told me he would welcome me any time I could come there and I took him up on it, perhaps sooner than he had expected. With his recommendation, Bowen's, and Code's, I was fortunate enough to get a Guggenheim Fellowship, which WARF and the Institute itself supplemented, to make my stay there possible.

The other visiting fellows in astrophysics that year were Anne Underhill and Su-shu Huang (both Yerkes PhD's like me), Hong-Yee Chiu, and Bob Christy, who I had known as a renowned theoretical physicist at Caltech, but who became interested in Cepheid-variable pulsation theory while in Princeton. We met once a week for an astrophysical lunch with the Princeton University group, which by then included Spitzer, Schwarzschild, Field, Rogerson, Bahng, Stuart Pottasch, and Bob Danielson. Woltjer also came to the Institute for the second semester and organized a symposium on interstellar matter in galaxies, in which Oort played a leading role. Robert Oppenheimer was the director of the Institute, and the only person I ever knew who had real Van Gogh and Gauguin paintings hanging on the walls of his home, to which we visitors and our spouses all were invited for dinner sometime during the year. He was interested in astrophysics and attended some of our colloquia and parts of the symposium, frequently asking questions—more, it seemed to me, to show his erudition than to seek information. Oppenheimer was a lonely, tragic figure, emaciated, a chain smoker, but a brilliant physicist, scientific leader, and loyal citizen who had been most unfairly charged, tried, and found guilty in a security hearing for his honest advice on the H-bomb issue.

I went to the Institute to work on the heating of the chromosphere and corona by the dissipation of magnetohydrodynamic waves generated as sound waves in the hydrogen convection zone. They become more and more magnetic in character ("Alfvén waves") as they propagate upward. My aim was to understand the sun, which we see close up in detail, first, and then to apply the same physics to all late-type stars, from dwarfs to supergiants. I hoped to understand the strong correlation between absolute magnitude and width of the Ca II H and K emission cores, which Olin Wilson had found empirically in these objects. He had been doing this research at Mount Wilson and Palomar while I was in Pasadena, and I had discussed it often with him. The corona and the chromosphere are low-density gases with many similarities to nebulae, so it was not so strange as it may seem for me to get into this subject.

That year I learned a lot about magnetohydrodynamic waves, how they are refracted and transformed, and how they dissipate; by the spring I could barely understand how the sun worked. I could not understand the chromosphere and I only had a crude understanding of the corona (Osterbrock 1961). Part of the problem undoubtedly was that I did not realize how extremely inhomogeneous the corona is. However, I knew by then that I would never be able to calculate the structure of the chromosphere and corona in other stars from first principles, so I had to abandon my more ambitious program. Nevertheless, my paper had an effect on the direction of solar research, and I was glad I had done it.

I also studied the escape of resonance-line radiation (like H I Lα) from a nebula in which the optical depth at the center of the line is very large. Resonance-line photons emitted in such a nebula are absorbed and re-emitted many times before they escape or are destroyed by another process, such as absorption by dust. The solution for central optical depths up to $t_0 \approx 10^4$, where only scattering in the Doppler core comes into play, was well known; I was able to extend it into the damping wings in an approximation based on the mean free path in frequency space and the resulting occasional shifts far enough from the center of the line for the photon to escape (Osterbrock 1962). This approximation was good enough to show that in general Lα photons would escape from idealized, dust-free nebulae but have a fairly high probability of being absorbed by dust in a real planetary or diffuse nebula (Osterbrock 1971), a process that leads to the observed infrared continuum radiation from these objects.

I was glad to have had the opportunity to spend a year working completely on theoretical problems, but I was glad to get back to the telescope and scanner at Wisconsin the next year, too.

LONDON

In 1968–69 I spent a year on leave from Wisconsin, working full-time on research at University College London (UCL) under a National Science Foundation fellowship. Nebular astrophysics (which includes the gas in Seyfert galaxies and quasars) is largely an application of atomic physics, and Seaton was by then the recognized world expert in the quantum mechanical calculations relevant to the atoms and ions of chief interest. After our joint paper, he had come to California where I had met him for the first time; later I saw him briefly at the Institute for Advanced Study, and still later he had visited us in Wisconsin and gone to the Walworth County Fair, near my wife's hometown, for a real taste of American life.

I understood fairly well in principle what the cross section (or "collision strength") calculations, which lead to excitation rates for nebular emission lines, are about, but I wanted to do some myself, to be sure I really did know, and to learn how to estimate their accuracy. Working conditions were ideal; I had a small office at UCL, very conveniently located in Bloomsbury. Mike, Dave Moores, Werner Eissner, Harry Nüssbaumer, and Hannelore Saraph gave me a crash course in their methods and computer programs, and I learned to specialize them to C^{++}, which emits the C III] λ1909 line. It is actually a close doublet, $\lambda\lambda$1907, 1909, strong in the ultraviolet spectra of quasars and, as we later learned, Seyfert galaxies. It was one of the few ions with strong observed lines for which Seaton and his collaborators had not yet calculated a cross section. There was a published estimate of it, from the U.S.S.R., which if correct made this cross section very small. However, we did not believe it. At the I. A. U. meeting in Prague in 1967 and at the subsequent planetary-nebula symposium in eastern Czechoslovakia, Seaton and I both tried to find out more about this estimate from the Russian astronomers and astrophysicists who could attend these meetings "behind the iron curtain" but could only very rarely get to Western Europe or the United States. None of them could tell us much about it, so we both thought that C^{++} would be a good case for me to tackle.

I did most of the calculations on the UCL electronic computer, which then used punched-card input of the program (which had to be "compiled" every time I made any change in any subprogram) and starting data, and I provided the output on cards and a hard-copy printout. My main problem was to find every single programming error, and then to speed up the calculation as much as possible. It was slow going, and since the English did not then assign computer time, but at UCL operated under a "gentlemen's understanding" that no one would use "too much" time, by spring the turnaround time was often two days. I thought I would never be able to finish my calculations and get the final results before I had to return to Wisconsin in September. Luckily, however, as summer approached, nearly all the UCL physicists went off "on holiday" or on working visits to the States, and the pressure for computer time decreased markedly. I completed my calculations and published the resulting cross sections and excitation rates (Osterbrock 1970a,b), which were used for several years by all workers in the field—although by now greatly improved values, calculated with much more powerful computers, have completely superseded them.

During our year in London I got to know several of Seaton's students well, especially David Flower (now at Durham University) and Malcolm Brocklehurst (who went on to an alternative career, applying his computer skills to programming the complicated British Rail train schedules). I attended the monthly Royal Astronomical Society meetings in London, where I became acquainted with many British astronomers, and I gave colloquia at Cambridge, Oxford, the Royal Greenwich Observatory (then at Herstmonceux), Leeds, and Manchester. I also took part in several atomic physics meetings in Britain, and went to France and the Netherlands to give talks on my work as well. Thus I made many friends, especially in Britain, and have followed and used their work for years. Soon after I returned to America, Nüssbaumer and I published a joint paper, written in correspondence. It showed that even such high-ionization emission lines as [Fe X], [Fe XI], and [Fe XIV], observed in some Seyfert galaxies, could be understood in terms of photoionization, contrary to the general opinion of that time. This paper was another result of my year at UCL (Nüssbaumer & Osterbrock 1971).

LICK OBSERVATORY

In 1973, I left the University of Wisconsin to be the director of Lick Observatory. It was a hard decision for me; the University of California astronomers had been after me (and probably one or two other potential candidates) with varying degrees of intensity for nearly five years, and Chancellor Dean E. McHenry, the head of the Santa Cruz campus where the Lick staff had moved in 1966, met me in

London on one occasion to look me over and to tell me he would get me some day!

I loved Wisconsin and Madison, as my wife and our three children did; we still consider it our real home, though Santa Cruz is nice too, especially in winter and spring. Art Code, John Mathis, Bob Bless, and I had been together for many years; they were (and are) among my closest friends. I hated to leave them and the University of Wisconsin, but the lure of the big telescope in the land of clear skies won out. I believed I had done most of what I could on nebulae with our 36-inch telescope at Pine Bluff Observatory. Lick's 120-inch reflector was then the second largest operating telescope in the world, superbly equipped with a digital image-dissector scanner unrivaled anywhere. A contributing, though much less important, factor in my decision was the turbulent atmosphere at UW in those years of the "student revolt," which culminated in the bombing of Sterling Hall, where our department was located, in September 1970. Bomb threats, strikes, coarse verbal attacks on university leaders, battles with the police, and fires had preceded it, and a physics postdoc I knew was killed in the bombing itself. After that, although support for the violence that always accompanied the protests fell off rapidly, I felt that the university administration and the Madison city government did not pursue, apprehend, and prosecute the bombers (who had struck and fled under cover of darkness) nearly vigorously enough. I still feel the same, a quarter of a century later, but not as intensely as I did when I finally decided to leave Madison. By the time I accepted the Lick post, in the spring of 1972, it was too late for me to abandon my students and other responsibilities at UW at the end of that academic year. Nevertheless, McHenry had me appointed director on July 1, but put me on leave (with the title but no salary!) until the summer of 1973. However, in the early fall of 1972, I flew out to Lick to inspect my new empire and to observe for parts of a few nights with Miller, my former PhD student at Wisconsin, and Joe Wampler, a Yerkes PhD about ten years after me and the builder (with Lloyd Robinson) of the image-dissector scanner. By then I was eager to do research on quasars and their "little siblings," radio galaxies. I felt that by applying the nebular diagnostics and methods I knew well to the emission-line spectra of radio galaxies, several of which are quite close, fairly bright, and reasonably well resolved, we could begin to understand their physical nature, and perhaps by extension, some aspects of quasars as well. Hence, on this first visit I concentrated on Cygnus A, the "colliding galaxy" that had been the breakthrough identification of a radio source made by Baade & Minkowski (1953) twenty years earlier. When I returned to Lick to stay the next year, I continued work on Cygnus A with Miller. We found it had a rich emission-line spectrum, covering a wide range of ionization, which could be understood as resulting from photoionization of gas clouds containing "normal" abundances of the elements by a "hard" input spectrum (Osterbrock & Miller 1975).

Lick Observatory was an interesting place and organization, even older than Yerkes. As part of the University of California, it had been almost strictly a research institution until 1966. All the astronomers had lived in a little astronomy village surrounding the observatory where they worked on Mount Hamilton, by 1973 less than an hour's drive from San Jose. They had no teaching duties except advising graduate students from Berkeley who were working on observational theses with the telescopes. But in the 1960s the university administration decided there should be no more campuses without classes or students, so when the new University of California, Santa Cruz campus opened, the Lick faculty, offices, library, computers, and everything but the telescopes, their operators, and others who maintained them moved there. Some, who had loved the more isolated life at Mount Hamilton, soon left; others were glad to be on a campus with students, physicists, other natural and social scientists, humanities professors, cultural events, and all kinds of activities.

UCSC was McHenry's vision of the Cambridge and Oxford college system transferred to an American state university. It was planned as a small campus, originally for undergraduate students only, but he quickly discovered that without grad-student TAs it would be impossible to meet the budget. All faculty members were supposed to spend half their time and effort teaching general courses (outside their specialty, for instance a general survey of civilization and science) in a college, and the other half, not in a department—for McHenry and the other advanced educational thinkers of his day were opposed to departments—but in a board of studies (for instance, in astronomy and astrophysics) which within a few years looked like a department, acted as a department, and quacked like a department, but was not called a department. Each college had a theme, like Cowell (an education with your friends) or Crown (natural sciences), to which most of the astronomers belonged. Others were more bizarre. Promotions and salary increases had to be recommended by both the board and the college. The Lick observational astronomers were exempt from having to teach in a college, but the theorists were not. This led to many tensions. I remember a scheduled joint meeting of the Lick and Kresge College (known informally as the "touchee-feelee" college) senior faculty, for which we were told we would have to take off our shoes before entering their conference room. It was famous for its thick rug which extended up all the walls! I declined to attend under these conditions, and the meeting never occurred, but we managed to hire the theorist we wanted, Stan Woosley, anyway.

The problem with McHenry's dream was that most professors, for example in the University of California, want to do research, and their primary loyalty is to their discipline, not to a single university or campus, as I had demonstrated by coming to Santa Cruz. If some faculty members wanted to stay at UCSC for the rest of their days, and the system there never changed, it would have worked out for them; but after five years or so of spending half their time teaching general science to freshmen and sophomores, or winemaking or Beethoven to advanced undergraduates, as some of our Lick faculty did once or twice though not as a way of life (all the specialized college courses were lumped as "basket-weaving" by some skeptics), no other university would ever want them.

However, I got along fine with McHenry; he was glad to have Lick and its instant recognition, several National Academy of Science members, a big research budget (but never big enough for its director), and an excellent astronomical library, which had become the nucleus of the UCSC science library. He recognized that all I wanted was to make Lick the best observatory in the world and to do research myself. All he asked me to do was attend his monthly administrative meetings (at 8:00 AM, not a good time for me), and he supported us as well as he could, within his budget constraints.

The main problem with UCSC as far as I was concerned was its instability in the upper echelons. When McHenry was sweet-talking me about all we could do together if only I would come to Lick, he neglected to mention that he intended to retire very soon, and he did so at the end of my first year there. He was nearly fifteen years older than me, and the student revolt, though muted at UCSC, and much later in arriving there than at Madison or Berkeley, had worn him down too. In my eight years as director, I worked for four different chancellors, beginning with McHenry, and reported through seven different vice chancellors (the early term) or deans (later) of natural sciences. Nobody wanted this job, but some scientist would be persuaded to take it, and then, when he found how really bad it was, would resign. I would just finish taking one new dean or chancellor to Mount Hamilton, to show him around and point out our needs, and he would leave the job.

However, I am glad to say the last chancellor I worked under, Bob Sinsheimer, a pioneering DNA biologist and former chairman of the division of biological sciences at Caltech, was excellent. He became chancellor in 1977 and lasted ten years, retiring at the mandatory age for top administrators in the University of California and going back to a biology lab at UC Santa Barbara. Sinsheimer turned Santa Cruz into a more conventional university campus, organized around departments, with the colleges surviving only as dormitories, eating halls, and centers for the students' extracurricular activities. They are excellent for providing an identity for younger students, and a sense of belonging to something smaller and more personal than an 11,000-student campus. Our current chancellor, M. R. C. Greenwood, a former food scientist and NSF administrator, has been very good, and our present dean of natural sciences, Dave Kliger, a chemist who does research and publishes papers, has remained in the post nine years and is excellent.

When I returned to Santa Cruz from Madison with my family in the fall of 1973, I acquired two graduate-student assistants—Alan Koski, who had completed his required course work, and Mark Phillips, a beginning student. We did a lot of spectroscopic observing of radio galaxies, which then had been little studied in the optical region. A year later, Rafael Costero, whom I had taught at UW, completed his master's degree there and came to Lick to work with me for a year before returning to Mexico. We found that radio galaxies' optical spectra were similar in many ways to those of Seyfert galaxies, but statistically there were differences between the two groups. We helped to confirm and strengthen the concept of a broad-line region (small, containing relatively high-density gas with a high internal-velocity distribution), surrounded by a much larger, lower-density, lower internal-velocity, narrow-line region (Osterbrock et al 1975, 1976; Osterbrock & Costero 1977).

I soon realized that the published quantitative astrophysical data on Seyfert galaxies was quite sparse, so we began observing them in earnest too, and measuring their physical properties. We found it was possible to regard any Seyfert-galaxy spectrum as a composite of a Seyfert 1 (broad permitted lines and narrow forbidden lines) and Seyfert 2 (relatively narrow emission lines only) (Osterbrock & Koski 1976, Osterbrock 1977). We measured the electron temperature in NGC 1052, one of the rare elliptical galaxies with strong emission lines which I had observed at Palomar years earlier, and found that it was, though somewhat elevated, consistent with heating primarily by photoionization as in Seyfert galaxies (Koski & Osterbrock 1976). Koski (1978) and Phillips (1977; 1978a,b) both wrote very good theses as part of this program, on Seyfert 2 galaxies and Seyfert 1 galaxies with strong Fe II emission, respectively.

Joel Tohline, who wanted to become a theorist but also wanted to see where observational data came from, worked with me one year. While reducing our spectra, he found a huge variation of the broad Hα emission line in the spectrum of NGC 7603, which had been a Seyfert 1 galaxy but became almost a Seyfert 2, and then came back again (Tohline & Osterbrock 1976). I had been skeptical of earlier reported variations in the spectra of other Seyferts, but no one could doubt this one. To me it clearly established that strong extinction by dust, either of the radiation from the ionizing source or, more likely, of the broad-line region (as seen from our direction), had occurred.

From the diagnostic spectral line-intensity ratios, it is straightforward to determine the mean density in each region, and from them and the dimensions, the filling factors. Others had done this previously, but we had more good data on more galaxies than any other group. I summarized our results in a symposium on active galactic nuclei in 1977 in Copenhagen at which Strömgren, who had returned to Denmark, presided (Osterbrock 1978a). After that summer, I went on leave for two quarters at the University of Minnesota, where as a visiting professor I could work nearly full-time on research. My good friend Ed Ney, the head of its astrophysics group, had invited me to come. Considering all the observational data, the pioneering paper by Woltjer (1959), and a recent specific model by Shields (1977), I realized that the small, dense broad-line regions, with their high internal-velocity fields, must be rapidly rotating and hence flattened, so that ionizing radiation from the central source could escape more readily along its axis than in the "equatorial plane." Based on these ideas, I sketched an observational model showing ionization cones in the surrounding more extended, lower-density, narrow-line region (Osterbrock 1978b). The next summer at another symposium in Copenhagen, this one in honor of Strömgren on his seventieth birthday, I refined and repeated these ideas, and included the concept of strong density fluctuations in both the broad- and narrow-line regions, as demanded by the observed filling factors. Each condensation or "cloud" is more highly ionized on the side facing the central radiation source, and may be neutral or nearly neutral on the opposite side, if sufficiently optically thick to the ionizing radiation (Osterbrock 1978c).

Much of my subsequent research, and that of the postdocs, graduate students, and others who worked with me, has been devoted to further extending, testing, and specifying more quantitatively this basic model. A few years later, Tohline, by then a theorist at Los Alamos, and I showed on observational and theoretical grounds that the axis of rotation of the central broad-line region need not at all be the same as that of the galaxy to which it belongs (Tohline & Osterbrock 1982). Today, with our knowledge of galactic mergers and interactions, that statement seems commonplace, but it was not widely accepted at the time. Since then, however, many high-angular-resolution radio observations have shown that the jets in radio galaxies, and at a fainter level in many Seyfert galaxies, which must be along the axis of the central black hole, can be highly inclined to their galactic "planes." I continued to emphasize and re-emphasize the importance of dust in all active galactic nuclei, as shown by the optical observations themselves, and later by better and better infrared measurements by other observers (Osterbrock 1979).

Weedman, my former student, along with Ed Khachikian had introduced the classification of Seyfert-galaxy spectra into types 1 and 2; Koski and I had further subdivided them by adding 1.2, 1.5, and 1.8, in order of decreasing relative strength of the broad components of the permitted lines (Khachikian & Weedman 1974, Osterbrock & Koski 1976). Our instrumental power was increasing, thanks to Robinson, who was continuously upgrading our digital data-taking and data-reduction capabilities, and to Miller, who pushed through to completion a new, much more stable and powerful, remotely controlled spectrometer. With it I was able to detect very faint wings on the Hα, and in some cases Hβ, narrow emission lines of "nearly Seyfert 2" objects (Osterbrock 1981). We called these Seyfert 1.8 or 1.9 (according to whether Hβ could or could not be seen to have wings), and in colloquia and talks I liked to joke that there were surely also Seyfert 1.95, 1.99, and $1.99 + \varepsilon$ objects if we could only look hard enough. All this indicated that dust was blocking our view of the central ionization source and the broad-line region. Ross Cohen (1983), in his thesis with me, measured quantitatively the narrow-line spectra of many of these "intermediate" Seyfert 1 galaxies, and showed that they were practically the same as those of Seyfert 2 galaxies, a natural prediction of this picture.

This was confirmed and greatly extended by Robert Antonucci & Miller (1985), who were able to observe the faint polarized flux from NGC 1068, a bright, nearby Seyfert 2 galaxy, which showed a weak Seyfert 1 spectrum in the polarized light. The plane of polarization was perpendicular to the weak radio jets in NGC 1068, which are inclined to the plane of the galaxy, indicating (as these authors stated) that the concepts of a tipped nucleus and of high obscuration by dust in its "equatorial plane" were quite correct for this object. They put forward the "hidden broad-line region" model: that most or all Seyfert 2 galaxies are identical with Seyfert 1's seen from a direction in which the nucleus cannot be directly observed. Much further work by Miller, Antonucci, Bob Goodrich, Laura Kay, and several other observers, including Hien Tran and André Martel (both of whom I was fortunate to have as assistants before they went on to do their theses with Miller on spectropolarimetry), has confirmed this picture for several more Seyfert 2's.

Rick Pogge (1988, 1989), in an outstanding thesis with me, obtained direct CCD images with interference filters chosen to isolate strong emission lines, which showed the ionization cones themselves in NGC 1068 and other Seyfert 2 galaxies. This further confirmed the observational model for most, but not necessarily all, Seyfert galaxies, as some of the Seyfert 2's observed did not show ionization cones. Postdoc Dick Shaw and I, and later Martel and I, tried to define the mean opening angle of the ionization cones statistically, by classifying all the Seyfert galaxies in well-defined complete samples of a limited, thoroughly searched region (Osterbrock & Shaw 1988) and in the brighter CfA sample covering the entire sky (Osterbrock & Martel 1993).

My last PhD thesis student, Sylvain Veilleux, was also outstanding. Before he began his thesis, he and I worked up a quantitative spectral-classification scheme for distinguishing Seyfert 2 galaxies from starburst and H II region galaxies, objects in which the primary photoionization source is radiation from hot stars. Others had earlier proposed similar methods, but ours was superior, we believed, in using only intensity ratios of emission lines close to one another in wavelength, thus minimizing the effects of reddening by dust and of errors in the calibration into energy units (Veilleux & Osterbrock 1987). In his thesis, Veilleux used line profiles and long-slit spectroscopy to draw up what was undoubtedly the best model of the velocity field in a Seyfert galaxy up to that time. It combined a central broad-line region and the surrounding narrow-line region, blending into interstellar matter farther out, with a gradual tilt (or a "warp") between the center, where the flow is outward along the axis, and the outer parts, where the flow is rotational with probably a slight inward component (Veilleux 1991a,b,c).

I was fortunate also in having three other postdocs work with me at Lick. The first, Steve Grandi, worked on the optical spectra of broad-line radio galaxies (Grandi & Osterbrock 1978) and on the highest-ionization Seyfert 1 galaxies (Grandi 1978). Later, Jim Shuder worked on the broad-line profiles in Seyfert 1 galaxies and QSOs. The results, necessarily indirect because none of these objects can be resolved, implied a mixture of rotational and radial (both inward and outward, fairly closely balanced) internal velocities, with the highest densities and the highest velocities concentrated to the center (Shuder & Osterbrock 1981, Shuder 1982, Osterbrock & Shuder 1982). Still later, Michael De Robertis worked on the narrow-line profiles in Seyfert 2 galaxies. They showed that in a general way the ionization and density both decreased outward, although with large overlaps (indicating strong density fluctuations) in both quantities (De Robertis & Osterbrock 1984, 1986). All my postdocs at Wisconsin and Lick also worked on related projects of their own, which I was glad to discuss with them, thus broadening my own knowledge. Unfortunately, there is little room to describe these projects here.

The main aim of active galactic nuclei (AGNs) research, after describing the AGNs physically, is to understand how they work and how they evolve. I quickly became convinced that the central radiation source is powered by a black hole, surrounded by what is basically an accretion disk, but probably much more complicated than any of the disk models yet worked out by theorists, on which potential, kinetic, and magnetic energy are converted to heat and radiation, along the lines first suggested by Donald Lynden-Bell and Martin Rees. Some of my contemporaries were skeptical of this interpretation for years—largely, as far as I could see, because it involves the general theory of relativity. To me it seemed obvious that none of the other suggested models could work, because we had known back when I was a student, years earlier, that gaseous "stars" and stellar objects with masses higher than $\sim 10^2$ M$_o$ are highly unstable.

To work, the "accretion disk" must be continuously refueled by mass arriving at the center, for times of order 10^7 or 10^8 yr, which requires that it have almost zero angular momentum on a galactic scale. Two Wisconsin students I taught (though they did not do their theses with me), Tom Adams and Sue Simkin, were pioneers in suggesting the accepted mechanism of today. Adams (1977), from early direct image-tube photographs he obtained at McDonald Observatory, noted that many Seyfert galaxies are highly distorted or have nearby companions, differentiating them from most other spiral galaxies. Simkin et al (1980) then postulated that the resulting non-circularly symmetric gravitational fields could perturb stars, and especially interstellar gas, into orbits—some of which could, after further interactions within the galaxy, deliver mass with nearly zero angular momentum at the center. Oved Dahari was the only thesis student I had who worked on this problem. He proposed, planned, and carried out an ambitious observational program to test statistically whether or not AGNs occur preferentially in galaxies that have "companions." He found that indeed they do, thus confirming this mechanism (Dahari 1984, 1985). This contradicted the conclusion of an earlier study, and Dahari's results were attacked, largely along the lines that his control sample (over which the Seyfert galaxies showed a significant excess of nearby "companions") was not sufficiently similar to the Seyferts, except for not having an active nucleus. I believe the critics were incorrect; there are very few spiral galaxies that have the same morphological type as most Seyferts (early-type spirals) which are not themselves Seyfert galaxies! Galaxies "just like" Seyfert galaxies, but which do not have broad emission lines covering a wide range in ionization, are relatively rare.

After nearly a decade of AGN research, I thought the time had come for me to revise and update my little blue book on nebulae, and to add a few chapters on the extensions to the higher-energy photons, which are relevant to Seyfert and radio galaxies. I began this task while on leave for one quarter as a research professor at Ohio State University, where Capriotti, my former student, was chair of the astronomy department. Once a week, I worked at my alternate visitor's office at Perkins Observatory, near Delaware, Ohio, just north of Columbus, because it had an excellent library. The 69-in Perkins telescope had been the first "big" reflector I had ever seen. My father took me there one cloudy, miserable Saturday, and I had been enthralled, in spite of the water dripping from it as fog condensed in the dome. By 1980, Armbruster, who had been my editor at W. H. Freeman for *Astrophysics of Gaseous Nebulae* (commonly abbreviated *AGN*), had his own company, University Science Books. Miller and I were his astronomy editors, and he published my new book, *Astrophysics of Gaseous Nebulae and Active Galactic Nuclei*, $(AGN)^2$, in 1984. Like the earlier book, it has been very widely used in graduate courses and by research workers in the field.

LICK DIRECTORSHIP

When I accepted its directorship, my aim was to make Lick Observatory an even more outstanding research institution than it already was, and to increase its role in training some of the best research workers of the next generation. I thought I should devote all my energies to these goals, and that the best way I could be a scientific leader was by example. In Madison I had become the letters editor of the *Astrophysical Journal* in 1971, when Chandra stepped down as managing editor after twenty years, and Helmut Abt succeeded him in editing the main journal. I liked being letters editor; it meant that I saw a lot of hot, new research results and dealt with many interesting scientists. But I resigned the editorship before I left for Lick, to devote my full time to its affairs.

At Santa Cruz, I had a separate science office where I could isolate myself for half of each day from administrative tasks, conferences, phone calls, and all the trivia that could easily fill whatever time I would allow. I instructed my secretary that I would not take calls in the morning from anyone below the chancellor or the president of the university, so I could work on my research and talk with my postdoc and graduate students. I attended colloquia regularly and made it clear that I expected all the other faculty members to do the same. I kept faculty meetings to a minimum, with well-defined agendas distributed in advance.

I was helped immensely by Joe Calmes, the assistant to the director who I was fortunate to inherit. He was fairly new on the job, and had previously been a theology student and an affirmative-action coordinator. Naturally I distrusted him at first! But he was extremely intelligent, loyal, and trustworthy. Joe soon learned the arcane University of California accounting rules and procedures backward and forward, and became a master at shifting allotted funds from what I considered unnecessary diversions to the research needs I thought essential. Joe was very sensitive in interpreting me to the Lick staff members and interpreting them to me.

I observed regularly, always with our telescope at Mount Hamilton, so I had a direct stake in making sure that everything was working at top efficiency. If I did not observe during a month, I tried to get up to the mountain for one day with Calmes, to see the daytime workers and learn about their problems. John Bumgarner, the mountain superintendent, was always glad to see me and was quite effective. After he retired, Don Reiterman, his successor, was a bit abrasive at first, but gradually learned to fit in and get the essential jobs done. Then Don left and we promoted Ron Laub, from within the mountain staff, to superintendent. He was excellent, and remained in the job for many years before moving on to the Keck Observatory in Hawaii.

A lot of our success with the telescope and instruments was due to the excellent Lick shops in Santa Cruz. Frank Melsheimer was in charge of the engineering and mechanical groups when I came; a Berkeley-trained engineer, he was outstanding in design and in updating the telescope into the digital age. But Frank was young and single-minded, resenting criticism of his ideas—especially by astronomers who, he thought, did not understand them. I hated to see him go, but I could not prevent it; some of my senior colleagues did not want him to stay, and I was too new myself to be able to change their minds. Frank's departure worked out fine for him, for he and his brother have made a great success of their DFM Engineering company, specializing in telescopes and advanced control systems. We did not fare so well; though our engineers were very good indeed, none of them wanted to head the group for long. Nevertheless, we got along, but at the cost of considerable time and effort by Dave Rank and Miller. Our machine shop, headed by Neal Jern during most of my time as director, and our electronics group, headed by Robinson, were very good, and Lick was well known for its excellent instruments and data-taking systems, which were highly advanced but reliable.

Although our Lick faculty in Santa Cruz was responsible for maintaining and operating the observatory, astronomers from other campuses—Berkeley, Los Angeles, and San Diego—could use it on an equal basis with us. This had caused some friction earlier, when it was first put into place, but most of the malcontents were gone by 1973, and sharing the telescope seemed quite natural to me. My friends Geoff Burbidge at UCSD and Abell at UCLA, as well as Charlie Townes at Berkeley, whom I did not know well then but greatly respected, had urged me to accept the Lick directorship, and I could work easily with them. Townes, Ford, Margaret Burbidge, Dan Popper, and Hy Spinrad, all from the other campuses, were among the best and most productive observers we had.

In 1977, Lick received a lucky, entirely unexpected bequest of $52,000 from the estate of Anna L. Nickel, a previously unknown San Francisco admirer of the observatory. Her gift was to be spent at the director's discretion for whatever we most needed. We had just acquired a 40-inch mirror blank, ordered before I came, which was intended to upgrade the old Crossley reflector. Rank and George Herbig convinced me instead to have a new telescope, dedicated to research, built around it with the Nickel funds. Howard Cowan ground and polished the disk into a paraboloid in our optical shop, and made the Cassegrain secondary for it. All the mechanical and electronic work was done in our shops, and we put the completed Nickel telescope in the dome previously used for the 12-inch refractor, which dated back to the 1880s. The 12-inch was an excellent visual telescope, praised by double-star observers from S. W. Burnham and Robert G. Aitken to Eggen, and I hated to take it out of action, but it was mothballed. I still hope that some day a generous donor will provide funds to erect it on the Santa Cruz campus for student use. The Nickel telescope, spearheaded by Rank, Robinson, and Jack Osborne of our engineering group, has a very good control system and auxiliary instrumentation; it has been extremely useful, especially for grad students' theses from throughout the UC system.

Even before I came, the Lick astronomers were hoping to raise much larger amounts of money for a "150-inch class" telescope, to be located at a better dark-sky site than Mount Hamilton, now light-polluted by San Jose and its suburbia. The rationale was that so much good research was being done by UC astronomers that we needed a second large telescope; there was no thought of closing Lick Observatory. The Lick group favored putting the proposed telescope on Junipero Serra Peak, in the Santa Lucia mountains of the Coast Range, well south of Monterey in a relatively dark area of California. Astronomers at the other campuses were not very favorable to that site; some members of the Berkeley group preferred White Mountain, and those at UCLA and UCSD, a Southern California peak. Nonetheless, we pushed ahead with plans for Junipero Serra, hoping the money could somehow be raised. Merle Walker played a key role in this planned project, but we never got the funds. It took me a long time to realize that no millionaire, no foundation, nor any group of politicians wants to provide money for "just another" moderately large telescope. George Ellery Hale got the money for four successive largest telescopes in the world, but Lick only once, in a rare combination of circumstances, succeeded in financing a "second-best" telescope, the 120-inch (3-meter) reflector, just after World War II.

Ultimately, Wampler and Rank came up with a plan for a radically new type of telescope, a 10-meter, thin-mirror "monolith." Jerry Nelson, then at the UC Lawrence Berkeley Laboratory, proposed an even more advanced 10-meter segmented telescope, its primary mirror made up of thirty-six 1.8-meter hexagons, their positions and tilts to be continually read out and corrected by a closed-loop servo system. There were proponents of each plan in the UC system, and debates raged about which design we should adopt and work toward realizing. We had several studies, and in the end I appointed a "Graybeards' Committee," composed of two senior astronomers from each of the four UC campuses with sizeable observational groups, to decide which plan to follow in attempting to raise funds to build what we called the "Ten-Meter Telescope." Wampler, Nelson, and their groups had already made several preliminary studies, which they presented to the Graybeards in written form and orally. By a narrow margin, Nelson's segmented mirror won out, five to three.

Not long after that, in fall 1981, I stepped down as director. I felt that after eight years (the median of the terms of the eight Lick directors before me) I had worked my passage, and could honorably continue as an active faculty member. It was the job I probably would have accepted without question years earlier, after the Lick faculty moved to Santa Cruz. By 1981 I was tired of the administrative burden, and no longer had the full support of my colleagues (as had inevitably happened to all my predecessors as well). Throughout my scientific career I enjoyed doing the best research I could with the means at my disposal. I did not want to spend any more time in planning a big telescope and heading the group that would build it, nor in raising the funds to do so; I had no experience nor interest in those branches of astronomy, although I realized that someone had to do them. Bob Kraft, who had been acting director for four years of the five just before I had come to Lick, had not wanted the job on a long-term basis back then. He had filled in as acting director each time I went on leave, and I felt that now he would be willing to accept the directorship, as indeed he did.

As I had finally foreseen, the idea of the biggest telescope in the world, based on a radically new concept, was much more exciting and interesting than another 3- or 4-meter telescope on a mountain in central California. Chancellor Sinsheimer, UC Presidents (successively) David Saxon and David Gardner, and many other UC administrators were far more enthusiastic about it, and willing to work hard to get it, than they had ever been for our Junipero Serra plan. Kraft and everyone else tried hard, but we never got the money to build it for the University of California alone. In the end we combined with Caltech, which we all agreed was the best partner we could possibly have; the 10-meter William M. Keck is now a reality on Mauna Kea, Hawaii, and is certainly the best ground-based telescope in the world. It has produced results to match those words, and has even spun off a twin, Keck II, also now in active operation. Kraft and Miller, his successor as director of what is now University of California Observatories/Lick Observatory, played very important roles in pushing them through.

After 1981, I continued active research at Lick, as sketched in the previous section. I visited the Institute for Advanced Study twice more, 1982–83 and 1989–90. John Bahcall had become its professor of astrophysics, and was even better than Strömgren, for he chose to have his office in the same building with all the astrophysics postdocs and visiting fellows, rather than in Fuld Hall with the other professors. Thus John was in immediate contact with what everyone was doing. Strömgren, though wonderful in discussions, had been a little remote from our much smaller group in 1960–61. In 1982–83, I worked on several papers on AGNs and completed my book on James E. Keeler, the pioneer American astrophysicist who was the second director of Lick Observatory. In my later years there, I worked on a long review article on AGNs and wrote most of my book on George Willis Ritchey and George Ellery Hale, the "pauper and the prince" of the early big-telescope era in American astronomy. One of the key ideas I emphasized in the review was the very large number of "little AGNs," LINERs and galactic nuclei with barely detectable activity, probably all with black holes that have not been refueled recently (Osterbrock 1991).

After I retired from the active faculty on December 31, 1992, I stopped observing entirely. I felt I had done most of what I could do at the telescope. I had had my full share of observing nights, if not more, and it was time to step aside for the younger astronomers who needed to get their data and test their theories. I still had about two years of spectra to measure, analyze, and discuss, but since publishing my last research paper on AGNs (Osterbrock & Fulbright 1996) and my last review article (Osterbrock 1997), I have confined myself to night-sky spectroscopy, using Keck data taken by others, and to the history of astronomy, my two current interests. They are stories for another time and place.

I had much good fortune in my scientific career, but the best was always working with outstanding colleagues, postdocs, grad students, and staff members. My education was almost entirely at government expense, for which I have always been grateful. All the universities at which I worked had outstanding astronomy libraries and were generous in providing physical facilities and financial support for my research, as were the National Science Foundation and the other foundations I have mentioned.

Visit the Annual Reviews home page at www.AnnualReviews.org

LITERATURE CITED

Abell GO. 1958. *Ap. J. Suppl.* 3:211

Adams TF. 1977. *Ap. J. Suppl.* 33:19

Antonucci RRJ, Miller JS. 1985. *Ap. J.* 297:621

Baade W, Minkowski R. 1953. *Ap. J.* 119:206

Capriotti ER. 1964. *Ap. J.* 139:225

Capriotti ER, Daub CT. 1960. *Ap. J.* 132:677

Cohen RD. 1983. *Ap. J.* 273:489

Dahari O. 1984. *Astron. J.* 89:966

Dahari O. 1985. *Ap. J. Suppl.* 57:643

De Robertis MM, Osterbrock DE. 1984. *Ap. J.* 286:171

De Robertis MM, Osterbrock DE. 1986. *Ap. J.* 301:727

Dufour RJ. 1975. *Ap. J.* 195:315

Federer CA. 1952. *Sky Telesc.* 11:138

Ford HC. 1970. PhD thesis. Univ. Wisc

Ford HC. 1971. *Bull. Am. Astron. Soc.* 3: 19

Gingerich O. 1994. *Phys. Today* 47:34

Grandi SA. 1978. *Ap. J.* 221:500

Grandi SA, Osterbrock DE. 1978. *Ap. J.* 220:783

Härm R, Schwarzschild M. 1955. *Ap. J. Suppl.* 1:319

Hoyle F, Schwarzschild M. 1955. *Ap. J. Suppl.* 2:1

Khachikian E, Weedman DW. 1974. *Ap. J.* 192:581

Koski AT. 1978. *Ap. J.* 223:56

Koski AT, Osterbrock DE. 1976. *Ap. J. Lett.* 203:L49

MacAlpine GM. 1972. *Ap. J.* 175:11

Mallik DCV. 1975. *Ap. J.* 197:355

Mathis JS. 1957a. *Ap. J.* 125:318

Mathis JS. 1957b. *Ap. J.* 125:328

Miller JS. 1968. *Ap. J.* 151:473

Minkowski R, Osterbrock DE. 1959. *Ap. J.* 129:523

Morgan WW. 1971. *Astron. J.* 76:1000

Morgan WW. 1988. *Annu. Rev. Astron. Astrophys.* 26:1

Morgan WW, Dreiser R. 1983. *Ap. J.* 269:438

Morgan WW, Osterbrock DE. 1969. *Ap. J.* 74:515

Morgan WW, Sharpless S, Osterbrock DE. 1952. *Ap. J.* 57:3

Naur P, Osterbrock DE. 1953. *Ap. J.* 117:306

Nüssbaumer H, Osterbrock DE. 1971. *Ap. J.* 161:811

O'Dell CR. 1962. *Ap. J.* 135:371

O'Dell CR, Collins GW, Daub CT. 1961. *Ap. J.* 133:471

Osterbrock DE. 1952. *Ap. J.* 116:164

Osterbrock DE. 1953. *Ap. J.* 118:529

Osterbrock DE. 1955. *Ap. J.* 122:235

Osterbrock DE. 1957. *Publ. Astron. Soc. Pac.* 69:227

Osterbrock DE. 1958a. *Publ. Astron. Soc. Pac.* 70:180

Osterbrock DE. 1958b. *Publ. Astron. Soc. Pac.* 70:399

Osterbrock DE. 1960. *Ap. J.* 132:325

Osterbrock DE. 1961. *Ap. J.* 134:347

Osterbrock DE. 1962. *Ap. J.* 135:196

Osterbrock DE. 1963. *Planet. Space Sci.* 11:621

Osterbrock DE. 1970a. *J. Phys. B* 3:149

Osterbrock DE. 1970b. *Ap. J.* 160:25

Osterbrock DE. 1971. *J. Quant. Spectrosc. Radiat. Transf.* 11:623

Osterbrock DE. 1977. *Ap. J.* 215:733

Osterbrock DE. 1978a. *Phys. Scr.* 17:285

Osterbrock DE. 1978b. *Proc. Natl. Acad. Sci. USA* 75:540

Osterbrock DE. 1978c. *Astronomical Papers Dedicated to Bengt Strömgren*, ed. A Reiz, T Anderson, p. 299. Copenhagen: Copenhagen Univ. Obs.

Osterbrock DE. 1979. *Astron. J.* 84:901

Osterbrock DE. 1981. *Ap. J.* 250:462

Osterbrock DE. 1991. *Rep. Prog. Phys.* 54:579

Osterbrock DE. 1997. *The Universe at Large: Key Problems in Astronomy and Cosmology*, ed. G Münch, A Mampaso, F Sanchez, p. 171. Cambridge, UK: Cambridge Univ. Press

Osterbrock DE, Capriotti ER, Bautz LP. 1963. *Ap. J.* 138:63

Osterbrock DE, Costero R. 1977. *Ap. J.* 211:675

Osterbrock DE, Flather E. 1959. *Ap. J.* 129:26

Osterbrock DE, Fulbright JP. 1996. *Publ. Astron. Soc. Pac.* 108:183

Osterbrock DE, Koski AT. 1976. *MNRAS* 176:61P

Osterbrock DE, Koski AT, Phillips MM. 1975. *Ap. J. Lett.* 197:L41

Osterbrock DE, Koski AT, Phillips MM. 1976. *Ap. J.* 206:898

Osterbrock DE, Martel A. 1993. *Ap. J.* 414:552

Osterbrock DE, Miller JS. 1975. *Ap. J.* 197:535

Osterbrock DE, O'Dell CR. 1962. *Ap. J.* 136:559

Osterbrock DE, Parker RAR. 1965. *Ap. J.* 141:892

Osterbrock DE, Parker RAR. 1966. *Ap. J.* 143:268

Osterbrock DE, Shaw RA. 1988. *Ap. J.* 327:89

Osterbrock DE, Shuder JM. 1982. *Ap. J. Suppl.* 49:149

Osterbrock DE, Stockhausen RE. 1960. *Ap. J.* 131:310

Osterbrock DE, Stockhausen RE. 1961. *Ap. J.* 133:2

Osterbrock DE, Stockhausen RE. 1965. *Ap. J.* 141:287

Pacholczyk AG, Weymann R. 1968. *Ap. J.* 73:836

Phillips MM. 1977. *Ap. J.* 215:746

Phillips MM. 1978a. *Ap. J. Suppl.* 38:18 7

Phillips MM. 1978b. *Ap. J.* 226:736

Pogge RW. 1988. *Ap. J.* 328:519

Pogge RW. 1989. *Ap. J.* 345:751

Sandage A, Schwarzschild M. 1952. *Ap. J.* 116:463

Seaton MJ. 1954. *Ann. Ap.* 17:74

Seaton MJ, Osterbrock DE. 1957. *Ap. J.* 125:66

Shields GA. 1977. *Ap. Lett.* 18:119

Shields GA. 1999. *Publ. Astron. Soc. Pac.* 111:661

Shuder JM. 1982. *Ap. J.* 259:48

Shuder JM, Osterbrock DE. 1981. *Ap. J.* 250:55

Simkin SM, Su HJ, Schwarz MP. 1980. *Ap. J.* 208:20

Strömgren B. 1948. *Ap. J.* 108:242

Tohline JE, Osterbrock DE. 1976. *Ap. J.* 210:117

Tohline JE, Osterbrock DE. 1982. *Ap. J. Lett.* 252:L49

Veilleux S. 1991a. *Ap. J. Suppl.* 75:357

Veilleux S. 1991b. *Ap. J. Suppl.* 75:383

Veilleux S. 1991c. *Ap. J.* 364:331

Veilleux S, Osterbrock D. 1987. *Ap. J. Suppl.* 63:295

Weedman DW. 1968. *Ap. J.* 153:49

Williams RE. 1967. *Ap. J.* 147:556

Woltjer L. 1959. *Ap. J.* 130:38

Annu. Rev. Astron. Astrophys. 2001. 38:35–77

STELLAR STRUCTURE AND EVOLUTION:
Deductions from Hipparcos

Yveline Lebreton

*Observatoire de Paris, DASGAL-UMR CNRS 8633, Place J. Janssen, 92195 Meudon,
France; e-mail: Yveline.Lebreton@obspm.fr*

Key Words   Stars: physical processes, distances, H-R diagram, ages, abundances,
Hipparcos

■ **Abstract**   During the last decade, the understanding of fine features of the structure and evolution of stars has become possible as a result of enormous progress made
in the acquisition of high-quality observational and experimental data, and of new
developments and refinements in the theoretical description of stellar plasmas. The
confrontation of high-quality observations with sophisticated stellar models has allowed many aspects of the theory to be validated, and several characteristics of stars
relevant to Galactic evolution and cosmology to be inferred. This paper is a review of
the results of recent studies undertaken in the context of the Hipparcos mission, taking
benefit of the high-quality astrometric data it has provided. Successes are discussed, as
well as the problems that have arisen and suggestions proposed to solve them. Future
observational and theoretical developments expected and required in the field are also
presented.

1.   INTRODUCTION

Stars are the main constituents of the observable Universe. The temperatures
and pressures deep in their interiors are out of reach for the observer, while the
description of stellar plasmas requires extensive knowledge in various domains of
modern physics such as nuclear and particle physics, atomic and molecular physics,
thermo- and hydrodynamics, physics of the radiation and of its interaction with
matter, and radiative transfer. The development of numerical codes to calculate
models of stellar structure and evolution began more than forty years ago with
the pioneering works of Schwarzschild (1958) and Henyey et al (1959). These
programs have allowed at least the qualitative study and understanding of numerous
physical processes that intervene during the various stages of stellar formation and
evolution.

During the last two decades, observational data of increasingly high accuracy
have been obtained as a result of 1) the coming of modern ground-based or space

telescopes equipped with high-quality instrumentation and with detectors giving access to almost any possible range of wavelengths and 2) the elaboration of various sophisticated techniques of data reduction. Ground-based astrometry has progressed, while space astrometry was initiated with Hipparcos. In the meantime, CCD detectors on large telescopes opened the era of high-resolution, high signal-to-noise ratio spectroscopy while multi-color filters were designed for photometry. New fields have appeared or are under development, such as helio- and asteroseismology or interferometry. On the other hand, stellar models have been enriched by a continuously improved physical description of the stellar plasma, while the use of increasingly powerful computers has led to a gain in numerical accuracy.

The confrontation of models with observations allows testing and even validation of the input physics of the models if numerous observations of high quality are available. Fundamental returns are expected in many domains that make use of quantitative results of the stellar evolution theory such as stellar, Galactic, and extragalactic astrophysics as well as cosmology. Because of their positions, movements, or interactions with the interstellar medium, stars are actors and tracers of the dynamical and chemical evolution of the Galaxy. Astrophysicists aim to determine their ages and chemical compositions precisely. For example, the firm determination of the ages of the oldest stars, halo stars or members of globular clusters, is a long-standing objective because it is one of the strongest constraints for cosmology.

Although great progress has been made, a number of observations cannot be reproduced by stellar models, which raises many questions regarding both the observations and the models. In the last few years, two scientific meetings have been explicitly devoted to unsolved problems in stellar structure and evolution (Noels et al 1995, Livio 2000). A major point of concern is that of transport processes at work in stellar interiors (transport of the chemical elements, angular momentum or magnetic fields by microscopic diffusion and/or macroscopic motions). Observations show that transport processes are indeed playing a role in stellar evolution but many aspects remain unclear (sometimes even unknown) and need to be better characterized. Another crucial point concerns the atmospheres, which link the stellar interior model to the interstellar medium and are the intermediate agent between the star and the observer. Uncertainties and inconsistencies in atmospheric descriptions generate errors in the analysis of observational data and in model predictions.

This paper is the third of the series in ARAA dedicated to the results of the Hipparcos mission; Kovalevsky (1998) presented the products of the mission and the very first astrophysical results obtained immediately after the release of the data, while Reid (1999) reviewed the implications of the Hipparcos parallaxes for the location of the main sequence (MS) in the Hertzsprung-Russell (H-R) diagram, the luminosity calibration of primary distance indicators, and the Galactic distance scale. Also, van Leeuwen (1997) presented the results of the mission, and Baglin (1999) and Lebreton (2000) discussed the impact of Hipparcos data on stellar structure and evolution.

Hipparcos has provided opportunities to study rather large and homogeneous samples of stars sharing similar properties for instance, in terms of their space location or chemical composition. I review studies based on Hipparcos observations which (1) confirmed several elements of stellar internal structure theory, (2) revealed some problems related to the development of stellar models, and (3) yielded more precise characteristics of individual stars and clusters. In Sections 2 and 3, I discuss the recent observational (including Hipparcos) and theoretical developments from which new studies could be undertaken. In Section 4, I concentrate on the nearest stars, observed with highest precision (A-K disk and halo single or binary field stars, and members of open clusters). In Section 5, I review recent results on variable stars, globular clusters and white dwarfs based on Hipparcos data. The stars considered are mostly of low or intermediate mass, and except for white dwarfs, the evolutionary stages cover the main sequence and subgiant branch. Throughout this paper, I emphasize that the smaller error bars on distances that result from Hipparcos make the uncertainties on the other fundamental stellar parameters more evident; fluxes, effective temperatures, abundances, gravities, masses, and radii have to be improved correspondingly, implying in many cases the need for progress in atmospheric description.

2. NEW HIGH-ACCURACY OBSERVATIONAL MATERIAL

This section presents a brief review of Hipparcos results and complementary ground-based or space observations which, if combined, provide very homogeneous and precise sets of data.

2.1 Space Astrometry with Hipparcos

The Hipparcos satellite designed by the European Space Agency was launched in 1989. The mission ended in 1993 and was followed by 3 years of data reduction. The contents of the Hipparcos Catalogue (Eur. Space Agency 1997) were described by Perryman et al (1997a). The data were released to the astrophysical community in June 1997. General information on the mission is given in van Leeuwen's (1997) and Kovalevsky's (1998) review papers.

Stars of various masses, chemical compositions and evolutionary stages located either in the Galactic disk or in the halo were observed; this was done systematically to a V-magnitude that depends on the galactic latitude and spectral type of the star, and more generally, with a limit of V $\sim$ 12.4 mag. The Hipparcos Catalogue lists positions, proper motions, and trigonometric parallaxes of 117 955 stars as well as the intermediate astrometric data, from which the astrometric solutions were derived; this allows alternative solutions for the astrometric parameters to be reconstructed according to different hypotheses (see van Leeuwen & Evans 1998).

A total of 12 195 double or multiple systems are resolved (among which about 25% were previously classified as single stars), and 8 542 additional stars are suspected to be non-single. Detailed information on multiple systems, as described by Lindegren et al (1997), can be found in The Double and Multiple System Annex of the Catalogue.

The median accuracy on positions and parallaxes (π) is typically $\sim$1 milliarcsecond (1 mas), whereas precisions on proper motions are about 1 mas per yr. Precisions become much higher for bright stars and worsen toward the ecliptic plane and for fainter stars. The astrometric accuracy and formal precision of Hipparcos data have been investigated by Arenou et al (1995) and Lindegren (1995), and discussed by van Leeuwen (1999a): for the Catalogue as a whole, the zero-point error on parallaxes is below 0.1 mas and the formal errors are not underestimated by more than 10%. After Hipparcos about 5 200 single stars and 450 double (or multiple) stars have parallaxes known with an accuracy σ_π/π better than 5%, 20 853 stars have σ_π/π lower than 10% and 49 333 stars have σ_π/π lower than 20% (Mignard 1997). Martin et al (1997, 1998) and Martin & Mignard (1998) determined the masses of 74 astrometric binaries with accuracies in the range 5–35%. Söderhjelm (1999) obtained masses and improved orbital elements for 205 visual binaries from a combination of Hipparcos astrometry and ground-based observations; among these, 12 (20) systems have mass-errors below 5 (7.5)%.

The Hipparcos Catalogue also includes detailed and homogeneous photometric data for each star, obtained from an average number of 110 observations per star. The broad-band Hipparcos (Hp) magnitude corresponding to the specific passband of the instrument spanning the wavelength interval $\sim$350–800 nanometers (see Figure 1 in van Leeuwen et al 1997) is provided with a median precision of 0.0015 mag for Hp < 9 mag. The Johnson V magnitude derived from combined satellite and ground-based observations is given with a typical accuracy of 0.01 mag. The star mapper Tycho had passbands close to the Johnson B and V bands and provided two-color B_T and V_T magnitudes (accuracies are typically 0.014 mag and 0.012 mag for stars with $V_T < 9$).

Hipparcos provided a detailed variability classification of stars (van Leeuwen et al 1997), resulting in 11 597 variable or possibly variable stars. Among these 2 712 stars are periodic variables (970 new) including 273 Cepheids, 186 RR Lyrae, 108 δ Scuti or SX Phoenicis stars, and 917 eclipsing binaries.

Hipparcos was planned more than fifteen years ago, and while its development proceeded, significant progress was made in the derivation of ground-based parallaxes using CCD detectors. Parallaxes with errors less than 1.4 mas have already been obtained for a few tens of stars, and errors are expected to drop to $\pm$0.5 mas in the years to come (Harris et al 1997, Gatewood et al 1998). In addition, the Hubble Space Telescope (*HST*) Fine Guidance Sensor observations can provide parallaxes down to V $\sim$ 15.8 with errors at the 1 mas level (Benedict et al 1994, Harrison et al 1999). However, the distances of a rather small number of stars will be measured by *HST* because of the limited observing time available for astrometry. The enormous advantage of Hipparcos resides in the large number of stars it dealt with, providing homogeneous trigonometric parallaxes that are essentially absolute.

2.2   Ground-Based Photometry and Spectroscopy

The fundamental stellar parameters (bolometric magnitude M_{bol}, effective temperature T_{eff}, surface gravity g, and chemical composition) can be determined from photometry and/or from detailed spectroscopic analysis. However, the determination largely relies on model atmospheres and sometimes uses results of interior models. Direct masses and radii can be obtained for stars belonging to binary or multiple systems. Interferometry combined with distances yields stellar diameters giving direct access to T_{eff}, but still for a very limited number of rather bright stars which then serve to calibrate other methods. The different methods (and related uncertainties) used to determine the fundamental stellar parameters mainly for A to K Galactic dwarfs and subgiants are briefly discussed, and improvements brought by Hipparcos are underlined.

Bolometric Magnitudes

Integration of UBVRIJHKL photometry gives access to the bolometric flux on Earth F_{bol}, at least for F-G-K stars where most energy is emitted in those bands and which are close enough not to be affected by interstellar absorption; the (small) residual flux, emitted outside the bands, is estimated from model atmospheres. Recently, Alonso et al (1995) applied the method to $\sim$100 F-K dwarfs and subdwarfs and obtained bolometric fluxes accurate to about 2% and, as a by-product, empirical bolometric corrections for MS stars.

If F_{bol} and distances are known, M_{bol} can be derived with no need for bolometric correction. The accuracy is then $\sigma_{M_{bol}} = \log e \ [(2.5 \frac{\sigma_{F_{bol}}}{F_{bol}})^2 + (5 \frac{\sigma_\pi}{\pi})^2]^{\frac{1}{2}}$, meaning that if $\frac{\sigma_{F_{bol}}}{F_{bol}} \sim 2\%$ then $\sigma_{M_{bol}}$ is dominated by the parallax error as soon as $\frac{\sigma_\pi}{\pi} > 1\%$. In other cases, when the distance is known, M_{bol} is obtained from any apparent magnitude m and its corresponding bolometric correction $BC(m)$, derived from empirical calibrations or from model atmospheres. Up to now Hipparcos magnitudes Hp have not been used extensively, despite their excellent accuracy (0.0015 mag), because of remaining difficulties in calculating BC(Hp) (Cayrel et al 1997a).

Effective Temperatures

The InfraRed Flux Method (IRFM; Blackwell et al 1990), applicable to A-K stars proceeds in two steps. First, the stellar angular diameter ϕ is evaluated by comparing the IR flux observed on Earth in a given band to the flux predicted by a model atmosphere calculated with the observed gravity and abundances and an approximate T_{eff} (the IR flux does not depend sensitively on T_{eff}). Then T_{eff} is obtained from the total (integrated) flux F_{bol} and ϕ. Iteration of the procedure yields a "definite" value of T_{eff}. Using IRFM, Alonso et al (1996a) derived temperatures of 475 F0-K5 stars (T_{eff} in the range 4000–8000 K) with internal accuracies of $\sim$1.5%. The zero-point of their T_{eff}-scale is based on direct interferometric measures by

Code et al (1976), and the resulting systematic uncertainty is ~1%. Accuracies of ~1% were obtained by Blackwell & Linas-Gray (1998), who applied IRFM to 420 A0-K3 stars, corrected for interstellar extinction using Hipparcos parallaxes. Both sets of results compare well, with differences below $0.12 \pm 1.25\%$ for the 93 stars in common.

The surface brightness method (Barnes et al 1978) was applied by Di Benedetto (1998) to obtain a (T_{eff}, V-K) calibration. The calibration is based on 327 stars with high-precision K-magnitudes from the Infrared Space Observatory (ISO), Hipparcos V-magnitudes and parallaxes (the latter to correct for interstellar extinction), and bolometric fluxes from Blackwell & Linas-Gray (1998). First, the visual surface brightness $S_V = V + 5 \log \phi$ is calibrated as a function of (V-K) using stars with precise ϕ from interferometry. Then for any star S_V is obtained from (V-K), ϕ from S_V and V, yielding in turn T_{eff} from F_{bol} and ϕ. From the resulting (T_{eff}, V-K) calibration, Di Benedetto derived T_{eff} values of 537 ISO A-K dwarfs and giants with $\pm 1\%$ accuracy. The method produces results in good agreement with those of IRFM and is less dependent on atmosphere models.

Multiparametric empirical calibrations of T_{eff} as a function of the color indices and eventually of metallicity [Fe/H] (logarithm of the number abundances of Fe to H relative to the solar value) and gravity can be derived from the empirical determinations of the effective temperatures of the rather nearby stars. In turn, the effective temperature of any star lying in the (rather narrow) region of the H-R diagram covered by a given calibration can easily be derived (see for example Alonso et al 1996b). Empirical calibrations also serve to validate purely theoretical calibrations based on model atmospheres; these latter have the advantage of covering the entire parameter space of the H-R diagram (i.e. wide ranges of color indices, metallicities and gravities; see Section 3.2 later in this article).

Spectroscopic determination of T_{eff} is based on the analysis of chosen spectral lines that are sensitive to temperature; for instance, the Balmer lines for stars with T_{eff} in the interval 5000–8000 K. Because of the present high quality of the stellar spectra, precisions of ± 50–80 K on T_{eff} that correspond to the adjustment of the theoretical line profile to the observed one are commonly found in the literature (Cayrel de Strobel et al 1997b, Fuhrmann 1998). This supposes that theoretical profiles are very accurate, and therefore neglects the model atmosphere uncertainties.

Popper (1998) used detached eclipsing binaries with rather good Hipparcos parallaxes, accurate radii, and measured V-flux to calibrate the radiative flux as a function of (B-V); he found good agreement with similar calibrations based on interferometric angular diameters. From the same data, Ribas et al (1998) derived effective temperatures (this required bolometric corrections) and found them to be in reasonable agreement (although systematically smaller by 2–3%) with T_{eff} derived from photometric calibrations. However, the stars are rather distant, which implies rather significant internal errors on M_{bol} and T_{eff} (a parallax error of 10% is alone responsible for a T_{eff}-error of 5%). In Ribas et al's sample, only a few systems have $\sigma_\pi / \pi < 10\%$, and because errors on radius, magnitudes, and BC also intervene, only 5 systems have T_{eff} determined to better than 3%.

Surface Gravities

If T_{eff} and M_{bol} are known, the radius of the star may be derived from the Stefan-Boltzmann law and the mass estimated from a grid of stellar evolutionary models, yielding in turn the value of g. This method has been applied to a hundred metal-poor subdwarfs and subgiants with accurate distances from Hipparcos (Nissen et al 1997, Fuhrmann 1998, Clementini et al 1999). Nissen et al showed that among the various sources of errors, the error on distance still dominates, but pointed out that if the distance error is lower than 20% then the error on log g may be lower than ± 0.20 dex.

On the other hand, g can be determined from spectroscopy. Different gravities produce different atmospheric pressures, modifying the profiles of some spectral lines. Two methods have been widely used to estimate g. The first method is based on the analysis of the equation of ionization equilibrium of abundant species, iron, for instance. The iron abundance is determined from FeI lines that are not sensitive to gravity, and then g is adjusted so that the analysis of FeII lines, which are sensitive to gravity, leads to the same value of the iron abundance. The accuracy in log g is in the range ± 0.1–0.2 dex (Axer et al 1994). The second method relies on the analysis of the wings of strong lines broadened by collisional damping, such as Ca I (Cayrel et al 1996) or the Mg Ib triplet (Fuhrmann et al 1997), leading to uncertainties smaller than 0.15 dex. The two methods often produce quite different results, with systematic differences of $\sim$0.2–0.4 dex, at least when ionization equilibria are estimated from models in local thermodynamical equilibrium (LTE).

Thévenin & Idiart (1999) have studied the effects of departures from LTE on the formation of FeI and FeII lines in stellar atmospheres, and found that modifications of the ionization equilibria resulted from the overionization of iron induced by significant UV fluxes. The nice consequence is that the gravities they inferred from iron ionization equilibrium for a sample of 136 stars spanning a large range of metallicities become very close to gravities derived either from pressure-broadened strong lines or through Hipparcos parallaxes.

Abundances of the Chemical Elements

The spectroscopic determination of abundances of chemical elements rests on the comparison of the outputs of model atmospheres (synthetic spectra, equivalent widths) with their counterpart in the observed spectra. This requires a preliminary estimate of T_{eff} and g. If high-resolution spectra are used, the line widths are very precise and the internal uncertainty in abundance determinations depends on uncertainties in g and T_{eff}, on the validity of the model atmosphere, and on the oscillator strengths. Error bars in the range ± 0.05–0.15 dex are typical (Cayrel de Strobel et al 1997b, Fuhrmann 1998). Also when different sets of [Fe/H] determinations are compared, the solar Fe/H ratio used as reference must be considered;

values differing by ∼0.15 dex used to be found in the literature (Axer et al 1994). This has resulted in long-standing difficulties in determining the solar iron abundance from FeI or FeII lines, because of uncertain atomic data. In a recent paper, Grevesse & Sauval (1999) reviewed the problem and opted for a "low" Fe-value, $A_{Fe} = 7.50 \pm 0.05$ ($A_{Fe} = \log(n_{Fe}/n_H) + 12$ is the logarithm of the number density ratio of Fe to H particles), in perfect agreement with the meteoritic value.

Furthermore, if abundances are estimated from model atmospheres in LTE, perturbations of statistical equilibrium by the radiation field are neglected. Thévenin & Idiart (1999) found that in metal-deficient dwarfs and subgiants, the iron overionization resulting from reinforced UV flux modifies the line widths. They obtained differential non-LTE/LTE abundance corrections increasing from 0.0 dex at [Fe/H] −0.0 to +0.3 dex at [Fe/H] = −3.0. These corrections are indeed supported by the agreement between spectroscopic gravities and "Hipparcos" gravities discussed previously.

Helium lines do not form in the photosphere of low-temperature stars which precludes a direct determination of helium abundance. The calibration of the solar model in luminosity and radius at solar age yields the initial helium content of the Sun (Christensen-Dalsgaard 1982), while oscillation frequencies give access to the present value in the convection zone (Kosovichev et al 1992). In other stars, it is common to use the well-known scaling relation $Y - Y_p = Z \frac{\Delta Y}{\Delta Z}$, which supposes that the helium abundance has grown with metallicity Z from the primordial value Y_p to its stellar birth value Y (Y and Z represent abundances in mass fraction); $\Delta Y / \Delta Z$ is the enrichment factor.

α-element abundances (O, Ne, Mg, Si, S, Ar, Ca, Ti) have now been widely measured in metal-deficient stars. Stars with [Fe/H] < ∼ −0.5 dex generally exhibit an α-element enhancement with respect to the Sun ([α/Fe]) quite independent of their metallicity (Wheeler et al 1989, Mc William 1997). Recent determinations of [α/Fe] in 99 dwarfs with [Fe/H] < −0.5 from high-resolution spectra by Clementini et al (1999) yield [α/Fe] = +0.26 ± 0.08 dex.

3. RECENT THEORETICAL AND NUMERICAL PROGRESS

Recent developments in the physical description of low and intermediate mass stars are briefly presented.

3.1 Microscopic Physics

The understanding of stellar structure benefited substantially from the complete reexamination of stellar opacities by two groups: the Opacity Project (OP, see Seaton et al 1994) and the OPAL group at Livermore (see Rogers & Iglesias 1992). Both showed by adopting different and independent approaches, that improved atomic physics lead to opacities generally higher than the previously almost "universally" used Los Alamos opacities (Huebner et al 1977). The opacity enhancements reach factors of 2-3 in stellar envelopes with temperatures in the range 10^5-10^6 K. With these new opacities, (1) a number of long-standing problems in stellar evolution have been solved or at least lessened and (2) finer tests of stellar structure could be undertaken. Since opacity is very sensitive to metallicity, any underlying uncertainty on metallicity may be problematic.

Great efforts have also been invested in the derivation of low-temperature opacities, including millions of molecular and atomic lines and grain absorption that are fundamental for the calculation of the envelopes and atmospheres of cool stars (Kurucz 1991, Alexander & Ferguson 1994).

OP and OPAL opacities have been shown to be in reasonable agreement (Seaton et al 1994, Iglesias & Rogers 1996); and very good agreement between OPAL and Alexander & Ferguson's or Kurucz's opacities is found in the domains where they overlap. Although some uncertainties remain that are difficult to quantify, the largest discrepancies between the various sets of tables do not exceed 20% and are generally well understood (Iglesias & Rogers), making opacities much more reliable today than they were ten years ago.

The re-calculation of opacities required appropriate equations of state (EOS). The MH&D EOS (see Mihalas et al 1988) is part of the OP project, while the OPAL EOS was developed at Livermore (Rogers et al 1996). In the meantime, another EOS was designed to interpret the first observations of very low-mass stars and brown dwarfs (Saumon & Chabrier 1991). OPAL and OP EOS are needed to satisfy the strong helioseismic constraints (Christensen-Dalsgaard & Däppen 1992).

3.2 Atmospheres

Atmospheres intervene at many levels in the analysis of observations (Section 2.2). They also provide external boundary conditions for the calculation of stellar structure and necessary relations to transform theoretical (M_{bol}, T_{eff}) H-R diagrams to color-magnitude (C-M) or color-color planes. Models have improved during the last two decades, and attention has been paid to the treatment of atomic and molecular line blanketing. The original programs MARCS of Gustafsson et al (1975) and ATLAS by Kurucz (1979) evolved toward the most recent ATLAS9 version, appropriate for O-K stars (Kurucz 1993) and NMARCS for A-M stars (see Brett 1995 and Bessell et al 1998). On the other hand, very low-mass stellar model atmospheres were developed; Carbon (1979) and Allard et al (1997) reviewed calculation details and remaining problems (such as incomplete opacity data, poor treatment of convection, neglect of non-LTE effects or assumption of plane-parallel geometry).

Color-Magnitude Transformations

Different sets of transformations (empirical or theoretical) were used to analyze the "Hipparcos" stars. Empirical transformations have been discussed in Section 2.2.

The most recent theoretical transformations are compiled by Bessell et al (1998), who used synthetic spectra derived from ATLAS9 and NMARCS to produce broadband colors and bolometric corrections for a very wide range of T_{eff}, g and [Fe/H] values. These authors found fairly good agreement with empirical relations except for the coolest stars (M dwarfs, K-M giants).

Interior/Atmosphere Interface

The external boundary conditions for interior models are commonly obtained from $T(\tau)$-laws (τ is the optical depth) derived either from theory or full atmosphere calculation. This method is suitable for low- and intermediate-mass stars (it is not valid for masses below ~ 0.6 $M_\odot$, Chabrier & Baraffe 1997). Morel et al (1994) and Bernkopf (1998) focused on the solar case where seismic constraints require a careful handling of external boundary conditions. Morel et al pointed out that homogeneous physics should be used in interior and atmosphere (opacities, EOS, treatment of convection) and showed that the boundary level must be set deep enough, in zones where the diffusion approximation is valid. Bernkopf discussed some difficulties in reproducing Balmer lines related to the convection treatment.

3.3 Transport Processes

Convection

3-D numerical simulations at current numerical resolution are able to reproduce most observational features of solar convection such as images, spectra, and helioseismic properties (Stein & Nordlund 1998). However, the "connection" with a stellar evolution code is not easy, and stellar models still mostly rely on 1-D phenomenological descriptions such as the mixing-length theory of convection (MLT, Böhm-Vitense 1958). The mixing-length parameter α_{MLT} (ratio of the mixing-length to the pressure scale height) is calibrated so that the solar model yields the observed solar radius at the present solar age. The question of the variations of α_{MLT} in stars of various masses, metallicities, and evolutionary stages remains a matter of debate (Section 4). As pointed out by Abbett et al (1997), the MLT can reproduce the correct entropy jump across the superadiabatic layer near the stellar surface, but fails to describe the detailed depth structure and dynamics of convection zones. Abbett et al found that the solar entropy jump obtained in 3-D simulations corresponds to predictions of the MLT for $\alpha_{MLT} \approx 1.5$. Ludwig et al (1999) calibrated α_{MLT} from 2-D simulations of compressible convection in solar-type stars for a broad range of T_{eff} and g-values. The solar α_{MLT} inferred from 3-D and 2-D simulations is close to what is obtained in solar model calibration. The α_{MLT} dependence with T_{eff} and g of Ludwig et al can be used to constrain the range of acceptable variations of α_{MLT} in stellar models (see Section 4).

Overshooting

Penetration of convection and mixing beyond the classical Schwarzschild convection cores (overshooting process) modifies the standard evolution model of stars of masses $M > \sim 1.2\ M_\odot$, in particular the lifetimes (see for instance Maeder & Mermilliod 1981, Bressan et al 1981). The extent of overshooting was estimated for the first time from the comparison of observed and theoretical MS widths of open clusters (Maeder & Mermilliod 1981), which yields an overshooting parameter $\alpha_{ov} \sim 0.2$ (ratio of overshooting distance to pressure scale height). As discussed in detail by Roxburgh (1997), α_{ov} is still poorly constrained despite significant efforts made to establish the dependence of overshooting with mass, evolutionary stage, or chemical composition (see Section 4). Andersen (1991) first pointed out that the simultaneous calibration of well-known binaries (masses and radii at 1–2%) may provide improved constraints for α_{ov}. A modeling of the sample of the best-known binaries indicates a trend for α_{ov} to increase with mass and suggests a decrease of α_{ov} with decreasing metallicity (Ribas 1999), although a larger sample would be desirable to confirm those trends. Further advances are expected from asteroseismology (Brown et al 1994, Lebreton et al 1995).

Diffusion of Chemical Elements

Various mixing processes may occur in stellar radiative zones (see Pinsonneault 1997). In low-mass stars, microscopic diffusion due to gravitational settling carries helium and heavy elements down to the center and modifies the evolutionary course as well as the surface abundances. It has been proved that microscopic diffusion can explain the low helium abundance of the solar convective zone derived from seismology (Christensen-Dalsgaard et al 1993). On the other hand, turbulent mixing (resulting, for instance, from hydrodynamical instabilities related to rotation, see Zahn 1992) probably inhibits microscopic diffusion. Richard et al (1996) did not find any conflict between solar models including rotation-induced mixing (to account for Li and Be depletion at the surface) and microscopic diffusion (to account for helioseismic data). More constraints are required to clearly identify (and quantify the effects of) the various candidate mixing processes; this will be illustrated in the following sections.

4.   STUDIES OF THE BEST-KNOWN OBJECTS

Stellar model results depend on a number of free input parameters. Some are observational data (mass, chemical composition and age, the latter for the Sun only), whereas others enter phenomenogical descriptions of poorly-known physical processes (mixing-length parameter for convection, overshooting, etc). The model outputs have to be compared with the best available observational data: luminosity, T_{eff} or radius, oscillation frequencies, etc. Numerous and precise observational constraints allow assessment of the input physics or give more precise values of the free parameters. They may reveal the necessity to include processes previously neglected, and in the best cases, to characterize them.

The model validation rests on (1) the nearest objects with the most accurate observations, (2) special objects with additional information such as stars belonging to binary systems, members of stellar clusters, or stars with seismic data, and (3) large samples of objects giving access to statistical studies.

4.1   Stars in Binary Systems

Masses are available for a number of stars belonging to binary systems, allowing their "calibration" under the reasonable assumption that the stars have the same age and were born with the same chemical composition (Andersen 1991, Noels et al 1991). A solution is sought which reproduces the observed positions in the H-R diagram of both stars. Andersen (1991) claimed that the only systems able to really constrain the internal structure theory are those with errors lower than 2% in mass, 1% in radius, 2% in T_{eff} and 25% in metallicity.

However, additional observations may sometimes cast doubts on an observed quantity previously determined with good internal accuracy. This occurred recently for the masses of stars in the nearest visual binary system α Centauri. The system has been widely modeled in the past (Noels et al 1991, Edmonds et al 1992, Lydon et al 1993, Fernandes & Neuforge 1995) with the objective of getting (among others) constraints on the mixing-length parameter. At that time, the astrometric masses were used (internal error of 1%) but the [Fe/H]-value was controversial, leading to various possibilities for α_{MLT}-values. Today the situation is still confused: metallicity is better assessed, but new radial velocity measurements yield masses higher than those derived from astrometry (by 6–7%, Pourbaix et al 1999). The higher masses imply a reduction in age by a factor of 2 and slightly different α_{MLT}-values for the two stars. However, the orbital parallax corresponding to the high-mass "option" is smaller than and outside the error bars, of both ground-based and Hipparcos parallax π_{Hipp}. Pourbaix et al noted the lack of reliability of π_{Hipp} given in the Hipparcos Catalogue, but since then it has been re-determined from intermediate data by Söderhjelm (1999) and is now close to (and in agreement with) the ground-based parallax. More accurate radial velocity measurements are therefore needed to assess the high-mass solution.

Possible variations of α_{MLT} have been investigated through the simultaneous modeling of selected nearby visual binary systems (Fernandes et al 1998, Pourbaix et al 1999, Morel et al 2000). Small variations of α_{MLT} (not greater than ≈ 0.2) in the two components of αCen (Pourbaix et al) and ιPeg (Morel et al) have been suggested. Fernandes et al, who calibrated 4 systems and the Sun with the same program and input physics, found that α_{MLT} is almost constant for [Fe/H] in the range [Fe/H]$_\odot \pm 0.3$ dex and masses between 0.6 and 1.3 $M_\odot$. In this mass range the sensitivity of models to α_{MLT} increases with mass (due to the increase with mass of the entropy jump across the superadiabatic layer) which makes the MS slope vary with α_{MLT}. Also, I estimate from my models that a change of α_{MLT} of ± 0.15 around 1 $M_\odot$ translates into a T_{eff}-change of ~ 40–55 K depending on the metallicity. On the other hand, with the solar-α_{MLT} value the MS slope of field stars and Hyades stars is well fitted (Section 4.3). It is therefore reasonable to adopt the solar-α_{MLT} value to model *solar-type stars*. For other stars, the situation is less clear. The calibration of α_{MLT} depends on the external boundary condition applied to the model, itself sensitive to the low-temperature opacities, and on the color transformation used for the comparison with observations. Chieffi et al (1995) examined the MS and red giant branch (RGB) in metal-deficient clusters and suggested a constancy of α_{MLT} from MS to RGB and a decrease with decreasing Z. They found variations of α_{MLT} with Z of $\approx$0.2–0.4, but these are difficult to assess considering uncertainties in the observed and theoretical cluster sequences. On the other hand, calibration of α_{MLT} with 2-D simulations of convection gives complex results (Freytag & Salaris 1999; Freytag et al 1999). In particular, (1) for solar metallicity, α_{MLT} is found to decrease when T_{eff} increases above solar T_{eff}, and to increase slightly when stars move toward the RGB (by $\approx$0.10–0.15) and, (2) α_{MLT} does not vary importantly when metallicity decreases at solar T_{eff}. More work is needed to go into finer details, and other calibrators of α_{MLT} are required, such as binary stars in the appropriate range of mass and with various chemical compositions.

The modeling of a large sample of binaries might give information on the variation of helium Y and age with metallicity Z, of great interest for Galactic evolution studies. The combined results for six binary systems and the Sun with the same program by Fernandes et al (1998) and Morel et al (2000) show a general trend for Y to increase with Z: Y increases from 0.25 to 0.30 ($\pm$0.02) when Z increases from 0.007 to 0.03 ($\pm$0.002). However, the Hyades appear to depart from this tendency (see Section 4.3).

The sample of binaries with sufficiently accurate temperatures and abundances is still too meager to allow full characterization of physical processes. Additional data are needed, such as observations of binaries in clusters (see Section 4.3) or asteroseismological measurements.

4.2   The Nearest Disk and Halo Stars

Fine Structure of the H-R Diagram

Highly accurate distances for a rather large number of stars in the solar neighborhood were provided by Hipparcos. This allowed the first studies of the fine structure of the H-R diagram and related metallicity effects to be undertaken.

Among an ensemble of "Hipparcos" F-G-K stars closer than 25 pc, with error on parallax lower than 5%, Lebreton et al (1997b) selected stars with [Fe/H] in the range [−1.0, +0.3] from detailed spectroscopic analysis ($\sigma_{[Fe/H]} \simeq 0.10$ dex, Cayrel de Strobel et al 1997b), F_{bol} and T_{eff} from Alonso et al (1995, 1996a) with $\frac{\sigma_{F_{bol}}}{F_{bol}} \sim 2\%$ and $\frac{\sigma_{T_{eff}}}{T_{eff}} \sim 1.5\%$ (see Section 2.2) and not suspected to be unresolved binaries. Figure 1 presents the H-R diagram of the 34 selected stars: the error bars are the smallest obtained for stars in the solar neighborhood ($\sigma_{M_{bol}}$ are in the range 0.031–0.095 with an average value $\langle \sigma_{M_{bol}} \rangle \simeq 0.045$ mag). The sample is compared with theoretical isochrones derived from standard stellar models in Figure 2. Models cover the entire [Fe/H]-range. They account for an α-element

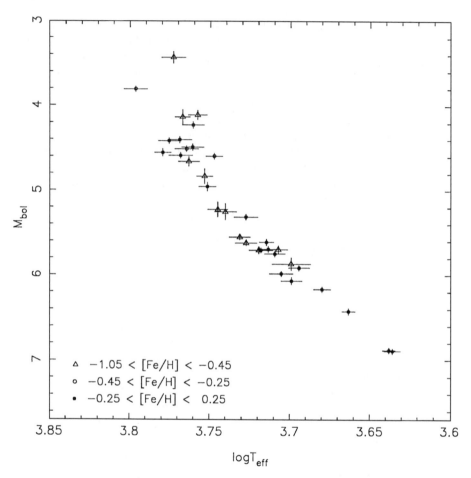

Figure 1 Hipparcos H-R diagram of the 34 best-known nearby stars. The parallax accuracies σ_π/π are in the range 0.003–0.041. Bolometric fluxes and effective temperatures are available from Alonso et al's (1995, 1996a) works ($\frac{\sigma_{F_{bol}}}{F_{bol}} \sim 2\%$ and $\frac{\sigma_{T_{eff}}}{T_{eff}} \sim 1.5\%$, see Section 2.2). Resulting $\sigma_{M_{bol}}$ are in the range 0.031–0.095 mag (from Lebreton et al 1999).

enhancement $[\alpha/Fe] = +0.4$ dex for $[Fe/H] \leq -0.5$ and, for non-solar $[Fe/H]$, have a solar-scaled helium content ($Y = Y_p + Z\,(\Delta Y/\Delta Z)_\odot$). The splitting of the sample into a solar metallicity sample and a moderately metal-deficient one (Figure 2a and b) shows that:

1. The slope of the MS is well reproduced with the solar α_{MLT}.
2. Stars of solar metallicity and close to it occupy the theoretical band corresponding to their (LTE) metallicity range, while for moderately metal deficient stars there is a poor fit.

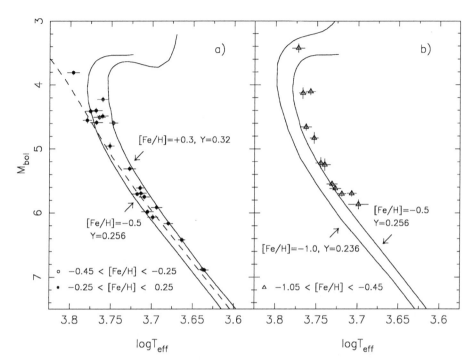

Figure 2   The sample of Figure 1 split into two metallicity domains. Figure 2*a*: shows stars with [Fe/H] close to solar ([Fe/H] ∈ [−0.45, +0.25]). Figure 2*b* shows moderately metal-deficient stars ([Fe/H] ∈ [−1.05, −0.45]). Theoretical isochrones are overlaid on the observational data. Figure 2*a*: the lower isochrone (10 Gyr) is for [Fe/H] = −0.5, $Y = 0.256$ and [α/Fe] = 0.4; the upper isochrone (8 Gyr) is for [Fe/H] = +0.3, $Y = 0.32$ and [α/Fe] = 0.0; the dashed line is a solar ZAMS ($\alpha_{MLT} = 1.65$, $Y_{\odot} = 0.266$ and $Z_{\odot} = 0.0175$). The brightest star is the young star γ Lep. Figure 2*b*: 2 isochrones (10 Gyr) with [α/Fe] = 0.4, the lower is for [Fe/H] = −1.0, $Y = 0.236$ and the upper for [Fe/H] = −0.5, $Y = 0.256$. All stars but one are sitting above the region defined by the isochrones. $(\Delta Y/\Delta Z)_{\odot} = 2.2$ is obtained with Balbes et al's (1993) primordial helium $Y_{p} = 0.227$ (from Lebreton et al 1999).

In general, stars have a tendency to lie on a theoretical isochrone corresponding to a higher metallicity than the spectroscopic (LTE) value. This trend was already noticed by Axer et al (1995) but it is now even more apparent because of the high accuracy of the data. Helium content well below the primordial helium value would be required to resolve the conflict.

This is exemplified by the star μCas A, the A-component of a well-known, moderately metal-deficient binary system that has a well-determined mass (error in mass of 8 per cent). The standard model (Figure 3) is more than 200 K hotter than the observed point and is unable to reproduce the observed T_{eff} even if (reasonable) error bars are considered (Lebreton 2000). On the other hand, the mass-luminosity properties of the star are well reproduced if the helium abundance is chosen to be

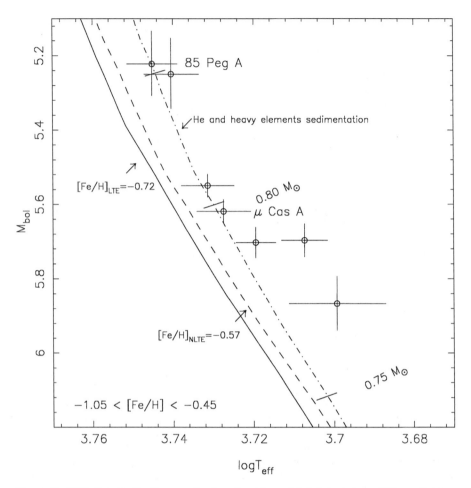

Figure 3   H-R diagram for the unevolved moderately metal deficient stars of Figure 2 (mean LTE metallicity $[Fe/H]_{LTE} = -0.72$, meaan non-LTE value $[Fe/H]_{NLTE} = -0.57$, see text). Full and dashed lines are standard isochrones (10 Gyr) computed with, respectively, the $[Fe/H]_{LTE}$ and $[Fe/H]_{NLTE}$ values. The dot-dashed isochrone (10 Gyr) includes He and heavy elements sedimentation: at the surface it has $[Fe/H]_{NLTE} = -0.57$ but the initial $[Fe/H]$ was ≈ -0.5. (from Lebreton et al 1999).

close to the primordial value, although the error bar in mass is somewhat too large to provide strong constraints.

Several reasons can be invoked to explain the poor fit at low metallicities:

1. *Erroneous temperature-scale.* 3-D model atmospheres could still change the T_{eff}-scale as a function of metallicity (Gustafsson 1998), but with the presently (1-D) available models it seems difficult to increase Alonso et al's (1996a) T_{eff} by as much as 200–300 K. As noted by Nissen (1998), this scale is already higher than other photometric scales, by as much as

100 K. Also, Lebreton et al (1999) verified that spectroscopic effective temperatures lead to a similar misfit.

2. *Erroneous metallicities*. As discussed in Section 2.2, the [Fe/H]-values inferred from model atmosphere analysis should be corrected for non-LTE effects. According to Thévenin & Idiart (1999) no correction is expected at solar metallicity, whereas for moderately metal-deficient stars the correction amounts to ~0.15 dex.

3. *Inappropriate interior models*. In low-mass stars, microscopic diffusion by gravitational settling can make helium and heavy elements sink toward the center, changing surface abundances as well as inner abundance profiles. In metal-deficient stars this process may be very efficient for three reasons: (1) densities at the bottom of the convection zone decrease with metallicity, which favors settling; (2) the thickness of the convection zones decreases with metallicity, making the reservoir easier to empty; and (3) metal-deficient generally means older, which implies more time available for diffusion.

The two latter reasons are attractive because they qualitatively predict an increasing deviation from the standard case when metallicity decreases. As shown in Figure 3, a combination of microscopic diffusion effects with non-LTE [Fe/H] corrections could remove the discrepancy noted for μCas A: an increase of [Fe/H] by 0.15 dex produces a rightward shift of 80 K of the standard isochrone, representing about one third of the discrepancy. Additionally, adding microscopic diffusion effects, according to recent calculations by Morel & Baglin (1999), provides a match to the observed positions. Moreover, the general agreement for solar metallicity stars (Figure 2a) should remain: (1) at solar metallicities non-LTE corrections are found to be negligible, and (2) at ages of ~5 Gyr chosen as a mean age for those (expectedly) younger stars, diffusion effects are estimated to be smaller than the error bars on T_{eff} (Lebreton et al 1999).

To conclude on this point, the high-level accuracy reached for a few tens of stars in the solar neighborhood definitely reveals imperfections in interior and atmosphere models. It casts doubts on abundances derived from model atmospheres in LTE, and favors models that include microscopic diffusion of helium and heavy elements toward the interior over standard models. Also, diffusion makes the surface [Fe/H]-ratio decrease by ~0.10 dex in 10 Gyr in a star like μCas (Morel & Baglin 1999), which is rather small and hidden in the observational error bars. In very old, very deficient stars, the [Fe/H] decrease is expected to be larger (Salaris et al 2000), which makes the relation between observed and initial abundances difficult to establish. In the future, progress will come from the study of enlarged samples reaching the same accuracies and of the acquisition of additional parameters to constrain the models. The knowledge of masses for several binaries in a narrow mass range but large metallicity range would help to constrain the helium abundances, while access to seismological data for at least one or two stars would help to better characterize mixing processes.

Statistical Studies

Complete H-R diagrams of stars of the solar neighborhood have been constructed by adopting different selection criteria, and have been compared to synthetic H-R diagrams based on theoretical evolutionary tracks.

- Schröder (1998) proposed diagnostics of MS overshooting based on star counts in the different regions of the "Hipparcos" H-R diagram of stars in the solar neighborhood (d < 50–100 pc). In the mass range 1.2–2 $M_\odot$, convective cores are small, and it is difficult to estimate the amount of overshooting with isochrone shapes. Schröder suggested using the number of stars in the Hertzsprung gap, associated with the onset of H-shell burning, as an indicator of the extent of overshooting around 1.6 $M_\odot$; the greater the overshooting on the MS, the larger the He-burning core, and in turn the longer the passage through the Hertzsprung gap. Actual star counts favor an onset of overshooting around $\sim$1.7 $M_\odot$ (no overshooting appears necessary below that mass), which is broadly consistent with other empirical calibrations (MS width, eclipsing binaries), but finer quantitative estimates would require more accurate observational parameters, mainly in T_{eff} and Z.

- Jimenez et al (1998) compared the red envelope of "Hipparcos" subgiants ($\sigma_\pi/\pi < 0.15$, $\sigma_{(B-V)} < 0.02$ mag) with isochrones to determine a minimum age of the Galactic disk of 8 Gyr, which is broadly consistent with ages obtained with other methods (white-dwarf cooling curves, radioactive dating, isochrones, or fits of various age-sensitive features in the H-R diagram). The fit is still qualitative: the metallicities of subgiants are unknown because of the inadequacy of model atmospheres in that region. For this reason, Jimenez et al investigated the isochrones in other regions, MS and clump (He core burning). They calculated the variations with mass of the clump position for a range of metallicities in the disk, and showed that stars with masses from 0.8 to 1.3 $M_\odot$ (ages from 2 to 16 Gyr) all occupy a well-defined vertical branch, the red-edge of the clump. The color of this border line is sensitive to metallicity, which makes it a good metallicity indicator in old metal-rich populations.

- Ng & Bertelli (1998) revised the ages of stars of the solar neighborhood and derived corresponding age-metallicity and age-mass relations. Fuhrmann (1998) combined the [Mg/H]-[Fe/H] relation with age and kinematical information to distinguish thin and thick disk stars. Several features seem to emerge from these studies: (1) no evident age-metallicity relation exists for the youngest (<8 Gyr) thin-disk stars; some of them are rather metal-poor, and super metal-rich stars appear to have been formed early in the history of the thin disk; (2) there is an apparent lack of stars in the age-interval 10–12 Gyr which is interpreted by Fuhrmann as a signature of the thin-disk formation; and (3) beyond 12 Gyr there is a slight decrease of metallicity with increasing age for stars of the thick disk; some of them are as old as halo stars. To assess these suggestions and to assist progress in the understanding of the Galactic evolution scenario (see Fuhrmann 1998 for details), enlarged stellar samples and further improvements on age determinations are of course required.

The Subdwarf/Subgiant Sequence

Hipparcos provided the very first high-quality parallaxes for a number of halo stars. Age determinations of the local halo could be undertaken, as well as comparisons with globular cluster sequences.

Among a large sample of Population II Hipparcos halo subdwarfs, Cayrel et al (1997b) extracted the best-known stars with criteria similar to those adopted by Lebreton et al (1999) for disk stars. Stars were corrected for reddening, excluding stars with E(B-V) > 0.05. Prior to Hipparcos, only 5 halo stars had parallax errors smaller than 10%; now there are 17, which represents sizeable progress. The halo stars are plotted in Figure 4; subdwarfs but also subgiants are present, delineating an isochrone-like shape with a turn-off region.

To make a first estimate of the age of the local halo, Cayrel et al kept 13 stars with the lowest error bars and spanning a narrow metallicity range ([Fe/H] = -1.5 ± 0.3), the most commonly found in the halo (Figure 5). They found that halo stars, like disk stars, are colder than the theoretical isochrone corresponding to their metallicity. The misfit was also noted by Nissen et al (1997) and Pont et al (1997) in larger samples of halo stars. The discrepancy amounts to 130 to 250 K depending on the metallicity, and comparisons indicate that it is independent of the particular set of isochrones used. Again, non-LTE corrections leading to increased [Fe/H]-values (Δ[Fe/H] = +0.2 for [Fe/H] ~ -1.5 according to Thévenin & Idiart 1999), added to the effects of microscopic diffusion, can be invoked to reduce the misfit. Figure 5a compares Cayrel et al's sample with standard isochrones by PA Bergbusch & DA VandenBerg (2000, in preparation), showing that the subdwarf main sequence cannot be reproduced by isochrones computed with the LTE [Fe/H]-value, but increasing the metallicity (to mimic non-LTE corrections) improves the fit. Figure 5b compares the halo sequence with Proffitt & VandenBerg's (1991) isochrones that include He sedimentation. Microscopic diffusion makes the isochrones redder, modifies their shape, and predicts a lower turn-off luminosity: the best fit with the observed sequence is achieved for an age smaller by 0.5–1.5 Gyr than that obtained without diffusion. Models by Castellani et al (1997) show that, if sedimentation of metals is also taken into account, including its effects on the matter opacity, the isochrone shift is smaller than the shift obtained with He diffusion only.

Cayrel et al (1997b) and Pont et al (1997) estimated the local halo to be 12–16 Gyr old (from standard isochrones). To improve the precision, more stars with accurate parallaxes are required. Subgiants are about 100 times rarer than subdwarfs, and we have only two subgiants with $\sigma_\pi/\pi < 12.5\%$ (and no subgiant with $\sigma_\pi/\pi < 5\%$). After Hipparcos the position of the subgiant branch is still

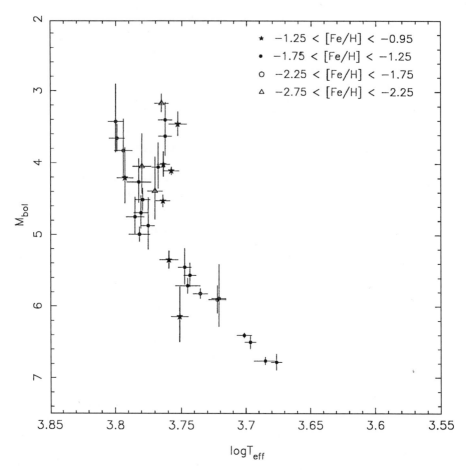

Figure 4 Hipparcos H-R diagram of the 32 halo stars with $\sigma_\pi/\pi < 0.22$ (the parallax accuracies σ_π/π are in the range 0.007–0.214). Bolometric fluxes and effective temperatures are available from Alonso et al's (1995, 1996a) works (see the caption for Figure 1). Resulting $\sigma_{M_{bol}}$ are in the range 0.03–0.48 mag. A bunch of subgiants emerges with an isochrone-like shape (from Cayrel et al 1997b).

poorly determined, which limits the accuracy on the age determination of the halo stars.

The ZAMS Positions

The sample made of Hipparcos disk and halo stars spans the whole Galactic metallicity range. Figure 6 shows the non-evolved stars ($M_{bol} > 5.5$) of Figure 1 and Figure 4 along with standard isochrones of various metallicities and solar-scaled helium ($(\Delta Y/\Delta Z)_\odot = 2.2$). It allows a discussion of the position of the zero

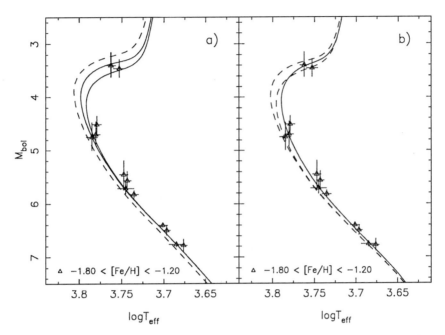

Figure 5 Hipparcos H-R diagram of 13 halo stars with $[Fe/H]_{LTE} = -1.5 \pm 0.3$ and $\sigma_\pi/\pi < 0.12$ (the parallax accuracies σ_π/π are in the range 0.01–0.12). Bolometric fluxes and effective temperatures are available from Alonso et al's (1995, 1996a) works (see the caption for Figure 1). Resulting $\sigma_{M_{bol}}$ are in the range 0.03–0.26 mag. All isochrones were kindly provided by DA Vandenberg. Figure 5a illustrates the effect of a non-LTE correction of +0.2 dex in [Fe/H] as inferred from PA Bergbusch & DA VandenBerg's (2000, in preparation) models (see text): the dashed line is a standard isochrone of 12 Gyr ($[\alpha/Fe] = +0.3$, $Y \simeq 0.24$) with $[Fe/H] = -1.54$ (LTE value), and the full lines are isochrones with $[Fe/H] = -1.31$ (non-LTE value) of 12 Gyr (upper line) and 14 Gyr (lower line). Figure 5b illustrates the effect of microscopic diffusion of He as inferred from Proffitt's & VandenBerg's (1991) models: isochrones ($[Fe/H] = -1.3$ and $[O/Fe] = 0.55$), of 12 Gyr with (full line) and without (dashed-line; upper line 12 Gyr, lower 14 Gyr) microscopic diffusion are plotted.

age main sequence (ZAMS) as a function of metallicity and implications for the unknown helium abundances.

- *MS width.* Although stars generally do not lie where predicted, in particular at low metallicities, the observational and theoretical MS widths are in reasonable agreement for $\Delta Y/\Delta Z = 2.2$. This qualitative agreement is broadly consistent with the $\Delta Y/\Delta Z$ ratio of $\simeq 3 \pm 2$ derived from similar measures of the lower MS width by Pagel & Portinari (1998) and the lower limit $\Delta Y/\Delta Z > \sim 2$ obtained by Fernandes et al (1996) from pre-Hipparcos MS. It also agrees with extragalactic determinations (see Izotov et al 1997) or nucleosynthetic predictions.

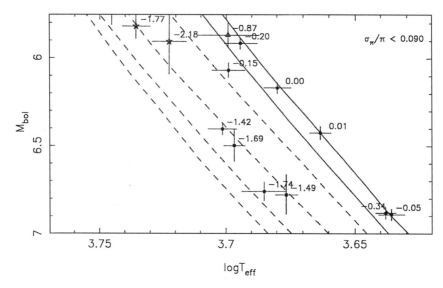

Figure 6 "Hipparcos" H-R diagram of non-evolved stars with $\sigma_\pi/\pi < 0.10$. Each star is labeled with its [Fe/H]-value. Standard isochrones are plotted with, from left to right, [Fe/H] $= -2.0$, -1.5, -1.0, -0.5, 0.0, 0.3 (from Lebreton 2000).

■ *Helium abundance at solar metallicities*. It can be noted from Figure 6 that there are 4 stars with Fe/H close to solar on the [Fe/H] $= 0.3$ isochrone. Non-LTE [Fe/H]-corrections are negligible at solar metallicity. Microscopic diffusion may produce a shift in the H-R diagram: for a 0.8 $M_\odot$ star of solar Fe/H at 5 Gyr the shift is small and comparable to the observational error bars (but it increases with age). These disk stars are not expected to be very old, and the shift could instead indicate that their He-content is lower than the solar-scaled value. Calibration of individual objects and groups with metallicities close to solar indicate an increase of helium with metallicity corresponding to $\Delta Y/\Delta Z \simeq 2.2$ from the Sun (Lebreton et al 1999) and $\Delta Y/\Delta Z \simeq 2.3 \pm 1.5$ from visual binaries (Fernandes et al 1998) but exceptions are found, such as in the (rather young) Hyades which, although metal-rich ([Fe/H] $= 0.14$), appear to have a solar or even slightly sub-solar helium content with $\Delta Y/\Delta Z \simeq 1.4$ (Perryman et al 1998). Going into finer resolution would clearly require more complete data including masses for enlarged samples of non-evolved stars.

■ *Position of metal-deficient stars*. Very few metal-deficient stars have accurate positions in the non-evolved part of the H-R diagram: a gap appears for [Fe/H] $\in [-1.4, -0.3]$ and only 4 subdwarfs are found below [Fe/H] ~ -1.4. The empirical dependence of the ZAMS location with metallicity is impossible to establish for these stars, which are expected to have practically primordial helium contents. This adds to difficulties in estimating the distances of globular clusters (Eggen & Sandage 1962; Sandage 1970, 1983; Chaboyer et al 1998).

4.3 Stars in Open Clusters

Hipparcos observed stars in all open clusters closer than 300 pc and in the richest clusters located between 300 and 500 pc providing valuable material for distance scaling of the Universe and for studies of kinematical and chemical evolution of the Galaxy. The absolute cluster sequences in the H-R diagram may be constructed directly from Hipparcos distances independently of any chemical composition consideration. Each sequence covers a large range of stellar masses and contains stars which can reasonably be considered to be born at the same time with similar chemical composition. Several clusters provide tests of the internal structure models for a wide range of initial parameters, in particular for different metallicities.

The Hyades

Obtaining high-quality astrometric data for the Hyades has been crucial, for it is the nearest rich star cluster, used to define absolute magnitude calibrations and the zero-point of the Galactic and extragalactic distance scales. Individual distances (mean accuracy of 5%) and proper motions were given by Hipparcos, providing a consistent picture of the Hyades distance, structure and dynamics (Perryman et al 1998). The recent determinations of the Hyades distance modulus $(m - M)$ are all in very good agreement while the internal accuracy was largely improved with Hipparcos:

- ground-based: $m - M = 3.32 \pm 0.06$ mag (104 stars, van Altena et al 1997b)
- *HST*: $m - M = 3.42 \pm 0.09$ mag (7 stars, van Altena et al 1997a)
- Hipparcos: $m - M = 3.33 \pm 0.01$ mag (134 stars within 10 pc of the cluster center, Perryman et al 1998)
- Statistical parallaxes based on Hipparcos proper motions: $m - M = 3.34 \pm 0.02$ mag (43 stars, Narayanan & Gould 1999a who also showed that the systematic error on the parallaxes toward the Hyades is lower than 0.47 mas).

Greatly improved precision is seen in the H-R diagrams built with Hipparcos data combined with the best ground-based observations (Perryman et al 1998):

- Figure 7 shows 40 stars with T_{eff} and Fe/H $= 0.14 \pm 0.05$ from detailed spectroscopic analysis delineating the lower part of the observational MS of the cluster (Cayrel de Strobel et al 1997a).
- Figure 8 is the whole H-R diagram in the $(M_V, B\text{-}V)$ plane for 69 cluster members. Known or suspected binaries, variable stars, and rapid rotators

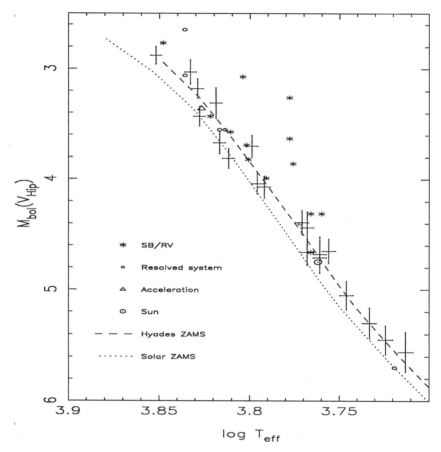

Figure 7 Hipparcos H-R diagram for 40 selected low MS stars in the Hyades. The mean [Fe/H] is 0.14 ± 0.05. Stars with error bars are not suspected to be double or variable. Internal errors on T_{eff} are in the range 50–75 K. Double or variable stars are also indicated: objects resolved by Hipparcos or known to be double systems are shown as circles, triangles denote objects with either detected photocentric acceleration or objects possibly resolved in photometry, and '*' means spectroscopic binary or radial velocity variable. Theoretical ZAMS loci are given for the Hyades (dashed line, $Y = 0.26$ $Z = 0.024$) and solar (dotted line, $Y = 0.266$ $Z = 0.0175$) chemical compositions (from Perryman et al 1998).

have been excluded (Perryman et al 1998). Also, Dravins et al (1997) derived dynamical parallaxes for the Hyades members from the relation between the cluster space motion, the positions and the projected proper motions; these parallaxes are more precise (by a factor of about 2) than those directly measured by Hipparcos, yielding in turn a remarkably well-defined MS sequence in the H-R diagram, narrower than that given in Figure 8 (see Figure 2 in Dravins et al 1997).

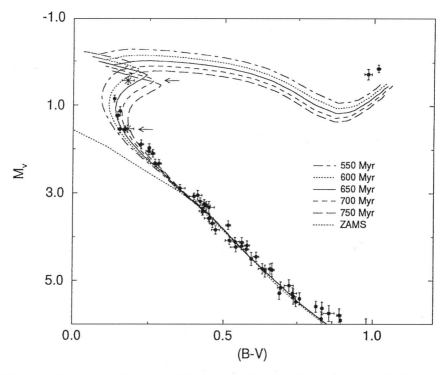

Figure 8 Hipparcos C-M diagram of the Hyades. V and B-V values are from the Hipparcos catalogue ($\sigma_{(B-V)} < 0.05$ mag). The loci of ZAMS and theoretical isochrones calculated with overshooting ($\alpha_{ov} = 0.2$) are indicated. Arrows indicate the position of the components of the binary system θ^2 Tau used for the age determination (from Perryman et al 1998).

- Figure 9 shows the empirical mass-luminosity (M-L) relation drawn from the (very accurate) masses of 5 binary systems (see caption). Nine of the stars are MS stars.

Comparisons with theoretical models yield some of the cluster characteristics (Lebreton et al 1997a, Perryman et al 1998, Lebreton 2000):

- The comparison of the lowest part of the MS (Figure 7), representing the non-evolved stars, with theoretical ZAMS corresponding to the mean observed [Fe/H] yields the initial cluster helium content $Y_H = 0.26 \pm 0.02$ and metallicity $Z_H = 0.024 \pm 0.04$. Metallicity is the dominant source of the uncertainty on Y.

- The comparison of the whole observed sequence with model isochrones yields the cluster age. Figure 8 shows that the optimum fit is achieved with an isochrone of 625 ± 50 Myr, $Y_H = 0.26$, $Z_H = 0.024$ and including overshooting. The turn-off region (which in the Hyades corresponds to the instability strip of δ Scuti stars) is rather well represented by the 625 Myr

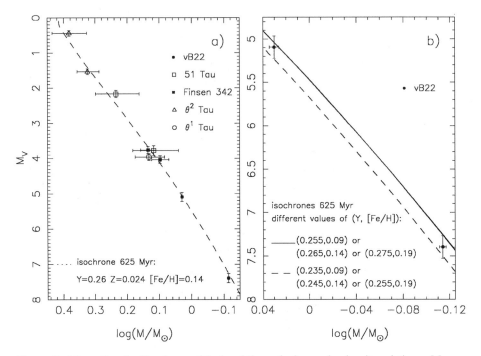

Figure 9 Figure 9*a*: the Hyades empirical and theoretical mass-luminosity relations. Masses are from Peterson & Solensky (1988, vB22), Torres et al (1997a,b,c, Finsen 342, θ^1, θ^2 Tau) and Söderhjelm (1999, 51 Tauri from Hipparcos data). The isochrone is for 625 Myr, taken from the same calculations used in Figure 8. Figure 9*b* illustrates how the precise positions of the members of vB22 allow to discriminate between different (*Y*, [Fe/H]) values.

isochrone (see also Antonello & Pasinetti Fracassini 1998). The quoted uncertainty on age only includes the contribution from visual fitting of the isochrones. Additional errors on age result from unrecognized binaries, rotating stars, color calibrations and bolometric corrections, and from theoretical models in particular through the parameterization of overshooting (Lebreton et al 1995). It is therefore reasonable to give an overall age uncertainty of at least 100 Myr.

■ In Figure 9*a* the observed M-L relation is compared with the theoretical isochrone of 625 Myr, $Y_H = 0.26$, $Z_H = 0.024$, showing an excellent agreement. The lower part of the relation is defined by the very accurate masses of the two components of vB22. This system gives additional constraints on the Y_H-value derived from ZAMS calibration. Figure 9*b* illustrates how the positions of the two vB22 components may be used to constrain Y_H in the whole metallicity range allowed by observations, [Fe/H] $= 0.14 \pm 0.05$ (see also Lebreton 2000).

Furthermore, in the turn-off region of the Hyades, 5 δScuti stars are found that are quite rapid rotators ($v_e \sin i$ in the range 80–200 km $\cdot$ s^{-1}, see Antonello & Pasinetti Fracassini 1998). From the measurement and analysis of their oscillation frequencies and the identification of the corresponding modes by means of models (of same age and chemical composition), we should be able to derive the inner rotation profile and learn about the size of convective cores and transport processes at work in the interiors (Goupil et al 1996, Michel et al 1999). For instance, the rotation profile is related to the redistribution of angular momentum by internal motions which could be generated by meridional circulation and shear turbulence in a rotating medium (see Zahn 1992). On the other hand, such motions might induce internal mixing, and as shown by Talon et al (1997), in the H-R diagram rotational effects "mimic" overshooting (for instance, in a star of 9 M$_\odot$, a rotational velocity of $\sim$100 km $\cdot$ s^{-1} is equivalent to an overshooting of $\alpha_{ov} \sim 0.2$).

The study and intercomparison of accurate observations of the non-pulsating and pulsating stars located in the instability strip should clearly provide deeper insight into the internal structure and properties of stars of the Hyades cluster. However, such analysis has to integrate the various complications related to rotation, such as the displacements in any photometric H-R diagram by amounts depending on the equatorial velocity and inclination (Maeder & Peytremann 1972, Pérez-Hernández et al 1999) or the splitting of oscillation frequencies, which has to be considered in the mode identification.

The Pleiades and Other Open Clusters

The membership of stars in nine clusters closer than 300 pc was carefully assessed by van Leeuwen (1999a) and Robichon et al (1999a). Robichon et al also studied nine rich clusters within 500 pc with more than 8 members and 32 more distant clusters. For clusters closer than 500 pc, the accuracy on the mean parallax is in the range 0.2–0.5 mas and the accuracy on the mean proper motions is of the order of 0.1 to 0.5 mas per year. Results from the two groups are in very good agreement. Platais et al (1998) looked for (new) star clusters in Hipparcos data and found one new, a nearby cluster in Carina (d = 132 pc) with 7 identified members.

In order to obtain an optimal mean parallax with correct error estimates, van Leeuwen (1999a) and Robichon et al (1999a) worked with the Hipparcos intermediate data corresponding to each cluster, parallax and proper motion of the cluster center, and position of each cluster member, instead of making a straight average of the parallaxes of the cluster members. Stars in open clusters are located within a few degrees on the sky and hence were often observed in the same field of view of the satellite. A combined solution can be obtained from intermediate data, which allows angular correlations to be taken into account and the resulting parallax errors to be minimized (van Leeuwen & Evans 1998; van Leeuwen 1999a,b; Robichon et al 1999a).

Mermilliod et al (1997), Robichon et al (1997), van Leeuwen (1999b), and Mermilliod (2000) compared the sequences of the various clusters in C-M diagrams derived from different photometric systems, and found puzzling results that are at odds with the common idea that differences in metallicity fully explain the relative positions of the non-evolved parts of the MS of different clusters:

- Some clusters have different metallicities but define the same main sequence in the (M_V, B-V) plane (Praesepe, Coma Ber, αPer, Blanco 1). For instance, Coma Ber has a quasi-solar metallicity while its sequence is similar to that of the Hyades, or of the metal-rich Praesepe.

- Some clusters sequences (Pleiades, IC 2391 and 2602) are abnormally faint with respect to others, for instance Coma Ber. The metallicity of the Pleiades as determined from spectroscopy is almost solar, and similar to that of Coma Ber, but the Pleiades sequence lies (unexpectedly) $\sim$0.3–0.4 mag below the Praesepe, Coma Ber, or Hyades sequence.

- Van Leeuwen (1999a,b) even suggested a possible (although unexpected) correlation between the age of a cluster and the position of the non-evolved part of its MS sequence in the H-R diagram.

Prior to Hipparcos, precise trigonometric parallaxes had not been obtained for clusters except the Hyades. Distances to open clusters were evaluated through the main sequence fitting technique: the non-evolved part of the (observed) cluster sequence was compared to the non-evolved part of the (absolute) lower MS (ZAMS) of either (1) theoretical isochrones, (2) field stars or (3) Hyades after a possible correction of chemical composition differences. The magnitude differences between absolute and apparent ZAMS directly yielded the distance modulus of the cluster.

The Hipparcos distances to the 5 closest open clusters (Hyades, Pleiades, αPer, Praesepe, and Coma Ber) can be compared to those recently derived from MS fitting by Pinsonneault et al (1998); they compared theoretical isochrones, translated into the C-M plane by means of Yale color calibrations, to observational data both in the (M_V, B-V) and (M_V, V-I) planes. The B-V color indice is more sensitive to metallicity than V-I (Alonso et al 1996b), so Pinsonneault et al derived as a by-product the value of the metallicity that gives the same distance modulus in the two planes, and compared it to spectroscopic determinations. They judged their distance modulii to be in good agreement with Hipparcos results except for the Pleiades and Coma Ber. For Coma Ber, the problem could result from the old VRI colors used. For the Pleiades the discrepancy with Hipparcos amounts to 0.24 mag, and the [Fe/H]-value derived from MS fitting in the two color planes agrees with the spectroscopic determination of Boesgaard & Friel (1999), [Fe/H] = -0.034 ± 0.024, although values in the range -0.03 to $+0.13$ can be found in the literature. In fact, with that metallicity the Hipparcos sequence of the Pleiades could be reproduced by classical theoretical models, provided they have a high helium content. The exact value depends on the model set and its input physics: Pinsonneault et al found $Y = 0.37$, Belikov et al (1998) found $Y = 0.34$ but for [Fe/H] $= 0.10$ and I find $Y \sim 0.31$. In any case, high helium content is only marginally supported by observations (Nissen 1976). Pinsonneault et al examined other possible origins of the discrepancy (erroneous metallicity, age-related effects, reddening) and concluded that none of them is likely to be responsible for the Pleiades discrepancy.

In parallel, Soderblom et al (1998) looked for young solar-type stars appearing as (anomalously) faint as the Pleiades. They found 50 field stars expected to be young (i.e. showing activity from Ca II H and K lines), but none of them lies significantly below the ZAMS. They also examined the subluminous stars observed by Hipparcos: they chose six stars among those lying well below the ZAMS, measured their spectroscopic metallicities, and found them to be metal-deficient with respect to the Sun with, in addition, kinematics typical of stars of a thick disk or halo population.

Soderblom et al and Pinsonneault et al concluded that, taking the Hipparcos results for the Pleiades at face value, it would be abnormal not to find stars similar to the Pleiades in the field. They inferred that the distance obtained from multi-color MS fitting is correct and accurate to about 0.05 mag, and concluded that the distance to Pleiades obtained from the analysis of Hipparcos data is possibly wrong at the 1 mas level, which is greater than the mean random error. They invoked statistical correlations between right ascension and parallax ($\rho^{\pi}_{\alpha \cos \delta}$) arising from the non-uniform distribution of Hipparcos observations over time (and in turn along the parallactic ellipse) which affects all stars, including clusters. Pinsonneault et al noted that in the Pleiades the brightest stars (1) are highly concentrated near the cluster center and are therefore subject to spatial correlations which gives them nearly the same parallax, (2) have smaller σ_π than fainter stars which gives them more weight in the mean parallax, and (3) are those which have the highest values of $\rho^{\pi}_{\alpha \cos \delta}$ and also the highest parallaxes in the Hipparcos Catalogue. They suggest that the "true" parallax (close to that obtained through MS fitting) is obtained if the brightest stars with high $\rho^{\pi}_{\alpha \cos \delta}$ are excluded from the calculation.

Narayanan & Gould (1999b) determined the parallaxes of the Pleiades stars by means of Hipparcos proper motions. The resulting distance modulus has a rather large error bar ($m - M = 5.58 \pm 0.18$ mag), but it is in disagreement with that derived directly from Hipparcos parallaxes ($m - M = 5.36 \pm 0.07$ mag), and in agreement with that obtained through MS fitting ($m - M = 5.60 \pm 0.05$ mag). Narayanan & Gould also argue that the differences between the Hipparcos trigonometric parallaxes and the parallaxes derived from Hipparcos proper motions reflect spatial correlations over small angular scales with an amplitude of up to 2 mas.

Robichon et al (1999a,b) and van Leeuwen (1999b) have subsequently derived more reliable distance estimates to these clusters and performed tests that do not support Pinsonneault et al's conclusion. The difference between Hipparcos and MS fitting distance moduli is small for the Hyades (0.01 mag), whereas for other clusters it ranges from -0.17 mag (αPer) to $+0.24$ mag (Pleiades). In fact, except for the Hyades, the difference is always larger than the error on MS fitting distance modulus (0.05 mag). Robichon et al showed that while the solution proposed by Pinsonneault et al improves the situation for the Pleiades, it would introduce new difficulties for Praesepe. By means of Monte-Carlo simulations of the Pleiades stars, they showed that the mean value of the Pleiades parallax does not depend on the correlations $\rho^{\pi}_{\alpha\cos\delta}$. They also carefully examined distant stars and clusters with high $\rho^{\pi}_{\alpha\cos\delta}$. Through these tests, Robichon et al made the Hipparcos distance to Coma Ber or Pleiades more secure, and did not find any obvious bias on the parallax resulting from a correlation between right ascension and parallax, either for stars within a small angular region or for the whole sky.

On the other hand, distances from MS fitting could be subject to higher error bars than quoted by Pinsonneault et al. They depend on reddening and on transformations from the (M_{bol}, T_{eff}) to the C-M plane if theoretical ZAMS are used as reference (or on metallicity corrections if empirical ZAMS are compared). Robichon et al (1999b) compared solar ZAMS from Pinsonneault et al to those I calculated both in the theoretical and in the (M_V, B-V) planes. They showed that while the two ZAMS are within 0.05 mag in the theoretical H-R diagram, they differ by 0.15–0.20 mag in the range B-V = 0.7–0.8 in the (M_V, B-V) plane, simply because different C-M transformations have been applied. Also, MS fitting often relies on rather old and inhomogeneous color sources (in the separate Johnson and Kron-Eggen RI systems) requiring transformations to put all data on the same (Cape-Cousins system) scale. It would therefore be worthwhile to verify the quality and precision of these data by making new photometric measurements of cluster stars.

Let us come back to the difficult question of metallicity. As pointed out by Mermilliod et al (1997), photometric and spectroscopic approaches may produce quite different results. Metallicities have been derived recently by M Grenon (1998, private communication) from large sets of homogeneous observations in the Geneva photometric system. He obtained [Fe/H] = -0.112 ± 0.025 for the Pleiades (quite different from published spectroscopic values) and [Fe/H] = 0.170 ± 0.010 and 0.143 ± 0.008 for Praesepe and the Hyades respectively (both in agreement with spectroscopy). The observed cluster sequences obtained with Hipparcos distances for the three clusters can be roughly reproduced by theoretical models computed with the photometric metallicities (and allowing for small variations of the helium content around the solar-scaled value) and transformed to the C-M plane according to the Alonso et al (1996b) and Bessell et al (1998) calibrations (Robichon et al 1999b).

In conclusion, we point out that a detailed study of the fine structure of the H-R diagram of the Pleiades (and other clusters) requires supplementary observations (colors and abundances) and further progress in model atmospheres. Today there is no obvious solid argument against the published Hipparcos distances. In order to identify and understand the remaining discrepancies with stellar models, the entire set of observed clusters has to be considered (van Leeuwen 1999a). Furthermore, not only the positions of the sequences in the H-R diagram but also the density of stars along them have to be intercompared. For instance, the luminosity function of young clusters exhibits a particular feature (local peak followed by dip) that is interpreted as a signature of pre-MS stars and might provide information on the initial mass function and stellar formation history (see the study of the Pleiades by Belikov et al 1998). On the other hand, since the error bars on luminosity are now small with respect to errors on color indices, stronger constraints are expected from the mass-luminosity relation, as in the Hyades. Observations of binaries in clusters are urgently needed and there is hope to detect them in the future, for, as pointed out by Soderblom et al, the difficult detection and measurement of visual binary orbits in the Pleiades is within the capabilities of experiments on board *HST*.

5. RARE, FAINT, SPECIAL, OR INACCESSIBLE OBJECTS

5.1 Globular Clusters Through Halo Stars

Globular clusters were beyond the possibilities of Hipparcos, but the knowledge of distances to nearby subdwarfs gave distance estimates to a few of them through the MS fitting technique (Sandage 1970; Reid 1997, 1998; Gratton et al 1997; Pont et al 1997, 1998; Chaboyer et al 1998), comparing the non-evolved part of the (absolute) subdwarf main-sequence to the non-evolved part of the (observed) globular cluster sequence. Although simple, the technique has to be applied with caution:

- Only halo subdwarfs and globular clusters with the most precise data should be retained. Abundances should be accurate and on a consistent scale. Globular cluster abundances are usually determined only for giants, while recent preliminary values have been obtained for subgiants in M92 (King et al 1998). Abundance comparisons between (1) field and cluster stars and (2) dwarfs and giants have shown sometimes puzzling differences (King et al 1998, Reid 1999). Questions have been raised as to whether they are primordial or appear during evolution, but definite answers clearly require better spectra for all types of stars as well as spherical model atmospheres with better treatment of convection. The globular cluster sequence should be determined from good photometry well below the MS turn-off, and the correction for interstellar reddening has to be well estimated. Very few halo stars have parallaxes accurate enough to fix precisely the position of the ZAMS (Section 4.2).

- Biases (see e.g. Lutz & Kelker 1973; Hanson 1979; Smith 1987) resulting from the selection of the sample in apparent magnitude, parallax, and metallicity, have to be corrected for (Pont et al 1998, Gratton et al 1997); alternatively, samples free of biases must be selected, which implies retaining the very nearby stars with highly accurate parallaxes (Chaboyer et al 1998, Brown et al 1997).

- Globular cluster sequence and halo sequence should (ideally) have similar *initial* chemical compositions. Because of the small number of subdwarfs in each interval of metallicity, it is not possible to properly establish the variation of the observed ZAMS position with metallicity, and to correct for chemical composition differences between globular clusters and subdwarfs empirically. Chaboyer et al (1998) found it safer to limit the method to globular clusters that have their equivalent in the field with the same [Fe/H] and [α/Fe] content. Gratton et al (1997) and Pont et al (1998) applied theoretical color corrections to the subdwarf data to account for metallicity differences with globulars. In addition, element sedimentation might introduce further difficulties, as already mentioned in Section 4.2 nearby. As pointed out by Salaris & Weiss (2000), the present surface chemical composition of field subdwarfs no more reflects the initial one if microscopic diffusion has been efficient during evolution, while, in globular cluster giants which have undergone the first dredge-up, the chemical abundances have been almost restored to the initial ones.

- Unresolved known or suspected binaries can introduce errors in the definition of the ZAMS position. Chaboyer et al and Gratton et al excluded them whereas Pont et al applied an average correction of 0.375 mag on their position, a procedure that has been criticized by Chaboyer et al and Reid (1998).

- Evolved stars have to be excluded (there is no certainty that globular clusters and halo dwarfs have exactly the same age). From theoretical models it is estimated that stars fainter than $M_V \simeq 5.5$ are essentially unevolved.

The number of globular clusters studied by the different authors varies because of the different criteria and techniques chosen to select the subdwarfs samples. Nevertheless they all agree on the general conclusion that globular cluster distances derived from MS fitting are larger by ∼5–7% than was previously found. Chaboyer (1998) calculated an average of distances to globular clusters obtained with different methods (MS fitting, astrometry, white dwarf sequence fitting, calibration of the mean magnitudes of RR Lyrae stars in the Large Magellanic Cloud, comparison with theoretical models of horizontal branch stars, statistical parallax absolute magnitude determinations of field RR Lyrae from the Hipparcos proper motions, etc) and noted that the distance scale is larger (by 0.1 mag) than his pre-Hipparcos reference.

It is worth pointing out that the statistical parallax method alone favors a shorter distance scale (by ≈0.3 mag with respect to MS fitting result). As reviewed by Layden (1998) and by Reid (1999), the statistical parallax method was applied by independent groups who found concordant results. However, the absolute magnitudes $M_V(RR)$ of the halo RR Lyrae derived from the statistical parallax method (on the basis of Hipparcos proper motions and of radial velocities) are also ≈0.3 mag fainter than the magnitudes obtained through a method directly based on Hipparcos parallaxes (Groenewegen & Salaris 1999); these latter being in turn in good agreement with the MS fitting result. As discussed thoroughly by Reid (1999), there are several difficulties related to the M_V(RR)-calibration and to its comparison with other distance calibrations that still hinder the coherent and homogeneous understanding of the local distance scale.

In Caputo's (1998) and Reid's (1999) reviews the ages of globulars are discussed. After Hipparcos, ages of globular clusters are reduced by typically 2–3 Gyr, because of both larger distances and improvements in the physics of the models, mainly in the equation of state and in the consideration of microscopic diffusion. Present ages are now in the range 10–13 Gyr, which can be compared to the previous interval of 13–18 Gyr (see VandenBerg et al's review, 1996).

Chaboyer et al (1998) claimed that the (theoretical) absolute magnitude M_V and lifetimes of stars at the MS turn-off (T-O) in globular clusters are now well understood, since the physics involved is very similar to that of the Sun, which is in turn well constrained by seismology. In particular, M_V(T-O) is quite insensitive to uncertainties related to model atmospheres or convection modeling (see also Freytag & Salaris 1999). Chaboyer et al suggested that no significant changes (more than $\sim 5\%$) in the derived ages of globular clusters are expected from future improvements in stellar models. Conversely, distance and abundance determinations are far from definite, and the quasi-verticalness of isochrones in the T-O region makes the determination of M_V(T-O) difficult (see Vandenberg et al 1996). Further revision of the ages is therefore not excluded. Also a global agreement between an entire globular cluster sequence and the corresponding model isochrone is far from being reached.

Age of the Universe   The ages of the oldest objects in the Galaxy, the most metal-poor halo or globular cluster stars, provide a minimum value for the age of the Universe T_U. Globular cluster ages (10–13 Gyr) from comparisons of isochrones with observed M_V(T-O) are presently the most reliable, but two independent methods look promising:

- Thorium (half-life 14.05 Gyr) has been detected and measured by Sneden et al (1996) in an ultra-metal-poor giant, too faint for observation by Hipparcos, and the star radioactive decay age is estimated to be 15 ± 4 Gyr. In the future, such observations of more stars and the possible detection of Rhenium and Uranium could provide strong constraints for T_U.

- Observations of (faint) white dwarfs (WD) in globular clusters are now within reach of experiments on board *HST*, and a lower limit to the age of WD in M4 of ~ 9 Gyr has been derived from a comparison with theoretical WD cooling curves (Richer et al 1997). Future access to cooler and fainter objects will better constrain T_U.

According to Sandage & Tammann (1997) and Saha et al (1999), the Hubble constant H_0 should be in the range 55 ± 5 km $\cdot$ s^{-1} $\cdot$ Mpc^{-1}, which implies $T_U =$

$\frac{2}{3}H_0^{-1} \approx 11$–$13.5$ Gyr, indicating that no strong discrepancy with the age of the oldest known stars remains.

5.2 Variable Stars

I shall not discuss the revisions of the distance scale based on pulsating stars (RR Lyrae, Cepheids, Miras, high-amplitude δScuti stars) because this topic has been extensively reviewed by Caputo (1998) and Reid (1999).

Both new insight as well as new questions about the physics governing pulsating stars have been generated from the combination of Hipparcos distances with asteroseismic data. When the magnitude of a star is modified and its error box reduced, the mass and evolutionary stage attributed to the star may be modified. For variable stars, a different evolutionary stage may give a drastically different eigenmode spectrum, and in turn may change the mode identification and asteroseismic analysis (see Liu et al 1997). Høg & Petersen (1997) showed that for the two double-mode, high-amplitude δScuti variables SX Phe and AI Vel, the masses derived on one hand from stellar envelope models and pulsation theory, and on the other hand from the position in the H-R diagram through stellar evolution models are in nice agreement if the Hipparcos parallaxes (accurate to 5–6%) are used. Further implications of Hipparcos distances on the understanding of δScuti stars, λBootis, and rapidly oscillating Ap stars have been discussed in several papers (see for instance Audard et al 1998, Viskum et al 1998, Paunzen et al 1998, Matthews et al 1999), whereas the physical processes relevant to the Asymptotic Giant Branch and pulsation modeling of Miras and Long Period Variables were examined by Barthès (1998, see also references therein).

5.3 White Dwarfs

The white dwarf (WD) mass-radius (M-R) relation was first derived by Chandrasekhar (1931) from the theory of stars supported by the fully degenerate electron gas pressure. It has been refined by Hamada & Salpeter (1961), who calculated zero-temperature (fully degenerate) WD models of different chemical composition (He, C, Mg, Si, S, and Fe) and by Wood (1995), who calculated WD models with carbon cores and different configurations of hydrogen and/or helium layers and followed the thermal evolution of WD as they cool. Although theoretical support is strong, it has long been difficult to confirm the relation empirically because of (1) the very few available WD with measures of masses and radii, (2) the size of the error bars and (3) the intrinsic mass distribution of the WD, which concentrates them in only a small interval around 0.6 $M_\odot$ (Schmidt 1996).

The M-R relation, assuming that WD have a carbon core, is a basic underlying assumption in most studies of WD properties. It serves to determine the mass of WD, and in turn their mass distribution and luminosity function. It is important because WD feature in many astrophysical applications such as the calibration of distances to globular clusters (Renzini et al 1996) or the estimate of the age of

Galactic disk and halo by means of WD cooling sequences (see Winget et al 1987, d'Antona & Mazzitelli 1990). The more precisely the M-R relation is defined by observations, the better the tests of theoretical models of WD interiors that can be undertaken. These include tests of the inner chemical composition of WD, thickness of the hydrogen envelope of DA WD, or the characterization of their strong inner magnetic fields.

Depending on the white dwarf considered, the empirical M-R relation can be obtained by different means:

1. *Surface brightness method.* If T_{eff} and log g are determined (generally from spectroscopy), then model atmospheres allow calculation of the energy flux at the surface of the star, which when compared to the flux on Earth yields the angular diameter ϕ (see Schmidt 1996). The radius R is obtained from the parallax and ϕ, M is deduced from R and log g. This method requires high-resolution spectra and largely depends on model atmospheres.

2. *Gravitational redshift.* The strong gravitational field at the surface of a WD causes a redshift of the spectral lines, the size of which depends on the gravitational velocity $v_{grs} = \frac{GM}{Rc}$ (c is the speed of light). If v_{grs} can be measured and the gravity is known, then M and R can easily be obtained independently of the parallax. v_{grs} can only be measured in WD members of binary systems, common proper motion pairs (CPM), or clusters because the radial velocity is required to distinguish the gravitational redshift from the line shift due to Doppler effect. Also, very high-resolution spectra are needed.

3. *WD in visual binary systems.* Masses may be derived directly from the orbital parameters through the Kepler's third law, provided the parallax is known. Radii are derived from the knowledge of T_{eff} and distances.

More than 15 years ago, when the Hipparcos project began, uncertainties on WD ground-based parallaxes were at least 10 mas. During the last 10 years, due to great instrumental progress, parallax determinations were improved by a factor of about 2, and more accurate atmospheric parameters T_{eff}, g and v_{grs} were obtained. In the meantime, Hipparcos observed 22 white dwarfs (11 field WD, 4 WD in visual binaries, and 7 in CPM systems) among which 17 are of spectral type DA. Although they are close to the faint magnitude limit of Hipparcos, the mean accuracy on their parallaxes is $\sigma_\pi \simeq 3.6$ mas (Vauclair et al 1997).

Vauclair et al (1997) and Provencal et al (1998) studied the whole sample of WD observed by Hipparcos. The M-R relation is narrower and most points are within 1σ of Wood's (1995) evolutionary models of WD with carbon cores and hydrogen surface layers. The theoretical shape is still difficult to confirm because of the lack of objects in the regions of either high or low mass. Furthermore, the error bars are still too large to distinguish fine features of the theoretical models, such as between evolutionary and zero-temperature sequences or thick and thin hydrogen envelopes (Vauclair et al 1997), except for some particular stars (Shipman et al

1997, Provencal et al 1998). Other effects such as alterations due to strong internal magnetic fields are not yet testable (Suh & Mathews 2000).

WD in Binary Systems   Prior to Hipparcos, Sirius B was the only star roughly located on the expected theoretical M-R relations; the others (Procyon B, 40 Eridani B and Stein 2051) were at least 1.5σ below the theoretical position (see Figure 1 in Provencal et al 1998). After Hipparcos, as shown by Provencal et al, the error on the radius is dominated by errors on flux and T_{eff}. On the other hand, the parallax error still dominates the error on mass, except for Procyon where the error on the component separation plays a major role.

- Sirius B is more precisely located on a Wood's (1995) M-R relation for a DA white dwarf of the observed T_{eff} with a thick H layer and carbon core (Holberg et al 1998). Also compatible with Wood's thick H layer models is the position of V471 Tau, a member of an eclipsing binary system for which the Hipparcos parallax supports the view that it is a member of the Hyades (Werner & Rauch 1997, Barstow et al 1997). The mass of 40 Eri B increased by 14%. The star is now back on Hamada & Salpeter's (1961) M-R relation for carbon cores, making it compatible with single star evolution (Figure 1 by Shipman et al 1997) and it does not appear to have a thick H layer. Sirius B (the most massive known WD) and 40 Eri B (one of the less massive) nicely anchor the high-mass and low-mass limits of the M-R relation.

- The case of Procyon B remains puzzling. The position is not compatible with models with carbon cores, and would be better accounted for with iron or iron-rich core models (Provencal et al 1997). Provencal et al (1998) also examined seven white dwarf members of CPM pairs (more distant and fainter than WD in visual binaries) with Hipparcos distances and gravitational redshift measurements. They showed that two of them also lie on theoretical M-R relations corresponding to iron cores. This is not predicted by current stellar evolution theory, and further work is required to clarify this problem.

In conclusion, better distances from Hipparcos and high-resolution spectroscopy has allowed better assessment of the theoretical M-R relation for white dwarfs, and has shown evidence for difficulties for a few objects that do not appear to have carbon cores. Future progress will come from further parallax improvements and from better T_{eff}, v_{grs}, magnitudes, and orbital parameters for visual binaries. A better understanding of the atmospheres is required: convection plays a role in the cooler WD, additional pressure effects due to undetectable helium affect the gravity determination, and incorrect H layer thickness estimates change the mass attributed to WD. Further information coming from asteroseismology or spectroscopy would help.

6.   FUTURE INVESTIGATIONS

Hipparcos has greatly enlarged the available stellar samples with accurate and homogeneous astrometric and photometric data. To fully exploit this new information, many studies have been undertaken (several hundred papers devoted to stellar studies based on or mentioning Hipparcos data can be found in the literature), therefore the present review could not be fully exhaustive.

The Hipparcos mission succeeded in clarifying our knowledge of nearby objects, and allowed first promising studies of rarer or farther objects. After Hipparcos, the theory of stellar structure and evolution is further anchored, and some of its physical aspects have been better characterized. For instance, new indications that the evolution of low-mass stars is significantly modified by microscopic diffusion have been provided by fine studies of the H-R diagram, and consequences for age estimates or surface abundance alterations have been further investigated. On the other hand, Hipparcos left us with intriguing results that raised new questions. For example, the unexpected position of the white dwarf Procyon B on a theoretical mass-radius relation corresponding to iron cores is still not understood.

Today, uncertainties on distances of nearby stars have been reduced significantly such that other error sources emerge to dominate, hindering further progress in the fine characterization of stellar structure. Progress on atmosphere modeling is worth being pursued, for it has implications for observational parameters (effective temperatures, gravities, abundances, bolometric corrections), theoretical models (outer boundary conditions), and color calibrations. A thorough theoretical description of transport processes (convection, diffusion) and related effects (rotation, magnetic fields) is needed to improve stellar models, as well as further improvements or refinements in microscopic physics (low-temperature opacities, nuclear reaction rates in advanced evolutionary stages).

What is now needed from the observational side is (1) enlarged samples of rare objects (distant objects, faint objects or objects undergoing rapid evolutionary phases), (2) an increased number of more "common" objects with extremely accurate data (including masses), and (3) a census over all stellar populations.

These goals should be (at least partially) achieved by future astrometric missions. The NASA Space Interferometry Mission (SIM), scheduled for launch in 2006, will have the capability to measure parallaxes to 4 μarcsecond and proper motions to 1–2 μarcsecond per year down to the 20th magnitude which represents a gain of three orders of magnitude with respect to Hipparcos (Peterson & Shao 1997). The ESA candidate mission GAIA is dedicated to the observation of about one billion objects down to $V \simeq 20$ mag (and typical $\sigma_\pi \sim 10$ μarcsecond at $V = 15$ mag). GAIA will also provide multi-color, multi-epoch photometry for each object, and will give access to stars of various distant regions of the Galaxy (halo, bulge, thin and thick disk, spiral arms). It is aimed to be launched in 2009 (Perryman et al 1997b).

Asteroseismology has already proved to be a unique tool to probe stellar interiors. Space experiments are under study. The first step will be the COROT mission (aimed to be launched in 2003), designed to detect and characterize oscillation modes in a few hundred stars, including solar-type stars and δ Scuti stars (see Baglin 1998).

ACKNOWLEDGMENTS

I enjoyed very much working with Ana Gómez, Roger and Giusa Cayrel, João Fernandes, Michael Perryman, Nöel Robichon, Annie Baglin, Jean-Claude Mermilliod, Marie-Nöel Perrin, Frédéric Arenou, Catherine Turon, and Corinne Charbonnel on the various subjects presented here. I particularly thank Roger Cayrel, Annie Baglin, Michael Perryman, and Catherine Turon for a careful reading of the original manuscript. I am also very grateful to Allan Sandage for his remarks and suggestions which helped improving the paper. I am especially grateful to Frédéric Thévenin, Hans-Günter Ludwig, Misha Haywood, Don VandenBerg, Marie-Jo Goupil, Pierre Morel, Jordi Carme and Franck Thibault for fruitful discussions and advice. The University of Rennes 1 is acknowledged for working facilities. This review has made use of NASA's Astrophysics Data System Abstract Service mirrored at CDS (Centre des Données Stellaires, Strasbourg, France).

Visit the Annual Reviews home page at www.AnnualReviews.org

LITERATURE CITED

Abbett WP, Beaver M, Davids B, Georgobiani D, Rathbun P, Stein RF. 1997. *Ap. J.* 480:395–99

Alexander DR, Ferguson JW. 1994. *Ap. J.* 437:879–91

Allard F, Hauschildt PH, Alexander DR, Starrfield S. 1997. *Annu. Rev. Astron. Astrophys.* 35:137–77

Alonso A, Arribas S, Martínez-Roger C. 1995. *Astron. Astrophys.* 297:197–215

Alonso A, Arribas S, Martínez-Roger C. 1996a. *Astron. Astrophys. Suppl. Ser.* 117:227–54

Alonso A, Arribas S, Martínez-Roger C. 1996b. *Astron. Astrophys.* 313:873–90

Andersen J. 1991. *Astron. Astrophys. Rev.* 3:91–126

Antonello E, Pasinetti Fracassini LE. 1998. *Astron. Astrophys.* 331:995–1001

Arenou F, Lindegren L, Frœschlé M, Gómez AE, Turon C, et al. 1995. *Astron. Astrophys.* 304:52–60

Axer M, Fuhrmann K, Gehren T. 1994. *Astron. Astrophys.* 291:895–909

Axer M, Fuhrmann K, Gehren T. 1995. *Astron. Astrophys.* 300:751–68

Audard N, Kupka F, Morel P, Provost J, Weiss WW. 1998. *Astron. Astrophys.* 335:954–58

Baglin A. 1999. *Highlights of Astronomy*, ed. J Andersen, *XXIIIrd IAU General Assembly*, 11B:555, Dordrecht: Kluwer

Baglin A and the COROT team 1998. In *New Eyes to See Inside the Sun and the Stars*, ed. F-L Deubner, J Christensen-Dalsgaard, D Kurtz, *IAU Symp 185*, p. 301. Dordrecht: Kluwer

Balbes MJ, Boyd RN, Mathews GJ. 1993. *Ap. J.* 418:229–34

Barnes TG, Evans DS, Moffett TJ. 1978. *MNRAS* 183:285–304

Barstow MA, Holberg JB, Cruise AM, Penny AJ. 1997. *MNRAS* 290:505–14

Barthès D. 1998. *Astron. Astrophys.* 333:647–57

Battrick B, ed. 1997. *HIPPARCOS Venice '97: Presentation of the Hipparcos and Tycho Catalogues and First Astrophysical Results of the Hipparcos Space Astrometry Mission.* Sci. Coord. MAC Perryman, PL Bernacca, ESA SP-402. Noordwijk, The Netherlands: ESA Publ. Div. ESTEC

Belikov AN, Hirte S, Meusinger H, Piskunov E, Schilbach E. 1998. *Astron. Astrophys.* 332:575–85

Benedict GF, McArthur BJ, Nelan E, Story D, Whipple AL, et al. 1994. *PASP* 106:327–36

Bernkopf J. 1998. *Astron. Astrophys.* 332:127–34

Bessell MS, Castelli F, Plez B. 1998. *Astron. Astrophys.* 333:231–50

Blackwell DE, Lynas-Gray AE. 1998. *Astron. Astrophys. Suppl. Ser.* 129:505–15

Blackwell DE, Petford AD, Arribas S, Haddock DJ, Selby MJ. 1990. *Astron. Astrophys.* 232:396–410

Boesgaard AM, Friel ED. 1990. *Ap. J.* 351:467–91

Böhm-Vitense E. 1958. *Zeitschrift f. Astrophysik* 45:108–43

Bressan AG, Bertelli G, Chiosi C. 1981 *Astron. Astrophys.* 102:25–30

Brett JM. 1995. *Astron. Astrophys.* 295:736–54

Brown AGA, Arenou F, van Leeuwen F, Lindegren L, Luri X. 1997. See Battrick 1997, pp. 63–68

Brown TM, Christensen-Dalsgaard J, Weibel-Mihalas B, Gilliland RL. 1994. *Ap. J.* 427:1013–34

Caputo F. 1998. *Astron. Astrophys. Rev.* 9:33–61

Carbon FC. 1979. *Annu. Rev. Astron. Astrophys.* 17:513–49

Castellani V, Ciacio F, Degl'Innocenti S, Fiorentini G. 1997. *Astron. Astrophys.* 322:801–6

Cayrel R, Castelli F, Katz D, van't Veer C, Gómez AE, Perrin M-N. 1997a. See Battrick 1997, pp. 433–35

Cayrel R, Faurobert-Scholl M, Feautrier N, Spielfieldel A, Thévenin F. 1996. *Astron. Astrophys.* 312:549–52

Cayrel R, Lebreton Y, Perrin M-N, Turon C. 1997b. See Battrick 1997, pp. 219–24

Cayrel de Strobel G, Crifo F, Lebreton Y. 1997a. See Battrick 1997, pp. 687–88

Cayrel de Strobel G, Soubiran C, Friel ED, Ralite N, François P. 1997b. *Astron. Astrophys. Suppl. Ser.* 124:299–305

Chaboyer B. 1998. In *Post-Hipparcos Cosmic Candles*, ed. A Heck, F Caputo, *Astrophys. Space Sci. Libr.* 237:111–25, Dordrecht: Kluwer

Chaboyer B, Demarque P, Kernan PJ, Krauss LM. 1998. *Ap. J.* 494:96–110

Chabrier G, Baraffe I. 1997. *Astron. Astrophys.* 327:1039–53

Chandrasekhar S. 1931. *MNRAS* 91:456

Chieffi A, Straniero O, Salaris M. 1995. *Ap. J.* 445:L39–42

Christensen-Dalsgaard J. 1982. *MNRAS* 199:735–61

Christensen-Dalsgaard J, Däppen W. 1992. *Astron. Astrophys. Rev.* 4:267–361

Christensen-Dalsgaard J, Proffitt CR, Thompson MJ. 1993. *Ap. J.* 403:L75–78

Clementini G, Gratton RG, Carretta E, Sneden C. 1999. *MNRAS* 302:22–36

Code AD, Davis J, Bless RC, Hanbury Brown R. 1976. *Ap. J.* 203:417–34

D'Antona F, Mazzitelli I. 1990. *Annu. Rev. Astron. Astrophys.* 28:139–81

Di Benedetto GP. 1998. *Astron. Astrophys.* 339:858–71

Dravins D, Lindegren L, Madsen S, Holmberg J. 1997. See Battrick 1997, pp. 733–38

Edmonds P, Cram L, Demarque P, Guenther DB, Pinsonneault MH. 1992. *Ap. J.* 394:313–19

Eggen OJ, Sandage A. 1962. *Ap. J.* 136:735–47

Eur. Space Agency. 1997. *The Hipparcos and Tycho Catalogues*, Eur. Space Agency Spec. Publ. 1200. Noordwijk, The Netherlands: ESA Publ. Div. ESTEC

Fernandes J, Lebreton Y, Baglin A 1996. *Astron. Astrophys.* 311:127–34

Fernandes J, Lebreton Y, Baglin A, Morel P. 1998. *Astron. Astrophys.* 338:455–64

Fernandes J, Neuforge C. 1995. *Astron. Astrophys.* 295:678–84

Freytag B, Ludwig HG, Steffen M. 1999. In: "Theory and Tests of Convection in Stellar Structure," ASP, Conference Series, Vol. 173, eds: A Giménez, EF Guinan, and B Montesinos, pp. 225–28

Freytag B, Salaris M. 1999. *Ap. J.* 513:L49–52

Fuhrmann K. 1998. *Astron. Astrophys.* 330: 626–30

Fuhrmann K, Pfeiffer M, Frank C, Reetz J, Gehren T. 1997. *Astron. Astrophys.* 323:909–22

Gatewood G, Kiewiet de Jonge J, Persinger T. 1998. *Astron. J.* 116:1501–3

Goupil M-J, Dziembowski WA, Goode PR, Michel E. 1996. *Astron. Astrophys.* 305:487–97

Gratton RG, Fusi Pecci F, Carretta E, Clementini G, Corsi CE, Lattanzi M. 1997. *Ap. J.* 491:749–71

Grevesse N, Sauval AJ. 1999. *Astron. Astrophys.* 347:348–54

Groenewegen MAT, Salaris M. 1999. *Astron. Astrophys.* 348:L33–36

Gustafsson B. 1998. In *Fundamental Stellar Properties: The Interaction between Observation and Theory*, ed. TR Bedding, AJ Booth, J Davis, *IAU Symp. 189*, p. 261, Dordrecht: Kluwer

Gustafsson B, Bell RA, Eriksson K, Nordlund Å. 1975. *Astron. Astrophys.* 42:407–32

Hamada T, Salpeter EE. 1961. *Ap. J.* 134:683–98

Hanson RB. 1979. *MNRAS* 186:875–96

Harris HC, Dahn CC, Monet DG. 1997. See Battrick 1997, pp. 105–11

Harrison TE, McNamara BJ, Szkody P, McArthur BE, Benedict GF, et al. 1999. *Ap. J.* 515:L93–96

Henyey LG, Wilets L, Böhm KH, LeLevier R, Levee RD. 1959. *Ap. J.* 129:628–36

Høg E, Petersen JO. 1997. *Astron. Astrophys.* 323:827–30

Holberg JB, Barstow MA, Bruhweiler FC, Cruise AM, Penny AJ. 1998. *Ap. J.* 497:935–42

Huebner WF, Merts AL, Magee NH, Argo MF. 1977. *Los Alamos Sci. Rep. LA-6760-M*

Iglesias CA, Rogers FJ. 1996. *Ap. J.* 464:943–53

Izotov YI, Thuan TX, Lipovetsky VA. 1997. *Ap. J. Suppl.* 108:1–39

Jimenez R, Flynn C, Kotoneva E. 1998. *MNRAS* 299:515–19

King JR, Stephens A, Boesgaard AM, Deliyannis CP. 1998. *Astron. J.* 115:666–84

Kosovichev AG, Christensen-Dalsgaard J, Däppen W, Dziembowski WA, Gough DO, Thompson MJ. 1992. *MNRAS* 259:536–58

Kovalevsky J. 1998. *Annu. Rev. Astron. Astrophys.* 36:99–129

Kurucz RL. 1979. *Ap. J. Suppl.* 40:1–340

Kurucz RL. 1991. In *Stellar Atmospheres: Beyond Classical Models*, ed. L Crivellari, I Hubeny, DG Hummer, *NATO ASI Ser.*, pp. 441–48, Dordrecht: Kluwer

Kurucz RL. 1993. In *Peculiar versus Normal Phenomena in A-type and Related Stars*, ed. MM Dworetsky, F Castelli, R Faraggiana, *IAU Colloq. 138*, pp. 87, San Francisco: Astron. Soc. Pac.

Layden AC. 1998. In *Post-Hipparcos Cosmic Candles*, ed. A Heck, F Caputo, *Astrophys. Space Sci. Lib.* 237:37–52, Dordrecht: Kluwer

Lebreton Y. 2000. See Livio, 2000, pp. 107

Lebreton Y, Gómez AE, Mermilliod J-C, Perryman MAC. 1997a. See Battrick 1997, pp. 231–36

Lebreton Y, Michel E, Goupil MJ, Baglin A, Fernandes J. 1995. In *Astronomical and Astrophysical Objectives of Sub-Milliarcsecond Optical Astrometry*, ed. E Høg, PK Seidelman, *IAU Symp. 166*, pp. 135–42, Dordrecht: Kluwer

Lebreton Y, Perrin M-N, Cayrel R, Baglin A, Fernandes J. 1999. *Astron. Astrophys.* 350:587–97

Lebreton Y, Perrin M-N, Fernandes J, Cayrel R, Baglin A, Cayrel de Strobel G. 1997b. See Battrick 1997, pp. 379–82

Lindegren L. 1995. *Astron. Astrophys.* 304:61–68

Lindegren L, Mignard F, Söderhjelm S, Badiali M, Bernstein HH, et al. 1997. *Astron. Astrophys.* 323:L53–56

Liu YY, Baglin A, Auvergne M, Goupil M-J, Michel E. 1997. See Battrick 1997, pp. 363–66

Livio M. ed. 2000. *Unsolved Problems in Stellar Evolution* STScI May 1998 Symp., Cambridge: Cambridge University Press

Ludwig HG, Freytag B, Steffen M. 1999. *Astron. Astrophys.* 346:111–24

Lutz TE, Kelker DH. 1973. *PASP* 85:573–78

Lydon TJ, Fox PA, Sofia S. 1993. *Ap. J.* 413:390–400

Maeder A, Mermilliod J-C. 1981. *Astron. Astrophys.* 93:136–49

Maeder A, Peytremann E. 1972. *Astron. Astrophys.* 21:279–84

Martin C, Mignard F. 1998. *Astron. Astrophys.* 330:585–99

Martin C, Mignard F, Frœschlé M. 1997. *Astron. Astrophys. Suppl. Ser.* 122:571–80

Martin C, Mignard F, Hartkopf WI, McAlister HA. 1998. *Astron. Astrophys. Suppl. Ser.* 133:149–62

Matthews JM, Kurtz DW, Martinez P. 1999. *Ap. J.* 511:422–28

Mc Williams A. 1997. *Annu. Rev. Astron. Astrophys.* 35:503–56

Mermilliod J-C. 2000. In *Very Low-Mass Stars and Brown Dwarfs in Stellar Clusters and Associations*, eds. R Rebolo, MR Zapatorio-Osorio, Cambridge: Cambridge University Press

Mermilliod J-C, Turon C, Robichon N, Arenou F, Lebreton Y. 1997. See Battrick 1997, pp. 643–50

Michel E, Hernández MM, Houdek G, Goupil M-J, Lebreton Y, et al. 1999. *Astron. Astrophys.* 342:153–66

Mignard F. 1997. See Battrick 1997, pp. 5–10

Mihalas D, Däppen W, Hummer DG. 1988. *Ap. J.* 331:815–25

Morel P, Baglin A. 1999. *Astron. Astrophys.* 345:156–62

Morel P, Morel Ch, Provost J, Berthomieu G. 2000. *Astron. Astrophys.* 354:636–44

Morel P, van't Veer C, Provost J, Berthomieu G, Castelli F, et al. 1994 *Astron. Astrophys.* 286:91–102

Narayanan VK, Gould A. 1999a. *Ap. J.* 515:256–64

Narayanan VK, Gould A. 1999b. *Ap. J.* 523:328–39

Ng YK, Bertelli G. 1998. *Astron. Astrophys.* 329:943–50

Nissen PE. 1976. *Astron. Astrophys.* 50:343–52

Nissen PE. 1998. In *The First MONS Workshop: Science with a Small Space Telescope*, eds. H Kjeldsen, TR Bedding, pp. 99–104, Aarhus: Aarhus Universitet

Nissen PE, Høg E, Schuster WJ. 1997. See Battrick 1997, pp. 225–30

Noels A, Fraipont D, Gabriel M, Grevesse N, Demarque P eds. 1995. *Stellar Evolution*: What Should Be Done? Proc. 32nd Liège International Astrophysical Coll., Liège: Univ. de Liège

Noels A, Grevesse N, Magain P, Neuforge C, Baglin A, Lebreton Y. 1991. *Astron. Astrophys.* 247:91–94

Pagel BEJ, Portinari L. 1998. *MNRAS* 298:747–52

Paunzen E, Weiss WW, Kuschnig R, Handler G, Strassmeier KG, et al. 1998. *Astron. Astrophys.* 335:533–38

Pérez-Hernández F, Claret A, Hernández MM, Michel E. 1999. *Astron. Astrophys.* 346:586–98

Perryman MAC, Brown AGA, Lebreton Y, Gómez AE, Turon C, et al. 1998. *Astron. Astrophys.* 331:81–120

Perryman MAC, Lindegren L, Kovalevsky J, Høg E, Bastian U, et al. 1997a. *Astron. Astrophys.* 323:L49–L52

Perryman MAC, Lindegren L, Turon C. 1997b. See Battrick 1997, pp. 743–48

Peterson DM, Shao M. 1997. See Battrick 1997, pp. 749–53

Peterson DM, Solensky R. 1988. *Ap. J.* 333:256–66

Pinsonneault MH. 1997. *Annu. Rev. Astron. Astrophys.* 35:557–605

Pinsonneault MH, Stauffer J, Soderblom DR, King JR, Hanson RB. 1998. *Ap. J.* 504:170–91

Platais I, Kozhurina-Platais V, van Leeuwen F. 1998. *Astron. J.* 116:2423–30

Pont F, Mayor M, Turon C, VandenBerg DA. 1998. *Astron. Astrophys.* 329:87–100

Pont F, Charbonnel C, Lebreton Y, Mayor M, Turon C, VandenBerg DA. 1997. See Battrick 1997, pp. 699–704

Popper DM. 1998. *PASP* 110:919–22

Pourbaix D, Neuforge-Verheecke C, Noels A. 1999. *Astron. Astrophys.* 344:172–76

Proffitt CR, VandenBerg DA. 1991. *AP. J. Suppl.* 77:473–514

Provencal JL, Shipman HL, Høg E, Thejll P. 1998. *Ap. J.* 494:759–67

Provencal JL, Shipman HL, Wesemael F, Bergeron P, Bond HE, et al. 1997. *Ap. J.* 480:777–83

Reid IN. 1997. *Astron. J.* 114:161–79

Reid IN. 1998. *Astron. J.* 115:204–28

Reid IN. 1999. *Annu. Rev. Astron. Astrophys.* 37:191–237

Renzini A, Bragaglia A, Ferraro FR, Gilmozzi R, Ortolani S, et al. 1996. *Ap. J.* 465:L23–26

Ribas I. 1999. PhD thesis. Univ. de Barcelona, Spain

Ribas I, Giménez A, Torra J, Jordi C, Oblak E. 1998. *Astron. Astrophys.* 330:600–4

Richard O, Vauclair S, Charbonnel C, Dziembowski WA. 1996. *Astron. Astrophys.* 312 1000–11

Richer HB, Fahlman GG, Ibata RA, Pryor C, Bell RA, et al. 1997. *Ap. J.* 484:741–60

Robichon N, Arenou F, Turon C, Mermilliod J-C, Lebreton Y. 1997. See Battrick 1997, pp. 567–70

Robichon N, Arenou F, Lebreton Y, Turon C, Mermilliod J-C. 1999b. In *Harmonizing Cosmic Distance Scales in a Post-Hipparcos Era*, ed. D Egret, A Heck. ASP Conf. Ser. 167, p. 72–77

Robichon N, Arenou F, Mermilliod J-C, Turon C. 1999a. *Astron. Astrophys.* 345:471–84

Rogers FJ, Iglesias CA. 1992. *Ap. J. Suppl.* 79:507–68

Rogers FJ, Swenson FJ, Iglesias CA. 1996. *Ap. J.* 456:902–8

Roxburgh IW. 1997. In *Solar Convection and Oscillations and their Relationship*, SCORe '96, ed. FP Pijpers, J Christensen-Dalsgaard, C Rosenthal, *Astrophys. Space Sci. Lib.* 225:23–50, Dordrecht: Kluwer

Saha A, Sandage A, Labhardt L, Tammann GA, Macchetto FD, Panagia N. 1999. *Ap. J.* 486:1–20

Salaris M, Groenewen MAT, Weiss A. 2000. *Astron. Astrophys.* 355:299–307

Sandage A. 1970. *Ap. J.* 162:841–70

Sandage A. 1983. *Astron. J.* 88:1569–78

Sandage A, Tammann GA. 1997. In *Critical Dialogues in Cosmology*, ed. N Turok, pp. 130, Singapore: World Scientific

Saumon D, Chabrier G. 1991. *Phys. Rev. A* 44:5122–41

Schmidt H. 1996. *Astron. Astrophys.* 311:852–57

Schröder K-P. 1998. *Astron. Astrophys.* 334:901–10

Schwarzschild M. 1958. *Structure and Evolution of the Stars*. Princeton: Princeton University Press

Seaton MJ, Yan Y, Mihalas D, Pradhan AK. 1994. *MNRAS* 266:805–28

Shipman HL, Provencal JL, Høg E, Thejll P. 1997. *Ap. J.* 488:L43–46

Smith H. 1987. *Astron. Astrophys.* 188:233–38

Sneden C, McWilliam A, Preston GW, Cowan JJ, Burris DL, Armosky BJ. 1996. *Ap. J.* 467:819–40

Soderblom DR, King JR, Hanson RB, Jones BF, Fischer D, et al. 1998. *Ap. J.* 504:192–99

Söderhjelm S. 1999. *Astron. Astrophys.* 341:121–40

Stein RF, Nordlund Å. 1998. *Ap. J.* 499:914–33

Suh I-S, Mathews GJ. 2000. *Ap. J.* 530:949–54

Talon S, Zahn J-P, Maeder A, Meynet G. 1997. *Astron. Astrophys.* 322:209–17

Thévenin F, Idiart TP. 1999. *Ap. J.* 521:753–63

Torres G, Stefanik RP, Latham DW. 1997a. *Ap. J.* 474:256–71

Torres G, Stefanik RP, Latham DW. 1997b. *Ap. J.* 479:268–78

Torres G, Stefanik RP, Latham DW. 1997c. *Ap. J.* 485:167–81

van Altena WF, Lee JT, Hoffleit ED. 1997b. *Baltic Astron* 6:27

van Altena WF, Lu C-L, Lee JT, Girard TM, Guo X, et al. 1997a. *Ap. J.* 486:L123–27

VandenBerg DA, Bolte M, Stetson PB. 1996. *Annu. Rev. Astron. Astrophys.* 34:461–510

van Leeuwen F. 1997. *Space Science Reviews* 81:201–409

van Leeuwen F. 1999a. In *Harmonizing Cosmic Distance Scales in a Post-Hipparcos Era*, ed. D Egret, A Heck, ASP Conf. Ser. 167, p. 52–71

van Leeuwen F. 1999b. *Astron. Astrophys.* 341:L71–74

van Leeuwen F, Evans DW. 1998. *Astron. Astrophys. Suppl. Ser.* 130:157–172

van Leeuwen F, Evans DW, Grenon M, Grossmann V, Mignard F, Perryman MAC. 1997. *Astron. Astrophys.* 323:L61–64

Vauclair G, Schmidt H, Koester D, Allard N. 1997. *Astron. Astrophys.* 325:1055–62

Viskum M, Kjeldsen H, Bedding TR, Dall TH, Baldry IK, et al. 1998. *Astron. Astrophys.* 335:549–60

Werner K, Rauch T. 1997. *Astron. Astrophys.* 324:L25–28

Wheeler JC, Sneden C, Truran JW. 1989. *Annu. Rev. Astron. Astrophys.* 27:279–349

Winget DE, Hansen CJ, Liebert J, Van Horn HM, Fontaine G, et al. 1987. *Ap. J.* 315:L77–81

Wood MA. 1995. In *Proc. 9th European Workshop on White Dwarfs*, ed. D Koester, K Werner, pp. 41–45, Berlin: Springer-Verlag

Zahn J-P. 1992. *Astron. Astrophys.* 265:115–32

Annu. Rev. Astron. Astrophys. 2000. 38:79–111

THE FIRST 50 YEARS AT PALOMAR, 1949–1999 ANOTHER VIEW: Instruments, Spectroscopy and Spectrophotometry and the Infrared

George Wallerstein

*Department of Astronomy, University of Washington, Seattle, Washington 98195-1580;
e-mail: wall@astro.washington.edu*

J. B. Oke

*National Research Council, Herzberg Institute for Astrophysics, Victoria, British
Columbia V8X 4M6, Canada; e-mail: bev.oke@hia.nrc.ca*

PROLOGUE

In volume 37 (1999) Allan Sandage gave an account of the research carried
out at Palomar in the first fifty years of the operation of the Hale telescope.
In that article he described much of the work carried out in the areas of
galaxies and cosmology as well as the early work on stellar evolution and
high-energy asrophysics. However, so much important research was carried
out at Palomar in the first half century of operation that it was impossible to
cover everything in a single article. Thus, I invited George Wallerstein and
J. B. Oke to write a second article covering many of the achievements that
were not described by Allan Sandage. Of course, there are still many
omissions. For example, the work of Fritz Zwicky and his collaborators in
the extragalactic field, much of it done with the Palomar Schmidt telescopes,
and the later work of Chip Arp on galaxies, radio sources and QSOs has not
been described in either article.

—Geoffery Burbidge, Editor

■ **Abstract** We review the research on a wide variety of topics using data obtained
with the 200-inch Hale telescope. Using state-of-the-art spectrographs, photometers,
spectrometers and infrared detectors, the Palomar astronomers investigated the spectra
of stars, interstellar matter, AGNs and quasars in great detail. Spectral resolutions
ranged from 1000 A for broad-band photometry to 0.04 A using interferometric
techniques.

0066-4146/00/0915-0079$14.00

INTRODUCTION

As the 200-inch Hale telescope came into full operation in the late 1940s and early 1950s, the stage was set for great advances in both stellar and extragalactic astronomy. By that time, much of the basic physics of stellar interiors had been established. For example, in 1955, Härm & Schwarzschild published a summary of their numerical integrations to model the stellar structure not only of the main sequence stars but also of red giants. They included some of the first applications of "large electronic computers" to this kind of study. Methods for modeling stellar atmospheres had been developed by S Chandrasekhar, VA Ambartsumian, and others and were summarized in Unsöld's monograph (Unsöld 1955).

By the early 1950s, research on stellar spectroscopy had moved from the era of line identification and radial velocity measurements into fields that were closely related to theoretical astrophysics and cosmology. Analysis of stellar atmospheres had been initiated during the pre-World War II era by individuals such as C Payne-Gaposhkin, HN Russell, A Unsöld, and many others. On the observational side, three astronomers, AH Joy, PW Merrill, and RF Sanford, had used the Mount Wilson telescopes and spectrographs to assemble a huge database of line identifications and radial velocities that proved to be of great value to the Palomar astronomers working on general problems of astrophysics. Two developments in 1951–1952 proved to be particularily auspicious for the future. One of these was the discovery by Merrill (1952a) of the unstable element technitium in S-type stars and soon after in some carbon stars. Because of its short half-life the presence of Tc in stars showed that nuclear reactions must be taking place in the interiors of stars in which Tc appeared. The other was the analysis by Chamberlain & Aller (1951) of two subdwarfs that they found to be metal poor by a factor 100 relative to the sun (Croswell 1995, Wallerstein 1999). One further important development was on the Caltech Campus, where W. A. Fowler and his associates were conducting experiments in the Kellogg Laboratory on the nuclear reactions first recognized by Bethe (1939) to be responsible for energy generation in the sun and stars.

In this review we have strongly concentrated on research with the 200-inch Hale telescope and have not covered topics studied primarily with the 48-inch and 18-inch Schmidt telescopes or the relatively new 60-inch reflector. Even within these limits it would be impossible to cover all areas of research carried out with the 200-inch telescope. Instead we have selected a limited, perhaps biased, number of areas to discuss.

INSTRUMENTATION

Optical Instrumentation

Not only was the 200-inch telescope by far the largest in the world, it also was equipped with state-of-the-art instrumentation. The spectrographs were designed by the director, IS Bowen, who had been a member of the Caltech physics department engaged in optical spectroscopy. They employed high-speed Schmidt cameras to provide maximum efficiency and speed with superb optical quality. They are described in detail by Bowen (1952). The Coudé spectrograph had a beam size of 12 inches, which was required if the full-potential light-gathering power of the 200-inch telescope was to be achieved. Such an unprecedented beam size was made possible by using a mosaic of four large plane reflection gratings rule by HW Babcock and very fast on-axis Schmidt cameras. This spectrograph provided dispersions ranging from 2.2 to 37 Å/mm. The very fast prime-focus spectrograph with solid Schmidt cameras provided dispersions 105–430 Å/mm for work on very faint stars and galaxies.

These first spectrographs, along with a prime-focus imaging system based on a Ross corrector, were intended for use with photographic plates, and these were used almost exclusively until the mid-1960s. Early on, a prime-focus photoelectric photometer designed to use cooled photomultiplier tubes was built. Conventional cold boxes, direct current amplifiers, and strip chart recorders were provided. One staff member, WA Baum, whose PhD was in physics rather than astronomy, developed a pulse-counting system for the photometer that was based on techniques used in physics laboratories, and which had many potential advantages (Baum 1955a,b). Nearly all subsequent work with photomultiplier tubes at Palomar used pulse-counting techniques.

To make use of the new photocathodes with their superior quantum efficiencies, two spectroscopic instruments were built in the mid-1960s. The first of these, completed in 1965, was the prime-focus photoelectric spectrum scanner, which used single photomultiplier tubes and which could measure absolute spectral energy distributions from 3,200 to 10,000 Å, with a resolution of 20–160 Å of very faint stars and galaxies. The second instrument, completed in 1967, was the Cassegrain image tube spectrograph (Dennison et al 1969), which used an image intensifier and reimaged the phosphor onto a photographic plate. This instrument, because of its high speed, essentially replaced the prime-focus nebular spectrograph. In 1968, the multichannel spectrometer (MCSP) (Oke 1969b) was completed. This instrument (a) was double, (b) separated the first- and second-order spectra with a dichroic mirror, and (c) had 32 photomultiplier tubes, which could cover the range from 3,100 to 11,000 Å simultaneously with a resolution of 20–360 Å and which could measure sky and object simultaneously. It could measure absolute spectral energy distributions of extremely faint objects with high accuracy and was limited only by photon statistics. The large digital output was stored on paper tape using a hardwired digital system.

The first computer system for telescope control and data acquisition was designed and installed in 1971 under the direction of Dr. EW Dennison.

Although pulse-counting technology was well in hand when using single photomultiplier tubes, the next step was clearly to combine the high quantum efficiency and low dark current of photocathodes with an area detection capability. Image intensifiers were mounted on the various cameras in the Coudé spectrograph to provide high speed with high dispersion. Another effort in this direction was the digital image recorder built in a collaboration between Princeton University and Caltech. This used a high-gain intensified Vidicon tube with enough gain so that individual photon events could be detected and recorded spatially. This device was used at Palomar in the Coudé spectrograph on numerous occasions (Lowrance et al 1972), but the cost of the special Vidicons made the program nonsustainable. A similar device, the image photon counting system, was developed in England by Boksenberg (1972) and began being used at Palomar in 1973 (Boksenberg & Sargent 1975).

In the late 1970s, experimental charge coupled devices (CCDs) began to appear. Caltech and the Jet Propulsion Laboratories began a collaboration to make these devices suitable for astronomical use. In 1977, Oke and coworkers constructed a cooler and electronics for a Texas Instruments 100×160 CCD and used it to obtain low-resolution far-red and near-infrared spectra of Seyferts and quasars (Oke 1978). At about the same time, JA Westphal and J Kristian were building a Dewar and electronics for a Texas Instruments 400×400 CCD. This was first used in 1977–1978 on the 200-inch telescope by Young et al (1978).

The first instrument designed specifically to use CCDs was the prime-focus universal extragalactic instrument designed and built in 1979 (Gunn & Westphal 1981). This used first a Texas Instruments 500×500 pixel CCD and somewhat later an 800×800 device. It was used only for direct imaging. The next major optical instrument was the double spectrograph, which was designed to go at the Cassegrain focus. It separated the light into two wide bands just below the input slit and the two bands went into two separate spectrographs, each of which had an 800×800 pixel CCD camera (Oke & Gunn 1982). A later addition allowed the use of a multislit, which could have up to eight simultaneous slits. Shortly after this spectrograph was completed, Gunn designed and built the four shooter (Gunn et al 1987), which produced four contiguous direct images on four 800×800 pixel CCDs. In addition, a small part of the field could feed a low-resolution CCD spectrograph; up to about 10 objects could be observed simultaneously.

In 1992, the Norris multi-object fiber spectrograph was completed (Hamilton et al 1993). It consisted of 178 fiber heads deployable onto a magnetic plate using an x-y positioner. In 1994, the Carnegie Observatories, which were part of the consortium running the Hale telescope, completed a prime-focus multislit and imaging camera called COSMIC (Kells et al 1998). In the imaging mode, COSMIC uses a variety of broad-band and narrow-band filters. In the spectroscopic mode, it can have as many as 50 slits in a single row.

Infrared Instrumentation

The first instrumentation used single-element detectors, PbS photoconductors, for the near-infrared and Germanium bolometers for the mid-infrared. The first mapping of the galactic center was done with these. In:Sb photovoltaic detectors soon replaced the PbS ones. Early work was done with the f/16 standard Casegrain secondary, but in 1973 this was replaced by an f/70 chopping secondary.

In 1989, a 64×64 element In:Sb array was introduced for use with the f/70 chopping secondary. A guider was added to allow accurate offsetting and long exposure times. This array was replaced by a 256×256 array in 1995. With a scale changer, it was possible to carry out speckle observations. This array is also used with two cooled spectrographs, one a single slit spectrograph with a resolution of 850–3000 and the other an integral field spectrometer (Murphy et al 1999) with comparable resolution. The secondary chopper was modified to allow motion in two directions to correct image wander.

Cornell, one of the partners who operate the Hale telescope, has built a 10 μm array camera for both imaging and spectrographic observations.

The Caltech infrared group made a millimeter bolometer camera that was used between 1973 and 1983 on the Hale telescope, which at the time was the largest millimeter dish in existence. Its purpose was to carry out millimeter-wavelength projects before the large dedicated millimeter dishes at very dry sites became available.

PLANETARY ATMOSPHERES

Following the early work on the spectra of planets by T Dunham and RS Richardson using the 100-inch Coudé to observe the atmospheres of planets, there was little research in this field by the Mt. Wilson and Palomar staffs. In the 1970s, however, G Münch of Caltech participated in the development of a Fabry-Perot spectrometer called PEPSIOS for the mapping of faint emission-line sources. They applied it to the [S II] nebula around Jupiter. The nebula consists of ions that originated from the volcanoes on Io and are trapped in Jupiter's magnetic field (Trauger, Münch & Roesler 1980).

Using the same PEPSIOS instrument, Münch, Trauger & Roesler (1977) searched unsuccessfully for the H_2 (3,0) S1 line in the spectrum of Titan.

Using both the multislit at the 200-inch and the Fabry-Perot, Münch & Bergstrahl (1977) mapped the distribution and its changes of Na I emission around Io. They related the distribution of the emission to the interaction of Jupiter's magnetic field with Io. The electron fluxes necessary to support the observed phenomena turned out to be an order of magnitude greater than the proton fluxes measured by the Pioneer 10 spacecraft.

By observing the integrated spectrum of Uranus Trauger, Roesler & Münch (1978) used a resolution of 0.035 Å (i.e. 140,000) and derived a rotational broadening of a reflected solar line. The resulting equatorial velocity so derived is 3.5 ± 0.4 km s^{-1}. Their measurement was only 0.5 km s^{-1} less than those of

Moore & Menzel (1930) and Lowell & Slipher (1912). The rotational period is then 13.0 ± 1.3 h.

RADIAL VELOCITIES

High-Precision Velocities

Beginning in 1971, RF Griffin and JE Gunn installed a velocity meter (Griffin 1967) on the 144-inch camera of the 200-inch telescope to measure radial velocities in star clusters. The purpose was both to derive orbits of members that are spectroscopic binaries and to investigate the velocity dispersion of the cluster. Their first target was the Hyades. A paper summarizing their results for 27 binaries (Griffin et al 1985) added greatly to our knowledge of the mass-luminosity relation along the Hyades main sequence. From the observation of more than 400 Hyades stars, Griffin et al (1988) and Gunn et al (1988) determined a new convergent point and hence a new distance modulus to this fundamental open cluster. Their value of $m - M = 3.23 \pm 0.10$ confirmed the suggestion by Hodge & Wallerstein (1966) that the then generally accepted modulus of 3.03 was significantly too small.

Working with Griffin et al and Gunn et al on open clusters, Mathieu et al (1986) measured velocities of 39 and 170 stars in M11 and M67, respectively. Some of the spectra were obtained with the facilities of the Center for Astrophysics at Harvard. In M11, 28 stars appear to be radial velocity members. The two single-lined binaries appear not to be members. In M67, 118 appear to be members.

Turning to the globular clusters, Gunn & Griffin (1979) determined the radial velocities of 111 stars in M3 with a typical uncertainty of 1.0 km s^{-1}. Except for two objects of very high velocity, the data may be accounted for by standard "King-Michie" models without any dark matter. Continuing their work on globulars, Lupton et al (1987) measured radial velocities of 147 stars in M13. The one-dimensional central velocity dispersion of the giants is about 7 km s^{-1} and the rotational velocity is less than 5 km s^{-1}. They used their models to estimate the mass function for the lower main sequence to below 0.5 M$_\odot$.

Velocities of Subdwarfs

Using both the Palomar 20-inch and Mt. Wilson 60-inch telescopes, Sandage (1964, 1969) initiated a program to discover new subdwarfs by obtaining UBV photometry of high proper motion stars. He then used the 18-inch camera on the 200-inch Coudé spectrograph on nonphotometric nights to obtain radial velocities of 112 of 300 photometric candidates. A full photometric catalogue of 1690 stars observed with the Mt. Wilson 60-inch and Palomar 20-inch telescopes was finally published by Sandage & Kowal (1986), accompanied by a radial velocity catalogue (Fouts & Sandage 1986).

WHITE DWARFS

The Hale 200-inch telescope and its auxilliary instruments made it possible to carry out a thorough search for and study of white dwarfs, which are intrinsically very faint objects.

The work was begun in the 1950s by OJ Eggen and JL Greenstein, with the first publication in 1965. They used primarily the proper motion catalogs of Giclas and Luyten to pick out candidate white dwarfs. Then photometric observations and spectra were obtained to verify that the candidates were white dwarfs. A series of 11 papers was produced, first by Eggen and Greenstein and then by Greenstein and various collaborators with data on over 500 white dwarfs. The eighth paper (Greenstein 1974) gives the references for the first six papers; the eleventh and last paper in the series was published in 1979 (Greenstein 1979). Some time after the MCSP became available (Oke 1969b), many of the discovered white dwarfs were observed with this instrument, which could produce absolute spectral energy distributions and sufficient resolution and sensitivity that even the faint broad spectral features could be studied quantitatively. These studies are in papers 9–11.

The data available from these papers included spectra and spectral classifications, colors and magnitudes, and spectral energy distributions (Sion et al 1983, Greenstein & Liebert 1990). The list provided for the first time a good sample of white dwarfs that were not only of spectral types DAs, but also of types DOs, DBs, DFs, etc, as well as some very peculiar objects. Greenstein's spectra of the cool white dwarf, Van Maanen 2 were analyzed in detail with model atmospheres by Weidemann (1960).

Because white dwarfs often have large parallaxes and are sometimes in binary systems, and because there is a well-known mass-radius relation, they are prime candidates for measuring the gravitational redshift, which can be in the range of 20–90 km sec^{-1} and which gives an additional relation between mass and radius.

An early study of the gravitational redshifts was by Greenstein & Trimble (1967). They measured the spectral line positions in 53 white dwarfs, 37 of which were of type DA and six of which were in the Hyades cluster. They carried out a statistical analysis, which gave a net redshift of 62.5 km s^{-1} when the Hyades white dwarfs were omitted. For the Hyades white dwarfs, the actual radial velocity was known and the resulting redshift was 53 km s^{-1}. Using the best available effective temperatures, they concluded that the average radius was 0.0107 R$_\odot$ and the average mass was 0.98 M$_\odot$. A small correction to the effective temperature scale led to a mean mass of 0.86 M$_\odot$.

The next such study was the daunting task of measuring the gravitational redshift of Sirius B (Greenstein et al 1971). Sirius B was then, at the time of observation in 1971, at a nearly maximum angular distance from Sirius A in its orbit, but it was necessary to apodize the aperture of the Hale telescope to move the diffraction spikes away from Sirius B. Excellent spectra were obtained with the Coudé spectrograph, which yielded a gravitational redshift of 89 ± 16 km s^{-1}. Furthermore, high-quality Balmer line profiles were derived, which, when compared with the best current model stellar atmospheres yielded an accurate effective temperature of 32000 K and a radius of 0.0078 $R_\odot$. The measured gravitational redshift agrees well with that predicted from the mass and radius.

Greenstein & Trimble (1972) found a gravitational redshift of 23 ± 5 km s^{-1} for 40 Eri B, which was in good agreement with an earlier determination of 21 km s^{-1} by Popper (1954) and that predicted by the mass and radius. Greenstein et al (1977b) obtained high-resolution spectra of 13 DA white dwarfs. They found that Hα has a very narrow core, and for the majority of the stars, they found rotational velocities of less than 40 km s^{-1}. For the remainder, the rotational velocities are less than 60–75 km s^{-1}.

With the completion first of the prime-focus scanner and then of the MCSP, it became possible to measure accurate absolute energy distributions for very faint stars. These instruments were particularly useful for white dwarfs, with their very broad and often shallow spectral lines.

Absolute spectral energy distributions from 3,200 to 10,000 Å were obtained for six white dwarfs, and these were compared with white dwarf model atmospheres to derive effective temperatures (Oke & Shipman 1971). They found that DAs have temperatures that range from 13,000 to 50,000 K, that DBs range from 15,000 to 25,000 K, that one DC was at 10,000 K, and that DF and DGs had effective temperatures ranging from 4500 to 9000 K. Similar energy distributions for 38 white dwarfs of various types were published by Oke (1974). Because the whole energy distribution could be measured, it was possible to set up bands to derive color mesurements using systms that were tailord to white dwarf spectra. This program was begun by Greenstein in mid-1975 and continued until mid-1985 and provided many insights into the nature of peculiar white dwarfs and information on effective temperature ranges for various classes of white dwarfs. For example, Greenstein (1982) found that the variable ZZ Ceti stars occurred in only a small effective temperature range, from 11,000 to 11,700 K. In the early 1980s, good model atmospheres became available for white dwarfs. Oke et al (1984) observed and analyzed all the DB white dwarfs that could be observed from Palomar. They found effective temperatures that ranged from 12000 to 29000 K with the vast majority between 12000 and 18000 K. The logarithm of the gravity ranged from 7.5 to 8.2. This delineated the strict range of temperatures that DB white dwarfs could have. They found a mean mass for DB white dwarfs of 0.55 ± 0.10 $M_\odot$. Using similar data, they studied the DB white dwarf GD358, which is the prototype variable DB star (Koester et al 1985) and found $T_e = 24000 \pm 1000$ K and log g = 8.0 ± 0.3.

With the large catalog of white dwarfs becoming available, it was found that a few of the objects were abnormal. Angel et al (1972) used the MCSP to measure the circular polarization as a function of wavelength of BD + 70° 8247. They found circular polarization of several percent, which changed abruptly at specific wavelengths. In 1974, Landstreet & Angel (1974) measured the circular polarization of GD229 and found structure of the same sort as in BD + 70° 8247. At the same time, Greenstein et al (1974) obtained spectra of GD229 and found absorption lines similar to those in BD + 70° 8247. These objects showed broad, multicomponent lines. Greenstein (1978) observed the very-high-velocity white dwarf G35-26 and identified H and He lines. The spectrum was consistent with a magnetic field of 8 MG and an effective temperature of 16000 K. Greenstein & Boksenberg (1978) looked for line variability in GD229 but found none; they inferred a magnetic field of 200 MG. They also observed Feige 7 and identified the Zeeman pattern of H and HeI lines, which suggested a magnetic field of 20–50 MG. Greenstein & Oke (1982) obtained absolute energy distributions for Feige 7 and BD + 70° 8247 both in the visual and in the ultraviolet down to 1200 Å using the IUE satellite, which they compared to black bodies. They found an effective temperature of 19,000–23,000 K for Feige 7 and 14,000 K for BD + 70° 8247. For the latter star, using its parallax, they found a very small radius corresponding to a mass of about 1 $M_\odot$. Greenstein & McCarthy (1985) observed the emission-line object GD356 and identified $H\alpha$ and $H\beta$ Zeeman triplets in emission, which corresponded to a dipole field of 20 MG, although a fit was also possible with small-scale random fields of 11 MG. They also observed the $H\alpha$ absorption line Zeeman triplet in G99-47, which gave a field of 14 MG. Following a suggestion by Angel (1978) that the lines in BD + 70° 8247 might be very highly shifted hydrogen components with a field of about 200 MG, Greenstein et al (1985) calculated the expected line profiles and verified this interpretation.

MAIN SEQUENCE STARS OF LOW LUMINOSITY

In a continuation of his work on stars of very low luminosity Greenstein began to assemble photometry, MCSP, and low resolution spectra of dM, sdM and sdk stars. In a summary paper (Greenstein 1989a,b) he presented his data along with proper motions and parallaxes to derive space motions and bolometric magnitudes. The latter reach $M_{bol} = 13.7$ mag and reach very close to the limit below which stars are not able to sustain their luminosities by their nuclear reactions. Some halo stars show speeds up to 425 km/sec confirming the high escape velocity, hence the high total mass of the Galaxy.

SPECTRA OF MIRA AND OTHER LATE-TYPE STARS

Prior to the opening of Palomar, Spitzer (1939) had found violet-displaced absorption lines in several M supergiants. They were described in detail by Adams (1956) using Mt. Wilson Coudé spectrograms. In a fundamental paper based on both Mt. Wilson and Palomar spectra, Deutsch (1956) showed that the violet-displaced components of various resonance lines were also visible as projected against the G0 III type companion of α Her A (spectral type M5 II), thus proving that the ejected material was being permanently lost by the primary. This was the definitive discovery of mass loss from giant stars. Deutsch monitored many supergiants and Mira stars at Palomar until his untimely death in 1969.

While IS Bowen was adjusting the 200-inch Coudé spectrograph, he worked on the optics during the daytime and took long exposures at night. It is not clear whether or not he slept! It was his practice to take long exposures, often of Mira stars near minimum light, and turn them over to PW Merrill for measurement because Merrill had reached compulsary retirement at Mt. Wilson and was no longer permitted to observe. Sample papers that were devoted to line identifications and radial velocity measurements are by Merrill (1952b–f). Merrill's (1953) study of χ Cyg near minimum revealed some 40 unidentified emission lines. Many of them were later identified by Herbig (1956) as due to inverse predissociation transitions of the AlH molecule. Merrill later teamed up with Deutsch (1959) to describe the bizare low excitation (probably chromospheric) emission-line spectrum of R Cyg during very low maxima. In another cooperation, Merrill & Greenstein (1956) published an extensive list of absorption lines in the Tc-rich S-type Mira, R And.

Keenan et al (1969) used Palomar spectra, mostly of 18 Å/mm dispersion, to study the ephemeral AlO bands in Mira stars. A number of stars show the bands in absorption, but at intensities that do not correlate with the TiO band strengths. Now and then the AlO bands go over to emission, but again not correlating with other observable parameters. Absorption bands of VO and ScO behave as expected, correlating with temperature and hence TiO strength. The reason for one molecule showing emission while others show absorption has never been explained.

Following the discovery of maser emission by the OH, H_2O, and SiO molecules in the radio spectra of a variety of M giants, supergiants, and Mira stars, Wallerstein (1975, 1977) investigated their optical spectra in great detail in an effort to determine where in their envelopes the maser emission comes from and to establish the velocity at which the cool gas is escaping from the star. For long-period variables and some other late M giants, velocities of departure from the photosphere of 6–17 km s^{-1} were noted. Such small velocities are well below the escape velocity, indicating that the escaping matter must be pushed continuously if it is to escape the star. Infrared observations by others showed that the absorption of visible light by dust almost certainly provides the necessary impetus. For the supergiant sources, including the extreme example of VY CMa, the stellar photospheric velocity falls midway between the two OH maser velocities and the expansion velocities that range from 15 to 40 km s^{-1}. By combining Palomar spectra taken over 20 years with spectra from other observatories, Wallerstein (1985) produced a kinematic model of the outer layers of the bright Mira variable χ Cyg.

STELLAR CHROMOSPHERES

Following work at Mt. Wilson by AH Joy & RE Wilson (1949), OC Wilson began to monitor the Ca II emission in G, K, and M stars in search of variations analogous to the solar cycle. Using spectra of 10.2 Å/mm from Mt. Wilson and 9.0 Å/mm from Palomar, Olin Wilson was not able to recognize intensity changes (see, however, Wilson 1978). In an outgrowth of that program, Wilson & Bappu

(1957) found a fundamental relationship between the width of the Ca II emission and the M_v of the star. The relationship was remarkably linear and was calibrated using the sun and the K giants in the Hyades cluster (Wilson 1959). A complete catalog of Wilson's measurements of the Ca II emission-line widths was published in 1976. The Ca II-M_v correlation was recalibrated by Wallerstein et al (1999) using accurate Hipparcos parallaxes. The new calibration overcame the error in the Wilson calibration that depended on an erroneous distance to the Hyades.

To investigate the behavior of the 10830-Å line of He I, formed in the chromospheres of cool stars, Vaughan & Zirin (1968) used an S-1 image tube at the 200-inch Coudé, usually with the 72-inch camera, which provided spectra of 15.6-Å/mm dispersion. Some spectra were exposed with the 144-inch and 36-inch cameras. A survey of 455 stars was finally published by Zirin (1982). With such a large database, some interesting correlations were recognized. Above all is the correlation of the intensity of the 10830 line with soft X-ray flux. The X-rays emitted by the corona ionize He I in the photosphere. Then recombination into the metastable ^{3}S state is followed by absorption at 10830 Å by a permitted transition to a ^{3}P^o state. Binaries with periods under 200 days showed an average equivalent width of the 10830-Å line that is three times the overall average, whereas longer period binaries showed no noticeable enhancement. A number of peculiar stars provided interesting results. Almost all RS CVn stars showed 10830-Å absorption. T Tau stars showed both absorption and emission. The symbiotic star with much surrounding nebulosity, R Aqr, was observed to be in emission on 10 occasions over 12 years, including the deep minimum around 1977.

That deep minimum was recognized by G Wallerstein, who recalled that AJ Deutsch of the Mt. Wilson and Palomar staff had told him that he, Deutsch, had looked at R Aqr every year in hope of finding its spectrum to be dominated by a bright blue continuum, as had happened in the early 1930s. The faint minimum, near $V = 12.0$, permitted a detailed investigation of the nebulosity close to the star (Wallerstein & Greenstein 1980). Soon thereafter, Willson et al (1981) used the 1978 deep minimum to suggest an orbital period of about 43.5 years.

STELLAR MAGNETIC FIELDS

Following his discovery of a 1500-G field in the A2p star 78 Vir with his circularly polarized analyzer, HW Babcock continued to measure magnetic fields in stars, first at Mt. Wilson and then at Palomar. A catalog of magnetic stars was published by Babcock in 1957. He continued to observe stellar magnetic fields, usually of a few to 10 kG, but for HD 215441 he found a field of 34 kG (Babcock 1960). Surveys of Ap stars for magnetic fields were continued by Preston (1971), who concentrated on the Sr-Cr-Eu subclass. He found a frequency distribution characterized by a maximum between 2 and 3 kG with a tail that extends beyond 10 kG. In a related survey of 256 late B-type stars, Wolff & Preston (1978) derived the distribution of rotational velocities and the percentage (16%) that are HgMn stars.

After Babcock became director of the combined Mt. Wilson and Palomar Observatories, GW Preston continued using the stellar magnetic analyzer, along lines similar to those he had started while at the Lick Observatory. One star of particular interest is HD 24712, a Sr Cr Eu star (Preston 1972). He found an effective field that varies from 4 to 12 kG without changing polarity. The Mg and Eu lines varied in anti-phase with the magnetic variations, with Mg strongest when the field was minimal and Eu strongest near maximum field strength. The data fit the oblique rotator model, which by then seemed to be the preferred theory, despite the lack of a field reversal.

While the oblique rotator model was being accepted as representative of the magnetic variations in the A0 stars, the origin of the enormous abundance anomalies remained a mystery. Efforts to explain them by surface nuclear reactions (Fowler, Burbidge & Burbidge 1955) and by thermonuclear reactions in the stellar interior (Fowler et al 1965) were not successful. Advised by WLW Sargent, Michaud (1970) developed a new explanation based on gravitational settling of species that appear to be underabundant and upward diffusion driven by radiation pressure of elements that are in excess on the stellar surface.

Using Palomar spectra of 4.5 Å/mm, White et al (1976) investigated the intriguing problem of the Hg isotopes in the Mn Hg stars and showed that the isotope distribution varies with T_{eff}, indicating that mass-dependent fractionation has occurred. Diffusion calculations by Michaud, Reeves & Charland (1974) seem to model the dependence, provided that certain assumptions are fulfilled.

INFRARED PHOTOMETRY OF GLOBULAR CLUSTER STARS

By the 1970s, infrared detectors had emerged from the domain of low-temperature physics and were available for use by astronomers. Continuing a program started at Kitt Peak and Cerro Tololo, JG Cohen, along with JA Frogel and SE Persson, obtained J, H, K, CO, and H_2O photometry of red giants in globular clusters. They used their data on 26 globular clusters to derive color-magnitude diagrams, color-color relations, reddenings, and effective temperatures (Frogel, Persson & Cohen 1983a,b). Their data provided valuable guidance to those who soon began to use CCD detectors to study the chemical composition of globular cluster stars, as discussed below.

CHEMICAL COMPOSITION OF STARS

The commissioning of the Palomar Coudé spectrograph for general use in 1953 came at the perfect time for the rapidly developing related fields of stellar chemical composition, stellar evolution, and nucleosynthesis. Many of these developments have been reviewed in the nontechnical books by Hearnshaw (1986) and Croswell (1995).

The rapid development of digital computers combined with the new and efficient method of solving the differential equations of stellar structure (Henyey et al 1959) permitted the calculation of stellar evolution from the main sequence to the red giants region and somewhat beyond. As mentioned above, the research on the rates of nuclear reactions at the Kellogg Laboratory of Caltech, headed by WA Fowler, provided further stimulus to investigations of the affects of nuclear reactions on the chemical composition of stars. Finally, the series of research contracts from the Air Force Office of Scientific Research to JL Greenstein enabled him to hire a series of postdoctoral fellows to conduct research on abundances in stars. For some time, the 200-inch telescope was unavailable to anyone except staff members of Caltech and Mt. Wilson, so Greenstein obtained spectra for his postdocs, who also observed with the excellent Coudé spectrograph on the 100-inch telescope at Mt. Wilson. For this reason it is not possible to describe progress in this field due to observations obtained at Palomar alone.

One of Greenstein's earliest projects was to obtain moderate dispersion (18 Å/mm in the blue) of stars in the globular clusters M92 and M13. Those clusters were selected because their color-magnitude diagrams were being obtained BDDDDDdage and Arp with the advice of Baade. The resulting paper by Helfer, Wallerstein & Greenstein (1959; see also Kraft 1999) showed a metal deficiency of 100 and 20 for M92 and M13, respectively. A field star, HDE 232078, which has the colors of a star on the red giant tip of a globular cluster, also appeared to be metal poor by a factor of 20. In M92, the s-process elements appeared to be deficient by a further factor of 3.

In an effort to confirm and extend the discovery by Chamberlain & Aller (1951) of the metal deficiency of field subdwarfs, Greenstein & Aller (1960) reobserved HD 140283 and HD 19445 and added another subdwarf, HD 219617, and a likely horizontal branch star, HD 161817. They found metal deficiencies of 100, 50, 25, and 4 for the four stars, respectively. Though the uncertainties for the heavy elements were large, an overdeficiency of the s-product species was indicated.

There were clear hints from the age and metallicity of the globular clusters, combined with Baade's ideas about stellar populations, that the general metal abundance in the Galaxy has been increasing. As a test of this, Wallerstein and Helfer, at the suggestion of Greenstein, used Mt. Wilson spectra to compare two G dwarfs in the Hyades with the sun. The similarity in temperature made it likely that the stellar atmospheres are similar (though we now know that the chromospheres of the Hyades stars are much more active than that of the sun), so that the curve-of-growth method should provide accurate relative abundances. Wallerstein & Helfer (1959a) found the metallicity of the Hyades to be the same as that of the sun within 25%. Using Palomar spectra of higher dispersion, Parker et al (1961) found an excess of metals in the Hyades compared with the sun of 0.1 dex, within the uncertainty of the previous study. More recent efforts by others have yielded an excess of 0.15 dex, but the age factor does not seem to dominate because the metallicity of the younger Pleiades is almost exactly solar. A solar metallicity was found for the G dwarf HD 115043 in the Ursa Major cluster by Helfer, Wallerstein & Greenstein (1960). It appeared that there must be a small intrinsic scatter among stars formed at any given time near the solar circle in our Galaxy. In a continuation of the G dwarf program, Wallerstein & Helfer (1959, 1961) found 85 Peg to be metal poor by a factor 3 and 20 LMi to be metal rich by a factor 2.

Following Arp's (1962) color-magnitude diagram of NGC 752, whose age is intermediate between the sun and the Hyades, Kraft and Gunn (1963) obtained high resolution spectra of two of its stars near the main sequence turn-off. They showed metallicities nearly equal to the solar value.

The bright K0 dwarf Groombridge 1830 has long been known to show a high space velocity and weak lines. Krishna Swamy (1966) used line profiles and model atmospheres to analyze Lick, Mt. Wilson, and Palomar spectra. He found that a model with $T_{eff} = 5000$ K and most metals depleted by a factor 20 reproduced the profiles of strong metallic lines. Na appeared to be depleted by an extra factor of two. It is interesting that the Hα profile could not be fitted with a scaled solar $T(\tau)$ relation even though the Hα profile for the standard K dwarf, ϵ Eridani, could be fitted without trouble. The Hα profile in Groombridge 1830 could be fit by reducing the radiative temperature gradient due to convection setting in at a higher level in the atmosphere than in stars of solar composition.

In other abundance work, Bonsack & Greenstein (1960) and Bonsack (1961) analyzed the Li resonance line at 6708 Å in 12 T Tau stars. They found a Li abundance 100 times that in the sun and comparable to the abundance in the Earth and meteorites. Apparently the T Tau stars, Earth, and Sun formed with similar Li abundances, but the Sun has depleted its Li by mixing its surface material to depths at which 99% of its Li was destroyed by proton capture.

Ever since their recognition by Morgan (1934) of the Ap stars, with their strange enhancements of the absorption lines of various elements, the origin of their surface abundance excesses has been an astrophysical mystery. In 1960, WP Bidelman found that the B5 star 3 Cen A showed strong lines of phosphorus, a new type of peculiarity related in some way to the Si stars, which have similar temperatures. In two papers, Sargent & Jugaku (1961) and Jugaku, Sargent & Greenstein (1961) found the surface abundances to be bizarre. Helium is deficient by a factor of six and most of the remaining He is ^{3}He rather than ^{4}He. The most drastic abundance excesses are those of P, Kr, and Ga, which are factors of 100, 1300, and 8000 above their normal ratios. Observations by Babcock failed to reveal a magnetic field.

Sargent et al (1969) conducted a search on 9-Å/mm plates for Ne I lines in the hotter Ap stars, most of which have temperatures of late B stars. Only four stars showed Ne I lines, and they all showed certain similarities to 3 Cen. Nine others are blue enough for Ne I lines to be expected, but they are not present. The same stars lack He I lines or have much weaker He I lines than expected.

The problem of the He/H ratio in stars with low metal abundance was a fundamental one when the steady state theory of cosmology was considered to be viable.

Steady state required the continuous creation of matter, presumably as protons and electrons. Helium and heavier elements were produced in stars by nuclear reactions in their interiors and redistributed throughout the Galaxy by supernovae and less-violent forms of mass loss. Hence, stars of very low metal content should also have a very low helium abundance. Unfortunately, the stars identified to have low metal content, such as field subdwarfs and globular cluster red giants, are too cool to show lines of He. However, hot horizontal branch stars in metal poor globulars ought to show lines of He I, so a number of authors investigated their spectra (Searle & Rodgers 1966, Greenstein & Münch 1966, Sargent & Searle 1967). Frequently the He I lines were found to be significantly weaker than expected, which would favor steady state. However, the spectra showed other peculiarities that cast doubt that the apparent He abundance in the stellar atmosphere represented the He abundance with which the star was born. However, at the low resolution required for stars in globular clusters, no other lines could be seen as tests of this conclusion. Higher-resolution spectra were obtained of hot stars at high galactic latitude that showed similar effective temperatures and surface gravities of hot horizontal branch stars. The analysis by Sargent & Searle (1967) of the halo star Feige 86 showed that its atmosphere resembles that of the peculiar population I B star 3 Cen; it showed weak HeI and PII lines along with many lines of Fe II and Si II. This showed that the 3 Cen syndrome can affect horizontal branch stars and give a false impression of an intrinsic He deficiency. Another B-type subdwarf, HD 4539, was analyzed by Baschek, Sargent & Searle (1972) and found to have $1/16$ of the normal He/H ratio. An excess of N by a factor of 3 and a similar deficiency of O were also found. Once more, they concluded that the low helium need not have been primordial.

Over a period of 20 years, Greenstein accumulated data on faint blue stars located at high galactic latitude. Target stars were discovered by a wide variety of techniques, starting with the survey of Humason & Zwicky (1947) with the original 18-inch Schmidt at Palomar. Spectra and colors for 189 stars were presented by Greenstein & Sargent (1974). Using colors and hydrogen line profiles, they derived T_{eff} and log g for most of the stars and examined the strengths of various spectral lines, especially those of helium. A wide variety of stars were found, including white dwarfs, normal population I B stars, horizontal branch stars, extended (i.e. very hot) horizontal branch stars, and some that appear to be population II, but lying above the horizontal branch. Among the hottest stars, a few were found to be nuclei of diffuse planetary nebulae. Some of the extreme horizontal branch stars showed weak He lines and lines from unusual elements, such as S III and P II, indicating radiative diffusion of certain elements, in agreement with the stars discussed above.

Greenstein made use of his contract funds to attract a number of European spectroscopists and experts in the structure of stellar atmospheres.

Weidemann (1960), who had recently completed his PhD at Kiel, analyzed Greenstein's spectra of the cool white dwarf Van Maanen 2. He found hydrogen to be deficient by a factor 100 and metals by a factor of 10^4. Van Maanen 2 appears to be a star with an atmosphere of almost pure helium from which heavier elements have settled out by gravitation diffusion. Using spectra taken by both Greenstein and Münch, Traving (1962) analyzed the atmospheres of two high-latitude B2 stars, one of which, Barnard 29, is a member of the globular cluster M13. Compared with the population I B star, τ Sco, B29 shows a H/He ratio reduced by a factor two. B29 was reinvestigated by Stoeckley & Greenstein (1968), who found a normal ratio of He/H and a reduction of heavy elements, C thru Si, of a factor 20, which is about the same as in the red giants of M13. Low C and O combined with high N indicate past CNO cycling, but there is no evidence of He burning. The field star BD $+ 33°$ 2642 showed a normal H/He ratio and a reduction of the heavier elements by a factor of 5–10.

Cayrel & Cayrel (1963) analyzed spectra taken by Bonsack of the G8 II–III star ϵ Virginis. They found the chemical composition to be the same as that of the Sun, with the possible exception of an excess of Na by a factor of 2. On the basis of their study, ϵ Vir has been used as a standard for the comparison of many G8-K2 giants.

Stawikowski & Greenstein (1964) used 18 Å/mm spectra of Comet Ikeya (1963a) to derive the $^{12}C/^{13}C$ ratio of 70 $\pm$ 15. Hence the comet shows a ratio that is the same as that on the Earth and Sun and far above the value near 4.0 found in some carbon stars. Koelbloed (1967), of the University of Amsterdam, used spectra taken by Greenstein to analyze the atmospheres of two high-velocity stars, HD 2665 and HD 6755. They proved to have metal deficiencies of 50 and 10 times compared with the Sun. Curiously, europium appeared with a small excess. The same phenomenon was noted in the extremely metal-poor red giant HD 165195. These early hints of an excess of Eu compared with the lighter rare earths have been strongly confirmed by many others to demonstrate an excess of r-process/s-process in stars with [Fe/H] $\lesssim -2.0$.

Following further work on G dwarfs based on Mt. Wilson spectra that primarily showed that the ratio of α/Fe (where α stands for elements whose nuclei consist of an integral number of α particles) increases to about two to three times the solar ratio as the iron abundance drops to about 1/10 solar, Wallerstein (1962) and Helfer, Wallerstein & Greenstein (1963) turned their attention to metal-poor red giants. They first set up the Hyades red giants to serve as standards (Helfer & Wallerstein 1964). Curiously, they found Na/Fe to be greater than in the Sun, just as noted by the Cayrels in ϵ Vir. Twenty-seven additional stars, all with galactic disk kinematics, were compared with the Hyades K giants from spectra taken at both Mt. Wilson and Palomar. The stars all show disk kinematics. Values of [Fe/H] from +0.25 to −0.85 were represented. The K giants with [Fe/H] > -0.4 showed the same small Na excess as did the Hyades, whereas stars with [Fe/H] between −0.4 and −0.9 showed the solar ratio of Na/Fe. Excesses of α/Fe of about 0.2 dex were seen in most of the more metal-poor stars. There appeared to be a group of stars with an excess of the odd iron-group species, V, Mn, and Co, relative to iron.

Moving on to extremely metal-poor giants, Wallerstein et al (1963, referred to by Pagel as "the thundering herd") analyzed Palomar spectra of three stars,

HD 122563, HD 165195, and HD 221170, selected by their extreme ultraviolet excesses. Metal deficiencies by factors of 800, 500, and 500 were derived, with a further deficiency of the heavy s-process species in HD 122563 by a factor of fifty.

Among the metal-poor red giants is a group of stars with very high space velocities, weak metallic lines, strong bands of CH, and weak bands of C_2 (Keenan 1942). They are naturally called CH stars. From Palomar spectra of 6.7 Å/mm in the yellow-red and 4.5 Å/mm in the blue, Wallerstein & Greenstein (1964) found HD 26 and HD 201626 to be metal poor by factors of 5 and 30 compared with ϵ Vir. Carbon is enhanced with C/Fe five times that in ϵ Vir, but ^{13}C is not detected. The heavy s-process species are enhanced by a factor of 20 in both stars. A qualitative scenario of He burning, CN cycling to produce ^{13}C, and its destruction by the reaction $^{13}C + \alpha \rightarrow {}^{16}O + n$ to release neutrons for s-processing was presented. Caughlan & Fowler (1964) made quantitative calculations of the above processes. However, the story of the CH stars does not end here. McClure & Woodsworth (1990) showed that the CH stars (as well as the Ba stars) are all binaries. Apparently the present atmospheres of these stars were once part of the now-defunct companion that produced abundance anomalies and dumped its atmospheres onto the currently observed CH and Ba stars.

Other carbon stars have attracted the attention of Palomar astronomers and their associates. Searle (1961) analyzed spectra of the H-deficient star R CrB taken by Greenstein, finding that the ratio of C/H is 10. Among the metals and s-process species, there were no anomalies exceeding a factor of 2. The composition parameters were found to be $X = 0.005$, $Y = 0.91$, and $Z = 0.09$, where X, Y, and Z are abundances of H, He, and heavier species by weight. Carbon dominates the heavies.

Following Herbig's (1965) discovery of the presence of Li in field F & G dwarfs in quantities much larger than in the Sun, there was a flurry of activity to survey stars for the presence of Li and Be in order to understand their origins. In their important review paper, Burbidge et al (1957) had listed D, ^3He, Li, Be, and B as produced by the "x-process," i.e. the unknown process. Danziger & Conti (1966) used Palomar spectra of Pleiades F dwarfs to compare with field F dwarfs. For the Pleiades stars they found Li excesses from 30–100 compared with the Sun, with a significant variation from star to star, a result similar to that in the Hyades (Wallerstein et al 1965). The problem of the variations of Li content within clusters was not clarified until the discovery of the "Li dip" in the Hyades by Boesgaard & Tripicco (1986). Observations of some fainter stars in the Pleiades and nine stars in the UMa cluster by Danziger (1967) helped to clarify the dependence of Li depletion by convective mixing in clusters with different ages.

Conti & Wallerstein (1969) conducted a search for Li and Be in 31 stars of type F and G Ib and 21 cepheids. Only upper limits could be derived for the stars. Evidently these stars have once been to the red giant tip of their evolution, where convection carried their Li and Be down to levels where it was destroyed by proton capture. Hence the vast majority of F-G Ib stars and cepheids are on the first "blue loop" after once being cool giants.

The third most abundant element in the atmospheres of normal stars is oxygen. Because of its high abundance, it is a significant contributor to the opacity in stellar interiors at depths below which H and He are fully ionized. Oxygen lines of various states of ionization are readily measurable in stars of types O to early G, but in cooler stars the permitted lines of OI arise from levels of excitation that are too high to be sufficiently populated. However, the [OI] lines at 6300 and 6363 Å arise from the ground level and are heavily populated. The transition probability of the 6300 Å line is 0.0069 $\sec^{-1}$ and is one third that value for the 6363 Å line. Based on Palomar, Lick, and Mt. Wilson spectra, Conti et al (1967) conducted a large survey of O in red giants. Equivalent widths were measured in 29 stars and eye estimates made for another 91 objects. Model atmosphere analyses were used to show that most low-velocity stars showed that same O/H ratio as the Sun. Among the metal-poor stars, O was less depleted than iron. Later work at various observatories showed that as [Fe/H] goes from 0 to -1.0, [O/Fe] rises from 0 to 0.4. For [Fe/H] between -0.1 and -2.5, the issue of whether [O/Fe] remains at $+0.4$ or rises to $+1.0$ is still open. (The Conti et al paper (1967) was originally written by the first four authors in New York dialect and translated into the King's—or rather the former King's—English by the fifth author.)

Evolving low-mass stars suffer the well-known helium-flash at the red giant tip and become restabilized on the horizontal branch. Horizontal-branch stars in globular clusters were too faint to analyze with high dispersion, but a few such objects have been recognized in the general field. Kodaira et al (1969) analyzed two horizontal-branch A stars, HD 86986 and HD 109995, using Palomar spectra of 9 Å/mm. Kodaira's (1964) work on HD 161817 was included. Metal deficiencies of -1.54, -1.76, and -1.16, respectively, were found for them. As with metal-poor red giants and G dwarfs, the α nuclei are least deficient and the s-process nuclei are somewhat more deficient. Both C and O are enhanced relative to iron, indicating that products of the He flash, including some α capture by ^{12}C to produce ^{16}O, have found their way to the surface in HD 86986 and HD 109995.

Although not luminous enough to be horizontal-branch stars, the λ Boo stars have long been known to have weak metallic lines for their temperatures. An analysis by Baschek & Searle (1969) of five such stars found that only three really satisfied the criteria for λ Boo star. They have metal deficiencies compared with hydrogen. The origin of these deficiencies in these low-velocity stars remained uncertain until Venn & Lambert (1990) showed that only elements that condense onto dust are deficient.

Another unusual A star, BD $+39°$ 4926, was found by Kodaira, Greenstein & Oke (1970) to have an outlandish chemical composition. Species from Na to Y show deficiencies ranging from -2.97 dex for Al and -3.42 dex for Sr to only -1.43 dex for Si and no deficiency at all for S. CNO are near normal. No nuclear astrophysics scheme could be found to cause such strange anomalies. The derived surface gravity, log g $= 1$, implies that $M_v = -3$, far above the horizontal branch. More recent work on stars with similar anomalies, including some RV Tau stars, has revealed similar abundance ratios. The most likely suggestion is that they are post-AGB stars whose atmospheric composition has been drastically altered by the loss of elements with high condensation temperatures (Mathis & Lamers 1992). Hence they have been blown away while stuck onto dust grains (as with the λ Boo stars discussed above).

Based on Lick and Palomar spectra of 4.5 Å/mm in the blue, Adelman (1973a,b) analyzed 21 cool Ap stars with rotational velocities less than 10 km s^{-1}. For some species, such as Sr and Eu, the range in atmospheric abundance reaches 4 dex within this sample of Sr Cr Eu stars. Remarkably, the abundance of carbon showed the least dispersion of the 23 elements studied. In another Caltech thesis, Oinas (1974) analyzed 26 K giants and dwarfs, 10 of which had been classified as "super-metal-rich (SMR)" by Spinrad & Taylor (1970) or Taylor, Spinrad & Schweizer (1972). He confirmed that the SMR stars show heavy element excesses of 0.2 to 0.3 dex.

Using primarily Palomar spectra, Smith & Wallerstein (1983) found the inter-combination line at λ5924.47 Å of the unstable element Tc to be present in one S star and four SC stars, but not in a cool Ba star. Their quantitative analysis was flawed, however, because they overlooked the unusually broad hyperfine structure of the Tc line. More accurate abundances of Tc relative to its neighboring species, such as Zr, have been published by Dominy & Wallerstein (1988) and by Vanture et al (1991), who showed that several late M stars without measurable excesses of other heavy elements show Tc lines.

With the acquisition of a CCD detector for the Coudé spectrograph of the 200-inch telescope, a new window of abundances in globular cluster stars was opened. The authors of this review took advantage of the opportunity presented by the large aperture and sensitive detector to investigate abundances in a number of clusters. We discuss here only one example because of both limited space and our effort (usually unsuccessful) to keep our own work to less than 50% of the references.

Early work on individual stars in M13 showed a spread in oxygen (Pilachowski, Wallerstein & Leep 1980) and sodium (Peterson 1980). Following this up, Brown, Wallerstein & Oke (1991) used the Palomar Coudé in the visual and near-infrared for the [OI] and red CN bands, as well as the Cassegrain double spectrograph for the CH and NH bands. A full analysis of six stars for abundances of C, N, O, and $^{12}C/^{13}C$ was carried out. Although the N abundance may be obtained from the CN bands, any errors in the C or O abundances propagate into the analysis of the CN bands. Hence, the NH bands near 3350 Å were also observed to confirm the N abundances. For six observed stars, oxygen abundances ranged from log $N_O =$ <6.55 to 7.6 [on the usual scale of log $N_H = 12.0$, a slight revision of Russell's (1929) convention]. The N abundances ranged from log $N_N = 7.6$ to 7.3 for the lowest and highest O values. When added together, $C + N + O$ ranged from ≥7.6 to 7.9, which may all be considered the same, within expected errors. This shows that the range of oxygen abundances was determined by the extent of CNO cycling, presumably in the star, rather than reflecting the initial composition. The $^{12}C/^{13}C$ ratios were 5 or 6 in the four stars in which the ratio was measurable. Both the presence of low O abundances and the low $^{12}C/^{13}C$ ratios indicate that most of the material currently on the surface of the stars has experienced CNO cycling. Not only has processed material been convected up from the H burning region, but material originally in the stellar atmosphere has been convected down so as to participate in the CNO cycle. Soon thereafter, Kraft and his collaborators showed that Na is in substantial excess in the O-deficient stars, having presumably been produced by the NeNa cycle operating at temperatures only a little above the CNO cycling region.

An interesting opportunity arose when the authors were observing with the Palomar Coudé and CCD. The type II cepheid, W Vir, whose [Fe/H] value is -1.1, was found to be at its phase of rising light during their run. Hence, it showed emission lines of both H and He, which could be analyzed to yield the H/He ratio. With the essential modeling of the shock ionization and postshock recombination, Raga, Wallerstein & Oke (1989) derived a ratio of He/H of 0.12 with limits from 0.08 to 0.18. At the low end, the He/H ratio is similar to that of unevolved metal-poor stars, whereas the upper end would indicate some enhancement in the He abundance because of the stellar evolution, mixing, and mass-loss.

CATACLYSMIC VARIABLES, X-RAY BINARIES, NOVAE

Spectroscopic observations of these objects became much easier with the 200-inch Hale telescope because the objects are faint and their periods are so short that exposures of only a few minutes are required.

Most of the early work was done by RP Kraft. Greenstein & Kraft (1959) obtained spectra of the old nova DQ Herculis and measured intensities and velocities of the emission lines. They found that the high members of the Balmer series and the HeII lines indicated the presence of a rotating hot disk that followed the orbital motion in the binary system. This paper was followed (Kraft 1959) by an interpretation of the data in terms of a disk at a temperature of about 40,000 K. Greenstein & Schneider (1979) obtained spectral energy distributions and slit spectra of this object. Although they could not clearly identify the faint star in the system, they were able to infer that its spectral type was probably M4 or later. Kraft (1962) obtained spectra of eight cataclysmic variables and showed that at least five of them were binaries, which suggests that all such stars are close binaries. In his study of 10 old novae, Kraft (1964) found that most of them were clearly close binaries with masses in the range of 0.1–3.0 solar masses and suggested that all novae were binaries. Using absolute spectral energy distributions, Wade (1979) found that the radiation came from a flat continuum radiated by an accretion disk and a faint M5-M7 dwarf star. Szkody & Wade (1981) measured the mass-loss rate in Z Cam as 6×10^{-10} M$_\odot$ per year. Thorstensen et al (1986) found that SU UMa had a period of 110 min and the emission lines had an amplitude of 58 km^{-1}: The mass function, however, is low. Using spectral energy distributions of AE Aqr obtained on timescales of seconds, Welsh et al (1993) found that the flickering must come from an object much smaller than the accretion disk.

The launching of X-ray satellites led to the discovery of a related class of close binaries, the so-called X-ray binaries. Sandage et al (1966) identified ScoX-1 and obtained a series of spectra that showed variability on a timescale of 1 day. Wallerstein (1967) found a minimum distance of 270 parsecs (pc) from the strengths of the interstellar lines. Westphal, Sandage & Kristian (1968) found rapid spectroscopic and photometric changes in ScoX-1. They found velocity changes of 100 km^{-1} and that the H and HeII lines had different phases, which suggested binary motion. Neugebauer et al (1969) obtained absolute energy distributions from 0.3 to 2.2 μm and were able to rule out synchrotron or free-free radiation. Sandage et al (1969) reported on photometric monitoring of ScoX-1. Other X-ray binaries have also been studied both spectroscopically and photometrically. Wade & Oke (1977) obtained absolute energy distributions and slit spectra of the transient X-ray binary A0535 + 26 and found that the main star was of type B0 whereas the X-ray source is in orbit about this star. Greenstein et al (1977) did a thorough spectroscopic study of AM Herculis whereas Priedhorsky et al (1978) did visual and infrared monitoring over four cycles. They found that at the principal minimum, the light comes from an M2V star.

Of particular interest was the X-ray nova A0620-00, which appeared on August 3, 1975, and which at its maximum was one of the brightest X-ray sources in the sky. Oke & Greenstein (1977) obtained optical spectral energy distributions in 1975–1976, which when combined with infrared measurements by G Neugebauer & EE Becklin showed a radiation spectrum like a black body at a temperature of 25,000–30,000 K produced by the accretion disk with a flow of matter onto a neutron star. When the nova had faded, Oke (1977) found evidence that the faint normal star was a K5-K7 dwarf and that the accretion disk could be modeled as a black body at 10,000 K with a radius of 1.7×10^9 cm and a cool 4000 K black body of radius 13.4×10^9 cm. Further spectroscopic observations by Johnston et al (1989) confirmed the earlier result by McClintock & Remillard (1986) that the invisible object had a mass of at least 3.2 M$_\odot$, which is close to the maximum possible mass of a neutron star. Using the velocity from lines formed at the outer edge of the emission-line region, they were able to show that the minimum mass was at least 4.0 and possibly as high as 9 M$_\odot$, indicating that the invisible object was a black hole.

PULSARS

With the optical identification of the crab pulsar by Cocke, Disney & Taylor (1969), efforts were made in many places, including Palomar, to study the object. The absolute spectral energy distribution from 3350 to 8000 Å of the pulses and of the interpulse radiation were measured by Oke (1969a). They were similar and not very different from the general radiation from the nearby crab nebula. Infrared measurements at 1.65 and 2.2 μm showed a smooth extension of the Oke curve (Neugebauer et al 1969). Further measurements at 2.2 and 3.5 μm by Becklin et al

(1973) confirmed that the light curve was the same as in the visible. Kristian et al (1970) measured the linear polarization around the cycle. The polarization changed smoothly from 25% to nearly 0%, whereas the plane of polarization changed from 150° to 60°.

INTERSTELLAR MATTER

With the commissioning of the 200-inch Coudé, G Münch began a series of observations of interstellar absorption lines. He first observed numerous O and B stars in the northern Milky Way to extend the data of Adams (1949) both in distance and galactic longitude, because Adams could not reach stars north of $\delta = 50°$ with the Mt. Wilson 100-inch Coudé. Dispersions of 4.5 and 9.0 Å/mm in the blue and 6.7 and 13.5 Å/mm in the yellow were used to observe Ca II and Na I lines. The effects of galactic rotation were clearly present, with components from local and distant gas showing differences of up to 30 km s^{-1}. Around distant OB associations, expanding motions of about 20 km s^{-1} were seen. Münch & Zirin (1961) then investigated interstellar lines in high-galactic-latitude B stars. Clouds out to $z = 1$ kpc were found. Such distances are greater than expected from the velocity distribution of clouds in the galactic plane. Münch & Unsöld (1962) found interstellar Ca II lines in α Oph, a star only 18 pc from the Sun and in nine other stars in the same general direction. In the direction of the cluster NGC 7822, Münch (1964) found a group of stars with surprisingly strong interstellar lines of the CN molecule.

In a different type of study, Wilson et al (1959) used specially designed multislits on the 72-inch and 36-inch cameras of the 200-inch Coudé. In this way a substantial area of the Orion nebula was covered and radial velocities of the emission lines of [OII] and [OIII] were mapped over the face of the nebula. A mean expansion of [OIII] with respect to the Trapezium stars was found. Velocity gradients and line splitting was observed in some areas. By imaging the Orion nebula in the OI 8446 Å line of OI, Münch & Taylor (1974) found a unique filamentary structure not seen in the 6300 Å line of [OI]. They explained the high level of fluctuations as due to the resonant excitation of $\lambda 8446$ Å by Lyβ. Line profiles were also obtained with a Fabry-Perot étalon. Following this up, Taylor & Münch (1978) found features of relatively high velocity near Θ^2 Ori that they ascribed to stellar winds that drive shocks into the ambient gas.

Using Mt. Wilson and Palomar spectra, the latter in the fourth-order ultraviolet, Wallerstein & Goldsmith (1974) observed 15 stars for their interstellar Ti II line at 3384 Å. Because Ti II and H I have almost exactly the same ionization potential, the ratio of Ti/H can be inferred directly from the column densities of the two species. The derived ratios show a Ti/H deficiency of 20–200 compared with the solar ratio. It seems that much of the Ti (and Ca as well) is locked up on grains, a conclusion reached a few months later for many species by Morton (1974) from Copernicus observations of ζ Ophiuchi.

Spectroscopic observations of nebulae began as soon as the spectrographs for the Hale telescope were available. With the high throughput and large collecting area, Bowen (1955) measured accurate wavelengths of forbidden lines in planetary nebulae and Aller et al (1955) measured intensities of 263 lines in the spectrum of the planetary nebula NGC7027. They found a wide range of ionization and 40 species of ions. Minkowski & Osterbrock (1960) measured the electron density in two planetaries using the ratio of [OII]λ3727 to [OII]λ3729. They found electron densities in the range of 120–840 cm^{-3}, with a systematic difference between the two objects.

Minkowski & Osterbrock (1959) and Osterbrock (1960) made a detailed study of the gas in elliptical galaxies. They found that both the density and the total mass of gas were small and that the ionization could probably be produced by blue stars in the galaxies.

The first results reported using the new millimeter wave detector were those of Westbrook et al (1976), who reported on 1-arcmin resolution observations of the central regions of molecular clouds associated with four HII regions. They were able to estimate that there were steep density gradients perhaps produced during the collapse of the central luminous object. Phillips et al (1977) first measured the flux from the $J = 3$ to $J = 2$ line of CO at 870 μm in an Orion molecular cloud. They found wide wings on the lines, which suggests high-velocity gas at or near the front face of the cloud.

Huggins et al (1979) measured molecular lines in HCN, HNC, and HCO$^+$. The HCN observations suggest a hydrogen density of 10^6 cm^{-3}. It was indicated that the HCN and HCO$^+$ lines came from different locations in the clouds.

THE GALACTIC CENTER

It was of course known that there was a strong radio source, Sagitarius A, which was thought to be at the galactic center. Optical observations were useless because the obscuration toward the galactic center was enormous. The only way to probe the galactic center was in the infrared. This was difficult because infrared detectors were single pixel devices. The problem was tackled by Becklin & Neugebauer (1968), who made observations between 1.65 and 3.4 μm by scanning areas of sky and chopping relative to a reference black body. They found that there was a point-like source centered on a much larger source, which was centered on Sagitarius A. They made further observations (Becklin & Neugebauer 1969) between 1.65 and 19.5 μm using 7.5- and 5-arcsec apertures. They found that the 10- and 20-μm radiation came from a source that was approximately 1 pc in diameter. The energy distribution was similar to that of the Seyfert galaxy NGC 1068. They noted that Sagitarius A was larger than the infrared source. Becklin & Neugebauer (1975) obtained 2.5-arcsec resolution maps of the central 1 arcmin between 2.2 and 10 μm. They found that most of the 2.2-μm radiation came from a discrete source in the central 2 pc with an absolute magnitude of -8. In all there were nine discrete sources. Further observations by Neugebauer et al (1976) showed that some of the discrete sources had CO absorption at 2.3 μm, indicating that they were M-type supergiants. The central source, identified with Sagitarius A, did not show CO absorption. Neugebauer et al (1978), working at Cerro Tololo, made Bγ observations of the ionized gas and found it concentrated in the central parsec of the galactic center. They also found CO absorption in the central 30 arcsec.

SUPERNOVAE AND SUPERNOVA REMNANTS

Observers on the Hale telescope were in a particularly good position to work on supernovae because Zwicky already had supernova searches in place with the Palomar 18-inch Schmidt and later the Palomar 48-inch Schmidt. As supernovae were discovered, finding charts would be made available almost immediately to the 200-inch observer. Spectra from the 200-inch were obtained in a piecemeal fashion starting in about 1960. References to papers where these spectra are displayed can be found in an article by Oke & Searle (1974). Greenstein & Minkowski (1973) published an atlas of spectra of 12 supernovae of different classes. Branch & Greenstein (1971) made some of the early models for type II supernovae and compared them with observations made mostly by Zwicky of supernova SN, 1961 in NGC 1058. They found that the redshifted emission lines were caused by blueshifted absorption. They found the atmospheres to contain at least 0.3 M$_\odot$ of material.

With the availability of the multichannel spectrometer, which could produce absolute spectral energy distributions at a spectral resolution adequate to study the supernova spectra, it was decided to make a determined effort to obtain such observations for many supernovae and also to observe how they changed with time. Shortly after this decision had been made, SN 1972e, the brightest supernova since that in IC4182 in 1939 (Minkowski 1939), was discovered by Kowal in NGC5253 with the Palomar 48-inch Schmidt. Both MCSP and infrared observations were obtained early (Kirshner et al 1973), and MCSP observations were continued until 720 days after maximum (Kirshner & Oke 1975). Available simple models indicated that effective temperatures were decreasing with time, from about 10,000 K to 7,000 K. At very late stages, [FeII] emission lines were observed that indicated the presence of at least 0.01 M$_\odot$ of iron in the expanding envelope. SN 1972e was also found to show interstellar lines from its parent galaxy (Wallerstein, Conti & Greenstein 1972) despite its large distance from its parent galaxy.

During this same time, seven other supernovae were observed with the MCSP, usually at several different time epochs (Oke et al 1973). For the type II supernovae, whose spectra are dominated by hydrogen emission lines, expansion velocities of 20,000 km s^{-1} were derived and the abundances were found to be near solar. For type I supernovae, it was shown that the most prominant P-Cygni–shaped features were the CaII H and K line and the CaII infrared triplet. Kirshner & Kwan (1975) used MCSP absolute energy distributions to model type II supernovae up to 400 days after maximum. They measured electron densities as a function of the time and again found essentially solar abundances. In 1985, Filippenko & Sargent (1985) obtained spectra of SN 1985f in NGC 4618. Although similar to type I supernovae in having no hydrogen lines, the spectrum was dominated by broad low excitation emission lines, including the sodium D lines and a very strong [OI]$\lambda\lambda$ 6300, 6364 blend. These were later given a classification as type Ib.

Infrared light curves of type I supernovae have been obtained by Elias et al (1985) for objects observed between 1971 and 1984. For type Ia supernovae, they found strong and variable absorption at 1.2 and also probably 3.5 μm. Type Ib supernovae have no such absorption. They also found a dispersion in color and absolute magnitude of 0.2 for type 1a supernovae.

The scale and stability of the 200-inch telescope made it ideal for studying the expansion of supernovae remnants using proper motions. Kamper & Van den Bergh (1976) summarized many years of observations of Cas A. They concluded from the proper motions that the supernova occurred about 1657 and that there is no remnant star brighter than $M_r = 8.0$. Trimble (1968) analyzed proper motions of filaments in the crab nebula. According to the proper motions, the expansion began about 1140, close to the observed date of the explosion. The expansion rate is 70 km s^{-1} or 0.01 arcsec year^{-1}.

QUASARS, AGNs, SEYFERTS

The very early work on quasars done with the 200-inch telescope in the early 1960s involved the identification and redshift determination of radio objects such as those in the 3C and 3CR catalogs. The first work is described in the article by Sandage (2000). The small group of emission-line galaxies first recognized by Seyfert (1943), which appeared to be low-luminosity counterparts of quasars were, of course, already known. In addition, there were the catalogs of peculiar galaxies of Markarian & Lipovetsky (1972), which had been found using the objective prism on their 40-inch Schmidt. The catalogs contained strong emission-line objects that were either giant HII regions or active galactic nuclei similar to quasars. Zwicky was also searching the Palomar Schmidt plates to find compact objects with bright nuclei (see Zwicky & Zwicky 1971). He was obtaining low-resolution spectra of a few of these and was also providing finding charts and magnitude information, particularly to Sargent and Oke, who incorporated these objects into their more general observational programs of these kinds of objects. Starting in 1986, Schmidt et al (1986a,b; 1987a,b) began a survey using a four shooter in its continuous readout mode to discover high redshift quasars optically.

Apart from the direct cosmological implications of QSOs and their very large distances, there was an immediate need to study the physical properties of these objects to understand their nature. Among the first studies were absolute flux measurements over wide wavelength ranges to make comparisons with the radio flux results. Of the many papers in this area, we mention only four. Using broad band

UBV photometry of 43 QSOs with various redshifts, Sandage (1966) constructed a mean continuum energy distribution from rest wavelengths of 1000 Å to 5500 Å, from which K corrections due to redshift were calculated for use in QSO Hubble diagrams. The 3000 Å hump was seen, ascribed now to an acretion disk.

Oke et al (1970) studied 26 quasars in the spectral range from 0.3 to 2.2 μm. They found no characteristic in this spectral range that distinguished radio-loud from radio-quiet quasars. They also found that the emission-line strength of $L\alpha$ was much too weak to be produced by an opaque Lyman continuum, i.e. under case B conditions. Neugebauer et al (1976) observed 18 Markarian galaxies in the visual and near-infrared and found that the energy distributions were consistent with a combined nonthermal power law and a stellar component. Ten of the objects were giant HII regions, with four of them having dust emission at 10 μm. Yee & Oke (1978) analyzed 26 3CR radio sources and found that the energy distributions could be matched by combining elliptical-galaxy and power-law energy distributions. They also found that the Balmer emission-line strengths were correlated with the ultraviolet power law flux, which suggests ionization and recombination.

Using both Lick and Palomar accurate energy distributions, Wampler & Oke (1967) identified several broad blended emission features with permitted lines of FeII. The absence of forbidden [FeII] lines indicated that the density of the emitting gas was at least 10^6–10^7 cm^{-3}. Oke & Lauer (1979) did a similar analysis of MKN 79 and 1Zw1, which have relatively sharp lines. They found that the FeII line profiles were similar to the broad H-line profiles.

The first study of the spectral changes in variable quasars was the work of Sandage et al (1966), who made observations during the outburst of 3C446. They found that the equivalent widths of the emission lines and the slope of the continuum both changed. The absolute intensities of the lines remained constant, indicating that only the continuum was changing rapidly. Similar results were obtained by Oke (1967) in both 3C446 and 3C279. Continuum changes of 20% were seen in 3C279 on timescales of 1 day. Oke et al (1967) measured optical and infrared changes in brightness of the two Seyfert galaxies, IZw1727 + 50 and 3C120, and concluded that the changes were due to the nonthermal component in the continuum. In 1967, Sandage (1967) and Oke (1967) reported changes in the brightness of the radio galaxy 3C371 by 1 mag over a period of 1 year and 0.10–0.15 mag in days. Yee & Oke (1981) studied the energy distributions and line intensities and profiles of the two variable objects 3C382 and 3C390.3 and found that the variability was due to changes in the nonthermal component of the continuum. They also found substantial changes in the profiles of the very broad Balmer lines and suggested that they were formed in an expanding shell or ring with a radius of a few light years. Further studies of the changes in the Balmer line profiles (Oke 1987) indicated that the emission lines could come from an accretion disk.

In 1973, a program was begun to detect and look for variability in quasars and AGNs at millimeter wavelengths. Twenty-three objects were observed by Elias et al (1978). They detected nine of them; seven appeared to be nonthermal whereas the other two appeared to be thermal radiators. BL Lac was found to vary by a factor 2 and the variations were correlated with radio variations. 3C84, 3C120, and 3C273 also were found to be variable. Further observations, until 1981, by Ennis et al (1982) confirmed and extended the variability observations in time. For the blazers, there was a good correlation between 1-mm and 2-cm fluxes. An outburst of 3C84 at 1 mm had not appeared at 2 cm a year later.

It was realized early on that quasars were not necessarily stellar-like sources because the earliest pictures of 3C48 showed an extension out of the nucleus. To study this "fuzz" was a formidable task because it was so faint compared with the very bright adjacent nucleus. One early application was to try to determine the redshift and the nature of BL Lac, which had a featureless spectrum. Using annuli as apertures, Oke & Gunn (1974) obtained the spectrum of the "fuzz" and determined that the spectrum was like that of a giant elliptical galaxy with a redshift of 0.07. Richstone & Oke (1977) were able to isolate emission lines at the same redshift as the quasar at a distance of 2 arcsec from the center of 3C249.1, as was done by Wampler et al (1975) for 3C48 and by Stockton (1976) for 4C37.43. In 1982, Boroson & Oke (1982) obtained spectra of the "fuzz" both north and south of the nucleus of 3C48 and showed that the light was being generated by hot stars with strong Balmer absorption lines. This showed conclusively that 3C48 is at a cosmological distance. In a series of papers, Boroson et al (1982), Boroson & Oke (1984), and Boroson et al (1985) studied the "fuzz" around QSOs. In the low-luminosity objects, they found a continuum produced almost certainly by stars. In the high-luminosity QSOs they found two groups of objects, one in which the "fuzz" was blue and where the emission lines were strong, and the other where the continuum was red and the emission lines faint. They concluded that there must be about 10^8 M$_\odot$ of gas in the "fuzz" around high-luminosity objects. A model was discussed in which QSOs had variable accretion rates onto a black hole. In subsequent years much better observations of the surroundings of QSOs was possible with the Canada-France-Hawaii Telescope and Hubble telescopes.

The first absorption lines in the spectra of QSOs were found in 3C191 (Burbidge et al 1966, Stockton & Lynds 1966). Additional such objects were observed soon after at Palomar Observatory. Arp et al (1967) found absorption lines at nearly the same redshift as the emission lines in PKS0237-23, whereas Greenstein & Schmidt (1967) found a second absorption system at a smaller redshift in the same object. Bahcall et al (1968) identified five systems with z between 1.36 and 2.2. Bahcall et al (1969) found three systems in Ton 1530, and Lowrence et al (1972) found at least four systems in PHL957. In 1975, a program was begun by Sargent and Boksenberg to study absorption-line systems in great detail using the enormously effective image photon counting system along with a moderately high-dispersion spectrograph. It was used first to observe PKS0237-23, in 1978, and a large number of absorption systems were found. Sargent and Boksenberg, along with a number of collaborators, accumulated and analyzed absorption spectra of a substantial number of QSOs. Using data from six QSOs, Sargent et al (1980) differentiated between Lα systems, which did not appear to arise from galaxy halos, and absorption line systems with metal-line absorption, which did appear to come from galaxy halos. Young et al (1979) found a strong correlation at the $L\alpha/L\beta$ wavelength ratio, as expected. In 1982, Young et al (1982) found evidence for evolution with z in the $L\alpha$ forest. Tytler et al (1987) studied the MgII aborption in 24 low redshift QSOs and found an increase in the density of systems with redshift. In the late 1980s and early 1990s, Sargent et al (1989) obtained spectra of large numbers of both low- and high-z QSOs. In a study of CIV absorption in 55 QSOs, Sargent et al (1989) found that the density of absorbers decreased with increasing z and suggested that this could be caused by a decrease in the abundance of CIV as z increases. In a sample of 59 QSOs, Sargent et al (1989) found 37 Lyman limit absorption systems, of which two thirds had associated heavy-element absorption, indicating that these systems were produced by absorption in galaxies. Steidel & Sargent (1992) studied again the MgII absorption in over 100 QSOs and found that the density decreased somewhat with z.

The discovery of very close pairs of radio sources and the possibility that they were lensed systems (Walsh et al 1979, Weymann et al 1979) led to a number of important papers based on Palomar observations. Young et al (1980, 1981) observed the double radio source Q0957+561 and found the lensing galaxy with a redshift of 0.36. The brightening of B by 0.3 mag led to an analysis of the system, which predicted time delays of up to 5 years in the light-travel time. The triple source Q1115+080 could be fitted with a quintuple image model (Young et al 1981). Schneider et al (1985) found a very red galaxy in the midst of the three radio-source positions. Hewett et al (1987) observed the source 0023+171, which is double with a separation of 5 arcsec. They concluded that it could be a lensed system or simply a close pair of radio sources.

Visit the Annual Reviews home page at www.AnnualReviews.org

LITERATURE CITED

Adams WS. 1949. *Ap. J.* 109:354–79

Adams WS. 1956. *Ap. J.* 123:189–200

Adelman SJ. 1973a. *Ap. J.* 183:95–120

Adelman SJ. 1973b. *Ap. J. Suppl.* 26:1

Aller LH, Bowen IS, Minkowski R. 1955. *Ap. J.* 122:62–71

Angel JRP. 1978. *Annu. Rev. Astron. Astrophys* 16:487–519

Angel JRP, Landstreet JD, Oke JB. 1972. *Ap. J.* 171:L11–15

Arp HC, 1962. *Ap. J.* 136:66–74

Arp HC, Bolton JG, Kinman TD. 1967. *Ap. J.* 147:840–45

Babcock HW. 1947. *Ap. J.* 105:105–19

Babcock HW. 1957. *Ap. J. Suppl.* 3:141

Babcock HW. 1960. *Ap. J.* 132:521–31

Bahcall J, Greenstein JL, Sargent WLW. 1968. *Ap. J.* 153:689–98

Bahcall JN, Osmer PS, Schmidt M. 1969. *Ap. J.* 156:L1–L5

Barker T, et al. 1971. *Ap. J.* 165:67–86

Baschek B, Sargent WLW, Searle L. 1972. *Ap. J.* 173:611–18

Baschek B, Searle L. 1969. *Ap. J.* 155:537

Baum WA. 1955a. *Sky Telesc.* 14:264–67

Baum WA. 1955b. *Sky Telesc.* 14:330–34

Becklin EE, Kristian J, Matthews K, Neugebauer G. 1973. *Ap. J.* 186:L137–39

Becklin EE, Neugebauer G. 1968. *Ap. J.* 151:145–61

Becklin EE, Neugebauer G. 1969. *Ap. J.* 157:L31–36

Becklin EE, Neugebauer G. 1975. *Ap. J.* 200:L71–74

Bethe H. 1939. *Phys. Rev.* 55:434–456

Bidelman WP. 1960. *Publ. Astron. Soc. Pac.* 72:24–28

Boesgaard AM, Tripicco MJ. 1986. *Ap. J.* 302:L49–53

Boksenberg A. 1972. *Aux. Inst. Large Telesc., ESO-CERN Conf.*, ed. S Lautsen, A Reiz. p. 295. ESO-CERN, Geneva, Switzerland

Boksenberg A, Sargent WLW. 1975. *Ap. J.* 198:31–43

Bonsack WK. 1961. *Ap. J.* 133:340–434

Bonsack WK, Greenstein JL. 1960. *Ap. J.* 131:83–98

Boroson TA, Oke JB. 1982. *Nature* 296:397–99

Boroson TA, Oke JB. 1984. *Ap. J.* 281:535–44

Boroson TA, Oke JB, Green RF. 1982. *Ap. J.* 263:32–42

Boroson TA, Persson SE, Oke JB. 1985. *Ap. J.* 293:120–31

Bowen IS. 1952. *Ap. J.* 116:1–7

Bowen IS. 1955. *Ap. J.* 121:306–11

Branch D, Greenstein JL. 1971. *Ap. J.* 167:89–100

Brown JA, Wallerstein G, Oke JB. 1991. *Astron. J.* 101:1693

Burbidge EM, Burbidge GR, Fowler WA, Hoyle F. 1957. *Rev. Mod. Phys.* 29:547

Burbidge EM, Lynds CR, Burbidge GR. 1966. *Ap. J.* 144:447–51

Caughlan GR, Fowler WA. 1964. *Ap. J.* 139:1180–94

Cayrel G, Cayrel R. 1963. *Ap. J.* 137:431–69

Chamberlain JW, Aller LH. 1951. *Ap. J.* 114:52–72

Cocke WJ, Disney MJ, Taylor DJ. 1969. *Int. Astron. Union* Circ. No. 2128

Conti PS, Greenstein JL, Spinrad H, Wallerstein G, Vardya MS. 1967. *Ap. J.* 148:105–27

Conti PS, Wallerstein G. 1969. *Ap. J.* 155:11–15

Crosswell K. 1995. *The Alchemy of the Heavens.* New York: Anchor Books

Danziger IJ. 1967. *Ap. J.* 150:733–35

Danziger IJ, Conti PS. 1966. *Ap. J.* 146:392–98

Dennison EW, Schmidt M, Bowen IS. 1969. *Adv. Electron. Electron Phys.* 28:767–71

Deutsch AJ. 1956. *Ap. J.* 123:210–27

Deutsch AJ, Merrill PW. 1959. *Ap. J.* 130:570

Dominy JF, Wallerstein G. 1988. *Ap. J.* 330:937

Eggen OJ. Greenstein JL. 1965. *Ap. J.* 141:83–108

Elias JH, et al. 1978. *Ap. J.* 220:25–41

Elias JH, Matthews K, Neugebauer G, Persson SE. 1985. *Ap. J.* 296:379–89

Ennis DJ, Neugebauer G, Werner MW. 1982. *Ap. J.* 262:451–59

Filipenko AV, Sargent WLW. 1985. *Nature* 316:407

Fouts G, Sandage A. 1986. *Astron J.* 91:1189–1208

Fowler WA, Burbidge EM, Burbidge GR, Hoyle F. 1965. *Ap. J.* 142:423

Fowler WA, Burbidge GR, Burbidge EM. 1955. *Ap. J. Suppl.* 2:167–94

Frogel JA, Persson SE, Cohen JG. 1983a. *Ap. J. Suppl.* 53:713–49

Frogel JA, Persson SE, Cohen JG. 1983b. *Ap. J.* 275:773–89

Greenstein JL. 1974. *Ap. J. Lett.* 189:L131–33

Greenstein JL. 1978. *Publ. Astron. Soc. Pac.* 90:303–6

Greenstein JL. 1979. *Ap. J.* 227:244–51

Greenstein JL. 1982. *Ap. J.* 258:661–73

Greenstein JL. 1989a. *Pub. Astron. Soc. Pac.* 101:787–810

Greenstein JL. 1989b. *Comments on Ap.* 13, 303

Greenstein JL, Aller LH. 1960. *Ap. J. Suppl.* 5:139

Greenstein JL, Boksenberg A. 1978. *MNRAS* 185:823–32

Greenstein JL, Boksenberg A, Carswell R, Shortridge K. 1977a. *Ap. J.* 212:186–97

Greenstein JL, Henry RJW, O'Connell RF. 1985. *Ap. J. Lett.* 289:L25–29

Greenstein JL, Kraft RP. 1959. *Ap. J.* 130:99–109

Greenstein JL, Liebert JW. 1990. *Ap. J.* 360:662–684

Greenstein JL, McCarthy JK. 1985. *Ap. J.* 289:732–47

Greenstein JL, Minkowski R. 1973. *Ap. J.* 182:225–43

Greenstein JL, Münch G. 1966. *Ap. J.* 146: 618–20

Greenstein JL, Oke JB. 1982. *Ap. J.* 252:285–95

Greenstein JL, Oke JB, Shipman HL. 1971. *Ap. J.* 169:563–66

Greenstein JL, Sargent AI. 1974. *Ap. J. Suppl.* 28:157–209

Greenstein JL, Sargent WLW, Boroson TA, Boksenberg A. 1977b. *Ap. J.* 218:L121–27

Greenstein JL, Schmidt M. 1967. *Ap. J.* 148:L13–15

Greenstein JL, Schmidt M, Searle L. 1974. *Ap. J. Lett.* 190:L27–28

Greenstein JL, Trimble V. 1967. *Ap. J.* 149: 283–98

Greenstein JL, Trimble V. 1972. *Ap. J.* 175: L1–5

Griffin RF. 1967. *Ap. J.* 148:465–76

Griffin RF, Gunn JE, Zimmerman BA, Griffin REM. 1988. *Astron. J.* 96:172–97

Griffin RF, Gunn JE, Zimmerman BA, Griffin REM. 1985. *Astron. J.* 90:609–42

Gunn JE, Carr M, Schneider DP, Zimmerman BA. 1987. *Opt. Eng.* 26:779–87

Gunn JE, Griffin RF. 1979. *Astron. J.* 84:752

Gunn JE, Griffin RF, Griffin REM, Zimmerman BA. 1988. *Astron. J.* 96:198–210

Gunn JE, Westphal JE. 1981. *Proc. Soc. Photo-Opt. Inst. Eng.* 290:16–23

Hamilton D, Oke JB, Carr MA, Cromer J, Harris FH, et al. 1993. *Publ. Astron. Soc. Pac.* 105:1308–21

Härm R, Schwarzschild M. 1955. *Ap. J. Suppl.* 1:319–430

Helfer HL, Wallerstein G. 1964. *Ap. J. Suppl.* 16:1–194

Helfer HL, Wallerstein G, Greenstein JL. 1959. *Ap. J.* 129:700–19

Helfer HL, Wallerstein G, Greenstein JL. 1960. *Ap. J.* 132:553–64

Helfer HL, Wallerstein G, Greenstein JL. 1963. *Ap. J.* 138:97–117

Henyey LG, Wilets L, Böhm KH, Lelevier R, Levee, RD. 1959. *Ap. J.* 129:628

Herbig GH. 1956. *Publ. Astron. Soc. Pac.* 68:204–10

Herbig GH. 1965. *Ap. J.* 141:588–609

Hearnshaw JB. 1986. *The Analysis of Starlight.* Cambridge, UK: Cambridge Univ. Press

Hewitt JN, et al. 1987. *Ap. J.* 321:706–13

Hodge PW, Wallerstein G. 1966. *Pub. Astron. Soc. Pac.* 78:411

Huggins PJ, Phillips TG, Neugebauer G, Werner MW, Wannier PG, Ennis D. 1979. *Ap. J.* 227:441–45

Humason ML, Zwicky F. 1947. *Ap. J.* 105: 85–91

Johnston HM, Kulkarni SR, Oke JB. 1989. *Ap. J.* 345:492–97

Joy AH, Wilson RE. 1949. *Ap. J.* 109:231–43

Jugaku J, Sargent WLW, Greenstein JL. 1961. *Ap. J.* 134:783

Kamper K, van den Bergh S. 1976. *Ap. J. Suppl.* 32:351–66

Keenan PC, Deutsch AJ, Garrison RF. 1969. *Ap. J.* 158:261

Kells W, et al. 1998. *Publ. Astron. Soc. Pac.* 110:1487–98

Kirshner RP, Kwan J. 1975. *Ap. J.* 197:415–24

Kirshner RP, Oke JB. 1975. *Ap. J.* 200:574–81

Kirshner RP, Oke JB, Penston M, Searle L. 1973. *Ap. J.* 185:303–22

Kirshner RP, Wilner SP, Becklin EE, Neugebauer G, Oke JB. 1973. *Ap. J. Lett.* 180:L97–100

Kodaira K. 1964. *Z. Astrophys.* 59:139–81

Kodaira K, Greenstein JL, Oke JB. 1969. *Ap. J.* 155:525–36

Kodaira K, Greenstein JL, Oke JB. 1970. *Ap. J.* 159:485–512

Koelbloed D. 1967. *Ap. J.* 149:299–315

Koester D, Vauclair G, Dolez N, Oke JB, Greenstein JL, Weidemann V, 1985. *Astron. Astrophys.* 149:423–28

Kraft RP. 1959. *Ap. J.* 130:110–22

Kraft RP. 1962. *Ap. J.* 135:408–23

Kraft RP. 1964. *Ap. J.* 139:457–75

Kraft RP. 1999. *Ap. J.* 525, Part 3:876–897
Kraft RP. Gunn, JE. 1963. *Ap. J.* 137:301–315
Krishna Swamy K. 1966. *Ap. J.* 145:174–94
Kristian J, Visvanathan N, Westphal JA, Snellen GH. 1970. *Ap. J.* 162:475–83
Landstreet JD, Angel JRP. 1974. *Ap. J. Lett.* 190:L25-L26
Lowell P, Slipher UM. 1912. *Lowell Obs. Bull.* 2:17
Lowrance JL, Morton DC, Zucchino P, Oke JB, Schmidt M. 1972. *Ap. J.* 171:233–51
Lupton RH, Gunn JE, Griffin RK. 1987. *Astron. J.* 93:1114–36
Markarian BY, Lipovetsky VA. 1972. *Astrofizika* 8:155–65
Mathieu RD, Latham DW, Griffin RF, Gunn JE. 1986. *Astron. J.* 92:1100–17
Mathis JS, Lamers HJGLM. 1992. *Astron. Astrophys.* 259:39
McClintock JE, Remillard RA. 1986. *Ap. J.* 308:110–22
McClure RD, Woodsworth AW. 1990. *Ap. J.* 352:709
Merrill PW. 1952a. *Ap. J.* 116:21–26
Merrill PW. 1952b. *Ap. J.* 116:18–20
Merrill PW. 1952c. *Ap. J.* 116:21
Merrill PW. 1952d. *Ap. J.* 116:337–43
Merrill PW. 1952e. *Ap. J.* 116:344–47
Merrill PW. 1952f. *Ap. J.* 116:523–24
Merrill PW. 1953. *Ap. J.* 118:453–58
Merrill PW, Greenstein JL. 1956. *Ap. J. Suppl.* 2:225–40
Michaud G, Reeves H, Charland Y. 1974. *Astron. Astrophys.* 37:313–24
Michland G. 1970. *Ap. J.* 160:641–58
Minkowski R. 1939. *Ap. J.* 89:156–217
Minkowski R, Osterbrock DE. 1959. *Ap. J.* 129:583–99
Minkowski R, Osterbrock DE. 1960. *Ap. J.* 131:537–40
Moore J, Menzel DH. 1930. *Pub. Astron. Soc. Pac.* 42:330
Morgan WW. 1934. *Publ. Yerkes Obs.* 7:III
Morton DC. 1974. *Ap. J.* 193:L35–39
Münch G. 1957. *Ap. J.* 125:42–65
Münch G. 1964. *Ap. J.* 140:107–11
Münch G, Bergstrahl JT. 1977. *Pub. Astron. Soc. Pac.* 89:232
Münch G, Persson SE. 1971. *Ap. J.* 165:241–58
Münch G, Taylor K. 1974. *Ap. J.* 192:L93–95
Münch G, Trauger JT, Roesler FL. 1977. *Ap. J.* 216:963–66
Münch G, Unsöld A. 1962. *Ap. J.* 135:711–14
Münch G, Zirin H. 1961. *Ap. J.* 133:11–28
Murphy TW Jr, Matthews K, Soifer BT. 1999. *Pub. Astron. Soc. Pac.* 111:1176–84
Neugebauer G, Becklin EE, Beckwith S, Matthews K, Wynn-Williams CG. 1976. *Ap. J.* 205:L139–41
Neugebauer G, Becklin EE, Kristian J, Leighton RB, Snellen G, Westphal JA. 1969. *Ap. J.* 156:L115–19
Neugebauer G, Becklin EE, Matthews K, Wynn-Williams CG. 1978. *Ap. J.* 220:149–55
Neugebauer G, Becklin EE, Oke JB, Searle L. 1976. *Ap. J.* 205:29–43
Neugebauer G, Oke JB, Becklin E, Garmire G. 1969. *Ap. J.* 155:1–9
Oinas V. 1974. *Ap. J. Suppl.* 27:405–14
Oke JB. 1967a. *Ap. J.* 147:901–7
Oke JB. 1967b. *Ap. J.* 150:L5–8
Oke JB. 1969a. *Ap. J.* 156:L49–53
Oke JB. 1969b. *Publ. Astron. Soc. Pac.* 81:11–22
Oke JB. 1974. *Ap. J. Suppl.* 27:21–35
Oke JB. 1977. *Ap. J.* 217:181–85
Oke JB. 1978. *J. R. Astron. Soc. Can.* 72:121–37
Oke JB. 1987. In *Superluminal Radio Sources*, ed. JA Zensus, TJ Pearson, pp. 267–72. Cambridge, UK: Cambridge Univ. Press
Oke JB, Becklin EE, Neugebauer G. 1970. *Ap. J.* 159:341–55
Oke JB, Greenstein JL. 1977. *Ap. J.* 211:872–80
Oke JB, Gunn JE. 1974. *Ap. J.* 189:L5–8
Oke JB, Gunn JE. 1982. *Publ. Astron. Soc. Pac.* 94:586–94
Oke JB, Lauer TR. 1979. *Ap. J.* 230:360–72
Oke JB, Sargent WLW, Neugebauer G, Becklin EE. 1967. *Ap. J.* 150:L173–76
Oke JB, Searle L. 1974. *Annu. Rev. Astron. Astrophys.* 12:315–29

Oke JB, Shipman HL. 1971. *Int. Astron. Union Symp.* 42:67–76

Oke JB, Weidemann V, Koester D. 1984. *Ap. J.* 281:276–85

Osterbrock DE. 1960. *Ap. J.* 132:325

Parker R, Greenstein JL, Helfer HL, Wallerstein G. 1961. *Ap. J.* 133:101

Peterson. RC. 1980. *Ap. J.* 237:L87–91

Phillips TG, Huggins PJ, Neugebauer G, Werner MW. 1977. *Ap. J.* 217:L161–64

Pilachowski CA, Wallerstein G, Leep EM. 1980. *Ap. J.* 236:508–21

Popper DM. 1954. *Ap. J.* 120:316–21

Preston G. 1972. *Ap. J.* 175:465–72

Preston GW. 1971. *Ap. J.* 164:309–15

Priedhorsky W, Matthews K, Neugebauer G, Werner M, Krzeminski W. 1978. *Ap. J.* 226:397–404

Raga A, Wallerstein G, Oke JB. 1989. *Ap. J.* 347:1107–13

Richstone DO, Oke JB. 1977. *Ap. J.* 213:8–14

Russell HN. 1929. *Ap. J.* 70:11–82

Sandage A. 1964. *Ap. J.* 139:442–50

Sandage A. 1966. *Ap. J.* 146:13

Sandage A. 1967. *Ap. J.* 150:L9–L12

Sandage A. 1969. *Ap. J.* 158:1115–36

Sandage A. 2000. *Annu. Rev. Astron. Astrophys.* 38:000–00

Sandage A, Kowal C. 1986. *Ap. J.* 91:1140

Sandage A, Westphal JA, Kristian J. 1969. *Ap. J.* 156:927–42

Sandage A, Westphal JA, Kristian J. 1969. *Ap. J.* 156:927–42

Sandage A, Westphal JA, Strittmatter PA. 1966. *Ap. J.* 146:322–25

Sandage A, et al. 1966. *Ap. J.* 146:316–22

Sargent AI, Greenstein JL, Sargent WLW. 1969. *Ap. J.* 157:757–68

Sargent WLW, Jugaku J. 1961. *Ap. J.* 134:777–82

Sargent WLW, Searle L. 1966. *Ap. J.* 145:652–54

Sargent WLW, Searle L. 1967. *Ap. J.* 150:L33–L37

Sargent WLW, Steidel CC, Boksenberg A. 1989. *Ap. J. Suppl.* 69:703–61

Sargent WLW, Young PJ, Boksenberg A, Tytler D. 1980. *Ap. J. Suppl.* 42:41–81

Schmidt M, Schneider DP, Gunn JE. 1986a. *Ap. J.* 306:411–27

Schmidt M, Schneider DP, Gunn JE. 1986b. *Ap. J.* 310:518–33

Schmidt M, Schneider DP, Gunn JE. 1987a. *Ap. J.* 316:L1–3

Schmidt M, Schneider DP, Gunn JE. 1987b. *Ap. J.* 321:L7–10

Schneider DP, Greenstein JL. 1979. *Ap. J.* 233:935–45

Schneider DP, Lawrence CR, Schmidt M, Gunn JE, Turner EL, et al. 1985. *Ap. J.* 294:66–69

Seyfert CK. 1943. *Ap. J.* 98:28

Searle L. 1961. *Ap. J.* 133:531

Searle L, Rodgers AW. 1966. *Ap. J.* 143:809–22

Sion EM, Greenstein JL, Landstreet JD, Liebert J, Shipman HL, Wegner GA. 1983. *Ap. J.* 269:253–257

Smith VV, Wallerstein G. 1983. *Ap. J.* 273:742–48

Spinrad H, Taylor BJ. 1970. *Ap. J. Suppl.* 22:177–248

Spitzer L Jr. 1939. *Ap. J.* 90:494–540

Stawikowski A, Greenstein JL. 1964. *Ap. J.* 140:1280–91

Steidel CC, Sargent WLW. 1992. *Ap. J. Suppl.* 80:1–108

Stockton A. 1976. *Ap. J. Lett.* 205:L113–16

Stockton AN, Lynds CR. 1966. *Ap. J.* 144:451–53

Stoekly R, Greenstein JL. 1969. *Ap. J.* 154:909–22

Szkody P, Wade RA. 1981. *Ap. J.* 251:201–4

Taylor BS, Spinrad H, Schweizer F. 1972. *Ap. J.* 173:619–630

Taylor K, Münch G. 1978. *Astron. Astrophys.* 70:359–66

Thorstensen JR, Wade RA, Oke JB. 1986. *Ap. J.* 309:721–31

Trauger JT, Münch G, Roesler FL. 1980. *Ap. J.* 236:1035–42

Trauger JT, Roesler F, Münch G. 1978. *Ap. J.* 219:1079–1083

Traving G. 1962. *Ap. J.* 135:439

Trimble V. 1968. *Astron. J.* 73:535–47

Tytler D, Boksenberg A, Sargent WLW, Young P, Kunth D. 1987. *Astrophys. J. Suppl.* 64:667–702

Unsöld A. 1955. *Physyk der Sternatmosphären.* Berlin, Germany: Springer

Vanture AD, Wallerstein G, Brown JA, Bozan G. 1991. *Ap. J.* 381:278–87

Vaughan AH, Zirin H. 1968. *Ap. J.* 152:123–39

Venn K, Lambert DL. 1990. *Ap. J.* 363:234

Wade RA. 1979. *Astron. J.* 84:562–69

Wade RA, Oke JB. 1977. *Ap. J.* 215:568–73

Wallerstein G. 1959. *Publ. Astron. Soc. Pac.* 71:316–20

Wallerstein G. 1962. *Ap. J. Suppl.* 6:407–79

Wallerstein G. 1967. *Astrophys. Lett.* 1:31–32

Wallerstein G. 1975. *Ap. J. Suppl.* 29:375–96

Wallerstein G. 1977. *Ap. J.* 211:170–77

Wallerstein G. 1985. *Publ. Astron. Soc. Pac.* 97:994–1000

Wallerstein G. 1999. *Ap. J.* 525, Part 3:426–449

Wallerstein G, Conti PS, Greenstein JL. 1972. *Astrophys. Lett.* 12:101–102

Wallerstein G, Goldsmith D. 1974. *Ap. J.* 187:237–42

Wallerstein G, Greenstein JL. 1964. *Ap. J.* 139:1163–79

Wallerstein G, Greenstein JL. 1980. *Pub. Astron. Soc. Pac.* 92:275–83

Wallerstein G, Greenstein JL, Parker R, Helfer HL, Aller LH. 1963. *Ap. J.* 137:280–300

Wallerstein G, Helfer HL. 1959a. *Ap. J.* 129: 347–55

Wallerstein G, Helfer HL. 1959b. *Ap. J.* 129: 720–23

Wallerstein G, Helfer HL. 1961. *Ap. J.* 133:562

Wallerstein G, Herbig GH, Conti PS. 1965. *Ap. J.* 141:610–16

Wallerstein G, Machado-Pelaez L, Gonzalez G. 1999. *Pub. Astron. Soc. Pac.* 111:335–41

Walsh D, Carswell RF, Weymann RJ. 1979. *Nature* 279:381–384

Wampler EJ, Oke JB. 1967. *Ap. J.* 148:695–704

Wampler EJ, Robinson LB, Burbidge EM, Baldwin JA. 1975. *Ap. J. Lett.* 198:L49–52

Weidemann V. 1960. *Ap. J.* 131:638–63

Welsh WF, Horne K, Oke JB. 1993. *Ap. J.* 406:229–33

Westbrook WE, Werner MW, Elias JH, Gezari DY, Hauser MG, et al. 1976. *Ap. J.* 209:94–101

Westphal JA, Sandage A, Kristian J. 1968. *Ap. J.* 154:139–56

Weymann RJ, Chafee FH, Davis M, Curleton NP, Walsh D, et al. 1979. *Ap. J. Lett.* 233: L43–L46

White RE, Vaughan AH, Preston GW, Swings JP. 1976. *Ap. J.* 204:131–40

Willson LA, Garnavich P, Mattei JA. 1981. *International Bulletin of Variable Stars* 1961:1

Wilson OC. 1959. *Ap. J.* 130:499–506

Wilson OC. 1976. *Ap. J.* 205:823–40

Wilson OC. 1978. *Ap. J.* 226:379–96

Wilson OC, Bappu MKV. 1957. *Ap. J.* 125: 661–83

Wilson OC, Münch G, Flather EM, Coffeen MF. 1959. *Ap. J. Suppl.* 4:199–56

Wilson OC, Münch G, Flather EM, Coffeen MF. 1960. *Ap. J. Suppl.* 4:199–256

Wolff SC, Preston GW. 1978. *Ap. J. Suppl.* 37:371–92

Yee HKC, Oke JB. 1978. *Ap. J.* 226:753–69

Yee HKC, Oke JB. 1981. *Ap. J.* 248:472–84

Young P, Deverill RS, Gunn JE, Westphal JA, Kristian J. 1981. *Ap. J.* 244:723–35

Young P, Gunn JE, Kristian J, Oke JB, Westphal JA. 1980. *Ap. J.* 241:507–20

Young P, Gunn JE, Kristian J, Oke JB, Westphal JA. 1981. *Ap. J.* 244:736–55

Young P, Sargent WLW, Boksenberg A. 1982. *Ap. J. Suppl.* 48:455–506

Young P, Sargent WLW, Boksenberg A, Carswell RF, Whelan JAJ. 1979. *Ap. J.* 229:891–908

Young PJ, Westphal JA, Kristian J, Wilson CP, Landauer FP. 1978. 221:721–730

Zarro DM, Zirin H. 1986. *Ap. J.* 304:365–70

Zirin H. 1982. *Ap. J.* 260:655–69

Zwicky F, Zwicky MA. 1971. *Catalogue of Compact Galaxies and of Post-Eruptive Galaxies.* Published by F. Zwicky, Guemligen, Switzerland

Annu. Rev. Astron. Astrophys. 2000. 38:113–41

COMMON ENVELOPE EVOLUTION OF MASSIVE BINARY STARS

Ronald E. Taam

Department of Physics & Astronomy, Northwestern University, Evanston, IL 60208;
e-mail: taam@apollo.astro.nwu.edu

Eric L. Sandquist

Department of Astronomy, San Diego State University, San Diego, CA 92182;
e-mail: erics@mintaka.sdsu.edu

Key Words close binaries, hydrodynamics

■ **Abstract** The common envelope phase of binary star evolution plays an essential role in the formation of short period systems containing a compact object. In this process, significant mass and angular momentum are lost, transforming a wide progenitor system into a close remnant binary. The pathways leading to this phase and the outcomes are described. Emphasis is placed on the conditions that are required for survival of the binary according to the results of three-dimensional hydrodynamics calculations. The evolution of high-mass systems containing neutron stars is discussed, including double neutron stars, binary pulsars, Thorne-Zytkow objects, and high- and low-mass X-ray binaries.

1. INTRODUCTION

Close binary stars have been important objects in stellar astronomy since the beginning because they are living proof that stars do not always go through their lives in solitude. Indeed, in a number of systems it is possible to observe the transmutation of stars—a process that can be heralded by transient and, at times, highly energetic activity. With the launches of space-based platforms providing broad wavelength coverage and high temporal resolution, studies of close binaries have come to the forefront of contemporary astrophysics. Binary systems containing compact objects are among the sources revealing the most energetic behavior and the quickest variability, and so have also received the greatest attention. However, this subset of close binaries also contains a contradiction. Because the interactions between the binary components are essential to explanations of the various observed phenomena, the stars must be in close proximity to each other. On the other hand, the creation of white dwarfs, neutron stars, and black holes requires that the parent star go through a giant or supergiant phase.

In a binary system, stars can evolve differently than they would in isolation because the gravitational influence of one of the components can limit the radius of the other. Mass beyond this radius can be transferred to the binary companion and/or lost from the system. This transfer or loss of mass and angular momentum is central to understanding all classes of close binary systems. The critical radius is referred to as the Roche radius for components in synchronous rotation and circular motion about their common center of mass. For a binary system of orbital separation A, the Roche radius R_L of a star of mass M_1 with a companion of mass M_2 is given to a good approximation by the equation:

$$R_L = A \frac{0.49q^{2/3}}{0.6q^{2/3} + \log(1 + q^{1/3})}$$

where q is the mass ratio ($\frac{M_1}{M_2}$) of the system (see Eggleton 1983). The orbital separation is related to the orbital angular momentum, J, of the system by

$$A = \frac{J^2 (M_1 + M_2)}{G M_1^2 M_2^2}.$$

Mass loss of the star in response to the imposition of such a radius constraint, coupled with the orbital evolution described by the preceding set of equations, determines the outcome of the system. For fundamental reviews of binary star evolution, see Paczynski (1971), Thomas (1977), and more recently Vanbeveren et al (1998). A general description of orbital variations in the presence of mass and angular momentum loss can be found in Soberman et al (1997).

Historically, the common envelope concept originated from attempts to understand the formation of cataclysmic variables. Ground-based optical observations revealed cataclysmic variables to be binary systems in which mass is transferred from a Roche lobe-filling main sequence-like star to its white dwarf companion. Because these systems are characterized by orbital periods of a few hours, early attempts to explain their origin relied on similarities between their properties and those of other classes of systems—for example, W Ursae Majoris binaries (Kraft 1962). In these systems, both components fill their respective Roche lobes (and are in contact), and are surrounded by a corotating common envelope. Although the similarity of their total masses and orbital periods to cataclysmic variables was suggestive, theoretical understanding of the evolution of contact binary systems was lacking. Little progress was made on the origin of cataclysmic variables until it was suggested by Paczynski (1976) and Ostriker (1975, in the context of massive binaries in which one member of the system is a neutron star) that these systems evolve from long period progenitor systems (required to incubate the white dwarf in the interior of a red giant) in which severe losses of mass and angular momentum occur (see also Ritter 1976). In this framework, a system characterized by low specific angular momentum could be produced from one of high specific angular momentum. It was envisioned that this transformation involved a phase in which the companion to a red giant or asymptotic red giant star (in a binary system characterized by an orbital separation comparable to an astronomical unit) is engulfed in the giant's differentially rotating envelope. The frictional torque associated with the gravitational interaction of the two stellar cores within this envelope was hypothesized to remove orbital energy, causing the orbit to shrink to separations comparable to a solar radius. Orbital angular momentum of the system is thus converted into the spin angular momentum of the common envelope, and the envelope mass is ejected using the energy released from the orbit.

The spiraling of a star into a red giant envelope was first discussed in the context of a white dwarf—red giant supernova model proposed by Sparks & Stecher (1974) following earlier work on tidal instability for binaries in circular and synchronous orbits, first by Darwin (1879) and then revived by Kopal (1972), Counselman (1973), and Hut (1980). The common envelope phase has now become the standard framework used to understand the formation of short period systems with a compact member. This scenario has applications to many branches of contemporary astrophysics, including X-ray astronomy (low-mass X-ray binaries, high-mass X-ray binaries), gravitational theory (binary neutron stars and black holes), supernova theory (double white dwarfs), and cataclysmic variable star research.

The purpose of this review is to summarize the current theory of the common envelope phase of binary star evolution and to discuss existing models for the formation of compact neutron star binary systems, updating earlier descriptions by Taam & Bodenheimer (1992) and Taam (1996). For a discussion of the evolution of the many subclasses of these systems, see Bhattacharya & Van den Heuvel (1991), Verbunt (1993), Verbunt & Van den Heuvel (1995), and Bhattacharya (1995, 1996). This review differs from an earlier review of common envelope binaries by Iben & Livio (1993) in that the focus here is on the evolution of binaries containing high-mass stars and the formation of short period systems containing neutron stars. Short period systems containing black holes are described in papers by Romani (1996), Portegies Zwart et al (1997), Ergma & Van den Heuvel (1998), Brown et al (1999), and Kalogera (1999). The recent review by Vanbeveren et al (1998) is complementary to the present review in concentrating on the details of massive close binary evolution (in the conservative and quasi-conservative modes of evolution) and its incorporation into massive star population synthesis models.

2.   EVOLUTION INTO THE COMMON ENVELOPE PHASE

The binaries that evolve into the common envelope stage typically have components differing significantly in both size and mass. The size difference is necessary for the formation of an evolved core (the progenitor of the compact component), whereas the mass difference is required to establish a differentially rotating common envelope. Since the components must interact, the more massive component must fill its Roche lobe at some time during its evolution, which implies that the initial orbital period of the system should be $\lesssim 10$ years. Provided that these conditions are realized, the binary system can be considered a candidate for forming a short period binary containing a compact object. The specific evolutionary path a primordial binary takes to the common envelope phase may involve one of two possibilities. [We will not discuss the possibility that a common envelope system may form as a result of the capture of stars during collisional interactions in dense stellar systems (see e.g. Bailyn 1988, Rasio & Shapiro 1991, Sigurdsson & Hernquist 1992).]

In the first case, the more massive component of the system initiates mass transfer to its companion while the binary is in a state of synchronous rotation. During mass exchange, the mean mass transfer rate for a primarily radiative star is given approximately by

$$\dot{M} \sim \frac{M}{\tau_{KH}}$$

where M is the mass of the more massive component and τ_{KH} is its Kelvin-Helmholtz timescale (see Paczynski 1971). Here,

$$\tau_{KH} = \frac{3 \times 10^7 M^2}{RL} \text{yr}$$

with the radius R, luminosity L, and mass expressed in solar units. For massive stars ($\gtrsim 12 \, M_\odot$), the mass transfer rates can exceed $10^{-3} \, M_\odot \, \text{yr}^{-1}$. The ultimate evolution of the system is dependent on its mass ratio and the evolutionary state of the more massive component.

During the core hydrogen burning phase and early in the hydrogen shell burning phase, the star has a radiative envelope, and so will contract upon mass loss since the specific entropy decreases from the surface into the interior. This contraction is less rapid than the shrinkage of its Roche lobe if total mass and angular momentum are conserved. For extreme mass ratios ($\gtrsim 5$–10), it is likely that the companion will be dragged into the envelope of the more massive star (see Vanbeveren et al 1998). For smaller mass ratios, the system may also evolve into contact because the main sequence accretor is unable to assimilate the accreted mass in a thermal equilibrium state. However, the system may avoid the common envelope stage if the timescale for mass transfer does not significantly differ from the timescale to establish corotation. If that is true, the matter surrounding the two components may expand to the outer Lagrangian point where mass can be lost with specific angular momentum greater than the average for the binary. This loss of both mass and angular momentum from the system would lead to shorter orbital periods than the purely conservative approximation would indicate. Such an evolution has been denoted as quasi-conservative (see Van den Heuvel 1992; Vanbeveren et al 1998).

The response of the accretor may differ from that just described if the less massive star is a neutron star (the remnant of what was originally the more massive star) instead of a main sequence star. In this case, the estimated mass transfer rates can easily exceed the Eddington rate, $\dot{M}_{Edd}$ (given by the condition that the radiation pressure gradient generated by the accretion luminosity is sufficient to balance the force of gravity), which can be expressed as

$$\dot{M}_{Edd} = \frac{4\pi c R}{\kappa}$$

where R is the radius of the neutron star and κ is the opacity of the accreting matter. For electron scattering opacity in a hydrogen-rich plasma accreting onto a neutron star of radius 10 km, the critical rate corresponds to $\sim 1.7 \times 10^{-8}\ M_\odot\ yr^{-1}$. Mass transferred in excess of the Eddington rate may be ejected. This follows from the fact that the energy released from a small amount of matter accreted at the neutron star surface can unbind a large amount of matter at large distances (see Blandford & Begelman 1999). Under the assumption that matter is blown away at the trapping radius, where the outward diffusive speed of the photons is balanced by the infall speed of the accreted matter, and that this radius lies within the Roche lobe of the neutron star, King & Begelman (1999) estimate that the common envelope phase can be initially avoided at the onset of mass transfer if the mass loss rates from a main sequence star are $\lesssim 10^{-3}\ M_\odot\ yr^{-1}$. However, for systems with extreme mass ratios the substantial losses of mass and angular momentum lead to such significant orbital shrinkage that the trapping radius lies outside the Roche lobe of the neutron star, and the common envelope phase cannot be avoided.

On the other hand, for a star on the red giant branch during thin shell hydrogen and/or helium burning, the envelope is in a state of convective equilibrium. Based on condensed polytrope models with point mass cores and convective envelopes, Hjellming & Webbink (1987) find that stars having core mass fractions less than 0.214 expand. Detailed studies of adiabatic mass loss from more realistic stellar models (Hjellming 1989) confirm this tendency, revealing that expansion takes place whenever the convective envelope exceeds 50% of the stellar mass. Hence, in thin shell burning stages, there is no tendency for the star to contract in response to mass loss (as the star's Roche lobe radius and orbital separation of the system do), so the mass loss proceeds more rapidly than in the case of stars with radiative envelopes. The timescale for mass loss is greater than the dynamical timescale, but smaller than the thermal timescale. Hence, the companion is engulfed in the envelope of the red giant, and corotation cannot be maintained. The inclusion of a wind from the neutron star cannot prevent the system from entering into the common envelope state.

The second route to common envelope evolution involves the initiation of mass transfer for a binary in which the mass-losing star is in a state of nonsynchronous rotation. This can occur if there is insufficient orbital angular momentum in the system that can be drawn on to bring the more massive star into a state of corotation. This situation can develop for systems that have orbital separations smaller than the critical separation at which the total angular momentum of a synchronous system is a minimum. This instability is tidal in origin and was originally discussed by Darwin (1879), and more recently by Kopal (1972), Counselman (1973), and

Hut (1980). It occurs when the spin angular momentum of the evolved star is more than one-third the orbital angular momentum of the binary system. Because the radius of gyration

$$r_g^2 = \frac{I}{MR^2}$$

(I is the moment of inertia) of evolved red giant stars is about 0.1 (De Greve et al 1975), this implies that the tidal instability occurs for mass ratios $\gtrsim$5–6. In this case, the companion will spiral into the envelope of the red giant. In earlier phases of evolution for massive stars, the radius of gyration ranges from $\sim$0.07 on the main sequence to $\sim$0.04 at the end of core hydrogen burning. This leads to a tidal instability for mass ratios $\gtrsim$25. In this context, a binary containing a 1.4 $M_\odot$ neutron star companion would only be of academic interest because such massive stars ($\gtrsim$35 $M_\odot$) lose a substantial amount of mass which can prevent the star from expanding significantly, thus avoiding Roche lobe overflow and mass transfer (see Vanbeveren et al 1998).

3.   MODEL OF THE INTERACTION AND EARLY NUMERICAL WORK

Because the interaction of the less massive star with the envelope of the more massive star is complex, the first models used the Bondi-Hoyle (1944) approximation. In this framework, the engulfed star is considered to be a point mass (i.e. finite size effects are neglected), and it is assumed to gravitationally interact with a uniformly moving homogeneous medium. Although the usefulness of such a formulation is limited because of the neglect of rotational effects associated with the binary motion (the stars rotating about the center of mass of the system), its use led to important insights. Initial studies carried out by Sparks & Stecher (1974) and Alexander et al (1976) were primarily concerned with the drag on the embedded star, and not the influence of the energy dissipation on the evolution of the red giant. Since the motion of the engulfed star is highly supersonic, the drag is produced by a detached bow shock in the direction of motion and by a gravitational wake (see Ruderman & Spiegel 1971) in the opposite direction. Based on dimensional arguments, the energy dissipation rate $\dot{E}$ can be estimated to be

$$\dot{E} \sim \pi R_A^2 \rho v^3$$

where R_A is the accretion radius, ρ is the local density within the envelope, and v is the velocity of the embedded star relative to the common envelope. The accretion radius is the gravitational interaction length scale based on the Bondi-Hoyle interpolation formula, and is given by

$$R_A = \frac{2GM}{v^2 + c^2}$$

where M is the mass of the embedded star and c is the local speed of sound in the common envelope. The finite size of the star in the envelope can also give rise to an additional contribution associated with the aerodynamic drag force exerted on the star. These interactions exert torques on the star, leading to the decay of its orbit and to the spin-up of the common envelope. Although not originally considered, the energy deposited in the envelope decreases its binding energy with respect to the binary system, thereby enhancing the possibility of its subsequent ejection.

Because the conversion of orbital energy into kinetic energy of mass loss is central to the formation of short period systems, the prescription of the energy generation in the common envelope is important. In the first approximation an angular velocity profile in the envelope must be specified to calculate the relative velocity, since the interaction leads to flows that transfer angular momentum from the orbit to the gas. As a simple description, the flow was assumed to establish an angular velocity profile consistent with marginal stability (constant specific angular momentum in space; see Taam et al (1978), uniform rotation (Taam 1979), or uniform rotation to a tidal radius beyond which the angular velocity took the form of a power law decreasing with radius from the engulfed star (Meyer & Meyer-Hofmeister 1979).

The early pioneering studies were carried out by Taam et al (1978), Taam (1979), and Delgado (1980) for the high-mass X-ray binary progenitors of binary radio pulsars. In a parallel investigation, Meyer & Meyer-Hofmeister (1979) studied the interaction of a red giant with a main sequence-like companion for the progenitors of pre-cataclysmic variable systems (see also Livio & Soker 1984 for planet interactions in the common envelope). In all these investigations, the evolution was treated in the simplest one-dimensional approximation. Notwithstanding the simplifying approximations used to make the problem tractable, a number of their qualitative conclusions are characteristic of interactions of this type. Specifically, the common envelope phase is very rapid (less than about 1,000 years) and hence, nearly adiabatic. Furthermore, it was shown that the removal of orbital energy is very effective, leading to reductions in binary separation by a factor of about 100. The numerical results are not sensitive to the prescription for the angular velocity profile because higher relative velocities (which result from uniform rotation) lead to lower densities and comparable energy generation rates (Taam 1979). The numerical results also indicate that some mass ejection would take place, but the calculations were insufficiently resolved to demonstrate conclusively that the entire common envelope would be ejected, or to identify the physical mechanism that causes the companion to stop its inward spiral and form an evolved short period binary remnant.

Subsequently, a number of simplifying assumptions were relaxed. The most notable is the treatment of the problem in one dimension—in particular, the assumptions inherent in the one-dimensional modeling (namely, the effect of rotation on the structure of the giant, the redistribution of angular momentum, and the spherical averaging process for energy dissipation) are no longer necessary. The first two-dimensional investigations of the common envelope phase were carried out by Bodenheimer & Taam (1984, 1986). For the case of supergiant-neutron star binaries, they showed that hydrodynamic mass ejection does develop, but more importantly the ejection of matter is confined to the equatorial plane of the binary system. This result reflects the spatial distribution of the energy deposited by interaction of the cores with the common envelope. In addition some matter is ejected with greater than the escape speed. This leads to a reduction in the efficiency of mass ejection from the envelope. Here the efficiency, α_{CE}, is given by

$$\alpha_{CE} = \frac{\Delta E_b}{\Delta E_{orb}}$$

where ΔE_b is the change in the binding energy of the ejected matter and ΔE_{orb} is the change in the orbital energy of the binary (see Iben & Tutukov 1984). In this description, the efficiency as determined from one-dimensional stellar models and the hydrodynamic calculations (not analytically as in Iben & Tutukov 1984) ranges from 40% to 50%. A similar range of efficiency was also found for the asymptotic giant branch progenitor systems of cataclysmic variables as well (Taam & Bodenheimer 1989, 1991). Although there is evidence that significant mass from the common envelope is lost from the progenitor binary systems (e.g. Yorke et al 1995), none of the two-dimensional simulations demonstrate conclusively that the entire envelope would be ejected.

4.   THREE-DIMENSIONAL MODELS

To provide a quantitative description of the common envelope phase, a further relaxation of the approximations to include the three-dimensional aspects of the evolution is necessary. The importance of this stems from the fact that the initial phases of the orbital evolution can occur on timescales that are shorter than or comparable to the orbital timescale of the stellar components. In addition, the gravitational interactions of the engulfed star and the supergiant's core with the common envelope can be modeled without adopting approximate prescriptions for the dissipation of orbital energy and for the gravitational torque.

With the development of sophisticated numerical algorithms and improved computational technology, the three-dimensional study of the common envelope phase has become practical. Because the length scales in the system can range from less than a solar radius to several astronomical units, and the timescales can range from hours to several years, the complete solution of the common envelope problem is still a substantial challenge. The application of numerical techniques such as smoothed particle hydrodynamics (SPH; see Lucy 1977; Gingold & Monaghan 1977; Monaghan 1985, 1992), which is adaptive in both space and time, and multigrid Eulerian hydrodynamics (see Berger & Oliger 1984, Berger & Colella 1989) has enabled researchers to make significant progress.

Although early three-dimensional numerical work was very limited in spatial resolution, investigations by de Kool (1987) (based on SPH calculations) and by Livio & Soker (1988) (based on fixed-grid Eulerian hydrodynamics), showed that nonaxisymmetric effects are important in the initial evolution. During this phase, gravitational tides are effective enough to drag the less massive companion into the red giant component on a timescale comparable to the dynamical timescale of the binary system. In addition, there is some indication that mass is ejected in the orbital plane of the system. Subsequent higher resolution SPH calculations carried out by Terman et al (1994) demonstrated that mass could be ejected in significant amounts with some preference for loss along the orbital plane. Because most of the mass ejection takes place when the two cores are deep in the gravitational potential well of the binary (where the orbital decay timescale is significantly longer than the orbital period), the global mass loss distribution is qualitatively similar to the earlier two-dimensional simulations, which confirmed the general outflow morphology obtained by Bodenheimer & Taam (1984) and Taam & Bodenheimer (1989, 1991).

In more recent three-dimensional studies, Rasio & Livio (1996) followed a binary somewhat beyond the early dynamical interaction of a red giant and a main sequence star for application to pre-cataclysmic binaries, and Terman et al (1995) followed the evolution for high-mass primaries with neutron star companions for application to short-period binary radio pulsars. In addition, further work was undertaken by Terman & Taam (1996) and Sandquist et al (1998) on pre-cataclysmics, and by Sandquist et al (2000) on pre-cataclysmics and double helium white dwarf systems.

In the following, we describe the general features of common envelope evolution with attention focused on the rapid decay and envelope ejection phases, and on the question of survival or merger. These phases are common to the evolution of a wide variety of binary systems that go through common envelope evolution. In particular, we focus our description on the evolution of systems with a neutron star component that may have been through a common envelope phase in the past or that will have such a phase in the future. Applications to the evolutionary history of specific classes of systems consisting of a neutron star and its companion will be deferred to the next section.

As the neutron star initially approaches the surface of its larger companion, the orbital separation decreases slowly under the action of gravitational torques. Although energy and angular momentum are transferred from the orbit to the gas, little matter is unbound and none is lost from the system at this point. The timescale of the orbital decay relative to the orbital period is found to be a function of the evolutionary state of the companion star with a more rapid spiral-in occurring for more evolved giant configurations. This is because a greater fraction of stellar matter is located near the surface for more evolved configurations, leading to greater gravitational torques on the neutron star (see Terman et al 1995). Once the neutron star nears the giant surface, the timescale for orbital decay is comparable to the orbital period of the binary for red supergiant stars and is significantly longer than the orbital period for less evolved stars. For example, the initial orbital decay timescale is of the order of 200 years for a star in its early core helium burning stage, whereas it is 3 years for a star in its late core helium burning stage (see Terman et al 1995). As a consequence of the rapid decay, the orbit changes from

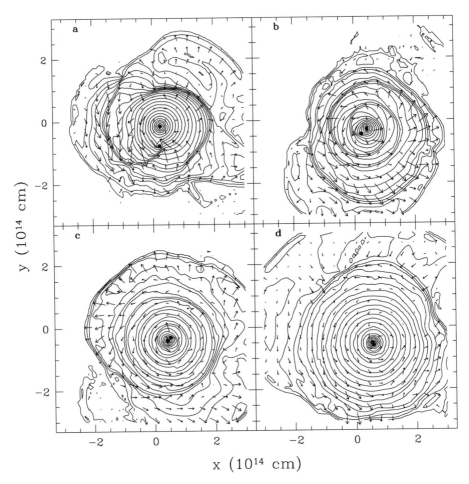

y (10^{14} cm)

x (10^{14} cm)

Figure 1   Density distribution in the orbital plane of the binary system at (*a*) 2610 d, (*b*) 3392 d, (*c*) 3814 d, and (*d*) 4655 d after the start of the common envelope evolution of a system consisting of a 20 $M_\odot$ red supergiant at the onset of core carbon burning and a 1.4 $M_\odot$ neutron star. The contour levels correspond to 5 per decade with a maximum of 6×10^{-7} gm cm^{-3}. The solid dots indicate the position of the stellar cores. The velocity field is on the same scale in all frames, with a maximum value of 60 km s^{-1}.

nearly circular to elliptical, reflecting the loss of a greater fraction of orbital angular momentum compared to energy. The subsequent evolution leads to a shrinkage in the orbital separation by nearly a factor of 100 (Terman et al 1995).

The rapid decay leads to the acceleration of gas relative to the neutron star, leading to supersonic velocities and to the generation of a shock near the neutron star. A similar structure develops about the core of the giant. From the time development of this structure, it can be seen (see Figure 1) to evolve into trailing

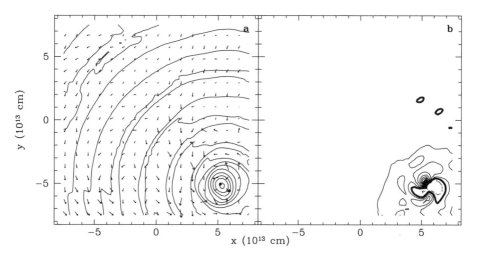

Figure 2   The left panel illustrates the density distribution and velocity field in the innermost region of the common envelope at a time (4,495 days) when matter in the vicinity of the two cores is spun up. The density contour levels correspond to 5 per decade and the maximum velocity vector corresponds to a value of 60 km s^{-1}. The core of the red supergiant is located at higher densities. The right panel plots the angular velocity contours, spaced every 2.5% of the Keplerian angular velocity of the binary with a maximum of 33%. The solid contour corresponds to no rotation and the dashed line corresponds to counter-rotation of 5%. Gas at higher angular velocity is situated in the vicinity of the core of the red supergiant at this time.

spiral shocks behind the two cores. A tightly wound spiral emerges as the orbital velocities of the two cores exceed the rotational velocity of the envelope gas. The shocks are essential for transferring the orbital angular momentum to the spin of the common envelope, and this conversion occurs quite early in the evolution. When the two stellar cores finally begin orbiting in denser portions of the common envelope, orbital energy is dissipated faster than orbital angular momentum, which tends to circularize the orbit. At the same time, the orbital decay timescale lengthens and eventually exceeds the binary orbital period. Matter in the vicinity of the two cores is eventually spun up (see Figure 2), and this facilitates the onset of mass loss from the system.

As a result of this spin-up of the common envelope in the neighborhood of the two cores, the effective gravity decreases, and mass loss driven by unbalanced pressure gradients develops. From the work of Terman et al (1995), it is evident that the major phase of mass loss is initiated when the two cores are within a few solar radii of each other. Mass is driven outward by the rapid energy input from the two cores ($\sim 10^{40}$ ergs s^{-1}) with the mass loss rates reaching about $1\,M_\odot$ yr^{-1} when the neutron star encounters the high-density region surrounding the evolved red supergiant core. At the same time, this outward flow mixes with matter at higher latitudes off the orbital plane to set up a circulation pattern with

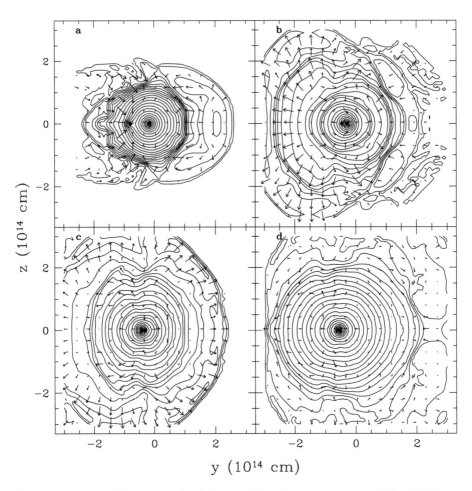

Figure 3   Density distributions and velocity fields in a plane perpendicular to the orbital plane corresponding to the times shown in Figure 1. Note the presence of circulatory flow from the orbital plane to the polar direction in (*b*) and the outward polar flow that merges with the equatorial flow in (*c*) and (*d*).

inflow along the polar direction and outflow along the equatorial direction. The inward polar flow results from pressure forces driving mass flow toward the region from which mass has been lost in the equatorial wind. This feedback prolongs the spiral-in phase because a fraction of the matter driven outward is not lost, but is recirculated (e.g. see Figure 3). Because of the significant decrease in the amount of mass in the vicinity of the two cores (within a distance where tidal effects are important—typically about 3 times the orbital separation; see

Taam et al 1994), the gravitational tidal torques are significantly reduced, and the timescale for orbital decay greatly increases. Provided the giant is sufficiently evolved and the companion is sufficiently massive, the timescale for ejection of the common envelope is shorter than the orbital decay timescale, which suggests that the entire common envelope can be ejected without causing the two cores to spiral together and coalesce. For less-evolved blue or yellow massive supergiant stars (i.e. during the early and the main core helium burning stages), and for low-mass stars on the faint end of the first red giant branch, the systems are likely to merge even if there is sufficient energy to unbind the envelope. This reflects the absence of a steep density gradient above the nuclear burning shells, which means that substantial mass is stored close to the core of the giant. As an example of the envelope structure, the mass-radius profiles for a 20 $M_\odot$ star are illustrated at various evolutionary phases in Figure 4. It is evident that the profiles are flattest for the most evolved configurations (where the star is in its core carbon burning phase).

The efficiency of the mass ejection process is of the order of 30%–50% for supergiant stars (this depends on the mass of the giant and the mass of the companion). The efficiencies for mass ejection from binaries consisting of giants on the first giant branch and the asymptotic giant branch are also less than 50%, but the actual values are less certain because spatial resolution has been insufficient to accurately determine it.

5. APPLICATIONS TO SPECIFIC CLASSES OF SYSTEMS

The results of the multi-dimensional numerical simulations point toward successful ejection of the common envelope and the survival of a short period binary system provided that the progenitor system contains an evolved red giant or red supergiant component. In these systems, the presence of a steep density gradient above the nuclear burning shell in the red giant configuration facilitates the deceleration and termination of the orbital decay. Merger of the system is virtually certain for binaries that enter into the common envelope phase at short orbital periods since the more massive component cannot be on the giant branch. In particular, for stars near the main sequence, the binding energy of the star's envelope is too large for energy from the orbital motion to completely eject it, whereas for yellow supergiants the system will merge even though the envelope binding energy is low enough. Because stars more massive than ~35 $M_\odot$ do not evolve to the red supergiant phase (as a result of their strong stellar winds), these stars would tend to evolve as if they were in isolation and would not contribute to the population of compact stars formed in common envelopes (see Van den Heuvel 1993). In the following section we consider these theoretical results while examining the evolutionary channels for the formation of systems that have a neutron star component.

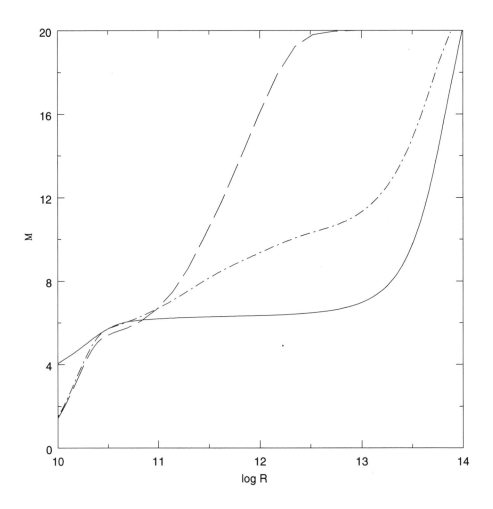

Figure 4   The distribution of mass (in units of $M_\odot$) as a function of radius for a $20\,M_\odot$ star in various phases of evolution. The dashed curve and dot-dashed curve correspond to the core helium burning phase with a central helium content of 0.5 and 0.34 respectively. The solid curve corresponds to the core carbon burning phase. Note the existence of a flat mass-radius profile for the most advanced stage of evolution where the mass contained within a spherical shell ranging from 10^{11} cm to 10^{12} cm is low ($\sim$0.16 $M_\odot$). This property of very evolved red supergiant stars facilitates the deceleration of the spiral-in of its companion, and hence favors the formation of a remnant short period binary system.

5.1 Double Neutron Stars

Among the known binary systems that contain compact objects, none better typifies the requirement of severe mass and angular momentum loss than systems containing a pair of neutron stars. The discovery of the double neutron star system PSR 1913+16 with an orbital period $P = 0.32$ days by Hulse & Taylor (1975) has opened a new window for studying general relativity in the weak field regime, while also raising many questions about their previous evolution. For these reasons, many surveys have been conducted to search for additional systems of this type. There are currently four candidate double neutron star systems in addition to PSR 1913+16; three are located in the Galactic disk [PSR 1534+12 ($P = 0.42$ days; Wolszczan 1991), PSR 1518+49 ($P = 8.6$ days; Nice et al 1996), PSR 1820–11 ($P = 357.8$ days; Lyne & McKenna 1989)], and one is located in the globular cluster M15 [$P = 0.34$ days; PSR 2127+11C (Anderson et al 1990)]. It is likely that two neutron stars are present in these systems since they share the characteristic of high orbital eccentricities (although see next paragraph). The high eccentricities of these systems can easily be explained by a supernova explosion of the observed pulsar's companion after its spin-up phase (see Srinivasan & Van den Heuvel 1982).

The nature of an additional double neutron star candidate PSR 2303+46 ($P = 12.34$ days; Stokes et al 1985) is currently uncertain. The possible detection of an optical counterpart of PSR 2303+46 by Van Kerkwijk & Kulkarni (1999) suggests that the pulsar companion in this system may be a massive white dwarf, which implies that high orbital eccentricity ($e = 0.66$) alone may be insufficient to classify a binary pulsar as a neutron star pair. In this case, the neutron star would have to form after the white dwarf star to impart the eccentricity to the system. This can occur if the originally less massive star had accreted sufficient mass to evolve to a neutron star end state. Thus, the evolution would involve the spiral of a white dwarf through the envelope of its more massive red supergiant companion. The successful ejection of the envelope leaves a remnant system consisting of the white dwarf and the evolved core. The subsequent evolution of the helium exhausted core ($\gtrsim 4\ M_\odot$) to the supernova phase and the kick imparted to the system leads to the observed state (see later in this section). The long orbital period of PSR 2303+46 may be difficult to accommodate within a common envelope framework, however, unless significant mass loss took place in the advanced red supergiant phase (during carbon/oxygen core burning) prior to the common envelope phase. Within this scenario, the neutron star should not have accreted mass nor been spun up. In agreement with this idea, the spin period of the pulsar (at 1.066 s) is similar to those observed for single radio pulsars (Bhattacharya et al 1992, Lorimer et al 1993). The long period highly eccentric pulsar PSR 1820–11 ($e = 0.795$) also may not belong to the double neutron star class. Phinney & Verbunt (1991) have suggested that the pulsar companion is a main sequence star (see also Portegies Zwart & Yungelson 1999), based partly on the fact that it is the only binary pulsar known with a spin period shorter than the rebirth line (indicating that it has not been recycled). A history in terms of independent evolution of the binary components for such a system has been suggested for its origin (see Terman et al 1996). In this interpretation, the system would be an immediate progenitor to an X-ray binary system, and its future evolution may give rise to a low-mass X-ray binary with a giant companion similar to GX 1+4 (Davidsen et al 1977, Chakrabarty & Roche 1997). Although PSR 2127+11C is also a possible double neutron star pair, its evolutionary history may differ from the other systems in the class because it is likely to have formed in an environment where stellar dynamical interactions were important (see Phinney & Sigurdsson 1991).

The origin of binary pulsars was first discussed in Flannery & Van den Heuvel (1975), Webbink (1975), and De Loore et al (1975), but the evolutionary channel involving a phase of common envelope for double neutron stars was suggested by Smarr & Blandford (1976) and further elaborated upon by Srinivasan & Van den Heuvel (1982), Van den Heuvel & Taam (1984), Taam & Bodenheimer (1992), and Terman & Taam (1995). In this picture, a high-mass X-ray binary system is the immediate progenitor. Unstable mass transfer resulting from the Roche lobe overflow of the OB star leads to the spiral-in of the neutron star into the envelope of its companion. Although many of these systems are expected to merge, the envelope can be successfully ejected provided that the evolved star is in late core helium burning or a more advanced stage (see Section 4). The remnant binary consists of the evolved core with its neutron star companion in an orbit that has a period as short as a fraction of a day (see Taam et al 1978, Terman et al 1995). Systems such as Cyg X-3 (with an orbital period of 4.8 hours) may be in such an evolutionary state (Van den Heuvel & De Loore 1973). The lack of observed systems of this type has been attributed to a number of observational selection effects (see Van den Heuvel & Bitzaraki 1995). We note, however, that the ejection of the common envelope is only successful during late core helium burning or carbon burning stages. Hence, the number of systems that are able to evolve to this stage is small compared to the number for which envelope ejection would occur in earlier stages, and their detection would also be made difficult by the short evolutionary timescale ($\lesssim 10^5$ years) of the remnant system.

The subsequent evolution of the remnant core-neutron star system does not necessarily lead to the formation of a neutron star pair—the outcome is dependent on the mass of the remnant core. For example, for massive core remnants ($\gtrsim 3.5\ M_\odot$) little envelope expansion takes place during the more advanced stages of evolution ($R \lesssim 10 R_\odot$) and the cores evolve to the Fe photodisintegration phase. As a consequence, a second supernova explosion will occur. For cores more massive than $4.2\ M_\odot$, the binary system would become unbound for a spherically symmetric explosion (Blaauw 1961). If an asymmetry in the explosion exists, the system can remain bound depending on the direction of the velocity kick. The systems that become unbound contribute to the high-velocity single radio pulsar population. Recent discussions of the origin of single radio pulsars and the necessity of kicks can be found in Van den Heuvel & van Paradijs (1997) and Portegies Zwart &

Van den Heuvel (1999; but see Iben & Tutukov 1996 for an alternative point of view). For the observed characteristics of the binary pulsars located in the Galactic disk, Fryer & Kalogera (1997) showed that they must have been formed with a kick exceeding 200 km s^{-1} imparted to the neutron star. Velocity kicks of this magnitude had already been suggested by Shklovskii (1969), Gunn & Ostriker (1970), and Dewey & Cordes (1987), and require asymmetries of the order of a few percent in the supernova explosion (Woosley 1987, Woosley & Weaver 1992). The common envelope phase is able to produce the systems at short periods (PSR 1913+16 at 7.75 hrs and PSR 1534+12 at 10.1 hr), but has difficulty in producing the long period system PSR 1518+49 at 8.63 days (see Bagot 1997, Portegies Zwart & Yungelson 1998) unless mass loss has reduced the envelope mass of the progenitor sufficiently. The best known class of systems that could be transformed into double neutron stars are the Be X-ray binaries (Rappaport & Van den Heuvel 1982, Van den Heuvel & Taam 1984, Van den Heuvel & Rappaport 1987, Van den Heuvel 1992). Because these systems have orbital periods ranging from ~15 days to ~500 days, they will enter the common envelope phase when the B star is sufficiently evolved. Taking into account the requirements for successful ejection of the envelope, the double neutron star binaries are likely to be descendants of the long period portion of this high-mass X-ray binary population. It is also possible that evolution involving components of very nearly equal mass (to within 4%) may also contribute to the double neutron star group (see Brown 1995, Wettig & Brown 1996, Portegies Zwart & Yungelson 1998). In this case, the immediate progenitor of the neutron star binary is a system consisting of two helium stars.

On the other hand, for remnant cores less massive than about 3.5 $M_\odot$, the system can interact further as a result of the expansion of the core in response to its nuclear evolution (see Arnett 1978, Habets 1986). Provided that stellar wind loss does not prevent this expansion (see Vanbeveren et al 1998), the system may enter a case BB mass exchange phase (occurring after the exhaustion of helium in the core) in which matter is transferred at rates ~10^{-4} $M_\odot$ yr^{-1} (Delgado & Thomas 1981; see also Law & Ritter 1983). If the envelope of the mass losing star is radiative, the mass transfer occurs on the relatively long thermal timescale, and the neutron star may eject the matter in the form of a super-Eddington wind, thereby avoiding the common envelope phase (see King & Begelman 1999). Provided that the remnant core after case BB mass exhange exceeds about 2.2 $M_\odot$ (Habets 1986), it will evolve to Fe photodisintegration and supernova. The resulting system may consist of a neutron star pair. On the other hand, if the envelope of the evolving star is convective, the system enters into a second common envelope phase. However, the system may not survive if the density gradients are insufficiently steep in the common envelope configuration to halt the orbital decay. In this case it is possible that the system may evolve into a Thorne-Zytkow object (see Section 5.3) or the neutron star may collapse to form a black hole (see Fryer & Kalogera 1997), perhaps surrounded by a massive helium-rich accretion disk. We note that the latter configuration has been hypothesized as the site where γ-ray bursts of a few seconds' duration are produced (see Popham et al 1999).

Recently Chevalier (1993) and Brown (1995) have suggested that the mass accretion rate (as estimated from a Bondi-Hoyle formalism) onto the neutron star during the common envelope phase may be so high that the Eddington rate is not the appropriate limit for accretion. In this situation, the temperatures produced in the accretion flow are sufficiently high that the loss of energy by neutrino cooling could be enough to allow the neutron star to accrete at rates significantly above the Eddington rate (see Zeldovich et al 1972). As a consequence, the accretion could lead to the collapse of the neutron star to a black hole, making this scenario for the formation of binary radio pulsars less viable. However, consideration of the angular momentum of the accreted gas (Chevalier 1996; see also Armitage & Livio 2000—see below) leads to the conclusion that the neutron star will accrete little mass in progenitor systems of long orbital period where survival of the remnant binary is likely (although see Bethe et al 1999 for an alternative viewpoint).

The future evolution of double neutron star binaries is of special interest because their merger under the action of angular momentum losses can lead to a burst of detectable gravitational wave radiation (Clark & Eardley 1977, Thorne 1996) and possibly to γ-ray bursts (Paczynski 1986, Goodman et al 1987, Eichler et al 1989). Their formation and merger rates have been simulated by Lipunov et al (1997) and by Portegies Zwart & Yungelson (1998), and vary over a wide range due to uncertainties in the mass ejection efficiency of the common envelope phase and especially in the kick velocity magnitude and distribution (e.g. Kalogera & Lorimer 1999). The merger rate may be comparable to the upper limit on the birth rate of coalescing neutron stars obtained by Bailes (1996) of 10^{-5} yr^{-1} from the statistics of normal pulsars (in neutron star binary pairs in comparison to recycled pulsars in these systems), by Van den Heuvel & Lorimer (1996) of $\sim 8 \times 10^{-6}$ yr^{-1}, and by Kalogera & Lorimer (2000) of $\sim 10^{-5}$ yr^{-1}.

5.2 Binary Pulsars

For a progenitor system composed of an intermediate mass asymptotic giant branch (AGB) star and a neutron star, a common envelope phase can result in the formation of a binary pulsar characterized by a short orbital period ($\lesssim 1$ day) and a high-mass white dwarf companion ($\gtrsim 0.6\,M_\odot$). The evolution into a common envelope phase is unavoidable in this case because the star fills its Roche lobe at a point in its evolution when it has a deep convective envelope. This ensures that the mass transfer is unstable because the orbital separation shrinks upon mass transfer and there is no tendency for the AGB star to contract. The mass loss from the AGB star reaches rates $\sim 10^{-3}-10^{-1}\,M_\odot$ yr^{-1}. Van den Heuvel & Taam (1984) suggested such an evolution for PSR 0655+64, a system that has an orbital period of 1.03 days, a white dwarf companion of $0.7\,M_\odot$, and a spin period of 196 ms. Based on three-dimensional numerical simulations of common envelope evolution for intermediate mass stars (Terman & Taam 1996, Sandquist et al 1998), the efficiency of the mass ejection process is expected to be $\sim 40\%$. An immediate progenitor system containing a star of $\sim 5\,M_\odot$ and a neutron star in a binary with an orbital period of $\sim$100 days will likely result in the successful formation of a binary radio pulsar with a final orbital period of $\sim$1 day.

The formation of binary pulsars can also involve a common envelope phase during the second stage of mass exchange for systems with an initial mass ratio less than 1.5. This type of evolution may be responsible for the formation of the possible white dwarf-neutron star systems PSR 2303+46 and PSR J1141–6545 (see above; also Portegies Zwart & Yungelson 1999 and Tauris & Sennels 2000). The first mass transfer phase is conservative because the mass ratio can be reversed soon after the initiation of Roche lobe overflow, meaning subsequent mass transfer leads to an increasing separation of the binary components. Further mass transfer leads to the depletion of the envelope of the initially more massive component and to the formation of a white dwarf. Provided that the initial masses of the stars are not significantly less than the lower limit for the formation of neutron stars, the accreting star will be massive enough to supernova (see Tutukov & Yungelson 1993; Portegies Zwart & Verbunt 1996). The common envelope phase occurs when the white dwarf spirals into its more massive companion as the latter expands to the red supergiant phase. Successful ejection of the envelope leads to the formation of a stellar remnant in its advanced core helium or carbon burning stage. The further nuclear evolution of the remnant leads to the formation of a neutron star (after the formation of the white dwarf) and to a system similar to PSR 2303+46. Based on population synthesis calculations, the birth rate of such systems in which the neutron star is born after the white dwarf has been found to be comparable to the birth rate of double neutron star binaries in the study of Portegies Zwart & Yungelson (1999), but about an order of magnitude larger in the study of Tauris & Sennels (2000). The uncertainties in these rates reflect uncertainties in the efficiency of mass ejection, i.e. α_{CE}, and the distribution and magnitude of the kick velocity.

The common envelope phase has also been discussed in the context of the formation of millisecond binary radio pulsars with short orbital periods (e.g. Tauris 1996). For these systems the neutron star has accreted sufficient mass ($\gtrsim$0.05 $M_\odot$) to spin up to millisecond periods. After mass transfer ends, the neutron star may appear as a radio pulsar. There are currently 35 systems known with orbital periods ranging from 2 hours to 175 days (see Taam et al 2000), with an absence of systems within periods between 22.7 and 56 days. It is not likely that the period gap is due to observational selection effects (Camilo 1995a,b, 1999; Bhattacharya 1996; Taam et al 2000), and if it is not a statistical fluctuation, its origin must reflect the previous evolutionary history of these systems. To explain this gap, Tauris (1996) assumed that a neutron star system that survives the common envelope phase can produce orbital periods as large as 20 days, thus providing the upper period cutoff in the short period population. For this to be true, the efficiency of mass ejection in the common envelope process (α_{CE}) was increased to $\sim$4 in order to accommodate the large final orbital separation ($\gtrsim$25 $R_\odot$) and period (12.2 days) of the system B1855+09. Although common envelope evolution can contribute to the short period population, there is great difficulty in producing systems at orbital periods as large as 20 days because there is insufficient energy in the orbit to unbind and eject the envelope. By definition, a mass ejection efficiency greater than 1 requires a source of energy in addition to the orbital motion of the binary. Sources such as recombination and nuclear burning have been suggested to facilitate mass ejection, although Sandquist et al (2000) conclude that these sources are not likely to be very effective. For example, the timescale for releasing nuclear energy is too long compared to the mass ejection timescale to be important. While ionization energy may assist in ejecting mass situated at large radii (although see Harpaz 1998), it is not important for ejecting the hot gas deep in the gravitational potential of the binary where most of the energy is required.

A more natural explanation of the apparent absence of systems between 23 and 56 days has been suggested by Taam et al (2000). Their interpretation involves mass transfer and orbital evolution taking place after core hydrogen exhaustion (Case B binary evolution). In this framework, the long period population is produced in part from binaries where the mass donor is a low-mass star on the red giant branch. Orbital expansion takes place in this situation because mass is transferred from the giant to the neutron star companion (low mass Case B evolution). For example, an initial orbital period of $\sim$2.5 day leads to the formation of a low-mass helium white dwarf ($\gtrsim 0.26\ M_\odot$) at long orbital period ($\gtrsim 60$ days), as originally suggested by Savonije (1983), Joss & Rappaport (1983), and Paczynski (1983), and more recently discussed by Tauris & Savonije (1999), following earlier work by Taam (1983) and Webbink et al (1983). Similar systems containing CO white dwarfs can be produced through early massive Case B evolution (see Kippenhahn & Weigert 1967) where mass transfer takes place in the Hertzsprung gap for a star with a radiative envelope and nondegenerate helium core ($M \gtrsim 2.25\ M_\odot$) and a mass ratio less than a critical value (King & Ritter 1999). This evolution is distinct from the low-mass case B evolution because the mass transfer rates are highly super-Eddington ($\sim 100\ \dot{M}_{Edd}$), making the evolution very nonconservative, but avoiding common envelope (King & Begelman 1999, Tauris et al 2000). The requirement of efficient mass loss from the system has been pointed out by King & Ritter (1999) and Podsiadlowski & Rappaport (2000) in constructing an evolutionary history of the low-mass X-ray binary Cyg X-2 consistent with the unusual evolutionary state of its optical companion (a low mass of $\sim 0.5 - 0.7 M_\odot$ and large radius of $\sim 7 R_\odot$).

The short period population descends in part from short period ($\lesssim 1$ day) low-mass Case B systems, which can produce low-mass helium white dwarfs ($\lesssim 0.2\ M_\odot$) provided that angular momentum losses are sufficiently effective in driving the system to shorter orbital periods (see Pylyser & Savonije 1989, Ergma et al 1998). In this case, the angular momentum loss associated with a magnetized stellar wind leads to the loss of spin angular momentum from the noncompact component of the system. The strong tidal coupling between the two components thus results in the loss of orbital angular momentum from the system, and hence facilitates mass transfer as the system evolves to shorter orbital periods (see Verbunt & Zwaan 1981). For greater initial orbital periods ($\gtrsim 1$ day), angular momentum losses are unimportant, and early massive Case B evolution (for period-dependent critical mass ratios) can lead to the formation of helium stars more massive than 0.35 $M_\odot$, which evolve to become CO white dwarfs. Common envelope evolution of long period binaries can also contribute short period binary pulsars with helium white dwarf companions in the mass range 0.25–0.5 $M_\odot$ if the binary interaction occurs on the red giant branch, and with CO white dwarfs $\gtrsim 0.6\,M_\odot$ if the interaction occurs on the asymptotic giant branch. In all these channels, there is significant mass and angular momentum loss during the evolution. The absence of systems in the period gap reflects the rarity of producing systems with periods greater than 30 days via early massive Case B evolution, and of producing systems less than 60 days via low-mass Case B evolution.

Thus, the orbital period distribution provides indirect evidence for nonconservative evolution in these systems. We note that mass and angular momentum can be lost without a common envelope phase for super-Eddington mass transfer from the companion (Kalogera & Webbink 1996, King & Ritter 1999, Tauris & Savonije 1999) or super-Eddington mass flow in the inner disk during an X-ray transient outburst (see King et al 1996, King et al 1997, Li et al 1998). This suggests that systematic differences may exist in the mass of the pulsar (Taam et al 1999) depending on the particular formation channel. If neutron stars are formed with identical initial masses, systems that have evolved through common envelope interaction or nonconservative early massive Case B mass transfer would have neutron stars that accreted little mass, whereas systems dominated by continuous angular momentum losses (causing evolution to shorter orbital periods) would have mass transfer rates significantly lower than the Eddington value and would thus be capable of accreting a non-negligible amount of mass onto the neutron star. Precise measurements of pulsar masses (e.g. Thorsett & Chakrabarty 1999) in these types of systems can provide an important test of such an evolutionary dependence.

5.3   Thorne-Zytkow Objects

For high-mass X-ray binary systems characterized by short orbital periods ($\lesssim 100$ days), the common envelope phase leads to the merger of the neutron star with its massive companion. Systems such as LMC X-4, Cen X-3, 4U1700–37, and SMC X-1 with orbital periods of 1.4, 2.1, 3.4, 3.89 days respectively fall into such a class. A possible fate of these systems is a star resembling a red supergiant, but with a neutron degenerate core. The structure of such objects was originally studied in pioneering investigations by Thorne & Zytkow (1975, 1977) and more recently by Eich et al (1989), Biehle (1991, 1994), Cannon et al (1992), and Cannon (1993). The detailed evolution and fate of these systems has been addressed by Bhattacharya & Van den Heuvel (1991) in the context of the formation of low-velocity radio pulsars, and by Podsiadlowski et al (1995) and Podsiadlowski (1996) in the context of soft X-ray transients containing a black hole. An assumption in all of these studies is that the accretion rate onto the neutron star is limited to the Eddington value. Although there is evidence of such a limitation from the observed luminosities of X-ray binaries, conditions in the supergiant envelope are sufficiently different from those in stable mass accreting systems that this assumption requires reexamination. Indeed, numerical work by Fryer et al (1996) based on spherically symmetric accretion (following analytic work by Chevalier 1993 and Brown 1995) suggests that accretion may occur onto a neutron star at much higher rates in a neutrino-dominated regime. In this case, the neutron star may be induced to collapse to a black hole, thus avoiding the Thorne-Zytkow phase. Recently, the angular momentum aspects of the problem have been considered by Chevalier (1996), who showed that the neutron star can avoid collapse within the envelope of giant-like companions, although it may not survive for less evolved companions.

The existence of Thorne-Zytkow objects is dependent on the rate at which the neutron star can accrete. In the common envelope framework, the mass accretion rates are based on a Bondi-Hoyle description and can exceed the Eddington rate by a large factor. Strict application of this theory to a common envelope binary system can be questioned, however, since the motion of the star about a common center of mass, the existence of a tidal radius, and the effect of significant density gradients within the gravitational interaction region are not incorporated in the Bondi-Hoyle picture. In particular, the possibility that gravitational torques can spin up the matter to near corotation can have an important impact on the hydrodynamical flow of matter in the common envelope.

In spite of the shortcomings of the theory, it is possible that accretion in the neutrino-dominated regime may take place at rates significantly higher than the Eddington value. Although the fully three-dimensional simulations of the common envelope phase for the progenitors of these objects have not been carried out, calculations in fewer dimensions may provide some insight into possible behaviors. In particular, the possibility exists that neutrinos produced close to the neutron star can heat the flow at larger radii, initiating a temporary phase in which accretion is interrupted by outflow (see Fryer et al 1996). In addition, it is possible that outflows from a disk can reduce the accretion rate onto the neutron star and deposit energy into the common envelope (Armitage & Livio 2000, King & Begelman 1999). How this evolution deep in the gravitational well of the neutron star influences the rotating common envelope at much larger radii is not known. Since it is possible that such effects could prevent the collapse of the neutron star to a black hole, the outcome may depend on the conditions within the common envelope, and hence the evolutionary state of the evolved massive star. Detailed simulations will be necessary to determine the range of conditions leading to the survival of a neutron star or production of a black hole in these merged binary systems.

5.4 Low-Mass X-Ray Binary Systems

These systems are the neutron star analogs of cataclysmic variables: a low-mass star fills its Roche lobe and transfers mass to its neutron star companion. They are also the progenitor systems to low-mass binary radio pulsars, as confirmed observationally by the recent discovery of the 2.5 ms X-ray pulsar SAX J1808.4-3658 by Wijnands & van der Klis (1998) and the discovery of its 2.01 hour orbital period by Chakrabarty & Morgan (1998). In the low-mass X-ray binary systems, the mass transfer can be driven by several mechanisms: nuclear expansion of an evolved low-mass companion (Taam 1983, Webbink et al 1983), angular momentum losses associated with magnetic braking in short period systems (Verbunt & Zwaan 1981), and X-ray irradiation of the companion by the accreting neutron star (Podsiadlowski 1991, Harpaz & Rappaport 1991, Tavani 1991, Frank et al 1992, Hameury et al 1993). For a review of the possible evolutionary channels to the low-mass X-ray binary stage, see Bhattacharya & Van den Heuvel (1991).

Common envelope evolution is a major channel leading to their formation. Binary progenitors of extreme mass ratios are required because the progenitor of the neutron star must be more massive than $\sim 10 \, M_\odot$. A common envelope phase occurring after the massive star has reached an advanced stage of evolution (helium shell burning or core carbon burning) can lead to successful ejection of the common envelope, leaving behind a remnant binary composed of the evolved core and the low-mass companion. If the exposed core can evolve to the supernova stage without significant interaction with its companion (see Section 5.1), a neutron star can be formed in a short period orbit. The effective tidal interaction between the components leads to the circularization of the orbits (see Zahn 1975, 1977) and to the eventual transfer of mass from the companion to the neutron star. Population synthesis studies (Terman et al 1996, Kalogera 1997, Kalogera & Webbink 1998; see also Brandt & Podsiadlowski 1995 and Sutantyo 1999) indicate that asymmetries in the supernova mass ejection are necessary for the binary to remain bound (however, see Iben & Tutukov 1996).

Routes that do not require a phase of common envelope have also been identified. One such mechanism is "direct supernova" for stars evolving independently in a very long period binary system ($\gtrsim 10$ years). Evolution of this kind has been suggested by Kalogera (1998; see also Terman et al 1996), who finds that a root mean square kick velocity of less than ~ 100 km s^{-1} is required for a significant binary population to survive. However, the mean pulsar velocities found by Lyne & Lorimer (1994: 450 km s^{-1}) and more recently by Hansen & Phinney (1997: 250–300 km s^{-1}) suggest that this evolutionary channel is not a major contributor to the low-mass X-ray binary population. A third mechanism (the accretion induced collapse model) involves mass transfer to a massive O-Ne-Mg white dwarf in a binary system (Flannery & Van den Heuvel 1975, Canal & Schatzman 1976, Canal et al 1990). The contribution to the population of Low-mass X-ray binary systems by this evolutionary channel has not been well determined because of uncertainties in the formation of massive white dwarfs and in binary evolution (e.g. Fryer et al 1999).

Evolutionary paths to the low-mass X-ray binary phase depend on the initial orbital period of the remnant system. For example, in short period ($\lesssim 1$ day) systems angular momentum losses may be sufficient to cause the companion star to fill its Roche lobe as a main sequence star and transfer mass to the neutron star (Pylyser & Savonije 1989, Ergma et al 1998). On the other hand, for long period systems, expansion of the star to the giant phase is necessary to initiate mass transfer (Taam 1983, Webbink et al 1983).

As described earlier, the successful ejection of the common envelope is far more likely for a star in an advanced evolutionary state because of lower envelope binding energy and the existence of a steep density gradient above the nuclear burning shells (especially for the carbon core burning phase of massive stars; see Habets 1986). For such progenitor systems, the exposed core naturally enters the supernova phase and forms a neutron star [provided that the mass of the core is not too large ($\lesssim 8\ M_\odot$)]. The critical orbital period for the survival of a detached binary system consisting of the remnant core and its low-mass companion can be estimated and is dependent on the masses of the progenitor system. For example, lower-mass companions and more massive evolved progenitors of the neutron star require longer initial orbital periods. Specifically, the critical period for a low-mass star of 1.4 $M_\odot$ (see Terman et al 1995) ranges from about 80 days for a massive companion of 12 $M_\odot$ to a period of 2 years for 24 $M_\odot$ (assuming an efficiency of mass ejection of 40% from common envelope computations). Given the successful ejection of the envelope, the evolution to the X-ray phase still does not necessarily follow because it requires the satisfaction of additional constraints. In particular, Webbink (1992), Terman et al (1996), and Kalogera & Webbink (1998) emphasize the importance of considering the timescale to evolve to the low-mass X-ray binary phase and the stability of mass transfer in determining the birth rate of the low-mass X-ray binary population. The theoretically estimated rates range from 10^{-6} to 10^{-5} yr^{-1}, but are uncertain because they depend on the poorly known frequency of primoridal binaries at long orbital periods and extreme mass ratios.

6. CONCLUSIONS

Substantial progress in our understanding of the common envelope phase has been achieved in the last decade as a result of numerical simulations of increasing complexity and realism. The most important step in this regard has been the introduction of three-dimensional computer simulations. The investigations have shown that the common envelope scenario is a viable evolutionary channel for producing short period systems with degenerate compact companions. In particular, it has been demonstrated for specific progenitor binary systems that the common envelope can be ejected without the companion spiraling into the core. The evolutionary state of the more massive component in the system has been identified as a very important factor in successful ejection of the common envelope. Stars in the red giant stage of evolution preferentially survive because the binding energy of the star's envelope is low and the envelope structure naturally has sharp density gradients above the hydrogen burning shell, where little mass is distributed over a large spatial extent. The former condition favors envelope ejection on energetic grounds, whereas the latter condition leads to the termination of the spiral-in phase by significantly reducing the magnitude of the gravitational torques transferring angular momentum from the orbit to the spin of the envelope.

In the future, high-resolution numerical calculations will be necessary for a quantitative calculation of the efficiency of the mass ejection process for a wide range of binary system parameters. Because the efficiency is likely to be a function of the masses of the binary components as well as their evolutionary states, such studies will provide the long-sought functional relationship required to relate the properties of compact systems to their pre-common envelope progenitor states. Population synthesis studies that determine the fraction of binary systems that evolve into these short period compact systems will enable theoretical determination of their birth rates, which will have far-reaching implications for studies in gravitational wave astronomy, γ-ray bursts, and Type I supernovae. With projected increases in computational capabilities and rapid advances in algorithm development in the next decade, the attainment of these goals seems within reach.

ACKNOWLEDGMENTS

This research was supported at Northwestern University by the National Science Foundation under Grant Nos. AST-9415423 and AST-9727875 and was written in part during visits to the Institute of Theoretical Physics, which is supported in part by the NSF under Grant No. PHY94-07194, and at the Aspen Center for Physics. The calculations were carried out at the Pittsburgh Supercomputing Center and at the Max Planck Institute for Astronomy. We also acknowledge the many contributions of Prof. Peter Bodenheimer and Dr. Andreas Burkert during the course of the investigations reported in this review.

Visit the Annual Reviews home page at www.AnnualReviews.org

LITERATURE CITED

Alexander ME, Chau WY, Henriksen RN. 1976. *Ap. J.* 204:879–88

Anderson SB, Gorham PW, Kulkarni SR, Prince TA, Wolszczan A. 1990. *Nature* 346:42–44

Armitage PJ, Livio M. 2000. *Ap. J.* In press

Arnett WD. 1978. In *Physics and Astrophysics of Neutron Stars and Black Holes*, ed. R Giacconi, R Ruffini, pp. 367–436. North Holland: Amsterdam

Bagot P. 1997. *Astron. Astrophys.* 322:533–44

Bailes M. 1996. In *Compact Stars in Binaries, IAU Symp., 165th, The Hague, 1994*, ed. J

van Paradijs, EPJ Van den Heuvel, E Kuulkers, pp. 213–23. Dordrecht: Kluwer

Bailyn CD. 1988. *Nature* 332:330–32

Berger MJ, Colella P. 1989. *J. Comput. Phys.* 82:64–84

Berger MJ, Oliger J. 1984. *J. Comput. Phys.* 53:484–512

Bethe HA, Brown GE, Lee CH. 1999. Preprint

Bhattacharya D. 1995. In *X-Ray Binaries*, ed. WHG Lewin, J van Paradijs, EPJ Van den Heuvel, pp. 233–51. Cambridge: Cambridge Univ. Press

Bhattacharya D. 1996. In *Compact Stars in Binaries, IAU Symp., 165th, The Hague, 1994*, ed. J van Paradijs, EPJ Van den Heuvel, E Kuulkers, pp. 243–56. Dordrecht: Kluwer

Bhattacharya D, Van den Heuvel EPJ. 1991. *Phys. Rep.* 203:1–124

Bhattacharya D, Wijers RAMJ, Hartman JW, Verbunt F. 1992. *Astron. Astrophys.* 254:198–212

Biehle GT. 1991. *Ap. J.* 380:167–84

Biehle GT. 1994. *Ap. J.* 420:364–73

Blaauw A. 1961. *Bull. Astron. Inst. Neth.* 15:265–90

Blandford RD, Begelman MC. 1999. *MNRAS* 303:L1–5

Bodenheimer P, Taam RE. 1984. *Ap. J.* 280:771–79

Bodenheimer P, Taam RE. 1986. In *The Evolution of Galactic X-Ray Binaries, Proc. NATO Adv. Res. Work., Rottach-Egern, 1985*, ed. J Truemper, WHG Lewin, W Brinkmann, pp. 13–24. Dordrecht: Reidel

Bondi H, Hoyle F. 1944. *MNRAS* 104:273–82

Brandt N, Podsiadlowski P. 1995. *MNRAS* 274:461–84

Brown GE. 1995. *Ap. J.* 440:270–79

Brown GE, Lee CH, Bethe HA. 1999. *New Ast.* 4:313–23

Camilo F. 1995a. *A search for millisecond pulsars.* PhD thesis. Princeton Univ. pp. 1–117

Camilo F. 1995b. In *The Lives of Neutron Stars, Proc. NATO Adv. Study Inst., Kemer, 1993* , ed. MA Alpar, U Kiziloglu, J van Paradijs, pp. 243–57. Dordrecht: Kluwer

Camilo F. 1999. In *Pulsar Timing, General Relativity, and the Internal Structure of Neutron Star*, ed. Z Arzoumanian, R van der Hooft, and EPJ van del Heuvel, pp. 115–23. Amsterdam: North Holland

Canal R, Isern J, Labay J. 1990. *Annu. Rev. Astron. Astrophys.* 28:183–214

Canal R, Schatzman E. 1976. *Astron. Astrophys.* 46:229–35

Cannon RC. 1993. *MNRAS* 263:817–38

Cannon RC, Eggleton PP, Zytkow AN, Podsiadlowski P. 1992. *Ap. J.* 386:206–14

Chakrabarty D, Morgan EH. 1998. *Nature* 394:346–48

Chakrabarty D, Roche P. 1997. *Ap. J.* 489:254–71

Chevalier RA. 1993. *Ap. J.* 411:L33–36

Chevalier RA. 1996. *Ap. J.* 459:322–29

Clark JPA, Eardley DM. 1977. *Ap. J.* 215:311–22

Counselman CC. 1973. *Ap. J.* 180:307–14

Darwin GH. 1879. *Proc. R. Soc. London* 29:168–81

Davidsen A, Malina R, Bowyer S. 1977. *Ap. J.* 211:866–71

De Greve JP, De Loore C, Sutantyo W. 1975. *Astrophys. Space Sci.* 38:301–12

de Kool M. 1987. *Models of interacting binary stars.* PhD thesis. Univ. Amsterdam. pp. 1–146

De Loore C, De Greve JP, De Cuyper JP. 1975. *Astrophys. Space Sci.* 36:219–25

Delgado AJ. 1980. *Astron. Astrophys.* 87:343–48

Delgado AJ, Thomas HC. 1981. *Astron. Astrophys.* 96:142–45

Dewey R, Cordes JM. 1987. *Ap. J.* 321:780–98

Eggleton PP. 1983. *Ap. J.* 268:368–69

Eich C, Zimmerman ME, Thorne KS, Zytkow AN. 1989. *Ap. J.* 346:277–83

Eichler D, Livio M, Piran T, Schramm DN. 1989. *Nature* 340:126–28

Ergma E, Sarna MJ, Antipova J. 1998. *MNRAS* 300:352–58

Ergma E, Van den Heuvel EPJ. 1998. *Astron. Astrophys.* 331:L29–32

Flannery BP, Van den Heuvel EPJ. 1975. *Astron. Astrophys.* 39:61–67

Frank J, King AR, Lasota JP. 1992. *Ap. J.* 385:L45–48

Fryer C, Benz W, Herant M. 1996. *Ap. J.* 460:801–26

Fryer C, Benz W, Herant M, Colgate SA. 1999. *Ap. J.* 516:892–99

Fryer C, Kalogera V. 1997. *Ap. J.* 489:244–53

Gingold RA, Monaghan JJ. 1977. *MNRAS* 181:375–89

Goodman J, Dar A, Nussinov S. 1987. *Ap. J.* 314:L7–10

Gunn JE, Ostriker JP. 1970. *Ap. J.* 160:979–1002

Habets GMHJ. 1986. *Astron. Astrophys.* 167:61–76

Hameury JM, King AR, Lasota JP, Raison F. 1993. *Astron. Astrophys.* 277:81–92

Hansen BMS, Phinney ES. 1997. *MNRAS* 291:569–77

Harpaz A. 1998. *Ap. J.* 498:293–95

Harpaz A, Rappaport S. 1991. *Ap. J.* 383:739–44

Hjellming MS. 1989. *Rapid mass transfer in binary systems.* PhD thesis. Univ. Ill. pp. 1–161

Hjellming MS, Webbink RF. 1987. *Ap. J.* 318:794–808

Hulse RA, Taylor JH. 1975. *Ap. J.* 195:L51–53

Hut P. 1980. *Astron. Astrophys.* 92:167–70

Iben I Jr, Livio M. 1993. *Publ. Astron. Soc. Pac.* 105:1373–406

Iben I Jr, Tutukov AV. 1984. *Ap. J. Supp.* 54:335–72

Iben I Jr, Tutukov AV. 1996. *Ap. J.* 456:738–49

Joss PC, Rappaport S. 1983. *Nature* 304:419–21

Kalogera V. 1997. *Publ. Astron. Soc. Pac.* 109:1394 (Abstr.)

Kalogera V. 1998. *Ap. J.* 493:368–74

Kalogera V. 1999. *Ap. J.* 521:723–34

Kalogera V, Lorimer DR. 2000. *Ap. J.* 530:890–95

Kalogera V, Webbink RF. 1996. *Ap. J.* 458:301–11

Kalogera V, Webbink RF. 1998. *Ap. J.* 493:351–67

King AR, Begelman MC. 1999. *Ap. J.* 519:L169–71

King AR, Kolb U, Burderi L. 1996. *Ap. J.* 464:L127–30

King AR, Kolb U, Szuszkiewicz E. 1997. *Ap. J.* 488:89–93

King AR, Ritter H. 1999. *MNRAS* 309:253–60

Kippenhahn R, Weigert A. 1967. *Z. Astrophys.* 65:251–73

Kopal Z. 1972. *Astrophys. Space Sci.* 16:3–51

Kraft RP. 1962. *Ap. J.* 135:408–23

Law WY, Ritter H. 1983. *Astron. Astrophys.* 123:33–38

Li XD, Van den Heuvel EPJ, Wang ZR. 1998. *Ap. J.* 497:865–69

Lipunov VM, Postnov KA, Prokhorov ME. 1997. *MNRAS* 288:245–59

Livio M, Soker N. 1984. *MNRAS* 208:763–81

Livio M, Soker N. 1988. *Ap. J.* 329:764–79

Lorimer DR, Bailes M, Dewey RJ, Harrison PA. 1993. *MNRAS* 263:403–15

Lucy L. 1977. *Astron. J.* 82:1013–24

Lyne AG, Lorimer DR. 1994. *Nature* 369:127–29

Lyne AG, McKenna J. 1989. *Nature* 340:367–69

Meyer F, Meyer-Hofmeister E. 1979. *Astron. Astrophys.* 78:167–76

Monaghan JJ. 1985. *Comp. Phys. Rep.* 3:71–124

Monaghan JJ. 1992. *Annu. Rev. Astron. Astrophys.* 30:543–74

Nice DJ, Sayer RW, Taylor JH. 1996. *Ap. J.* 466:L87–90

Ostriker JP. 1975. Presented at The Structure and Evolution of Close Binary Systems, IAU Symp., 73rd, Cambridge

Paczynski B. 1971. *Annu. Rev. Astron. Astrophys.* 9:183–208

Paczynski B. 1976. In *The Structure and Evolution of Close Binary Systems, IAU Symp., 73rd, Cambridge*, ed. P Eggleton, S Mitton, J Whelan, pp. 75–80. Dordrecht: Reidel

Paczynski B. 1983. *Nature* 304:421–22

Paczynski B. 1986. *Ap. J.* 308:L43–46

Phinney ES, Sigurdsson S. 1991. *Nature* 349:220–23

Phinney ES, Verbunt F. 1991. *MNRAS* 248:21p–23p

Podsiadlowski P. 1991. *Nature* 350:136–38

Podsiadlowski P. 1996. In *Compact Stars in Binaries, IAU Symp., 165th, The Hague, 1994*, ed. J van Paradijs, EPJ Van den Heuvel, E Kuulkers, pp. 29–41. Dordrecht: Kluwer

Podsiadlowski P, Cannon RC, Rees MJ. 1995. *MNRAS* 274:485–90

Podsiadlowski P, Rappaport S. 2000. *Ap. J.* 529:946–51

Popham R, Woosley SE, Fryer C. 1999. *Ap. J.* 518:356–74

Portegies Zwart SF, Van den Heuvel EPJ. 1999. *New Ast.* 4:355–63

Portegies Zwart SF, Verbunt F. 1996. *Astron. Astrophys.* 309:179–96

Portegies Zwart SF, Verbunt F, Ergma E. 1997. *Astron. Astrophys.* 321:207–12

Portegies Zwart SF, Yungelson LR. 1998. *Astron. Astrophys.* 332:173–88

Portegies Zwart SF, Yungelson LR. 1999. *MNRAS* 309:26–30

Pylyser EHP, Savonije GJ. 1989. *Astron. Astrophys.* 208:52–62

Rappaport S, Van den Heuvel EPJ. 1982. In *Be Stars, IAU Symp., 98th, Munich, 1981*, ed. M Jaschek, HG Groth, pp. 327–46. Dordrecht: Reidel

Rasio FA, Livio M. 1996. *Ap. J.* 471:366–76

Rasio F, Shapiro SL. 1991. *Ap. J.* 377:559–80

Ritter H. 1976. *MNRAS* 175:279–95

Romani R. 1996. In *Compact Stars in Binaries, IAU Symp., 165th, The Hague, 1994*, ed. J van Paradijs, EPJ Van den Heuvel, E Kuulkers, pp. 93–103. Dordrecht: Kluwer

Ruderman MA, Spiegel EA. 1971. *Ap. J.* 165:1–15

Sandquist EL, Taam RE, Burkert A. 2000. *Ap. J.* 533 In Press

Sandquist EL, Taam RE, Chen X, Bodenheimer P, Burkert A. 1998. *Ap. J.* 500:909–22

Savonije GJ. 1983. *Nature* 304:422–23

Shklovskii IS. 1969. *Astr. Zh.* 46:715–20

Sigurdsson S, Hernquist L. 1992. *Ap. J.* 401:L93–96

Smarr LL, Blandford RD. 1976. *Ap. J.* 207:574–88

Soberman GE, Phinney ES, Van den Heuvel EPJ. 1997. *Astron. Astrophys.* 327:620–35

Sparks WM, Stecher TP. 1974. *Ap. J.* 188:149–53

Srinivasan G, Van den Heuvel EPJ. 1982. *Astron. Astrophys.* 108:143–47

Stokes GH, Taylor JH, Dewey RJ. 1985. *Ap. J.* 294:L21–24

Sutantyo W. 1999. *Astron. Astrophys.* 344:505–10

Taam RE. 1979. *Ap. L.* 20:29–35

Taam RE. 1983. *Ap. J.* 270:694–99

Taam RE. 1996. In *Compact Stars in Binaries,*
IAU Symp., 165th, The Hague, 1994, ed. J van Paradijs, EPJ Van den Heuvel, E Kuulkers, pp. 3–15. Dordrecht: Kluwer

Taam RE, Bodenheimer P. 1989. *Ap. J.* 337:849–57

Taam RE, Bodenheimer P. 1991. *Ap. J.* 373:246–49

Taam RE, Bodenheimer P. 1992. In *X-Ray Binaries and Recycled Pulsars,* ed. EPJ Van den Heuvel, SA Rappaport, pp. 281–291. Dordrecht: Kluwer

Taam RE, Bodenheimer P, Ostriker JP. 1978. *Ap. J.* 222:269–80

Taam RE, Bodenheimer P, Rozyczka M. 1994. *Ap. J.* 431:247–53

Taam RE, King AR, Ritter H. 2000. *Ap. J.* In press

Tauris TM. 1996. *Astron. Astrophys.* 315:453–62

Tauris TM, Savonije GJ. 1999. *Astron. Astrophys.* 350:928–44

Tauris TM, Sennels T. 2000. *Astron. Astrophys.* 35:236–44

Tauris TM, Vanden Heurel EPJ, Savonije GJ. 2000. *Ap. J.* 530:L93–96

Tavani M. 1991. *Ap. J.* 366:L27–31

Terman JL, Taam RE. 1995. *MNRAS* 276:1320–26

Terman JL, Taam RE. 1996. *Ap. J.* 458:692–98

Terman JL, Taam RE, Hernquist L. 1994. *Ap. J.* 422:729–36

Terman JL, Taam RE, Hernquist L. 1995. *Ap. J.* 445:367–76

Terman JL, Taam RE, Savage CO. 1996. *MNRAS* 281:552–64

Thomas HC. 1977. *Annu. Rev. Astron. Astrophys.* 15:127–51

Thorne KS. 1996. In *Compact Stars in Binaries, IAU Symp., 165th, The Hague, 1994*, ed. J van Paradijs, EPJ Van den Heuvel, E Kuulkers, pp. 153–83. Dordrecht: Kluwer

Thorne KS, Zytkow AN. 1975. *Ap. J.* 199:L19–24

Thorne KS, Zytkow AN. 1977. *Ap. J.* 212:832–58

Thorsett SE, Chakrabarty D. 1999. *Ap. J.* 512:288–99

Tutukov AV, Yungelson LR. 1993. *Astron. Rep.* 37:411–31

Vanbeveren D, De Loore C, Van Rensbergen W. 1998. *Astron. Astrophys. Rev.* 9:63–152

Van den Heuvel EPJ. 1992. In *X-Ray Binaries and Recycled Pulsars*, ed. EPJ Van den Heuvel, SA Rappaport, pp. 233–56. Dordrecht: Kluwer

Van den Heuvel EPJ. 1993. *Sp. Sci. Rev.* 66:309–22

Van den Heuvel EPJ, Bitzaraki O. 1995. In *The Lives of Neutron Stars*, ed. MA Alpar, U Kiziloglu, J van Paradijs, pp. 421–48. Dordrecht:Kluwer

Van den Heuvel EPJ, De Loore C. 1973. *Astron. Astrophys.* 25:387–95

Van den Heuvel EPJ, Lorimer DR. 1996. *MNRAS* 283:L37–40

Van den Heuvel EPJ, Rappaport S. 1987. In *Physics of Be Stars, Proc. IAU Colloq., 92nd, Boulder, 1986*, ed. A Slettebak, TP Snow, pp. 291–308. Cambridge, UK: Cambridge Univ. Press

Van den Heuvel EPJ, Taam RE. 1984. *Nature* 309:235–37

Van den Heuvel EPJ, van Paradijs J. 1997. *Ap. J.* 483:399–401

Van Kerkwijk MH, Kulkarni SR. 1999. *Ap. J.* 516:L25–28

Verbunt F. 1993. *Annu. Rev. Astron. Astrophys.* 31:93–127

Verbunt F, Van den Heuvel EPJ. 1995. In *X-Ray Binaries*, ed. WHG Lewin, J van Paradijs, EPJ Van den Heuvel, pp. 457–94. Cambridge, UK: Cambridge Univ. Press

Verbunt F, Zwaan C. 1981. *Astron. Astrophys.* 100:L7–9

Webbink RF. 1975. *Astron. Astrophys.* 41:1–8

Webbink RF. 1992. In *X-Ray Binaries and Recycled Pulsars*, ed. EPJ Van den Heuvel, SA Rappaport, pp. 269–80. Dordrecht: Kluwer

Webbink RF, Rappaport S, Savonije GJ. 1983. *Ap. J.* 270:678–93

Wettig T, Brown GE. 1996. *New Astr.* 1:17–34

Wijnands R, van der Klis, M. 1998. *Nature* 394:344–46

Wolszcan A. 1991. *Nature* 350:688–90

Woosley SE. 1987. In *The Origin and Evolution of Neutron Stars, IAU Symp., 125th, Nanjing, 1986*, ed. D Helfand, JH Huang, pp. 255–70. Dordrecht: Reidel

Woosley SE, Weaver TA. 1992. In *The Structure and Evolution of Neutron Stars*, ed. D Pines, R Ramagaki, S Tsuruta, pp. 235–49. Redwood City, CA: Addison Wesley

Yorke HW, Bodenheimer P, Taam RE. 1995. *Ap. J.* 451:308–13

Zahn JP. 1975. *Astron. Astrophys.* 41:329–44

Zahn JP. 1977. *Astron. Astrophys.* 57:383–94

Zeldovich YaB, Ivanova J, Nadyozhin DK. 1972. *Sov. Astron.* 16:209–18

Annu. Rev. Astron. Astrophys. 2000. 38:143–90

THE EVOLUTION OF ROTATING STARS

André Maeder and Georges Meynet

Geneva Observatory, CH–1290 Sauverny, Switzerland;
e-mail: andre.maeder@obs.unige.ch, georges.meynet@obs.unige.ch

Key Words   stellar rotation, stellar evolution, mass loss, mixing, chemical abundances

■ **Abstract**   In this article we first review the main physical effects to be considered in the building of evolutionary models of rotating stars on the Upper Main-Sequence (MS). The internal rotation law evolves as a result of contraction and expansion, meridional circulation, diffusion processes, and mass loss. In turn, differential rotation and mixing exert a feedback on circulation and diffusion, so that a consistent treatment is necessary.

We review recent results on the evolution of internal rotation and the surface rotational velocities for stars on the Upper MS, for red giants, supergiants, and W-R stars. A fast rotation enhances the mass loss by stellar winds and, conversely, high mass loss removes a lot of angular momentum. The problem of the breakup or Ω-limit is critically examined in connection with the origin of Be and LBV stars. The effects of rotation on the tracks in the HR diagram, the lifetimes, the isochrones, the blue-to-red supergiant ratios, the formation of Wolf-Rayet stars, and the chemical abundances in massive stars as well as in red giants and AGB stars are reviewed in relation to recent observations for stars in the Galaxy and Magellanic Clouds. The effects of rotation on the final stages and on the chemical yields are examined, along with the constraints placed by the periods of pulsars. On the whole, this review points out that stellar evolution is not only a function of mass M and metallicity Z, but of angular velocity Ω as well.

1.   INTRODUCTION

Stellar rotation is an example of an astronomical domain that has been studied for several centuries and in which the developments are rather slow. A short historical review since the discovery of the solar rotation by Galileo Galilei is given by Tassoul (1978). Few of the early works apply to real stars, because in general gaseous configurations were not considered, and no account was given to radiative energy transport. The equations of rotating stars in radiative equilibrium were first considered by Milne (1923), von Zeipel (1924) and Eddington (1925); see also Tassoul (1990) for a more recent history. From the early days of stellar evolution, the studies of rotation and evolution have been closely associated. Soon after the first models showing that Main-Sequence (MS) stars move farther into the giant

0066-4146/00/0915-0143$14.00

and supergiant region (Sandage & Schwarzschild 1952), rotation was used as a major test for the evolution. Oke & Greenstein (1954) and Sandage (1955) found that the observed rotational velocities were consistent with the proposed evolutionary sequence.

Stellar evolution, like other fields of science, proceeds using as a guideline the principle of Occam's razor, which says that the explanation relying on the smallest number of hypotheses is usually the one to be preferred. Thus, as a result of the many well-known successes of the theory of stellar evolution, rotation was and is still generally considered only a second order effect. Over recent years, however, a number of serious discrepancies between current models and observations have been noticed. They particularly concern the helium and nitrogen abundances in massive O- and B-type stars and in giants and supergiants, as well as the distribution of stars in the HR diagram at various metallicities. The observations show that the role of rotation has been largely overlooked. All the model outputs (tracks in the HR diagram, lifetimes, actual masses, surface abundances, nucleosynthetic yields, supernova precursors, etc) are greatly influenced by rotation; thus, it turns out that stellar evolution is basically a function of mass M, metallicity Z, and angular velocity Ω.

A number of reviews exist concerning stellar rotation—for example, Strittmatter (1969), Fricke & Kippenhahn (1972), Tassoul (1978, 1990), Zahn (1983, 1994) and Pinsonneault (1997). Here, we focus on rotation in Upper MS stars, where the effects are likely the largest ones. The consequences for blue, yellow, and red supergiants and Wolf-Rayet (W-R) stars, as well as for red giants and Asymptotic Giant Branch (AGB) stars, are also examined. The rotation of low-mass stars, where spin-down resulting from magnetic coupling between the wind and the central body is important, has been treated in a recent review (Pinsonneault 1997); rotation and magnetic activity were also reviewed by Hartmann & Noyes (1987). The role of rotation in pre-Main Sequence evolution with accretion disks has been discussed by Bodenheimer (1995).

2.   BASIC PHYSICAL EFFECTS OF ROTATION

2.1 Hydrostatic Effects

In a rotating star, the centrifugal forces reduce the effective gravity according to the latitude and also introduce deviations from sphericity. The four equations of stellar structure need to be modified. The idea of the original method devised by Kippenhahn & Thomas (1970) and applied in most subsequent works (Endal & Sofia 1976; Pinsonneault et al 1989, 1990, 1991; Fliegner & Langer 1995; Heger et al 2000) is to replace the usual spherical eulerian or lagrangian coordinates with new coordinates characterizing the equipotentials. This method applies when the effective gravity can be derived from a potential, i.e. when the problem is conservative, which occurs for solid body rotation or for constant rotation on cylinders centered on the axis of rotation. If so, the structural variables P, T, ρ, ... are constant on an equipotential $\Psi = \Phi + \frac{1}{2}\Omega^2 r^2 \sin^2 \vartheta$, where Φ is the gravitational potential, ϑ is the colatitude, and Ω is the angular velocity. Thus the problem can be kept one-dimensional, which is a major advantage. However, the internal rotation generally evolves toward rotation laws that are non-conservative; in this case the above method is not physically consistent. Unfortunately, it has been and is still used by most authors.

A particularly interesting case of differential rotation is that with Ω constant on isobars (Zahn 1992). This case is called "shellular rotation," and it is often approximated by $\Omega = \Omega(r)$, which is valid at low rotation. It is supported by the study of turbulence in the Sun and stars (Spiegel & Zahn 1992, Zahn 1992). Such a law results from the fact that the turbulence is very anisotropic, with a much stronger, geostrophic-like transport in the horizontal direction than in the vertical one, where stabilization is favored by the stable temperature gradient. The horizontal turbulence enforces an essentially constant rotation rate on isobars, thus producing the preceding rotation law. The star models are essentially one-dimensional, which enormously simplifies the computations. Shellular rotation is likely to occur in fast as well as in slow rotators.

The equations of stellar structure can be written consistently for a differentially rotating star, if the rotation law is shellular. Then, the isobaric surfaces satisfy the same equation $\Psi = const.$ as the equipotentials of the conservative case (Meynet & Maeder 1997), and thus the equations of stellar structure can be written as a function of the coordinate of an isobar, either in the lagrangian or eulerian form. Let us emphasize that in general the hydrostatic effects of rotation have only very small effects of the order of a few percent on the internal evolution (Faulkner et al 1968, Kippenhahn & Thomas 1970). Recent two-dimensional models including the hydrostatic effects of rotation confirm the smallness of these effects (Shindo et al 1997).

The above potential Ψ describes the shape of the star in the conservative case for the so-called Roche model, where the distortions of the gravitational potential are neglected. However, the stellar surface deviates from a surface given by $\Psi = const.$ in the case of non-conservative rotation law (Kippenhahn 1977).

2.2   The von Zeipel Theorem

The von Zeipel (1924) theorem is essential for predicting the distribution of temperature at the surface of a rotating star. It applies to the conservative case and states that the local radiative flux $\vec{F}$ is proportional to the local effective gravity $\vec{g}_{\text{eff}}$, which is the sum of the gravity and centrifugal force, if the star is not close to the Eddington limit,

$$\vec{F} = -\frac{L(P)}{4\pi G M_\star(P)} \, \vec{g}_{\text{eff}}, \tag{1}$$

with $M_*(P) = M(1 - \frac{\Omega^2}{2\pi G \bar{\rho}})$; $L(P)$ is the luminosity on an isobar, and $\bar{\rho}$ is the mean internal density. Thus, the local effective temperature on the surface of a rotating star varies like $T_{\text{eff}}(\vartheta) \sim g_{\text{eff}}(\vartheta)^{\frac{1}{4}}$. This shows that the spectrum of a rotating star is in fact a composite spectrum made up of local atmospheres of different gravity and T_{eff}. If it is meaningful to define an average, a reasonable choice is to take $T_{\text{eff}}^4 = L/(\sigma S(\Omega))$, where σ is Stefan's constant and $S(\Omega)$ is the total actual stellar surface. In the case of non-conservative rotation law, the corrections to the von Zeipel theorem depend on the opacity law and on the degree of differential rotation, but the corrections are likely to be small, i.e. $\leq 1\%$ in current cases of shellular rotation (Kippenhahn 1977, Maeder 1999a). There are some discussions (Langer 1999; Langer et al 1999a) as to whether von Zeipel must really be used; we would like to emphasize that this is a mere and inescapable consequence of Newton's law and basic thermodynamics (see Section 5.3).

2.3 Transport of Angular Momentum and Chemical Elements

Inside a rotating star the angular momentum is transported by convection, turbulent diffusion, and meridional circulation. The equation of transport was derived by Jeans (1928), Tassoul (1978), Chaboyer & Zahn (1992), and Zahn (1992). For shellular rotation, the equation of transport of angular momentum in the vertical direction is (in lagrangian coordinates)

$$\rho \frac{d}{dt}(r^2 \Omega)_{M_r} = \frac{1}{5r^2}\frac{\partial}{\partial r}(\rho r^4 \Omega U(r)) + \frac{1}{r^2}\frac{\partial}{\partial r}\left(\rho D r^4 \frac{\partial \Omega}{\partial r}\right). \tag{2}$$

$\Omega(r)$ is the mean angular velocity at level r, $U(r)$ is the vertical component of the meridional circulation velocity, and D is the diffusion coefficient resulting from the sum of the various turbulent diffusion processes (see Section 2.5). The factor $\frac{1}{5}$ comes from the integration in latitude. If both $U(r)$ and D are zero, we just have the local conservation of the angular momentum $r^2\Omega = const.$ for a fluid element in case of contraction or expansion. The solution of Equation 2 gives the *nonstationary solution* of the problem. The rotation law is not arbitrarily chosen, but is allowed to evolve with time as a result of transport by meridional circulation, diffusion processes, and contraction or expansion. In turn, the differential rotation built up by these processes generates some turbulence and meridional circulation, which are themselves functions of the rotation law. This coupling provides feedback, and the self-consistent solution for the evolution of $\Omega(r)$ must be found (Zahn 1992).

Some characteristic times can be associated to both the processes of meridional circulation and diffusion:

$$t_{\text{circ}} \simeq \frac{R}{U}, \qquad t_{\text{diff}} \simeq \frac{R^2}{D}. \tag{3}$$

These timescales are essential quantities, because the comparison with the nuclear timescales will show the relative importance of the transport processes in the considered nuclear phases. Equation 2 also admits a *stationary solution* when one of the above characteristic times is short with respect to the nuclear evolution time, a situation that only occurs at the beginning of the MS:

$$U(r) = -\frac{5D}{\Omega}\frac{\partial\Omega}{\partial r}. \tag{4}$$

This equation (Randers 1941, Zahn 1992) expresses that the (inward) flux of angular momentum transported by meridional circulation is equal to the (outward) diffusive flux of angular momentum. As a matter of fact, this solution is equivalent to considering local conservation of angular momentum (Urpin et al 1996).

Instead of Equation 2, the transport of angular momentum by circulation is often treated as a diffusion process (Endal & Sofia 1978; Pinsonneault et al 1989, 1990; Langer 1991a; Fliegner & Langer 1995; Chaboyer et al 1995a,b; Heger et al 2000). We see from Equation 2 that the term with U (advection) is functionally not the same as the term with D (diffusion). Physically, advection and diffusion are quite different. Diffusion brings a quantity from where there is a lot of this quantity to other places where there is little. This is not necessarily the case for advection; for example, the circulation of money in the world is not a diffusive process, but an advective one. Let us make it clear that circulation with a positive value of $U(r)$, i.e. rising along the polar axis and descending at the equator, is in fact making an *inward* transport of angular momentum. If this process were treated as a diffusive function of $\frac{\partial\Omega}{\partial r}$, even the sign of the effect may be wrong.

A differential equation like Equation 2 is subject to boundary conditions at the edge of the core and at the stellar surface. At both places, this condition is usually $\frac{\partial\Omega}{\partial r} = 0$, with in addition the assumptions of solid body rotation for the convective core and $U = 0$ at the surface (Talon et al 1997, Denissenkov et al 1999). If there is magnetic coupling at the surface (Hartmann & Noyes 1987; Pinsonneault et al 1989, 1990; Pinsonneault 1997) or mass loss by stellar winds, the surface condition must be modified accordingly (Maeder 1999a). Various asymptotic regimes for the angular momentum transport can be considered (Zahn 1992) depending on the presence of a wind with or without magnetic coupling.

2.3.1 Transport of Chemical Elements   The transport of chemical elements is also governed by a diffusion–advection equation like Equation 2 (Endal & Sofia 1978; Schatzman et al 1981; Langer 1991a, 1992; Heger et al 2000). However, if the horizontal component of the turbulent diffusion is large, the vertical advection of the elements (and not that of the angular momentum) can be treated as a simple diffusion (Chaboyer & Zahn 1992) with a diffusion coefficient D_{eff},

$$D_{\mathrm{eff}} = \frac{|rU(r)|^2}{30D_h}, \tag{5}$$

where D_h is the coefficient of horizontal turbulence, for which an estimate is $D_h = |rU(r)|$ (Zahn 1992). Equation 5 expresses that the vertical advection of chemical elements is severely inhibited by the strong horizontal turbulence characterized by D_h. Thus, the change of the mass fraction X_i of the chemical species i is simply

$$\left(\frac{dX_i}{dt}\right)_{M_r} = \left(\frac{\partial}{\partial M_r}\right)_t \left[(4\pi r^2 \rho)^2 D_{\text{mix}} \left(\frac{\partial X_i}{\partial M_r}\right)_t\right] + \left(\frac{dX_i}{dt}\right)_{\text{nucl}}. \tag{6}$$

The second term on the right accounts for composition changes resulting from nuclear reactions. The coefficient D_{mix} is the sum $D_{\text{mix}} = D + D_{\text{eff}}$, where D is the term appearing in Equation 2 and D_{eff} accounts for the combined effect of advection and horizontal turbulence. The characteristic time for the mixing of chemical elements is therefore

$$t_{\text{mix}} \simeq \frac{R^2}{D_{\text{mix}}}. \tag{7}$$

Noticeably, the characteristic time for chemical mixing is not t_{circ} given by Equation 3, as has been generally assumed (Schwarzschild 1958). This makes the mixing of the chemical elements much slower, since D_{eff} is very much reduced. In this context, we recall that several authors have reduced by arbitrary factors (up to 30 or 100) the effect of the transport of chemicals in order to better fit the observed surface compositions (Pinsonneault et al 1989, 1991; Chaboyer et al 1995a,b; Heger et al 2000). This is no longer necessary with the more appropriate expressions given above.

2.4 Meridional Circulation

Meridional circulation arises from the local breakdown of radiative equilibrium in a rotating star (Vogt 1925, Eddington 1925). In a uniformly rotating star, the equipotentials are closer to each other along the polar axis than along the equatorial axis. Thus, according to von Zeipel's theorem, the heating on an equipotential is generally higher in the polar direction than in the equatorial direction, which thus drives a large-scale circulation rising at the pole and descending at the equator. This problem has been studied for about 75 years (see reviews by Tassoul 1978 or Zahn 1983). The classical formulation (Sweet 1950; Mestel 1953, 1965; Kippenhahn & Weigert 1990) for rigid rotation predicts a value of the vertical velocity of the Eddington-Sweet circulation

$$U_{ES} = \frac{8}{3}\omega^2 \frac{L}{gM} \frac{\gamma - 1}{\gamma} \frac{1}{\nabla_{\text{ad}} - \nabla} \left(1 - \frac{\Omega^2}{2\pi G\rho}\right), \tag{8}$$

with $\omega^2 = \frac{\Omega^2 r^3}{GM_r}$ the local ratio of centrifugal force to gravity and γ the ratio of the specific heats C_P/C_V. The term $\frac{\Omega^2}{2\pi G\rho}$, often called the Gratton-Öpik term (Tassoul

1990), predicts that U_{ES} becomes negative at the stellar surface because of the presence of the term $1/\rho$. This means an inverse circulation, i.e. descending at the pole and rising at the equator. The dependence in $\frac{1}{\rho}$ also makes $U(r)$ diverge at the surface. This has led to some controversies on what is limiting $U(r)$ (Tassoul & Tassoul 1982, 1995; Zahn 1983; Tassoul 1990). The timescale for circulation mixing defined in Equation 3 becomes, with the above Eddington-Sweet velocity,

$$t_{ES} \simeq t_{KH} \frac{g}{\Omega^2 R}, \qquad (9)$$

where g is the surface gravity and R is the stellar radius. Even for modest rotation velocities, t_{ES} is much shorter than the MS lifetime (Schwarzschild 1958; see also Denissenkov et al 1999), so that most stars should be mixed, if this timescale were applicable. However, the presence of μ-gradients was not taken into account in the above expressions. When this is done, rotation is found to allow circulation above only a certain rotation limit that depends on the value of the μ-gradient (Mestel 1965, Kippenhahn 1974, Kippenhahn & Weigert 1990). The balance of rotation and μ-gradients has been considered in Vauclair (1999).

The velocity of the meridional circulation in the case of shellular rotation was derived by Zahn (1992), who considered the effects of the latitude-dependent μ-distribution (Mestel 1953, 1965). Even more important are the effects of the vertical μ-gradient ∇_μ and of the horizontal turbulence (Maeder & Zahn 1998). Contrary to the conclusions of the previous works, the μ-gradients were shown not to introduce a velocity threshold for the existence of the circulation, but to progressively reduce the circulation when ∇_μ increases. We then have

$$U(r) = \frac{P}{\rho g C_P T \left[\nabla_{ad} - \nabla + (\varphi/\delta)\nabla_\mu \right]} \left\{ \frac{L}{M_\star} (E_\Omega + E_\mu) \right\}, \qquad (10)$$

where P is the pressure, C_P is the specific heat, and E_Ω and E_μ are terms depending on the Ω- and μ-distributions respectively, up to the third-order derivatives. Because the derivative of $U(r)$ appears in Equation 2, we see that the consistent solution to the problem is of fourth order (Zahn 1992). This makes the numerical solution difficult (Talon et al 1997, Denissenkov et al 1999, Meynet & Maeder 2000). Whereas the classical solution predicts an infinite velocity at the interface between a radiative and a semiconvective zone with an inverse circulation in the semiconvective zone, Equation 10 gives a continuity of the solution with no change of sign. In evolutionary models, the term ∇_μ in Equation 10 may be one or two orders of magnitude larger than $\nabla_{ad} - \nabla$ in some layers, so that $U(r)$ may be reduced by the same ratio. This considerably increases the characteristic time t_{circ} with respect to the classical estimate t_{ES}.

2.5   Instabilities and Transport

The subject of the instabilities in moving plasmas is a field in itself. Here we limit ourselves to a short description of the main instabilities currently considered influential in the evolution of Upper MS stars (Endal & Sofia 1978; Zahn 1983, 1992; Pinsonneault et al 1989; Heger et al 2000).

2.5.1 Convective and Solberg-Høiland Instability   In a rotating star, the Ledoux or Schwarzschild criteria for convective instability should be replaced by the Solberg–Høiland criterion (Kippenhahn & Weigert 1990). This criterion accounts for the difference of the centrifugal force for an adiabatically displaced fluid element; the condition for dynamical stability is

$$N^2 + \frac{1}{s^3} \frac{d(s^2\Omega)^2}{ds} \geq 0. \tag{11}$$

The Brunt-Väisälä frequency N^2 is given by $N^2 = \frac{g\delta}{H_p}(\nabla_{ad} - \nabla + \frac{\varphi}{\delta}\nabla_\mu)$, where the various symbols have their usual meaning (Kippenhahn & Weigert 1990), and the term s is the distance to the rotation axis. For no rotation, the Ledoux criterion is recovered. If the thermal effects are ignored, we just recover Rayleigh's criterion for stability, which says that the specific angular momentum $s^2\Omega$ must increase with the distance to the rotation axis. For displacements parallel to the rotation axis, convective instability occurs when $N^2 < 0$. For displacements perpendicular to the rotation axis, the stability is in general reinforced. Thus, criterion (Equation 11) is sensitive to the type of axisymmetric displacements considered (cf. Ledoux 1958). In the absence of rotation, a zone located beween the places where $\nabla = \nabla_{ad} + \nabla_\mu$ and $\nabla = \nabla_{ad}$ is called semiconvective. There, non-adiabatic effects can drive growing oscillatory instabilities (Kato 1966, Kippenhahn & Weigert 1990). An appropriate diffusion coefficient describing the transport in such zones was derived by Langer et al (1983). However, there is no diffusion coefficient yet available for semiconvective mixing in the presence of rotation.

The assumption of solid body rotation is generally made in convective regions, owing to the strong turbulent coupling. However, the collisions or scattering of convective blobs influence the rotation law in convective regions (Kumar et al 1995): Solid body rotation only occurs for an isotropic scattering. For some forms of anisotropic scattering, an outward rising rotation profile such as that observed in the Sun can be produced. Two-dimensional models of rotating stars (Deupree 1995, 1998) also show that the angular velocity in convective cores is not uniform, but it decreases with distance from the center and is about constant on cylinders. A considerable overshoot is obtained by Deupree (1998) and amounts to about 0.35 H_P, where H_P is the local pressure scale height. Similar conclusions were obtained by Toomre (1994), who also found penetrative convection at the edge of the convective core of rotating A-type stars. The very large Reynolds number characterizing stellar turbulence prevents direct numerical simulations of the Navier-Stokes equation, so that some new specific methods have been proposed to study convective turbulence (Canuto 1994, Canuto et al 1996, Canuto & Dubovikov 1997) and its interplay with differential rotation (Canuto et al 1994, Canuto 1998); these last results have not yet been applied to evolutionary models in rotation.

2.5.2 Shear Instabilities: Dynamical and Secular

In a radiative zone, shear caused by differential rotation is likely to be a very efficient mixing process. Indeed, shear instability grows on a dynamical timescale that is of the order of the rotation period (Zahn 1992, 1994). Stability is maintained when the Richardson number Ri is larger than a critical value Ri_{cr}

$$Ri = \frac{N^2}{\left(\frac{dV}{dz}\right)^2} > Ri_{cr} = \frac{1}{4}, \tag{12}$$

where V is the horizontal velocity and z is the vertical coordinate. Equation 12 means that the restoring force of the density gradient is larger than the excess energy $\frac{1}{4}\left(\frac{dV}{dz}\right)^2$ present in the differentially rotating layers (Chandrasekhar 1961). In Equation 12, heat exchanges are ignored and the criterion refers to the dynamical shear instability (Endal & Sofia 1978, Kippenhahn & Weigert 1990).

When thermal dissipation is significant, the restoring force of buoyancy is reduced and the instability occurs more easily (Endal & Sofia 1978); the timescale is longer, however, because it is the thermal timescale. This case is sometimes referred to as secular shear instability. For small thermal effects ($Pe \ll 1$), a factor equal to Pe appears as multiplying N^2 in Equation 12 (Zahn 1974). The number Pe is the ratio of the thermal cooling time to the dynamical time, i.e. $Pe = \frac{v\ell}{K}$, where v and ℓ are the characteristic velocity and length scales, and $K = (4acT^3)/(3C_P\kappa\rho^2)$ is the thermal diffusivity. Pe varies typically from 10^9 in deep interiors to 10^{-2} in outer layers (Cox & Giuli 1968). For general values of Pe, a more general expression of the Richardson's criterion can be found (Maeder 1995, Maeder & Meynet 1996); it is consistent with the case of low Pe treated by Zahn (1974). The problem of the Richardson criterion has also been considered by Canuto (1998), who suggests that for $Pe > 1$, i.e. negligible radiative losses $Ri_{cr} \sim 1$, and for $Pe < 1$, i.e. important radiative losses $Ri_{cr} \sim Pe^{-1}$. Thus, similar dependences with respect to Pe are obtained, but Canuto (1998) finds that turbulence may exist beyond the $\frac{1}{4}$ limit in Equation 12.

Many authors have shown that the μ-gradients appear to inhibit the mixing too much with respect to what is required by the observations (Chaboyer et al 1995a,b; Meynet & Maeder 1997). Changing Ri_{cr} from $\frac{1}{4}$ to 1 does not solve the problem, since the difference is a matter of one or two orders of magnitude. For example, instead of using a gradient ∇_μ, some authors write $f_\mu \nabla_\mu$ with an arbitrary factor $f_\mu = 0.05$ or even smaller (Heger et al 2000; see also Chaboyer et al 1995a,b). Most of the zone external to the convective core, where the μ-gradient inhibits mixing, is in fact semiconvective and is thus subject to thermal instability anyway. This has led to the hypothesis that the excess energy in the shear is degraded by turbulence on the thermal timescale, which changes the entropy gradient and consequently the μ-gradient (Maeder 1997). This gives a diffusion coefficient D_{shear}, which tends toward the diffusion coefficient for semiconvection by Langer et al (1983) when shear is negligible and toward the value $D_{shear} = (K/N^2)(\Omega\frac{d\ln\Omega}{d\ln r})^2$ given by Zahn (1992) when semiconvection is negligible. Another proposition was made by Talon & Zahn (1997), who took into account the homogeneizing effect of the horizontal diffusion on the restoring force produced by the μ-gradient. This also reduces the excessive stabilizing effect of the μ-gradient. Both of the above suggestions lead to an acceptable amount of mixing in view of the observations. We stress that the Reynolds condition $D_{\text{shear}} \geq \frac{1}{3}\nu Re_c$ must be satisfied when the medium is turbulent, where Re_c is the critical Reynolds number (Zahn 1992, Denissenkov et al 1999) and ν is the total viscosity (radiative + molecular).

Globally, we may expect the secular instability to work during the MS phase, where the Ω-gradients are small and the lifetimes are long, whereas the dynamical instability could play a role in the advanced stages.

2.5.3 Other Instabilities   Baroclinicity, i.e. the non-coincidence of the equipotentials and surface of constant ρ, generates various instabilities in the case of non-conservative rotation laws. Some instabilities are axisymmetric, like the GSF instability (Goldreich & Schubert 1967, Fricke 1968, Korycansky 1991). The GSF instability is created by fluid elements displaced between the directions of constant angular momentum and of the rotational axis. Stability demands a uniform or constant rotation on cylinders, which is incompatible with shellular rotation. The GSF instability thus favors solid-body rotation, on a timescale of the order of that of the meridional circulation (Endal & Sofia 1978). This instability is inhibited by the μ-gradients (Knobloch & Spruit 1983), nevertheless Heger et al (2000) find that it plays a role near the end of the helium-burning phase. Another axisymmetric instability is the ABCD instability (Knobloch & Spruit 1983). Fluid elements displaced between the surfaces of constant P and T create the ABCD instability, a kind of horizontal convection. The ABCD instability is oscillatory, and its efficiency is difficult to estimate for now. Non-axisymmetric instabilities, such as salt-fingers, may also occur (Spruit & Knobloch 1984). They are not efficient when rotation is low; however, in the case of fast rotation they may occur everywhere in rotating stars, so that one-dimensional models are likely to be an unsatisfactory idealization in this case.

The study of the transport of angular momentum by gravity waves has been stimulated by the finding of an almost solid body rotation for most of the radiative interior of the Sun (Schatzman 1993, Montalban 1994, Kumar & Quataert 1997, Zahn et al 1997, Talon & Zahn 1998). Gravity waves are supposed to transport angular momentum from the external convective layers to the radiative interior. However, Ringot (1998) has recently shown that quasi-solid rotation of the radiative zone of the Sun cannot be a direct consequence of the action of gravity waves. Thus, even for the Sun the importance of gravity waves remains uncertain.

In Upper MS stars, we could expect gravity waves to be generated by turbulent motions in the convective core (Denissenkov et al 1999). The momentum will be deposited where the Doppler shift of the waves resulting from differential rotation is equal to the initial wave frequency. From the work by Montalban & Schatzman (1996), we know that in general the deposition of energy decreases very quickly away from the boundaries of a convective zone. The same was found by Denissenkov et al (1999), who show that uniform rotation sustained by gravity waves is limited to the very inner radiative envelope; the size of the region of uniform rotation enforced by gravity waves likely increases with stellar mass. Only angular momentum may be directly transported by gravity waves and not chemical elements. Nonetheless, the transport of momentum by waves, which reduces the differential rotation, could also indirectly influence the distribution of chemical elements in stars.

2.6   Mass Loss and Rotation

Mass loss by stellar winds is a dominant effect in the evolution of Upper MS stars (Chiosi & Maeder 1986). The mass loss rates currently applied in stellar models are based on the observations of de Jager et al 1988, Lamers & Cassinelli 1996. A significant growth of the mass flux of OB stars with rotation, i.e. by 2–3 powers of 10, was found by Vardya (1985). Nieuwenhuijzen & de Jager (1988) suggested that the correlation found by Vardya mainly reflects the distributions of the mass loss rates $\dot{M}$ and of the rotation velocities v_{rot} over the HR diagram. After trying to disentangle the effects of L, T_{eff}, and v_{rot}, they found that the $\dot{M}$-rates seem to increase only slightly with rotation for O- and B-type stars. The result by Vardya might not be incorrect; when the data for OB stars by Nieuwenhuijzen & de Jager (1988) are considered, a correlation of the mass fluxes with v_{rot} is noticeable. These authors also point out that the equatorial $\dot{M}$-rates of Be stars are larger by a factor 10^2. Since Be stars are essentially B stars with fast rotation, a single description of the large changes of the $\dot{M}$-rates from the low to the high values of v_{rot} should be considered.

On the theoretical side, Pauldrach et al (1986) and Poe & Friend (1986) found a very weak change of the $\dot{M}$-rates with v_{rot} for O-stars: The increase amounts to about 30% for $v_{rot} = 350$ km/sec. Friend & Abbott (1986) found an increase of the $\dot{M}$-rates, which can be fitted by the relation (Langer 1998, Heger & Langer 1998)

$$\dot{M}(v_{rot}) = \dot{M}(v_{rot} = 0) \left(\frac{1}{1 - \frac{v_{rot}}{v_{crit}}} \right)^{\xi}, \qquad (13)$$

with $\xi = 0.43$; this expression is often used in evolutionary models.

The previous wind models of rotating stars are incomplete because they do not account for the von Zeipel theorem. The gravity darkening at the equator leads to a reduction of the equatorial mass flux (Owocki et al 1996; Owocki & Gayley 1997, 1998). This leads to very different predictions for the wind morphology than those of the current wind-compressed disk model by Bjorkman & Cassinelli (1993), which is currently advocated to explain disk formation. Equatorial disks may form quite naturally around rotating stars, however. The theory of radiative winds, with revised expressions of the von Zeipel theorem and of the Eddington factor, has been applied to rotating stars (Maeder 1999a). There are two main sources of wind anisotropies: (*1*) the g_{eff}-effect, which favors polar ejection, since the polar caps of a rotating star are hotter; and (2) the opacity or κ-effect, which favors an equatorial ejection when the opacity is large enough at the equator as a result of an opacity law that increases rapidly with decreasing temperature. In O-type stars, because opacity results mainly from the T-independent electron scattering, the g_{eff}-effect is likely to dominate and to raise a fast, highly ionized polar wind. In B-type and later stars, where a T-growing opacity is present in the external layers, the opacity effect should favor a dense equatorial wind and ring formation, with low terminal velocities and low ionization.

The so-called B[e] stars (Zickgraf 1999) are known to show both a fast, highly ionized polar wind and a slow, dense, low ionized equatorial ejection, and they may be a template of the g_{eff}- and κ-effects. At some values of T, the ionization equilibrium of the stellar wind changes rather abruptly, as do the opacity and the force-multipliers that characterize the opacities (Kudritzki et al 1989, Lamers 1997, Lamers 1999). Such transitions, which Lamers calls the bi-stability of stellar winds, may favor strong anisotropies of the winds, and even create some symmetrical rings at the latitude where an opacity peak occurs on the T-varying surface of the star.

It could be thought at first sight that wind anisotropies have no direct consequences for stellar evolution. This is not at all the case. Like magnetic coupling for low-mass stars, the anisotropic mass loss selectively removes the angular momentum (Maeder 1999a) and influences further evolution. Winds through polar caps, as are likely in O-stars, remove very little angular momentum, whereas equatorial mass loss removes considerable angular momentum from the stellar surface.

3.   MAIN SEQUENCE EVOLUTION OF ROTATING STARS

3.1   Evolution of Internal Rotation

As a result of transport processes, contraction, and expansion, the stars should be differentially rotating, with a strong horizontal turbulence enforcing a rotation law of the form Ω constant on isobars (Zahn 1992). The whole problem must be treated self-consistently, because differential rotation in turn determines the behaviors of the meridional circulation and turbulence, which themselves contribute to differential rotation. There are various approximations to treat this physical problem (Pinsonneault et al 1989, 1991; Chaboyer et al 1995a,b; Langer 1998; Heger & Langer 1998). In some works, rigid rotation is assumed, whereas in other works advection is treated as a diffusion with the risk that even the sign of the effect is the wrong one! Some authors, in order to fit the observations, introduce several efficiency factors such as f_μ, f_c, and so on. The problem is that the sensitivity of the results to these many efficiency factors is as large, or even larger, than the sensitivity to rotation.

A simplification was applied by Urpin et al (1996), who assumed equilibrium between the outward transport of angular momentum by diffusion and the inward transport by circulation, as was also suggested by Zahn (1992). This is the stationary case discussed in Section 2.3 for which Equation 4 applies. The values of $U(r)$ are always positive, and the circulation has only one loop. Urpin et al (1996) point out that the stationary distribution of Ω arranges itself so as to reduce $U(r)$ to a minimum value over the bulk of the star. This makes the values of $U(r)$ of the order of $10^{-5}-10^{-6}$ cm/s, which is quite insufficient to produce any efficient mixing in a 20 $M_\odot$ star.

The initial nonstationary approach to equilibrium in 10 and 30 $M_\odot$ stars has been studied by Denissenkov et al (1999). In a very short time—about 1% of the MS lifetime t_{MS}—$\Omega(r)$ converges towards a profile with a small degree of differential rotation and with very small values of $U(r)$ (Urpin et al 1996). The circulation shows two cells, an internal one rising along the polar axis and an external one descending at the pole. The evolution is not calculated, but it is noted that the timescale t_{circ} (which behaves like Ω^{-2}) is very short with respect to the MS lifetime for most Upper MS stars, and that the ratio $\frac{t_{\text{mix}}}{t_{MS}} \geq 1$ (Equation 7). It is noticeable that $t_{\text{circ}} \ll t_{\text{mix}}$, so that no efficiency factors are needed to reduce the mixing of chemical elements compared with the transport of angular momentum. This reduction naturally results from the effect of the horizontal turbulence.

The full evolution of the rotation law has been studied with the nonstationary scheme for a 9 $M_\odot$ star by Talon et al (1997), and for stars from 5 to 120 $M_\odot$ (including the effects of mass loss) by Meynet & Maeder (2000). The very fast initial convergences of $\Omega(r)$ and of $U(r)$ are confirmed. However, after convergence the asymptotic state of $U(r)$ does not correspond to the stationary approximation. In the full solution, $U(r)$ changes sign in the external region and thus transports some angular momentum outward, which is not the case in the stationary solution. Also, contrary to the classical result of the Eddington-Sweet circulation (Equation 8), it is found that $U(r)$ depends very little on the initial rotation.

Figure 1 shows the evolution of $\Omega(r)$ during MS evolution of a 20 $M_\odot$ star. Mass loss at the stellar surface removes a substantial fraction of the total angular momentum, which makes $\Omega(r)$ decrease with time everywhere in the star. The outer zone with inverse circulation progressively deepens during MS evolution, because a growing part of the outer layers has lower densities. This inverse circulation contributes to the outward transport of angular momentum. The deepening of the inverse circulation also has the consequence that the stationary and nonstationary solutions differ more and more as the evolution proceeds, because no inverse circulation is predicted by the stationary solution. This shows that the stationary solutions are too simplified and that outward and inward transport never reach exact equilibrium, contrary to the initial expectations.

Figure 2 shows the various diffusion coefficients inside a 20 $M_\odot$ star when the hydrogen mass fraction at the center is equal to 0.20. We notice that in general $K \geq D_h \geq D_{\text{shear}} \geq D_{\text{eff}}$. This confirms the basic hypothesis of a large D_h necessary to validate the assumption of shellular rotation.

The characteristic time t_{mix} of the mixing processes is of the same order as the lifetime t_H of the H-burning phase for the Upper MS stars (Maeder 1987). Indeed,

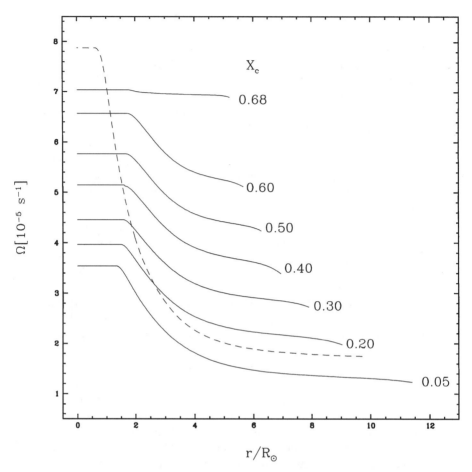

Figure 1 Evolution of the angular velocity Ω as a function of the distance to the center in a 20 $M_\odot$ star with an initial $v_{rot} = 300$ km/s. X_c is the hydrogen mass fraction at the center. The broken line shows the profile when the He-core contracts at the end of the H-burning phase.

if shear mixing is the dominant mixing process, the timescale is $t_{mix} \simeq \frac{R^2}{D_{shear}}$, and for a given degree of differential rotation it behaves like $t_{mix} \simeq K^{-1}$, which itself goes like $M^{-1.7}$. The timescale t_H behaves like $M^{-0.7}$ for $M \geq 15$ $M_\odot$ (Maeder 1998). Thus, for larger masses t_{mix} tends to decrease much faster than t_H, and mixing processes grow in importance. From the end of MS evolution when $X_c \leq 0.05$, central contraction starts dominating the evolution of the central $\Omega(r)$, which grows quickly until core collapse. During these post-MS phases, the average value of t_{mix} will be longer than the nuclear lifetimes. Thus, the rotational mixing processes during these phases may be globally unimportant (Heger et al 2000, Meynet & Maeder 2000). Nonetheless, it is likely that in some regions of a rotating star in the

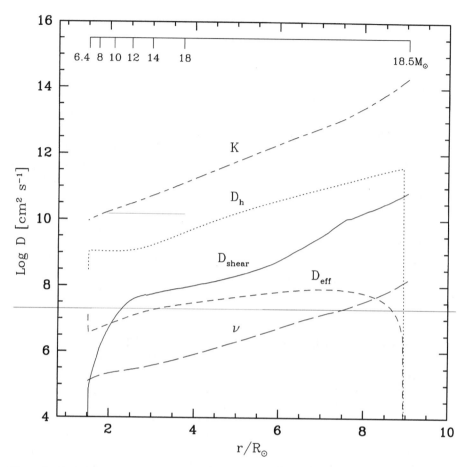

Figure 2   Internal values of K the thermal diffusivity, D_h the coefficient of horizontal turbulence, D_{shear} the shear diffusion coefficient, D_{eff} the effective diffusivity (see text) and ν the total viscosity (radiative+molecular) in the radiative envelope of a 20 $M_\odot$ star with an initial $v_{rot} = 300$ km/s. The lagrangian mass coordinate is given on the upper scale. Here, the hydrogen mass fraction at the center $X_c = 0.20$.

advanced stages, the Ω-gradient may become so large as a result of extreme central contraction that some other local instabilities develop, leading to fast mixing.

3.2   Evolution of V_{rot}: The Case of Be Stars

The evolution of the surface rotational velocities v_{rot} at the equator is a consequence of the processes discussed in the preceding section. Let us first consider the two extreme cases of coupling and no coupling between adjacent layers, first examined in the early works by Oke & Greenstein (1954) and Sandage (1955): (*1*) the case of rigid rotation and (*2*) the case of local conservation. (*1*) For rigid rotation, v_{rot}

remains nearly constant during the MS phase (Fliegner & Langer 1995). This is because the effects of core contraction and envelope expansion nearly compensate for one another. (*2*) For local conservation of the angular momentum we have $v_{\text{rot}} = \Omega R \sim R^{-1}$, whereas the critical velocity changes like $v_{\text{crit}} \sim R^{-\frac{1}{2}}$. Thus, the inflation of the stellar radius R during evolution makes rotation less and less critical. The opposite effect may occur during a blueward crossing of the HR diagram, and then critical rotation may be reached (Section 5.2). Endal & Sofia (1979) have shown that the ratio $v_{\text{rot}}(case\ 1)/v_{\text{rot}}(case\ 2) \simeq 1.8$ before the crossing of the HR diagram for intermediate mass stars. The truth generally lies between the two cases, closer to the rigid case during the MS phase because transport processes have more time to proceed. In post-MS phases, and in particular during the fast crossings of the HR diagram, the evolution timescale is short, so that little transport occurs and the evolution of v_{rot} closely resembles that of local conservation.

Stars in solid body rotation may reach the breakup limit before the end of the MS phase even for moderate initial v_{rot} (Sackmann & Anand 1970, Langer 1997). However, this is a consequence of the simplified assumption of solid body rotation. With diffusion and transport, it is less easy for the star to reach the breakup limit. For a 20 $M_{\odot}$ model without mass loss (dotted line in Figures 3 and 4), v_{rot} grows and the ratio $\frac{\Omega}{\Omega_{\text{crit}}}$ may become close to 1 before the end of the MS phase. For stars with $M < 15\ M_{\odot}$, where mass loss is small, the ratio $\frac{\Omega}{\Omega_{\text{crit}}}$ also increases during the MS phase and the breakup or Ω-limit may be reached during the MS phase. If so, the mass loss should then become very intense, until the velocity again becomes subcritical. It is somehow paradoxical that small or no mass loss rates during the bulk of the MS phase (as at low metallicity Z) may lead to very high mass loss at the end of the MS evolution for rotating stars. This may be the cause of some ejection processes as in Luminous Blue Variables (LBV), B[e], and Be stars. We may wonder whether the higher relative number of Be stars observed in lower Z regions (Maeder et al 1999) is simply a consequence of the lower average mass loss in lower metallicity regions, or whether this is related to star formation.

If mass loss is important during MS evolution, then v_{rot} decreases substantially (Figures 3 and 4). Therefore it is not surprising that Be stars, which are likely close to breakup (Slettebak 1966), form not among O-type stars, but mainly among B-type stars, with a relative maximum at type B3. Indeed, to form a Be star it is probably not necessary that the breakup limit be reached exactly. The conditions for an equatorial ejection responsible for the Be spectral features occur when the κ-effect is important (Section 2.6), which requires that the equatorial regions of the rapidly rotating star be below the bi-stability limit. Of course, the higher the rotation, the higher the equatorial mass loss will be.

Any magnetic coupling between the star and the wind would dramatically reduce v_{rot}. However, such a coupling does not seem to be important in general, except for Bp and Ap stars. MacGregor et al (1992) show that even in the presence of a small magnetic field of 100 G, the rotation velocities of OB stars should be much lower than observed. This result agrees with that of Mathys (1999), who finds no detectable magnetic field in hot stars.

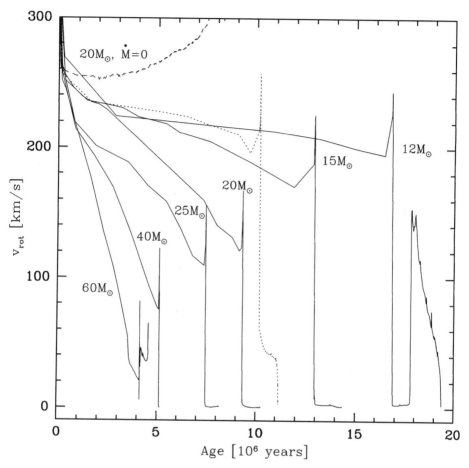

Figure 3   Evolution of the surface equatorial velocity as a function of time for stars of various initial stellar masses (Meynet & Maeder 2000). All models have an initial velocity of 300 km/s. The continuous lines refer to solar metallicity models, the dotted line corresponds to a 20 $M_\odot$ star with Z = 0.004. The dashed line corresponds to a 20 $M_\odot$ star without mass loss.

3.2.1 Effects of Mass Loss on Rotation   Mass loss by stellar winds drastically reduces v_{rot} during evolution (Packet et al 1980, Langer & Heger 1998; Figures 3 and 4). Even if isotropic, the stellar winds carry away considerable angular momentum, and this is even more important in the case of equatorial mass loss. The new surface layers then have a lower v_{rot} as a result of expansion and redistribution.

With the simplified assumption of solid body rotation for a 60 $M_\odot$ model with mass loss, Langer (1997, 1998) found a convergence of v_{rot} toward the critical value (the Ω-limit), before the end of the MS phase, for all initial velocities above 100 km/s. The overall result is that the final velocities are the same (all are at the critical

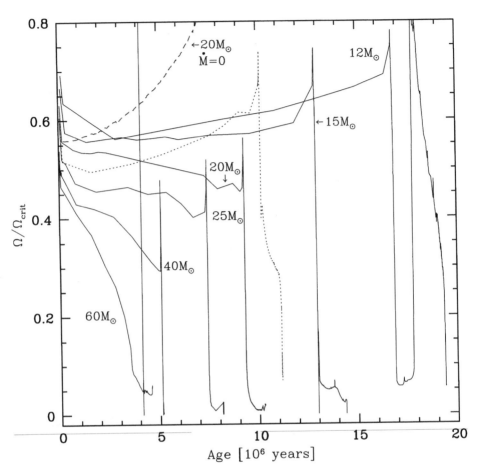

Figure 4 Same as Figure 3 for the ratio Ω/Ω_{crit} of the angular velocity to the break-up velocity at the stellar surface.

limit), whereas the final MS masses strongly depend on the initial rotation. This convergence toward the Ω-limit clearly results also from the simplified assumption of rigid rotation, which grossly exagerates the coupling of the surface layers.

Figure 3 illustrates the decrease in v_{rot} during evolution when the various transport mechanisms are followed in detail. The decrease in v_{rot} is much greater for larger initial stellar masses because mass loss is greater for them. The same is true for $\frac{\Omega}{\Omega_{crit}}$ (Figure 4). For $M \geq 40\,M_\odot$ the velocities v_{rot} will remain largely subcritical for all initial velocities, except during the overall contraction phase at the end of MS evolution. For a given initial mass, the resulting scatter of v_{rot} should be smaller at the end of the MS phase, but this is not a convergence toward the Ω-limit, as occurs for models with solid body rotation (Langer 1997, 1998). The final masses are of course lower for greater rotation; because mass loss is enhanced by rotation, this results in a large scatter of masses and v_{rot} at a given luminosity.

3.2.2 Comparison with Observations   Only a few comparisons between the observed v_{rot} and the model predictions have been made until now. Conti & Ebbets (1977) found that v_{rot} in O-type giants and supergiants is not as low as they expected from models with rotation conserved in shells. This fact and the relative absence of low rotators among the evolved O-stars led them to conclude that another line-broadening mechanism, such as macroturbulence, should be present in these objects. The same conclusion based on similar arguments was supported by Penny (1996) and Howarth et al (1997). This conclusion needs to be further verified, since the decrease predicted by the new models is not as rapid as for local conservation.

It is well known that the average value of v_{rot} increases from the early O-type to the early B-type stars (Slettebak 1970). This may be the signature of the effect of higher losses of mass and angular momentum in the more massive stars (Penny 1996), which leads to a lower average rotation in the course of the MS phase. We notice that the increase of v_{rot} from O to B stars is larger for the stars of class IV than for the stars of class V (Fukuda 1982), which is expected because the difference resulting from mass loss is more visible near the end of the MS phase (Figure 3). The stars of luminosity class I (Fukuda 1982) show a fast decline of the average v_{rot} from the O-type to the B-type stars. Because the supergiants of class I originate from the most massive stars, which evolve at approximately constant luminosity, this last effect could be related to the fact that for a given initial mass, v_{rot} declines strongly as the star moves away from the MS (Langer 1998).

3.3 The HR Diagram, Lifetimes, and Isochrones

As always in stellar evolution, the shape of the tracks is closely related to the internal distribution of the mean molecular weight μ. All results show that the convective cores are slightly increased by rotation (Maeder 1987, 1998; Langer 1992; Talon et al 1997; Meynet 1998, 1999; Heger et al 2000). The height of the μ- discontinuity at the edge of the core is reduced, and there is a mild composition gradient built up from the core to the surface, which may then be slightly enriched in helium and nitrogen. For low or moderate rotation the convective core shrinks as usual during MS evolution, whereas for high masses ($M \geq 40\,M_\odot$) and large initial rotations ($\frac{\Omega}{\Omega_{crit}} \geq 0.5$) the convective core grows in mass during evolution. These behaviors, i.e. reduction or growth of the core, determine whether the star will follow respectively the usual redward MS tracks in the HR diagram, or whether it will bifurcate to the blue (cf Maeder 1987, Langer 1992) toward the classical tracks of homogeneous evolution (Schwarzschild 1958) and likely produce W-R stars (Section 5.4). Also, for O- and B-type stars, fast rotation increases the He content of the envelope, and the decrease of the opacity also favors a blueward track.

Figure 5 shows the overall HR diagram for rotating and nonrotating stars. The atmospheric distortions produce a shift to the red in the HR diagram by several 0.01 in (B-V), with only a small change of luminosity, on average (Maeder & Peytremann 1970, Collins & Sonneborn 1977). During MS evolution, the luminosity of the rotating stars grows faster and the tracks extend farther away from the ZAMS, as in the case of a moderate overshooting (Maeder 1987, Langer 1992, Sofia et al 1994, Talon et al 1997). This effect introduces a significant scatter in the mass-luminosity relation (Meynet 1998), in the sense that fast rotators are overluminous with respect to their actual masses. This may explain some of the discrepancies between the evolutionary masses and the direct mass estimates in some binaries (Penny et al 1999). In this context, we recall that for a decade a severe mass discrepancy beween spectroscopic and evolutionary masses was claimed by some authors (Groenewegen et al 1989, Kudritzki et al 1992, Herrero et al 1992). Most of the problem has collapsed and was shown to be a result of the proximity of O-stars to the Eddington limit (Lamers & Leitherer 1993, Herrero et al 1999) and of the large effect of metal line blanketing not usually accounted for in the atmosphere models of massive stars (Lanz et al 1996).

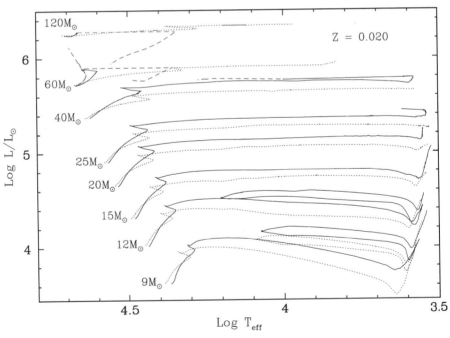

Figure 5 Part of the evolutionary tracks for non-rotating (dotted lines) and rotating (continuous lines) models with solar metallicity (Meynet & Maeder 2000). The rotating models have an initial velocity v_{rot} of 300 km/s. The long dashed track corresponds to a very fast rotating star ($v_{rot} \sim 400$ km/s) of 60 $M_\odot$, which follows a nearly homogeneous evolution. The short dashed tracks indicate the beginning of the W-R phases.

There is little difference between tracks with $v_{rot} = 200$ or 300 km/s (Meynet & Maeder 2000; see also Talon et al 1997). If the effects behaved like v_{rot}^2, larger differences would exist. This saturation effect occurs because outward transport of angular momentum by shears is larger when rotation is larger; also, a larger rotation produces more mass loss, which further reduces rotation during evolution.

For the model of 12 $M_\odot$, a blue loop appears when rotation is included. This results from the smaller core in the rotating model, because in this model the growth of the core is prevented by a very large external convective zone. Thus, rotation not only modifies the mass-luminosity relation for Cepheids, but also it increases the maximum possible luminosity for the occurrence of Cepheids.

3.3.1 Lifetimes and Isochrones   The lifetimes t_H in the H-burning phase grow only moderately because more nuclear fuel is available, but at the same time the luminosity is larger. The net result is an increase by about 20% to 30% for an initial velocity of 200 km/s (Talon et al 1997, Meynet & Maeder 2000). This influences the isochrones and age determinations. For example, for $v_{rot} = 200$ km/s, the isochrone of log age $= 7.0$ is the same as that of log age $= 6.90$ without rotation (Meynet 2000). Thus, accounting for rotation could lead to ages larger by about 25% for O- and early B-type stars. However, since the cluster ages are generally determined on the basis of the blue envelope of the observed sequence, where most low rotators lie (Maeder 1971), it is likely that the effect of rotation in current age determinations is rather small. If a blueward track occurs, the larger core and mixing lead to much longer lifetimes in the H-burning phase. In this case, the fitting of timelines becomes hazardous.

4.   ROTATION AND CHEMICAL ABUNDANCES

4.1   Observations

Chemical abundances are a very powerful test of internal evolution, and they give strong evidence in favor of some additional mixing processes in O- and B-type stars, in supergiants, and in red giants of lower masses.

4.1.1 Abundances in MS O- and B-Type Stars   Much evidence of He and N excesses in O-type and early B-type stars have been reported over the last decade (Gies & Lambert 1992; Herrero et al 1992, 1998; Kilian 1992; Kendall et al 1995, 1996; Lyubimkov 1996, 1998). We can extract the following main points:

1. The OBN stars show significant He and N excesses. OBN stars are more frequent among stars above 40 $M_\odot$ (Walborn 1988, Schönberner et al 1988, Herrero et al 1992).

2. All fast rotators among the O-stars show some He excesses (Herrero et al 1992, 1998, 1999; Lyubimkov 1996).

3. Although rather controversial initially, there seems to be an increase of the He and N abundances with relative age (i.e. the fraction t/t_{MS} of the MS lifetime spent) for the early B-type stars (Lyubimkov 1991, 1996; Gies & Lambert 1992, see note added in proof; Denissenkov 1994). Lyubimkov (1996) suggests a sharp rise from $(He/H) = 0.08$–0.10 to 0.20 in number for O-stars when $t/t_{MS} \geq 0.5$–0.7, whereas for B-type stars the corresponding value rises to 0.12–0.14. As for nitrogen, its abundance is estimated to rise to about 3 times for a 14 $M_\odot$ and 2 times for a 10 $M_\odot$ star. An increase of the N abundance by a factor of 2–3 for an O-star with the average v_{rot} of 200 km/s is the order of magnitude typically considered as a constraint for recent stellar models (Heger et al 2000).

4. The boron abundances in five B-stars on the MS were found to be smaller than the cosmic meteoritic value by a factor of at least 3 or 4 (Venn et al 1996). Boron depletion occurs in stars that also show N excess; this supports the idea that rotational mixing occurs throughout the star (Fliegner et al 1996).

4.1.2 *Abundances in Supergiants*   The main observations are as follows:

1. He and N excesses seem to be the rule among OB supergiants (Walborn 1988). According to Walborn, only the small group of the "peculiar" OBC stars has the normal cosmic abundances. An excess of He, sometimes called the "helium discrepancy," and corresponding excesses of N have been found by a number of authors (Voels et al 1989; Lennon et al 1991; Gies & Lambert 1992; Herrero et al 1992, 1999; Smith & Howarth 1994; Venn 1995a,b; Crowther 1997; McEarlean 1998; McEarlean et al 1999). As shown by these last authors, the determination of the helium abundance also depends on the adopted value for microturbulence. However, Villamariz & Herrero (1999) and Herrero et al (1999) point out that the helium discrepancy is only reduced, not solved, when microturbulence is accounted for.

2. Evidence of highly CNO-processed material is present for B supergiants in the range 20–40 $M_\odot$, (McEarlean et al 1999). Values of [N/H] (i.e. the difference in log with respect to the solar values) amounting to 0.6 dex have been found for B supergiants around 20 $M_\odot$ (Venn 1995a,b). Such values agree with the enrichments found in the ejecta of SN 1987A (Fransson et al 1989).

3. The values of [N/H] for galactic A-type supergiants around 12 $M_\odot$ lie between 0 and 0.4 dex (Venn 1995a,b, 1999). All of these values are globally consistent with the preceding results from Lyubimkov (1996). Takeda & Takada-Hidai (1995) have suggested that these excesses are greater for larger masses, a result in agreement with theory and also recently confirmed by McEarlean et al (1999). For the A-type supergiants

in the SMC, the N/H excesses are much larger, spanning a range of [N/H] = 0 to 1.2 dex (Venn 1998; Venn 1999).

4. Na excesses have been found in yellow supergiants (Boyarchuk & Lyubimkov 1983), and the overabundances seem to be larger for higher-mass stars (Sasselov 1986). Two different explanations have been proposed—one based on the reaction $^{22}Ne(p,\gamma)^{23}Na$ (Denissenkov 1994), and the other one based on $^{20}Ne(p,\gamma)^{21}Na$ (Prantzos et al 1991), with some additional mixing processes in both cases. The latter reaction seems to have a too-low rate, whereas the first one may work (Denissenkov 1994). The important point is that the observed Na excesses imply some mixing from the deep interior to the surface.

5. Very few abundance determinations exist in yellow and red supergiants. Some excesses of N with respect to C and O have been found by Luck (1978). Barbuy et al (1996) found both N enrichments and normal compositions among the slow rotating F-G supergiants. Isotopic ratios $^{13}C/^{12}C$, $^{17}O/^{16}O$, and $^{18}O/^{17}O$ for red supergiants would provide very useful information. However, the dilution factor in the convective envelope of red giants and supergiants is so large that it is not possible from the rare data available to draw any conclusions about the presence of additional mixing (Maeder 1987, Dearborn 1992, Denissenkov 1994, El Eid 1994).

4.2 Comparisons of Models and Observations

4.2.1 Massive Stars in MS and Post-MS Phases   Let us first recall that from the comparison of t_{mix} and t_H, it is clear that mixing processes are more efficient in more massive stars. For the intermediate-mass stars of the B and A types, no global mixing is currently predicted. Often, the comparisons with the observed abundance excesses for O- and B-type stars are used to adjust some efficiency factors in the models (Pinsonneault et al 1989, Weiss et al 1988, Weiss 1994, Chaboyer et al 1995a,b, Heger et al 2000). Although not fully consistent, these approaches help us appreciate the importance of the various possible effects. The old prescriptions of Zahn (1983) were applied by Maeder (1987), Langer (1992), and Eryurt et al (1994), and led to some surface He and N enrichments.

The prescriptions by Zahn (1992) were applied to the evolution of a 9 $M_\odot$ star (Talon et al 1997). These last authors found essentially no He enrichment and a moderate enhancement (factor $\sim$2) of N at the stellar surface, for an initial velocity of 300 km/s. Figure 6 illustrates the changes of the N/H ratios from the ZAMS to the red supergiant stage for 20 and 25 $M_\odot$ stars (Meynet & Maeder 2000). For non-rotating stars, the surface enrichment in nitrogen occurs only when the star reaches the red supergiant phase; there, CNO elements are dredged up by deep convection. For rotating stars, N excesses occur already during the MS phase, and they are larger for high rotation and initial stellar masses. At the end of the MS phase, for solar metallicity Z = 0.02, the predicted excesses

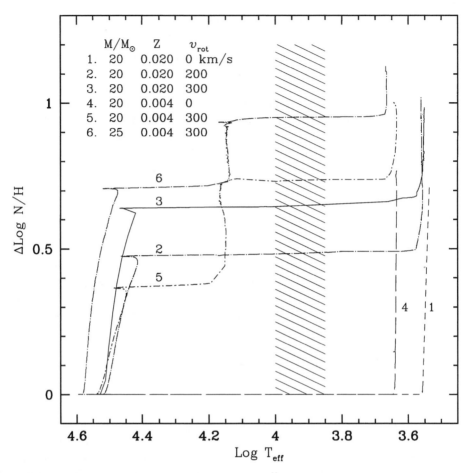

Figure 6   Evolution as a function of log T_{eff} of $\Delta \log \frac{N}{H} = \log(N/H) - \log(N/H)_i$, where N and H are the surface abundances (in number) of nitrogen and hydrogen respectively, the index i indicates initial values. The initial masses, metallicities and rotational velocities are indicated. The shaded area corresponds to the range of observed values (Venn 1999) for A-type supergiants in the SMC.

amount to factors of 3 and 4 for initial $v_{rot} = 200$ and 300 km/s, respectively. At lower metallicity, the N enrichment during the MS phase is smaller, probably because of lower mass loss; however, there is a very large increase (up to a factor of ∼10) for late B-type supergiants, because the star spends a lot of time in the blue phase, and mixing processes have time to work. The predictions of Figure 6 agree with the observed excesses for galactic B- and A-type supergiants (Venn 1995a,b, Venn 1998). Also, the very large excesses observed for A-type supergiants in the SMC (Venn 1998, 1999) are remarkably well accounted for.

4.2.2 Questions About Nitrogen Many studies of galactic halo stars, blue compact galaxies, and highly redshifted galaxies have revealed the need for initial production of primary N in addition to the currently accepted mechanism of secondary nitrogen (cf Edmunds & Pagel 1978, Matteucci 1986, Pettini et al 1995, Thuan et al 1995, Centurion et al 1998, Pilyugin 1999, Henry & Worthey 1999). The early production of N in the evolution of galaxies seems to imply that some N is produced in massive stars (Matteucci 1986, Thuan et al 1995). The problem is that the usual stellar models do not show such a production without ad hoc assumptions. Models of rotating stars allow us to clearly identify the conditions for the production of primary N: The star must have an He-burning core and a thick and long-lived H-burning shell, and then diffusion and transport of new ^{12}C from the core to the shell may generate some primary ^{14}N. W-R stars do not seem favorable because the H-shell does not live long enough—it is quickly extinguished and removed by mass loss. Low-metallicity supergiants that have not suffered large mass loss are very favorable sites, especially if rotation is faster at low metallicities.

4.3 Red Giants and AGB Stars

MS stars in the range of 1.5 to $\sim$10 $M_\odot$ show no evidence of extra-mixing; however, there are interesting indications for red giants. The study of ^{12}C/^{13}C in cluster red giants (Gilroy 1989) shows that stars between 2.2 $M_\odot$ and 7 $M_\odot$ have ratios close to the standard predictions without mixing. However, red giants below 2.2 $M_\odot$ show ^{12}C/^{13}C ratios much lower than the predictions, indicating some extra-mixing (Gilroy 1989; see also Harris et al 1988); the lower the mass, the higher the mixing. On the red giant branch of M67, there are indications of extra-mixing for stars brighter than $\log L/L_\odot \simeq 1.0$, where the first dredge-up occurs (Gilroy & Brown 1991).

From the data on M67, it appears that extra-mixing is only efficient when the H-burning shell reaches the μ-discontinuity left by the inward progression of the outer convective zone (Charbonnel 1994). Prior to this stage, the μ-gradient created by the first dredge-up acts as a barrier to any mixing below the convective envelope (Charbonnel et al 1998). The μ-gradient necessary to prevent mixing is found to be in agreement with that expected to stop the meridional circulation. For stars with $M \geq 2.2$ $M_\odot$, helium ignition occurs in nondegenerate cores, i.e. early enough so that the H-shell does not reach the border left by the outer convective zone and there is always a μ-gradient high enough to prevent mixing (Charbonnel et al 1998). Boothroyd & Sackmann (1999) confirm that extra-mixing and the associated CNO nuclear processing (cool bottom processing, CBP) occur when the H-burning shell erases the μ-barrier established by the first dredge-up, and they predict that the effects of the CBP behave like M^{-2} and Z^{-1}. Further observations of red giants with various masses are very much needed to confirm the relative absence of enrichment for masses ≥ 2.2 $M_\odot$.

For the more advanced stages of intermediate mass stars, the critical questions concern nucleosynthesis and the processes leading to the production of the s-elements in AGB stars (Iben 1999). Rotation appears to allow the formation of larger degenerate cores (Sackmann & Weidemann 1972, Maeder 1974); then, the C/O core mass is further increased during the early AGB phases (Dominguez et al 1996). The large Ω-gradients between the bottom of the convective envelope and the H-burning shell can drive mixing, mainly through the GSF instability and to a lesser extent through shear and meridional circulation between the H- and ^{12}C-rich layers during the third dredge-up in AGB stars (Langer et al 1999b). The neutron production by ^{13}C$(\alpha,n)^{16}$O between the thermal pulses favors the production of s-elements. In further studies on the role of rotation in AGB stars, the exact treatment of the instabilities in regions of steep Ω- and μ-gradients will play a crucial role.

5.   POST-MS EVOLUTION WITH ROTATION

The post-MS evolution of rotating stars differs from that of nonrotating stars for three main reasons: (*1*) The structure at He-ignition is different because of the rotationally induced mixing during the previous H-burning phase; in rotating stars, the He cores are more massive (Sreenivasan & Wilson 1985b, Sofia et al 1994) and the radiative envelope is enriched with CNO-burning products (Maeder 1987, Heger et al 2000, Meynet & Maeder 2000); (*2*) The mass loss rates are increased by rotation (Friend & Abbott 1986, Langer 1998; Section 2.6); and (*3*) Rotational transport mechanisms may also operate in the interior during the post-MS phases. In massive stars, however, the timescales for mixing and circulation, t_{mix} and t_{circ}, are much larger than the evolutionary timescale by one to two orders of magnitude during the He-burning phase (Endal & Sofia 1978) and even larger in the post–He-burning phases (Heger et al 2000). Thus, these processes will have small global effects during these stages. However, because of very high angular velocity gradients occuring locally, some instabilities may appear on much smaller timescales (Endal & Sofia 1978, Deupree 1995).

5.1   Internal Effects

The fast contraction of the core and expansion of the envelope that follow the end of the MS phase produce an acceleration of Ω in the inner regions and a slowing down in the outer layers (Kippenhahn et al 1970, Endal & Sofia 1978, Talon et al 1997, Heger et al 2000, Meynet & Maeder 2000). Typically, for a 20 $M_\odot$ model with an initial $v_{\text{rot}} = 300$ km/s (Figure 5), the ratio of the central to surface angular velocity never exceeds 5 during the MS phase, whereas it increases to 10^5 or 10^6 during the He-burning phase (Section 5.5).

At the beginning of the He-burning phase, the He cores in rotating models are more massive by about 15% for $v_{\text{rot}} = 300$ km/s. The further evolution of the convective core depends on the adopted criterion for convection. If the Ledoux criterion is used, the growth of the convective core is prevented by the μ-barriers and remains small. Above it, several small convective zones appear, each separated by semiconvective layers (e.g. Langer 1991c). If the Schwarzschild criterion is used, the convective core simply grows in mass. For both cases, the rotational effects depend on their sensitivity to the μ-gradients. For example, models with the Ledoux criterion show that when rotational mixing is artificially made insensitive to μ-gradients, the shear mixing operates efficiently in semiconvective regions and considerably enlarges the final C/O core masses (Heger et al 2000). Typically, the C/O core mass increases from a value of 1.77 $M_\odot$ in the 15 $M_\odot$ nonrotating model to a value of 3.4 $M_\odot$ for an initial $v_{rot} = 200$ km/s. For a similar rotating model using the Schwarzschild criterion and incorporating the inhibiting effect of the μ-gradients, Meynet & Maeder (2000) obtained a C/O core mass of 2.9 $M_\odot$, to be compared with the value of 2 $M_\odot$ obtained in the nonrotating model. The treatment of the μ-gradient is thus critical, because the C/O core masses play a key role in determining the stellar remnants as well as the chemical yields (Section 5.5).

5.2   Evolution in the HR Diagram, Lifetimes, and Rotational Velocities

Rotating as well as nonrotating models with initial masses between 9 and 40 $M_\odot$ at solar metallicity evolve toward the red supergiant (RSG) stage after the MS phase (Kippenhahn et al 1970, Sofia et al 1994, Heger 1998, Meynet & Maeder 2000). Because of the larger He cores, the rotating stars have higher luminosities, as long as mass loss is not too large. The initial distribution of the rotational velocities will thus introduce some scatter in the luminosities of the supergiants originating from the same initial mass. The lower the sensitivity of the mixing processes to the μ-gradients, the greater the scatter. For initial v_{rot} between 0 and 300 km/s, the difference will be of the order of 0.25 mag (Figure 5).

In rotating stars of initial mass between 20 and 40 $M_\odot$, because of the large cores, the quantity of nuclear fuel is larger, but the luminosities are also higher and thus the He-burning lifetimes change slightly; as an example, for initial $v_{rot} =$ 200–300 km/s, the changes are less than 5%. The ratios t_{He}/t_H of the He to H-burning lifetimes are not very sensitive to rotation and remain around 10% (Heger 1998, Meynet & Maeder 2000).

5.2.1 The Number Ratio of Blue to Red Supergiants   The variation with metallicity Z of the number of blue and red supergiants (RSG) is important in relation to the nature of the supernova progenitors in different environments (cf Langer 1991b,c) and population synthesis (e.g. Cervino & Mas-Hesse 1994, Origlia et al 1999). The observations show that the number ratio (B/R) of blue to red supergiants increases steeply with Z. Cowley et al (1979) examined the variation of the B/R ratio across the Large Magellanic Cloud (LMC) and found that it increases by a factor of 1.8 when the metallicity is larger by a factor of 1.2. For M_{bol} between -7.5 and -8.5, the B/R ratio is up to 40 or more in inner Galactic regions and only about 4 in the SMC (Humphreys & McElroy 1984). A difference in the B/R ratio of an order of magnitude between the Galaxy and the Small Magellanic Cloud (SMC) was also found from cluster data (Meylan & Maeder 1982). Langer & Maeder (1995) compared different stellar models with the observations and concluded that most massive star models have problems reproducing this observed trend.

Part of this difficulty certainly arises from the fact that supergiants are often close to a neutral state between a blue and a red location in the HR diagram. Even small changes in mass loss, in convection or other mixing processes, greatly affect the evolution and the balance between the red and the blue locations (Stothers & Chin 1973, 1975, 1979, 1992a,b; Maeder 1981; Brunish et al 1986; Maeder & Meynet 1989; Arnett 1991; Chin & Stothers 1991; Langer 1991b,c, 1992; Salasnich et al 1999). As stated by Kippenhahn & Weigert (1990), "the present phase is a sort of magnifying glass, revealing relentlessly the faults of calculations of earlier phases."

The choice of the criterion for convection plays a key role, particularly when the mass loss rates are small. Models with the Ledoux criterion, with or without semiconvection, predict at low metallicity (Z between 0.002 and 0.004) both red and blue supergiants. However, when the metallicity increases, the B/R ratio decreases in contradiction with the observed trend (Stothers & Chin 1992a, Brocato & Castellani 1993, Langer & Maeder 1995). Models with the Schwarzschild criterion, with or without overshooting, can more or less reproduce the observed B/R ratio in the solar neighborhood. However, they predict very few or no red supergiants at the metallicity of the SMC, whereas many are observed (Brunish et al 1986, Schaller et al 1992, Bressan et al 1993, Fagotto et al 1994).

When the mass loss rates are low (i.e. at low Z), a large intermediate convective zone forms in the vicinity of the H-burning shell, homogenizing part of the star and maintaining it as a blue supergiant (Stothers & Chin 1979, Maeder 1981). For larger mass loss rates, the intermediate convective zone is drastically reduced, and the formation of RSG is favored. A further increase of the mass loss rates may bring the star back to the blue. When the He core encompasses more than some critical mass fraction q_c of the total mass (Chiosi et al 1978, Maeder 1981), the star moves to the blue and becomes either a blue supergiant or a W-R star (e.g. Schaller et al 1992, Salasnich et al 1999, Stothers & Chin 1999). The critical mass fraction q_c is equal to 67% at 60 $M_\odot$, 77% at 30 $M_\odot$, and 97% at 15 $M_\odot$ (Maeder 1981). This agrees with the investigations made for lower masses by Giannone et al (1968).

Rotation mainly affects the B/R ratio through its effect on the interior structure and on the mass loss rates. Maeder (1987), Sofia et al (1994), and Talon et al (1997) showed that the effect of additional mixing caused by rotational instabilities in some respect mimics that of a small amount of convective overshoot, which does not favor the formation of RSG at low Z. However, fast rotation also implies higher mass loss rates by stellar winds, and in general this favors the formation of RSG. In view of these two opposite effects, it is still uncertain whether rotation may solve the B/R problem. Because of the initial distribution of rotational velocities, one expects a scatter of the mass loss rates and therefore different evolutionary scenarios for a given initial mass star (Sreenivasan & Wilson 1985a). However, by producing high mass loss rates even at low metallicity, rotation may help resolve the B/R problem.

Is rotation responsible for the observed characteristics of the blue progenitor of the SN 1987A? The presence of the ring structures around SN 1987A (Burrows et al 1995, Meaburn et al 1995), which likely result from axisymmetric inhomogeneities in the stellar winds ejected by the progenitor (Eriguchi et al 1992, Lloyd et al 1995, Martin & Arnett 1995), and the high level of nitrogen enhancements in the circumstellar material (Fransson et al 1989, Panagia et al 1996, Lundqvist & Fransson 1996) are features that may be explained at least in part by rotation. Woosley et al (2000) suggest that any mechanism that reduces the helium core while simultaneously increasing helium in the envelope would favor a blue supernova progenitor.

5.2.2 Evolution of Surface Velocities   As we already stated in Section 3.2, when the star evolves from the blue toward the RSG stage, surface velocities quickly decrease (Endal & Sofia 1979, Langer 1998). For the stars shown in Figure 5, velocities between 20 and 50 km/s are obtained when $\log T_{\text{eff}} = 4.0$. Values of the order of 1 km/s are reached at the RSG stage. Observations confirm this rapid decline of surface velocities (Rosendhal 1970, Fukuda 1982). The values at $\log T_{\text{eff}} = 4.0$ are in good agreement with Verdugo et al's (1999) recent determinations of rotational velocities for galactic A-type supergiants.

When the star evolves back to the blue from the RSG stage, as is the case for the rotating 12 $M_\odot$ model shown in Figure 5, the rotational velocity approaches the breakup velocity (Heger & Langer 1998, Meynet & Maeder 2000). This behavior results from the stellar contraction which concentrates a large fraction of the angular momentum of the star (previously contained in the extended convective envelope of the RSG) in the outer few hundredths of a solar mass. At the maximum extension of the blue loop, the equatorial velocity at the surface of the 12 $M_\odot$ star (Figure 5) reaches values as high as 150 km/s. As the star evolves back toward the RSG stage, the surface velocity declines again to about 2 km/s. When the star crosses the Cepheid instability strip, its surface velocity is between 10 and 20 km/s, well inside the observed range (Kraft et al 1959, Kraft 1966, Schmidt-Kaler 1982).

The increase in the surface velocity occurs every time a star leaves the Hayashi line to hotter zones of the HRD (Heger & Langer 1998). In most cases, there is observational evidence for axisymmetric circumstellar matter: disks around T Tauri stars during pre-Main-Sequence evolution (e.g. Guilloteau & Dutrey 1998), bipolar planetary nebulae (e.g. Garcia-Segura et al 1999), structures around SN 1987A (e.g. Meaburn et al 1995), and rings around W-R stars (e.g. Marston 1997).

5.3 The Ω, Γ Limits and the LBVs

In recent years, the problem of the very luminous stars close to the Eddington limit and reaching the breakup limit (Langer 1997, 1998) has been discussed in relation to the Luminous Blue Variables (LBVs) and their origin (see Davidson et al 1989, de Jager & Nieuwenhuijzen 1992, Nota & Lamers 1997). The LBVs, also called the Hubble-Sandage Variables and S Dor Variables, are extreme OB supergiants with $\log L/L_\odot \simeq 6.0$ and T_{eff} between about 10,000 and 30,000 K (Humphreys 1989). Only a few of them are known in the Milky Way; among them is η Carinae (Davidson & Humphreys 1997, Davidson et al 1997). LBVs experience giant outbursts with shell ejections. Often they show surrounding bipolar nebulae (Nota et al 1997). Many models and types of instabilities have been proposed to explain the LBV outbursts (de Jager & Nieuwenhuijzen 1992; Stothers & Chin 1993, 1996, 1997; Nota & Lamers 1997).

5.3.1 Physics of the Breakup Limit The first problem concerns the expression of the breakup limit for stars close to the Eddington limit, i.e. for the brightest supergiants. When the radiation field is strong, the radiative acceleration $\vec{g}_{\text{rad}}$ must be accounted for in the total acceleration

$$\vec{g}_{\text{tot}} = \vec{g}_{\text{grav}} + \vec{g}_{\text{rot}} + \vec{g}_{\text{rad}} = \vec{g}_{\text{eff}} + \vec{g}_{\text{rad}}, \tag{14}$$

with a modulus $g_{\text{rad}} = (\kappa L/4\pi c R^2)$, where R is the equatorial radius. Thus, the breakup velocity obtained when $\vec{g}_{\text{tot}} = 0$ is found to be $v_{\text{crit}}^2 = \frac{GM}{R}(1 - \Gamma)$ (Langer 1997, 1998; Langer & Heger 1998; Lamers 1997), where Γ is the Eddington factor $\Gamma = \kappa L/(4\pi c GM)$. For the most luminous stars, $\Gamma \to 1$, and thus the critical velocity tends toward zero. This has led Langer (1997, 1998) to conclude that for any initial rotation, the critical limit is reached before the Eddington limit. Therefore, Langer claims that we should speak of an Ω-limit for LBV stars, rather than a Γ-limit.

Glatzel (1998) suggested that the Ω-limit is an artifact resulting from the absence of von Zeipel's relation in the expression of g_{rad}. Indeed, with von Zeipel's relation the radiative flux tends toward zero when the resulting gravity is zero. Thus, the critical velocity is just $v_{\text{crit}}^2 = \frac{GM}{R}$, while rotation reduces (up to 40%; Glatzel 1998) the limiting luminosity. Stothers (1999) also considered that fast rotation reduces the limiting luminosity.

For stars close to the Eddington limit, convection may develop in the outer layers (Langer 1997, Glatzel & Kiriakidis 1998). This is not an objection to the application of the von Zeipel theorem, however, since most of the flux is carried by radiation at the surface. A possible objection though (Langer et al 1999a) is that, according to a generalization of the von Zeipel theorem by Kippenhahn (1977), the radiative flux at the equator may be reduced or increased depending on the internal rotation law. However, the deviations from von Zeipel's theorem are negligible in the current cases of models with shellular rotation (Maeder 1999a). Thus, a study of the physical conditions, the critical velocity, and the instabilities in rotating stars close to the Eddington limit is still necessary.

5.3.2 The Evolution of LBVs   According to the evolutionary models at high masses (Schaller et al 1992, Stothers & Chin 1996, Salasnich et al 1999), there are three possible ways for very massive stars to reach the Ω-limit in the HR diagram:

1. Stars with very large initial mass and high rotation, especially if their v_{rot} is increased by blueward evolution during the MS phase, may reach the Ω-limit in the blue part of the HR diagram. Some fast rotators may reach the breakup limit during the overall contraction phase at the end of the MS, as shown for the 60 $M_\odot$ (Figure 4). If the mass loss for O-stars is mainly bipolar (Maeder 1999a), the reduction of v_{rot} during the MS phase may be smaller. For smaller mass loss rates, as in lower-metallicity galaxies, v_{crit} could possibly be reached earlier in evolution. The star η Carinae shows evidence that the Ω-limit is reached in the blue, and it is likely at the end of its MS phase or beyond in view of its surface composition (Davidson et al 1986, Viotti et al 1989).

2. After the end of the MS phase, when the star evolves redward in the HR diagram, the value of $\frac{\Omega}{\Omega_{crit}}$ becomes quite small because the star evolves with essentially local conservation of angular momentum. Thus, rotation is less important. However, the Γ-limit without rotation lies at a much lower luminosity there (Lamers 1997, Ulmer & Fitzpatrick 1998), so the Ω-limit may be reached by the very massive stars during their redward crossing of the HR diagram.

3. When stars leave the red supergiant phase, either on blue loops or evolving toward the W-R stage, the ratio $\frac{\Omega}{\Omega_{crit}}$ increases significantly. This is caused by conservation of angular momentum in retreating convective envelopes, which contributes to accelerate the rotation of the blueward evolving stars (Heger & Langer 1998). Thus, the Ω-limit may also be reached from the red side. This possibility is particularly interesting because the observed CNO abundances in some nebulae around LBV stars are the same as in red supergiants, which suggests that some LBV may originate from red supergiants (Smith 1997). A similar conclusion was obtained by Waters (1997), who found evidence in some LBV nebulae of crystalline forms of silicates, with composition similar to that of red supergiants.

5.3.3 The Nebulae Around LBVs: Signature of Rotation?   Almost all nebulae around LBV stars show a bipolar structure (Nota et al 1995, Nota & Clampin 1997), which may be related to binarity (Damineli 1996, Damineli et al 1997). Most models invoke collisions of winds of different velocities and densities, emitted at different phases of their evolution. In some cases, an equatorial density enhancement is assumed before the outburst (Frank et al 1995; see also Nota et al 1995), whereas other models assume rather arbitrary nonspherical winds or a ring-like structure interacting with a previous spherical wind (Dwarkadas & Balick 1998, Frank et al 1998). The models by Garcia-Segura et al (1996, 1997) and Langer et al (1999) consider three phases in the formation of the nebula for η Carinae. In both the pre- and post-outburst phases, the star has the spherical fast and low density wind typical of a blue supergiant. At the breakup limit, the star is assumed to have a slow dense wind concentrated in the equatorial plane. The bipolar structure then arises because the shell ejected in the third phase expands more easily into the lower density at the pole. Langer et al (1999) assumed that the equatorial enhancement during the outburst results from the wind-compressed disk model (Bjorkman & Cassinelli 1993), which does not apply if the von Zeipel theorem is used (Owocki & Gayley 1997, 1998). Nevertheless, we note that because of the κ-effect at the breakup limit (Maeder 1999a), a strong equatorial ejection occurs quite naturally and is characterized by a high density and a low velocity, as is required by the colliding wind model of Langer et al (1999).

5.4 Rotation and W-R Star Formation

5.4.1 Generalities   Recent reviews on the Wolf-Rayet (W-R) phenomenon have been presented by Abbott & Conti (1987), van der Hucht (1992), Maeder & Conti (1994), and Willis (1999). Wolf-Rayet stars are bare cores of initially massive stars (Lamers et al 1991). Their original H-rich envelope has been removed by stellar winds or through a Roche lobe overflow in a close binary system. Observationally, most W-R stars appear to originate from stars initially more massive than about $40\,M_\odot$ (Conti et al 1983, Conti 1984, Humphreys et al 1985, Tutukov & Yungelson 1985); however, a few stars may originate from initial masses as low as 15–$25\,M_\odot$ (Thé et al 1982, Schild & Maeder 1984, Hamann et al 1993, Hamann & Koesterke 1998a, Massey & Johnson 1998). The stars enter the W-R phase as WN stars, i.e. with surface abundances representative of equilibrium CNO processed material. If the peeling off proceeds deep enough, the star may enter the WC phase, during which the He-burning products appear at the surface.

Many observed features are well reproduced by current stellar models. Typically, good agreement is obtained between the observed and predicted values for the surface abundances of WN stars (Crowther et al 1995, Hamann & Koesterke 1998a). This indicates the general correctness of our understanding of the CNO cycle and of the relevant nuclear data (Maeder 1983), but is not a test of the model structure. For WC stars, comparisons with observed surface abundances also generally show good agreement (Willis 1991, Maeder & Meynet 1994). In particular, the strong surface Ne enrichments predicted by the models of WC stars have been confirmed by ISO observations (Willis et al 1997, 1998; Morris et al 1999; Dessart et al 1999).

The star number ratios W-R/O, W-R/RSG, and WN/WC show a strong correlation with metallicity (Azzopardi et al 1988, Smith 1988, Maeder 1991, Maeder & Meynet 1994, Massey & Johnson 1998). For instance, the W-R/O number ratio increases with the metallicity Z of the parent galaxy. Despite many other claims (Bertelli & Chiosi 1981, 1982; Garmany et al 1982; Armandroff & Massey 1985; Massey 1985; Massey et al 1986), the main cause is metallicity Z, which through stellar winds influences stellar evolution and thus the W-R lifetimes (Smith 1973, Maeder et al 1980, Moffat & Shara 1983). The higher the metallicity, the stronger the mass loss by stellar winds, and thus the earlier the entry in the W-R phase for a given star; also, the minimum initial mass for forming a W-R star is lowered.

5.4.2 Remaining Problems with W-R Stars   Despite the successes discussed previously, observations indicate some remaining problems:

1. It is possible to reproduce the W-R/O and WN/WC number ratios observed in the Milky Way and in various galaxies of the Local Group, but only by using models with mass loss rates enhanced by a factor of 2 during the MS and WNL phases (Maeder & Meynet 1994). The relative populations of WN and WC stars observed in young starburst regions are also better reproduced when models with high mass loss rates are used (Meynet 1995, Schaerer et al 1999). This is not satisfactory, because clumping in the winds of hot stars tends to reduce the observed mass loss rates by a factor of 2 to 3 (Nugis et al 1998; Hamann & Koesterke 1998b).

2. The lower limit for the luminosities of WN stars (around $\log L/L_{\odot} \sim 5.0$; Hamann & Koesterke 1998a) is fainter than that predicted by standard evolutionary tracks. Massey & Johnson (1998) found that the presence of luminous red supergiants (RSG) and W-R stars is well correlated for the OB associations in M31 and M33, which suggests that some stars with mass ≥ 15 $M_{\odot}$ go through both the RSG and W-R phases.

3. For WN stars, there is a continuous transition from high H-surface abundances (0.4–0.5 in mass fraction) to hydrogen-free atmospheres, whereas standard models predict an abrupt transition (Langer et al 1994, Hamann & Koesterke 1998a; see Figure 7 below).

4. Smith & Maeder (1998) showed that, besides the mass, a second parameter affecting the mass loss rates and terminal velocities of the wind is necessary to characterize the hydrogen-free WN stars.

5. Standard models do not reproduce the observed number of stars in the transition WN/WC phase, characterized by spectra with both H- and He-burning products. These models indeed predict an abrupt transition from WN to WC stars, because the He core is growing and thus building up a steep chemical discontinuity at its outer edge (e.g. Schaller et al 1992). Thus, almost no (<1%) stars with intermediate characteristics of WN and WC stars are predicted. However, 4–5% of the W-R stars are in such a transition phase (Conti & Massey 1989, van der Hucht 1999), which shows that some extra-mixing is at work (Langer 1991b).

5.4.3 Rotation and Formation of W-R Stars   Rotation may affect the formation and properties of W-R stars in several ways (Sreenivasan & Wilson 1982, 1985a; Maeder 1987; Fliegner & Langer 1995; Maeder 1999b; Meynet 1999):

1. Surface abundances characteristic of the WNL stars may appear in a rotating star, not only as a result of the mass loss, which uncovers inner layers, but also as a result of mixing in radiative zones. The same remark applies to the entry into the WC phase.

2. Rotation may imply different evolutionary scenarios. Before becoming a W-R star, the nonrotating 60 $M_\odot$ model at solar metallicity is likely to go through a short LBV phase after the H exhaustion in its core. In the case of fast rotation, the star may enter the W-R phase while still burning hydrogen in its core (Maeder 1987, Fliegner & Langer 1995, Meynet 1999), thus skipping the LBV phase and spending more time in the W-R phase.

3. Rotation favors the formation of W-R stars from lower initial mass, through its effects on both the mass loss rates (Sreenivasan & Wilson 1982; Section 2.6) and the mixing. Typically, for the nonrotating models shown in Figure 5, the minimum mass for W-R star formation is between 35 and 40 $M_\odot$. It decreases to about 25 $M_\odot$ for initial $v_{rot} = 300$ km/s. This effect may help to explain the low luminous WN stars reported by Hamann & Koesterke (1998a). It also favors entry into the W-R phase from the RSG stage.

4. During the WN phase, the surface abundances are different. Indeed, as a consequence of the first point in this list, the N/C, N/O ratios obtained at the surface of the rotating WN models may not yet have reached the full nuclear equilibrium, in contrast with the nonrotating case where nuclear equilibrium is reached as soon as the star enters the WN phase. However, the CNO ratios are close to the equilibrium values (see Figure 7). During the transition WN/WC phase, nitrogen enhancements can be observed simultaneously with carbon and neon enhancements. After this transition phase, the ^{22}Ne enhancement reaches more or less the same high equilibrium level regardless of the initial angular velocity, which agrees with determinations of the neon abundance at the surface of WC stars (Willis et al 1997).

5. Higher rotational velocities lead to longer W-R lifetimes. As an example, for a 60 $M_\odot$ model (Figure 7), the W-R lifetime is increased by more than a factor of 3 when rotation is included. The durations of the WN and of the transition WN/WC phases are increased. The ratio of the lifetimes of the WC to the WN phase is reduced.

6. High rotations lead to less luminous WC stars. This is because a rapidly rotating star enters the W-R stage earlier in its evolution and thus begins to lose large amounts of mass early. Therefore, fast rotators enter the WC phase with a small mass and a low luminosity; the final masses are also smaller.

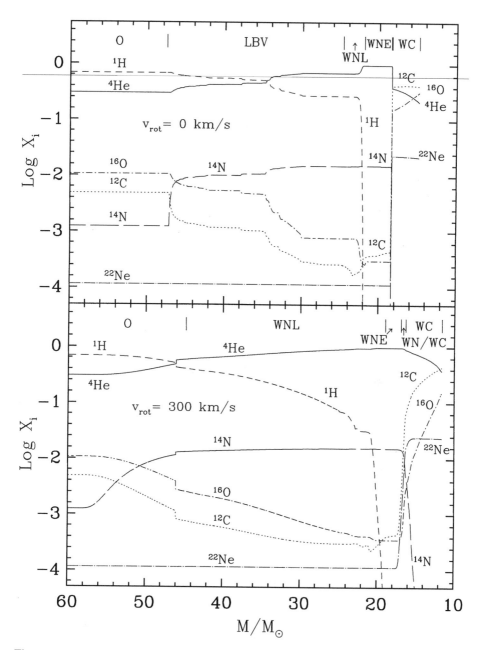

Figure 7   Evolution of the abundances at the surface of a 60 $M_\odot$ star as a function of the remaining stellar mass for different initial rotational velocities v_{rot}. The parts of the evolution during which the star may be considered as an O-type star, a LBV and a W–R star are indicated. During the W–R phase, the WN, the transition "WN/WC" and the WC phases are distinguished.

Rotation could thus remove or at least alleviate the problems mentioned above. The need to enhance the mass loss rates to reproduce the observed W-R/O number ratios no longer appears necessary. Rotation also implies effects that cannot be reproduced by an increase in the mass loss rate. In particular, mixing induced by rotation produces milder chemical gradients and leads to a more progressive decrease of the hydrogen abundance at the surface of WN stars (Figure 7).

5.4.4 Are W-R Stars Fast or Slow Rotators?   Direct attempts to measure the rotational velocity of W-R stars have been performed only for a few cases. Massey (1980) and Koenigsberger (1990) obtained $v \sin i \sim 500$ km/s for WR138; however, the binary nature of this object (Annuk 1991) blurs this picture, because the origin of this high velocity might be the O-type companion. The second case, WR3 with $v \sin i \sim 150$–200 km/s, looks more promising (Massey & Conti 1981) because the broadened absorption lines move in phase with the W-R emission lines (Moffat et al 1986).

Indirect evidence points toward the existence of some axisymmetric features around W-R stars (see Drissen et al 1992, Marchenko 1994). For instance, Arnal (1992) has mapped the environment of six W-R stars at a frequency of 1.42 MHz and found that all the HI cavities around these have an elongated shape with a mean major-to-minor axis ratio of about 2.2. Other evidence of anisotropies was found by Schulte-Ladbeck et al (1992) and Miller & Chu (1993). According to Harries et al (1998), about 15% of W-R stars have anisotropic winds. They suggest that the main cause of the wind anisotropy is equatorial density enhancements produced by fast rotation rates, and estimate the rotational velocities to be about 10–20% of the breakup velocity.

The surface velocities of W-R stars depend mainly on the initial velocity and on the amount of angular momentum lost during the previous stages. This amount will depend on the exact evolutionary sequence followed; in particular, the questions are whether the star has passed through the RSG stage and what were the anisotropies of the stellar winds. Some other effects may also intervene, for instance the possible presence of a magnetic field (e.g. Cassinelli 1992, Sreenivasan & Wilson 1982; Section 3.2).

For the 60 $M_\odot$ model shown in Figure 7 with $v_{rot} = 300$ km/s, computed assuming spherically symmetric winds and no magnetic fields, the surface velocity ranges between 20 and 40 km/s, i.e. between 3% and 6% of the breakup velocity during most of the WNL phase. At the beginning of the core He-burning phase, the He core contracts, the small H-rich envelope expands, and the surface velocity reaches the breakup limit. Huge mass loss rates ensue, ejecting about 3 $M_\odot$ of material and forming an anisotropic nebula with abundances characteristic of CNO equilibrium. When the star has lost sufficient angular momentum, it drops below the breakup limit and pursues its evolution with a nearly constant surface velocity around 40 km/s. The contraction of the W-R star at the very end of the He-burning phase may again increase v_{rot}, but it remains far below the breakup limit ($\Omega / \Omega_{crit} \sim 5\%$).

5.5   Late Stages, Remnants, and Chemical Yields

5.5.1 The Post–He-Burning Phases   The masses of the C/O cores are larger in rotating stars, which do not evolve through a W-R phase (Sofia et al 1994, Heger et al 2000). For example, at the end of the He-burning phase, the C/O core mass in a rotating 20 $M_\odot$ model with an initial $v_{rot} = 300$ km/s is 5.7 $M_\odot$ (Meynet & Maeder 2000). The value in a nonrotating model is 3.8 $M_\odot$. Thus, a rotating 20 $M_\odot$ star will have a behavior during the late stages similar to that of a nonrotating 25 $M_\odot$ star.

It is also interesting to notice that because of the larger He cores, the $^{12}C(\alpha, \gamma)^{16}O$ reaction is more active at the end of the He-burning phase; therefore, the fraction of carbon left in the C/O core decreases with respect to that in the nonrotating model (by a factor of about 2.5 in the preceding example). This leads to an increase in the oxygen yield. Moreover, since the carbon burning phase is considerably reduced, the stellar core has less time to remove its entropy through heavy neutrino losses, thus favoring the formation of black holes (Woosley 1986). According to Fryer (1999) the lower mass limit for black hole formation is likely lowered by rotation (see also Fryer & Heger 1999).

After central He exhaustion, the He-shell ignites and the layers above it expand, leading to a decrease in the strength of the H-burning shell. The smaller thermal gradient near the H-shell favors the mixing of chemical elements there (Heger et al 2000). Protons are brought from the H-shell down into the underlying He-rich layers where they will be engulfed by the convective He-burning shell. At the same time, because of the contraction of the C/O core, the temperature at the bottom of the He-shell increases and the overlying convective zone extends in mass, engulfing the region that is rotationally enriched in protons and nitrogen. These species then burn very rapidly, on a timescale shorter than the convective turnover time; thus, they have almost completely disappeared before reaching the bottom of the He-convective zone (Heger 1998). The large extension of the He-burning shell has important consequences for nucleosynthesis. This extension is greater for larger core masses, i.e. for large initial mass and/or for large rotational velocities.

When evolution proceeds further, the rotating core speeds up more and more, possibly becoming unstable with respect to nonaxisymmetric perturbations (Kippenhahn et al 1970, Ostriker & Bodenheimer 1973, Tassoul 1978). However, the results obtained by Heger et al (2000) for stars with initial v_{rot} between 200 and 300 km/s suggest that, before the core collapse, the ratio of rotational to potential energy is lower than that required for such instabilities to occur.

5.5.2 Rotational Periods of Pulsars   According to the models by Heger et al (2000), at their birth neutron stars (NS) should have rotation periods of about 0.6 ms, since they are nearly at the breakup rate. What is striking is that these periods are much smaller than the measured periods for young pulsars, which are around 20–150 ms (Marshall et al 1998). This means that the models have between $\sim$20 and 100 times more specific angular momentum than is found in the young neutron stars. Various effects may be responsible for this excess rotation in the models. The efficiency of some rotationally induced mixing processes may have been underestimated, or some important transport mechanism may still be missing. Kippenhahn et al (1970) speculated about the possibility for rapidly rotating dense cores to shed some mass into the envelope at its equator, in a way similar to rapidly rotating stars shedding mass into the circumstellar envelope. The equatorial mass loss by anisotropic stellar winds heavily modifies the surface boundary conditions and may remove a huge amount of angular momentum (Maeder 1999a). Other braking mechanisms, such as the removal of angular momentum from the convective core by gravity waves (Denissenkov et al 1999; Section 2.5) or through a magnetic field (Spruit & Phinney 1998), may also be invoked. As pointed out by Fricke & Kippenhahn (1972), the coupling of the core and envelope cannot be complete because, with solid body rotation at all times, the core would rotate too slowly ($P \simeq 650$ ms) to form pulsars with the observed periods.

The evacuation of the excess angular momentum could also have occurred during the formation of a neutron star. The NS could also have been born spinning at breakup velocity and could have been very efficiently slowed down during the first years. However, as discussed by Hardorp (1974), some arguments suggest that NS have never been near breakup (Rudermann 1972); the study of the Crab pulsar supports this view (Trimble & Rees 1970). Indeed, in this case, the release of such an important amount of rotational energy, if not emitted in the form of a γ-ray burst or through gravitational waves, would have shown up in the expansion energy of the nebula and in the optical light during historical times, which was not reported. The original rotation period of the Crab pulsar at birth is estimated to be 5 ms, still one order of magnitude from the breakup period (Hardorp 1974). Therefore, the stellar core must have been spinning slowly before its collapse.

If a large angular momentum is embarrassing at present when we explain the observed rotating periods of young pulsars, it may give some support to the collapsar model proposed by Woosley (1993; MacFadyen & Woosley 1999) for the γ-ray bursts. A collapsar is a black hole formed by the incomplete explosion of a rapidly rotating massive star. The rapid rotation is necessary to allow the formation of an accretion disk outside the black hole. The accretion disk efficiently transforms the gravitational binding energy into heat which can then power a highly relativistic jet. The burst and its afterglow in various wavelengths are attributed to the jet and its interactions with the external medium. The models by Heger et al (2000) and Meynet & Maeder (2000) have enough angular momentum to support matter in a stable disk outside a black hole, and thus could offer interesting progenitors for this kind of evolution, if it exists.

5.5.3 Chemical Yields   Rotation affects chemical yields in many ways. The larger He cores obtained in rotating models at core collapse imply larger production of helium and other α-nuclei elements (Heger 1998). This is by far the most important effect of rotation on chemical yields. Also, by enhancing the mass loss rates and by making the formation of W-R stars easier, rotation favors the enrichment of the interstellar medium by stellar winds (Maeder 1992). Indeed, the stronger the winds, the richer the ejecta in helium and carbon and the lower the ejecta in oxygen.

The rotational diffusion during the H-burning phase enriches the outer layers in CNO-processed elements (Maeder 1987, Fliegner & Langer 1995, Meynet 1998, Heger 1998). Some ^{14}N is extracted from the core and saved from further destruction. The same can be said for ^{17}O and ^{26}Al, a radioisotope with a half-life of 0.72 M. The mixing in the envelope of rotating stars also leads to faster depletion of the temperature-sensitive light isotopes, for instance lithium and boron (Fliegner et al 1996).

The presence of ^{26}Al in the interstellar medium is responsible for the diffuse galactic emission observed at 1.8 MeV (e.g. Oberlack et al 1996). If the nucleosynthetic sites of this element appear to be the massive stars (Prantzos & Diehl 1996), it is still not clear how the production is shared between the supernovae and the W-R stars, and how it is affected by rotation and binarity. For stellar masses between 12 and 15 $M_\odot$, the lifetimes are much longer than that of ^{26}Al, and therefore most of the ^{26}Al produced during central H-burning and partially mixed in the envelope has decayed at the time of the supernova explosion. Thus, for this mass range, rotation does not seem to effect important changes (Heger 1998). However, when the star is massive enough to go through the W-R phase, the stellar winds may remove ^{26}Al-enriched layers at a much earlier stage. In this case, rotation may substantially increase the quantity of ^{26}Al injected in the interstellar medium (Langer et al 1995).

The convective zone associated with the He-shell in rotating models transports H-burning products to the He-burning shell (Heger 1998). The injection of protons and nitrogen into a He-burning zone opens new channels of nucleosynthesis (Jorissen & Arnould 1989). In particular, this enhances the s-process and the formation of ^{14}C, ^{18}O, and ^{19}F. Because these elements are produced just before the core collapse, they can survive until the supernova explosion. The injection of protons into a He-burning zone may also be responsible for primary ^{14}N production through ^{12}C(p,γ)^{13}N(β^+)^{13}C(p,γ)^{14}N, but this primary nitrogen is rapidly destroyed to produce ^{18}O. As we discussed in Section 4.2, stars with an He-burning core and a thick and long-lived H-burning shell seem to be more favorable sites for primary ^{14}N production. Another important effect of the growth of the He-shell in these late stages is the production of some ^{15}N shortly before core collapse. In nonrotating models this element is destroyed, whereas in rotating models it is synthesized (Heger 1998). The very low ^{14}N/^{15}N ratios measured in star-forming regions of the LMC and in the core of the (post-) starburst galaxy NGC 4945 (Chin et al 1999) support an origin of ^{15}N in massive stars.

6. PERSPECTIVES

We hope to have shown that rotation is indispensable for a proper understanding and modeling of the evolution for the Upper MS stars.

Further progress requires more studies of the physical effects of rotation, in particular of the various instabilities that can produce mixing of the elements and transport of angular momentum, both in the early and advanced phases of evolution. The influence of rotation on mass loss is also critical. In this respect, the shape and composition of the asymmetric nebulae observed around many massive stars provide interesting constraints on the models of rotating stars.

ACKNOWLEDGMENTS

We express our thanks to Dr. Laura Fullton for her most useful advice on the manuscript.

Visit the Annual Reviews home page at www.AnnualReviews.org

LITERATURE CITED

Abbott DC, Conti PS. 1987. *Annu. Rev. Astron. Astrophys.* 25:113–50

Annuk K. 1991. In *Wolf-Rayet Stars and Interrelations with Other Massive Stars in Galaxies,* IAU Symp. 143, ed. KA van der Hucht, B Hidayat, pp. 245–50. Dordrecht: Kluwer

Armandroff TE, Massey P. 1985. *Ap. J.* 291:685–92

Arnal EM. 1992. *Astron. Astrophys.* 250:171–78

Arnett D. 1991. *Ap. J.* 383:295–307

Azzopardi M, Lequeux J, Maeder A. 1988. *Astron. Astrophys.* 189:34–38

Barbuy B, De Medeiros JR, Maeder A. 1996. *Astron. Astrophys.* 305:911–19

Bertelli G, Chiosi C. 1981. In *The Most Massive Stars,* ed. S d'Odorico, D Baade, K Kjär, pp. 211–13. ESO Workshop, ESO Publ, Garchingb. München

Bertelli G, Chiosi C. 1982. In *Wolf–Rayet Stars: Observations, Physics, Evolution,* IAU Symp. 99, ed. CWH de Loore, AJ Willis, pp. 359–63. Dordrecht: Reidel

Bjorkman JE, Cassinelli JP. 1993. *Ap. J.* 409:429–49

Bodenheimer P. 1995. *Annu. Rev. Astron. Astrophys.* 33:199–238

Boothroyd AI, Sackmann IJ. 1999. *Ap. J.* 510:232–50

Boyarchuk AA, Lyubimkov LS. 1983. *Izv. Krym. Astrofiz. Obs.* 66:130

Bressan A, Fagotto F, Bertelli G, Chiosi C. 1993. *Astron. Astrophys.* 100:647–64

Brocato E, Castellani V. 1993. *Ap. J.* 410:99–109

Brunish WM, Gallagher JS, Truran JW. 1986. *Astron. J.* 91:598–601

Burrows CJ, Krist J, Hester JJ, Sahai R, Trauger JT, et al. 1995. *Ap. J.* 452:680–84

Canuto VM. 1994. *Ap. J.* 428:729–52

Canuto VM. 1998. *Ap. J.* 508:767–79

Canuto VM, Dubovikov M. 1997. *Ap. J. Lett.* 484:161–63

Canuto VM, Goldman I, Mazzitelli I. 1996. *Ap. J.* 473:550–59

Canuto VM, Minotti FO, Schilling O. 1994. *Ap. J.* 425:303–25

Cassinelli JP. 1992. See Drissen et al 1992, pp. 134–44

Centurion M, Bonifacio P, Molaro P, Vladilo G. 1998. *Ap. J.* 509:620–32

Cervino M, Mas-Hesse JM. 1994. *Astron. Astrophys.* 284:749–63

Chaboyer B, Demarque P, Pinsonneault MH. 1995a. *Ap. J.* 441:865–75

Chaboyer B, Demarque P, Pinsonneault MH. 1995b. *Ap. J.* 441:876–85

Chaboyer B, Zahn JP. 1992. *Astron. Astrophys.* 253:173–77

Chandrasekhar S. 1961. In *Hydrodynamic and Hydromagnetic Stability*, Oxford: Clarendon Press, p. 491

Charbonnel C. 1994. *Astron. Astrophys.* 282:811–20

Charbonnel C, Brown JA, Wallerstein G. 1998. *Astron. Astrophys.* 332:204–14

Chin CW, Stothers RB. 1991. *Ap. J. Suppl.* 77:299–316

Chin Y, Henkel C, Langer N, Mauersberger R. 1999. *Ap. J.* 512:L143–46

Chiosi C, Maeder A. 1986. *Annu. Rev. Astron. Astrophys.* 24:329–75

Chiosi C, Nasi E, Sreenivasan SR. 1978. *Astron. Astrophys.* 63:103–24

Collins GW, Sonneborn GH. 1977. *Ap. J. Suppl.* 34:41–94

Conti PS. 1984. In *Observational Tests of the Stellar Evolution Theory*, IAU Symp. 105, ed. A Maeder, A Renzini, pp. 233–54. Dordrecht: Reidel

Conti PS, Ebbets D. 1977. *Ap. J.* 213:438–47

Conti PS, Garmany CD, de Loore C, Vanbeveren D. 1983. *Ap. J.* 274:302–12

Conti PS, Massey P. 1989. *Ap. J.* 337:251–71

Cowley AP, Dawson P, Hartwick FDA. 1979. *Publ. Astron. Soc. Pac.* 91:628–31

Cox JP, Giuli RT. 1968. *Stellar Structure.* New York: Gordon & Breach. p. 307

Crowther PA. 1997. See Wolf et al 1999, pp. 51–57

Crowther PA, Hillier DJ, Smith LJ. 1995. *Astron. Astrophys.* 293:403–26

Damineli A. 1996. *Ap. J. Lett.* 460:49–52

Damineli A, Conti PS, Lopes DF. 1997. *New Astron.* 2:107–17

Davidson K, Dufour RJ, Walborn NR, Gull TR. 1986. *Ap. J.* 305:867–79

Davidson K, Ebbets D, Johansson S, Morse JA, Hamann FW. 1997. *Astron. J.* 113:335–45

Davidson K, Humphreys RM. 1997. *Annu. Rev. Astron. Astrophys.* 35:1–32

Davidson K, Moffat AFJ, Lamers HJGLM. 1989. In *Physics of Luminous Blue Variables*, IAU Coll. 113, ed. A. Nota, HJGLM Lamers, Dordrecht: Kluwer. 404 pp.

Dearborn DSP. 1992. *Phys. Rep.* 210:367

de Jager C, Nieuwenhuijzen H. 1992. In *Instabilities in Evolved Super- and Hypergiants*, ed. C. de Jager, H. Nieuwenhuijzen. Amsterdam: North-Holland. 199 pp.

de Jager C, Nieuwenhuijzen H, van der Hucht KA. 1988. *Astron. Astrophys. Suppl.* 72:259–89

Denissenkov PA. 1994. *Astron. Astrophys.* 287:113–30

Denissenkov PA, Ivanova NS, Weiss A. 1999. *Astron. Astrophys.* 341:181–89

Dessart L, Willis AJ, Crowther PA, Morris PW, Hillier DJ. 1999. See van der Hucht et al 1999, pp. 233–34

Deupree RG. 1995. *Ap. J.* 439:357–64

Deupree RG. 1998. *Ap. J.* 499:340–47

Dominguez I, Straniero O, Tornambé A. 1996. *Ap. J.* 472:783–88

Drissen L, Leitherer C, Nota A. 1992. *Nonisotropic and Variable Outflows from Stars, ASP Conf. Ser.* San Francisco: Astron. Soc. Pacific 22:408

Dwarkadas VV, Balick B. 1998. *Astron. J.* 116:829–39

Eddington AS. 1925. *The Observatory* 48:73–75

Edmunds MG, Pagel BEJ. 1978. *MNRAS* 185:77–80

El Eid MF. 1994. *Astron. Astrophys.* 285:915–28

Endal AS, Sofia S. 1976. *Ap. J.* 210:184–98

Endal AS, Sofia S. 1978. *Ap. J.* 220:279–90

Endal AS, Sofia S. 1979. *Ap. J.* 232:531–40

Eriguchi Y, Yamaoka H, Nomoto K, Hashimoto M. 1992. *Ap. J.* 392:243–48

Eryurt D, Kirbiyik H, Kiziloglu N, Civelek R,

Weiss A. 1994. *Astron. Astrophys.* 282:485–92

Fagotto F, Bressan A, Bertelli G, Chiosi C. 1994. *Astron. Astrophys. Suppl. Ser.* 105:39–45

Faulkner J, Roxburgh IW, Strittmatter PA. 1968. *Ap. J.* 151:203–16

Fliegner J, Langer N. 1995. In *Wolf-Rayet Stars: Binaries, Colliding Winds, Evolution,* IAU Symp. 163, ed. KA van der Hucht, PM Williams, pp. 326–28. Dordrecht: Kluwer

Fliegner J, Langer N, Venn KA. 1996. *Astron. Astrophys.* 308:L13–16

Frank A, Balick B, Davidson K. 1995. *Ap. J.* 441:L77–80

Frank A, Ryu D, Davidson K. 1998. *Ap. J.* 500:291–301

Fransson C, Cassatella A, Gilmozzi R, Kirshner RP, Panagia N, Sonneborn G, Wamsteker W. 1989. *Ap. J.* 336:429–41

Fricke KJ. 1968. *Zeitschr. Ap.* 68:317–44

Fricke KJ, Kippenhahn R. 1972. *Annu. Rev. Astron. Astrophys.* 10:45–72

Friend DB, Abbott DC. 1986. *Ap. J.* 311:701–7

Fryer CL. 1999. *Ap. J.* 522:413–18

Fryer CL, Heger A. 1999. astro-ph /9907433

Fukuda I. 1982. *Publ. Astron. Soc. Pac.* 92:271–84

Garcia-Segura G, Langer N, Mac Low MM. 1997. See Nota & Lamers 1997, pp. 121–27.

Garcia–Segura G, Langer N, Rózyczka M, Franco J. 1999. *Ap. J.* 517:767–81

Garcia–Segura G, Mac Low MM, Langer N. 1996. *Astron. Astrophys.* 305:229–44

Garmany CD, Conti PS, Chiosi C. 1982. *Ap. J.* 263:777–90

Giannone P, Kohl K, Weigert A. 1968. *Z. Astrophys.* 68:107–29

Gies DR, Lambert DL. 1992. *Ap. J.* 387:673–700

Gilroy KK. 1989. *Ap. J.* 347:835–48

Gilroy KK, Brown JA. 1991. *Ap. J.* 371:578–83

Glatzel W. 1998. *Astron. Astrophys.* 339:L5–8

Glatzel W, Kiriakidis M. 1998. *MNRAS* 295:251–56

Goldreich P, Schubert G. 1967. *Ap. J.* 150:571–87

Groenewegen MAT, Lamers HJGLM, Pauldrach A. 1989. *Astron. Astrophys.* 221:78–88

Guilloteau S, Dutrey A. 1998. *Astron. Astrophys.* 339:467–76

Hamann WR, Koesterke L. 1998a. *Astron. Astrophys.* 333:251–63

Hamann WR, Koesterke L. 1998b. *Astron. Astrophys.* 335:1003–8

Hamann WR, Koesterke L, Wessolowski U. 1993. *Astron. Astrophys.* 274:397–414

Hardorp J. 1974. *Astron. Astrophys.* 32:133–36

Harries TJ, Hillier DJ, Howarth ID. 1998. *MNRAS* 296:1072–88

Harris MJ, Lambert DL, Smith VV. 1988. *Ap. J.* 325:768–75

Hartmann LW, Noyes RW. 1987. *Annu. Rev. Astron. Astrophys.* 25:271–01

Heger A. 1998. PhD thesis. Max-Planck-Institut für Astrophysik. *MPA* 1120

Heger A, Langer N. 1998. *Astron. Astrophys.* 334:210–20

Heger A, Langer N, Woosley SE. 2000. *Ap. J.* 528:368–396

Henry RBC, Worthey G. 1999. *Publ. Astron. Soc. Pac.* 111:919–45

Herrero A, Corral LJ, Villamariz MR, Martin EL. 1999. *Astron. Astrophys.* 348:542–52

Herrero A, Kudritzki RP, Vilchez JM, Kunze D, Butler K, Haser S. 1992. *Astron. Astrophys.* 261:209–34

Herrero A, Villamariz MR, Martin EL. 1998. See Howarth 1998, pp. 159–68

Howarth ID. 1998. *Boulder–Munich II: Properties of Hot, Luminous Stars,* ASP Conf. Ser. San Francisco, 131:456. Astron. Soc. Pacific

Howarth ID, Siebert KW, Hussain GAJ, Prinja RK. 1997. *MNRAS* 284:265–85

Humphreys RM. 1989. In *Physics of Luminous Blue Variables,* ed. K Davidson, AFJ Moffat, HJGLM Lamers, pp 1–14. Dordrecht: Kluwer

Humphreys RM, McElroy DB. 1984. *Ap. J.* 284:565–77

Humphreys RM, Nichols M, Massey P. 1985. *Astron. J.* 90:101–8

Iben I. 1999. See Le Bertre et al 1999, p. 591

Jeans JH. 1928. In *Astronomy and Cosmogony,* p. 273. Cambridge, UK: Cambridge Univ. Press

Jorissen A, Arnould M. 1989. *Astron. Astrophys.* 221:161–79

Kato S. 1966. *Publ. Astron. Soc. Pac.* 18:374–83

Kendall TR, Dufton PL, Lennon DJ. 1996. *Astron. Astrophys.* 310:564–76

Kendall TR, Lennon DJ, Brown PJF, Dufton PL. 1995. *Astron. Astrophys.* 298:489–504

Kilian J. 1992. *Astron. Astrophys.* 262:171–87

Kippenhahn R. 1974. In *Late Stages of Stellar Evolution,* IAU Symp. 66, ed. RJ Tayler, JE Hesser, pp. 20–40. Dordrecht: Reidel

Kippenhahn R. 1977. *Astron. Astrophys.* 58:267–71

Kippenhahn R, Meyer-Hofmeister E, Thomas HC. 1970. *Astron. Astrophys.* 5:155–61

Kippenhahn R, Thomas HC. 1970. See Slettebak 1970, pp. 20–29

Kippenhahn R, Weigert A. 1990. In *Stellar Structure and Evolution.* Berlin: Springer Verlag. 468 pp.

Knobloch E, Spruit HC. 1983. *Astron. Astrophys.* 125:59–68

Koenigsberger G. 1990. *Rev. Mex. Astron. Astrofis.* 20:85–112

Korycansky DG. 1991. *Ap. J.* 381:515–25

Kraft RP. 1966. *Ap. J.* 144:1008–15

Kraft RP, Camp DC, Fernie JD, Fujita C, Hughes WT. 1959. *Ap. J.* 129:50–8

Kudritzki RP, Hummer DG, Pauldrach A, Puls J, Najarro F, Imhoff J. 1992. *Astron. Astrophys.* 257:655–62

Kudritzki RP, Pauldrach A, Puls J, Abbott DC. 1989. *Astron. Astrophys.* 219:205–18

Kumar P, Narayan R, Loeb A. 1995. *Ap. J.* 453:480–94

Kumar P, Quataert EJ. 1997. *Ap. J. Lett.* 475:43–46

Lamers HJGLM. 1997. See Nota & Lamers 1997, pp. 79–82.

Lamers HJGLM. 1999. See Wolf et al 1999, pp. 159–68

Lamers HJGLM, Cassinelli JP. 1996. See Leitherer et al 1996, pp. 162–73

Lamers, HJGLM, Leitherer C. 1993. *Ap. J.* 412:471–91

Lamers HJGLM, Maeder A, Schmutz W, Cassinelli JP. 1991. *Ap. J.* 368:538–44

Langer N. 1991a. *Astron. Astrophys.* 243:155–59

Langer N. 1991b. *Astron. Astrophys.* 248:531–37

Langer N. 1991c. *Astron. Astrophys.* 252:669–88

Langer N. 1992. *Astron. Astrophys.* 265:L17–20

Langer N. 1997. See Nota & Lamers 1997, pp. 83–89

Langer N. 1998. *Astron. Astrophys.* 329:551–58

Langer N. 1999. See Wolf et al 1999, pp. 359–67

Langer N, Braun H, Fliegner J. 1995. *Astrophys. Space Sci.* 224:275–78

Langer N, Fricke KJ, Sugimoto D. 1983. *Astron. Astrophys.* 126:207–8

Langer N, Garcia-Segura G, Mac Low MM. 1999a. *Ap. J. Lett.* 520:49–53

Langer N, Hamann WR, Lennon M, Najarro F, Pauldrach AWA, Puls J. 1994. *Astron. Astrophys.* 290:819–33

Langer N, Heger A. 1998. See Howarth 1998, pp. 76–84

Langer N, Heger A, Wellstein S, Herwig F. 1999b. *Astron. Astrophys.* 346:L37–40

Langer N, Maeder A. 1995. *Astron. Astrophys.* 295:685–92

Lanz T, de Koter A, Hubeny I, Heap SR. 1996. *Ap. J.* 465:359–62

Le Bertre T, Lèbre A, Waelkens C. 1999. *Asymptotic Giant Branch Stars, IAU Symp. 191, Publ. Astron. Soc. Pac* 632 pp.

Ledoux P. 1958. Stellar Stability, in *Handbuch der Physik, 51: Astrophysik II: Sternaufbau.,* ed. S. Flügge. pp. 605–688. Berlin: Springer Verlag. 830 pp.

Leitherer C, Fritze-von Alvensleben U, Huchra T. 1996. *From Stars to Galaxies: The Impact of Stellar Physics on Galaxy Evolution, ASP Conf. Ser. 98,* pp. 624: Astron. Soc Pacific, San Francisco

Lennon DJ, Becker ST, Butler K, Eber F, Groth HG, et al. 1991. *Astron. Astrophys.* 252:498–507

Lloyd HM, O'Brien TJ, Kahn FD. 1995. *MNRAS* 273:L19–23

Luck RE. 1978. *Ap. J.* 219:148–64

Lundqvist P, Fransson C. 1996. *Ap. J.* 464:924–42

Lyubimkov LS. 1991. See Michaud & Tutukov 1991, pp. 125–35

Lyubimkov LS. 1996. *Astroph. Space Sci.* 243:329–49

Lyubimkov LS. 1998. *Astron. Rep.* 42:53–9

MacFadyen AI, Woosley SE. 1999. *Ap. J.* 524:262–289

MacGregor KB, Friend DB, Gilliland RL. 1992. *Astron. Astrophys.* 256:141–47

Maeder A. 1971. *Astron. Astrophys.* 10:354–61

Maeder A. 1974. *Astron. Astrophys.* 34:409–14

Maeder A. 1981. *Astron. Astrophys.* 102:401–10

Maeder A. 1983. *Astron. Astrophys.* 120:113–35

Maeder A. 1987. *Astron. Astrophys.* 178:159–69

Maeder A. 1991. *Astron. Astrophys.* 242:93–111

Maeder A. 1992. *Astron. Astrophys.* 264:105–20

Maeder A. 1995. *Astron. Astrophys.* 299:84–88

Maeder A. 1997. *Astron. Astrophys.* 321:134–44

Maeder A. 1998. See Howarth 1998, pp. 85–95

Maeder A. 1999a. *Astron. Astrophys.* 347:185–93

Maeder A. 1999b. See van der Hucht et al 1999, pp. 177–86

Maeder A, Conti P. 1994. *Annu. Rev. Astron. Astrophys.* 32:227–75

Maeder A, Grebel E, Mermilliod JC. 1999. *Astron. Astrophys.* 346:459–64

Maeder A, Lequeux J, Azzopardi M. 1980. *Astron. Astrophys.* 90:L17–20

Maeder A, Meynet G. 1989. *Astron. Astrophys.* 210:155–73

Maeder A, Meynet G. 1994. *Astron. Astrophys.* 287:803–16

Maeder A, Meynet G. 1996. *Astron. Astrophys.* 313:140–44

Maeder A, Peytremann E. 1970. *Astron. Astrophys.* 7:120–32

Maeder A, Zahn JP. 1998. *Astron. Astrophys.* 334:1000–6

Marchenko SV. 1994. *Astrophys. Space Sci.* 221:169–80

Marshall FE, Gotthelf EV, Zhang W, Middleditch J, Wang QD. 1998. *Ap. J.* 499:L179–82

Marston AP. 1997. *Ap. J.* 475:188–93

Martin CL, Arnett D. 1995. *Ap. J.* 447:378–90

Massey P. 1980. *Ap. J.* 236:526–35

Massey P. 1985. *Publ. Astron. Soc. Pac.* 97:5–24

Massey P, Armandroff TE, Conti PS. 1986. *Astron. J.* 92:1303–33

Massey P, Conti PS. 1981. *Ap. J.* 244:173–78

Massey P, Johnson O. 1998. *Ap. J.* 505:793–827

Mathys G. 1999. See Wolf et al 1999, pp. 95–102

Matteucci F. 1986. *MNRAS* 221:911–21

McEarlean ND. 1998. See Howarth 1998, pp. 148–52

McEarlean ND, Lennon DJ, Dufton PL. 1999. *Astron. Astrophys.* 349:553–72

Meaburn J, Bryce M, Holloway AJ. 1995. *Astron. Astrophys.* 299:L1–4

Mestel L. 1953. *MNRAS* 113:716–45

Mestel L. 1965. In *Stellar Structure,* ed. LH Aller, DB McLaughlin, pp. 465–97. Chicago: Univ. Chicago Press

Meylan G, Maeder A. 1982. *Astron. Astrophys.* 108:148–56

Meynet G. 1995. *Astron. Astrophys.* 298:767–83

Meynet G. 1998. See Howarth 1998, pp. 96–107

Meynet G. 1999. See Wolf et al 1999, pp. 377–80

Meynet G. 2000. In *Spectrophotometric Dating of Stars and Galaxies,* ASP Conf. Ser., ed. I Hubeny, S Heap, R Cornett. In press

Meynet G, Maeder A. 1997. *Astron. Astrophys.* 321:465–76

Meynet G, Maeder A. 2000. *Astron. Astrophys.* In press

Michaud G, Tutukov AV. 1991. *Evolution of Stars: The Photospheric Abundance Connection, IAU Symp. 145,* Dordrecht: Kluwer Acad. Publ., pp. 487

Miller GJ, Chu YH. 1993. *Ap. J. Suppl.* 85:137–43

Milne EA. 1923. *MNRAS* 83:118–47

Moffat AFJ, Lamontagne R, Shara MM, McAlister HA. 1986. *Astron. J.* 91:1392–99

Moffat AFJ, Shara MM. 1983. *Ap. J.* 273:544–61

Montalban J. 1994. *Astron. Astrophys.* 281:421–32

Montalban J, Schatzman E. 1996. *Astron. Astrophys.* 305:513–18

Morris PW, van der Hucht KA, Willis AJ, Dessart L, Crowther PA, Williams PM. 1999. See van der Hucht et al 1999, pp. 77–79

Nieuwenhuijzen H, de Jager C. 1988. *Astron. Astrophys.* 203:355–60

Nota A, Clampin M. 1997. See Nota & Lamers 1997, pp. 303–9

Nota A, Lamers HJGLM. 1997. *Luminous Blue Variables: Massive Stars in Transition.* ASP Conf. Ser. 120. San Francisco: Astron. Soc. of Pacific, 404 pp.

Nota A, Livio M, Clampin M, Schulte-Ladbeck R. 1995. *Ap. J.* 448:788–96

Nota A, Smith L, Pasquali A, Clampin M, Stroud M. 1997. *Ap. J.* 486:338–54

Nugis T, Crowther PA, Willis AJ. 1998. *Astron. Astrophys.* 333:956–69

Oberlack U, Bennett K, Bloemen H, Diehl R,

Dupraz C, et al. 1996. *Astron. Astrophys. Suppl.* 120:311–14

Oke JB, Greenstein JL. 1954. *Ap. J.* 120:384–90

Origlia L, Goldader JD, Leitherer C, Schaerer D, Oliva E. 1999. *Ap. J.* 514:96–108

Ostriker JP, Bodenheimer P. 1973. *Ap. J.* 180:171–80

Owocki SP, Cranmer SR, Gayley KG. 1996. *Ap. J. Lett.* 472:115–18

Owocki SP, Gayley KG. 1997. See Nota & Lamers 1997, pp. 121–27

Owocki SP, Gayley KG. 1998. See Howarth 1998, pp. 237–44

Packet W, Vanbeveren D, De Greve JP, de Loore C, Sreenivasan R. 1980. *Astron. Astrophys.* 82:73–78

Panagia N, Scuderi S, Gilmozzi R, Challis PM, Garnavich PM, Kirshner RP. 1996. *Ap. J.* 459:L17–21

Pauldrach A, Puls J, Kudritzki RP. 1986. *Astron. Astrophys.* 164:86–100

Penny LR. 1996. *Ap. J.* 463:737–46

Penny LR, Gies DR, Bagnuolo WG. 1999. See van der Hucht et al 1999, pp. 86–89

Pettini M, Lipman K, Hunstead RW. 1995. *Ap. J.* 451:100–10

Pilyugin LS. 1999. *Astron. Astrophys.* 346:428–31

Pinsonneault M. 1997. *Annu. Rev. Astron. Astrophys.* 35:557–605

Pinsonneault MH, Delyannis CP, Demarque P. 1991. *Ap. J.* 367:239–52

Pinsonneault MH, Kawaler SD, Demarque P. 1990. *Ap. J. Suppl.* 74:501–50

Pinsonneault MH, Kawaler SD, Sofia S, Demarque P. 1989. *Ap. J. Suppl.* 338:424–52

Poe CH, Friend DB. 1986. *Ap. J.* 311:317–25

Prantzos N, Coc A, Thibaud JP. 1991. *Ap. J.* 379:729–33

Prantzos N, Diehl R. 1996. *Phys. Rep.* 267:1–69

Randers G. 1941. *Ap. J.* 94:109–23

Ringot O. 1998. *Astron. Astrophys.* 335:L89–92

Rosendhal JD. 1970. *Ap. J.* 159:107–18

Rudermann M. 1972. *Annu. Rev. Astron. Astrophys.* 10:427–76

Sackmann IJ, Anand SPS. 1970. *Ap. J.* 162:105–12

Sackmann IJ, Weidemann V. 1972. *Ap. J.* 178:427–32

Salasnich B, Bressan A, Chiosi C. 1999. *Astron. Astrophys.* 342:131–52

Sandage AR. 1955. *Ap. J.* 122:263–70

Sandage AR, Schwarzschild M. 1952. *Ap. J.* 116:463–76

Sasselov DD. 1986. *Publ. Astron. Soc. Pac.* 98:561–71

Schaerer D, Contini T, Kunth D. 1999. *Astron. Astrophys.* 341:399–417

Schaifers K, Voigt HH. 1982. *Astronomy and Astrophysics–stars and star clusters* Landolt–Bornstein New Series. Berlin: Springer Verlag 2b, 456 pp.

Schaller G, Schaerer D, Meynet G, Maeder A. 1992. *Astron. Astrophys. Suppl. Ser.* 96:269–331

Schatzman E. 1993. *Astron. Astrophys.* 279:431–46

Schatzman E, Maeder A, Angrand F, Glowinsky R. 1981. *Astron. Astrophys.* 96:1–16

Schild H, Maeder A. 1984. *Astron. Astrophys.* 136:237–42

Schmidt-Kaler Th. 1982. See Schaifers & Voigt, p. 1

Schönberner D, Herrero A, Becker S, Eber F, Butler K, et al. 1988. *Astron. Astrophys.* 197:209–22

Schulte-Ladbeck RE, Meade MR, Hillier DJ. 1992. See Drissen et al 1992, pp. 118–29

Schwarzschild M. 1958. In *Structure and Evolution of the Stars,* Princeton: Princeton Univ. Press, p. 183

Shindo M, Hashimoto M, Eriguchi Y, Müller E. 1997. *Astron. Astrophys.* 326:177–86

Slettebak A. 1966. *Ap. J.* 145:126–29

Slettebak A. 1970. In *Stellar Rotation,* IAU Coll. 4, ed. A Slettebak, pp. 3–8. Dordrecht: Reidel

Smith KC, Howarth ID. 1994. *Astron. Astrophys.* 290:868–74

Smith LF. 1973. In *W-R and High-Temperature Stars,* IAU Symp. 49, ed. MKV Bappu, J Sahade, pp. 15–41. Dordrecht: Reidel

Smith LJ. 1988. *Ap. J.* 327:128–38

Smith LJ. 1997. See Nota & Lamers 1997, pp. 310–15

Smith LJ, Maeder A. 1998. *Astron. Astrophys.* 334:845–56

Sofia S, Howard JM, Demarque P. 1994. In *Pulsation; Rotation; and Mass Loss in Early-Type Stars,* IAU Symp. 162, ed. LA Balona, et al, pp. 131–44. Dordrecht: Kluwer

Spiegel EA, Zahn JP. 1992. *Astron. Astrophys.* 265:106–114

Spruit HC, Knobloch E. 1984. *Astron. Astrophys.* 132:89–96

Spruit HC, Phinney ES. 1998. *Nature* 393:139–41

Sreenivasan SR, Wilson WJF. 1982. *Ap. J.* 254:287–96

Sreenivasan SR, Wilson WJF. 1985a. *Ap. J.* 290:653–59

Sreenivasan SR, Wilson WJF. 1985b. *Ap. J.* 292:506–10

Stothers RB. 1999. *Ap. J.* 513:460–63

Stothers RB, Chin CW. 1973. *Ap. J.* 179:555–68

Stothers RB, Chin CW. 1975. *Ap. J.* 198:407–17

Stothers RB, Chin CW. 1979. *Ap. J.* 233:267–79

Stothers RB, Chin CW. 1992a. *Ap. J.* 390:136–43

Stothers RB, Chin CW. 1992b. *Ap. J.* 390:L33–35

Stothers RB, Chin CW. 1993. *Ap. J.* 408:L85–88

Stothers RB, Chin CW. 1996. *Ap. J.* 468:842–50

Stothers RB, Chin CW. 1997. *Ap. J.* 489:319–30

Stothers RB, Chin CW. 1999. *Ap. J.* 522:960–64

Strittmatter PA. 1969. *Annu. Rev. Astron. Astrophys.* 7:665–84

Sweet PA. 1950. *MNRAS* 110:548–58

Takeda Y, Takada-Hidai M. 1995. *Publ. Astron. Soc. Jpn.* 47:169–88

Talon S, Zahn JP. 1997. *Astron. Astrophys.* 317:749–51

Talon S, Zahn JP. 1998. *Astron. Astrophys.* 329:315–18

Talon S, Zahn JP, Maeder A, Meynet G. 1997. *Astron. Astrophys.* 322:209–17

Tassoul JL. 1978. In *Theory of Rotating Stars,* Princeton: Princeton Univ. Press

Tassoul JL. 1990. In *Angular Momentum and Mass Loss for Hot Stars,* ed. LA Willson, R Stalio, pp. 7–32. Dordrecht: Kluwer

Tassoul JL, Tassoul M. 1982. *Ap. J. Suppl.* 49:317–50

Tassoul JL, Tassoul M. 1995. *Ap. J.* 440:789–809

Thé PS, Arens M, van der Hucht KA. 1982. *Astrophys. Lett.* 22:109–18

Thuan TX, Izotov YJ, Lipovetsky VA. 1995. *Ap. J.* 445:108–23

Toomre L. 1994. *AAS* 184.5002

Trimble V, Rees MJ. 1970. *Astrophys. Lett.* 5: 93

Tutukov AV, Yungelson LR. 1985. *Sov. Astron.* 29:352

Ulmer A, Fitzpatrick EL. 1998. *Ap. J.* 504:200–6

Urpin VA, Shalybkov DA, Spruit HC. 1996. *Astron. Astrophys.* 306:455–463

van der Hucht KA. 1992. *Astron. Astrophys. Rev.* 4:123–59

van der Hucht KA. 1999. See van der Hucht et al 1999, pp. 13–20

van der Hucht KA, Koenigsberger G, Eenens PRJ. 1999. *Wolf-Rayet Phenomena in Massive Stars and Starburst Galaxies,* IAU Symp. 193. San Francisco: Astron. Soc. Pacific, 788 pp.

Vauclair, S. 1999. *Astron. Astrophys.* 351:973–80

Vardya MS. 1985. *Ap. J.* 299:255–64

Venn KA. 1995a. *Ap. J.* 449:839–62

Venn KA. 1995b. *Ap. J. Suppl.* 99:659–92

Venn KA. 1998. See Howarth 1998, pp. 177–87

Venn KA. 1999. *Ap. J.* 518:405–21

Venn KA, Lambert DL, Lemke M. 1996. *Astron. Astrophys.* 307:849–59

Verdugo E, Talavera A, Gómez de Castro AI. 1999. *Astron. Astrophys.* 346:819–30

Villamariz MR, Herrero A. 2000. *Astron. Astrophys.* In press

Viotti R, Rossi L, Cassatella A, Altmore A, Baratta GB. 1989. *Ap. J. Suppl.* 71:983–1009

Voels SA, Bohannan B, Abbott DC, Hummer DG. 1989. *Ap. J.* 340:1073–190

Vogt H. 1925. *Astron. Nachr.* 223:229

von Zeipel H. 1924. *MNRAS* 84:665–701

Walborn NR. 1988. In *Atmospheric Diagnostics of Stellar Evolution,* IAU Coll. 108, ed. K Nomoto, pp. 70–78. New York: Springer

Waters LBFM. 1997. See Nota & Lamers 1997, pp. 326–31

Weiss A. 1994. *Astron. Astrophys.* 284:138–44

Weiss A, Hillebrandt W, Truran JW. 1988. *Astron. Astrophys.* 197:L11–14

Willis AJ. 1991. See Michaud & Tutukov 1991, pp. 195–207

Willis AJ. 1999. See van der Hucht et al 1999, pp. 1–12

Willis AJ, Dessart L, Crowther PA, Morris PW, Maeder A, Conti PS, van der Hucht KA. 1997. *MNRAS* 290:371–79

Willis AJ, Dessart L, Crowther PA, Morris PW, van der Hucht KA. 1998. *Astrophys. and Space Sc.* 255:167–8

Wolf G, Stahl O, Fullerton AW. 1999. *Variable and Non-Spherical Winds in Luminous Hot Stars,* IAU Coll. 169, Lecture Notes in Physics 523. Heidelberg: Springer Verlag 424 pp.

Woosley SE. 1986. In *Nucleosynthesis and Chemical Evolution,* 16th Saas-Fee Course, ed. A Maeder, B Hauck, G Meynet, pp. 1–195. Geneva: Geneva Observatory

Woosley SE. 1993. *Ap. J.* 405:273–77

Woosley SE, Heger A, Weaver TA, Langer N. 2000. In *SN 1987A: Ten Years After,* ed. MM Phillips, NB Suntzeff, Fifth CTIO/ESO/LCO Workshop. In press

Zahn JP. 1974. In *Stellar Instability and Evolution,* IAU Symp. 59, ed. P Ledoux, et al, pp. 185–195. Dordrecht: Reidel

Zahn JP. 1983. In *Astrophysical Processes in*

Upper Main Sequence Stars, ed. B Hauck, A Maeder, Saas Fee Course, pp. 253–329. Geneva: Geneva Observatory

Zahn JP. 1992. *Astron. Astrophys.* 265:115–32

Zahn JP. 1994. *Space Sci. Rev.* 66:285–97

Zahn JP, Talon S, Matias J. 1997. *Astron. Astrophys.* 332:320–28

Zickgraf FJ. 1999. See Wolf et al 1999, pp. 40–48

Annu. Rev. Astron. Astrophys. 2000. 38:191–230

TYPE IA SUPERNOVA EXPLOSION MODELS

Wolfgang Hillebrandt[1] and Jens C. Niemeyer[2]

[1]*Max-Planck-Institut für Astrophysik, Karl-Schwarzschild-Str. 1, 85740 Garching, Germany; e-mail: wfh@MPA-Garching.mpg.de*
[2]*University of Chicago, Enrico-Fermi Institute, 5640 South Ellis Ave., Chicago, Illinois 60637; e-mail: j-niemeyer@uchicago.edu*

Key Words   stellar evolution, hydrodynamics

■ **Abstract**   Because calibrated light curves of type Ia supernovae have become a major tool to determine the local expansion rate of the universe and also its geometrical structure, considerable attention has been given to models of these events over the past couple of years. There are good reasons to believe that perhaps most type Ia supernovae are the explosions of white dwarfs that have approached the Chandrasekhar mass, $M_{chan} \approx 1.39\,M_\odot$, and are disrupted by thermonuclear fusion of carbon and oxygen. However, the mechanism whereby such accreting carbon-oxygen white dwarfs explode continues to be uncertain. Recent progress in modeling type Ia supernovae as well as several of the still open questions are addressed in this review. Although the main emphasis is on studies of the explosion mechanism itself and on the related physical processes, including the physics of turbulent nuclear combustion in degenerate stars, we also discuss observational constraints.

1.   INTRODUCTION

Changes in the appearance of the night sky, visible with the naked eye, have always called for explanations (and speculations). But, although "new stars," i.e. novae and supernovae, have been observed by humans for thousands of years, the modern era of supernova research began only about a century ago, on August 31, 1885, when Hartwig discovered a "nova" near the center of the Andromeda galaxy, which became invisible about 18 months later. In 1919, Lundmark estimated the distance of M31 to be about 7×10^5 lyear, and by that time it became obvious that Hartwig's nova had been several 1000 times brighter than a normal nova (Lundmark 1920). It was also Lundmark (1921) who first suggested an association between the supernova observed by Chinese astronomers in 1054 and the Crab nebula.

A similar event as S Andromeda was observed in 1895 in NGC 5253 ("nova" Z Centauri), and this time the "new star" appeared to be five times brighter than the entire galaxy. But it was not before 1934 that a clear distinction between classical novae and supernovae was made (Baade & Zwicky 1934). Systematic searches, performed predominantly by Zwicky, lead to the discovery of 54 supernovae in the years up to 1956 and, owing to improved observational techniques, 82 further supernovae were discovered in the years from 1958 to 1963, all of course in external galaxies (e.g. Zwicky 1965).

Until 1937, spectrograms of supernovae were rare, and what was known seemed to be not too different from common novae. This changed with the very bright ($m_V \simeq 8.4$) supernova SN1937c in IC 4182, which had spectral features different from any object that had been observed before (Popper 1937). All the other supernovae discovered in the following years showed little dispersion in their maximum luminosity, and their postmaximum spectra looked similar at any given time. Based on this finding, Wilson (1939) and Zwicky (1938a) suggested supernovae be used as distance indicators.

In 1940 it became clear, however, that there exist at least two distinctly different classes of supernovae. SN 1940c in NGC 4725 had a spectrum different from all other previously observed supernovae for which good data were available at that time, leading Minkowski (1940) to introduce the names type I for those with spectra like SN 1937c and type II for SN 1940c–like events, representing supernovae without and with Balmer lines of hydrogen near maximum light.

Whether or not the spectral differences also reflect a different explosion mechanism was not known. In contrast, the scenario originally suggested by Zwicky (1938b), that a supernova occurs as the transition from an ordinary star to a neutron star and gains its energy from the gravitational binding of the newly born compact object, was for many years the only explanation. Hoyle & Fowler (1960) were the first to discover that thermonuclear burning in an electron-degenerate stellar core might trigger an explosion and (possibly) the disruption of the star. Together with the idea that the light curves could be powered by the decay-energy of freshly produced radioactive ^{56}Ni (Truran et al 1967, Colgate & McKee 1969), this scenario is now the generally accepted one for a subclass of all type I supernovae called type Ia. It is amusing to note that all supernovae (besides the Crab nebula) on which Zwicky had based his core-collapse hypothesis were in fact of type Ia and most likely belonged to the other group, whereas the first core-collapse supernova, SN1940c, was observed only about a year after he published his paper.

To be more precise, supernovae that do not show hydrogen lines in their spectra but a strong silicon P Cygni feature near maximum light are named type Ia (Wheeler & Harkness 1990). They are believed to be the result of thermonuclear disruptions of white dwarfs, either consisting of carbon and oxygen with a mass close to the Chandrasekhar mass, or of a low-mass C + O core mantled by a layer of helium, the so-called sub-Chandrasekhar mass models [see the recent reviews by Woosley (1997b), Woosley & Weaver (1994a, 1994b) and Nomoto et al (1994b, 1997)]. The main arguments in favor of this interpretation include: (*a*) the apparent lack of neutron stars in some of the historical galactic supernovae (e.g. SN1006, SN 1572, SN1604); (*b*) the homogeneous appearance of this subclass; (*c*) the excellent fits to the light curves, which can be obtained from the simple assumption that a few tenths of a solar mass of ^{56}Ni is produced during the explosion; and (*d*) the good agreement with the observed spectra of typical type Ia supernovae. Several of these observational aspects are discussed in some detail in Section 2, together with their cosmological implications. Models of light curves and spectra are reviewed in Section 3, and questions concerning the nature of the progenitor stars are addressed in Section 4.

But having good arguments in favor of a particular explosion scenario does not mean that this scenario is indeed the right one. Besides that, one would like to understand the physics of the explosion, the fact that the increasing amount of data also indicates that there is a certain diversity among the type Ia supernovae seems to contradict a single class of progenitor stars or a single explosion mechanism. Moreover, the desire for using them as distance indicators makes it necessary to search for possible systematic deviations from uniformity. Here, again, theory can make important contributions. In Section 5, therefore, we discuss the physics of thermonuclear combustion, its implementation into numerical models of exploding white dwarfs, and the results of recent computer simulations. A summary and conclusions follow in Section 6.

2. OBSERVATIONS

The efforts to systematically obtain observational data of SNe Ia near and far have gained tremendous momentum in recent years. This is primarily a result of the unequaled potential of SNe Ia to act as "standardizable" candles (Branch & Tammann 1992, Riess et al 1996, Hamuy et al 1995, Tripp 1998) for the measurement of the cosmological expansion rate (Hamuy et al 1996b, Branch 1998) and its variation with look-back time (Perlmutter et al 1999, Schmidt et al 1998, Riess et al 1998). For theorists, this development presents both a challenge to help understand the correlations among the observables and an opportunity to use the wealth of new data to constrain the zoo of existing explosion models. There exist a number of excellent reviews about SNe Ia observations in general (Filippenko 1997b), their spectral properties (Filippenko 1997a), photometry in the infrared (IR) and optical bands (Meikle et al 1996, 1997), and their use for measuring the Hubble constant (Branch 1998). Recent books that cover a variety of observational and theoretical aspects of type Ia supernovae are Ruiz-Lapuente et al (1997) and Niemeyer & Truran (2000). Below, we highlight those aspects of SN Ia observations that most directly influence theoretical model building at the current time.

2.1 General Properties

The classification of SNe Ia is based on spectroscopic features: the absence of hydrogen absorption lines, distinguishing them from type II supernovae, and the presence of strong silicon lines in the early and maximum spectrum, classifying them as type Ia (Wheeler & Harkness 1990).

The spectral properties, absolute magnitudes, and light curve shapes of the majority of SN Ia are remarkably homogeneous, exhibiting only subtle spectroscopic and photometric differences (Branch & Tammann 1992, Hamuy et al 1996c,

Branch 1998). It was believed until recently that approximately 85% of all observed events belonged to this class of normal ["Branch-normal" (Branch et al 1993)] SNe Ia, represented for example by SNe 1972E, 1981B, 1989B, and 1994D. However, the peculiarity rate can be as high as 30% as suggested by Li et al (2000).

The optical spectra of normal SN Ia contain neutral and singly ionized lines of Si, Ca, Mg, S, and O at maximum light, indicating that the outer layers of the ejecta are mainly composed of intermediate mass elements (Filippenko 1997b). Permitted Fe II lines dominate the spectra roughly 2 weeks after maximum when the photosphere begins to penetrate Fe-rich ejecta (Harkness 1991). In the nebular phase of the light curve tail, beginning approximately 1 month after peak brightness, forbidden Fe II, Fe III, and Co III emission lines become the dominant spectral features (Axelrod 1980). Some Ca II remains observable in absorption even at late times (Filippenko 1997a). The decrease of Co lines (Axelrod 1980) and the relative intensity of Co III and Fe III (Kuchner et al 1994) give evidence that the light curve tail is powered by radioactive decay of ^{56}Co (Truran et al 1967, Colgate & McKee 1969).

The early spectra can be explained by resonant scattering of a thermal continuum with P Cygni–profiles whose absorption component is blueshifted according to ejecta velocities of up to a few times 10^4 km s^{-1}, rapidly decreasing with time in the early phase (Filippenko 1997a). Different lines have different expansion velocities (Patat et al 1996), which suggests a layered structure of the explosion products.

Photometrically, SN Ia rise to maximum light in a period of approximately 20 days (Riess et al 1999b) reaching

$$M_B \approx M_V \approx -19.30 \pm 0.03 + 5\log(H_0/60) \qquad (1)$$

with a dispersion of $\sigma_M \leq 0.3$ (Hamuy et al 1996b). It is followed by a first rapid decline of about three magnitudes in a matter of 1 month. Later, the light curve tail falls off in an exponential manner at a rate of approximately one magnitude per month. In the I-band, normal SNe Ia rise to a second maximum approximately 2 days after the first maximum (Meikle et al 1997).

It is especially interesting that the two most abundant elements in the universe, hydrogen and helium, have not been unambiguously detected in SN Ia spectra (Filippenko 1997a; for a possible identification of He, see Meikle et al 1996), and there are no indications for radio emission of SNe Ia. Cumming et al (1996) failed to find any signatures of H in the early time spectrum of SN 1994D and used this fact to constrain the mass accretion rate of the progenitor wind (Lundqvist & Cumming 1997). The later spectrum of SN 1994D also did not exhibit narrow Hα features (Filippenko 1997b). Another direct constraint for the progenitor system accretion rate comes from the nondetection of radio emission from SN 1986G (Eck et al 1995), used by Boffi & Branch (1995) to rule out symbiotic systems as a possible progenitor of this event.

2.2 Diversity and Correlations

Early suggestions (Pskovskii 1977, Branch 1981) that the existing inhomogeneities among SN Ia observables are strongly intercorrelated are now established beyond doubt (Hamuy et al 1996a, Filippenko 1997a). Branch (1998) offers a recent summary of correlations between spectroscopic line strengths, ejecta velocities, colors, peak absolute magnitudes, and light curve shapes. Roughly speaking, SNe Ia appear to be arrangeable in a one-parameter sequence according to explosion strength, wherein the weaker explosions are less luminous, are redder, and have a faster declining light curve and slower ejecta velocities than the more energetic events (Branch 1998). The relation between the width of the light curve around maximum and the peak brightness is the most prominent of all correlations (Pskovskii 1977, Phillips 1993). Parameterized either by the decline rate Δm_{15} (Phillips 1993, Hamuy et al 1996a), a "stretch parameter" (Perlmutter et al 1997), or a multi-parameter nonlinear fit in multiple colors (Riess et al 1996), it was used to renormalize the peak magnitudes of a variety of observed events, substantially reducing the dispersion of absolute brightnesses (Riess et al 1996, Tripp 1998). This correction procedure is a central ingredient of all current cosmological surveys that use SNe Ia as distance indicators (Perlmutter et al 1999, Schmidt et al 1998).

SN 1991bg and SN 1992K are well-studied examples for red, fast, and subluminous supernovae (Filippenko et al 1992a, Leibundgut et al 1993, Hamuy et al 1994, Turatto et al 1996). Their V-, I-, and R-band light curve declined unusually quickly, skipping the second maximum in I, and their spectrum showed a high abundance of intermediate mass elements (including Ti II) with low expansion velocities but only little iron (Filippenko et al 1992a). Models for the nebular spectra and light curve of SN 1991bg consistently imply that the total mass of ^{56}Ni in the ejecta was very low ($\sim 0.07\ M_\odot$) (Mazzali et al 1997a). On the other side of the luminosity function, SN 1991T is typically mentioned as the most striking representative of bright, energetic events with broad light curves (Phillips et al 1992, Jeffery et al 1992, Filippenko et al 1992b, Ruiz-Lapuente et al 1992, Spyromilio et al 1992). Rather than the expected Si II and Ca II, its early spectrum displayed high-excitation lines of Fe III but returned to normal a few months after maximum (Filippenko et al 1992b).

Peculiar events like SN 1991T and SN 1991bg were suggested as belonging to subgroups of SNe Ia different from those of the normal majority, created by different explosion mechanisms (Mazzali et al 1997a, Filippenko et al 1992b, Fisher et al 1999). Until recently, the overall SN Ia luminosity function seemed to be very steep on the bright end (Vaughan et al 1995), implying that "normal" events are essentially the brightest whereas the full class may contain a large number of undetected subluminous SNe Ia (Livio 1999). New results (Li et al 2000) indicate, however, that the luminosity function may be shallower than anticipated.

There is also mounting evidence that SN Ia observables are correlated with the host stellar population (Branch 1998). SNe Ia in red or early type galaxies show, on average, slower ejecta velocities and faster light curves, and they are dimmer by ≈ 0.2–0.3 mag than those in blue or late-type galaxies (Hamuy et al 1995, 1996a; Branch et al 1996). The SN Ia rate per unit luminosity is nearly a factor of 2 higher in late-type galaxies than in early type ones (Cappellaro et al

1997). In addition, the outer regions of spirals appear to give rise to similarly dim SNe Ia as ellipticals whereas the inner regions harbor a wider variety of explosion strengths (Wang et al 1997). When corrected for the difference in light curve shape, the variation of absolute magnitudes with galaxy type vanishes along with the dispersion of the former. This fact is crucial for cosmological SN Ia surveys, making the variations with stellar population consistent with the assumption of a single explosion strength parameter (Perlmutter et al 1999, Riess et al 1998).

2.3 Nearby and Distant SNe Ia

Following a long and successful tradition of using relatively nearby [$z \leq 0.1$, comprised mostly of the sample discovered by the Calán/Tololo survey (Hamuy et al 1996a)] SNe Ia for determining the Hubble constant (Branch 1998), the field of SN Ia cosmology has recently seen a lot of activity, expanding the range of observed events out to larger redshift, $z \approx 1$. Systematic searches involving a series of wide-field images taken at epochs separated by 3–4 weeks, in addition to prescheduled follow-up observations to obtain detailed spectroscopy and photometry of selected events, have allowed two independent groups of observers—the Supernova Cosmology Project (SCP) (Perlmutter et al 1999) and the High-z Supernova Search Team (Schmidt et al 1998)—to collect data of more than 50 high-redshift SNe. Extending the Hubble diagram out to $z \approx 1$, one can, given a sufficient number of data points over a wide range of z, determine the density parameters for matter and cosmological constant, Ω_M and Ω_Λ, independently (Goobar & Perlmutter 1995), or, in other words, constrain the equation of state of the universe (Garnavich et al 1998). Both groups come to a spectacular conclusion (Riess et al 1998, Perlmutter et al 1999): The distant SNe are too dim by ≈ 0.25 mag to be consistent with a purely matter-dominated, flat or open Friedmann-Robertson-Walker universe. Interpreted as being a consequence of a larger-than-expected distance, this discrepancy can be resolved only if Ω_Λ is non-zero, implying the existence of an energy component with negative pressure. In fact, the SN Ia data is consistent with a spatially flat universe made up of two parts vacuum energy and one part matter.

Both groups discuss in detail the precautions that were taken to avoid systematic contaminations of the detection of cosmological acceleration, including SN Ia evolution, extinction, and demagnification by gravitational lensing. All of these effects would, in all but the most contrived scenarios, give rise to an increasing deviation from the $\Omega_\Lambda = 0$ case for higher redshift, whereas the effect of a non-zero cosmological constant should become less significant as z grows. Thus, the degeneracy between a systematic overestimation of the intrinsic SN Ia luminosity and cosmological acceleration can be broken when sufficiently many events at $z \geq 0.85$ are observed (Filippenko & Riess 1999). Meanwhile, the only way to support the cosmological interpretation is by "...adding to the list of ways in which they are similar while failing to discern any way in which they are different"

(Riess et al 1999a). This program has been successful until recently: The list of similarities between nearby and distant SNe Ia includes spectra near maximum brightness (Riess et al 1998) and the distributions of brightness differences, light curve correction factors, and $B - V$ color excesses of both samples (Perlmutter et al 1999). Moreover, although the nearby sample covers a range of stellar populations similar to the one expected out to $z \approx 1$, a separation of the low-z data into subsamples arising from different progenitor populations shows no systematic shift of the distance estimates (Filippenko & Riess 2000). However, a recent comparison of the rise times of more than 20 nearby SNe (Riess et al 1999b) with those determined for the SCP high-redshift events gives preliminary evidence for a difference of roughly 2.5 days. This result was disputed by Aldering et al (2000) who conclude that the rise times of local and distant supernovae are statistically consistent.

2.4 Summary: Observational Requirements for Explosion Models

To summarize the main observational constraints, any viable scenario for the SN Ia explosion mechanism has to satisfy the following (necessary but probably not sufficient) requirements:

1. *Agreement of the ejecta composition and velocity with observed spectra and light curves.* In general, the explosion must be sufficiently powerful (i.e. produce enough ^{56}Ni) and produce a substantial amount of high-velocity intermediate mass elements in the outer layers. Furthermore, the isotopic abundances of "normal" SNe Ia must not deviate significantly from those found in the solar system.

2. *Robustness of the explosion mechanism.* In order to account for the homogeneity of normal SNe Ia, the standard model should not give rise to widely different outcomes, depending on the fine-tuning of model parameters or initial conditions.

3. *Intrinsic variability.* Although the basic model should be robust with respect to small fluctuations, it must contain at least one parameter that can plausibly account for the observed sequence of explosion strengths.

4. *Correlation with progenitor system.* The explosion strength parameter must be causally connected with the state of the progenitor white dwarf in order to explain the observed variations as a function of the host stellar population.

3. LIGHT CURVE AND SPECTRA MODELING

Next we discuss the problem of coupling the interior physics of an exploding white dwarf to what is finally observed, namely light curves and spectra, by means of radiative transfer calculations. For many astrophysical applications, this problem is not solved, and SN Ia are no exceptions. In fact, radiation transport is even more complex in type Ia than for most other cases.

A rough sketch of the processes involved can illustrate some of the difficulties (see, e.g. Mazzali & Lucy 1993, Eastman & Pinto 1993). Unlike most other objects we know in astrophysics, SN Ia do not contain any hydrogen. Therefore the opacities are always dominated either by electron scattering (in the optical) or by a huge number of atomic lines [in the ultraviolet (UV)]. In the beginning, the supernova is an opaque expanding sphere of matter into which energy is injected from radioactive decay. This could happen in an inhomogeneous manner, as is discussed later. As the matter expands, diffusion times eventually get shorter than the expansion time and the supernova becomes visual. However, because the star is rapidly expanding, the Doppler shift of atomic lines causes important effects. For example, a photon emitted somewhere in the supernova may find the surrounding matter more or less transparent until it finds a line Doppler shifted such that it is trapped in that line and scatters many times. As a consequence, the spectrum might look thermal although the photon "temperature" has nothing in common with the matter temperature.

It is also obvious that radiation transport in SN Ia is nonlocal and that the methods used commonly in models of stellar atmospheres need refinements. As a consequence, there is no agreement yet among the groups modeling light curves and spectra as to what the best approach is. Therefore it can happen that even if the same model for the interior physics of the supernova is inserted into one of the existing codes for modeling light curves and spectra, the predictions for what should be "observed" could be different—again an unpleasant situation. Things get even worse because all such models treat the exploding star as being spherically symmetric, an assumption that is at least questionable, given the complex combustion physics discussed below.

In the following subsections we outline some of the commonly used numerical techniques and also discuss their predictions for SN Ia spectra and light curves. For more details on the techniques used by the various groups, we refer readers to the articles by Eastman (1997), Blinnikov (1997), Pinto (1997), Baron et al (1997), Mazzali et al (1997b), Höflich et al (1997), and Ruiz-Lapuente (1997).

3.1 Radiative Transfer in Type Ia Supernovae

In principle, the equations that have to be solved are well known, either in the form of the Boltzmann transport equation for photons or as a transport equation for the monochromatic intensities. However, to solve this time-dependent, frequency-dependent radiation transport problem, including the need to treat the atoms in non-local thermodynamic equilibrium (NLTE), is expensive, even in spherically symmetric situations. Therefore, approximations of various kinds are usually made which give rise to controversial discussions.

Conceptually, it is best to formulate and solve the transport equation in the co-moving (Lagrangian) frame (cf. Mihalas & Weibel Mihalas 1984). This makes the transport equation appear simpler, but it causes problems in calculating the "co-moving" opacity, in particular if the effect of spectral lines on the opacity of an expanding shell of matter is important, as in the case of SN Ia (Karp et al 1977).

There are different ways to construct approximate solutions of the transport equation. One can integrate over frequency and replace the opacity terms by appropriate means, leaving a single (averaged) transport equation. Unfortunately, in order to compute the flux-mean opacity, one has to know the solution of the transport equation. Frequently the flux mean is replaced, for example, by the Rosseland mean, allowing for solutions, but at the expense of consistency (see, e.g., Eastman 1997).

Another way out is to replace the transport equation by its moment expansion, introducing, however, the problem of closure. In its simplest form, the diffusion approximation, the radiation field is assumed to be isotropic, the time rate of change of the flux is ignored, and the flux is expressed in terms of the gradient of the mean intensity of the radiation field. Replacing the mean intensity by the Planck function and closing the moment expansion by relating the radiation energy density and pressure via an Eddington factor (equal to one third for isotropic radiation) finally leads to a set of equations that can be solved (Mihalas & Weibel Mihalas 1984).

Again, this simple approach has several obvious shortcomings. First, the transition from an optically thick to thin medium at the photosphere requires a special treatment mainly because the radiation field is no longer isotropic. One can compensate for this effect by putting in either a flux limiter or a variable Eddington factor to describe the transition from diffusion to free streaming, but both approaches are not fully satisfactory because it is difficult to calibrate the newly invented parameters (e.g. Kunasz 1984, Fu 1987, Blinnikov & Nadyoshin 1991, Mair et al 1992, Stone et al 1992, Yin & Miller 1995).

Alternatively, one can bin frequency space into groups and solve the set of fully time-dependent coupled monochromatic transport equations for each bin. In this approach, the problem remains of computing average opacities for each frequency bin. Moreover, because of computer limitations, in all practical applications the number of bins cannot be large, which introduces considerable errors, given the strong frequency dependence of the line-opacities (Blinnikov & Nadyoshin 1991, Eastman 1997) (see also Figure 1).

Finally, in order to get synthetic spectra one might apply Monte-Carlo techniques, as was done by Mazzali et al (1997b) and Lucy (1999). Here the assumption is that the supernova envelope is in homologous spherical expansion and that the luminosity and the photospheric radius are given. The formation of spectral lines is then computed by considering the propagation of a wave packet emitted from the photosphere subject to electron scattering and interaction with lines. Line formation is assumed to occur by coherent scattering, and the line profiles and escape probabilities are calculated in the Sobolev approximation. Although this approach

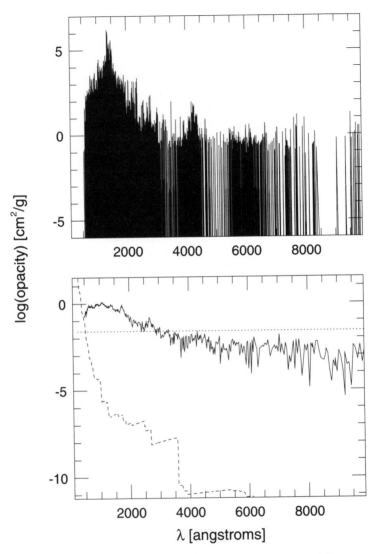

Figure 1 Mass opacity in a hot (23,000 K) plasma of cobalt at a density of 10^{-12} g/cm^3. The upper plot shows the line opacities (calculated by Iglesias, Rogers & Wilson (1987) with OPAL), the lower one the bound-free (*long dashed*), electron scattering (*short dashed*), and the "line expansion opacity" (*solid curve*). (Courtesy of R. Eastman 1999, personal communication)

appears to be a powerful tool to get synthetic spectra, it lacks consistency because the properties of the photosphere have to be calculated be other means.

But having a numerical scheme at hand to solve the transport equation is not sufficient. It is even more important to have accurate opacities. The basic problem, namely that at short wavelengths the opacity is dominated by a huge number of weak lines, was mentioned before. In practice this means that because the list included in anyone's code is certainly incomplete and the available information may not always be accurate, it is difficult to estimate possible errors. Moreover, there is no general agreement among the different groups calculating SN Ia light curves and spectra on how to correct the opacities for Doppler shifts of the lines, caused by the expansion of the supernova. The so-called "expansion opacity" (see Figure 1) that should be used in approaches based on the diffusion equation as well as on moment expansions of the transport equation is still controversial (Pauldrach et al 1996, Blinnikov 1997, Baron et al 1997, Eastman 1997, Hoeflich et al 1997, Mazzali et al 1997b, Pinto 1997).

Other open questions include the relative importance of absorption and scattering of photons in lines, and whether one can calculate the occupation numbers of atomic levels in equilibrium or whether it has to be calculated by means of the Saha-equation (Pauldrach et al 1996, Nugent et al 1997, Hoeflich et al 1998b).

3.2 Results of Numerical Studies

Despite the problems discussed in the previous subsection, radiation hydrodynamic models have been used widely as a diagnostic tool for SN Ia. These studies include computations of γ-ray (Burrows & The 1990, Müller et al 1991, Burrows et al 1991, Shigeyama et al 1993, Ruiz-Lapuente et al 1993b, Timmes & Woosley 1997, Hoeflich et al 1998a, Watanabe et al 1999), UV and optical rays (Branch & Venkatakrishna 1986; Ruiz-Lapuente et al 1992; Nugent et al 1995, 1997; Pauldrach et al 1996; Hoeflich et al 1997; Hatano et al 1998; Hoeflich et al 1998b; Lentz et al 1999a,b; Fisher et al 1999; Lucy 1999) and of infrared light curves and spectra (Spyromilio et al 1994, Hoeflich 1995, Wheeler et al 1998). All studies are based on the assumption, that the explosion remains on average spherically symmetric, an assumption that is questionable, as is discussed in Section 5. Although spherical symmetry might be a good approximation for temperatures, densities, and velocities, the spatial distribution of the products of explosive nuclear burning is expected to be nonspherical, and it is the distribution of the heavier elements, both in real and in velocity space, that determines to a large extent light curves and spectra.

With the possible exception of SN 1991T, where a 2–3σ detection of the ^{56}Co decay lines at 847 keV and 1238 keV has been reported (Morris et al 1997; but see also Leising et al 1995), only upper limits on γ-ray line emission from SN Ia are known. On the basis of the models this is not surprising because the flux limits of detectors such as COMPTEL on GRO [10^{-5} photons per cm^2 and second (Schoenfelder et al 1996)] allows detections out to distances of about 15 Mpc in the most favorable cases, i.e. delayed-detonation models producing lots of ^{56}Ni in the outer parts of the supernova (Timmes & Woosley 1997). In fact, the tentative detection of decay lines from SN 1991T at a distance of about 13 Mpc can be explained by certain delayed-detonation models and was even predicted by some of them (Müller et al (1991), see also Section 5).

Synthetic (optical and UV) spectra of hydrodynamic models of SN Ia have been computed by several groups (Hoeflich & Khokhlov 1996, Nugent et al 1997) and have been compared with the observations. The bottom line of these investigations is that Chandrasekhar-mass deflagration models are in good agreement with observations of Branch-normals such as SN 1992A and SN 1994D (Hoeflich & Khokhlov 1996, Nugent et al 1997), and delayed detonation are equally good. The reason is that in both classes of models, the burning front starts by propagating out slowly, giving the star some time to expand. The front then speeds up to higher velocities, i.e. to a fair fraction of the sound velocity for deflagration models and to supersonic velocity for detonations, which is necessary to match the obseved high velocities of the ejecta. But as far as the amount of radioactive Ni is concerned, the predictions of both classes of models are not too different (Nugent et al 1997). It also appears that sub-Chandrasekhar models cannot explain the observed UV flux and the colors of normal SNe Ia (Khokhlov et al 1993). Moreover, although sub-Chandrasekhar models eject considerable amounts of He, according to the synthetic spectra, He lines should not be seen, eliminating them as a tool to distinguish between the models (Nugent et al 1997).

In the IR, SN Ia do show nonmonotonic behavior (Elias et al 1985), and as for the bolometric light curves, a correlation between peak brightness and light curve shape seems to exist (Contardo & Leibundgut 1998, Contardo 1999). Therefore calculations of IR light curves and spectra are of importance, and they might prove to be a good diagnostic tool. Broad-band IR light curves have been computed by Hoeflich et al (1995) with the result that the second IR peak can be explained as an opacity effect. Although the fits were not perfect, the general behavior, again, was consistent with both the deflagration and the delayed-detonation models. Detailed early IR spectra have been calculated only recently (Wheeler et al 1998), and the models provide a good physical understanding of the spectra. Again, a comparison between several of the delayed-detonation models and SN 1994D gave good agreement, but one might suspect that certain deflagrations would do equally well. However, in principle, synthetic IR spectra are sensitive to the boundary between explosive C and O and between complete and incomplete Si burning (Wheeler et al 1998) and should provide some information on the progenitors and the explosion mechanism.

In conclusion, models of SN Ia light curves and spectra can fit the observations well, but so far, their predictive power is limited. The fact that multidimensional effects are ignored and that the opacities as well as the radiative-transfer codes have obvious shortcomings makes it difficult to derive strong constraints on the explosion mechanism. It appears, however, that although it seems to be difficult to distinguish between pure deflagrations and delayed detonations on the basis of synthetic light curves and spectra, sub-Chandrasekhar models cannot fit normal SN Ia equally well.

4.   PROGENITOR SYSTEMS

In contrast to supernovae from collapsing massive stars, for which in two cases the progenitor star was identified and some of its properties could be inferred directly from observations [SN1987A in the Large Magellanic Cloud (Blanco 1987, Gilmozzi 1987, Gilmozzi et al 1987, Hillebrandt et al 1987) and SN1993J in M81 (Benson et al 1993, Schmidt et al 1993, Nomoto et al 1993, Podsiadlowski 1993)], there is not a single case known where we have this kind of information for the progenitor of a SN Ia. This is not too surprising given the fact that their progenitors are most likely faint compact dwarf stars and not red or blue supergiants. Therefore we must rely on indirect means to determine their nature.

The standard procedure is to eliminate all potential candidates if some of their properties disagree with either observations or physical principles, and to hope that a single and unique solution is left. Unfortunately, for the progenitors of type Ia supernovae, this cannot be done unambiguously, the problem being the lack of strong candidates that pass all possible tests beyond doubt.

In this section, we first repeat the major constraints that must be imposed on the progenitor systems and then discuss in some detail the presently favored candidates, Chandrasekhar-mass C + O white dwarfs and low-mass C + O white dwarf cores embedded in a shell of helium. It is shown, however, that even if we could single out a particular progenitor system, this would narrow the parameter space for the initial conditions at the onset of the explosion but might not determine them sufficiently well, in particular if we are aiming at a quantitative understanding. Some of the discussion given below follows recent reviews of Renzini (1996) and Livio (1999).

4.1   Observational Constraints on Type Ia Progenitors

As discussed in Section 2, SNe Ia are (spectroscopically) defined by the absence of emission lines of hydrogen and the presence of a (blueshifted) Si II absorption line with a rest wavelength of 6355-Å near-maximum light. The first finding requires that the atmosphere of the exploding star contains no or at most $0.1\,M_\odot$ of hydrogen, and the second one indicates that some nuclear processing takes place and that products of nuclear burning are ejected in the explosion. Mean velocities of the ejecta, as inferred from spectral fits, are around $5000\ \mathrm{km\ s^{-1}}$ and peak velocities exceeding $20{,}000\,\mathrm{km\ s^{-1}}$ are observed, which is consistent with fusing about $1\,M_\odot$ of carbon and oxygen into Fe group elements or intermediate-mass elements such as Si or Ca. The presence of some UV flux, the width of the peak of the early light curve, and the fact that radioactive-decay models (^{56}Ni $\rightarrow$ ^{56}Co $\rightarrow$ ^{56}Fe) can fit the emission very well all point toward compact progenitor stars with radii of less than about 10,000 km.

After about 2 weeks, the typical SN Ia spectrum changes from being dominated by lines of intermediate-mass nuclei to being dominated by Fe II. Because a Co III feature is identified at later stages, this adds evidence to the interpretation that they are indeed thermonuclear explosions of compact stars, leaving the cores of stars with main sequence masses near 6–8 $M_\odot$ or white dwarfs as potential candidates. Moreover, the energetics of the explosion and the spectra seem to exclude He white dwarfs (Nomoto & Sugimoto 1977, Woosley et al 1986), mainly because such white dwarfs would undergo very violent detonations.

Next one notes that most SNe Ia, of the order of 85%, have similar peak luminosities, light curves, and spectra. The dispersion in peak blue and visual brightness is only of the order of 0.2–0.3 mag, calling for a homogeneous class of progenitors. It is mainly this observational fact that seems to single out Chandrasekhar-mass white dwarfs as their progenitors. Because the ratio of energy to mass determines the velocity profile of the exploding star, the homogeneity would be explained in a natural way. However, as discussed in Section 2, there exist also significant differences among the various SNe Ia, which may indicate that this simple interpretation is not fully correct. The difference in peak brightness, ranging from subluminous events such as SN 1991bg in NGC 4374 ($B_{max} = -16.54$) (Turatto et al 1996), compared with the mean of the Branch-normals of $B_{max} \simeq -19$ (Hamuy et al 1996c) to bright ones such as SN 1991T, which was about 0.5 mag brighter in B than a typical type Ia in the Virgo cluster (Mazzali et al 1995), is commonly attributed to different ^{56}Ni masses produced in the explosion. They range from about 0.07 $M_\odot$ for SN 1991bg (see, e.g., Mazzali et al 1997a) to at least 0.92 $M_\odot$ for SN 1991T (Khokhlov et al 1993; but see also Fisher et al 1999), with typically 0.6 $M_\odot$ for normal SNe Ia (Hoeflich & Khokhlov 1996, Nugent et al 1997). It is hard to see how this large range can be accommodated in a single class of models.

The stellar populations in which SNe Ia show up include spiral arms as well as elliptical galaxies, with some weak indication that they might be more efficiently produced in young populations (Bartunov et al 1994). Again, if we insist on a single class of progenitors, the very fact that they do occur in ellipticals would rule out massive stars as potential candidates. On the other hand, the observations may tell us that there is not a unique class of progenitors. In particular, the fact that the bright and slowly declining ones (such as SN 1991T) are absent in an elliptical and S0 galaxies may point toward different progenitor classes (Hamuy et al 1996c).

All in all, the observational findings summarized so far are consistent with the assumption that type Ia supernovae are the result of thermonuclear disruptions of white dwarfs, C + O white dwarfs being the favored model. The diversity among them must then be attributed to the history and nature of the white dwarf prior to the explosion and/or to the physics of thermonuclear burning during the event. The

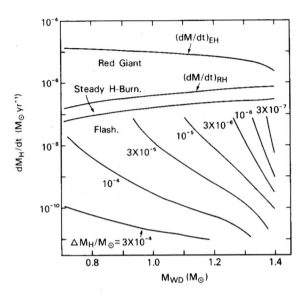

Figure 2 The likely outcome of hydrogen accretion onto white dwarfs of different masses is shown. From Kahabka and Van Den Heuvel (1997).

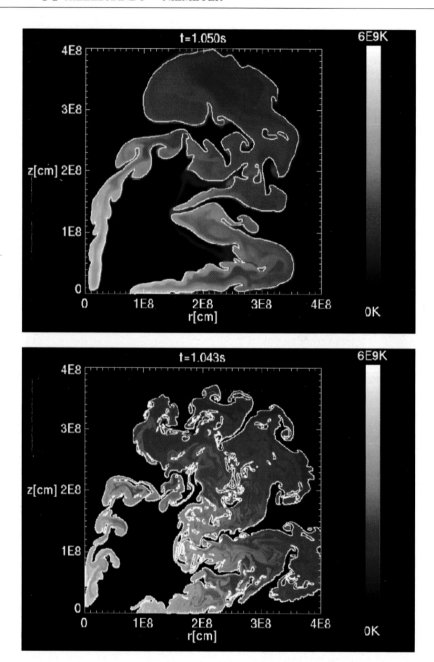

Figure 3 Snapshots of the temperature and the front geometry in a Chandrasekhar-mass deflagration model at 1.05 s (from Reinecke et al 1999c). Shown are a model with "low" resolution (256^2) (*upper figure*) and one with three times higher resolution, respectively. Because of the larger surface area of the better-resolution model, it exploded, whereas the other one remained marginally bound.

possibility cannot be excluded, however, that at least some SNe Ia have a different origin, such as accretion-induced collapse of massive O-Ne-Mg (or O-Ne) white dwarfs for SN 1991bg–like objects (Nomoto et al 1994a, 1995, 1996; Fryer et al 1999). Also it is not clear whether there is a clear-cut distinction between type Ib/c supernovae, defined by the absence of the Si II feature, and the (faint) SNe Ia. The former are believed to reflect the core collapse of a massive star, its hydrogen-rich envelope being pealed off because of mass loss in a binary system. For example, SN1987K started out as a SN II with H lines in its spectrum but changed into a SN Ib/c–like spectrum after 6 months (Filippenko 1988), supporting this interpretation. It should be noted that SN 1991bg–like objects are not often observed, but that this may well be a selection effect. Suntzeff (1996), for example, argues that up to 40% of all type Ia's could perhaps belong to that subgroup.

4.2 Presupernova Evolution of Binary Stars

In spite of all these uncertainties, it is the current understanding and belief that the progenitors of SNe Ia are $C + O$ white dwarfs in binary systems evolving to the stage of explosion by mass overflow from the companion (single-degenerate scenario) or by the merger of two white dwarfs (double-degenerate scenario). Binary evolution of some sort is necessary because $C + O$ white dwarfs are typically born with a mass around 0.6 $M_\odot$ (Homeier et al 1998) but need either to be near the Chandrasekhar mass or to accumulate a shell of helium in order to explode. In this subsection, we summarize the arguments in favor and against both scenarios.

Double-degenerates as potential type Ia progenitors had many ups and downs in the past, beginning with the classic papers of Iben & Tutukov (1984) and Webbink (1984). The arguments in favor are that such binaries should exist as a consequence of stellar evolution, they would explain very naturally the absence of hydrogen, and they could, in principle, be an easy way to approach a critical mass. In fact, several candidate systems of binary white dwarfs have recently been identified, but most of the short-period ones (at present eight systems are known with orbital periods of less than half a day), which could merge in a Hubble time because of the emission of gravitational radiation, have a mass less than M_{chan} (Saffer et al 1998; for a recent review, see Livio 2000). There is only one system known [KPD 0422 + 5421 (Koen et al 1998)] with a mass which, within the errors, could exceed M_{chan}, a surprisingly small number. Nonetheless, it is argued that from population synthesis one could arrive at about the right frequency of sufficiently massive mergers (Livio 2000).

Besides the lack of convincing direct observational evidence for sufficiently many appropriate binary systems, the homogeneity of "typical" SNe Ia may be an argument against this class of progenitors. It is not easy to see how the merging of two white dwarfs of (likely) different mass, composition, and angular momentum with different impact parameters, etc, will always lead to the same burning conditions and, therefore, the production of a nearly equal amount of ^{56}Ni. Moreover, some investigations of white dwarf mergers seem to indicate that an off-center ignition will convert carbon and oxygen into oxygen, neon, and magnesium, leading to gravitational collapse rather than a thermonuclear disruption (Nomoto & Iben 1985; Woosley & Weaver 1986a; Saio & Nomoto 1985, 1998; Mochkovitch & Livio 1990). Finally, based on their galactic chemical evolution model, Kobayashi et al (1998) claim that double-degenerate mergers lead to inconsistencies with the observed O/Fe as a function of metallicity, but this statement is certainly model dependent. In any case, mergers might, if they are not responsible for the bulk of the SNe Ia, still account for some peculiar ones, such as the superluminous SN 1991T–like explosions.

Single-degenerate models are in general favored today. They consist of a low-mass white dwarf accreting matter from the companion star until either it reaches M_{chan} or a layer of helium has formed on top of its $C + O$ core that can ignite and possibly drive a burning front into the carbon and oxygen fuel. This track to thermonuclear explosions of white dwarfs was first discussed by Whelan & Iben (1973), Nomoto (1982a), Iben & Tutukov (1984) and Paczynski (1985). The major problem of these models has always been that nearly all possible accretion rates can be ruled out by strong arguments (Nomoto 1982a, Munari & Renzini 1992, Cassisi et al 1996, Tutukov & Yungelson 1996, Livio et al 1996, King & Van Teeseling 1998, Kato & Hachisu 1999, Cassisi et al 1998). In short, it is believed that white dwarfs accreting hydrogen at a low rate undergo nova eruptions and lose more mass in the outburst than they have accreted prior to it (e.g. Beer 1974, Gehrz et al 1993). At moderate accretion rates, a degenerate layer of helium is thought to form which might flash and could give rise to sub-Chandrasekhar explosions (which have other problems, as is discussed later). Next, still higher accretion rates can lead to quiet hydrostatic burning of H and He, but these systems should be so bright that they could easily be detected. However, it is not clear beyond doubt that they coincide with any of the known symbiotic or cataclysmic binaries. Very high accretion rates, finally, would form an extended H-rich red giant envelope around the white dwarf with debris not seen in the explosions (Nomoto et al 1979) (see also Figure 2). Therefore, it is uncertain whether white dwarfs accreting hydrogen from a companion star can ever reach the M_{chan} (Cassisi et al 1998).

Some of these arguments may be questioned, however. Firstly, a class of binary systems has recently been discovered, the so-called Supersoft X-ray Sources, which are best interpreted as white dwarfs accreting hydrogen-rich matter at such a high rate that H burns steadily (Truemper et al 1991, Greiner et al 1991, Van Den Heuvel et al 1992, Southwell et al 1996, Kahabka & Van den Heuvel 1997). It appears that if these white dwarfs could retain the accreted gas, they might be good candidates for SN Ia progenitors. In principle, they could accrete a few tenths of a solar mass with a typical accretion rate of a few 10^{-7} $M_\odot$/year over the estimated lifetime of such systems of several 10^9 years. Because most of them are heavily extinct, their total number might be sufficiently high (Di Stefano & Rappaport 1994; see also Livio 1996, Yungelson et al 1996), although this statement is certainly model dependent. However, some of the Supersoft X-ray Sources are known to be variable in X-rays (but not in the optical wave bands) on timescales of weeks (Pakull et al 1993), too short to be related with the H-burning shell, possibly indicating substantial changes in the accretion rates. It therefore may not be justified to assume that the accretion rates we see now are sustained over several 10^9 years. But their very existence provides a first and strong case for the single-degenerate scenario.

Secondly, the minimum accretion rate at which hydrogen burns quietly without a nova outburst is uncertain. All models that compute this rate ignore important pieces of physics, and therefore, their predictions could be off by orders of magnitude. For example, classical nova outbursts require that the accreted hydrogen-rich envelope of the white dwarf also be heavily enriched in C and O from the white dwarf's core (see, e.g., Starrfield et al 1972, 1978; Sparks et al 1976; Truran 1982). One possible explanation has been that convective mixing and dredge-up might happen during the thermonuclear runaway, but recent numerical simulations indicate that this mechanism is insufficient (Kercek et al 1999). In contrast to spherically symmetric models, their three-dimensional (3D) simulations lead to a phase of quiet H-burning for accretion rates as low as $5 \times 10^{-9} \, M_\odot$/year for a white dwarf of 1 $M_\odot$ rather than a nova outburst with mass loss from the core. Other shortcomings include the assumption of spherical accretion with zero entropy, the neglect of magnetic fields, etc. So the dividing line between steady hydrogen burning and nova eruptions might leave some room for SN Ia progenitors.

Finally, it has been argued that the interaction of a wind from the white dwarf with the accretion flow from lobe-filling, low-mass red giant may open a wider path to type Ia supernovae. In a series of papers Hachisu et al (1996, 1999a, 1999b) discuss the effect that when the mass accretion rate exceeds a certain critical value, the envelope solution on the white dwarf is no longer static but corresponds to a strong wind. The strong wind stabilizes the mass transfer and limits the accretion rate in such a way that wind-loss rates and accretion rates become nearly equal. Consequently, the radius of the white dwarf does not increase with time and accretion rates leading to SNe Ia seem to be possible. However, their model assumes spherical accretion onto and a spherical wind from the white dwarf, which seem to be contradicting assumptions. But the idea should certainly be followed up.

4.3 Evolution to Ignition

In what follows, we assume most of the time that SN Ia progenitors are Chandrasekhar-mass C + O white dwarfs because, as discussed in the previous sections, this class of models seems to fit best the "typical" or "average" type Ia. In this subsection also, we do not discuss models in which two degenerate white dwarfs merge and form a critical mass for the ignition of carbon, mainly because the merging process will, in reality, be complex, and it is difficult to construct realistic explosion models (although with increasing computational resources it may be possible in the future).

But even if we consider only Chandrasekhar-mass white dwarfs as progenitor candidates, the information that is needed in order to model the explosion cannot be obtained easily. In particular, the thermal structure and the chemical composition are uncertain. The C/O-ratio, for example, has to be known throughout the white dwarf, but this ratio depends on the main sequence mass of its progenitor and the metallicity of the gas from which it formed (Umeda et al 1999, Wellstein & Langer 1999). It was found that, depending on the main sequence mass, the central C/O can vary from 0.4 to 0.6, considerably less than assumed in most supernova models.

Next, the thermal structure of a white dwarf on its way to an explosion depends on the (convective) URCA process (Paczynski 1973; Iben 1978, 1982; Barkat & Wheeler 1990; Mochkovitch 1996). The URCA pairs $A = 21, 23$, and 25 (such as, e.g., $^{21}Ne / ^{21}F, \ldots$) can lead to either heating or cooling, and possibly even to a temperature inversion near the center of the white dwarf. The abundances of the URCA pairs depends again on the initial metallicity, which could, thus, affect the thermal structure of the white dwarf. Unfortunately, the convection in the degenerate star is likely to be nonlocal, time-dependent, 3D, and subsonic, but it needs to be modeled over very long (secular) timescales. It is not likely that in the near future we will be able to model these processes in a realistic manner, even on supercomputers.

Because of these difficulties, numerical studies of the explosion rely on ad hoc assumptions fixing the initial conditions, which are usually chosen to be as simple as possible. Realistic simulations have to be multidimensional, as is explained in the next section, and therefore numerical studies can only investigate a small fraction of the available parameter space. The failure or success of a particular model to explain certain observational results may, therefore, not be conclusive.

5.   EXPLOSION MODELING

Numerical models are needed to provide the density, temperature, composition, and velocity fields of the supernova ejecta that result from the thermonuclear explosion of a white dwarf, accepted by most researchers as the "standard model" for SNe Ia (Sections 2, 4). This information can then be used to compute the resulting light curve and spectra with the help of radiation transport codes (Section 3) or to compare the relative distribution of isotopes with the observed solar abundances.

To a very good approximation, the exploding white dwarf material can be described as a fully ionized plasma with varying degrees of electron degeneracy, satisfying the fluid approximation. The governing equations are the hydrodynamical equations for mass, species, momentum, and energy transport, including gravitational acceleration, viscosity, heat and mass diffusion (Landau & Lifshitz 1995), and nuclear energy generation (Arnett 1996). They must be supplemented by an equation of state for an ideal gas of nuclei, an arbitrarily relativistic and degenerate electron gas, radiation, and electron-positron pair production and annihilation (Cox & Giuli 1968). The gravitational potential is calculated with the help of the Poisson equation. In numerical simulations that fully resolve the relevant length scales for dissipation, diffusion, and nuclear burning, it is possible to obtain the energy generation rate from a nuclear reaction network (for a recent overview, see Timmes 1999) and the diffusion coefficients from an evaluation of the kinetic transport mechanisms (Nandkumar & Pethick 1984). If, on the other hand, these scales are unresolved—as is usually the case in simulations on scales of the stellar radius—subgrid-scale models are required to compute (or parameterize) the effective large-scale transport coefficients and burning rates, which are more or less unrelated to the respective microphysical quantities (Khokhlov 1995, Niemeyer & Hillebrandt 1995b).

Initial conditions can be obtained from hydrostatic spherically symmetric models of the accreting white dwarf or—for Chandrasekhar mass progenitors—from the Chandrasekhar equation for a fully degenerate, zero-temperature white dwarf (Kippenhahn & Weigert 1989). Given the initial conditions and symmetries specifying the boundary conditions, the dynamics of the explosion can in principle be determined by numerically integrating the equations of motion. Müller (1998) gives a detailed account of some current numerical techniques used for modeling supernovae.

Until the mid-1990s, most work on SN Ia explosions was done studying 1D, spherically symmetric models. This approach inherently lacks some important aspects of multidimensional thermonuclear burning relevant for M_{chan}-explosion models, e.g. off-center flame ignition, flame instabilities, and turbulence, which have to be mimicked by means of a spherical flame front with an undetermined turbulent flame speed, e.g. Nomoto et al (1976, 1984), Woosley & Weaver (1986a), Woosley (1990). In spite of these caveats, 1D models still represent the only reasonable approach to combine the hydrodynamics with detailed nucleosynthesis calculations and to carry out parameter studies of explosion scenarios. In fact, most of the phenomenology of SN Ia explosions and virtually all of the model predictions for spectra and light curves are based on spherically symmetric models. Several recent articles (Woosley 1990, Nomoto et al 1996, Hoeflich & Khokhlov 1996, Iwamoto et al 1999) describe the methodology and trends observed in these studies, as well as their implications regarding the cosmological supernova surveys (Hoeflich et al 1998b, Ruiz-Lapuente & Canal 1998, Umeda et al 1999, Sorokina et al 1999).

Following the pioneering work of Müller & Arnett (1982, 1986), some groups have explored the dynamics of 2D (Livne 1993; Arnett & Livne 1994a,b; Niemeyer & Hillebrandt 1995b; Niemeyer et al 1996; Arnett 1997; Reinecke et al 1999a) and 3D (Khokhlov 1994, 1995; Bravo & Garcia-Senz 1997; Benz 1997) explosion models, triggering the development of numerical algorithms for representing thin propagating surfaces in large-scale simulations (Khokhlov 1993a, Niemeyer & Hillebrandt 1995b, Bravo & Garcia-Senz 1995, Arnett 1997, Garcia-Senz et al 1998, Reinecke et al 1999b). It has also become possible to perform 2D and 3D direct numerical simulations, i.e. fully resolving the relevant burning and diffusion scales, of microscopic flame instabilities and flame-turbulence interactions (Niemeyer & Hillebrandt 1995a, Khokhlov 1995, Niemeyer & Hillebrandt 1997, Niemeyer et al 1999).

5.1 Chandrasekhar Mass Explosion Models

Given the overall homogeneity of SNe Ia (Section 2.1), the good agreement of parameterized 1D M_{chan} models with observed spectra and light curves, and their reasonable nucleosynthetic yields, the bulk of normal SNe Ia is generally assumed to consist of exploding M_{chan} C + O white dwarfs (Hoyle & Fowler 1960, Arnett 1969, Hansen & Wheeler 1969). In spite of three decades of work on the hydrodynamics of this explosion mechanism (beginning with Arnett (1969)), no clear consensus has been reached as to whether the star explodes as a result of a subsonic nuclear deflagration that becomes strongly turbulent (Ivanova et al 1974; Buchler & Mazurek 1975; Nomoto et al 1976, 1984; Woosley et al 1984), or whether this turbulent flame phase is followed by a delayed detonation during the expansion (Khokhlov 1991a,b; Woosley & Weaver 1994a) or after one or many pulses (Khokhlov 1991b; Arnett & Livne 1994a,b). Only the prompt detonation mechanism is agreed to be inconsistent with SN Ia spectra, as it fails to produce sufficient amounts of intermediate mass elements (Arnett 1969, Arnett et al 1971).

This apparently slow progress is essentially a consequence of the overwhelming complexity of turbulent flame physics and deflagration-detonation transitions (DDTs) (Williams 1985, Zeldovich et al 1985) that makes first-principle predictions based on M_{chan} explosion models nearly impossible. The existence of an initial subsonic flame phase is, it seems, an unavoidable ingredient of all M_{chan} models (and only those) where it is required to preexpand the stellar material prior to its nuclear consumption in order to avoid the almost exclusive production of iron-peaked nuclei (Nomoto et al 1976, 1984; Woosley & Weaver 1986a).

Guided by parameterized 1D models that yield estimates for the values for the turbulent flame speed S_t and the DDT transition density ρ_{DDT} (e.g. Hoeflich & Khokhlov 1996), much work has been done recently on the physics of buoyancy-driven, turbulent thermonuclear flames in exploding M_{chan} white dwarfs. The close analogy with thin chemical premixed flames has been exploited to develop a conceptual framework that covers all scales from the white dwarf radius to the microscopic flame thickness and dissipation scales (Khokhlov 1995, Niemeyer & Woosley 1997). In the following discussions of nuclear combustion, flame ignition, and the various scenarios for M_{chan} explosions characterized by the sequence of combustion modes, we emphasize the current understanding of physical processes rather than empirical fits of light curves and spectra.

5.1.1 Flames, Turbulence, and Detonations   Owing to the strong temperature dependence of the nuclear reaction rates, $\dot{S} \sim T^{12}$ at $T \approx 10^{10}$ K (Hansen & Kawaler 1994:247), nuclear burning during the explosion is confined to microscopically thin layers that propagate either conductively as subsonic deflagrations ("flames") or by shock compression as supersonic detonations (Courant & Friedrichs 1948, Landau & Lifshitz 1995:Ch. 14). Both modes are hydrodynamically unstable to spatial perturbations, as can be shown by linear perturbation analysis. In the nonlinear regime, the burning fronts either are stabilized by forming a cellular structure or become fully turbulent—either way, the total burning rate increases as a result of flame surface growth (Lewis & von Elbe 1961, Williams 1985, Zeldovich et al 1985). Neither flames nor detonations can be resolved in explosion simulations on stellar scales and therefore must be represented by numerical models.

When the fuel exceeds a critical temperature T_c where burning proceeds nearly instantaneously compared with the fluid motions [for a suitable definition of T_c, see Timmes & Woosley (1992)], a thin reaction zone forms at the interface between burned and unburned material. It propagates into the surrounding fuel by one of two mechanisms allowed by the Rankine-Hugoniot jump conditions: a deflagration ("flame") or a detonation (cf Figure 2.5 in Williams 1985).

If the overpressure created by the heat of the burning products is sufficiently high, a hydrodynamical shock wave forms that ignites the fuel by compressional heating. A self-sustaining combustion front that propagates by shock heating is called a detonation. Detonations generally move supersonically and therefore do not allow the unburned medium to expand before it is burned. Their speed depends mainly on the total amount of energy released per unit mass, ϵ, and is therefore more robustly computable than deflagration velocities. A good estimate for the velocity of planar strong detonations is the Chapman-Jouget velocity (Lewis & von Elbe 1961, Zeldovich et al 1985, Williams 1985, and references therein). The nucleosynthesis, speed, structure, and stability of planar detonations in degenerate $C+O$ material was analyzed by Imshennik & Khokhlov (1984), by Khokhlov (1988, 1989, 1993b), and recently by Kriminski et al (1998) and Imshennik et al (1999), who claim that $C+O$ detonations are one-dimensionally unstable and therefore cannot occur in exploding white dwarfs above a critical density of $\sim 2 \times 10^7$ g cm^{-3} (Kriminski et al 1998) (cf Section 5.1.3).

If, on the other hand, the initial overpressure is too weak, the temperature gradient at the fuel-ashes interface steepens until an equilibrium between heat diffusion (carried out predominantly by electron-ion collisions) and energy generation is reached. The resulting combustion front consists of a diffusion zone that heats up the fuel to T_c, followed by a thin reaction layer where the fuel is consumed and energy is generated. It is called a deflagration or simply a flame and moves subsonically with respect to the unburned material (Landau & Lifshitz 1995). Flames, unlike detonations, may therefore be strongly affected by turbulent velocity fluctuations of the fuel. Only if the unburned material is at rest, a unique laminar flame speed S_l can be found, which depends on the detailed interaction of burning and diffusion within the flame region (e.g. Zeldovich et al 1985). According to Landau & Lifshitz (1995), it can be estimated by assuming that in order for burning and diffusion to be in equilibrium, the respective timescale timescales, $\tau_b \sim \epsilon/\dot{w}$ and $\tau_d \sim \delta^2/\kappa$, where δ is the flame thickness and κ is the thermal diffusivity, must be similar: $\tau_b \sim \tau_d$. Defining $S_l = \delta/\tau_b$, one finds $S_l \sim (\kappa \dot{w}/\epsilon)^{1/2}$, where $\dot{w}$ should be evaluated at $T \approx T_c$ (Timmes & Woosley 1992). This is only a crude estimate due to the strong T dependence of $\dot{w}$. Numerical solutions of the full equations of hydrodynamics, including nuclear energy generation and heat diffusion, are needed to obtain more accurate values for S_l as a function of ρ and fuel composition. Laminar thermonuclear carbon and oxygen flames at high to intermediate densities were investigated by Buchler et al (1980), Ivanova et al (1982), and Woosley & Weaver (1986b), and, using a variety of different techniques and nuclear networks, by Timmes & Woosley (1992). For the purpose of SN Ia explosion modeling, one needs to know the laminar flame speed $S_l \approx 10^7 \ldots 10^4$ cm s^{-1} for $\rho \approx 10^9 \ldots 10^7$ g cm^{-3}, the flame thickness $\delta = 10^{-4} \ldots 1$ cm (defined here as the width of the thermal preheating layer ahead of the much thinner reaction front), and the density contrast between burned and unburned material $\mu = \Delta\rho/\rho = 0.2 \ldots 0.5$ [all values quoted here assume a composition of $X_C = X_O = 0.5$, (Timmes & Woosley 1992)]. The thermal expansion parameter μ reflects the partial lifting of electron degeneracy in the burning products and is much lower than the typical value found in chemical, ideal gas systems (Williams 1985).

Observed on scales much larger than δ, the internal reaction-diffusion structure can be neglected and the flame can be approximated as a density jump that propagates locally with the normal speed S_l. This "thin flame" approximation allows a linear stability analysis of the front with respect to spatial perturbations. The result shows that thin flames are linearly unstable on all wavelengths. It was discovered first by Landau (1944) and Darrieus (1944), and is hence called the Landau-Darrieus (LD) instability. Subject to the LD instability, perturbations grow until a web of cellular structures forms and stabilizes the front at finite perturbation amplitudes (Zeldovich 1966). The LD instability therefore does not, in general, lead to the production of turbulence. In the context of SN Ia models, the nonlinear LD instability was studied by Blinnikov & Sasorov (1996), using a statistical approach based on the Frankel equation, and by Niemeyer & Hillebrandt (1995a) employing 2D hydrodynamics and a one-step burning rate. Both groups concluded that the cellular stabilization mechanism precludes a strong acceleration of the burning front as a result of the LD instability. However, Blinnikov & Sasorov (1996) mention the possible breakdown of stabilization at low stellar densities (i.e. high μ), which is also indicated by the lowest density run of Niemeyer & Hillebrandt (1995a)—this may be important in the framework of active turbulent combustion (see below). The linear growth rate of LD unstable thermonuclear flames with arbitrary equation of state was derived by Bychkov & Liberman (1995a). The same authors also found a 1D, pulsational instability of degenerate C + O flames (Bychkov & Liberman 1995b), which was later disputed by Blinnikov (1996).

The best-studied and probably most important hydrodynamical effect for modeling SN Ia explosions is the Rayleigh-Taylor (RT) instability (Rayleigh 1883, Chandrasekhar 1961) resulting from the buoyancy of hot, burned fluid with respect to the dense, unburned material. Several groups have investigated the RT instability of nuclear flames in SNe Ia by means of numerical hydrodynamical simulations (Müller & Arnett 1982, 1986; Livne 1993; Khokhlov 1994, 1995; Niemeyer & Hillebrandt 1995b). After more than five decades of experimental and numerical work, the basic phenomenology of nonlinear RT mixing is fairly well understood (Fermi 1951, Layzer 1955, Sharp 1984, Read 1984, Youngs 1984): Subject to the RT instability, small surface perturbations grow until they form bubbles (or "mushrooms") that begin to float upward while spikes of dense fluid fall down. In the nonlinear regime, bubbles of various sizes interact and create a foamy RT mixing layer whose vertical extent h_{RT} grows with time t according to a self-similar growth law, $h_{RT} = \alpha g(\mu/2)t^2$, where α is a dimensionless constant ($\alpha \approx 0.05$) and g is the background gravitational acceleration (Sharp 1984, Youngs 1984, Read 1984).

Secondary instabilities related to the velocity shear along the bubble surfaces (Niemeyer & Hillebrandt 1997) quickly lead to the production of turbulent velocity fluctuations that cascade from the size of the largest bubbles ($\approx 10^7$ cm) down to the microscopic Kolmogorov scale, $l_k \approx 10^{-4}$ cm, where they are dissipated (Niemeyer & Hillebrandt 1995b, Khokhlov 1995). Because no computer is capable of resolving this range of scales, one must resort to statistical or scaling approximations of those length scales that are not properly resolved. The most prominent scaling relation in turbulence research is Kolmogorov's law for the cascade of velocity fluctuations, stating that in the case of isotropy and statistical stationarity, the mean velocity v of turbulent eddies with size l scales as $v \sim l^{1/3}$ (Kolmogorov 1941). Knowledge of the eddy velocity as a function of length scale is important to classify the burning regime of the turbulent combustion front (Niemeyer & Woosley 1997, Niemeyer & Kerstein 1997, Khokhlov et al 1997). The ratio of the laminar flame speed and the turbulent velocity on the scale of the flame thickness, $K = S_l/v(\delta)$, plays an important role: If $K \gg 1$, the laminar flame structure is nearly unaffected by turbulent fluctuations. Turbulence does, however, wrinkle and deform the flame on scales l, where $S_l \ll v(l)$, i.e. above the Gibson scale l_g defined by $S_l = v(l_g)$ (Peters 1988). These wrinkles increase the flame surface area and therefore the total energy generation rate of the turbulent front (Damköhler 1940). In other words, the turbulent flame speed, S_t, defined as the mean overall propagation velocity of the turbulent flame front, becomes larger than the laminar speed S_l. If the turbulence is sufficiently strong, $v(L) \gg S_l$, the turbulent flame speed becomes independent of the laminar speed, and therefore of the microphysics of burning and diffusion, and scales only with the velocity of the largest turbulent eddy (Damköhler 1940, Clavin 1994):

$$S_t \sim v(L). \tag{2}$$

Because of the unperturbed laminar flame properties on very small scales, and the wrinkling of the flame on large scales, the burning regime where $K \gg 1$ is called the corrugated flamelet regime (Pope 1987, Clavin 1994).

As the density of the white dwarf material declines and the laminar flamelets become slower and thicker, it is plausible that at some point turbulence significantly alters the thermal flame structure (Khokhlov et al 1997, Niemeyer & Woosley 1997). This marks the end of the flamelet regime and the beginning of the distributed burning, or distributed reaction zone, regime (e.g. Pope 1987). So far, modeling the distributed burning regime in exploding white dwarfs has not been attempted explicitly because neither nuclear burning and diffusion nor turbulent mixing can be properly described by simplified prescriptions. Phenomenologically, the laminar flame structure is believed to be disrupted by turbulence and to form a distribution of reaction zones with various lengths and thicknesses. In order to find the critical density for the transition between both regimes, we need to formulate a specific criterion for flamelet breakdown. A criterion for the transition between both regimes is discussed by Niemeyer & Woosley (1997), Niemeyer & Kerstein (1997) and Khokhlov et al (1997):

$$l_{\text{cutoff}} \leq \delta. \tag{3}$$

Inserting the results of Timmes & Woosley (1992) for S_l and δ as functions of density, and using a typical turbulence velocity $v(10^6 \text{cm}) \sim 10^7$ cm s^{-1}, the transition from flamelet to distributed burning can be shown to occur at a density of $\rho_{\text{dis}} \approx 10^7$ g cm^{-3} (Niemeyer & Kerstein 1997).

The close coincidence of ρ_{dis} and the preferred value for ρ_{DDT} (Hoeflich & Khokhlov 1996, Nomoto et al 1996) inspired some authors (Niemeyer & Woosley 1997, Khokhlov et al 1997) to suggest that both are related by local flame quenching and reignition via the Zeldovich induction time gradient mechanism (Zeldovich et al 1970), whereby a macroscopic region with a uniform temperature gradient can give birth to a supersonic spontaneous combustion wave that steepens into a detonation (Woosley 1990, and references therein). In the context of the SN Ia explosion mechanism, this effect was first analyzed by Blinnikov & Khokhlov (1986, 1987). Whether or not the gradient mechanism can account for DDTs in the delayed detonation scenario for SNe Ia is still controversial: Khokhlov et al (1997) conclude that it can, whereas Niemeyer (1999)—using arguments based on incompressible computations of microscopic flame-turbulence interactions by Niemeyer et al (1999)—states that thermonuclear flames may be too robust with respect to turbulent quenching to allow the formation of a sufficiently uniform temperature gradient.

Assuming that the nonlinear RT instability dominates the turbulent flow that advects the flame, the passive-surface description of the flame neglects the additional stirring caused by thermal expansion within the flame brush itself, accelerating the burnt material in random directions. Both the spectrum and cutoff scale may be affected by "active" turbulent combustion (Kerstein 1996, Niemeyer & Woosley 1997). Although the small expansion coefficient μ indicates that the effect is weak compared with chemical flames, a quantitative answer is still missing.

Finally, we note that some authors also studied the multidimensional instability of detonations in degenerate C + O matter (Boisseau et al 1996, Gamezo et al 1999), finding unsteady front propagation, the formation of a cellular front structure, and locally incomplete burning in multidimensional C + O detonations. These effects may have interesting implications for SN Ia scenarios involving a detonation phase.

5.1.2 Flame Ignition   As the white dwarf grows close to the Chandrasekhar mass $M_{chan} \approx 1.4 M_\odot$, the energy budget near the core is governed by plasmon neutrino losses and compressional heating. The neutrino losses increase with growing central density until the latter reaches approximately 2×10^9 g cm^{-3} (Woosley & Weaver 1986a). At this point, plasmon creation becomes strongly suppressed while electron screening of nuclear reactions enhances the energy generation rate until it begins to exceed the neutrino losses. This "smoldering" of the core region marks the beginning of the thermonuclear runaway (Arnett 1969, 1971; Woosley & Weaver 1986a). During the following ~ 1000 years, the core experiences internally heated convection with progressively smaller turnover timescale timescales τ_c. Simultaneously, the typical timescale for thermonuclear burning, τ_b, drops even faster as a result of the rising core temperature and the steep temperature dependence of the nuclear reaction rates.

During this period, the entropy and temperature evolution of the core is affected by the convective URCA process, a convectively driven electron capture-beta decay cycle leading to neutrino-antineutrino losses. It was first described in this context by Paczynski (1972), who argued it would cause net cooling and therefore delay the runaway. Since then, the convective URCA process was revisited by several authors (e.g. Bruenn 1973, Iben 1982, Barkat & Wheeler 1990, Mochkovitch 1996), who alternately claimed that it results in overall heating or cooling. The most recent analysis (Stein et al 1999) concludes that although the URCA neutrinos carry away energy, they cannot cool the core globally but instead slow down the convective motions.

At $T \approx 7 \times 10^8$ K, τ_c and τ_b become comparable, indicating that convective plumes burn at the same rate as they circulate (Nomoto et al 1984, Woosley & Weaver 1986a). Experimental or numerical data describing this regime of strong reactive convection is not available, but several groups are planning to conduct numerical experiments at the time this article is written. At $T \approx 1.5 \times 10^9$ K, τ_b becomes extremely small compared with τ_c, and carbon and oxygen virtually burn in place. A new equilibrium between energy generation and transport is found on much smaller length scales, $l \approx 10^{-4}$ cm, where thermal conduction by degenerate electrons balances nuclear energy input (Timmes & Woosley 1992). The flame is born.

The evolution of the runaway immediately prior to ignition of the flame is crucial for determining its initial location and shape. Using a simple toy model, Garcia-Senz & Woosley (1995) found that under certain conditions, burning bubbles subject to buoyancy and drag forces can rise a few hundred kilometers before flame formation, which suggests a high probability for off-center ignition at multiple, unconnected points. As a consequence, more material burns at lower densities, thus producing higher amounts of intermediate mass elements than a centrally ignited explosion. In a parameter study, Niemeyer et al (1996) and Reinecke et al (1999a) demonstrated the significant influence of the location and number of initially ignited spots on the final explosion energetics and nucleosynthesis.

5.1.3 Prompt Detonation   The first hydrodynamical simulation of an exploding M_{chan} white dwarf (Arnett 1969) assumed that the thermonuclear combustion commences as a detonation wave, consuming the entire star at the speed of sound. Given no time to expand prior to being burned, the C + O material in this scenario is transformed almost completely into iron-peak nuclei and thus fails to produce significant amounts of intermediate mass elements, in contradiction to observations (Filippenko 1997a,b). It is for this reason that prompt detonations are generally considered ruled out as viable candidates for the SN Ia explosion mechanism.

In addition to the empirical evidence, the ignition of a detonation in the high-density medium of the white dwarf core was argued to be an unlikely event. In spite of the smallness of the critical mass for detonation at $\rho \approx 2 \times 10^9$ g cm^{-3} (Niemeyer & Woosley 1997, Khokhlov et al 1997) and the correspondingly large number of critical volumes in the core ($\sim 10^{18}$), the stringent uniformity condition for the temperature gradient of the runaway region (Blinnikov & Khokhlov 1986, 1987) was shown to be violated even by the minute amounts of heat dissipated by convective motions (Niemeyer & Woosley 1997). A different argument against the occurrence of a prompt detonation in C + O white dwarf cores was given by Kriminski et al (1998), who found that C + O detonations may be subject to self-quenching at high material densities ($\rho > 2 \times 10^7$ g cm^{-3}) (see also Imshennik et al 1999).

5.1.4 Pure Turbulent Deflagration   Once ignited (Section 5.1.2), the subsonic thermonuclear flame becomes highly convoluted as a result of turbulence produced by the various flame instabilities (Section 5.1.1). It continues to burn through the star until it either transitions into a detonation or is quenched by expansion. The key questions with regard to explosion modeling are the following: (*a*) What is the effective turbulent flame speed S_t as a function of time, (*b*) is the total amount of energy released during the deflagration phase enough to unbind the star and produce a healthy explosion, and (*c*) does the resulting ejecta composition and velocity agree with observations?

By far the most work has been done on 1D models, ignoring the multidimensionality of the flame physics and instead parameterizing S_t in order to answer the second and third questions (for reviews, see Woosley & Weaver 1986a, Nomoto et al 1996). One of the most successful examples, model W7 of Nomoto et al (1984), clearly demonstrates the excellent agreement of "fast" deflagration models with SN Ia spectra and light curves. S_t has been parameterized differently by different authors, for instance as a constant fraction of the local sound speed (Hoeflich & Khokhlov 1996, Iwamoto et al 1999), using time-dependent convection theory (Nomoto et al 1976, 1984; Buchler & Mazurek 1975; Woosley et al 1984), or with a phenomenological fractal model describing the multiscale character of the wrinkled flame surface (Woosley 1990, 1997b). All these studies essentially agree that very good agreement with the observations is obtained if S_t accelerates up to roughly 30% of the sound speed. There remains a problem with the overproduction of neutron-rich iron-group isotopes in fast deflagration models (Woosley et al 1984, Thielemann et al 1986, Iwamoto et al 1999), but this may be alleviated in multiple dimensions (see below). Turning this argument around, Woosley (1997a) argues that ^{48}Ca can only be produced by carbon burning in the very-high-density regime of a M_{chan} white dwarf core, providing a clue that a few SNe Ia need to be M_{chan} explosions igniting at $\rho \geq 2 \times 10^9$ g cm^{-3}. A slightly different approach to 1D SN Ia modeling was taken by Niemeyer & Woosley (1997), who employed the self-similar growth rate of RT mixing regions (Section 5.1.1) to prescribe the turbulent flame speed. Here, all the free parameters are fixed by independent simulations or experiments. The result shows a successful explosion, albeit short on intermediate mass elements, which suggests that the employed flame model is still too simplistic.

A number of authors have studied multidimensional deflagrations in exploding M_{chan} white dwarfs using a variety of hydrodynamical methods (Livne 1993, Arnett & Livne 1994a, Khokhlov 1995, Niemeyer & Hillebrandt 1995b, Niemeyer et al 1996, Reinecke et al 1999a). The problem of simulating subsonic flames in large-scale simulations has two aspects: the representation of the thin, propagating surface separating hot and cold material with different densities, and the prescription of the local propagation velocity $S_t(\Delta)$ of this surface as a function of the hydrodynamical state of the large-scale calculation with numerical resolution Δ. The former problem has been addressed with artificial reaction-diffusion fronts in codes based on the Piecewise Parabolic Method (PPM) (Khokhlov 1995, Niemeyer & Hillebrandt 1995b, Niemeyer et al 1996) and Smoothed Particle Hydrodynamics (SPH) (Garcia-Senz et al 1998), in a PPM-specific flame-tracking technique (Arnett 1997), and in a hybrid flame-capturing/tracking method based on level sets (Reinecke et al 1999b) (see Figure 3, color insert). Regarding the flame speed prescription, some authors assigned the local front propagation velocity assuming that the flame is laminar on unresolved scales $l < \Delta$ (Arnett & Livne 1994a), by postulating that $S_t(\Delta)$ is dominated by the terminal rise velocity of Δ-sized bubbles (Khokhlov 1995), or by using Equation 2 together with a subgrid-scale model for the unresolved turbulent kinetic energy providing $v(\Delta)$ (Niemeyer & Hillebrandt 1995b, Niemeyer et al 1996, Reinecke et al 1999a).

In most multidimensional calculations on stellar scales to date, the effective turbulent flame speed stayed below the required 30% of the sound speed. The detailed outcome of the explosion is controversial: Whereas some calculations show that the star remains gravitationally bound after the deflagration phase has ceased (Khokhlov 1995), others indicate that S_t may be large enough to produce a weak but definitely unbound explosion (Niemeyer et al 1996). These discrepancies can probably be attributed to differences in the description of the turbulent flame and to numerical resolution effects that plague all multidimensional calculations.

Niemeyer & Woosley (1997) and Niemeyer (1999) speculate about additional physics that can increase the burning rate in turbulent deflagration models, in particular multipoint ignition and active turbulent combustion (ATC), i.e. the generation of additional turbulence by thermal expansion within the turbulent flame brush. ATC can, in principle, explain the acceleration of S_t up to some fraction of the sound speed (Kerstein 1996), but its effectiveness is unknown. Multipoint ignition, on the other hand, has already been shown to significantly increase the total energy release compared with single-point ignition models (Niemeyer et al 1996, Reinecke et al 1999a). Furthermore, it allows more material to burn at lower densities, thus alleviating the nucleosynthesis problem of 1D fast deflagration models (Niemeyer et al 1996).

We conclude the discussion of the pure turbulent deflagration scenario with a checklist of the model requirements summarized in Section 2.4. Assuming that some combination of buoyancy, ATC, and multipoint ignition can drive the effective turbulent flame speed to $\sim 30\%$ of the sound speed—which is not evident from multidimensional simulations—one can conclude from 1D simulations that pure deflagration models readily comply with all observational constraints. Most authors agree that S_t decouples from microphysics on large enough scales and becomes dominated by essentially universal hydrodynamical effects, making the scenario intrinsically robust. A noteworthy exception is the location and number of ignition points that can strongly influence the explosion outcome and may be a possible candidate for the mechanism giving rise to the explosion strength variability. Other possible sources of variations include the ignition density and the accretion rate of the progenitor system (Umeda et al 1999, Iwamoto et al 1999). All these effects may potentially vary with composition and metallicity and can therefore account for the dependence on the progenitor stellar population.

5.1.5 Delayed Detonation

Turbulent deflagrations can sometimes be observed to undergo spontaneous transitions to detonations [deflagration-detonation transitions (DDTs)] in terrestrial combustion experiments (e.g. Williams 1985:217–19). Thus inspired, it was suggested that DDTs may occur in the late phase of a M_{chan} explosion, providing an elegant explanation for the initial slow burning required to preexpand the star, followed by a fast combustion mode that produces large amounts of high-velocity intermediate mass elements (Khokhlov 1991a, Woosley & Weaver 1994a). Meanwhile, many 1D simulations have demonstrated the capability of the delayed detonation scenario to provide excellent fits to SN Ia spectra and light curves (Woosley 1990, Hoeflich & Khokhlov 1996), as well as reasonable nucleosynthesis products with regard to solar abundances (Khokhlov 1991b, Iwamoto et al 1999). In the best-fit models, the initial flame phase has a slow velocity of roughly 1% of the sound speed and transitions to detonation at a density of $\rho_{DDT} \approx 10^7$ g cm^{-3} (Hoeflich & Khokhlov 1996, Iwamoto et al 1999). The transition density was also found to be a convenient parameter to explain the observed sequence of explosion strengths (Hoeflich & Khokhlov 1996).

Various mechanisms for DDT were discussed in the early literature on delayed detonations (see Niemeyer & Woosley 1997, and references therein). Recent investigations have focussed on the induction time gradient mechanism (Zeldovich et al 1970, Lee et al 1978), analyzed in the context of SNe Ia by Blinnikov & Khokhlov (1986, 1987). It was realized by Khokhlov et al (1997) and Niemeyer

& Woosley (1997) that a necessary criterion for this mechanism is the local disruption of the flame sheet by turbulent eddies or, in other words, the transition of the burning regime from flamelet to distributed burning (Section 5.1.1). Simple estimates (Niemeyer & Kerstein 1997) show that this transition should occur at roughly 10^7 g cm^{-3}, providing a plausible explanation for the delay of the detonation.

The critical length (or mass) scale over which the temperature gradient must be held fixed in order to allow the spontaneous combustion wave to turn into a detonation was computed by Khokhlov et al (1997) and Niemeyer & Woosley (1997); it is a few orders of magnitude thicker than the final detonation front and depends very sensitively on composition and density.

The virtues of the delayed detonation scenario can again be summarized by completing the checklist of Section 2.4. It is undisputed that suitably tuned delayed detonations satisfy all the constraints given by SN Ia spectra, light curves, and nucleosynthesis. If ρ_{DDT} is indeed determined by the transition of burning regimes—which in turn might be composition dependent (Umeda et al 1999)—the scenario is also fairly robust and ρ_{DDT} may represent the explosion strength parameter. Note that in this case, the variability induced by multipoint ignition needs to be explained away. If, on the other hand, thermonuclear flames are confirmed to be almost unquenchable, the favorite mechanism for DDTs becomes questionable (Niemeyer 1999). Moreover, should the mechanism DDT rely on rare, strong turbulent fluctuations, one must ask about those events that fail to ignite a detonation following the slow deflagration phase, which, on its own, cannot give rise to a viable SN Ia explosion. They might end up as pulsational delayed detonations or as unobservably dim, as yet unclassified explosions. Multidimensional simulations of the turbulent flame phase may soon answer whether the turbulent flame speed is closer to 1% or 30% of the speed of sound and hence decide whether DDTs are a necessary ingredient of SN Ia explosion models.

5.1.6 Pulsational Delayed Detonation

In this variety of the delayed detonation scenario, the first turbulent deflagration phase fails to release enough energy to unbind the star that subsequently pulses and triggers a detonation upon recollapse (Nomoto et al 1976; Khokhlov 1991b). This model was studied in 1D by Hoeflich & Khokhlov (1996) and Woosley (1997b) (who calls it "pulsed detonation of the first type") and in 2D by Arnett & Livne (1994b). Hoeflich & Khokhlov (1996) report that it produces little ^{56}Ni but a substantial amount of Si and Ca and may therefore explain very subluminous events, such as SN 1991bg. Using a fractal flame parameterization, Woosley (1997b) also considered "pulsed deflagrations," i.e. reignition occurs as a deflagration rather than a detonation, and "pulsed detonations of the second type," in which the burning also reignites as a flame but later accelerates and touches off a detonation. This latter model closely resembles the standard delayed detonation, whereas the former may or may not produce a healthy explosion, depending on the prescribed speed of the rekindled flame (Woosley 1997b).

Obtaining a DDT by means of the gradient mechanism is considerably more plausible after one or several pulses than during the first expansion phase (Khokhlov et al 1997), as the laminar flame thickness becomes macroscopically large during the expansion, allowing the fuel to be preheated, and turbulence is significantly enhanced during the collapse.

The "checklist" for pulsational delayed detonations looks similar to that of simple delayed detonations (see above), with somewhat less emphasis on the improbability for DDT. Some fine-tuning of the initial flame speed is needed to obtain a large enough pulse in order to achieve a sufficient degree of mixing, while avoiding to unbind the star in a very weak explosion (Niemeyer & Woosley 1997). Again, these "fizzles" may be very subluminous and may have escaped discovery. We finally note that all pulsational models are in conflict with multidimensional simulations that predict an unbound star after the first deflagration phase.

5.2   Sub-Chandrasekhar Mass Models

C + O white dwarfs below the Chandrasekhar mass do not reach the critical density and temperature for explosive carbon burning by accretion and therefore need to be ignited by an external trigger. Detonations in the accreted He layer were suggested to drive a strong enough shock into the C + O core to initiate a secondary carbon detonation (Weaver & Woosley 1980, Nomoto 1980, 1982b; Woosley et al 1980, Sutherland & Wheeler 1984, Iben & Tutukov 1984). The nucleosynthesis and light curves of sub-M_{chan} models, also known as helium ignitors or edge-lit detonations, were investigated in 1D (Woosley & Weaver 1994b, Hoeflich & Khokhlov 1996) and 2D (Livne & Arnett 1995) and found to be superficially consistent with SNe Ia, especially subluminous ones (Ruiz-Lapuente et al 1993a). Their ejecta structure is characterized almost inevitably by an outer layer of high-velocity Ni and He above the intermediate-mass elements and the inner Fe / Ni core.

These models are favored mostly by the statistics of possible SN Ia progenitor systems (Yungelson & Livio 1998, Livio 1999) and by the straightforward explanation of the one-parameter strength sequence in terms of the white dwarf mass (Ruiz-Lapuente et al 1995). However, they appear to be severely challenged both photometrically and spectroscopically: Owing to the heating by radioactive ^{56}Ni in the outer layer, they are somewhat too blue at maximum brightness and their light curve rises and declines too steeply (Hoeflich & Khokhlov 1996, Nugent et al 1997, Hoeflich et al 1997). Perhaps even more stringent is the generic prediction of He ignitors to exhibit signatures of high-velocity Ni and He, rather than Si and Ca, in the early and maximum spectra, which is in strong disagreement with observations (Nugent et al 1997, Hoeflich et al 1997).

With respect to the explosion mechanism itself, the most crucial question is whether and where the He detonation manages to shock the C + O core sufficiently to create a carbon detonation. By virtue of their built-in spherical symmetry, 1D models robustly (and unphysically) predict a perfect convergence of the inward propagating pressure wave and subsequent carbon ignition near the core (Woosley

& Weaver 1994b). Some 2D simulations indicate that the C + O detonation is born off-center but still due to the convergence of the He-driven shock near the symmetry axis of the calculation (Livne 1990, Livne & Glasner 1991) whereas others find a direct initiation of the carbon detonation along the circle where the He detonation intersects the C + O core (Livne 1997, Arnett 1997, Wiggins & Falle 1997, Wiggins et al 1998). Using 3D SPH simulations, Benz (1997) failed to see carbon ignition in all but the highest-resolution calculations, where carbon was ignited directly at the interface rather than by shock convergence. Further, C ignition is facilitated if the He detonation starts at some distance above the interface, allowing the build-up of a fully developed pressure spike before it hits the carbon (Benz 1997). This result was confirmed by recent 3D SPH simulations (Garcia-Senz et al 1999) that also examined the effect of multiple He ignition points, finding enhanced production of intermediate mass elements in this case. Hence, multidimensional SPH and PPM simulations presently confirm the validity of He-driven carbon detonations, in particular by direct ignition, but they also demonstrate the need for very high numerical resolution in order to obtain mutually consistent results (Arnett 1997, Benz 1997).

To summarize, sub-M_{chan} models are most severely constrained by their prediction of an outer layer of high-velocity Ni and He. Should further research conclude that spectra, colors, and light curves are less contaminated by this layer than presently thought, they represent an attractive class of candidates for SNe Ia, especially subluminous ones, from the point of view of progenitor statistics and the one-parameter explosion strength family. Note, however, that the SN Ia luminosity function in this scenario is directly linked to the distribution of white dwarf masses, predicting a more gradual decline on the bright side of the luminosity function than indicated by observations (Vaughan et al 1995, Livio 1999). The explosion mechanism itself appears realistic, at least in the direct carbon ignition mode, but more work is needed to firmly establish the conditions for ignition of the secondary carbon detonation.

5.3   Merging White Dwarfs

The most obvious strength of the merging white dwarfs, or double-degenerate, scenario for SNe Ia (Webbink 1984, Iben & Tutukov 1984, Paczynski 1985) is the natural explanation for the lack of hydrogen in SN Ia spectra (Livio 1999) (cf Section 2.1). Furthermore, in contrast to the elusive progenitor systems for single-degenerate scenarios, there is some evidence for the existence of double-degenerate binary systems (Saffer et al 1998) despite earlier suspicions to the contrary (e.g. Bragaglia 1997). These systems are bound to merge as a consequence of gravitational wave emission with about the right statistics (Livio 1999) and give rise to some extreme astrophysical event, albeit not necessarily a SN Ia.

Spherically symmetric models of detonating merged systems, parameterized as C + O white dwarfs with thick envelopes, were analyzed by Hoeflich et al (1992), Khokhlov et al (1993) and Hoeflich & Khokhlov (1996), giving reasonable agreement with SN Ia light curves. SPH simulations (3D) of white dwarfs mergers (Benz et al 1990, Rasio & Shapiro 1995, Mochkovitch et al 1997) show the disruption of the less-massive star in a matter of a few orbital times, followed by the formation of a thick hot accretion disk around the more-massive companion. The further evolution hinges crucially on the effective accretion rate of the disk: In case $\dot{M}$ is larger than a few times 10^{-6} $M_\odot$ year^{-1}, the most likely outcome is off-center carbon ignition leading to an inward propagating flame that converts the star into $O + Ne + Mg$ (Nomoto & Iben 1985, Saio & Nomoto 1985, Kawai et al 1987, Timmes et al 1994, Saio & Nomoto 1998). This configuration, in turn, is gravitationally unstable owing to electron capture onto ^{24}Mg and will undergo accretion-induced collapse to form a neutron star (Saio & Nomoto 1985, Mochkovitch & Livio 1990, Nomoto & Kondo 1991). A recent reexamination of Coulomb corrections to the equation of state of material in nuclear statistical equilibrium indicates that accretion-induced collapse in merged white dwarf systems is even more likely than previously anticipated (Bravo & Garcia-Senz 1999).

Dimensional analysis of the expected turbulent viscosity due to magneto-hydrodynamical (MHD) instabilities (Balbus et al 1996) suggests that it is very difficult to avoid such high accretion rates (Mochkovitch & Livio 1990, Livio 1999). Even under the unphysical assumption that angular momentum transport is dominated entirely by microscopic electron-gas viscosity, the expected life time of $\sim 10^9$ years (Mochkovitch & Livio 1990, Mochkovitch et al 1997) and high UV luminosity of these accretion systems would predict the existence of $\sim 10^7$ such objects in the Galaxy, none of which have been observed (Livio 1999).

A possible solution to the collapse problem is to ignite carbon burning as a detonation rather than a flame immediately during the merger event, either in the core of the more massive star (Shigeyama et al 1992) or at the contact surface (D Arnett & PA Pinto, private communication). This alternative clearly warrants further study.

To summarize, the merging white dwarf scenario must overcome the crucial problem of avoiding accretion-induced collapse before it can be seriously considered as a SN Ia candidate. Its key strengths are a plausible explanation for the progenitor history, yielding reasonable predictions for SN Ia rates, the straightforward explanation of the absence of H and He in SN Ia spectra, and the existence of a simple parameter for the explosion strength family (i.e. the mass of the merged system).

6.   SUMMARY

In this review we have outlined our current understanding of type Ia supernovae, summarizing briefly the observational constraints, but putting more weight on models of the explosion. From the tremendous amount of work carried out over the past couple of years, it has become obvious that the physics of SNe Ia is complex, ranging from the possibility of different progenitors to the complexity of the physics leading to the explosion and the complicated processes that couple the interior physics to observable quantities. None of these problems is fully understood yet, but what one is tempted to state is that, from a theorist's point of view, it appears to be a miracle that all the complexity seems to average out in a mysterious way to make the class so homogeneous. In contrast, as it stands, a safe prediction from theory seems to be that SNe Ia should get more diverse with increasing observed sample sizes. If, however, homogeneity would continue to hold, this would certainly add support to the Chandrasekhar-mass single-degenerate scenario. On the other hand, even an increasing diversity would not rule out Chandrasekhar-mass single-degenerate progenitors for most of them. In contrast, there are ways to explain how the diversity is absorbed in a one-parameter family of transformations, such as the Phillips relation or modifications of it. For example, we have argued that the size of the convective core of the white dwarf prior to the explosion might provide a physical reason for such a relation.

As far as the explosion/combustion physics and the numerical simulations are concerned, significant recent progress has made the models more realistic (and reliable). Thanks to ever increasing computer resources, 3D simulations that treat the full star with good spatial resolution and realistic input physics have become feasible. Already the results of 2D simulations indicate that pure deflagrations waves in Chandrasekhar-mass $C + O$ white dwarfs can lead to explosions, and one can expect that going to three dimensions, because of the increasing surface area of the nuclear flames, should add to the explosion energy. If confirmed, this would eliminate pulsational detonations from the list of potential models. On the side of the combustion physics, the burning in the distributed regime at low densities needs to be explored further, but it is not clear anymore whether a transition from a deflagration to a detonation in that regime is needed for successful models. In fact, according to recent studies, such a transition appears to be unlikely.

Finally, sub–Chandrasekhar-mass models seem to face problems, both from the observations and from theory, leaving us with the conclusion that we seem to be lucky and Nature was kind to us and singled out from all possibilities the simplest solution, namely a Chandrasekhar-mass $C + O$ white dwarf and a nuclear deflagration wave, to make a type Ia supernova explosion.

Visit the Annual Reviews home page at www.AnnualReviews.org

LITERATURE CITED

Aldering G, Knop R, Nugent P. 2000. *Astron J.* In press

Arnett WD. 1969. *Astrophys. Space Sci.* 5:180–212

Arnett WD. 1971. *Ap. J.* 169:113

Arnett WD. 1996. *Supernovae and Nucleosyn-* *thesis.* Princeton: Princeton Univ. Press

Arnett WD. 1997. In Ruiz-Lapuente, Canal, & Isern 1997, pp. 405–40

Arnett WD, Livne E. 1994a. *Ap. J.* 427:315–29

Arnett WD, Livne E. 1994b. *Ap. J.* 427:330–41

Arnett WD, Truran JW, Woosley SE. 1971. *Ap. J.* 165:87

Axelrod TS. 1980. *Late time optical spectra from the Ni-56 model for Type 1 supernovae.* Ph.D. thesis, Univ. Calif., Santa Cruz

Baade W, Zwicky F. 1934. *Phys. Rev.* 46:76

Balbus SA, Hawley JF, Stone JM. 1996. *Ap. J.* 467:76

Barkat Z, Wheeler JC. 1990. *Ap. J.* 355:602–16

Baron E, Hauschildt PH, Mezzacappa A. 1997. In Ruiz-Lapuente, Canal, & Isern 1997, pp. 627–46

Bartunov OS, Tsvetkov DY, Filimonova IV. 1994. *PASP* 106:1276–84

Beer A. 1974. *Vistas in Astronomy,* vol. 16. Oxford: Pergamon

Benson PJ, Little-Marenin IR, Herbst W, Salzer JJ, Vinton G, et al. 1993. *Am. Astron. Soc. Meet.,* vol. 182, p. 2916. Washington DC: Am. Astron. Soc.

Benz W. 1997. In Ruiz-Lapuente, Canal, & Isern 1997, pp. 457–74

Benz W, Cameron AGW, Press WH, Bowers RL. 1990. *Ap. J.* 348:647–67

Blanco VM. 1987. In *ESO Workshop SN 1987 A,* ed. IF Danziger, pp. 27–32. Garching, Ger: Europ. South. Obs.

Blinnikov SI. 1996. In *Proc. 8th Workshop on Nuclear Astrophysics,* eds. W Hillebrandt, E Müller, pp. 25–9. Garching, Ge: Max-Planck-Inst. Astrophys.

Blinnikov SI. 1997. In Ruiz-Lapuente, Canal, & Isern 1997, pp. 589–95

Blinnikov SI, Khokhlov AM. 1986. *Sov. Astron. Lett.* 12:131

Blinnikov SI, Khokhlov AM. 1987. *Sov. Astron.* 13:364

Blinnikov SI, Nadyoshin DK. 1991. The shock breakout in SN1987a modelled with the time-dependent radiative transfer. *Tech. Rep. N92–1293303–90,* Sternberg Astron. Inst., Moscow

Blinnikov SI, Sasorov PV. 1996. *Phys. Rev. E* 53:4827

Boffi FR, Branch D. 1995. *PASP* 107:347

Boisseau JR, Wheeler JC, Oran ES, Khokhlov AM. 1996. *Ap. J. Lett.* 471:L99

Bragaglia A. 1997. In Ruiz-Lapuente, Canal, & Isern 1997, pp. 231–37

Branch D. 1981. *Ap. J.* 248:1076–80

Branch D. 1998. *Annu. Rev. Astron. Astrophys.* 36:17–56

Branch D, Fisher A, Nugent P. 1993. *Astron J.* 106:2383–91

Branch D, Romanishin W, Baron E. 1996. *Ap. J.* 465:73

Branch D, Tammann GA. 1992. *Annu. Rev. Astron. Astrophys.* 30:359–89

Branch D, Venkatakrishna KL. 1986. *Ap. J. Lett.* 306:L21–3

Bravo E, Garcia-Senz D. 1995. *Ap. J. Lett.* 450:L17

Bravo E, Garcia-Senz D. 1997. *Proc. IAU Symp. 187,* p. E3

Bravo E, Garcia-Senz D. 1999. *MNRAS* 307:984–92

Bruenn SW. 1973. *Ap. J. Lett.* 183:L125

Buchler JR, Colgate SA, Mazurek TJ. 1980. *J. Phys.* 41:2–159

Buchler JR, Mazurek TJ. 1975. *Mem. Soc. R. Sci. Liege* 8:435–45

Burrows A, Shankar A, van Riper KA. 1991. *Ap. J. Lett.* 379:L7–11

Burrows A, The LS. 1990. *Ap. J.* 360:626–38

Bychkov VV, Liberman MA. 1995a. *Astron. Astrophys.* 302:727

Bychkov VV, Liberman MA. 1995b. *Ap. J.* 451:711

Cappellaro E, Turatto M, Tsvetkov D, Bartunov O, Pollas C, et al. 1997. *Astron. Astrophys.* 322:431–41

Cassisi S, Castellani V, Tornambe A. 1996. *Ap. J.* 459:298

Cassisi S, Iben I. J, Tornambe A. 1998. *Ap. J.* 496:376

Chandrasekhar S. 1961. *Hydrodynamic and Hydromagnetic Stability.* Oxford: Oxford Univ. Press

Clavin P. 1994. *Annu. Rev. Fluid Mech.* 26:321

Colgate SA, McKee C. 1969. *Ap. J.* 157:623

Contardo G. 1999. In *Future Directions in Supernova Research: Progenitors to Remnants.* In press

Contardo G, Leibundgut B. 1998. In *Proc. 9th*

Workshop Nucl. Astrophys., eds. W Hillebrandt, E Müller, p. 128. Garching, Ger: Max-Planck-Inst. Astrophys.

Courant. R, Friedrichs KO. 1948. *Supersonic Flow and Shock Waves.* New York: Springer

Cox JP, Giuli RT. 1968. *Principles of Stellar Structure,* vol. II, pp. 183–261. New York: Gordon & Breach

Cumming RJ, Lundqvist P, Smith LJ, Pettini M, King DL. 1996. *MNRAS* 283:1355–60

Damköhler G. 1940. *Z. Elektrochem.* 46:601

Darrieus G. 1944. *La Tech. Mod.* Unpublished

Di Stefano R, Rappaport S. 1994. *Ap. J.* 437: 733–41

Eastman RG. 1997. In Ruiz-Lapuente, Canal, & Isern 1997, pp. 571–88

Eastman RG, Pinto PA. 1993. *Ap. J.* 412:731–51

Eck CR, Cowan JJ, Roberts DA, Boffi FR, Branch D. 1995. *Ap. J. Lett.* 451: L53

Elias JH, Matthews K, Neugebauer G, Persson SE. 1985. *Ap. J.* 296:379–89

Fermi E. 1951. In *Collected Works of Enrico Fermi,* ed. E Segre, pp. 816–21. Chicago: Univ. Chicago Press

Filippenko AV. 1988. *Astron. J.* 96:1941–8

Filippenko AV. 1997a. *Annu. Rev. Astron. Astrophys.* 35:309–55

Filippenko AV. 1997b. In Ruiz-Lapuente, Canal, & Isern 1997, pp. 1–32

Filippenko AV, Richmond MW, Branch D, Gaskell M, Herbst W, et al. 1992a. *Astron. J.* 104:1543–56

Filippenko AV, Richmond MW, Matheson T, Shields JC, Burbidge EM, et al. 1992b. *Ap. J. Lett.* 384:L15–8

Filippenko AV, Riess AG. 2000. See Niemeyer & Truran 2000

Fisher A, Branch D, Hatano K, Baron E. 1999. *MNRAS* 304:67–74

Fryer C, Benz W, Herant M, Colgate SA. 1999. *Ap. J.* 516:892–9

Fu A. 1987. *Ap. J.* 323:227–42

Gamezo VN, Wheeler JC, Khokhlov AM, Oran ES. 1999. *Ap. J.* 512:827–42

Garcia-Senz D, Bravo E, Serichol N. 1998. *Ap. J. suppl.* 115:119

Garcia-Senz D, Bravo E, Woosley SE. 1999. *Astron. Astrophys.* 349:177–88

Garcia-Senz D, Woosley SE. 1995. *Ap. J.* 454:895

Garnavich PM, Jha S, Challis P, Clocchiatti A, Diercks A, et al. 1998. *Ap. J.* 509:74–9

Gehrz RD, Truran JW, Williams RE. 1993. In *Protostars and Planets,* ed. E Levy, pp. 75–95. Tucson: Univ. Arizona Press

Gilmozzi R. 1987. In *ESO Workshop SN 1987 A,* ed. IF Danziger, pp. 19–24. Garching, Ger: Europ. South. Obs.

Gilmozzi R, Cassatella A, Clavel J, Gonzalez R, Fransson C. 1987. *Nature* 328:318–20

Goobar A, Perlmutter S. 1995. *Ap. J.* 450:14

Greiner J, Hasinger G, Kahabka P. 1991. *Astron. Astrophys.* 246:L17–20

Hachisu I, Kato M, Nomoto K. 1996. *Ap. J. Lett.* 470:L97

Hachisu I, Kato M, Nomoto Ki. 1999a. *Ap. J.* 522:487–503

Hachisu I, Kato M, Nomoto Ki, Umeda H. 1999b. *Ap. J.* 519:314–23

Hamuy M, Philips MM, Maza J, Suntzeff NB, Della Valle M, et al. 1994. *Astron. J.* 108:2226–32

Hamuy M, Phillips M, Suntzeff NB, Schommer RA, Maza J, et al. 1996a. *Astron. J.* 112:2391

Hamuy M, Phillips MM, Maza J, Suntzeff NB, Schommer RA, et al. 1995. *Astron. J.* 109:1–13

Hamuy M, Phillips MM, Suntzeff NB, Schommer RA, Maza J, et al. 1996b. *Astron. J.* 112:2398

Hamuy M, Phillips MM, Suntzeff NB, Schommer RA, Maza J, et al. 1996c. *Astron. J.* 112:2438

Hansen CJ, Kawaler SD. 1994. *Stellar Interiors.* p. 247 New York: Springer

Hansen CJ, Wheeler JC. 1969. *Astrophys. Space Sci.* 3:464

Harkness RP. 1991. In *Supernova 1987A and Other Supernovae, Conf. Proc. 37,* eds. IJ Danziger, K Kjär, pp. 447–56. Garching, Ger: Europ. South. Obs.

Hatano K, Branch D, Baron E. 1998. In *Am. Astron. Soc. Meet.*, vol. 193, p. 4703

Hillebrandt W, Hoeflich P, Weiss A, Truran JW. 1987. *Nature* 327:597–600

Hoeflich P. 1995. *Ap. J.* 443:89–108

Hoeflich P, Khokhlov A. 1996. *Ap. J.* 457:500

Hoeflich P, Khokhlov A, Müller E. 1992. *Astron. Astrophys.* 259:549–66

Hoeflich P, Khokhlov A, Wheeler JC, Nomoto K, Thielemann FK. 1997. In Ruiz-Lapuente, Canal, & Isern 1997, pp. 659–79

Hoeflich P, Khokhlov AM, Wheeler JC. 1995. *Ap. J.* 444:831–47

Hoeflich P, Wheeler JC, Khokhlov A. 1998a. *Ap. J.* 492:228

Hoeflich P, Wheeler JC, Thielemann FK. 1998b. *Ap. J.* 495:617

Homeier D, Koester D, Hagen HJ, Jordan S, Heber U, et al. 1998. *Astron. Astrophys.* 338:563–75

Hoyle F, Fowler WA. 1960. *Ap. J.* 132:565

Iben I. J. 1978. *Ap. J.* 219:213–25

Iben I. J. 1982. *Ap. J.* 253:248–59

Iben I. J, Tutukov AV. 1984. *Ap. J. suppl.* 54:335–72

Imshennik VS, Kal'Yanova NL, Koldoba AV, Chechetkin VM. 1999. *Astron. Lett.* 25:206–14

Imshennik VS, Khokhlov AM. 1984. *Sov. Astron. Lett.* 10:262

Ivanova LN, Imshennik VS, Chechetkin VM. 1974. *Astrophys. Space Sci.* 31:497–514

Ivanova LN, Imshennik VS, Chechetkin VM. 1982. *Pisma Astron. Zhurnal* 8:17–25

Iwamoto K, Brachwitz F, Nomoto KI, Kishimoto N, Umeda H, et al. 1999. *Ap. J. suppl.* 125:439–62

Jeffery DJ, Leibundgut B, Kirshner RP, Benetti S, Branch D, et al. 1992. *Ap. J.* 397:304–28

Kahabka P, Van Den Heuvel EPJ. 1997. *Annu. Rev. Astron. Astrophys.* 35:69–100

Karp AH, Lasher G, Chan KL, Salpeter EE. 1977. *Ap. J.* 214:161–78

Kato M, Hachisu I. 1999. *Ap. J. Lett.* 513:L41–4

Kawai Y, Saio H, Nomoto KI. 1987. *Ap. J.* 315:229–33

Kercek A, Hillebrandt W, Truran JW. 1999. *Astron. Astrophys.* 345:831–40

Kerstein AR. 1996. *Combust. Sci. Technol.* 118:189

Khokhlov AM. 1988. *Astrophys. Space Sci.* 149:91–106

Khokhlov AM. 1989. *MNRAS* 239:785–808

Khokhlov AM. 1991a. *Astron. Astrophys.* 245:114–28

Khokhlov AM. 1991b. *Astron. Astrophys.* 245:L25–28

Khokhlov AM. 1993a. *Ap. J. Lett.* 419:L77

Khokhlov AM. 1993b. *Ap. J.* 419:200

Khokhlov AM. 1994. *Ap. J. Lett.* 424:L115–17

Khokhlov AM. 1995. *Ap. J.* 449:695

Khokhlov AM, Müller E, Hoeflich P. 1993. *Astron. Astrophys.* 270:223–48

Khokhlov AM, Oran ES, Wheeler JC. 1997. *Ap. J.* 478:678

King AR, Van Teeseling A. 1998. *Astron. Astrophys.* 338:965–70

Kippenhahn R, Weigert A. 1989. *Stellar Structure and Evolution.* Berlin: Springer

Kobayashi C, Tsujimoto T, Nomoto KI, Hachisu I, Kato M. 1998. *Ap. J. Lett.* 503:L155

Koen C, Orosz JA, Wade RA. 1998. *MNRAS* 300:695–704

Kolmogorov AN. 1941. *Dokl. Akad. Nauk SSSR* 30:299

Kriminski SA, Bychkov VV, Liberman M. 1998. *New Astron.* 3:363–77

Kuchner MJ, Kirshner RP, Pinto PA, Leibundgut B. 1994. *Ap. J. Lett.* 426:L89

Kunasz PB. 1984. *Ap. J.* 276:677–90

Landau LD. 1944. *Acta Physicochim. URSS* 19:77

Landau LD, Lifshitz EM. 1995. *Fluid Mechanics.* London: Pergamon

Layzer D. 1955. *Ap. J.* 122:1

Lee JHS, Knystautas R, Yoshikawa N. 1978. *Acta Astronaut.* 5:971

Leibundgut B, Kirshner RP, Phillips MM, Wells LA, Suntzeff N, et al. 1993. *Astron. J.* 105:301–13

Leising MD, Johnson WN, Kurfess JD,

Clayton DD, Grabelsky DA, et al. 1995. *Ap. J.* 450:805

Lentz EJ, Baron E, Branch D, Hauschildt PH, Nugent PE. 1999. In *Am. Astron. Soc. Meet.*, vol.194, p. 8606

Lentz EJ, Baron E, Branch D, Hauschildt PH, Nugent PE. 2000. *Ap. J.* 530:966–76

Lewis B, von Elbe G. 1961. *Combustion, Flames, and Explosions of Gases.* New York: Academic

Li WD, Filippenko AV, Riess AG, Hu JY, Qiu YL. 2000. In *Proc. 10th Ann. Oct. Astrophys. Conf. Cosmic Explosions.* Univ. Maryland. In press

Livio M. 1996. In *Supersoft X-ray Sources, Proc. Int. Workshop*, ed. J Greiner, vol. 472, pp. 183–91. Garching, Ger: Max-Planck-Inst. Extraterr. Phys.

Livio M. 2000. In Niemeyer & Truran 2000

Livio M, Branch D, Yungelson LB, Boffi F, Baron E. 1996. In *IAU Colloq. 158: Cataclysmic Variables and Related Objects*, pp. 407–15

Livne E. 1990. *Ap. J. Lett.* 354:L53–5

Livne E. 1993. *Ap. J. Lett.* 406:L17–20

Livne E. 1997. In Ruiz-Lapuente, Canal, & Isern 1997, pp. 425–40

Livne E, Arnett D. 1995. *Ap. J.* 452:62

Livne E, Glasner AS. 1991. *Ap. J.* 370:272–81

Lucy LB. 1999. *Astron. Astrophys.* 345:211–20

Lundmark K. 1920. *Svenska Vetenkapsakad. Handl.* 60:8

Lundmark K. 1921. *Publ. Astron. Soc. Pac.* 33:234

Lundqvist P, Cumming RJ. 1997. In *Advances in Stellar Evolution*, pp. 293–6. Cambridge, UK: Cambridge Univ. Press

Mair G, Hillebrandt W, Hoeflich P, Dorfi A. 1992. *Astron. Astrophys.* 266:266–82

Mazzali PA, Chugai N, Turatto M, Lucy LB, Danziger IJ, et al. 1997a. *MNRAS* 284:151–71

Mazzali PA, Danziger IJ, Turatto M. 1995. *Astron. Astrophys.* 297:509

Mazzali PA, Lucy LB. 1993. *Astron. Astrophys.* 279:447–56

Mazzali PA, Turatto M, Cappellaro E, Della Valle M, Benetti S, et al. 1997b. In Ruiz-Lapuente, Canal, & Isern 1997, pp. 647–58

Meikle WPS, Bowers EJC, Geballe TR, Walton NA, Lewis JR, et al. 1997. In Ruiz-Lapuente, Canal, & Isern 1997, pp. 53–64

Meikle WPS, Cumming RJ, Geballe TR, Lewis JR, Walton NA, et al. 1996. *MNRAS* 281:263–80

Mihalas D, Weibel Mihalas B. 1984. *Foundations of Radiation Hydrodynamics.* New York: Oxford Univ. Press

Minkowski R. 1940. *Publ. Astron. Soc. Pac.* 52:206

Mochkovitch R. 1996. *Astron. Astrophys.* 311:152–4

Mochkovitch R, Guerrero J, Segretain L. 1997. In Ruiz-Lapuente, Canal, & Isern 1997, pp. 187–204

Mochkovitch R, Livio M. 1990. *Astron. Astrophys.* 236:378–84

Morris DJ, Bennett K, Bloemen H, Diehl R, Hermsen W, et al. 1997. In *AIP Conf. Proc. 410. Proc. 4th Compton Symp.*, p. 1084

Müller E. 1998. In *Computational Methods for Astrophysical Fluid Flow*, eds. O Steiner, A Gautschy, pp. 343–494. Berlin, New York: Springer

Müller E, Arnett WD. 1982. *Ap. J. Lett.* 261:L109–15

Müller E, Arnett WD. 1986. *Ap. J.* 307:619–43

Müller E, Hoeflich P, Khokhlov A. 1991. *Astron. Astrophys.* 249:L1–4

Munari U, Renzini A. 1992. *Ap. J. Lett.* 397:L87–90

Nandkumar R, Pethick CJ. 1984. *MNRAS* 209:511–24

Niemeyer JC. 1999. *Ap. J. Lett.* 523:L57–60

Niemeyer JC, Bushe WK, Ruetsch GR. 1999. *Ap. J.* 524:290

Niemeyer JC, Hillebrandt W. 1995a. *Ap. J.* 452:779

Niemeyer JC, Hillebrandt W. 1995b. *Ap. J.* 452:769

Niemeyer JC, Hillebrandt W. 1997. In Ruiz-Lapuente, Canal, & Isern 1997, pp. 441–56

Niemeyer JC, Hillebrandt W, Woosley SE. 1996. *Ap. J.* 471:903

Niemeyer JC, Kerstein AR. 1997. *New Astron.* 2:239–44

Niemeyer JC, Truran JW, eds. 2000. *Type Ia Supernovae: Theory and Cosmology,* Cambridge: Cambridge Univ. Press

Niemeyer JC, Woosley SE. 1997. *Ap. J.* 475:740

Nomoto K. 1980. *Space Sci. Rev.* 27:563–70

Nomoto K. 1982a. *Ap. J.* 257:780–92

Nomoto K. 1982b. *Ap. J.* 253:798–810

Nomoto K, Iben I. J. 1985. *Ap. J.* 297:531–7

Nomoto K, Iwamoto K, Thielemann F, Brachwitz F, et al. 1997. In Ruiz-Lapuente, Canal, & Isern 1997, pp. 349–78

Nomoto K, Iwamoto K, Yamaoka H, Hashimoto M. 1995. In *ASP Conf. Ser. 72: Millisecond Pulsars. A Decade of Surprise,* p. 164

Nomoto K, Nariai K, Sugimoto D. 1979. *Publ. Astron. Soc. Jpn.* 31:287–98

Nomoto K, Sugimoto D. 1977. *Publ. Astron. Soc. Jpn.* 29:765–80

Nomoto K, Sugimoto D, Neo S. 1976. *Astrophys. Space Sci.* 39:L37–42

Nomoto K, Suzuki T, Shigeyama T, Kumagai S, Yamaoka H, et al. 1993. *Nature* 364:507–9

Nomoto K, Thielemann FK, Yokoi K. 1984. *Ap. J.* 286:644–58

Nomoto K, Yamaoka H, Pols OR, Van den Heuvel EPJ, Iwamoto K, et al. 1994a. *Nature* 371:227

Nomoto K, Yamaoka H, Shigeyama T, Iwamoto K. 1996. In *Supernovae and Supernova Remnants. Proc. IAU Colloq. 145. Xian, China,* eds. R McCray, Z Wang, pp. 49–68. Cambridge: Cambridge Univ. Press

Nomoto K, Yamaoka H, Shigeyama T, Kumagai S, Tsujimoto T. 1994b. In *Supernovae, Les Houches Session LIV,* eds. J Audouze, S Bludman, R Mochovitch, J Zinn-Justin, pp. 199–249. Amsterdam: Elsevier

Nomoto KI, Kondo Y. 1991. *Ap. J. Lett.* 367:L19–22

Nugent P, Baron E, Branch D, Fisher A, Hauschildt PH. 1997. *Ap. J.* 485:812

Nugent P, Baron E, Hauschildt PH, Branch D. 1995. *Ap. J. Lett.* 441:L33–6

Paczynski B. 1972. *Ap. J.* 11:L53

Paczynski B. 1973. *Astron. Astrophys.* 26:291

Paczynski B. 1985. In Cataclysmic variables and low-mass X-ray binaries. *Proc. 7th North Am. Workshop,* pp. 1–12. Dordrecht: Reidel

Pakull MW, Moch C, Bianchi L, Thomas HC, Guibert J, et al. 1993. *Astron. Astrophys.* 278:L39–42

Patat F, Benetti S, Cappellaro E, Danziger IJ, Della Valle M, et al. 1996. *MNRAS* 278:111–24

Pauldrach AWA, Duschinger M, Mazzali PA, Puls J, Lennon M, et al. 1996. *Astron. Astrophys.* 312:525–38

Perlmutter S, Aldering G, Goldhaber G, Knop RA, Nugent P, et al. 1999. *Ap. J.* 517:565–86

Perlmutter S, Gabi S, Goldhaber G, Goobar A, Groom DE, et al. 1997. *Ap. J.* 483:565

Peters N. 1988. In *Symp. Int. Combust., 21st,* p. 1232. Pittsburgh: Combust. Inst.

Phillips MM. 1993. *Ap. J. Lett.* 413:L105–8

Phillips MM, Wells LA, Suntzeff NB, Hamuy M, Leibundgut B, et al. 1992. *Astron. J.* 103:1632–37

Pinto PA. 1997. In Ruiz-Lapuente, Canal, & Isern 1997, pp. 607–27

Podsiadlowski P. 1993. *Space Sci. Rev.* 66:439

Pope SB. 1987. *Annu. Rev. Fluid Mech.* 19:237–70

Popper DM. 1937. *Publ. Astron. Soc. Pac.* 49:283

Pskovskii IP. 1977. *Sov. Astron.* 21:675–82

Rasio FA, Shapiro SL. 1995. *Ap. J.* 438:887–903

Rayleigh LK. 1883. *Proc. London Math. Soc.* 14:170

Read KI. 1984. *Physica D* 12:45

Reinecke M, Hillebrandt W, Niemeyer JC. 1999a. *Astron. Astrophys.* 347:739–47

Reinecke M, Hillebrandt W, Niemeyer JC, Klein R, Gröbl A. 1999b. *Astron. Astrophys.* 347:724–33

Renzini A. 1996. In *Supernovae and Supernova Remnants. Proc. IAU Colloq. 145. Xian, China,* eds. R McCray, Z Wang, p. 77. Cambridge: Cambridge Univ. Press

Riess AG, Filippenko AV, Challis P, Clocchiatti A, Diercks A, et al. 1998. *Astron. J.* 116:1009–38

Riess AG, Filippenko AV, Li W, Schmidt BP. 1999a. *Astron. J.* 118:2668–74

Riess AG, Filippenko AV, Li W, Treffers RR, Schmidt BP, et al. 1999b. *Astron. J.* 118:2675–88

Riess AG, Press WH, Kirshner RP. 1996. *Ap. J.* 473:88

Ruiz-Lapuente P. 1997. In Ruiz-Lapuente, Canal, & Isern 1997, pp. 681–704

Ruiz-Lapuente P, Burkert A, Canal R. 1995. *Ap. J. Lett.* 447:L69

Ruiz-Lapuente P, Canal R. 1998. *Ap. J. Lett.* 497:L57

Ruiz-Lapuente P, Canal R, Isern J, eds. 1997. *Thermonuclear Supernovae,* Dordrecht. Kluwer

Ruiz-Lapuente P, Cappellaro E, Turatto M, Gouiffes C, Danziger IJ, et al. 1992. *Ap. J. Lett.* 387:L33–6

Ruiz-Lapuente P, Jeffery DJ, Challis PM, Filippenko AV, Kirshner RP, et al. 1993a. *Nature* 365:728

Ruiz-Lapuente P, Lichti GG, Lehoucq R, Canal R, Casse M. 1993b. *Ap. J.* 417:547

Saffer RA, Livio M, Yungelson LR. 1998. *Ap. J.* 502:394

Saio H, Nomoto K. 1985. *Astron. Astrophys.* 150:L21–3

Saio H, Nomoto KI. 1998. *Ap. J.* 500:388–97

Schmidt BP, Kirshner RP, Eastman RG, Grashuis R, Dell'Antonio I, et al. 1993. *Nature* 364:600–2

Schmidt BP, Suntzeff NB, Phillips MM, Schommer RA, Clocchiatti A, et al. 1998. *Ap. J.* 507:46–63

Schoenfelder V, Bennett K, Bloemen H, Diehl R, Hermsen W, et al. 1996. *Astron. Astrophys. Suppl.* 120:C13

Sharp DH. 1984. *Physica D* 12:3

Shigeyama T, Kumagai S, Yamaoka H, Nomoto K, Thielemann FK. 1993. *Astron. Astrophys. Suppl.* 97:223–4

Shigeyama T, Nomoto Ki, Yamaoka H, Thielemann FK. 1992. *Ap. J. Lett.* 386:L13–6

Sorokina EI, Blinnikov SI, Bartunov OS. 1999. In *Proc. Astron. at the Eve of the New Century*

Southwell KA, Livio M, Charles PA, O'Donoghue D, Sutherland WJ. 1996. *Ap. J.* 470:1065

Sparks WM, Starrfield S, Truran JW. 1976. *Ap. J.* 208:819–25

Spyromilio J, Meikle WPS, Allen DA, Graham JR. 1992. *MNRAS* 258:53P–56P

Spyromilio J, Pinto PA, Eastman RG. 1994. *MNRAS* 266:L17

Starrfield S, Truran JW, Sparks WM. 1978. *Ap. J.* 226:186–202

Starrfield S, Truran JW, Sparks WM, Kutter GS. 1972. *Ap. J.* 176:169

Stein J, Barkat Z, Wheeler JC. 1999. *Ap. J.* 523:381–5

Stone JM, Mihalas D, Norman ML. 1992. *Ap. J. suppl.* 80:819–45

Suntzeff NB. 1996. In *Supernovae and Supernova Remnants. Proc. IAU Colloq. 145. Xian, China,* eds. R McCray, Z Wang, p. 41. Cambridge: Cambridge Univ. Press

Sutherland PG, Wheeler JC. 1984. *Ap. J.* 280:282–97

Thielemann FK, Nomoto K, Yokoi K. 1986. *Astron. Astrophys.* 158:17–33

Timmes FX. 1999. *Ap. J. Suppl.* 124:241

Timmes FX, Woosley SE. 1992. *Ap. J.* 396:649–67

Timmes FX, Woosley SE. 1997. *Ap. J.* 489:160

Timmes FX, Woosley SE, Taam RE. 1994. *Ap. J.* 420:348–63

Tripp R. 1998. *Astron. Astrophys.* 331:815–20

Truemper J, Hasinger G, Aschenbach B, Braeuninger H, Briel UG. 1991. *Nature* 349:579–83

Truran JW. 1982. In *Essays in Nuclear Astrophysics,* eds. C Barnes, DD Clayton, DN Schramm, WA Fowler, p. 467. Cambridge: Cambridge Univ. Press

Truran JW, Arnett D, Cameron AGW. 1967. *Canad. J. Physics* 45:2315–32

Turatto M, Benetti S, Cappellaro E, Danziger IJ, Della Valle M, et al. 1996. *MNRAS* 283:1–17

Tutukov A, Yungelson L. 1996. *MNRAS* 280:1035–45

Umeda H, Nomoto K, Kobayashi C, Hachisu I, Kato M. 1999. *Ap. J. Lett.* 522:L43–7

Van den Heuvel EPJ, Bhattacharya D, Nomoto K, Rappaport SA. 1992. *Astron. Astrophys.* 262:97–105

Vaughan TE, Branch D, Miller DL, Perlmutter S. 1995. *Ap. J.* 439:558–64

Wang L, Hoeflich P, Wheeler JC. 1997. *Ap. J. Lett.* 483:L29

Watanabe K, Hartmann DH, Leising MD, The LS. 1999. *Ap. J.* 516:285–96

Weaver TA, Woosley SE. 1980. In *AIP Conf. Proc. 63. Supernovae Spectra. La Jolla, CA*, p. 15

Webbink RF. 1984. *Ap. J.* 277:355–60

Wellstein S, Langer N. 1999. *Astron. Astrophys.* 350:148–62

Wheeler JC, Harkness RP. 1990. *Rep. Prog. Phys.* 53:1467–557

Wheeler JC, Hoeflich P, Harkness RP, Spyromilio J. 1998. *Ap. J.* 496:908

Whelan J, Iben I. J. 1973. *Ap. J.* 186:1007–14

Wiggins DJR, Falle SAEG. 1997. *MNRAS* 287:575–82

Wiggins DJR, Sharpe GJ, Falle SAEG. 1998. *MNRAS* 301:405–13

Williams FA. 1985. *Combustion Theory.* Menlo Park, CA: Benjamin/Cummings

Wilson OC. 1939. *Ap. J.* 90:634

Woosley SE. 1990. In *Supernovae*, ed. AG Petschek, pp. 182–210. Berlin: Springer-Verlag

Woosley SE. 1997a. *Ap. J.* 476:801

Woosley SE. 1997b. In Ruiz-Lapuente, Canal, & Isern 1997, pp. 313–37

Woosley SE, Axelrod TS, Weaver TA. 1984.

In *Stellar Nucleosynthesis*, eds. C Chiosi, A Renzini, p. 263. Dordrecht: Kluwer

Woosley SE, Taam RE, Weaver TA. 1986. *Ap. J.* 301:601–23

Woosley SE, Weaver TA. 1986a. *Annu. Rev. Astron. Astrophys.* 24:205–53

Woosley SE, Weaver TA. 1986b. In *Radiation Hydrodynamics in Stars and Compact Objects, Lect. Notes Phys.*, eds. D Mihalas, KHA Winkler, vol. 255, p. 91. Berlin: Springer-Verlag

Woosley SE, Weaver TA. 1994a. In *Supernovae, Les Houches Session LIV*, eds. J Audouze, S Bludman, R Mochovitch, J Zinn-Justin, p. 63. Amsterdam: Elsevier

Woosley SE, Weaver TA. 1994b. *Ap. J.* 423:371–79

Woosley SE, Weaver TA, Taam RE. 1980. In *Texas Workshop Type I Supernovae*, ed. J Wheeler, pp. 96–112. Austin, TX: Univ. Texas

Yin WW, Miller GS. 1995. *Ap. J.* 449:826

Youngs DL. 1984. *Physica D* 12:32

Yungelson L, Livio M. 1998. *Ap. J.* 497: 168

Yungelson L, Livio M, Truran JW, Tutukov A, Fedorova A. 1996. *Ap. J.* 466:890

Zeldovich YB. 1966. *J. Appl. Mech. and Tech. Phys.* 7:68

Zeldovich YB, Barenblatt GI, Librovich VB, Makhviladze GM. 1985. *The Mathematical Theory of Combustion and Explosions.* New York: Plenum

Zeldovich YB, Librovich VB, Makhviladze GM, Sivashinsky GL. 1970. *Astronaut. Acta* 15:313

Zwicky F. 1938a. *Phys. Rev.* 55:726

Zwicky F. 1938b. *Ap. J.* 88:522

Zwicky F. 1965. In *Stellar Structure*, eds. LH Aller, DB McLaughlin, pp. 367–424. Chicago: Univ. of Chicago Press

Annu. Rev. Astron Astrophys 2000. 38:231–88

EXTREME ULTRAVIOLET ASTRONOMY

Stuart Bowyer,[1] Jeremy J. Drake,[2] and Stéphane Vennes[3]

[1]*Space Sciences Laboratory, University of California, Berkeley, CA, 94720-7450;*
e-mail: bowyer@ssl.berkeley.edu
[2]*Harvard-Smithsonian Center for Astrophysics, MS-3, 60 Garden Street, Cambridge, MA 02138; e-mail: jdrake@cfa.harvard.edu*
[3]*Astrophysical Theory Centre, School of Mathematical Sciences, Australian National University, Canberra, ACT 0200, Australia; e-mail: vennes@wintermute.anu.edu.au*

Key Words planets, interstellar medium, stars, extragalactic

■ **Abstract**   Astronomical studies in the extreme ultraviolet (EUV) band of the spectrum were dismissed during the early years of space astronomy as impossible, primarily because of the mistaken view that radiation in this band would be absorbed by the interstellar medium. Observations in the 1980s from sounding rockets and limited duration orbital spacecraft began to show the potential of this field and led to the deployment of two spacecraft devoted to EUV astronomy: the UK Wide Field Camera and the Extreme Ultraviolet Explorer. The instrumentation in these missions, although quite limited in comparison with instrumentation in other fields of space astronomy, provided unique and far-reaching results. These included new information on solar system topics, stellar chromospheres and corona, white dwarf astrophysics, cataclysmic variables, the interstellar medium, galaxies, and clusters of galaxies. We summarize these findings herein.

CONTENTS

0066-4146/00/0915-0231$14.00   **231**

1.   INTRODUCTION

For many years extreme ultraviolet (EUV) astronomy, defined as astronomy covering the decade of wavelength from 90 to 912 Å, was dismissed as astronomy that could not be done even in principle. The average density of the interstellar medium in the Galactic plane is about one hydrogen atom per cm^3, which at 150 Å translates to an optical depth of one at a distance of only 3 pc. This dire prophecy was wrong because of the complex nature of the interstellar medium; in fact, the Sun resides in a large region in which most of the hydrogen is ionized.

An additional problem for the development of this field was that for many years the technology for focusing, dispersing, and detection of EUV radiation did not exist. Inexpensive grazing incidence mirrors made of metal (Bowyer & Green 1988), variable line space gratings (Hettrick & Bowyer 1983), and imaging photon counting EUV detectors (Siegmund et al 1986) were developed in response to this need. Finally, there was a general reluctance among astronomers to reconsider the longstanding belief that EUV astronomy was impossible. Observations made during the APOLLO-SOYUZ mission in 1975 forced a reconsideration. "The Berkeley EUV telescope flown on that mission detected several discrete EUV sources, the discovery of which essentially launched the subject of EUV astronomy" (Pounds et al 1993).

At this point in time, EUV astronomy has produced substantial results in a wide variety of fields. Over 1100 sources have been identified, and spectroscopic studies of many of these sources have been carried out. All of this has been accomplished with photometers with less than 10 cm^2 effective area and spectrometers with a few cm^2 effective area. In the following we summarize many of the results obtained in this field. We apologize to those authors whose work we have not included because of lack of space.

2.   ALL SKY SURVEYS

A number of Extreme Ultraviolet surveys have been carried out, and over 1100 EUV sources have been identified.

2.1   UK Wide Field Camera Surveys

The UK Extreme Ultraviolet Wide Field Camera (WFC) was launched as a piggyback instrument attached to the German ROSAT X-ray satellite. It carried out the first EUV All Sky Survey during 1990 and 1991. Only a portion of the EUV band was surveyed; data was obtained in two bands: 60–140 Å and 120–200 Å. The first results were reported by Pounds et al (1991), who identified 35 EUV sources from a "mini-survey" covering 1800 square degrees of sky. An analysis of the entire all sky survey data was provided by Pounds et al (1993), who listed 383 EUV sources. Almost 90% of these sources had reasonable optical counterparts. Log N – log S distributions were provided for hot white dwarfs and active late type stars. These suggested that the detections of late type stars were luminosity limited, while observations of the intrinsically more luminous hot white dwarfs were limited by the patchy interstellar medium.

Pye et al (1995) carried out a definitive analysis of the WFC All Sky Survey data and found 479 EUV sources. Optical identifications, $\log N - \log S$ relations and source variability analyses were provided. Kreysing et al (1995) provided a catalog of EUV sources detected by the WFC during the pointed phase of the ROSAT mission. A total of 328 sources were identified in these pointings, including 113 new sources not detected in the all sky survey data. White dwarfs, assumed to be constant flux sources, were repeatedly observed. This allowed for the correction of a substantial degradation of the WFC sensitivity over the course of the observations, a feature not included in the other WFC catalogs.

2.2   Extreme Ultraviolet Explorer Surveys

The Extreme Ultraviolet Explorer (EUVE) carried out an extensive survey of celestial EUV emission. The instruments on EUVE allowed for two distinct data sets. The primary data set was an all-sky survey carried out in four bands centered at approximately 100, 200, 400, and 600 Å. The 10% transmission points of these bandpasses were 58–60, 170–230, 370–600, and 500–740 Å. A second data set with substantially more integration time was obtained with the Deep Survey Telescope. This instrument obtained data over a strip of sky $2° \times 180°$ along the plane of the ecliptic. Data in this survey were obtained in two bandpasses: 65–180 and 170–360 Å. All-sky data were obtained between July 1992 and January 1993, and missing sky regions were observed over the following year. Approximately 97% of the sky was surveyed with exposures ranging from a minimum of 20 s per field of view to 20,000 s per field of view, depending upon the sky location. Selected white dwarfs were observed through the course of the mission; these showed that all the EUVE instruments were stable to the limit of the photon statistics ($\sim 4\%$) over the entire duration of the mission.

A number of searches for EUV emission were carried out using the EUVE data set. A Bright Source Catalog was produced by Malina et al (1994). This survey utilized a preliminary (and subsequently discarded) source extraction procedure to search for sources in pre-selected sky locations. Some 356 sources were reported. Of these, 128 had already been identified in the WFC catalogs (Pounds et al 1991, 1993), and of the 228 new sources claimed to have been found, almost 20% were not confirmed in later, more definitive searches.

Bowyer et al (1994) developed an eight-stage processing chain to extract point sources from the EUVE data set with rigorously defined criteria uniformly applied to all observed sky locations. The results were published in the First EUVE Source Catalog. This catalog provided the first all-sky survey data at EUV wavelengths longer than 200 Å. In total, 410 sources were listed; 372 had plausible optical, ul-traviolet, or X-ray counterparts, which were also presented. A number of statistical diagnostics for the results were provided, including source counts and positional error uncertainties.

TABLE 1   Percentage of sources by class in the Second
EUVE and EUVE/ROSAT catalogs

Object type	EUVE Cat 2 %	EUVE/ROSAT %
Late-type stars	55	77
Early-type stars	5	8
White Dwarfs	21	5
Dwarf novae and CVs	13	1
Other or no type	8	9
Extragalactic	8	0

The Second EUVE Catalog was developed by Bowyer et al (1996). A new detection method was employed that provided improved sensitivity and reliability. Additional data from sky locations not in the First EUVE Catalog were included. A total of 734 sources were detected in one or more of the bands surveyed. A wide variety of stellar objects were reported. Three lists were provided: all-sky survey detections, deep survey detections, and sources detected during other phases of the mission.

Lampton et al (1997) searched for objects detected jointly in the EUVE 100 Å band all-sky survey and in the ROSAT X-ray Telescope 0.25 keV band. This joint detection requirement allowed less intense EUV sources to be identified in this band without compromising the integrity of the survey. A total of 534 sources were found, including 166 new sources that were not in either the Second EUVE Source Catalog or in the WFC catalogs.

The percentage of the sources detected by class in the Second EUVE Catalog and the joint EUVE/ROSAT catalog is shown in Table 1.

Additional sources of EUV emission were derived from the EUVE Right Angle Program. This program used the all-sky survey telescopes, whose fields of view were at right angles to the Deep Survey and Spectrometer Telescope, to obtain photometric data simultaneously with long spectrographic observations. These observations had substantially higher sensitivity than the all sky survey observations because of their longer integration times, but only a fifth of the sky was observed. The first catalog of sources found in this data set (McDonald et al 1994) included 99 new sources. The Second Right-Angle Program catalog (Christian et al 1999) listed 169 new sources. The categories of objects detected were consistent with those found in the all-sky surveys.

2.3 Optical Identifications

Mason et al (1995) provided optical finding charts for the unidentified WFC sources and reported on optical studies of objects in these fields. Optical identifications for 195 of the unidentified sources in the WFC catalogs were obtained. A variety of information for these sources, including the V and B magnitudes, spectral type, and information on interesting lines in the spectra, was provided.

Shara et al (1997) extracted 7 × 7 degree optical images centered on 540 EUVE source locations from the Space Telescope Science Institute digitized sky archives, and they provided these images as finder charts to aid observers in searching for unidentified EUV sources or to characterize known sources.

Craig et al (1995) found the optical counterparts for two hot DA white dwarfs, and six late-type stars. Vennes et al (1997c) identified two previously unrecognized hot white dwarfs and four active late type stars. Identifications for 30 previously unidentified sources in the southern hemisphere were reported by Craig et al (1997b). Twenty-three were identified as late-type (dKe and dMe) stars, and four were identified as DA white dwarfs. A signature of chromospheric activity (emission in the Balmer series and Ca H and K lines) was found in all spectra of sources identified as late-type stars. Polomski et al (1997) identified 21 EUV faint sources; 14 were late-type stars, 3 were white dwarfs, and 4 were AGN. A detailed description was provided for each of these sources. Twenty identifications of northern hemisphere faint EUV sources were made by Craig et al (1999). Eighty percent of the sources identified were late-type stars.

At this point virtually all of the brighter EUV sources have secure optical counterparts, and less than 25% of the total set of reported EUV sources are missing optical counterparts.

3.   LATE-TYPE STELLAR OUTER ATMOSPHERES

3.1   Preview

The EUV and X-ray emission from stellar outer atmospheres is widely interpreted in terms of a stellar analogy to solar coronal emission, in which hot plasma is trapped within closed magnetic structures generated by dynamo action (e.g., Vaiana & Rosner 1978). Progress in understanding this hot plasma emission and the physical processes giving rise to it has come from both photometric and spectroscopic EUV observations, though the latter has arguably provided the major breakthroughs. Some limited EUV spectroscopy of stellar coronae was obtained by the earlier EXOSAT satellite that was equipped with a grating spectrograph with EUV as well as soft X-ray capability, albeit at fairly low resolution. A small handful of bright coronae were observed before its premature demise (Lemen et al 1989). The advent of the astrophysical spectroscopic capability of EUVE, with its comparatively high spectral resolution, truly represents a landmark in the study of stellar outer atmospheres; this resolution is sufficient to resolve individual spectral lines and has provided the stellar astronomer with the wealth of plasma diagnostics lying in the EUV. Progress in EUV spectroscopy of stellar coronae has been reviewed by Drake (1996a), Jordan (1996), and Laming (1998) (see also Doschek 1991 and Feldman et al 1992 for solar perspectives).

3.2 Detections and Survey Studies of Late-Type Stars

More than half of the EUV sources detected have been identified with late-type stars. Some of the most active, rapidly rotating single late-type stars known have been identified following their detection at EUV wavelengths (e.g. Matthews et al 1994; Jeffries et al 1994), together with previously unknown RS CVn-type and BY Dra-type binaries (Jeffries et al 1993; Jeffries & Bromage 1993). Hodgkin & Pye (1994) and Wood et al (1994) have examined luminosity functions of late-type stars detected in the WFC survey. Hodgkin & Pye found a larger number of K dwarfs than expected based on volume-limited samples of nearby stars, this excess likely comprising a population of pre-zero-age main sequence stars. Both studies used filter photometry with bands covering wavelengths from 60–140 Å and 112–200 Å to confirm the trend of increasing coronal temperature with increasing activity level found from earlier X-ray observations. Jeffries & Jewel (1993) have examined the detailed kinematics of WFC late-type stellar sources, confirming the sample as being kinematically young, with a mean age of 1–2 Gyr. They identified components likely due to the Pleiades-age Local Association, including the excess of K dwarfs found in luminosity functions by Hodgkin & Pye (1994).

Mathioudakis et al (1994) searched for EUV emission among the nearby, slowest rotating, low activity stars. Of the 19 stars studied, only one—HD 4268 (dK4)—was detected. This star was not detected in the ROSAT PSPC all-sky survey, and in order to reconcile these two results, Mathioudakis et al suggested the coronal temperature must be 10^6 K or less—a temperature consistent with that predicted for a corona heated acoustically rather than through dissipation of magnetic energy. EUV emission was searched for in a sample of 104 RS CVn-type binaries by Mitrou et al (1997), resulting in 38 significant detections and adding 11 and 8 new sources to the respective EUVE and WFC source catalogs.

3.3 Temperature Structure of Stellar Coronae

3.3.1 Two-Temperature Coronae   Analyses of early EINSTEIN imaging proportional counter observations of stellar coronae revealed three very general results. Most important, the observed X-ray luminosity was well-correlated with stellar rotation rate but not with spectral type (Pallavicini et al 1981; Walter & Bowyer 1981; Vaiana et al 1981; the analogous correlation of EUV emission with rotation was verified by Mathioudakis et al 1995). A correlation with rotation rate, together with the observations of resolved magnetic loop structures on the Sun and indirect evidence that hot coronae of active stars must be confined by forces other than gravity (Walter et al 1980), contradicted ideas of acoustically heated coronae. Instead pointed to magnetic dynamo action was indicated as the main controlling influence on coronal emission; the faster rotating stars having more active dynamos. Second, coronal temperatures were well-correlated with coronal luminosity, with the most luminous coronae having temperatures an order of magnitude higher than that of the Sun. Third, it was found that very low resolution spectra could generally be well-approximated by a combination of two isothermal optically thin plasmas (e.g. Swank et al 1981; Schmitt et al 1990), hailing the birth of the concept of the *two-temperature corona*.

3.3.2 Emission Measure Distributions

The ability of EUVE to resolve spectral lines from different ions formed over a wide range of coronal temperatures offered the potential to discern temperature structure in stellar coronae, if appreciable structure over and above the two-temperature parameterization were to exist.

The low densities of stellar coronae ($N_e \sim 10^{12} - 10^{13}$ cm^{-3}) render emission essentially optically thin and collision-dominated. Under these conditions, the observed intensity of a spectral line can be written (e.g. Pottasch 1963, and later refinement of Craig & Brown 1976)

$$I_{ul} = A K_{ul} \int_{\Delta T_{ul}} G_{ul}(T) \overline{N_e^2(T)} \frac{dV(T)}{d \log T} d \log T \text{ erg cm}^{-2} \text{ s}^{-1}. \qquad (1)$$

where A is the elemental abundance, K_{ul} is a known constant that includes the wavelength of the transition and the stellar distance, and $G_{ul}(T)$ is the "contribution" function of the line containing all the relevant atomic physics parameters (parent ion population and collisional excitation rates). The source term $N_e^2(T)\frac{dV(T)}{d \log T}$ is known as the *differential emission measure* (because different authors use different definitions of the DEM or do not use the differential form, we hereafter use the more general term "EM distribution").

At least in principle, through inversion of Equation 1 the fluxes in the various spectral lines can yield information on the EM distribution source term. However, inverting this equation (a Fredholm equation of the first type) to extract the actual source EM distribution from observed spectral line fluxes is actually a mathematically ill-conditioned problem and can result in ambiguity (e.g., Sylwester et al 1980; Judge et al 1997; Kashyap & Drake 1998). Despite this, EM distributions do provide some first order information on the gross coronal temperature structure. Early analyses of EUVE spectra appeared to confirm the results from low resolution X-ray observations concerning trends of increasing coronal temperature with increasing luminosity and coronal activity. Many analyses of spectra of different sources did not, however, reveal evidence for significant bimodality in EM distributions in the RS CVn-type binaries (e.g. Dupree et al 1993; Schrijver et al 1995), in active M dwarfs (e.g. Monsignori-Fossi et al 1995b; Stern & Drake 1996), in active and moderately active G and K single stars (e.g. Schrijver et al 1995; Laming et al 1996; Schmitt et al 1996c), or in the low activity stars Procyon and α Cen AB (e.g. Drake et al 1995a, 1996c; Mewe et al 1995). Controversy also arose based on the methods used to derive EM distributions. Methods that employed global model fitting of radiative loss models to the whole EUVE spectrum tended to produce a "hot tail"—a large rise in the derived EM at $\log T \geq$ 7.5. The "hot tails" gave rise to continuum flux in the EUV but not line flux, the ion formed at the highest temperature and exhibiting strong lines in the EUVE bandpass being FeXXIV. X-ray measurements tended to argue against the reality of such hot material (Schmitt et al 1996b), and explanations based on low metal abundances or resonance scattering in strong lines were advanced to explain the low derived line-to-continuum ratios (Schrijver et al 1994). Drake et al (1997) and Schmitt et al (1996b) argued against both explanations, concluding instead that the radiative loss models employed were missing a large number of lines (see also Jordan 1996). The missing lines would be from Fe ions with $n = 3$ ground state electrons (the "M-shell") and $n = 2$ ground states of other abundant elements such as Mg, Si, S, and Ar. This hypothesis has now been confirmed by Beiersdorfer et al (1999), who have examined laboratory spectra of Fe ions with $n = 3$ ground states.

These early results prompted Drake (1996a) to propose a general trend of EM distribution with increasing activity: a minimum in EM occurring somewhere between about log $T \sim 5$–6 that monotonically increases up to a peak, and at $T > T_{max}$ the EM falls off fairly sharply. This peak in EM moves about a decade in temperature with increasing coronal luminosity from a solar-like level to the very active RS CVn stars (log $T_{max} \sim 6.2$–7.2). Drake concluded that the two temperature models are then simply parameterizations of continuous EM distributions, which fit adequately the limited temperature discrimination of the low resolution X-ray instruments.

Since these early results, more EM distribution studies have been performed, in particular for dwarf stars of intermediate activity such as κ^1 Ceti (G5 V; $P_{rot} = 9.3d$; Landi et al 1997), ξ Boo A (G8V; $P_{rot} = 6.4d$; Laming & Drake 1999), and ϵ Eri (K2 V; $P_{rot} = 11.3d$; Laming et al 1996; Schmitt et al 1996c), and in giants such as β Cet and 31 Com (K0III; Ayres et al 1998). More active RS CVn-type binaries V711 Tau (HR 1099; K1IV+G5V; $P = 2.8d$), AR Lac (G2 IV-V+K0IV; $P = 2.0d$), II Peg (K2IV-V with unseen companion; $P = 6.7d$) and UX Ari (G5V+K0IV; $P = 6.4d$) have also been investigated in some detail (Drake & Kashyap 1998; Griffiths & Jordan 1998; Mitrou et al 1996; Mewe et al 1997; Güdel et al 1999), as has the only Algol-type binary accessible to EUVE spectrometers—Algol itself—(Stern et al 1995a; 1998), together with the rapidly rotating, putatively single stars AB Dor (K0-K2 IV-V; $P_{rot} = 0.51d$; Rucinski et al 1995; Mewe et al 1996), EK Dra (G0V; $P_{rot} = 2.75d$; Güdel et al 1997), and HD 35850 (F8V; $P_{rot} \sim 1d$; Mathioudakis & Mullan 1999; Gagné et al 1999). At the extremes of binary rotation, the contact or quasi-contact W UMa-type binaries 44i Boo (G8V+K2V; $P = 0.27d$), VW Cep (K0-K2, with an uncertain secondary type; $P_{rot} = 0.28d$), and ER Vul (G1V+G1V; $P_{rot} = 0.70d$) have been studied by Brickhouse & Dupree (1998) and Rucinski (1998). Analysis of later spectral types include AU Mic (M2 Ve; Monsignori-Fossi & Landini 1994; Schrijver et al 1995; Monsignori-Fossi et al 1996), EQ Peg (dM4e+dM5e; Monsignori-Fossi et al 1995b), AT Mic (dM4.5e+dM4.5e; Monsignori-Fossi et al 1995a), the well-known dM3e flare star AD Leo (Cully et al 1997), and the late-type BY Dra-type binaries FK Aqr (dM2e+dM2e; $P = 4.1d$), BF Lyn (K2V+dK; $P = 3.8d$), and DH Leo [(K0V+K7V)+K5V; $P = 1.1d$; Stern & Drake 1996].

One important conclusion from this work is that the EM distributions of the most active stars all look similar to one another and peak at temperatures of log $T \sim 7$, regardless of spectral type, binarity, and whether the components of a binary are

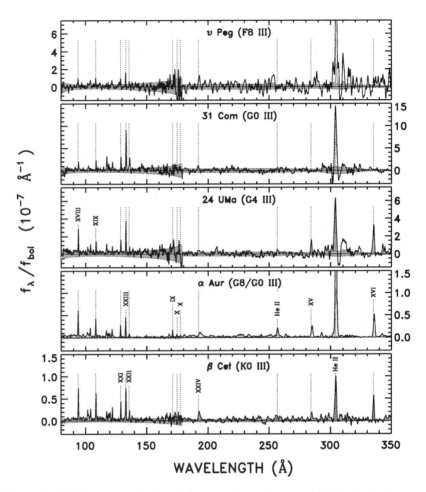

Figure 1 EUVE spectra of selected giant stars from Ayres et al 1996, depicted in normalized flux density and corrected for interstellar absorption. Shading indicates $\pm 2\sigma$ photometric errors. Important features are identified explicitly; bare Roman numerals indicate ionization stages of iron. Structure in vicinity of HeII 304 results from subtraction of strong geocoronal background. (Reprinted from *Ap. J.*)

tidally synchronized or not. Some evidence for bimodality in the EM distributions has been found (e.g. Griffiths & Jordan 1998; Gagné et al 1999), although the true significance of this structure has not been examined in any detail. Differences are to be found in EM distributions of giant stars; Ayres et al (1998) provide an interesting comparison of the EM distributions of intermediate mass giants at different evolutionary phases. They point out that the Herztsprung gap giant 31 Com shows a peak in its EM distribution at $\log T \sim 7.2$, whereas the clump giant β Ceti is more "Capella-like" with a peak in EM distribution at $\log T \sim 6.8$; these differences are actually visible in the EUVE short wavelength spectra themselves (Figure 1). The

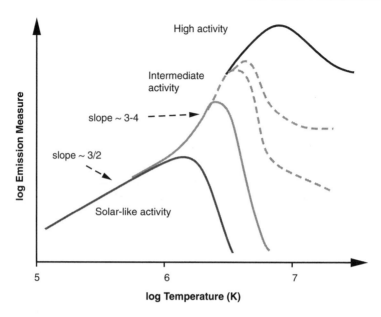

Figure 2   Rough sketch of the general trend of EM distribution changes as a function of stellar activity level. The peak in the solar-like distribution gradually moves to higher temperatures, while a hot extension toward $\log T > 10^7$ K to the distribution begins to develop and then dominate in stars of intermediate and higher activity.

Capella binary (G1III+G8III) illustrates the peril of taking an EM distribution at face value; as one of the brightest stars in the EUV sky, it has attracted the attention of several different workers (e.g. Dupree et al 1993; Brickhouse et al 1995; Brickhouse 1996; Ayres 1996; Jordan 1996; Schrijver et al 1995), and different EM distributions have been derived by each. All the analyses resulted in a peak in the EM distribution at $\log T \sim 6.8$ which is unambiguously indicated from strong FeXVIII and XIX lines in the EUVE SW spectrum.

There is some evidence from the trends of EM distributions in intermediate activity stars to suggest small modifications of the Drake (1996a) picture. The peak in the solar-like low-activity EM distribution gradually shifts to higher and higher temperatures with increasing stellar activity. However, in the highest activity stars it appears that this peak might only reach temperatures of $\log T \sim 6.8$–7, regardless of whether or not the actual full EM distribution peaks at higher temperatures. At some point when this peak is close to $\log T \sim 6.6$, there also begins to appear increasing quantities of plasma at temperatures extending to $\log T \sim 7$ and higher; it is this plasma that gives rise to the EM peak in the most active stars. This revised picture is illustrated in Figure 2. Of considerable interest in this scenario is the nature of the plasma in the peak and the hot "extension." The former could be produced in the active stellar analogy of solar bright active regions, whereas the latter could be produced from a superposition of continuous flaring.

3.4   Coronal Plasma Densities

The ability to determine electron densities from *EUVE* spectra represents a major advance in the field because it provides us with direct clues regarding coronal structure: armed with an EM—essentially the product $N_e^2 V$—knowledge of the electron density yields the emitting volume V. Density sensitivity of spectral lines arises in transitions that are influenced by metastable levels, which have relatively long radiative decay timescales that allow level populations to be excited or de-excited by collisions. Diagnostics for $\sim 10^6$ K can be found in ions FeX, XII, XIII, and XIV and are generally only detectable in stars with lower activity levels. Procyon (F5IV) and the α Cen AB binary (G2V+K0V) present the highest quality spectra for these purposes. Schmitt et al (1994, 1996a), Drake et al (1995a), and Young & Mason (1996) have derived densities for Procyon of about 5×10^9 cm^{-3}—similar to typical solar active region densities. These densities indicate the corona must be magnetically confined and are therefore inconsistent with the otherwise appealing picture of an acoustically heated corona for Procyon, as suggested by Mullan & Cheng (1994). Using static loop models, Schmitt et al (1994) estimate a filling factor of $\sim 6\%$, and solar-like loop lengths of a few 10,000 km. Similar densities of $\sim 10^9$ cm^{-3} have been found for α Cen AB (Mewe et al 1995; Drake et al 1997; Young & Mason 1996). The α Cen EUV spectrum is very similar in appearance to that of the Sun, showing prominent lines of FeIX-XVI; the density estimates also indicate structural similarity. Stars of intermediate activity show some evidence for slightly higher densities. Estimates for ϵ Eri and ξ Boo A based on FeXIII–XIV lines suggest values closer to 10^{10} cm^{-3} (Schmitt et al 1996c, Laming et al 1996, Laming & Drake 1999).

At higher coronal temperatures, FeXXI provides some density-sensitive ratios, but only for relatively high ranges of density $N_e \gtrsim 10^{11}$ cm^{-3}. These density diagnostics are discussed by Laming (1998). Additional density-sensitive ratios are available from FeXIX, FeXX (e.g. Brickhouse & Dupree 1998), and FeXXII, though again these are only sensitive for densities in the range $N_e \gtrsim 5 \times 10^{12}$ cm^{-3}. Early estimates of the electron density based on FeXXI line ratios in the EUVE spectrum of Capella by Dupree et al (1993) indicated values in the range $N_e \sim 10^{11} - 10^{13}$ cm^{-3}. These high densities have been seen in solar spectra only during flares. An analysis (Schrijver et al 1995) and a re-analysis of Capella spectra co-added from different observations (Brickhouse 1996), indicated densities of $\sim 10^{12}$ cm^{-3} from FeXIX-XXII line ratios, but $N_e \sim 10^9$ cm^{-3} from FeXII-XIV lines. If the estimates are reliable, the difference between them—two orders of magnitude or more—implies that the dominant emission at $T \sim 1-2 \times 10^6$ K is from regions physically distinct to the regions dominating the $T \sim 10^7$ K emission.

Density estimates based on spectra of other active stars such as the RS CVn σ Gem (Schrijver et al 1995), 44*i* Boo (Brickhouse & Dupree 1998) the M dwarf flare star AU Mic (Brown 1996; Monsignori-Fossi et al 1996), the active system ξ UMa (Schrijver et al 1995), and the BY Dra-type binary FK Aqr (Stern & Drake 1996) suggest that high (up to $\sim 10^{13}$ cm^{-3}) densities might be common on the very active stars during periods of relative quiescence as well as (possibly) during flares. However, Griffiths & Jordan (1998) arrived at somewhat lower values. Using the ratio of the observed FeXXI 128.7 Å line flux to the global emission measure based on lines of different Fe ions, densities of $\sim 1.6 \times 10^{11}$ cm^{-3} were inferred for V711 Tau, AR Lac, and II Peg. They dismissed higher densities of 2–3 $\times 10^{12}$ cm^{-3} based on the FeXXI 102.3/128.7 ratio as uncertain due to the weakness of the FeXXI 102.3 Å line. A similar density of 5 $\times 10^{12}$ cm^{-3} was derived by Güdel et al (1999) for UX Ari based on the same FeXXI ratio. In the case of the active F dwarf HR 1817, only upper limits of $N_e \lesssim 10^{12}$ cm^{-3} and $N_e \lesssim 10^{12}$ cm^{-3} were derived from FeXXI lines by Mathioudakis & Mullan (1999) and Gagné et al (1999), respectively. The latter also derived $N_e > 10^{13}$ cm^{-3} based on the FeXXII 114.4/117.2 ratio, but discarded this in the light of the likely more reliable FeXXI diagnostics.

Clearly, density estimates for the hotter plasma are prone to substantial uncertainties, largely from the flux measurement of the weaker line in a density-sensitive pair. Taken at face value, the high densities imply small emitting volumes confined by strong magnetic fields. The static loop models of Rosner et al (1978) suggest loop lengths of $\sim$100–1000 km for $N_e \sim 10^{13}$ cm^{-3} and $T \sim 10^7$ K and confining magnetic field strengths of $\sim$1 kG. Although these estimates are obviously very rough (the applicability of quasi-static models to these coronae might also be questioned, e.g. Drake et al 2000), they are to be compared with the equivalent solar loop parameters for $T \sim 2 \times 10^6$ K and active region densities of $N_e \sim 10^9 - 10^{10}$ cm^{-3}; a typical loop length using the same static model is $\sim$20,000 km—about two orders of magnitude larger—with a much weaker confining magnetic field of the order of 10 G. Also interesting are the filling factors such high densities imply. For an active RS CVn primary with radius $R \sim 2R_\odot$ and a volume EM of $\sim 10^{54}$ cm^{-3} at 2 $\times 10^7$ K and with $N_e \sim 10^{13}$, the implied surface filling factor is small—about 1% of the stellar surface. For static loop models, the filling factor varies roughly as $1/N_e$, so that a lower $N_e \sim 10^{12}$ still implies filling factors of only 10% or so. One possible way out of the rather puzzlingly small filling factors and ultra-compact coronae the high densities imply might be filamentary or other small-scale structure within the loops themselves. The loops could be larger and more extended, but the plasma within the loops must then be highly inhomogeneous. Evidence does exist for low filling factors within flaring loops on the Sun (e.g. Phillips et al 1996).

It does appear that there is a general pattern in the different density estimates when viewed as a whole. The general increase in EUV and X-ray flux (and therefore in EM) in the more active (rapidly rotating) stars probably arises as a result of an increase in N_e rather than in V; coronae (at least the parts contributing significantly to the observed emission) do not become more spatially extended but confine more plasma in a similar volume with increasing activity.

3.5   Chemical Compositions of Stellar Coronae

Evidence that the composition of the solar corona differs from that of the photosphere is now well established (Meyer 1985a,b, 1996; Feldman 1992). Similar to the abundance patterns seen in cosmic rays, elements with low First Ionization Potential (FIP; <10 eV; e.g. Mg, Si, Fe) show average enhancements of a factor $\sim$4 when compared with elements with high FIP ($\geq$10 eV; e.g. O, Ne, Ar). The mechanisms and sites responsible for this "FIP-effect" have not been firmly identified and are not yet understood.

The launch of the EUVE and ASCA satellites in the early 1990s enabled for the first time the study of stellar coronal abundances; the stellar field has been reviewed by Drake (1996a), Drake et al (1996a), Jordan et al (1998), and, from the standpoint of low-resolution X-ray studies, S. A. Drake (1996b). As in the case of investigations of coronal temperature structure, there are two basic methods that have been employed to determine element abundances using EUVE spectra: detailed line-by-line analyses and global radiative loss model fitting. Coronal abundance studies appeared to fall into two categories (Drake 1996a): stars that exhibit a solar-like FIP-effect (e.g. Drake et al 1997; Laming et al 1996; Laming & Drake 1999), with an enhancement of low FIP species over high FIP species compared with photospheric values; and stars that exhibit metal poor coronae, which are apparently metal poor relative to the Sun and might be depleted in metals relative to photospheric values (e.g. Stern et al 1995a; Rucinski et al 1995; Schmitt et al 1996d; see also Schrijver et al 1995; ASCA results have been reviewed by S.A. Drake 1996b). Schmitt et al (1996d), on finding a coronal Fe abundance for the RS CVn-type binary CF Tuc five to ten times lower than that of the solar photosphere, dubbed the phenomenon the *metal abundance deficiency* (MAD) *syndrome*. A third example, the F5 sub-giant Procyon, was analyzed in detail by Drake et al (1995a,b), who failed to find evidence for a difference between photospheric and coronal abundances.

The division between the MAD coronae and those characterized by a FIP-effect appeared to be one of stellar activity, with the more active stars appearing to have low coronal abundances. However, there are few detailed photospheric abundance studies with which to compare coronal values, and for some stars available photospheric estimates appear to match coronal values (e.g. Schmitt et al 1996d; see also Drake & Kashyap 1998). In other cases—those of AB Dor and II Peg being the most pronounced—photospheric abundances appeared significantly higher than the coronal values (Rucinski et al 1995; Mewe et al 1996; Mewe et al 1997). Ottmann et al (1998) concluded that active stars did appear to have anomalously metal-poor coronae based on a study of photospheric abundances for a handful of active stars for which coronal estimates based on ASCA studies were available. However, it has been pointed out (Jordan et al 1998) that existing abundance studies based on global model fits to EUVE and low resolution X-ray spectra have not taken into account the true uncertainties involved in the analyses, such as errors and omissions in the radiative loss models involved, and uncertainties in instrument calibration.

If the general division of active stellar coronae being MAD and solar-like coronae being low FIP element enriched were correct, then one would expect a transition at some point in activity level from coronae characterized by a FIP-effect to MAD coronae. Drake & Kashyap (2000) and Laming & Drake (1999) have searched for such a boundary in the intermediate activity G8 dwarf ξ Boo A, but found it to fall into the FIP category. At higher activity levels, some more recent studies have also failed to find evidence for MAD coronae, with abundances apparently more in keeping with photospheric results (e.g. Güdel et al 1997; Brickhouse & Dupree 1998; Gagné et al 1999).

There is one further anomaly that we do not discuss here in any depth, and that is an apparent increase in coronal abundances during large flare events on RS CVn-type systems found mostly from analyses of low resolution X-ray spectra but also including joint EUV and ASCA observations (e.g. Stern et al 1992; Mewe et al 1997; Güdel et al 1999). In addition to the obvious interpretation of this effect in terms of abundances, further investigation of the possible breakdown in the underlying assumptions, such as that of ionization equilibrium, would be well-motivated.

Some of the proposed mechanisms by which coronal abundances might differ from those in the solar photosphere according to the observed FIP bias have been discussed by Meyer (1996) and Feldman (1992). FIP-biased compositions are thought to be rooted in processes that occur in the chromosphere where low FIP species are largely ionized but high FIP species are neutral. Both electric and magnetic forces have been invoked, and variants on both concepts have been developed (e.g. Peter 1998; Steinitz & Kunoff 1999). In the case of MAD coronae, gravitational settling within what must be stable loop systems has been suggested by van den Oord et al (1997), whereas Drake (1998) proposed that the continua of the EUV and X-ray spectra could be enhanced through super-solar He abundances rather than metals being depleted.

3.6 Variability in Coronal EUV Emission

3.6.1 Flare Activity   Acton et al (1992) based on a review of *Yohkoh* observations of the Sun stated: "The X-ray corona is never static. *Yohkoh* observes it to be in a constant state of brightening, fading and reconfiguration. The dividing line between what is a flare and what is a coronal change blurs as observations improve." A careful study of EUV photometry leads to the same conclusion. EUV light curves are invaluable for the study of coronal variability and flaring because the length of pointed observations—from several days up to one month—is typically much longer than has been obtained with X-ray instruments.

Coronae of solar-like to intermediate activity in disk-integrated light sampled by the EUV photometric instruments appear either fairly constant (e.g. Vedder et al 1993; Drake et al 1995a) or show relatively small, slow variations amounting to only 10% or so of the total output and with infrequent if any small, short ($\lesssim 1000s$) flare-like brightenings (Laming & Drake 1999). In contrast, EUV light curves of the most active single stars, RS CVn–type and BY Dra–type active binaries, exhibit many flares with typical decay timescales from 1–10 ks or so and, much more rarely, large flares with decay timescales of a day or more (e.g. Cully et al 1994; Christian et al 1998; Christian & Vennes 1999; Gagné et al 1999; Drake et al 1994b; Drake & Kashyap 1998; Mewe et al 1997; Monsignori-Fossi et al 1995a,b; Schmitt et al 1996d; Stern & Drake 1996; Walter 1996; White et al 1996). The most comprehensive investigation of flares in EUV light curves is the recent study of EUVE photometric observations of 16 RS CVn-type binaries by Osten & Brown (1999). They find 30 out of 31 observations to show statistically significant variability, and that, on average, RS CVn-type binaries spend about 40% of the time in a flaring state. The RS CVn light curves show not only strong impulsive flares but also many gradual events with amplitudes of up to 50–100%, showing the ambiguity of what constitutes a flare as opposed to a "coronal change". EUV light curves of normal giant stars, where coronae are characterized by a 10^7 K plasma, do not appear to exhibit flares (e.g. Ayres 1996).

The time-dependent spectroscopic capability of EUVE offers a method of diagnosing the temperature and density of flaring plasma as a function of time, and thereby yields new information on the physics of the flare process at work. Unfortunately, only the largest, most energetic flares are sufficiently bright to allow such time-dependent studies. Two flares have been studied in detail in this way: the 1992 July 15–16 event on AU Mic, and the 1993 March 1–3 event on AD Leo. The first of these had an estimated energy of $\sim 5 \times 10^{35}$ erg (Cully et al 1994) and the latter, $\sim 2 \times 10^{33}$ erg (Hawley et al 1995). This is to be compared with energies of the largest solar flares of $\sim 10^{32}$ erg. Hawley et al (1995) applied a flare loop model to the AD Leo event, finding a loop half-length of 4×10^{10} cm^3—larger than the stellar radius. EUV and U-band data provided evidence for the "Neupert effect": the EUV luminosity during the impulsive phase appears to follow the time-integrated UV emission $L_{EUV}(t) \propto \int_0^t L_U(t)dt'$. Following the standard *thick target* solar flare model analogy, this relation suggests that the EUV emission is a result of material evaporated from the chromosphere and heated by electrons (and perhaps also protons) accelerated in the flaring loop.

At typical flare densities ($N_e \sim 10^{11}$–10^{12} cm^{-3}), the radiative cooling time is of the order of 30 min or less—much shorter than the observed decay times of either the AU Mic or AD Leo flares (the cooling time also depends heavily on the assumed plasma composition and density). Cully et al (1994) interpreted the AU Mic flare in terms of a coronal mass ejection (CME), in which rapid expansion decreased the density sufficiently quickly to avoid catastrophic radiative cooling. Conversely, others have argued that substantial postflare heating is more likely (e.g. Drake et al 1994a; Katsova et al 1999), supported in part by uncertain density estimates of $N_e \sim 10^{12}$–10^{13} cm^{-3} during the flare (Monsignori-Fossi et al 1996). Katsova et al (1999) propose a model similar to large flares seen on the Sun in which the source of emission is a system of high coronal loops that form after a CME, fed by additional postflare heating through reconnection in a vertical current sheet.

3.7   Rotational Modulation and Eclipses

We know from direct EUV and X-ray imaging that the solar corona comprises structures with typical scale-heights of up to $0.1R_\odot$ or so. By direct analogy, it is probably reasonable to assume that other solar-like stellar coronae share similar morphologies; estimates of coronal densities from EUVE spectra support this view (§3.4). However, the most active stars can be up to four orders of magnitude brighter than the Sun in the EUV and X-rays. On such active stars it is not immediately obvious what the spatial distribution of the emitting plasma might be, and one has to resort to indirect methods of studying their spatial structure, such as rotational or eclipse modulation. However, all methods relying on a modulation of emission through geometric variation suffer from the problem that active coronae can vary strongly on the same timescales as the eclipse or modulation by rotation, leading to possible ambiguity in interpretation of variations in terms of geometric obscuration or instrinsic source variability.

The young, rapidly rotating single star AB Dor has a period of only 12.4 h and, being relatively nearby (~25 pc), represents one of the best candidates for studying rotational modulation. White et al (1996) observed AB Dor over a period of nearly 7 days in November 1993, covering more than 13 stellar rotations, but found no obvious modulation on the rotational period (see also the analysis of ROSAT survey data by Pagano et al 1993). A subsequent study of 5 years of ROSAT monitoring of AB Dor found that only 10–20% of the modulation of X-ray flux could be attributed to rotation (Kürster et al 1997); moreover, no long-term variability over the 5-year period was found. Kürster et al (1997) suggest that these observations support the idea of a "saturated" but compact corona, almost uniformly covering the underlying star.

At least three RS CVn-type binaries—V711 Tau, AR Lac, and CF Tuc—have presented some evidence for rotational modulation, the latter two being partially eclipsing systems (Drake et al 1994b; Walter 1996; Christian et al 1996a; Schmitt et al 1996d; Gunn et al 1997). In contrast to these results, Osten & Brown (1999) did not find evidence for eclipses in the eclipsing RS CVn-type binary ER Vul. The strongest evidence for rotationally modulated coronal EUV emission was found for the W UMa eclipsing binary 44 i Boo, with a periodicity in the EUV being longer by nearly 2% than the optical period, suggestive of differential rotation between the corona and photosphere (Brickhouse & Dupree 1998). To explain the morphology of the light curve, within the constraint of a very high estimate of the plasma density of $N_e = 2 \times 10^{13}$ cm^{-3}, Brickhouse & Dupree (1998) suggest that a compact, localized "hot spot" is present near the pole of the G0 primary. Evidence for rotational modulation of coronal emission on stars of later type was presented for the BY Dra-type binary DH Leo by Stern & Drake (1996).

The conclusion one can draw from these studies is that even on the most active stars a significant fraction of the observed coronal EUV emission must come from structures that are of a scale similar to or smaller than the stellar radius.

3.8   Understanding Coronal Structure
from EUV Observations

The challenge for understanding stellar coronae is to take what are by necessity integrated, full-disk spectra and photometry and attempt to reconstruct from these limited data the spatial, temporal, and thermodynamic workings of the coronal plasma and its origins in the underlying star.

First, we note that coronae of completely different types of the very active stars—from F dwarfs to M dwarfs, from pre-zero-age main sequence stars to evolved giants—appear, based on their spectra, to be similar. Because convection zones and rotation (differential or otherwise)—the stellar characteristics thought to engender the magnetic dynamo action that is thought to feed coronal activity—differ substantially even across the range of higher activity, it is perhaps disappointing not to see obvious signs of such differences in spectra. Until accurate and reliable density diagnostics are available for a range of stars, such differences in coronal structure will remain difficult to discern.

The lack of strong differences is either due to active stars all having, to first order, similar coronae, or to their coronae arriving at fairly similar temperature structures, despite having diverse morphologies and perhaps different heating characteristics. The observable, though fairly subtle, differences in the temperature structure of the coronae of intermediate mass giant stars pointed out by Ayres et al (1998) is an interesting example that provides palpable evidence for significant structural changes with evolutionary phase. They sketch a picture in which the remnant field from the main-sequence B- and A-type stars plays a possible role in the coronal emission that develops as they become convective and spin down through magnetic breaking. Based on a lack of long-term changes in the observed coronal emission from the very late M dwarf VB8, Drake et al (1996d) have suggested that the coronae of active stars—including the supposedly "fully-convective" late M-dwarfs—could be dominated by a turbulent dynamo process, as discussed by Durney et al (1993; see also Vaiana & Rosner 1978; Stern et al 1995b), rather than a large-scale field dynamo that dominates solar activity but which has doubtful application to very late types. At first sight the turbulent, or distributive, dynamo would give rise to small-scale coronal structures and so could be identified through a definitive absence of extended coronal structure. However, it has been suggested that, provided stellar rotation is sufficiently rapid to influence convection and give it a definite handedness, large-scale fields could result from turbulence-driven dynamo processes (D. Hughes, private communication). Large, solar-scale coronal structures generated by a turbulent dynamo are then not inconceivable. The possible stellar regimes in which such a situation might prevail remain to be investigated.

In terms of quantitative modeling of EUVE observations based on energy balance and other physical considerations, progress has been relatively slow. Griffiths (1999), based on the Griffiths & Jordan (1998) EM distributions for the coronae of RS CVn-type binaries, concluded that two families of loop structures were adequate to explain the observed EM distributions—a result similar to those of other studies of EUVE and X-ray spectra of different types of stars (e.g. Schmitt et al 1996c; van den Oord et al 1997; Ventura et al 1998 and references therein). Griffiths (1999) claimed that the photospheric area coverage required by these loops is consistent with observed spot distributions, and that the magnetic field required for heating is consistent with measured values. The ratios of conductive and radiative fluxes in these models were found to be close to the value for minimum energy loss.

It has been pointed out that the EM distributions of the intermediate activity stars ξ Boo A and ϵ Eri are too steep to be explained by quasi-static, constant cross-section loops (Drake et al 2000). Similar conclusions were reached by van den Oord et al (1997) in the case of other stars and indeed were reached earlier in the study of EXOSAT spectra. van den Oord et al (1997) adopted loops with expanding geometries to account for this and pointed out that no unique solutions of loop model parameters, such as loop length and base conductive flux, can be derived from the observations.

The study of Drake et al (2000) also pointed out the similarity between the EM distributions and their peak temperatures for the brightest solar active regions observed by *Yohkoh* and the EM distributions and peak temperatures of the EM distributions of ξ Boo A and ϵ Eri. They concluded that the intermediate activity stars are likely covered in such active regions, whereas still more active coronae are either radially more extended or more dense, with plasma at temperatures higher than $\log T \sim 7.0$ probably originating from continuous flaring. The possibility that the very active coronae could be largely or wholly heated by stellar flares was raised by Güdel (1997), following earlier work in both solar and stellar contexts (e.g. Collura et al 1988). EUVE was the first satellite to obtain sufficiently long exposure times to begin to examine the flare heating question in detail. One diagnosis of whether or not flares might contain sufficient energy to power the observed coronal output lies in the relationship between the flare energy, E, and the frequency of occurrence N. For the Sun, this flare frequency distribution conforms to a power law $dN/dE = kE^{-\alpha}$; the observed value of α then yields, when integrated over all flare energies, the total energy dissipated by flares. For $\alpha \geq 2$, there is potentially very large energy at small flare energy scales. Audard & Güdel (1999) have examined the EUVE light curves of the active solar analogues 47 Cas and EK Dra to determine the value of α and have found values in the range $\alpha \approx 2.2 \pm 0.2$—strongly suggestive of a flare-dominated corona.

4.   EARLY-TYPE STARS

EUV emission from early-type (OB) stars is expected to arise from two different processes. At longer EUV wavelengths, the hot stellar photospheres of O-type and early B-type stars produce copious EUV flux, although detection of this in a large number of stars was not expected because of intervening interstellar absorption. Emission at shorter EUV wavelengths (100 Å or so) is thought most likely to arise from shock-heated, high velocity winds, at least for earlier spectral types. Analogous to the emission observed in soft X-rays (e.g. Cohen et al 1997), this emission should also, at least in principle, be detectable.

Detections of short wavelength EUV flux from early-type stars from all-sky survey observations, whether arising from shock-heated wind emission or other mechanisms, are hampered by the finite transmissions of the relevant filters to UV radiation. While early-type stars are seen in survey data, it is difficult to unambiguously assign the detection to a UV leak or to actual EUV flux. A handful of early-type stars were listed among the WFC survey sources (Pounds et al 1993). Based on the EUVE survey, Bowyer et al (1996) identified ϵ CMa and β CMa as copious emitters of EUV radiation. Other early-type stars that were apparently valid EUV detections were β Per (K0IV+B8V), ζ Pup (B5), and θ Hya (B9.5V); the first of these is the prototype of the Algol close binaries whose EUV (and X-ray) emission is dominated by the corona of the late-type component rather than by the early-type star. The star θ Hya has been shown to be harboring a hot white dwarf companion that is responsible for the EUV emission (Burleigh & Barstow 1999), and the star ζ Pup has also been shown to harbor a hot white dwarf companion that is producing the EUV emission (Vennes et al 1997a; Burleigh & Barstow 1998). This is perhaps not altogether surprising based on the results of the soft X-ray survey of Cohen et al (1997), who found that the X-ray flux, and by proxy the EUV flux, of near-main-sequence B stars sharply declines for types later than B1.5.

Prior to the launch of EUVE, the detection of photospheric emission at longer EUV wavelengths was predicted for only a small handful of early-type stars (Kudritzki et al 1991). Two of these were actually detected in the longer wavelength EUVE bandpasses (345–605 Å and 500–740 Å): the B giants ϵ CMa (B2II) and β CMa (B1II-III). Both were also detected using the ROSAT WFC 520–730 Å P2 filter (Hoare et al 1993). As reported by Hoare et al (1993), while the observed flux from β CMa was within expected bounds, ϵ CMa was much brighter than expected. Vallerga et al (1993) noted that the observed flux exceeded that of the hot white dwarf HZ43, previously thought to be the brightest nonsolar EUV source, by a factor of 30.

Spectroscopic observations of ϵ CMa obtained by EUVE were analyzed by Cassinelli et al (1995). The H Lyman continuum and the HeI continuum shortward of 504 Å were found to be stronger than predictions of both empirical and theoretical models by a factor of up to 30. Cassinelli et al (1995) suggested additional heating of the outer atmospheric layers, possibly from back-warming by the shocked wind, might be responsible. Both photospheric and wind emission and absorption features were identified in the spectrum, including Bowen-fluoresced OIII 374 Å and Fe emission lines due to ions Fe^{+8}–Fe^{+15}. Many of these features were examined in more detail based on ORFEUS-SPAS II observations by Cohen et al (1998), the higher resolution of this instrument revealing wind-broadening and blue-shifts of up to 800 km s^{-1}.

In the case of β CMa, Cassinelli et al (1996) reported that observed EUV fluxes and fluxes predicted by both LTE and non-LTE models were discrepant by up to a factor of 5—much less severe than for ϵ CMa and ameliorated by uncertainty in the appropriate effective temperature. Absorption and emission line spectra of ϵ CMa and β CMa were otherwise found to be superficially similar. The latter star is a β Cephei variable, and Cassinelli et al (1996) were able to measure a pulsation-driven EUV magnitude variation of 0.1 mag. This variation is largely a result of small photospheric temperature changes, and variability in the EUV range is consequently much larger than that at optical wavelengths. The dominant pulsing has a period of approximately 0.25 days but was observed to have a beat frequency. Pulse periods derived using a three component fit agreed with those based on optical and UV observations. This agreement implies that the pulsations propagate between the deeper layers where optical and UV continua are formed and the higher layers where emergent EUV flux is formed. Cassinelli et al (1996) speculate that deposition of some of this pulsational energy in the outer photosphere might be heating these layers, perhaps explaining the flux excess compared to current models. Further progress, given that there is little prospect for future EUV observations, lies on the modeling side.

5.   WHITE DWARF ASTROPHYSICS

5.1   Preview

White dwarf stars fall into two spectroscopic groups, dominated either by hydrogen or by helium. Short diffusion timescales in high-gravity atmospheres were believed to lead to a stratified structure with the lightest element floating on top (Schatzman 1958), thereby defining two main spectral classes: DA for hydrogen-dominated white dwarfs and DB/DO for helium-dominated white dwarfs (see Wesemael et al 1993). Trace abundance of elements heavier than the main constituent (either H or He), as estimated from optical spectra (Strittmatter & Wickramasinghe 1971), supported the view that the bimodal abundance pattern may be altered by accretion of material from the interstellar medium on timescales much shorter than the evolutionary timescales. The effects of the accretion/diffusion sequence may be further altered by competing actions of convective mixing, stellar magnetic field, and selective radiation pressure (Strittmatter & Wickramasinghe 1971; Vauclair et al 1979; Fontaine & Michaud 1979). By itself, the bimodal spectroscopic classification has its origin within the first few million years of white dwarf evolution and is the result of the phases of evolution characterized by temperatures near or exceeding 10^5 K.

With some remarkable insights into things to come, the white dwarf HZ 43, observed during the Apollo-Soyuz mission, became known as the first nonsolar EUV source (Lampton et al 1976). Observations of the white dwarf Feige 24, obtained during the same mission, led Margon et al (1976) to conclude that a "reasonable fit to the observations at $\lambda > 170$ Å is obtained at 60,000 K," but that "additional sources of opacity may permit agreement with the shorter wavelength upper limits." The detection of HZ 43 and upper limits from Sirius B (Shipman et al 1977) posed similar challenges to previous models. Hydrogen-rich atmospheres with traces of helium [$n(\text{He})/n(\text{H}) \sim 10^{-4}$] offered the best explanation for the spectral energy distributions of HZ 43 (Auer & Shipman 1977) and Sirius B (Shipman et al 1977). Apollo-Soyuz observations indicated that both HZ 43 and Feige 24 were among the hottest, hence youngest, members of their class.

EXOSAT low-energy (LE) telescope observations of $\sim$20 young white dwarf stars offered some evidence that, as a class, young DA white dwarfs hide photospheric contaminants—contaminants as yet detected only in the EUV spectral range—and that their abundance may correlate with temperature, i.e. with age. Jordan et al (1987) and Paerels & Heise (1989) described a possible correlation between an hypothetical homogeneous helium abundance in the photosphere and effective temperature. Based upon diffusion calculations, Vennes et al (1988) rejected that homogeneous model in favor of a hydrogen /helium stratified model. Model atmosphere calculations, incorporating the effect of the pure diffusive equilibrium of hydrogen and helium, showed a correlation between the hydrogen layer thickness and effective temperature (Koester 1989; Vennes & Fontaine 1992).

Several non-DA white dwarfs of the PG 1159 class were also surveyed by the EXOSAT LE revealing soft photospheric emission, albeit of extremely high temperatures. That emission implicated probable carbon-oxygen enrichments in white dwarfs' atmospheres (Nousek et al 1986; Vennes et al 1989b; Barstow & Holberg 1990; Barstow & Tweedy 1990).

EXOSAT transmission grating spectrometer observations of Sirius B, HZ 43, and Feige 24 (sources initially studied during the Apollo-Soyuz mission) answered some questions but also raised new ones. Paerels et al (1988) measured a relatively low temperature for Sirius B, which explained its nondetection at long wavelengths and established the existence of line-blanketing and back-warming effects in white dwarf atmospheres (Wesemael et al 1980). Heise et al (1988) determined an upper limit to the helium abundance in HZ 43, whereas Paerels et al (1986) unveiled the complex EUV spectrum of Feige 24. Vennes et al (1989a) proposed that the EXOSAT spectrum of Feige 24 indicated a host of trace heavy elements in an abundance consistent with calculations of selective radiation pressure.

5.2 First Extensive EUV Photometric Surveys of White Dwarfs: Population, Chemical Composition, Binarity

The first large-scale samples of EUV-selected white dwarfs were extracted from the ROSAT WFC all-sky survey (Pounds et al 1991; Pye et al 1995) and the EUVE all-sky surveys (Bowyer et al 1994, 1996). Additional EUV detections of white dwarfs have been made through combined EUVE and ROSAT observations

(Lampton et al 1997) and detection during EUV pointed observations (Kreysing et al 1995; McDonald et al 1994; Christian et al 1999). Highlights include:

1. Homogeneous samples of hot DA white dwarfs (more than 100 objects) have been constructed (Vennes et al 1996b; Marsh et al 1997a; Vennes et al 1997d; Finley et al 1997). It is now possible to present a complete phenomenological description of the early cooling sequence of white dwarf stars.

2. Twelve of the 15 white dwarfs with masses exceeding $M_\odot$ 1.2 were discovered through EUV studies (Vennes et al 1996b; Marsh et al 1997a; Vennes et al 1997d; Finley et al 1997; Vennes 1999). The origins of ultra-massive white dwarfs is a profound mystery. Suggestions include that they are the result of double-degenerate mergers (Segretain et al 1997), or they are "missed" neutron stars with ONeMg cores (Nomoto 1984).

3. Magnetic fields have been discovered in association with massive white dwarfs ($\geq$1.1 $M_\odot$) in the double-degenerate systems RE J0317$-$853/EUVE J0317$-$855 (Barstow et al 1995b; Ferrario et al 1997) and EUVE J1439+750 (Vennes et al 1999a). The isolated magnetic white dwarf EUVE J0823$-$254/RE J0823$-$254 (Ferrario et al 1998) has a mass and magnetic field similar to that of another EUV-selected massive magnetic white dwarf, PG 1658+441 (Schmidt et al 1992). Overall, some 25% of EUV-selected massive white dwarfs show detectable magnetic fields, in contrast to a mere 4% proportion found in the overall white dwarf population (Schmidt & Smith 1995).

4. RE 1738+665 (T_{eff} = 90$-$95 $\times$ 10^3 K; Barstow et al 1994a; Finley et al 1997) is possibly the hottest hydrogen-rich white dwarf detected in the ROSAT WFC/PSPC all-sky survey. A low helium and heavy element abundance is implicated, as well as the probable existence of a direct hydrogen-rich formation channel, next to a helium-rich channel (DO/DAO stars).

5. Helium, in particular, suppresses EUV emission below 227 Å in DO white dwarfs cooler than about 100,000 K. Nonetheless, two relatively cool helium-rich white dwarfs have also been detected, one during the all-sky survey (MCT 0501$-$2858/EUVE J0503$-$288/RE 0503$-$289; Vennes et al 1994; Barstow et al 1994c) and the other during a pointed observation (HD 149499B, see §5.3). MCT 0501$-$2858, at T_{eff} $\sim$ 70,000 K, has a carbon abundance between that of ordinary DO white dwarfs and the PG 1159 stars (Dreizler & Werner 1996).

6. Three ultrahot white dwarfs of the PG 1159 class and several central stars of planetary nebulae (Fruscione et al 1995) have been detected in EUV all-sky surveys. Optical spectroscopic investigation revealed carbon/oxygen enrichment in a usually helium-dominated atmosphere (Dreizler & Heber 1998, and references therein).

Multiwavelength studies (optical spectra constraining the effective temperature and surface gravity, EUV photometry constraining trace element abundances) have resulted in strict determinations of abundance patterns in hot DA white dwarfs. Sample-wide, metallicity appears to diminish with cooling ages (0.1 Myr to 30 Myr). Barstow et al (1993) confirmed this trend in a sample of 30 DA white dwarfs from the ROSAT WFC sample. Based on goodness-of-fit arguments, Barstow et al (1993) also rejected helium as the primary source of opacity. Using new Opacity Project data (Pradhan 1996), Vennes et al (1996b) defined a metallicity index relative to G191 B2B ($I_Z^{DA}/I_Z^{G191B2B}$), which, applied to the DA stars LB 1628 and GD 246, resulted in abundance ratios $I_Z^{LB1628} : I_Z^{GD246} : I_Z^{G191B2B} = 0.30 : 0.06 : 1.00$. Extreme abundance variations were found to take place early-on in the cooling sequence (age $\leq 10^6$ years).

EUV photometric studies were carried out on 89 objects in the ROSAT WFC sample, confirming the same trend. Most objects below 40,000 K were found to be apparently devoid of trace elements, while objects over 55,000 K all appeared to be contaminated, presumably with iron and other elements from the iron group (Marsh et al 1997b). Many objects with intermediate temperatures (40,000–55,000 K), such as HZ 43, also appeared to be devoid of trace elements.

Using results of a model-atmosphere analysis of EUV spectroscopic and photometric data, Finley (1996) established a temperature/gravity (or age) effect on the heavy element abundance (see Figure 3). Ages have been derived from evolutionary models of Wood (1995) using stellar parameters obtained from analyses of Balmer line profiles and heavy element abundance constrained with EUV photometric measurements. These data, along with a comparison with diffusion theory predictions (Chayer et al 1995a,b), showed gradual depletion of the heavy element supply over a timescale of a few million years, with the exception of the silicon-rich DA white dwarf GD 394 and the ultramassive white dwarf GD 50. Neither of these white dwarfs, each having temperatures near 40,000 K, is expected to hold an appreciable heavy element abundance; both, however, are evidently contaminated.

While numerous questions regarding the exact nature of young white dwarf stars were being raised, an EUV spectroscopic survey was initiated with EUVE's SW, MW, and LW spectrometers (see §5.3), complemented by IUE high-dispersion spectroscopy of EUV-selected white dwarfs. The predominance of heavy elements in hot DA white dwarfs was initially established with the detection of iron and nickel in the white dwarfs Feige 24 and G191 B2B (Vennes et al 1992; Holberg et al 1994; Werner & Dreizler 1994). These results were confirmed by similar detections in the two EUV-selected DA white dwarfs RE 0623−377/ EUVE J0623−374 and RE 2214−492/ EUVE J2214−492 (Holberg et al 1993).

Studies of binarity and close binary evolution have also been conducted with EUV samples. To a great extent, samples of EUV-selected white dwarfs do not discriminate for the presence of non-interacting companions; moreover, strong EUV emission associated with normal or weakly active stars is often indicative of the presence of an unseen companion, i.e. a white dwarf star.

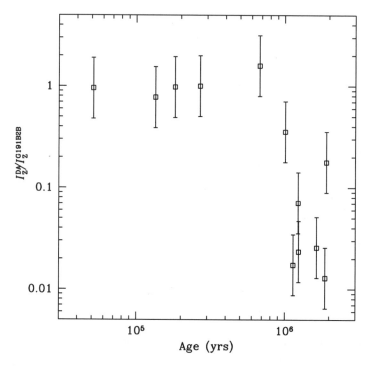

Figure 3 Heavy element abundance in a selection of twelve objects relative to the DA white dwarf G191 B2B. Set against temperature alone, no clear pattern would emerge, but set against age a correlation is apparent with the abundance dropping by at least an order of magnitude after ~10^6 years (Vennes 1999, private communication).

Analyses of likely candidates of EUV sources have resulted in a deeper understanding of the stellar content in the Solar neighborhood. White dwarf companions to bright stars (e.g. those stars listed in the Harvard-Revised and Henry-Drapper catalogs) were identified based upon their unambiguous EUV signatures. Using follow-up International Ultraviolet Explorer low-dispersion FUV spectroscopy, Hodgkin et al (1993), Wonnacott et al (1993), and Landsman et al (1993) confirmed the presence of white dwarf companions to the F stars HD 33959 C and HD 15638, the A star HR 8210, and the K0 IV star HR 1608. Additional detections were reported by Barstow et al (1994a), Vennes et al (1995b), Génova et al (1995), Christian et al (1996b), Burleigh et al (1997), and Vennes et al (1997b).

A new ultramassive white dwarf companion to the A8 V star HR 8210 was identified (Figure 4; see Landsman et al 1993). Harper (1927) suspected the presence of a $1.2 M_\odot$ underluminous companion orbiting the A star in 21.7 days, leading to speculations about its nature (Trimble & Thorne 1969).

Vennes et al (1998a) presented a comprehensive study of stellar and system properties, concluding that 3 out of 12 binaries with a white dwarf primary

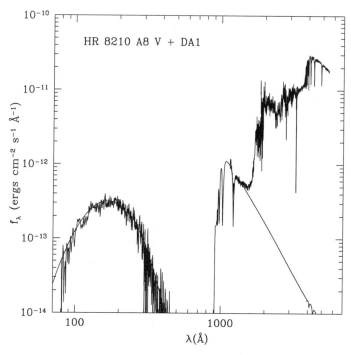

Figure 4   Optical to EUV spectral energy distribution of the A8 V plus DA binary HR 8210. A pure-hydrogen white dwarf model illustrates the expected contribution from the white dwarf throughout this spectral range (from Vennes et al 1998a; reprinted with permission of *Ap. J.*).

and a luminous secondary show notable radial velocity variations (HR 1608, HD 33959 C, HR 8210). However, HD 33858C is now considered a triple system with its third component, most likely a late-type star, responsible for both the hard X-ray emission detected by ROSAT and the radial velocity variations of the bright F star. Vennes et al (1998b) determined an average mass of $\sim 0.7 M_\odot$ for the white dwarfs in these systems, which is comparable to average mass for the general population of white dwarfs.

Conversely, several white dwarfs (both known and newly identified) have faint late-type companions, and at least 25% of the EUV-selected population of white dwarf stars appears to be composed of multiple systems. Three new DA+dMe close binaries were identified in the EUV-selected sample of white dwarfs, each of which showed traces of helium and carbon in optical and HST GHRS spectra (Vennes et al 1999b, and references therein). Orbital studies showed periods of ~ 1 day; theory (e.g. King et al 1994) suggests that because of its relatively large secondary mass and its mass-ratio close to unity, EUVE J0720−317/RE J0720−314 may come into contact within a Hubble timescale and may initiate unstable mass transfer.

The large carbon and helium abundance also suggests that the white dwarfs in these three systems may be accreting mass-loss material from their close late-type companions. Known systems such as Feige 24 ($P \sim 4.2$ days) and V471 Tauri ($P \sim 12.5$ h) are also part of the EUV selection of close binaries. The white dwarf Feige 24 is a spectroscopic twin of the isolated white dwarf G191 B2B and does not allow a conclusive test of the accretion hypothesis.

5.3 First Extensive Ultraviolet Spectroscopic Studies of DA White Dwarf Stars: From the Near to the Extreme Ultraviolet

Spectroscopy allows us to disentangle the effects of stellar parameters on photospheric EUV emission in white dwarf stars. Effective temperature and chemical structure, and to a lesser extent surface gravity, contribute to the shaping of EUV spectra. Early studies showed that white dwarf and ISM properties are inextricably coupled in EUV photometric measurements. The analysis of EUV spectroscopy thus follows two main lines of work: first, the identification of elements and full characterization of abundance pattern in individual objects, and second, the definition of class characteristics.

Initially set forth on theoretical grounds (Vennes et al 1988), structural stratification with ultrathin hydrogen layers did not leave an obvious spectroscopic signature in any EUV white dwarf spectra. The overwhelming effects of a host of trace elements detected in all DA white dwarfs foreseen as stratified have nearly hidden the potential effects of photospheric H / He stratification.

Ionized helium continuum absorption observed in the spectrum of the hot DA white dwarf GD 246 was found to originate in the ISM without any evidence of photospheric absorption (Vennes et al 1993). The effect of heavy elements is apparent at $\lambda \leq 200$ Å. As a class, hot DA white dwarfs are contaminated by a host of trace elements. Dupuis et al (1995) utilized high-metallicity model atmospheres based on Opacity Project atomic data (Pradhan 1996), establishing a heavy element blanketing effect on the EUV spectra of the white dwarfs Feige 24, G191 B2B, and MCT 0455−2812. The mysterious EUV opacity in the photosphere of the DA white dwarf GD 394 has been identified with heavy elements such as C, N, O, Si (Barstow et al 1996), and Fe (Wolff et al 1998).

Lanz et al (1996) proposed a comprehensive FUV/EUV non-LTE model of the hot DA white dwarf G191 B2B (qualitatively similar to that of Dupuis et al 1995) involving homogeneously distributed heavy elements. However, Barstow & Hubeny (1998) have revived the hydrogen / helium stratified model, including homogeneously mixed heavy element opacities as an explanation for the EUV spectrum of that star. Finally, Dreizler & Wolff (1999) have suggested that a better representation of G191 B2B may involve an inhomogeneous distribution of heavy elements as a function of depth.

Chayer et al (1997) and Wolff et al (1998) established the presence of heavy element opacities in several objects from the EUVE spectroscopic survey, characterizing such opacities as a general property of the youngest members of the DA

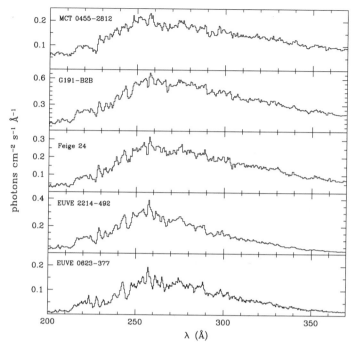

Figure 5 EUVE mini-spectroscopic survey of high-metallicity DA white dwarfs (from Chayer et al 1997; reprinted with permission of Kluwer).

white dwarf class and not simply of the hot DA white dwarf Feige 24, as originally modeled by Vennes et al (1989a). Figure 5 depicts a comparison of 5 hot DA white dwarfs and a detailed correspondence between all spectroscopic features. Figure 6 shows possible identifications of many lines and blends in the EUVE spectrum of G191 B2B (Chayer et al 1997).

These studies illustrate that this field of research has matured from mere qualitative characterization into the realm of detailed physics. The implications of these findings are fundamental to the questions of the spectral evolution of white dwarf stars and the origin of the bimodal chemical structure.

At the other extreme, several white dwarfs are devoid of heavy elements. A pure hydrogen atmosphere appears to set in early in a white dwarf's life. An EUV spectroscopic survey of 10 DA white dwarfs (Dupuis & Vennes 1996) and a largely overlapping survey of 13 white dwarfs (Barstow et al 1997) showed very little departure from pure hydrogen atmospheres in these objects. Most remarkably, the hot DA white dwarf HZ 43 has a helium abundance no greater than log He/H $\sim$ -7 (Barstow et al 1995a). These findings led to the conclusion that mass loss effectively reduces the heavy element content in hot DA white dwarf atmospheres on a timescale of 10^6 years or less (Chayer et al 1995b). Direct evidence of this

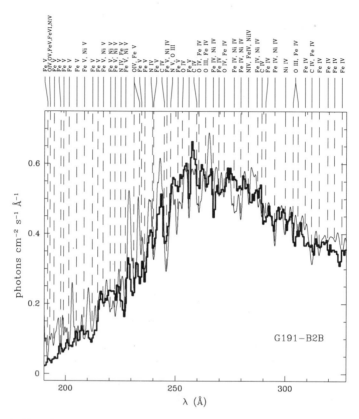

Figure 6   Tentative heavy element (C, N, O, Fe, Ni) line identifications in the EUVE spectrum of G191 B2B (from Chayer et al 1997; reprinted with permission of *Ap. J.*).

mass loss is yet to be observed in DA white dwarfs, but such evidence has already been established in the case of the PG 1159 stars (Koesterke & Werner 1998).

The presence of a companion in a wide orbit may not influence a white dwarf's chemical structure; moreover, the peculiar abundance pattern in these binaries mimics the pattern noted in isolated white dwarfs. The DA white dwarfs in the binary HD 15638, HR 8210 (Barstow et al 1994b), and EUVE J1925−563 (Vennes et al 1998a) are modeled with pure hydrogen atmospheres, but the DA white dwarf in the (triple?) system HD 33959C, despite being somewhat cooler than the white dwarf in HD 15638, shows heavy element traces (Vennes et al 1998b). The white dwarf in HD 33959C appears similar to the DA white dwarf GD 394, and the much hotter white dwarf in HD 223816 is demonstrably a spectroscopic twin of the hot isolated white dwarf MCT 0455−2812 (Vennes et al 1998b). In summary, white dwarfs in long-period binaries are phenomenologically similar to isolated white dwarfs.

However, this does not appear to be the case for white dwarfs in close non-interacting binaries. The close binary Feige 24 and the older white dwarfs in the close binaries EUVE J0720−317, 1016−053, and 2013+400 are characterized by large helium abundance and trace heavy element abundance, which are uncharacteristic of the abundance noted in isolated white dwarfs with similar effective temperatures and surface gravities. Vennes et al (1996b) noted surface abundance inhomogeneities in EUV photometric measurements and evidence of reduced solar-like abundance in the EUV photosphere of the white dwarf EUVE J1016−053. These variations appear to be coupled with a white dwarf rotation period of ∼57 min. Similar findings are reported for the binary EUVE J0720−317 (Dupuis et al 1997; Dobbie et al 1999), which appears to rotate with a period of 11.03 h. Abundance studies have also been initiated in HST/GHRS spectroscopic measurements (Vennes et al 1999b), which have revealed large concentrations of helium and carbon.

Perhaps the most remarkable finding of the EUV spectroscopic surveys of white dwarf stars is the discovery of helium in the ultramassive ($M \sim 1.2\ M_\odot$) white dwarf GD 50 (Vennes et al 1996a). Traces of helium appear homogeneously mixed in the atmosphere (although no evidence of surface abundance variations has been uncovered); these helium traces largely exceed the abundance level predicted by theory. An EUV study of another ultramassive white dwarf, EUVE J1746−706 (Dupuis & Vennes 1997), has placed an upper limit to its helium content slightly below the abundance measured in GD 50. Ultramassive white dwarfs are not expected to emerge from single-star evolution but are possible postmerger survivors. The trace helium identified may be a relic of a CO + He merger process (Segretain et al 1997). A high rotation rate in GD 50 is now considered improbable (Vennes 1999). In sum, the nature of ultramassive white dwarfs remains a mystery.

EUVE spectroscopy has established the presence of a white dwarf companion to the B5V star HR 2875 showing that main sequence stars with masses up to at least 5 $M_\odot$ will likely evolve into white dwarfs (Vennes et al 1997a; Burleigh & Barstow 1998).

Located alongside the hydrogen-rich cooling sequence, the DO white dwarfs show possible evidence of a highly dynamic process involving the ongoing separation of hydrogen, helium, and carbon into overlying strata (see Unglaub & Bues 1997, 1998). Detailed EUV and FUV spectroscopic studies of two of these objects, MCT 0501−2858 (Vennes et al 1994; Barstow et al 1994c; Dreizler & Werner 1996; Vennes et al 1998b) and HD 149499B (Jordan et al 1997), illustrate the possible effect of time on the hydrogen and carbon abundance relative to helium. Relatively high carbon abundance in MCT 0501−2858 (Figure 7) is attributable to its young cooling age and its affiliation with the PG 1159 predegenerate stars. Increasing concentrations of hydrogen in HD 149499B may be due to its advanced stage of evolution and its eventual metamorphosis into a DA white dwarf. Migration of hydrogen from deeper layers or accretion from circumstellar environment (see MacDonald & Vennes 1991, and references therein) has been proposed to explain the complex spectral evolution of white dwarf stars.

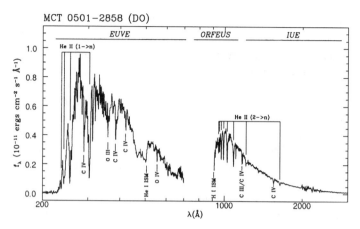

Figure 7 Complete spectral energy distribution of the DO white dwarf MCT 0501−2858 (from Vennes et al 1998b) revealing, from the extreme ultraviolet to the far ultraviolet, a large carbon concentration in a helium-rich atmosphere. (Reprinted with permission of *Ap. J.*)

Spectroscopic investigation of predegenerate stars extended the study of the chemical abundance structure of white dwarf stars to the formation stage. From central stars of planetary nebulae such as NGC 1360 (Hoare et al 1996) to PG 1159 stars with old planetary nebulae such as PG 1520+525 (Werner et al 1996) and the bare nucleus H1504+65 (Werner & Wolff 1999), large concentrations of oxygen have been linked to core material. In particular, the high abundance of neon measured in the EUVE spectrum of H1504+65 provides a deeper understanding of CO core formation. EUV spectroscopic analysis of the PG 1520+525 led Werner et al (1996) to delineate the boundaries of the GW Vir instability strip.

6.   CATACLYSMIC VARIABLES

6.1   Preview

Cataclysmic variables (CVs) consist of a white dwarf (WD) primary and a lower main-sequence secondary that fills its Roche lobe and transfers matter onto the more massive primary. There are four main types of CV: classical novae, novalikes, dwarf novae, and magnetic CVs (see Warner 1995 for a detailed treatise covering the entire field of CVs). In magnetic CVs the accreted material is channeled into a stream that funnels the material to one, or both, poles. Magnetic CVs are divided into two classes: AM Herculis-type stars, or polars, with field strengths from ~10 to 80 MG, and DQ Herculis-type stars, or intermediate polars, with field strengths from ~1 to 10 MG.

TABLE 2 EUV sources identified as CVs in original catalogs

WFC survey[a]	2nd EUVE Cat[b]	EUVE RAPI[c]	WFC pointed[d]
7	13	3	17

[a]Pye et al. 1995
[b]Bowyer et al. 1996
[c]McDonald et al. 1994
[d]Kreysing et al. 1995

A substantial number of dwarf nova and the majority of the currently identified magnetic CVs were found in the EUV all sky surveys and in follow-up pointed studies. In Table 2, we list the number of CVs initially identified in published EUV catalogs.

6.2 Polars

A number of discrepancies between theoretical models of these systems and the available X-ray data were identified early on. Perhaps the most serious of these discrepancies was the detection of far more soft X-ray emission than was predicted; this was dubbed "the soft X-ray puzzle." The preferred solution was that blobs of material in the accretion column pushed through the shocked boundary layer that was producing the X-rays and penetrated the WD photosphere where it produced large amounts of EUV emission but no additional X-ray flux (Kuijpers & Pringle 1982). This seemed a satisfactory solution to the problem but was unproven.

A few studies of the EUV emission of polars had been carried out using EXOSAT, which had EUV capabilities. Osborne et al (1986) used EXOSAT to study the polar E2003+225 (QQ Vul). They found a complex EUV light curve that could partially be attributed to an eclipse of the emitting region by the accretion stream. An EUV spectrum obtained with the transmission grating was reasonably fit by blackbody models with temperatures between 18 and 29 eV. A comparison of the EUV luminosity with the thermal bremsstrahlung flux obtained simultaneously with the medium energy X-ray detectors revealed an EUV excess of >4.5.

Osborne et al (1987) obtained additional EXOSAT observations of QQ Vul approximately two years after the first observation discussed above. The average EUV count rate had doubled from that of the previous observation. The EUV light curve was vastly different than that obtained previously; it showed two similar peaks and two unequal minima separated by half a period. This was modeled by a changing projected area of the polar cap emission as the system rotated, in combination with obscuration by the accretion column.

Mittaz et al (1992) studied RE 1149+28 and identified the object as a polar based on optical spectroscopy, and Buckley et al (1993) reported the discovery of the polar RE 1938−461, a very bright EUV source in the WFC survey.

Ramsey et al (1994) used the lowest energy soft X-ray/EUV channel of the ROSAT proportional counter detector to obtain spectral parameters for 18 polars.

They found the amount of the excess EUV emission was correlated with the magnetic field strength of the WD and concluded the EUV flux excess was the product of blobs in the accreting stream of the system and that the magnetic field of the system influenced the proportion of these blobs in the stream.

Howell et al (1995) used EUVE to study the EUV light curve variations in the polar RE 1149+28. These data yield an orbital period of 90.14 ± 0.015 m. Persistent dips were seen near the time of EUV flux maximum.

Warren et al (1995) used EUVE to study EUV light curves of UZ Fornacis. These data strongly constrained the emission and absorption geometry of this system. A dip in the EUV accretion spot radiation was shown to be absorption by the accretion stream; this result was in conflict with previous models based on optical data (Schwope et al 1990; Bailey & Cropper 1991). Warren et al (1995) showed the ratio of the accretion region to the surface area of the WD was <0.0005 and that this region was raised above the surface by about 5% of the WD radius.

Vennes et al (1995b) determined properties of the polar VV Puppis using spectroscopy and photometry obtained with EUVE. A WD model atmosphere suggested the effective temperature of the accretion region was 270,000 < Teff < 360,000 K. They provided evidence for OVI absorption features in the EUV spectrum and found that the EUV emitting region had a vertical extension of 1.3% of the WD radius.

Beardmore et al (1995) studied the polar QQ Vul with the WFC, ROSAT, and Ginga. They found the temperature of the best-fit blackbody was 28 eV and the X-ray bremsstrahlung component was 18 keV. They estimated the size of the low temperature emitting region to be $\sim 10^{16}$ cm^2 and found an EUV excess of >2.5, which they attributed to the effects of blobs of material in the accretion flow.

Ramsay et al (1996) used the soft X-ray/EUV channel of the proportional counter on ROSAT to carry out a detailed study of the spectral variability of the polars AA Her, VV Pup, BL Hyi, and UX For. They found significant variations in the low energy flux in four of these systems and significant variations in the ratio of the EUV to X-ray flux in three of these systems. They attributed these effects as the product of dense blobs of material in the accretion stream plunging through the WD photosphere; the amount of the material in these blobs was influenced by the magnetic field of the system.

Rosen et al (1996) studied the polar ZS Tel with EUVE, WFC, and ROSAT. The EUVE data showed a double-peaked light curve separated by about half a cycle, indicating two accretion sites. This was dramatically different from that seen in an observation of this system with the WFC during the survey mode of ROSAT, where only one site was observed. The emergence of a second emission site was attributed to a rather larger accretion rate during this later observation. These authors favored enhanced accretion in the form of blobs in the accretion stream as the underlying mechanism for this increase. A deep narrow dip was present in both the EUVE and ROSAT flux maxima. These were attributed to an occultation of the emission site by the accretion column. A blackbody fit to the EUV spectrum yielded a temperature of ~15 eV. Tentative evidence was found for an ionization edge at 85 Å and absorption lines at 98 and 116 Å. The EUV luminosity was found to be more than 15 times larger than the hard X-ray flux.

Paerels et al (1996) obtained EUVE spectroscopy of the polar AM Her. They detected ionization edges and absorption lines from NeVI and NeVIII but did not see the OVI edges apparently seen in the lower resolution EXOSAT transmission-grating spectrometer data (Paerels et al 1994). There was evidence for limb brightening of the spectrum, indicating the presence of a temperature inversion in the photosphere. They interpreted this result as the effect of hard X-ray irradiation of the WD atmosphere by an accretion shock, as opposed to mechanical heading of the WD photosphere by blobs of material plunging into the star.

Szkody et al (1999) obtained EUVE photometry and spectroscopy of the polar BL Hydri. The EUV photometry showed a broad hump during those phases in which both poles, known to produce X-ray emission, were visible. This suggested that the EUV emission covered a substantial area. The EUV spectrum was parameterized by a 17 eV blackbody. The total EUV energy was about equal to the hard X-ray and cyclotron energy, which suggests that this system had a conventional accretion column and that the WD polar cap was primarily heated from above by radiation rather than from below by mechanical heating from blobs that penetrated the photosphere.

Szkody et al (1997) studied the very high magnetic field polar AR Ursae Majoris with EUVE, ASCA, RXTE, and ground-based optical instruments. The X-ray luminosity during the high accretion phase was low and showed no obvious modulation. The high state EUV flux was highly modulated but never entirely disappeared, indicating at least some of the heated area was visible at all times. The EUV spectra were not well fit by either a blackbody or a standard stellar atmosphere, but using blackbody parameters the temperatures of the fainter phases were ~215,000 K, and during the dip phase the temperature was 320,000 K.

Sirk & Howell (1998) studied 10 polars observed with EUVE and developed a geometrical model of the emission region. A schematic drawing of an edge-on view of the emission region in their model is shown in Figure 8.

The model employs spot sizes that average 10^{-3} of the WD's surface area, with average spot heights of 0.023 of the WD radius. The emitting region is modeled as a sector of rotation, and the observed brightness is directly proportional to the cross-sectional area visible at a given phase. With the addition of a brightness scale factor and emission from the accretion column above the spot that decreases exponentially with height, the earliest and latest phase emission can be explained. The dips in the light curves are described by absorption due to a cylindrical accretion column.

The model is singularly successful in explaining the complex light curves of the 10 systems studied, which include a wide variety of observational orientations. It seems very likely that the model is broadly descriptive of the character of the emitting region in all polars.

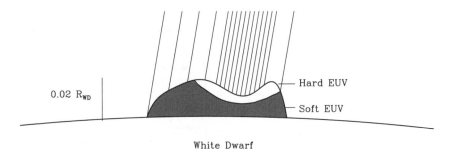

Figure 8   Schematic diagram of the accretion spot with the accretion column tilted 10 degrees away from the magnetic pole. The densest portion of the stream is offset from the center of the spot in the direction away from the magnetic pole. (Reprinted with permission of *Ap. J.*)

Mauche (1999) analyzed the EUV spectra of 11 polars with good quality data. Three models were studied: blackbody, a pure hydrogen stellar atmosphere with $\log g = 8$, and a solar abundance stellar atmosphere. The fits to these models were extraordinarily poor. Szkody et al (1999) pointed out that this is probably because any emission close to a magnetic WD would show strong energy level displacements, and Zeeman effects would distort the continuum from that of a simple blackbody or atmosphere model.

Mauche argued that despite their inadequacies, blackbody fits were broadly descriptive and the models were smooth. Hence any residuals from the fits would reveal features such as lines and edges. With the exception of AM Her, there are no sources with positive residuals $>3\ \sigma$ that would correspond to emission lines. However, nearly all the sources show negative residuals $>3\ \sigma$ corresponding to absorption lines and edges. The features with the highest significance were found in those spectra with the highest signal-to-noise ratio, which suggests that similar features could be detected in all of the sources if better data were obtained. Features detected and their probable identifications included NeVIII at 98 Å and NeVII at 116.5 Å; other unidentified features were found at 108.7 Å, 92.9 Å, 96.5 Å, and 94.1 Å.

6.3   Intermediate Polars

Hurwitz et al (1997) observed the intermediate polar EX Hydrae with EUVE. Both the photometric data and the EUV spectrum were analyzed. The EUV flux was strongly modulated at both the binary period of 98 min and the WD spin period of 67 min. An eclipse attributed to the disc bulge had an energy dependence consistent with photoelectric absorption through a neutral column of 10^{20} cm^{-2}. The shape of the light curve was well described by a simple geometric absorption model. The EUV emission was shown not to arise from the accretion disc or the secondary star but to emanate from a region covering less than 4% of the WD surface, which was located within 0.5 WD radii (in the orbital plane) and 4.5

WD radii (perpendicular to the plane) relative to the center of the star. Further arguments placed this emission in the main stream of the accretion flow.

The EUV spectrum showed many narrow emission lines characteristic of a plasma at 10^7 K, with some features indicating temperatures as low as 10^6 K. The volume emission measure of the EUV emission is substantially less than that derived by Singh & Swank 1993 using X-ray data, but their interstellar medium column, if valid, would have precluded any possibility of observing EUV radiation from the system and was clearly incorrect.

Howell et al (1997) obtained EUV photometry and spectroscopy and ground-based optical data of the (peculiar) intermediate polar PQ Geminorum. The EUV light curve showed the WD spin period, with a clear absorption dip that had previously been observed by Duck et al (1994) using the low energy channel of the proportional counter on ROSAT. No significant EUV modulation was found with either the beat or the binary orbital period. The EUV spectrum was unremarkable and showed only a few weak emission lines suggesting emission from $\sim 10^6$ K plasma.

6.4   Dwarf Novae in Outburst

Standard models of DN developed in the 1970s and 1980s associated the quiescent phase with periods of low mass transfer onto the WD. Simple models of the outburst phase (Pringle 1977) and more sophisticated radiative transfer calculations (Kley 1989; Popham & Narayan 1995) were based on the premise that the outbursts were the result of high mass transfer from the inner disk boundary layer to the WD. At these times this boundary layer would become optically thick as this material was thermalized. The effective temperature of this layer would be much cooler than during the quiescent phase and would be in the range of 20 to 30 eV. The resulting emission would be in the EUV.

EXOSAT light curves of OU Car (Naylor et al 1988) and ROSAT light curves of the nova-like variable UX Uma (Wood 1995) showed the low-level EUV emission was not eclipsed in these systems. This suggests the EUV emission of the inner boundary layer was always obscured from view. The EUV emission that was observed was attributed to a corona above the disk or to radiation from wind driven material.

Polidan et al (1990) obtained upper limits to the EUV flux from VW Hyrdi in normal and super outbursts. Mauche et al 1991 combined these upper limits with EXOSAT EUV data and determined the temperature of the boundary layer to be ~ 10.5 eV. They found the ratio of the luminosity of the boundary layer to the luminosity of the disk was ~ 0.04, as opposed to theoretical predictions that this ratio would be ~ 1. Van Teeseling et al (1993) argued that these results suggested the boundary layer remained optically thin at X-ray energies during outburst. Wheatley et al (1996a) used Voyager EUV data with X-ray data from ROSAT, Ginga, and EXOSAT to study an ordinary outburst of VW Hydri. Their analysis indicated that the boundary layer also remained optically thin at X-ray energies during a normal outburst.

Ponman et al (1995) observed SS Cygni during a decline from outburst with the WFC, the ROSAT proportional counter, and the high energy proportional counters on Ginga. The EUV luminosity was much larger than the X-ray luminosity but appeared to be less than the optical and far UV flux from the accretion disk. The observed EUV flux declined at least twice as fast as the optical flux. However, given the limited EUV bandpass covered, a drop in the emission temperature could have reduced the observed flux even if the luminosity of the source was unchanged. The co-temporally obtained X-ray data showed the additional energy of the outburst was not diverted into X-ray emission.

Mauche et al (1995) studied SS Cygni in outburst with EUVE as the optical magnitude increased from V∼12 to V∼8. The EUV flux increase of a factor of 300 initially lagged the optical outburst by ∼3 days, but this delay decreased to zero as the two light curves approached maximum. Parameterizing the EUV emission as blackbody emission suggested a temperature of 20–30 eV. While theoretical models suggested the boundary layer and accretion disk luminosity should be roughly equal, these data showed the boundary layer emission was 1/15 of the disk emission. The EUV emission spectrum was virtually constant over a change in EUV emission of greater than two orders of magnitude, which seems impossible to reconcile with the standard picture of boundary layer emission.

Wheatley et al (1996b) observed Z Cam in outburst with the WFC and obtained coincident X-ray data with the ROSAT X-ray detectors. They found the quiescent X-ray emission decreased by a factor of 10 during the rise to peak optical outburst, while the EUV emission increased to at least an order of magnitude larger than the quiescent X-ray emission. The EUV emission then declined more rapidly than the optical flux; this could have been the result of a change in the mass transfer rate, or a drop in the effective temperature of the emission moving the flux out of the WFC bandpass.

Long et al (1996) studied U Geminorum in outburst with EUVE. They found the EUV continuum emission in U Gem was eclipsed, but the lines were not eclipsed, implying they were produced in a region that is larger than the boundary layer. The EUV spectra showed NeVI-VIII, MgVI-VII, and FeVII-X emission lines superposed on a ∼12 eV blackbody continuum. The plasma temperature is <160,000 K, and the gas is photoionized. They concluded the emission features are produced by resonant scattering of photoionized material in massive winds at large distances from the WD. The continuum source is much smaller and is partially occulted near a location where calculations suggest the disk and/or stream should be thickest. In this system, the boundary layer emission is comparable to the UV/optical luminosity of the disk, and there is no boundary layer problem.

Mauche (1998) studied OY Carmia in super outburst with EUVE. No eclipses of the EUV flux were seen, implying that the EUV emitting region was significantly extended. The EUV spectrum obtained simultaneously showed substantial line emission superposed on a weak continuum. The spectrum suggested that resonant scattering of the boundary layer continuum rather than thermal emission from a shocked wind was the source of the EUV flux.

Howell et al (1999) carried out a multiwavelength study of a superoutburst in T Leonis. The X-ray flux at optical peak was at least a factor of 5 less than the quiescent X-ray emission. The EUV emission emanated from a region that was a few times the size of the white dwarf and was attributed to emission from the boundry layer. This flux was reasonably fit with a black body temperature of $\sim$90,000 K.

At this point major discrepancies exist between observational data and theoretical predictions of the boundary layer during outburst. Many, but not all, of the boundary layer luminosities are far lower than that predicted. The effective temperature of the boundary layer is predicted to be $\sim$20–30 eV; however, in many of the systems studied a lower temperature is indicated.

7.   SOLAR SYSTEM

7.1   The Moon

EUV observations of the Moon are of considerable interest since theoretical studies suggested that lunar EUV emissions would be dominated by L- and M-shell fluorescence and hence would provide useful diagnostics of surface elemental abundances.

Gladstone et al (1994) derived EUV lunar albedos of the first and last quarter Moon from data obtained in the all-sky survey portion of the EUVE mission. Average geometric albedos for the four survey bandpasses were all low, increasing from less than 0.13% for the 100 Å bandpass to 3.5% for the 600 Å bandpass. These albedos were well fit by scaled reflectivities of SiO_2 and Al_2O_3 and were consistent with reflected sunlight rather than fluorescence, but line emission could not be ruled out given the broad bandpasses of the survey filters.

Flynn et al (1998) obtained pointed EUVE full Moon data with the 200, 400, and 600 Å filters in addition to obtaining spectrometer data over the entire EUV range of the spectrometers. The whole disc lunar albedos were systematically lower than the first and last quarter Moon images obtained by Gladstone et al 1994, which suggests that the lunar EUV phase function is more shallow at high phase angles than it is in the FUV and visible. A high degree of surface contrast was found at 200 Å with albedo reversal compared with visible light, whereas the 600 Å image showed no detectable contrast. It was believed that the difference between these two images was compositional in nature. Reflectivities derived from the continuum spectrum from 140–760 Å showed the lunar reflectivity decreases by a factor of about 60 over this range. No spectral lines were detected despite the earlier theoretical predictions. The upper limits for the florescent line flux of the calcium 488 Å line was a factor of 10 below the predicted level. The reasons for this nondetection are not clear.

7.2 Venus

An EUV spectrum of Venus was obtained with the Galileo spacecraft during a flyby in 1990. A HeI 584 Å flux of $200 \pm 60R$ and a OII 834 Å flux of $180 \pm 60R$ were reported (Hord et al 1991). Gladstone et al (2000) reported EUVE observations of the Venus dayglow. A brightness of $145 \pm 35R$ was found for the HeI 584 Å emission and $40 \pm 8R$ for the combined HeI 537 Å and OII 539 Å emissions. Detailed modeling showed the $\sim$538 Å feature was $\sim$30% resonantly scattered solar HeI 537 Å flux from the Venus corona and $\sim$70% OII 539 Å flux from solar EUV photoionization excitation of oxygen atoms in the atmosphere.

7.3 Mars

A number of theorists developed models to predict the amount of helium in the Martian atmosphere and its escape rate; the results were in wide variance. As was pointed out by Krasnopolsky & Gladstone (1996), the "detection of any new species in Mars atmosphere is a rare and important event, and this is especially true for helium, which has implications for Mars' evolution." Krasnopolsky et al (1993) used the spectrometer on EUVE in an early attempt to detect neutral helium in Mars' atmosphere through a measurement of the 584 Å flux produced by resonance scattering of solar HeI. Only upper limits to this flux were obtained. The first detection of helium on Mars was made using an improved method of separating Mars' resonantly scattered helium emission from the geocoronal background (Krasnopolsky 1994). The helium escape rate was then calculated from the measured helium-mixing ratio.

Barabash et al (1995) used data from a space plasma detector on the Phobos 2 Mars Orbiter to obtain the escape rate of HeII and then modeled the helium-mixing ratio. They obtained a value that was 50 times higher than that derived from the EUVE data.

Krasnopolsky & Gladstone (1996) used an improved value for the solar EUV flux and an improved estimate of the geocoronal HeI line profile with the EUVE measured EUV flux to rederive the helium escape rate. They found a value of 7.2×10^{23} s^{-1} for this parameter; this was somewhat larger than the original estimate based on the EUVE data. Equally important, they reanalyzed the results of Barabash et al and corrected several errors in that work. This corrected value was within a factor of two of the EUVE derived results and was consistent with that value given the difference in solar EUV flux at the differing times of the measurements. Krasnopolsky & Gladstone provided an extended discussion of the implications of these measurements for the relative abundances of radioactive elements in the Martian interior.

7.4 Jupiter and the Io Torus

A fundamental property of planetary atmospheres is the eddy diffusion coefficient in the vicinity of the homopause. McConnell et al (1981) used measurements of the resonantly scattered solar HeI line obtained during the Voyager flyby of Jupiter to derive this parameter. Substantial modeling was required because of the complexity of the Jovian atmosphere and because of the low (30 Å) resolution of the Voyager spectrometer, and the uncertain background of this instrument. These authors concluded the value of the eddy diffusion was between 10^6 and 10^7 cm^2 s^{-1}, which was in reasonable agreement with values obtained by other methods.

Shemansky (1985) questioned some of the approaches used by McConnell et al (1981) and challenged their conclusion that there was a longitudinal variation in the He 584 brightness on Jupiter. Vervack et al (1995) carried out a much more sophisticated analysis of the HeI 584 Å Voyager data and concluded there was insufficient evidence to claim a longitudinal variation in this parameter. They were, however, able to narrow the range for the value of the eddy diffusion coefficient to $10^6 - 4 \times 10^6$ cm^2 s^{-1}.

Extensive observations of the Jovian system were made with EUVE before, during, and after the impact of the fragments of comet Shoemaker-Levy 9 (Gladstone et al 1995). Two to four hours after the impact of several of the larger fragments, the HeI emission at 584 Å temporarily increased by a factor of ten. A variety of explanations for this enhancement were explored and were rejected. The most likely explanation for this enhancement is resonant scattering of solar HeI 584 Å radiation from widespread neutral helium in the high altitude impact plumes, although this explanation requires an unexpectedly large amount of material ejected from deeper levels in the Jovian atmosphere.

EUV observations of the Io torus are of special importance because this emission can only come from species located in the higher temperature plasma near the center of the torus. Also, many of the strong lines that dominate the radiative energy loss from the torus occur in this spectral band. Results from the EUV spectrometer on the Voyager spacecraft in the late 1970s suggested that the Io torus composition was dominated by oxygen and sulfur ions (Shemansky & Smith 1981). Moos et al (1991) detected OII emission in the EUV portion of the spectrum of Io obtained with the Hopkins Ultraviolet Telescope spectrometer. This "confirmed the conclusion drawn from the analysis of the Voyager UVS torus data which has always had some uncertainty because of the low spectral resolution and the necessary spectral modeling" (Moos et al 1991). Hall et al (1994) used the spectrometers on EUVE to obtain the EUV spectrum of the Io torus and found several lines of OI, OII, OIII and SiII, III, and IV. They obtained an improved value of the oxygen sulfur ion ratio of about two.

Hall et al (1995) analyzed the extensive EUVE observations of the Io torus obtained in the period before and after the comet Shoemaker-Levy 9 impacts. These data showed that no new spectral features appeared in the 400–700 Å spectral region because of the impact, nor were the known lines enhanced. The brightest feature detected was SiIII at 680 Å. They estimated that roughly 20% of the power radiated from the torus was emitted in the 650–710 Å spectral interval. The fact that no enhancement in the torus emission was observed contradicted some theoretical predictions but confirmed the prediction of Dessler & Hill (1994) that because of the unfavorable geometry of the comet tail traversal, the mass loading rate of the material into the torus would not exceed a few percent of the average.

7.5 Saturn

Sandel et al (1982) used Voyager flyby data of Saturn's HeI 584 resonantly scattered dayglow combined with the methane density profile deduced from stellar occultation data to obtain the eddy diffusion coefficient for this planet; they concluded this parameter was within the range 6×10^6 to 2×10^8 cm^2 s^{-1}.

Parkinson et al (1998) derived the eddy diffusion coefficient for Saturn using measurements of the HeI 584 Saturnian emission obtained in the Voyager flybys, radiative transfer models with partial frequency redistribution, and inhomogeneous atmospheric models. They found values for this parameter were greater than 10^9, which, they acknowledged, was unreasonably large. A variety of possibilities were explored as means of reducing this value, but they found that in all reasonable scenarios this parameter was greater than 10^8.

A possible solution to the quandary is that the Voyager data are incorrect. The reduction of these data is quite complex, and the resultant signal is only 0.3% of the total counts detected. The remainder are eliminated as background as determined in part from ground (rather than in-flight) calibrations. A suggestion that this may be the case is provided by EUVE observations of Saturn. No 584 Å flux was detected with upper limits that were more than a factor of 2 below the Voyager detections (Moses et al 2000).

7.6 Comets

Intense EUV and soft X-ray emission from comets was a totally unexpected phenomena discovered with the ROSAT proportional counter and the UK Wide Field Camera (Lisse et al 1996), and shortly thereafter, by EUVE (Mumma et al 1997). Spatial images from these instruments showed the maximum emission was displaced sunward of the nucleus.

Prompted by this discovery, Dennerl et al (1997) searched the archival ROSAT All-Sky Survey database for evidence of soft X-ray emission from other comets that were observed by chance. This search resulted in the retrospective discovery of X-ray emission from comets C/1990 K1 (Levy), C/1990 N1 (Tsuchiya-Kiuchi), 45P (Honda-Mrkos-Pajdusakova), and C/1991 (Arail) in the lowest energy channels of the proportional counter. They established that greater than 80% of the detected energy was at energies less than 0.4 keV. They compared their findings with a variety of suggested emission mechanisms and concluded the most likely production mechanism was charge exchange between solar wind heavy ions and cometary gas followed by radiative de-excitation. Primary arguments against another suggested mechanism, scattering of solar X-rays by very small cometary particles (so-called "attogram" dust with a mean mass of about 1×10^{-19} gm), were the lack of emission from the dust tails and the requirement that substantial quantities of attogram dust must be present at distances greater than 10^6 km from the cometary core.

Given this surprising result, over a dozen mechanisms were suggested for the origin of this flux. Krasnopolsky (1997) provided a listing of most of these mechanisms and showed that only two—scattering of solar X-rays by attogram dust and charge transfer of solar wind heavy ions to cometary gas followed by radiative de-excitation—were tenable source mechanisms. Krasnopolsky et al (1997) observed comet Hale-Bopp (C/1995 O1) in September 1996 using EUVE. Both an image of the comet in the 100 Å bandpass and an EUV spectrum were obtained. An extended emission feature was found that did not correlate with the dust jets seen in the optical image, which argues against the dust scattering source model. Helium 584 Å line emission was detected as was a line near 538 Å. Neon in Hale-Bopp was found to be depleted in the cometary ice relative to the solar abundance by a factor of more than 25, and it was depleted relative to the sum of ice and dust by a factor of over 200.

Mumma et al (1997) used EUVE to study EUV emission from four comets: 6P/d'Arrest, C/1995 Q1 Bradfield, Hyakutake (C/1996 B2), and Hale-Bopp. Both spatial imagery and spectra were obtained. The comets' dust and gas production rates were correlated with their EUV emission. The authors argued that the brightness offsets, production rates, and maximum brightness of the comets suggested charge transfer of solar wind heavy ions to cometary neutrals was the primary EUV emission mechanism.

Owens et al (1998) studied the EUV and soft X-ray emission from Hale-Bopp with BeppoSAX. They did not detect carbon or oxygen soft X-ray line emission, which argued against charge transfer of heavy solar wind ions to cometary neutrals as the underlying source mechanism. This, in combination with a variety of other observational results, led these authors to conclude, contrary to Mumma et al (1997), that the primary production mechanism in Hale-Bopp was scattering by dust.

The source of the EUV and X-ray emission in Hyakutake was studied by Krasnopolsky (1998), who used the differing energy bandpasses of the ROSAT High Resolution Imager and the Wide Field Camera for his analysis. Bremsstrahlung and electron impact excitation were shown to be insufficient for the production of the EUV emission by two orders of magnitude. Krasnopolsky argued that these data suggested charge transfer of heavy solar wind ions to cometary neutrals was the most likely source of the EUV emission in Hyakutake.

Krasnopolsky et al (1999) detected EUV emission in a pointed post perihelion observation of Hale-Bopp. They also searched the EUV archives and found EUV emission from comet Borrelly. A reanalysis of the total set of comet observations led these authors to suggest that the regular EUV emission is primarily produced by attogram dust scattering, while enhanced emissions are produced by gas release and enhanced charge transfer, but they acknowledged limitations with this hypothesis.

It is clear that EUV and soft X-ray emission is a general property of comets. However, our understanding of the production mechanisms for this emission is not yet complete.

8.   INTERSTELLAR MEDIUM

8.1   Introduction

The interstellar medium (ISM) produces significant absorption in virtually all observations in the EUV. While this limits the depth of field in many view directions, it provides an advantage in that the resultant absorption signature can be studied to provide unique information on the intervening material. The EUV bandpass is especially useful in this regard because both the HeI Lyman edge at 504 Å and the HeII Lyman edge at 228 Å fall in this band. When these absorption edges are observed in a continuum spectrum, they provide a direct measurement of the column density of these species. An additional useful EUV spectral feature pointed out by Rumph et al (1994) is the neutral helium autoionization feature at 206 Å. This feature is at a higher energy than the 504 Å HeI edge, and it allows a determination of the intervening neutral helium in more heavily absorbed sources. In this review, we begin with studies near the Sun and progress outward into the Galaxy.

8.2   The EUV Radiation Field

The EUV radiation field is a major component of the ionization balance of the local ISM, yet this parameter was known only to a factor of 20 before the launch of EUVE (Cox & Reynolds 1987). Vallerga (1996, 1998) derived an estimate of ionizing the EUV background flux by summing the stellar spectra of stars observed with EUVE. The resultant sum was estimated to be more than 90% complete longward of 200 Å. Using this data, Vallerga (1998) developed a model of the hydrogen and helium ionization in the Local Interstellar Cloud. He found the helium ionization stays roughly constant throughout the cloud, while the hydrogen ionization strongly increases in the direction of the cloud edge. This result is in agreement with existing data on the Local Interstellar Cloud, but this radiation field cannot explain the helium ionization in the ISM beyond the cloud. Vallerga predicted a gradient would be found in the hydrogen ionization of the ISM in the general direction of Canis Majoris because of the strong ionization produced by B stars in this direction.

8.3   Neutral and Ionized Hydrogen and Helium in the ISM

Before the advent of EUV astronomy, estimates of neutral hydrogen columns to nearby stars had been obtained through the study of hydrogen Lyman series lines seen in asbsorption and in a few cases using the EUV portion of the Voyager spectrometer to measure the continuum absorption in the spectra of hot white dwarfs. There were no estimates of neutral helium in the local ISM, only global values were available. Estimates of the ionization of hydrogen and helium in the ISM were based on indirect and/or theoretical arguments. These estimates covered all possible combinations of these parameters, and the true ionization state was, in fact, unknown.

With EUV observations, the neutral hydrogen column density could be determined through the attenuation of the spectrum in the Lyman continuum of an appropriate star (typically a DA white dwarf). The strength of the HeI absorption edge or autoionization feature provided the neutral helium column through the same material. These values could be combined to obtain a direct measurement of the ionization of hydrogen and helium in the ISM.

Green et al (1990) were the first to obtain direct observational data relevant to this issue. They obtained the EUV spectrum of the white dwarf G191-B2B with an instrument flown on a sounding rocket. The neutral hydrogen and helium columns were obtained in the manner described above. Comparison of these values with their relative cosmic abundance provided an estimate of the relative ionization of the two species. Kimble et al (1993) obtained a more precise measurement of these same parameters through a longer observation of G191-B2B with the Hopkins Ultraviolet Telescope flown on the space shuttle. Both results were in agreement and showed that both hydrogen and helium were ionized; this contradicted the then-widely-held view that hydrogen was far more ionized than helium.

The first direct detection of ionized helium in the local ISM was made with observations of the white dwarf GD 246. Vennes et al (1993) obtained a spectrum of this object with EUVE, including data on both sides of the HeI 228 Å edge. The helium ionization fraction observed was $\sim$25%, and the upper limit to the hydrogen ionization fraction was 27%. Because the hydrogen column in the direction of this star is roughly a factor of ten higher than the average value for the Local Interstellar Cloud, this line of sight must sample other diffuse clouds, but the helium ionization fraction obtained was consistent with the value for the Local Interstellar Cloud. This demonstrated that helium must be partially ionized in warm diffuse clouds in the local ISM.

Dupuis et al (1995) studied six DA white dwarfs to better determine the ionization state of the local ISM. The neutral hydrogen and helium columns were obtained as above. Because of limitations in the signal-to-noise ratio, only upper limits to the amount of HeII were obtained. The sum of the HeI and the upper limits on HeII set an upper limit on the total amount of helium, because HeIII is essentially nonexistent in the Galaxy (Heiles et al 1996). Multiplying the total helium by 10 gives the total hydrogen, which can then be compared with the measured HI; the difference is HII. Using these results, Dupuis et al showed that helium is more ionized than hydrogen in much of the local ISM.

Barstow et al (1997) carried out a detailed study of EUV spectra of 13 DA white dwarfs. Their choice of objects and long observation times resulted in an especially clean separation of ISM and possible stellar atmosphere effects. Direct evidence of ionized helium was found in the line of sight to 5 of these stars.

They found weighted mean ionization fractions of HII ~0.27 and HeII ~0.35. All the data were consistent with these values being representative of the local ISM.

Vallerga et al (1993) using EUVE and Hoare et al (1993) using the Wide Field Camera reported the discovery of intense EUV emission from Beta Canis Majoris, a B2II star at a distance of about 190 pc, in the 600 Å bandpass of these instruments. Beta Canis Majoris at 210 pc was also detected in this bandpass (Hoare et al 1993). These detections at the longest EUV wavelengths could only be possible with a very low column of intervening neutral hydrogen. This was estimated to be ~0.002 cm^{-3} in this direction. These stars are the dominant stellar source of local hydrogen ionization and have a substantial impact on the local ISM, as discussed elsewhere in this section.

Wolff et al (1999) studied EUV spectra of 29 white dwarfs and carried out an extensive analysis of the ionization state of the ISM. They were able to directly determine the ionization of both hydrogen and helium in 17 of these objects. They found that the helium ionization has an average value of ~40% and shows no trend with distance. However, the hydrogen ionization does vary with distance and shows an increase in the general direction of the Canis Majoris ionized cavity. This provided observational evidence for the ionization gradient predicted by Vallerga (1998). However, Wolff et al pointed out that the photoionization that they observed cannot itself explain the large ionization of helium, and this provides observational evidence for a nonequilibrium state for the ISM.

8.4 The Size and Physical Conditions within the Local Bubble

EUV observations have provided unique information on the extent of the extended ionization region surrounding the Sun, the so-called "Local Bubble," and upon the character of the conditions within this region. The McKee-Ostriker model (McKee & Ostriker 1977) has been considered the standard description of the ISM from its inception. This model assumes thermal pressure equilibrium between the warm and hot phases of the ISM. Breitschwerdt & Schmutzler (1994) have advanced an alternate model for the large-scale structure and ionization state of the ISM; this model assumes adiabatically cooling nonequilibrium plasma. In regards to the character of the clouds themselves and the space surrounding these clouds, hydrodynamic models of cloud formation/evaporation that do not invoke thermal pressure equilibrium have been developed by Ballestrons-Pardes et al (1999). Both of these alternative models are consistent with existing data but are not yet established.

Bowyer et al (1995) and Berghoefer et al (1998) utilized shadows in the EUV background to obtain a direct measurement of the thermal pressure of the hot phase of the local ISM. Measurements were obtained in three different view directions. The pressures in all these directions were identical and were more than eight times larger than the pressure of the warm cloud surrounding the Sun, directly contradicting the McKee-Ostriker model.

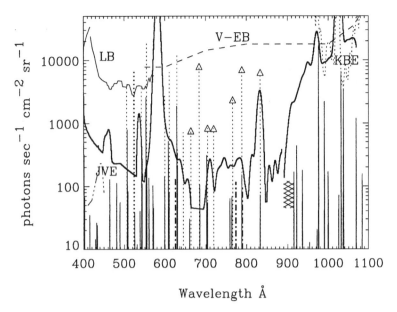

Figure 9 Two sigma upper limits to the diffuse EUV cosmic background are shown by the heavy line. Previous limits are shown by the curves marked LB, V-EB, and KBE. The vertical lines are emission expected from various models of the ISM.

Measurements of the EUV background have provided challenges to other long-held views of the ISM. The highly ionized phase of the ISM is generally considered to be produced by a 10^6 X-ray emitting plasma (Cox & Reynolds 1987). Jelinsky et al (1995) used an extensive data set obtained with the EUVE spectrometer to search for lines that should be present if the region surrounding the Sun is primarily composed of material at this temperature. These lines were not found. Several escapes from this dilemma are possible, including nonstandard abundances or nonequilibrium ionic abundances, but this result is nonetheless disturbing.

Edelstein et al (1999) analyzed data from a diffuse EUV spectrometer flown on the Spanish MiniSat satellite to study the local ISM. This instrument covers the wavelength band from 350–1100 Å with ~6 Å resolution (Bowyer et al 1997). Measurements of diffuse radiation in this bandpass were obtained that are ~100 times more sensitive than previous upper limits, and lines from either a 10^6 thermal plasma or a nonequilibrium gas adiabatically cooling from ~10^6 should have been observed. Unexpectedly, no lines were detected. The current upper limits to this radiation, along with expected emission lines from a variety of models, are shown in Figure 9.

Welsh et al (1999) analyzed the amount of absorption to 450 EUV sources with reliable identifications and distance estimates. They then compared these results with the boundary of the ionized region surrounding the Sun, or Local

Bubble, derived by Sfeir et al (1999). They found that the vast majority of the EUV sources detected are within this ionized region. However, at high galactic latitudes, 20 extra-galactic sources have been detected along lines of sight with low neutral hydrogen column densities. They interpret this result as showing that the local interstellar cavity is open ended in the direction of the galactic poles. They note that this open region corresponds with regions of enhanced X-ray emission and suggest the Local Bubble is, in fact, a Local Tube or Chimney.

8.5   Decaying Neutrinos in the Local ISM

Sciama, in an extensive series of papers (see Sciama 1998 for a partial listing), showed that relic neutrinos, if massive with a radiative decay lifetime of $\sim 10^{24}$ s, could explain a large number of otherwise puzzling astronomical phenomena, both in the local ISM and in intergalactic space. A number of efforts were made to detect the radiation from these decaying neutrinos, but none was definitive. Bowyer et al (1999b) searched for this decay line predicted to be at a wavelength between 890 and 912 Å, using data from the diffuse EUV spectrometer on MiniSat. The intensity of the expected emission was determined from neutrinos decaying in the ionized cavity, as defined by Sfeir et al (1999) from observations of interstellar NaI. This emission was then absorbed by the Local Interstellar Cloud, as determined by Redfield & Linsky (2000), to give a predicted observational flux. Upper limits to the diffuse EUV flux in the relevant band were decidedly below this predicted emission, ruling out this intriguing hypothesis.

9.   EXTRAGALACTIC SOURCES OF EUV EMISSION

9.1   Galaxies

Given the very high opacity of the ISM for EUV radiation and a correspondingly limited horizon, the discovery of a substantial number of extragalactic sources emitting EUV radiation was quite surprising. These detections were possible because of the relatively low-column density of neutral hydrogen in selected regions near the galactic poles. Pounds et al (1993) listed seven AGN detected with the Wide Field Camera. Marshall et al (1995) reported 13 AGN using a preliminary version of the EUV All-Sky Survey data processing routines. Fruscione (1996) searched the EUVE All-Sky Survey archival data set and found 20 galaxies detected with significance greater than 4 σ. Sixty-eight additional galaxies were detected with significance between three and four σ, but a substantial number of these sources were believed to be spurious. Additional sources have been identified through optical studies (Craig & Fruscione 1996; Génova et al 1995).

Kaastra et al (1995) observed the Seyfert 1 galaxy NGC 5548 with the EUVE spectrometer and identified a number of emission line features that were identified as a NeVII/NeVIII blend at 88 Å and a SiVII line at 70 Å. A fit to a thermal plasma code yielded a temperature for the plasma of 6×10^5 K. Marshall et al

(1997) measured the light curve and spectrum of NGC 5548 over 20 days as part of a coordinated observation with HST and IUE. The EUV light curve showed several factors of 2 variation on 12-hour timescales and a factor of 4 variation over 2 days. The EUV and UV variations were simultaneous; the EUV amplitudes were twice those in the UV. These authors did not detect any emission lines in the EUV spectrum and concluded the findings of Kaastra et al were the product of fixed pattern noise in the detector that had not been properly taken into account. They concluded that a thermal model best described the EUV spectral shape and its variations.

Halpren & Marshall (1996) monitored the Seyfert 1 galaxy RX J0437.4−4711 for 20 days with the EUVE Deep Survey Telescope and Short Wavelength Spectrometer. The object was detected over the 70 to 110 Å range. The EUV emission showed variability by a factor of four over the twenty-day period with a minimum doubling time of 5 h. No emission lines were found. Ramos et al (1997) observed 3C 273 as part of a multiwavelength campaign. The EUV data showed variability at the 99% confidence level.

Vennes et al (1995a) reported intense EUV emission from the Seyfert 1 galaxy TON S180 and obtained the broadband energy distribution of the galaxy from the IR through X-rays. They showed this galaxy had a large EUV to soft X-ray flux ratio.

Hwang & Bowyer (1997) obtained EUVE spectra of three Seyfert galaxies: MRK 279, MRK 478, and TON S180. The variability of the light curves provided an upper limit to the size of the EUV emitting region that excluded the BLR and diffuse intercloud regions as a source of this emission. All three of these Seyferts exhibited discrete line features; in the case of MRK 478, the lines were found to vary on a timescale of less than a day. The most striking feature was an emission line at ∼79 Å in the rest frame of the galaxies; this feature was seen in the spectra in each of these galaxies. The summed spectrum is shown in Figure 10. These authors suggested the feature could be a fluorescent transition of neutral Fe that was related to the 6.4 keV K-alpha iron emission line often observed in the X-ray spectra of Seyfert galaxies, but the true source of this line is uncertain.

Konigl et al (1995) obtained a spectrum in the EUV of the BL Lacertae object PKS 2155-304. They demonstrated that this spectrum was best attributed to Doppler-smeared lines of magnesium, neon, and iron. They showed that both the EUV spectra and the unexplained X-ray absorption in this object (Canizares & Kruper 1984) were due to absorption by clouds accelerated to about 0.1 c that intercept our line of sight within the beamed emission cone of the relativistic jet. The inferred ionization parameters and densities of these clouds were comparable to those of nuclear QSO clouds. A schematic picture of the configuration is shown in Figure 10. Kartje et al (1997) carried out a long EUVE observation of the BL Lacertae object MRK 421 and found an absorption spectrum similar to that observed in PKS 2155-304. They identified the lines as mostly L- and N-shell transitions of magnesium and neon occurring in clouds with a total column density of about 5×10^{21} cm^{-2}, which had been accelerated in the core of the object. The

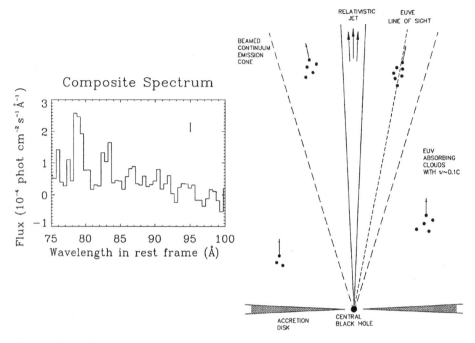

Figure 10 Left: The summed rest frame EUV spectra of MRK 279, MRK 478, and TON S180. (Reprinted with permission from *Ap. J.*) Right: The configuration of the absorbing clouds producing the spectral features in PKS 2155-304.

velocity range spanned by these clouds was relatively small, from about 0.05 c to 0.1 c. They argued that these results were best explained by a scenario in which the clouds were initially accelerated to 0.05 c by a magnetized outflow from a nuclear accretion disc with radiation pressure accelerating them to higher velocities after they entered the beamed emission cone of the jet.

9.2   Clusters of Galaxies

Extreme ultraviolet emission in excess of that produced by the well-studied X-ray emitting gas in clusters of galaxies was reported in five clusters of galaxies: the Virgo cluster (Lieu et al 1996b), the Coma cluster (Lieu et al 1996a), Abell 1795 (Mittaz et al 1998), Abell 2199 (Lieu et al 1999b), and Abell 4038 (Bowyer et al 1998). These results were all obtained with the Deep Survey Telescope of EUVE. The effective bandpass of these observations is defined by the intrinsic response of the telescope combined with the absorption of the intervening galactic ISM. In these observations, the bandpass has a peak response at 80 Å with 10% transmission points at ~65 and 100 Å. A wide variety of instrumental effects that might have explained these results were suggested by a variety of workers, but these were shown to be incapable of explaining the data (Bowyer et al 1998). A particular ionization state of the galactic ISM could have produced these results (Fabian 1996), but a detailed analysis by Bowyer et al (1996) showed that the required ionization was not, in fact, realized.

A number of suggestions were made as to the source of this EUV emission. Initial work focused on additional components of "warm" gas (about 10^6 K) in these clusters (Lieu et al 1996a,b; Fabian 1997). The problem with this hypothesis is that gas at this temperature is near the peak of the cooling curve, and substantial energy would be needed to supply the energy continually radiated away. Several authors (Hwang 1997; Ensslin & Biermann 1998) suggested that the EUV flux in the Coma cluster was inverse Compton emission. The intensity of the EUV emission was consistent with that expected from an extrapolation of the population of electrons producing the radio emission interacting with the 3 K blackbody background. However, Bowyer & Berghoefer (1998) showed that the spatial distribution of the flux would be inconsistent with the observed spatial distribution of the EUV emission in the Coma cluster, and a new population of cosmic rays will be required if inverse Compton emission is the source of the EUV excess.

Lieu et al (1999b) suggested that the Coma cluster contains a cosmic ray population that is producing the 25–80 keV emission seen by BeppoSAX via inverse Compton emission. They proposed this same cosmic ray population extrapolated to lower energies would produce the observed EUV flux by inverse Compton scattering. However, these authors did not address the fact that these electrons will produce a spatial distribution of the EUV flux that is inconsistent with spatial data reported by Bowyer & Berghoefer (1998).

Ensslin et al (1999) also explored processes that might explain both the EUV excess and the high-energy X-ray emission in the Coma cluster by inverse Compton emission. After a detailed analysis, they concluded that scattering of 3 K blackbody photons by a low-energy tail of the electron population producing the radio halo in the cluster was possible but problematic, and this population could not explain the high-energy X-ray emission. Inverse Compton scattering of starlight photons from galaxies in the cluster against 4–10 meV electrons could produce the EUV flux but could not produce the high-energy X-ray emission.

Sarazin & Lieu (1998) developed an evolutionary scenario for cosmic ray production and decay in clusters of galaxies and showed that all clusters might contain a relic population of electrons that would be unobservable in the radio. This population could produce the excess EUV flux by inverse Compton scattering against the 3 K blackbody background. Lieu et al (1999a) provided an analysis that claimed to show that the distribution of the EUV and soft X-ray emission in both Abell 2199 and 1795 were vastly different, and hypothesized that this result could be explained as due to emission lines in the EUV.

Bowyer et al (1999a) carried out a detailed analysis of the EUV emission from clusters of galaxies and found that the explanation for the apparent universality of excess EUV emission in these clusters was, in fact, due to the use of an incorrect theoretically derived instrument background in the reduction of the EUVE data. When the correct observationally derived background was employed, the excess EUV emission was not found to be present in Abell 1795, 2199, and 4038. However, excess EUV emission was confirmed to be present in the Coma cluster (Bowyer et al 1999a) and in the Virgo cluster (Berghoefer et al 1999).

The source of the excess EUV emission in the Coma and Virgo clusters is unknown. However, Bowyer et al (1999a) and Berghoefer et al (1999) showed that the spatial distribution of the EUV flux in each of these clusters was inconsistent with that of a gravitationally bound gas, and hence the underlying mechanism must be nonthermal in origin.

It is also not understood why some clusters display this emission, whereas some do not. The fact that both the Virgo cluster and the Coma cluster show current activity while the clusters that do not have excess EUV emission are quiescent may be suggestive.

10. CONCLUDING COMMENTS

The information obtained from an "astronomy that cannot be done even in principle" has turned out to be quite diverse and in most cases unique. Most of this information has been obtained with telescopes with very small effective collecting areas. It is clear that EUV instruments with even modestly improved capabilities for exploiting this unique window will yield even richer insights into the character of the universe.

Visit the Annual Reviews home page at www.AnnualReviews.org

LITERATURE CITED

Acton LW, Tsuneta S, Ogawara Y, Bentley R, Bruner M, et al. 1992. *Science.* 258:618–625

Audard M, Güdel M. 1999. *Ap. J.* 513:L53–L56

Auer LH, Shipman HL. 1977. *Ap. J.* 211:L103

Ayres TR. 1996. *Astrophysics in the Extreme Ultraviolet*, Proc. IAU Collection. 152, ed. S Bowyer, RF Malina, 113–120. Kluwer: Dordrecht.

Ayres TR, Simon TR, Stern RA, Drake SA, Wood BE, et al. 1998. *Ap. J.* 496:428–448

Bailey J, Cropper M. 1991. *MNRAS.* 253:27

Ballestrons-Pardes J, Vazquez-Semadenie E, Scalo J. 1999. *Ap. J.* 515:286–303

Barabash S, Kallio E, Lundin, R; Koskinen H. 1995. *J. Geophys. Res.* 100:21, 307–16

Barstow MA, Dobbie PD, Holberg JB, et al. 1997. *MNRAS.* 286:58

Barstow MA, Fleming TA, Diamond CJ, Finley DS, Samson AE, et al. 1993. *MNRAS.* 264:16

Barstow MA, Holberg JB. 1990. *MNRAS.* 245:370

Barstow MA, Holberg JB, Fleming TA, et al. 1994a. *MNRAS.* 270:499

Barstow MA, Holberg JB, Hubeny I, et al. 1996. *MNRAS.* 279:1120

Barstow MA, Holberg JB, Koester D. 1994b. *MNRAS.* 270:516

Barstow MA, Holberg JB, Koester D. 1995a. *MNRAS.* 274:L31

Barstow MA, Holberg JB, Werner, K, Buckley DAH, Stobie RS. 1994c. *MNRAS.* 267: 653

Barstow MA, Hubeny I. 1998. *MNRAS.* 299:379

Barstow MA, Jordan S, O'Donoghue D,

Burleigh MR, Napiwotzki R, et al. 1995b. *MNRAS.* 277:971

Barstow MA, Tweedy RW. 1990. *MNRAS.* 242:484

Beardmore A, Ramsay G, Osborne J, Mason K, Nousek J, et al. 1995. *MNRAS.* 273:742–750

Beiersdorfer P, Lepson JK, Brown GV, Utter SM, et al. 1999. *Ap. j. Lett.* 519:L185–L188

Berghoefer T, Bowyer S, Korpela E. 1999. *Ap. J.* 526:592–598

Berghoefer T, Bowyer S, Lieu R, Knude J. 1998. *Ap. J.* 500:838–846

Bowyer S, Berghoefer T. 1998. *Ap. J.* 506:502–508

Bowyer S, Berghoefer T, Korpela E. 1999b. *Ap. J.* 526:592–598

Bowyer S, Edelstein J, Lampton M. 1997. *Ap. J.* 485:523–532

Bowyer S, Green J. 1988. *Appl. Opt.* 27:1414–1422

Bowyer S, Korpela E, Edelstein J, Lampton M, Morales C, et al. 1999b. *Ap. J.* 526:10–13

Bowyer S, Lampton M, Lewis J, Wu X, Jelinsky P, et al. 1996. *Ap. J. Suppl.* 102:129–160

Bowyer S, Lieu R, Lampton M, Lewis J, Wu X, et al. 1994. *Ap. J. Suppl.* 93:569–587

Bowyer S, Lieu R, Mittaz J. 1998. In *The Hot Universe,* Proc. IAU Symp. 153, ed. K Koyama et al, pp. 185–188. Dordrecht: Kluwer

Bowyer S, Lieu R, Sidher S, Lampton M, Knude J. 1995. *Nature.* 375:212

Breitschwerdt D, Schmutzler T. 1994. *Nature.* 371:744–776

Brickhouse NS. 1996. In *Astrophysics in the Extreme Ultraviolet,* Proc. IAU Coll. 152, ed. S Bowyer, RF Malina, 105–112. Kluwer: Dordrecht

Brickhouse NS, Dupree AK. 1998. *Ap. J.* 502:918–931

Brickhouse NS, Raymond JC, Smith BW. 1995. *Ap. J. Suppl.* 97:551–570

Brown A. 1996. In *Astrophysics in the Extreme Ultraviolet,* Proc. IAU Coll. 152, ed. S Bowyer, RF Malina, 89–96. Kluwer: Dordrecht

Buckley D, O'Donoghue D, Hassall B, Kellett B, Mason K, et al. 1993. *MNRAS.* 262:93–108

Burleigh MR, Barstow MA. 1998. *MNRAS.* 295:L15

Burleigh MR, Barstow MA. 1999. *Astron. Astrophys.* 341:795–798

Burleigh MR, Barstow MA, Fleming TA. 1997. *MNRAS.* 287:381

Canizares C. Kruper J. 1984. *Ap. J. Lett.* 278:L99–L102

Cassinelli JP, Cohen DH, MacFarlane JJ, Drew JE, Lynas-Gray AE, et al. 1995. *Ap. J.* 438:932–949

Cassinelli JP, Cohen DH, MacFarlane JJ, Drew JE, Lynas-Gray AE, et al. 1996. *Ap. J.* 460:949–963

Chayer P, Fontaine G, Wesemael F. 1995a. *Ap. J. Suppl.* 99:189

Chayer P, Vennes S, Dupuis J, Thejll P, Pradhan AK. 1997. In *White Dwarfs,* ed. J Isern, M Hernanz, E Garcia-Berro, p. 273. Dordrecht: Kluwer

Chayer P, Vennes S, Pradhan AK, Thejll P, Beauchamp A, et al. 1995b. *Ap. J.* 454:429

Christian DJ, Craig N, Cahill W, Roberts B, Malina R. 1999. *Astron. J.* 117:2466–2484.

Christian DJ, Drake JJ, Mathioudakis M. 1998. *Astron. J.* 115:316–324

Christian DJ, Drake JJ, Patterer RJ, Vedder PW, Bowyer S. 1996a. *Astron. J.* 112:751–760

Christian DJ, Vennes S. 1999. *Astron. J.* 117:1852–1856

Christian DJ, Vennes S, Thorstensen JR, Mathioudakis M. 1996b. *Astron. J.* 112:258

Cohen DH, Cassinelli JP, MacFarlane JJ. 1997. *Ap. J.* 487:867–884

Cohen DH, Hurwitz M, Cassinelli JP, Bowyer S. 1998. *Ap. J.* 500:L51–L54

Collura A, Pasquini L, Schmitt JHMM. 1988. *Astron. Astrophys.* 205:197–206

Cox D, Reynolds R. 1987. *Annu. Rev. Astron. Astrophys.* 25:303–344

Craig I, Brown J. 1976. *Astron. Astrophys.* 49:239–250

Craig N, Christian D, Dupuis J, Roberts B. 1997b. *Astron. J.* 114:1–14

Craig N, Christian D, Roberts B. 1999. *ASP Conference Series 54.* The Tenth Cambridge Workshop on Cool Stars, Stellar Systems, and the Sun, ed. R Donahue, J Bookbinder, pp. 1009–1020. San Francisco: ASP

Craig N, Fruscione A. 1996. *Astron. J.* 114:1356–1364

Craig N, Fruscione A, Dupuis J, Mathioudakis M, Drake J, et al. 1995. *Astron. J.* 110:1304–1316

Cully SL, Fisher G, Abbott MJ, Siegmund OHW. 1994. *Ap. J.* 435:449–463

Cully SL, Fisher GH, Hawley SL, Simon T. 1997. *Ap. J.* 491:910–924

Dennerl K, Engelhausen J, Truemper J. 1997. *Science.* 277:1625–1630

Dessler A, Hill T. 1994. *Geophys. Res. Lett.* 21:1043–1046

Dobbie PD, Barstow MA, Burleigh MR, Hubeny I. 1999. *Astron. Astrophys.* 346:163

Doschek GA. 1991. In *Extreme Ultraviolet Astronomy*, ed. S Bowyer, RF Malina, pp. 94–112. Pergamon: Oxford

Drake JJ. 1996a. In *Cool Stars, Stellar Systems and the Sun.* 9th Cambridge Workshop, ASP Conf. Ser. ed. R Pallavicini, AK Dupree, pp. 203–214. San Francisco: ASP

Drake JJ. 1998. *Ap. J. Lett.* 496:L33–L37

Drake JJ, Brown A, Bowyer S, Jelinsky P, Malina RF, et al. 1994a. In *Cool Stars, Stellar Systems and the Sun,* The 8th Cambridge Workshop, ed. ASP Conf. Ser., 64, Caillaut J-P, pp. 35–37. San Francisco: ASP

Drake JJ, Brown A, Patterer RJ, Vedder PW, Bowyer S, et al. 1994b. *Ap. J.* 421:L43–L46

Drake JJ, Kashyap V. 1998. In *Cool Stars, Stellar Systems and the Sun,* 10th Cambridge Workshop, ASP Conf. Ser., ed. R Donahue, J Bookbinder, p. 1014. San Francisco: ASP.

Drake JJ, Kashyap V. 2000. *Ap. J.* Accepted

Drake JJ, Laming JM, Widing KG. 1995a. *Ap. J.* 443:393–415

Drake JJ, Laming JM, Widing KG. 1996a. *Astrophysics in the Extreme Ultraviolet*, Proc. IAU Coll. 152, ed. S Bowyer, RF Malina, pp. 97–112. Dordrecht: Kluwer

Drake JJ, Laming JM, Widing KG. 1996c. In *UV and X-Ray Spectroscopy of Astrophysical and Laboratory Plasmas*, ed. K Yamashita, T Watanabe, pp. 267–270. Tokyo: Univ. Acad. Press

Drake JJ, Laming JM, Widing KG. 1997. *Ap. J.* 478:403–416

Drake JJ, Laming JM, Widing KG, Schmitt JHMM, Haisch B, et al. 1995b. *Science.* 267:1470–1473

Drake JJ, Peres G, Orlando S, Laming JM, Maggio A. 2000. *Ap. J.* Accepted

Drake JJ, Stern RA, Stringfellow G, Mathioudakis M, Laming JM, et al. 1996d. *Ap. J.* 469:828–833

Drake SA. 1996b. In *Cosmic Abundances,* ASP Conf. Ser., 99, eds. SS Holt, G Sonneborn, pp. 215–226. San Francisco: ASP

Dreizler S, Heber U. 1998. *Astron. Astrophys.* 334:618

Dreizler S, Werner K. 1996. *Astron. Astrophys.* 314:217

Dreizler S, Wolff B. 1999. *Astron. Astrophys.* 348:189

Duck S, Rosen S, Ponman T, Norton A, Watson M, et al. 1994. *MNRAS.* 271:372–384

Dupree AK, Brickhouse NS, Doschek GA, Green JC, Raymond JC. 1993. *Ap. J. Lett.* 418:L41–L44

Dupuis J, Vennes S. 1996. In *Astrophysics in the Extreme Ultraviolet IAU Collection*, 152, ed. S Bowyer, R Malina, pp. 217–222. Dordrecht: Kluwer

Dupuis J, Vennes S. 1997. *Ap. J.* 475:L131

Dupuis J, Vennes S, Bowyer S. 1997. In *White Dwarfs*, ed. J Isern, M Hernanz, E Garcia-Berro, p. 277. Dordrecht: Kluwer

Dupuis J, Vennes S, Bowyer S, Pradhan AK, Thejll P. 1995. *Ap. J.* 455:574–589

Durney BR, De Young DS, Roxburgh IW. 1993. *Solar Phys.* 145:207–225

Edelstein J, Bowyer S, Korpela E, Lampton, M, Trapero J, et al. 1999. In *Advances in Spac. Res.* ed. A Gimenez. Dordrecht: Kluwer. Accepted

Ensslin R, Biermann P. 1998. *Astron. Astrophys.* 330:96–107

Ensslin T, Lieu R, Biermann P. 1999. *Astron. Astrophys.* 344:409–21

Fabian A. 1996. *Science.* 271:1244–45

Fabian A. 1997. *Science.* 275:48–49

Feldman U. 1992. *Phys. Scripta.* 46:202–220

Feldman U, Mandelbaum P, Seely JF, Doschek GA, Gursky H. 1992. *Ap. J. Suppl.* 81:387–408

Ferrario L, Vennes S, Wickramasinghe DT. 1998. *MNRAS* 299:L1

Ferrario L, Vennes S, Wickramasinghe DT, Bailey JA, Christian DJ. 1997. *MNRAS* 292:205

Finley D. 1996. In *Astrophysics in the Extreme Ultraviolet,* Proc. IAU Coll. 152, eds. S Bowyer, R Malina, pp. 223–228. Dordrecht: Kluwer.

Finley D, Koester D, Basri G. 1997. *Ap. J.* 488:375

Flynn B, Vallerga J, Gladstone G, Edelstein J. 1998. *Geophys. Res. Lett.* 25:3253–3256

Fontaine G, Michaud G. 1979. *Ap. J.* 231:826

Fruscione A. 1996. *Ap. J.* 459:509–19

Fruscione A, Drake J, McDonald K, Malina RF. 1995. *Ap. J.* 441:726

Gagné M, Valenti JA, Linsky JL, Tagliaferri G, Covino S, et al. 1999. *Ap. J.* 515:423–434

Génova R, Bowyer S, Vennes S, Lieu R, Henry P, et al. 1995. *Astron. J.* 110:788–793

Gladstone G, Hall D, Waite J. 1995. *Science.* 268:1595–1597

Gladstone G, McDonald J, Boyd W, Bowyer S. 1994. *Geophys. Res. Lett.* 21:461–464

Gladstone G, Stern S, Slater D, Paxton L. 2000. Submitted to *J. Geoph. Res.* 397:L99–L102

Green J, Jelinsky P, Bowyer S. 1990. *Ap. J.* 359:499–505

Griffiths NW. 1999. *Ap. J.* 518:873–889

Griffiths NW, Jordan C. 1998. *Ap. J.* 497:883–895

Güdel M. 1997. *Ap. J.* 480:L121–L124

Güdel M, Guinan EF, Mewe R, Kaastra JS, Skinner SL. 1997. *Ap. J.* 479:416–426

Güdel M, Linsky JL, Brown A, Nagase F. 1999. *Ap. J.* 511:405–421

Gunn A, Migenes V, Doyle JG, Spencer RE,

Mathioudakis M. 1997. *MNRAS.* 287:199–210

Hall D, Gladstone GR, Moos H, Bagenal F, Clark J, et al. 1994. *Ap. J.* 426:L51–54

Hall D, Gladstone GR, Herbert F, Lieu R, Thomas N. 1995. *Geophys. Res. Lett.* 22:3441–3444

Halpren J, Marshall H. 1996. *Ap. J.* 464:760–764

Harper WE. 1927. *Publ. Dom. Astrophys. Obs. Victoria.* 4:161

Hawley SL, Fisher GH, Simon T, Cully SL, Deustua SE, et al. 1995. *Ap. J.* 453:464–479

Heilas C, Koo B, Levenson N, Reach W. 1996. *Ap. J.* 462:326–338

Heise J, Paerels FBS, Bleeker JAM, Brinkman AC. 1988. *Ap. J.* 334:958

Hettrick M, Bowyer S. 1983. *Appl. Opt.* 1983. 22:3921–3924

Hoare M, Drew J, Denby M. 1993. *MNRAS.* 262:L19–L22

Hoare MG, Drake JJ, Werner K, Dreizler S. 1996. *MNRAS.* 283:830

Hodgkin ST, Barstow MA, Fleming TA, Monier R, Pye JP. 1993. *MNRAS* 263:229

Hodgkin ST, Pye JP. 1994. *MNRAS.* 267:840–870

Holberg JB, Barstow MA, Buckley DAH, Chen A, et al. 1993. *Ap. J.* 416:806–819

Holberg JB, Hubeny I, Barstow MA, Lanz T, Sion EM, et al. 1994. *Ap. J.* 425:L105

Hord C, Barth C, Esposito L, McClintock W, Pryor W, et al. 1991. *Science* 253:1548–1550

Howell S, Ciardi D, Szkody P, van Pardijs J, Kuulkers E, et al. 1999. *Publish. Astron. Soc. Pacific* 111:342–355

Howell S, Sirk M, Ramsay G, Cropper M, Potter S, et al. 1997. *Ap. J.* 485:333–340

Howell SB, Sirk MN, Malina RF, Mittaz J, Mason K. 1995. *Ap. J.* 439:991–995

Hurwitz M, Sirk M, Bowyer S, Ko Y-K. 1997. *Ap. J.* 477:390–397

Hwang C-Y. 1997. *Science.* 278:1917–1918

Hwang C-Y, Bowyer S. 1997. *Ap. J.* 475:552–556

Jeffries RD, Bromage GE. 1993. *MNRAS.* 260:132–140

Jeffries RD, Elliot KH, Kellett BJ, Bromage GE. 1993. *MNRAS.* 265:81–87

Jeffries RD, James DJ, Bromage GE. 1994. *MNRAS.* 271:476–480

Jeffries RD, Jewel S. 1993. *MNRAS.* 264:106–120

Jelinsky P, Vallerga J, Edelstein J. 1995. *Ap. J.* 442:653–661

Jordan C. 1996. In *Astrophysics in the Extreme Ultraviolet*, Proc. IAU Coll. 152, eds. S Bowyer, R Malina, pp. 81–88. Dordrecht: Kluwer

Jordan C, Doschek GA, Drake JJ, Galvin AB, Raymond JC. 1998. In *Cool Stars, Stellar Systems, and the Sun*, 10th Cambridge Workshop, ASP Conf. Ser., ed. R. Donahue, J. Bookbinder, pp. 91–109. San Francisco: ASP

Jordan S, Koester D, Wulf-Mathies C, Brunner H. 1987. *Astron. Astrophys.* 185:253

Jordan S, Napiwotzki R, Koester D, Rauch T. 1997. *Astron. Astrophys.* 318:461

Judge PG, Hubeny V, Brown, JC. 1997. *Ap. J.* 475:275–290

Kaastra J, Roos N, Mewe R. 1995. *Astron. Astrophys.* 300:25–30

Kartje J, Konigl A, Hwang C-Y, Bowyer S. 1997. *Ap. J.* 474:630–638

Kashyap V, Drake JJ. 1998. *Ap. J.* 503:450–466

Katsova MM, Drake JJ, Livshits A. 1999. *Ap. J.* 510:986–998

Kimble R, Davidsen A, Blain W, Bowers C, Dixon WVD, et al. 1993. *Ap. J.* 404:663–672

King AR, Kolb U, de Kool M, Ritter H. 1994. *MNRAS.* 269:907

Kley W. 1989. *Astron. Astrophys.* 222:141

Koester D. 1989. *Ap. J.* 342:999

Koesterke L. Werner K. 1998. *Ap. J.* 502:858

Konigl A, Karje S, Bowyer S, Kahn S, Hwang C-Y. 1995. *Ap. J.* 446:598–601

Krasnopolsky V. 1997. *Icarus.* 128:368–385.

Krasnopolsky V. 1998. *J. Geophys. Res.* 103:2069–2075.

Kranopolsky V, Bowyer S, Chakrabarti S, Gladstone R, McDonald J. 1994. *Icarus.* 109:337–351

Kranopolsky V, Chakrabarti S, Gladstone R. 1993. *J. Geophys. Res.* 98:15,061–15,068

Kranopolsky V, Gladstone R. 1996. *J. Geophys. Res.* 101:15,765–15,772

Krasnopolsky V, Mumma M, Abbott M. 1999. *Icarus.* Accepted.

Krasnopolsky V, Mumma M, Abbott BC, Flynn KJ, Meech DK, et al. 1997. *Science.* 277:1488–1491

Kreysing H, Brunner H, Staubert R. 1995. *Astron. Astrophys. Suppl.* 114:465–485.

Kudritzki RF, Puls J, Gabler R, Schmitt JHMM. 1991. In *Extreme Ultraviolet Astronomy*, ed. RF Malina, S Bowyer, pp. 130–152. London: Pergamon

Kuijpers J, Pringle J. 1982. *Astron. Astrophys.* 114:L4

Kürster M, Schmitt JHMM, Cutispoto G, Dennerl K. 1997. *Astron. Astrophys.* 320:831–839

Laming JM. 1998. In *Cool Stars, Stellar Systems and the Sun, ASP Conf. Ser.,* 10th Cambridge Workshop, ed. R Donahue, J Bookbinder, pp. 447–461. San Francisco: ASP

Laming JM, Drake JJ. 1999. *Ap. J.* 516:324–334

Laming JM, Drake JJ, Widing KG. 1996. *Ap. J.* 462:948–959

Lampton M, Lieu R, Schmitt J, Bowyer S, Voges W, et al. 1997. *Ap. J. Suppl.* 108:545–557

Lampton M, Margon B, Paresce F, Stern R, Bowyer S. 1976. *Ap. J.* 203:L71–L74

Landi E, Landini M, Del Zanna G. 1997. *Astron. Astrophys.* 324:1027–1035

Landsman W, Simon T, Bergeron P. 1993. *PASP.* 105:841

Lanz T, Barstow MA, Hubeny I, Holberg JB. 1996. *Ap. J.* 473:1089

Lemen JR, Mewe R, Schrijver CJ, Fludra A. 1989. *Ap. J.* 341:474–483

Lieu R, Bonamente M, Mittaz J. 1999a. *Ap. J. Letters.* 517:L91–L95

Lieu R, Ip W-H, Axford I, Bonamente M. 1999b. *Ap. J.* 510:L25–28

Lieu R, Mittaz J, Bowyer S, Breen J, Lockman F, et al. 1996a. *Science.* 274:1335–1340

Lieu R, Mittaz L, Bowyer S, Lockman F, Hwang C-Y, et al. 1996b. *Ap. J.* 458:L5–L7

Lisse C, et al. 1996. *Science.* 274:205–209

Long K, Mauche C, Raymond J, Szkody P, Mattei J. 1996. *Ap. J.* 469:841–853

MacDonald J, Vennes S. 1991. *Ap. J.* 371:719

Malina R, Marshall H, Antia B, Christian C, Dobson C, et al. 1994. *Astron. J.* 107:751–765

Margon B, Lampton M, Bowyer S, Stern R, Paresce F. 1976. *Ap. J.* 210:L79–L82

Marsh MC, Barstow MA, Buckley DA, Burleigh MR, Holberg JB, et al. 1997a. *MNRAS.* 286:369

Marsh MC, Barstow MA, Buckley DA, Burleigh MR, Holberg JB, et al. 1997b. *MNRAS.* 287:705

Marshall H, Carone T, Peterson B, Clavel J, Crenshaw D, et al. 1997. *Ap. J.* 479:222–230

Marshall H, Fruscione A, Carone T. 1995. *Ap. J.* 439:90–97

Mason KO, Hassall BJM, Bromage GE, Buckley DAH, Naylor T, et al. 1995. *MNRAS.* 274:1194–1218

Mathioudakis M, Drake JJ, Vedder PW, Schmitt JHMM, Bowyer S. 1994. *Astron. Astrophys.* 291:517–520

Mathioudakis M, Fruscione F, Drake JJ, McDonald K, Bowyer S, et al. 1995. *Astron. Astrophys.* 300:775–782

Mathioudakis M, Mullan DJ. 1999. *Astron. Astrophys.* 342:524–530

Matthews L, et al. 1994. *MNRAS.* 266:757–760

Mauche C. 1998. In *Proceedings of Wild Stars in the Old West*, ed. S Howell, E Kuulkers, C Woodward, pp. 113–121, ASP Conf. Ser. San Francisco: ASP

Mauche C. 1999. In *Proceedings of Annapolis Workshop on Magnetic Cataclysmic Variables*, ed. C Hellier, K Mukai, pp. 157–168, ASP Conf. Ser. San Francisco: ASP

Mauche C, Raymond J, Mattei J. 1995. *Ap. J.* 446:842–851

McConnell J, Sandel B, Broadfoot A. 1981. *Planet. Space Sci.* 29:283–292

Mauche C, Wade R, Polidan R, van der Woerd H, Paerels F. 1991. *Ap. J.* 372:659–663

McDonald K, Craig N, Sirk MM, Drake JJ, Fruscione A, et al. 1994. *Astron. J.* 108:1843–1853

McKee C, Ostriker J. 1977. *Ap. J.* 218:148–169

Mewe R, Kaastra JS, Schrijver CJ, van den Oord GHJ, Alkemade FJM. 1995. *Astron. Astrophys.* 296:477–498

Mewe R, Kaastra JS, van den Oord GHJ, Vink J, Tawara Y. 1997. *Astron. Astrophys.* 320:147–158

Mewe R, Kaastra JS, White SM, Pallavicini R. 1996. *Astron. Astrophys.* 315:170–178

Meyer J-P. 1985a. *Ap. J. Suppl.* 57:151–171

Meyer J-P. 1985b. *Ap. J. Suppl.* 57:172–204

Meyer J-P. 1996. In *Proc. Cosmic Abundances, 6th Ann. Astrophysics Conf. in Maryland*, ed. SS Holt, G Sonneborn, pp. 127–146. San Francisco: ASP

Mitrou CK, Doyle JG, Mathioudakis M, Antonopoulou B. 1996. In *Cool Stars, Stellar Systems, and the Sun. The Ninth Cambridge Workshop*, ASP Conf. Ser. ed. R Pallavicini, AK Dupree, pp. 275–276. San Francisco: ASP

Mitrou CK, Mathioudakis M, Doyle JG, Antonopoulou E. 1997. *Astron. Astrophys.* 317:776–785

Mittaz J, Lieu R, Lockman F. 1998. *Ap. J.* 498:L17–L20

Mittaz J, Rosen S, Mason K, Howell S. 1992. *MNRAS.* 258:277–284

Monsignori-Fossi BC, Landini M. 1994. *Astron. Astrophys.* 284:900–906

Monsignori-Fossi BC, Landini M, Del Zanna G, Bowyer S. 1996. *Ap. J.* 466:427–436

Monsignori-Fossi BC, Landini M, Drake JJ, Cully, SL. 1995a. *Astron. Astrophys.* 302:193–201

Monsignori-Fossi BC, Landini M, Fruscione A, Dupuis J. 1995b. *Ap. J.* 449:376–385

Moos H, Feldman P, Durrance S, Blair W, Bowers C, et al. 1991. *Ap. J.* 382:L105–108

Moses J, Bezaud B, Lellouch E, Gladstone R, Feuchgruben et al. 2000. *Icarus.* In press

Mullan DJ, Cheng QQ. 1994. *Ap. J.* 435:435–448

Mumma M, Krasnopolsky V, Abbott M. 1997. *Ap. J.* 491:L125–128

Naylor T, Bath P, Chalres P, Hassall B. 1988. *MNRAS.* 231:237–255

Nomoto K. 1984. *Ap. J.* 277:791

Nousek JA, Shippman HL, Holberg JB, Liebert J, Pravdo SH et al. 1986. *Ap. J.* 309:230–240.

Osborne J, Beuermann K, Charles P, Maraschi L, Mukai K, et al. 1987. *Ap. J.* 315:L123–L127

Osborne J, Bonnet-Bidand J-M, Bowyer S, Charles P, Chiappetti L, et al. 1986. *MNRAS.* 221:823–838

Osten RA, Brown A. 1999. *Ap. J.* 515:746–761

Ottman R, Pfeiffer MJ, Gehren T. 1998. *Astron. Astrophys.* 338:661–673

Owens A, Parmar A, Oosterbrock T, Orr A, Antonelli L, et al. 1998. *Ap. J.* 493:L47–L51

Paerels F, Bleeker J, Brinkman AC, Heise J. 1986. *Ap. J.* 309:L33

Paerels F, Bleeker J, Brinkman AC, Heise J. 1988. *Ap. J.* 329:849

Paerels F, Heise J. 1989. *Ap. J.* 339:1000

Paerels F, Heise J, Teeseling A. 1994. *Ap. J.* 426:313–326

Paerels F, Hur M, Mauche C, Heise J. 1996. *Ap. J.* 464:884–889

Pagano I, et al. 1993. In *Physics of Solar and Stellar Coronae*, ed. JF Linsky, S Serio, pp. 457–462. Dordrecht: Kluwer

Pallavicini R, Golub L, Vaiana GS, Ayres T, Linsky JL. 1981. *Ap. J.* 247:692–702

Parkinson C, Griffioen E, McConnell J, Gladstone G, Sandel B. 1998. *Icarus.* 133:210–220

Peter H. 1998. *Astron. Astrophys.* 335:691–702

Phillips KJH, Bhatia AK, Mason HE, Zarro DM. 1996. *Ap. J.* 466:549–560

Polidan R, Mauche C, Wade, R. 1990. *Ap. J.* 356:211–222

Polomski E, Vennes S, Thorstensen JR, Mathioudakis M, Falco EE. 1997. *Ap. J.* 486:179–196

Ponman T, Belloni T, Duck S, Verbunt F, Watson M, et al. 1995. *MNRAS.* 276:495–504

Popham R, Narayan R. 1995. *Ap. J.* 442:337

Pottasch SR. 1963. *Ap. J.* 137:945–966

Pounds K, Abbey M, Barstow M, Bentley R, Bewick A, et al. 1991. *MNRAS.* 253:364–368

Pounds K, Allen D, Barber C, Barstow M, Bertram D, et al. 1993. *MNRAS.* 260:77–102

Pradhan AK. 1996. In *Astrophysics in the Extreme Ultraviolet*, Proc. IAU Coll. 152, ed. S Bowyer, R Malina, pp. 569–576. Dordrecht: Kluwer

Pringle J. 1977. *MNRAS* 178:195–202

Pye J, McGale P, Allen D, Barber C, Bertram D, et al. 1995. *MNRAS.* 274:1165–1193

Ramos E, Kafatos M, Fruscione A, Bruhweiler F, McHardy I, et al. 1997. *Ap. J.* 482:167–172

Ramsay G, Cropper M, Mason K. 1996. *MNRAS.* 278:285–293

Ramsay G, Mason K, Cropper M, Watson M, Clayton K. 1994. *MNRAS.* 270:692–702

Redfield S, Linsky J. 2000. *Ap. J.* To appear in May 10

Rosen S, Mittaz J, Buckley D, Layden A, Clayton K, et al. 1996. *MNRAS.* 280:1121–1142

Rosner R, Tucker WH, Vaiana GS. 1978. *Ap. J.* 220:643–665

Rucinski SM. 1998. *Astron. J.* 115:303–315

Rucinski SM, Mewe R, Kaastra JS, Vilhu O, White SM. 1995. *Ap. J.* 449:900–908

Rumph T, Bowyer S, Vennes S. 1994. *Ap. J.* 107:2108–2115

Sandel B, McConnell J, Strobel D. 1982. *J. Geophys. Res. Lett.* 9:1077–1081

Sarazin C, Lieu R. 1998. *Ap. J.* 494:L177–180

Schatzman, E. 1958. *White Dwarfs.* Amsterdam: North Holland

Schmidt GD, Bergeron P, Liebert J, Saffer RA. 1992. *Ap. J.* 394:603

Schmidt GD, Smith PS. 1995. *Ap. J.* 448:305

Schmitt JHMM, Collura A, Sciortino S, Vaiana GS, Harnden FR Jr, et al. 1990. *Ap. J.* 365:704–728

Schmitt JHMM, Drake JJ, Haisch BM, Stern RA. 1996a. *Ap. J.* 467:841–850

Schmitt JHMM, Drake JJ, Stern RA. 1996b. *Ap. J.* 465:L51–L54

Schmitt JHMM, Drake JJ, Stern RA, Haisch B. 1996c. *Ap. J.* 457:882–891

Schmitt JHMM, Haisch BM, Drake JJ. 1994. *Science.* 265:1420–1422

Schmitt JHMM, Stern, RA, Drake JJ, Kürster M. 1996d. *Ap. J.* 464:898–909

Schrijver CJ, Mewe R, van den Oord GHJ, Kaastra JS. 1995. *Astron. Astrophys.* 302:438–456

Schrijver CJ, van den Oord GHJ, Mewe R. 1994. *Astron. Astrophys.* 289:L23–L26

Schwope A, Beuermann K, Thomas H. 1990. *Astron. Astrophys.* 230:120

Sciama D. 1998. *Astron. Astrophys.* 335:12–18

Segretain L, Chabrier G, Mochkovitch R. 1997. *Ap. J.* 481:355

Sfeir D, Lallement R, Crifo F, Welsh B. 1999. *Astron. Astrophys.* 346:785–797

Shara M, Bergeron L, Christian C, Craig N, Bowyer S. 1997. *PASP.* 109:998–1056

Shemansky D. 1985. *J. Geophys. Res.* 90:2673–2694.

Shemansky D, Smith G. 1981. *J. Geophys. Res.* 86:9179–9192

Shipman HL, Margon B, Bowyer S, Lampton M, Paresce F, Stern R. 1977. *Ap. J.* 213:L25–L28

Siegmund O, Lampton M, Bixler J, Chakrabarti S, Vallerga J, et al. 1986. *J. Opt. Soc. Am-A.* 3:2139–2145

Singh J, Swank J. 1993. *MNRAS.* 262:1000–1006

Sirk M, Howell S. 1998. *Ap. J.* 506:824–841

Steinitz R, Kunoff E. 1999. *Ap. J.* 510:L77

Stern RA, Drake JJ. 1996. In *Astrophysics in the Extreme Ultraviolet,* ed. S Bowyer, RF Malina, pp. 135–140. Dordrecht: Kluwer

Stern RA, Lemen JR, Schmitt JHMM, Pye JP. 1995a. *Ap. J.* 444:L45–L48

Stern RA, Schmitt JHMM, Kahabka PT. 1995b. *Ap. J.* 448:683–704

Stern RA, Uchida Y, Tsuneta S, Nagase F. 1992. *Ap. J.* 400:321–329

Stern RA, Lemen JR, Antunes S, Drake SA, Nagase F, et al. 1998. In *Cool Stars, Stellar Systems and the Sun,* 10th Cambridge Workshop, ASP Conf. Ser., ed. R Donahue, J

Bookbinder. pp. 1166–1174. San Francisco: ASP

Strittmatter PA, Wickramasinghe DT. 1971. *MNRAS.* 152:47

Swank JH, White NE, Holt SS, Becker RH. 1981. *Ap. J.* 246:208–214

Sylwester J, Schrijver J, Mewe R. 1980. *Solar Phys.* 67:285–309

Szkody P, Vennes S, Schmidt G, Wagner R, Mark R. et al. 1999. *Ap. J.* 520:841–848

Szkody P, Vennes S, Sion E, Long K, Howell S, 1997. *Ap. J.* 487:916–920

Trimble VL, Thorne KS. 1969. *Ap. J.* 156:1013

Unglaub K, Bues I. 1997. *Astron. Astrophys.* 321:485

Unglaub K, Bues I. 1998. *Astron. Astrophys.* 338:75

Vaiana GS, Rosner R. 1978. *Annu. Rev. Astron. Astrophys.* 16:393–428

Vaiana GS, et al. 1981. *Ap. J.* 245:163–182

Vallerga J. 1996. *Space Sci. Rev.* 78:277–288

Vallerga J. 1998. *Ap. J.* 497:921–927

Vallerga J, Vedder P, Welsh B. 1993. *Ap. J.* 414:L65–L67

van den Oord GHJ, Schrijver CJ, Camphens M, Mewe R, Kaastra JS. 1997. *Astron. Astrophys:* 326:1090–1102

Van Teeseling A, Verbunt F, Heise J. 1993. *Astron. Astrophys.* 270:159–164

Vedder PW, Patterer RJ, Jelinsky P, Brown A, Bowyer S. 1993. *Ap. J. Lett.* 414:L61–L64

Vennes S. 1999. *Ap. J.* 525:995–1008

Vennes S, Berghoefer TW, Christian DJ. 1997a. *Ap. J.* 491:L85

Vennes S, Bowyer S, Dupuis J. 1996a. *Ap. J.* 461:L103–106

Vennes S, Chayer P, Fontaine G, Wesemael F. 1989a. *Ap. J.* 336:L25

Vennes S, Chayer P, Thorstensen JR, Bowyer CS, Shipman, HL. 1992. *Ap. J.* 392:L27

Vennes S, Christian DJ, Mathioudakis M, Doyle JG. 1997b. *Astron. Astrophys.* 318:L9

Vennes S, Christian DJ, Thortsensen JR. 1998a. *Ap. J.* 502:763

Vennes S, Dupuis J, Chayer P, Polomski EF, Dixon WVD, et al. 1998b. *Ap. J.* 500:L41

Vennes S, Dupuis J, Rumph T, Drake J, Bowyer S, et al. 1993. *Ap. J.* 410:L119–122

Vennes S, Dupuis J, Bowyer S, Fontaine G, Wiercigroch A, et al. 1994. *Ap. J.* 421:L35

Vennes S, Ferrario L, Wickramasinghe DT. 1999a. *MNRAS.* 302:L49

Vennes S, Fontaine G. 1992. *Ap. J.* 401:288

Vennes S, Fontaine G, Wesemael F. 1989b. In *White Dwarfs*, ed. G. Wegner, p. 363. Berlin: Springer.

Vennes S, Korpela E, Bowyer S. 1997c. *Astron. J.* 114:1567–1572.

Vennes S, Pelletier C, Fontaine G, Wesemael F. 1988. *Ap. J.* 331:876

Vennes S, Polomski E, Bowyer S, Thorstensen J. 1995a. *Ap. J. Lett.* 448:L9–12

Vennes S, Szkody P, Sion E, Long K. 1995b. *Ap. J.* 445:921–926

Vennes S, Thejll PA, Génova-Galvan R, Dupuis J. 1997d. *Ap. J.* 480:714

Vennes S, Thejll PA, Wickramasinghe DT, Bessell MS. 1996b. *Ap. J.* 467:782

Vennes S, Thorstensen JR, Polomski EF. 1999b. *Ap. J.* 523:386

Ventura R, Maggio A, Peres G. 1998. *Astron. Astrophys.* 334:188–200

Vervack R Jr, et al. 1995. *Icarus.* 114:163–173

Walter FM. 1996. In *Astrophysics in the Extreme Ultraviolet*, ed. S Bowyer, RF Malina, pp. 129–133. Dordrecht: Kluwer

Walter FM, Bowyer S. 1981. *Ap. J.* 245:671–676

Walter FM, Cash W, Charles PA, Bowyer S. 1980. *Ap. J.* 236:212–218

Warner B. 1995. *Cataclysmic Variable Stars.* Cambridge: Cambridge University Press

Warren J, Sirk M, Vallerga J. 1995. *Ap. J.* 445:909–920

Welsh B, Sfeir D, Sirk M, Lallement R. 1999. *Astron. Astrophys.* 352:308–316

Werner K, Dreizler S. 1994. *Astron. Astrophys.* 286:L31

Werner K, Dreizler S, Heber U, Rauch T. 1996. In *Astrophysics in the Extreme Ultraviolet*, Proc. IAU Coll. 152, ed. S Bowyer, R Malina, pp. 229–234. Dordrecht: Kluwer

Werner K, Wolff B. 1999. *Astron. Astrophys.* 347:L9

Wesemael F, Auer LH, Van Horn HM, Savedoff MP. 1980. *Ap. J. Suppl.* 43:159

Wesemael F, Greenstein JL, Liebert J, Lamontagne R, Fontaine G, et al. 1993. *PASP.* 105:761

Wheatley P, van Teeseling A, Waton M, Verbunt F, Pfefferman E. 1996b. *Monthly Notices Roy. Ast. Soc.* 283:101–106

Wheatley P, Verbunt F, Belloni T, Watson M, Naylor T, et al. 1996a. *Astron. Astrophys.* 307:137–148

White S, Lim J, Rucinski S, Roberts D. 1996. In *Astrophysics in the Extreme Ultraviolet*, Proc. IAU Coll. 152, ed. S Bowyer, RF Malina, pp. 165–169. Dordrecht: Kluwer

Wolff B, Koester D, Dreizler S, Hass S. 1998. *Astron. Astrophys.* 329:1045

Wolff B, Koester D, Lallement R. 1999. *Astron. Astrophys.* 346:969–978

Wonnacott D, Kellett BJ, Stickland DJ. 1993. *MNRAS.* 262:277

Wood BE, Brown A, Linsky JL, Kellett BJ, Bromage GE, et al. 1994. *Ap. J. Suppl.* 93: 287–307

Wood M. 1995. In *White Dwarfs*, ed. D Koester, K Werner, p. 41. Berlin: Springer

Young PR, Mason HE. 1996. In *Cool Stars, Stellar Systems, and the Sun,* 9th Cambridge Workshop, ASP Conf. Ser., ed. R Pallavicini, AK Dupree, pp. 301–302. San Francisco: ASP

Annu. Rev. Astron. Astrophys. 2000. 38:289–335

X-RAY PROPERTIES OF GROUPS OF GALAXIES

John S. Mulchaey

*The Observatories of the Carnegie Institution of Washington, 813 Santa Barbara St.,
Pasadena, CA 91101; e-mail: mulchaey@ociw.edu*

Key Words   intragroup medium, temperature, metallicity, masses, dark matter

■ **Abstract**   ROSAT observations indicate that approximately half of all nearby groups of galaxies contain spatially extended X-ray emission. The radial extent of the X-ray emission is typically 50–500 h_{100}^{-1} kpc or approximately 10–50% of the virial radius of the group. Diffuse X-ray emission is generally restricted to groups that contain at least one early-type galaxy. X-ray spectroscopy suggests the emission mechanism is most likely a combination of thermal bremsstrahlung and line emission. This interpretation requires that the entire volume of groups be filled with a hot, low-density gas known as the intragroup medium. ROSAT and ASCA observations indicate that the temperature of the diffuse gas in groups ranges from approximately 0.3 keV to 2 keV. Higher temperature groups tend to follow the correlations found for rich clusters between X-ray luminosity, temperature, and velocity dispersion. However, groups with temperatures below approximately 1 keV appear to fall off the cluster L_X-T relationship (and possibly the L_X-σ and σ-T cluster relationships, although evidence for these latter departures is at the present time not very strong). Deviations from the cluster L_X-T relationship are consistent with preheating of the intragroup medium by an early generation of stars and supernovae.

There is now considerable evidence that most X-ray groups are real, physical systems and not chance superpositions or large-scale filaments viewed edge-on. Assuming the intragroup gas is in hydrostatic equilibrium, X-ray observations can be used to estimate the masses of individual systems. ROSAT observations indicate that the typical mass of an X-ray group is $\sim 10^{13} h_{100}^{-1}$ $M_\odot$ out to the radius to which X-ray emission is currently detected. The observed baryonic masses of groups are a small fraction of the X-ray determined masses, which implies that groups are dominated by dark matter. On scales of the virial radius, the dominant baryonic component in groups is likely the intragroup medium.

1.   INTRODUCTION

Redshift surveys of the nearby universe indicate that most galaxies occur in small groups (e.g. Holmberg 1950, Humason, Mayall & Sandage 1956, de Vaucouleurs 1965, Materne 1979, Huchra & Geller 1982, Geller & Huchra 1983, Tully 1987, Nolthenius & White 1987). Despite diligent work in this area over the last two decades, the nature of poor groups is still unclear. Dynamical studies of groups are generally hampered by small number statistics: a typical group contains only a few luminous galaxies. For this reason, the dynamical properties of any individual group are always rather uncertain. In fact, many cataloged groups may not be real physical systems at all (e.g. Hernquist et al 1995, Frederic 1995, Ramella et al 1997), but rather chance superpositions or large-scale structure filaments viewed edge-on. Given the small number of luminous galaxies in a group, the prospects for uncovering the nature of these systems from studying the galaxies alone seem rather bleak.

The discovery that many groups are X-ray sources has provided considerable new insight into these important systems. X-ray observations indicate that about half of all poor groups are luminous X-ray sources. In many cases, the X-ray emission is extended, often beyond the optical extent of the group. X-ray spectroscopy suggests the emission mechanism is a combination of thermal bremsstrahlung and line emission from highly ionized trace elements. The spatial and spectral properties of the X-ray emission suggest the entire volume of groups is filled with hot, low-density gas. This gas component is referred to as the intragroup medium, in analogy to the diffuse X-ray emitting intracluster medium found in rich clusters (e.g. Forman & Jones 1982).

To first order, groups can be viewed as scaled-down versions of rich clusters. Many of the fundamental properties of groups, such as X-ray luminosity and temperature, are roughly what one expects for a "cluster" with a velocity dispersion of several hundred kilometers per second. However, some important physical differences exist between groups and clusters. The velocity dispersions of groups are comparable to the velocity dispersions of individual galaxies. Therefore, some processes such as galaxy-galaxy merging are much more prevalent in groups than in clusters. Other mechanisms that are important in the cluster environment, such as ram-pressure stripping and galaxy harassment, are not expected to be important in groups. The spectral nature of the X-ray emission is also somewhat different in groups than in clusters. At the typical temperature of the intracluster medium, almost all abundant elements are fully ionized, and the X-ray emission is dominated by a thermal bremsstrahlung continuum. At the lower temperatures of groups, most of the trace elements retain a few atomic electrons, and line emission dominates the observed X-ray spectrum. Thus, while the cluster analogy is a useful starting point, detailed studies of groups as a class are also important. Although no strict criterion exists for separating groups from poor clusters, for the context of this article I focus on systems with velocity dispersions less than about 500 km/s.

The idea that poor groups might contain diffuse hot gas dates back to the classic Kahn & Woltjer (1959) paper on the "timing mass" of the Local Group. Kahn & Woltjer (1959) found that the mass of the Local Group far exceeded the visible stellar mass and suggested the bulk of the missing mass was in the form of a warm, low-density plasma. Although it is now generally believed that the Local Group is dominated by dark matter, Kahn & Woltjer's estimates for the properties of the intragroup medium are remarkably similar to more recent estimates. More than a decade after Kahn & Woltjer, the idea of diffuse gas in the Local Group and other groups was revisited by Oort (1970), Ruderman & Spiegel (1971), Hunt & Sciama (1972), and Silk & Tarter (1973).

The earliest claims for X-ray detections of groups came from the non-imaging X-ray telescopes Uhuru, Ariel 5, and HEAO 1 in the 1970s. Cooke et al (1978) produced a catalog (known as the 2A) of 105 bright X-ray sources from the Leicester Sky Survey Instrument on Ariel 5. Based on positional coincidences, Cooke et al (1978) suggested the identification of seven X-ray sources in the 2A catalog as groups of galaxies. Subsequent observations showed that several of these X-ray sources were variable, indicating they were actually active galaxies within the group (Ricker et al 1978, Ward et al 1978, Griffiths et al 1979). However, several of the remaining objects in Cooke et al (1978) were later shown to be poor clusters (Schwartz et al 1980).

X-ray studies of lower-mass systems received a major boost with the launch of the Einstein Observatory in November 1978. Einstein observations firmly established that some poor clusters with bright central galaxies (i.e. MKW and AWM clusters; Morgan et al 1975, Albert et al 1977) were X-ray sources (Kriss et al 1980, 1983, Burns et al 1981, Price et al 1991, Dell'Antonio et al 1994). The X-ray luminosities of these poor clusters range from several times 10^{41} ergs s^{-1} h_{100}^{-2} up to several times 10^{43} ergs s^{-1} h_{100}^{-2}. The X-ray emission in these poor clusters was shown to be extended (out to radii as great as $0.5 \, h_{100}^{-1}$ Mpc) with temperatures in the range T $\sim$1–5 keV. Although most of these systems are somewhat richer than the typical groups considered in this review, these Einstein observations clearly demonstrated that diffuse X-ray emission was not restricted to rich clusters.

Several attempts were also made to study even poorer galaxy systems with Einstein. Biermann and collaborators detected extended emission in two nearby elliptical-dominated groups (Biermann et al 1982; Biermann & Kronberg 1983). In both cases, the X-ray emission was centered on the dominant galaxy. For the NGC 3607 group, Biermann et al (1982) concluded that the X-ray emission most likely originated from a hot, intergalactic gas because it was extended on scales larger than the galaxy. (Biermann et al estimate a Gaussian width for the X-ray emission of $4.7' \approx 13 \, h_{100}^{-1}$ kpc.) From a rough fit to the X-ray spectrum, a temperature of $\approx 5 \times 10^6$ K and an X-ray luminosity of $2 \times 10^{40} \, h_{100}^{-2}$ ergs s^{-1} was found. Following their discovery of X-ray emission in the NGC 3607 group, Biermann & Kronberg (1983) found a similar component in the NGC 5846 group. The Einstein Observatory was also used to study the X-ray properties of compact groups. Bahcall et al (1984) studied five compact groups, including four from Hickson's (1982) catalog. Three of the compact groups were detected with Einstein. The Einstein exposure times for these groups were very short, resulting in only $\sim$20–60 net counts in the X-ray detected cases. Bahcall et al (1984) noted that the X-ray luminosities of two of the groups were of order $\sim 10^{42}$ ergs s^{-1} h_{100}^{-2}, much higher than the X-ray emission expected from the member galaxies alone. The emission was also extended in these two groups, and in the case of Stephan's

Quintet, the shape of the X-ray spectrum was unlike that expected from individual galaxies. These X-ray properties led Bahcall et al (1984) to conclude that the X-ray emission likely originated in a hot intragroup gas in at least two of the five groups they studied. Thus, although it was not possible to unambiguously separate a diffuse component from galaxy emission with Einstein, there were strong indications that intragroup gas was likely present in some groups.

2.   X-RAY TELESCOPES

While there were hints from Einstein observations that some groups of galaxies might contain a hot intragroup medium, it was not until the 1990s that the presence of diffuse gas in groups was firmly established. Group studies were aided by the launch of two important X-ray telescopes, ROSAT (the ROentgen SATellite) and ASCA (Advanced Satellite for Cosmology and Astrophysics). Both of these telescopes were capable of simultaneous X-ray imaging and spectroscopy in the energy range appropriate for poor groups. In addition, the field of view for both telescopes was large enough that nearby groups could effectively be studied.

2.1   ROSAT

ROSAT consisted of two telescopes. The X-ray telescope (Aschenbach 1988) was sensitive to photons in the energy range of 0.1–2.4 keV, whereas the Wide Field Camera (Wells et al 1990) covered the energy range 0.070–0.188 keV. The relatively high luminosity of the X-ray background combined with the strong effects of absorption by the Galaxy limited the study of diffuse extragalactic gas with the Wide Field Camera. Therefore, this instrument was not useful for studies of groups and will not be discussed further. Two different kinds of detectors were used with the X-ray telescope: the Position Sensitive Proportional Counter (PSPC) and the High Resolution Imager (HRI). ROSAT was flown with two nearly identical PSPC detectors (Pfeffermann et al 1988). The low internal background, large field of view, and good sensitivity to soft X-rays made the PSPC detectors ideal for studying X-ray emission from groups. The PSPC detectors also had modest energy resolution, allowing the spectral properties of the X-ray emission to be studied. Although the ROSAT HRI provided higher spatial resolution than the PSPC detectors ($\sim 5''$ versus $\sim 25''$ for an on-axis source), the internal background of the HRI was high enough that the low surface brightness diffuse emission found in groups could in general not be studied with this instrument. Therefore, most ROSAT studies of groups were performed with the PSPC.

The ROSAT mission consisted of two main scientific phases. The first was a six-month, all-sky survey (Voges 1993) performed with one of the PSPC detectors (until that detector was destroyed during an accidental pointing at the Sun in January 1991). The mean exposure time for the all-sky survey was approximately 400 seconds. Following the completion of the survey, ROSAT was operated in so-called "pointed mode"—that is, with longer pointings at individual targets. Typical exposure times during the pointed mode of the mission were in the range 5000 to 25,000 seconds, or roughly 10 to 50 times longer than the all-sky survey exposures. Although the pointed mode of the ROSAT mission lasted until early 1999, the second PSPC detector ran out of gas in late 1994, effectively ending studies of diffuse emission in groups.

2.2 ASCA

ASCA, a joint Japanese–United States effort, was launched in early 1993. ASCA consists of four identical grazing-incident X-ray telescopes each equipped with an imaging spectrometer (Tanaka et al 1994). The focal plane detectors are two CCD cameras (known as the Solid-State Imaging Spectrometers, or SIS; Gendreau 1995) and two gas scintillation imaging proportional counters (Gas Imaging Spectrometer, or GIS; Ohashi et al 1996). The SIS detectors have superior energy resolution, whereas the GIS detectors provide a larger field of view. The angular resolution of ASCA is considerably worse than that of ROSAT, with a half power diameter of approximately $3'$. However, ASCA's spectral resolution is much higher than that of the ROSAT PSPC ($E/\Delta E \sim 20$ for the SIS at 1.5 keV versus $E/\Delta E \sim 3$ for the PSPC), so this instrument has primarily played a role in the study of the spectral properties of the intragroup gas. Although the detectors aboard ASCA have undergone serious degradation, this mission is expected to remain operational until sometime in the year 2000.

3.   PROPERTIES OF THE INTRAGROUP MEDIUM

3.1   First ROSAT Results

The great potential of ROSAT for group studies was demonstrated in early papers by Mulchaey et al (1993) and Ponman & Bertram (1993). Each of these papers presented a detailed look at the X-ray properties of an individual group. Mulchaey et al (1993) studied the NGC 2300 group, a poor group dominated by an elliptical-spiral pair. The X-ray emission in the NGC 2300 group is not centered on any particular galaxy, but is instead offset from the elliptical galaxy NGC 2300 by several arcminutes. The X-ray emission can be traced to a radius of at least $\sim$150 h_{100}^{-1} kpc ($\sim$25$''$). Ponman & Bertram (1993) studied Hickson Compact Group 62 (HCG 62). In this case, the X-ray emission is extended to a radius of at least 210 h_{100}^{-1} kpc ($\sim$18$''$). Although the presence of intragroup gas had been suggested by earlier Einstein observations, these ROSAT PSPC results were the first to unambiguously separate a diffuse component related to the group from emission associated with individual galaxies. The intragroup medium interpretation was also supported by the ROSAT PSPC spectra, which are well-fit by a thermal model with a temperature of approximately 1.0 keV ($\sim$10^7 K). The ROSAT PSPC spectrum of HCG 62 contained enough counts that Ponman & Bertram (1993) could also derive a temperature profile for the gas. Ponman & Bertram (1993) found evidence for cooler gas near the center of the group, which they interpreted as evidence for a cooling flow. Many of the X-ray properties of the NGC 2300 group and HCG 62 are consistent with the idea of these systems being scaled-down versions of more massive clusters.

The early ROSAT observations of groups also provided some surprises. For both groups, the gas metallicity derived from the X-ray spectra was much lower than the value found for rich clusters ($\sim$6% solar for NGC 2300 and $\sim$15% solar for HCG 62, compared with $\sim$20–30% solar found for clusters; Fukazawa et al 1998). The X-ray data were also used to estimate the total masses of the groups. In each case, the mass of the group is approximately 10^{13} h_{100}^{-1} $M_{\odot}$. Comparing the total mass as measured by the X-ray data with the total mass in observed baryons, Mulchaey et al (1993) and Ponman & Bertram (1993) concluded that the majority of mass in these groups is dark. In the case of the NGC 2300 group, Mulchaey et al (1993) estimated a baryon fraction that was low enough to be consistent with $\Omega = 1$ and the baryon fraction predicted by standard big bang nucleosynthesis. However, subsequent analysis of the ROSAT PSPC data suggests the true baryon fraction is higher in this group (David et al 1995, Pildis et al 1995, Davis et al 1996).

3.2 ROSAT Surveys of Groups

Unfortunately, the results of Mulchaey et al (1993) and Ponman & Bertram (1993) came late enough in the lifetime of the ROSAT PSPC that large systematic follow-up surveys of groups were not carried out with this instrument. However, the ROSAT PSPC observed many galaxies during its lifetime, and because most galaxies occur in groups, many groups were observed serendipitously. Furthermore, the field of view of the PSPC was large enough that many groups were also observed when the primary target was a star, an active galaxy, or a QSO. In the end, over 100 nearby groups were observed by the ROSAT PSPC during its lifetime, and most of our current understanding of the X-ray properties of groups comes from this dataset.

The existence of an excellent data archive has led to many X-ray surveys of groups using ROSAT PSPC data (Pildis et al 1995, David et al 1995, Doe et al 1995, Saracco & Ciliegi 1995, Mulchaey et al 1996a, Ponman et al 1996, Trinchieri et al 1997, Mulchaey & Zabludoff 1998, Helsdon & Ponman 2000; Mulchaey et al 2000). These surveys indicate that not all poor groups contain an X-ray—emitting intragroup medium. The exact fraction of groups that contain hot intragroup gas has been difficult to quantify because of biases in the sample selection. For example, many of the samples used in archival surveys contain groups that were a priori known to be bright X-ray sources or were likely to be bright X-ray sources based on morphological selection (such as a high fraction of early-type galaxies). These samples are almost certainly not representative of poor groups in general. Furthermore, the term "X-ray detected" has a variable meaning in the literature; some authors use this term only when a diffuse, extended X-ray component (i.e. intragroup medium) is present, whereas others also include cases when emission is associated primarily with the individual galaxies.

There has been considerable interest in the Hickson Compact Groups (HCGs; Hickson 1982; for a review see Hickson 1997). The short crossing times implied for these systems has led some authors to suggest the HCGs are chance alignments of unrelated galaxies within looser systems (Mamon 1986, Walke & Mamon 1989), bound configurations within loose groups (Diaferio et al 1994, Governato et al 1996) or filaments viewed edge-on (Hernquist et al 1995). X-ray observations can potentially help distinguish between these various scenarios (Ostriker et al 1995; Diaferio et al 1995). Ebeling et al (1994) detected eleven HCGs in the ROSAT All-Sky Survey (RASS) data. For some of the detections, the X-ray emission was clearly extended and thus consistent with hot intragroup gas. However, in other cases the sensitivity of the RASS was not good enough to determine the nature of the X-ray emission. Still, Ebeling et al's sample was the first to suggest a correlation between the presence of X-ray emission and a high fraction of early-type galaxies in groups. Pildis et al (1995) and Saracco & Ciliegi (1995) each analyzed ROSAT pointed-mode observations of 12 HCGs (there was considerable overlap in these two samples). Both surveys found that approximately two-thirds of the HCGs were X-ray detected, although in many cases the X-ray emission could not be unambiguously attributed to intragroup gas. (Note also that many of the X-ray detections in these two surveys overlapped with Ebeling et al's earlier RASS detections.) A much more complete study of the HCGs was presented by Ponman et al (1996). This survey combined pointed ROSAT PSPC observations with ROSAT All-Sky Survey data to search for diffuse gas in 85 HCGs. These authors detected extended X-ray emission in ~26% (22 of 85 groups) of the systems studied and inferred that ~75% of the HCGs contain a hot intragroup medium (when one corrects for the detection limits of the observations). Although this is intriguing, some caution must be expressed regarding the Ponman et al (1996) results. Given the compactness of these groups, the nature of the X-ray emission in some of the detected HCGs is far from clear. For example, although Stephan's Quintet (HCG 92) is extended in the ROSAT PSPC data (Sulentic et al 1995), a higher-resolution ROSAT HRI image suggests that most of the extended emission is associated with a shock feature and not with a smooth intragroup gas component (Pietsch et al 1997). Thus, some of the detections in the Ponman et al (1996) survey may not be related to an intragroup medium at all.

Many of the problems inherent to the study of compact groups can be avoided with loose groups. Helsdon & Ponman (2000) studied a sample of 24 loose groups from the catalog of Nolthenius (1993) and found that half of the systems contain intragroup gas. Mulchaey et al (2000) detected diffuse gas in 27 of 57 groups selected from redshift surveys (including the Nolthenius catalog). Both of these studies relied on fairly deep ROSAT pointings and therefore are sensitive to gas down to low X-ray luminosities ($\sim 5 \times 10^{40}\, h_{100}^{-2}$ ergs s^{-1}). The majority of the groups in both Helsdon & Ponman (2000) and Mulchaey et al (2000) were observed serendipitously with the ROSAT PSPC. Based on their velocity dispersions and morphological composition, these samples are fairly representative of groups in nearby redshift surveys. Therefore, these surveys suggest that $\sim$50% of nearby optically-selected groups contain a hot X-ray–emitting intragroup medium.

ROSAT All-Sky Survey (RASS) data have also played an important role in our understanding of the X-ray properties of groups. While the RASS observations are generally not very deep, the nearly complete coverage of the sky allows for larger samples to be studied than is possible with the pointed mode data alone. Henry et al (1995) used the RASS data in the region around the north ecliptic pole to define the first X-ray selected sample of poor groups. The survey by Henry et al (1995) was sensitive to all groups more luminous than $\sim$2.3 × 10^{41} h_{100}^{-2} ergs s^{-1}. Although their sample was rather small (8 groups), Henry et al (1995) were able to show that X-ray–selected groups lie on the smooth extrapolation of the cluster X-ray luminosity and temperature functions. The X-ray selected groups also have lower spiral fractions than typical optically- selected groups, which may suggest that X-ray selection produces a more dynamically evolved sample of groups (Henry et al 1995).

The RASS data have also been used to study optically-selected group samples. Burns and collaborators have devoted considerable effort into studying the X-ray properties of the WBL poor clusters and groups (White et al 1999), which were selected by galaxy surface density. One of the more important results from these studies is the derivation of the first X-ray luminosity function for an optically selected sample of groups and poor clusters (Burns et al 1996). The luminosity function derived by Burns et al (1996) is a smooth extrapolation of the rich cluster X-ray luminosity function and is consistent with the luminosity function Henry et al (1995) derived from their X-ray–selected sample of groups. Follow-up work on some of the brighter sources in the WBL catalog indicates that many of these objects are more massive than typical groups with gas temperatures of 2–3 keV (Hwang et al 1999). These systems are important because they represent the transition objects between poor groups and rich clusters.

Mahdavi et al (1997, 2000) used the RASS database to study the X-ray properties of a large sample of groups selected from the CfA redshift survey (Ramella et al 1995). After accounting for selection effects, Mahdavi et al (2000) estimate that $\sim$40% of the groups are extended X-ray sources. From these detections, the authors derive a relationship between X-ray luminosity and velocity dispersion that is much shallower than is found for rich clusters (see Section 4.2). They suggest that this result is consistent with the X-ray emission in low velocity dispersion groups being dominated by intragroup gas bound to the member galaxies as opposed to the overall group potential. Unfortunately, the RASS observations of groups typically contain very few counts, so detailed spatial studies of the emission are not possible with this dataset. A much deeper X-ray survey of an optically-selected group sample like the one used in Mahdavi et al (2000) would be very useful and should be a priority for future X-ray missions.

3.3   Spatial Properties of the Intragroup Medium

3.3.1 X-Ray Morphologies   The morphology of the X-ray emission can provide important clues into the nature of the hot gas. There is a considerable range in the observed X-ray morphologies of groups. X-ray luminous ($L_X > 10^{42} h_{100}^{-2}$ erg s^{-1}) groups tend to have somewhat regular morphologies (see Figure 1). The total extent of the X-ray emission in these cases is often beyond the optical extent of the group as defined by the galaxies. The peak of the X-ray emission is usually coincident with a luminous elliptical or S0 galaxy, which tends to be the most optically luminous group member (Ebeling et al 1994, Mulchaey et al 1996a, Mulchaey & Zabludoff 1998). The position of the brightest galaxy is also indistinguishable from the center of the group potential, as defined by the mean velocity and projected spatial centroid of the group galaxies (Zabludoff & Mulchaey 1998). Therefore, the brightest elliptical galaxy lies near the dynamical center of the group. There is also a tendency for the diffuse X-ray emission to roughly align with the optical light of the galaxy in many cases (Mulchaey et al 1996a, Mulchaey & Zabludoff 1998). These morphological characteristics are similar to those found for rich clusters containing cD galaxies (e.g. Rhee et al 1992, Sarazin et al 1995, Allen et al 1995).

At lower luminosities, more irregular X-ray morphologies are often found (see Figure 2). In these cases, the X-ray emission is not centered on one particular galaxy, but rather is distributed around several galaxies. Low X-ray luminosity groups also tend to have lower gas temperatures. Dell'Antonio et al (1994) and Mahdavi et al (1997) suggested that the change in X-ray morphologies at low X-ray luminosities indicates a change in the nature of the X-ray emission. They proposed a "mixed-emission" scenario where the observed diffuse X-ray emission originates from both a global group potential and from intragroup gas in the potentials of individual galaxies. In this model, the latter component becomes dominant in low-velocity dispersion systems. This model is consistent with the fact that the X-ray emission is distributed near the luminous galaxies in many of the low-luminosity systems. Another possible source of diffuse X-ray emission in the low-luminosity systems might be gas that is shock-heated to X-ray temperatures by galaxy collisions and encounters within the group environment. This appears to be the case in HCG 92, where the diffuse X-ray emission comes predominantly from an intergalactic feature also detected in radio continuum maps (Pietsch et al 1997). Given that many of the groups with irregular X-ray morphologies are currently experiencing strong galaxy-galaxy interactions (e.g. HCG 16, HCG 90), shocks may be important in many cases. Regardless of the exact origin of the gas, the clumpy X-ray morphologies suggest that the X-ray gas may not be virialized in these cases.

3.3.2 Spatial Extent   To estimate the extent of the hot gas, the usual method is to construct an azimuthally-averaged surface brightness profile and determine at what radial distance the emission approaches the background value. For most rich

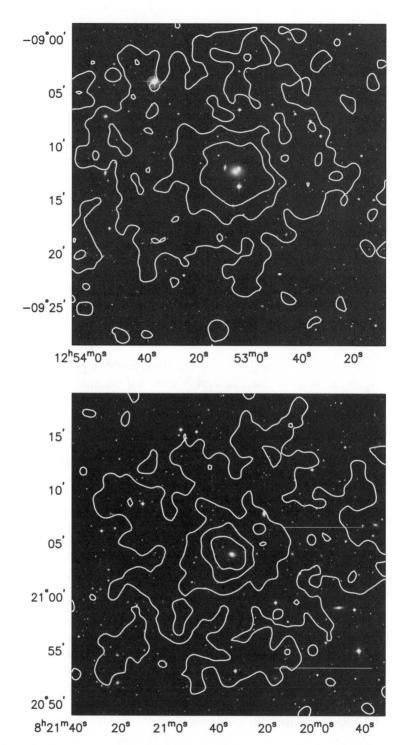

Figure 1 Contour map of the diffuse X-ray emission as traced by the ROSAT PSPC in HCG 62 (*top*) and the NGC 2563 group (*bottom*) overlayed on the STScI Digitized Sky Survey. The X-ray data have been smoothed with a Gaussian profile of width 30″. The coordinate scale is for epoch J2000.

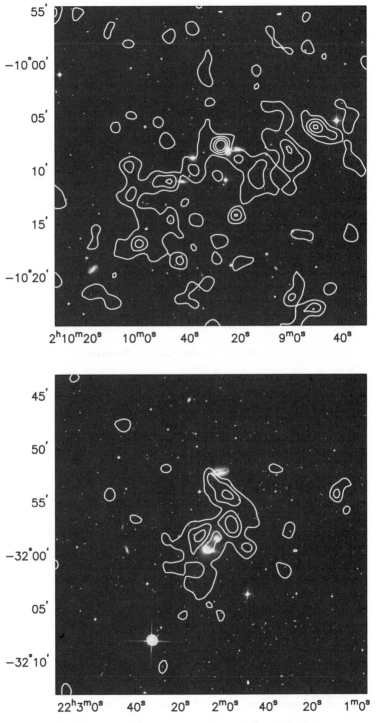

Figure 2 Contour map of the diffuse X-ray emission as traced by the ROSAT PSPC in HCG 16 (*top*) and HCG 90 (*bottom*) overlayed on the STScI Digitized Sky Survey. The X-ray data have been smoothed with a Gaussian profile of width 30″. The coordinate scale is for epoch J2000.

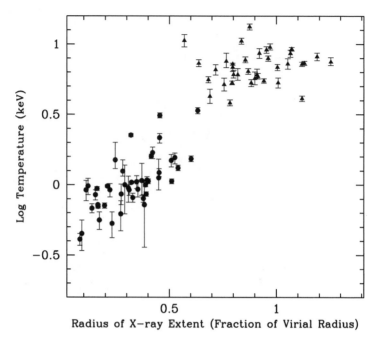

Figure 3 Total radius of X-ray extent plotted as a fraction of the virial radius of each system versus the logarithm of the temperature for a sample of groups (circles) and rich clusters (triangles). The groups were taken from Mulchaey et al (1996a), Hwang et al (1999) and Helsdon & Ponman (2000). The clusters plotted are a redshift-selected subset of the clusters in White (2000). The virial radius for each system was calculated assuming $r_{virial}(T) = 1.85$ $(T/10$ keV$)^{0.5}$ $(1+z)^{-1.5}$ h_{100}^{-1} Mpc (Evrard et al 1996).

clusters, the central surface brightness of the intracluster medium is several orders of magnitude higher than the surface brightness of the X-ray background. Not surprisingly, the central surface brightness of less massive systems like groups tends to be much lower. In fact, in many of the X-ray weakest groups, the central surface brightness of the intragroup gas is just a few times higher than that of the background. Therefore, the measured extent of the X-ray emission in groups is usually much less than that of rich clusters. When comparing groups and clusters, it is useful to normalize the radial extent of the X-ray gas by the mass of the system. Figure 3 plots X-ray extent normalized by the virial radius (R_{virial}) of each system versus temperature for a sample of groups and clusters. Figure 3 indicates that many rich clusters are currently detected to approximately R_{virial}, whereas groups are typically detected to a small fraction of R_{virial}. In some cases, the group X-ray extents are less than 10% of the virial radius. There is also a strong correlation between the radius of detection in virial units and the temperature of the gas in groups: cool groups are detected to a smaller fraction of their virial radius than hot groups. This correlation is important because it suggests that a smaller fraction of the gas mass, and thus, X-ray luminosity, is detected in low temperature systems. Therefore, it is very important to account for this effect when one compares X-ray properties of systems spanning a large range in temperature (i.e. mass). Unfortunately, this has generally not been done in the literature.

3.3.3 The Beta Model Traditionally, a hydrostatic isothermal model has been used to describe the surface brightness profiles of rich clusters (e.g. Jones & Forman 1984). By analogy to the richer systems, this model is usually adopted for poor groups. The hydrostatic isothermal model assumes that both the hot gas and the galaxies are in hydrostatic equilibrium and isothermal. These assumptions appear to be valid for groups with regular X-ray morphologies, but are likely incorrect for groups with irregular X-ray morphologies (although this model is often applied even in these cases). With King's (1962) analytic approximation to the isothermal sphere, the X-ray surface brightness at a projected radius R is given by:

$$S(R) = S_o \, (1 + (R/r_c)^2)^{-3\beta+0.5}$$

where r_c is the core radius of the gas distribution. This model is often referred to as the standard beta model in the literature. The parameter β is the ratio of the specific energy in galaxies to the specific energy in the hot gas:

$$\beta \equiv \mu m_p \sigma^2 / kT_{gas}$$

where μ is the mean molecular weight, m_p is the mass of the proton, σ is the one-dimensional velocity dispersion, and T_{gas} is the temperature of the intragroup medium. For high-temperature systems such as clusters, the X-ray emissivity is fairly independent of temperature over the energy range observed by ROSAT ($\sim$0.1–2 keV). Therefore, the gas density profile can be derived from the surface brightness profile even if the gas temperature varies somewhat within the cluster. However, at the temperatures more typical of groups, the X-ray emissivity is a strong function of temperature. Thus, to invert the observed surface brightness profiles of groups to a gas density profile, the gas must be fairly isothermal.

Based on fits to ROSAT PSPC data, most authors have derived β values of around $\sim$0.5 for groups (Ponman & Bertram 1993, David et al 1994, Pildis et al 1995, Henry et al 1995, Davis et al 1995, David et al 1995, Doe et al 1995, Mulchaey et al 1996a). This number is somewhat lower than the typical value found for clusters (e.g. $\sim$0.64; Mohr et al 1999). However, simulations of clusters indicate that the β value derived from a surface brightness profile depends strongly on the range of radii used in the fit (Navarro et al 1995, Bartelmann & Steinmetz 1996). In particular, β values derived on scales much less than the virial radius tend to be systematically low. As most groups are currently detected to a much smaller fraction of the virial radius than rich clusters, a direct comparison between group and cluster β values may not be particularly meaningful.

Although the hydrostatic isothermal model has almost universally been used for groups, in most cases it provides a poor fit to the data. In general, the central regions of groups exhibit an excess of emission above the extrapolation of the beta model to small radii. This steepening of the profile is often accompanied by a drop in the gas temperature, which has led some authors to suggest that the central deviations are related to a cooling flow (Ponman & Bertram 1993, David et al 1994, Helsdon & Ponman 2000). Alternatively, the excess flux could be emission associated with the central elliptical galaxy (Doe et al 1995, Ikebe et al 1996, Trinchieri et al 1997, Mulchaey & Zabludoff 1998).

Mulchaey & Zabludoff (1998) have shown that the surface brightness profiles in many groups can be adequately fit using two separate beta models. Although the various parameters are not well-constrained with the two-component models, Mulchaey & Zabludoff (1998) found a systematic trend for the β values to be larger with this model than in the case of a single beta model. Similar behavior has been found for rich clusters of galaxies (Ikebe et al 1996, Mohr et al 1999). Mohr et al (1999) suggest that the effect is a consequence of the strong coupling between the core radius (r_c) and β in the fitting procedure; a beta model with a large core radius and high β value can produce a profile similar to that of a beta model where both parameters are lower. Therefore, the presence of a central excess drives the core radius (and thus β) to lower values in the single beta model fits. While Helsdon & Ponman (2000) verified the need for multiple components in groups, they did not derive systematically higher β values. The likely explanation is that the argument in Mohr et al (1999) applies exclusively to systems where the extended component (i.e. the group/cluster gas) dominates the central component. In many of the lower-luminosity systems in Helsdon & Ponman's sample, however, the central component is dominant.

Helsdon & Ponman (2000) also compared the β values of groups and rich clusters and found a trend for β to decrease as the temperature of the system decreases. A similar trend had previously been found in samples of poor and rich clusters (e.g. David et al 1990, White 1991, Bird et al 1995, Mohr & Evrard 1997, Arnaud & Evrard 1999). Mohr et al (1999) reexamined the effect in clusters and found that it disappears when the surface brightness profiles are properly modeled using the two-component beta models. This explanation does not appear to work for poor groups, however, because Helsdon & Ponman (2000) used two-component beta models in their study. The lower β values in groups may be an indication that non-gravitational heating has played a more important role in low-mass systems (David et al 1995, Knight & Ponman 1997, Horner et al 1999, Helsdon & Ponman 2000). However, as noted above, simulations indicate that the derived β value depends strongly on the radii over which the surface brightness fit is performed. Thus, given the strong correlation between system temperature and X-ray extent (Figure 3), conclusions about how β varies with temperature (i.e. mass) may be premature.

3.4   Spectral Properties

X-ray spectral studies of groups have followed the techniques previously used for other diffuse X-ray sources such as elliptical galaxies and rich clusters. The observed data from X-ray instruments such as ROSAT or ASCA do not give the actual spectrum of the source but a convolution of the source spectrum with the instrument response. In general, it is not possible to uniquely invert the convolution and obtain the input spectrum. The usual solution is to adopt a model spectrum with a few adjustable parameters and to find the best fit to the observed data. By analogy to rich clusters, it has generally been assumed that the dominant emission mechanism in groups is thermal emission from diffuse, low-density gas. Many authors have calculated the spectrum emitted by a hot, optically thin plasma. The most popular models are that of Raymond & Smith (1977) and Mewe and collaborators (the so-called MEKAL model; Mewe et al 1985, Kaastra & Mewe 1993, Liedahl et al 1995). For simplicity, single-temperature (i.e. isothermal) models are usually assumed. The free parameters of interest in the isothermal plasma models include the gas temperature and metal abundance. For very hot systems, such as rich clusters, the X-ray emission in the isothermal model is dominated by the free-free continuum from hydrogen and helium. For the temperatures more typical of groups ($\sim 10^7$ K), much of the flux is found in line emission and bound-free continuum.

3.4.1 Gas Temperature   In general, isothermal plasma models provide good fits to the ROSAT PSPC spectra of groups. The derived gas temperatures are in the range ~ 0.3–1.8 keV (see Figure 3), which is roughly what is expected given the range of observed velocity dispersions for groups (e.g. Ponman et al 1996, Mulchaey et al 1996a, Mulchaey & Zabludoff 1998, Helsdon & Ponman 2000). There is generally good agreement in the literature on the temperature of the gas; multiple authors have derived temperature values within 10% of each other, even when temperatures were derived over vastly different physical apertures (e.g. Mulchaey et al 1996a). The temperatures derived from the different plasma models (i.e. Raymond-Smith, MekaL) are also fairly consistent with each other (e.g. Mulchaey & Zabludoff 1998). Furthermore, there is very good agreement between gas temperatures determined by the ROSAT PSPC and ASCA for systems with temperatures less than about 2 keV. For higher-temperature gas (i.e. clusters), the ROSAT data appear to underestimate the true gas temperature by approximately 30% (Hwang et al 1999). All these observations suggest that the derived temperatures for the intragroup medium are fairly robust.

For some of the groups observed by ROSAT, it is possible to measure temperature profiles for the hot gas (Ponman & Bertram 1993, David et al 1994, Doe et al 1995, Davis et al 1996, Trinchieri et al 1997, Mulchaey & Zabludoff 1998, Helsdon & Ponman 2000, Buote 2000b). These profiles suggest that the gas is not strictly isothermal, but rather follows a somewhat universal form: the gas temperature is at a minimum at the center of the group, rises to a temperature maximum in the inner $\sim$50–75 h_{100}^{-1} kpc, and drops gradually at large radii. The temperature minimum in the inner regions of the group is coincident with the sharp rise in the X-ray surface brightness profile. This behavior is consistent with that expected from a "cooling flow" (cf Fabian 1994). The temperature drop at larger radii is often based on lower-quality spectra, and in most cases is not statistically significant. Even if this latter effect is present, the gas temperature at large radii is usually within 10–15% of the temperature maximum. Therefore, isothermality is not a bad assumption over most of the group, as long as the central regions are excluded. However, when global gas temperatures are quoted for groups in the literature, the central regions are almost always included. Because the central regions dominate the total counts in the spectrum, the temperatures found in the literature may underestimate the global temperatures in many cases.

3.4.2 β_{spec} Although most authors have estimated the ratio of specific energy of the galaxies to the specific energy of the gas (i.e. the β parameter) from surface brightness profiles (see Section 3.3.3), β can in principle be determined by directly measuring σ and T_{gas}. Unfortunately, because σ is usually derived from only a few velocity measurements, this method is often not very robust. Detailed membership studies have been made for a few X-ray groups (i.e. Ledlow et al 1996, Zabludoff & Mulchaey 1998, Mahdavi et al 1999), and in these cases the velocity dispersion estimates are more reliable. Using such estimates, Mulchaey & Zabludoff (1998) found $\beta_{\mathrm{spec}} \sim$1 for most of the groups in their sample. Helsdon & Ponman (2000) found a similarly high value for β_{spec} for groups with temperatures of $\sim$1 keV, but noted a trend for β_{spec} to decrease in the lower-temperature systems. However, almost all of the low-temperture groups in the Helsdon & Ponman (2000) sample have velocity dispersions determined from a small number of galaxies. Thus, while the current data suggest a trend for β_{spec} to decrease as the temperature of the group decreases, detailed spectroscopy of cool groups will be required to verify this result.

The $\beta \sim$1 values derived for hot groups from the direct measurement of temperature and velocity dispersion (β_{spec}) are significantly higher than the values of β often derived from surface brightness profile fits (β_{fit}). This so-called β-discrepancy problem has been discussed extensively for rich clusters (e.g. Mushotzky 1984, Sarazin 1986, Edge & Stewart 1991, Bahcall & Lubin 1994). Based on simulations, Navarro et al (1995) concluded that β_{fit} is biased low in galaxy clusters because of the limited radial range used in the X-ray profiles. This explanation may also explain the discrepancy found for groups, which are typically detected to a much smaller fraction of the virial radius than their rich cluster counterparts. Therefore, the β-discrepancy in groups may be an indication that the current derived β_{fit} values underestimate the true β values in many cases.

3.4.3 Gas Metallicity   In addition to measuring gas temperatures, ROSAT PSPC and ASCA observations of groups have been used to estimate the metal content of the intragroup medium. As noted earlier, X-ray spectra of groups are dominated by emission line features. The strongest emission lines are produced when an electron in a highly ionized atom is collisionally excited to a higher level and then radiatively decays to a lower level. The most important features in the X-ray spectra of groups include the K-shell (n = 1) transitions of carbon through sulfur and the L-shell (n = 2) transitions of silicon through iron. Particularly important is the Fe L-shell complex in the spectral range ∼0.7–2.0 keV (Liedahl et al 1995). The wealth of line features in the soft X-ray band potentially provides powerful diagnostics of the physical conditions of the gas, including the excitation mechanism and the elemental abundance (Mewe 1991, Liedahl et al 1990).

Unfortunately, the X-ray telescopes flown to date have not had high enough spectral resolution to resolve individual line complexes. Still, many attempts have been made to estimate the elemental abundance of the gas. For groups, this method primarily measures the iron abundance in the gas, because lines from this element dominate the spectra. Spectral fits to both ROSAT and ASCA data suggest that the metallicity of the intragroup medium varies significantly from group to group; some systems are very metal-poor (∼10–20% solar), whereas others are more enriched (∼50–60% solar; Mulchaey et al 1993; Ponman & Bertram 1993; David et al 1994; Davis et al 1995; Saracco & Ciliegi 1995; Davis et al 1996; Ponman et al 1996; Mulchaey et al 1996a; Fukazawa et al 1996, 1998; Mulchaey & Zabludoff 1998; Davis et al 1999; Finoguenov & Ponman 1999; Hwang et al 1999; Helsdon & Ponman 2000). The low metallicities measured in some groups are surprising because the ratio of stellar mass to gas mass is higher in groups than in clusters. Consequently, one would naively expect the metallicities of the gas to be higher in groups than in rich clusters.

Several potential problems have been noted with the low metallicity measurements for the intragroup medium. Ishimaru & Arimoto (1997) pointed out that most X-ray studies have adopted the old photospheric value for the solar Fe abundance (Fe/H ∼ 4.68×10^{-5}), whereas the commonly accepted "meteoritic" value is significantly lower (Fe/H ∼ 3.24×10^{-5}). (Note that more recent estimates of the photospheric Fe abundance in the sun are consistent with the meteoritic value; see McWilliam 1997.) Thus, essentially all the Fe measurements in the X-ray literature should be increased by a factor of ∼1.44 to renormalize to the meteoritic value. This is particularly important when comparing the X-ray metallicities to chemical-evolution models, which usually adopt the meteoritic Fe solar abundance. The ability of ROSAT data to properly measure the gas abundance has also been questioned. Bauer & Bregman (1996) measured metallicities with the ROSAT PSPC for stars with known metallicities close to the solar value, and found the ROSAT metallicities were typically a factor of five lower than the optical measurements. Bauer & Bregman (1996) suggested several possible explanations for the discrepancy, including instrumental calibration uncertainties, problems with the plasma codes and possible differences in the photospheric and coronal abundances of stars. Instrumental uncertainties with the ROSAT PSPC are unlikely to be the major source of the problem because ASCA spectroscopy of groups also indicates low gas metallicities (Fukazawa et al 1996, 1998; Davis et al 1999; Finoguenov & Ponman 1999; Hwang et al 1999). The possibility that the plasma models are inaccurate or incomplete has been a major concern. While abundance measurements for rich clusters are derived primarily from the well-understood Fe K-α line, group measurements rely on the much more complicated Fe L-shell physics. Problems with the plasma models were in fact identified by early ASCA observations of cooling flow clusters (Fabian et al 1994). Liedahl et al's (1995) revision to the standard MEKA thermal emission model likely accounts for the largest problems in the earlier plasma codes. However, fits to ASCA spectra of groups with the revised model still require very low metal abundances. Hwang et al (1997) have shown that for clusters with sufficient Fe L and Fe K emission (i.e. clusters with temperatures in the range $\sim$2–4 keV), the metallicities derived from the Fe L line complex are consistent with the values derived from the better understood Fe K complex (see also Arimoto et al's 1997 analysis of the Virgo cluster). Unfortunately, it is not clear that the reliability of the Fe L diagnostics implied from $\sim$2–4 keV poor clusters necessarily extends down to lower temperature groups, since other Fe lines dominate the spectrum below $\sim$1 keV (Arimoto et al 1997). Therefore, some problems with the plasma models may still exist.

Another potentially important problem is that the usually assumed isothermal model may be inappropriate for groups (Trinchieri et al 1997; Buote 1999, 2000a). There is clear evidence for temperature gradients in groups, particularly in the inner $\sim$50 h_{100}^{-1} kpc. In fact, the surface brightness profiles of ROSAT PSPC data suggest the presence of at least two distinct components in groups (Mulchaey & Zabludoff 1998). Mixing of multiple-temperature components is particularly an issue for ASCA data because separating out the central component from more extended emission is not possible with the ASCA point spread function. Buote (1999, 2000a) has studied this problem in detail for both elliptical galaxies and groups, and finds that in general single-temperature models provide poor fits to the ASCA spectra. By adopting a two-temperature model, one can obtain better fits, and the metallicities derived are substantially higher. For a sample of 12 groups, Buote (2000a) derives an average metallicity of $Z = 0.29 \pm 0.12\,Z_\odot$ for the isothermal model and $Z = 0.75 \pm 0.24\,Z_\odot$ for the two-temperature model (a single metallicity is assumed for the gas in these models). Buote (2000a) also finds that a multiphase cooling flow model provides a good description of the data. This model also requires higher metallicities ($Z = 0.65 \pm 0.17\,Z_\odot$). Buote (2000a) finds a trend for the metallicities to be lowest in those groups for which the largest extraction apertures were used. This result is consistent with metallicity gradients in groups (see also Buote 2000c). Alternatively, it may simply reflect that the relative contribution of the "group" gas component increases as one adopts a larger aperture. In fact, given the results of the ROSAT surface brightness profile fits, emission from the central elliptical galaxy may dominate the flux in the typical ASCA aperture and thus likely dominates the metallicity measurement. Therefore, the ASCA measurements may not be providing an accurate gauge of the global metal content of the group gas. Regardless, the work of Buote (1999, 2000a) is an important reminder that the properties derived from X-ray spectroscopy are very sensitive to the choice of the input model.

Matsushita et al (2000) also considered multi-temperature models for a large sample of early-type galaxies observed with ASCA. In contrast to Buote (1999, 2000a), Matsushita et al (2000) concluded that the poor spectral fits to ASCA data were not caused by incorrect modeling of multi-temperature emission. Furthermore, the multi-temperature models used by Matsushita et al (2000) produced relatively small increases in the overall abundance in many cases. Matsushita et al (2000) suggested that the strong coupling between the abundance of the so-called α-elements (i.e. O, Ne, Mg, Si, S) and the abundance of Fe hampers a unique determination of the overall metallicity. By fixing the abundance of the α-elements, Matsushita et al (2000) found that the derived metallicities are approximately solar. Although Matsushita et al (2000) restricted their analysis to early-type galaxies, these results may be applicable to groups, which have X-ray properties very similar to those of X-ray luminous ellipticals.

Although the dominant line features for the intragroup medium are produced by iron, strong lines are also expected from elements such as oxygen, neon, magnesium, silicon, and sulfur. The relative abundance of these various elements provides strong constraints on the star formation history of the gas. Some authors have attempted to fit the ASCA spectra with an isothermal model where the α-elements are varied together and separately from the iron abundance (Fukazawa et al 1996, 1998; Davis et al 1999; Finoguenov & Ponman 1999; Hwang et al 1999). In general, these studies find that the α-element to iron ratio is approximately solar in groups. Unfortunately, the determination of this ratio is very sensitive to the spectral model adopted (Buote 2000a) and if the isothermal assumption is not valid, these determinations are not particularly meaningful.

In summary, despite the great potential of X-ray spectroscopy to provide clues into the enrichment history of the intragroup medium, it is not possible at the present time to make strong conclusions about the metal content of the hot gas. Until we have higher resolution X-ray spectra and more complete plasma codes, the metallicity of the intragroup medium will remain an open issue.

3.4.4 Absorbing Column   The soft X-ray band is sensitive to low-energy photoabsorption by gas both within the source and along the line of sight. This absorption must be included in the X-ray spectral fits. It is usually assumed that the X-ray flux is diminished by:

$$A(E) = \exp(-N_H \sigma(E))$$

where N_H is the hydrogen column density and $\sigma(E)$ is the photo-electric cross section (solar abundances are almost universally assumed for the absorbing gas). The cross sections in Morrison & McCammon (1983) are commonly adopted for X-ray analysis. The standard procedure is to allow N_H to be a free parameter in the spectral fit. If the best-fit spectral model returns a value of N_H significantly higher than the Galactic value, this is taken as evidence for excess absorption intrinsic to the group or central galaxy. The ROSAT and ASCA spectra of groups are often not of high enough quality to adequately constrain the absorbing column. Therefore, many authors have chosen to fix N_H to the Galactic value for spectral fits. For a few groups, however, column densities above the Galactic value have been inferred (Fukazawa et al 1996; Davis et al 1999; Buote 2000a,b). Buote (2000b) undertook the most ambitious study of absorption in groups, measuring N_H as a function of radius in a sample of 10 luminous systems observed by the ROSAT PSPC. Buote (2000b) found that the value of N_H derived depends strongly on the bandpass used in the X-ray analysis and suggested the bandpass-dependent N_H values are consistent with additional absorption in the group from a collisionally ionized gas. This excess absorption manifests itself primarily as a strong oxygen edge feature at $\sim$0.5 keV. Buote (2000b) found that within the central regions of the groups, the estimated masses of the absorbers are consistent with the matter deposited by a cooling flow over the lifetime of the flow. If a warm absorber exists in groups, as suggested by Buote (2000b), it should be verified by the next generation of X-ray telescopes.

3.4.5 X-Ray Luminosity   For a thermal plasma, the X-ray luminosity is a rough measure of the total mass in gas. Therefore, the total X-ray luminosity of a group provides a potentially interesting probe of a group's properties. In almost all cases in the literature, the total flux or luminosity quoted is out to the radius to which X-ray emission is detected. In this sense, quoted X-ray luminosities should be thought of as "isophotal luminosities." The measured luminosity is also sensitive to the exact techniques used in the X-ray analysis. For example, the total radial extent of the X-ray emission (and thus the total X-ray luminosity) is strongly dependent on the assumed background level (Henriksen & Mamon 1994, Davis et al 1996). Because of this, different authors often derive vastly different X-ray luminosities for the same group using the same ROSAT observation (Mulchaey et al 1996a).

It is a common practice to quote bolometric luminosities in the literature. The bolometric correction is estimated by extrapolating the spectral model for the gas beyond the limited bandpass of the particular telescope and by making a correction for any absorption along the line of sight. In the case of ROSAT observations, these corrections can easily double the luminosity of the source. The bolometric correction is also somewhat sensitive to uncertainties in the spectral model such as gas metallicity. For very shallow observations, such as those based on ROSAT All-Sky Survey data, a spectral model must usually be assumed to estimate the total X-ray luminosity. The bolometric luminosities of groups are typically in the range several times 10^{40} h_{100}^{-2} to nearly 10^{43} h_{100}^{-2} (Mulchaey et al 1996a,

Ponman et al 1996, Helsdon & Ponman 2000). Thus, the X-ray luminosities of groups can be several orders of magnitude lower than the X-ray luminosities of rich clusters (cf Forman & Jones 1982).

Finally, it is worth noting that because X-ray emission is usually traced only to a fraction of the virial radius in groups, it is likely that the isophotal measurements significantly underestimate the true luminosities of the hot gas. This is particularly true for the coolest groups. Helsdon & Ponman (2000) have attempted to account for the missing luminosity by extrapolating the gas density profile models out to the virial radius. A comparison of the observed isophotal luminosities to the corrected virial luminosities in the Helsdon & Ponman sample indicate that in many cases, over half of the luminosity could occur beyond the radius to which X-ray emission is currently detected.

4. CORRELATIONS

There has been considerable interest in how the X-ray and optical properties of groups differ from those of richer clusters. Such comparisons are often limited by the poorly determined group properties. Most optical properties of groups are derived from existing redshift surveys, which typically include only the most luminous group members. Consequently, global properties such as velocity dispersion and morphological composition are subject to small number uncertainties. The properties of the hot gas also tend to be more uncertain in poorer systems than in clusters because of the lower X-ray fluxes of groups. It should also be remembered that the X-ray properties of groups and clusters are often derived over very different gas density contrasts, which further complicates the comparisons of these systems. Despite these potential problems, group and cluster comparisons have provided considerable insight into the nature of X-ray groups.

4.1 T-σ Relation

Because both the temperature of the intragroup medium and the velocity dispersion of the galaxies provide a measure of the gravitational potential strength, a correlation between these two quantities is expected. Although there is considerable scatter in the data, ROSAT observations are consistent with such a correlation (Figure 4; Ponman et al 1996, Mulchaey & Zabludoff 1998, Helsdon & Ponman 2000). High-temperature groups (T $\sim$ 1 keV) appear to follow the extrapolation of the trend found for rich clusters; the ratio of specific energy in the galaxies to specific energy in the gas is approximately one (i.e. $\beta \sim 1$ and T $\propto \sigma^2$; Mulchaey & Zabludoff 1998, Helsdon & Ponman 2000). Ponman et al (1996) and Helsdon & Ponman (2000) have claimed that the T-σ relation becomes much steeper for cooler groups. However, Figure 4 suggests that given the large scatter, evidence for a systematic deviation from the cluster relationship is at this point rather scarce.

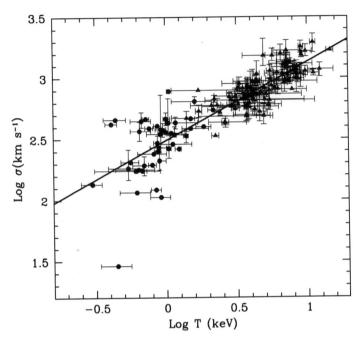

Figure 4 Logarithm of the X-ray temperature versus logarithm of optical velocity dispersion for a sample of groups (circles) and clusters (triangles). The group data are taken from the literature compilation of Xue & Wu (2000), with the addition of the groups in Helsdon & Ponman (2000). The cluster data are taken from Wu et al (1999). The solid line represents the best-fit found by Wu et al (1999) for the clusters sample (using an orthogonal distance regression method). Within the large scatter, the groups are consistent with the cluster relationship.

4.2 L_X-σ and L_X-T Relations

Strong correlations are also found between X-ray luminosity and both velocity dispersion and gas temperature in groups. However, there is considerable disagreement in the literature over the nature of these correlations. Figure 5 shows the L_X-σ relationship for all the groups observed by the ROSAT PSPC in pointed-mode and a sample of clusters observed with various X-ray telescopes (Wu et al 1999). The solid line shows the best-fit relationship Wu et al (1999) derived from the cluster sample alone. Figure 5 shows that for the most part, groups are consistent with the cluster relationship, although there is considerable scatter particularly among the lowest luminosity groups. This conclusion was reached by Mulchaey & Zabludoff (1998), who found that a single relationship fit their sample of groups and rich clusters. Ponman et al (1996) and Helsdon & Ponman (2000) also found that the L_X-σ for groups was basically consistent with the cluster relationship,

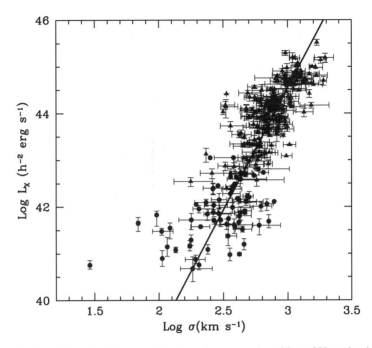

Figure 5 Logarithm of optical velocity dispersion versus logarithm of X-ray luminosity for a sample of groups (circles) and clusters (triangles). The data are taken from the same sources cited in Figure 4. The solid line represents the best-fit found by Wu et al (1999) for the clusters sample (using an orthogonal distance regression method).

although both studies noted that the relationship may become somewhat flatter for low velocity dispersion systems. (Within the errors, the slopes derived by Mulchaey & Zabludoff (1998), Ponman et al (1996) and Helsdon & Ponman (2000) are indistinguishable; $L_X \propto \sigma^{4.3}$, $\sigma^{4.9}$ and $\sigma^{4.5}$, respectively). Therefore, there is fairly good agreement among the ROSAT studies based on pointed-mode data. However, Mahdavi et al (1997) derived a significantly flatter slope from their ROSAT All Sky Survey data ($L_X \propto \sigma^{1.56}$) and suggested that for low velocity dispersion systems the X-ray emission is dominated by hot gas clumped around individual galaxies. More recently, Mahdavi et al (2000) presented X-ray luminosities for a much larger sample of loose groups. In agreement with their earlier result, they find a much flatter L_X-σ for groups than for rich clusters. Mahdavi et al (2000) modeled the L_X-σ relationship as a broken power law, with a very flat slope ($L_X \propto \sigma^{0.37}$) for systems with velocity dispersion less than 340 km s^{-1} and a cluster-like value ($L_X \propto \sigma^{4.0}$) for higher velocity dispersion systems. However, a visual inspection of Mahdavi et al's (2000) L_X-σ relationship (see Figure 4 of their paper) reveals that the need for a broken power law fit is driven by the one or two lowest velocity dispersion groups (out of a total sample of 61 detected groups.) Furthermore, nearly all the L_X upper limits derived by Mahdavi et al (2000) fall below their broken power law relationship (and therefore require a "steeper" relationship). Thus, the case for deviations from the L_X-σ cluster relationship is far from compelling. It is also worth noting that the velocity dispersions of the groups that appear to deviate the most from the cluster relationship are often based on very few velocity measurements (for example the most "deviant" system in Figures 4 and 5 has a velocity dispersion based on only four velocity measurements.) Zabludoff & Mulchaey (1998) have found that when velocity dispersions are calculated for X-ray groups from a large number of galaxies, as opposed to just the four or five brightest galaxies, the velocity dispersion is often significantly underestimated. Therefore, more detailed velocity studies of low velocity dispersion groups could prove valuable in verifying deviations from the cluster L_X-σ relation.

There is also considerable disagreement in the literature about the relationship between X-ray luminosity and gas temperature. Mulchaey & Zabludoff (1998) found that a single L_X-T relationship could describe groups and clusters ($L_X \propto T^{2.8}$). However, both Ponman et al (1996) and Helsdon & Ponman (2000) found much steeper relationships for groups ($L_X \propto T^{8.2}$ and $L_X \propto T^{4.9}$, respectively). These differences might be attributed to the different temperature ranges included in the studies. Mulchaey & Zabludoff's (1998) sample was largely restricted to hot groups (i.e. $\sim$1 keV), whereas Ponman and collaborators have included much cooler systems (down to $\sim$0.3 keV). Indeed, Helsdon & Ponman (2000) found that the steepening of the L_X-T relationship appears to occur below about 1 keV. Figure 6 suggests that the deviation of the cool groups from the cluster relationship is indeed significant. The fact that the L_X-σ relationship for groups appears to be similar to the relationship found for clusters, while the relationships involving gas temperature significantly depart from the cluster trends, may be an indication that non-gravitational heating is important in groups (Ponman et al 1996, Helsdon & Ponman 2000). However, the group X-ray luminosities may be biased somewhat low because groups are detected to a smaller fraction of their virial radius than richer systems, and if comparisons are made at the same mass over-density level, groups would likely fall closer to the cluster relation.

4.3 Galaxy Richness and Optical Luminosity

Most authors have found little or no correlation between X-ray luminosity and the number of luminous galaxies in a group or the total optical luminosity of the group (Ebeling et al 1994, Doe et al 1995, Mulchaey et al 1996a, Ponman et al 1996). The lack of correlation between X-ray luminosity and number of group members is not too surprising because galaxy-galaxy merging is likely prevalent in groups, and thus the number of galaxies in a group is likely not conserved in time (Ponman et al 1996). The fact that there is no relationship between optical and X-ray luminosity is important because it suggests that the X-ray emission is

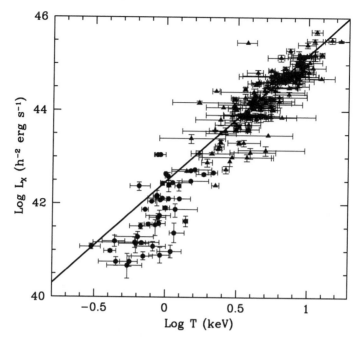

Figure 6 Logarithm of the X-ray temperature versus logarithm of X-ray luminosity for a sample of groups (circles) and clusters (triangles). The data are taken from the same sources cited in Figure 4. The solid line represents the best-fit found by Wu et al (1999) for the clusters sample (using an orthogonal distance regression method). The observed relationship for groups is somewhat steeper than the best-fit cluster relationship.

not associated with individual galaxies for most of the samples studied (Ponman et al 1996).

Mahdavi et al (1997) came to a very different conclusion with their RASS survey of optically-selected groups: They found a strong correlation between X-ray luminosity and optical luminosity. The differences between Mahdavi et al's (1997) results and those of other authors suggests that the groups in Mahdavi et al (1997) may be systems dominated by X-ray emission from individual galaxies and not intragroup gas.

4.4 Morphological Content

Correlations between the presence of X-ray emission and the morphological composition of groups were suggested from the earliest ROSAT studies. Ebeling et al (1994) were the first to claim such an effect, noting that all but one of the X-ray detected HCGs in the ROSAT All-Sky Survey data had spiral fraction less than 50%. Subsequent studies of small samples appeared to support this trend (Henry et al 1995, Pildis et al 1995, Mulchaey et al 1996a). However, Ponman et al

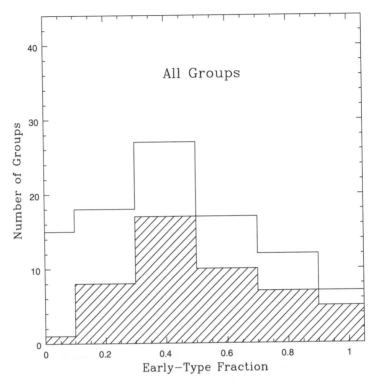

Figure 7 Distribution of early-type fraction for all groups (*open histogram*) and groups with diffuse X-ray emission (*shaded histogram*). The top panel shows the result for all published PSPC pointed-mode observations, whereas the lower panel contains only groups selected from optical redshift surveys. (*continued*)

(1996) came to a very different conclusion based on their much larger survey of the HCGs. They detected several groups with high spiral fractions, including the extreme example HCG 16, a compact group that contains only spirals.

Figure 7 shows the distribution of early-type fraction for all the groups with published pointed observations with ROSAT. For the purposes of this plot, a group is considered "X-ray detected" only if there is evidence for an extended intragroup medium component. As is apparent from this figure, a significant number of spiral-rich groups do contain diffuse X-ray emission, which confirms the conclusion of Ponman et al (1996). In fact, in contrast to the earlier studies, the distribution of early-type fractions is surprisingly flat for the X-ray detected groups. The apparent contradiction with the earlier results can be explained by the fact that the majority of the groups in the current sample were selected from optical redshift surveys and were serendipitously observed by ROSAT (Helsdon & Ponman 2000, Mulchaey et al 2000), whereas the earlier studies were biased toward X-ray luminous groups, which tend to have higher early-type fractions (Mulchaey & Zabludoff 1998).

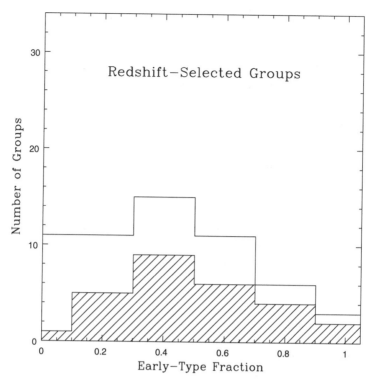

Figure 7   (*Continued*)

A closer examination of Figure 7 reveals that while many spiral-rich systems are X-ray sources, spiral-only groups tend not to contain a diffuse X-ray component. The one exception in Figure 7 is HCG 16. However, the true nature of the X-ray emission in HCG 16 is unclear. The ROSAT image of the group indicates that the emission is very clumpy and concentrates around the brightest group members (see Figure 2, top). Some authors have attributed all of the X-ray emission to individual galaxies (Saracco & Ciliegi 1995; see also an earlier Einstein observation by Bahcall et al 1984), whereas others have claimed the existence of intragroup gas (Ponman et al 1996). Dos Santos & Mamon (1999) have reanalyzed the ROSAT PSPC data on HCG 16, paying special attention to the removal of emission associated with galaxies. Although Dos Santos & Mamon (1999) derived a lower luminosity for the diffuse gas than Ponman et al (1996), they still found evidence for some diffuse gas. However, the presence of diffuse emission does not necessarily mean that HCG 16 contains a diffuse intragroup medium. One possibility is that the emission is related to the unusually high number of active galaxies in the group (HCG 16 contains one Seyfert galaxy, two LINERs, and three starburst galaxies; Ribeiro et al 1996). The X-ray to infrared luminosity ratio of this system is much higher than one would expect if the X-ray emission is related to the galaxies' activity, however (Ponman, private communication). Alternatively, the X-ray emission may be associated with shocked gas, as appears to be the case in Stephan's Quintet (Pietsch et al 1997).

With the possible exception of HCG 16, all X-ray detected groups studied to date contain at least one early-type galaxy. There are several possible explanations for why spiral-only groups do not contain diffuse X-ray emission. One possibility is that all spiral-only groups are chance superpositions and not real, physical systems. This possibility seems unlikely, given the existence of our own spiral-only Local Group (see Section 5.10 for a discussion of the intragroup medium in the Local Group). Another possibility is that the intragroup gas in spiral-only groups is too cool to produce appreciable amounts of X-ray emission (Mulchaey et al 1996b). Based on velocity dispersions, the virial temperatures of spiral-only groups do tend to be lower than those of their early-type dominated counterparts (Mulchaey et al 1996b). While a cool (i.e. several million degrees K) intragroup medium would be difficult to detect in X-ray emission, such gas might produce prominent absorption features in the far-ultraviolet or X-ray spectra of background quasars (Mulchaey et al 1996b, Perna & Loeb 1998, Hellsten et al 1998). In fact, several such groups may have already been detected as OVI $\lambda\lambda$1031.93,1037.62 Å absorption systems (Bergeron et al 1994, Savage et al 1998). A third possibility is that the gas densities in spiral-only groups are too low to be detected in X-rays. Low gas densities in spiral-only groups are in fact consistent with recent prediction of preheating models for groups (Ponman et al 1999; see Section 5.9).

5.   COSMOLOGICAL IMPLICATIONS OF X-RAY GROUPS

5.1   The Physical Nature of Groups

Simulations of local large-scale structure suggest that a significant fraction of the groups identified in redshift surveys are not real, bound systems (Frederic 1995, Ramella et al 1997). The existence of diffuse X-ray emitting gas is often cited as evidence that a group is real. This is not necessarily the case, however. Hernquist et al (1995) noted that primordial gas may be shock-heated to X-ray emitting temperatures along filaments. When these filaments are viewed edge-on, a "fake" group with an X-ray halo could be observed. Ostriker et al (1995) proposed a test of the Hernquist et al (1995) filament model by defining an observable quantity Q, that is proportional to the axis ratio of the group. Applying this test to the early ROSAT observations of HCGs, Ostriker et al (1995) found that the Q values for most HCGs are consistent with their being frauds. However, the low Q values for groups can also be explained if the ratio of gas mass to total mass is smaller in groups than in rich clusters. Both ROSAT observations (David et al 1995, Pildis et al 1995, Mulchaey et al 1996a) and simulations (Diaferio et al 1994, Pildis et al 1996) of X-ray groups are in fact consistent with this idea, suggesting that the Ostriker et al test may in the end not be very useful.

Several arguments support the idea that at least some X-ray groups are real, bound systems and that the X-ray gas is virialized. In the most X-ray luminous groups, the diffuse gas extends on scales of hundreds of kiloparsecs and appears smooth. This is consistent with what one expects for a "smooth" group potential. The gas temperature in these cases agrees fairly well with the temperature expected, based on the velocity dispersion of the groups. Furthermore, most of these groups show evidence for cooling flows in their centers, suggesting that the gas is in an equilibrium state and has probably existed for at least several gigayears.

Ironically, perhaps the best evidence for the reality of the X-ray luminous groups has come from optical studies of these systems. Zabludoff & Mulchaey (1998) used multifiber spectroscopy to study the faint galaxy population in a small sample of groups and found large differences in the number of faint galaxies in X-ray detected and non-detected groups. All of the X-ray detected groups in the Zabludoff & Mulchaey (1998) sample contain at least 20–50 group members (down to magnitudes as faint as $M_B \sim -14 + 5 \log_{10} h_{100}$). Even down to these relatively faint magnitude limits, many of the X-ray detected groups have very high early-type fractions (nearly 60% in some cases). The large number of group galaxies argue that these X-ray groups must be real, physical systems and not radial superpositions. There are also strong correlations between dynamical measures of the gravitational potential (i.e. velocity dispersion/gas temperature) and the early-type fraction of the group (Zabludoff & Mulchaey 1998, Mulchaey et al 1998). These correlations imply either that galaxy morphology is set by the local potential at the time of galaxy formation (Hickson et al 1988) or that the potential grows as the group evolves (Diaferio et al 1993). Either scenario requires that most X-ray luminous groups be real, bound systems.

However, it is likely that some X-ray detected groups are not virialized systems. In particular, low-luminosity, low-temperature groups tend to have irregular X-ray morphologies with the X-ray emission distributed in the immediate vicinity of individual galaxies. These X-ray morphologies suggest that these groups are still dynamically evolving. In some cases, such as HCG 92, gas has apparently reached X-ray emitting temperatures by other mechanisms such as shocks. Therefore, X-ray detection alone does not indicate that a system is virialized.

5.2 Mass Estimates

One of the most important applications of X-ray observations of groups has been mass estimates. Prior to ROSAT, mass determinations for groups were largely based on application of the virial theorem to the group galaxies. For a typical cataloged group with only four or five velocity measurements, the virial method can be unreliable (e.g. Barnes 1985, Diaferio et al 1993).

The method used to estimate group masses from X-ray data is analogous to the technique developed for rich clusters (e.g. Fabricant et al 1980, 1984; Fabricant & Gorenstein 1983; Cowie et al 1987). The fundamental assumption is that the hot gas is trapped in the potential well of the group and is in rough hydrostatic equilibrium. This assumption is probably a reasonable one for most groups, given the short sound-crossing times in these systems. A further assumption is that the only source of heating for the gas is gravitational, i.e. that the gas temperature is a direct measure of the potential depth and therefore of the total mass. This assumption may not be strictly true for some groups. In particular, the fact that the heavy metal abundance of the intragroup medium is non-zero suggests that some of the gas has been reprocessed in the stars in galaxies and ejected by supernovae-driven winds. In addition to polluting the intragroup gas with metals, such winds also provide additional energy to the gas. It has generally been assumed in the literature that the energy contribution of such winds is negligible. Semi-analytic models suggest that this assumption is fair as long as the temperature of the system is greater than about 0.8 keV (Balogh et al 1999, Cavaliere et al 1999). Thus, for many groups, the hydrostatic mass estimator should be valid.

With the further assumption of spherical symmetry, the mass interior to radius R is given by (Fabricant et al 1984):

$$M_{\text{total}}(<R) = \frac{kT_{\text{gas}}(R)}{G\mu m_p}\left[\frac{d\log\rho}{d\log r} + \frac{d\log T}{d\log r}\right]R$$

where k is Boltzmann's constant, $T_{\text{gas}}(R)$ is the gas temperature at radius R, G is the gravitational constant, μ is the mean molecular weight, m_p is the mass of the proton, and ρ is the gas density. In principle, all of the unknowns in this equation can be calculated from the X-ray data. Typically, the gas temperature is measured directly from the X-ray spectrum and the gas density profile is determined by fitting the standard beta model to the surface brightness profile. Unfortunately, it is often necessary to make a further assumption that the gas is isothermal (i.e. $\frac{d\log T}{d\log r} = 0$). For a few groups, the temperature profile can be directly measured. The resulting mass estimates suggest that the isothermal assumption generally results in an error in the mass of no more than about 10% (e.g. David et al 1994, Davis et al 1996). With the isothermal assumption, $M_{\text{total}}(<R) \propto T_{\text{gas}}\beta R$ (as long as R is much larger than the core radius in the beta model. Therefore, if β is underestimated from the surface brightness profile fits by a factor of $\sim$2 (see Sections 3.3.3 and 3.4.3), then the mass estimates are also too small by a factor of $\sim$2.)

ROSAT measurements indicated a small range of total group masses with nearly all of the systems clustered around $10^{13}\,h^{-1}\,M_{\odot}$ (see Figure 8; Mulchaey et al 1993, Ponman & Bertram 1993, David et al 1994, Pildis et al 1995, Henry et al 1995, Mulchaey et al 1996a). The narrow range of group masses is not too surprising, given that nearly all the groups in these surveys have temperatures of $\sim$1 keV.

The X-ray mass estimates can generally be applied only to a radius of several hundred kiloparsecs. Beyond that, the gas density profile is not well-constrained. Because the virial radius for a 1 keV group is approximately $\sim$0.5 h_{100}^{-1} Mpc, the X-ray method measures only a fraction of the total mass (Ponman & Bertram 1993; David et al 1995; Henry et al 1995). Simply extrapolating out to the virial radius, the total group masses are a factor of approximately two to three times larger than those implied from the X-ray studies (Mass $\propto$ R). However, if

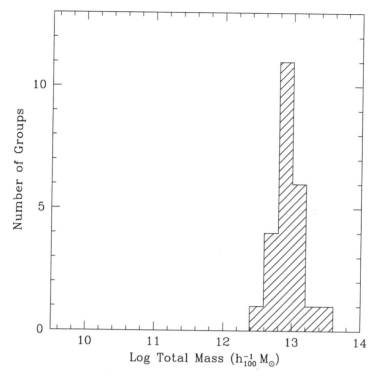

Figure 8 Distribution of X-ray–determined total group masses. In each case, the masses are determined out to the radius to which the X-ray emission is detected. The sample is based on the compilation given in Mulchaey et al 1996a, with the addition of a few groups with more recent X-ray mass estimates in the literature.

non-gravitational heat is important in groups, the extrapolation out to the virial radius is more uncertain (Loewenstein 2000).

Because of their relatively large masses, X-ray groups make a substantial contribution to the mass density of the universe (Mulchaey et al 1993, Henry et al 1995, Mulchaey et al 1996a). Based on their X-ray-selected group sample, Henry et al (1995) estimate that X-ray luminous groups contribute $\Omega \sim 0.05$. However, their sample contained only the most luminous, elliptical-rich groups. When one corrects for the groups missing from Henry et al's (1995) sample (assuming a similar mass density), groups might contribute as much as $\Omega \sim 0.25$. These estimates are comparable to the numbers found for richer clusters, which verifies the cosmological significance of poor groups.

5.3 Baryon Fraction

The ratio of baryonic to total mass in groups and clusters can provide interesting constraints on cosmological models (e.g. Walker et al 1991, White et al

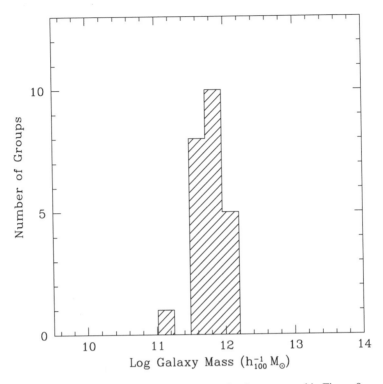

Figure 9   Distribution of galaxy mass for the sample of groups used in Figure 8.

1993). The two known baryonic components in groups are the galaxies and the hot gas. The total mass in galaxies can be estimated by measuring the total galaxy light and assuming an appropriate mass-to-light ratio for each galaxy based on its morphological type. While ideally the luminosity function of each group should be used to measure the total light, generally most authors have included only the contribution of the most luminous galaxies. Fortunately, these galaxies account for nearly all the light in the group. The mass-to-light ratios of X-ray groups are generally in the range $M/L_B \sim 120$–$200\ h_{100}\ M_\odot/L_\odot$ (Mulchaey et al 1996a), which is comparable to the mass-to-light ratios found in rich clusters. However, these estimates are made out to the radius of X-ray detection, so the values out to the virial radius could be larger. Assuming standard mass-to-light ratios for ellipticals and spirals, the mass in galaxies in X-ray groups is typically in the range 3×10^{11}–$2 \times 10^{12}\ h_{100}^{-1}\ M_\odot$ (Figure 9).

The mass in the intragroup medium can be estimated from the model fit to the surface brightness profile. The gas-mass estimates depend both on the radius out to which X-rays are detected (Henriksen & Mamon 1994) and on the spectral properties assumed (for example, the gas metallicity; Pildis et al 1995). For these reasons, different authors have derived significantly different gas masses for the

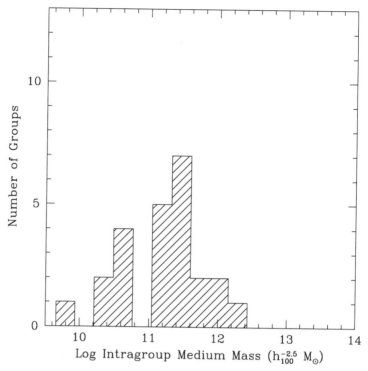

Figure 10 Distribution of intragroup medium mass for the sample of groups used in Figure 8.

same systems (cf Mulchaey et al 1996a). For most groups, the gas mass is in the range $\sim 2 \times 10^{10}$–$10^{12} h_{100}^{-5/2} M_\odot$ (Figure 10). This is somewhat less than or comparable to the mass in galaxies. Note, however, that the gas mass is much more strongly dependent on H_o, and for more realistic (i.e. lower) values of H_o, the gas mass can be somewhat higher than the galaxy mass. The observed gas mass–to–stellar mass ratio tends to decrease as the temperature of the system decreases. This trend extends from rich clusters to individual elliptical galaxies. David et al (1995) estimate that the gas-to–total mass fraction is approximately 2% in ellipticals, 10% in groups and 20–30% in rich clusters. However, the hot gas in groups is detected to a much smaller fraction of the virial radius than in rich clusters, so comparisons made at the current level of X-ray detection may not accurately reflect the global gas fractions (Loewenstein 2000). In fact, much of the intragroup gas probably lies beyond the current X-ray detection limits, and on more global scales, groups may not be gas-poor compared to clusters. Consequently, the total gas masses of groups may be severely underestimated by ROSAT observations. On scales of the virial radius, the intragroup medium is likely the dominant baryonic component in these systems. In fact, Fukugita et al (1998) estimated that diffuse gas in groups is the

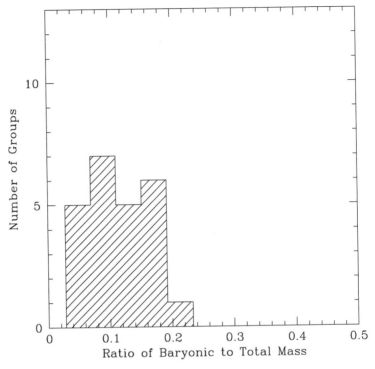

Figure 11   Distribution of total observed baryonic mass to total group mass for the sample of groups used in Figure 8. The low "baryonic fractions" derived for groups indicate that these systems are dominated by dark matter.

dominant baryon component in the nearby universe. A fundamental assumption in Fukugita et al's calculation is that all groups contain an intragroup medium and that the absence of X-ray detections in many groups is primarily a result of lower virial temperature rather than the absence of plasma. Regardless of whether this assumption is valid or not, it is now clear that intragroup gas is an important baryonic constituent of the local universe.

Adding up the baryons in galaxies and intragroup gas and comparing to the total mass, one finds that the known baryonic components typically account for only 10–20% of the total mass that is derived using the X-ray data (Figure 11; Mulchaey et al 1993; Ponman & Bertram 1993; David et al 1994; Pildis et al 1995; David et al 1995; Doe et al 1995; Davis et al 1995, 1996; Mulchaey et al 1996a; Pedersen et al 1997). This provides some of the strongest evidence to date that small groups of galaxies are dominated by dark matter. The ratio of mass in observed baryonic components to total mass (i.e. the "baryon fraction") in general is smaller in groups than in rich clusters (David et al 1995, David 1997). However, the lower observed baryon fractions of groups may largely reflect the fact that much of the hot gas occurs beyond the radius of current X-ray detection. Even if the observed baryon fractions of groups are representative of the global values, the baryon fractions in X-ray groups are still too high to be consistent with the low baryon fractions required for $\Omega = 1$ and standard big bang nucleosynthesis (cf White et al 1993).

5.4 Large-Scale Structure

Redshift surveys of the nearby universe indicate that groups of galaxies are good tracers of large-scale structure (e.g. Ramella et al 1989). The presence of a hot intragroup medium in many groups suggests that X-ray observations can also be used to map out the distribution of mass in the universe. Recent ROSAT results demonstrate the great potential of large area X-ray surveys. Mullis et al (2000) have recently completed an optical follow-up survey of the $\sim$500 X-ray sources detected in the ROSAT All-Sky Survey in a 9×9 square-degree region around the north ecliptic pole. They identify 65 galaxy systems, $\sim$30% of which are poor groups. Remarkably, some 23% of the galaxy systems found in this field belong to a single wall-like structure at $z = 0.088$. Although a supercluster consisting of six Abell clusters had previously been identified in this region (Batuski & Burns 1985), the X-ray data reveal that this supercluster is significantly larger than implied by the optical data alone. Furthermore, the X-ray data show that the massive Abell clusters are linked together by groups and poor clusters. The supercluster spans the entire area surveyed by Mullis et al (2000), suggesting that the true extent of this structure could be larger still. Numerical simulations imply that future X-ray missions such as CHANDRA and XMM will be able to map out even lower-density regions such as filaments (Pierre et al 2000). Such X-ray studies will be very important because many current models suggest that the majority of baryons occur in these filaments (Miralda-Escude et al 1996, Cen & Ostriker 1999).

5.5 Moderate Redshift Groups

Despite the cosmological significance of groups, remarkably little is known about these systems at high redshift. Optical studies of high redshift groups have been limited because low galaxy densities make groups difficult to recognize even at moderate redshifts. X-ray emission from the intragroup medium provides a potentially useful method for finding groups at high redshift. A number of searches for faint, extended X-ray sources have been performed in recent years using deep ROSAT PSPC observations (e.g. Rosati et al 1995, Griffiths et al 1995, Scharf et al 1997, Burke et al 1997, Jones et al 1998, Schmidt et al 1998, Vikhlinin et al 1998, Zamorani et al 1999). Although the goal of these surveys is often to find rich clusters of galaxies at high redshift, many X-ray groups at redshifts $z = 0.1$–0.6 have also been found. Unfortunately, the ROSAT observations of these groups generally contain very few counts, so it is not possible to determine the temperature or the metallicity of the gas with the existing data. However, studies of the spectral properties of the intragroup medium out to $z \sim 0.3$ will be possible with both XMM and CHANDRA. Furthermore, deep images with these telescopes will likely uncover X-ray groups at even higher redshifts. Therefore, the first studies of the evolution of the intragroup medium should be possible within the next decade.

5.6   Gravitational Lensing

The efficiency of a massive system to act as a gravitational lens is a function of both the mass-density profile and the source-lens-observer geometry (cf Blandford & Narayan 1992). Given their relatively high mass densities, X-ray groups at moderate ($z > 0.2$) redshifts are expected to be efficient lenses (Mendes de Oliveira & Giraud 1994, Montoya et al 1996). Unfortunately, because very few samples of galaxy groups at moderate redshift exist in the literature, systematic searches for lensing in these objects have not been carried out. However, several of the well-studied, multiple-image QSO systems are lensed by galaxies that belong to spectroscopically-confirmed poor groups (Kundic et al 1997a,b; Tonry 1998; Tonry & Kochanek 2000). Although the primary lens in each of these cases is an individual galaxy, the group potential also contributes to the observed lensing. The presence of an extended group potential acts as a source of external sheer (Keeton et al 1997; Kundic et al 1997a,b). To properly model the lensing system, the group potential must be included. Most authors have attempted to measure the velocity dispersion of the group and then assume a form for the potential. Unfortunately, these dispersions are based on only a few velocity measurements and are subject to the large uncertainties that have plagued optical studies of nearby groups. Still, good fits to the lensing data are often obtained. In the case of the quadruple lens PG 1115+080, the measured velocity dispersion of the group (Kundic et al 1997a; Tonry 1998) is consistent with the value predicted earlier from the lensing data (Schecter et al 1997). Obtaining an adequate model for the group potential is also necessary to derive cosmological parameters like the Hubble Constant (H_o) from lensing experiments. Future X-ray observations may be the key to such techniques. High-resolution X-ray images taken with CHANDRA and XMM should allow the potential of the lensing groups to be mapped in detail. A better determination of the lensing potential will result in tighter constraints on cosmological parameters.

5.7   Cooling Flows

Galaxy groups display many of the signatures of cooling flows that have previously been observed in rich clusters and elliptical galaxies (Fabian 1994). The surface brightness profiles of the X-ray emission are sharply peaked, indicating that the gas density is rising rapidly towards the center of the group. In addition, at least half of all groups with measured temperature profiles show direct evidence for cooler gas in the central regions (Ponman & Bertram 1993; David et al 1994; Trinchieri et al 1997; Mulchaey & Zabludoff 1998; Helsdon & Ponman 2000). In some cases, the central gas is cooler than the mean gas temperature by nearly 50%. Cooling flow models also appear to provide a better fit to the ASCA spectra of groups than an isothermal plasma model (Buote 2000a). While these observations are consistent with the cooling flow interpretation, there are other possibilities. For example, Mulchaey & Zabludoff (1998) noted that the above features could also be explained if there is a distinct X-ray component associated with the central elliptical galaxy.

Perhaps the strongest case for a cooling flow in a low-mass system is the NGC 5044 group. David et al (1994) obtained a very deep ROSAT PSPC observation of this system that allowed the construction of a detailed temperature profile. They found evidence for a cooling flow with an essentially constant mass accretion rate from approximately 20 h_{100}^{-1} kpc out to the cooling radius ($\sim$50–75 h_{100}^{-1} kpc). This suggests a nearly homogeneous cooling flow. In contrast, the cooling flows in rich clusters tend to be inhomogeneous; a significant amount of the gas cools out at large radii (cf Fabian 1994). David et al (1994) suggest that gravitational heating is more important in the NGC 5044 group than in clusters because in groups the temperature of the hot gas is comparable to the virial temperature of the central galaxy, whereas for rich clusters the gas temperature is significantly higher. Therefore, most of the observed X-ray emission in the cooling flow region can be provided by the gravitational energy in groups, whereas mass deposition dominates in rich clusters.

5.8 Fossil Groups

Because of their relatively low velocity dispersions and high galaxy densities, groups of galaxies provide ideal sites for galaxy-galaxy mergers. Numerical simulations suggest that the luminous galaxies in a group will eventually merge to form a single elliptical galaxy (Barnes 1989, Governato et al 1991, Bode et al 1993, Athanassoula et al 1997). The merging timescales for the brightest group members ($M \approx M^*$) are typically a few tenths of a Hubble time for an X-ray detected group (Zabludoff & Mulchaey 1998). Therefore, by the present day some groups have likely merged into giant ellipticals. Outside of the high-density core, the cooling time for the intragroup medium is longer than a Hubble time; thus, while the luminous galaxies in some groups have had enough time to merge into a single object, the large-scale X-ray halo of the original groups should remain intact. This means that a merged group might appear today as an isolated elliptical galaxy with a group-like X-ray halo (Ponman & Bertram 1993).

Using the ROSAT All-Sky Survey data, Ponman et al (1994) found the first such "fossil" group candidate. The RXJ1340.6+4018 system has an X-ray luminosity comparable to a group, but $\sim$70% of the optical light comes from a single elliptical galaxy (Jones et al 2000). The galaxy luminosity function of RXJ1340.6+4018 indicates a deficit of galaxies at approximately M^*. The luminosity of the central galaxy is consistent with it being the merger product of the missing M^* galaxies. Jones et al (2000) have studied the central galaxy in detail and find no evidence for spectral features implying recent star formation, which indicates the last major merger occurred at least several gigayears ago.

Several other fossil group candidates are now known. Mulchaey & Zabludoff (1999) discovered a large X-ray emitting halo around the optically- selected isolated elliptical NGC 1132. Although the NGC 1132 system contains no other luminous galaxies, there is evidence for an extensive dwarf galaxy population clustered around the central galaxy. The dwarfs in NGC 1132 are comparable in number and distribution to the dwarfs found in X-ray groups (Zabludoff & Mulchaey 1998). The existence of a clustered dwarf population in fossil groups is not surprising because the galaxy-galaxy merger and dynamical friction timescales for faint galaxies in groups are significantly longer than the timescales for the luminous galaxies (Zabludoff & Mulchaey 1998). Hence, the dwarf galaxy population, like the X-ray halo, will remain long after the central elliptical has formed.

Vikhlinin et al (1999) have found four potential fossil groups in their large-area ROSAT survey of extended X-ray sources. (Their sample includes RXJ1340.6+ 4018 and two X-ray sources detected in earlier Einstein surveys but not previously recognized as potential group remnants.) Given the large surface area they covered in their survey, Vikhlinin et al were able to estimate the spatial density of X-ray fossil groups for the first time and found that these objects represent $\sim$20% of all clusters and groups with an X-ray luminosity greater than $5 \times 10^{42} h_{100}^{-2}$ ergs s^{-1}. The number density of fossil groups is comparable to the number density of field ellipticals, so most, if not all, luminous field ellipticals may be the product of merged X-ray groups.

Although the X-ray and optical properties of some luminous, isolated elliptical galaxies are consistent with the merged group interpretation, another possibility is that these systems may have simply formed with a deficit of luminous galaxies (Mulchaey & Zabludoff 1999). Distinguishing between these two scenarios will be difficult, if not impossible. Regardless, these objects are massive enough and found in large enough numbers that they are cosmologically important. Vikhlinin et al (1999) estimated that the contribution of fossil groups to the mass density of the universe is comparable to the contribution of massive clusters. These objects are also an important reminder that galaxies are not always good tracers of mass and large-scale structure: optical group catalogs would miss these large mass concentrations.

5.9   The Origin and Evolution of the Intragroup Medium

The presence of heavy elements in the intragroup medium indicates that a substantial fraction of the diffuse gas must have passed through stars. The presence of iron is particularly important because it suggests that supernovae played an important role in the enrichment of the gas. In principle, X-ray spectroscopy can provide detailed constraints on the stars responsible for the enrichment. For example, the relative abundance of the α-burning elements to iron is a measure of the relative importance of Type II to Type 1a supernovae (Renzini et al 1993, Renzini 1997, Gibson et al 1997). For the gas temperatures characteristic of groups ($\sim$1 keV), strong emission lines are expected for many of the α elements including oxygen, neon, magnesium, silicon and sulfur. Although most ASCA studies suggest that the α/Fe ratio is approximately solar in groups, this result is somewhat inconclusive at present because of uncertainties in the spectral modeling.

Renzini and collaborators have used the concept of iron mass-to-light ratio to study the history of the hot gas in groups and clusters (Renzini et al 1993, Renzini 1997). They find that the X-ray emitting gas in rich clusters contains ~ 0.01 $h^{-1/2}$ $M_\odot$ of iron for each $L_\odot$ of blue light. The iron mass-to-light ratio is effectively constant for clusters with temperature between ~ 2 and 10 keV. However, this ratio is typically a factor of ~ 50 lower in X-ray groups (Renzini et al 1993, Renzini 1997, Davis et al 1999). The iron mass-to-light ratios of groups are lower than those of clusters because both the overall iron abundance and the gas-to–stellar mass ratio are lower in groups than in clusters (Renzini 1997). The low iron mass-to-light ratios may be evidence that a significant amount of mass has been lost in groups. The escape velocities of groups are comparable to the escape velocities of individual galaxies. Thus, material that is ejected from galaxies may also escape the group. Several mechanisms have been proposed to eject material from groups, including galactic winds and outflows powered by supernovae or nuclear activity (Renzini 1997). The material lost from groups may have contributed significantly to the enrichment of the intergalactic medium (Davis et al 1999).

The iron mass-to-light ratios of groups could be somewhat underestimated if the true iron abundances are higher than the sub-solar values usually derived from isothermal model fits. However, the gas-mass estimates are less sensitive to the iron abundance assumed and uncertainties in the iron abundances likely lead to inaccuracies in the gas-mass estimates of at most $\sim 50\%$ (Pildis et al 1995). A potentially bigger problem is that many groups are detected to a much smaller fraction of the virial radius than their rich clusters counterparts. Thus, the true gas masses in some groups may be significantly underestimated from the existing X-ray data. In fact, it is possible that the differences in the iron mass-to-light ratios of groups and clusters may largely be a result of this effect and not necessarily evidence for mass loss.

The mechanisms responsible for producing metals may also inject energy into the gas. Numerical simulations indicate that in the absence of such non-gravitational heating, the density profiles of groups and clusters are nearly identical (Navarro et al 1997). There is now considerable evidence for departures from such uniformity. In the standard hierarchical clustering models, the X-ray luminosity is expected to scale with temperature as $L_X \propto T^2$ (e.g. Kaiser 1991). The observed relationship is considerably steeper, especially for small groups (see Figure 6). Furthermore, the ratio of specific energy of the galaxies to specific energy of the gas (i.e. the β parameter) is less than one for low-mass systems. (However, see Section 3.3.3 for a discussion of why the observed β values for groups may be biased low.) Both of these observations suggest that the gas temperature may not be a good indicator of the virial temperature in poor groups. Entropy profiles for groups and clusters indicate that the entropy of the group gas is also higher than can be achieved through gravitational collapse alone (David et al 1996, Ponman et al 1999, Lloyd-Davies et al 2000). All of these observations are consistent with preheating models for the hot gas (Kaiser 1991; Evrard & Henry 1991; Metzler & Evrard 1994; Knight & Ponman 1997; Cavaliere et al 1997, 1998, 1999; Arnaud & Evrard 1999; Balogh et al 1999; Tozzi et al 2000; Loewenstein 2000; Tozzi & Norman 2000). Such preheating leads to a more extended gas component in groups than in rich clusters (i.e. lower central gas densities and shallower density slopes). Moreover, without preheating, groups appear to over-produce the X-ray background (Wu et al 2000).

Ponman and collaborators have estimated the excess entropy associated with the preheating in groups and find that it corresponds to a temperature of $\sim$0.3 keV (Ponman et al 1999, Lloyd-Davies et al 2000). The preheating temperature can be combined with the excess entropy to estimate the electron density of the gas into which the energy was injected. The resulting value ($n \sim 4 \times 10^{-4} h_{100}^{0.5}$ cm^{-3}) implies that the heating occurred prior to the cluster collapse but after a redshift of $z \sim 10$ (Lloyd-Davies et al 2000). The current estimates for the entropy associated with the preheating have been based on rather small samples of groups and clusters, and these techniques will undoubtably improve with the next generation of X-ray telescopes. Already it is clear that such research can provide considerable insight into the history of the gas and group formation.

5.10 The Local Group

Finally, it is interesting to consider the implications X-ray observations of other groups have for our own Local Group. The idea that the Local Group might contain a hot intragroup medium dates back to the work of Kahn & Woltjer (1959). The X-ray detection of other groups has led to renewed interest in this idea. Suto et al (1996) proposed that a hot halo around the Local Group with a temperature of $\sim$1 keV and column density $N_H \sim 10^{21}$ cm^{-2} could explain the observed excess in the X-ray background below 2 keV. The X-ray halo would also generate temperature anisotropies in the microwave background via the Sunyaev-Zeldovich effect. There is no evidence for such anisotropies in the COBE MDR maps, however (Banday & Górski 1996). Furthermore, the gas temperature and column density assumed by Suto et al (1996) are probably overestimated, given the ROSAT observations of other groups (Pildis & McGaugh 1996). In fact, the strong trend for spiral-only groups not to be X-ray detected suggests that the Local Group is unlikely to produce appreciable amounts of X-ray emission (Pildis & McGaugh 1996, Mulchaey et al 1996b).

Although the Local Group is probably not X-ray bright, a significant gas component may exist at cooler temperatures (Mulchaey et al 1996b, Fields et al 1997). Given the expected virial temperature of the Local Group ($\sim$0.2 keV), the detection of this gas in emission would be exceedingly difficult. However, an enriched collisionally ionized gas at these temperatures is expected to produce prominent absorption features in the far-UV region. The strongest features result from lithium-like ions O VI, Ne VIII, Mg X and Si XII (Verner et al 1994). Lines of sight to hot stars in the Magellanic Clouds are known to show O VI absorption features, but it is not clear whether this gas is associated with intragroup gas or gas in our own Galaxy. There may be other ways to infer the presence of warm gas in the Local Group. Wang & McCray (1993) found evidence in the soft X-ray background for a thermal component with temperature $\sim$0.2 keV, which could be due to a warm intragroup medium in the Local Group (see, however, Sidher et al 1999, who argue that the X-ray halo of the Galaxy dominates). Maloney & Bland-Hawthorn (1999) have recently considered the ionizing flux produced by warm intragroup gas and find that it is unlikely to dominate over the cosmic background or the ultraviolet background produced by the luminous members of the Local Group. Still, encounters between the intragroup gas and the Magellanic Stream may be responsible for the strong Hα emission detected by Weiner & Williams (1996).

The existence of an intragroup medium in the Local Group may also be relevant to the H I high velocity clouds (HVCs; for a review see Wakker & van Woerden 1997). Recently, Blitz et al (1999) revived the idea that many of the HVCs may be dark-matter dominated structures falling onto the Local Group. In this scenario, some of the HVCs collide near the center of the Local Group and produce a warm intragroup medium. If the Blitz et al (1999) scenario is correct, one would expect to find similar H I clouds in other nearby groups. Blitz et al (1999) suggested that several HVC analogs have indeed been found. However, Zwaan & Briggs (2000) completed a H I strip survey of the extragalactic sky with Arecibo and detected no objects resembling the HVCs in other groups. The failure of the Arecibo survey to detect H I does not necessarily rule out the Blitz et al (1999) model. One possibility is that the groups in the Zwaan & Briggs (2000) survey contain an X-ray-emitting intragroup gas and that the H I clouds do not survive this hostile environment. Unfortunately, the X-ray properties of the Zwaan & Briggs (2000) groups are currently unknown. The conclusions of Zwaan & Briggs (2000) are also sensitive to the masses assumed for the H I clouds. Braun & Burton (2000) argued for a lower HVC H I mass and concluded that the sensitivity and coverage of Zwaan & Briggs's (2000) survey was not sufficient to detect analogs of the HVCs in other groups. A more serious problem may be the number statistics of moderate redshift Mg II and Lyman limit absorbers, which appear to be inconsistent with a Local Group origin for the HVCs (Charlton et al 2000). Regardless, it is clear that future H I surveys of X-ray detected and X-ray–non-detected groups could provide important insight into the relationship between hot and cold gas in galaxy groups.

6. FUTURE WORK

X-ray telescopes launched in the 1990s have firmly established the presence of a hot X-ray-emitting intragroup medium in nearby groups of galaxies. X-ray observations suggest that many groups are real, physical systems. The masses of X-ray groups are substantial and make a significant contribution to the mass density of the universe. Although most of the mass in groups appears to be in dark matter, the intragroup medium may be the dominant baryonic component in the nearby universe.

While we have made significant progress towards understanding groups in the last decade, there are still many outstanding issues. Ambiguities about the proper spectral model for the gas and our inability to detect gas to a large fraction of the virial radius are particularly troubling because the resulting uncertainties propagate into cosmological applications. Furthermore, the contribution of individual galaxies to the observed X-ray emission remains a point of contention. Our ability to understand the intragroup medium has largely been limited by the poor spatial and spectral resolution of the X-ray instruments. This situation is about to change drastically, however, with the availability of new powerful X-ray telescopes. Recently, NASA successfully launched CHANDRA (formerly known as AXAF). This telescope will produce high-resolution X-ray images of groups ($\sim 1''$) that will allow the relative contribution of galaxies and diffuse gas to be quantified. In late 1999, the European Space Agency (ESA) launched XMM-Newton. Although the spatial resolution of XMM-Newton is poorer than that of CHANDRA, the collecting area of this telescope is much greater. Therefore, XMM-Newton will obtain the deepest X-ray exposures ever of nearby groups and will extend the studies of the group environment to higher redshifts. The combination of CHANDRA and XMM-Newton will probably answer many of the questions raised by the recent generation of X-ray telescopes.

ACKNOWLEDGMENTS

I would like to thank my collaborators and colleagues, particularly Arif Babul, Dave Burstein, David Buote, David Davis, Steve Helsdon, Pat Henry, Lawrence Jones, Lori Lubin, Gary Mamon, Bill Mathews, Kyoko Matsushita, Chris Mullis, Richard Mushotzky, Gus Oemler, Trevor Ponman, Matthias Steinmetz, Jack Sulentic, Ben Weiner, and Ann Zabludoff for useful discussions on X-ray groups. I would also like to thank David Davis, Richard Mushotzky and Allan Sandage for comments on the manuscript. This work was supported in part by NASA under grant NAG 5-3529.

Visit the Annual Reviews home page at http://www.AnnualReviews.org

LITERATURE CITED

Albert CE, White RA, Morgan WW. 1977. *Ap. J.* 211:309–10

Allen SW, Fabian AC, Edge AC, Böhringer H, White DA. 1995. *MNRAS* 275:741–54

Arimoto N, Matsushita K, Ishimaru Y, Ohashi T, Renzini A. 1997 *Ap. J.* 477:128–143

Arnaud M, Evrard AE. 1999. *MNRAS* 305:631–40

Aschenbach B. 1988. *Appl. Optics* 27:1404–13

Athanassoula E, Makino J, Bosma A. 1997. *MNRAS* 286:825–38

Bahcall NA, Harris DE, Rood HJ. 1984. *Ap. J.* 284:L29–33

Bahcall NA, Lubin LM. 1994. *Ap. J.* 426:513–15

Balogh ML, Babul A, Patton DR. 1999. *MNRAS* 307:463–79

Banday AJ, Górski KM. 1996. *MNRAS* 283:L21–25

Barnes JE. 1985. *MNRAS* 215:517–36

Barnes JE. 1989. *Nature* 338:123–26

Bartelmann M, Steinmetz M. 1996. *MNRAS* 283:431–46

Batuski DJ, Burns JO. 1985. *Astron. J.* 90:1413–24

Bauer F, Bregman JN. 1996. *Ap. J.* 457:382–89

Bergeron J, Petitjean P, Sargent WLW, Bahcall JN, Boksenberg A, et al. 1994. *Ap. J.* 436:33–43

Biermann P, Kronberg PP. 1983. *Ap. J.* 268:L69–73

Biermann P, Kronberg PP, Madore BF. 1982. *Ap. J.* 256:L37–40

Bird CM, Mushotzky RF, Metzler CA. 1995. *Ap. J.* 453:40–47

Blandford RD, Narayan R. 1992. *Annu. Rev. Astron. Astrophy.* 30:311–58

Blitz L, Spergel DN, Teuben PJ, Hartmann D, Burton WB. 1999. *Ap. J.* 514:818–43

Bode PW, Cohn HN, Lugger PM. 1993. *Ap. J.* 416:17–25

Braun R, Burton WB. 2000. *Astron. Astrophys.* submitted

Buote DA. 1999. *MNRAS* 309:685–714

Buote DA. 2000a. *MNRAS* 311:176–200

Buote DA. 2000b. *Ap. J.* in press

Buote DA. 2000c. *Ap. J.* in press

Burke DJ, Collins CA, Sharples RM, Romer AK, Holden BP, et al. 1997. *Ap. J.* 488:L83–86

Burns JO, Gregory SA, Holman GD. 1981. *Ap. J.* 250:450–63

Burns JO, Ledlow MJ, Loken C, Klypin A, Voges W, et al. 1996. *Ap. J.* 467:L49–52

Cavaliere A, Menci N, Tozzi P. 1997. *Ap. J.* 484:L21–24

Cavaliere A, Menci N, Tozzi P. 1998. *Ap. J.* 501:493–508

Cavaliere A, Menci N, Tozzi P. 1999. *MNRAS* 308:599–608

Cen R, Ostriker JP. 1999. *Ap. J.* 514:1–6

Charlton JC, Churchill CW, Rigby JR. 2000. *Ap. J.* submitted

Cooke BA, Ricketts MJ, Maccacaro T, Pye JP, Elvis M, et al. 1978. *MNRAS* 182:489–515

Cowie LL, Henriksen M, Mushotzky R. 1987. *Ap. J.* 317:593–600

David LP. 1997. *Ap. J.* 484:L11–15

David LP, Arnaud KA, Forman W, Jones C. 1990. *Ap. J.* 356:32–40

David LP, Jones C, Forman W. 1995. *Ap. J.* 445:578–90

David LP, Jones C, Forman W. 1996. *Ap. J.* 473:692–706

David LP, Jones C, Forman W, Daines S. 1994. *Ap. J.* 428:544–54

Davis DS, Mulchaey JS, Mushotzky RF. 1999. *Ap. J.* 511:34–40

Davis DS, Mulchaey JS, Mushotzky RF, Burstein D. 1996. *Ap. J.* 460:601–11

Davis DS, Mushotzky RF, Mulchaey JS, Worrall DM, Birkinshaw M, Burstein D. 1995. *Ap. J.* 444:582–89

Dell'Antonio IP, Geller MJ, Fabricant DG. 1994. *Astron. J.* 107:427–47

de Vaucouleurs G. 1965. in *Stars and Stellar Systems*, ed. A. Sandage, M. Sandage and J. Kristian (Chicago: Univesity of Chicago Press)

Diaferio A, Geller MJ, Ramella M. 1994. *Astron. J.* 107:868–79

Diaferio A, Geller MJ, Ramella M. 1995. *Astron. J.* 109:2293–2303

Diaferio A, Ramella M, Geller MJ, Ferrari A. 1993. *Astron. J.* 105:2035–46

Doe SM, Ledlow MJ, Burns JO, White RA. 1995. *Astron. J.* 110:46–67

Dos Santos S, Mamon GA. 1999. *Astron. Astrophys.* 352:1–18

Ebeling H, Voges W, Böhringer H. 1994. *Ap. J.* 436:44–55

Edge AC, Stewart GC. 1991. *MNRAS* 252:428–41

Evrard AE, Henry JP. 1991. *Ap. J.* 383:95–103

Evrard AE, Metzler CA, Navarro JF. 1996. *Ap. J.* 469:494–507

Fabian AC. 1994. *Annu. Rev. Astron. Astrophys.* 32:277–318

Fabian AC, Arnaud KA, Bautz MW, Tawara Y. 1994. *Ap. J.* 436:L63–66

Fabricant D, Gorenstein P. 1983. *Ap. J.* 267:535–46

Fabricant D, Lecar M, Gorenstein P. 1980. *Ap. J.* 241:552–60

Fabricant D, Rybicki G, Gorenstein P. 1984. *Ap. J.* 286:186–95

Fields BD, Mathews GJ, Schramm DN. 1997. *Ap. J.* 483:625–37

Finoguenov A, Ponman TJ. 1999. *MNRAS* 305:325–37

Forman W, Jones C. 1982. *Annu. Rev. Astron. Astrophys.* 20:547–85

Frederic JJ. 1995. *Ap. J. Suppl.* 97:259–74

Fukazawa Y, Makishima K, Matsushita K, Yamasaki N, Ohashi T, et al. 1996. *Publ. Astron. Soc. Japan* 48:395–407

Fukazawa Y, Makishima K, Tamura T, Ezawa H, Xu H, et al. 1998. *Publ. Astron. Soc. Japan* 50:187–93

Fukugita M, Hogan CJ, Peebles PJE. 1998. *Ap. J.* 503:518–30

Geller MJ, Huchra JP. 1983. *Ap. J. Suppl.* 52:61–87

Gendreau KC. 1995. *X-ray CCDs for Space Applications: Calibration, Radiation Hardness, and Use for Measuring the Spectrum of the Cosmic X-ray Background.* PhD thesis. Massachusetts Institute of Technology.

Gibson BK, Loewenstein M, Mushotzky RF. 1997. *MNRAS* 290:623–28

Governato F, Bhatia R, Chincarini G. 1991. *Ap. J.* 371:L15–18

Governato F, Tozzi P, Cavaliere A. 1996. *Ap. J.* 458:18–26

Griffiths RE, Georgantopoulos I, Boyle BJ, Stewart GC, Shanks T, Della Ceca R. 1995. *MNRAS* 275:77–88

Griffiths RE, Schwartz DA, Schwarz J, Doxsey RE, Johnson MD, et al. 1979. *Ap. J.* 230: L21–25

Hellsten U, Gnedin NY, Miralda-Escude J. 1998. *Ap. J.* 509:56–61

Helsdon SF, Ponman TJ. 2000. *MNRAS* in press

Henriksen MJ, Mamon GA. 1994. *Ap. J.* 421:L63–66

Henry JP, Gioia IM, Huchra JP, Burg R, McLean B, et al. 1995. *Ap. J.* 449:422–30

Hernquist L, Katz N, Weinberg DH. 1995. *Ap. J.* 442:57–60

Hickson P. 1982. *Ap. J.* 255:382–91

Hickson P. 1997. *Annu. Rev. Astron. Astrophys.* 35:357–88

Hickson P, Huchra J, Kindl E. 1988. *Ap. J.* 331:64–70

Holmberg E. 1950. *Medd. Lunds Obs. Ser. 2* 128:1–56

Horner DJ, Mushotzky RF, Scharf CA. 1999. *Ap. J.* 520:78–86

Huchra JP, Geller MJ. 1982. *Ap. J.* 257:423–37

Humason ML, Mayall NU, Sandage AR. 1956. *Ap. J.* 61:97–162

Hunt R, Sciama DW. 1972. *MNRAS* 157:335–48

Hwang U, Mushotzky RF, Burns JO, Fukazawa Y, White RA. 1999. *Ap. J.* 516:604–18

Hwang U, Mushotzky RF, Loewenstein M, Markert, TH, Fukazawa Y, Matsumoto H. 1997. *Ap. J.* 476:560–71

Ikebe Y, Ezawa H, Fukazawa Y, Hirayama M, Izhisaki Y, et al. 1996. *Nature* 379:427–29

Ishimaru Y, Arimoto N. 1997. *PASJ* 49:1–8

Jones C, Forman W. 1984. *Ap. J.* 276:38–55

Jones LR, Ponman TJ, Forbes DA. 2000. *MNRAS* 312:139–50

Jones LR, Scharf C, Ebeling H, Perlman E, Wegner G, et al. 1998. *Ap. J.* 495:100–14

Kaastra JS, Mewe R. 1993. *Astron. Astrophys. Suppl.* 97:443–82

Kahn FD, Woltjer L. 1959. *Ap. J.* 130:705–17

Kaiser N. 1991. *Ap. J.* 383:104–111

Keeton CR, Kochanek CS, Seljak U. 1997. *Ap. J.* 482:604–20

King IR. 1962. *Astron. J.* 67:471–85

Knight PA, Ponman TJ. 1997. *MNRAS* 289: 955–72

Kriss GA, Canizares CR, McClintock JE, Feidelson ED. 1980. *Ap. J.* 235:L61–65

Kriss GA, Cioffi DF, Canizares CR. 1983. *Ap. J.* 272:439–48

Kundic T, Cohen JG, Blandford RD, Lubin LM. 1997a. *Astron. J.* 114:507–10

Kundic T, Hogg DW, Blandford RD, Cohen JG, Lubin LM, et al. 1997b. *Astron. J.* 114:2276–83

Ledlow MJ, Loken C, Burns JO, Hill JM, White RA. 1996. *Astron. J.* 112:388–406

Liedahl DA, Kahn SM, Osterheld AL, Goldstein WH. 1990. *Ap. J.* 350:L37–40

Liedahl DA, Osterheld AL, Goldstein WH. 1995. *Ap. J.* 438:L115–118

Lloyd-Davies EJ, Ponman TJ, Cannon DB. 2000. *MNRAS* In press

Loewenstein M. 2000. *Ap. J.* In press

Mahdavi A, Böhringer H, Geller MJ, Ramella M. 1997. *Ap. J.* 483:68–76

Mahdavi A, Böhringer H, Geller MJ, Ramella M. 2000. *Ap. J.* in press

Mahdavi A, Geller MJ, Böhringer H, Kurtz MJ, Ramella M. 1999. *Ap. J.* 518:69–93

Maloney PR, Bland-Hawthorn J. 1999. *Ap. J.* 522:L81–84

Mamon GA. 1986. *Ap. J.* 307:426–30

Materne J. 1979. *Astron. Astrophys.* 74:235–43

Matsushita K, Ohashi T, Makishima K. 2000. *PASJ* in press

McWilliam A. 1997. *Annu. Rev. Astron. Astrophys.* 35:503–56

Mendes de Oliveira C, Giraud E. 1994. *Ap. J.* 437:L103–106

Metzler CA, Evrard AE. 1994. *Ap. J.* 437:564–83

Mewe R. 1991. *Astron. & Astrophys. Review* 3:127–68

Mewe R, Gronenschild EHBM, van den Oord GHJ. 1985. *Astron. Astrophys. Suppl.* 62:197–254

Miralda-Escude J, Cen R, Ostriker JP, Rauch M. 1996. *Ap. J.* 471:582–616

Mohr JJ, Evrard AE. 1997. *Ap. J.* 491:38–44

Mohr JJ, Mathiesen B, Evrard AE. 1999. *Ap. J.* 517:627–49

Montoya ML, Dominguez-Tenreiro R, Gonzalez-Casado G, Mamon GA, Salvador-Sole E. 1996. *Ap. J.* 473:L83–86

Morgan WW, Kayser S, White RA. 1975. *Ap. J.* 199:545–48

Morrison R, McCammon D. 1983. *Ap. J.* 270:119–22

Mulchaey JS, Davis DS, Mushotzky RF, Burstein D. 1993. *Ap. J.* 404:L9–12

Mulchaey JS, Davis DS, Mushotzky RF, Burstein D. 1996a. *Ap. J.* 456:80–97

Mulchaey JS, Davis DS, Mushotzky RF, Burstein D. 2000. *Ap. J.* in preparation

Mulchaey JS, Mushotzky RF, Burstein D, Davis DS. 1996b. *Ap. J.* 456:L5–8

Mulchaey JS, Zabludoff AI. 1998. *Ap. J.* 496:73–92

Mulchaey JS, Zabludoff AI. 1999. *Ap. J.* 514:133–37

Mullis CR, Henry JP, Gioia IM, Böhringer H, Briel UG. 2000. *Ap. J.* in preparation.

Mushotzky RF. 1984. *Physica Scripta* T7:157–62

Navarro JF, Frenk CS, White SDM. 1995. *MNRAS* 275:720–40

Navarro JF, Frenk CS, White SDM. 1997. *Ap. J.* 490:493–508

Nolthenius R. 1993. *Ap. J. Suppl.* 85:1–25

Nolthenius R, White SDM. 1987. *MNRAS* 225:505–30

Ohashi T, Ebisawa K, Fukazawa Y, Hiyoshi K, Horii M, et al. 1996. *PASJ* 48:157–70

Oort JH. 1970. *Astron. Astrophys.* 7:381–404

Ostriker JP, Lubin LM, Hernquist L. 1995. *Ap. J.* 444:L61–64

Pedersen K, Yoshii Y, Sommer-Larsen J. 1997. *Ap. J.* 485:L17–20

Perna R, Loeb A. 1998. *Ap. J.* 503:L135–138

Pfeffermann E, Briel UG, Hippmann H, Kettenring G, Metzner G, et al. 1988. *Proc. SPIE* 733:519–32

Pierre M, Bryan G, Gastaud R. 2000. *Astron. Astrophys.* Submitted

Pietsch W, Trinchieri G, Arp H, Sulentic JW. 1997. *Astron. Astrophys.* 322:89–97

Pildis RA, Bregman JN, Evrard AE. 1995. *Ap. J.* 443:514–26

Pildis RA, Evrard AE, Bregman JN. 1996. *Astron. J.* 112:378–87

Pildis RA, McGaugh SS. 1996. *Ap. J.* 470:L77–79

Ponman TJ, Allan DJ, Jones LR, Merrifield M, McHardy IM, et al. 1994. *Nature* 369:462–64

Ponman TJ, Bertram D. 1993. *Nature* 363:51–54

Ponman TJ, Bourner PDJ, Ebeling H, Böhringer H. 1996. *MNRAS* 283:690–708

Ponman TJ, Cannon DB, Navarro JF. 1999. *Nature* 397:135–137

Price R, Duric N, Burns JO, Newberry MV. 1991. *Astron. J.* 102:14–29

Ramella M, Geller MJ, Huchra JP. 1989. *Ap. J.* 344:57–74

Ramella M, Geller MJ, Huchra JP, Thorstensen JR. 1995. *Astron. J.* 109:1458–75

Ramella M, Pisani A, Geller MJ. 1997. *Astron. J.* 113:483–91

Raymond JC, Smith BW. 1977. *Ap. J. Suppl.* 35:419–39

Renzini A. 1997. *Ap. J.* 488:35–43

Renzini A, Ciotti L, D'Ercole A, Pellegrini S. 1993. *Ap. J.* 419:52–65

Rhee G, van Haarlem M, Katgert P. 1992. *Astron. J.* 103, 1721–28

Ribeiro ALB, De Carvalho RR, Coziol R, Capelato HV, Zepf SE. 1996. *Ap. J.* 463:L5–8

Ricker G, Doxsey RE, Dower RG, Jernigan JG, Delvailee JP, et al. 1978. *Nature* 271:35–37

Rosati P, Della Ceca R, Burg R, Norman C, Giacconi R. 1995. *Ap. J.* 445:L11–14

Ruderman MA, Spiegel EA. 1971. *Ap. J.* 165:1–15

Saracco P, Ciliegi P. 1995. *Astron. Astrophys.* 301:348–58

Sarazin CL. 1986. *Review Modern Physics* 58:1–115

Sarazin CL, Burns JO, Roettiger K, McNamara BR. 1995. *Ap. J.* 447:559–71

Savage BD, Tripp TM, Lu L. 1998. *Astron. J.* 115:436–50

Scharf CA, Jones LR, Ebeling H, Perlman E, Malkan M, et al. 1997. *Ap. J.* 477:79–92

Schechter PL, Bailyn CD, Barr R, Barvainis R, Becker CM, et al. 1997. *Ap. J.* 475:L85–88

Schmidt M, Hasinger G, Gunn, J, Schneider D, Burg R, et al. 1998. *Astron. Astrophys.* 329:495–503

Schwartz DA, Schwarz J, Tucker W. 1980. *Ap. J.* 238:L59–62

Sidher SD, Sumner TJ, Quenby JJ. 1999. *Astron. Astrophys.* 344:333–41

Silk J, Tarter J. 1973. *Ap. J.* 183:387–410

Sulentic JW, Pietsch W, Arp H. 1995. *Astron. Astrophys.* 298:420–26

Suto Y, Makishima K, Ishisaki Y, Ogasaka Y. 1996. *Ap. J.* 461:L33–36

Tanaka Y, Inoue H, Holt SS. 1994. *Publ. Astron. Soc. Japan* 46, L37–41

Tonry JL. 1998. *Astron. J.* 115:1–5

Tonry JL, Kochanek CS. 2000. *Astron. J.* 117:2034–38

Tozzi P, Norman C. 2000. *Ap. J.* in press

Tozzi P, Scharf C, Norman. C. 2000. *Ap. J.* in press

Trinchieri G, Fabbiano G, Kim D-W. 1997. *Astron. Astrophys.* 318:361–75

Tully RB. 1987. *Ap. J.* 321:280–304

Verner DA, Tytler D, Barthel PD. 1994. *Ap. J.* 430:186–90

Vikhlinin A, McNamara BR, Forman W, Jones C, Quintana H, et al. 1998. *Ap. J.* 502:558–81

Vikhlinin A, McNamara BR, Hornstrup A, Quintana H, Forman W, et al. 1999. *Ap. J.* 520:L1–4

Voges W. 1993. *Adv. Space Res.* 13:12391–97

Wakker BP, van Woerden H. 1997. *Annu. Rev. Astron. Astrophys.* 35:217–66

Walke DG, Mamon GA. 1989. *Astron. Astrophys.* 225:291–302

Walker TP, Steigman G, Kang HS, Schramm DM, Olive KA. 1991. *Ap. J.* 376:51–69

Wang QD, McCray R. 1993. *Ap. J.* 409:L37–40

Ward MJ, Wilson AS, Penston MV, Elvis M, Maccacaro T, et al. 1978. *Ap. J.* 223:788–97

Weiner BJ, Williams TB. 1996. *Astron. J.* 111:1156–63

Wells A, Abbey AF, Barstow MA, Cole RE, Pye JP, et al. 1990. *Proc. SPIE* 733:519–532

White DA. 2000. *MNRAS* 312:663–88

White RA, Bliton M, Bhavsar SP, Bornmann P, Burns JO, et al. 1999. *Astron. J.* 118:2014–37

White RE. 1991. *Ap. J.* 367:69–77

White SDM, Navarro JF, Evrard AE, Frenk CS. 1993. *Nature* 366:429–33

Wu KKS, Fabian AC, Nulsen PEJ. 2000. *MNRAS* in press

Wu X-P, Xue Y-J, Fang L-Z. 1999. *Ap. J.* 524:22–30

Xue Y, Wu X-P. 2000. *Ap. J.* in press

Zabludoff AI, Mulchaey JS. 1998. *Ap. J.* 496:39–72

Zamorani G, Mignoli M, Hasinger G, Burg R, Giacconi R, et al. 1999. *Astron. Astrophys.* 346:731–52

Zwaan MA, Briggs FH. 2000. *Ap. J.* 530:L61–64

Annu. Rev. Astron. Astrophys. 2000. 38:337–77

THEORY OF LOW-MASS STARS AND SUBSTELLAR OBJECTS

Gilles Chabrier and Isabelle Baraffe

*Centre de Recherche Astrophysique de Lyon (UMR CNRS 5574),
Ecole Normale Supérieure de Lyon, 69364 Lyon Cedex 07, France;
e-mail: chabrier@ens-lyon.fr, ibaraffe@ens-lyon.fr*

Key Words   stars: fundamental parameters; stars: low mass, brown dwarfs; stars: cataclysmic variables; stars: luminosity function, mass function; stars: planetary systems; Galaxy: stellar content

■ **Abstract**   Since the discovery of the first bona-fide brown dwarfs and extra-solar planets in 1995, the field of low-mass stars and substellar objects has progressed considerably, both from theoretical and observational viewpoints. Recent developments in the physics entering the modeling of these objects have led to significant improvements in the theory and to a better understanding of these objects' mechanical and thermal properties. This theory can now be confronted with observations directly in various observational diagrams (color-color, color-magnitude, mass-magnitude, mass-spectral type), a stringent and unavoidable constraint that became possible only recently with the generation of synthetic spectra. In this paper we present the current state-of-the-art general theory of low-mass stars and sub stellar objects, from one solar mass to one Jupiter mass, regarding primarily their interior structure and evolution. This review is a natural complement to the previous review by Allard et al (1997) on the atmosphere of low-mass stars and brown dwarfs. Special attention is devoted to the comparison of the theory with various available observations. The contribution of low-mass stellar and sub stellar objects to the Galactic mass budget is also analyzed.

1.   INTRODUCTION

Interest in the physics of objects at the bottom of and below the Main Sequence (MS) originated in the early demonstration by Kumar (1963) that hydrogen-burning in a stellar core no longer occurs below a certain mass, and that below this limit hydrostatic equilibrium against gravitational collapse is provided by electron degeneracy. Simple analytical arguments, based on the balance between the classical ionic thermal pressure and the quantum electronic pressure, yielded for this H-burning minimum mass (HBMM) $m_{HBMM} \approx 0.1 \, M_\odot$, whereas the first detailed evolutionary calculations gave $m_{HBMM} \approx 0.085 \, M_\odot$ (Grossman et al 1974). Tarter

0066-4146/00/0915-0337$14.00

(1975) proposed the term "brown dwarfs" (BD) for objects below this H-burning limit. D'Antona & Mazzitelli (1985) first pointed out that the luminosity below m_{HBMM} would stretch by a few orders of magnitude over a few hundredths of a solar mass, making the observation of BDs a tremendously difficult task. Subsequent benchmarks in low-mass star (LMS) and BD theory came primarily from VandenBerg et al (1983), D'Antona & Mazzitelli (1985), Dorman et al (1989), Nelson et al (1986, 1993a) and the Tucson group (Lunine et al 1986, 1989; Burrows et al 1989, 1993). In spite of this substantial theoretical progress, all of these models failed to reproduce the observations at the bottom of the MS and thus could not provide a reliable determination of the characteristic properties (mass, age, effective temperature, luminosity) of low-mass stellar and substellar objects. A noticeable breakthrough came from the first calculations by a few groups of synthetic spectra and atmosphere models characteristic of cool ($T_{eff} \lesssim 4000$ K) objects (Allard 1990, Brett & Plez 1993, Saumon et al 1994, Allard & Hauschildt 1995, Brett 1995, Tsuji et al 1996, Hauschildt et al 1999). This allowed the computation of consistent non-grey evolutionary models (Saumon et al 1994, Baraffe et al 1995) and the direct confrontation of theory and observation in photometric passbands and color-magnitude diagrams, thus avoiding dubious transformations of observations into theoretical L-T_{eff} Hertzsprung-Russell diagrams.

In the meantime, the search for faint (sub)stellar objects has bloomed over the past few years. Several BDs have now been identified since the first discoveries of bona-fide BDs (Rebolo et al 1995, Oppenheimer et al 1995), either in young clusters (see Basri, this volume, and Martín 1999 for reviews and references therein) or in the Galactic field (Ruiz et al 1997, Delfosse et al 1997, Kirkpatrick et al 1999a). The steadily increasing number of identified extra-solar giant planets (EGPs) since the discovery of 51PegB (Mayor & Queloz 1995; see Marcy & Butler 1998 for a review) has opened up a new era in astronomy. Ongoing and future ground-based and space-based optical and infrared surveys of unprecedented faintness and precision are likely to reveal several hundred more red dwarfs, brown dwarfs, and giant planets, which will necessitate the best possible theoretical foundation. The correct understanding of the physical properties of these objects bears major consequences for a wide range of domains of physics and astrophysics—namely, dense matter physics, planet and star formation and evolution, galactic evolution, and missing mass.

A general outline of the basic physics entering the structure and the evolution of BDs can be found in the excellent reviews of Stevenson (1991) and Burrows & Liebert (1993). It is the aim of this review to summarize the most recent progress realized in the theory of LMS, BDs, and EGPs. We focus mainly on the internal structure and the evolution of these objects, since a comprehensive review on the atmosphere of LMS and BDs appeared recently (Allard et al 1997). We also consider the implications of LMS and BDs in a more general galactic context and evaluate their contribution to the Galactic mass budget.

2.   THE PHYSICS OF DENSE OBJECTS

2.1   Interior Physics

2.1.1 Equation of State   Central conditions for LMS—hereafter identified generically as objects below a solar mass—and for substellar objects (SSO) for solar composition range from a maximum density $\rho_c \simeq 10^3$ gcm^{-3} at the hydrogen-burning limit ($m \approx 0.07\ M_\odot$, see below) to $\rho_c \simeq 10$ gcm^{-3} for Saturn ($= 5 \times 10^{-4}$ $M_\odot$) at 5 Gyr, and from $T_c \simeq 10^7$ K for the Sun to $T_c \simeq 10^4$ K for Saturn at the same age, spanning several orders of magnitudes in mass, density, and temperature. Effective temperatures range from $\sim$6000 K to $\sim$2000 K in the stellar domain and extend down to $\sim$100 K for Saturn. Molecular hydrogen and other molecules become stable for $kT \lesssim 3 \times 10^{-2}$ Ryd ($T \approx 5 \times 10^3$ K), a condition encountered in the atmosphere, or even deeper layers, of most of these objects. Under these conditions, the interior of LMS and SSO is essentially a fully ionized H$^+$/He^{++} plasma characterized by coupling parameters $\Gamma_i = \langle Z^{5/3} \rangle e^2 / a_e kT \approx 0.1 - 50$ for classical ions and $r_S = \langle Z \rangle^{-1/3} a/a_0 \approx 0.1 - 1$ for degenerate electrons. With $\Gamma_i = \langle Z^{5/3} \rangle (2.693 \times 10^5\ \text{K}/T) n_{24}^{1/3}$, $r_S = 1.39/(\rho/\mu_e)^{1/3} = 1.172\, n_{24}^{-1/3}$, $n_{24} \equiv n_e/10^{24}$ cm$^{-3} \approx (\rho/1.6605\,\text{gcm}^{-3})\mu_e^{-1}$ the electron number-density, $a = (\frac{3}{4\pi}\frac{V}{N_i})^{1/3} = a_e \langle Z \rangle^{1/3}$ is the mean interionic distance, $\mu_e^{-1} = \langle Z \rangle / \langle A \rangle$ is the electron mean molecular weight, ρ is the mass-density, A is the atomic mass, a_0 is the electronic Bohr radius, and $\langle \rangle$ denotes the average number-fraction $\langle X \rangle = \sum_i x_i X_i$ ($x_i = N_i / \sum_i N_i$). The temperature is of the order of the electron Fermi temperature T_F so that the degeneracy parameter $\psi = kT/kT_F \approx 3.314 \times 10^{-6}\ T\ (\mu_e/\rho)^{2/3}$ is of the order of unity. The classical (Maxwell-Boltzman) limit corresponds to $\psi \to +\infty$, whereas $\psi \to 0$ corresponds to complete degeneracy. The aforementioned thermodynamic conditions yield $\psi \approx 2$–0.05 in the interior of LMS and BDs, which implies that finite-temperature effects for the electrons must be included to accurately describe the thermodynamic properties of the correlated, partially degenerate electron gas. Moreover, the Thomas-Fermi wavelength $\lambda_{TF} = \left(kT_F/(6\pi n_e e^2)\right)^{1/2}$ is of the order of a, so that the electron gas is polarized by the external ionic field, and electron-ion coupling must be taken into account in the plasma hamiltonian. Last but not least, the electron average binding energy can be of the order of the Fermi energy $Ze^2/a_0 \sim \epsilon_F$ so that *pressure*-ionization takes place along the internal profile. Figure 1 illustrates central characteristic quantities for LMS and SSOs from the Sun to Jupiter.

Above $\sim 0.4\ M_\odot$, the structure evolves slowly with increasing mass from a $n = 3/2$ towards a $n = 3$ polytrope, which yields the correct P_c for the Sun, as a result of the growing central radiative core. This leads to increasing central pressures and densities for increasing mass (increasing polytropic index) in this mass range. Below ~ 0.3–$0.4\ M_\odot$, the core becomes entirely convective and follows the behavior of a $n = 3/2$ polytrope. Because the gas is still in the classical regime ($\psi \gtrsim 1$), $m \propto R$ and the central density increases with decreasing mass,

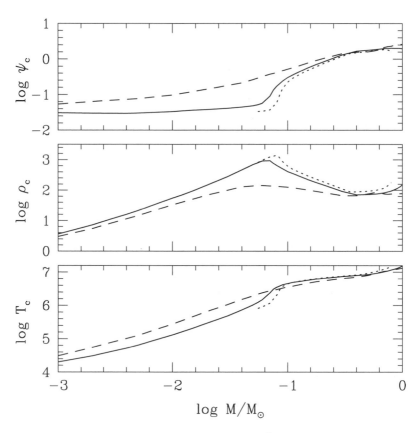

Figure 1   Central temperature (in K), density (in g cm^{-3}), and degeneracy parameter along the LMS-SSO mass range for objects with $Z = Z_\odot$ at 5 Gyr (*solid line*) and 10^8 yr (*dashed line*), and with a metallicity $Z = 10^{-2} Z_\odot$ at 5 Gyr (*dotted-line*).

$\rho_c \propto m^{-2}$. Below the H-burning limit, electron degeneracy becomes dominant ($\psi \lesssim 0.1$), so that one approaches the relation $m \propto R^{-3}$ (for $\psi = 0$), and density decreases again with decreasing mass, $\rho_c \propto m^2$. These various effects yield a non-monotomic behavior of the central density and pressure with mass, with a minimum in the stellar regime around 0.4 $M_\odot$ and a maximum near the H-burning limit.

The equation of state (EOS) of low-mass objects thus requires a detailed description of strongly correlated, polarizable, partially degenerate classical and quantum plasmas, plus an accurate treatment of pressure partial ionization, a severe challenge for dense matter physicists. Several steps toward the derivation of such an accurate EOS for dense astrophysical objects have been realized since the pioneering work of Salpeter (1961), with major contributions by Fontaine et al (1977), Magni & Mazzitelli (1979), Marley & Hubbard (1988) (for objects with $m < 0.2\ M_\odot$), Hümmer & Mihalas (1988) (primarily devoted to conditions characteristic of solar-type stellar envelopes), and Saumon & Chabrier (Chabrier 1990; Saumon & Chabrier 1991, 1992; Saumon et al 1995; hereafter SCVH). We refer the reader to Saumon (1994) and SCVH for a detailed review and an extensive comparison of these different EOSs in the domain of interest. The SCVH EOS presents a consistent treatment of pressure ionization and includes significant improvements with respect to previous calculations in the treatment of the correlations in dense plasmas. It compares well with available high-pressure shock-wave experiments in the molecular domain, and with Monte-Carlo simulations in the fully-ionized, metallic domain (see references above for details). Recently, laser-driven shock-wave experiments on D_2 have been conducted at Livermore (Da Silva et al 1997, Collins et al 1998); these experiments were the first to reach the pressure-dissociation and ionization domain and thus directly probe the thermodynamic properties of dense hydrogen under conditions characteristic of BDs and giant planets (GP). The relevance of these experiments for the interior of SSOs can be grasped from Figure 1 of Collins et al (1998). These experiments have shown good agreement between the predictions of the Saumon-Chabrier model and the actual data (although there is certainly room for improvement). In particular, the strong compression factor arising from hydrogen pressure-dissociation and ionization observed in the experiment ($\rho/\rho_i \sim 5.8$) agrees well with the predicted theoretical value (see Figure 3 of Collins et al 1998). Any EOS devoted to the description of the interior of dense astrophysical objects must now be confronted with these available data.

The SCVH EOS is a pure hydrogen and helium EOS, based on the so-called additive-volume law between the pure components (H and He). The accuracy of the additive-volume-law has been examined in detail by Fontaine et al (1977). The invalidity of this approximation for accurately describing the thermodynamic properties of the mixture is significant only in the *partial ionization* region (see SCVH), which concerns only a few percent of the stellar mass under LMS and BD conditions. The effect of metals on the structure and the evolution of these objects has been examined in detail by Chabrier & Baraffe (1997; Section 2.1). These authors show that because of their negligible number-abundance ($\sim 0.2\%$), metals do not contribute significantly to the EOS and barely modify the structure and evolution ($\sim 1\%$ in T_{eff} and $\sim 4\%$ in L) of these objects. In the low-density limit characteristic of the atmosphere, the SCVH EOS recovers the perfect gas limit, and thermal contributions from various atomic or molecular species can be added within the aforementioned additive-volume law formalism, which is exact in this regime. Only in the (denser) metal-depleted atmospheres ($Z \lesssim 10^{-2}\ Z_\odot$) of the densest objects ($\lesssim 0.1\ M_\odot$) does one find a slight departure from ideality, with a 1% to 4% effect on the adiabatic gradient ∇_{ad} (Chabrier & Baraffe 1997).

2.1.2 Nuclear Rates. Screening Factors   Although the complete PP chain is important for nucleosynthesis, the thermonuclear processes relevant from the energetic viewpoint under LMS and BD conditions are given by the PPI chain (see e.g. Burrows & Liebert 1993, Chabrier & Baraffe 1997):

$$p + p \rightarrow d + e^{+} + \nu_{e}; \; p + d \; \rightarrow \; {}^{3}He + \gamma; \; {}^{3}\text{He} + {}^{3}\text{He} \rightarrow {}^{4}\text{He} + 2p. \quad (1)$$

Below $\sim 0.7\, M_{\odot}$, the PPI chain contributes to more than 99% of the energy generation on the zero-age Main Sequence (ZAMS), and the PPII chain to less than 1%. The destruction of ^{3}He by the above reaction is important only for $T > 6 \times 10^{6}$ K, i.e masses $m \gtrsim 0.15\, M_{\odot}$ for ages $t < 10$ Gyr, for which the lifetime of this isotope against destruction becomes eventually smaller than a few Gyr.

As examined in Section 4.6, the abundance of light elements (D, Li, Be, B) provides a powerful diagnostic for identifying the age and/or mass of SSOs. The rates for these reactions (see e.g. Nelson et al 1993b) in the vacuum, or in an almost perfect gas where kinetic energy largely dominates the interaction energy, are given by Caughlan & Fowler (1988) and Ushomirsky et al (1998) for updated values. The reaction rate R_0 (in cm$^{-3}s^{-1}$) in the vacuum is given by $R_0 \propto e^{-3\epsilon_0/kT}$, where ϵ_0 corresponds to the Gamow-peak energy for non-resonant reactions (Clayton 1968). However, as we mentioned in the previous section, non-ideal effects dominate in the interior of LMS and BDs and lead to polarization of the ionic and electronic fluids. These polarization effects resulting from the surrounding particles yield an enhancement of the reaction rate, as was first recognized by Schatzman (1948) and Salpeter (1954). The distribution of particles in the plasma reads:

$$n(r) = \bar{n}e^{-Ze\phi(r)/kT}, \quad \text{with} \quad \phi(r) = \frac{Ze}{r} + \psi(r), \quad (2)$$

where $\psi(r)$ is the induced mean field potential resulting from the polarization of the surrounding particles. This induced potential lowers the Coulomb barrier between the fusing particles and thus yields an enhanced rate in the plasma $R = E \times R_0$, where

$$E = \lim_{r \to 0} \left\{ g_{12}(r) exp \left(\frac{Z_1 Z_2 e^2}{rkT} \right) \right\} \quad (3)$$

is the enhancement (screening) factor and $g_{12}(r)$ is the pair-distribution function of particles in the plasma. Under BD conditions, ion screening must be considered as well as electron screening, i.e. $E = E_i \times E_e$. Both effects are of the same order ($E_i \sim E_e \sim$ a few) and must be included in the calculations for a correct estimate of the light element-depletion factor (see Section 4.6). Figure 2 portrays the evolution of the central temperature for objects respectively above, at the limit of, and below the hydrogen-burning minimum mass, with lines indicating the hydrogen, lithium, and deuterium burning temperatures in the plasma.

2.1.3 Transport Properties

Energy in the interior of LMS below $\sim 0.4\, M_{\odot}$, BDs, and GPs is transported essentially by convection (see e.g. Stevenson 1991). According to the mixing length theory (MLT), the convective flux reads (Cox

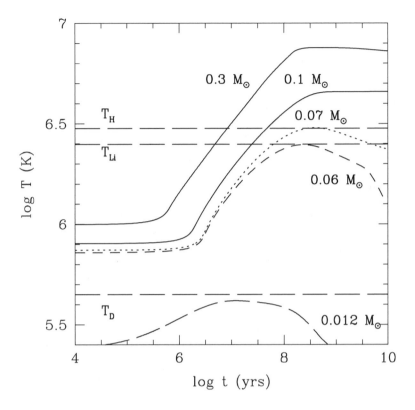

Figure 2 Central temperature as a function of age for different masses. T_H, T_{Li}, and T_D indicate the hydrogen, lithium, and deuterium burning temperatures, respectively.

& Giuli 1968):

$$F_{conv} \propto \rho v_{conv} C_P \, \delta T \propto \left(Q^{1/2}/\Gamma_1^{1/2}\right)\left(\frac{l}{H_P}\right)^2 \rho C_P c_S T (\nabla - \nabla_e)^{3/2}, \quad (4)$$

where $Q = -(\frac{\partial \ln \rho}{\partial \ln T})_P$ is the volume expansion coefficient, C_P is the specific heat at constant pressure, $c_S = (\Gamma_1 P/\rho)^{1/2}$ is the adiabatic speed of sound, δT is the temperature excess between the convective eddy and the surrounding ambient medium, ∇ is the temperature gradient, v_{conv} is the convective velocity, l and H_P denote the mixing length and the pressure scale height, respectively, and $\nabla_e \sim \nabla_{ad}$ is the eddy temperature gradient. The last term in parentheses on the righthand side of Equation 4 defines the fluid superadiabaticity, i.e. the fractional amount by which the real temperature gradient exceeds the adiabatic temperature gradient. Below a certain mass, the inner radiative core vanishes and the star becomes entirely convective (VandenBerg et al 1983, D'Antona & Mazzitelli 1985, Dorman et al 1989). This minimum mass, calculated with consistent non-grey atmosphere models and with the most recent OPAL radiative opacities (Iglesias & Rogers 1996) for the interior, is found to be $m_{conv} = 0.35\ M_\odot$ within the metallicity range $10^{-2} \leq Z/Z_\odot \leq 1$ (Chabrier & Baraffe 1997; Section 3.2). The Rayleigh number is defined as $Ra = \frac{gQH_P^3}{\xi\nu}(\nabla - \nabla_{ad})$, where $g \sim 10^3 – 10^5$ cm s^{-2} is the surface gravity, ξ is the thermal diffusivity (conductive or radiative diffusivity), and ν is the kinematic viscosity. In the interior of LMS, $Ra \sim 10^{25}$ so that convection is almost perfectly adiabatic and the MLT provides a fairly reasonable description of this transport mechanism. Variation of the MLT parameter $\alpha = l/H_P$ between 1 and 2 in the interior is found to be inconsequential below $\sim 0.6\ M_\odot$ (see Baraffe et al 1997).

However, convection can be affected or even inhibited by various mechanisms. Maximum rotation velocities for LMS and BDs are usually in the range $\sim 20–30$ km s^{-1} (see e.g. Delfosse et al 1998a) with values as large as 50 km s^{-1} and 80 km s^{-1} in the extreme case of Kelu 1 (Ruiz et al 1997, Basri et al 2000). This corresponds to an angular velocity $\Omega \sim 5 \times 10^{-4}$ rad s^{-1} for a characteristic radius $R \sim 0.1\ R_\odot$ ($\Omega_{Jup} = 1.76 \times 10^{-4}$ rad s^{-1}). Within most (>95% in mass) of the interior of LMS and BDs, v_{conv} is of the order of 10^2 cm s^{-1} ($\ll c_S$) and $H_P \sim 10^9$ cm. Thus, the Rossby number $Ro = v_{conv}/\Omega l \approx 10^{-3}$, where $l \sim H_P$, and convection in principle can be inhibited by rotation. However, the fact that lithium is not observed in objects with $m \gtrsim 0.06\ M_\odot$ (see Section 4.6) suggests that some macroscopic transport mechanism (e.g. meridional circulation, turbulence, or convection) remains efficient throughout the star. The Reynolds number $Re = v_{conv} l/\nu$ remains largely above unity in stellar and SSO interiors, so that convection is not inhibited by viscosity. Magnetic inhibition requires a magnetic velocity $v_A = (B^2/4\pi\rho)^{1/2} > v_{conv}$ (Stevenson 1979, 1991), i.e. $B \gtrsim 10^4$ Gauss under the conditions of interest, which is only slightly above the predicted fields in these objects (see Section 2.3). However, Stevenson (1979) proposed that in a rapidly rotating fluid with a significant magnetic field, the Proudman-Taylor theorem no longer applies, so that vertical convective motions could be possible. Last, convection efficiency can be diminished by the presence of a density gradient. As shown by Guillot (1995), this may occur in the interior of gaseous planets because of a gradient of heavy elements, thus leading to inefficient convective transport in the envelope of these objects.

The treatment of convection in the outer (molecular) layers, above and near the photosphere, is a more delicate question. The Rayleigh number in this region is $Ra \sim 10^{15}$, a rather modest value by the usual standards in turbulence. Since the MLT, by definition, applies to the asymptotic regime $Ra \to \infty$, it is no longer valid near the photosphere. Much work has been devoted to the improvement of this formalism and to the derivation of a non-local treatment of convection. One of the most detailed attempts came from Canuto and Mazzitelli (Canuto & Mazzitelli 1991; CM). The original CM formalism was based on a linear stability analysis, whereas energy transport by turbulence is a strongly non-linear process. It was improved recently by the inclusion to some extent of non-linear modes in the energy rate (Canuto et al 1996), but it yields results similar to the initial CM model. The formalism, however, still requires the calibration of a free parameter, which represents a characteristic "mixing" scale and in this sense resembles the MLT. A more severe shortcoming of the CM formalism is that the predicted outermost limit of the convection zone for the Sun and the amount of superadiabaticity in this zone disagree significantly, both quantitatively and qualitatively, with the results obtained from coupled hydrodynamic-radiation 3D simulations (Demarque et al 1999, Nordlund & Stein 1999). The 3D simulations yield excellent agreement with the observational data at several diagnostic levels: the thermal structure (and the inferred depth of the convection zone), the dynamical structure (vertical velocity amplitude and spectral line synthesis), and the p-mode frequencies and amplitudes for the Sun, without the use of free parameters (see e.g. Spruit & Nordlund 1990 for a review, also Nordlund & Stein 1999). This now provides a high degree of confidence in 3-D hydrodynamic models of stellar surface layers and the inferred transition from convective to radiative energy transport.

The thermal structure, for example, is a very robust property of the numerical models, since it depends relatively little on the level of turbulence (and thus on numerical resolution) (Stein & Nordlund 2000, Nordlund & Stein 1999). These results illustrate the fact that any formalism based on a homogeneous description of convection (e.g. MLT, CM) cannot accurately describe this strongly inhomogeneous process. Interestingly enough, the standard MLT is found to compare reasonably well with these simulations, at least for the thermal profile (see aforementioned references), and thus seems to offer a reasonable (or least worst!) overall description of convective transport, even in the small-efficiency convective regions. Indeed, in contrast to 1-D models computed with the CM formalism, the 3-D models do not have a steeper and more narrow superadiabatic structure for the Sun than was obtained with the classical Böhm-Vitense mixing length recipe (Nordlund & Stein 1999). It it thus fair to say that, in the absence of a correct non-local treatment of convection in LMS and BD interiors, the standard MLT is probably the most reasonable choice—at least for the present objects, LMS and SSOs. Clearly, the development of 3-D hydrodynamic models of the atmosphere of these objects and the calibration of the mixing length from these simulations, as is done for solar-type stars (Ludwig et al 1999), represents one of the next major challenges in the theory.

Another possible mechanism of energy transport in stellar interiors is conduction. Its efficiency can be estimated as follows: The distance l over which the temperature changes significantly is $l \sim (\chi t)^{1/2}$, where χ is the thermal diffusivity and t is the time during which the temperature change occurs. For $\chi \sim 10^{-1}$ cm^2 s^{-1}, characteristic of metals, and $t \sim 10^9$ yr, $l \sim 10^2$ km. Thus, heat can possibly be transported by conduction only over a limited range of the interior, providing that the density is high enough and the temperature low enough for electron conductivity to become important. Indeed, below the HBMM $\sim 0.07\ M_\odot$, the interior becomes degenerate enough during cooling that the conductive flux

$F_{\rm cond} = K_{\rm cond} \nabla T$, where $K_{\rm cond} = \frac{4ac}{3} \frac{T^3}{\rho \kappa_{\rm cond}}$ and $\kappa_{\rm cond}$ is the conductive opacity, becomes larger than the convective flux. Old enough BDs in the mass-range 0.02–$0.07\ M_\odot$ become degenerate enough to develop a conductive core, which slows down the cooling $L(t)$ (Chabrier et al 2000).

2.2 Atmosphere

Knowledge of the atmosphere is needed for two reasons: (1) as a boundary condition for the interior profile in the optically-thick region, and (2) as a description of the emergent radiative flux. A comprehensive review of the physics of the atmosphere of LMS and BDs can be found in Allard et al (1997). Only a general outline of the main characteristics of these atmospheres is mentioned in the present review.

2.2.1 Spectral Distribution
Surface gravity $g = Gm/R^2$ for LMS and SSOs range from $\log g \simeq 4.4$ for a main sequence $1.0\ M_\odot$ star to $\log g \simeq 3.4$ for Jupiter, with a maximum $\log g \simeq 5.5$ at the H-burning limit, for solar metallicity. This yields $P_{ph} \sim g/\bar\kappa \sim 0.1 - 10$ bar and $\rho_{ph} \sim 10^{-6}$–10^{-4} g cm^{-3} at the photosphere. Collision effects become significant under these conditions and induce molecular dipoles on e.g. H_2 or He-H_2, yielding so-called collision-induced absorption (CIA) between roto-vibrational states of molecules that otherwise would have only quadrupolar transitions (Linsky 1969, Borysow et al 1997). At first order this CIA coefficient scales as $\kappa_{CIA} \propto n_{H_2}^2 \times \kappa_{H_2 H_2}$ and thus becomes increasingly important as soon as H_2 molecules become stable. The CIA of H_2 suppresses the flux longward of 2 μm in the atmosphere of LMS, BDs, and GPs. This is one of the reasons (the main one for metal-depleted objects) for the redistribution of the emergent radiative flux toward shorter wavelengths in these objects (Lenzuni, Chernoff & Salpeter 1991, Saumon et al 1994, Allard & Hauschildt 1995, Baraffe et al 1997). The CIA of H_2 and H_2-He and the bound-free and free-free opacities of H$^-$ and H_2^- provide the main continuum opacity sources below ~ 5000 K for objects with metal-depleted abundances. Below $T_{\rm eff} \lesssim 4000$ K, most of the hydrogen is locked into H_2, and most of the carbon in CO. Excess oxygen is bound in molecules such as TiO, VO and H_2O, with some amounts also in OH and O (see e.g. Figure 1 of Fegley & Lodders 1996). Metal oxides and metal hydrides (FeH, CaH, MgH) are also present. The energy distribution of solar-abundance M-dwarfs is thus entirely governed by the line absorption of TiO and VO in the optical, and H_2O and CO in the infrared, with no space for a true continuum. The VO and TiO band strength indexes are used to classify M-dwarf spectral types (see e.g. Kirkpatrick et al 1991). The strongly frequency-dependent absorption coefficient resulting from line transitions of these molecules, along with Rayleigh scattering ($\propto \nu^4$) shortward of $\sim 0.4\ \mu$m, yield a strong departure from a black-body energy distribution (see e.g. Figure 5 of Allard et al 1997).

At $T_{\rm eff} \approx 2000$ K, near the H-burning limit, signatures of metal oxides and hydrides (TiO, VO, FeH, CaH bands) disappear from the spectral distribution (although some amount of TiO remains present in the atmosphere), as observed e.g. in GD 165B (Kirkpatrick et al 1999b). Alkalies are present under their atomic form. The disappearance of TiO-bands prompted astronomers to suggest a new spectral type classification, the "L"-type, for objects with T_{eff} below the afore-mentioned limit (Martín et al 1997, 1998, Kirkpatrick et al 1999a). Below a local temperature $T \approx 1300-1500$ K, P $\sim$ 3–10 bar for a solar abundance-distribution, carbon monoxide CO was predicted to dissociate and the dominant equilibrium form of carbon was predicted to become CH_4 (Allard & Hauschildt 1995, Tsuji et al 1995, Fegley & Lodders 1996). Note that this transition occurs gradually, with some of the two elements present in the stability field of the other (see e.g. Fegley & Lodders 1996, Lodders 1999), so that cool stars could contain some limited amount of methane, and CO may be visible in objects with $T_{eff} < 1800$ K as indeed detected in G1229B (Noll et al 1997; Oppenheimer et al 1998). This prediction has been confirmed by the spectroscopic observation of Gliese229B and the identification of methane absorption features at 1.7, 2.4, and 3.3 μm (Oppenheimer et al 1995). The presence of methane in its spectrum unambiguously confirmed its sub-stellar nature (Oppenheimer et al 1995, Allard et al 1996, Marley et al 1996). Objects like Gl229B, characterized by the strong signature of methane absorption in their spectra, define a new spectral class of objects, the "methane" (sometimes called "T") BDs. Methane absorption in the H and K bands yields a steep spectrum at shorter wavelengths and thus blue near infrared colors, with $J - K \lesssim 0$, but $I - J \gtrsim 5$ (Kirkpatrick et al 1999a, Allard 1999). Although the transition temperature between "L" and "methane" dwarfs is not precisely determined yet, it should lie in the range $1000 \lesssim T_{eff} \lesssim 1700$ K (see Section 4.5.2).

Below $T_{eff} \sim 2800$ K, complex O-rich compounds condense in the atmosphere, slightly increasing the carbon:oxygen abundance ratio (see e.g. Tsuji et al 1996, Fegley & Lodders 1996, Allard et al 1997). Different constituents will condense at a certain location in the atmosphere with an abundance determined by chemical equilibrium conditions (although non-equilibrium material may form, depending on the time-scale of the reactions, as mentioned below) between the gas phase and the condensed species. The formation of condensed species depletes the gas phase of a number of molecular species (e.g. VO, TiO, which will be sequestered into perovskite $CaTiO_3$), significantly modifying the emergent spectrum (see e.g. Fegley & Lodders 1994 for the condensation chemistry of refractory elements in Jupiter and Saturn). The equilibrium abundances can be determined from the Gibbs energies of formation either by minimization of the total Gibbs energy of the system (Sharp & Huebner 1990, Burrows & Sharp 1999), or by computing the equilibrium pressures of each grain species (Grossman 1972, Alexander & Ferguson 1994, Allard et al 1998). At each temperature, the fictitious pressure P_C of each condensed phase under consideration is calculated from the partial pressures of the species i which form the condensate (e.g. Al and O for corundum Al_2O_3) $P_i = {}^{\cdot}N_i \mathcal{R} T$, where N_i is the number of moles of species i and $\mathcal{R}$ is the gas constant, determined by the vapour phase equilibria. This fictitious pressure

P_C is compared to the equilibrium pressure P_{eq}, calculated from the Gibbs energy of formation of the condensate. The abundance of a condensed species is determined by the condition that this species be in equilibrium with the surrounding gas phase, $P_C \geq P_{eq}$ (Grossman 1972). The opacities of the grains are calculated from Mie theory. As suggested by Fegley & Lodders (1996) and Allard et al (1998), refractory elements Al, Ca, Ti, Fe, and V are removed from the gaseous atmosphere by grain condensation at about the corundum (Al_2O_3) or perovskite ($CaTiO_3$) condensation temperature $T \lesssim 1800$ K. Rock-forming elements (Mg, Si, Fe) condense as iron and forsterite (Mg_2SiO_4) or enstatite ($MgSiO_3$) within about the same temperature range (depending on P). Therefore the spectral features of all these elements will disappear gradually for objects with T_{eff} below these temperatures (see e.g. Fegley & Lodders 1994, 1996; Lodders 1999; Burrows & Sharp 1999 for detailed calculations). For jovian-type effective temperatures ($T_{eff} \sim 125$ K), H_2O and NH_3 condense near and below the photosphere, and water and ammonia bands disappear completely for $T_{eff} \lesssim 150$ K and 80 K, respectively (Guillot 1999).

The gas abundance strongly depends on pressure and temperature, so that the abundances of various species vary significantly with the mass (and T_{eff}) of the astrophysical body. As shown e.g. in Figure 2 of Lodders (1999), the condensation point of the dominant clouds lies much closer to the photosphere for M-dwarfs than for Gl229B or—worse—for Jupiter. In other words, the location of a given grain condensation lies deeper in Jupiter than in Gl229B. For late M-dwarfs and for massive and/or young BDs, the main cloud formation is predicted to occur very near the photosphere. This is consistent with the fact that all the DENIS and 2MASS objects discovered near the bottom of and below the MS exhibit strong thermal heating and very red colors (Delfosse et al 1998b, Kirkpatrick et al 1999a; see Section 4.5.2). Indeed, the atmospheric heating resulting from the large grain opacity (the so-called greenhouse or backwarming effect), along with the resulting enhanced H_2O dissociation, yield a redistribution of the IR flux, as proposed initially by Tsuji et al (1996).

A key question for the grain formation process is the size of the grains (see e.g. Alexander & Ferguson 1994). The suppression of the flux in the optical in Gl229B suggests grain sizes of ~ 0.1 μm as a source of continuum opacity (Jones & Tsuji 1997, Griffith et al 1998), although the wings of alkali resonance lines (KI, NaI) are also a source of absorption in this region (Tsuji et al 1999, Burrows et al 2000). Recent calculations assume a grain-size distribution in the submicron range (Allard 1999). Inclusion of the grain absorption with this size distribution in the atmosphere of objects near the limit of the MS successfully reproduces the observed colors of GD 165B, Kelu-1 and of DENIS objects (Leggett et al 1998, Allard 1999, Goldman et al 1999, Chabrier et al 2000). However, the spectrum of GL229B shows no indication for dust in its IR spectrum from 1 to 5 μm (Allard et al 1996, Marley et al 1996, Tsuji et al 1996, Oppenheimer et al 1998, Schultz et al 1998). This suggests ongoing dynamical processes such as grain settling in SSO atmospheres. Indeed, as noted by Chabrier et al (2000), although convection

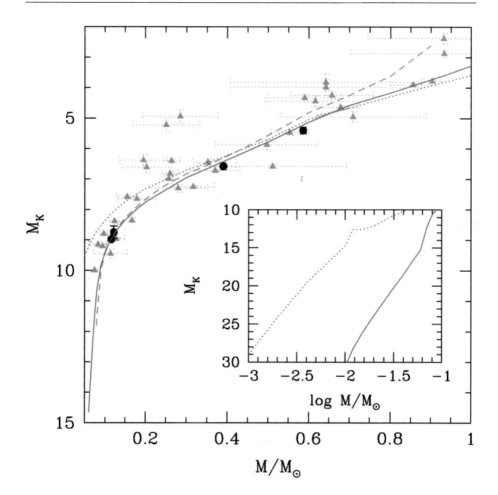

Figure 9 Mass-M_K relationship. Observationally determined masses are indicated by filled tri-angles (Henry & McCarthy 1993) and circles (Delfosse et al 1999c, Forveille et al 1999). Solid line: [M/H] = 0, $t = 5 \times 10^9$ yr; dotted line: [M/H] = 0, $t = 10^8$ yr; dash-line: [M/H] = -0.5, $t = 10^{10}$ yr. Note the log-scale in the inset.

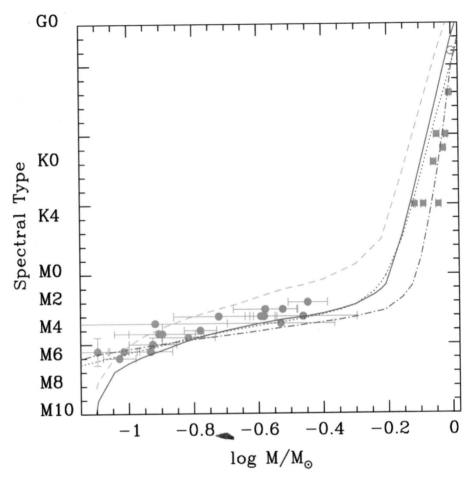

Figure 10 Mass-Spectral type relationships. Observations are from: Kirkpatrick & McCarthy (1994) for disk M-dwarfs (circles) and Clausen et al. (1999) for K- and G-dwarfs (squares). Note that the latter objects are likely to be young. The Sun is included at $Sp = G2$. The theoretical models are from Baraffe et al (1998). Solid line: $[M/H] = 0$, $t = 5 \times 10^9$ yr; dotted line: $[M/H] = 0$, $t = 10^8$ yr; dot-dash-line: $[M/H] = 0$, $t = 3 \times 10^7$ yr; dash-line: $[M/H] = -0.5$, $t = 5 \times 10^9$ yr.

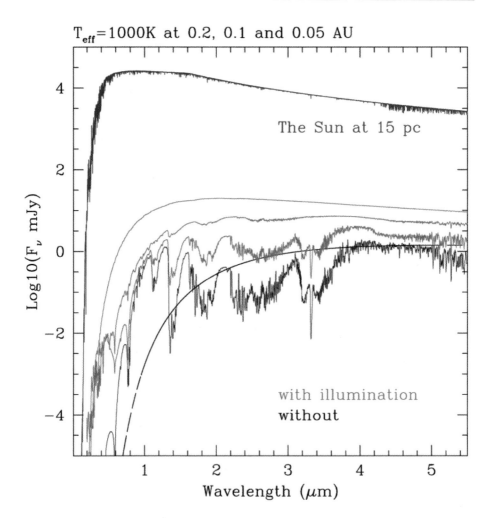

Figure 11 Spectrum of a young EGP (T_{eff} = 1000 K, defined as the effective temperature of the non-irradiated object) irradiated by a G2V primary at 0.2, 0.1, and 0.05 AU distance (bottom to top). The plot shows the incident spectrum (topmost curve), the spectrum emitted by the irradiated planet (magenta curves with more flux and shallower water and methane bands) and non-irradiated spectrum of the EGP (bottom full black curve) assuming a distance of 15 pc to Earth. The dashed black line is the blackbody spectrum for T_{eff} = 1000 K. The EGP model includes dust opacities. (Courtesy of F. Allard and P.H. Hauschildt).

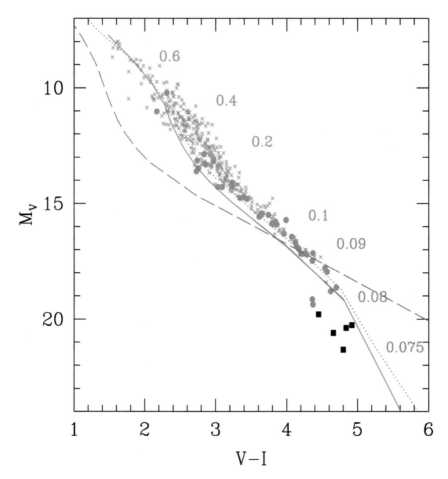

Figure 12 M_V vs (V-I) diagram for different ages and metallicites: DUSTY models (see text) for [M/H] = 0 for 10^8 yr (dotted line) and 5×10^9 yr (solid line); grainless models for [M/H] = –2 and t = 10 Gyr (dashed line). The crosses, circles and squares correspond to observations from Dahn et al (1995), Monet et al (1992) and Dahn et al (1999), respectively. The indicated masses (in $M_\odot$) correspond to the 5 Gyr [M/H] = 0 isochrone.

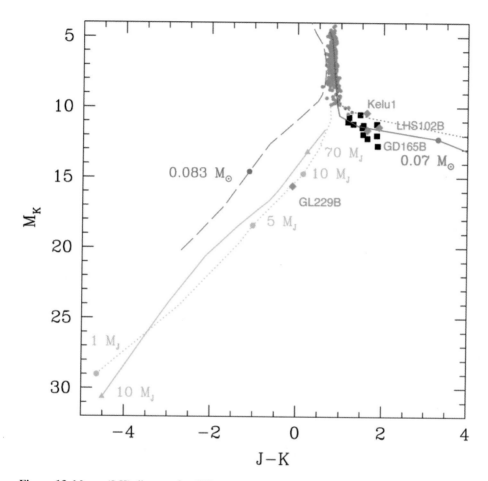

Figure 13 M_K vs (J-K) diagram for different ages and metallicites; [M/H] = 0 for 10^8 yr (dotted lines) and 5×10^9 yr (solid lines); [M/H] = −2, t = 10 Gyr (dashed line). The red curves on the right correspond to the DUSTY models for [M/H] = 0. The blue curves on the left correspond to the COND models for [M/H] = 0 (see text, §4.5.2). Filled circles and triangles on the isochrones indicate masses either in $M_\odot$ or M_J (1 M_J ≈ 10^{-3} $M_\odot$). Small green circles: MS stars from Leggett (1992) and Leggett et al (1996); black squares: Dahn et al (1999). Some identified BDs are also indicated (green diamonds).

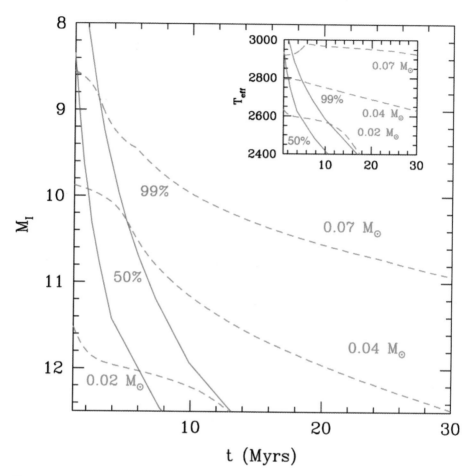

Figure 14 Deuterium-depletion curves as a function of age. The dashed lines correspond to different SSO masses, while the solid lines correspond to the 50% and 99% D-depletion limit, respectively. The inset displays the corresponding $T_{eff}(t)$.

occurs only in optically-thick regions ($\tau > 1$) for L-dwarfs, the top of the convection zone lies only about one pressure-scale height or less below the photosphere ($\tau \sim 1$), close to the region where most grains condense for objects near the bottom of the MS. This can have important consequences on the mixing and the settling of atmospheric grains. Indeed although the temperature at the top of the convective zone is found to be generally above the condensation temperature of all grains, rotation-induced advection or turbulent diffusion could efficiently bring material upward to the region of condensation and maintain small-grain layers, which otherwise will settle gravitationally. In general, grain formation process involves a balance between various dynamical timescales, such as the condensation, evaporation, coagulation, coalescence, and convection timescales (see e.g. Rossow 1978), not to mention large-scale hydrodynamical instabilities, which are typical of weather conditions on Earth. See for example Marley et al (1999) for an early attempt at cloud modelling for BD and EGP atmospheres. Moreover, although to an excellent first approximation the atmosphere of the solar jovian planets are in thermodynamical equilibrium, non-equilibrium species with chemical equilibrium timescales larger than the convection timescale can be dredged up to the photosphere by convection, as e.g. CO, PH_3, GeH_4 in Jupiter (Fegley & Lodders 1994). A correct understanding and a reliable description of this complex grain formation process represent a major challenge for theorists, and the observation of spectra of SSOs over a significant temperature range is necessary to provide guidance through this maze.

2.2.2 Transport Properties

As mentioned previously, below ~ 5000 K—i.e. $m \sim 0.6$–$0.8\,M_\odot$ depending on the metallicity—hydrogen atoms recombine, and n_{H_2} increases, as does the opacity through H_2 CIA. The radiative opacity κ increases by several orders of magnitude over a factor 2 in temperature, which in turn decreases the radiative transport efficiency ($\mathcal{F}_{rad} \propto \frac{\nabla}{\kappa}$). On the other hand, the presence of molecules increases the number of internal degrees of freedom (vibration, rotation) and thus the molar specific heat C_p, which in turn decreases the adiabatic gradient $(dT/dP)_{ad} = \frac{\mathcal{R}}{C_p}\frac{T}{P}$. Both effects strongly favor the onset of convection in the optically-thin ($\tau < 1$) atmospheric layers (Copeland et al 1970, Allard 1990, Saumon et al 1994, Baraffe et al 1995) with a maximum radial extension for the convection zone around $T_{eff} \sim 3000$ K. The combination of relatively large densities, opacities, and specific heat in layers where H_2 molecules are present contributes to a very efficient convective transport. Convection is found to be adiabatic almost up to the top of the convective region, and the extension of the superadiabatic layers is very small compared to solar-type stars (Brett 1995, Allard et al 1997). Thus a variation of the mixing length between H_P and $2H_P$ barely affects the thermal atmospheric profile (Baraffe et al 1997; Section 3). The presence of convection in the optically thin layers precludes radiative equilibrium in the atmosphere ($\mathcal{F}_{tot} \neq \mathcal{F}_{rad}$) and requires the resolution of the transfer equation for radiative-convective atmosphere models for objects below $\sim 0.7\,M_\odot$. This and the aforementioned strong frequency-dependence of the molecular absorption coefficients exclude grey model atmosphere and/or the use of radiative $T(\tau)$ relationships to determine the outer boundary condition to the interior structure. As demonstrated by Saumon et al (1994) and Baraffe et al (1995, 1997, 1998), such outer boundary conditions overestimate the effective temperature and the HBMM and yield erroneous m-T_{eff} and m-L relationships, two key relations for the calibration of the temperature scale and for the derivation of mass functions (see Chabrier & Baraffe 1997, Section 2.5 for a complete discussion). Correct evolutionary models for low-mass objects require (1) the connection between the non-grey atmospheric (P, T) profile, characterized by a given $(\log g, T_{eff})$ and the interior (P, T) profile at a given optical depth τ, preferably large enough for the atmospheric profile to be adiabatic; and (2) consistency between the atmospheric profiles and the synthetic spectra used to determine magnitudes and colors. For a fixed mass and composition, only one atmosphere profile matches the interior profile for the aforementioned boundary condition. This determines the complete stellar model (mass, radius, luminosity, effective temperature, colors) for this mass and composition (Chabrier & Baraffe 1997; Burrows et al 1997 for objects below 1300 K in their Figure 1).

2.3 Activity

The Einstein and Rosat surveys of the solar neighborhood and of several open clusters have allowed the identification of numerous M-dwarfs as faint X-ray sources, with typical luminosities of the order of the average solar luminosity $L_X \sim 10^{27}$ erg s^{-1}, reaching up to $\sim 10^{29}$ erg s^{-1} in young clusters (see e.g. Randich, 1999 and references therein for a recent review). This suggests that late M-dwarfs, as coronal emitters, are as efficient as other cool stars in terms of L_X/L_{bol}, which expresses the level of X-ray activity. Moreover, about 60% of M-dwarfs with spectral-type $>M5$ ($M_{bol} \simeq 12$) show significant chromospheric activity with $\log(L_{H\alpha}/L_{bol}) \simeq -3.9$ (Hawley et al 1996). These observations provide important empirical relationships between age, rotation, and activity. The L_X/L_{bol} ratio seems to saturate above a rotational velocity threshold at a limit $\log(L_X/L_{bol}) \simeq -3$, implying that the intrinsic coronal emission of LMS and BDs is quite low and decreases with L_{bol}, and thus mass, and does not increase with increasing rotation above the threshold limit. This suggests a saturation relation in terms of a Rossby number, i.e. the ratio of the rotational period over the convective turnover time $Ro = \frac{P}{t_{conv}}$ (see Section 2.1.3). Because t_{conv} increases for lower-mass stars, they will saturate at progressively larger periods, i.e. lower rotational velocities. The velocity threshold for a 0.4 $M_\odot$ star is estimated around 5–6 km s^{-1} (Stauffer et al 1997). Comparison of the Pleiades and the Hyades samples shows a steep decay of the X-ray activity in the Hyades. This result is at odds with a simple, monotonic rotation-activity relationship. The spin-down timescales for M-dwarfs are much longer than for solar-type stars. Hyades

M-dwarfs do indeed show moderate or even rapid rotation ($v \sin i \geq 10$ km s^{-1}) and thus should show strong X-ray emission. The fainter L_X in the Hyades than in the Pleiades thus reinforces the saturation scenario.

Delfosse et al (1998a) obtained projected rotational velocities and fluxes in the H$_\alpha$ and H$_\beta$ lines for a volume-limited sample of 118 field M-dwarfs with spectral-type M0–M6. They found a strong correlation between rotation and both spectral type (measured by $R - I$ colors) and kinematic population: All stars with measurable rotation are later than M3.5 and have kinematic properties typical of the young disk population, or intermediate between young disk and old disk. They interpret this correlation as evidence for a spin-down timescale that increases with decreasing mass, and this timescale is a significant fraction of the age of the young disk (~ 3 Gyr) at spectral type M4 ($\sim 0.15\, M_\odot$). These data confirm the saturation relation inferred previously for younger or more massive stars: L_X/L_{bol} and L_{H_α}/L_{bol} both correlate with $v \sin i$ for $v \sin i \lesssim 4$–5 km s^{-1} and then saturate at $10^{-2.5}$ and $10^{-3.5}$, respectively. Recently, Neuhäuser & Comeron (1998) detected X-ray activity in young BDs with $m \approx 0.04\, M_\odot$ in the Chamaleon star-forming region, with $\log L_X = 28.41$ and $\log L_X/L_{bol} = -3.44$. Coronal activity in BDs has been confirmed by Neuhäuser et al (1999), with all the objects belonging to young clusters or star-forming regions.

All these observations show that there is no drop in L_X/L_{bol} in field stars up to a spectral type of $M7$, which is well below the limit $m \sim 0.3\, M_\odot$ where stellar interiors are predicted to become fully convective. However, recent observations of the 2MASS objects (Kirkpatrick et al, 1999a: Figure 15a) seem to show a significant decrease of activity for L-type objects, which confirms the decline of activity suggested previously for very late M-spectral types ($\gtrsim$M8) (Hawley et al 1996, Tinney & Reid 1998). In any event, X-ray emission does not disappear in fully convective stars, and even at least young SSOs can support magnetic activity. A fossil field can survive only over a timescale $\tau_d \approx R^2/\eta \sim$ a few years for fully convective stars ($\eta \equiv \eta_t$ is the turbulent magnetic diffusivity and R the stellar radius), so that a dynamo process is necessary to generate the magnetic field. The data suggest that (1) the dynamo-generation believed to be at work in the Sun does not apply for very-low-mass objects, (2) dynamo generation in fully convective stars is as efficient as in stars with a radiative core, and (3) whatever the dynamo mechanism is, it is bounded by the saturation condition at least for late-type stars.

It is generally admitted that magnetic activity in the Sun results from the generation of a large-scale toroidal field by the action of differential rotation on a poloidal field at the interface between the convective envelope and the radiative core—where differential rotation is strongest—known as the tachocline (Spiegel & Weiss 1980, Spiegel & Zahn 1992). In this region, a sufficiently strong magnetic field is stable against buoyancy. In this case the shear dominates the helicity; this is the so-called $\alpha\Omega$ dynamo-generation, which predicts a correlation between activity and rotation, as observed in solar-type stars.

As shown initially by Parker (1975), buoyancy prevents the magnetic field generated by the turbulent motions in a convective zone to become global in character. For this reason, the absence of a radiative core—i.e. of a region of weak buoyancy and strong differential rotation—precludes in principle the generation of a large-scale magnetic field. A dynamo generated by a turbulent velocity field that would generate chaotic magnetic fields in the absence of rotation has been proposed as an alternative process for fully convective stars (Durney et al 1993). They found that a certain level of magnetic activity can be maintained without the generation of a large-scale field. The turbulent velocity field can generate a self-maintained *small-scale* magnetic field, providing that the magnetic Reynolds number $Re = (vl/\eta)^2$ is large enough. The scale of this field is comparable to that of turbulence and is in rough energy equipartition with the fluid motions, i.e. $\frac{1}{8\pi}\langle(B_P^2 + B_\phi^2)\rangle \approx \frac{1}{2}\langle\rho v_{conv}^2\rangle$. Rotation is not essential in this case, but it increases the generation rate of the field.

Recently, Küker and Chabrier (in preparation) explored an other possibility, namely the generation of a *large-scale* field by a pure α^2-effect. In the α^2-dynamo, helicity is generated by the action of the Coriolis force on the convective motions in a rotating, stratified fluid. The α^2-effect strongly depends on the Rossby number, or the equivalent Coriolis number $\Omega^\star = 2t_{conv}\Omega = 4\pi/Ro$. In low-mass objects, the convective turnover time is longer than the rotation period (see Section 2.1.3), so that $\Omega^\star \gg 1$. They find that the α^2-dynamo is clearly supercritical. It generates a large-scale, non-axisymmetric steady (co-rotating) field that is symmetric with respect to the equatorial plane. Equipartition of energy yields field strengths of several kiloGauss. The possible decrease of activity in very late-type stars, as observed for example with the small chromospheric activity of the M9.5 BD candidate BRI 0021-0214 in spite of its fast rotation ($v \sin i = 40\,\text{km s}^{-1}$; Tinney et al 1998), is a more delicate issue that requires the inclusion of dissipative processes. Indeed, these calculations show that α^2-dynamo can efficiently generate a large-scale magnetic field in the interior of fully convective stars. Since conductivity decreases in the outermost layers of the star, however, no current would be created by the field in these regions, and thus no dissipative process and no activity (see e.g. Meyer & Meyer-Hofmeister 1999).

The observed continuous transition in rotation and activity at the fully convective boundary suggests in fact that the α^2-dynamo is already at work in the convection zones of the more massive stars. An interesting possibility for verifying the present theory would be Doppler imaging of fast-rotating LMS and BDs. Turbulent dynamo is likely to yield a spatially uniform chromospheric activity, whereas the large-scale α^2 process suggested by Küker & Chabrier generates asymmetry. Moreover, non-axisymmetric fields can propagate in longitudinal directions without any cyclic variation of the total field energy, whereas dynamo waves generated by $\alpha\Omega$ processes propagate only along the lines of constant rotation rate (Parker 1955). Therefore, we do not expect cycles for uniformly rotating (fully convective) stars.

3.   MECHANICAL AND THERMAL PROPERTIES

3.1 Mechanical Properties

Figure 3 portrays the mass-radius behavior of LMS and isolated SSOs from the Sun to Jupiter for $t = 6 \times 10^7$ (*dot-dash line*) and 5×10^9 yr (*solid line*) for $Z = Z_\odot$ and $Z = 10^{-2} \times Z_\odot$ (*dashed-line*). The time required to reach the ZAMS, arbitrarily defined as the time when $L_{\mathrm{nuc}} = 95\%\ L_{\mathrm{tot}}$, for solar metallicity LMS is given in Table 1.

The general *m-R* behavior reflects the physical properties characteristic of the interior of these objects, as inferred from Figure 1. For $m \gtrsim 0.2\ M_\odot$, $\psi > 1$ for all ages, so that the internal pressure is dominated by the classical perfect

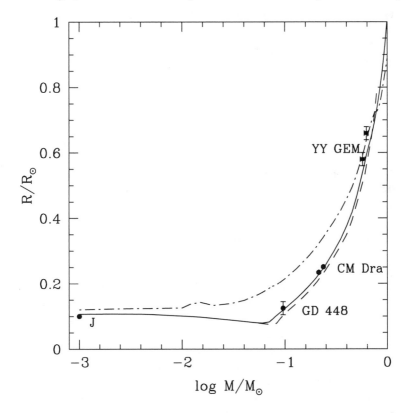

Figure 3   Mass-radius relationship for LMS and SSOs for two ages, $t = 6 \times 10^7$ yr (*dot-dash line*), 5×10^9 yr (*solid line*) for $Z = Z_\odot$, and $t = 5 \times 10^9$ yr for $Z = 10^{-2} \times Z_\odot$ (*dashed line*). The HBMM is 0.075 $M_\odot$ for $Z = Z_\odot$ and 0.083 $M_\odot$ for $Z = 10^{-2} \times Z_\odot$. Also indicated are the observationally-determined radii of various objects (see text) and the position of Jupiter radius (J). The bump on the 6×10^7 yr isochrone illustrates the initial D-burning phase.

TABLE 1   ZAMS age for solar metallicity LMS (after Baraffe et al 1998)

$m/M_\odot$	0.075	0.08	0.09	0.1	0.2	0.4	0.6	0.8	1
t_{ZAMS} (Gyr)	3	1.7	1	0.7	0.33	0.20	0.12	0.06	0.04

gas ion+electron contribution ($P = \rho kT/\mu m_H$, neglecting the molecular rotational/vibrational excited level contributions), plus the correcting Debye contribution arising from ion and electron interactions ($P_{DH} \propto -\rho^{3/2}/T^{1/2}$) and $R \propto m$ in first order from hydrostatic equilibrium. When the density becomes large enough during pre-MS contraction so that $\psi < 1$, which occurs at t $\gtrsim$ 50 Myr for $m = 0.15\,M_\odot$ and at t $\gtrsim$ 10 Myr for the HBMM $m = 0.075\,M_\odot$, the EOS starts to be dominated by the contribution of the degenerate electron gas ($P \propto \rho^{5/3}$). This yields a minimum in the m-R relationship $R_{min} \approx 0.08 R_\odot$ for $m \sim 0.06$–$0.07\,M_\odot$. Full degeneracy ($\psi \simeq 0$) would yield the well-known zero-temperature relationship $R \propto m^{-1/3}$, as in white dwarf interiors, but partial degeneracy and the non-negligible contribution arising from the (classical) ionic Coulomb pressure (which implies $R \propto m^{+1/3}$) combine to yield a smoother relation $R = R_0 m^{-1/8}$ at t = 5 Gyr, where $R_0 \simeq 0.06\,R_\odot$ for 0.01 $M_\odot \lesssim m \lesssim 0.07\,M_\odot$, i.e. an almost constant radius. For the age and metallicity of the solar system, this radius is of the order of the Jupiter radius $R_J \simeq 0.10 \times R_\odot$. At 5 Gyr, the radius reaches a maximum $R \simeq 0.11\,R_\odot$ for $m \simeq 4 \times M_J$ (where $M_J = 9.5 \times 10^{-4}\,M_\odot$ is the Jupiter mass) (Zapolsky & Salpeter 1969, Hubbard 1994, Saumon et al. 1996). Below this limit, degeneracy saturates (see Figure 1) and the classical ionic pressure contribution becomes important enough so that we recover a nearly classical behavior. Figure 3 also displays the astrophysically determined radii of the two eclipsing binary systems YY-Gem (Leung & Schneider 1978) and CM-Dra (Metcalfe et al 1996), and of the white dwarf companion GD448 (Maxted et al 1998). The bump on the 60 Myr isochrone near $m \sim 0.01 M_\odot$ results from initial deuterium burning (see Section 3.2). The m-R relationship is essentially determined by the EOS and is weakly sensitive to the outer boundary condition and the structure of the atmosphere, since the latter represents at most (for 1 $M_\odot$) a few percent of the total radius of LMSs and SSOs (Dorman et al 1989, Chabrier & Baraffe 1997). The effect of metallicity Z on the m–R relation remains modest. A decrease in metallicity yields a slight decrease in the radius at a given age or at the same stage of nuclear burning (i.e. the same H content). Indeed, lower metallicity yields a larger T_{eff} at a given mass (see Section 3.2). This in turns implies an increase in the nuclear energy production to reach thermal equilibrium, and thus a larger central temperature T_c. Since $R \propto \mu m/T$ from hydrostatic equilibrium, where μ is the mean molecular weight, determined essentially by hydrogen and helium, a lower-metallicity star contracts more to reach thermal equilibrium. As shown by Beuermann et al (1998), the variation of R with Z predicted by the models of Baraffe et al (1998) is in agreement with the radii deduced from observations.

3.2   Thermal Properties

Figure 4 exhibits the mass-effective temperature relationships for representative ages and metallicities. The arrows indicate the onset of formation of H_2 near the photosphere (Auman 1969, Copeland et al 1970, Kroupa et al 1990). As discussed in Section 2.2.2, molecular recombination favors convective instability in the atmosphere. Convection yields a smaller T-gradient ($\nabla \sim \nabla_{ad}$) and thus a cooler structure in the deep atmosphere (Brett 1995, Allard & Hauschildt 1995, Chabrier & Baraffe 1997; Figure 5). Therefore, a model with atmospheric convection corresponds to a larger T_{eff} since the (P, T) interior-atmosphere boundary is fixed for a given mass (Section 2.2.2). Furthermore, the adiabatic gradient in the regions of H_2 recombination decreases to a minimum value $\nabla_{ad} \sim 0.1$, compared to $\nabla_{ad} \sim 0.4$ for an ideal monoatomic gas (Copeland et al 1970, Saumon et al 1995: Figure 17). Therefore, even if the atmosphere were already convective without

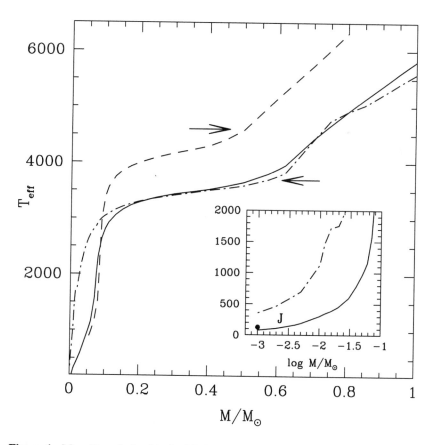

Figure 4   Mass-T_{eff} relationship for LMS and SSOs. Same legend as in Figure 3. The arrows indicate the onset of H_2 formation near the photosphere. Note the log scale in the inset.

H_2, molecular recombination yields a flatter inner temperature gradient in the atmosphere and thus enhances the former effect, i.e. a larger T_{eff} for a given mass. Formation of H_2 in the atmosphere occurs at higher T_{eff} for decreasing metallicity, because of the denser (more transparent) atmosphere (see Figure 4). The relation between the central and effective temperatures for LMS and SSOs can be inferred from Figures 1 and 4 and is commented on by Chabrier & Baraffe (1997; Section 4.2) and Baraffe et al. (1998; Section 2). As these authors demonstrate, a grey approximation or the use of a radiative $T(\tau)$ relationship significantly overestimates the effective temperature (from ~50 to 300 K depending on the atmospheric treatment).

A representative T_c-ρ_c diagram is shown in Figure 5 (see also Figure 8 of Burrows et al 1997) and illustrates the different evolutionary paths for LMS and SSOs in this diagram. The bumps appearing on the isochrones between 10^6 and 10^8 yr and log $T_c \sim 5.4$–5.8 result from initial deuterium burning. For

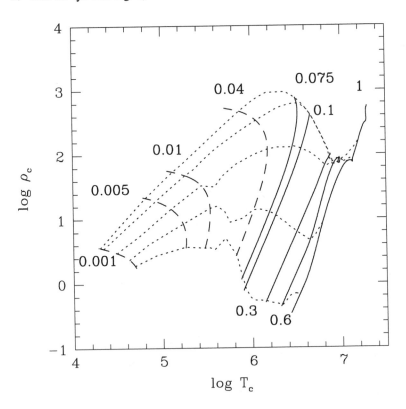

Figure 5 T_c-ρ_c relationship (in cgs) for LMS (*solid lines*) and SSOs (*dashed lines*) from 1 $M_\odot$ to 0.001 $M_\odot$ (masses in $M_\odot$ indicated on the curves). (*Dotted lines*) represent 10^6, 10^7, 10^8, 10^9, and 5 10^9 yr isochrones from bottom to top. The bumps on the 10^6–10^8 yr isochrones at log $T_c \sim 5.4$–5.8 correspond to the initial deuterium burning phase.

stars, T_c and ρ_c always increase with time until they reach the ZAMS. For BDs, T_c first increases for $\sim 10^7$–10^9 yr, for masses between ~ 0.01 and $\sim 0.07\,M_\odot$, respectively. Then, when degeneracy becomes dominant, T_c reaches a maximum and decreases. For objects with m $\lesssim 5 \times 10^{-3}\,M_\odot$, T_c always decreases for $t \gtrsim 1$ Myr. In the substellar regime there is no steady nuclear energy generation, by definition, and the evolution is governed by the change of internal energy $\int_M \frac{dE}{dt}\,dm$ and the release of contraction work $\int_M \frac{P}{\rho^2}\frac{d\rho}{dt}\,dm$. Even when degeneracy becomes important, SSOs keep contracting, though very slowly, within a Hubble time.

4.   EVOLUTION

4.1   Evolutionary Tracks

Figure 6 exhibits $T_{\mathrm{eff}}(t)$ obtained from consistent non-grey calculations for several masses, for $[M/H]^1 = 0$ (helium mass fraction $Y = 0.275$) and $[M/H] = -2.0$ ($Y = 0.25$), respectively. Initial deuterium burning (with an initial mass fraction $[D]_0 = 2 \times 10^{-5}$) for masses m $\gtrsim 0.013\,M_\odot$ proceeds quickly at the early stages of evolution and lasts about $\sim 10^6$–10^8 years. Objects below this limit are not hot enough to fuse deuterium in their core (see Figure 2). For masses above $0.07\,M_\odot$ for $[M/H] = 0$ and $0.08\,M_\odot$ for $[M/H] = -2.0$, the internal energy provided by nuclear burning quickly balances the gravitational contraction energy, and after a few Gyr the lowest-mass star reaches complete thermal equilibrium ($L = \int \epsilon_{\mathrm{nuc}}\,dm$, where ϵ_{nuc} is the nuclear energy rate) for both metallicities. The lowest mass for which thermal equilibrium is reached defines the HBMM and the related hydrogen-burning minimum luminosity HBML. Stars with $m \geq 0.4\,M_\odot$ develop a convective core near the ZAMS for a relatively short time (depending on the mass and metallicity), which results in the bumps for 0.6 and 1 $M_\odot$ at $\sim 10^8$ yr and $\sim 3 \times 10^7$ yr, respectively (Chabrier & Baraffe 1997; Section 3.2). The dotted lines portray $T_{\mathrm{eff}}(t)$ for objects with solar abundance when grain opacity is taken into account in the atmosphere (see Section 4.5.2). The Mie opacity resulting from the formation of refractory silicate grains produces a blanketing effect that lowers the effective temperature and luminosity at the edge of the main sequence, an effect first noticed by Lunine et al (1989). However, as a whole, grain formation only moderately affects the evolution near and below the bottom of the main sequence and thus the HBMM and HBML. Models with grainless atmospheres yield $m = 0.072\,M_\odot$, $L = 5 \times 10^{-5}\,L_\odot$, and $T_{\mathrm{eff}} = 1700$ K at the H-burning limit, whereas models with grain opacity give $m = 0.07\,M_\odot$, $L = 4 \times 10^{-5}\,L_\odot$ and $T_{\mathrm{eff}} = 1600$ K, for solar composition (Chabrier et al 2000). As shown in Figure 6, young and massive BDs can have the same effective temperature (or luminosity) as older, very-low-mass stars, a possible source of contamination for

---

[1]$[M/H] = \log\,(Z/Z_\odot)$.

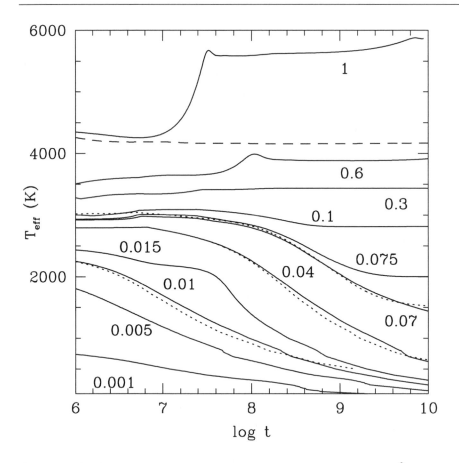

Figure 6   Effective temperature versus time (yr) for objects from 1 $M_\odot$ to 10^{-3} $M_\odot$ (masses indicated in $M_\odot$). (*Solid lines*): $Z = Z_\odot$, no dust opacity; (*dotted lines*): $Z = Z_\odot$, dust opacity included, shown for 0.01, 0.04, and 0.07 $M_\odot$; (*dashed line*): $Z = 10^{-2} \times Z_\odot$ (only for 0.3 $M_\odot$).

the determination of the local stellar luminosity function at the bottom of the MS (see Section 5).

Note the quick decrease of T_{eff} (and L) with time for objects below the HBMM, $L \propto t^\alpha$ with $\alpha \sim -5/4$ (Stevenson 1991; Burrows et al 1994, 1997), with a small dependence of α on the presence of grains. Slightly below ~ 0.072 $M_\odot$ (resp. 0.083 $M_\odot$) for $[M/H] = 0$ (resp. $[M/H] \leq -1$), nuclear ignition still takes place in the central part of the star, but cannot balance the ongoing gravitational contraction (see Figure 2). Although these objects are sometimes called "transition objects," we prefer to consider them as massive BDs, because strictly speaking they will never reach thermal equilibrium. Indeed, just below the HBMM, the contributions from the nuclear energy source $\int \epsilon_{\text{nuc}} \, dm$ and the entropy source $\int T \frac{dS}{dt} \, dm$ are comparable, and cooling proceeds at a much slower rate than mentioned above.

Below about 0.07 $M_\odot$ (resp. 0.08 $M_\odot$) for [M/H] = 0 (resp. [M/H] ≤ -0.5), the energetic contribution arising from hydrogen-burning, though still present for the most massive objects, is orders of magnitude smaller than the internal heat content, which provides essentially all the energy of the star ($\epsilon_{\mathrm{nuc}} \ll T|\frac{dS}{dt}|$). Once on the ZAMS the radius for stars is essentially constant, whereas for BDs the contraction slows down when $\psi \lesssim 0.1$ (see Figures 1 and 3).

The effects of metallicity on the atmosphere structure (Brett 1995, Allard and Hauschildt 1995) and on the evolution (Saumon et al 1994, Chabrier & Baraffe 1997) have been discussed extensively in the previously cited references and can be apprehended with simple arguments: the lower the metallicity Z, the lower the mean opacity $\bar{\kappa}$ and the more transparent the atmosphere, so that the same optical depth lies at deeper levels and thus higher pressure ($\frac{dP}{d\tau} = \frac{g}{\bar{\kappa}}$). Therefore, for a given mass (log g) the (T, P) interior profile matches, for a given optical depth τ, an atmosphere profile with larger T_{eff} (dashed-line on Figure 4), and thus higher luminosity L since the radius depends only weakly on the metallicity. The consequence is a larger HBMM for lower Z because a larger L requires more efficient nuclear burning to reach thermal equilibrium, and thus a larger mass.

A standard way to display evolutionary properties is a theoretical Hertzsprung-Russell diagram (HRD). Such an HRD for LMS and SSOs is shown in Figure 7 for several masses and isochrones from 1 Myr to 5 Gyr, whereas Figure 8 shows evolutionary tracks in a log g-T_{eff} diagram (see also Burrows et al 1997). These figures allow the determination of the mass and age of an object from the gravity and effective temperature inferred from its spectrum.

4.2   Mass-Magnitude

One of the ultimate goals of stellar theory is an accurate determination of the mass of an object for a given magnitude and/or color. Figure 9 (see color insert; see also Baraffe et al 1998) shows the comparison between theory and observationally-determined masses in the K-band. The solid line corresponds to a 5×10^9 yr isochrone, for which the lowest-mass stars have settled on the MS, for a solar metal-abundance ([M/H] = 0), whereas the dotted line corresponds to a 10^8 yr isochrone for the same metallicity, and the dashed line corresponds to a 10^{10} yr isochrone for [M/H] = -0.5, which is representative of the thick-disk population. A striking feature is the weak metallicity-dependence in the K-band, compared with the strong dependence in the V-band (see Figure 3 of Baraffe et al 1998). As we discuss in this chapter, this stems from two different effects. On one hand, the increasing opacity in the optical, dominated by TiO and VO lines, and the decreasing H_2 opacity in the K-band with increasing metallicity shift the peak of the flux toward larger wavelengths. Thus, for fixed T_{eff} the V-flux decreases and the K-flux increases with increasing [M/H]. On the other hand, for a given mass, the total flux (and T_{eff}) decreases with increasing metallicity, as we mentioned in Section 4.1. These two effects add up in the V-band and yield an important variation of the V-flux with metallicity. In the K-band, they cancel and yield similar fluxes

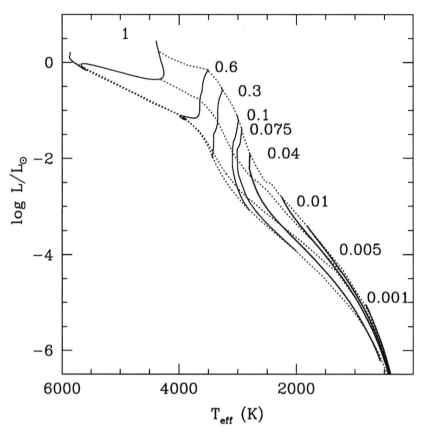

Figure 7   Theoretical H-R diagram for various masses (labeled in $M_\odot$). The weak dependence of radius on mass for SSOs yields the merging of the tracks for the lowest-mass objects. (*Dotted lines*) represent 10^6, 10^7, 10^8, and $5\ 10^9$ yr isochrones from right to left. The models are calculated with a mixing length $l = 1.9\ H_P$.

for a given mass below $\sim 0.4\ M_\odot$ ($T_{eff} \sim 3500$–4300 K, depending on [M/H]) at different metallicities. These arguments remain valid as long as H_2 CIA does not significantly depress the K-band, as it does for very metal-depleted objects.

4.3   Mass-Spectral Type

The knowledge of spectral type (Sp) is extremely useful for analyzing objects with unknown distance or with colors altered by reddening, as in young clusters, for example. The determination of a mass-Sp relationship provides a powerful complement to the mass-magnitude relationship for assigning a mass to an object. Based on spectroscopic observations of low-mass nearby composite systems, Kirkpatrick & McCarthy (1994) have determined an empirical m-Sp relationship for M-dwarfs, restricted to M2–M6 spectral types. A theoretical relation has been

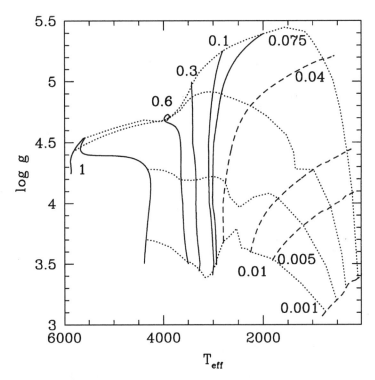

Figure 8   log g (in cgs) versus T_{eff} (in K) for LMS (*solid curves*) and SSOs (*dashed curves*) from 1 $M_\odot$ to 0.001 $M_\odot$ (masses in $M_\odot$ indicated on the curves). *Dotted lines* represent 10^6, 10^7, 10^8, and $5 \ 10^9$ yr isochrones from bottom to top.

derived by Baraffe & Chabrier (1996) from M0 to M10, emphasizing the non-linear behavior of this relationship near the bottom of the MS, for >M6. Figure 10 (see color insert) shows the m-Sp relationship for M-, K-, and G-dwarfs based on the Baraffe et al (1998) models. The spectral type for K- and G-dwarfs is derived from the model synthetic color $(I - K)$ through the empirical $Sp(I - K)$ relation recently derived by Beuermann et al (1998). As Baraffe & Chabrier (1996) stress, the m-Sp relationship depends crucially on age and metallicity. At ~1 Gyr, objects with Sp later than M10 are below the HBMM and can be considered as bona-fide BDs. This limit decreases to ~M7 at 100 Myr and ~M6 at 10 Myr. A difficulty arises when we're dealing with very young objects ($t \lesssim 10$ Myr) because of gravity effects. No reliable Sp-color relationship is presently available for the objects for which spectroscopic and photometric properties are found to be intermediate between giants and dwarfs (Luhman 1999). The present m-Sp relationships should be extended in the future to dwarfs cooler than M9.5–M10, the so-called L-dwarfs and methane-dwarfs. Note, however, that SSOs evolve at different rates through a series of spectral types as they cool, so we cannot associate a given spectral type with a specific mass for these objects.

4.4 Irradiated Planets

The mass-radius relation and evolutionary sequences described above reflect the relations for isolated objects. After a rapid initial accretion phase and subsequent hydrodynamical collapse, planets orbiting stars will evolve differently. As shown by Hubbard (1977) illumination from a parent star will yield thermal expansion of the less massive (low gravity) objects, toward an asymptotic temperature T_{eq} set only by the thermalized photons from the parent star and toward a (larger) asymptotic equilibrium radius R_p, with $R_p = 2a(L_{eq}/L_\star)^{1/2}[1/(1-A)]^{1/2}$ and $T_{eq} = (1-A)^{1/4}(R_\star/2a)^{1/2}T_{eff_\star}$, where a is the orbital distance, A is the Bond albedo, $L_\star = 4\pi\sigma R_\star^2 T_{eff_\star}$ is the parent star luminosity, effective temperature, and radius, and $L_{eq} = 4\pi\sigma R_p^2 T_{eq}^4$ denotes the equilibrium luminosity (see Saumon et al 1994). As noted in Guillot et al (1996) and Guillot (1999), EGPs in close orbits are heated substantially by their parent star, and their atmosphere cannot cool substantially. They develop an inner radiative region as a result of this stellar heating and contract at almost constant T_{eff}, at a much smaller rate than if they were not heated by the star. Planets whose luminosity is larger than the absorbed stellar flux of the parent star just evolve along a fully convective Hayashi track for ~10^6 yr before reaching the aforementioned equilibrium state.

The flux from an irradiated planet includes two contributions: the intrinsic thermal emission and the reflected starlight (the Albedo contribution) (see e.g. Seager & Sasselov 1998, Marley et al 1999):

$$F_\nu = \left(\frac{R_p}{d}\right)^2 \mathcal{F}_\nu^p + \left(\frac{A}{4}\right)P(\phi)\left(\frac{R_\star}{d}\right)^2\left(\frac{R_p}{a}\right)^2 \mathcal{F}_\nu^\star, \qquad (5)$$

where d is the distance of the system to Earth, and $P(\phi)$ is the dependence of the reflected light upon the phase angle between the star, the planet, and Earth ($P = 1$ if the light reflected by the planet is redistributed uniformly over 4π steradians). Figure 11 (see color insert) illustrates the flux from a young EGP (assuming the irradiation is isotropic) with $T_{eff} = 1000$ K orbiting a Sun-like star at 15 pc for various orbital distances.

4.5 Color-Magnitude Diagrams

Thanks to the recent progress both on the observational side, with the development of several ground-based and space-based observational surveys of unprecedented sensitivity in various optical and infrared passbands, and on the theoretical side, with the derivation of synthetic spectra and consistent evolutionary calculations based on non-grey atmosphere models, it is now possible to confront theory with observation directly in the observational planes, i.e. in color-color and color-magnitude diagrams (CMD). This avoids dubious color-T_{eff} or color-M_{bol} transformations and allows more accurate determinations of the intrinsic properties (m, L, T_{eff}, R) of an object from its observed magnitude and/or color.

In this section we examine the behavior of LMS and SSOs in various CMDs characteristic of different populations in terms of age and metallicity. These diagrams capture the essence of the observational signatures of the very mechanical and thermal properties of these objects, described in the previous sections.

4.5.1 Pre-Main Sequence and Young Clusters

4.5.1 Pre-Main Sequence and Young Clusters   Numerous surveys devoted to the detection of SSOs have been conducted in young clusters with ages spanning from $\sim$1–10 Myr to $\gtrsim 10^8$ yr for the Pleiades or the Hyades (see Martín 1999 and Basri, this volume, for reviews). Observations of young clusters present two important advantages, namely (1) all objects in the cluster are likely to be coeval within a reasonable range, except possibly in star-forming regions, where the spread in ages for cluster members can be comparable with the age of the cluster; and (2) young objects are brighter for a given mass (see Figure 6), which makes the detection of very-low mass objects easier. Conversely, they present four major difficulties: (1) extinction caused by the surrounding dust modifies both the intrinsic magnitude and the colors of the object; (2) accurate proper motion measurements are necessary to assess whether the object belongs to the cluster; (3) gravity affects both the spectrum and the evolution; and (4) the evolution and spectrum of very young objects ($t \lesssim 1$ Myr) may still be affected by the presence of an accretion disk or circumstellar material residual from the protostellar stage. Young multiple systems remove some of these difficulties and provide excellent tests for models at young ages—for example, the quadruple system GG TAU (White et al 1999), with components covering the whole mass-range of LMS and BDs from 1 $M_\odot$ to $\sim$0.02 $M_\odot$. Models based on non-grey atmospheres (Baraffe et al 1998) are the only ones consistent with these observations (White et al 1999, Luhman 1999). The comparison of $\sim$10^6 yr isochrones for SSOs with observations in a near-IR CMD is shown in Zapatero Osorio et al (1999) for the young cluster σ-Orionis (their Figure 1).

4.5.2 Disk Field Stars

4.5.2 Disk Field Stars   Figure 12 and 13 (see color insert) display the CMDs for LMS and SSOs in the optical and the infrared for various ages and metallicities (see also Burrows et al 1997). Various sets of models correspond to calculations where (1) grain formation is included in the atmosphere EOS but not in the transfer equations, mimicking a rapid settling below the photosphere (COND models); (2) grain formation is included in both the EOS and the opacity (DUSTY models) (see Allard et al 1999); and (3) grainless models. Models (1) and (2) represent extreme cases that bracket the complex grain processes at play in these atmospheres (see Section 2.2.1). The optical sequence shows a monotonic magnitude-color behavior with three changes of slope around $M_V \sim 10$, 13, and 19 for $Z = Z_\odot$, respectively, which correspond to 0.5 $M_\odot$, $T_{eff} = 3600$ K; 0.2 $M_\odot$, $T_{eff} = 3300$ K, and 0.08 $M_\odot$, $T_{eff} = 2330$ K on the 5 Gyr isochrone. They reflect, respectively, the onset of convection in the atmosphere, degeneracy in the core, and grain absorption near the photosphere, as described in the previous sections. The latter feature, which yields a steep increase of M_V at $(V - I) \sim 5$, does not appear in dust-free models (cf Baraffe et al 1998), where TiO is not depleted by grain formation in the atmosphere and thus absorbs the flux in the optical. It is interesting that this steepening is observed in the sample collected recently by Dahn et al (1999), which includes some DENIS and 2MASS L-dwarf parallaxes. This effect is also observed in the $(R - I)$ CMD of the Pleiades (Bouvier et al 1998), in agreement with the theoretical predictions (Chabrier et al 2000). This clearly illustrates the effect of grain formation in the atmosphere, which affects the spectra of LMS and SSOs below $T_{eff} \lesssim 2400$ K. In the near-infrared color $(J-K)$ (Figure 13, see color insert), grain condensation yields very red colors for the DUSTY models, whereas COND and grainless models loop back to short wavelengths (blueward) below $m \sim 0.11\, M_\odot$, $M_K \sim 10$ for [M/H] $= -2$, and $m \sim 0.072\, M_\odot$ (resp. $\sim$0.04 $M_\odot$) at 5 Gyr (resp. 100 Myr) for solar metallicity, i.e. $M_K \sim 12$. As described in Section 2.2, this stems from (1) the onset of H_2 CIA for low-metallicity objects and (2) the formation of CH_4 at the expense of CO when the peak of the Planck function moves in the wavelength range characteristic of the absorption bands of this species, for solar-like abundances.

In terms of colors, there is a competing effect between grain and molecular opacity sources for objects at the bottom and just below the MS. The backwarming effect resulting from the large grain opacity destroys molecules such as e.g H_2O, one of the main sources of absorption in the near-IR (cf Section 2.2.1), and yields the severe reddening near the bottom of the MS ($T_{eff} \lesssim 2300$–2400 K), as illustrated by the DENIS (Tinney et al 1998) and 2MASS (Kirkpatrick et al 1999a) objects and by GD165B (Kirkpatrick et al 1999b). As the temperature decreases, grains condense and settle at deeper layers below the photosphere, while methane absorption in the IR increases (see Section 2.2) so that the peak of the flux will move back to shorter wavelengths and the color sequence will go from the DUSTY one to the COND one. The exact temperature at which this occurs is still uncertain because it involves the complex grain thermochemistry and dynamics outlined in Section 2.2, but it should lie between $T_{eff} \sim 1000$ K, for Gl229B, and $\sim$1700 K, corresponding to J-$K \sim 2$ for the presently reddest observed L-dwarf with known parallax LHS102B (Goldman et al 1999, Basri et al 2000).

In spite of the aforementioned strong absorption in the IR, BDs around 1500 K radiate nearly 90% (99% with dust) of their energy at wavelengths longward of 1 μm, and infrared colors are still preferred to optical colors (at least for solar metal abundance), with J, Z, H as the favored bands for detection, and $M_M \sim M_{L'} \sim$ 10–11, $M_K \sim M_J \sim 11$–12 at the H-burning limit, at 1 Gyr (see Chabrier et al 2000).

4.5.3 Halo Stars. Globular Clusters

Observations of LMS belonging to the stellar halo population of our Galaxy (also called "spheroid" to differentiate it from the $\rho(r) \propto 1/r^2$ dark halo) or to globular clusters down to the bottom of the MS have been rendered possible with the HST optical (WFPC2) and IR (NICMOS) cameras and with parallax surveys at very faint magnitudes (Leggett

1992, Monet et al 1992, Dahn et al 1995). This provides stringent constraints for our understanding of LMS structure and evolution for metal-depleted objects. The reliability of the LMS theory outlined in Section 2 has been assessed by the successful confrontation to various observed sequences of globular clusters ranging from $[M/H] \simeq -2.0$ to -1.0^2 (Baraffe et al 1997). Figure 8 in Baraffe et al (1997) portrays several tracks for different metallicities, corresponding to the MS of various globular clusters. The tracks are superposed with the subdwarf sequence of Monet et al (1992), which is identified as stellar halo objects from their kinematic properties. As shown in this figure, the pronounced variations of the slope in the CMD around $\sim 0.5\, M_\odot$ and $\sim 0.2\, M_\odot$, which stem from the very physical properties of the stars—namely the onset of molecular absorption in the atmosphere and of degeneracy in the core (see Section 2)—are well reproduced by the theory at the correct magnitudes and colors. The predicted blue loop in IR colors caused by the ongoing CIA absorption of H_2 for the lowest-mass (coolest) objects (see Section 2.2.1 and Figure 13 for $[M/H] = -2$), has also been confirmed by observations with the NICMOS camera (Pulone et al 1998).

4.6   The Lithium and Deuterium Tests

One ironclad certification of a BD is a demonstration that hydrogen fusion has not occurred in its core. As Figure 2 illustrates, lithium burning through the $Li^7(p, \alpha)\, He^4$ reaction occurs at a lower temperature than is required for hydrogen fusion. The timescale for the destruction of lithium in the lowest-mass stars is about 100 Myr. Moreover, the evolutionary timescale in LMS/BDs is many orders of magnitude larger than the convective timescale (see e.g. Bildsten et al 1997), so core abundances and atmospheric abundances can be assumed identical. Therefore, the retention of lithium in a fully mixed object older than 10^8 yr signifies the lack of hydrogen burning. This provides the basis of the so-called "lithium-test" first proposed by Rebolo et al (1992) to identify bona-fide BDs. Evolutionary models based on non-grey atmospheres and the screening factors described in Section 2.1.2 yield a lithium-burning minimum mass $m_{Li} \simeq 0.06\, M_\odot$ (Chabrier et al 1996). Note from Figure 2 the strong age dependence of the lithium test: Young stars at $t \lesssim 10^8$ yr (depending on the mass) will exhibit lithium (which in passing precludes the application of the lithium test for the identification of SSOs in star forming regions), whereas massive BDs within the mass range $[0.06$–$0.07\, M_\odot]$ older than $\sim 10^8$ yr will have burned lithium.

By observing a young cluster older than $\sim 10^8$ yr, one can look for the boundary-luminosity below which lithium has not yet been depleted. Objects fainter than

---

[2]Globular clusters with an observed $[Fe/H]$ must be compared with theoretical models with the corresponding metallicity $[M/H] = [Fe/H] + [O/Fe]$ in order to take into account the enrichment of α-elements (see Baraffe et al 1997).

this limit will definitely be in the substellar domain. Conversely, the determination of the lithium depletion edge yields the age of the cluster, as was first proposed by Basri et al (1996) (see also e.g. Stauffer et al 1998). Indeed, the lithium test provides a "nuclear" age that may be even more powerful a diagnostic than the conventional nuclear age from the upper main sequence turn-off, since the evolution of the present fully convective low-mass objects does not depend on ill-constrained parameters such as mixing length or overshooting. As noted by Stauffer et al (1998), the age-scale for open clusters based on the (more reliable) lithium depletion boundary has important implications for stellar evolution: The ages of several clusters (Pleiades, α-Per, IC 2391) are all consistent with a small but non-zero amount of overshooting to be included in the evolutionary models at the turn-off mass in order to yield similar ages. See Basri (this issue) for a more detailed discussion of the lithium test. The deuterium-test can be used in a similar manner, extending the lithium test to smaller masses and younger ages, typically $t \lesssim 10^7$ yr (Béjar et al 1999). The D-burning minimum mass is predicted as $m_D \simeq 0.013\ M_\odot$ (see Figure 2) and has been proposed (Shu et al 1987) as playing a key role in the formation of isolated star-like objects, in contrast to objects formed in a protoplanetary disk. Deuterium-depletion is illustrated in Figure 14 (see color insert). This figure displays the evolution of SSOs above m_D in the I-band. The left and right diagonal solid lines correspond to 50% D-depletion ($[D]/[D]_0 = 1/2$) and 99% D-depletion, respectively ($[D]_0 = 2 \times 10^{-5}$ is the initial mass fraction). The inset displays the corresponding curves in T_{eff}. Spectroscopic signatures of deuterium include absorption lines of deuterated water HDO between 1.2 and 2.1 μm (Toth 1997). The observation of deuterated methane CH_3D cannot be used for the deuterium test as an age-indicator; at the temperature methane forms ($\lesssim$1800 K), SSOs above the D-burning minimum mass are old enough for all deuterium to have been burned (see Figure 6).

4.7   Low-Mass Objects in Cataclysmic Variables

The study of cataclysmic variables (CVs) is closely related to the theory of LMS and SSOs regarding their inner structure and the magnetic field generation. CVs are semi-detached binaries with a white dwarf (WD) primary and a low-mass stellar or substellar companion ($m \lesssim 1\ M_\odot$) that transfers mass to the WD through Roche-lobe overflow (see e.g. King 1988 for a review). In the current standard model, mass transfer is driven by angular momentum loss caused by magnetic wind braking for predominantly radiative stars, and by gravitational wave emission for fully convective objects. The Roche lobe filling the secondary's mean density $\bar{\rho}$ determines almost entirely the orbital period P, with $P_h = k/\bar{\rho}^{1/2}$ (where $k \simeq 8.85$ is a weak function of the mass ratio, P_h is the orbital period in hr, and $\bar{\rho} = (m_2/R_2^3)/(M_\odot/R_\odot^3)$ is the mean density of the secondary in solar units) (see King 1988). For objects obeying classical mechanics, i.e. in the stellar regime, $m \propto R$ (see Section 3.1 and Figure 3) so that the mean density increases with decreasing mass ($\bar{\rho} \propto 1/m^2$) and the orbital period decreases along evolution as the secondary loses mass $P_h \propto 1/\bar{\rho}^{1/2} \propto m_2$. The situation reverses near and below the HBMM when electron degeneracy dominates. In that case, the radius increases very slightly with decreasing mass (Figure 3), so that the mean density is essentially proportional to the mass ($\bar{\rho} \propto m$) and the orbital period increases along evolution $P_h \propto \sim m_2^{-1/2}$. Then the secular evolution of the orbital period reverses when the donor becomes a BD. The analysis of CV orbital evolution thus provides important constraints on LMS and BD internal structure.

The major puzzle of CVs is the observed distribution of orbital periods, namely the dearth of systems in the 2–3 h period range and the minimum period P_{min} at 80 min (see King 1988). The most popular explanation for the period gap is the disrupted magnetic braking scenario (Rappaport et al 1983, Spruit & Ritter 1983), and most evolutionary models including this process do reproduce the observed period distribution. However, the recent progress realized in the field of LMS and SSOs now allows the confrontation of the theoretical predictions with the observed atmospheric properties (colors, spectral types) of the secondaries (Beuermann et al 1998, Clemens et al 1998, Kolb & Baraffe 1999a). The aforementioned standard period-gap model may be in conflict with the observed spectral type of some CV secondaries (Beuermann et al 1998, Kolb & Baraffe 1999b). Although alternative explanations exist for the period gap, none of them has been proven as successful as the disrupted magnetic braking scenario. The most recent alternative suggestion, based on a characteristic feature of the mass-radius relationship of LMS (Clemens et al 1998), has been shown to fail to reproduce the observed period distribution around the period gap (Kolb et al 1998). On the other hand, as we discussed in Section 2.3, no change in the level of activity is observed in isolated M-dwarfs along the transition between partially and fully convective structures at $m \sim 0.35\,M_\odot$ and $Sp \sim$ M2–M4, which indicates that a magnetic field is still generated in the mass-range corresponding to the CV gap. In this case, the abrupt decrease of angular momentum losses by magnetic braking at the upper edge of the period gap could result from a rearrangement of the magnetic field when stars become fully convective, without necessarily implying a sudden decline in the magnetic activity, as suggested by Taam & Spruit (1989; see also Spruit 1994).

Finally, an ~10% discrepancy still remains between the theoretical and the observed minimum period $P_{min} = 80$ min, even when we include the most recent improvements in stellar physics (see Kolb & Baraffe 1999a for details). Residual shortcomings in the theory, either in the EOS or in the atmosphere, cannot be ruled out as the cause of this discrepancy. Alternatively, an additional driving mechanism to gravitational radiation in fully convective objects can reconcile predicted and observed P_{min} (Kolb & Baraffe 1999a). Because magnetic activity is still observed in fully convective, late spectral type, M-dwarfs, magnetic braking could operate in CV secondaries even down to P_{min}, but with a weaker efficiency than above the period gap. These open questions certainly require a better understanding of magnetic field generation and dissipation in LMS and BDs, a point already stressed in Section 2.3.

5.   GALACTIC IMPLICATIONS

5.1   Stellar Luminosity Function and Mass Function

It is now well established that visible stars are not numerous enough to account for the dynamics of our Galaxy—the so-called galactic dark-matter problem. A precise determination of their density requires the correct knowledge of the luminosity function (LF) down to the H-burning limit, and a correct transformation into a mass-function (MF). The latter issue has been improved significantly with models that describe more accurately the color-color and color-magnitude diagrams of LMS and BDs for various metallicities, and most importantly, that provide mass-magnitude relationships in good agreement with the observations (see Section 4). The former issue, however, is not completely settled at present, and significant differences still exist among various determinations of the nearby LF and between the nearby LF and the photometric LF determined from the ground and with the HST, as shown later in this section.

The main problem with the nearby sample is the limited number of stars at faint magnitudes. Although the sample is complete to 20 pc for stars with $M_V < 9.5$, it is severely incomplete beyond 5 pc for $M_V > 12$ (Henry et al 1997). Kroupa (unpublished) derived a nearby LF Φ_{near} by combining Hipparcos parallax data (which is essentially complete for $M_V < 12$ at r = 10 pc) and the sample of nearby stars (Dahn et al 1986) with ground-based parallaxes for $V > 12$ to a completeness distance r = 5.2 pc. The nearby LF determined by Reid & Gizis (1997) is based on a volume sample within 8 pc. Most of the stars in this survey have parallaxes. For all the late K and M dwarfs, however, trigonometric parallaxes are not available, and these authors use a spectroscopic TiO-index vs M_V^{TiO} relation to estimate the distance (Reid et al 1995). This sample was revised recently with Hipparcos measurements and new binary detections in the solar neighborhood (Delfosse et al 1999a) and leads to a revised northern 8-pc catalogue and nearby LF (Reid et al 1999). These authors argue that their sample should be essentially complete for $M_V < 14$. However, the analysis of completeness limits by Henry et al (1997, their Figure 1) shows that the known stellar census becomes substantially incomplete for distances larger than 5 pc. About 35% of the systems in the Reid et al sample are multiple, and ~45% of all stars have a companion in binary or multiple systems.

The photometric LFs, Φ_{phot}, based on large observed volumes via deep pencil-beam surveys, avoid the limitation because of small statistics but introduce other problems such as Malmquist bias and unresolved binaries (see e.g. Kroupa et al 1993, Kroupa 1995), yielding only the determination of the stellar *system* LF. For the HST LF (Gould et al 1997), with $I \lesssim 24$, the Malmquist bias is negligible

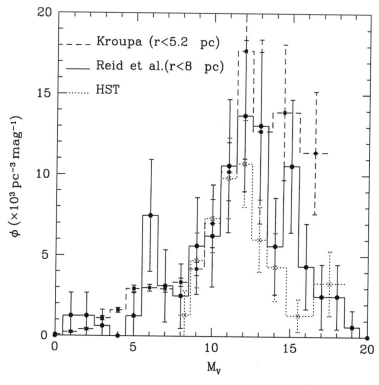

Figure 15   Luminosity function in the V-band from the 5-pc sample (*dashed-line*) (Kroupa, unpublished), the 8-pc sample (*solid-line*) (Reid et al 1999) and from the HST (*dotted-line*) (Gould et al 1998).

because all stars down to $\sim 0.1\,M_\odot$ are seen through to the edge of the thick disk. A major caveat of photometric LFs, however, is that the determination of the distance relies on a photometric determination from a color-magnitude diagram. In principle this requires the determination of the metallicity of the stars, since colors depend on metallicity (see Figure 12, see color insert). The illustration of the disagreement among the three aforementioned LFs, Φ_{5pc}, Φ_{8pc}, and Φ_{HST}, at faint magnitude is apparent in Figure 15. The disagreement between the two nearby LFs, in particular, presently has no robust explanation. It might come from incompleteness of the 8 pc sample at faint magnitudes. Note also that in the spectroscopic relation used to estimate the distance, M_V can be uncertain by ~ 1 mag (see Figure 3 of Reid et al 1995), certainly a source of Malmquist bias, even though the number of stars without parallax in this sample is small ($\sim 10\%$) and a Malmquist bias on this part should not drastically affect the results. On the other hand, as noted in Section 4.1, the end of the 5-pc sample might be contaminated by a statistically significant number of young BDs, or by stars with slightly under-solar abundances.

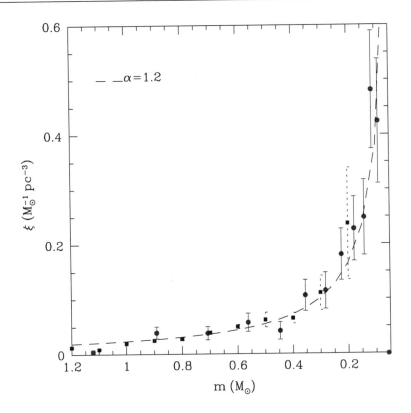

Figure 16   Mass function derived from the 5-pc LF (*squares*, last bin out of the figure) and the 8-pc LF (*circles*). *Dashed-line*: $dn/dm = 0.024\,m^{-1.2}$ (see text).

Figure 16 compares the MFs obtained from the two aforementioned nearby LFs, using the mass-magnitude relations described in Section 4.2. In practice, the observed sample constitutes a mixture of ages and metallicities. However, the Hipparcos CMD indicates that $\sim$90% of disk stars have $-0.3 < [M/H] < 0.1$ (Reid 1999), so the spread of metallicity is unlikely to significantly affect the derivation of the MF through the mass-magnitude relation. Moreover, the choice of m-M_K minimizes metallicity effects; see Figure 9, also Figure 3 of Baraffe et al 1998. The MS lifetimes for stars with m $\lesssim$ 1 $M_\odot$ are longer than a Hubble time, and age variations in the mass-magnitude relation will affect only objects $\lesssim$0.15 $M_\odot$ younger than $\sim$0.5 Gyr (see Figure 9).

Superposed is a power-law MF normalized at 0.8 $M_\odot$ on the Hipparcos sample $dn/dm = 0.031(m/0.8)^{-\alpha} = 0.024\,m^{-\alpha}\,M_\odot^{-1}\,pc^{-3}$ with $\alpha = 1.2$. As shown in the Figure 16, this provides a very reasonable description of the low-mass part of the MF, within a variation of $\sim$10% for α, thus confirming the previous analysis of Kroupa et al (1993)—except for the very last bin obtained from Kroupa's LF, which predicts about twice as many stars at 0.1 $M_\odot$ (out of the figure). Similar results are obtained from the bolometric and M_K-LFs for the 8-pc sample. Integration of this MF yields very-low-mass stars ($m \leq 0.8\,M_\odot$), number- and mass-densities $n_{VLMS} \simeq 7.0 \times 10^{-2}$ pc^{-3} and $\rho_{VLMS} \approx 2.0 \times 10^{-2}$ $M_\odot$pc^{-3}, respectively. Adding up the contribution from more massive stars, $\rho_{>0.8} \approx 1.4 \times 10^{-2}$ $M_\odot$pc^{-3} (Miller & Scalo 1978) and stellar remnants, $\rho_{WD+NS} \approx 3.0 \times 10^{-3}$ $M_\odot$pc^{-3} yields a stellar density $\rho_\star \approx 3.4 \times 10^{-2}$ $M_\odot$pc^{-3} in the Galactic disk, i.e. a surface density $\sum_\star \approx 22 \pm 2$ $M_\odot$pc^{-2} (assuming a scale height $h = 320$ pc). A Salpeter MF, $dn/dm \propto m^{-2.35}$, all the way down to the bottom of the MS would overestimate the density at 0.1 $M_\odot$ by more than a factor of 10.

For the Galactic spheroid, the question is more settled. The LFs determined from nearby surveys (Dahn et al 1995) and from the HST (Gould et al 1998) are comparable within about 2 sigmas, and yield a MF with $\alpha \lesssim 1$ (Graff & Freese 1996, Chabrier & Méra 1997, Gould et al 1998) and a stellar number-and mass-density $n_\star < 10^{-3}$ pc^{-3} and $\rho_\star < 4.0 \times 10^{-5}$ $M_\odot$pc^{-3}, i.e. an optical depth $\tau \sim 10^{-9}$, about 1% of the value measured toward the LMC. At present the dark halo MF is completely unknown, but both the Hubble Deep Field observations and the narrow-range of the observed time-distribution of the microlensing events toward the LMC strongly suggest an IMF different from a Salpeter IMF below ~ 1 $M_\odot$ (Chabrier 1999). Assuming a homogeneous distribution, both constraints yield a mass-density for LMS in the dark halo about two orders of magnitude smaller than the afore-mentioned spheroid one (Chabrier & Méra 1997). This means that essentially no stars formed below about a solar mass in the dark halo.

5.2   Brown Dwarf Mass Function

Even though the possibility that BDs could make up for the Galactic missing mass is now clearly excluded, a proper census of the number of BDs has significant implications for our understanding of how stars and planets form. The determination of the BD MF is a complicated task. By definition, BDs never reach thermal equilibrium, and most of the BDs formed at the early stages of the Galaxy will have dimmed to very low luminosities ($L \propto 1/t$). Thus, observations will be biased toward young and massive BDs. This age-undetermination is circumvented when we study the BD MF in clusters because objects in this case are likely to be coeval. The Pleiades cluster has been extensively surveyed, and several BDs have been identified down to ~ 0.04 $M_\odot$ (Martín et al 1998, Bouvier et al 1998, Hambly et al 1999). A single power-law function from ~ 0.4 to 0.04 $M_\odot$ seems to adequately reproduce the observations (not corrected for binaries) with some remaining uncertainties in the exponent: $\alpha \sim 0.6$–1.0. However, stellar objects in very young clusters ($\lesssim 10^6$ yr) might still be accreting, whereas older clusters may have already experienced significant dynamical evolution, mass segregation in the core, and/or evaporation in the outer regions (see e.g. Raboud & Mermilliod 1998). In this case, the present-day mass function does not reflect the initial mass function. For these reasons, it is probably premature to claim a robust determination of the MF in young clusters.

The DENIS (Delfosse et al 1999b) and 2MASS (Kirkpatrick et al 1999b, Burgasser et al 1999) surveys, which have covered an area of several hundred square degrees and are complete to $K = 13.5$ and 14.5 respectively, have revealed about 20 field L-dwarfs and 4 field methane-BDs. This yields an L-dwarf number-density $n_L \approx 0.03 \pm 0.01$ sq deg^{-1} for $K < 14.5$ and a methane-BD number-density $n_{CH_4} \approx 0.002$ sq deg^{-1} for $J < 16$. Although these numbers correspond to small statistics and should be considered with caution, they provide the first observational constraints on the substellar MF in the Galactic disk and are consistent with a slowly rising BD MF with $\alpha \simeq 1-2$ (Reid et al 1999; Chabrier, in preparation), assuming a constant formation rate.

Independent, complementary information on the stellar and substellar MF comes from microlensing observations. Indeed, the time distribution of the events provides a (model-dependent) determination of the mass-distribution and thus of the minimum mass of the dark objects: $dN_{ev}/dt_e = E \times \epsilon(t_e) \times d\Gamma/dt_e \propto P(m)/\sqrt{m}$, where E is the observed exposure (i.e. the number of star × years), ϵ is the experimental efficiency, Γ is the event rate, and $P(m)$ is the mass probability distribution. The analysis of the published 40 MACHO (Alcock et al 1997) + 9 OGLE (Udalsky et al 1994) events toward the bulge is consistent with a rising MF at the bottom of the MS, whereas a decreasing MF below 0.2 $M_\odot$ seems to be excluded at a high (>90%) confidence level (Han & Gould 1996, Méra et al 1998). Although the time distribution might be affected by various biases (e.g. blending) and robust conclusions must wait for larger statistics, the present results suggest that, in order to explain both star counts and the microlensing experiments, a substantial number of BDs must be present in the Galactic disk. Extrapolation of the stellar MF determined in Section 5.1 into the BD domain down to 0.01 $M_\odot$ yields for the Galactic disk a BD (*number*) density comparable to the stellar one, $n_{BD} \approx 0.1\ pc^{-3} \simeq n_\star$, and a mass-density $\rho_{BD} \approx 3.0 \times 10^{-3}\ \mathrm{M_\odot pc^{-3}}$, i.e. $\sum_{BD} \approx 2\ \mathrm{M_\odot pc^{-2}}$.

For the spheroid, extrapolation of the previously determined stellar MF yields $n_{BD} \lesssim 10^{-3}\ pc^{-3}$, $\rho_{BD} \lesssim 10^{-5}\ \mathrm{M_\odot pc^{-3}}$, less than 0.1% of the required dynamical density. For the dark halo the density is about two orders of magnitude smaller, as mentioned before.

It is obviously premature to try to infer the mass distribution of exoplanets. This will first require a clear theoretical and observational distinction between planets and brown dwarfs. However, an interesting preliminary result comes from the observed mass distribution of the companions of G and K stars. As shown in Figure 4 of Mayor et al (1998), there is a strong discontinuity in the mass distribution at $m_2/\sin i \approx 5\ M_J$, with a clear peak below this limit. This suggests that planet formation in a protoplanetary disk is a much more efficient mechanism than BD formation as star companions, at least around G and K stars (see e.g. Marcy & Butler 1998). It also suggests that the MF for BD stellar companions differs from the MF in the field. Ongoing observations around M-stars will tell us whether such a mass distribution still holds around low-mass stars.

6. CONCLUSIONS

This review has summarized the significant progress achieved within the past few years in the theory of cool and dense objects at the bottom of and beyond the main sequence: low-mass stars, brown dwarfs, and gaseous planets. The successful confrontation of the theory with the numerous detections of low-mass stellar and substellar objects allows a better understanding of their structural and thermal properties, and allows reliable predictions about their evolution. This in turn brings confidence in the predicted characteristic properties of these objects, a major issue in terms of search strategies for future surveys.

Important problems remain to be solved to improve the theory. A non-exhaustive list includes, for example: (1) a better determination of the EOS in the pressure-ionization region, with the possibility of a first-order phase transition; (2) the study of phase separation of elements in SSO interiors; (3) an improved treatment of convection in optically-thin regions; (4) a precise description of the dynamics of grain formation and sedimentation in SSO atmospheres; (5) the derivation of an accurate mass-$T_{\rm eff}$-age scale for SSOs and young objects; and (6) a correct understanding of magnetic-field generation and dissipation in active LMS and BDs.

The increasing number of observed LMS and SSOs, together with the derivation of accurate models, eventually will allow a robust determination of the stellar and substellar mass functions, of the minimum mass for the formation of star-like objects, and thus of the exact density of these objects in the Galaxy. As we discussed in Section 5, present determinations in various Galactic regions point to a slowly rising MF near and below the H-burning limit, with a BD number-density comparable to the stellar one and an MF truncated below $\sim 1\,M_\odot$ in the dark halo. A more precise determination must await confirmation from future observations. At last, the amazingly rapid pace of exoplanet discoveries should yield the determination of the planetary MF and maximum mass, and eventually the direct detection of such objects, a future formidable test for the theory.

ACKNOWLEDGMENTS

This review has benefited from various discussions with our colleagues F Allard, PH Hauschildt, D Alexander, K Lodders, A Burrows, J Lunine, B Fegley, T Forveille, X Delfosse, J Bouvier, P Kroupa, A Nordlund, T Guillot, D Saumon, and U Kolb. Our profound gratitude goes to these individuals, who helped by improving the original manuscript. We are also very indebted to P Kroupa, N Reid, HC Harris, and C Dahn for providing various data prior to publication.

Visit the Annual Reviews home page at www.AnnualReviews.org

LITERATURE CITED

Alcock C, Allsman RA, Alves D, Axelrod TS, Bennett DP, et al. 1997. *Ap. J.* 479: 119

Alexander DR, Ferguson JW. 1994. *Ap. J.* 437:879

Allard F. 1990. PhD thesis. Univ. Heidelberg, Germany. 160 pp.

Allard, F. 1999. *Proc. Euroconference, La Palma*

Allard F, Alexander DR, Tamanai A, Hauschildt P. 1998. *Proc. ASP Conf. Series* 134:438

Allard F, Hauschildt PH. 1995. *Ap. J.* 445:433

Allard F, Hauschildt PH, Alexander DR, Starrfield S. 1997. *Annu. Rev. Astron. Astrophys.* 35:137

Allard F, Hauschildt PH, Alexander DR, Tamanai A, Ferguson J. 2000. *Ap. J.* In press

Allard F, Hauschildt PH, Baraffe I, Chabrier G. 1996. *Ap. J.* 465:L123

Auman J. 1969. *Low Luminosity Stars.* New York: Gordon and Breach, 539 pp.

Baraffe I, Chabrier G. 1996. *Ap. J.* 461:L51

Baraffe I, Chabrier G, Allard F, Hauschildt PH. 1995. *Ap. J.* 446:L35

Baraffe I, Chabrier G, Allard F, Hauschildt PH. 1997. *Astron. Astrophys.* 327:1054

Baraffe I, Chabrier G, Allard F, Hauschildt PH. 1998. *Astron. Astrophys.* 337:403

Basri G, Marcy GW, Graham JR. 1996. *Ap. J.* 458:600

Basri G, Mohanty S, Allard F, Hauschildt PH, Delfosse X, et al. 2000. *Ap. J.* In press

Béjar VJS, Zapatero Osorio MR, Rebolo R. 1999. *Ap. J.* 521:671

Beuermann K, Baraffe I, Kolb U, Weichhold M. 1998. *Astron. Astrophys.* 339:518

Bildsten L, Brown EF, Matzner CD, Ushomirsky G. 1997. *Ap. J.* 482:442

Borysow A, Jorgensen UG, Zheng C. 1997. *Astron. Astrophys.* 324:185

Bouvier J, Stauffer JR, Martín EL, Barrado y Navascues B, Wallace B, Béjar VJS. 1998. *Astron. Astrophys.* 336:490

Brett C. 1995. *Astron. Astrophys.* 295:736

Brett C, Plez B. 1993. *Astron. Soc. Aust. Proc.* 10:250

Burgasser AJ, Kirkpatrick JD, Brown NE, Reid IN, et al. 1999. *Ap. J.* 522:65

Burrows A, Hubbard WB, Lunine JI. 1989. *Ap. J.* 345:939

Burrows A, Hubbard WB, Saumon D, Lunine JI. 1993. *Ap. J.* 406:158

Burrows A, Hubbard WB, Lunine JI. 1994. *Cool Stars, Stellar Systems and the Sun. ASP.* 64:528

Burrows A, Liebert J. 1993. *Rev. Mod. Phys.* 65:301

Burrows A, Marley M, Hubbard WB, Lunine JI, Guillot T, et al. 1997. *Ap. J.* 491:856

Burrows A, Marley M, Sharp CM. 2000. *Ap. J.* 531:43

Burrows A, Sharp CM. 1999. *Ap. J.* 512:843

Canuto VM, Goldman I, Mazzitelli I. 1996. *Ap. J.* 473:550

Canuto VM, Mazzitelli I. 1991. *Ap. J.* 370: 295

Caughlan GR, Fowler WA. 1988. *Atom. Data Nucl. Data Tables* 40:283

Chabrier G. 1990. *J. Phys.* 51:1607

Chabrier G. 1999. *Ap. J.* 513:L103

Chabrier G, Baraffe I. 1997. *Astron. Astrophys.* 327:1039

Chabrier G, Baraffe I, Allard F, Hauschildt PH. 2000. *Ap. J.* In press

Chabrier G, Baraffe I, Plez B. 1996. *Ap. J.* 459:L91

Chabrier G, Méra D. 1997. *Astron. Astrophys.* 328:83

Clausen JV, Helt BE, Olsen EH. 1999. *Theory and tests of convection in stellar structure.* ASP Conf. Series 173, *Am. Soc. Phys.*

Clayton DD. 1968. *Principles of Stellar Evolution and Nucleosynthesis.* Chicago: Chicago Press, 612 pp.

Clemens JC, Reid IN, Gizis JE, O'Brien MS. 1998. *Ap. J.* 496:392

Collins GW, Da Silva LB, Celliers P, Gold DM, Foord ME, et al. 1998. *Science* 281: 1178

Copeland H, Jensen JO, Jorgensen HE. 1970. *Astron. Astrophys.* 5:12

Cox JP, Guili RT. 1968. *Principles of Stellar Structure.* New York: Gordon and Breach 1305 pp.

Dahn CC, Guetter HH, Harris HC, Henden AA, Luginbuhl CB, et al. 1999. The evolution of galaxies on cosmological timescales. *ASP Conf. Series*

Dahn CC, Liebert J, Harris HC, Guetter HH. 1995. The bottom of the main sequence and beyond. *ESO Astrophys. Symp.*, p. 239. Berlin/Heidelberg: Springer Verlag

Dahn CC, Liebert J, Harrington RS. 1986. *Astron. J.* 91:621

Da Silva, et al. 1997. *Phys. Rev. Lett.* 78:483

D'Antona F, Mazzitelli I. 1985. *Ap. J.* 296:502

Delfosse X, Forveille T, Beuzit JL, Udry S, Mayor M, Perrier C. 1999a. *Astron. Astrophys.* 344:897

Delfosse X, Forveille T, Perrier C, Mayor M. 1998a. *Astron. Astrophys.* 331:581

Delfosse X, Forveille T, Tinney CG, Epchtein N. 1998b. See Allard et al. 1998, p. 67

Delfosse X, Tinney CG, Forveille T, Epchtein N, Bertin E, et al. 1997. *Astron. Astrophys.* 327:L25

Delfosse X, Tinney CG, Forveille T, Epchtein N, Borsenberger J, et al. 1999b. *Astron. Astrophys. Suppl.* 135:41

Delfosse X, Forveille T, Udry S, Beuzit JL, Mayor M, Perrier C. 1999c. *Astron. Astrophys. Lett.* 350:L39

Demarque P, Guenther DB, Kim Y-C. 1999. *Ap. J.* 517:510

Dorman B, Nelson LA, Chau WY. 1989. *Ap. J.* 342:1003

Durney B, De Young DS, Roxburgh IW. 1993. *Solar Phys.* 145:207

Fegley B, Lodders K. 1994. *Icarus* 110:117

Fegley B, Lodders K. 1996. *Ap. J.* 472:L37

Fontaine G, Graboske HC, VanHorn HM. 1977. *Ap. J. Suppl.* 35:293

Forveille T, Beuzit JL, Delfosse X, Segransan D, Beck F, et al. 1999. *Astron. Astrophys.* 351:619

Goldman B, Delfosse X, Forveille T, Afonso C, Alard C, et al. 1999. *Astron. Astrophys.* 351:L5

Gould A, Flynn C, Bahcall JN. 1998. *Ap. J.* 503:798

Gould A, Bahcall JN, Flynn C. 1997. *Ap. J.* 482:913

Graff D, Freese K. 1996. *Ap. J.* 456:49

Griffith CA, Yelle RV, Marley MS. 1998. *Science* 282:2063

Grossman AS, Hays D, Graboske HC. 1974. *Astron. Astrophys.* 30:95

Grossman L. 1972. *Geochim. Cosmochim. Acta* 36:597

Guillot T. 1995. *Science* 269:1697

Guillot T. 1999. *Science* 286:72

Guillot T, Burrows A, Hubbard WB, Lunine JI, Saumon D. 1996. *Ap. J.* 459:L35

Hambly NC, Hodgkin ST, Cossburn MR, Jameson RF. 1999. *MNRAS* 303:835

Han C, Gould A. 1996. *Ap. J.* 467:540

Hauschildt PH, Allard F, Baron E. 1999. *Ap. J.* 512:377

Hawley SL, Gizis JE, Reid IN. 1996. *Astron. J.* 112:2799

Henry TJ, McCarthy DW. 1993. *Astron. J.* 106:773

Henry TJ, Ianna PA, Kirkpatrick D, Jahreiss H. 1997. *Astron. J.* 114:388

Hubbard WB. 1977. *Icarus* 30:305

Hubbard WB. 1994. The equation of state in astrophysics. *Proc. IAU Coll., Saint-Malo, 1993.* 147:443. Cambridge, UK: Cambridge University Press

Hümmer DG, Mihalas D. 1988. *Ap. J.* 331:794

Iglesias CA, Rogers FJ. 1996. *Ap. J.* 464:943

Jones HRA, Tsuji T. 1997. *Ap J.* 480:L39

Kirkpatrick JD, Allard F, Bida T, Zuckerman B, Becklin EE, et al. 1999b. *Ap. J.* 519:834

Kirkpatrick JD, Henry TJ, McCarthy DW. 1991. *Ap. J. Suppl.* 77:417

Kirkpatrick JD, McCarthy DW. 1994. *AJ.* 107:333

Kirkpatrick JD, Reid IN, Liebert J, Cutri RM, Nelson B, et al. 1999a. *Ap. J.* 519:802

King AR. 1988. *Q. J. R. Astron. Soc.* 29:1

Kolb U, Baraffe I. 1999a. *MNRAS* 309:1034

Kolb U, Baraffe I. 1999b. Annapolis workshop on magnetic cataclysmic variables. *ASP Conf. Series.* 197:273

Kolb U, King AR, Ritter H. 1998. *MNRAS* 298:L29

Kroupa P. 1995. *Ap. J.* 453:350

Kroupa P, Tout CA, Gilmore G. 1990. *MNRAS* 244:76

Kroupa P, Tout CA, Gilmore G. 1993. *MNRAS* 262:545

Kumar SS. 1963. *Ap. J.* 137:1121

Leggett S. 1992. *Ap. J. Suppl.* 82:351

Leggett S, Allard F, Berriman G, Dahn CC, Hauschildt PH. 1996. *Ap. J. Suppl.* 104:117

Leggett S, Allard F, Hauschildt PH. 1998. *Ap. J.* 509:836

Lenzuni P, Chernoff DF, Salpeter EE. 1991. *Ap. J. Suppl.* 76:759

Leung KC, Schneider D. 1978. *Astron. J.* 83:618

Linsky J. 1969. *Ap. J.* 156:989

Lodders K. 1999. *Ap. J.* 519:793

Ludwig HG, Freytag B, Steffen M. 1999. *Astron. Astrophys.* 346:111

Luhman KL. 1999. *Ap. J.* 525:466

Lunine JI, Hubbard WB, Burrows A, Wang YP, Garlow K. 1989. *Ap. J.* 338:314

Lunine JI, Hubbard WB, Marley MS. 1986. *Ap. J.* 310:238

Magni G, Mazzitelli I. 1979. *Astron. Astrophys.* 72:134

Marcy GW, Butler RP. 1998. *Annu. Rev. Astron. Astrophys.* 36:57

Marley MS, Gelino C, Stephens D, Lunine JI, Freedman R. 1999. *Ap. J.* 513:879

Marley MS, Hubbard WB. 1988. *Icarus* 73:536

Marley MS, Saumon D, Guillot T, Freedman RS, Hubbard WB, et al. 1996. *Science* 272:1919

Martín E. 1999. See Allard 1999

Martín EL, Basri G, Delfosse X, Forveille T. 1997. *Astron. Astrophys.* 327:L29

Martín EL, Basri G, Zapatero-Osorio MR, Rebolo R, Lopez RJ. 1998. *Ap. J.* 507:L41

Maxted PFL, Marsh TR, Moran C, Dhillon VS, Hilditch RW. 1998. *MNRAS* 300:1225

Mayor M, Queloz D. 1995. *Nature* 378:355

Mayor M, Queloz D, Udry S. 1998. See Allard et al. 1998, p. 140

Méra D, Chabrier G, Schaeffer R. 1998. *Astron. Astrophys.* 330:937

Metcalfe TS, Mathieu RT, Latham DW, Torres G. 1996. *Ap. J.* 456:356

Meyer F, Meyer-Hofmeister E. 1999. *Astron. Astrophys.* 341:L23

Miller G, Scalo J. 1978. *PASP.* 90:506

Monet D, Dahn CC, Vrba FJ, Harris HC, Pier JR, et al. 1992. *Astron. J.* 103:638

Nelson LA, Rappaport S, Joss PC. 1986. *Ap. J.* 311:226

Nelson LA, Rappaport S, Chiang E. 1993b. *Ap. J.* 413:364

Nelson LA, Rappaport S, Joss PC. 1993a. *Ap. J.* 404:723

Neuhäuser R, Briceno C, Comeron F, Hearty T, Martín EL, et al. 1999. *Astron. Astrophys.* 343:883

Neuhäuser R, Comeron F. 1998. *Science* 282:83

Noll KS, Geballe TR, Marley MS, 1997. *Ap. J.* 489:L87

Nordlund AA, Stein RF. 1999. See Clausen, et al. 1999, p. 91

Oppenheimer BR, Kulkarni SR, Matthews K, Nakajima T. 1995. *Science* 270:1478

Oppenheimer BR, Kulkarni SR, Matthews K, van Kerkwijk MH. 1998, *Ap. J.* 502:932

Parker EN. 1955. *Ap. J.* 122:293

Parker EN. 1975. *Ap. J.* 198:205

Pulone L, De Marchi G, Paresce F, Allard F. 1998. *Ap. J.* 492:L41

Raboud D, Mermilliod JC. 1998. *Astron. Astrophys.* 333:897

Randich S. 1999. See Allard 1999

Rappaport S, Verbunt F, Joss PC. 1983. *Ap. J.* 275:713

Rebolo R, Martín EL, Magazzù A. 1992. *Ap. J.* 389:L83

Rebolo R, Zapatero Osorio MR, Martín EL. 1995. *Nature* 377:129

Reid IN. 1999. *Annu. Rev. Astron. Astrophys.* 37:191

Reid IN, Gizis JE. 1997. *Astron. J.* 113:2246

Reid IN, Hawley SL, Gizis JE. 1995. *Astron. J.* 110:1838

Reid IN, Kirkpatrick JD, Liebert J, Burrows A, Gizis JE, et al. 1999. *Ap. J.* 521:613

Rossow WB. 1978. *Icarus* 36:1

Ruiz MT, Leggett SK, Allard F. 1997. *Ap. J.* 491:L107

Salpeter EE. 1954. *Aust. J. Phys.* 7:373

Salpeter EE. 1961. *Ap. J.* 134:669

Saumon D. 1994. See Hubbard 1994, p. 306

Saumon D, Bergeron P, Lunine JI, Hubbard WB, Burrows A. 1994. *Ap. J.* 424:333

Saumon D, Chabrier G. 1991. *Phys. Rev. A.* 44:5122

Saumon D, Chabrier G. 1992. *Phys. Rev. A.* 46:2084

Saumon D, Chabrier G, VanHorn HM. 1995. *Ap. J. Suppl.* 99:713

Saumon D, Hubbard WB, Burrows A, Guillot T, Lunine JI, Chabrier G. 1996. *Ap. J.* 460:993

Schatzman E. 1948. *J. Phys. Rad.* 9:46

Schultz AB, Allard F, Clampin M, McGrath M, Bruhweiler FC, et al. 1998. *Ap. J.* 492:L181

Seager S, Sasselov DD. 1998. *Ap. J.* 502:L157

Sharp CM, Huebner WF. 1990. *Ap. J. Suppl.* 72:417

Shu FH, Adams FC, Lizano S. 1987. *Annu. Rev. Astron. Astrophys.* 25:23

Spiegel EA, Weiss NO. 1980. *Nature* 287:616

Spiegel EA, Zahn J-P. 1992 *Astron. Astrophys.* 265:106

Spruit H. 1994. Cosmical magnetism. *NATO ASI Series C.* 422:33–44

Spruit H, Nordlund AA. 1990. *Annu. Rev. Astron. Astrophys.* 28:263

Spruit H, Ritter H. 1983. *Astron. Astrophys.* 124:267

Stauffer JR, Hartmann LW, Prosser CF, Randich S, Balachandran S, et al. 1997. *Ap. J.* 479:776

Stauffer JR, Schultz G, Kirkpatrick JD. 1998. *Ap. J.* 499:L199

Stein RF, Nordlund. 2000. *Solar Phys.* 192:91

Stevenson DJ. 1979. *Geophys. Astrophys. Fluid Dyn.* 12:139

Stevenson DJ. 1991. *Annu. Rev. Astron. Astrophys.* 29:163

Taam RE, Spruit HC. 1989. *Ap. J.* 345:972

Tarter J. 1975. PhD thesis. Univ. Calif., Berkeley. 120 pp.

Tinney CG, Delfosse X, Forveille T, Allard F. 1998. *Astron. Astrophys.* 338:1066

Tinney CG, Reid IN. 1998. *MNRAS* 301:1031

Toth RA. 1997. *J. Mol. Spect.* 186:276

Tsuji T, Ohnaka K, Aoki W. 1995. See Dahn et al. 1995, p. 45

Tsuji T, Ohnaka K, Aoki W. 1996. *Astron. Astrophys.* 305:L1

Tsuji T, Ohnaka K, Aoki W. 1999. *Ap. J.* 520:L119

Udalsky A, Szymanski M, Stanek KZ, Kaluzny J, Kubiak M, et al. 1994. *Act. Astrono.* 44:165

Ushomirsky G, Matzner CD, Brown EF, Bildsten L, Hilliard VG, Schroeder P. 1998. *Ap. J.* 253:266

VandenBerg DA, Hartwick FDA, Dawson P, Alexander DR. 1983. *Ap. J.* 266:747

White RJ, Ghez AM, Reid IN, Schultz G. 1999. *Ap. J.* 520:811

Zapatero Osorio MR, Bejar VJS, Rebolo R, Martín EL, Basri G. 1999. *Ap. J.* 524:L115

Zapolsky HS, Salpeter EE. 1969. *Ap. J.* 158:809

Annu. Rev. Astron. Astrophys. 2000. 38:379–425

GAMMA-RAY BURST AFTERGLOWS

Jan van Paradijs,[1*] Chryssa Kouveliotou,[2]
and Ralph A. M. J. Wijers[3]

[1]*Astronomical Institute 'Anton Pannekoek', University of Amsterdam, Kruislaan 403,
1098 SJ Amsterdam, The Netherlands and Department of Physics, University of Alabama
in Huntsville, Huntsville, AL 35899*
[2]*Universities Space Research Association, NASA Marshall Space Flight Center,
SD–50, Huntsville, AL 35812; e-mail: chryssa.kouveliotou@iss.msfc.nasa.gov*
[3]*Department of Physics & Astronomy, State University of New York, Stony Brook,
NY 11794-3800; e-mail: rwijers@astro.sunysb.edu*

Key Words   gamma-ray bursts, compact objects, cosmology, particle acceleration,
shocks, supernovae

■ **Abstract**   The discovery of counterparts in X-ray and optical to radio wavelengths
has revolutionized the study of γ-ray bursts, until recently the most enigmatic of
astrophysical phenomena. We now know that γ-ray bursts are the biggest explosions in
nature, caused by the ejection of ultrarelativistic matter from a powerful energy source
and its subsequent collision with its environment. We have just begun to uncover
a connection between supernovae and γ-ray bursts, and are finally constraining the
properties of the ultimate source of γ-ray burst energy. We review here the observations
that have led to this breakthrough in the field; we describe the basic theory of the fireball
model and discuss the theoretical understanding that has been gained from interpreting
the new wealth of data on γ-ray bursts.

Ik zie een ster	*I see a star*
en onder mij	*and beneath me*
zakt de aarde langzaam weg.	*the Earth slowly falls away.*

1.   INTRODUCTION

The very first model of γ-ray bursts was proposed by Colgate (1968, 1974) even
before γ-ray bursts had been discovered; the model invoked γ-ray emission by
accelerated particles at the breakout of a shock from a supernova progenitor's
photosphere.

---

*deceased

Cosmic γ-ray bursts (GRBs) were first reported in 1973 by Klebesadel et al (1973), who in their discovery paper pointed out the lack of evidence for a connection between GRBs and supernovae as proposed by Colgate. It is an interesting twist of history that a quarter-century later, observations of the low-energy afterglows of GRBs have provided evidence that at least some GRBs originate from a probably rare type of supernova. The return to a discarded GRB model that preceded the discovery of GRBs is part of the flood of results that have become available since the initial discoveries of X-ray (Costa et al 1997b), optical (van Paradijs et al 1997), and radio (Frail et al 1997c) afterglows of GRBs. The review of these results is the subject of this paper.

Cavallo & Rees (1978) and Schmidt (1978) realized that the very small sizes (implied by a short variability time, δt) and high fluxes of GRBs ($F_\gamma \lesssim 10^{-4}$ erg/cm^2 s) implied a high photon density at the source. This, in turn, implied that photon-photon scattering would prevent emission of the observed MeV photons for any GRB population more distant than ~1 kpc (unless the source were expanding ultrarelativistically). Cavallo & Rees (1978) formalized the salient source properties into a single *compactness parameter* given by

$$\theta_* = \frac{L\sigma_T}{m_p c^3 R} \simeq \frac{F_\gamma d^2 \sigma_T}{m_p c^4 \delta t} \sim 1 F_{\gamma-4} d_{kpc}^2 / (\delta t)_{ms} \sim 10^{12} F_{\gamma-4} d_{Gpc}^2 / (\delta t)_{ms}, \quad (1)$$

where $F_{\gamma-4} = F_\gamma / 10^{-4}$ erg cm^{-2} s^{-1}, σ_T is the Thompson electron scattering cross section, m_p is the proton mass; L, d are the luminosity and distance of the GRB source, respectively; and $R = c\delta t$ is the radius of the emitting region. Barring relativistic expansion, $\theta_* \gtrsim 1$ implies that the source is optically thick for $\gamma\gamma$ interactions, and cannot be a copious emitter of high-energy photons. The preceding numerical values therefore suggest that nonrelativistic GRB sources would have to be in the Galaxy, and conversely, extragalactic GRBs would have to come from very relativistic sources.

Early indications against a Euclidean space distribution of GRBs was the apparent lack of very weak GRBs (Fishman et al 1978). This result anticipated by about a decade the discussion on the GRB distance scale in the early and mid-1990s. A detailed discussion of this issue, in particular the extent to which this result reflected instrumental effects, was given by Hurley (1986).

In spite of this uncertainty, by the mid-1980s there was a general belief that GRBs originate from galactic neutron stars—a belief that was supported by the cyclotron lines in the spectra of some GRBs reported by the Ginga team (Murakami et al 1988, Fenimore et al 1988), and by line features near 500 keV reported by the Konus team (Mazets et al 1981).

A galactic disc population of neutron stars was firmly excluded as the source of GRBs, based on the results of the first 6 months of BATSE observations, which showed that the GRB sky distribution is isotropic and that there is a strong lack of very faint GRBs (Meegan et al 1992). With the increasing number of GRBs detected with BATSE, the isotropy has become strongly established (Briggs et al 1996). The lack of very faint GRBs shows up as a turnover in the cumulative log $N(>P)$ distribution (here P is the GRB peak flux); for the brightest bursts, the slope of the distribution follows the Euclidean value of $-3/2$, but below $P \sim 5$ ph/cm^2 s ($\sim 10^{-6}$ ergs/cm^2 s) the slope decreases and becomes as low as ~ -0.6 at the faint end of the distribution (see Paciesas et al 1999 for a review; also Figure 19).

The statistical properties of GRBs are naturally explained if their typical distances are measured in Gpc (the cosmological distance scale; Paczyński 1986, Usov & Chibisov 1975), across which inhomogeneity in the distribution of luminous matter is averaged out (see Condon 1999 for a sky map of extragalactic radio sources nicely illustrating this point). The turnover in the log $N(>P)$ is the direct consequence of the effect of cosmological redshift on the observed GRB *rates* (downward shift of the low P end of the curve) and the GRB *peak fluxes* (leftward shift of the low P end of the curve).

However, there was a countervailing view that GRBs arise from neutron stars in a large galactic corona (Shklovskii & Mitrofanov 1985, Brainerd 1992, Podsiadlowski et al 1995, Lamb 1995, Bulik et al 1998), whose size had to be several 100 kpc to mask the offset of the Earth from the galactic center (Shklovskii & Mitrofanov 1985), but not much larger to avoid seeing excess GRBs from M31 (Hakkila et al 1994).

For several years, the GRB distance was the subject of a lively debate (see e.g. Lamb 1995, Paczyński 1995), but this debate did not lead to a consensus. It was generally felt that setting the distance scale would require the identification of possibly long-lived GRB counterparts in other wavebands (Fishman & Meegan 1995).

Independent of which of the two contending distance scales were correct, the compactness problem was severe, and by analogy to e.g. BL Lac objects and other superluminal AGN sources, appeared to require relativistic outflow of the GRB source (Cavallo & Rees 1978). Reviving the idea already hinted at by Cavallo & Rees, Rees & Mészáros (1992) suggested that γ-ray bursts can be produced if part of a relativistic bulk flow is converted back into high-energy photons through particle acceleration in a relativistic shock between the outflow and the circumsource medium. This basic mechanism has been developed into the fireball model, which provides the background of current discussions on GRBs.

Early attempts to find GRB counterparts employed several approaches, some of which are as follows:

1. Archival plate searches for optical transients in small GRB error boxes obtained from the IPN network of satellites through triangulation (Hurley et al 2000a,b). Schaefer et al (1981, 1984) reported the discovery of such transients in the error boxes of several GRBs after extensive searches in the Harvard plate archives. However, Żytkow (1990) and Greiner et al (1990) have argued, on the basis of an analysis of the 3-D distribution of the image in the plate emulsions, that these events are likely plate defects (but see Schaefer 1990).

2. Deep imaging observations (optical, X-ray, and radio) of small error boxes. The aim of these observations was to find out if particular objects appear in

these error boxes with a statistically significant excess, which might qualify them as possible GRB counterparts. These searches were not successful, i.e. they did not lead to a convincing detection of a source population connected to GRBs. For example, some error boxes were conspicuously devoid of host galaxies (Schaefer 1992, Band et al 1999), which was taken as evidence against extragalactic origin of GRBs or as a problem with specific extragalactic models (Schaefer 1992, Band et al 1999)—though correlations with extragalactic objects were also claimed (e.g. Larson et al 1996, Kolatt & Piran 1996) but disputed (Schaefer 1998, Hurley et al 1997).

3. Simultaneous sky coverage using wide-field optical meteor search cameras (see e.g. Hudec et al 1999, Greiner et al 1993, Krimm et al 1996). In none of the simultaneous photographic images was an optical event detected; the corresponding magnitude limit (for optical flashes lasting as long as the GRB) is ~5th magnitude (McNamara et al 1995).

Extensive summaries of the early attempts to find GRB counterparts have been given by McNamara et al (1995) and Frail & Kulkarni (1995). In hindsight, the lack of success of these early attempts is now understood as a result of these observations being either too late or not sensitive enough.

A breakthrough in the search for GRB counterparts occurred in early 1997, as the result of the first rapidly available (typically within a few hours) accurate (of order several arcminutes) GRB error boxes produced by the two Wide Field Cameras (WFCs; Jager et al 1995) on the Italian-Dutch satellite BeppoSAX (Boella et al 1997). These WFCs are coded-mask cameras with a full field of view of $40° \times 40°$ and a resolution of several arcminutes. GRBs are detected as transient events in the WFC ratemeter (2–20 keV). Trigger information that a GRB occurred is usually provided by the Gamma Ray Burst Monitor (GRBM) on BeppoSAX (Feroci et al 1997), and occassionally by other instruments (e.g. BATSE). In several cases, scans of a BATSE GRB error box with the Proportional Counter Array (PCA) of the Rossi X-ray Timing Explorer (RXTE) have produced relatively accurate GRB error boxes (Takeshima et al 1998; see also Section 5).

This review of the rapid increase in our understanding of GRBs during the last two and a half years thus describes to a large extent the success story of BeppoSAX. To many, the impact of BeppoSAX came as a surprise; however, the direct use of the WFCs for accurate GRB locations was an integral part of the scientific goals of this instrument from the very beginning (see e.g. Hurley 1986). Rapid follow-up observations of these error boxes have unambiguously shown that γ-ray bursts originate from the high-redshift universe.

The review is structured as follows. In Section 2 we provide a brief description of the developments in our understanding of the pure γ-ray aspects of GRBs. In Section 3 we give a very brief description of a simple version of the fireball model, because this provides the current framework of virtually all discussions of GRBs. This is done in the hope that it will guide the description of the observations, although we realize that in a sense we are prejudicial. Yet we feel the overall evidence for the baseline model is strong enough to proceed this way. In Section 4 we describe the main results of low-energy afterglow studies in the form of highlights. Numerical information and a very short narrative on individual GRB afterglow observations are collected in Section 5, in tabular form. In Section 6 we contrast the results of the afterglow observations with those expected from the simplest fireball models, and discuss the complexities of the GRB emitter that may explain the various discrepancies between the two. In Section 7 we briefly discuss the connection of GRB to host galaxies and cosmological aspects of γ-ray burst follow-up observations, among them a possible connection with the star formation rate history. We briefly mention some implications of the GRB association with galaxies and supernovae for the central engine. In Section 8 we close our review with some comments regarding the impact of GRB follow-up observations on our understanding of the origin of γ-ray bursts, and likely developments in the near future.

2. PROMPT γ-RAY EMISSION

A comprehensive review on γ-ray bursts was written by Fishman & Meegan in 1995. In the five years that have elapsed, BATSE has recorded an additional 1000 GRBs, with a grand total of over 2704 events. The isotropy and inhomogeneity of their distribution is now firmly established, as well as the GRB hardness-duration correlation and bimodal duration distribution (Paciesas et al 1999). The line detection paucity is still a hard fact (Briggs 1999), as is the non-detection of any GRB repeater (Paciesas et al 1999).

With almost 3000 events at hand, the next step in the pure γ-ray field was obtaining elaborate statistics, i.e. statistics on GRB properties that required the use of advanced (sometimes copious) analyses and/or model simulations and fits. This section describes some of the new results that have appeared in the last three years in the literature. Selection criteria applied were the broader impact of the result and its significance in improving our understanding of the physical mechanisms responsible for the production of the prompt γ-ray emission.

2.1 Faint (Untriggered) GRBs

Kommers et al (2000a,b) recently completed a search of six years of archival BATSE data for GRBs that were too faint to activate the real-time on-board burst trigger system (untriggered bursts). They found 873 untriggered events, 551 of which were faint—i.e. below the BATSE detection threshold (their detection efficiency falls below 50% at peak fluxes of 0.16 ph/cm^2 s). The events thus collected have peak fluxes a factor of $\sim$2 lower than those detected with the nominal BATSE trigger criteria.

The latest BATSE $\log N - \log P$ flattens below $P = 0.6$ ph/cm^2 s (Paciesas et al 1999; see Figure 19). The efficiency calculations of Pendleton et al (1998) show that this is a real effect and not an instrumental artifact. The combined faint + BATSE GRB cumulative $\log N - \log P$ distribution also exhibits a dramatic flattening consistent with the triggered burst results, an indication that few faint bursts have remained undetected with BATSE in the triggered or untriggered mode. However, Stern et al (1999) find no indication of a turnover in a similar study; the discrepancy appears to lie in the trigger efficiency calculation. Kommers et al (2000a) extended the GRB peak flux distribution ($\log N - \log P$) to $\sim$0.18 ph/cm^2 s and fitted it with several cosmological models with power-law luminosity distributions. Their results favor models in which the redshift distribution of the GRB rate approximately traces the star formation rate of the Universe.

2.2 A Hubble Relationship for GRBs?

Norris et al (2000) analyzed two samples of GRBs: (1) seven events (six of which have known redshifts) observed with BATSE and BeppoSAX that also have optical or radio counterparts, and (2) the 174 brightest long duration (over 2 s) GRBs. In particular, they computed cross-correlation lags between low (25–50 keV) and high (100–300 keV and >300 keV) energy bands and examined their dependence on burst γ/X peak flux ratio and peak luminosity. They find that the spectral lags and the burst peak intensities are *anticorrelated* in both samples. For the bursts with known redshifts, the connection is well fitted by a power-law, $L_{53} = 1.3 \times (\tau/0.01s)^{-1.15}$ (Figure 1).

A similar claim is made by Ramirez-Ruiz & Fenimore (1999), who have found a relationship between burst variability and absolute burst luminosity. If these results stand the test of time, they may provide a unique way of determining burster distances using the prompt γ-ray data, thus significantly expanding the distance

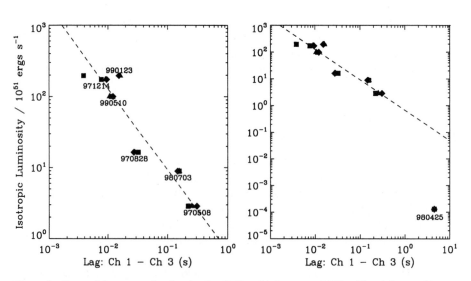

Figure 1 Spectral lag versus luminosity for GRBs with known redshifts. The right panel has an expanded luminosity range to include GRB 980425 (Norris et al 2000).

sample and enabling deep cosmological probing. However, the interpretation of the duration differences between bright and dim bursts previously attributed to cosmological time dilation (Norris et al 1994) has now become very unclear.

2.3 Prompt Gamma-ray "Afterglows"?

Costa et al (1997b) have shown that the BeppoSAX NFI X-ray afterglow light curve of GRB 970228 smoothly joins with the WFC prompt emission, indicating a gradual evolution between prompt and afterglow emission. Similarly, Connors & Hueter (1998) report distinct X-ray afterglow emission from GRB 780506, starting ~2 min after the main event and lasting for about 30 min. Burenin et al (1999) found a power-law decay in one GRANAT/SIGMA burst, albeit with a considerably flatter decay index ($\delta = -0.7$) than a normal ($\delta = -1.0$) afterglow.

Giblin et al (1999) have found evidence in the BATSE data for a prompt high-energy (25–300 keV) afterglow component from GRB 980923. After 40 s of variable emission, the γ-ray light-curve decays with a power-law index, $\delta = -1.81(2)$, in a smooth tail lasting ~400 s. An abrupt change in spectral shape is found when the tail becomes noticeable (Figure 2). More important, Giblin et al (1999) showed that the spectral evolution in the tail mimics that of a cooling synchrotron spectrum, similar to the spectral evolution of the low-energy afterglows of GRBs. Currently, Connaughton (2000) is working on a statistical analysis of GRB tail emission; her results indicate that high-energy tails are prevalent in long GRBs, albeit at low intensity levels. If confirmed, these results will provide evidence for a continuation of the emission during the GRB, and will constrain the emission models.

2.4 TeV Prompt Emission?

Atkins et al (2000) detected photons with energies greater than a few hundred GeV in one out of 54 BATSE searched events with the Milagrito detector, GRB 970417a. The excess has a chance probability of 2.8×10^{-5} of being a background fluctuation, and 1.5×10^{-3} of being a chance coincidence. If this result stands, this would be the highest energy emission ever associated with a GRB—a result that could provide a theoretical challenge for some GRBs because the fluence of GRB 970417a above 50 GeV is an order of magnitude higher than its sub-MeV fluence.

3. BASIC THEORY OF FIREBALLS AND BLAST WAVES

Here we give a brief overview, starting from first principles, of the evolution of spherical blast waves, to set the context for the subsequent discussion of the observations. More complex effects are discussed later (Section 6) in broad terms. Much more about theoretical developments, including many of the technical details, can be found in the review by Piran (1999).

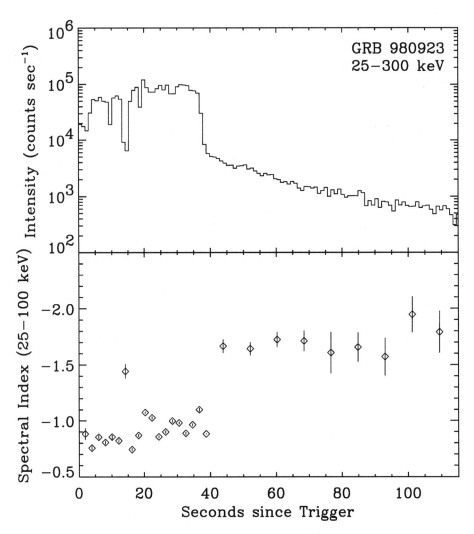

Figure 2   Time history of GRB 980923 (25–300 keV) plotted logarithmically to indicate the first part of the 400 s long tail (*upper panel*). The high-energy photon spectral index as a function of time is shown in the *bottom panel*. Note the abrupt change of the spectral index when the tail begins (Giblin et al 1999).

3.1 Dynamics

As the simplest possible model of gamma-ray burst dynamics, we consider the release of a large amount of energy into a small volume, and follow the resulting explosion. This gives a scenario much like a normal supernova, except that we choose parameters so that the exploding fireball is ultrarelativistic; consequently,

Lorentz contractions play an important role in setting the typical size, duration, and characteristic photon energy to values very different from those of a supernova.

Let us consider the release of an energy $E = 10^{52} E_{52}$ erg into a sphere with radius r_{in}. A rest mass M_0 of baryons is entrained in the volume, and the energy-to-mass ratio in the initial fireball is thus $\eta = E/M_0 c^2$. The evolution from these initial conditions in the context of gamma-ray bursts was pioneered by Cavallo & Rees (1978). It depends on one other parameter, the optical depth, τ, of the fireball. For energy primarily in MeV photons, this will be dominated by photon-photon scattering and pair production, so that $\tau \sim E\sigma_T/r_{in}^2 m_p c^2$ (σ_T is the Thompson cross section; see Cavallo & Rees 1978). Given energies appropriate for cosmological GRBs, the optical depth is very high. This means the internal energy can only be converted into kinetic energy, and an adiabatically expanding explosion is initiated.[1] A phase of acceleration now begins, and we can derive the rest-frame temperature T' and bulk Lorentz factor Γ of the exploding fireball from thermodynamics and energy conservation. Adiabatic expansion dictates that $T'V^{\gamma_a - 1} = const.$, so that the temperature decreases with fireball radius as $T' \propto R^{-1}$ (V is the source volume and γ_a the adiabatic index of the gas; $\gamma_a = 4/3$ for an ultrarelativistic gas). The total internal plus kinetic energy in the frame of an external observer equals $E = \Gamma M_0 (kT'/m_p + c^2)$. For relativistic temperatures, the first term dominates, so that $E \propto \Gamma T' = $ const. Combined with the thermodynamic relation, we then find that $\Gamma \propto R$. The bulk Lorentz factor of the gas thus increases linearly with radius, until it saturates at a value $\Gamma_0 \sim \eta$, at a radius $r_c \sim \eta r_{in}$. Beyond r_c, the now cold shell coasts along at constant Lorentz factor. Because all the matter in it has moved with $v \simeq c$ since the beginning, it is all piled up in a thin shell with thickness R/Γ^2 near the leading edge (Mészáros & Rees 1993).

As with supernovae, the coasting (or ballistic) phase ends when the energy contained in material swept up by the shell becomes a significant fraction of the total energy. The shell drives a blast wave into the ambient medium, with Lorentz factor $\gamma = \Gamma\sqrt{2}$ (Blandford & McKee 1976); the qualitative evolution of this system is easily understood in the case where the shocked ambient gas does not radiate. While the first material is swept up, the shock moves with Lorentz factor $\gamma \sim \gamma_0 \gg 1$. The shock jump conditions (Blandford & McKee 1976) imply that the rest-frame thermal energy of the mass, m, swept through the shock is γmc^2. In our frame, this energy is blueshifted, so that $E_s \sim \gamma^2 mc^2$. This means that about half the initial explosion energy, $E \sim \gamma_0 M_0 c^2$, resides in the swept-up mass when $m \sim m_{dec} = M_0/\gamma_0$, at which point the kinetic energy loss of the shell begins to be significant and the shell deceleration starts in earnest. This is much sooner than in the non-relativistic case, where $m_{dec} \sim M_0$. Well beyond this point, the shocked gas dominates the mass and energy of the expanding system, and thus

---

[1] Ironically, Cavallo & Rees already recognized and discussed the evolution in this opaque limit. At the time, however, GRBs were believed to be very nearby, and consequently they emphasized the behavior of far lower-energy Galactic versions of the model.

$E \simeq E_s \propto \gamma^2 m = const.$, so γ slowly decreases as $\gamma \propto m^{-1/2}$. For the simplest case of a uniform ambient medium, we then have the standard result for a self-similar relativistic blast wave: $\gamma \propto r^{-3/2}$.

To get the corresponding observer times, we need to introduce the kinematic relation between radius and observer time for a relativistically approaching shell (Rees 1964):

$$dt = dr/2\gamma^2 c. \tag{2}$$

Before serious deceleration, we may set $\gamma = \gamma_{0,2}$ and omit the differentials. For a pure hydrogen medium with number density $n = \rho/m_p$ we then get the radius and observer time of the onset of deceleration (Rees & Mészáros 1992):

$$r_{dec} = \left(3E_0/4\pi\gamma_0^2 nm_p c^2\right)^{1/3} = 3.8 \times 10^{16}\,(E_{52}/n)^{1/3}\,\gamma_{0,2}^{-2/3}\ \text{cm} \tag{3}$$

$$t_{dec} = r_{dec}/2\gamma_0^2 c = 63\,(E_{52}/n)^{1/3}\,\gamma_{0,2}^{-8/3}\ \text{s}. \tag{4}$$

Compared with the supernova case, deceleration takes place at somewhat smaller radii, but most starkly within seconds rather than centuries; this is caused by the higher speed of the ejecta, but even more by the two Lorentz factors in Equation 2. The relativistic phase ends and turns into the usual supernova behavior (Sedov 1969 Ch. IV, Taylor 1950) when the energy per particle approaches $m_p c^2$. At that time the Lorentz factor corrections disappear, and we may estimate the size simply as ct, hence

$$t_{NR} = \left(3E_0/4\pi nm_p c^5\right)^{1/3} = 1.2\,(E_{52}/n)^{1/3}\ \text{yr}. \tag{5}$$

As we see from the preceding qualitative discussion, the relativistic self-similar phase can span many orders of magnitude in time. We now derive the evolution for the blast wave more precisely and generally, following Huang et al (1999; see also Katz & Piran 1997). We assume that material passing through the shock quickly radiates a fraction ϵ of the post-shock thermal energy, and retains the rest as thermal energy. The total (kinetic plus thermal) energy of the burst is then $E = (\gamma - 1)(M_0 + m)c^2 + (1 - \epsilon)\gamma U$, and the radiated energy is $dE_{rad} = \epsilon\gamma(\gamma-1)c^2\,dm$ (Panaitescu & Mészáros 1998a, Blandford & McKee 1976). From the shock jump conditions, one finds $U = (\gamma - 1)mc^2$, which in combination with $dE = dE_{rad}$ gives

$$\frac{d\gamma}{dm} = -\frac{\gamma^2 - 1}{M_0 + \epsilon m + 2(1 - \epsilon)\gamma m}. \tag{6}$$

For fully radiative ($\epsilon = 1$) and fully adiabatic ($\epsilon = 0$) blast waves, one can write analytic solutions for the expansion. Equation 6 differs somewhat from the sometimes quoted

$$\frac{d\gamma}{dm} = -\frac{\gamma^2 - 1}{M}, \tag{7}$$

in which M is the total rest-frame mass (inclusive of mass equivalent of internal energy; Chiang & Dermer 1999, Katz 1994b). For $\epsilon = 1$, the forms are equivalent, but the adiabatic form of Equation 7 is only valid in the ultrarelativistic case: its solution near $\gamma = 1$ should yield the classical Sedov-Taylor solution but does not. Equation 6 does yield the Sedov-Taylor solution in the non-relativistic limit.

3.1.1 Fully Radiative Case, $\epsilon = 1$

The equation of motion is easily integrated to give (Blandford & McKee 1976, Huang et al 1999, Katz & Piran 1997)

$$\left(\frac{\gamma - 1}{\gamma + 1}\right)\left(\frac{\gamma_0 + 1}{\gamma_0 - 1}\right) = \left(\frac{M_0 + m_0}{M_0 + m}\right)^2, \qquad (8)$$

where $\gamma_0 \sim \eta/2$, $m_0 \sim M_0/\eta$ are the initial conditions of the deceleration. The ultrarelativistic limit is obtained for $M_0 \gg m$, since the deceleration starts in that limit, and the large radiative losses mean that the shock comes to a halt after having swept up only a few times m_0. Expanding Equation 8, we obtain $\gamma m = M_0$, i.e. $\gamma \propto m^{-1}$. For a uniform ambient medium, this gives the well-known $\gamma \propto r^{-3}$. Inserting this into the kinematic relation between radius and observer time (Equation 2), we find that $\gamma \propto t^{-3/7}$ and $r \propto t^{1/7}$: the Lorentz factor decreases substantially at very slowly changing radius. In the non-relativistic limit, the solution becomes $\beta \propto m^{-1}$, which for a uniform medium is the snowplow (radiative) phase of supernova remnant expansion.

3.1.2 Adiabatic Case, $\epsilon = 0$

The equation of motion for this case is easily integrated to give $\gamma M_0 + (\gamma^2 - 1)m = const$. To interpret this result, we subtract M_0 and multiply by c^2, and note that the result is the sum of thermal and bulk kinetic energies of the blast wave, hence

$$(\gamma - 1)M_0 c^2 + (\gamma^2 - 1)mc^2 = E_{K0}, \qquad (9)$$

where E_{K0} is a constant. In the ultrarelativistic limit ($m \gg M_0/\gamma$ and $\gamma \gg 1$), this implies $\gamma^2 m = const$. For a uniform ambient medium, this then gives $\gamma \propto r^{-3/2}$ as derived qualitatively above. With the kinematic $r-t$ relation, this then gives $\gamma \propto t^{-3/8}$ and $r \propto t^{1/4}$. This demonstrates that although the self-similar relativistic case can cover 6–7 orders of magnitude in time, the range in radii it covers is not nearly as large. In the non-relativistic limit ($\gamma \sim 1$, $m \gg M_0$) we get $\beta \propto r^{-3/2}$, which is the proper Sedov-Taylor solution.

The above relations define (the limiting cases of) the dynamical behavior of the blast wave. Unfortunately, the great distances make the blast waves unresolved on the sky, so we cannot directly test the relations between r, t, and β, γ. [The scintillation size determination of GRB 970508 (Frail et al 1997c) is the one exception to this.] Therefore, we will have to describe the flux and spectrum of radiation from the expanding shock before we can find direct tests of our model from data.

3.2 Radiation

The material behind the shock has relativistic temperatures; because energy transfer between particles in two-body collisions becomes less efficient with increasing temperature, many common emission mechanisms fare poorly in the shock-heated gas. The one mechanism that does well with relativistic particles is synchrotron radiation—provided a significant magnetic field is present. These efficiency considerations made synchrotron emission a favored model early on. By now, its dominant importance has been confirmed for the afterglow phase of the burst, and we will develop the theory for this case. The origin of the burst emission is less clear; although it could be synchrotron emission, its nature is still the subject of lively debate (Section 6.4).

3.3 The Afterglow Emission

In the afterglow, rest frame densities of particles and photons are quite small, and the electron scattering optical depth will typically be 10^{-5} or less. This means that optical depth effects can be neglected for now, and that two-body encounters are not efficient in producing radiation. Let us therefore consider synchrotron emission, which, as we will see, provides very good fits to the afterglow spectra and light curves. The first precise prediction of this emission was made by Mészáros & Rees (1997). The relations between the observable synchrotron spectra and shock parameters have been derived in a number of papers (Rees & Mészáros 1992, Paczyński and Rhoads 1993, Waxman 1997c, Granot et al 1999b, Wijers & Galama 1999). Here we omit details, and follow the notation and coefficients of Wijers & Galama (1999).

The formation of magnetic fields and acceleration of particles at and in the wake of a relativistic shock front are not yet well understood. For now, we summarize our ignorance of the post-shock parameters in as few parameters as possible. The post-shock conditions are related to the pre-shock ones by jump conditions (Blandford & Mckee 1976), which in the ultrarelativistic limit ($\gamma \gg 1$) can be approximated as

$$n' = 4\gamma n; \qquad U' = 4\gamma^2 n m_\mathrm{p} c^2, \tag{10}$$

where primed quantities are in the rest frame of the shock, n is the number density, and U is the energy density. (Strictly speaking, if the medium is not pure hydrogen, n denotes the nucleon density.) We assume that the energy density in magnetic field and relativistic electrons are a fixed fraction, ϵ_B and ϵ_e, respectively, of the post-shock total energy density. For the magnetic field, this means

$$B' = \sqrt{8\pi \epsilon_B U'} = \gamma c \sqrt{32\pi \epsilon_B n m_\mathrm{p}}. \tag{11}$$

For the electrons, we have to specify a shape of the energy distribution as well as a total energy. In many acceleration mechanisms, the electron Lorentz factor distribution becomes a power law above some minimum Lorentz factor γ_m and extends to much higher energies. For a power-law index p we can integrate this simple distribution and relate the Lorentz factor distribution to the total energy in relativistic electrons. We can then relate the minimum electron Lorentz factor to the energy fraction in these electrons:

$$\gamma_m = \frac{2}{1+X}\frac{m_p}{m_e}\frac{p-2}{p-1}\epsilon_e\gamma, \tag{12}$$

where X is the hydrogen fraction by mass and the first term is (to good approximation) the electron-to-nucleon number ratio in the ambient gas.

The calculation of the synchrotron radiation spectrum given the electron energy and magnetic field is lengthy, but standard [see Rybicki & Lightman (1979), Ch. 7 for a general treatment, and Wijers & Galama (1999) for an example of application to GRBs]. Figure 3 illustrates the schematic shape of the spectrum from radio to X rays for the typical adiabatic case from hours to weeks into the afterglow. The spectrum at late times is divided up into four regions. At very low frequencies the spectrum is self-absorbed and follows a blackbody shape, $F \propto \nu^2$. This region ends at the self-absorption frequency, ν_a, characteristically a few GHz, above which we find the standard low-frequency synchrotron slope $F \propto \nu^{1/3}$, up to the peak frequency ν_m. This frequency corresponds to the minimum-energy electrons; the flux at this point, from which the spectrum is usually anchored, is denoted as F_m. Above this, the slope of the spectrum depends on the electron energy index p in the usual way: $F \propto \nu^{-(p-1)/2}$. Finally, at very high energy, the injected electrons cool more rapidly than the characteristic time of the source (the expansion time), and thus the spectrum becomes steeper by a power $1/2$ at a third characteristic frequency ν_c, the cooling frequency. Schematically, for a uniform ambient medium, we find:

$$\nu_a = f_a(\epsilon_e, \epsilon_B, n, E, p, z)\,\text{Hz}$$

$$\nu_m = f_m(\epsilon_e, \epsilon_B, n, E, p, z)\,t^{-3/2}\,\text{Hz}$$

$$\nu_c = f_c(\epsilon_e, \epsilon_B, n, E, p, z)\,t^{-1/2}\,\text{Hz} \tag{13}$$

$$F_m = f_F(\epsilon_e, \epsilon_B, n, E, p, z)\,\text{Hz}$$

The detailed forms of the coefficients can be derived (Wijers & Galama 1999b, Granot et al 1999b) but are somewhat uncertain. The main reasons are that the derivations usually assume a uniform medium behind the shock, as opposed to the true self-similar shock structure (Blandford & McKee 1976). Also, emission from the shell is strongly forward-beamed within an angle $1/\gamma$ of the radial direction, which means we need to average the spectrum properly over the shell rather than just take the point on our line of sight as representative. All these effects, along with proper radiation transport, can be accounted for with enough patience, and some numerical work has been done on parts of the problem. Generally, they lead to much smoother changes between regimes, and to changes of order unity in the coefficients of Equation 13. There are also some true physical uncertainties—e.g. some assumption must be made about the pitch angle distribution of electrons relative to the field (here we take them to be isotropic, and to remain so even during cooling). These depend on the mechanisms of field formation and electron acceleration in the shock, about which little is yet known (but see later sections). The time dependence of the frequencies means that at very early times, typically minutes after the burst, we can have $\nu_c < \nu_m$, in which case the shape and evolution of the spectrum are different (Cohen et al 1998). Likewise, at very late times we may have $\nu_m < \nu_a$, which changes the evolution and shape of the radio spectrum. This may be so late that other effects, such as the blast turning non-relativistic, also play a role, preventing a clear-cut observation of either effect by itself. The bulk of current afterglow data are hours to weeks after the burst, however, where the spectrum and time evolution of Figure 3 and Equation 13 apply.

Despite some uncertainties in the coefficients, a number of very important inferences and tests of the model can be obtained from these relations. First, we note that for a burst with known redshift Equation 13 has four measurable quantities and four unknowns (p also follows from the measured spectrum). Therefore, the blast wave properties can be derived from a complete spectrum, and this has been accomplished in a few cases (Section 4). Also, because the spectrum evolves to

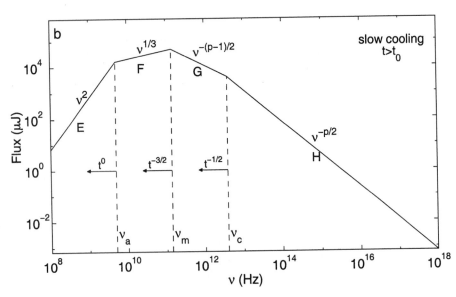

Figure 3 The piecewise power-law schematic shape of blast wave synchrotron spectra for later afterglow evolution (Sari et al 1998). The characteristic break frequencies and their time evolution are indicated, as is the spectral slope in each regime. This can be directly compared with the observed spectrum of GRB 970508 (Figure 12).

lower frequencies with time, one can sample in a given waveband and see the spectral breaks pass by. In this way, Frail et al (2000) have been able to measure the physical parameters of GRB 980703.

Second, if ϵ_e and/or ϵ_B were not constant, then the time dependencies of the measurable variables would change. Hence the observation that many afterglows follow the simple model predictions seems to justify the a priori uncertain ansatz that the electron and field energy densities scale like equipartition values—i.e. they are constant fractions of the total shock energy density. Given that sometimes $\epsilon_B \ll 1$ (e.g. in GRB 980703: Vreeswijk et al 1999, Bloom et al 1998b; and in GRB 990123: Galama et al 1999), the apparent good validity of this scaling is somewhat remarkable.

Third, now that we have the shape of the spectrum and the evolution of the breaks, we can compute the time evolution in any fixed observed waveband as well (Mészáros et al 1994, Wijers et al 1997, Sari et al 1998, Mészáros & Rees 1997, Mészáros et al 1998). Below ν_m, the spectral shape is independent of the electron spectrum, and its evolution has no free parameters; flux should increase as $t^{1/2}$ at all these frequencies. Above ν_m, the a priori unknown p enters. Recent theoretical calculations of particle acceleration in relativistic shocks (Gallant & Achterberg 1999, Gallant et al 1999) give $p \sim 2.3$–2.4, in good agreement with the observed range of 2–2.5. But observational precision is often much better, so the theoretical prediction is not yet accurate enough. Because observations often give both the spectral index and the time decay rate over a range of frequencies and times, however, this is only one parameter for two measured power-law indices. Therefore, this still leaves one test of the model. In practice, what one does is fit a model of the form $F(\nu, t) \propto \nu^{-\beta} t^{-\delta}$ to the data, and determine β and δ. Since $\beta = \beta(p)$ and $\delta = \delta(p)$, we then have a theoretical relation $\beta = \beta(\delta)$, which we may test with the fit to the data. A good summary of the scaling relations in the various spectral regimes for spherical fireballs is given by Sari, Piran, and Narayan (1998).

It must be noted first, though, that the relation is different for each dynamical evolution model of the blast wave, and also depends on whether our observed frequency is above or below ν_c. Therefore, unless we have other data from which to decide on the appropriate regime, our prediction will in fact consist of a few possible relations. In practice, the various possible predictions of β for a given δ differ only by 0.1–0.3; hence it is necessary to measure the spectral and temporal slopes of the afterglow to a few percent accuracy in order to obtain a stringent test of the model. Let us take the case of the adiabatic spherical fireball as an example, and imagine we have measured an optical temporal decay index $\delta = 1.0 \pm 0.1$. We have also measured a $B - R$ color, and from this deduced that the optical spectral index is $\beta = 0.70 \pm 0.03$. Is the optical band above or below ν_c? Well, if ν_c were below the optical, we would expect $\beta = (2\delta + 1)/3 = 1.00 \pm 0.07$, whereas if it were above the optical, we would expect $\beta = 2\delta/3 = 0.67 \pm 0.07$. Thus we see that the color measurement establishes that the latter is true with fair certainty, but for the somewhat larger errors in β and δ that are often obtained in practice the question is not resolvable.

4.   HIGHLIGHTS OF AFTERGLOW OBSERVATIONS

In late 1996 the BeppoSAX Team, following the derivation of an accurate (but not rapidly available) WFC error box for GRB 960720 (In't Zand et al 1997, Piro et al 1998b), was ready to quickly produce arcminute-sized WFC error boxes for subsequent follow-up observations. The first opportunity to apply this capability occurred on January 11, 1997.

The initial 10′ radius WFC error box (Costa et al 1997a), partly cut by the BeppoSAX-Ulysses IPN annulus, contained several faint X-ray sources detected with the BeppoSAX Narrow Field Instruments (NFI), some of which are also present in the ROSAT Sky survey (Butler et al 1997, Voges et al 1997, Frontera et al 1997, Feroci et al 1998). One of these sources coincided with a variable radio source (Frail et al 1997b, Frail et al 1997a). However, an improved (3′ radius) WFC error box (In't Zand et al 1997) obtained some three weeks after the event contained none of these objects. Likewise, optical observations that started less than a day after the burst gave no evidence for a candidate GRB counterpart (Gorosabel et al 1998).

4.1   Solving the GRB Riddle

The next opportunity, which would lead to a breakthrough, arose on February 28, 1997. After the hard X rays from the burst were discovered with a Wide Field Camera on BeppoSAX, pointed BeppoSAX NFI X-ray observations were made $\sim$8 hours later, and revealed the presence of an unknown soft X-ray point source with a 2–10 keV flux of several 10^{-12} erg/cm^2 s in the WFC error box. Four days later, the flux of this X-ray source had decreased by a factor $\sim$20; the first X-ray afterglow of a GRB had been detected (Costa et al 1997b; Figure 4, see color insert).

In the meantime, GRB 970228 had also become the first γ-ray burst for which an optical counterpart was found, independently from the soft X-ray afterglow detection. From a comparison of (V and I band) images made with the William Herschel and Isaac Newton Telescopes 21 hours and a week after the burst, Groot et al (1997b) discovered a decaying 21st magnitude object at a position consistent with all positional information on the γ-ray burst (Van Paradijs et al 1997; Figure 5). Subsequent deep images made with the ESO New Technology Telescope (Groot et al 1997a) and the Keck Telescope (Metzger et al 1997c) showed an $\sim$1″ extended object at the location of the optical transient, likely the host galaxy of the γ-ray burst (implying that the burst came from a distance of order Gpc).

HST observations were made in late March and early April 1997 (Sahu et al 1997) and in September 1997 (Fruchter et al 1999). These observations showed the presence of a point source whose brightness decayed according to a power law located near the edge of an extended object (diameter $\approx 0.8''$; Figure 6, see color insert). Later observations established its redshift as 0.695 (Djorgovski et al 1999), confirming that the host of GRB 970228 is a distant galaxy.

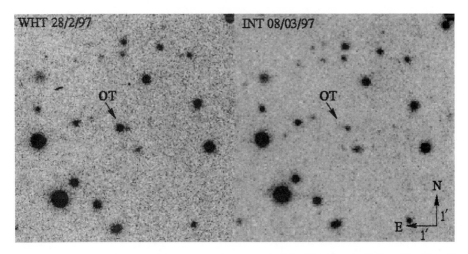

Figure 5   Discovery images of the optical afterglow of GRB 970228 at La Palma (Van Paradijs et al 1997).

The optical light curve of the afterglow of GRB 970228 showed a significant deviation from a pure power law (Galama et al 1997) about 1–2 weeks after the burst. Recently this has been interpreted as possible evidence for the presence of a supernova component in the light curve (Reichart 1999, Galama et al 2000; Figure 7), following an original suggestion for a similar deviation in GRB 980326 (Bloom et al 1999a). (See Section 4.3 for further discussion on the connection between γ-ray bursts and supernovae.)

The exponents of the power-law decay X-ray (Figure 8) and optical light curves were $\delta_X = -1.33^{+0.12}_{-0.11}$ and $\delta_{opt} = -1.46 \pm 0.16$. The exponents of the (assumed power law) spectrum of the afterglow were $\beta_X = -1.06 \pm 0.24$ and $\beta_{X\text{-opt}} = -0.78 \pm 0.02$ for the X-ray waveband and for the X-ray to R band interval, respectively. The relation between δ and β is consistent with that expected from a very simple version of the fireball model (see Section 3), and gave initial support to this model (Wijers et al 1997). However, after subtraction of the SN contribution, the temporal slope steepens and the light curve may be better fit by a blast wave propagating in a stellar wind (Chevalier & Li 1999).

The next big step forward was made with GRB 970508, whose optical counterpart (Bond 1997) initially increased in brightness from the first to the second night after the γ-ray burst and reached a maximum magnitude R $\simeq$ 19.8 two days after the burst (Pedersen et al 1998). It then started a power-law decline that could be followed for several hundred days (Galama et al 1998a, Djorgovski et al 1997, Sokolov et al 1998, Garcia et al 1998), and eventually flattened off, revealing the presence of a host (Zharikov et al 1998, Bloom et al 1998a, Fruchter et al 2000).

Optical spectroscopy obtained with the Keck telescope revealed the presence of absorption lines of Mg II, Fe II, and Mg I (Metzger et al 1997b; Figure 9) which are

light curves of GRB 970228

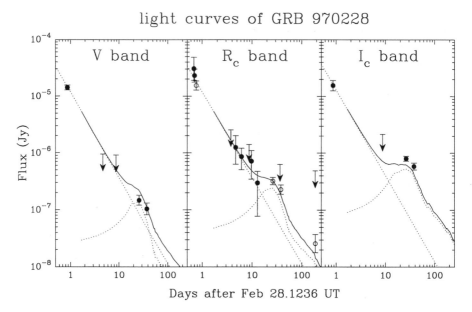

Days after Feb 28.1236 UT

Figure 7 The *VRI* light curves of the afterglow of GRB 970228, showing the evidence of a supernova component superposed on the power-law decline at late times (Galama et al 2000).

often found in quasar spectra (Steidel & Sargent 1992), redshifted by $z = 0.835$ (Metzger 1997). The subsequent discovery of [O II] and [Ne III] emission lines in the spectrum at the same $z = 0.835$ (Metzger et al 1997a) established the presence of an underlying host galaxy. HST observations revealed that the OT was at the center of a blue, actively star-forming dwarf galaxy (Pian et al 1998, Fruchter et al 2000, Natarajan et al 1997; Figure 10, see color insert) of $L_B = 0.12L_*$ and with a star formation rate of ≤ 1 $M_\odot$/yr (Bloom et al 1998a). This result unambiguously established that GRBs originate at cosmological distances, and terminated the discussion on the GRB distance scale: it established γ-ray bursts as the most luminous photon emitters in the entire universe, with peak luminosities of order $L\gamma \simeq 10^{52}$ erg/sec. The optical afterglow of GRB 970508 reached an absolute magnitude $M_V \simeq -24$, i.e. in optical emission it became two orders of magnitude brighter than a type Ia supernova.

GRB 970508 also has the distinction of being the first γ-ray burst for which radio afterglow was detected (Frail & Kulkarni 1997). During the first month the radio flux underwent strong irregular variations, around an average value of 0.6 mJy (8.5 GHz), which damped out after about a month (Figure 11). These variations are caused by interstellar scintillation in our Galaxy. The damping of the fluctuations reflects the increase in the size of the radio emitter (analogous to the absence of twinkling of planets). From the known source distance and the properties of the interstellar medium along the line of sight, Frail et al (1997c,

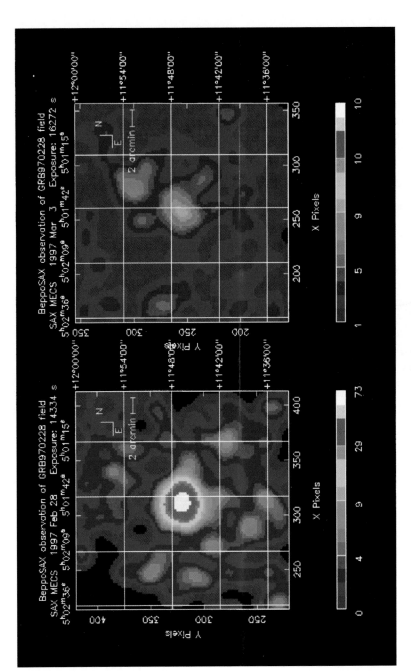

Figure 4 Discovery images of the X-ray afterglow of GRB 970228 with the BeppoSAX NFI (Costa et al 1976).

Figure 6 The host galaxy of GRB 970228 (*center*), imaged with HST. The six-month-old afterglow is still visible (*bright pixel*) at the top right edge of the host (Fruchter et al 1999).

Figure 10 HST/STIS image of the OT of GRB 970508 (Pian et al 1998). The faint host only became visible after the OT had faded.

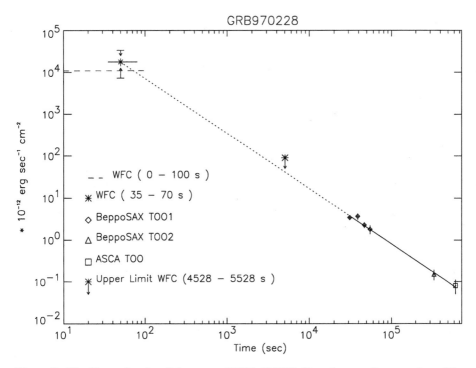

Figure 8 The X-ray afterglow light curve of GRB 970228. Note the smooth connection of the afterglow with the prompt X-ray flux from the GRB (Costa et al 1997b).

1997) inferred a value for the source size at the time the fluctuations disappeared. They concluded that the radio emitter expanded with a velocity consistent with that of light, which gives strong support to the relativistic fireball model for GRBs and their afterglows.

GRB 970508 was the first γ-ray burst for which the afterglow was observed all the way from X-rays, via optical/near IR, to mm and low-frequency radio waves (see Section 5). This allowed Galama et al (1998d) to reconstruct the X-ray to radio spectral energy distribution of the afterglow, as observed 12 days after the γ-ray burst (Figure 12). This spectrum consists of piecewise connected power-law distributions, with three clearly recognizable spectral breaks at the frequencies at which the different power laws are connected. At the low-frequency side we recognize the $\nu^{1/3}$ synchrotron low-frequency limit, which at even lower frequencies turns over because of self-absorption. The high-frequency end of the $\nu^{1/3}$ part of the spectrum connects, at the peak of the spectrum, to a power-law part whose slope depends on the power, p, of the power-law electron energy (or Lorentz factor) distribution and on particulars of the fireball synchrotron emitter (see Section 3). In the near infrared a third break is seen; this can be unambiguously identified with the cooling break, whose frequency corresponds to an electron

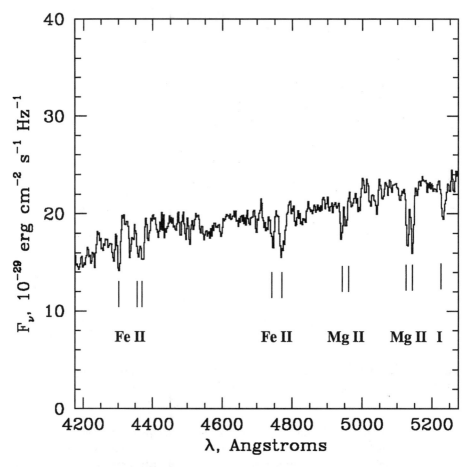

Figure 9   The spectrum of the OT of GRB 970508, showing Fe and Mg absorption lines at z = 0.835 and z = 0.77 (Metzger et al 1997b).

energy above which the synchrotron loss time is smaller than the flow timescale of the system. The identification is based on two observed facts: (1) the change in spectral slope equals 0.5 (Galama et al 1998d; Figure 12), and (2) the frequency, ν_c, of the cooling break changed with time since the burst as $\nu_c \propto t^{-1/2}$, as was evident from the progression of the spectral slope change (by 0.5) from the optical to the near-infrared passbands (Galama et al 1998a; Figure 13). These results show that relatively simple versions of the fireball model provide a reasonable description of the observed afterglow spectrum, and provide a strong argument for the idea that GRB afterglows are powered by the synchrotron emission of electrons accelerated in a relativistic shock.

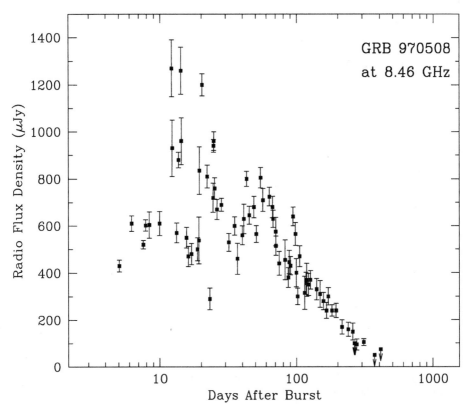

Figure 11   The 8.46 GHZ VLA light curve of the afterglow of GRB 970508 (Kulkarni et al 2000). Note the large scintillation fluctuations in the first month and their later absence, indicating that the source expanded (Frail et al 1997c).

4.2   Dark GRBs

The relative optical response in the afterglow of a γ-ray burst can vary enormously from one burst to another (for the X-ray afterglows the variations are more moderate; see Section 5). A very good example is provided by GRB 970828, a fairly strong γ-ray event for which optical observations were made within four hours after the burst and continued for eight consecutive nights after the burst. No potential optical counterpart was found to vary by more than 0.2 mag down to R = 23.8 (Groot et al 1998a). Compared to GRB 970508 the peak optical flux in the afterglow of GRB 970828, normalized to the fluence of the γ-ray burst, was at least a factor of 10^3 smaller. Absorption in the GRB host galaxy, by at least five magnitudes in the R band, may explain their large difference in optical afterglows. Note that for moderate redshifts ($z \sim 1$) the absorbed photons have wavelengths of $\sim$3000 Å, at which the interstellar extinction is a factor of $\sim$2.5 larger than in the

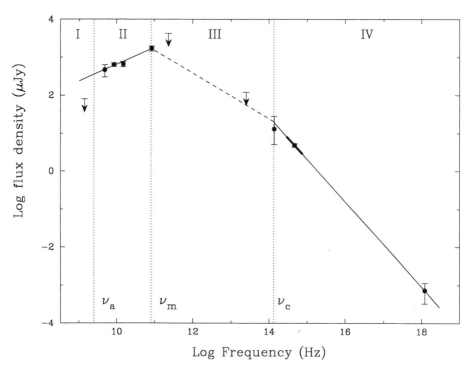

Figure 12 The radio to X-ray spectrum of the afterglow of GRB 970508, 12.1 days after trigger, showing all the characteristics of synchrotron emission (Galama et al 1998d).

R band. Therefore, with a normal dust-to-gas ratio in the host, a modest column density of $N_H \sim 10^{21}$ cm^{-2} would be sufficient to provide the large extinction. Recent results support this suggestion: at the position of GRB 970828 there is a galaxy with $z = 0.96$ (Djorgovski et al 2000) and $N_H = 4.1^{+2.1}_{-1.6} \times 10^{21}$ cm^{-2} (as measured in our rest frame by Yoshida et al 1999).

4.3 The Supernova–GRB Connection

The total energy budgets of γ-ray bursts corresponding to their cosmological distances are roughly of the same order of magnitude as those of supernovae; in γ-ray bursts, however, the energy is emitted much more rapidly than in supernovae. The reason is that in supernovae the energy (except that in neutrinos) is thermalized by a large amount of mass (several solar masses) and converted into heat and expansion at a speed of typically 10^4 km/s. In a γ-ray burst the energy cannot be shared with more than $\sim 10^{-3}$ M$_\odot$ in baryons, in order not to lose the required high Lorentz factors ($\Gamma \geq 10^2$). The similar amount of total energy (10^{53} ergs) involved in both phenomena kept the possibility of their connection open, despite the very large difference in the way the energy is emitted. For instance, in one of

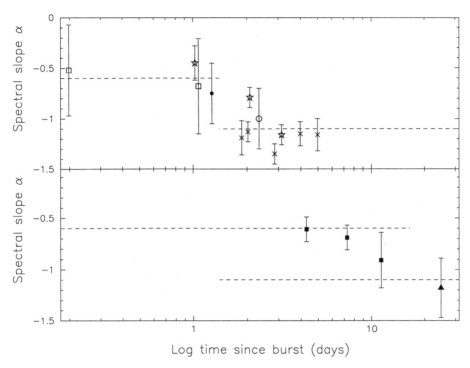

Figure 13   The evolution of the optical spectral index (top: $V - R$ and bottom: $R - K$) of the afterglow of GRB 970508, showing the passage of a break consistent with a cooling break (Galama et al 1998a).

the main γ-ray burst models, the collapsar or failed supernova model (Woosley 1993, Paczyński 1998), a γ-ray burst is the result of a massive core collapse, with extreme parameters (e.g. extreme mass or rotation of the progenitor). However, until April 1998, direct evidence for a relation between γ-ray bursts and supernovae was totally lacking.

It therefore came as something of a surprise when Galama et al (1998c, 1998b) found that the WFC error box of GRB 980425 contained the supernova SN 1998bw (Figure 14). This supernova is located in a spiral arm of the nearby galaxy ESO 184–G82, at a redshift of 2550 km/s, corresponding to a distance of 40 Mpc. On the basis of very conservative assumptions regarding the error box and the time window in which the supernova occurred, Galama et al (1998c) determined that the probability that any supernova with peak optical flux a factor of 10 below that of SN 1998bw would be found in the error box by chance coincidence is 10^{-4}; this provides strong evidence for a physical relation between the γ-ray burst and the supernova. The WFC error box also contains two weak X-ray sources, one of which, 1SAX J1935.0-5248, coincides with SN 1998bw and GRB 980425 (Piro et al 1998a, Pian et al 1999); the second source initially faded

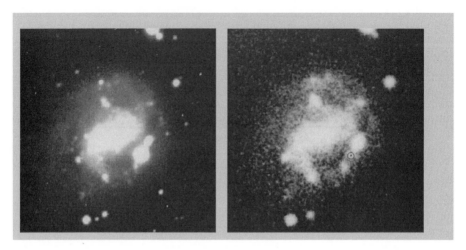

Figure 14   Image of the galaxy ESO 184-G82 with (*left*) and without (*right*) SN 1998bw (Galama et al 1998c).

but was subsequently redetected at a flux similar to the first observation (Pian et al 1999), which excludes it as a viable GRB afterglow. The case for a physical relation between the supernova and the GRB is therefore a strong one.

With respect to its apparent properties (peak flux, duration, burst profile) GRB 980425 was not remarkable. Of course, at its distance of 40 Mpc, its total energy $(8 \times 10^{47}$ erg/s) is some five orders of magnitude smaller than that of normal γ-ray bursts (Galama et al 1998c). The total energy in GRB 980425 is remarkably close to the value envisaged in Colgate's (1974) model.

Independent of its connection with a γ-ray burst SN 1998bw is extraordinary for its very high radio luminosity near the peak of the SN light curve (Kulkarni et al 1998b). According to the analysis of Kulkarni et al (1998b) the radio light curve requires the presence of a mildly relativistic ($\Gamma \sim 2$) outflow, which may account for the γ-ray burst emission. An analysis of the optical light curve (Galama et al 1998c; Figure 15) and its early spectra (Iwamoto et al 1998, Woosley et al 1999, Branch 1999) showed that SN 1998bw was an extremely energetic event [total explosive energy in the range $(2-6) \times 10^{52}$ erg, i.e. a factor of $\sim$30 higher than is typical for an Ib/c supernova], in which an extraordinarily large amount of ^{56}Ni $(0.5-0.7\ M_\odot)$ was ejected. The early expansion speed was as high as $\sim$60,000 km/s. According to Iwamoto et al (1998; see also Iwamoto 1999a, Iwamoto 1999b), the remnant mass of the core collapse exceeded $3\ M_\odot$, and a black hole was likely formed in SN 1998bw. [Note that by allowing for asymmetry in the supernova explosion, Höflich et al (1999) derive somewhat more moderate but still very large values for the energetics of this supernova.] GRB 980425 is the only γ-ray burst (out of more than 2000) for which the evidence of a connection with a supernova appears convincing. Attempts to search for further associations have not

SN 1998 bw light curves

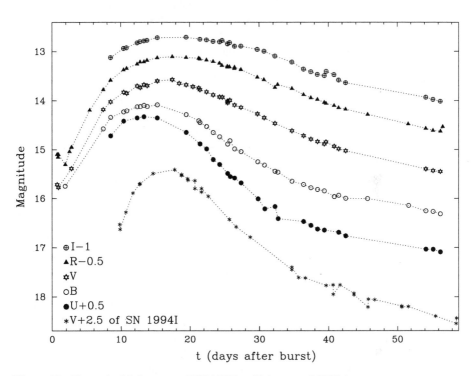

Figure 15   The optical light curve of SN 1998bw (Galama et al 1998c).

led to other strong candidates (Wang & Wheeler 1998, Woosley et al 1999, Bloom et al 1998c, Kippen et al 1998). [We consider the proposed connections between SN1997cy and GRB 970514 (Germany et al 1999) and between SN1999eb and GRB 991002 (Terlevich et al 1999) not convincing.]

Because the sampling volume for low-luminosity events such as GRB 980425 is smaller than that of the normal γ-ray bursts by a factor $\sim 10^6$, the rate (per galaxy) of the former events may well exceed those of the latter by a large factor. Because of their small distances they are expected to contribute a $P^{-3/2}$ component to the log N ($>P$) distribution. From the absence of a turn-up at the flux limit ($P \simeq 0.2$), Kommers et al (2000a) inferred that such a Euclidean component can contribute at most 10% to the observed BATSE burst sample (99% confidence limit). With a normal GRB rate of $\sim 10^{-8}$ per galaxy per year (Wijers et al 1998), the corresponding limit on events like SN 1998bw is thus a few 10^{-4} per galaxy per year. With an observed rate for type Ib/c supernovae of a few times 10^{-3} per galaxy per year (van den Bergh & Tammann 1991), this rather weak limit serves to show that at most a fraction of the SN Ib/c produce γ-ray bursts.

The observational basis for a connection between γ-ray bursts and supernovae was greatly enriched with the discovery by Bloom et al (1999a) of a late component superposed on the power law optical light curve of GRB 980326, which they argue reflects an underlying supernova (Figure 16). A similar interpretation has been proposed by Reichart (1999) and Galama et al (2000) for the long-known deviation from a pure power-law decay of the optical afterglow of GRB 970228 (Galama et al 1997).

The optical light curve of GRB 980326 showed an initial rapid decay ($\delta_{opt} = -2.0$; Groot et al 1998b); the light curve flattened after ~ 10 days to a constant value $R = 25.5 \pm 0.5$. Such flattening has been seen in the light curves of other afterglows as well, and has been interpreted as the signature of an underlying host galaxy. Observations by Bloom et al (1999a) made ~ 3 weeks after the burst revealed a surprising brightening of the afterglow, to a flux level 60 times above that expected from an extrapolation of the power-law decay. At the same time, the spectral energy distribution became very red. Observations made ~ 9 months after the burst showed that any host galaxy is fainter than $R = 27.3$. Using the multicolor light curve of SN 1998bw (Galama et al 1998c) as a template, Bloom et al (1999a) found that they can reproduce the observed optical afterglow light curve of GRB 980326 by a combination of a power-law (exponent -2.0) and a bright supernova at a redshift $z \sim 1$.

Reichart (1999) and Galama et al (2000) made the same decomposition of the optical light curve of GRB 970228 and found that this provides a good fit to the data. These results support the idea that at least a fraction of the γ-ray bursts originate from the collapse of a massive star. This confirms the models proposed by Woosley (1993) and Paczyński (1998), under various names, such as the failed supernova model or collapsar model, in which it is assumed that a black hole surrounded by a fairly massive torus is formed. It is still unclear what particular circumstances give rise to the GRB (e.g. a very high mass, rapid rotation, a particular evolutionary history), but it seems virtually certain that strong collimation of the outflow is required to accommodate both the extremely high Lorentz factor flow required for the γ-ray burst and the more sluggish flow connected with the supernova photospheric emission.

4.4 Prompt Optical Emission

Based on the fireball model, prompt optical emission simultaneous with the γ-ray burst is expected, with apparent V magnitudes ranging between 9 and 18 for typical GRB distances (Mészáros & Rees 1993, Mészáros & Rees 1997, Sari & Piran 1999, Katz 1994a). After many years of unsuccessful attempts to catch the optical signal of a γ-ray burst in progress (see e.g. McNamara et al 1995, Krimm et al 1996, Hudec & Soldan 1995, Lee et al 1997, Park et al 1997), the first such simultaneous detection was made of GRB 990123 by Akerlof et al (1999). The robotic camera ROTSE, triggered by BATSE, started a sequence of optical

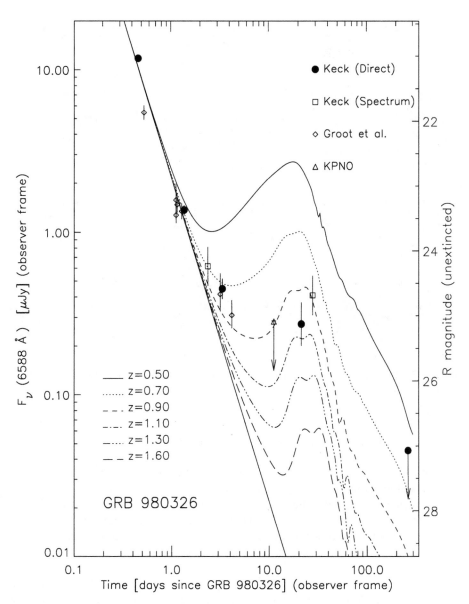

Figure 16   The *R* light curve of GRB 980326, with model curves representing the power-law decay of a relativistic afterglow superposed with the light curve of SN 1998bw, shifted to various redshifts (Bloom et al 1999a).

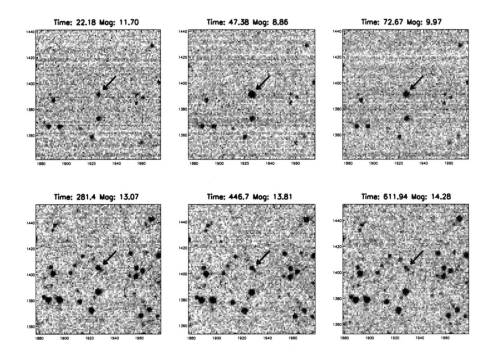

Figure 17   The ROTSE discovery images of the prompt optical emission from GRB 990123, from 22 to 800 s after trigger (Akerlof et al 1999).

images of the BATSE error box 22 seconds after the start of the burst (Figure 17). An optical transient was detected; ~50 seconds after the start of the burst, the transient reached magnitude 8.9 and afterward decayed to ~15th mag in ~10^3 s. The slope of the prompt light curve changed after ~300 s, and merged smoothly with the later afterglow light curve (Figure 18; Kulkarni et al 1999a, Galama et al 1999, Castro-Tirado et al 1999a). During the peak of the prompt optical emission the γ-ray burst reached an absolute magnitude $M_v \approx -36.5$, i.e. the event was then for a brief time interval 10 million times brighter than a Type Ia supernova.

The prompt optical emission is not proportional to the γ-ray flux; neither can it be understood as the low-energy extrapolation of the (variable) γ-ray burst spectrum (Galama et al 1999, Briggs et al 1999; see also insert of Figure 18). This indicates that the prompt optical emission and the γ rays originate from different regions in the fireball. It has been popular practice to ascribe the origin of the γ rays to internal shocks (Rees & Mészáros 1994, Kobayashi et al 1997)

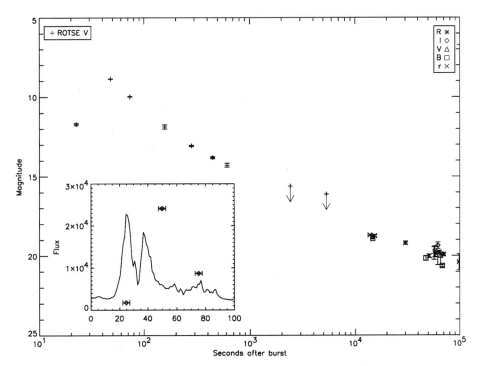

Figure 18   The R light curve of GRB 990123, showing the early emission rising above the back-extrapolated afterglow and then merging smoothly with the later afterglow light curve (Akerlof et al 1999). The γ-ray light curve of ARB 990123; the three points indicate the times during which prompt optical emission was detected with ROTSE (in an arbitrary intensity scale).

and the long-term afterglow to the external shock[2]. It has therefore been natural to ascribe the prompt optical emission to the reverse shock (Sari & Piran 1999, Mészáros & Rees 1999), which is observed only during a time interval of order of the burst duration (i.e. comparable to the time it takes for the reverse shock to travel through the ejecta). The radio afterglow properties of GRB 990123 are peculiar in that radio emission was only seen during a $\sim$ 1-day interval, about a day after the burst (Kulkarni et al 1999b, Galama et al 1999). This brief radio event has been interpreted by Kulkarni et al (1999b) as reverse-shock emission. Galama et al (1999) ascribe it to emission from the forward shock and interpret the peculiar nature of the radio emission as the result of a very low value of the

---

[2]Recent theoretical calculations, however, have reopened the issue: Dermer & Mitman (1999) have presented a plausible external-shock model for highly variable prompt gamma-ray emission, and Fenimore & Ramirez-Ruiz (1999) have shown that previous objections to external-shock models for the prompt emission can be circumvented.

synchrotron peak frequency one day after the burst. They suggest that differences in the afterglow properties (peak frequency, cooling frequency) reflect differences in the magnetic field strength in the afterglow-emitting regions.

4.5   GRB Polarimetry

Synchrotron radiation is highly polarized, with typical degrees of (linear) polarization for ordered magnetic fields of ~60% (Hughes & Miller 1991); one may therefore expect measurable amounts of polarization in afterglow emission. By analogy to AGNs, one might expect up to 10–20% polarization if the shock emission takes place in a collimated outflow. The strong intrinsic polarization is lowered by averaging over the unresolved source (Gruzinov 1999, Gruzinov & Waxman 1999, Medvede & Loeb 1999, Loeb & Perna 1998).

For GRB 990123, (Hjonth et al 1999) reported an upper limit to R band afterglow polarization of 2.3% (95% confidence level). The first positive detection of polarization, however, was made for GRB 990510 (Covino et al 1999, Wijers et al 1999), with polarization $\rho = 1.7 \pm 0.2$ % and 1.6 $\pm$ 0.2%, 0.77 and 0.86 days after the burst, respectively. An uncertain measurement made 1.8 days after the burst is consistent with these values (Wijers et al 1999). The angle of polarization remained constant during these observations. The rather small observed values of the polarization (compared to the high intrinsic values in the synchrotron process) may be the result of a highly tangled structure of the magnetic field, or of very symmetric field geometries.

5.   TABLE OF GRB, COUNTERPART, AND HOST PROPERTIES

We have assembled here a master table of all GRBs (Table 1) that have been rapidly followed up since the launch of BeppoSAX until August 1999 (including, for completeness, GRB 960720). The reason for this selection became apparent upon collecting the relevant literature: In the last year, most of the literature is in the form of IAU and GCN circulars, making for a cumbersome and confusing literature display. For the same reason, whenever all available data were diligently collected into one or more publications, these were preferentially referenced as the main sources of literature. However, we have always referenced the discovery announcements in each wavelength.

We have used the values of the BATSE durations, fluxes, and fluences for the prompt γ-ray emission throughout the table for consistency. These were substituted by the BeppoSAX/GRBM whenever the GRB was occulted for BATSE. The prompt X-ray values were supplied predominantly by the BeppoSAX WFC, and occasionally by the RXTE/ASM; the X-ray afterglow values were consistently supplied by the BeppoSAX NFI.

Detection in a wavelength was posted when generally accepted by the community. In cases that were arguable, we preferred erring on the conservative side, as

TABLE 1: GRB, counterpart, and host properties

Name	UT trig	$(\alpha,\delta)_{2000}$		(ℓ,b)	detected $\lambda\lambda$
other names of the transient					

pr: duration	errbox	F_γ^{peak}	S_γ	F_X^{peak}	S_X	prompt-refs
X: ΔUT_X	F_X^a	$F_X^a[\mu\text{Jy}]$	β_X	δ_X	N_H	X-refs
O: ΔUT_O	mag	F_O	β_O	δ_O	A_V	O-refs
IR: ΔUT_{IR}	mag	F_{IR}	β_{IR}	δ_{IR}		IR-refs
mm: ΔUT_{mm}	ν_{mm}	F_{mm}^a	β_{mm}	δ_{mm}		mm-refs
ra: ΔUT_{ra}	ν_{ra}	F_{ra}^a	β_{ra}	δ_{ra}		ra-refs
ho: z	host photometry (up to 3 bands)				OT offset	host-refs
Telegram-style narrative						

960720 0.48395[321]	17^{h}30^{m}36^s +49°5′49″ (3′)[206, 326]			(75.8,+33.4)	γ,X

pr: 12.5(2.5)[13]	~ 28	0.63(6)	0.29(5)[250]	0.025	0.008(2)[108]	326
X: 45.2	<0.04	<0.002[320, 277]				27, 148
O: —						
IR: —						
mm: —						
ra: 49.5	1.43	<2[99]				
ho: —						

First burst with WFC signal detected during an off-line analysis 45 days after it was recorded[326]. Final ($\rho = 3'$) error box contains[148, 249, 206, 149, 326] radio-loud QSO 4C 49.29. Relation with GRB very uncertain; quoted[149, 326] probability (2×10^{-4}) does not include number of trials. No optical, mm, IR observations.

970111 0.406[60]	15^{h}28^{m}11^s +19°35′54″ (1.8′)[186, 84, 125]			(29.6,+53.4)	γ,X

pr: 31.5(2)[231]	~ 3	4.4(2)	5.8(3)[125]	0.041(7)	0.16(1)[84]	107
X: 0.6875	<0.16	<0.008[84]				206
O: 0.7917	R>22.6	<2.8[146]				125
IR: —						
mm: 30.0	86.4	<10[367]				
ra: 1.2	1.43	<0.5[96]				125
ho: —						

First attempts (unsuccessful) to find low-energy afterglows in *rapid* follow-up observations of BeppoSAX WFC error box. Initial 10′ radius[60] error box contained candidate X-ray[35, 400, 110], radio[97, 125], and optical[237] counterparts, but none of these was included in the improved ($\rho = 3'$)[206] WFC error box[96, 125]. Deep BeppoSAX NFI observations of the further improved WFC error box ($\rho = 1.8'$)[186, 84], led to upper limit to X-ray flux only[84] (weak NFI source not included in the combined IPN/NFI error box). No variable optical afterglow detected within 19 hours of the GRB[146] with $\Delta R < 0.5$ mag at $R > 22.6$ (see also [153, 125]). No mm-wave detection[367].

970228 0.123620[58]	05^{h}01^{m}46.66^s +11°46′53.9″ (2)[399]			(188.9,−17.9)	γ,X,X^a,O
RX J050146+1146.9[112], 1SAX J0501.7+1146[62]					

pr: 80*[62]	4[200]	3.7(1)	1.1(1)^a[112]	0.14(1)	0.22[112]	299
X: 0.3445	2.8(4)	0.15(2)[112]	1.06(24)[112]	$1.33^{+0.13}_{-0.11}$[112]	$3.5^{+3.3}_{-2.3}$[112]	62, 423, 28
O: 0.8679	I=20.6(1)	13.7[399]	0.78(2)[136]	$1.73^{+0.09}_{-0.12}$[136]	0.78(12)[357]	159, 312, 167
IR: 18.7	J>21.5(2)	<4.1[136]				230, 373
mm: 7.9	86.4	<1.2[367]				358
ra: 3.9	4.63	<0.07[101]				124
ho: 0.695(2)[71]	V=25.8(2)[136] R=25.2(2)[136] K=22.9(3)[136]			0.3″[399]	158, 269, 120

First GRB for which X-ray[59, 62] and optical[159, 399] afterglow was detected, in BeppoSAX NFI images taken 0.3 and 4 days, and in (V, I) images taken at La Palma $\sim$1 day and $\sim$1 week after the GRB, respectively. X-ray flux decays as a power-law[62, 423, 112]. Backward extrapolation of X-ray afterglow smoothly joins the late X-ray emission in the GRB tail[62]. Optical afterglow located at the edge of an extended $(0.8'')$ optical source[399, 269, 348], which is a host galaxy at redshift 0.695[71]. Host galaxy among bluest in its magnitude range[120] suggestive of ongoing star formation with SFR$\sim$ 0.4 $M_\odot$/yr and $L_B \sim 0.05 - 0.1 L_*$[71]. Optical light curve[399, 122, 136] deviates in detail from a power-law decay; interpreted as a relatively weak supernova component superposed on the GRB afterglow[136, 338]. No optical counterpart in plate archives[142]. No radio, IR or mm detection[101, 358, 124, 367].

970402 0.9303[85]		$14^h 50^m 6^s$ $-69°20'0''$ $(50'')$[325]		(313.1,$-$8.8)		γ, X, X^a
1SAX J1450.1$-$6920[325]						
pr:	150*[284]	~ 28[183]	0.24(7)	0.82(9)$^{a\,284}$	0.007(2)	0.040(4)[284]	108
X:	0.35	0.22(6)	0.014(3)[284]	0.7[284]	1.57(3)[284]	<20[284]	325, 108
O:	0.77	R>21.0	<12.02[156]				303
IR:	—						
mm:	—						
ra:	—						
ho:	—						

BeppoSAX WFC ($\rho = 3'$)[183] and NFI afterglow[325] detection. Afterglow decays as a power-law[284]; spectral slope consistent with simple fireball model[284]. No variable optical counterpart within 18.5 hours of the GRB[156] with $\Delta R < 0.3$ mag at $R = 21$ (see also [303]). No radio ($\Delta UT = 0.703$d, $\nu =1.43$ GHz) detection[284]; observations with ISO ($\Delta UT = 2.3$d) result in upper limits of 140 and 350 μJy in 12 and 174 μm, respectively[45].

970508 0.904[61]		$6^h 53^m 49.45131^s$ $+79°16'19.51312''$ (15)[388]		(135.0,+26.7)		$\gamma, X, X^a, O, IR, mm, ra$
1SAX J0653.8+7916[323], VLA J065349.4+791619[98]						
pr:	25(4)[231]	~ 3.1[323]	0.61(17)	0.37(4)[231]	0.059(6)	0.033(5)[316, 108]
X:	0.2335	0.70(7)	0.036(4)[316]	1.1(6)[108]	1.1(1)[316]	~ 0[322]	293
O:	0.236	V=21.49(21)	9.1[29, 347]	1.11(6)[126]	1.141(14)[126]	<0.01[134]	81, 140, 309, 39, 315, 377, 426
IR:	4.3	K=18.2(1)	33(4)[53]	0.6(2)[53]	1.2(1)[53]		271, 134, 114, 315
mm:	10.1	86.24	2.38(51)[32]				366, 370, 359, 358, 165
ra:	5.056	8.46	0.43(3)[91, 98]				387, 388, 333, 133, 134
ho:	0.835(1)[267]	B=26.77(35)[16], R=25.7(2)[16], V=25.40(15)[117]			< 0.01''[117]		266, 268, 114, 140, 315, 283, 40, 98, 378

Optical counterpart[29, 347] became brighter during the first two days following the GRB. Afterglow spectrum shows redshifted interstellar absorption lines (z =0.835), unambiguously settling the GRB cosmological distance scale[268, 267]; [OII]λ 3727 detected at same redshift[266, 16] shows the absorption comes from the host galaxy. Optical afterglow coincides to within $\sim 0.01''$ with the center of the host galaxy[117] (exponential scale length $0.046''$). Host is active star-forming dwarf galaxy, of $M_B = -18.55$, $L_B = 0.12L_*$ and star formation rate < 1 $M_\odot$ yr^{-1} (see [16, 378]). Two-day bump in the optical light curve[81, 126, 309, 377, 426] is reflected in the X-ray afterglow[316]. Strong Fe K_α line emission in X-ray afterglow spectrum[322]. First radio counterpart detected[98, 388, 32, 134] shows interstellar scintillation variability, which confirms the relativistically expanding fireball model; first measured linear size of the fireball (R< 10^{17} cm)[98]. Detected in IR (2.16μm)[53] and mm[32] wavelengths. No sub-mm detection[370].

970616 0.7568[56]		$1^h 18^m 54^s$ $-5°30'00''$ $(21' \times 36')$[254]		(141.0,$-$67.4)		γ, X
pr:	66.2(5)[13]	~ 80[204]	5.78(49)	4.15(19)[231]			
X:	0.1667	11.0	0.57[254]				275, 151
O:	—						
IR:	—						
mm:	—						
ra:	—						
ho:	—						

Search of BATSE 2° error box[56] with the RXTE/PCA 4 hours after the GRB finds an X-ray source[254]. Combined RXTE/PCA and IPN error box[204] contains four ASCA X-ray sources[275] and 11 ROSAT sources[151]. None of the X-ray sources is a convincing GRB counterpart. Optical, radio and near-IR follow-up observations focus on the four ASCA sources and do not lead to detection of transient counterpart[143, 294, 413, 154]. $2'$ away from RXTE/IPN error box is a massive cluster of galaxies, probably unrelated to the GRB[143].

970815	0.5049[364]	$16^h 8^m 43^s$ +81°30′36″ ()[364]		(115.3,+32.5)	γ, X

pr:	150(4)[13]	~165[364]	2.0(3)	2.2(3)[231]	0.05	[364]
X:	3.2	<0.1	<0.005[278]			[147]
O:	0.6	$V>21.5$	<9.0[155]			[180, 381, 1, 238, 41]
IR:	1.5	$K>18$	<42.5[41]			[180]
mm:	—					
ra:	1.14	4.6	<0.35[238]			
ho:	—					

First RXTE/ASM detection[364]. Error box does not contain ASCA sources[278]; steady ASCA source is found slightly outside the RXTE/ASM error box[278]. ROSAT/HRI field of view contains 10 sources[147], one of which lies just at the border of the error box; maybe related to the GRB. No optical[155, 180, 381, 238, 1, 41], IR[180, 41], or radio[238] detection. No mm observations.

970828	0.73931[340]	$18^h 8^m 31.7^s$ +59°18′50″ (10″)[150]		(88.2,+28.5)	γ, X, X^a, ra
RX J1808.5+5918[150]					

pr:	160*[340]	~8[360]	5.5(5)	9.4(2)[231]	0.02	0.16[1340]	[196]
X:	0.15	11.5	0.6[252]	1[424]	1.44(7)[424]	$4.1^{+2.1}_{-1.6}$[424]	[280, 279, 150, 293]
O:	0.168	$R>23.8$	<0.91[157]				[150, 291, 43, 422, 379, 166]
IR:	12.1	$K>20$	<6.73[228]				
mm:	—						
ra:	3.5	8.46	0.15[103]				[69]
ho:	0.96[69]	$R = 24.2$[69]					

RXTE/PCA scan of RXTE/ASM error box[340, 368] within 3.6 hours[252], leading to detection of X-ray afterglow. ROSAT/HRI field contains 15 X-ray sources[150], one of which is inside the RXTE/IPN error box and coincides with an ASCA variable source[280, 424]. In the week following the GRB no variable optical counterpart detected in the ROSAT error box down to $R = 23.8$[157] (see also [291]). Very short-lived radio source detected within the 10″ ROSAT error box[103, 69]. Host galaxy detected at $z = 0.96$ with SFR $\simeq 1$ M$_\odot$/yr[69]. Most dusty host so far ($A_V = 2.2$). No IR detection[228]. No mm observations.

971024	0.4816[363]	$18^h 24^m 58^s$ +49°28′55″ ()[363]		(77.8,+24.4)	γ, X

pr:	132(8)[231]	~13.4[363]	0.15(7)	6.5(1.5)[231]	0.001	1[363]
X:	—					
O:	0.583	$R>19$	<76[274]			[406]
IR:	—					
mm:	—					
ra:	—					
ho:	—					

RXTE/ASM detection[363] of a very weak BATSE burst. Error box is large (~13.4 arcmin²); no X-ray follow-up and the optical follow-up is very scanty and does not lead to a counterpart detection[406, 274]. No IR, radio and mm observations.

971214 0.97272^{184} $11^h56^m26.4^s$ $+65°12'0.5''$ $(0.75)^{178}$ $(132.0,+50.9)$ γ,X,X^a,O,IR

1SAX J1156.4+6513^8

pr:	$31(1)^{13}$	$\sim 3.2^8$	$0.8(1)$	$1.1(1)^{231}$	0.02^{184}	$0.019(2)^{108}$	222
X:	0.2778	0.4	0.02^8	$1.03^{+0.51}_{-0.22}{}^{293}$	$1.20(2)^{293}$	$4.3^{+6.1}_{-2.5}{}^{293}$	
O:	0.4973	$I{=}21.30(4)$	$7.2^{175,\,178}$	$2.13(4)^{337}$	$1.20(2)^{68}$		234, 67, 341, 36
IR:	0.4673	$J{=}20.47^{+0.21}_{-0.19}$	10.7^{68}				141, 385, 337, 145
mm:	—						
ra:	0.6973	8.46	$<0.05^{92}$				337
ho:	$3.418(10)^{234}$	$R{=}25.60(17)^{234}$	$B{>}26.8^{234}$ $K{>}22.5^{337}$			$0.14(7)''^{290}$	232, 239, 290

So far the highest redshift measured ($z = 3.418$) for a host galaxy[234]. Host spectrum shows Lyα emission line; it is consistent with a star forming galaxy with SFR$>5M_\odot$ yr^{-1}, $M_B \sim -20.9$ (see [234]), L~ 0.2L$_*$, radius$= 0.16(2)''$ and e-folding scale length $0.15'' - 0.19''$ (see [290]). The burst flux[222] combined with the redshift implies an energy release of 3×10^{53} erg in γ rays, comparable to the entire energy available in a coalescing neutron star merger. Estimates[234, 290] of chance coincidence between the host and the OT are of the order of 10^{-3}. GRB detected also with the RXTE/ASM at the 470(140) mCrab level[222]. EUVE observations[26] 20 hours after the GRB provide a flux upper limit of 1.7×10^{-13} erg/cm^{2}s in 10nm; no sub-mm detection[368, 370]. No mm observations.

971227 0.34938^{55} $12^h57^m17.2^s$ $+59°24'0.2''$ $(1.5')^{327,\,9}$ $(121.6,+57.7)$ γ,X,X^a

1SAX J1257.3+5924^{327}

pr:	$7(2)^{13}$	$\sim 7^{327}$	$0.83(4)$	$1.14(15)^{231}$	0.04	0.01^9	55
X:	0.5833	0.3	0.016^{327}		$1.12^{+0.08}_{-0.05}{}^9$		
O:	0.89	$R{>}22.8$	$<2.3^{162}$				42, 123, 30, 205, 264, 336,
							300, 417, 168
IR:	12.75						227
mm:	—						
ra:	—						
ho:	—						

BATSE/BeppoSAX-WFC and NFI detections[419, 55, 327], but error box includes two BeppoSAX-NFI sources, one of which is a possible variable X-ray afterglow[327, 9]. The error box of the non-variable X-ray source contains 3 radio sources. List of optical afterglow limits and a debatable afterglow identification[42, 168] is given in[417]. For clarifications on the confusion about the optical counterparts see [30, 337, 77]. LOTIS observations starting 14 s after the GRB trigger show no simultaneous optical emission of $R > 13.2(2)$ (10 s integration time)[300, 417]. No IR detection[227]. No mm, radio observations.

980109 0.05024^{208} $0^h25^m56^s$ $-63°1'26''$ $(10')^{208}$ $(307.8,-53.9)$ γ,X

pr:	$31(4)^{231}$	$\sim314^{208}$	$0.74(14)$	$0.5(1)^{231}$	0.02	208
X:	—					
O:	2.99	$I{>}20$	$<23.8^{349}$			397
IR:	—					
mm:	—					
ra:	—					
ho:	—					

No BeppoSAX/NFI follow-up of WFC $\rho \simeq 10'$ error box[208]. No variable optical source (>0.4 mag) seen one day [397] and three days[349] after the GRB, down to $I = 21$ and 20 mags, respectively. No IR, mm, radio observations.

980326 0.888125^{51} $8^h36^m34.28^s$ $-18°51'23.9''$ $(0.4'')^{160}$ (242.4,13.0) γ,X,O

pr:	5*[34]	~76[202]	0.52(4)	0.22(4)[231]	0.1	0.05[51]	160
X:	0.34	<1.6	<0.08[253]				
O:	0.5319	R=21.2(1)	10.0[161,160]	0.8(4)[22]	2.0(1)[22]		163,164,83,398,38
IR:	—						
mm:	—						
ra:	—						
ho:	≃1[22]	R>27.3[21]					160,73,22

Optical transient found ~ 0.5d after the GRB trigger[161,160]. Light curve shows fast decline[160,73,22,38]. Saturation of optical light curve initially interpreted as host galaxy at $R = 25.5(5)$[160,73], but later observations show that $R_{host} > 27.3$[21,22]. Light curve contains a bump, about 3–4 weeks after the GRB, which is consistent with the presence of a $z \simeq 1$ type Ic supernova component[22]. The SN+PL interpretation naturally fits the collapsar model. No sub-mm detection (< 2.7mJy @ $850\mu m$, < 30mJy @ $450\mu m$)[370]; no IR, radio observations.

980329 0.1559^{111} $7^h2^m38.0217^s$ $+38°50'44.017''$ $(0.05'')^{390}$ (178.1,+18.7) γ,X,X^a,O,IR,mm,ra
1SAX J0702.6+3850[207], VLA J0702+3850[390]

pr:	18.9(3)[231]	~3.1[207]	5.9(2)	5.02(8))[231]	0.14	0.07(1)[108]	111
X:	0.2931	0.4	0.02[207]	1.4(4)[209]	1.35(3)[209]	10(4)[209]	293,152
O:	3.84	R=25.7(3)	0.16[79]		$1.21^{+0.13}_{-0.12}$[339]	0.24[339]	169,296,298,144,351, 80,229,304
IR:	3.84	K=20.7(2)	3.5[243]				335,242,265,298,339, 54,298
mm:	7.0	350	5(1.5)[369]		3.0[370]		394
ra:	2.94	8.4	0.248(16)[390]	+0.9[394]			
ho:	<3.9[69]	R=26.3(2)[69]					351,54

Variable radio afterglow[390,394] led the way to a decaying R to K band counterpart[79,296,298,229,54,243,265]. Optical/near IR light curve analysis[298,339,144] and host galaxy properties restrict the redshift $z < 3.9$[69] and indicate host maybe in a molecular cloud[241]; alternative models available[82,412].

980425 0.90915^{372} $19^h35^m3.316^s$ $-52°50'44.75''$ $(0.07'')^{236}$ (345.0,−27.7) γ,X,X^a,O,IR,mm,ra
SN1998bw,1SAX J1935.0−5248[315],J195303.3−525045[236]

pr:	23(1)[231]	~201[372]	0.30(3)	0.44(4)[231]	0.026	0.18(3)[313]	372,314
X:	0.425	0.30(4)	0.016[313]			0.4[313]	315,318,282
O:	2.49	R=15.7(1)	1585[128]		1.18[262]	0.20[128]	128,262,302,216,301,31
IR:	10.49	J=11.5	41446[246]				
mm:	11.8	150	39(11)[236]				
ra:	2.82	4.8	9[414]	0.75[236]			236,415,213,212,245
ho:	0.00841(5)[396,219]						128,236

BeppoSAX initial ($\rho = 8'$) error box[372] contains a bright supernova SN1998bw[128] of a peculiar Type Ib/c, located in an HII region in a spiral arm of the face-on barred spiral galaxy ESO 184−G82[128]. Probability of chance coincidence $< 10^{-4}$(see [128]). Redshift of the host is $z = 0.00841(5)$[396,219]; linear polarization of 0.53(8)%, $\theta = 49(3)°$ detected in the spectrum of the SN[219]. SN 1998bw spectra are very unusual[302,301]. BeppoSAX/NFI observations reveal two X-ray sources[315,314,313] one of which (1SAX J1935.0−5248) is found later to be consistent with the SN, after final corrections of the NFI position[318]. Very strong radio emission, indicating mildly relativistic afterglow ($\Gamma \sim 2$)[236]; IR detection[246]. Modeling of the optical light curves[128,262] indicates SN 1998bw is extremely energetic[214,420,31] and may have left behind a black hole[214]. Upper limit on gravitational wave detection[4].

980515 0.708495^{319} $21^{h}17^{m}25^{s}$ $-67°15'18''$ $(4')^{343}$ $\qquad$ $(326.5,-38.7)$ $\quad \gamma,X$

							ref
pr:	15*[343]	∼50[343]			0.035		343
X:	—						87
O:	—						
IR:	—						
mm:	—						
ra:	—						
ho:	—						

BeppoSAX GRBM and WFC trigger[318]. NFI follow-up observations[87] find a new X-ray source (1SAX J2116.8−6712) of $F_x = 1.6(4) \times 10^{-13}$ ergs/cm²s. The source is offset by 4.9′ from the centroid of the refined[343] WFC position and does not show variability[87]; it is likely not associated with the GRB. No optical, IR, mm, radio observations.

980519 0.51410^{273} $23^{h}22^{m}21.50^{s}$ $+77°15'43.25''$ $(0.1'')^{100}$ $\qquad$ $(118.0,+15.3)$ $\quad \gamma,X,X^{a},O,R$

1SAX J2322.3+7716[286], VLA J232221.5+771543[100]

							ref
pr:	30(2)[231]	∼2.2[286]	2.4(1.4)	0.01(1)[231]	0.01	0.18[210]	273
X:	0.4059	0.38(6)	0.02[286]	$1.8^{+0.6}_{-0.5}$[285]	1.8(3)[285]	$5.6^{+10.2}_{-5.1}$[293]	293
O:	0.3559	$I=18.5(1)$	95.0[215]	1.2(3)[176]	2.05(4)[176]	0.481[170]	402
IR:	—						
mm:	—						
ra:	2.79	8.3	0.102(19)[106]				103, 100
ho:		$R = 26.05(22)$[374]					24

BATSE/BeppoSAX-WFC GRB with one of the fastest ($\simeq 2$) temporal power-law decays recorded[285, 293, 402, 176]. For a complete photometry of the optical light curve see [176, 402]. Radio source detected 2.8 days after trigger with large light curve variations due to ISS[100]. The fast decline and the radio data (radio source very compact of size $< 1\mu$arcsec) indicate that the afterglow emission may originate either from a jet[176, 100] or from a blast wave propagating into the dense wind of the progenitor star[100]. No constraining mm observations[370]; no IR observations. Deep optical observations two months after the GRB detected a faint object at the OT position, presumably the host galaxy[374, 24].

980611 0.034^{13} $18^{h}20^{m}7.2^{s}$ $+54°4'58.8''$ $(7° \times 3.5')^{194}$ $\qquad$ $(82.6,+26.1)$ $\quad \gamma$

							ref
pr:	9.3(5)[13]	∼1500[194]		1.1(1)[13]			
X:	—						
O:	—						
IR:	—						
mm:	—						
ra:	—						
ho:	—						

BATSE GRB (# 6816). RXTE/PCA follow-up scan finds two X-ray sources; both are shown to be substantially outside the BATSE/Ulysses IPN annulus[194] and are very likely unrelated to the GRB.

980613 0.20215^{324} $10^{h}17^{m}57.82^{s}$ $+71°27'25.5''$ $(0.6'')^{171}$ $\qquad$ $(138.0,+40.8)$ $\quad \gamma,X,X^{a},O$

1SAX J1017.9+7127[57]

							ref
pr:	50*[371]	∼ 2.2[57]		0.17(3)[418]	0.014		371
X:	0.3597	0.11(3)	0.006[57]				172
O:	0.6979	$R=22.9(2)$	2.1[188]		1.3[171]	0.27[70]	49, 50, 288, 172, 66
IR:	—						263, 49, 50
mm:	—						
ra:	—						103
ho:	1.0964(3)[70]	$B=24.4(2)$[70]	$V=24.1(2)$[70]	$R=23.9(2)$[70]			74, 75, 76

BeppoSAX GRBM and WFC triggered GRB[324]; BATSE did not trigger because of prior trigger due to intense solar flare[418]. NFI follow-up detected 2 X-ray sources, one of which is variable[57]. Optical counterpart detected 0.69 days after trigger[188]. Host galaxy was found at the OT position with $z = 1.0964$[70]; spectrum shows strong emission line interpreted as [OII] 3727 and very blue featureless continuum indicating star-forming galaxy (SFR$\simeq$ 3.9 $M_\odot$/yr)[70]. No IR[263, 49, 50], radio[103] detections; no mm observations.

980703	0.18247[244]	$23^h59^m6.6661^s$ $+8°35'7.0939''$ $(0.0005'')$[389]			(101.6, −52.1)	γ, X, X^a, O, IR, R
1SAX J2359.1+0835[127]						

pr:	370(10)[231]	∼30.5[365]	1.9(1)	5(2)[231]	0.04	1[244]	
X:	0.9135	0.75	0.04[127]	1.51(32)[130]	<0.91[407]	0.34[407]	
O:	1.3	R=21.2	10.0[90]	1.01(1)[407]	1.61[407]	0.188[78]	425, 18, 47
IR:	5.4	K=18.8(1)	20.3[18]		1.43(11)[47]		187, 407
mm:	7.34	220	<5.2[18]				370
ra:	1.2	4.86	0.135(26)[90]				389
ho:	0.9662(2)[78]	$R=22.58^{+0.06}_{-0.05}$[407]	$V=23.04(8)$[407]	$K=19.62^{+0.12}_{-0.11}$[407]	0.21(12)$''$[18]		74, 375, 389, 376

RXTE/ASM[244] and BATSE GRB; X-ray source detected with the BeppoSAX NFI[127, 130]. Optical counterpart detected[90, 425] following detection of a radio source[90]. Host galaxy spectrum contains emission and absorption lines allowing redshift determination at $z = 0.9662(2)$[78]. Host is the brightest so far[18, 407] with $M_B = -21.2$[18]; star-forming galaxy with SFR$\gtrsim$ 10 $M_\odot$/yr[78, 376]. VLBI detects source unresolved at < 0.3 mas[389]. IR detection[407, 18]; no mm detection[18, 370].

980706	0.6665[13]	$11^h0^m33^s$ $+57°23'6''$ $(36' \times 1.56')$[255]		(148.2, +54.0)	γ

pr:	26(2)[13]	∼56[255]	5.12(5)[13]		
X:	—				
O:	—				
IR:	—			255	
mm:	—				
ra:	—				
ho:	—				

RXTE/PCA scan 2.7 hours after trigger of BATSE GRB finds a 2mCrab X-ray source that is not detected at the next scan 1.5 hours later[255]. It is unclear whether the source has faded below the PCA detection limit (implying a decay index of >3) or it is not located at IPN/PCA 'best position'. Most likely, PCA source is not related to the GRB. No BeppoSAX/NFI follow-up, no optical, IR, mm, radio observations.

981220	0.91135[361]	$3^h42^m33.8^s$ $+17°9'0''$ $(2.4' \times 4.5')$[199]		(171.0, −29.3)	γ, X

pr:	15*[89]	∼11[199]	2.4(4)	1.0(2)[1, 109]	0.012(1)[361]	
X:	—					
O:	2.12	V>22	<5.7[352]			401, 173, 270, 410, 12
IR:	—					
mm:	—					370
ra:	2.1	1.43	<0.2[131]			93, 95, 391
ho:	—					

RXTE/ASM, BeppoSAX, Ulysses and KONUS GRB[361, 89, 199, 109]. No optical counterpart found[352, 401, 410, 270, 12, 171]. Unusual radio variable source[131, 93] associated with a faint, slowly variable optical source[15] was later found to lie outside the refined IPN error box[201] and has a core-jet morphology in VLBA observations, strongly suggestive of a highly variable background intra-day variable (IDV) source[391]. No X-ray follow-up observations; no IR,mm observations.

981226 0.40793[65] 23^{h}29^{m}37.21^s −23°55′53.8″ (0.4″)[95]   (38.2,−71.3) γ,X,X^a,R
1SAX J2329.6−2356[113], VLA 232937.2−235553[102]

pr:	20*[65]	~3.2[113]			0.006	65
X:	0.4701	0.30(7)	0.016[113]			
O:	0.412	R>23	<1.9[247]			
IR:	—					95, 132, 421
mm:	3.74	350	<0.6(3.8)[370]			44
ra:	8.54	8.46	0.169(28)[95]		2.0(4)[95]	132
ho:		R=24.85(6)[95]				

BeppoSAX/WFC error box contains decaying NFI X-ray source[113]. Several suggestions for optical and IR counterpart made[132, 44, 421, 226], but none acceptable[342, 19, 356, 247]. Radio counterpart found 8.5 days after trigger[95] leads to potential host galaxy[95]. No mm detection[370].

990123 0.40759[330] 15^{h}25^{m}30.31^s +44°45′59.24″ (0.15″)[235]   (73.1,+54.6) γ,X,X^a,O,IR,R
1SAX J1525.5+4446[181], VLA J152530.3+444559[235]

pr:	63.3(3)[135]	~2.2[181]	23.7(2.3)	27.1(5)[231]	0.08	88	33
X:	0.245	11.0	0.57[329]		1.44(7)[233]	1.2[181]	276, 46
O:	0.169	R=18.65(4)	94.0[289]	0.75(23)[135]	1.52(15)[116]	0.053[135]	3, 2, 46, 189, 346, 292, 260
IR:	1.23	K=18.29(4)	32.5[20]	0.8(1)[233]	1.12(11)[233]		46
mm:	1.27	222.0	<2.41[235]				334
ra:	1.24	8.46	0.260(32)[94]	+1.4(7)[135]	<0.8[235]		
ho:	1.6004(8)[220]	V=24.20(15)[118], R=23.63(5)[118], K=21.65(30)[25]				0.6(1)″[25]	189, 233, 119, 72, 116, 376, 193, 7

First detection of prompt optical emission: BATSE trigger activates the robotic telescope ROTSE, which detects the OT in 6 images between 22 s and 10 min after the burst, with V between 8.9 and 14.3 mag[2, 3]. Prompt γ emission among top 0.4% in fluence[33]; gamma-ray and optical signals do not track one another[135, 33]. Later follow-up optical observations detected the OT at R ~18.6 and smoothly declining (see [233, 135, 46, 7, 120] and references therein). Optical afterglow light-curve composed of smoothly joined power-laws; break in the light curve at $t \sim 1$ d first interpreted as beaming, but is not achromatic[233, 120]. The combined γ/optical/radio light curve[2, 135, 233, 235, 46] indicates importance of internal and external (both outward and reverse) shocks. Optical polarization not detected (<2.3%)[189]. Redshifted metal absorption and lack of the Lyα forest limit the host redshift to $1.6 < z < 2.05$[220, 233, 189]. Irregular host morphology may indicate merging galaxies[120]; spectrum indicates relatively blue, star-forming galaxy of $M_B = -20.4$, $L \simeq 0.7L_*$ and SFR>6.0 M$_\odot$/yr[25]. Three active star forming regions identified[193]. Initial detection of nearby galaxies spawned suggestions of gravitational lensing that were later refuted (see GCN Circs 219,221,234-6,238,241,242-3). First detection of radio flare 1.24 d after the trigger[94, 235]. IR detection[20, 233, 46]; no mm detection[235].

990217 0.224618[386] 3^{h}2^{m}52^s −53°5′36″ (3′)[386]   (268.9,−54.5) γ,X

pr:	25*[6]	~28[386]	0.11(3)	0.13(2)^a[6]	0.016	6	386
X:	0.27	<0.1	<0.005[317]		>1.6[317]	
O:	0.28	R>23.5	<1.8[297]			353, 311
IR:	—					
mm:	—					
ra:	1.09	4.8	<0.28[411]			
ho:	—					

BeppoSAX GRBM/WFC detection with $\rho = 3'$ error box[386]; NFI follow-up[317] provides only upper limit for an X-ray afterglow of 10^{-13} erg/cm^2 s. No variable optical[297, 353, 311], or radio[411] counterpart. No IR, mm, observations.

pr: $106(12)^{355}$	$\sim317^{362}$	$0.63(14)$	$0.6(1)^{231}$	0.009	[362]
X: —					
O: 0.1367	$R=18.14(6)$	246^{355}	$0.38(25)^{355}$	$1.2(1)^{355}$	[417]
IR: —					
mm: —					
ra: 102.0	8.5	$<0.258^{355}$			
ho:	$R>25.7^{355}$ $K>23.3^{355}$				

Near-simultaneous LOTIS and Super-LOTIS observations of BATSE error box do not reveal afterglow coincident with GRB, with $V > 12.0$ (10s integration, starting 132s after trigger), $V > 13.4$ (10 min) for LOTIS, and $V > 15.3$ for Super-LOTIS[417]. Comparison of images taken 3.28 hours and 82 days after burst show likely optical counterpart[355], which decays by 0.08 ± 0.08 mag in 7.7 min. No radio counterpart[355]. No IR, mm observations. No evidence for a host galaxy in images taken 103 days after the burst ($R > 25.7$, $K > 23.3$)[355].

VLA J115450.1−2640.6[392]

pr: $131.3(2)^{223}$	$\sim30^{203}$	$7.6(6)$	$17.6(4)^{231}$		
X: 0.126	0.35	1.8^{251}		$1.9(6)^{384}$	
O: 0.616	$R>23$	$<1.9^{257}$			403, 354, 427, 404, 310, 350
IR: —					
mm: —					
ra: 1.73	8.4	$0.54^{393,\ 392}$			
ho:	$R=24.8(2)^{17}$				307

RXTE/PCA scanning of BATSE trigger #7549[223] detects a previously unknown X-ray source of 1.5 mCrab, which decayed by a factor of 2.4 in subsequent observations[251, 384]. The source is within the refined (30 arcminsq) IPN error box[203]. No evidence for optical afterglow to $R = 23$ mag[257, 310, 403, 404, 354, 427, 350]. Radio observations detect 4 sources within the PCA/IPN error box[391], one of which fades below the VLA detection limit (0.035 mJy) sometime beween 2−16 days after the burst trigger[393]. Optical observations[17] of the radio source ~36 days after the GRB reveal a faint, extended galaxy with irregular and possibly interacting morphology, potentially the host of the GRB. No IR, mm observations.

1SAX J1338.1−8030[240]

pr: $68(2)^{224}$	$\sim24^{195}$	$4.37(13)$	$2.53(9)^{231}$	0.14	4	64
X: 0.333	331				240
O: 0.35	$R=19.2(3)$	92.8^{408}	$0.61(12)^{380}$	$2.41(2)^{380}$	0.67^{179}	179, 211, 14, 190
IR: —					
mm: —					
ra: 3.3	8.7	$0.227(30)^{179}$			
ho: $1.619(2)^{409}$	$R>27.6^{14}$, $V>28^{115}$				211

X-ray and optical counterparts are found[331, 240, 408] in the combined WFC/IPN error box[328, 64, 195]. Later recalibration of R band measurement[218] indicates that the OT had been detected 3.5 hours after trigger[10] at $R = 17.75$ mag. Optical light curve shows clear achromatic break after ~1.5 d, that may signify beaming; first index is -0.8 and second is $-2.4^{179,\ 380}$. Radio counterpart detected in 8.7 GHz[179]. Metal absorption lines in the spectrum[409] limit the redshift above $z > 1.619(2)$; absence of Lyα forest places an upper limit $z < 2.0^{14}$. First burst with detection (0.86 d after trigger) of optical linear polarization[63, 416] of $1.6(2)\%$ and polarization angle $96(4)°$, that remains constant over about three days. No host galaxy detection; no IR, mm observations.

990520 0.08539^{137} $8^h35^m56^s$ $+51°18'36''$ $(3')^{137}$ $(167.5,+36.9)$ γ,X
1SAX J0835.9+5118[137]

pr:	8[225]	~28[137]		0.08(3)[225]	0.023	[137]
X:	—					
O:	0.735	$R>21.5$	<11.2[248]			256, 306, 305, 37, 177
IR:	—					
mm:	—					
ra:	0.605	4.8	<0.125[103]			104, 105
ho:	$R=20.75(7)^{177}$ $V=21.08(5)^{256}$ $B=21.58(8)^{177}$					23

Brief WFC transient ($\sim$ 10 s) barely seen in GRBM data[137]. WFC error box ($\rho = 3'$) contains VLA source[104, 105], coincident with a non-variable optical object, which may be a galaxy[23], but could also be a point source[37, 248]. The BATSE team reports on May 24.7 UT a faint untriggered event coincident with the BeppoSAX transient, that was recorded while the onboard trigger was disabled[225]. It is uncertain whether the BeppoSAX transient and the VLA source/optical galaxy are related, but it is likely that the BeppoSAX transient is associated with a GRB. No IR, mm observations.

990527 0.58182^{197} $22^h50^m27^s$ $-20°53'26''$ $(6')^{197}$ $(39.1,-61.7)$ γ

pr:	20*[197]	~100[197]		$0.8^{a\,197}$
X:	—			
O:	1.75	$R>22.0$	<7.0[308]	
IR:	—			
mm:	—			
ra:	—			
ho:	—			

GRB detected with Ulysses/Konus/NEAR[197]. Optical imaging of 100 arcmin^2 error box $\sim$1.75 d after the burst trigger does not reveal a counterpart[308] ($R > 22.0$). No X-ray follow-up, radio, IR, mm observations.

990627 0.20894^{272} $1^h48^m27^s$ $-77°4'36''$ $(1')^{272}$ $(298.8,-39.6)$ γ,X,X^a
1SAX J0148.5-7704[287]

pr:	50*[272]	~28[272]			0.007	[272]
X:	0.331	0.35	0.02[287]			
O:	0.929	$R>21.0$	<17.7[344]			
IR:	—					
mm:	—					
ra:	1.46	4.8	<0.125[383]			103
ho:	—					

BeppoSAX WFC error box[272] contains 4 radio sources[383], one of which coincides with a fading NFI source[287], which is a likely GRB counterpart. No variable optical counterpart is found at the $R > 21.0$ limit[344]. No IR, mm observations.

990704 0.7294^{332} $12^h19^m27.3^s$ $-3°50'22''$ $(1')^{86}$ $(287.7,+58.1)$ γ,X,X^a
1SAX J1219.5-0350[86]

pr:	40*[182]	~154[139]			0.14	[182]
X:	0.335	0.44	0.02[86]			281
O:	0.439	$R>21.2$	<14.7[405]			48, 261, 259, 217
IR:	—					
mm:	—					
ra:	1.021	4.88	<0.065[345]			
ho:	—					

BeppoSAX/WFC error box ($\rho = 7'$)[332, 139, 182] contains new fading X-ray source, which is the likely counterpart[86, 281]. Two radio sources in the WFC error box[345] unlikely related to GRB. Initial optical counterpart[261] later retracted[259]. No IR, mm observations.

990705   0.66765[52] $5^{\rm h}9^{\rm m}54.52^{\rm s}$ $-72°7'53.1''$ (0.3")[258]   (283.5, −33.4) $\gamma, X, X^{\rm a}, O, IR$

pr:	45*[52]	~3.5[198]		0.09	[52]	
X:	0.4744		[52,5]		0.72[258]	
O:	0.7324	$V=22.0(4)$	5.7[295]	>1[258]	0.40[258]	174
IR:	0.272	$H=16.57(5)$	252.0[295]	1.68(10)[258]		
mm:	—					
ra:	0.662	4.8	<0.1[382]			103
ho:		$V=23.80(15)$[258]				

BeppoSAX/WFC detection; radio observations[382] revealed 3 sources in the WFC ($\rho = 3'$) error box[52], which are not, however, in the BeppoSAX/Ulysses/NEAR IPN/WFC intersection[198]. NFI follow-up detects a new X-ray source[138]; detection contaminated by radiation from LMC X-2, which lies 52' off the NFI center[5]. First detection of an afterglow in NIR [295, 258] ($H = 16.57$), which decays with a power-law of index -1.68[258]. Elongated irregular, fuzzy object of $V = 23.8$ (size $2.4'' \times 0.8''$) partially superposed to the OT is suggested as the GRB host[258]. No neutrino detection ±48 hours from the GRB trigger[121]. No mm observations.

990712   0.69655[185] $22^{\rm h}31^{\rm m}53.1^{\rm s}$ $-73°24'29''$ (1")[11]   (315.3, −40.2) γ, X, O

pr:	90*[185]	~12.6[185]			
X:	—				
O:	0.1235	$R=19.4$	52.5[11]	1.05[221]	395, 191
IR:	—				
mm:	—				
ra:	—				
ho:	0.430(5)[129]	$R=21.79$[192]			221, 191

BeppoSAX WFC error box ($\rho = 2'$)[185] contains optical counterpart[11]; its spectrum shows pronounced emission lines and absorption lines both consistent with a redshift of $z = 0.430(5)$[129]. Evidence for a host galaxy at $R = 21.76$ from the saturation of the optical light curve[191, 192, 221]. No BeppoSAX NFI follow-up; no IR, radio or mm observations.

TABLE NOTES

General notation	
NNN	reference for the value just to the left of it
(MM)	error in the last digits of the number just to the left of it
*	the value is not determined in the standard way adopted for this entry; see reference for details
N	the value is not determined in the standard way, but in a frequent deviation explained below in these notes

Comments on specific columns	
Name	GRB name in YYMMDD format
UT trig	time of trigger, in fractional days since start of the UT day of the burst
duration	duration, defaults to the BATSE T90, unless marked by star*.
errbox	area of prompt emission or NFI error box, in square arcmin
F_γ^{peak}	prompt gamma-ray peak flux in units of $10^{-6}\,\mathrm{erg\,cm^{-2}\,s^{-1}}$. Defaults to BATSE flux above 20 keV
S_γ	Fluence of prompt gamma-ray emission in units of $10^{-5}\,\mathrm{erg\,cm^{-2}}$. Defaults to BATSE fluence above 20 keV. Reference and note are for both flux and fluence. Notes for frequent deviations: (a) SAX GRBM 40–700 keV.
F_X^{peak}	prompt 2–10 keV X-ray peak flux in units of $10^{-6}\,\mathrm{erg\,cm^{-2}\,s^{-1}}$.
S_X	Fluence of prompt 2–10 keV X-ray emission in units of $10^{-5}\,\mathrm{erg\,cm^{-2}}$. Reference and note are for both flux and fluence
$\Delta \mathrm{UT}_X$	time since trigger, in days, of X-ray afterglow discovery or limit; ditto other wavelengths
F_X^{a}	2–10 keV X-ray afterglow discovery flux or limit, first in units of $10^{-12}\,\mathrm{erg\,cm^{-2}\,s^{-1}}$ and then in μJy
$F_{\mathrm{O},\ldots,\mathrm{ra}}$	afterglow discovery fluxes or limits, in μJy for O and IR and in mJy for mm and radio.
β_X	X-ray afterglow flux spectral slope; ditto other wavelengths
δ_X	X-ray afterglow flux temporal slope; ditto other wavelengths
N_H	H column density, from X rays, in $10^{21}\,\mathrm{cm^{-2}}$
mag	optical ($UBVRI$) and near-IR (JHK) discovery magnitude, with band; all else being equal, R and K were preferred
A_V	visual absorption in mags. Foreground, or total, depending on reference and method.
ν_{mm}	millimeter frequency in GHz. All else being equal, 90 GHz was the preferred mm frequency
ν_{ra}	radio frequency in GHz; All else being equal, 5 GHz was the preferred radio frequency
z	redshift; in order of preference (i) an emission-line redshift, or (ii) the redshift of the presumed host, or (iii) the highest absorption redshift in the OT spectrum. Limits sometimes from (absence of) Ly break.

Table references

[1] Adams, MT et al. 1997 IAUC 6725.
[2] Akerlof, C et al. 1999 Nat 398:400.
[3] Akerlof, CW & McKay, TA 1999 GCN 205.
[4] Amati, L et al. 1999 A&AS 138:605.
[5] Amati, L et al. 1999 GCN 384.
[6] Amati, L et al. 1999 GCN 317.
[7] Andersen, MI et al. 1999 Science 283:2075.
[8] Antonelli, LA et al. 1997 IAUC 6792.
[9] Antonelli, LA et al. 1999 A&AS 138:435.
[10] Axelrod, T et al. 1999 GCN 315.
[11] Bakos, G et al. 1999 IAUC 7225.
[12] Bartolini, C et al. 1998 GCN 175.
[13] BATSE Team. Gamma-ray burst data. http://gammaray.msfc.nasa.gov/batse/ 2000.
[14] Beuermann, K et al. 1999 A&A 352:L26.
[15] Bloom, JS et al. 1999 GCN 196.
[16] Bloom, JS et al. 1998 ApJ 507:L25.
[17] Bloom, JS et al. 1999 GCN 351.
[18] Bloom, JS et al. 1998 ApJ 508:L21.
[19] Bloom, JS et al. 1998 GCN 182.
[20] Bloom, JS et al. 1999 GCN 240.
[21] Bloom, JS & Kulkarni, SR 1998 GCN 161.
[22] Bloom, JS et al. 1999 Nat 401:453.
[23] Bloom, JS et al. 1999 GCN 335.
[24] Bloom, JS et al. 1998 GCN 149.
[25] Bloom, JS et al. 1999 ApJ 518:L1.
[26] Boer, M et al. 1997 IAUC 6795.
[27] Boller, T & Voges, W 1996 IAUC 6469.
[28] Boller, T et al. 1997 IAUC 6580.
[29] Bond, HE 1997 IAUC 6654.
[30] Bond, H et al. 1998 GCN 22.
[31] Branch, D. In Livio, M, editor, *Supernovae and Gamma Ray Bursts* in press (astro–ph/9906168) 1999.
[32] Bremer, M et al. 1998 A&A 332:L13.
[33] Briggs, MS et al. 1999 ApJ 524:82.
[34] Briggs, MS et al. 1998 IAUC 6856.
[35] Butler, RC et al. 1997 IAUC 6539.
[36] Castander, FJ et al. 1997 IAUC 6791.
[37] Castro-Tirado, A et al. 1999 GCN 336.
[38] Castro-Tirado, AJ & Gorosabel, J 1999 A&AS 138:449.
[39] Castro-Tirado, AJ et al. 1998 Science 279:1011.
[40] Castro-Tirado, AJ et al. 1998 IAUC 6848.
[41] Castro-Tirado, AJ et al. 1997 IAUC 6744.
[42] Castro-Tirado, AJ et al. 1997 IAUC 6800.
[43] Castro-Tirado, AJ et al. 1997 IAUC 6730.
[44] Castro-Tirado, AJ et al. 1998 GCN 173.
[45] Castro-Tirado, AJ et al. 1998 A&A 330:14.
[46] Castro-Tirado, AJ et al. 1999 Science 283:2069.
[47] Castro-Tirado, AJ et al. 1999 ApJ 511:L85.
[48] Castro-Tirado, A et al. 1999 GCN 362.
[49] Castro-Tirado, AJ et al. 1998 GCN 102.
[50] Castro-Tirado, AJ et al. 1998 GCN 103.
[51] Celidonio, G et al. 1998 IAUC 6851.
[52] Celidonio, G et al. 1999 IAUC 7218.
[53] Chary, R et al. 1998 ApJ 498:L9.
[54] Cole, DM et al. 1998 IAUC 6866.
[55] Coletta, A et al. 1997 IAUC 6796.
[56] Connaughton, V et al. 1997 IAUC 6683.
[57] Costa, E et al. 1998 IAUC 6939.
[58] Costa, E et al. 1997 IAUC 6572.
[59] Costa, E et al. 1997 IAUC 6576.
[60] Costa, E et al. 1997 IAUC 6533.
[61] Costa, E et al. 1997 IAUC 6649.
[62] Costa, E et al. 1997 Nat 387:783.
[63] Covino, S et al. 1999 A&A 348:L1.
[64] Dadina, M et al. 1999 IAUC 7160.
[65] di Ciolo, L et al. 1998 IAUC 7074.
[66] Diercks, A et al. 1997 GCN 108.
[67] Diercks, A et al. 1997 IAUC 6791.
[68] Diercks, AH et al. 1998 ApJ 503:L105.
[69] Djorgovski, SG & et al. Hosts of GRB 970828 and GRB 980329. in preparation 2000.
[70] Djorgovski, SG et al. 1998 GCN 189.
[71] Djorgovski, SG et al. 1999 GCN 289.
[72] Djorgovski, SG et al. 1999 GCN 256.
[73] Djorgovski, SG et al. 1998 GCN 57.
[74] Djorgovski, SG et al. 1998 GCN 139.
[75] Djorgovski, SG et al. 1998 GCN 114.
[76] Djorgovski, SG et al. 1998 GCN 117.
[77] Djorgovski, SG et al. 1998 GCN 25.
[78] Djorgovski, SG et al. 1998 ApJ 508:L17.
[79] Djorgovski, SG et al. 1998 GCN 41.
[80] Djorgovski, SG et al. 1998 GCN 38.
[81] Djorgovski, SG et al. 1997 Nat 387:876.
[82] Draine, BT 2000 ApJ 532:273.
[83] Eichelberger, AC et al. 1998 GCN 33.
[84] Feroci, M et al. 1998 A&A 332:L29.
[85] Feroci, M et al. 1997 IAUC 6610.
[86] Feroci, M et al. 1999 IAUC 7217.
[87] Feroci, M et al. 1998 IAUC 6909.
[88] Feroci, M et al. 1999 IAUC 7095.
[89] Feroci, M et al. 1998 GCN 159.
[90] Frail, DA et al. 1998 GCN 128.
[91] Frail, DA & Kulkarni, SR 1997 IAUC 6662.
[92] Frail, DA & Kulkarni, SR 1997 GCN 7.
[93] Frail, DA & Kulkarni, SR 1998 GCN 170.
[94] Frail, DA & Kulkarni, SR 1999 GCN 211.
[95] Frail, DA et al. 1999 ApJ 525:L81.
[96] Frail, DA et al. 1997 ApJ 483:L91.
[97] Frail, DA et al. 1997 IAUC 6545.
[98] Frail, DA et al. 1997 Nat 389:261.
[99] Frail, DA et al. 1996 IAUC 6472.
[100] Frail, DA et al. 2000 ApJ 534:559.
[101] Frail, DA et al. 1998 ApJ 502:L119.

[102] Frail, DA et al. 1999 GCN 269.
[103] Frail, DA et al. In Kippen, RM et al., editors, *Gamma-Ray Bursts* in press 2000.
[104] Frail, DA et al. 1999 GCN 334.
[105] Frail, DA et al. 1999 GCN 337.
[106] Frail, DA et al. 1998 GCN 89.
[107] Frontera, F et al. In Meegan, C et al., editors, *Gamma-Ray Bursts* 446 New York 1998. AIP.
[108] Frontera, F et al. 2000 ApJS 127:59.
[109] Frontera, F et al. 1998 GCN 167.
[110] Frontera, F et al. 1997 IAUC 6567.
[111] Frontera, F et al. 1998 IAUC 6853.
[112] Frontera, F et al. 1998 A&A 334:L69.
[113] Frontera, F et al. 1998 IAUC 7078.
[114] Fruchter, A et al. 1997 IAUC 6674.
[115] Fruchter, A et al. 1999 GCN 386.
[116] Fruchter, A et al. 1999 GCN 354.
[117] Fruchter, AS et al. 2000 ApJ submitted (astro-ph/9903236).
[118] Fruchter, AS et al. 1999 ApJ 519:L13.
[119] Fruchter, A et al. 1999 GCN 255.
[120] Fruchter, AS et al. 1999 ApJ 516:683.
[121] Fulgione, W 1999 GCN 390.
[122] Galama, T et al. 1997 Nat 387:479.
[123] Galama, T et al. 1997 GCN 21.
[124] Galama, T et al. 1997 IAUC 6574.
[125] Galama, TJ et al. 1997 ApJ 486:L5.
[126] Galama, TJ et al. 1998 ApJ 497:L13.
[127] Galama, TJ et al. 1998 GCN 127.
[128] Galama, TJ et al. 1998 IAUC 6895.
[129] Galama, TJ et al. 1999 GCN 388.
[130] Galama, TJ et al. 1998 GCN 145.
[131] Galama, TJ et al. 1998 GCN 168.
[132] Galama, TJ et al. 1998 GCN 183.
[133] Galama, TJ et al. 1998 ApJ 500:L101.
[134] Galama, TJ et al. 1998 ApJ 500:L97.
[135] Galama, T et al. 1999 Nat 398:394.
[136] Galama, T et al. 2000 ApJ 536:185.
[137] Gandolfi, G et al. 1999 IAUC 7174.
[138] Gandolfi, G 1999 GCN 373.
[139] Gandolfi, G 1999 GCN 366.
[140] Garcia, MR et al. 1998 ApJ 500:L105.
[141] Garcia, MR et al. 1997 IAUC 6792.
[142] Gorosabel, J & Castro-Tirado, AJ 1998 A&A 333:417.
[143] Gorosabel, J et al. 1999 A&AS 138:455.
[144] Gorosabel, J et al. 1999 A&A 347:L31.
[145] Gorosabel, J et al. 1998 A&A 335:L5.
[146] Gorosabel, J et al. 1998 A&A 339:719.
[147] Greiner, J 1997 IAUC 6742.
[148] Greiner, J et al. 1996 IAUC 6487.
[149] Greiner, J et al. 1997 IAUC 6570.
[150] Greiner, J et al. 1997 IAUC 6757.
[151] Greiner, J et al. 1997 IAUC 6722.
[152] Greiner, J et al. 1998 GCN 59.
[153] Groot, PJ et al. 1997 IAUC 6574.
[154] Groot, PJ et al. 1997 IAUC 6723.
[155] Groot, PJ et al. 1997 IAUC 6723.
[156] Groot, PJ et al. 1997 IAUC 6616.
[157] Groot, PJ et al. 1998 ApJ 493:L27.
[158] Groot, PJ et al. 1997 IAUC 6588.
[159] Groot, PJ et al. 1997 IAUC 6584.
[160] Groot, PJ et al. 1998 ApJ 502:L123.
[161] Groot, PJ et al. 1998 IAUC 6852.
[162] Groot, P et al. 1997 GCN 27.
[163] Grossan, B et al. 1998 GCN 34.
[164] Grossan, B et al. 1998 GCN 35.
[165] Gruendl, RA et al. In Meegan, C et al., editors, *Gamma-Ray Bursts* 576 New York 1998. AIP.
[166] Guarnieri, A et al. 1997 IAUC 6733.
[167] Guarnieri, A et al. 1997 A&A 328:L13.
[168] Guarnieri, A et al. 1999 A&AS 138:457.
[169] Guarnieri, A et al. 1998 IAUC 6855.
[170] Hakkila, J et al. 1997 AJ 114:2043.
[171] Halpern, J & Fesen, R 1998 GCN 134.
[172] Halpern, J et al. 1998 GCN 106.
[173] Halpern, J et al. 1998 GCN 163.
[174] Halpern, J et al. 1999 GCN 381.
[175] Halpern, J et al. 1997 IAUC 6788.
[176] Halpern, JP et al. 1999 ApJ 517:L105.
[177] Halpern, JP et al. 1999 GCN 343.
[178] Halpern, JP et al. 1998 Nat 393:41.
[179] Harrison, FA et al. 1999 ApJ 523:L121.
[180] Harrison, TE et al. 1997 IAUC 6721.
[181] Heise, J et al. 1999 IAUC 7099.
[182] Heise, J et al. 1999 IAUC 7217.
[183] Heise, J et al. 1997 IAUC 6610.
[184] Heise, J et al. 1997 IAUC 6787.
[185] Heise, J et al. 1999 IAUC 7221.
[186] Heise, J et al. In Meegan, C et al., editors, *Gamma-Ray Bursts* 397 New York 1998. AIP.
[187] Henden, AA et al. 1998 GCN 140.
[188] Hjorth, J et al. 1997 GCN 109.
[189] Hjorth, J et al. 1999 Science 283:2073.
[190] Hjorth, J et al. 1999 GCN 320.
[191] Hjorth, J et al. 1999 GCN 389.
[192] Hjorth, J et al. 1999 GCN 403.
[193] Holland, S & Hjorth, J 1999 A&A 344:L67.
[194] Hurley, K 1998 GCN 100.
[195] Hurley, K & Barthelmy, S 1999 GCN 309.
[196] Hurley, K et al. 1997 IAUC 6728.
[197] Hurley, K & Cline, T 1999 GCN 347.
[198] Hurley, K et al. 1999 GCN 380.
[199] Hurley, K et al. 1998 GCN 160.
[200] Hurley, K et al. 1997 ApJ 485:L1.
[201] Hurley, K & Feroci, M 1999 GCN 270.
[202] Hurley, K et al. 1998 GCN 53.
[203] Hurley, K et al. 1999 GCN 298.
[204] Hurley, K et al. 1997 IAUC 6687.
[205] Ilovaisky, SA & Chevalier, C 1998 IAUC 6803.
[206] In 't Zand, J et al. 1997 IAUC 6569.
[207] in 't Zand, J et al. 1998 IAUC 6854.
[208] in 't Zand, J et al. 1998 IAUC 6805.

[209] in 't Zand, JJM et al. 1998 ApJ 505:L119.
[210] in 't Zand, JJM et al. 1999 ApJ 516:L57.
[211] Israel, GL et al. 1999 A&A 348:L5.
[212] Iwamoto, K 1999 ApJ 517:L67.
[213] Iwamoto, K 1999 ApJ 512:L47.
[214] Iwamoto, K et al. 1998 Nat 395:672.
[215] Jaunsen, AO et al. 1998 GCN 78.
[216] Jeffery, D 1999 astro-ph/ 9907015.
[217] Jensen, BL et al. 1999 GCN 371.
[218] Kałuzny, J et al. 1999 IAUC 7164.
[219] Kay, LE et al. 1998 IAUC 6969.
[220] Kelson, DD et al. 1999 IAUC 7096.
[221] Kemp, J & Halpern, J 1999 GCN 402.
[222] Kippen, MR et al. 1997 IAUC 6789.
[223] Kippen, RM 1999 GCN 306.
[224] Kippen, RM 1999 GCN 322.
[225] Kippen, RM et al. 1999 GCN 344.
[226] Klose, S 1998 GCN 186.
[227] Klose, S 1998 GCN 28.
[228] Klose, S et al. 1997 IAUC 6756.
[229] Klose, S et al. 1998 IAUC 6864.
[230] Klose, S et al. 1997 IAUC 6611.
[231] Koshut, T 2000 ApJ in preparation.
[232] Kulkarni, SR et al. 1998 GCN 29.
[233] Kulkarni, SR et al. 1999 Nat 398:389.
[234] Kulkarni, SR et al. 1998 Nat 393:35.
[235] Kulkarni, SR et al. 1999 ApJ 522:L97.
[236] Kulkarni, SR et al. 1998 Nat 395:663.
[237] Kulkarni, SR et al. 1997 IAUC 6559.
[238] Kulkarni, SR et al. 1997 IAUC 6723.
[239] Kulkarni, SR et al. 1998 GCN 27.
[240] Kuulkers, E et al. 1999 GCN 326.
[241] Lamb, DQ et al. 1999 A&AS 138:479.
[242] Larkin, J et al. 1998 GCN 44.
[243] Larkin, J et al. 1998 GCN 51.
[244] Levine, A et al. 1998 IAUC 6966.
[245] Li, ZY & Chevalier, RA 1999 ApJ 526:716.
[246] Lidman, C et al. 1998 IAUC 6895.
[247] Lindgren, B et al. 1999 GCN 190.
[248] Luginbuhl, C et al. 1999 GCN 341.
[249] Luginbuhl, C et al. 1996 IAUC 6526.
[250] Mallozzi, R. priv. commun. 2000.
[251] Marshall, F & Takeshima, T 1999 GCN/RXTE_PCA burst position notice Thu 06 May 99 17:27:37 UT.
[252] Marshall, FE et al. 1997 IAUC 6727.
[253] Marshall, FE & Takeshima, T 1998 GCN 58.
[254] Marshall, FE et al. 1997 IAUC 6683.
[255] Marshall, FE et al. 1998 GCN 138.
[256] Masetti, N et al. 1999 GCN 345.
[257] Masetti, N et al. 1999 GCN 327.
[258] Masetti, N et al. 2000 A&A 354:473.
[259] Maury, A 1999 IAUC 7217.
[260] Maury, A et al. 1999 GCN 220.
[261] Maury, A et al. 1999 IAUC 7214.
[262] McKenzie, EH & Schaefer, BE 1999 PASP 111:964.
[263] McMahon, RG et al. 1998 GCN 101.
[264] Mendez, J et al. 1998 IAUC 6806.
[265] Metzger, MR 1998 IAUC 6874.
[266] Metzger, MR et al. 1997 IAUC 6676.
[267] Metzger, MR et al. 1997 Nat 387:879.
[268] Metzger, MR et al. 1997 IAUC 6655.
[269] Metzger, MR et al. 1997 IAUC 6588.
[270] Metzger, MR et al. 1999 GCN 191.
[271] Morris, M et al. 1997 IAUC 6666.
[272] Muller, JM et al. 1999 IAUC 7211.
[273] Muller, JM et al. 1998 IAUC 6910.
[274] Munn, JA et al. 1997 GCN 4.
[275] Murakami, T et al. 1997 IAUC 6687.
[276] Murakami, T et al. 1999 GCN 228.
[277] Murakami, T et al. 1996 IAUC 6481.
[278] Murakami, T et al. 1997 IAUC 6722.
[279] Murakami, T et al. 1997 IAUC 6729.
[280] Murakami, T et al. 1997 IAUC 6732.
[281] Murakami, T et al. 1999 GCN 372.
[282] Nakamura, T 1999 ApJ 522:L101.
[283] Natarajan, P et al. 1997 New Astron. 2:471.
[284] Nicastro, L et al. 1998 A&A 338:L17.
[285] Nicastro, L et al. 1999 A&AS 138:437.
[286] Nicastro, L et al. 1998 IAUC 6912.
[287] Nicastro, L et al. 1999 IAUC 7213.
[288] Odewahn, S et al. 1998 GCN 105.
[289] Odewahn, SC et al. 1999 IAUC 7094.
[290] Odewahn, SC et al. 1998 ApJ 509:L5.
[291] Odewahn, SC et al. 1997 IAUC 6735.
[292] Offutt, W 1999 IAUC 7098.
[293] Owens, A et al. 1998 A&A 339:L37.
[294] Pahre, MA et al. 1997 IAUC 6691.
[295] Palazzi, E et al. 1999 GCN 377.
[296] Palazzi, E et al. 1998 GCN 48.
[297] Palazzi, E et al. 1999 GCN 262.
[298] Palazzi, E et al. 1998 A&A 336:L95.
[299] Palmer, DM et al. In Meegan, C et al., editors, Gamma-Ray Bursts 304–308 New York 1998. AIP.
[300] Park, HS et al. 1997 GCN 19.
[301] Patat, F et al. 1999 IAUC 7215.
[302] Patat, F et al. 1998 IAUC 7017.
[303] Pedersen, H et al. 1997 IAUC 6628.
[304] Pedersen, H et al. 1998 GCN 52.
[305] Pedersen, H et al. 1999 GCN 348.
[306] Pedersen, H et al. 1999 GCN 340.
[307] Pedersen, H et al. 1999 GCN 352.
[308] Pedersen, H et al. 1999 GCN 349.
[309] Pedersen, H et al. 1998 ApJ 496:311.
[310] Pedersen, H et al. 1999 GCN 342.
[311] Pedersen, K et al. 1999 GCN 267.
[312] Pedichini, F et al. 1997 A&A 327:L36.
[313] Pian, E et al. 1999 A&AS 138:463.
[314] Pian, E et al. 1998 GCN 69.
[315] Pian, E et al. 1998 ApJ 492:L103.
[316] Piro, L et al. 1998 A&A 331:L41.
[317] Piro, L et al. 1999 IAUC 7111.
[318] Piro, L et al. 1998 GCN 155.
[319] Piro, L & Costa, E 1998 GCN 99.

[320] Piro, L et al. 1996 IAUC 6480.
[321] Piro, L et al. 1996 IAUC 6467.
[322] Piro, L et al. 1999 ApJ 514:L73.
[323] Piro, L et al. 1997 IAUC 6656.
[324] Piro, L & Feroci, M 1998 GCN 72.
[325] Piro, L et al. 1997 IAUC 6617.
[326] Piro, L et al. 1998 A&A 329:906.
[327] Piro, L et al. 1997 IAUC 6797.
[328] Piro, L 1999 GCN 304.
[329] Piro, L 1999 GCN 203.
[330] Piro, L 1999 GCN 199.
[331] Piro, L 1999 GCN 311.
[332] Piro, L 1999 GCN 360.
[333] Pooley, G & Green, D 1997 IAUC 6670.
[334] Pooley, G 1999 GCN 244.
[335] Quashnock, JM et al. 1998 IAUC 6860.
[336] Ramaprakash, AN et al. 1998 GCN 24.
[337] Ramaprakash, AN et al. 1998 Nat 393:43.
[338] Reichart, DE 1999 ApJ 521:L111.
[339] Reichart, DE et al. 1999 ApJ 517:692.
[340] Remillard, R et al. 1997 IAUC 6726.
[341] Rhoads, J & Halpern, J 1997 IAUC 6793.
[342] Rhoads, J et al. 1998 GCN 181.
[343] Ricci, R et al. 1998 IAUC 6910.
[344] Rol, E et al. 1999 GCN 358.
[345] Rol, E et al. 1999 GCN 374.
[346] Sagar, R et al. 1999 Bull. Astron. Soc. India 27:3.
[347] Sahu, K et al. 1997 ApJ 489:L127.
[348] Sahu, KC et al. 1997 Nat 387:476.
[349] Sahu, KC & Sterken, C 1998 IAUC 6808.
[350] Sargent, WLW et al. 1999 GCN 297.
[351] Schaefer, BE 1998 IAUC 6865.
[352] Schaefer, BE 1998 GCN 165.
[353] Schaefer, BE 1999 GCN 259.
[354] Schaefer, BE 1999 GCN 293.
[355] Schaefer, BE et al. 1999 ApJ 524:L103.
[356] Schaefer, BE et al. 1998 GCN 185.
[357] Schlegel, DJ et al. 1998 ApJ 500:525.
[358] Shepherd, DS et al. 1998 ApJ 497:859.
[359] Shepherd, DS et al. 1997 IAUC 6664.
[360] Smith, D et al. 1997 IAUC 6728.
[361] Smith, DA 1998 GCN 159.
[362] Smith, DA et al. 1999 GCN 275.
[363] Smith, DA et al. 1999 ApJ 526:683.
[364] Smith, DA et al. 1997 IAUC 6718.
[365] Smith, DA et al. 1998 GCN 126.
[366] Smith, IA & Gruendl, RA 1997 IAUC 6663.
[367] Smith, IA et al. 1997 ApJ 487:L5.
[368] Smith, IA & Tilanus, RPJ 1997 GCN 15.
[369] Smith, IA & Tilanus, RPJ 1998 IAUC 6868.
[370] Smith, IA et al. 1999 A&A 347:92.
[371] Smith, MJS et al. 1998 IAUC 6938.
[372] Soffitta, P et al. 1998 IAUC 6884.
[373] Soifer, B et al. 1997 IAUC 6619.
[374] Sokolov, V et al. 1998 GCN 148.
[375] Sokolov, V et al. 1998 GCN 147.

[376] Sokolov, VV et al. 2000 Bull. Special Astrophys. Obs. submitted (astro-ph/0001357).
[377] Sokolov, VV et al. 1998 A&A 334:117.
[378] Sokolov, VV et al. 1999 A&A 344:43.
[379] Stanek, KZ et al. 1997 IAUC 6735.
[380] Stanek, KZ et al. 1999 ApJ 522:L39.
[381] Stanek, KZ et al. 1997 IAUC 6721.
[382] Subrahmanyan, R et al. 1999 GCN 376.
[383] Subrahmanyan, R et al. 1999 GCN 357.
[384] Takeshima, T & Marshall, F 1999 GCN/RXTE_PCA burst position notice Fri 07 May 99 02:31:42 UT.
[385] Tanvir, N et al. 1997 IAUC 6796.
[386] Tarei, G et al. 1999 IAUC 7110.
[387] Taylor, GB et al. 1997 IAUC 6670.
[388] Taylor, GB et al. 1997 Nat 389:263.
[389] Taylor, GB et al. 1998 GCN 152.
[390] Taylor, GB et al. 1998 GCN 40.
[391] Taylor, GB et al. 1999 GCN 168.
[392] Taylor, GB et al. 1999 GCN 350.
[393] Taylor, GB et al. 1999 GCN 308.
[394] Taylor, GB et al. 1998 ApJ 502:L115.
[395] Thompson, I et al. 1999 GCN 391.
[396] Tinney, C et al. 1998 IAUC 6896.
[397] Udalski, A. priv. comm. 1998.
[398] Valdes, F et al. 1998 GCN 56.
[399] van Paradijs, J et al. 1997 Nat 386:686.
[400] Voges, W et al. 1997 IAUC 6539.
[401] Vrba, FJ 1999 GCN 194.
[402] Vrba, FJ et al. 2000 ApJ 528:254.
[403] Vrba, FJ et al. 1999 GCN 294.
[404] Vrba, FJ et al. 1999 GCN 300.
[405] Vrba, FJ et al. 1999 GCN 365.
[406] Vrba, FJ & Munn, JA 1997 GCN 2.
[407] Vreeswijk, PM et al. 1999 ApJ 523:171.
[408] Vreeswijk, P et al. 1999 GCN 310.
[409] Vreeswijk, P et al. 1999 GCN 324.
[410] Wagner, RM & Starrfield, S 1998 GCN 162.
[411] Wark, R et al. 1999 GCN 266.
[412] Waxman, E & Draine, BT 2000 ApJ 537:796.
[413] Wheeler, JC et al. 1997 IAUC 6697.
[414] Wieringa, M et al. 1998 IAUC 6896.
[415] Wieringa, MH et al. 1999 A&AS 138:467.
[416] Wijers, RAMJ et al. 1999 ApJ 523:L33.
[417] Williams, GG et al. 1999 ApJ 519:L25.
[418] Woods, P et al. 1998 GCN 112.
[419] Woods, PM et al. 1997 IAUC 6798.
[420] Woosley, SE et al. 1999 ApJ 516:788.
[421] Wozniak, PR 1998 GCN 177.
[422] Yanagisawa, K et al. 1997 IAUC 6731.
[423] Yoshida, A et al. 1997 IAUC 6593.
[424] Yoshida, A et al. 1999 A&AS 138:433.
[425] Zapatero Osorio, MR et al. 1998 IAUC 6967.
[426] Zharikov, SV et al. 1998 A&A 337:356.
[427] Zhu, J & Zhang, HT 1999 GCN 295.

pointed out in the relevant GRB narrative. Narratives are collections of the salient points of each event, as well as depositories of counterpart peculiarities without a place in the tabular form.

We have attempted to collect all available literature on each GRB, from trigger to manifold publication. Nevertheless, given the wealth of communication media and journals it is inevitable that we have missed some—which is unintentional, but all the same irritating for the colleagues at the other end. It would improve future versions of this table if such omissions were brought to our attention, so we would be thankful for any gentle reminders of omissions.

6. COMPLICATIONS AND EXTENSIONS OF MODELS

Although the basic principles of relativistic blast waves are now well established and have a good grounding in the body of afterglow data collected thus far, it is clear from the previous section that we can already see beyond the basic spherical adiabatic model. Many afterglows deviate so significantly from the basic predictions that we know extensions or modifications of the basic model are required. However, we seldom have enough information on a given afterglow to pin down which of the many possible modifications applies to the afterglow at hand (if there is ever only one). Therefore, different models are often found for a given afterglow by different groups, and sober evaluation of the evidence shows that the difference cannot be resolved observationally. We thus omit here most of the technical details behind the more complex models. Instead, we focus on qualitative aspects of each of the models, paying particular attention to the relevance of more complex models to the broader issues in the field, such as what causes GRBs and what their true rate is.

6.1 Collimation or jets

The importance of collimation goes beyond influencing the detailed shape of the afterglow light curve: It reveals something about the central engine and affects our estimates of how many GRB progenitors there need to be.

Consider the simplest possible jet model: two cones, of opening angle θ_c around the z-axis, have outflowing material in them. The outflows have the same Lorentz factor, γ, everywhere within the cone, and no outflow exists outside it. The shock front formed is just the part of the previous spherical shock that lies within the cones; the outflow is collimated. As long as the flow is relativistic, the emission from each part of the shock front is strongly concentrated (beamed) within an angle $\theta_b \simeq 1/\gamma$. This means that an observer only sees emission from the material that flows within an angle θ_b of her line of sight. This effect is purely relativistic, unlike collimation, which can affect any flow.[3] Beaming also affects spherical gamma-ray bursts, so in those we also see no more than a small part of the

---

[3] One often finds the word 'beaming' used both for true beaming and for collimation in the current GRB literature, which occasionally causes confusion.

outflow. This means that we expect at least one transition to occur in a collimated outflow; as long as $\theta_b \sim 1/\gamma$ is less than θ_c, the observer sees a smaller part of the outflow than the cone, and therefore cannot distinguish between a spherical and a collimated flow. When $\theta_b = \theta_c$, which happens when the Lorentz factor becomes low enough ($\gamma \lesssim \gamma_b = 1/\theta_b$), the observer begins to see the edge of the cone and thus becomes aware of the collimated nature of the flow. When this happens, the falloff of the light curve becomes steeper; at early times, the decline of the afterglow is a balance between a very steep drop of the surface brightness of the shock and an increase in the observed emitting area proportional to $1/\gamma^2 \propto t^{3/4}$. When $\gamma < 1/\theta_b$, the emitting area, limited by the size of the cone, stays constant; hence, a drop in the exponent of the power law decline by $t^{-3/4}$ is expected.

Because the pressure is very high behind the shock, this decreases the likeliness of a pressure on the edge able to confine it, so one may well ask how long the flow can stay collimated. Here, relativity comes to the rescue: The jet expands sideways no faster than its internal speed of sound ($c/\sqrt{3}$ in the ultrarelativistic limit). Because to an external observer the apparent sideways expansion of the jet is superluminal with speed γc, the angular size of a region of the jet that is causally connected is only of order $1/\gamma$, and the angle by which the jet expands sideways is similar. Therefore, sideways expansion of the jet is unimportant as long as the total jet opening angle is larger than $\theta_b > 1/\gamma$. When the sideways expansion does become important, simple scaling suggests that the radius of the shock front stops expanding, and the energy is lost in one place. More detailed considerations show that in fact the shock Lorentz factor decreases exponentially with radius (Rhoads 1999). However, changes in the area and relativistic kinematics also take place at the same time, and the net result is that an external observer still sees a power-law decline of the brightness, albeit much steeper.

Since the critical angles for the end of collimation and the fanning-out of the emission are of the same order, it is unclear whether the collimation break of $t^{-3/4}$ (Mészáros & Rees 1999) would ever be seen. The detailed calculations by Rhoads (1999) only show a very broad transition between the initial spherical evolution and the steep (t^{-p}) late-time behavior, in which the beaming break is presumably hidden (see also Panaitescu & Mészáros 1999). More recent detailed calculations even suggest that complicating effects cause the collimation break to be much weaker (Kumar & Panaitescu 2000) or even absent (Huang et al 2000), though a jet may manifest itself via a steep break in the transition to non-relativistic evolution. An important aspect of the break resulting from a jet collimation transition is that it is expected to be achromatic, i.e. of the same strength and occurring at the same time in all wavelengths.

The first claim for a possible collimated burst came for GRB 980519 (Halpern et al 1999), which had a very sharply declining afterglow. In this burst, however, the break was not seen (it would have occurred before the first observation), and the data are also consistent with a spherical afterglow expanding into a $1/r^2$ stellar wind (Halpern et al 1999). When a break was seen in the afterglow of GRB 990123, starting about a day after the burst, this was again attributed to a jet (Kulkarni et al

1999a,b). However, closer investigation shows that the break is not nearly strong enough to be caused by a jet, even if only the beaming break occurs and is not seen in K band, which is contrary to expectation. Also, the optical data of Castro-Tirado et al (1999) indicate an achromatic decline possibly consistent with a cooling break moving through optical frequencies between days 1 and 3. The short radio afterglow of this source has also been advanced as support for a jet (Kulkarni et al 1999b), but can be interpreted without using a jet as well (Galama et al 1999). The afterglow of GRB 990510 appears to be the first in which the evidence for beaming is strong and has not been the subject of controversy: All optical data from U to K can be fit with a single transition time and the same asymptotic power-law indices, perhaps with some evidence of deviations in B. The values of the pre- and post-break power-law agree with the calculations of Rhoads (1999). In addition, polarization has been detected in this source (Wijers et al 1999; see Section 6.2), which may be related to its jet-like character.

In summary, some clear predictions from collimated outflows are strong enough to rule out certain sources as being beamed, and make at least one source a good candidate. Also, in some well-studied cases significant collimation is ruled out (e.g. GRB 970228 and GRB 970508). The implications for the physics of GRB are considerable: If a burst has a collimation angle of even $10°$ (very wide by the standard of AGN jets), then only 1.5% of the sky is illuminated by the burst. This means that when making the usual assumption of isotropy, we would overestimate the energy requirements of the central engine by a factor of 100, and at the same time underestimate the formation rate of GRB progenitors by that same factor of 100. This establishes the investigation of the reality of collimation in GRB, and any correlations between collimation and brightness, as one of the more pressing challenges in afterglow research.

6.2 Rings and Polarization

The beaming of the GRB emission has further consequences for the appearance of the afterglow. Because the Lorentz factor of the burst is continually declining, the surface we see at a given observer time is not perfectly elliptical as in the constant-γ case (Rees 1964), nor is it uniform in surface brightness. The point on the line of sight is closest to us, which means that among all the points we see at any given time, the light from that particular point left the surface the latest, so that point is oldest in the frame of the afterglow. Therefore, it has the lowest γ and lowest surface brightness. Consequently, the point approaching us is less bright than those immediately around it. Furthermore, the edge of the afterglow in our observer frame is only of order $1/\gamma$ away in angle from the center. The net result is that the observed surface brightness has a maximum away from the line of sight (by about an angle of $1/\gamma$), i.e. it appears as a ring (Panaitescu & Mészáros 1998b, Sari 1998, Waxman 1997a). At very low frequencies, the less steep dependence of the surface brightness on the Lorentz factor makes this effect nearly go away (Granot et al 1999a).

Because the angular size of the ring is usually too small to resolve and does not change any major scalings, the observable effects are small at best. However, it does offer the chance of an asymmetry that brings about net polarization of the afterglow.

Since the afterglow is synchrotron radiation, its intrinsic polarization will be 60–70%; therefore, rather than asking why it is polarized, we should ask why it is not. Two reasons have been advanced: First, the magnetic field is generated by some instability, and thus should be highly tangled in nature. If the coherence length of the field is much less than the size of the observable afterglow surface, then we expect greatly reduced polarization, to a value of about $60/\sqrt{N}\%$ for N independent patches (Gruzinov & Waxman 1999). Second, there could be net direction to the magnetic field even if it is generated by instability, especially if one accounts for aberration effects in the ring of emission (Medvedev & Loeb 1999, Gruzinov 1999, Ghisellini & Lazzati 1999). If the ring is perfect, the symmetry ensures zero net polarization, but any imperfections would give a net polarization.

In the latter case, beaming and collimation may combine to give a net polarization that varies with time. Initially, when a collimated outflow still has a very high Lorentz factor, we see a complete ring, and the symmetry precludes any net polarization. At very late times, we see the entire outflow, of which the symmetry once again precludes net polarization. At intermediate times, when the beaming cone is similar to the collimation angle and the afterglow light curve is breaking to a steeper decline, we can see part of a ring if our line of sight is offset from the center of the outflow. During this phase the polarization does not average to zero, since we do not get emission from the complete ring.

After a first attempt by Hjorth et al (1999) on GRB 990123, which set an upper limit of 2.3% on polarization, an actual detection was made with the ESO VLT for GRB 990510 (Wijers et al 1999, Covino et al 1999). The polarization was measured as 1.7% around the time of the jet break. The data are consistent both with a symmetry-breaking origin of the polarization and with the random-patch model. Much earlier and later data, during the power-law parts of the light curve, would be needed to distinguish between the two (Wijers et al 1999). Sari (1999) has calculated a toy model of the polarization in the jet case. He showed that the period of maximum polarization near the jet break contains considerable fine structure, with a few minima possible as a result of polarization sign changes. The other burst with measured polarization to date is GRB 990712 (Rol et al 2000). Curiously, this burst does not show any signs of a beaming break, and the middle of three polarization measurements has the lowest value, whereas the polarization angle is constant. It is possible that a beaming break is hidden by the effects of a bright host in this burst, but even so, a minimum in the polarization is not easy to obtain without changes in the polarization angle. Likewise, the random-patch model would predict large angle changes in the polarization as the value changes, so neither model provides a convincing interpretation of this event. The observational difficulties of improving the situation are clear: To detect polarization of 1%, one needs the object to be 5 or 6 magnitudes above the detection limit, and therefore measurements of polarization can only be made within the first few days after trigger.

6.3 External Density Profiles and Late Evolution

Depending on the type of progenitor, a burst may occur in more or less average interstellar medium, or it may be surrounded by a large amount of circumstellar medium. Consequently, it is not trivial that the ambient medium should have a uniform density. Specifically, the stellar-wind case of a $1/r^2$ density falloff has received some attention since the discovery of a GRB-supernova association (Sections 4 & 7).

Aside from providing smooth changes, circumstellar media also offer the possibility of strong inhomogeneities. For example, a wind has a termination shock where it encounters the older wind pressed up against the interstellar medium, and a forward shock driven into the ISM by the wind pressure. When the blast wave meets these, sudden density changes will lead to jumps and non–self-similar behavior in the afterglows. The maximum in the light curve of GRB 970508 after 1.5 days could represent such a situation. At a shock interface, instabilities may also lead to finger formation and other small-scale irregularity. Dermer & Mitman (1999) and Dermer et al (1999) have suggested that the irregular light curves of gamma-ray bursts may be caused by an encounter of the forward shock with these irregularities. Most recent work has attributed the gamma-ray burst proper to internal shocks in the relativistic outflow (Rees & Mészáros 1994, Kobayashi et al 1997); however, objections to the external shock model seem to be circumvented by this new model, so this has again become an open question (Fenimore et al 1999, Fenimore & Ramirez-Ruiz 1999).

What density structures an observable afterglow will meet depends on its energy; not long after it turns non-relativistic, it declines fast enough to become unobservable, so in broad terms we can define the afterglow phase as lasting from t_{dec} to t_{NR} (Section 3). The non-relativistic phase starts when a mass equal to E/c^2 has been swept up, since at that point the energy per particle is comparable to the rest mass energy, i.e. $M_{NR} = 0.05 E_{52} M_\odot$. This is small compared to the total wind mass ejected by a massive star, so indeed massive stars may produce GRB whose afterglows are entirely within their old wind (Chevalier & Li 1999). However, most of that wind is in a shocked, nearly uniform bubble, so it does not follow that all massive-star GRBs have rapidly fading afterglows characteristic of a $1/r^2$ density profile. This is even more true if the star moves with even a few tens of km/s, because then most of the ejected mass during its life is left far behind and plays no role when the star's life ends.

6.4 Finite Optical Depths, Prompt Emission, and Lines

Except at radio wavelengths, the optical depth of the afterglow to its own emission is negligible. Nonetheless, a variety of observations or other considerations have inspired observers to look for effects of finite optical depth. The oldest concern is that the peak frequency in a synchrotron model should be very strongly dependent on the Lorentz factor of the blast wave. Combined with the narrow range of observed peak frequencies in GRBs, this raises concerns about the required narrowness of the Lorentz factor distribution. Brainerd et al (1998) has proposed his Compton attenuation model specifically for this: Compton scattering by the external, non-moving medium imposes a signature at a fixed source-frame photon energy of 0.5 MeV. A problem with this model is the large required optical depth, which implies even larger energies for GRB than more conventional models, but also makes it difficult to understand why we generally see little or modest reddening in the optical afterglow. More recent suggestions achieve a signature of a non-moving medium in the spectrum by interaction of the GRB flux with previously emitted and scattered GRB photons, or with external sources of soft radiation (Madau & Thompson 2000, Madau et al 2000, Dermer et al 1999).

The optical depth to MeV photons during the burst has also been rediscussed recently, after the long-held belief that it must be small lest the spectrum become thermalized (the compactness problem; Section 1). It appears that the effects must be mostly small (Lazzati et al 2000, Mészáros & Rees 2000; but see Liang et al 1999), but some unexpected effects did turn up: Granot et al (2000) discovered that the optical depth at the earliest times in a burst could be large enough to cause self-absorption in X-rays. This may answer a long-standing issue that a significant fraction of burst spectra, especially early in the burst, rises more steeply than optically thin synchrotron spectra can account for (Preece et al 1998, Tavani et al 2000). It is important to note that low optical depths do not necessarily imply that any other emission process is energetically unimportant. For example, inverse Compton scattering may have a luminosity as large as $\gamma_e^2 \tau$ times the primary luminosity, whereas the random electron Lorentz factor, γ_e, can be hundreds of times the already large shock Lorentz factor. Therefore, even at $\tau \sim 10^{-6}$ this could still produce an important fraction of the energy output (Waxman 1997b, Mészáros et al 1994).

Another effect of finite optical depth (or rather, finite emission measure) is the possibility of emission and absorption lines in the spectra of GRB. An iron line was reported in BeppoSAX NFI X-ray spectra of GRB 970508 (Piro et al 1999). Since this line is not highly blueshifted, it must come from material that does not participate in the relativistic outflow. It may therefore signal the presence of high-density colder material that is being illuminated by the energetic radiation from the GRB, such as the remains of an exploding star (Böttcher 1999, Lazzati et al 1999).

7. HOSTS, COSMOLOGY, AND CENTRAL ENGINE

It has become clear that GRBs lie in star-forming galaxies and are associated with supernovae, and that their great brightness allows us in principle to observe them at high redshifts, perhaps up to $z = 20$ (Wijers et al 1998, Lamb & Reichart 2000). This has greatly increased the interest in gamma-ray bursts as tools for cosmology and laboratories of high- energy astrophysics, so we will briefly touch upon these subjects here.

7.1 The Association of GRB with Star Formation and Host Galaxies

Prior to 1997 there appeared to be a systematic dearth of sufficiently bright galaxies in the best determined error boxes of γ-ray bursts. This became known as the no host problem (e.g. Band et al 1999). However, as the first afterglows were found, it rapidly became clear that almost all detected counterparts lie in a host galaxy. Furthermore, the large energies required also pointed to source models involving stellar collapses and mergers. This prompted a number of attempts to associate the GRB rate in the universe with the star formation rate (Totani 1997, Wijers et al 1998). It was shown that the observed peak flux distribution of gamma-ray bursts (Figure 19) is consistent with the assumption that the GRB rate in the universe is directly proportional to the star formation rate. Recent discoveries of supernovae associated with GRBs (Section 4) have lent further support to this conclusion.

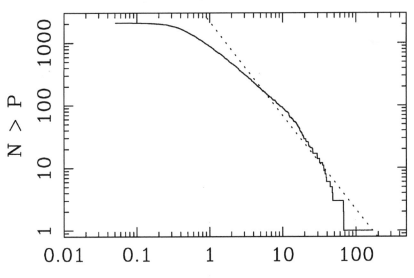

Figure 19 The cumulative peak flux distribution of gamma-ray bursts detected with BATSE. The dotted line (a $-3/2$ power-law) indicates the expected distribution for a uniform density of GRBs in a Euclidean space (M Briggs, private communication).

In star formation-related models, the GRB rate is only 10^{-8} per galaxy per year in the universe at present but was much higher at $z \sim 1$, and the characteristic peak luminosity is 10^{52} erg/s (Wijers et al 1998). Typical GRBs at the BATSE threshold would be at $z \simeq 4$. These results are quite different from those of earlier fits to the peak flux distribution, which assumed standard candles and no evolution of the GRB rate; they typically placed the dimmest BATSE bursts at $z \sim 1$ (e.g. Fenimore et al 1993, Paczyński 1992; but see Fenimore & Bloom 1995).

With a small dozen redshifts and a somewhat greater number of hosts known, it has become clear that GRB luminosities in all wavelengths range widely (Section 5), so the results of standard-candle fits to the flux distribution should be taken with a grain of salt (see e.g. Kommers et al 2000a, Krumholz et al 1998, Schmidt 1999). However, the bursts with known OTs are a much brighter group than the BATSE bursts as a whole, and their median redshift is about 1, making it likely that the dimmest GRBs are very far away indeed. This opens the prospect of using GRBs to study the early universe, e.g. by investigating absorption line forests in their spectra, as is done with quasars up to $z = 5$ (Wijers et al 1998, Lamb & Reichart 2000).

In Figure 20 we have assembled images of the known GRB hosts; measured properties of the hosts are assembled in the summary table (Section 5). They are rather diverse in nature, but share some important characteristics: all are blue, indicating the presence of an abundant number of young stars. In virtually all cases, the OT does not coincide with the center of the galaxy, but does lie within its detectable light distribution (e.g. Bloom et al 1999b). Many of the hosts are subluminous, but the wide range of values includes L_* galaxies (e.g. Hogg & Fruchter 1999). Star formation rates in several hosts have been estimated (Section 5). While they are not particularly high in many cases, the star formation rate per unit luminosity in some is quite substantial (e.g. in the small host of GRB 970508; Natarajan et al 1997). These average properties support the notion that GRBs occur where massive stars are born and die in the Universe.

7.2 Progenitors and Central Engines

The association of GRBs with supernovae and blue host galaxies, as well as the supernova-like energies, clearly suggest an origin of GRBs in some type of stellar death. The most popular among these have been mergers of neutron stars (Paczyński 1986, Goodman et al 1987, Eichler et al 1989, Mochkovitch et al 1993) and massive-star collapses (Woosley 1993, Paczyński 1998). The location of GRB counterparts within the blue parts of galaxies argues against high-velocity progenitors, such as merging neutron stars (Bloom et al 1999c, Bulik et al 1999).

Figure 20   The known host galaxies of GRBs, imaged with HST (0228, 0508, 1214, 0123, 0510), Keck (0828, 0326, 0329, 0519, 0613, 0703, 1226, 0506; courtesy Caltech GRB collaboration), and NTT (0425). Images are $14''$ on a side except 0828, 0123, and 1214, which are $7''$, and 0425 ($2'$ on a side).

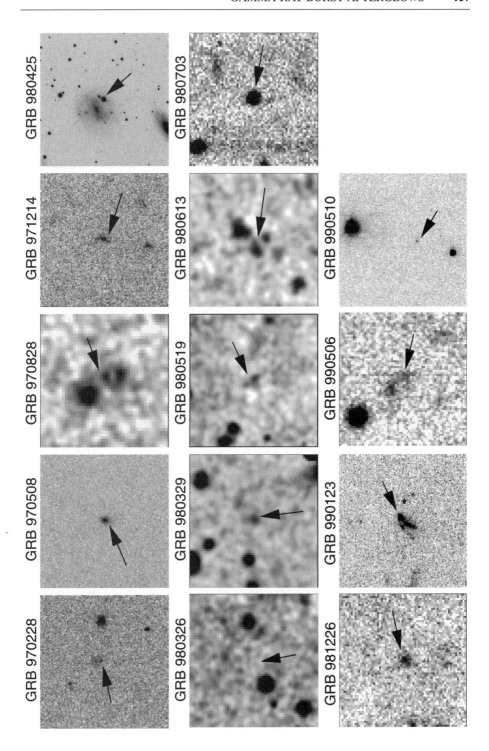

While this is true for the bursts thus far located accurately, it should be noted that those are all long-duration bursts. The short bursts (Kouveliotou et al 1993) have not yet been followed up, so it is possible that these represent another type of central engine (e.g. Fryer et al 1999).

The energies provided by many possible central engines are quite similar, since all eventually lead to the formation of a rotating compact object surrounded by debris (Mészáros et al 1999). For some bursts, such as GRB 971214 and GRB 990123, the implied isotropic energy is large $\sim 10^{54}$ erg (Kulkarni et al 1998a, Ramaprakash et al 1998, Halpern et al 1998, Kulkarni et al 1999a, Galama et al 2000). While still within the realm of the possible for the stellar-death models, the efficiency of converting the original energy to gamma rays could be low, so some collimation and beaming of the outflows may be necessary (Kumar 1999, Kumar & Piran 1999).

Two mechanisms have been suggested for the extraction of energy from the central engine. Both use a disk-like configuration around a compact object, and therefore naturally lead to some amount of collimation. First, neutrino annihilation can provide a large energy input while the central object is still hot and accreting rapidly. Because it depends steeply on the neutrino luminosity, it is not expected to last for more than a few to ten seconds (see e.g. Ruffert & Janka 1999). This suffices to push a jet through a helium-star envelope in a collapsar model (MacFadyen & Woosley 1999), but not to power bursts at the long end of the duration distribution (100–1000 s). Second, electromagnetic extraction of rotation energy from a central black hole (Blandford & Znajek 1977) has been proposed. This mechanism has the potential of lasting much longer and extracting somewhat higher energies. Its efficacy is not yet universally accepted (Li 1999, Livio et al 1999), but has recently been discussed in detail by Lee et al (2000a,b), who conclude that it is a viable central-engine model.

8.   CONCLUDING REMARKS

In the last three years we have witnessed a tremendous observational breakthrough in our understanding of GRBs. This, in turn, has led to a new generation of GRB models, ironically both based on very old initial concepts: the 1978 fireball model (Cavallo & Rees 1978) and the 1968 and 1974 supernova model (Colgate 1968, Colgate 1974). Over 30 GRBs have provided believable afterglows, and in at least a dozen of these a galaxy host has been clearly identified: The GRB cosmological distance scale is established beyond reasonable doubt. Also among the hard GRB afterglow facts, one should count the following (see Section 5): (1) temporal and spectral power-law decays for all wavelengths, varying between -1.1 and -2.1, and between -0.8 and -2.0, respectively; (2) initial source sizes of the order of μarcsecs (defined by VLA radio scintillation observations); (3) the existence of dark afterglows, i.e. cases where we observe the X-ray but not the optical counterpart; they are thought to lie in dense molecular clouds or have much steeper decay rates; (4) in some cases, an association between a GRB and a peculiar type of supernova has been established; (5) in almost all cases where a host has been identified, it is a rather blue and actively star-forming galaxy; and (6) given their peak luminosities and their distance scales, GRBs are the most powerful photon emitters in the Universe.

We are still seeking the answers to some major questions on gamma-ray bursts. What is the meaning of dark, afterglow-less bursts? What is the true energy output of GRBs, and how does the central engine deliver it? How high a redshift can we go to in chasing these cosmic explosions? In addition, the mystery of how the prompt gamma-ray emission is precisely produced is still with us.

As the GRB afterglow stamp collection grows, new evidence will emerge and fill in the puzzle. As we pointed out in the introduction to this article, none of this would have been achieved without the dedication of the scientific team of BeppoSAX; they deserve a major part of the credit. Another generous part of the credit should go to the GRB hunters, the tireless observers who scan enormous amounts of data for the elusive counterpart detection, quite often without reward.

The situation will dramatically change with the advent of the new GRB satellites, HETE-2 and SWIFT. Through them, fast and accurate GRB positions will be delivered automatically to the ground for subsequent follow-ups. Although the nature of the hunt may change, following up already identified counterparts instead of searching for them, the existing shortage of observing facilities will become more severe. A good ground-based observatory infrastructure therefore needs to be built and maintained to cope with the (predicted) future deluge of GRB observations. Monitoring, however, should be done in all wavelengths and with high-resolution capability, to further advance the field. GRBs have by now brought together multiple astrophysical disciplines, including early star formation and cosmology. It is a very healthy sign in a field when the excitement of discovery alternates between theorists and observers.

ACKNOWLEDGMENTS

We would like thank the many people whose invaluable help made the completion of this review possible, despite the unusually difficult circumstances. Ed van den Heuvel in Amsterdam and Jerry and Nancy Fishman in Huntsville provided unlimited support despite their own grief. In Amsterdam, Erica Veenhof and Jane Ayal helped with typing the text, and Paul Vreeswijk, Paul Groot, and Evert Rol gave invaluable practical support. We thank Astrid Havinga for keeping our bodies and souls together during the final writing, and Geoff Burbidge for his great patience and encouragement. The compilation of the enormous amount of literature would not have been possible without the authors of the IAU Circular Service, NASA's Astrophysics Data System, the Los Alamos Preprint Server, and Scott Barthelmy's GCN Circular Service. Jochen Greiner's web page (http://www.aip.de:8080/People/JGreiner/grbgen.html) served as an invaluable cross-check reference for Section 5. We are indebted to many colleagues for their helpful advice and cooperation—especially George Djorgovski, Dale Frail, Andy Fruchter, Tom Koshut, and Shri Kulkarni for sharing unpublished results with us, and Tim Giblin and Michael Briggs for preparing figures for this paper. We wish to thank Valerie Connaughton, Titus Galama, Ed van den Heuvel, Chip Meegan, Peter Mészáros, Elena Pian, Alan Sandage and Re'em Sari for valuable comments on the manuscript. To Titus Galama and Josh Bloom, who prepared one of the most beautiful figures of this review (host galaxies), we would like to say thank you for a job well done.

Visit the Annual Reviews home page at www.AnnualReviews.org

LITERATURE CITED

Akerlof C, Balsano R, Barthelmy S, Bloch J, Butterworth P, et al. 1999. *Nature* 398:400–2

Atkins R, Benbow W, Berley D, Chen ML, Coyne DG, et al. 2000. *Ap. J.* 533:L119–22

Band DL, Hartmann DH, Schaefer BE. 1999. *Ap. J.* 514:862–68

BATSE Team. 2000. URL http://gammaray.msfc.nasa.gov/batse/

Blandford RD, McKee CF. 1976. *Phys. Fluids* 19:1130–38

Blandford RD, Znajek RL. 1977. *MNRAS* 179:433–56

Bloom JS, Djorgovski SG, Kulkarni SR, Frail DA. 1998a. *Ap. J.* 507:L25–28

Bloom JS, Frail DA, Kulkarni SR, Djorgovski SG, Halpern JP, et al. 1998b. *Ap. J.* 508:L21–24

Bloom JS, Kulkarni SR, Djorgovski SG, Eichelberger AC, Cote P, et al. 1999a. *Nature* 401:453–56

Bloom JS, Kulkarni SR, Harrison F, Prince T, Phinney ES, Frail DA. 1998c. *Ap. J.* 506:L105–8

Bloom JS, Odewahn SC, Djorgovski SG, Harrison SRKFA, Koresko C, et al. 1999b. *Ap. J.* 518:L1–4

Bloom JS, Sigurdsson S, Pols OR. 1999c. *MNRAS* 305:763–69

Boella G, Butler RC, Perola GC, Piro L, Scarsi L, Bleeker JAM. 1997. *Astron. Astrophys. Suppl. Ser.* 122:299–307

Bond HE. 1997. *IAU Circ.* 6654

Böttcher M. 1999. *Ap. J.* In Press (astro-ph/9912030)

Brainerd JJ. 1992 *Nature* 355:552–4

Brainerd JJ, Preece RD, Briggs MS, Pendleton GN, Paciesas WS. 1998. *Ap. J.* 501:325–38

Branch D. 1999. In *Supernovae and Gamma Ray Bursts*, ed. M Livio. In Press (astro-ph/9906168)

Briggs MS. 1999. In *Gamma-Ray Bursts: The First Three Minutes*, ed. J Poutanen, R Svensson, p. 133–50. San Francisco: ASP

Briggs MS, Band DL, Kippen RM, Preece RD, Kouveliotou C, et al. 1999. *Ap. J.* 524:82–91

Briggs MS, Paciesas WS, Pendleton GN, Meegan CA, Fishman GJ, et al. 1996. *Ap. J.* 451:40–63

Bulik T, Belczynski K, Zbijewski W. 1999. *Astron. Astrophys. Suppl. Ser.* 138:483–84

Bulik T, Lamb DQ, Coppi PS. 1998. *Ap. J.* 505:666–87

Burenin RA, Vikhlinin AA, Gilfanov MR, Terekhov OV, Tkachenko AY, et al. 1999. *Astron. Astrophys.* 344:L53–56

Butler RC, Piro L, Costa E, Feroci M, Frontera F, et al. 1997. *IAU Circ.* 6539

Castro-Tirado AJ, Rosa Zapatero-Osorio M, Caon N, Marina Cairos L, Hjorth J, et al. 1999a. *Science* 283:2069–73

Castro-Tirado AJ, Zapatero-Osorio MR, Gorosabel J, Greiner J, Heidt J, et al. 1999b. *Ap. J.* 511:L85–88

Cavallo G, Rees MJ. 1978. *MNRAS* 183:359–65

Chevalier RA, Li ZY. 1999. *Ap. J.* 520:L29–32

Chiang J, Dermer CD. 1999. *Ap. J.* 512:699–710

Cohen E, Piran T, Sari R. 1998. *Ap. J.* 509:717–27

Colgate SA. 1968. *Can. J. Phys.* 46:S476–80

Colgate SA. 1974. *Ap. J.* 187:333–36

Condon JJ. 1999. *Proc. Nat. Acad. Sci. USA* 96:4756–58

Connaughton V. 2000, in preparation

Connors A, Huether GJ. 1998 *Ap. J.* 501:307–24

Costa E, Feroci M, Piro L, Frontera F, Zavattini G, et al. 1997a. *IAU Circ.* 6533

Costa E, Frontera F, Heise J, Feroci M, in't Zand J, et al. 1997b. *Nature* 387:783–85

Covino S, Lazzati D, Ghisellini G, Saracco P, Campana S, et al. 1999. *Astron. Astrophys.* 348:L1–4

Dermer CD, Böttcher M, Chiang J. 1999. *Ap. J.* 515:L49–52

Dermer CD, Mitman KE. 1999. *Ap. J.* 513:L5–8

Djorgovski SG, et al. 2000. in preparation

Djorgovski SG, Kulkarni SR, Bloom JS, Frail DA. 1999. GCN 289

Djorgovski SG, Metzger MR, Kulkarni SR, Odewahn SC, Gal RR, et al. 1997. *Nature* 387:876–78

Eichler D, Livio M, Piran T, Schramm DN. 1989. *Nature* 340:126–28

Fenimore EE, Bloom JS. 1995. *Ap. J.* 453:25–36

Fenimore EE, Conner JP, Epstein RI, Klebesadel RW, Laros JG, et al. 1988. *Ap. J.* 335:L71–74

Fenimore EE, Epstein RI, Ho C, Klebesadel RW, Lacey C, et al. 1993. *Nature* 366:40–42

Fenimore EE, Ramirez-Ruiz E. 1999. *Ap. J.* pp submitted (astro–ph/9909299)

Fenimore EE, Ramirez-Ruiz E, Wu B. 1999. *Ap. J.* 518:L73–76

Feroci M, Antonelli LA, Guainazzi M, Muller JM, Costa E, et al. 1998. *Astron. Astrophys.* 332:L29–33

Feroci M, Frontera F, Costa E, dal Fiume D, Amati L, et al. 1997. *Proceedings of SPIE* 3114:186–97

Fishman GJ, Meegan CA. 1995. *Annu. Rev. Astron. Astrophys.* 33:415–58

Fishman GJ, Meegan CA, Watts JW, Derrickson JH. 1978. *Ap. J.* 223:L13–15

Frail DA, Bloom JS, Kulkarni SR, Sari R, Taylor GB. 2000. in preparation

Frail DA, Kulkarni SR. 1995. *Astrophys. Space Sci.* 231:277–80

Frail DA, Kulkarni SR. 1997. *IAU Circ.* 6662

Frail DA, Kulkarni SR, Costa E, Frontera F, Heise J, et al. 1997a. *Ap. J.* 483:L91–94

Frail DA, Kulkarni SR, Nicastro L, dal Fiume D, Orlandini M, et al. 1997b. *IAU Circ.* 6545

Frail DA, Kulkarni SR, Nicastro L, Feroci M, Taylor GB. 1997c. *Nature* 389:261–63

Frontera F, Costa E, Piro L, Antonelli LA, Voges W, et al. 1997. *IAU Circ.* 6567

Fruchter AS, Pian E, Gibbons R, Thorsett SE, Ferguson H, et al. 2000. *Ap. J.* submitted (astro-ph/9903236)

Fruchter AS, Pian E, Thorsett SE, Bergeron LE, González RA, et al. 1999. *Ap. J.* 516:683–92

Fryer CL, Woosley SE, Hartmann DH. 1999. *Ap. J.* 526:152–77

Galama T, Briggs M, Wijers RAMJ, Vreeswijk PM, Rol E, et al. 1999. *Nature* 398:394–99

Galama T, Groot PJ, van Paradijs J, Kouveliotou C, Robinson CR, et al. 1997. *Nature* 387:479–81

Galama T, Tanvir N, Vreeswijk P, Wijers R, Groot P, et al. 2000. *Ap. J.* 536:185–94

Galama TJ, Groot PJ, van Paradijs J, Kouveliotou C, Strom RG, et al. 1998a. *Ap. J.* 497:L13–16

Galama TJ, Vreeswijk PM, Pian E, Frontera F, Doublier V, Gonzalez JF. 1998b. *IAU Circ.* 6895

Galama TJ, Vreeswijk PM, van Paradijs J, Kouveliotou C, Augusteijn T, et al. 1998c. *Nature* 395:670–72

Galama TJ, Wijers RAMJ, Bremer M, Groot PJ, Strom RG, et al. 1998d. *Ap. J.* 500:L97–100

Gallant YA, Achterberg A. 1999. *MNRAS* 305:L6–10

Gallant YA, Achterberg A, Kirk JG. 1999. *Astron. Astrophys. Suppl. Ser.* 138:549–50

Garcia MR, Callanan PJ, Moraru D, McClintock JE, Tollestrup E, Willner SP, et al. 1998. *Ap. J.* 500:L105–8

Germany LM, Reiss DJ, Sadler EM, Schmidt BP, Stubbs CW. 1999. *Ap. J.* 533:320–28

Ghisellini G, Lazzati D. 1999. *MNRAS* 309:L7–11

Giblin T, VanParadijs J, Kouveliotou C, Connaughton V, Wijers RAMJ, Fishman GJ. 1999. *Ap. J.* 524:L41–50

Goodman J, Dar A, Nussinov S. 1987. *Ap. J.* 314:L7–10

Gorosabel J, Castro-Tirado AJ, Wolf C, Heidt J, Seitz T, et al. 1998. *Astron. Astrophys.* 339:719–28

Granot J, Piran T, Sari R. 1999a. *Ap. J.* 513:679–89

Granot J, Piran T, Sari R. 1999b. *Ap. J.* 527:236–46

Granot J, Piran T, Sari R. 2000. *AP. J.* 534:L163–6

Greiner J, Wenzel W, Degel W. 1990. *Astron. Astrophys.* 234:251–61

Greiner J, Wenzel W, Hudec R, Pravec P, Rezek T, et al. 1993. In *Compton Gamma-Ray Observatory*, ed. M Friedlander, N Gehrels, DJ Macomb, pp. 828–32. New York: AIP

Groot PJ, Galama TJ, van Paradijs J, Kouveliotou C, Wijers RAMJ, et al. 1998a. *Ap. J.* 493:L27

Groot PJ, Galama TJ, van Paradijs J, Melnick G, vander Steene G, et al. 1997a. *IAU Circ.* 6588

Groot PJ, Galama TJ, van Paradijs J, Strom RG, Telting J, et al. 1997b. *IAU Circ.* 6584

Groot PJ, Galama TJ, Vreeswijk PM, Wijers RAMJ, Pian E, et al. 1998b. *Ap. J.* 502:L123–26

Gruzinov A. 1999. *Ap. J.* 525:L29–31

Gruzinov A, Waxman E. 1999. *Ap. J.* 511:852–61

Hakkila J, Meegan CA, Pendleton GN, Fishman GJ, Wilson RB, et al. 1994. *Ap. J.* 422:659–70

Halpern JP, Kemp J, Piran T, Bershady MA. 1999. *Ap. J.* 517:L105–8

Halpern JP, Thorstensen JR, Helfand DJ, Costa E. 1998. *Nature* 393:41–43

Hjorth J, Bjornsson G, Andersen MI, Caon N, Marina Cairos L, et al. 1999. *Science* 283:2073

Höflich P, Wheeler JC, Wang L. 1999. *Ap. J.* 521:179–89

Hogg DW, Fruchter AS. 1999. *Ap. J.* 520:54–58

Huang YF, Dai ZG, Lu T. 1999. *MNRAS* 309:513–16

Hudec R, Ceplecha Z, Spurny P, Florián J, Kolá A, et al. 1999. *Astron. Astrophys. Suppl. Ser.* 138:591–92

Hudec R, Soldan J. 1995. *Astrophys. Space Sci.* 231:311–14

Hughes PA, Miller L. 1991. In *Beams and Jets in Astrophysics*, ed. PA Hughes, 19:1–51, Cambridge:CUP

Hurley K. 1986. In *Gamma-ray Burst and Neutrino Star Physics*, ed. EP Liang, V Petrosian, 141:1. New York: AIP

Hurley K, Cline T, Mazets E, Aptekar R, Golenetskii S, et al. 2000a. *Ap. J.* 534:L23–5

Hurley K, Feroci M, Cinti M, Costa E, Preger B, et al. 2000b. *Ap. J.* 534:258–64

Hurley K, Hartmann D, Kouveliotou C, Fishman G, Laros J, et al. 1997. *Ap. J.* 479:L113–16

In't Zand, J, Heise J, Hoyng P, Jager R, Piro L, et al. 1997. *IAU Circ.* 6569

Iwamoto K. 1999a. *Ap. J.* 517:L67–67

Iwamoto K. 1999b. *Ap. J.* 512:L47–50

Iwamoto K, Mazzali PA, Nomoto K, Umeda H, Nakamura T, et al. 1998. *Nature* 395:672–74

Jager R, Heise J, In't Zand J, Brinkman AC. 1995. *Adv. Space Res.* 13:315–18

Katz JI. 1994a. *Ap. J.* 432:L107–9

Katz JI. 1994b. *Ap. J.* 422:248–59

Katz JI, Piran T. 1997. *Ap. J.* 490:772

Kippen RM, Briggs MS, Kommers JM, Kouveliotou C, Hurley K, et al. 1998. *Ap. J.* 506:L27–30

Klebesadel RW, Strong IB, Olson RA. 1973. *Ap. J.* 182:L85–88

Kobayashi S, Piran T, Sari R. 1997. *Ap. J.* 490:92

Kolatt T, Piran T. 1996. *Ap. J.* 467:L41–44

Kommers JM, Lewin WHG, Kouveliotou C, van Paradijs J, Pendleton GN, et al. 2000a. *Ap. J.* 533:696–709

Kommers JM, Lewin WHG, Kouveliotou C, van Paradijs J, Pendleton GN, et al. 2000b. *Ap. J.* In preparation

Kouveliotou C, Meegan CA, Fishman GJ, Bhat NP, Briggs MS, et al. 1993. *Ap. J.* 413:L101–4

Krimm HA, Vanderspek RK, Ricker GR. 1996. *Astron. Astrophys. Suppl. Ser.* 120:251

Krumholz MR, Thorsett SE, Harrison FA. 1998. *Ap. J.* 506:L81–84

Kulkarni SR, Berger E, Bloom JS, Chaffee F, Diercks A, et al. 2000. In *Gamma-Ray Bursts*, ed. Kippen RM, RS Mallozzi, GJ Fishman. In press

Kulkarni SR, Djorgovski SG, Odewahn SC, Bloom JS, Gal RR, et al. 1999a. *Nature* 398:389–94

Kulkarni SR, Djorgovski SG, Ramaprakash AN, Goodrich R, Bloom JS, et al. 1998a. *Nature* 393:35–39

Kulkarni SR, Frail DA, Moriarty-Schieven GH, Shepherd DS, Udomprasert P, et al. 1999b. *Ap. J.* 522:L97–100

Kulkarni SR, Frail DA, Wieringa MH, Ekers RD, Sadler EM, et al. 1998b. *Nature* 395:663–69

Kumar P. 1999. *Ap. J.* 523:L113–16

Kumar P, Panaitescu A. 2000. *Ap. J. Lett.* Submitted (astro-ph/0003264)

Kumar P, Piran T. 2000. *Ap. J.* 535:152–57

Lamb DQ. 1995. *Publ. Astron. Soc. Pac.* 107:1152

Lamb DQ, Reichart D.2000. *Ap. J.* 536:1–18

Larson SB, McLean IS, Becklin EE. 1996. *Ap. J.* 460:L95–98

Lazzati D, Campana S, Ghisellini G. 1999. *Astron. Astrophys. Suppl. Ser.* 138:547–48

Lazzati D, Ghisellini G, Celotti A, Rees MJ. 2000. *Ap. J.* 529:L17–20

Lee B, Akerlof C, Band D, Barthelmy S, Butterworth P, et al. 1997. *Ap. J.* 482:L125

Lee HK, Brown GE, Wijers RAMJ. 2000a. *Ap. J.* 536:416–19

Lee HK, Wijers RAMJ, Brown GE. 2000b. *Phys. Rep.* 325:83–114

Li LX. 1999. *Phys. Rev. D* 61:084016

Liang EP, Crider A, Böttcher M, Smith IA. 1999. *Ap. J.* 519:L21–24

Livio M, Ogilvie GI, Pringle JE. 1999. *Ap. J.* 512:100–4

Loeb A, Perna R. 1998. *Ap. J.* 495:597–603

MacFadyen AI, Woosley SE. 1999. *Ap. J.* 524:262–89

Madau P, Blandford RD, Rees MJ. 2000. *Ap. J.* In press

Madau P, Thompson C. 2000. *Ap. J.* 534:239–47

Mazets EP, Golenetskii SV, Aptekar RL, Gurian IA, Ilinskii VN. 1981. *Nature* 290:378–82

McNamara BJ, Harrison TE, Williams CL. 1995. *Ap. J.* 452:L25–28

Medvedev MV, Loeb A. 1999. *Ap. J.* 526:697–706

Meegan CA, Fishman GJ, Wilson RB, Paciesas WS, Pendleton GN, et al. 1992. *Nature* 355:143–45

Mészáros P, Laguna P, Rees MJ. 1993. *Ap. J.* 415:181–90

Mészáros P, Rees MJ. 1993. *Ap. J.* 418:L59–62

Mészáros P, Rees MJ. 1997. *Ap. J.* 476:232–37

Mészáros P, Rees MJ. 1999. *MNRAS* 306:L39–43

Mészáros P, Rees MJ. 2000. *Ap. J.* 530:292–98

Mészáros P, Rees MJ, Papathanassiou H. 1994. *Ap. J.* 432:181–93

Mészáros P, Rees MJ, Wijers RAMJ. 1998. *Ap. J.* 499:301–8

Mészáros P, Rees MJ, Wijers RAMJ. 1999. *New Astron.* 4:303–12

Metzger MR, Cohen JG, Chaffee FH, Blandford RD. 1997a. *IAU Circ.* 6676

Metzger MR, Djorgovski SG, Steidel CC, Kulkarni SR, Adelberger KL, Frail, DA. 1997b. *IAU Circ.* 6655

Metzger MR, Kulkarni SR, Djorgovski SG, Gal R, Steidel CC, Frail DA. 1997c. *IAU Circ.* 6588

Mochkovitch R, Hernanz M, Isern J, Martin X. 1993. *Nature* 361:236–38

Murakami T, Fujii M, Hayashida K, Itoh M, Nishimura J. 1988. *Nature* 335:234–5

Natarajan P, Bloom JS, Sigurdsson S, Johnson RA, Tanvir NR, et al. 1997. *New Astron.* 2:471–75

Norris JP, Marani GF, Bonnell JT. 2000. *Ap. J.* 534:248–57

Norris JP, Nemiroff RJ, Scargle JD, Kouveliotou C, Fishman GJ, et al. 1994. *Ap. J.* 424:540–545

Paciesas WS, Meegan CA, Pendleton GN, Briggs MS, Kouveliotou C, et al. 1999. *Astrophys. J. Suppl. Ser.* 122:465–95

Paczyński B. 1986. *Ap. J.* 308:L43–46

Paczyński B. 1992. *Nature* 355:521

Paczyński B. 1995. *Publ. Astron. Soc. Pac.* 107:1167

Paczyński B. 1998. *Ap. J.* 494:L45–48

Paczyński B, Rhoads JE. 1993. *Ap. J.* 418:L5–8

Panaitescu A. Mészáros P. 1998a. *Ap. J.* 501:772–9

Panaitescu A. Mészáros P. 1998b. *Ap. J.* 493:L31–34

Panaitescu A. Mészáros P. 1999. *AP. J.* 526:707–15

Park HS, Ables E, Band DL, Barthelmy SD, Bionta RM, et al. 1997. *Ap. J.* 490:99–108

Pedersen H, Jaunsen AO, Grav T, Ostensen R, Andersen MI, et al.1998. *Ap. J.* 496:311–5

Pendleton GN, Hakkila J, Meegan CA. 1998. In *Gamma-Ray Bursts*, ed. C Meegan, R Preece, T Koshut, p. 899. New York: AIP

Pian E, Amati L, Antonelli LA, Butler RC, Costa E, et al.1999. *Astron. Astrophys. Suppl. Ser.* 138:463–64

Pian E, Fruchter AS, Bergeron LE, Thorsett SE, Frontera F, et al. 1998. *Ap. J.* 492:L103–6

Piran T. 1999. *Phys. Rep.* 314:575–667

Piro L, Butler R, Fiore F, Antonelli A, Pian E. 1998a. *GCN 155*

Piro L, Costa E, Feroci M, Frontera F, Amati L, et al. 1999. *Ap. J.* 514:L73–77

Piro L, Heise J, Jager R, Costa E, Frontera F, et al. 1998b. *Astron. Astrophys.* 329:906–10

Podsiadlowski P, Rees MJ, Ruderman M. 1995. *MNRAS* 273:755–71

Preece RD, Briggs MS, Mallozzi RS, Pendleton GN, Paciesas WS, Band DL. 1998. *Ap. J.* 506:L23–26

Ramaprakash AN, Kulkarni SR, Frail DA, Koresko C, Kuchner M, et al. 1998. *Nature* 393:43–46

Ramirez-Ruiz E, Fenimore EE. 2000. In *Gamma-Ray Bursts* ed. RM Kippen, RS Malliozzi, GJ Fishman. In press

Rees MJ. 1964. PhD thesis. University of Cambridge

Rees MJ, Mészáros P. 1992. *MNRAS* 258:L41–43

Rees, MJ Mészáros P. 1994. *Ap. J.* 430:L93–96

Reichart DE. 1999. *Ap. J.* 521:L111–15

Rhoads JE. 1999. *Ap. J.* 525:737–49

Rol E, Wijers RAMJ, Vreeswijk PM, Galama TJ, Van Paradijs J, Kouveliotou C, et al. 2000. *Ap. J.* In press

Ruffert M, Janka HT. 1999. *Astron. Astrophys.* 344:573–606

Rybicki GB, Lightman AP. 1979. *Radiative Processes in Astrophysics.* New York: Wiley & Sons

Sahu KC, Livio M, Petro L, Macchetto FD, van Paradijs J, et al. 1997. *Nature* 387:476–78

Sari R. 1998. *Ap. J.* 494:L49–52

Sari R. 1999. *Ap. J.* 524:L43–46

Sari R, Piran T. 1999. *Ap. J.* 517:L109–12

Sari R, Piran T, Narayan R. 1998. *Ap. J.* 497:L17–20

Schaefer BE. 1981. *Nature* 294:722–24

Schaefer BE. 1990. *Ap. J.* 364:590–600

Schaefer BE. 1992. In *Gamma-Ray Bursts— Observations, Analyses and Theories.* ed. C Ho, RI Epstein, EE Fenimore, pp. 107–12. Cambridge, UK: Cambridge Univ. Press

Schaefer BE. 1998. In *Gamma-Ray Bursts*, ed. Meegan CA, Preece RD, Koshut TM. pp. 595–99. New York: AIP

Schaefer BE, Bradt HV, Barat C, Hurley K, Niel M, Vedrenne G, et al. 1984. *Ap. J.* 286:L1–L4

Schmidt M. 1999. *Ap. J.* 523:L117–20

Schmidt WKH. 1978. *Nature* 271:525–27

Sedov L. 1969. *Similarity and Dimensional Methods in Dynamics.* New York: Academic

Shklovskii IS, Mitrofanov IG. 1985. *MNRAS* 212:545–51

Sokolov VV, Kopylov AI, Zharikov SV, Feroci M, Nicastro L, Palazzi E. 1998. *Astron. Astrophys.* 334:117–23

Steidel CC, Sargent WLW. 1992. *Astrophys. J. Suppl. Ser.* 80:1–108

Stern B, Tikhomirova Y, Stepanov M, Kompaneets D, Berezhnoy A, Svensson R. 1999. *Astron. Astrophys. Suppl. Ser.* 138:413–14

Takeshima T, Marshall FE, Corbet RHD, Cannizzo JK, Valinia A, et al. 1998. In *Gamma-Ray Bursts.* ed. CA Meegan, RD Preece, TM Koshut, pp. 414–19. New York: AIP

Tavani M, Band D, Ghirlanda G. 2000. In *Gamma-Ray Bursts.* ed. RM Kippen, RS Mallozzi, GJ Fishman. In press

Taylor GB, Frail DA, Beasley AJ, Kulkarni SR. 1997. *Nature* 389:263–65

Taylor GI. 1950. *Proc. R. Soc. London, Ser. A* 201:159

Terlevich R, Fabian A, Turatto M. 1999. *IAU Circ.* 7269

Totani T. 1997. *Ap. J.* 486:L71–74

Usov VV, Chibisov GV. 1975. *Sov. Astron.* 19:115–16

van den Bergh S, Tammann GA. 1991. *Annu. Rev. Astron. Astrophys.* 29:363–407

van Paradijs J, Groot PJ, Galama T, Kouveliotou C, Strom, RG, et al. 1997. *Nature* 386: 686–89

Voges W, Boller T, Greiner J. 1997. *IAU Circ.* 6539

Vreeswijk PM, Galama TJ, Owens A, Oosterbroek T, Geballe TR, et al. 1999. *Ap. J.* 523:171–76

Wang L, Wheeler JC. 1998. *Ap. J.* 504:L87–90

Waxman E. 1997a. *Ap. J.* 491:L19–22

Waxman E. 1997b. *Ap. J.* 489:L33–36

Waxman E. 1997c. *Ap. J.* 485:L5–8

Wijers RAMJ, Bloom JS, Bagla JS, Natarajan P. 1998. *MNRAS* 294:L13–17

Wijers RAMJ, Galama TJ. 1999. *Ap. J.* 523:177–86

Wijers RAMJ, Rees MJ, Mészáros P. 1997. *MNRAS* 288:L51–56

Wijers RAMJ, Vreeswijk PM, Galama TJ, Rol E, van Paradijs J, et al. 1999. *Ap. J.* 523:L33–36

Woosley SE. 1993. *Ap. J.* 405:273–77

Woosley SE, Eastman RG, Schmidt BP. 1999. *Ap. J.* 516:788–96

Yoshida A, Namiki M, Otani C, Kawai N, Murakami T, et al. 1999. *Astron. Astrophys. Suppl. Ser.* 138:433–34

Zharikov SV, Sokolov VV, Baryshev YV. 1998. *Astron. Astrophys.* 337:356–62

Żytkow AN. 1990. *Ap. J.* 359:138–54

Annu. Rev. Astron. Astrophys. 2000. 38:427–83

ORGANIC MOLECULES IN THE INTERSTELLAR MEDIUM, COMETS, AND METEORITES: A Voyage from Dark Clouds to the Early Earth

Pascale Ehrenfreund[1] and Steven B. Charnley[2]

[1]*Raymond and Beverly Sackler Laboratory for Astrophysics at Leiden Observatory, P.O. Box 9513, 2300 RA Leiden, The Netherlands; e-mail: pascale@strw.leidenuniv.nl*
[2]*Space Science Division, NASA Ames Research Center, Moffett Field, CA 94305; e-mail: charnley@dusty.arc.nasa.gov*

Key Words   molecular clouds, solar system, carbon chemistry, origin of life, astrobiology

■ **Abstract**   Our understanding of the evolution of organic molecules, and their voyage from molecular clouds to the early solar system and Earth, has changed dramatically. Incorporating recent observational results from the ground and space, as well as laboratory simulation experiments and new methods for theoretical modeling, this review recapitulates the inventory and distribution of organic molecules in different environments. The evolution, survival, transport, and transformation of organics is monitored, from molecular clouds and the diffuse interstellar medium to their incorporation into solar system material such as comets and meteorites. We constrain gas phase and grain surface formation pathways to organic molecules in dense interstellar clouds, using recent observations with the Infrared Space Observatory (ISO) and ground-based radiotelescopes. The main spectroscopic evidence for carbonaceous compounds in the diffuse interstellar medium is discussed (UV bump at 2200 Å, diffuse interstellar bands, extended red emission, and infrared absorption and emission bands). We critically review the signatures and unsolved problems related to the main organic components suggested to be present in the diffuse gas, such as polycyclic aromatic hydrocarbons (PAHs), fullerenes, diamonds, and carbonaceous solids. We also briefly discuss the circumstellar formation of organics around late-type stars.

In the solar system, space missions to comet Halley and observations of the bright comets Hyakutake and Hale-Bopp have recently allowed a reexamination of the organic chemistry of dust and volatiles in long-period comets. We review the advances in this area and also discuss progress being made in elucidating the complex organic inventory of carbonaceous meteorites. The knowledge of organic chemistry in molecular clouds, comets, and meteorites and their common link provides constraints for the processes that lead to the origin, evolution, and distribution of life in the Galaxy.

0066-4146/00/0915-0427$14.00   **427**

1.   INVITATION TO THE VOYAGE

The interstellar medium (ISM) consists of gas and dust between the stars, which accounts for 20–30% of the mass of our galaxy. Much of this material has been ejected by old and dying stars. The ISM contains different environments showing large ranges in temperature ($10–10^4$ K) and densities ($100–10^8$ H atoms cm^{-3}). It is filled mainly with hydrogen gas, about 10% helium atoms, and $\sim$0.1% of atoms such as C, N, and O. Other elements are even less abundant. Roughly 1% of the mass is contained in microscopic (micron-sized) dust grains. The interstellar medium represents the raw material for forming future generations of stars, which may develop planetary systems like our own. Comets and carbonaceous chondrites may be witnesses of the processes occurring during the gravitational collapse of molecular clouds to form stars and protoplanets. There is evidently a link between chemical processes in dark embedded clouds and cometary volatiles; a detailed elucidation of the connection between interstellar, cometary, and meteoritic dust can provide constraints on the formation of the solar system and the early evolution of life on Earth (Irvine 1998).

Astronomical observations of the ISM and solar system bodies, in combination with laboratory investigations of meteoritic samples, have shown the presence of numerous organic molecules. Well over one hundred different molecules have been identified in interstellar and circumstellar regions, most of them are organic in nature (see Table 1; see also http://www.cv.nrao.edu/~awootten/allmols.html). Larger carbon-bearing species such as polycyclic aromatic hydrocarbons (PAHs) and fullerenes may also be present in interstellar gas or incorporated in dust and ice mantles. A large amount of the cosmic carbon is locked in carbonaceous solids on dust grains.

Dense clouds are characterized by very low temperatures (10–30 K) and high densities (10^{4-8} hydrogen atoms cm^{-3}). Cold gas phase chemistry can efficiently form simple species such as CO, N_2, O_2, C_2H_2 and C_2H_4, HCN, and simple carbon chains (Herbst 1995). Efficient accretion of atoms and molecules in such environments and subsequent reactions on the grain surface lead to the formation of molecules such as CO_2 and CH_3OH, which are later returned to the interstellar gas (Tielens & Hagen 1982, Schutte 1999). Evaporated ices can drive gas phase reactions by acting as precursors for the larger organics that are observed in hot cores (Charnley et al 1992). The largest organic molecule that has been unambiguously identified in the interstellar gas is $HC_{11}N$ (Bell et al 1999). The tentative detection of diethyl ether $[(C_2H_5)_2O]$ has also been recently reported (Kuan et al 1999). The chemistry in interstellar clouds and the production of organic molecules can be enriched by thermal and energetic processing, such as ultraviolet (UV) irradiation and cosmic rays. Although the UV radiation field of a young protostar is strongly attenuated within its close environment, cosmic rays can penetrate throughout the cloud and ionize H_2 molecules; the energetic electrons from these molecules can in turn excite H_2 to higher electronic states. These excited H_2 molecules subsequently

TABLE 1 Interstellar and circumstellar molecules as compiled by Al Wootten (see text)

2	3	4	5	6	7	8	9	10	11	13
H_2	C_3	$c\text{-}C_3H$	C_5	C_5H	C_6H	CH_3C_3N	CH_3C_4H	$CH_3C_5N?$	HC_9N	$HC_{11}N$
AlF	C_2H	$l\text{-}C_3H$	C_4H	$l\text{-}H_2C_4$	CH_2CHCN	$HCOOCH_3$	CH_3CH_2CN	$(CH_3)_2CO$		
AlCl	C_2O	C_3N	C_4Si	C_2H_4	CH_3C_2H	$CH_3COOH?$	$(CH_3)_2O$	$NH_2CH_2COOH?$		
C_2	C_2S	C_3O	$l\text{-}C_3H_2$	CH_3CN	HC_5N	C_7H	CH_3CH_2OH			
CH	CH_2	C_3S	$c\text{-}C_3H_2$	CH_3NC	$HCOCH_3$	H_2C_6	HC_7N			
CH^+	HCN	C_2H_2	CH_2CN	CH_3OH	NH_2CH_3		C_8H			
CN	HCO	$CH_2D^+?$	CH_4	CH_3SH	$c\text{-}C_2H_4O$					
CO	HCO^+	HCCN	HC_3N	HC_3NH^+						
CO^+	HCS^+	$HCNH^+$	HC_2NC	HC_2CHO						
CP	HOC^+	HNCO	HCOOH	NH_2CHO						
CSi	H_2O	HNCS	H_2CHN	C_5N						
HCl	H_2S	$HOCO^+$	H_2C_2O							
KCl	HNC	H_2CO	H_2NCN							
NH	HNO	H_2CN	HNC_3							
NO	MgCN	H_2CS	SiH_4							
NS	MgNC	H_3O^+	H_2COH^+							
NaCl	N_2H^+	NH_3								
OH	N_2O	SiC_3								
PN	NaCN									
SO	OCS									
SO^+	SO_2									
SiN	$c\text{-}SiC_2$									
SiO	CO_2									
SiS	NH_2									
CS	H_3^+									
HF										

Note that observations suggest the presence of large PAHs and fullerenes in the interstellar gas (Tielens et al 1999, Foing & Ehrenfreund 1997).

decay by emitting UV photons. The cosmic ray induced UV radiation field is approximately 10^3 photons s^{-1} cm^{-2} with an energy of ~ 10 eV (Prasad & Tarafdar 1983, Gredel et al 1989). The overall geometry of protostellar sources—including outflows, shock regions, and hot cores, as well as the filamentary and clumpy structure of interstellar clouds—strongly influence the general line-of-sight conditions and the strength of the UV radiation field. Therefore, only combined information from astronomical observations (infrared and radio), interferometry, laboratory spectroscopy, and supporting theoretical models allows us to understand the formation and distribution of organics in the dense ISM.

More than 50 years ago, diatomic molecules such as CH, CH^+, and CN were reported from visible absorption of diffuse clouds (Swings & Rosenfeld 1937, McKellar 1940, Douglas & Herzberg 1941). Much larger molecules are now expected to be present in the interstellar gas and on dust grains in such environments. Among the large organic molecules observed or suspected in diffuse clouds are PAHs, fullerenes, carbon-chains, diamonds, amorphous carbon (hydrogenated and bare), and complex kerogen-type aromatic networks. The formation and distribution of large molecules in the gas and solid state is far from being understood. In the envelopes of carbon-rich late-type stars, carbon is mostly locked in CO and C_2H_2. C_2H_2 molecules are precursors for soot formation, where PAHs might act as intermediates (Frenklach & Feigelson 1989). The ubiquitous presence of aromatic structures in the ISM and in external galaxies has been well documented by numerous observations with the Infrared Space Observatory, or ISO (see special issue of *Astron. Astrophys.* 315, 1996). Evidence for carbon-chains and fullerenes arises from the characterization of the Diffuse Interstellar Bands (DIBs) (Freivogel et al 1994, Tulej et al 1998, Foing & Ehrenfreund 1994, 1997). Diamonds were recently proposed to be the carriers of the 3.4 and 3.5 μm emission bands (Guillois et al 1999) observed in planetary nebulae. Hydrogenated amorphous carbon (HAC) seems to be responsible for the 2200 Å bump in the interstellar extinction curve (Mennella et al 1998). A variety of complex aromatic networks are likely to be present on carbonaceous grains (Henning & Salama 1998).

The cosmic carbon abundance is an important criterion when considering the inventory of organics in the universe. Although the determination of solar carbon values indicates 355 ± 50 atoms per 10^6 H atoms (ppm) (Grevesse & Noels 1993), recent measurements using the Goddard High Resolution Spectrograph (GHRS) on the HST (Hubble Space Telescope) indicate interstellar carbon abundances that are only 2/3 of solar values (Cardelli et al 1996). A total carbon abundance of 225 ± 50 ppm is currently adopted (Snow & Witt 1995). The mean gas phase abundance of carbon is estimated to be $\sim 140 \pm 20$ ppm and is stable among different lines of sight, indicating no significant exchange between the dust and gas (Cardelli et al 1996). These new results provide strong constraints on carbonaceous dust models.

Comets are an agglomerate of frozen gases, ices, and rocky debris, and are likely the most primitive bodies in the solar system (Whipple 1950a,b). Comets were formed in the region of giant planets and beyond, from remnant planetesimals that were not assembled into planets. Such comet nuclei have been ejected out of the solar system by gravitational interactions. A large number of comets formed in the zone of the giant planets are stored in the so-called Oort cloud at ~50,000 AU from the solar system. Comet nuclei that were formed in the trans-Neptunian region reside in the Edgewood-Kuiper belt (Jewitt et al 1998). When comets are perturbed and enter the solar system, the solar radiation heats the icy material and forms a gaseous cloud, the coma. During this sublimation process, "parent" volatiles can be subsequently photolysed and produce radicals and ions, the so-called "daughter" molecules. In 1986, an international fleet of spacecraft (Giotto, Vega 1, Vega 2, Suisei, Sakigake, and ICE) encountered comet 1/P Halley. In situ investigations showed that Halley contained a large amount of dust and organic material (Kissel & Krueger 1987). Recent observations of two exciting bright comets, Hyakutake and Hale-Bopp, allowed astronomers to identify more than 25 parent species and strongly improved our understanding of cometary outgassing (Mumma 1997; Biver et al 1997, 1999; Lis et al 1997; Crovisier & Bockelée-Morvan 1999; Irvine et al 2000; Bockelée-Morvan et al 2000). The composition of comets encodes information on their origin and can be used as a tracer for processes that were predominant in the protosolar nebula.

A large number of organic species has been identified in meteorites (Cronin & Chang 1993). Samples of carbonaceous meteorites are the only pristine extraterrestrial material that can currently be studied on Earth, until future cometary missions allow in situ measurements and sample returns. The isotopic composition of extraterrestrial matter—in particular, D-enrichments and the $^{13}C/^{12}C$ ratio—provide evidence that interstellar matter has been incorporated in meteoritic material (Kerridge 1999). Cometary and meteoritic impacts on Earth in the first 100 million years may have delivered large amounts of organic molecules (Chyba & Sagan 1992). The richness of extraterrestrial organics may have allowed a jump start to the development of life—a topic of great interest for the emerging field of astrobiology (see http://www.astrobiology.com/).

In the following sections we review the formation of organic molecules in various environments of molecular clouds, and compare and contrast this inventory with that of solar system material. In Section 2.1, we discuss the formation of organic molecules in dense clouds by gas phase and grain surface reactions. In Section 2.2, we describe our current knowledge on the formation, evolution, and distribution of organics in diffuse interstellar clouds, and the formation of complex organics in circumstellar environments. Data on cometary organics from space missions to comet Halley, along with recent remote sensing data of volatiles in the comae of comets Hale-Bopp and Hyakutake, are reviewed in Section 3. In Section 4 we discuss organics measured in carbonaceous meteorites. The link between organic molecules in dark clouds and in the solar system and their implications for the origin of life are described in Section 5.

2.   ORGANIC MOLECULES IN THE INTERSTELLAR MEDIUM

The ISM is composed mainly of three types of clouds: dark clouds ($A_V > 5$ mag), translucent clouds (1 mag $< A_V < 5$ mag), and diffuse clouds (visual extinction, $A_V \leq 1$ mag) (Spaans & Ehrenfreund 1999). Interstellar clouds are neither uniform nor dynamically quiescent on long timescales. They display "clumpy" structures, are continually evolving as new stars form, and are enriched by material ejected from dying stars that was formed during stellar nucleosynthesis. Observations of atoms and molecules in interstellar clouds have been previously examined by Winnewisser & Herbst (1993), and Herbst (1995).

2.1   Organic Molecules in Dense Clouds

Dense interstellar clouds are the birth sites of stars of all masses and their planetary systems. Interstellar molecules and dust become the building blocks for protostellar disks, from which planets, comets, asteroids, and other macroscopic bodies eventually form (Weaver & Danly 1989, Levy & Lunine 1993, Mannings et al 2000). Observations at infrared, radio, millimeter, and submillimeter frequencies show that a large variety of gas phase organic molecules are present in the dense interstellar medium (Irvine et al 1987, Ohishi & Kaifu 1998, Winnewisser & Kramer 1999). These include organic classes such as nitriles, aldehydes, alcohols, acids, ethers, ketones, amines, and amides, as well as many long-chain hydrocarbon compounds. Such species display large compositional variations between quiescent dark clouds and star-forming regions, as well as strong abundance gradients on small spatial scales within each type of cloud (Irvine et al 1987, Olano et al 1988, van Dishoeck et al 1993, van Dishoeck & Blake 1998).

In the following sections we describe the solid state and gas phase inventories of molecular clouds, as revealed by ground-based and space-based observations. Table 2 summarizes the most abundant species observed in different regions of dense clouds and in the comae of bright comets, and lists their abundances relative to H_2O.

2.1.1 Organic Molecules Formed on the Grain Surface

The dust particles present in cold molecular clouds are important chemical catalysts. They can acquire icy grain mantles through the efficient but slow (typically one per day) accretion and reaction of atoms and molecules from the gas (Greenberg 1986, Whittet 1993, Schutte 1996, Tielens & Whittet 1997, Tielens & Charnley 1997). At low gas temperatures, sticking coefficients are generally expected to be close to unity for heavy atoms and molecules (Schmitt 1994), and their efficient removal from the gas phase occurs on the depletion timescale of $\tau \sim 3 \times 10^9/n_H$ yr, where n_H is the hydrogen nucleon number density. For $n_H > 10^4 \, \text{cm}^{-3}$, $\tau < 10^6$ yr, which is less than the expected lifetime of a dense core (Blitz 1993); therefore, regulatory mechanisms such as selective desorption and mantle explosion are needed to maintain organic molecules in the gas phase (Léger et al 1985, Schutte & Greenberg 1991, Schmitt 1994, Willacy & Millar 1998).

An active chemistry takes place on grain surfaces that, in principle, can produce many complex interstellar molecules. At 10 K only H, D, C, O, and N atoms have sufficient mobility to scan the grain surface (by quantum tunneling or thermal hopping) and find a reaction partner (Tielens & Hagen 1982). Formation of the simplest mantle molecules (H_2O, NH_3, CH_4, etc.) can be explained by elementary exothermic hydrogen addition reactions, which do not possess activation-energy barriers (Allen & Robinson 1977, d'Hendecourt et al 1985, Brown & Charnley 1990, Hasegawa et al 1992). However, the presence of complex molecules such as CH_3OH and various unidentified organics indicates that other reaction processes also operate on and in the ice mantles. Although grain surface reactions of many neutrals, such as CO, possess activation barriers, H and D atoms can penetrate them by quantum tunneling, and can open up bonds for subsequent activation-less atom additions (Tielens & Hagen 1982). As dust grains with these simple ice mantles are transported from a cold quiescent phase to warm, dense, and active protostellar regions, other processes (such as ultraviolet irradiation, cosmic ray bombardment, and temperature variations) may become important and alter the grain mantle composition. Laboratory experiments show that irradiation processes of ice analogs lead to the formation of radicals, complex molecules, and even organic refractory material (Allamandola et al 1997, Moore & Hudson 1998, Kaiser & Roessler 1998, Greenberg 1999). Therefore, grain chemistry may become even more complex in high-energy environments.

Icy grain mantles have been recently studied with the Short Wavelength Spectrometer (SWS) on-board ISO. Observations of interstellar ices prior to ISO have been extensively reviewed by Whittet (1993) and Schutte (1996). In Figure 1 we display the spectrum toward the massive protostar W33A (Gibb et al 2000, Langer et al 2000). ISO allowed us for the first time observations of the complete range between 2.5 and 200 μm, free of any telluric contamination. The most recent abundances of interstellar ices near high-mass, low-mass, and field stars are listed in Table 2. H_2O is the major component of interstellar grains, but many features observed in the near- and mid-infrared are of organic nature. ISO-SWS and ground-based data show that the major organic species in interstellar ices are CO, CO_2, and CH_3OH (Table 2; Whittet et al 1996, Chiar et al 1998a, Dartois et al 1999b, Gerakines et al 1999). Organic species such as OCS, H_2CO, HCOOH, CH_4, and OCN^- are observed toward massive protostars and are characterized by abundances of a few percent relative to water ice (Whittet et al 1996, d'Hendecourt et al 1996, Schutte & Greenberg 1997, Ehrenfreund et al 1997a, Gibb et al 2000). Upper limits have been determined from ISO spectra for C_2H_2, C_2H_5OH, and C_2H_6 (Boudin et al 1998).

The stretching mode of CO is widely observed toward high- and low-mass protostars as well as in the quiescent medium (Tielens et al 1991, Chiar et al 1995, 1998a). The existence of hydrogen-rich ices (polar ices) and hydrogen-poor ices

TABLE 2 Millennium abundances of molecules in interstellar gas and ice and cometary volatiles

Molecule	W33A high	NGC7538 IRS9/high	Elias29 low	Elias16 field	Orion hot core	Comet Halley	Comet Hyakutake	Comet Hale-Bopp
H_2O	100	100	100	100	>100	100	100	100
CO	9	16	5.6	25	1000	15	6–30	20
CO_2	14	20	22	15	2–10	3	2–4	6–20
CH_4	2	2	<1.6	—	—	0.2–1.2	0.7	0.6
CH_3OH	22	5	<4	<3.4	2	1–1.7	2	2
H_2CO	1.7–7	5	—	—	0.1–1	0–5	0.2–1	1
OCS	0.3	0.05	<0.08	—	0.5	—	0.1	0.5
NH_3	15	13	<9.2	<6	8	0.1–2	0.5	0.7–1.8
C_2H_6	—	<0.4	—	—	—	—	0.4	0.3
HCOOH	0.4–2	3	—	—	0.008	—	—	0.06
OCN^-	3	1	<0.24	<0.4	—	—	—	—
HCN	<3	—	—	—	4	0.1	0.1	0.25
HNC	—	—	—	—	0.02	—	0.01	0.04

HNCO	—	—	—	0.06	—	0.07	0.06–0.1
C$_2$H$_2$	—	—	—	3–10	—	0.5	0.1
CH$_3$CN	—	—	—	0.2	—	0.01	0.02
HCOOCH$_3$	—	—	—	0.1	—	—	0.06
HC$_3$N	—	—	—	0.04	—	—	0.02
NH$_2$CHO	—	—	—	0.002	—	—	0.01
H$_2$S	—	—	—	1	0.04	0.8	1.5
H$_2$CS	—	—	—	0.01	—	—	0.02
SO	—	—	—	0.5	—	—	0.2–0.8
SO$_2$	—	—	—	0.6	—	—	0.1

Abundances are normalized to H$_2$O for interstellar ices and volatiles in comets Halley (1P/Halley), Hyakutake, (C/1996 B2) and Hale-Bopp (C/1995 O1). Abundances for the Orion hot core are normalized to CO. Recent ISO-SWS observations have allowed the ice abundances to be determined in the high-mass protostellar objects W33A (Gibb et al 2000, Keane et al 2000), NGC7538:IRS9 (Whittet et al 1996, Schutte 1999, Ehrenfreund & Schutte 2000a, Keane et al 2000), in the low-mass protostellar object Elias 29 (Boogert et al, Ehrenfreund & Schutte 2000a), and the field star Elias 16 (Whittet et al 1998, Schutte 1999). For W33A, the range given for each of HCOOH and H$_2$CO reflects the uncertainty in the H$_2$O ice column density (Gibb et al 2000, Keane et al 2000). Note that Keane et al (2000) recently estimated the values for NH$_3$ in the protostars W33A and NGC7538 to be below 9% relative to H$_2$O. Data for the Orion hot core are taken from vanDishoeck & Blake (1998) and Irvine et al (1999). Measurements for cometary volatiles are compiled from Crovisier & Bockelée-Morvan (1999), Mumma et al (1993), and Rodgers (1998). Molecules observed in comets, such as CO, H$_2$CO, HNC, OCS, and SO may have extended sources. For detailed information on the observations of individual species, the reader is referred to van Dishoeck & Blake (1998), Ehrenfreund & Schutte (2000a), Irvine et al (1999), Crovisier & Bockelée-Morvan (1999) and Cottin et al (1999).

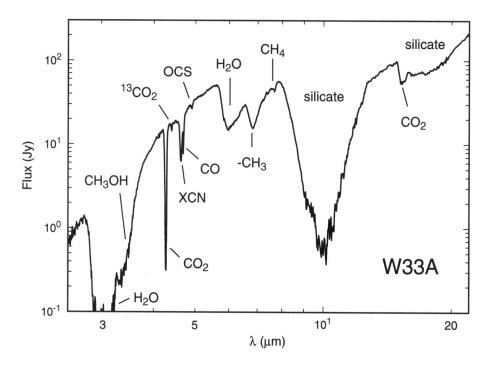

Figure 1   The ISO-SWS spectrum toward the protostar W33A (taken from Gibb et al 2000, Langer et al 2000). This cold and dense line of sight is characterized by large column densities of interstellar ices. Dominant absorption bands are visible at 3.0 μm (H_2O ice), 4.27 μm (CO_2 ice), 4.67 μm (CO ice), and 10 μm (silicates). The spectrum between 2 and 25 μm shows several additional species, many of them organic in nature.

(apolar ices) has been inferred from the CO band profile with the help of laboratory spectra (Sandford et al 1988, Tielens et al 1991). Polar ices, dominated by H_2O ice, in general evaporate around 90 K under astrophysical conditions and can therefore survive in higher-temperature regions close to the star (Tielens & Whittet 1997). Apolar ices, composed of molecules with high volatility (evaporation temperatures of <20 K) such as CO, O_2, and N_2 (Ehrenfreund et al 1998a), can only be formed and survive in cold, dense regions. During their passage through different cloud environments, grain mantles may accrete layers of polar and apolar ice.

The ubiquitous presence of CO_2 is one of the major discoveries of ISO. This molecule was observed through its stretching and bending vibrations at 4.27 and 15.2 μm, respectively (de Graauw et al 1996, d'Hendecourt et al 1996, Gürtler et al 1996, Whittet et al 1998, Gerakines et al 1999). Observations of $^{13}CO_2$ led to an estimate of the isotopic $^{13}C/^{12}C$ ratio in the galaxy (Boogert et al 2000). It has been shown that most of the CO_2 ice is present in annealed (hot) form on grain mantles and locked in intermolecular complexes with CH_3OH ice (Ehrenfreund et al 1998b, Dartois et al 1999a,b). The formation of CO_2 in dense clouds in such large quantities ($\sim$20% relative to water ice; de Graauw et al 1996, Gerakines et al 1999) remains a mystery. It may be formed by grain surface reactions, namely by oxidation of CO. Energetic processing (UV irradiation or cosmic rays) or gas phase production in shocks and subsequent accretion on grains are alternative formation routes. Abundances of gaseous CO_2 are surprisingly low (van Dishoeck et al 1996, Dartois et al 1998), probably a result of destruction of CO_2 in shocks (Charnley & Kaufman 2000). Like CO_2, CH_3OH may be formed either by grain surface reactions (hydrogenation of CO; Hiraoka et al 1994) or by energetic processing. Recent ground-based observations indicate large amounts of CH_3OH in the line of sight toward some high massive protostars, but rather low CH_3OH ice abundances are observed toward low-mass protostars (Dartois et al 1999a). Recent ISO results also show extensive ice segregation of CO_2 and CH_3OH in the vicinity of protostars (Ehrenfreund et al 1998b, 1999a).

ISO gave the first high-quality observations of the strong 6.0 μm absorption band that is generally seen toward embedded protostars. Previously this band was ascribed to the deformation mode of H_2O. However, new data clearly showed excess absorption on the blue as well as the red wing of this feature (Schutte et al 1996, Keane et al 2000), which indicates the presence of the simplest organic acid, HCOOH (Schutte et al 1996), and of aromatic moieties (Schutte et al 1998), respectively. The presence of H_2CO, displaying a strong feature at 5.81 μm, is not yet confirmed, but a comparison with laboratory data indicates that H_2CO could be present in interstellar ices with an abundance of a few percent relative to water ice (d'Hendecourt et al 1996, Gibb et al 2000, Keane et al 2000). ISO was able to confirm the presence of CH_4 at 7.68 μm toward several objects, and showed an abundance relative to water ice of $\sim$2–4% (Boogert et al 1996, Dartois et al 1998). Gas phase abundances of CH_4 are generally lower than solid state abundances (Dartois et al 1998, van Dishoeck 1998, Boogert et al 1998). OCS is currently the only sulphur-containing species identified in interstellar ices, and shows an abundance of maximal 0.2% relative to water ice (Palumbo et al 1995, 1997; d'Hendecourt et al 1996). The tentative identification of the famous interstellar XCN band (Tegler et al 1995, Pendleton et al 1999) with OCN^- indicates that ions can be present and that acid-base reactions may occur on interstellar grain surfaces (Grim & Greenberg 1987, Schutte & Greenberg 1997, Demyk et al 1998). A further indication of more complex organic molecules present in interstellar ices is the weak absorption band at 3.25 μm detected in a number of massive YSOs (Brooke et al 1999, Sellgren et al 1994), which is currently assigned to the CH stretching mode of frozen PAHs. The presence of diamonds in dense clouds has been proposed by Allamandola et al (1992). Figure 2 shows a schematic outline of processes that can lead to the chemical diversity of interstellar ices and the formation of complex molecules in ice and gas (Ehrenfreund & Schutte 2000a).

No theoretical studies have examined detailed chemical reactions that occur by energetic processing, though much work has been done on modeling the molecular

THE CYCLE OF ICE AND GAS IN DENSE CLOUDS

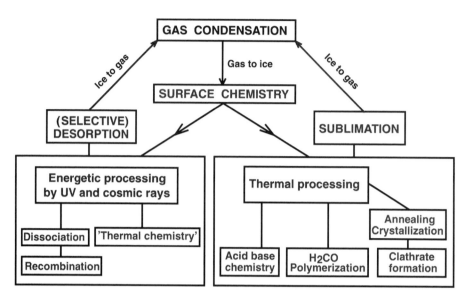

Figure 2   This diagram shows the cycle of gas and dust in dense clouds and the processes that lead to the formation of large complex molecules (Ehrenfreund & Schutte 2000a). For a detailed understanding, we refer to Schutte et al (1993) (formaldehyde polymerization), Gerakines et al (1996), Bernstein et al (1995), Briggs et al (1992) (UV processing), Ehrenfreund et al (1999a) (clathrate formation), Moore & Hudson (1992, 1998), Palumbo et al (1999) (cosmic ray processing), Schutte et al (1999) (acid-base reactions), Léger et al (1985) (selective desorption), and Sandford & Allamandola (1993) (sublimation).

complexity that could arise from grain surface reactions (Allen & Robinson 1977, d'Hendecourt et al 1985, Brown 1990, Hasegawa et al 1992, Hasegawa & Herbst 1993, Caselli et al 1993). However, progress in this area has been slow because most models neglect the stochastic nature of surface reactions and treat the surface populations as continuous deterministic variables (Pickles & Williams 1977). Even very simple surface reaction networks give different and implausible results when treated deterministically as opposed to stochastically (Tielens & Charnley 1997). Complex molecule abundances derived from deterministic models are therefore suspect. Despite recent attempts to reconcile the two approaches (Caselli et al 1998, Shalabiea et al 1998), at the present time a full stochastic simulation of both gas and grain chemistries appears to be required (Charnley 1998, Herbst 2000). The low solid and gas phase abundances of molecular oxygen found respectively by ISO and SWAS (Vandenbussche et al 1999, Melnick 2000) place stringent constraints on the efficiency of even simple surface reactions such as the combination of two oxygen atoms. Further experimental and theoretical effort is clearly necessary for understanding the chemistry that can occur on dust grains and in the ice mantles that cover them, both in cold clouds and in the energetic conditions of star-forming regions.

2.1.2 Organic Molecules in Cold Dark Clouds

Prior to the onset of star formation, dust grains effectively shield molecules from interstellar UV photons; however, cosmic rays can penetrate throughout and drive a rich ion-molecule chemistry, supplemented by neutral-neutral processes, in which many complex organic species may be produced (Herbst 1987, Prasad et al 1987, Millar et al 1997, Herbst & Leung 1989). These reactions, along with grain surface processes, account for the high observed D/H ratios in interstellar molecules (Tielens 1983, Millar et al 1989).

The dark cloud TMC-1 contains a distinctive suite of many unsaturated carbon chain molecules (Ohishi & Kaifu 1998). These include the cyanopolyynes ($HC_{2n+1}N$, $n = 1 - 5$), various cumulene carbenes (H_2C_n, $n = 3, 4, 6$), and chain radicals (HC_n, $n = 1-8$); C_nN, $n = 1, 3, 5$), as well as some methylated molecules such as methylcyanoacetylene (CH_3CCCN) and methyldiacetylene (CH_3CCCCH). Various small molecules (CCO, CCCO, CCS, CCCS) are observed; their higher homologues and new homologous series may also be present (McCarthy et al 1997, Langer et al 1997, Guélin et al 1997, 1998; Thaddeus et al 1998, Bell et al 1999). Curiously, these organic molecules show spatial abundance gradients—they have their peak emission in a region of the cloud, the so-called "cyanopolyyne peak," which is physically distinct from other emission peaks in ammonia and DCO^+ (Gúelin et al 1982, Olano et al 1988, Pratap et al 1997). The challenge of the organic chemistry of TMC-1 involves both identifying the formation routes and explaining the spatial abundance gradients; most studies have concentrated on the former.

Early attempts at modeling the TMC-1 chemistry were based on ion-molecule pathways (Huntress & Mitchell 1979, Leung et al 1984, Millar & Freeman 1984, Millar & Nejad 1985, Millar et al 1987, Herbst & Leung 1989). It was subsequently realized that, for HC_3N, production in the neutral process involving CN and C_2H_2 could be important (Herbst & Leung 1990). Cherchneff & Glassgold (1993) showed that neutral-neutral reactions involving higher cyanopolyynes and polyacetylenes could consistently explain the abundances of these compounds in the carbon-rich envelope of the red giant IRC+10216. However, inclusion of these processes in models of interstellar chemistry tended to inhibit the production of complex species, primarily because of the high abundances of O, N, and C atoms present (Herbst et al 1994, Bettens et al 1995). Support for cold phase organic synthesis by neutral-neutral reactions comes from their rapid low-temperature rates (Sims et al 1993, Chastaing et al 1998), the fact that many of them are exothermic and have no energy barriers (Fukuzawa et al 1998), and the fact that observations of various ^{13}C isotopomers in TMC-1 confirm production of HC_3N by the reaction of CN with acetylene (Takano et al 1998).

Simple chemical models of TMC-1 predict that the gas phase organics should peak at very early times ($\sim 10^5$ years), when atomic carbon is abundant. This has led to the view that the various emission peaks in the ridge are at different evolutionary states, with the "cyanopolyyne peak" being the youngest (Hirahara et al 1992). Alternatively, Ruffle et al (1997) have shown that after an accretion time ($\sim 10^6$ years) organic abundances can peak as a result of the more rapid loss of destructive oxygen atoms; however, the predicted large depletions of CO and $N_2(N_2H^+)$ are not seen in TMC-1.

The fact that a few oxygen-bearing grain mantle molecules (see Section 2.1.1) are present at the "cyanopolyyne peak" led Markwick et al (2000) to construct a model of a transient gas phase chemistry where magneto-hydrodynamic waves from the embedded IRAS source induce grain-grain streaming, which leads to mantle explosions. Here, the "cyanopolyyne peak" chemistry is driven by acetylene molecules, which were formed and accreted earlier in the evolution of the ridge gas; the calculated abundance gradients along the ridge agree well with those determined by Pratap et al (1997).

In conclusion, it appears that neutral-neutral reactions are important for the production of many observed carbon chain molecules. The discovery and mapping of more cold sources similar to TMC-1 would allow further testing of this scenario.

2.1.3 Organic Molecules in Hot Molecular Cores

Star formation has a profound effect on the chemistry of the surrounding medium (van Dishoeck & Blake 1998, Wyrowski et al 1999, Langer et al 2000, van Dishoeck & van der Tak 2000). In the early phases of protostellar evolution, the surrounding gas and dust experience radiative heating and shock waves. Hot molecular cores are regions where evaporation and sputtering of icy grain mantles has recently occurred (Millar & Hatchell 1998); gas phase observations of these cores allows ice mantle composition to be probed to lower abundance levels than is possible by IR absorption spectroscopy. Hot cores are particularly rich in complex organic molecules and are typified by large abundances of water and ammonia (see Table 2) and enhanced deuterium fractionation ratios (e.g. Blake et al 1987, van Dishoeck et al 1993, Sutton et al 1995, Tielens & Charnley 1997, Irvine et al 2000).

Molecules observed in hot cores can have three possible origins. First, cold gas phase chemistry, combined with accretion onto dust during core collapse, can account for the presence of CO, HCN, and acetylene (Lahuis & van Dishoeck 2000). Second, grain surface reactions produce simple saturated compounds such as water and ammonia (Brown et al 1988). Both of these processes, which occur in preexisting cold material, can also explain the high deuterium fractionation ratios observed in specific molecules, such as HDO, NH_2D, DCN, D_2CO, and CH_2DOH (Tielens 1983, Brown & Millar 1989, Turner 1990, Jacq et al 1993, Millar et al 1995, Charnley et al 1997). For some time it was accepted that many of the complex organic molecules must be the products of grain surface chemistry. However, some cores appear to be dominated by nitrogen-bearing molecules, and other adjacent ones by oxygen-bearing organic species (Blake et al 1987, Wyrowksi et al 1999). This can be explained by the third mechanism for producing hot core molecules: Evaporation of simple molecular mantles drives gas phase complexity (Charnley et al 1992). In this picture, the gas phase chemical differentiation results from small compositional differences between the ejected mantles; models have been proposed as to how this could occur (Caselli et al 1993).

Chemical models of hot cores can be used to identify molecules for which grain-surface production is necessary—e.g. ethanol, ketene, acetaldehyde, propynal, isocyanic acid, and formamide. This points to a surface chemistry seeded by CO molecules (Tielens & Hagen 1982). Hydrogen atom addition to CO leads to HCO (Hiraoka et al 1994, 1998), and subsequent H, C, N, and O atom additions lead to formaldehyde, ketene, isocyanic acid, and formic acid. These molecules can be further processed; for example, acetaldehyde and ethanol may be produced by consecutive hydrogen additions to ketene (Charnley 1997). When they are protonated, mantle-formed molecules such as methanol and ethanol (and perhaps higher alcohols) offer a very specific route to molecular complexity in hot cores through alkyl cation transfer to a neutral base, $R'X$,

$$ROH_2^+ + R'X \longrightarrow R'XR^+ + H_2O.$$

A large neutral organic molecule then results upon dissociative recombination with an electron. These reactions have been extensively studied in the laboratory (Mautner & Karpas 1986, Karpas & Mautner 1989). Evaporation of ices rich in methanol and ethanol should produce various pure and mixed ethers (Charnley et al 1995). The observed correlation of their distributions in star-forming regions (Minh et al 1993, Ikeda 1998) supports the view that dimethyl ether originates from the self-methylation of methanol. Figure 3 shows that several molecules could also be formed in this way and could predict the presence of many new, potentially detectable ones. Kuan et al (1999) have recently reported the tentative detection of diethyl ether. Recent calculations (Charnley & Rodgers 2000) suggest that evaporating ice mantles containing ammonia, as well as methanol and ethanol, could be the origin of the nitrogen-bearing organics (including CH_3CH_2CN, CH_2CHCN, CH_3CN, CH_3NH_2) that are seen in star-forming regions (Miao et al 1995, Mac-Donald et al 1996, Kuan et al 1996).

Future millimeter and submillimeter interferometers (cf. Bally et al 1995) such as ALMA will lead to the detection of complex molecules in protostellar cores with abundances almost a factor of a hundred below current detection limits. An exciting recent development is the demonstrated ability to detect simple organic species (e.g. HCN, HNC, CCH, H_2CO, CH_3OH) in the disks around low-mass stars (Dutrey et al 1997, Goldsmith et al 1999). Hence, an entire interstellar organic chemistry, including perhaps amino acids and other prebiotic molecules (Snyder 1997), may await discovery in objects that are prototypical of our own protosolar nebula.

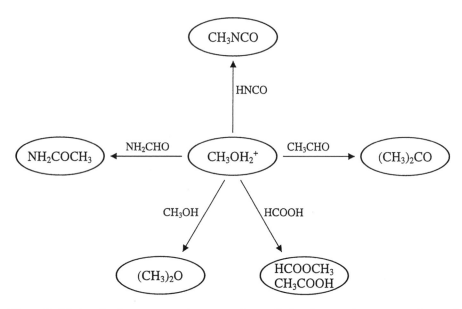

Figure 3   Methanol, when protonated, can transfer an alkyl cation to other evaporated neutral mantle products. In this figure, the step involving the electron dissociative recombination of the intermediate organic molecular ion has been suppressed for clarity.

2.2   Organic Molecules in the Diffuse Interstellar Medium

Diffuse clouds have moderate extinctions (<1 mag) and densities of roughly 100–300 cm^{-3}. They are characterized by an average temperature of ~ 100 K and a UV radiation field of approximately $\sim 10^8$ photons cm^{-2} s^{-1} (Mathis et al 1983). Since the initial discovery of simple diatomic molecules in interstellar space, CH, CN and CH^+ (Swings & Rosenfeld 1937, McKellar 1940, Douglas & Herzberg 1941), many more molecules have been detected in the photon-dominated diffuse medium, although at lower abundances than found in dense clouds. Species detected include: HCO^+, CO, OH, C_2, HCN, HNC, CN, CS and H_2CO (Lucas & Liszt 1997). Lucas & Liszt (2000) have recently detected C_2H and c-C_3H_2 in several extragalactic diffuse clouds and obtained upper limits on the column densities of l-C_3H and C_4H. The C_2H abundance varies little from diffuse to dense clouds whereas the c-C_3H_2 abundance is markedly higher in dense clouds. Reactions of the neutral molecules CH and CH_2 with C^+ can lead to the formation and build-up of polyatomic hydrocarbons. However, the presence of large carbon-bearing species is strongly dependent on their formation and survival rate because the diffuse medium is controlled by photochemistry. The larger carbonaceous molecules that enter the diffuse interstellar gas are detected in circumstellar envelopes around late-type stars (see Section 2.2.2). Evidence for carbonaceous dust particles is given by spectroscopy, polarization measurements, and theoretical models (Henning & Schnaiter 1999, Henning & Salama 1998).

About 1% per mass of the interstellar medium is in the form of solid dust grains, which may be carbon or silicon-based. Dust grains act as catalytic surfaces throughout the interstellar medium, and in general show dimensions on the submicron scale. The starlight is absorbed and scattered by dust grains and reaches the observer dimmed, a process referred to as extinction. The extinction-curve of the interstellar medium represents a superposition of the wavelength-dependent extinction properties of different dust particles. Different dust components that contribute to the line of sight extinction-curve are discussed by Dorschner & Henning (1995), Jenniskens & Greenberg (1993), Li & Greenberg (1997), and Vaidya & Gupta (1999). Dust particles in diffuse clouds and circumstellar envelopes can be composed of silicates, amorphous carbon (AC), hydrogenated amorphous carbon (HAC), diamonds, organic refractories, and carbonaceous networks such as coal, soot, graphite, quenched-carbonaceous condensates (QCC), and others. The dust size distribution could be inferred from astronomical observations in the UV, VIS, and IR (Mathis et al 1977, Dwek et al 1997). A three-component model of interstellar dust proposed by Desert et al (1990) suggests the coexistence of big grains (silicates with refractory mantles), very small grains (VSG, carbonaceous) and polycyclic aromatic hydrocarbons (PAHs). Li & Greenberg (1997) have modeled the interstellar extinction and polarization on the basis of a similar trimodal dust model, using the latest observational data and laboratory measurements.

Dust grains form in the cool expanding circumstellar environment of evolved stars (Sedlmayr 1994). Stellar winds inject dust into the ambient interstellar medium, which is then distributed by supernovae shock waves over large scales through the ISM. During this period, dust particles cycle several times through dense and diffuse clouds, which allows efficient mixing and processing of interstellar dust (Greenberg 1986, Mathis 1990). UV irradiation and cosmic rays, together with processes such as grain-grain collisions, sputtering, and grain growth, alter and destroy dust in interstellar and circumstellar regions (Jones et al 1996). Therefore, grains probably retain only traces of their origin, as evidenced by the isotopically anomalous composition of presolar grains found in meteorites (Zinner 1998). The life cycle of dust and interstellar depletions was recently discussed by Tielens (1998). Recent observations suggest that the abundance of carbon in the interstellar medium is only two-thirds of its solar value (Snow & Witt 1995, 1996; Cardelli et al 1996). This poses problems for many recent dust models, because only a limited amount of carbon is available for the dust phase (Snow & Witt 1995, Henning & Schnaiter 1999). In order to solve the so-called interstellar carbon crisis, Mathis (1996) proposed fluffy grain structures. Other solutions for the carbon abundance problem are discussed by Dwek (1997).

Organic molecules present in the diffuse medium can originate in one of three ways: (1) via gas phase reactions where model calculations of dispersing clouds allow the formation of molecules up to 64 C atoms with ion-molecule reactions and neutral-neutral reactions (Bettens & Herbst 1996); among them are unsaturated C-chains, single- and triple-ring structures, and fullerenes (Herbst 1995, Ruffle et al 1999); (2) by reactions in circumstellar envelopes and subsequent mixing into the diffuse medium (only very photoresistant species; see Section 2.2.2); and

(3) from carbonaceous dust by photochemical reactions and grain collisions. Grain surface reactions have also been suggested, but these are only expected to produce simple hydrides (e.g. NH; Wagenblast et al 1993).

2.2.1 Spectroscopic Evidence for Carbonaceous Compounds in the DISM

A large and diverse suite of organic compounds is believed to reside and evolve in the diffuse medium (Papoular et al 1996, Henning & Salama 1998). In this section we first list the main spectroscopic features observed in diffuse clouds, the carriers of which remain unknown, but are likely of a carbonaceous nature. The spectroscopic puzzles in the ultraviolet, optical, and infrared range are the ultraviolet bump at 2200 Å, the diffuse interstellar bands (DIBs), the extended red emission (ERE), the 3.4 μm absorption, and the infrared emission features between 3 and 15 μm (UIR). We then discuss the most important organic species such as PAHs, fullerenes, and diamonds, as well as solid organic compounds possibly present in the DISM.

The UV Bump at 2200 Å   By far the strongest feature in the interstellar extinction curve is the ultraviolet bump at 2200 Å. It is characterized by a stable position, but its band width changes according to the interstellar environment (Fitzpatrick & Massa 1986). The carrier of the ultraviolet bump has not yet been identified. Among the numerous proposed candidates are graphite grains (Stecher & Donn 1965, Annestad 1992), graphitic onions (Wright 1988), hydrogenated amorphous carbon (HAC) (Mennella et al 1996), core particles (made of silicates or graphite) with a mantle of PAHs (Mathis 1994, Duley & Seahra 1998), and coal-like material (Papoular et al 1996). An important contribution of a mixture of PAHs to the UV bump has been proposed by Joblin et al (1992). Molecular aggregates with aromatic double-ring structures produced in plasma discharge are discussed by Beegle et al (1997). Recent theoretical calculations and laboratory data propose onion particles as possible carriers (Henrard et al 1997, Wada et al 1999). Isolated nanometer-sized C particles show a UV bump as narrow as the interstellar feature and a good fit in wavelength position (Schnaiter et al 1998). Mennella et al (1998) attribute the UV bump to a population of nano-sized, UV-processed hydrogenated amorphous carbon grains. The latter authors propose a class of material whose optical properties vary according to the processing degree in different regions of the interstellar medium. All recent results point strongly toward a carbonaceous carrier for the UV bump.

The Diffuse Interstellar Bands (DIBs) between 4000 and 10,000 Å   The Diffuse Interstellar Bands (DIBs) are absorption lines detected in the spectra of nearby stars. They are observed in a spectral range extending from 0.4 to 1.3 μm and exhibit a strong diversity of profiles (Full Width at Half Maximum, or FWHM, ranging from 0.8 to 30 Å) (Herbig 1995). The detection of DIBs in almost all ISM regions

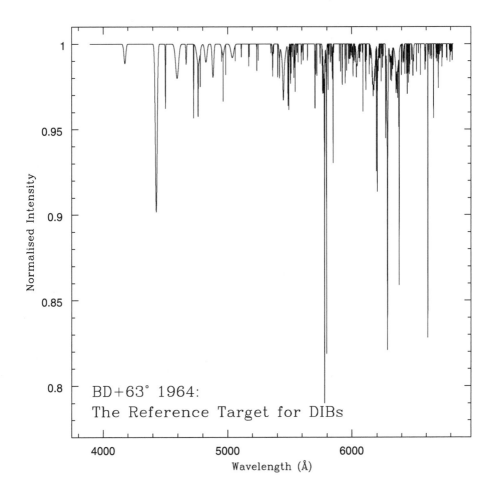

Figure 4 The synthetic spectrum of the hot B0II star BD+63 1964. This object, characterized by an $E_{(B-V)} = 1$, is currently used as a reference target of diffuse bands, displaying more than 250 DIBs in the range 3900–7000 Å and a significant enhancement of narrow DIBs (Ehrenfreund et al 1997b, O'Tuairisg et al 2000).

indicates chemically stable carriers. In the past 75 years, a number of surveys were initiated to constrain better the nature of the DIB carriers. The DIB number rose to around 200 in 1994 (Jenniskens & Desert 1994, Krelowski et al 1995). Presently, more than 250 DIBs are reported (O'Tuairisg et al 2000). Figure 4 shows the synthetic DIB spectrum of the hot star BD+63 1964 (Ehrenfreund et al 1997b, O'Tuairisg et al 2000).

As DIB strength correlates with reddening, dust grains and impurities embedded in dust particles have been among the proposed carriers in the past (Herbig

1995). However, the solid state theory did not explain the ubiquitous presence and the high wavelength stability of DIBs. The DIBs' strength is dependent on the environmental conditions (Herbig 1975, Snow et al 1995, Krelowski et al 1998). Observations show that the DIB strength progressively decreased in cloud interiors (Adamson et al 1991). This "skin effect" has recently found further support (Herbig 1993, Cami et al 1997, Sonnentrucker et al 1997).

Many results seem to point toward carrier molecules that are of gas phase origin, chemically stable, and carbonaceous in nature (Herbig 1995). Cami et al (1997) completed the first survey of DIB correlations over 4000 Å, which showed that most of the DIB carriers are undergoing photoionization and that all measured DIBs do originate from different carriers. The detection of DIBs in emission in the Red Rectangle (Scarrott et al 1992) and the detection of intrinsic substructures in the profile of several DIBs indicate the molecular nature of some DIB carriers (Sarre ET AL 1995, Ehrenfreund & Foing 1996, Krelowski & Schmidt 1997, Le Coupance et al1999).

Recent laboratory studies point toward PAHs, carbon chains, and carbon rings or fullerenes as potential carriers for some DIBs (Salama et al 1996, Freivogel et al 1994, Tulej et al 1998, Ehrenfreund & Foing 1996, Foing 1996, Foing & Ehrenfreund 1997). Theoretical models of large carbonaceous carrier molecules constrain the formation and destruction rate of such species in the diffuse medium (Allain et al 1996a,b, Bettens & Herbst 1997, Jochims et al 1999, Ruffle et al 1999). The search for specific bands of PAH cations in highly reddened stars shows some coincidences with small PAHs measured in the gas phase or in neon matrices (Salama et al 1999). In the near future, measurements using larger PAH molecules may show better agreement. Carbon chains measured in the gas phase show coincidences with the band position of some DIBs (Freivogel et al 1994). In particular, the anion C_7^- has been suggested as a potential DIB carrier (Tulej et al 1998). However, high-resolution astronomical data indicate that the band ratios for this species are variable in several stars (Galazutdinov et al 1999). Additionally, the abundance of carbon chains in the diffuse medium is estimated to be very low— 10^{-10} per H atom, a factor 1000 lower than the estimated PAH abundance. This strongly suggests a limited contribution of carbon chains to the DIB spectrum (Allamandola et al 1999a, Motylewski et al 2000). Foing & Ehrenfreund (1997) observed two DIBs at 9577 and 9632 Å as first evidence for C_{60}^+, the largest molecule ever detected in space. This conclusion was based on the laboratory spectrum of C_{60}^+, in a neon matrix obtained by Fulara et al (1993) and has to be confirmed by a comparison with the gas phase spectrum of the C_{60} cation.

At present, no definitive identification of the carriers of more than 250 bands exists. For a correct identification of the DIB carriers, gas phase spectroscopy is an important tool of the present and future (Freivogel et al 1994, Tulej et al 1998, Brechignac & Pino 1999, Romanini et al 1999). The complexity of interstellar processes, the wealth of DIBs and potential carriers, and technical limitations may account for the lack of success in the past. Attempts were made to correlate the DIB strength variations with extinction curve parameters, grain properties, the interstellar radiation field, and other interstellar atoms and molecules (Herbig 1993, Adamson et al 1991, Krelowski et al 1995, Desert et al 1995, Krelowski et al 1999, Sonnentrucker et al 1999). Although constraints could be evaluated for a few DIBs, a consistent picture remains elusive (Snow 1997). The recent search for DIBs in the solar system, namely in the coma of Hale-Bopp, was not successful (Herbig & McNally 1999). The identification of the DIB carriers will remain an important problem in astronomy in the new millennium.

The Extended Red Emission (ERE)   The extended red emission has been detected in many dusty astrophysical objects and appears to be a general characteristic of dust (Gordon et al 1998). ERE is a broad, structureless emission band that appears between 5400–9400 Å and is attributed to the photoluminescence of particular dust grains. The central wavelength of the ERE varies from object to object, and their spectral nature and photon conversion efficiency requires an abundant carrier in the ISM, such as carbon or silicon-based material (Witt et al 1998). The carbonaceous candidates proposed as carriers for the ERE are the same substances that have been proposed for the UV bump, namely HAC (Duley 1985), QCCs (Sakata et al 1992), coal (Papoular et al 1996), small carbon particles of mixed sp^2/sp^3 hybridized bonding (Seahra & Duley 1999), PAHs (d'Hendecourt et al 1986), and fullerenes (Webster 1993). The recent comparison of photoluminescence properties of carbon- and silicon-based material through observations indicates that the ERE carrier might be silicon-based (Witt et al 1998, Ledoux et al 1998). Zubko et al (1999) conclude that the intrinsic photon conversion efficiency of the photoluminescence by silicon nanoparticles must be near 100%, if they are the source of the ERE.

The 3.4 μm Feature of Aliphatic Hydrocarbons   The spectra of several sources located close to the Galactic center, and Wolf Rayet stars, show the signature of organic structures on grains at 3.4 μm (Pendleton et al 1994, Sandford et al 1991, 1995). This feature is attributed to the symmetric and asymmetric C–H stretching modes of aliphatic hydrocarbons. The detection of the 3.4 μm absorption toward the Seyfert galaxy NGC1068 (which has an active nucleus obscured by dust) and the IRAS galaxy 08572+3915 offers exciting future perspectives (Wright et al 1996). The strong consistency in spectral signatures in different galaxies suggests the preferential formation of a common carbonaceous structure. The 3.4 μm feature is also observed in infrared spectra of solar system material, such as captured interplanetary dust particles (IDPs) and organic extracts of the meteorites Orgueil and Murchison (Ehrenfreund et al 1991, Pendleton et al 1994; see also Figure 5). The C–H stretching mode of hydrocarbons is a rather unspecific feature that can arise in many different kinds of carbonaceous material. Therefore numerous fits of carbon-containing material have been proposed as identification of the 3.4 μm feature (see Pendleton et al 1994 for a review). It is generally assumed that the 3.4 μm band may result from aliphatic chains that act as bridges in a carbonaceous network (as is found in kerogens and coals), residing on interstellar grains.

A new 2.8–3.8 μm spectrum of the carbon-rich protoplanetary nebula CRL 618 (Chiar et al 1998b) confirms the previous detection of a circumstellar 3.4 μm absorption feature in this object (Lequeux & Jourdain de Muizon 1990). The observations imply that the carriers of the interstellar 3.4 μm feature are produced, at least in part, in the circumstellar environment by post-processing of carbon grains during the transition from the AGB to the planetary nebula phase (Chiar et al 1998b, Mennella et al 1999). This raises doubts as to whether this material is produced by the processing of interstellar ices in dense interstellar clouds, as was previously proposed (Greenberg et al 1995). Schnaiter et al (1999) tried to match the 3.4 μm profile of CRL618 with carbon particles condensed from carbon vapor in a hydrogen-rich quenching atmosphere. Mennella et al (1999) studied the interaction of atomic H with nano-sized carbon grains, the 3.4 μm feature produced in this way also fits the absorption band seen toward the Galactic center, as well as toward CRL618.

The absence of the 3.4 μm feature in AGB stars, the progenitor stars for CRL618 (Chiar et al 1998b), and in dense interstellar clouds (Allamandola et al 1992) may give clues to its origin. Recent laboratory studies of hydrocarbons covered with ice, simulating dense cloud conditions, suggest that the 3.4 μm band is not present, because in such an environment hydrocarbons are efficiently dehydrogenated through interaction with H_2O ice and UV photons (Munoz et al 2000). Spectropolarimetry of the 3.4 μm aliphatic C–H stretch feature, in the line of sight from the Galactic center source IRS 7, showed an unpolarized feature not associated with the aligned silicate component of interstellar dust (Adamson et al 1999). The simplest explanation is that the 3.4 μm carrier resides in a population of small, nonpolarizing carbonaceous grains, physically segregated from the silicates. To summarize, the current results indicate that a mechanism able to form C–H bonds must be active, and that carbon grains produced in carbon-rich AGB stars may be hydrogenated in the diffuse medium (Mennella et al 1999). Figure 5 shows a comparison of the 3.4 μm feature in an external galaxy, our Galactic center, and an extract of the Murchison meteorite (Pendleton 1997), which indicates that the formation process of such material is widespread in the Universe.

The Infrared Emission Bands between 3 and 15 μm   The infrared emission bands at 3.3, 6.2, 7.7, 8.6, 11.2, and 12.7 μm are observed in the diffuse interstellar medium, in circumstellar environments, and even in external galaxies. These bands are attributed to the C–H stretching and bending vibrations and C=C stretching modes in aromatic molecules (Léger & Puget 1984, Allamandola et al 1987, Puget & Léger 1989). In previous years, many components have been proposed to be responsible for these features. Among them are PAHs, hydrogenated amorphous carbon (HAC), quenched carbonaceous composites (QCCs), and coal (see Tielens et al 1999 for a recent review). The ISO satellite has revisited the IR emission bands and showed their ubiquitous presence in our own as well as external galaxies (see special issue of *Astron. Astrophys.* 315, 1996). High-resolution infrared spectra of different interstellar environments display (apart from the main bands)

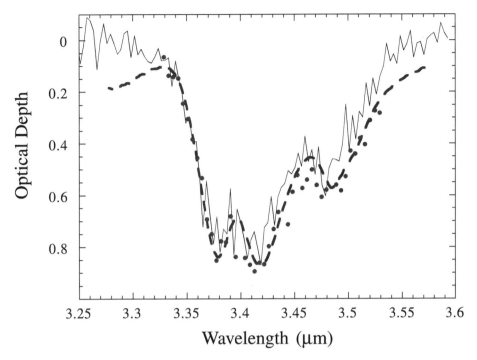

Figure 5   The 3.4 μm absorption signature of C–H stretching modes in aliphatic hydrocarbons is shown as observed toward the extragalactic source (IRAS 08572+3915) (*solid line*) and our Galactic center (IRS6) (*points*), and measured for an organic residue of the Murchison meteorite (*dashed line*) (taken from Pendleton 1997). The similarity of these spectra indicates pathways of carbon chemistry that are selected throughout the universe.

new additional weak features, shoulders, and subfeatures, as well as plateaus and a varying underlying continuum (Lemke et al 1998, Klein et al 1999, Tielens et al 1999). Figure 6 shows infrared spectra of various objects in the diffuse ISM. Observations of the infrared emission bands in cirrus clouds indicate the presence of PAHs in the gas phase. In such environments the radiation field is so low that there is no alternative mechanism to the transient heating of small particles by absorption of a single photon (Mattila et al 1996, Boulanger et al 1998). Modeling of the infrared dust bands by Lorentz profiles produced perfect fits to spectra obtained with ISO-CAM (on-board ISO). Those profiles indicate particles of a few hundred atoms (Boulanger et al 1998). Theoretical calculations of the relative emission strengths of neutral and ionized PAHs (Langhoff 1996) and laboratory studies have allowed the infrared emission bands to be successfully modeled by combining laboratory spectra of neutral and positively charged PAHs (Hudgins & Allamandola 1999a,b, Allamandola et al 1999b). The particle size of these fits indicates a PAH size distribution between 20 and 90 C atoms. Tielens et al (1999) argue for the presence of PAH with ∼50 C atoms. Apparently there is not

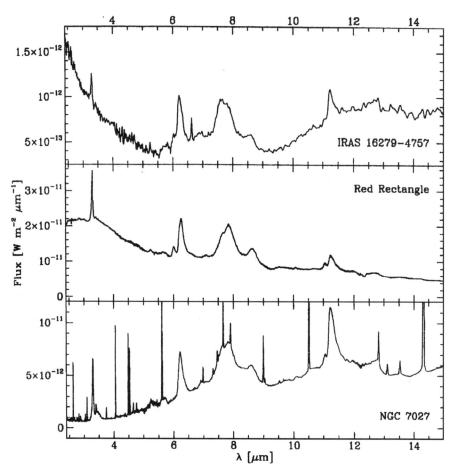

Figure 6   ISO-SWS 3–15 μm emission spectra of the post AGB objects, IRAS 16279-4757, the Red Rectangle, and the planetary nebula NGC 7027. The different objects display a rich and variable spectrum of infrared emission bands (taken from Tielens et al 1999).

yet a consensus on the PAH size distribution in the interstellar medium, but it is likely that free individual PAH molecules, PAH clusters, and particles composed of PAH subunits are responsible for the overall interstellar emission spectrum (Allamandola et al 1999b). The amount of cosmic carbon locked up in PAH molecules is a substantial fraction of ~20% (Dwek et al 1997). ISO spectra showed new bands in the region between 10 and 15 μm, which may be attributed to phenyl groups, and a plateau between 16 and 20 μm, which is attributed to larger PAH clusters, including five-ring structures such as fluorathenes (van Kerckhoven et al 2000). PAHs might show strong diversity because of various attached side groups. Organometallic molecules formed by efficient reactions between aromatics and transition metals have been proposed by Serra et al (1992).

The observed 2–13 μm emission spectra of carbon-rich post-AGB stars, including both bands and continuum, could be well matched with the computed infrared emission spectra of semi-anthracite coal grains (Guillois et al 1996), which provides evidence for the presence of solid aromatic compounds (Papoular et al 1996). Although the nature of molecules responsible for the infrared emission bands is no doubt aromatic, the individual components will have to be identified in future laboratory experiments.

2.2.2 Organic Molecules in Circumstellar Envelopes

The massive circumstellar envelopes (CSEs) of late-type, post-AGB stars are pivotal to the evolution of the ISM because they provide the dust and refractory species for the diffuse medium, and yet require the existence of the dense medium to beget the stars. The bulk of the stardust originates in oxygen-rich M giants, radio-luminous OH/IR stars, supergiants, and carbon stars (e.g. Gail & Sedlmayr 1987). Grains are formed near the photosphere and drive the circumstellar wind. For stars with O/C < 1, the carbon not in CO is primarily contained in acetylene and is available for incorporation into organic molecules, carbonaceous grains, and refractory molecules such as SiC (Blanco et al 1998, Mutschke et al. 1999). Carbon stars show strong PAH emission, and it has been proposed that PAH formation is the key step to the production of carbon dust (soot) (Frenklach & Feigelson 1989, Cherchneff et al 1992). Alternative chemical pathways to soot formation are based on neutral radicals, ions, C-chains, and fullerenes as intermediates (see Tielens & Charnley 1997). Pure carbon dust can exist only in hydrogen-deficient stars, therefore carbon-rich WR stars with envelopes of He, C, and O may be a source of interstellar fullerenes (Cherchneff & Tielens 1995). Recent ISO results showed that the chemical evolution of carbonaceous dust from aliphatic to aromatic structures can rapidly take place as the stars evolve (Kwok et al 1999).

In oxygen-rich CSEs, the available carbon is in CO and does not contribute to formation of grains, which are silicates. However, many objects show the anomalous presence of simple organic molecules such as HNC, HCN, CN, CS, and H_2CO (e.g. Omont 1993, Bujarrabal et al 1994). Several explanations have been advanced to explain these anomalies, but at present this problem is unresolved (Nejad & Millar 1988, Nercessian et al 1989, Beck et al 1992, Charnley & Latter 1997, Willacy & Millar 1997, Duari et al 1999). Interestingly, weak PAH emission in combination with silicate absorption has been found in the shells of several O-rich supergiants (Voors 1999, Voors et al 1999) and O-bearing molecules have been detected in carbon-rich proto-planetary nebulae (Herpin & Cernicharo 2000).

The carbon-rich envelope of IRC+10216 contains a large variety of organic molecules (Glassgold 1996) and is dominated by the presence of many carbon chain molecules, including polyacetylenes (e.g. C_8H; Cernicharo & Guélin 1996), cyanopolyynes (e.g. HC_9N; Bell et al 1992) and sulfuretted chains (CCS, CCCS, and possibly CCCCCS; Cernicharo et al 1987, Bell et al 1993). Structural isomerism has also been found in carbon chain molecules (e.g. HCCNC versus HC_3N;

Gensheimer 1997). Many of these molecules are also detected in the dark cloud TMC-1 (see Section 2.1.2). Molecular observations of IRC+10216 at IR and millimeter wavelengths have recently been reviewed by Cernicharo (2000) and Guélin (2000). In the outer regions of the envelope, outflowing gas is photodissociated and photoionized to produce ions and radicals. Neutral-neutral reactions are believed to be crucial to the formation of cyanopolyynes and polyacetylenes (Cherchneff & Glassgold 1993, Cherchneff et al 1993, Millar & Herbst 1994) as well as several organo-silicon compounds (Howe & Millar 1990, MacKay & Charnley 1999). Alternative explanations for the observed prevalence of carbon chains involve either direct disruption of carbonaceous grains by UV photons or grain collisions (Jura & Kroto 1990) or production on grain surfaces followed by UV photodesorption (Guélin et al 1993).

Red giant CSEs thus provide the bulk of the raw material for Galactic organic chemistry. Except for the most photochemically resilient species (PAHs, fullerenes, dust grains), molecules formed in the red giant CSEs are destroyed in the very outermost layers as they leave the circumstellar envelope and enter the diffuse ISM.

2.2.3 Evidence for Specific Organic Molecules

The current abundance estimates (per H atom) of large carbonaceous molecules contributing to the interstellar carbon budget are 10^{-7} for PAHs, 10^{-8} and 10^{-10} for carbon chains in dense and diffuse clouds, respectively, and 10^{-8} for fullerenes (Tielens et al 1999, Allamandola et al 1999a). A large amount of carbon (up to ~50% of the total cosmic carbon) may be locked in carbonaceous solids (e.g. O'Donnell & Mathis 1997, Li & Greenberg 1997).

Polycylic Aromatic Hydrocarbons (PAHs)   PAHs are believed to be the most abundant free organic molecules in space (Puget & Léger 1989, d'Hendecourt & Ehrenfreund 1997). Electron delocalization over their carbon skeleton makes them remarkably stable. As discussed in Section 2.2.2, PAH molecules are produced partly in the outer atmospheres of carbon stars or formed by shock fragmentation of carbonaceous solid material. PAHs may eventually also form in the diffuse interstellar gas by neutral-neutral reactions (Bettens & Herbst 1996) or by energetic processing of specific ices in dense clouds (Kaiser & Roessler 1998). PAHs play a central role in the gas phase chemistry (Bakes & Tielens 1994, 1998). The environmental conditions and the local ultraviolet radiation field determine their charge and hydrogenation state. Theoretical calculations imply that PAHs with <50 C atoms can be destroyed or fragmented in UV-dominated regions (Allain et al 1996a). In their evolutionary cycle, PAH molecules may freeze out onto grains in dense clouds, an idea that is supported by recent ISO observations toward M17 (Verstraete et al 1996). PAHs are ubiquitously distributed in the interstellar medium and are also seen in traces in solar system material (Clemett et al 1993). PAHs in the interstellar gas are likely responsible for the infrared emission bands and may contribute to the DIB spectrum.

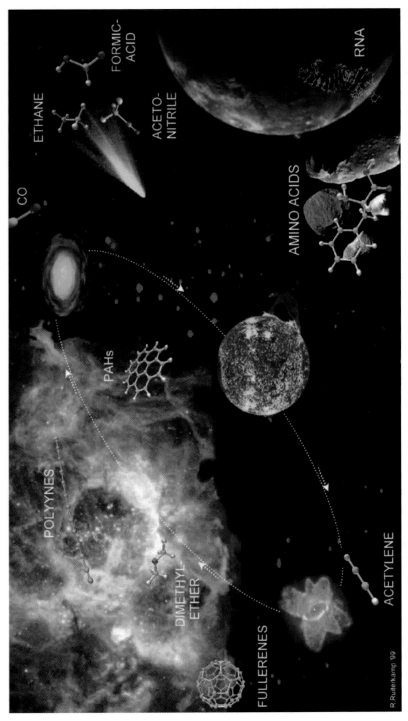

Figure 11 The cycle of organic molecules in the Universe. Interstellar organics are formed in the interstellar gas, in stellar outflows and on dust grains. This organic material is integrated in the Solar System, is partly chemically processed and/or destroyed. In the final stage of stars, dust and elements are returned to the interstellar medium. Organic molecules formed during this dust cycle may have seeded the early Earth.

Fullerenes   The polyhedral geometry of C_{60} was discussed for the first time by Kroto et al (1985). In 1990, C_{60} was efficiently synthesized in macroscopic quantities by Kraetschmer et al (1990). The presence of soot material in carbon-rich stars, along with the spontaneous formation and remarkable stability of the fullerene cage suggest the presence of fullerene compounds in interstellar space (see Ehrenfreund & Foing 1997 for a review). Fullerenes may be formed in small amounts in envelopes of R Coronae Borealis stars (Goeres & Sedlmayr 1992). Theoretical models show the possible formation of fullerenes in the diffuse interstellar gas via the buildup of C_2–C_{10} chains from C^+ insertion, ion-molecule reactions, and neutral-neutral reactions (Bettens & Herbst 1996, 1997). Recent fingerprints of the C_{60}^+ ion were discovered in the near-infrared, which indicate that fullerenes can play an important role in interstellar chemistry (see Section 2.2.1; Foing & Ehrenfreund 1994, 1997). The abundance of C_{60}^+ was inferred from optical measurements to be 0.3–0.9% of the cosmic carbon (Foing & Ehrenfreund 1997). The search for vibrational transitions (in the mid-IR) in emission of C_{60} and C_{60}^+ in the reflection nebula NGC 7023 was negative, and only uncertain upper limits could be determined (Moutou et al 1999). Garcia-Lario et al (1999) attribute the 21 μm dust feature observed in the C-rich protoplanetary nebula IRAS 16594-4656 to fullerenes, which may be formed during dust fragmentation. However, many different carrier species have been proposed for this broad strong emission band (e.g. Hill et al 1998). C_{60}^+ may be abundant in the ISM, as well as endo- and exohedral fullerene compounds (Kroto & Jura 1992). Interstellar hydrogenated fulleranes have been discussed by Webster (1995). Fullerenes—and in particular, higher fullerenes—have recently been detected in the meteorites (see Section 4; Becker & Bunch 1997, Becker et al 1999a). Fullerenes have also been detected in an impact crater on the LDEF spacecraft (di Brozzolo et al 1994).

Diamonds in Space   A main route of synthesis of diamonds in the interstellar medium could be the growth by chemical vapor deposition mechanism (CVD) in the expanding envelopes of Type II supernovae (Clayton et al 1995). But diamonds might also be formed in stellar environments by high velocity grain-grain collisions behind supernova shock waves (Tielens et al 1987, Blake et al 1988). Different methods for gas phase synthesis of nanodiamonds are discussed by Hahn (1997). The infrared band at 3.42 μm detected in the spectra of massive protostars falls near the position of C–H stretching vibrations in tertiary carbon atoms and suggests a carrier with a diamond-like structure (Allamandola et al 1992, 1993). Interstellar nanodiamonds have been proposed to be the carriers of the unidentified 21 μm infrared emission feature observed in the dust shells around some carbon-rich protoplanetary nebulae (Hill et al 1998). Small crystallites of diamonds have been recently identified in dusty envelopes by Guillois et al (1999), who attributed the two peculiar unidentified infrared emission bands at 3.43 and 3.53 μm to the vibrational modes of hydrogen-terminated crystalline facets of diamonds (see Figure 7). The intensities of these two features as measured in the Herbig Ae/Be star HD97048 correspond to a mass of 10^{-10}–10^{-9} solar masses of diamond dust

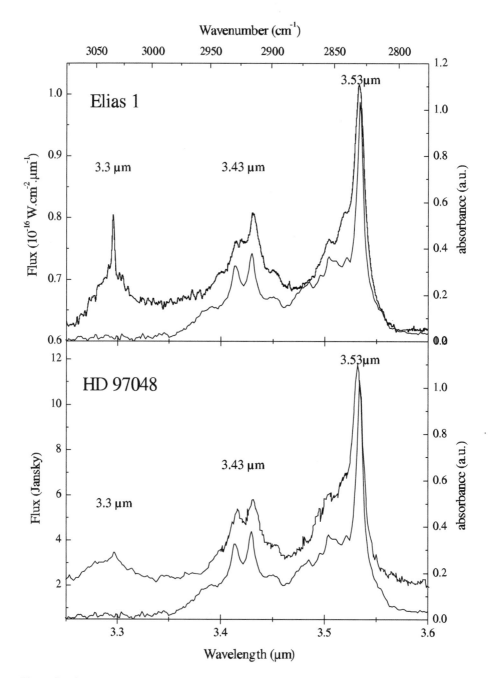

Figure 7 Spectroscopic comparison between the emission spectra of Elias 1 (top) and HD97048 (bottom) and the absorbance spectra of diamond nanocrystals (lower trace) measured at 300 K (from Guillois et al 1999).

at an equilibrium temperature of 1000 K. Atomic hydrogen recombination lines observed in this object indicate strong UV fluxes and a high temperature, which eventually anneal preexisting diamond grains (Guillois et al 1999). It has been recently demonstrated that carbon onions act as nanoscopic pressure cells and can be converted to diamonds during thermal annealing and irradiation with electrons (Banhart & Ajayan 1996).

Carbonaceous Solids   Carbonaceous dust in the interstellar medium may show strong diversity and may include amorphous carbon (AC), hydrogenated amorphous carbon (HAC), coal, soot, quenched-carbonaceous condensates (QCC), diamonds, and other compounds. The coexistence of PAHs and fullerenes together with complex carbonaceous dust suggests a common link and an evolutionary cycle that is dominated by energetic processing (Jenniskens et al 1993, Scott et al 1997). A detailed comparison of solid-state carbonaceous models of cosmic dust has been summarized by Papoular et al (1996). Solid carbon material such as graphite, diamonds, fullerite solids, and carbynes is discussed by Henning & Salama (1998). Figure 8 displays the chemical structure of some carbon compounds that are likely present in space.

Enormous progress has been made in the characterization of carbonaceous solids through laboratory studies (Henning & Schnaiter 1999), and new data show a strong link between the many different carbon phases (e.g. Ugarte 1992, 1995; Wdowiak et al 1995, Banhart & Ajayan 1996, Scott et al 1997). The improvement

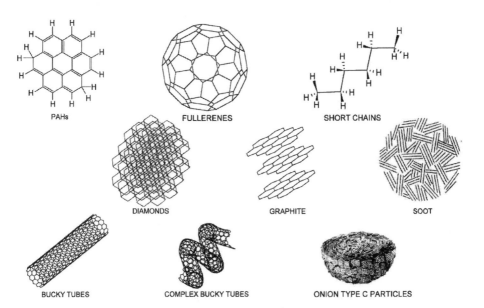

Figure 8   Some of the various forms of carbon that are likely present in gaseous and solid state in the ISM and in solar system material.

in instruments and new techniques, used to investigate material relevant to the astronomical solid state, reveal a rich diversity of carbonaceous material on the nanoscale (Herlin et al 1998, Jaeger et al 1998, Wada et al 1999). Carbon structures in soot material were recently investigated by electron microscopy, and displayed different well-defined structures (Rotundi et al 1998). Future progress in this field will strongly influence the characterization of carbonaceous matter in space, and will help to resolve the carbon crisis and the remaining spectroscopic mysteries such as the UV bump, the ERE, the DIBs, and the UIR bands.

3.   ORGANIC MOLECULES IN COMETS

The modern view of the composition of comets began with the "dirty Snowball" model of Whipple (1950a,b). This model proposed that cometary nuclei were predominantly made of water ice with an admixture of small rock particles. An evaluation of the most important issues in cometary astronomy during the last 4 decades is reviewed by Festou et al (1993a,b). Today this model has been refined, and the nuclei of comets are thought to be composed of ice, rock, and large organic entities (commonly referred to as CHON particles), since these nuclei are composed primarily of carbon, hydrogen, oxygen, and nitrogen (Mumma 1997, Irvine et al 2000). The chemical composition of a comet nucleus has been recently constrained by Greenberg (1998), who proposes that 26% of the mass is incorporated in silicates, 23% in refractory material, 9% in small carbonaceous molecules, and $\sim$30% in H_2O ice (with small contributions of CO, CO_2, CH_3OH, and other simple molecules). Remote observations of cometary gas and dust from ground-based and space telescopes throughout the electromagnetic spectrum are used to constrain the formation and composition of comets. Recent results have shown that the volatile component in comets is dominated by water, followed by CO and CO_2, with trace amounts of other chemical species such as CH_3OH, CH_4, C_2H_2, etc (see Table 2). The trace molecules are important, however, because they give clues as to the conditions under which comets were formed. CO is the most abundant carbon species in the coma, but it has an extended source; only less than 7% CO relative to water ice is assumed to originate from the nucleus (Irvine et al 2000). In Hale-Bopp 50% of the CO originated from the nucleus (DiSanti et al 1999).

The origin of comets and the extent to which they have been altered during the formation process is far from being understood. Three models have been considered for the origin of comets: (1) the interstellar model, which suggests that interstellar grains agglomerated to form cometary nuclei in the cold outer solar nebula far from the protosun (Greenberg 1982); (2) the complete chemical equilibrium model, in which presolar material is altered and chemically equilibrated (Lunine et al 1991); and (3) an intermediate model in which presolar material has been chemically and physically processed (Chick & Cassen 1997, Fegley 1999). It is currently assumed that comets are a mixture of interstellar and nebular material (Mumma et al 1993,

Irvine 1999, Ehrenfreund & Schutte 2000b). The original composition of comets may differ according to their place of origin.

Comets appear to originate from two distinct sources: the Oort cloud, the source of long-period comets (periods P > 20 years); and the Kuiper Belt, the source of short-period comets (P < 20 years) (Cochran et al 1995, Weissmann 1995). A better discriminant between cometary sources is the Tisserand invariant; short-period comets generally have a value greater than 2 (e.g. Weissmann 1999). Evidence for two classes of comets is found from their whole coma scattering properties at large phase angles (Levasseur-Regourd et al 1996). Comets with higher maximum of polarization seem to be dust-rich (Bockelée-Morvan et al 1995). The properties of 85 comets showed a dust/gas ratio that does not vary with the dynamical age of the comet (A'Hearn et al 1995) and indicates that the composition of comets may be related to their place of formation. For example, the presence of carbon-chain molecules (C_2, C_3) shows a dramatic decline as the Tisserand invariant increases above 2 (A'Hearn et al 1995). New dynamical studies suggest that the source of the Oort cloud comets was the entire giant planet region from Jupiter to Neptune. It seems that comets are formed in different nebular environments and probably experienced thermal and collisional processing before their ejection to the Oort cloud (Weissmann 1999). Processing and dynamical exchange of icy planetesimals in the giant planet zone may have "homogenized" cometary nuclei. Future observations of short-period comets, which have a different evolution (Levison & Duncan 1997), and dynamically new long-period comets will allow us to address the question of the diversity of comets.

The only way to measure the nuclear composition directly is via in situ measurements by a space probe, such as the GIOTTO mission to comet Halley (Keller et al 1987). However, by observing the coma one can in principle deduce the molecular inventory of the nucleus. Remote sensing by gas phase molecular spectroscopy is less ambiguous than mass spectrometry. IR observations can directly sample molecules coming off the nucleus (Mumma et al 1996, Mumma 1997, Crovisier 1998). The predominant chemical processes occurring in the coma are photodissociation and photoionization of parent molecules as they are exposed to the solar radiation field. The first simple analytical model of coma chemistry was given by Haser (1957), who considered only photodissociation of parent molecules to form "daughter" and "granddaughter" molecules. It is now appreciated that in the high-density inner coma ($\sim 10^{13}$ particles cm^{-3}), ion-molecule and neutral-neutral reactions are also likely to be important (e.g. Schmidt et al 1988).

In this section we summarize current knowledge concerning the organic inventory of the comae of recent comets, both refractory species and volatile molecules, and compare it with that found in the interstellar medium.

3.1 Cometary Dust and Refractory Organics

The cometary nucleus is a porous aggregate of ices and refractory material. In situ measurements on past space missions and observations strongly enhanced our knowledge about the structure and composition of comets. Several spacecraft performed a close flyby of Comet Halley in 1986. The particle composition was studied with mass spectrometers PUMA-1, PUMA-2, and PIA, flown on VEGA-1, VEGA-2, and GIOTTO, respectively. Time-of-flight mass spectra were recorded during the impact of dust particles on metal targets. All three instruments obtained a large collection of mass spectra of individual cometary dust particles in the mass range $3 \times 10^{-16} - 3 \times 10^{-10}$ g. These data revealed an overall mass ratio between 2 and 1 of siliceous to organic materials in Comet Halley dust (Fomenkova & Chang 1993).

In situ measurements made at Halley showed that about 70% of the dust grains comprised a mixed phase of organic material and refractory silicates; the remaining grains contained no organics (e.g. Jessberger et al 1988, Mumma et al 1993). An important discovery made by these encounters was the detection of CHON particles—that is, dust composed of carbon, hydrogen, oxygen, and nitrogen (Clark et al 1987, Langevin et al 1987, Jessberger et al 1988). Reanalysis of space-recorded data shows that CHON particles and silicate components are interdispersed at the submicrometer scale (Lawler & Brownlee 1992). Silicates, which comprise a major fraction of the refractory part of comets, appear to be a mix of high-temperature crystalline enstatite, forsterite, pyroxene, and glassy amorphous carbon (Hanner et al 1994, Crovisier et al 1997, Wooden et al 1999, Hanner 1999). Recent ISO observations showed crystalline silicates in comet Hale-Bopp as well as around young and evolved stars (Waelkens et al 1996, Waters et al 1996). The strong spectroscopic similarity of silicate material in all these objects has been reported (Malfait et al 1998). As the physical properties of disks around young stars and red giants are similar, Molster et al (1999) suggest that low temperature crystallization of silicate grains may occur in protoplanetary systems.

The data gathered by the PUMA mass spectrometer on the more refractory organic material existing in the coma of P/Halley constitute a unique data set (Kissel & Krueger 1987). Until future space missions are completed (e.g. STARDUST and ROSETTA), any review is limited to the Halley results (Kissel et al 1997, Cottin et al 1999). The PUMA data allowed the tentative, but not unambiguous, identification of many distinct organic molecules emanating from the carbonaceous dust (Kissel & Krueger 1987). The organic inventory has been summarized by Kissel & Krueger (1987), Cottin et al (1999), and Fomenkova (1999). The majority of grains contain heteropolymers and different mixtures of carbon phases and compounds including alcohols, aldehydes, ketones, and acids. Other organic species identified include PAHs, highly branched aliphatic hydrocarbons, and unsaturated hydrocarbon chains, including alkynes, dienes, various large nitriles, and analogous species with imino and amino end-groups. Several nitrogen heterocycles may also have been identified; these include pyrrole, pyrroline, pyridine, pyrimidine, imidazole, and perhaps purine and adenine.

Erosion or fragmentation of the mixed-phase particles may give rise to complex molecules in the coma (Festou 1999). This mechanism has been invoked to explain the existence of extended coma sources for various molecules: CO, H_2CO,

CN, C_2, C_3, and NH_2 (Mumma et al 1993). Extended sources may not reflect nuclear ice abundances (DiSanti et al 1999). This provides more indirect evidence that extremely complex organic molecules could be released from cometary dust particles. These include polyoxymethylene (POM), a polymer of H_2CO, whose decomposition was originally proposed by Huebner et al (1987) as an explanation of the extended formaldehyde source around comet Halley. Other complex molecules are probably being fragmented as well (Huebner et al 1989) and may include HCN-polymers (Matthews & Ludicky 1992). Eberhardt (1999) argues that photodissociation of H_2CO into CO can account for most of the extended CO source observed in comet Halley. Recently, Bernstein et al (1995) suggested hexamethylenetetramine (HMT, $C_6H_{12}N_4$) as a candidate component of cometary organics. In laboratory experiments on interstellar/cometary ice analogues, HMT is formed by strong UV photolysis and warming of ammonia/methanol-rich ices. Various POM-related compounds are also formed, but not at the concentrations seen in ammonia/formaldehyde-rich ices where thermally promoted reactions are predominant (e.g. Allamandola et al 1997). These latter reactions are also candidates for the origin of the POM identified in Halley; it is therefore likely that the CHON particles contain a mixture of polymers (Eberhardt 1999). The claimed identification of a polycylic aromatic hydrocarbon in Halley's comet (Moreels et al 1994) raises the possibility that other polyaromatic organics are also present in cometary dust. The 3.4 μm emission feature in comets, which shows a profile like the interstellar diffuse cloud absorption (see Section 2.2.1), was originally attributed to hydrocarbons in dust grains (Knacke et al 1986). Reanalysis of this feature showed that most of this emission can be attributed to CH_3OH (Reuter 1992, Bockelée-Morvan et al 1995) and other molecular organic species (DiSanti et al 1995, Mumma et al 1996). The exact composition of cometary refractories will be investigated by current and future space missions such as STARDUST and ROSETTA (Huntress 1999).

3.2 Cometary Volatiles

3.2.1 Interstellar Organics?

The recent apparitions of comets Hyakutake and Hale-Bopp have revolutionized our understanding of the volatile chemical inventory of comets and the interstellar-comet connection. Observational and theoretical work on the nucleus and coma of Hale-Bopp is extensively summarized in the proceedings of the International Conference on Comet Hale-Bopp published in volumes 77–79 of *Earth, Moon and Planets* (1999). Here we discuss selected results in the wider context of organic astrochemistry.

As a comet approaches the sun, solar radiation heats the surface layers of the nucleus and induces sublimation of the molecular ices. The gaseous molecules then stream away from the nucleus and carry the entrained dust grains with them. Remote sensing of the bright target Hale-Bopp showed that at large heliocentric distance the activity is driven by sublimation of CO. As the comet approaches the sun, the CO outgassing rate increases roughly as r_h^{-2}. Closer to the sun, crystallization of amorphous water ice is predicted to occur, and the release of trapped CO can then become the major CO source. At about 2.5 AU from the sun, sublimation of water ice from the surface layer becomes more and more efficient, and water outgassing becomes the major driver of the comet's activity (Biver et al 1997, Enzian et al 1998, Huebner & Benkhoff 1999). Figure 9 shows the evolution of the theoretical CO and water outgassing rates (Enzian 1999) compared to the observed rates (Biver et al 1999).

The model confirms that sublimation of CO ice at large heliocentric distance produces a gradual increase in the comet's activity as it approaches the Sun. Crystallization of amorphous water ice is predicted to begin at 7 AU from the Sun,

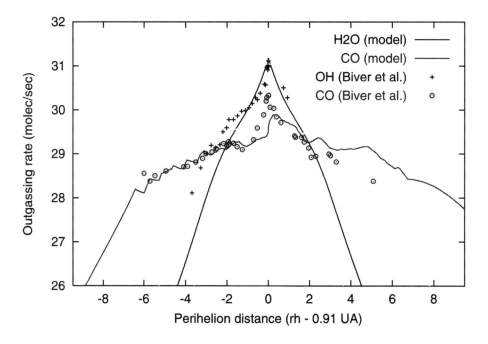

Figure 9   Evolution of the modeled CO and water outgassing rates (Enzian 1999) of comet Hale-Bopp compared with measured outgassing rates from astronomical observations (Biver et al 1999). The modeled comet nucleus has an albedo of 0.04 and a radius of 35 km; the radius is suggested by CCD observations (Weaver & Lamy 1999). The initial nucleus composition contains dust (50% by mass), amorphous water ice (40%), CO trapped in amorphous water ice (5%), and an additional small amount of CO ice (5%) in an independent phase. The thermal evolution is computed for two apparitions from 50 AU pre-perihelion to 50 AU post-perihelion. During the first orbit (which is shown above) the spin axis of the model comet nucleus is perpendicular to the orbital plane, whereas in the second orbit (which is shown above) the spin axis lies in the orbital plane and points toward the Sun at perihelion (from Enzian 1999).

but no outbursts were found. Seasonal effects and thermal inertia of the nucleus material lead to larger CO outgassing rates as the comet recedes from the Sun. Radio observations of comet Hale-Bopp by Biver et al (1997) showed that between 1.4 and 4 AU, the production rates of many simple molecules exhibited different dependencies on the heliocentric distance.

Many new cometary molecules were discovered or better constrained by IR and radio observations of comets Hyakutake and Hale-Bopp (Mumma et al 1996, Irvine et al 1996, Brooke et al 1996, Crovisier et al 1997, Bird et al 1997, Biver et al 1997, 1999, Lis et al 1997, Mumma 1997, Dello Russo et al 1998, Bockelée-Morvan et al 1998, DiSanti et al 1999, Magee-Sauer et al 1999, Dello Russo et al 2000, Crovisier & Bockelée-Morvan 1999, Bockelée-Morvan et al 2000). To understand the differences between cometary and interstellar material, we must first make a detailed abundance comparison between cometary volatiles and the molecular ices found in various regions of the dense ISM (Despois 1992, Mumma et al 1993, Mumma 1997, Crovisier 1998, Ehrenfreund et al 1997c, Ehrenfreund & Schutte 2000b, Irvine 1999, Irvine et al 2000). Table 2 shows such a comparison, including the gas phase abundances of selected molecules found in hot molecular cores where icy grain mantles have been evaporated. Generally, there are fewer unambiguous detections of large organics in cometary comae than in hot cores. This is due to difficulties in making radio observations of molecules in the more diffuse coma and the fact that abundances of large molecules probably decrease with increasing complexity. Despite this, significant upper limits have been obtained in Hale-Bopp for many organics (Crovisier 1998, Crovisier & Bockelée-Morvan 1999). Cometary spectra also contain many unidentified lines (e.g. Wyckoff et al 1999), so more organics may await discovery at shorter wavelengths.

It appears that interstellar and cometary species have similar compositions. The presence of HCOOH, HNCO, and NH_2CHO, as well as formaldehyde and methanol, is consistent with an origin on interstellar grains where hydrogen atom additions to CO, followed by C, O, and N atom additions, dominated the chemistry (see Section 2.1.1). The discovery of C_2H_6 (Mumma et al 1996) also lends support to the idea that the nuclear ices were formed in cold molecular clouds, since it could result from the reduction of acetylene on grains. Other molecules, if formed under interstellar conditions, appear to require gas phase synthesis prior to incorporation into ices; these include C_2H_2, CH_3CN, HC_3N, $HCOOCH_3$, and HCN (see Section 2.1.3)—although there is some debate as to whether $HCOOCH_3$ forms by gas phase reactions (MacDonald et al 1996, Helmich & van Dishoeck 1997).

Some notable exceptions in the abundance of interstellar ices and cometary volatiles are the large amounts of methanol and ammonia recently observed from the ground toward certain protostars (Dartois et al 1999a, Lacy et al 1998). Interstellar abundances of CH_3OH, which are consistent with cometary data, are currently observed predominantly in low-mass protostars. More thorough observations of such objects with the new class of 8+ m telescopes (VLT, KECK, SUBARU etc.) are needed in the future to investigate these discrepancies. OCS seems generally more abundant in comets than in interstellar ices toward YSOs, whereas the reverse holds true for CO_2, HCOOH, and HNCO (Ehrenfreund & Schutte 2000b). There are also significant differences between the cometary abundances of trace molecules when compared with abundances of hot core gas: some are markedly higher (HCOOH, NH_2CHO, HNCO), whereas others are much lower (CH_3CN, C_2H_2, HCN). This indicates that some cometary ice molecules are the result of a significant amount of processing. It is unclear whether these abundance differences reflect a different chemistry in the Sun's parent interstellar cloud or modifications induced as this material became incorporated into the protosolar disk. A further possibility is that some of these molecules (e.g. CH_3CN, HC_3N, $HCOOCH_3$) could be produced in the coma by gas phase reactions (see Section 3.2.2). Fully quantifying ISM-comet abundance differences will probably require detailed chemical modeling of the infall of interstellar gas and dust through the accretion shock into a protostellar disk, and of the chemical evolution followed in the comet-forming regions (e.g. Aikawa & Herbst 1999a). Such work may also shed light on the compositional differences in trace species to be expected as future comets are subjected to detailed chemical scrutiny.

3.2.2 The Origin of HNC

Simple organic molecules may also provide important probes of cometary composition and origin. Cold interstellar clouds have large HNC/HCN ratios ~0.1–1 (Ohishi & Kaifu 1998). The detection of HNC in comets and the associated HNC/HCN ratios (0.06–0.2) (Irvine et al 1996, Biver et al 1997, Hirota et al 1999, Irvine et al 1999) therefore strongly suggested an ion-molecule origin for the HNC in a cold molecular cloud. However, in the dense inner coma, chemical reactions could occur that synthesize HNC from HCN; the HNC/HCN ratio in comets has been viewed from this perspective (Rodgers & Charnley 1998, Irvine et al 1998a,b). Ion-molecule reactions can produce some HNC from HCN but cannot produce the observed HNC/HCN ratio in Hyakutake. The prodigious molecule production rate in comet Hale-Bopp suggests that isomerization of HCN by suprathermal hydrogen atoms could reproduce both the observed HNC/HCN ratio and its variation with heliocentric distance (Rodgers & Charnley 1998); however, this mechanism fails to produce the HNC/HCN ratio in Hyakutake's coma (0.06), which suggests another source for the HNC in this comet. Comet Lee has a water production rate similar to that of Hyakutake but an HNC/HCN ratio of about 0.12 (Crovisier 2000). Hence, the above gas phase reactions would also fail to reproduce the comet's HNC/HCN ratio. These studies suggest that there may be two sources of cometary HNC: nuclear sources and coma-synthesized sources. Indeed, interferometric observations of Hale-Bopp by Blake et al (1999) indicate two HNC components: one corresponding to nuclear emission and another corresponding to more extended or jet emission. It is interesting that the HNC/HCN ratio of the nuclear component is similar to those of comets Hyakutake and Lee, whereas the jet component is almost two to three times as large. Hence, in comets with large water production rates, the nuclear HNC component may be overwhelmed by that due to coma synthesis. Alternatively, Blake et al (1999) favor an explanation based on trapping of a large (interstellar) nuclear HNC/HCN ratio, which is then modified during sublimation. More measurements of this important ratio are clearly needed from more comets.

3.2.3 Isotopic Ratios

Enhanced isotopic fractionation ratios provide strong evidence that cometary material can retain a chemical memory of an origin at the cold temperatures of interstellar clouds (Crovisier 1998). Comets Halley, Hale-Bopp, and Hyakutake had water D/H ratios of $\sim 3 \times 10^{-4}$ (Bockelée-Morvan et al 1998, Irvine et al 2000), similar to those of hot cores (Gensheimer et al 1996). Blake et al (1999) have proposed that the nuclear ice HDO/H_2O ratio in Hale-Bopp is ~ 1.5–2.5×10^{-3}, and there is recent evidence that interstellar ices could have significantly higher HDO/H_2O ratios (Teixeira et al 1999); however, it is difficult to reconcile these with the lower values seen in hot core gas. Meier et al (1998) measured DCN/HCN $\sim 2 \times 10^{-3}$, which is lower than that found in interstellar clouds ($\sim 2 \times 10^{-2}$). Hence, although these D/H ratios support an interstellar origin, it appears that the material either was processed sufficiently to lower these values prior to comet assembly, or was simply formed in higher-temperature interstellar gas (Millar et al 1989). Recently, Aikawa & Herbst (1999b) showed that cometary D/H ratios can be produced in the outer regions of protostellar disks, which supports the view that cometary volatiles are a mixture of interstellar matter and matter that is processed in situ. Finally, other isotopic ratios in oxygen, nitrogen, sulfur, and carbon have been measured in simple cometary organics and found to be consistent with cosmic (solar) values (see Crovisier & Bockelée-Morvan 1999).

By characterizing the organic composition, isotopic fractionation, and spin ratios of future comets, it may eventually become possible to identify the formation sites of individual comets and to quantify the degree to which their nuclei differ from pristine interstellar material.

4.   ORGANIC MOLECULES IN METEORITES

It was established over a century ago that some meteorites contain carbonaceous material. Organic compounds that have been identified in carbonaceous C1 and C2 chondrites include amines and amides; alcohols, aldehydes, and ketones (Cronin & Chang 1993); aliphatic and aromatic hydrocarbons (Hahn et al 1988, Kerridge et al 1987, Gilmour & Pillinger 1994, Messenger et al 1998, O'D Alexander et al 1998, Sephton et al 1998, Stephan et al 1998); sulfonic and phosphonic acids (Cooper et al 1992); amino, hydroxycarboxylic, and carboxylic acids (Cronin & Pizzarello 1990, Cronin et al 1993, Peltzer et al 1984); purines and pyrimidines (Stoks & Schwartz 1979, 1981); and kerogen-type material (Becker et al 1999b).

The study of presolar grains in meteorites allows us to trace the circumstellar origin of some dust species (Bernatowicz & Zinner 1996, Bernatowicz 1997). The isotopic composition of presolar graphite grains, diamonds, and SiC grains isolated from meteoritic samples indicates their interstellar origin and a mass ratio of graphite:SiC:diamonds = 1:5:400 (Ott 1993). Nanodiamonds are present in primitive meteorites in abundances of up to ∼1400 ppm (Huss & Lewis 1995). These grains contain an anomalous Xe isotopic component (Xe-HL) considered to be characteristic of nucleosynthetic processes in supernovae (Lewis et al 1987). The ^{15}N depletion and the low C/N ratio in such grains is consistent with carbon-rich stellar environments. Only low abundances are observed for graphite in meteorites (Nuth 1985); however, the greatest concentrations of noble gases are measured in the poorly graphitized carbon (PGC) that makes up the bulk of the carbon material in these meteorites. SiC is present in many carbon stars and some PN, but has not been observed in the ISM. Conversely, presolar micron-size SiC grains are found in abundance in carbonaceous chondrites. Elemental and isotopic ratios of noble gases in most meteoritic SiC grains show excellent agreement with predictions for material enriched with s-process nucleosynthesis dredged up from the helium shells of AGB stars during thermal pulses (Ott 1993). GEMS (Glasses with Embedded Metal and Sulfide) are a subgroup of polyphase grains occurring in interplanetary dust particles, whose depletion and inclusions indicate their interstellar origin (Bradley 1994, Bradley et al 1999).

Through the use of laser desorption mass spectrometry (LDMS), PAHs, kerogen-like material, fullerenes, and fulleranes have been detected in meteorites (Becker 1999). The presence of carbon onions in acid residues of the Allende meteorite suggested that higher fullerenes or nanotubes may be present in meteorites. This has been recently confirmed by LDMS measurements of the higher fullerenes (C_{100} to C_{400}) in the Allende meteorite (see Figure 10; Becker et al 1999a). In addition, the higher fullerenes have also been isolated from the Murchison carbonaceous residue, and measurements of noble gases (helium, neon, and argon) in both the Murchison and Allende fullerenes indicate that these molecules are indeed extraterrestrial in origin (Becker et al 2000). The meteoritic PGC material retains the newly detected fullerene component; fullerenes may therefore represent an important carrier phase for noble gases that has been previously overlooked (Becker et al 2000).

Many classes of organic molecules have been discovered in the Murchison carbonaceous chondrite (Cronin & Chang 1993). One explanation for the origin of these compounds is that they were formed from organic matter residing on interstellar dust grains that survived entry into the region of the protosolar nebula where comets and other icy bodies formed and that ultimately became incorporated into the parent bodies of carbonaceous meteorites (Bunch & Chang 1980). The Murchison organic inventory displays a large deuterium enrichment, particularly in the amino and hydroxycarboxylic acids. This constitutes one of the most convincing pieces of evidence that the original carbon reservoir was interstellar (Kerridge & Chang 1985, Zinner 1988) because in cold clouds, enhanced molecular D/H ratios can easily be attained by gas phase and grain-surface processes (Millar et al 1989, Tielens 1983). In the "interstellar-parent body" scenario, melting of the parent

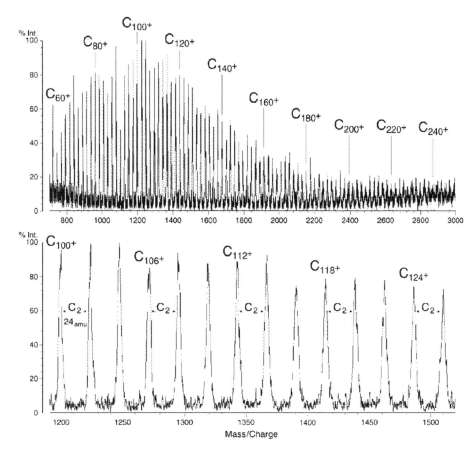

Figure 10   The detection of higher fullerenes in the Allende meteorite was recently reported by Becker et al (1999a). The figure shows laser desorption time-of-flight mass spectra of the meteoritic extract. A small peak for C_{60}^+ and remarkably stable larger carbon clusters between C_{100}^+ and C_{250}^+ (*top*) are displayed, as well as a closer examination of the larger clusters (*bottom*).

body water ice resulted in the relatively pristine interstellar material undergoing a variety of organic reactions in solution ("aqueous alteration"), leading to the formation of a second generation of more complex organics. Therefore, many of the organics identified in Murchison have been postulated to be either truly interstellar components (e.g. alcohols, aldehydes, and amines) or altered parent body products, such as urea and guanylurea and amino acids, which were derived from various interstellar precursors (NH_3, HNCO, NH_2CN, acetone, acrolein, vinyl cyanide).

Detailed analyses of Murchison extracts have shown very marked trends within each class of organic compound, as well as some characteristics that are quite general (Cronin & Chang 1993, Cronin et al 1995). For a given number of carbon atoms there is complete structural diversity within most classes of organic compounds, which indicates formation mechanisms involving free C-bearing radicals. All stable isomeric forms are present, and branched-chain isomers are the most abundant. Within each homologous series, the molecular concentrations decline with the increasing number of carbon atoms, which is consistent with chain growth by single-carbon additions. These characteristics are particularly evident in the large number of amino acids identified in the extracts obtained from carbonaceous chondrites, such as Murchison. An extraterrestrial (though not necessarily interstellar) origin is suggested for many of these amino acids because, for example, α-aminoisobutyric acid (AIB) and isovaline are present in appreciable abundances (e.g. Cronin & Pizzarello 1983) but are particularly rare on Earth (Brinton et al 1998); other amino acids are unknown in terrestrial biochemistry.

Strecker-cyanohydrin synthesis of meteoritic α-amino and α-hydroxy acids during aqueous alteration of a parent body is an appealing scenario (Cronin & Chang 1993). It accounts for the strong structural correspondence between amino and hydroxy acids at a given carbon number, and the fact that both exhibit high isotopic fractionation. This latter facet strongly supports the hypothesis that interstellar molecules were the simple precursors required for Strecker-cyanohydrin processes (Cronin 1989). There are two problems with this scenario, however (Cronin & Chang 1993, Kerridge 1999). First, the ^{13}C and D fractionation data appear to be inconsistent with a common (aldehyde or ketone) precursor—the amino acids are far more highly enriched in D and ^{13}C. This places doubt on there being a common precursor for each type of acid (Cronin et al 1993). Second, imino acids should also be produced in this synthesis but are not abundant in Murchison (Kerridge 1999). The connection between meteoritic amino acids and interstellar chemistry would be strengthened with the discovery of interstellar glycine (see Snyder 1997). If alkyl cation transfer reactions with glycine can proceed, driven by evaporated alcohol-rich ice mantles, (see section 2.1.3), they could lead to other meteoritic amino acids in hot cores; these include α-alanine, α-aminoisobutyric acid, and 2-aminobutyric acid (Charnley 2000). These molecules may be detectable with future millimeter-wave arrays. Recently it has been claimed that equilibrium (Strecker) reactions could produce amino acids and hydroxy acids in interstellar molecular gas (Chakrabarti & Chakrabarti 2000). The reactions and rates employed in that particular study included those from a model of amino acid synthesis in submarine hydrothermal systems of the Earth's early oceans (Schulte & Shock 1995). It is inappropriate to extrapolate data from equilibrium chemistry in liquids to interstellar plasmas, where the chemistry is always far from equilibrium, and so the claim of Chakrabarti & Chakrabarti (2000), that interstellar glycine, alanine, and also the DNA base adenine, could be formed by these reactions is most probably incorrect.

Recent meteoritic organic studies have also specifically addressed the question of their relation to the development of life. Enantiomeric excesses of 7–9% have recently been measured in several Murchison amino acids known to be of extraterrestrial origin (e.g. 2-amino-2,3-dimethylpentanoic acid; Cronin & Pizzarello

1997). This fact suggests an extraterrestrial provenance for the homochirality found in protein amino acids, or at least a primordial L/D excess that could subsequently be amplified by further reactions. The PAH component of meteorites (e.g. Krishnamurthy et al 1992, McKay et al 1996, Clemett et al 1998) has been invoked as an integral part of the claim that the Martian meteorite ALH84001 contains extinct microbial life—a claim that is currently the subject of intense debate (Jull et al 1998, Bada et al 1998, Becker et al 1999b).

In conclusion, carbonaceous meteoritic material represents a sample of interstellar matter—albeit a highly processed one. Future studies will undoubtedly expand the catalogue of molecules found in this material, along with its connection to comets and the interstellar medium.

5. THE VOYAGE

The life cycle of interstellar dust is tied to the evolution of stars and planets in the Galaxy. Molecules with up to 15 carbon atoms have been observed in cold clouds and hot cores in the ISM. What degree of complexity can be reached by reactions in molecular cloud gas is not evident, but might be explored by future millimetric observations aimed to search for larger species. Hydrogenation and oxidation reactions lead to simple organic species on the grain surface (Tielens & Hagen 1982, Charnley 1997, Schutte 1999). Energetic processing and heating of complex ice mixtures in laboratory experiments revealed the presence of complex molecules such as alcohols, ketones, polyoxymethylene (POM), and HMT ($C_6H_{12}N_4$) (Bernstein et al 1995, 1997). Because energetic processing in dense clouds is not well constrained, those laboratory experiments may however, only partly reflect the conditions in dense clouds.

In diffuse clouds and circumstellar shells, a variety of carbon compounds are observed. Among the most abundant organic molecules are PAHs (Tielens et al 1999), which are also identified in meteorites (Krishnamurthy et al 1992), and interplanetary dust particles (IDPs) (Clemett et al 1993), possibly in comets (Moreels et al 1994) and in the atmosphere of Titan and Jupiter (Sagan et al 1993). Other important carbon species likely present in molecular clouds are fullerenes and diamonds (Foing & Ehrenfreund 1997, Guillois et al 1999). A large fraction of the cosmic carbon seems to be tied up in carbonaceous solids that may be present in many different structures (Henning & Salama 1998).

The study of cometary nuclei and their comae is of fundamental importance because comets are thought to be the most pristine members of the solar system (e.g. Delsemme 1982). Carbonaceous chondrites are unique with respect to their content of carbon in the form of organic compounds. A central question is: How much of the organic material present in primitive bodies, such as comets, asteroids, and meteorites, is pristine interstellar material, and to what extent is the chemical inventory of these objects a result of chemical processing within the nebula (Fegley & Prinn 1989, Mumma et al 1993, Irvine et al 2000)? While in some models comet material is assumed to be basically unprocessed interstellar ice and dust (Greenberg 1982), other models state that the original interstellar ices were processed in the solar nebula before becoming incorporated into the cometary body (Engel et al 1990, Lunine et al 1991, Chick & Cassen 1997, Fegley 1999). To address this problem in a quantitative manner, we need a detailed understanding of the relevant chemical reactions and processing that interstellar gas and dust experiences as it evolves from the diffuse ISM into cold molecular clouds, and thereafter into star-forming cores. Additionally, we need to obtain evidence showing the extent to which organics have been preserved in solar system bodies (Sandford 1996, Mumma 1997, Irvine et al 2000).

Cometary volatiles and their relative proportions are more compatible with interstellar solid state abundances (Mumma 1997, Ehrenfreund et al 1997c). A comparison of interstellar and cometary ices using recent ISO data has revealed a general correspondence in the composition of cometary and interstellar ices, but discrepancies are also apparent (see Table 2; Ehrenfreund & Schutte 2000b). If interstellar icy grains are incorporated unaltered into comets, ice species should be observed in comets at varying heliocentric distances because of their different sublimation temperatures (Ehrenfreund 1999a). A comparison of volatiles in the cometary coma and interstellar abundances shows that certain aspects of cometary activity are consistent with interstellar ice chemistry and gas species in hot cores. However, there is strong evidence that cometary ices are a mixture of original interstellar material with material that has been moderately to heavily processed in the presolar nebula (Irvine 1999, Ehrenfreund & Schutte 2000b, Ehrenfreund 2000, Irvine et al 2000).

The early Earth may have obtained most of its volatile organic material from the arrival of meteorites and comets at its surface. Some of this matter was accumulated in the form of surviving organic compounds (Anders 1989, Chyba et al 1990, Maurette et al 1995, Pierazzo & Chyba 1999). Determining the most likely distribution of meteoritic and cometary organic molecules that could seed primitive planets is a key astrobiological objective because it sets the initial conditions for at least part of their phase of prebiotic chemical evolution (Wood & Chang 1985, Cronin & Chang 1993, Oro & Lazcano 1997). Precursor molecules for life, including HCN, H_2CO, amino acids, and purines, have been identified in astronomical observations of the interstellar gas, in comets, and in laboratory investigations of extraterrestrial material. Recent laboratory experiments lend support to the idea that PAHs may have been important intermediates in the cHl pathways that led from space to the origin of life on Earth (Bernstein et al 1999). In H_2O-PAH icy mixtures, biogenic compounds were formed by UV radiation and may have found their way to the early Earth (Bernstein et al 1999, Ehrenfreund 1999b). The recent detection of higher fullerenes in meteorites shows that those compounds may have played an important role on the early Earth, and perhaps other planets, by providing not only a source of carbon but also the volatiles that contributed to the development of planetary atmospheres (Becker et al 2000).

6.   FUTURE STUDIES OF ORGANICS IN THE UNIVERSE

Current and future space missions are crucial to enhance our knowledge of organic compounds in the Universe. Although the analysis of ISO data will provide us with important information on organic molecules in space for at least a decade, the next IR satellite SIRTF is already on the start ramp. Further IR missions planned for the next millenium are the airborne observatory SOFIA and the FIRST satellite. Future ground-based IR facilities and (sub)millimeter interferometers, such as VLT, KECK and ALMA, will lead to the detection of complex molecules with abundances almost a factor of a hundred below current detection limits.

Milestones are expected in the search for extrasolar planets and in the planetary exploration of our own solar system. STARDUST is on its journey to comet Wild-2, and will capture comet dust and volatiles by impact into an ultra-low-density aerogel at a low speed of $6.1 \, \text{km s}^{-1}$ and an encounter distance of 150 km from the nucleus. The collected material will be dropped off in a reentry capsule that will parachute to Earth in 2006, and will allow the first analysis of cometary samples in Earth laboratories (http://stardust.jpl.nasa.gov/top.html). CASSINI-HUYGENS is on its way to meet Saturn and Titan in 2004. The HUYGENS probe will be released to parachute through the atmosphere of Titan with an entry speed of 20,000 km/h. Aerobraking will allow a 2.5-hour descent of the probe; meanwhile, six instruments will measure the chemical properties of Titan's atmosphere, which is known to contain organic molecules that reflect prebiotic conditions (http://sci.esa.int). The ROSETTA comet rendezvous mission will be launched in 2003 with the Ariane 5 rocket for a rendezvous maneuver with comet Wirtanen in 2011–2013. More than 20 instruments on the orbiter and the lander will obtain data on cometary origin and the interstellar-comet connection, which will broaden our insight into the origin of our solar system (http://sci.esa.int). Future Mercury missions (MESSENGER, BEPICOLOMBO) and lunar missions (SMART1, SE-LENE) will also investigate the presence of ice deposits (and eventually organics) in the permanently shadowed polar areas. Other important missions include the search for organics and extinct/extant life on Mars (MARS-EXPRESS/BEAGLE-2, Mars Sample Return 2005) and missions to Jupiter's moon Europa (EUROPA ORBITER) and the farthest planet in our solar system, Pluto (PLUTO-KUIPER EXPRESS).

In the life cycle of cosmic dust, organic molecules formed in the ISM are later incorporated in solar system material, and probably supplied organic material that seeded the early Earth (see Figure 11, color insert). The interdisciplinary effort combining astronomical observations, laboratory studies, and theoretical models in the 1990's has shown significant progress in our understanding of the ISM-planetary connection (Owen & Bar-Nun 1998, Cruikshank 1997, 1999, Ehrenfreund et al 1999b, McCord et al 1997, Irvine et al 2000, Lunine et al 2000) and will certainly continue to do so in the new millennium. The study of the formation, stability, and evolution of large carbon molecules in the universe therefore supports our understanding of the physical and chemical processes in the ISM and in solar systems including ours, and so may shed some light on the emergence of life and our own existence.

ACKNOWLEDGMENTS

Special thanks go to R Ruiterkamp and S Rodgers for graphic art and their help in compiling the tables. We thank E Deul for computer support. The authors especially want to thank M Mumma and M Fomenkova for discussion on cometary dust, and L Becker for discussion on organic molecules in meteorites. We are grateful to L Becker, A Enzian, O Guillois, Y Pendleton, A Tielens, and D Whittet, who allowed us to include figures of their recent work. We further thank our colleagues (in alphabetical order) for stimulating discussions: A Enzian, BH Foing, JM Greenberg, Y Pendleton, S Rodgers, WA Schutte, M Spaans and E van Dishoeck. We thank the referee, F Shu, for his comments.

This work was supported by the Netherlands Research School for Astronomy (NOVA). Theoretical astrochemistry at NASA Ames is supported by NASA's Origins of Solar Systems and Exobiology Programs.

Visit the Annual Reviews home page at www.AnnualReviews.org

LITERATURE CITED

Adamson AJ, Whittet DC, Chrysostomou A, Hough JH, Aitken DK, et al. 1999. *Ap. J.* 512:224–29

Adamson AJ, Whittet DC, Duley WW. 1991. *MNRAS* 252:234–45

A'Hearn MF, Millis RL, Schleicher DG, Osip DJ, Birch PV. 1995. *Icarus* 118:223–70

Aikawa Y, Herbst E. 1999a. *Astron. Astrophys.* 351:233–46

Aikawa Y, Herbst E. 1999b. *Ap. J.* 526:314–26

Allain T, Leach S, Sedlmayr E. 1996a. *Astron. Astrophys.* 305:602–15

Allain T, Leach S, Sedlmayr E. 1996b. *Astron. Astrophys.* 305:616–30

Allamandola JM, Sandford SA, Tielens AGGM, Herbst TM. 1992. *Ap. J.* 399:134–46

Allamandola LJ, Bernstein MP, Sandford SA. 1997. In *Astronomical & Biochemical Origins and the Search for Life in the Universe*, ed. CB Cosmovici, S Bowyer, D Wertheimer, pp. 23–47. Bologna: Editrice Compositori

Allamandola LJ, Hudgins DM, Bauschlicher W, Langhoff SR. 1999a. *Astron. Astrophys.* 352:659–64

Allamandola LJ, Hudgins DM, Sandford SA. 1999b. *Ap. J.* 511:L115–19

Allamandola LJ, Sandford SA, Tielens AGGM, Herbst TM. 1993. *Science* 260:64–66

Allamandola LJ, Tielens AGGM, Barker JR. 1987. In *Interstellar Processes*, ed. DJ Hollenbach, HA Thronson, pp. 471–90. Dordrecht: Reidel

Allen M, Robinson GW. 1977. *Ap. J.* 212:396–415

Anders E. 1989. *Nature* 342:255–57

Annestad PA. 1992. *Ap. J.* 386:627–34

Bada JL, Glavin DP, McDonald GD, Becker L. 1998. *Science* 279:362–65

Bakes ELO, Tielens AGGM. 1994. *Ap. J.* 427:822–38

Bakes ELO, Tielens AGGM. 1998. *Ap. J.* 499:258–66

Bally J, Black JH, Blake GA, Churchwell E, Knapp G, et al. 1995. In *Report of Galactic Molecular Clouds and Astrochemistry*

Working Group, NRAO Science Workshop, National Radio Astronomy Observatory, 9 pages

Banhart F, Ajayan PM. 1996. *Nature* 382:433–35

Beck HKB, Gail HP, Henkel R, Sedlmayr E. 1992. *Astron. Astrophys.* 265:626–42

Becker L. 1999. In *Laboratory Astrophysics and Space Research*, ed. P Ehrenfreund, K Krafft, H Kochan, V Pirronello, pp. 377–98. Dordrecht: Kluwer

Becker L, Bunch TE. 1997. *Meteoritics* 32:479–87

Becker L, Bunch TE, Allamandola LJ. 1999a. *Nature* 400:228–29

Becker L, Popp B, Rust T, Bada J. 1999b. *Earth Planet. Sci. Lett.* 167:71–79

Becker L, Poreda R, Bunch T. 2000. *Proc. Nat. Acad. Sci.* 97(7):2979–83

Beegle LW, Wdowiak TJ, Robinson MS, Cronin JR, McGehee MD, et al. 1997. *Ap. J.* 487:976–82

Bell MB, Avery LW, Feldman PA. 1993. *Ap. J.* 417:L37–40

Bell MB, Avery LW, MacLeod JM, Matthews HE. 1992. *Ap. J.* 400:551–55

Bell MB, Feldman PA, Watson JKG, McCarthy MC, Travers MJ, et al. 1999. *Ap. J.* 518:740–47

Bernatowicz TJ. 1997. In *From Stardust to Planetesimals*, ed. Y Pendleton, AGGM Tielens, 122:227–52. Provo, Utah: Astron. Soc. Pac.

Bernatowicz TJ, Zinner E. 1996. *Astrophysical Implications of the Laboratory Study of Presolar Materials.* New York: AIP Press, 749 pp.

Bernstein M, Sandford SA, Allamandola LJ. 1997. *Ap. J.* 476:932–42

Bernstein M, Sandford SA, Allamandola LJ, Chang S, Scharberg MA. 1995. *Ap. J.* 454:327–44

Bernstein M, Sandford SA, Allamandola LJ, Gillette JS, Clemett SJ, et al. 1999. *Science* 283:1135–38

Bettens RP, Herbst E. 1996. *Ap. J.* 468:686–93

Bettens RP, Herbst E. 1997. *Ap. J.* 478:585–93

Bettens RPA, Lee HH, Herbst E. 1995. *Ap. J.* 443:664–74

Bird MK, Huchtmeier WK, Gensheimer P, Wilson TJ, Janardhan P, et al. 1997. *Astron. Astrophys.* 325:L5–8

Biver N, Bockelée-Morvan D, Colom P, Crovisier J, Davies KJ, et al. 1997. *Science* 275:1915–18

Biver N, Bockelée-Morvan D, Crovisier J, Davies JK, Matthews HE, et al. 1999. *Astron. J.* 118(4):1850–72

Blake GA, Qi C, Hogerheijde MR, Gurwell MA, Muhlman DO. 1999. *Nature* 398:213–16

Blake GA, Sutton EC, Masson CR, Phillips TG, 1987. *Ap. J.* 315:621–45

Blake DF, Freund F, Krishnan KFM, Echer CJ, Shipp R. 1988. *Nature* 332/6165:611–13

Blanco A, Borghesi A, Fonti S, Orofino V. 1998. *Astron. Astrophys.* 330:505–14

Blitz L, 1993. *Nature* 364:757–58

Bockelée-Morvan D, Brooke TY, Crovisier J. 1995. *Icarus* 116:18–39

Bockelée-Morvan D, Gautier D, Lis D, Young K, Keene J, et al. 1998. *Icarus* 133:147–62

Bockelée-Morvan D, Lis DC, Wink JE, Despois D, Crovisier J, et al. 2000. *Astron. Astrophys.* 353:1101–14

Boogert ACA, Ehrenfreund P, Gerakines PA, Tielens AGGM, Whittet DCB, et al. 2000. *Astron. Astrophys.* 353:349–62

Boogert ACA, Helmich FP, van Dishoeck EF, Schutte WA, Tielens AGGM, et al. 1998. *Astron. Astrophys.* 336:352–58

Boogert ACA, Schutte WA, Tielens AGGM, Whittet DCB, Helmich FP, et al. 1996. *Astron. Astrophys.* 315:L377–80

Boudin N, Schutte WA, Greenberg JM. 1998. *Astron. Astrophys.* 331:749–59

Boulanger F, Boissel P, Cesarsky D, Ryter C. 1998. *Astron. Astrophys.* 339:194–200

Bradley JP. 1994. *Science* 265:925–29

Bradley JP, Keller LP, Snow T, Hanner M, Flynn GJ, et al. 1999. *Science* 285:1716–18

Brechignac Ph, Pino Th, 1999. *Astron. Astrophys.* 343:L49–52

Briggs R, Ertem G, Ferris JP, Greenberg JM,

McCain PJ, et al. 1992. *Orig. Life Evol. Biosph.* 22:287–307

Brinton KL, Engrand C, Glavin DP, Bada JL, Maurette M. 1998. *Orig. Life Evol. Biosph.* 28:413–24

Brooke TY, Sellgren K, Geballe TR. 1999. *Ap. J.* 517:883–900

Brooke TY, Tokunaga AT, Weaver HA, Crovisier J, Bockelée-Morvan D, Crisp D. 1996. *Nature* 383:606–8

Brown PD. 1990. *MNRAS* 43:65–71

Brown PD, Charnley SB. 1990. *MNRAS* 244:432–43

Brown PD, Charnley SB, Millar TJ. 1988. *MNRAS* 231:409–17

Brown PD, Millar TJ. 1989. *MNRAS* 237:661–71

Bujarrabal V, Fuente A, Omont A. 1994. *Astron. Astrophys.* 285:247–71

Bunch T, Chang S. 1980. *Geochim. Cosmochim. Acta.* 44:1543–77

Cami J, Sonnentrucker P, Ehrenfreund P, Foing BH. 1997. *Astron. Astrophys.* 326:822–30

Cardelli J, Meyer DM, Jura M, Savage BD. 1996. *Ap. J.* 467:334–40

Caselli P, Hasegawa TI, Herbst E. 1993. *Ap. J.* 408:548–58

Caselli P, Hasegawa TI, Herbst E. 1998. *Ap. J.* 495:309–16

Cernicharo J. 2000. In *Astrochemistry: From Molecular Clouds to Planetary Systems*, IAU Symposium 197, ed. YC Minh, EF van Dishoeck, pp. 375–90. Sogwipo: Astron. Soc. Pac.

Cernicharo J, Guélin M. 1996. *Astron. Astrophys.* 309:L27–30

Cernicharo J, Guélin M, Hein H, Kahane C. 1987. *Astron. Astrophys.* 181:L9–12

Chakrabarti S, Chakrabarti SK. 2000. *Astron. Astrophys.* 354:L6–8

Charnley SB. 1997. In *Astronomical & Biochemical Origins and the Search for Life in the Universe*, ed. CB Cosmovici, S Bowyer D Wertheimer, pp. 89–96. Bologna: Editrice Compositori

Charnley SB. 1998. *Ap. J.* 509:L121–24

Charnley SB. 2000. In *The Bridge Between the Big Bang and Biology*, ed. F. Giovannelli, Rome: Consiglio Nazionale delle Ricerche. In press

Charnley SB, Kaufman MJ. 2000. *Ap. J. Lett.* 529:L111–14

Charnley SB, Kress ME, Tielens AGGM, Millar TJ. 1995. *Ap. J.* 448:232–39

Charnley SB, Latter WB. 1997. *MNRAS* 287:538–42

Charnley SB, Rodgers SD. 2000. *Ap. J.* Submitted

Charnley SB, Tielens AGGM, Millar TJ. 1992. *Ap. J. Lett.* 399:L71–74

Charnley SB, Tielens AGGM, Rodgers SD. 1997. *Ap. J. Lett.* 482:203–6

Chastaing D, James PL, Sims IR, Smith IWM. 1998. *Faraday Discussions* 109:165–82. Cambridge, UK: R. Soc. Chem.

Cherchneff I, Barker JR, Tielens AGGM. 1992. *Ap. J.* 401:269–87

Cherchneff I, Glassgold AE. 1993. *Ap. J.* 419:L41–44

Cherchneff I, Glassgold AE, Mamon GA. 1993. *Ap. J.* 410:188–201

Cherchneff I, Tielens AGGM. 1995. In *Wolf-Rayet Stars*, ed. KA van der Hucht, PM Williams, pp. 346–54. Dordrecht: Kluwer

Chiar JE, Adamson AJ, Kerr TH, Whittet DCB. 1995. *Ap. J.* 455:234–43

Chiar JE, Gerakines PA, Whittet DCB, Pendleton YJ, Tielens AGGM, et al. 1998a. *Ap. J.* 498:716–27

Chiar JE, Pendleton YJ, Geballe TR, Tielens AGGM. 1998b. *Ap. J.* 507:281–86

Chick KM, Cassen P. 1997. *Ap. J.* 477:398–409

Chyba C, Sagan C. 1992. *Nature* 355:125–32

Chyba C, Thomas PJ, Brookshaw L, Sagan C. 1990. *Science* 249:366–73

Clark BC, Mason LW, Kissel J. 1987. *Astron. Astrophys.* 187:779–84

Clayton DD, Meyer BS, Sanderson CI, Russell SS, Pillinger CT. 1995. *Ap. J.* 447:894–905

Clemett SJ, Dulay MT, Gillette JS, Chillier XDF, Mahajan TB, et al. 1998. *Faraday*

Discussions 109:417–36. Cambridge, UK: R. Soc. Chem.

Clemett SJ, Maechling CR, Zare RN, Swan PD, Walker RM. 1993. *Science* 262:721–25

Cochran AL, Levison HF, Stern AS, Duncan MJ. 1995. *Ap. J.* 455:342–46

Cooper G, Unwo WM, Cronin J. 1992. *Geochim. Cosmochim. Acta.* 56:4109–15

Cottin M, Gazeau MC, Raulin F. 1999. *Planet. Space Sci.* 47:1141–62

Cronin J. 1989. *Adv. Space. Res.* 9(2):59–64

Cronin J, Cooper G, Pizzarello S. 1995. *Adv. Space. Res.* 15(3):91–97

Cronin J, Pizzarello S. 1983. *Adv. Space. Res.* 3(9):5–18

Cronin J, Pizzarello S. 1990. *Geochim. Cosmochim. Acta.* 54:2859–68

Cronin J, Pizzarello S. 1997. *Science* 275:951–55

Cronin J, Pizzarello S, Epstein S, Krishnamurthy RV. 1993. *Geochim. Cosmochim. Acta.* 57:4745–52

Cronin JR, Chang S. 1993. In *The Chemistry of Life's Origins*, ed. JM Greenberg, CX Mendoza-Gomez, V Pirronello, pp. 209–58. Dordrecht: Kluwer

Crovisier J. 1998. In *Faraday Discussions* 109:437–52. Cambridge, UK: R. Soc. Chem.

Crovisier J. 2000. In *Astrochemistry: From Molecular Clouds to Planetary Systems*, IAU Symposium 197, ed. YC Minh, EF van Dishoeck, pp. 461–470. Sogwipo: Astron. Soc. Pac.

Crovisier J, Bockelée-Morvan D. 1999. *Space Science Reviews* 90:19–32

Crovisier J, Leech K, Bockelée-Morvan D, Brooke TY, Hanner MS. 1997. *Science* 279:1904–7

Cruikshank D. 1997. In *From Stardust to Planetesimals*, ed. Y Pendleton, AGGM Tielens, 122:315–334. Provo, Utah: Astron. Soc. Pac.

Cruikshank D. 1999. In *Laboratory Astrophysics and Space Research*, ed. P Ehrenfreund, K Krafft, H Kochan, V Pirronello, pp. 37–68. Dordrecht: Kluwer

Dartois E, Demyk K, d'Hendecourt L,

Ehrenfreund P. 1999b. *Astron. Astrophys.* 351:1066–74

Dartois E, d'Hendecourt L, Boulanger F, Jourdain de Muizon M, Breitfellner M, et al. 1998. *Astron. Astrophys.* 331:651–60

Dartois E, Schutte WA, Geballe TR, Demyk K, Ehrenfreund P, et al. 1999a. *Astron. Astrophys.* 342:L32–35

de Graauw T, Whittet DCB, Gerakines PA, Bauer OH, Beintema DA, et al. 1996. *Astron. Astrophys.* 315:L345–48

Dello Russo N, DiSanti MA, Mumma MJ, Magee-Sauer K, Rettig TW. 1998. *Icarus* 135:377–88

Dello Russo N, Mumma MJ, DiSanti MA, Magee-Sauer K, Novak R, Rettig TW. 2000. *Icarus* 143:324–337

Delsemme AH. 1982. In *Comets*, ed. LL Wilkening, pp. 85–130. Tucson: Univ. Ariz. Press

Demyk K, Dartois E, d'Hendecourt L, Jourdain de Muizon M, Heras MA, Breitfellner M. 1998. *Astron. Astrophys.* 339:553–60

Desert FX, Boulanger F, Puget JL. 1990. *Astron. Astrophys.* 237:215–36

Desert FX, Jenniskens P, Dennefeld M. 1995. *Astron. Astrophys.* 303:223–32

Despois D. 1992. In *The Astrochemistry of Cosmic Phenomena*, IAU Symposium 150, ed. PD Singh, pp. 451–58. Dordrecht: Kluwer

d'Hendecourt L, Ehrenfreund P. 1997. *Adv. Space Res.* 19:1023–32

d'Hendecourt L, Jourdain de Muizon M, Dartois E, Breitfellner M, Ehrenfreund P, et al. 1996. *Astron. Astrophys.* 315:L365–68

d'Hendecourt L, Léger A, Olofsson G, Schmidt W. 1986. *Astron. Astrophys.* 170:91–96

d'Hendecourt LB, Allamandola LJ, Greenberg JM. 1985. *Astron. Astrophys.* 152:130–50

di Brozzolo FR, Bunch TE, Fleming RH, Macklin J. 1994. *Nature* 369:37–40

DiSanti MA, Mumma MJ, Dello Russo N, Magee-Sauer K, Novak R, Rettig TW. 1999. *Nature* 399:662–665

DiSanti MA, Mumma MJ, Geballe TR, Davies JK. 1995. *Icarus* 116:1–17

Dorschner J, Henning T. 1995. *Astron. Astrophys. Rev.* 6(4):271–333

Douglas AE, Herzberg G. 1941. *Ap. J.* 94:381–81

Duari D, Cherchneff I, Willacy K. 1999. *Astron. Astrophys.* 341:L47–50

Duley WW. 1985. *MNRAS* 215:259–63

Duley WW, Seahra S. 1998. *Ap. J.* 507:874–88

Dutrey A, Guilloteau S, Guélin M. 1997. *Astron. Astrophys.* 31:L55–58

Dwek E. 1997. *Ap. J.* 484:779–84

Dwek E, Arendt RG, Fixsen DJ, Sodroski TJ, Odegard N, et al. 1997. *Ap. J.* 475:565–79

Eberhardt P. 1999. *Space Science Reviews* 90:45–52

Ehrenfreund P. 1999a. *Space Science Reviews* 90:233–38

Ehrenfreund P. 1999b. *Science* 283:1123–24

Ehrenfreund P. 2000. In *Cometary Nuclei in Space and Time*, ed. M A'Hearn. IAU Coll 168: ASP Conf. Series. In press

Ehrenfreund P, Boogert A, Gerakines P, Schutte WA, Thi W, et al. 1997a. In *First ISO Workshop on Analytical Spectroscopy*, 419:3–10. Noordwijk: ESA Publ. Div.

Ehrenfreund P, Boogert A, Gerakines P, Tielens AGGM. 1998a. *Faraday Discussions* 109:463–74. Cambridge, UK: R. Soc. Chem.

Ehrenfreund P, Cami J, Dartois E, Foing BH. 1997b. *Astron. Astrophys.* 318:L28–31

Ehrenfreund P, Dartois E, Demyk K, d'Hendecourt L. 1998b. *Astron. Astrophys.* 339:L17–20

Ehrenfreund P, d'Hendecourt LB, Dartois E, Jourdain de Muizon M, Breitfellner M, et al. 1997c. *Icarus* 130:1–15

Ehrenfreund P, Foing BH. 1996. *Astron. Astrophys.* 307:L25–28

Ehrenfreund P, Foing BH. 1997. *Adv. Space Res.* 19(7):1033–42

Ehrenfreund P, Kerkhof O, Schutte WA, Boogert ACA, Gerakines P, et al. 1999a. *Astron. Astrophys.* 350:240–53

Ehrenfreund P, Krafft K, Kochan H, Pir-

ronello V. 1999b. *Laboratory Astrophysics and Space Research.* Dordrecht: Kluwer. 687 pp

Ehrenfreund P, Robert F, d'Hendecourt L, Behar F. 1991. *Astron. Astrophys.* 252:712–17

Ehrenfreund P, Schutte WA. 2000a. In *Astrochemistry: From Molecular Clouds to Planetary Systems*, IAU Symposium 197, ed. YC Minh, EF van Dishoeck, pp. 135–46. Sogwipo: Astron. Soc. Pac.

Ehrenfreund P, Schutte WA. 2000b. *Adv. Space Res.* 25(11):2177–88

Engel S, Lunine JI, Lewis JS. 1990. *Icarus* 85:380–93

Enzian A. 1999. *Space Science Reviews* 90:131–39

Enzian A, Cabot H, Klinger J. 1998. *Planet. Space Sci.* 46:851–58

Fegley B. 1999. *Space Science Reviews* 90:239–52

Fegley B, Prinn RG. 1989. In *The Formation and Evolution of Planetary Systems*, ed. HA Weaver, L Danly, pp. 171–211. Cambridge: Cambridge University Press

Festou MC. 1999. *Space Science Reviews.* 90:53–67

Festou MC, Rickman H, West RM. 1993a. *Astron. Astrophys. Rev.* 4:363–447

Festou MC, Rickman H, West RM. 1993b. *Astron. Astrophys. Rev.* 5:37–163

Fitzpatrick EL, Massa D. 1986. *Ap. J.* 307:286–94

Foing BH, Ehrenfreund P. 1994. *Nature* 369:296–98

Foing BH, Ehrenfreund P. 1997. *Astron. Astrophys.* 317:L59–62

Fomenkova MN, 1999. *Space Science Reviews* 90:109–14

Fomenkova MN, Chang S. 1993. *Lun. Planet. Sci. Conf.* 24:501–2

Freivogel P, Fulara J, Maier JP. 1994. *Ap. J.* 431:L151–54

Frenklach M, Feigelson ED. 1989. *Astron. Astrophys.* 341:372–84

Fukuzawa K, Osamura Y, Schaeffer HF. 1998. *Ap. J.* 505:278–85

Fulara J, Jakobi M, Maier JP. 1993. *Chem. Phys. Lett.* 211:227–34

Gail HP, Sedlmayr E. 1987. In *Physical Processes in Interstellar Clouds*, ed. GE Morfill, M Scholer, pp. 275–303. Dordrecht: Kluwer

Galazutdinov GA, Krelowski J, Musaev FA. 1999. *MNRAS.* 310:1017–22

Garcia-Lario P, Manchado A, Ulla A, Manteiga M. 1999. *Ap. J.* 513:941–46

Gensheimer P. 1997. *Ap. J. Lett.* 479:L75–79

Gensheimer PD, Maursberger R, Wilson TL. 1996. *Astron. Astrophys.* 314:281–94

Gerakines PA, Schutte WA, Ehrenfreund P, van Dishoeck EF. 1996. *Astron. Astrophys.* 312:289–305

Gerakines PA, Whittet DCB, Ehrenfreund P, Boogert ACA, Tielens AGGM, et al. 1999. *Ap. J.* 522:357–77

Gibb E, Whittet DCB, Schutte WA, Chiar J, Ehrenfreund P, et al. 2000. *Ap. J.* 536:347–56

Gilmour I, Pillinger CT. 1994. *MNRAS* 269:235–40

Glassgold AE. 1996. *Annu. Rev. Astron. Astrophys.* 34:241–78

Goeres A, Sedlmayr E. 1992. *Astron. Astrophys.* 265:216–36

Goldsmith PF, Langer WD, Velusamy T. 1999. *Ap. J. Lett.* 519:L173–76

Gordon KD, Witt AN, Friedmann BC. 1998. *Ap. J.* 498:522–40

Gredel R, Lepp S, Dalgarno A, Herbst E. 1989. *Ap. J.* 347:289–293

Greenberg JM. 1982. In *Comets,* ed. LL Wilkening, pp. 131–62. Tucson: Univ. Ariz. Press

Greenberg JM. 1986. *Astrophys. Space Sci.* 128:17–29

Greenberg JM. 1998. *Astron. Astrophys.* 330:375–80

Greenberg JM. 1999. In *Formation and Evolution of Solids in Space*, ed. JM Greenberg, A Li, pp. 53–76. Dordrecht: Kluwer

Greenberg JM, Li A, Mendoza-Gomez C, Schutte WA, Gerakines P, de Groot M. 1995. *Ap. J.* 455:L177–80

Grevesse N, Noels A. 1993. In *Origin and Evolution of the Elements*, ed. N Prantzos, et al., pp. 15–25. Cambridge: Cambridge Univ. Press

Grim RJA, Greenberg JM. 1987. *Ap. J.* 321:L91–96

Guélin M, 2000. In *Astrochemistry: From Molecular Clouds to Planetary Systems*, IAU Symposium 197, ed. YC Minh, EF van Dishoeck, pp. 365–74. Sogwipo: Astron. Soc. Pac.

Guélin M, Cernicharo J, Travers MJ, McCarthy MC, Gottlieb CA, et al. 1997. *Astron. Astrophys.* 317:L1–4

Guélin M, Langer WD, Wilson RW. 1982. *Astron. Astrophys.* 107:107–127

Guélin M, Lucas R, Cernicharo J. 1993. *Astron. Astrophys.* 280:L19–22

Guélin M, Neininger N, Cernicharo J. 1998. *Astron. Astrophys.* 335:L1–4

Guillois O, Ledoux G, Reynaud C. 1999. *Ap. J. Lett.* 521:L133–36

Guillois O, Nenner I, Papoular R, Reynaud C. 1996. *Ap. J.* 464:810–17

Gürtler J, Henning TH, Koempe C, Pfau W, Kraetschmer W, et al. 1996. *Astron. Astrophys.* 315:L189–92

Hahn H. 1997. In *Nano Structured Materials*, Vol. 9, pp. 3–12. Amsterdam: Elsevier Sci. Ltd.

Hahn JH, Zenobi R, Bada J, Zare RN. 1988. *Science* 239:1523–25

Hanner M. 1999. *Space Science Reviews* 90:99–108

Hanner MS, Lynch DH, Russell RW. 1994. *Ap. J.* 425:274–85

Hasegawa T, Herbst E. 1993. *MNRAS* 261:83–102

Hasegawa T, Herbst E, Leung CM. 1992. *Ap. J. Suppl.* 82:167–95

Haser L. 1957. *Bull. Acad. R. Belg.* 43:740. Classe de Sciences

Helmich FP, van Dishoeck EF. 1997. *Astron. Astrophys. Suppl.* 124:205–53

Henning TH, Salama F. 1998. *Science* 282:2204–10

Henning TH, Schnaiter M. 1999. In *Laboratory*

Astrophysics and Space Research, ed. P Ehrenfreund, K Krafft, H Kochan, V Pirronello, pp. 249–78. Dordrecht: Kluwer

Henrard L, Lambin P, Lucas AA. 1997. *Ap. J.* 487:719–27

Herbig GH. 1975. *Ap. J.* 196:129–60

Herbig GH. 1993. *Ap. J.* 407:142–56

Herbig GH. 1995. *Annu. Rev. Astron. Astrophys.* 33:19–74

Herbig GH, McNally D. 1999. *MNRAS* 304:951–56

Herbst E. 1987. In *Interstellar Processes*, ed. DJ Hollenbach, HA Thronson, pp. 611–30. Dordrecht: Reidel

Herbst E. 1995. *Annu. Rev. Phys. Chem.* 46:27–53

Herbst E. 2000. In *Astrochemistry: From Molecular Clouds to Planetary Systems*, IAU Symposium 197, ed. YC Minh, EF van Dishoeck, pp. 147–59. Sogwipo: Astron. Soc. Pac.

Herbst E, Lee H-H, Howe DA, Millar TJ. 1994. *MNRAS* 268:335–49

Herbst E, Leung CM. 1989. *Ap. J.* 69:271–300

Herbst E, Leung CM. 1990. *Astron. Astrophys.* 233:177–80

Herlin N, Bohn I, Reynaud C, Cauchetier M, Galvez A, Rouzaud J. 1998. *Astron. Astrophys.* 330:1127–35

Herpin F, Cernicharo J. 2000. *Ap. J. Lett.* 530:L129–32

Hill HGM, Jones AP, d'Hendecourt LB. 1998. *Astron. Astrophys.* 336:L41–44

Hirahara Y, Suzuki H, Yamamoto S, Kawaguchi K, Kaifu N, et al. 1992. *Ap. J.* 394:539–51

Hiraoka K, Miyagoshi T, Takayama T, Yamamoto K, Kihara Y. 1998. *Ap. J.* 498:710–15

Hiraoka K, Ohashi N, Kihara Y, Yamamoto K, Sato T, et al. 1994. *Chem. Phys. Lett.* 229:408–14

Hirota T, Yamamoto S, Kawaguchi K, Sakamoto A, Ukita N. 1999. *Ap. J.* 520:895–900

Howe DA, Millar TJ. 1990. *MNRAS* 244:444–49

Hudgins DM, Allamandola LJ. 1999a. *Ap. J.* 516:L41–44

Hudgins DM, Allamandola LJ. 1999b. *Ap. J.* 513:L69–73

Huebner W, Benkhoff J. 1999. *Space Science Reviews* 90:117–30

Huebner WF, Boice DC, Korth A. 1989. *Adv. Space Res.* 9:29–34

Huebner WF, Boice DC, Sharp CM. 1987. *Ap. J.* 320:L149–52

Huntress WT. 1999. *Space Science Reviews* 90:329–40

Huntress WT, Mitchell GF. 1979. *Ap. J.* 231:456–67

Huss GR, Lewis RS. 1995. *Geochim. Cosmochim. Acta* 59:115–60

Ikeda M. 1998. An Observational Study of the Chemical Composition of Massive Star-Forming Regions. *Doctoral Thesis.* Grad. Univ. Adv. Studies. 104 pp.

Irvine WM. 1998. *Orig. Life Evol. Biosph.* 28:365–83

Irvine WM. 1999. *Space Science Reviews* 90:203–18

Irvine WM, Bergin EA, Dickens JE, Jewitt D, Lovell AJ, et al. 1998a. *Nature* 393:547–50

Irvine WM, Bockelée-Morvan D, Lis DC, Matthews HE, Biver N, et al. 1996. *Nature* 383:418–20

Irvine WM, Dickens JE, Lovell AJ, Schloerb FP, Senay M, et al. 1998b. *Faraday Discussions* 109:475–92. Cambridge, UK: R. Soc. Chem.

Irvine WM, Dickens JE, Lovell AJ, Schloerb FP, Senay M, et al. 1999. *Earth Moon Planets* 78:29–35

Irvine WM, Goldsmith PF, Hjalmarson AA. 1987. In *Interstellar Processes*, ed. DJ Hollenbach, HA Thronson, pp. 561–610. Dordrecht: Reidel

Irvine WM, Schloerb FP, Crovisier J, Fegley B, Mumma MJ. 2000. In *Protostars and Planets IV*, ed. V Mannings, A Boss, S Russell. Tucson: Univ. Ariz. Press. p. 1159

Jacq T, Walmsley CM, Maursberger R, Anderson T, Herbst E, De Lucia F. 1993. *Astron. Astrophys.* 271:276–81

Jaeger C, Mutschke H, Henning TH. 1998. *Astron. Astrophys.* 332:291–99

Jenniskens P, Baratta GA, Kouchi A, De Groot MS, Greenberg JM, Strazzulla G. 1993. *Astron. Astrophys.* 273:583–600

Jenniskens P, Desert X. 1994. *Astron. Astrophys. Suppl.* 106:39–78

Jenniskens P, Greenberg, JM. 1993. *Astron. Astrophys.* 274:439–50

Jessberger EK, Christoforidis A, Kissel J. 1988. *Nature* 332:691–95

Jewitt D, Luu J, Trujillo C. 1998. *Astron. J.* 115:2125–35

Joblin C, Léger A, Martin P. 1992. *Ap. J.* 393:L79–82

Jochims HW, Baumgartel H, Leach S. 1999. *Ap. J.* 512:500–10

Jones A, Tielens AGGM, Hollenbach DJ. 1996. *Ap. J.* 469:740–64

Jull AJT, Courtney C, Jeffrey DA, Beck JW. 1998. *Science* 279:366–68

Jura M, Kroto HW. 1990. *Astron. Astrophys.* 351:222–29

Kaiser RI, Roessler K. 1998. *Ap. J.* 503:959–75

Karpas Z, Mautner M. 1989. *J. Phys. Chem.* 93:1859–63

Keane J, Tielens AGGM, Boogert ACA, Schutte WA, Whittet DCB. 2000. *Astron. Astrophys.* Accepted

Keller HU, Delamere WA, Huebner WF, Reitsema HJ, Schmidt HU, et al. 1987. *Astron. Astrophys.* 187:807–23

Kerridge JF. 1999. *Space Science Reviews* 90:275–88

Kerridge JF, Chang S, Shipp R. 1987. *Geochim. Cosmochim. Acta.* 51:2527–40

Kerridge JF, Chang S. 1985. In *Protostars and Planets II*, ed. DC Black, MS Matthews, pp. 738–54. Tucson: Univ. Ariz. Press

Kissel J, Krueger FR. 1987. *Nature* 3(26)755–60

Kissel J, Krueger FR, Roessler K. 1997. In *Comets and the Origins of Life*, ed. PJ Thomas, CF Chyba, CP McKay, pp. 69–109. New York: Springer-Verlag

Klein R, Henning Th, Cesarsky D. 1999. *Astron. Astrophys.* 343:L53–56

Knacke RF, Brooke TY, Joyce RR. 1986. *20th ESLAB Symp.* 2:95–99

Kraetschmer W, Lamb L, Fostiropoulos K, Huffman DR. 1990. *Nature* 347:354–58

Krelowski J, Ehrenfreund, P, Foing B, Snow T, Weselak T, et al. 1999. *Astron. Astrophys.* 347:235–42

Krelowski J, Galazutdinov GA, Musaev FA. 1998. *Ap. J.* 493:217–21

Krelowski J, Schmidt M. 1997. *Ap. J.* 477:209–17

Krelowski J, Sneden C, Hiltgen D. 1995. *Planet. Space Sci.* 43:1195–203

Krishnamurthy RV, Epstein S, Cronin J, Pizzarello S, Yuen GU. 1992. *Geochim. Cosmochim. Acta.* 56:4045–58

Kroto HW, Heath JR, O'Brien SC, Curl RF, Smalley RE. 1985. *Nature* 318:162–63

Kroto HW, Jura M. 1992. *Astron. Astrophys.* 263:275–80

Kuan Y-J, Charnley SB, Wilson TL, Ohishi M, Huang HC, Snyder L. 1999. *BAAS* 194:942

Kuan Y-J, Mehringer DM, Snyder LE. 1996. *Ap. J.* 459:619–26

Kwok S, Volk K, Hrivnak BJ. 1999. *Astron. Astrophys.* 350:L35–38

Lacy JH, Faraji H, Sandford SA, Allamandola LJ. 1998. *Ap. J.* 501:L105–9

Lahuis F, van Dishoeck EF. 2000. *Astron. Astrophys.* 355:699–712

Langer WD, van Dishoeck EF, Bergin EA, Blake GA, Tielens AGGM, Velusamy T, Whittet DCB. 2000. In *Protostars and Planets IV*, ed. V Mannings, A Boss, S Russell. Tucson: Univ. Ariz. Press. p. 29

Langer WD, Velusamy T, Kuiper TB, Peng R, McCarthy MC, et al. 1997. *Ap. J. Lett.* 480:L63–66

Langevin Y, Kissel J, Bertaux JL, Chassefiere E. 1987. *Astron. Astrophys.* 187:761–66

Langhoff SR. 1996. *J. Phys. Chem.* 100:2819–41

Lawler ME, Brownlee DE. 1992. *Nature* 359:810–12

Le Coupanec P, Rouan D, Moutou C, Léger A. 1999. *Astron. Astrophys.* 347:669–75

Ledoux G, Ehbrecht M, Guillois O, Huisken

F, Kohn B, et al. 1998. *Astron. Astrophys.* 333:L39–42

Léger A, Jura M, Omont A. 1985. *Astron. Astrophys.* 144:147–60

Léger A, Puget JL. 1984. *Astron. Astrophys.* 137:L5–8

Lemke D, Mattila K, Lethinen K, Laureijs RJ, Liljestrom T, et al. 1998. *Astron. Astrophys.* 331:742–48

Lequeux J, Jourdain de Muizon M. 1990. *Astron. Astrophys.* 240:L19–22

Leung CM, Herbst, E, Huebner WF. 1984. *Astron. Astrophys. Suppl.* 56:231–56

Levasseur-Regourd AC, Hadamcik E, Renard JB. 1996. *Astron. Astrophys.* 313:327–33

Levison HF, Duncan MJ. 1997. *Icarus* 127:13–32

Levy EH, Lunine JI. 1993. *Protostars and Planets III.* Tucson: Univ. Ariz. Press. 1596 pages

Lewis RS, Ming T, Wacker JF, Anders E, Steel E. 1987. *Nature* 326:160–62

Li A, Greenberg JM. 1997. *Astron. Astrophys.* 323:566–84

Lis DC, Keene J, Young K, Phillips TG, Bockelée-Morvan D, et al. 1997. *Icarus* 130:355–72

Lucas R, Liszt HS. 1997. In *Molecules in Astrophysics: Probes and Processes*, ed. EF van Dishoeck, pp. 421–30. Dordrecht: Kluwer

Lucas R, Liszt HS. 2000. *Astron. Astrophys.* 358:1069–76

Lunine J, Engel S, Rizk B, Horanyi M. 1991. *Icarus* 94:333–44

Lunine JI, Owen TC, Brown RH. 2000. In *Protostars and Planets IV*, ed. V Mannings, A Boss, S Russell. Tucson: Univ. Ariz. Press. p. 1055

MacDonald GH, Gibb AG, Habing RJ, Millar TJ. 1996. *Astron. Astrophys. Suppl.* 119:333–67

MacKay DDSM, Charnley SB. 1999. *MNRAS* 302:793–800

Magee-Sauer K, Mumma MJ, DiSanti MA, Dello Russo N, Rettig TW. 1999. *Icarus* 142:498–508

Malfait K, Waelkens C, Waters LBFM, Van-denbussche B, Huygen E, et al. 1998. *Astron. Astrophys.* 332:L25–28

Mannings V, Boss A, Russell S. 2000. In *Protostars and Planets IV*, ed. V Manning, A Boss, S Russell. Tucson: Univ. Ariz. Press. 535:256–65

Markwick A, Millar TJ, Charnley SB. 2000. *Ap. J.* 535:256–65

Mathis JS. 1990. *Annu. Rev. Astron. Astrophys.* 28:37–70

Mathis JS. 1994. *Ap. J.* 422:176–86

Mathis JS. 1996. *Ap. J.* 472:643–55

Mathis JS, Metzger PG, Panagia N. 1983. *Astron. Astrophys.* 128:212–29

Mathis JS, Rumpl W, Nordsieck KH. 1977. *Ap. J.* 217:425–33

Matthews CN, Ludicky RA. 1992. *Adv. Space Res.* 12(4):21–32

Mattila K, Lemke D, Haikala LK, Laureijs RJ, Léger A, et al. 1996. *Astron. Astrophys.* 315:L353–56

Maurette M, Brack A, Kurat G, Perreau M, Engrand C. 1995. *Adv. Space Res.* 15:113–26

Mautner M, Karpas Z. 1986. *J. Phys. Chem.* 90:2206–10

McCarthy MC, Travers MJ, Kovacs A, Gottlieb CA, Thaddeus P. 1997. *Ap. J. Suppl.* 113:105–20

McCord TB, Hansen GB, Clark RN, Martin PD, Hibbitts CA, et al. 1997. *Science* 278:271–75

McKay DS, Gibson EK, Thomas-Keprta KL, Vali H, Romanek CS, et al. 1996. *Science* 273:924–30

McKellar A. 1940. *Publ. Astr. Soc. Pacific* 52:187–92

Meier R, Owen TC, Jewitt DC, Matthews HE, Senay M, et al. 1998. *Science* 279:1707–10

Melnick G. 2000. In *Astrochemistry: From Molecular Clouds to Planetary Systems*, IAU Symposium 197, ed. YC Minh, EF van Dishoeck, pp. 161–74. Sogwipo: Astron. Soc. Pac.

Mennella V, Brucato JR, Colangeli L, Palumbo P. 1999. *Ap. J.* 524:L71–74

Mennella V, Colangeli L, Bussoletti E, Palumbo

P, Rotundi A. 1998. *Ap. J.* 507:L177–80

Mennella V, Colangeli L, Palumbo P, Rotundi A, Schutte W, et al. 1996. *Ap. J.* 464:191–94

Messenger S, Amari S, Gao X, Walker RM, Clemett S, et al. 1998. *Ap. J.* 502:284–95

Miao Y, Mehringer DM, Kuan Y-J, Snyder LE. 1995. *Ap. J. Lett.* 445:L59–62

Millar TJ, Bennett A, Herbst E. 1989. *Ap. J.* 340:906–20

Millar TJ, Farquhar PRA, Willacy K. 1997. *Astron. Astrophys. Suppl.* 121:139–85

Millar TJ, Freeman A. 1984. *MNRAS* 207:405–23

Millar TJ, Hatchell J. 1998. *Faraday Discussions* 109:15–30. Cambridge, UK: R. Soc. Chem.

Millar TJ, Herbst E. 1994. *Astron. Astrophys.* 288:561–71

Millar TJ, Leung CM, Herbst E. 1987. *Astron. Astrophys.* 183:109–17

Millar TJ, Nejad LAM. 1985. *MNRAS* 217:507–22

Millar TJ, Roueff E, Charnley SB, Rodgers SD. 1995. *Intl. J. Mass Spec. Ion Proc.* 149:389–402

Minh YC, Ohishi M, Roh DG, Ishiguro M, Irvine WM. 1993. *Ap. J.* 411:773–77

Molster FJ, Yamamura I, Waters LBFM, Tielens AGGM, de Graauw Th. 1999. *Nature* 401:563–565

Moore MH, Hudson RL. 1992. *Ap. J.* 401:353–60

Moore MH, Hudson RL. 1998. *Icarus* 135:518–27

Moreels G, Clairemidi J, Hermine P, Brechignac P, Rousselott P. 1994. *Astron. Astrophys.* 282:643–56

Motylewski T, Linnartz H, Vaizert O, Maier JP, Galazutdinov GA. 2000. *Ap. J.* 531:312–20

Moutou C, Sellgren K, Verstraete L, Léger A. 1999. *Astron. Astrophys.* 347:949–56

Mumma MJ. 1997. In *From Stardust to Planetesimals*, ed. Y Pendleton, AGGM Tielens, 122:369–96. Provo, Utah: Astron. Soc. Pac.

Mumma MJ, DiSanti MA, Dello Russo N,

Fomenkova M, Magee-Sauer K, et al. 1996. *Science* 272:1310–14

Mumma MJ, Weissmann PR, Stern SA. 1993. In *Protostars and Planets III*, ed. EH Levy, JI Lunine, pp. 1177–252. Tucson: Univ. Ariz. Press

Munoz G, Mennella V, Ruiterkamp R, Schutte WA, Greenberg JM, et al. 2000. In *Thermal Emission Spectroscopy and Analysis of Dust, Disks, and Regolith*, ed. ML Sitko, AL Sprague, D Lynch, pp. 333–45. Houston: Astron. Soc. Pac.

Mutschke H, Andersen AC, Clement D, Henning Th, Peiter G. 1999. *Astron. Astropyhys.* 345:187–202

Nejad LAM, Millar TJ. 1988. *MNRAS* 230:79–86

Nercessian E, Omont A, Benayoun JJ, Guiloteau S. 1989. *Astron. Astrophys.* 210:225–35

Nuth JA. 1985. *Nature* 318:166–68

O'D Alexander CM, Russell SS, Areden JW, Ash RD, Grady MM, et al. 1998. *M&PS* 33:603–22

O'Donnell JE, Mathis JS. 1997. *Ap. J.* 479:806–17

Ohishi M, Kaifu N. 1998. *Faraday Discussion,* 109:205–16. Cambridge, UK: R. Soc. Chem.

Olano C, Walmsley CM, Wilson TL. 1988. *Astron. Astrophys.* 196:194–200

Omont A. 1993 *J. Chem. Soc. Faraday Trans.* 89:2137–45

Oro J, Lazcano A. 1997. In *Comets and the Origins of Life*, ed. PJ Thomas, CF Chyba, CP McKay, pp. 3–19. New York: Springer-Verlag

Ott U. 1993. *Nature* 364:25–33

O'Tuairisg S, Cami J, Foing BH, Sonnentrucker P, Ehrenfreund P. 2000. *Astron. Astrophys. Suppl.* 142:225–238

Owen T, Bar-Nun A. 1998. *Faraday Discussions* 109:453–62. Cambridge, UK: R. Soc. Chem.

Palumbo ME, Castorina AC, Strazzulla G. 1999. *Astron. Astrophys.* 342:551–62

Palumbo ME, Geballe TR, Tielens AGGM, et al. 1997. *Ap. J.* 479:839–44

Palumbo ME, Tielens AGGM, Tokunaga AT. 1995. *Ap. J.* 449:674–80

Papoular R, Conard J, Guillois O, Nenner I, Reynaud C, Rouzaud JN. 1996. *Astron. Astrophys.* 315:222–36

Peltzer ET, Bada JL, Schlesinger G, Miller SL. 1984. *Adv. Space Res.* 4:69–74

Pendleton Y. 1997. *Orig. Life Evol. Biosph.* 27:53–78

Pendleton Y, Sandford S, Allamandola LJ, Tielens AGGM, Sellgren K. 1994. *Ap. J.* 437:683–96

Pendleton YJ, Tielens AGGM, Tokunaga, AT, Bernstein MP. 1999. *Ap. J.* 513:294–304

Pickles J, Williams DA. 1977. *Astrophys. Space Sci.* 52:443–52

Pierazzo E, Chyba CF. 1999. *M&PS* 34(6)909–18

Prasad SS, Tarafdar SP. 1983. *Ap. J.* 267:603–9

Prasad SS, Tarafdar SP, Villere KR, Huntress WT. 1987. In *Interstellar Processes*, ed. DJ Hollenbach, HA Thronson, pp. 631–66. Dordrecht: Reidel

Pratap P, Dickens JE, Snell RL, Miralles MP, Bergin EA, et al. 1997. *Ap. J.* 486:862–85

Puget JL, Léger A. 1989. *Annu. Rev. Astron. Astrophys.* 27:161–98

Reuter DC. 1992. *Ap. J.* 386:330–335

Rodgers SD. 1998. *A Comparison of Cometary and Interstellar Ices.* PhD thesis. Univ. Manchester Inst. Sci. Tech. 210 pp.

Rodgers SD, Charnley SB. 1998. *Ap. J. Lett.* 501:L227–30

Romanini D, Biennier L, Salama F, Katchanov A, Allamandola LJ, et al. 1999. *Chem. Phys. Lett.* 303:165

Rotundi A, Rietmeijer FJM, Colangeli L, Mennella V, Palumbo P, Bussoletti E, et al. 1998. *Astron. Astrophys.* 329:1087–96

Ruffle DP, Bettens RP, Terzieva R, Herbst E. 1999. *Ap. J.* 523:678–82

Ruffle DP, Hartquist TW, Taylor SD, Williams DA. 1997. *MNRAS* 291:235–40

Sagan C, Khare BN, Thompson WR, McDonald GD, Wing MR, et al. 1993. *Ap. J.* 414:399–405

Sakata A, Wada S, Narisawa T, Asano Y, Iijima Y, et al. 1992. *Ap. J.* 393:L83–86

Salama F, Bakes ELO, Allamandola LJ, Tielens AGGM. 1996. *Ap. J.* 458:621–36

Salama F, Galazutdinov GA, Krelowski J, Allamandola LJ, Musaev FA. 1999. *Ap. J.* 526:265–73

Sandford S. 1996. *Meteoritics* 31:449–76

Sandford S, Allamandola LJ, Tielens AGGM, Sellgren K, Tapia M, Pendleton Y. 1991. *Ap. J.* 371:607–20

Sandford SA, Allamandola LJ. 1993. *Ap. J.* 417:815–25

Sandford SA, Allamandola LJ, Tielens AGGM, Valero LJ. 1988. *Ap. J.* 329:498–510

Sandford SA, Pendleton Y, Allamandola LJ. 1995. *Ap. J.* 440:697–705

Sarre PJ, Miles JR, Kerr TH, Hibbins RE, Fossey SJ, Somerville WB. 1995. *MNRAS* 277:L41–43

Scarrott SM, Watkin S, Miles JR, Sarre PJ. 1992 *MNRAS* 255:L11–16

Schmitt B. 1994 In *Molecules and Grains in Space,* ED. I Nenner, pp. 735–57 New York: AIP Press.

Schmidt HU, Wegmann R, Huebner WF, Boice DC. 1998 In *Comp. Phys. Comm.* 49:17–59

Schnaiter M, Henning TH, Mutschke H, Kohn B, Ehbrecht M, et al. 1999. *Ap. J.* 519:687–96

Schnaiter M, Mutschke H, Dorschner J, Henning TH, Salama F. 1998. *Ap. J.* 498:486–96

Schulte M, Shock E. 1995. *Orig. Life Evol. Biosph.* 25:161–73

Schutte WA. 1996. In *The Cosmic Dust Connection,* ed. JM Greenberg, pp. 1–42. Dordrecht: Kluwer

Schutte WA. 1999. In *Laboratory Astrophysics and Space Research,* ed. P Ehrenfreund, K Krafft, H Kochan, V Pironello, pp. 69–104. Dordrecht: Kluwer

Schutte WA, Allamandola LJ, Sandford SA. 1993. *Icarus* 104:118–37

Schutte WA, Boogert ACA, Tielens AGGM, Whittet DCB, Gerakines PA, et al. 1999. *Astron. Astrophys.* 343:966–76

Schutte WA, Greenberg JM. 1991. *Astron. Astrophys.* 244:190–204

Schutte WA, Greenberg JM. 1997. *Astron. Astrophys.* 317:L43–46

Schutte WA, Tielens AGGM, Whittet DCB, Boogert ACA, Ehrenfreund P, et al. 1996. *Astron. Astrophys.* 315:L333–36

Schutte WA, Van der Hucht KA, Whittet DCB, Boogert ACA, Tielens AGGM, et al. 1998. *Astron. Astrophys.* 337:261–74

Scott A, Duley WW, Pinho GP. 1997. *Ap. J.* 489:L193–95

Seahra SS, Duley WW. 1999. *Ap. J.* 520:719–23

Sedlmayr E. 1994. In *Molecules in the Stellar Environment*, ed. UG Jorgensen, p. 163. Berlin: Springer-Verlag

Sellgren K, Smith RG, Brooke TY. 1994. *Ap. J.* 433:179–86

Sephton MA, Pillinger CT, Gilmour I. 1998. *Geochim. Cosmochim. Acta.* 62:1821–28

Serra G, Chaudret B, Saillard Y, Le Beuze A, Rabaa H, et al. 1992. *Astron. Astrophys.* 260:489–93

Shalabiea O, Caselli P, Herbst E. 1998. *Ap. J.* 502:652–60

Sims IR, Queffelec J-L, Travers D, Rowe BR, Herbert LB, Karth J, Smith IWM. 1993. *Chem. Phys. Lett.* 211:461–68

Snow T. 1997. In *From Stardust to Planetesimals*, ed. Y Pendleton, AGGM Tielens, 122:147–57. Provo, Utah: Astron. Soc. Pac.

Snow T, Bakes ELO, Buss RH, Seab CG. 1995. *Astron. Astrophys.* 296:L37–40

Snow T, Witt A. 1995. *Science* 270:1455–60

Snow T, Witt A. 1996. *Ap. J. Lett.* 468:L65–68

Snyder L. 1997. *Orig. Life Evol. Biosph.* 27:115–33

Sonnentrucker P, Cami J, Ehrenfreund P, Foing BH. 1997. *Astron. Astrophys.* 327:1215–21

Sonnentrucker P, Foing BH, Breitfellner M, Ehrenfreund P. 1999. *Astron. Astrophys.* 346:936–46

Spaans M, Ehrenfreund P. 1999. In *Laboratory Astrophysics and Space Research*, ed.

P Ehrenfreund, K Krafft, H Kochan, V Pirronello, pp. 1–36. Dordrecht: Kluwer

Stecher TP, Donn B. 1965. *Ap. J.* 142:1681–83

Stephan T, Rost D, Jessberger EK, Greshake A. 1998. *M&PS* 32:A149–50

Stoks PG, Schwartz A. 1979. *Nature* 282:709–10

Stoks PG, Schwartz A. 1981. *Geochim. Cosmochim. Acta.* 45:563–69

Sutton EC, Peng R, Danchi WC, Jaminet PA, Sandell G, Russell APG. 1995. *Ap. J. Suppl.* 97:455–96

Swings P, Rosenfeld L. 1937. *Ap. J.* 86:483–88

Takano S, Masuda A, Hirahara Y, Suzuki H, Ohishi M, et al. 1998. *Astron. Astrophys.* 329:1156–69

Tegler SC, Weintraub DA, Rettig TW, Pendleton YJ, Whittet DCB, Kulesa CA. 1995. *Ap. J.* 439:279–87

Teixeira TC, Devlin JP, Buch V, Emerson JP. 1999. *Astron. Astrophys.* 347:L19–22

Thaddeus P, McCarthy MC, Travers MJ, Gottlieb CA, Chen W. 1998. *Faraday Discussions* 109:121–136. Cambridge, UK: R. Soc. Chem.

Tielens AGGM. 1983. *Astron. Astrophys.* 119:177–84

Tielens AGGM. 1998. *Ap. J.* 499:267–72

Tielens AGGM, Charnley SB. 1997. *Orig. Life Evol. Biosph.* 27:23–51

Tielens AGGM, Hagen W. 1982. *Astron. Astrophys.* 114:245–60

Tielens AGGM, Hony S, van Kerckhoven C, Peeters E. 1999. In *The Universe as seen by ISO*, 427:579–88. Noordwijk: ESA Publ. Div.

Tielens AGGM, Seab CG, Hollenbach DJ, McKee CK. 1987. *Ap. J. Lett.* 319:103–107

Tielens AGGM, Tokunaga AT, Geballe TR, Baas F. 1991. *Ap. J.* 381:181–99

Tielens AGGM, Whittet DCB. 1997. In *Molecules in Astrophysics: Probes and Processes*, ed. EF van Dishoeck, pp. 45–60. Dordrecht: Kluwer

Tulej M, Kirkwood DA, Pachkov M, Maier JP. 1998. *Ap. J.* 506:L69–73

Turner BE. 1990. *Ap. J. Lett.* 362:L29–32

Ugarte D. 1992. *Nature* 359:707–9

Ugarte D. 1995. *Ap. J.* 443:L85–88

Vaidyia DB, Gupta R. 1999. *Astron. Astrophys.* 348:594–99

Vandenbussche B, Ehrenfreund P, Boogert ACA, van Dishoeck EF, Schutte WA, et al. 1999. *Astron. Astrophys.* 346:L57–60

van Dishoeck E. 1998. *Faraday Discussions* 109:31–46. Cambridge, UK: R. Soc. Chem.

van Dishoeck EF, Blake GA. 1998. *Annu. Rev. Astron. Astrophys.* 36:317–68

van Dishoeck EF, Blake GA, Draine BT, Lunine JI. 1993. In *Protostars and Planets III*, ed. EH Levy, JI Lunine, pp. 163–244. Tucson: Univ. Ariz. Press

van Dishoeck EF, Helmich FP, de Graauw T, Black JH, Boogert ACA, et al. 1996. *Astron. Astrophys.* 315:L349–52

van Dishoeck EF, van der Tak FS. 2000. In *Astrochemistry: From Molecular Clouds to Planetary Systems*, IAU Symposium 197, ed. YC Minh, EF van Dishoeck, pp. 97–112. Sogwipo: Astron. Soc. Pac.

van Kerckhoven C, Hony S, Peeters E, Tielens AGGM, Allamandola LJ, et al. 2000. *Astron. Astrophys.* 357:1013–19

Verstraete L, Puget JL, Falgarone E, Drapatz S, Wright CM, Timmermann R. 1996. *Astron. Astrophys.* 315:L337–40

Voors RHM. 1999. *Infrared Studies of Hot Stars with Dust*. PhD thesis. Univ. Utrecht. 190 pages

Voors RHM, Waters LBFM, Morris PW, Trams NR, de Koter A, Bouwman J. 1999. *Astron. Astrophys.* 341:L67–70

Wada S, Kaito C, Kimura S, Ono H, Tokunaga AT. 1999. *Astron. Astrophys.* 345:259–64

Waelkens C, Waters LBFM, de Graauw MS, Huygen E, Malfait K, et al. 1996. *Astron. Astrophys.* 315:L245–48

Wagenblast R, Williams DA, Millar TJ, Nejad LAM. 1993. *MNRAS* 260:420–24

Waters LBFM, Molster FJ, de Jong T, Beintema DA, Waelkens C, et al. 1996. *Astron. Astrophys.* 315:L361–64

Wdowiak TJ, Wei L, Cronin J, Beegle LW, Robinson MS. 1995. *Planet. Space Sci.* 43:1175–82

Weaver HA, Danly L. 1989. *The Formation and Evolution of Planetary Systems*. Cambridge: Cambridge Univ. Press

Weaver H, Lamy P. 1999. *Earth, Moon, and Planets* 79:17–33

Webster A. 1993. *MNRAS* 264:L1–2

Webster A. 1995. In *The Diffuse Interstellar Bands*, ed. A Tielens, T Snow, pp. 349–58. Dordrecht: Kluwer

Weissmann P. 1999. *Space Science Reviews* 90:301–11

Weissmann PR. 1995. *Science* 269:1120

Whipple FL. 1950a. *Ap. J.* 111:375–94

Whipple FL. 1950b. *Ap. J.* 113:464–74

Whittet DCB. 1993. In *Dust and Chemistry in Astronomy*, ed. TJ Millar, DA Williams, pp. 9–35. Bristol: IOP Publ. Ltd.

Whittet DCB, Gerakines PA, Tielens AGGM, Adamson AJ, Boogert ACA, et al. 1998. *Ap. J.* 498:L159–63

Whittet DCB, Schutte WA, Tielens AGGM, Boogert ACA, de Graauw T, et al. 1996. *Astron. Astrophys.* 315:L357–60

Willacy K, Millar TJ. 1997. *Astron. Astrophys.* 324:237–48

Willacy K, Millar TJ. 1998. *MNRAS* 298:562–68

Witt AN, Gordon KD, Furton DG. 1998. *Ap. J.* 501:L111–15

Winnewisser G, Herbst E. 1993. *Rep. Prog. Phys.* 56:1209–73

Winnewisser G, Kramer C. 1999. *Space Science Reviews* 90:181–202

Wright EL. 1988. *Nature* 336:227–28

Wright GS, Geballe T, Bridger A, Pendleton YJ. 1996. In *Changing Perceptions of the Morphology, Dust Content, and Dust-Gas Ratios in Galaxies*, ed. DL Block, JM Greenberg, pp. 143–150. Dordrecht: Kluwer

Wood JA, Chang S. 1985. In *The Cosmic History of the Biogenic Elements and Compounds*. Washington, DC: Sci. & Tech. Info. Branch, NASA SP-476.

Wooden DH, Harker DE, Woodward CE,

Butner HM, Koike C, Witteborn FC, Mc-Murthy CW. 1999. *Ap. J.* 517:1034–58

Wyckoff S, Heyd RS, Fox R. 1999. *Ap. J. Lett.* 512 L73–76

Wyrowski F, Schilke P, Walmsley CM, Menten KM. 1999. *Ap. J. Lett.* 514:43–46

Zinner E. 1988. In *Meteorites and the Early Solar System*, ed. J Kerridge, M Shapley Mathews, pp. 956–83. Tucson: Univ. Ariz. Press

Zinner E. 1998. *Annu. Rev. Earth Planet. Sci.* 26:147–88

Zubko VG, Smith TL, Witt AN. 1999. *Ap. J.* 511:L57–60

Annu. Rev. Astron. Astrophys. 2000. 38:485–519

OBSERVATIONS OF BROWN DWARFS

Gibor Basri

*Astronomy Dept. (MC 3411), University of California, Berkeley,
California 94720; e-mail: basri@soleil.berkeley.edu*

Key Words   substellar, low-mass stars, mass function, binaries, young clusters

■ **Abstract**   The brown dwarfs occupy the gap between the least massive star and the most massive planet. They begin as dimly stellar in appearance and experience fusion (of at least deuterium) in their interiors. But they are never able to stabilize their luminosity or temperature and grow ever fainter and cooler with time. For that reason, they can be viewed as a constituent of baryonic "dark matter." Indeed, we currently have a hard time directly seeing an old brown dwarf beyond 100 pc. After 20 years of searching and false starts, the first confirmed brown dwarfs were announced in 1995. This was due to a combination of increased sensitivity, better search strategies, and new means of distinguishing substellar from stellar objects. Since then, a great deal of progress has been made on the observational front. We are now in a position to say a substantial amount about actual brown dwarfs. We have a rough idea of how many of them occur as solitary objects and how many are found in binary systems. We have obtained the first glimpse of atmospheres intermediate in temperature between stars and planets, in which dust formation is a crucial process. This has led to the proposal of the first new spectral classes in several decades and the need for new diagnostics for classification and setting the temperature scale. The first hints on the substellar mass function are in hand, although all current masses depend on models. It appears that numerically, brown dwarfs may well be almost as common as stars (though they appear not to contain a dynamically interesting amount of mass).

1.   INTRODUCTION

The least massive star has 75 times the mass of Jupiter. What about objects of intermediate mass? What are their properties and how do they compare with those of stars and planets? How many of these objects are there? These questions take us into the realm of the newly discovered "brown dwarfs." Although theories discussing such objects go back to Kumar (1963) the quest for an observation of an incontrovertible brown dwarf was frustrating. There was a series of proposed candidates over a 20-year period, each of which failed further confirmation. There were several unrelated breakthroughs in 1995, followed rapidly by detection of many further convincing cases. By now the number of truly confirmed brown dwarfs has passed 20, with over 100 very likely detections. There have been several

0066-4146/00/0915-0485$14.00

recent conferences and workshops whose proceedings contain valuable reviews on this and related topics. Of particular note is *Brown Dwarfs and Extrasolar Planets* (Rebolo et al 1998a) and *From Giant Planets to Cool Stars* (Griffith & Marley 2000). Other reviews that are useful to consult are Allard et al (1997), Jameson & Hodgkin (1997), Kulkarni (1997), and Oppenheimer et al (2000).

1.1 What Is a Brown Dwarf?

Before we follow the story of discovery, let us sharpen the definitions of "star," "brown dwarf" (BD), and "planet." The defining characteristic of a star is that it will stabilize its luminosity for a period of time by hydrogen burning. A star derives 100% of its luminosity from fusion during the main sequence phase, whereas the highest-mass BD always has gravitational contraction as at least a small part of its luminosity source. The BD is brightest when it is born and continuously dims and cools (at the surface) after that. There can be some hydrogen fusion in the higher-mass BDs, and all objects down to about 13 Jupiter masses (jupiters) will at least fuse deuterium (Saumon et al 1996). The lower-mass limit of the main sequence lies at about 0.072 times the mass of the Sun (or 75 jupiters) for an object with solar composition. The limit is larger for objects with lower metallicity, reaching about 90 jupiters for zero metallicity (Saumon et al 1994). I refer you for details to the article by G. Chabrier in this volume that describes the theory of the structure and evolution of these objects.

Amazingly, astronomers are currently somewhat undecided on just how to define "planet." At the low-mass end of planets, an example of the difficulty is provided by the recent controversy over Pluto. At the high-mass end of planets, we are now aware of extrasolar "giant planets" (Marcy & Butler 1998) ranging up to more than 10 jupiters. At what point in mass should these be more properly called brown dwarfs? The traditional line of thinking holds that brown dwarfs form like stars—through direct collapse of an interstellar cloud into a self-luminous object. As this object forms, the material with higher angular momentum will settle into a disk of gas and dust around it. The dust in the disk can coagulate into planetesimals (kilometers in radius), and these can crash together to eventually form rock/ice cores. When a core reaches 10–15 earth masses, and if the gas disk is still present, it can begin to rapidly attract the gas and build up to a gas giant planet. Because of the nature of this process, one naïvely expects the planet to be in an almost circular orbit. The layout of our solar system also suggests that a massive enough core can only be produced if icy planetesimals are widely available, which occurs at about the distance of Jupiter (the "ice boundary").

This traditional picture (based on our own solar system) has been seriously challenged by the discovery of the extrasolar planetary systems. All of these that are not tidally circularized by being too close to the star have eccentric orbits. They are all inside the ice boundary (though this is largely an observational selection effect). Some are very close to the star, where formation of a giant planet seems nearly impossible. These facts led Black (1997) to claim that most of the extrasolar planets found so far are really BDs, because objects found by Doppler searches only have lower limits on their masses. Such a claim is unsupportable because of the statistics of these lower limits (Marcy & Butler 1998); if they reflected a population of BDs, then many others would show up closer to their true masses because of the random inclination of orbits, and they would be even easier to detect. It is possible that neither the size nor shape of the orbits reflect their initial values. This makes it difficult to distinguish giant planets from BDs on an orbital basis.

Stars often form with a companion. This process involves formation in disks (both cirmcumstellar and circumbinary), as does the formation of a lone star. Binary star formation leads to companions at any separation, with eccentric orbits. The difference between the formation of binary stellar companions and planets is thought to be the lack of a need for stars or BDs to first form a rock/ice core. Unfortunately, there is no current method for determining whether there was such an initial core in extrasolar objects. Thus, formation in a disk does not by itself distinguish star from planet formation, and apparently neither do orbital eccentricity or separation. It is possible that giant planets form both by gas accretion onto a rocky core and by more direct forms of gravitational collapse in gaseous disks (Boss 1997). Even the requirement that a planet be found orbiting a star is now thought overly restrictive; when several giant planets form in a system, it is easy for one or more to be ejected by orbital interactions and end up freely floating. For a much more detailed discussion of formation issues, see *Protostars and Protoplanets IV* (Mannings et al 2000).

Given these difficult issues, there is a rising school of thought that the definition of brown dwarfs should have a basis more similar to the definition of stars (based on interior physics). One intuitive difference between stars and planets is that stars experience nuclear fusion, whereas planets do not. We can therefore define the lower mass limit for BDs on that basis. Because significant deuterium fusion does not occur below 13 jupiters (Saumon et al 1996), that is the proposed lower mass limit for BDs. It is also thought to be near the lower limit for direct collapse of an interstellar cloud. With this definition, one must only determine its mass to classify an object. We can avoid the observational and theoretical uncertainties associated with a formation-based definition by using the mass-based definition, and that is what I advocate. Nonetheless, there actually is some evidence that most planets and most BDs form by different mechanisms. It is much more probable to find a planet rather than a BD as a companion to a solar-type star (see Section 5.2). It may turn out that substellar objects form in more than one way, but at least we'll know what to call them.

2.   THE SEARCH FOR BROWN DWARFS

The search for brown dwarfs can be divided into three qualitatively different arenas. The most obvious is the search for old visible BDs, whose temperature and luminosity obviously lie below the minimum possible value for stars. This can be done by looking for stellar companions, or looking in the field. The second search arena is for dynamical BDs, where orbital information suggests a mass below the minimum stellar mass. Here, we don't need to actually see the BD; its effect on a stellar companion can reveal it. If the orbit is spatially resolved, the actual masses of the components can be found; otherwise, only lower limits can be placed because of the unknown orbital inclination.

The third arena involves searching for young BDs, which are visible and at their brightest. Although these are the easiest to see, it is more difficult to verify that they are really BDs. At early ages, the BDs occupy the same region of temperature and luminosity as very low-mass stars (VLMS). One can trade off mass and age to infer either a BD or a VLMS at a given observed value of luminosity or temperature. For isolated objects in the field, this is a particularly acute problem because their age is not generally known. Even in a cluster, the mere fact that an object occupies a position in an HR or color-magnitude diagram, where theory tells us to expect BDs at the age of the cluster, has not proved convincing by itself. This is partly because the theory that converts observational quantities to mass is still being refined, and partly because other factors may invalidate the conclusion. Among these factors is the possibility that the object may not actually be a member of the cluster, or that the age of the cluster may have a large spread or may not have been correctly determined.

2.1 A Brief History of the Searches

A review of early observational efforts can be found in Oppenheimer et al (2000). One of the first efforts to directly image BDs as companions to nearby stars was made by McCarthy et al (1985). Using an infrared speckle technique, they reported a companion to VB8, with inferred properties that would guarantee its substellar status. This was the highlight of the first conference on brown dwarfs (Kafatos 1986). Unfortunately, their result was never confirmed. Later surveys (e.g. Skrutskie et al 1989, Henry & McCarthy 1990) did not find good BD candidates (but did find several VLMS companions). In a survey of white dwarfs, Becklin & Zuckerman (1988) turned up a very red and faint companion, GD 165B, whose spectrum was quite enigmatic. Kirkpatrick et al (1999a) argue that this is probably a BD.

The next good candidate came from a radial velocity survey. Latham et al (1989) were conducting a survey of about 1,000 stars with 0.5 km s^{-1} precision. Among their roughly 20 radial velocity standards, HD 114762 exhibited periodic variations just at the limit of detectability. This orbit has been confirmed by the precision radial velocity groups and implies a lower mass limit for the companion of about 11 jupiters. The difficulty is that the orbital inclination is not known. It would not be too surprising to find a very low inclination stellar companion in a sample of 1,000 stars, but much more surprising in a sample of 20. This argument remains unsettled, though subsequent surveys have shown a real dearth of companions to solar-type stars in the BD mass range (see Sections 5.2, 6.2). Until the actual orbital inclination for this object is measured (by a space interferometer?), it must remain unconfirmed but tantalizing.

During the early 1990s, there were a number of surveys aimed at finding BDs in young clusters. Forrest et al (1989) announced a number of candidates in Taurus-Aurigae, which were later shown to be background giants (Stauffer et al 1991). Surveys of star-forming regions (e.g. Williams et al 1995) also found objects that might well be substellar, but there is no obvious way to confirm them. Hambly et al (1993; HHJ) conducted a deep proper motion survey of the Pleiades and found a number of objects that models suggested should be substellar. Stauffer et al (1994) were also conducting a survey for BDs in this cluster, working from color-magnitude diagrams. Both surveys went substantially deeper than before and uncovered interesting objects. This set the stage for the next (ultimately successful) effort to find cluster BDs. Nonetheless, we should remember that at the ESO Munich conference on "The Bottom of the Main Sequence—and Beyond" (Tinney 1994), there was a palpable sense of frustration at the failure of many efforts to confirm a single BD.

Working from the new Pleiades lists, Basri and collaborators were finally able to announce at the June 1995 meeting of the AAS (Science News 147, p. 389) the first successful application of the lithium test for substellarity (Section 3.1). This was the first public declaration of a BD that is currently still solid. The object, PPl 15, would have an inferred mass well below the substellar limit, except that concurrently the age of the Pleiades was revised substantially upward (Section 3.2.1). This moved the mass of PPl 15 just under the substellar limit. Along with community unfamiliarity with the lithium test, this delayed acceptance of PPl 15 as a true BD (though there is no question about it now; see Section 5.3). Given this fact, any fainter Pleiades members should automatically be BDs. In September, Rebolo et al (1995) announced the discovery of such an object: Teide 1. Any remaining doubt could be removed by confirming lithium in it; this was accomplished by Rebolo et al (1996). These two objects are now accepted as undeniable BDs (along with many subsequently discovered faint Pleiades members). Their masses are in the 55–70 jupiter range.

2.2   The First Incontrovertible Brown Dwarf: Gl 229B

Only a month after the publication of Teide 1, any debate over the existence of brown dwarfs was ended by the announcement in Florence (at the Tenth Cambridge Cool Stars Workshop) of the discovery of a very faint companion to a nearby M star. Its temperature and luminosity are well below the minimum main sequence values. With the additional revelation at the same session of the first extrasolar planet, it was suddenly very clear that Nature has no problem manufacturing substellar objects.

Gl 229B was found in a coronographic survey of nearby low-mass stars (Naka-jima et al 1995). The survey was originally chosen to be biased toward younger M stars (though not strictly so). It ended up as a complete survey of stars to 8pc (almost 200 targets; Oppenheimer 1999). Of these, only Gl 229 shows a substellar companion. The companion was first detected in 1994, but the group showed commendable forbearance in waiting for proper motion confirmation that it was physically associated with the primary (allowing the known parallax of the primary to be applied to find its luminosity). They also obtained a spectrum that confirmed the remarkably low temperature implied by its luminosity (Oppenheimer et al 1995). In particular, the spectrum contains methane bands at 2 microns—features that had previously been detected only in planetary atmospheres and that are not expected in any main sequence star.

The mass of Gl 229B is still somewhat uncertain. Its large separation from the primary means we will have to wait a few decades to find a dynamical mass from the orbit. The primary, though a member of the young disk population kinematically, is not a particularly active star. The uncertainty in age translates directly to a possible mass range. There is only a weak constraint on the gravity from atmospheric diagnostics. The allowed mass is from about 20–50 jupiters; 40 jupiters is a reasonable value to take for now (given the inactivity of the primary, which implies an older age). A number of BDs have been found since that have masses lower than Gl 229B, which is distinguished by being the coolest BD (and therefore the oldest). This was a watershed discovery in the search for BDs; the next example of a similar object was not found until 1999.

3.   DISTINGUISHING YOUNG BROWN DWARFS FROM STARS

Stars and BDs can have identical temperatures and luminosities when they are young (though the star would have to be older than the BD). "Young" in this context extends up to several gigayears. We therefore require a more direct test of the substellar status of a young BD candidate before it can be certified. Because the difference between BDs and VLMS lies in the nuclear behavior of their cores, it is natural to look for a nuclear test of substellarity. For this we can use a straightforward diagnostic that is fairly simple, both theoretically and observationally: the "lithium test." In addition to verifying substellar status, observations of lithium can be used to assess the age of stars in clusters, which is helpful in the application of the lithium test itself. Lithium observations of very cool objects can be useful in constraining the nature of BD candidates in clusters, in the field, and in star-forming regions.

3.1   The Principle of the Lithium Test

In simplest terms, stars will burn lithium in a little over 100 Myr (megayears) at most, whereas most BDs will never reach the core temperature required to do so. This stems from the fact that even before hydrogen burning commences, core temperatures in a star reach values that cause lithium to be destroyed. On the other hand, in most BDs the requisite core temperature is never reached because of core degeneracy (see Chabrier, this volume). Furthermore, at masses near and below the substellar boundary the objects are all fully convective, so that surface material is efficiently mixed to the core. Finally, the surface temperatures of young candidates are favorable for observation of the neutral lithium resonance line, which is strong and occurs in the red. Some subtleties should be considered in the application of the test, as discussed later in this section. A more comprehensive review of this subject is provided by Basri (1998a).

The idea behind the lithium test was implicit in calculations of the central temperature of low-mass objects by D'Antona & Mazzitelli (1985) and others. They found that the minimum lithium burning temperature was never reached in the cores of objects below about 60 jupiters. On the other hand, all M stars on the main sequence are observed to have destroyed their lithium. The first formal proposal to use lithium to distinguish between substellar and stellar objects was made by Rebolo et al (1992). This induced Nelson et al (1993) to provide more explicit calculations useful in the application of the lithium test.

The theory of lithium depletion in VLM objects is comparatively simple. Because the objects are fully convective, their central temperature is simply related to their luminosity evolution. The semi-analytic study of lithium depletion by Bildsten et al (1997) is a particularly revealing exposition of the heart of the problem. The physical complications in VLM objects, including partially degenerate equations of state and very complicated surface opacities, do not obscure the basic relation between the effective temperature and lithium depletion. The complications of mixing theory, which lead to many fascinating effects in the observations of surface lithium in higher-mass stars, are simply not relevant for fully convective objects.

Pavlenko et al (1995) studied lithium line formation at very cool temperatures. Their basic result, that the lithium line should be quite strong in the 1,500–3,000 K range, is confirmed by observations. NLTE effects and the effect of chromospheric activity have been considered by them and by Stuik et al (1997) and found to be of secondary importance. The strength of the resonance line means that it does not begin to desaturate until more than 90% of the initial lithium has been depleted. The timescale over which the lithium line disappears is about 10 Myr, which is roughly 10% of the age at which it occurs in substellar objects. However, the observational disappearance of the line occurs even more rapidly (after desaturation).

Based on the clear possibility of using the lithium test to confirm substellar status, the group at the IAC embarked on an effort to apply it to the best existing BD candidates. They used 4-m class telescopes at spectral resolutions of 0.05 nm, for a brighter initial sample (Magazzù et al 1993), and 0.2–0.4 nm (Martín et al 1995). This latter resolution is lower than ideal, but the observations are very difficult owing to the faintness of VLM objects. The group was unable to detect lithium in any of the candidates. For most targets (since the ages are unknown), this implied a lower mass limit greater than 60 jupiters but did not resolve the question of whether they are BDs.

The results were puzzling for their Pleiades candidates. These were drawn from the Hambly et al (1993) list of very faint proper motion objects, and those authors had already suggested BD candidacy based on the color-magnitude position of the objects compared to evolutionary tracks for the age of the Pleiades (thought to be 70 Myr). Martín et al (1995) realized that there was an inconsistency between the inferred mass of these Pleiades members and the lack of lithium. The situation was even more striking in the results of Marcy et al (1994), who observed, using the newly commissioned Keck 10-m telescope, a yet fainter Pleiades member (HHJ 3) with better upper limits on the lithium line.

3.2   The Lithium Test in Young Clusters

The first application of the lithium test to a BD candidate with a positive result came in the study of PPL 15 by Basri et al (1996). PPL 15 is an object only slightly fainter than HHJ 3, and was the faintest known Pleiades member at the time of the study. Basri et al reported a detection of the lithium line, but apparently weaker than expected for undepleted lithium in an M6.5 star (based on high-resolution model spectra). At the same time, they confirmed that PPL 15 had the right radial velocity and Hα strength to be a cluster member [it was discovered by Stauffer et al (1994) in a photometric, rather than proper motion, survey]. More recently, Hambly et al (1999) have also confirmed that it is a proper motion member of the cluster.

To explain how lithium could appear in PPL 15 but not in HHJ 3, Basri et al used an empirical bolometric correction to convert to luminosity. The solution becomes apparent in a luminosity-age diagram, with the lithium depletion region displayed (e.g. Figure 1, see color insert). This shows that the lithium test is more subtle than was presented above. One wrinkle is that it takes stars a finite amount of time to deplete their lithium. Thus, if an object is sufficiently young, it will show lithium despite having a mass above the hydrogen-burning limit (giving the possibility of a false positive in the test). On the other hand, the minimum mass for lithium destruction is below the minimum mass for stable hydrogen burning. Thus, if we wait long enough, the high-mass BDs will deplete their lithium too (giving the possibility of a false negative in the test).

Basri et al resolved the problem of the non-detection of lithium in HHJ 3 and its presence in PPL 15 by suggesting that the Pleiades is substantially older than was previously thought. They showed that with an age of 115 Myr (rather than the classical age of 70 Myr), the behavior of both stars makes sense. The inferred mass of VLM Pleiades members is thereby raised (since they have longer to cool to the observed temperatures), with PPL 15 just about at the substellar boundary. The prediction was that any cluster members that are fainter than PPL 15 would show strong lithium.

This prediction was tested in short order on Teide 1, a fainter M8 Pleiades member with apparently good cluster membership credentials. Field M8 stars are quite unlikely to be young enough to show lithium. Rebolo et al (1996) used the Keck telescope to confirm strong lithium in both Teide 1 and a very similar object (Calar 3). Because these are well below PPL 15 in luminosity, they must be considered ironclad BDs in the cluster. They have masses in the range of 55–60 jupiters (given the new age for the cluster; they would be substantially lower using the classical age).

3.2.1 The Age Scale for Young Clusters The work of Basri et al (1996) suggested that a new method of determining ages of clusters has been found: lithium dating. Stauffer et al (1999b) pursued such a program for the Pleiades and obtained very clear confirmation of the lithium boundary found by Basri et al. They agreed that the explanation is that the cluster is more than 50% older (125 Myr) than its classical age. Further progress has occurred for several clusters. Basri & Martín (1999) found lithium in a (previously known) member of the α Per cluster and determined that the classical age of α Per should be corrected substantially upward. More objects were needed to pin down the lithium boundary, and Stauffer et al (1999a) provide them. They conclude that the age of α Per is about 85 Myr, rather than the classical age of 50 Myr (a similar correction as in the Pleiades). Barrado y Navascués et al (1999) also find that the younger cluster IC 2391 needs a correction of less than 50% to its classical age (50 Myr old instead of 35 Myr) on the basis of lithium dating.

Lithium dating is fundamentally a nuclear age calibrator. In that sense, it is like the upper-main sequence turnoff age, which is the "classical" means of assessing cluster ages. There is good reason to regard the lithium ages as more reliable than the classical method for young clusters. This is because the stars turning off the main sequence in young clusters are massive enough that they have convective nuclear burning cores. The issue of convective overshoot is then quite crucial—the more there is, the more hydrogen from the convectively stable envelope that can be enlisted into the main sequence phase. This increases the main sequence lifetime of the star, and thus the age inferred from the turnoff. Stellar evolution theory had already been grappling with this problem; a review of the topic in this context can be found in Basri (1998b). The treatment of convective overshoot is quite uncertain, and the problem must be inverted to find observational constraints to what is otherwise an essentially free parameter.

In lithium dating, on the other hand, the details of convection are rendered unimportant by the fully convective nature of the objects (which are then forced to adiabatic temperature gradients). The precision of lithium dating is limited by the width of the depletion boundary, errors in the conversion of magnitudes to luminosities (owing to bolometric corrections and cluster distances), and possible corrections to the age scale because of opacity issues in very cool objects. But it probably has similar precision to, and greater accuracy than, classical dating methods. Indeed, this may prove one of the most powerful methods to finally provide a value for the convective overshoot in high-mass stars. Lithium dating can only work up to about 200 Myr, when the lowest-mass object that can deplete lithium will have done so. Furthermore, the correction for core convective overshoot can only apply for clusters younger than about 2 Gyr; stars leaving the main sequence in older clusters have radiative cores.

As a cluster gets older, the luminosity of the lithium depletion boundary gets fainter. Thus, while the Hyades is one-third the distance of the Pleiades, its lithium boundary is at fainter apparent magnitudes. Searches for BDs here have been less successful (cf. Reid & Hawley 1999). Although α Per is farther away, its youth means that the apparent magnitude of the lithium boundary is similar to the Pleiades. Given a correct age, the luminosity of the substellar boundary can then be inferred from models. This will not be coincident with the depletion boundary in general (only at the age of the Pleiades). Once the boundary is established, the search for BDs can proceed to fainter objects using cluster membership as the sole criterion.

3.3   The Lithium Test in the Field

Can the lithium test be used for field objects, given that one generally does not know the age of an object? Clearly it works to distinguish main sequence M stars from BDs less massive than 60 jupiters (that was the original idea). Basri (1998a) refined the discussion of how to apply the lithium test in the field. Figure 1 shows that the lithium depletion region, taken with the observed luminosity or temperature of the object, provides a lower bound to the mass and age (jointly) if lithium is not seen. Conversely, it provides an upper bound to the mass and age if lithium is seen. The temperature at which an object at the substellar limit has just depleted lithium sets a crucial boundary. It is the temperature below which the object must automatically be substellar if lithium is observed. More massive (stellar) objects will have destroyed lithium before they can cool to this temperature. A substellar mass limit of 75 jupiters implies a temperature limit of about 2,700 K for lithium detection, which roughly corresponds to a spectral type of M6. Thus, *any object M7 or later that shows lithium must be substellar.* This form of the test is easier to apply than that employing luminosity, which requires one to know the distance and extinction to an object. Otherwise, the two forms are equivalent.

For instance, the spectral type of the object 269A (Thackrah et al 1997) is M6, so one cannot be sure it is a BD even though it shows lithium (though it certainly lies in the region where it might be a BD; the age would have to be known to be sure). A more definitive case is provided by LP 944-20 (Tinney 1998). It is sufficiently cool (M9), so the fact that lithium is detected guarantees it is a BD, even though we know little about its age (the lithium detection provides an upper limit on the age). This is also true for the enigmatic object PC0025+0447 (M9.5), which displays prodigious Hα emission. Martín et al (1999a) claim a lithium detection for it during a less active state, which would imply that it is a (probably very young) BD. The objects in Hawkins et al (1998) were originally suggested to have luminosities around 10^{-4} solar. If they were confirmed to be below that level, then they would be BDs independent of a lithium observation (since that is the minimum main sequence luminosity). They are cool enough so that if they showed lithium, they would definitely be BDs. Unpublished observations by Basri and Martín find that the brightest of them does not show lithium, and recent work by Reid (1999) makes it unlikely that these are actually BDs (they are apparently farther away and thus more luminous). As discussed in Section 4, the lithium test has been applied quite successfully to objects that are cooler than M; those that show lithium (about a third of them, so far) must certainly be substellar.

3.4   Brown Dwarfs in Star-Forming Regions

The lithium test is less obviously useful in a star-forming region (SFR). Even clear-cut stars have not had time to deplete lithium yet. Nonetheless, there have been numerous reports of BDs in SFRs. They are identified as BDs on the basis of their position in color-magnitude or HR diagrams, using pre-main sequence evolutionary tracks. One must worry about whether the pre-main sequence tracks for these objects are correct, or if there are residual effects of the accretion phase. If one of these candidates doesn't show lithium, it can be immediately eliminated as being a non-member of the SFR. The lithium test as applied in the field still works: If a member of a SFR is cooler than about M7 (here we should be mindful that the pre-main sequence temperature scale might be a little different) and the object shows lithium, then it must be substellar. Indeed, for an object to be so cool at such an early age pushes it very comfortably into the substellar domain.

Good BD candidates have now been found in a number of SFRs, including Taurus (Briceño et al 1998), Chameleon (Neuhauser & Comeron 1999), ρ Oph (Williams et al 1995, Wilking et al 1999), the Trapezium cluster (Hillenbrand 1997), IC 348 (Luhman et al 1997), the σ Ori cluster (Bejar et al 1999), and others. Some very faint/cool objects have been found whose substellar status seems relatively firm (if they are members). The lowest of these may be as small as 10–15 jupiters (Tamura et al 1998). Obtaining spectroscopic confirmation of these candidates is imperative (recent unpublished observations by Martín and Basri show that some of these objects are not substellar). Spectroscopic confirmation has been obtained for a BD near the deuterium-burning boundary in σ Ori (Zapatero-Osorio et al 1999a). Such observations indicate that the substellar mass function may extend right down through the lowest-mass BDs. It is natural to wonder how far it goes below that, since there is no obvious reason why it should stop where we have defined the boundary between BDs and planets.

4.   OBJECTS COOLER THAN M STARS

Although we cannot be fully certain of the substellar nature of GD 165B, it deserves mention as the first known object of the new "L" spectral type. Its spectrum was mysterious until recently (Kirkpatrick et al 1999a). It is very red, suggesting that it is very cool, but it does not show the TiO and VO molecular features in the optical and near infrared (NIR) that characterize the M stars. Even before other such objects were finally discovered, work on model atmospheres was showing convincingly that such a spectrum arises because of the onset of photospheric dust formation (Tsuji et al 1996a, Allard 1998).

Dust actually begins to form in mid M stars. The TiO bands are saturated, then weaken, as one moves to the latest M types. Because they are the defining features of the M spectral class, it was suggested by Kirkpatrick (1998) that we really should have another spectral class for cooler objects (which were being called unsatisfactory names like "M10+" or "≫M9"). Martín et al (1997) proposed "L" as an appropriate choice, bearing the same relation to M that A does to B at hotter spectral classes. I should emphasize that not all L stars are BDs, nor are all BDs L stars (and let us agree that "star" in this context is not to be taken literally). Whether or not a BD is an L star depends on both its mass and its age. A BD generally starts in the mid to late M spectral types and then cools through the L spectral class as it ages (eventually becoming a "methane dwarf"). We do not know at which L subclass the minimum main sequence star resides; estimates of its temperature lie in the 1,800–2,000 K range (probably somewhere in the L2–L4 region).

4.1   The Discovery of Field Brown Dwarfs

The discovery of BDs in the field was somewhat impractical until the advent of wide-field CCD cameras or infrared all-sky surveys. Of particular note are the 2MASS and DENIS surveys. These American and European efforts are the first comprehensive, deep looks at the sky in the NIR, and these surveys are producing many new faint red objects in the solar neighborhood. Recently they have been joined by the SDSS optical survey, which can detect a similar volume of such objects. BDs lay beyond the sensitivity of older surveys such as the Palomar Sky Survey because of their extremely red color and faintness. Even the coolest M subclasses were very sparsely known until recently. Discoveries of BDs in the field were preceded by both cluster and companion BD discoveries. The first announcements were made in 1997, from two very different searches.

One of these was the culmination of a long search for faint red objects with high proper motion (the Calan-ESO survey). A red spectrum of a candidate was obtained in March 1997 (Ruiz et al 1997). This spectrum shows the features now associated with the L dwarfs: broad potassium lines, hydrides, and a lack of TiO bands (Figure 2). Equally striking, it showed the lithium line. As discussed above, this guarantees substellar status for all L dwarfs. The team dubbed the object "Kelu-1" (a Chilean native word for "red").

At about the same time, the DENIS BD team led by Delfosse and Forveille was studying three objects that were as red or redder. They obtained NIR spectra of these objects and showed them also to be L dwarfs (though both discoveries pre-date the introduction of the "L" terminology). There was a suggestion that the coolest of them might show methane (Delfosse et al 1997), but this has not

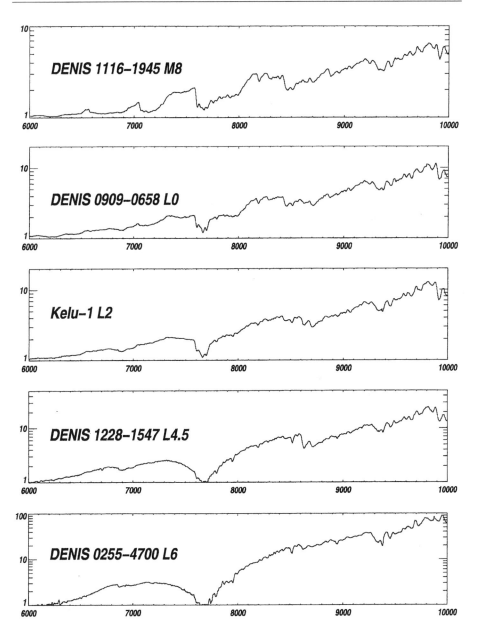

Figure 2   Low-resolution optical spectra of very cool stars. Spectra taken with LRIS on Keck in the red optical range. The dips at 6600 and 7100 in the M8 spectrum are due to TiO; note how they disappear in the L stars. The potassium doublet is best visible in the L0 spectrum at 7700; it then causes the broad depression there in the later L types. The CsI line is also most visible at 8500 in the coolest objects. The molecular features at 8600–8700 are from CrH and FeH; redder features are mostly water.

been confirmed. These objects and Kelu-1 were discussed at the workshop on *Brown Dwarfs and Extrasolar Planets* held in Tenerife in March 1997 (Rebolo et al 1998a). This was the first meeting at which the new discoveries of substellar objects were summarized and discussed in detail.

The DENIS objects and Kelu-1 were studied in the optical at high resolution by Martín et al (1997) and Basri et al (1998). They confirmed the lithium in Kelu-1 and also found lithium in DENIS-P J1228-1547. Lithium detection can be used to place good limits on the mass and age of the objects. They also confirmed that the potassium lines are responsible for the exceptionally strong absorption near 770 nm in these objects. Finally, they found that all the objects are rotating rapidly. Lithium in the DENIS object was quickly confirmed by Tinney et al (1997), who also presented the first suite of low-resolution optical observations of L stars.

The 2MASS survey was also under way and soon greatly surpassed the first few objects with a continuing flood of late M and L stars. The early discoveries are summarized by Kirkpatrick et al (1999b), who present a detailed low-resolution spectral analysis of 25 objects and propose a scheme for the L spectral subclasses. Seven of their objects also show lithium (it is still very strong at L5), so they are definite BDs. It is clear that the lithium test works down to the minimum main sequence temperature, below which all objects are automatically BDs. Concerns about whether such very cool objects are still fully convective (probably not) are irrelevant, partly because they are so cool, and partly because they were fully convective at the time they were depleting lithium (when they resembled the Pleiades BDs). A very substantial fraction of the L objects are substellar. The discovery of objects by all-sky surveys has continued apace, and the number of such objects known is rapidly approaching 100. I discuss their numbers further in Section 6.3.

4.2 Definition of the L Spectral Class

A good compilation of the temperature scale for all spectral classes can be found in DeJager & Nieuwenhuijzen (1987). The temperature ranges spanned by the traditional spectral classes are not uniform; they reflect historical ignorance and old observing techniques, as well as diverse effects of temperature on the appearance of different spectral ranges. The OB spectral classes cover large ($>10,000$ K) temperature ranges. The A class covers almost 3,000 K, and the rest are between 1,000 K and 1,500 K (the shortest range is for G stars). The M stars span a range of 1,500 K.

Although the temperature scale attached to late M stars is still not fully settled, there is general agreement that it ends a little above 2,000 K. This dictates the beginning of the L spectral class. Where to place the cool end of the L class is not obvious from purely spectral considerations. The main optical/NIR spectral characteristics of L stars are the dominance of hydrated molecules and the strong neutral alkali atomic lines. The Cs I lines are still visible in Gl 229B, and the Na I and K I line wings are a dominant opacity source in the optical spectra. The conversion of CO to CH_4 is similar to the conversion of other oxides to hydrides that happens at the beginning of the L class. It is not even settled whether we should use the CCD red or NIR ranges for spectral classification.

Nevertheless, the community seems agreed that Gl 229B (a "methane dwarf") deserves yet another spectral class (on the basis of its strikingly different NIR spectrum). Kirkpatrick has suggested spectral class "T" for methane dwarfs, and this has already received wide usage (Martín et al 1999c prefer "H"). We do not know how close the coolest currently known L dwarfs are to showing methane, nor is the appearance of methane a logically necessary end for the L spectral class (there is still weak TiO in early L stars). Indeed, the appearance of methane depends on which band we're talking about. The strongest (but observationally more difficult) 3.5 micron band is predicted to appear at about 1,600 K. The two micron bands seen in Gl 229B probably appear below 1,500 K and become very strong by 1,200 K, where the optical methane bands are just becoming visible.

Delfosse et al (1999) display a sequence of NIR spectra of L stars. Tokunaga & Kobayashi (1999) find a well-behaved color index in the NIR, but neither set of authors defines a subclass scheme. Kirkpatrick et al (1999b) provide a classification scheme for L stars founded primarily on the optical appearance or disappearance of various molecules. Based on model predictions about these molecules (but not on detailed model fitting), they suggest that L0 begin just above 2,000 K and that L8 begin at about 1,400 K. Martín et al (1999c) present another large set of optical observations and propose a subclass designation similar in temperature to that of Kirkpatrick et al. Theirs is based primarily on optical color band indices, and its temperature scale is informed by the detailed model fitting of alkali line strengths by Basri et al (2000). They make the more specific suggestion that L0 be at 2,200 K, and that each subclass be 100 K cooler. This means that L9 would occur at 1,300 K, consistent with the Kirkpatrick et al scale. The two schemes agree on the spectral appearance of L0–L4 objects.

There is disagreement between the two groups about the actual temperature of the coolest 2MASS objects, however. Based on the weakening of CrH, Kirkpatrick et al believe their coolest object is about 1,400 K. Based on fitting of the Cs I and Rb I line profiles, Basri et al assign it a temperature closer to 1,700 K. An additional fact in favor of the hotter temperature is that methane is not detected in similar DENIS objects (Tokunaga & Kobayashi 1999, Noll et al 1999), whereas it should be observable at the lower temperature. This is only important because one of the classification schemes will need adjustment to assign the appropriate subclass for the coolest currently known L dwarfs. The community will have to settle this question after a full range of ultra-cool objects is discovered and studied in both the CCD and near-IR spectral ranges and the models are improved.

4.3 Atmospheres of Very Cool Objects

The behavior of VLM stars and BDs in color-magnitude and color-color plots has been defined both observationally (e.g. Leggett et al 1998a) and theoretically

(Chabrier, this volume). Since it is well discussed in the latter reference, I concentrate here on the appearance of the spectrum. What distinguishes L stars from M stars is that they are so cool that Ti has been captured in refractory grains, and is not visible in the red in molecular bands (especially at low spectral resolution). The only atomic features visible in the optical are lines of neutral alkaline metals, such as Na and K, as well as the much rarer Cs and Rb (and of course sometimes Li). In the CCD range commonly observed (650–900 nm), the most striking is the resonance doublet of K at 766,770 nm, which merge together and become a very broad bowl-shaped feature covering more than 10 nm as one moves to the cooler L objects (Tinney et al 1998). The NaD lines are an even more spectacular source of opacity, but most spectra do not have the sensitivity to show such broadly depressed flux. Ruiz et al (1997) and Tinney et al (1998) have shown the first comparisons of model atmosphere calculations to low-resolution optical spectra of L stars. The models are generally (but not completely) successful. The molecular bands visible in CCD spectra include some VO (in early L stars) and hydrides like FeH, CrH, and CaH.

In the near infrared, steam bands become increasingly strong (Figure 3), along with H_2 and CO (Allard et al 1997). A good compilation of NIR spectra can be found in Delfosse et al (1999). A few atomic lines are seen, particularly lines of Na I. The ordering of objects by temperature as deduced from NIR spectra agrees well with that from optical spectra. A detailed discussion of a spectrum and modeling for an L star is in Kirkpatrick et al (1999a). The best-fitting models there, as well as in Leggett et al (1998a,b), include both dust formation and dust opacities (although the distribution of grain shapes and sizes is unknown). These do much better, in particular, than models in which dust formation has not been considered. Dust is known to play a strong role even in the late M stars (Tsuji et al 1996b, Jones & Tsuji 1997, Allard et al 1997).

From the first observation of strong alkali lines in a cool dwarf, Basri & Marcy (1995) suggested they could be important spectral diagnostics for very cool stars. They had already been observed in very cool giants, and modelers were aware of their potential utility. It is now clear that Cs I resonance lines can serve as a spectral diagnostic with simple behavior throughout the L spectral range (Figure 4) and extending to the methane dwarfs. One scenario that is fairly successful in modeling the optical line profiles allows the dust to form (and deplete elements like Ti from the molecular source list) but does not use the dust opacity that might

Figure 3   Infrared model spectra of very cool stars. Spectra from the "Dusty" models of the Lyon group. The three humps at 1.2, 1.7, and 2.2 microns are caused by water absorption in the objects (the same transitions help define the J, H, and K bands in the Earth's atmosphere). Note the reddening of the spectrum at the shortest wavelengths for cooler objects, whereas the objects actually get bluer in J-K color. A feature at 2.3 microns is due to CO; alkali lines become strong at 1.65 and 2.2 microns in the coolest object. (Thanks to France Allard for these spectra.)

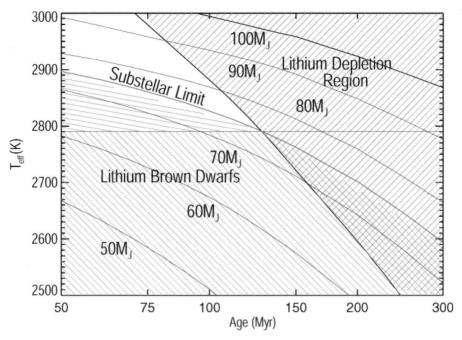

Figure 1 The lithium test (color). The effective temperature vs. age of low mass objects, from models by Baraffe and Chabrier. The solid lines labelled with mass are coolin tracks in the relevant age range. The substellar limit at 75 jupiters is noted in blue. The region beyond which lithium depletion has proceeded to 99% (where it could be easily noted spectroscopically) is marked with the red hatching. The horizontal line marks the temperature at which the substellar boundary crosses the depletion region. Below this line, in the green hatched region, an observation of lithium in an object guarantees that it is substellar. In the red/green region the lithium test for substellarity will give a false negative, while in the blue hatched region it does not distinguish between stars and brown dwarfs (unless the age is known).

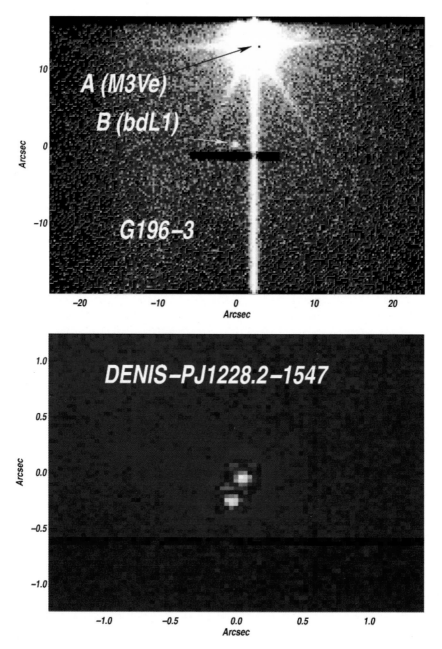

Figure 6 Brown dwarf companions (color). a) An image of G196-3 (a low mass brown dwarf companion to an active M star). It was taken with the HIRES finder camera; the central dark horizontal feature is the spectrograph slit. b) An image of the double brown dwarf DENIS-P J1228-1547 obtained with the NICMOS camera on HST. Note the small scale of the image (the rings around the stars are optical effects of the instrument).

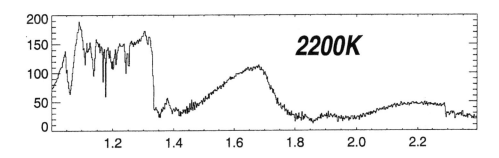

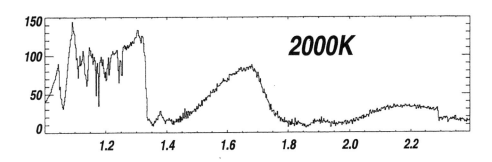

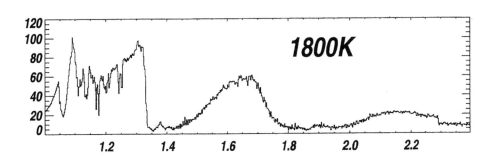

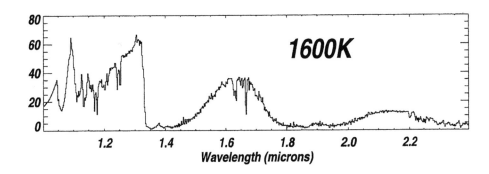

Wavelength (microns)

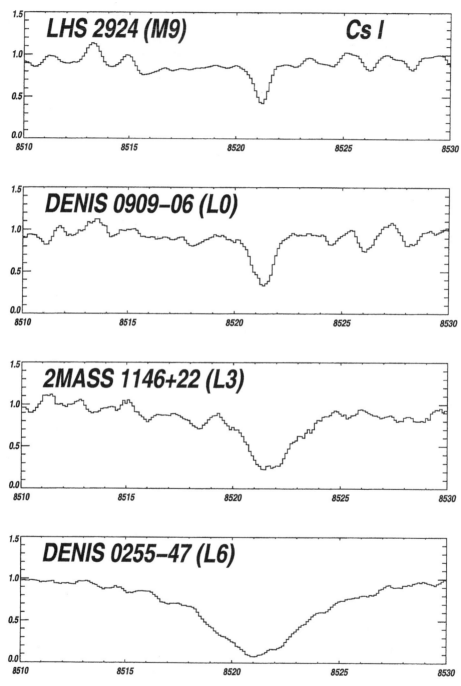

Figure 4 High-resolution spectra of the Cs I line in L stars. Spectra taken with the HIRES echelle at Keck. The line grows in strength as the objects get cooler. The sharpness of molecular features yields the rotational velocity (there are stronger features elsewhere). Molecular features here are smoothed out by the line wings of the coolest object.

result from that. The physical situation mimicked by such "cleared dust" models is condensation of dust followed by gravitational settling of the grains below the line formation region. Tinney et al (1998) find that low-resolution optical spectra are better fit by cleared dust models. Basri et al (2000) show that such models are also successful in explaining observations of the Cs I and Rb I line profiles. They derive a temperature scale for the L stars using these models.

The influence of atmospheric convection, cloud formation, and particle suspensions remains to be properly treated (see also Section 4.4.1). It is very likely that the discrepancies above arise because we do not yet understand the formation and disposition of dust in L star atmospheres. One possibility is that the dust in the upper cooler layers condenses to large enough size to settle down to where it still influences the infrared but not the optical. Such a model has been discussed by Tsuji et al (1999) in the context of Gl 229B, but it may well apply to warmer objects. It is worth recalling that there is a range of temperatures in the atmospheres of these objects; in particular, they are substantially cooler than the effective temperature in the upper layers. The Lyon group is also working on new "settled dust" models. As we discover more very cool objects, this will be an active area of investigation for the next several years.

4.4 The "Methane" or "T" Dwarfs

At extremely cool temperatures (<1,400 K, which only BDs can attain), methane becomes an increasingly important form for carbon molecules. The infrared spectral energy distribution for Gl 229B was first presented by Nakajima et al (1995). While such objects are very red in a color like I-J, they become bluer in colors like J-K because of methane absorption in the K band (see Chabrier, this volume). There has been a good deal of follow-up work on Gl 229B: Oppenheimer et al (1995, 1998), Matthews et al (1996), Geballe et al (1996), Golimowski et al (1998), Schulz et al (1998), and Leggett et al (1999).

After the discovery of Gl 229B, there was a gap of four years before the next methane dwarf was announced. This is largely due to the faintness of such objects; the current surveys can only see them out to about 10 pc. In fact, most BDs in the Galaxy should be methane dwarfs, since they will cool to the required low temperatures within 1–2 gigayears. There was a burst of discoveries in 1999 (announced at the June AAS meeting). The SDSS team (Strauss et al 1999) found two and the 2MASS team (Burgasser et al 1999) announced four. At first glance they appear very similar to Gl 229B in the NIR, though more careful analysis implies they lie between 1,000–1,300 K. These are all apparently single objects in the field (though of course some could be close BD pairs not yet resolved).

4.4.1 The Atmosphere of Gl 229B   The first analysis of the spectrum of Gl 229B was by Oppenheimer et al (1995). This was followed by more detailed papers (Allard et al 1996, Marley et al 1996). They, along with Oppenheimer et al

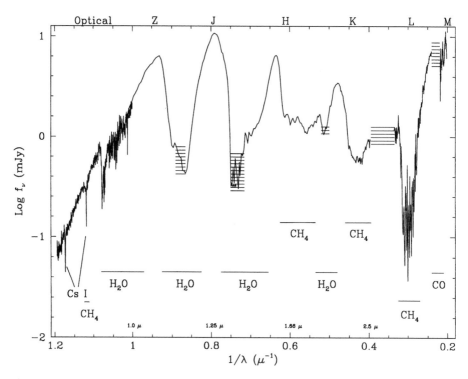

Figure 5   The spectrum of Gl 229B from the optical through the infrared. A full compilation
of the spectrum from the work of Ben Oppenheimer. Water and methane dominate the infrared
features, whereas the Cs I lines are still visible in the optical. Note the strong methane features in
the H and K bands that are not seen in L stars; these define the proposed T spectral type.

(1998), generally conclude that it is not matched by an atmosphere containing
dust, but the dust-free models fit. The effective temperature of Gl 229B is about
900 K. Features resulting from H_2O dominate the spectrum (Figure 5). Methane
(CH_4) is now also a dominant producer of molecular absorption, particularly in the
K band (and presumably at 3.5 microns as well). CO is also seen (Noll et al 1997,
Oppenheimer et al 1998), and that is surprising for such a cool object. This has been
interpreted to mean that there is some convective overshoot that passes through the
subphotospheric radiative zone predicted by models (e.g. Burrows et al 1997) and
brings up species from the hotter interior. The chemical equilibrium of species
is quite complicated in the methane dwarfs. It has been discussed with varying
degrees of sophistication by Fegley & Lodders (1996), Burrows et al (1997),
Lodders (1999), and Griffith & Yelle (1999), among others (see also Chabrier, this
volume). It is important that calculations be done in the context of self-consistent
radiative/convective equilibrium models, or the temperature structure and mixing
will be incorrectly treated and will produce misleading results.

The presence of strong alkali lines (e.g. Cs I; Oppenheimer et al 1998) is indicative that they have either not yet formed molecules or are dredged up from below. The optical colors of Gl 229B seemed to require some sort of broadband opacity in excess of the dust-free models, which are substantially brighter than observed (e.g. Golimowski et al 1998). This was taken to indicate that a proper treatment of dust, hazes, and aerosols in the atmosphere might be important (Griffith et al 1998, Burrows & Sharp 1999). Recently, however, Burrows et al (1999) and Tsuji et al (1999) have suggested that the missing opacity in the 700–950 nm range is actually just the enormous damping wings of K I and Na I (apparently not treated properly in the initial calculations). This has very recently been confirmed spectroscopically.

Tsuji et al also reconsider the question of where the dust might be and show that hybrid models with the dust settled below a certain (currently arbitrary) layer do a better job of matching the spectrum. Basri et al (2000) were led independently to a similar suggestion for the L stars, so this issue will be important to pursue. Optical flux is blocked by the dust in the inner photosphere (where it is cool enough to form dust but not hot enough to evaporate it) and reprocessed to the infrared. The dust is more transparent at longer wavelengths, of course. Then, above a certain layer, the grains may become large enough to settle out, and the optical opacity is freed of the dust (above the infrared photosphere but in the optical line-forming region).

Gl 229B provided us the first opportunity to test our understanding of atmospheres intermediate between stars and the giant planets in our Solar System. Because methane dwarfs are brighter than cold planets, it is likely that the first extrasolar planets whose spectra are recorded will be in this temperature range (planets begin as L stars when very young). The discovery of Gl 229B has stimulated a resurgence in the work on opacities, chemistry, and the atmospheric structure of such objects. It is clear that the discovery of more methane dwarfs covering a range of temperatures will now greatly advance this effort.

4.5 Rotation and Activity in Very Low Mass Objects

It is now possible to draw the first conclusions about the nature of magnetic activity and angular momentum evolution for objects near and below the substellar boundary. Among convective solar-type stars, there is a well-known connection between the rotation of an object and the amount of magnetic activity at its surface. The more rapid the rotation, the more active the object, leading to emission in spectral lines like CaII K or Hα, or in coronal X-rays. This in turn leads to a magnetized wind from the object that carries away angular momentum and spins it down (reducing the level of activity). The field is generated by a dynamo, which in solar-type stars is thought to arise primarily at the bottom of the convective zone. Recent thinking is that the non-cyclical half of the Sun's flux might arise in a turbulent dynamo throughout the convective zone (Title & Schrijver 1998). The fraction contributed by the turbulent dynamo probably increases with the depth of the convection zone, until it takes over when the star becomes fully convective. That would explain why there is no obvious change in stellar activity passing through early M stars (Giampapa et al 1996).

The first indications that something else might happen near the substellar boundary came from observation of an M9.5 star at high spectral resolution by Basri & Marcy (1995). They found that this (old field) star, BRI 0021, has an amazingly high spin rate and virtually no Hα emission. Later, however, an Hα flare was seen on this star (Reid et al 1999a). This suggests that it had never had much magnetic braking, and that the connection between rotation and activity does not apply to VLMS. Delfosse et al (1998) surveyed a complete sample of nearby early and mid M stars and found that the fraction of fast (>5 km s^{-1}) rotators is quite low until M4 or so (the boundary for fully convective stars) and then begins to increase rapidly. Basri et al (1996, 2000) and Tinney & Reid (1998) have found that rapid rotation becomes ubiquitous later than M7 or so (despite the effect of equatorial inclination on $v \sin i$). These rapid rotators are characterized by moderate to very weak Hα emission, and all the rotators above 20 km s^{-1} have weak Hα emission (less than 5A equivalent width).

Most of the DENIS and 2MASS L dwarfs show no Hα emission. There are a few earlier than L4 that show a little Hα emission (Leibert et al 2000), but the implied surface fluxes are extremely low. Because of the extremely cool photospheres, Hα can only show up if there is chromospheric or coronal heating. It is also the case that a given value of emission equivalent width (say 5A) represents a dramatically weakening surface flux as we move into the late M and L stars. The continuum, which defines the normalization of equivalent width, is dropping very quickly with temperature (Hα now occurs in the Wien part of the Planck function). There cannot be a corona in the stars showing no Hα because it would create a chromosphere by photoionization (Cram 1982) that would easily show up. Basri et al (2000) find that most of the L dwarfs have $v \sin i$ corresponding to rotation periods of at most a few hours. Thus it is quite clear that for older BDs and VLMS, the usual rotation-activity connection is completely broken and may even be reversed (since the late M stars showing stronger emission tend to be the slower rotators).

There are several possible explanations for these results. One is that the ionization levels in the photosphere may have become so low that there is insufficient conductivity to allow coupling of the magnetic field to the gas. Then gas motions do not twist up the fields, and there is no dissipation to heat the upper atmosphere. This has to be true even in the face of ambipolar diffusion, which couples small numbers of ions to the neutrals fairly effectively (as in T Tauri disks). The alkali metals that are the last suppliers of electrons are becoming quite neutral in the L stars. A possible counterexample to this hypothesis is provided by the detection of (non-flaring) Hα emission in a methane dwarf (Liebert et al 2000).

All low-mass objects should have turbulent dynamos, which are driven by convective motions. Rotation can enhance production of fields, and the amplitude of convective velocities also does. But convective overturn times scale with luminosity in these objects. At the bottom of the main sequence they can increase to months, while typical spin periods are dropping to hours. The traditional rotation-activity connection may arise because activity increases with decreasing Rossby number (the ratio of rotation period to convective overturn time). Activity levels increase steadily from a Rossby number of unity down to 0.1. They saturate between 0.1 and 0.01, with a hint of a downturn at 0.01 (Randich 1998). The BDs have Rossby numbers in the range from 0.01 to 0.001. I speculate that the dynamo may be unable to operate efficiently at such low levels, perhaps because rotation organizes the flows too much. A possible counterexample to this hypothesis is provided by the very rapid rotator Kelu-1, which exhibits a persistent (though very weak) $H\alpha$ emission line.

A related possibility is that the field is not actually quenched by rapid rotation but instead takes on a relatively stable, large-scale character (see Chabrier, this volume) like that of Jupiter. In that case, the field might be sufficiently quiet (especially in conjunction with the low atmospheric conductivity) that it does not suffer the dissipative configurations that power stellar activity. To the extent that acoustic or magneto-acoustic heating play a role, the low convective velocities in these objects will reduce it. Thus, the objects might still have strong fields but no stellar activity.

This could be tested in principle using Zeeman diagnostics. Valenti et al (2000) have suggested using FeH for objects in this temperature range and shown that it can work in late M stars. Occasional flaring does occur on some of these objects. Flares have been seen in objects that seem otherwise quite quiescent, such as VB10 (Linsky et al 1995) and 2MASSW J1145572+231730 (Leibert et al 1999). Another possibility is to search for rotational periodicities (photometrically or spectroscopically). These traditionally indicate the presence of magnetic spots. Some very cool objects have shown such behavior (Martín et al 1996, Bailer-Jones & Mundt 1999), but many have not. A possible complication arises if dust clouds condense inhomogeneously in the atmospheres of these objects. One might then detect rotational modulation due to "weather" (Basri et al 1998, Tinney & Tolley 1999). There is no confirmation of this yet; one will have to very carefully distinguish between the two possible sources of variability (spots or clouds) by showing that opacity rather than temperature is the cause (they will cause different effects in different spectral features).

The only BDs that seem to show strong magnetic activity are the very young ones (e.g. Neuhauser et al 1999 for X-rays; many examples of $H\alpha$ emission in SFRs and young clusters). These are sufficiently luminous objects that are hot enough and/or perhaps not rotating too fast. In the youngest cases, there may be an added contribution due to accretion phenomena. They all eventually become relatively inactive as the convection weakens and the atmosphere cools. Apparently most objects near or below the substellar boundary are rapid rotators because they have not experienced much magnetic braking.

5.   BROWN DWARFS IN BINARY SYSTEMS

5.1   Visible Brown Dwarf Companions

Many of the original searches for BDs were imaging or radial velocity surveys for companions to nearby stars. That these searches were unsuccessful or had very low yields caused some of the pessimism about finding BDs before 1995. This pessimism was codified in the phrase "brown dwarf desert" (e.g. Marcy & Butler 1998). One must remember that while it is convenient to search around stars, this covers only a subset of possible places to find BDs. The search that discovered GD 165B included several hundred white dwarf primaries, and that which uncovered Gl 229B tested several hundred M dwarf primaries. There have been numerous searches from the ground and with HST that came up empty around solar neighborhood G-M stars (e.g. Forrest et al 1988, Henry & McCarthy 1990, Simons et al 1996), and Hyades low-mass stars (Macintosh et al 1996, Reid & Gizis 1997, Patience et al 1998). These have been pretty successful at finding VLMS companions, but not clearcut BDs. We can conclude that there is a relatively low (<1%) fraction of stars with well-separated visible BD companions.

Recent searches have had slightly better luck. LHS 102B (Goldman et al 2000) is an L-type companion to an early M star (found in a proper motion study of EROS observations), although it fails the lithium test. That does not exclude it from being a BD, but it must have a mass greater than 60 jupiters. This is also true for GD 165B; unfortunately, we cannot quite be sure that these are not VLMS unless their precise ages can be determined (or if it turns out they are just under the minimum main sequence temperature). In a survey careful to examine only young systems, Rebolo et al (1998b) found a BD companion (G196-3) after searching only 60 M primaries (Figure 6, see color insert). This is one of the lowest-mass confirmed BDs (about 20 jupiters, based on the high activity of its primary, which implies a Pleiades age). The companion to GG Tau B (White et al 1999) is an even younger example of such a system. The coolest current T dwarf has recently been found as a wide member of the Gl 570 system (Burgasser et al 2000). Both this and GG Tau are quadruple systems. Two other very wide systems with more massive primaries have been identified recently by Kirkpatrick et al (in prep.). Although there is an observational selection effect against finding faint companions to brighter primaries, the results in the next subsection indicate this is not the main reason for the lack of companions around higher-mass primaries.

5.2   Radial Velocity Brown Dwarf Candidates

Sensitive radial velocity surveys of G-M stars do not suffer from the fading of BDs with age or brightness contrast. They examine a separation range closer in than that of the imaging surveys. Precision radial velocity (PRV) searches will automatically find BDs more easily than planets. It is possible that the first BD was in fact found this way. HD114762 was detected as a companion to a solar-type star by Latham et al (1989; Section 2.1). It has been generally referred to as an extrasolar planet, but the minimum mass for it (11 jupiters) is quite near the planet/BD boundary, so that the inclination correction is likely to push it into the BD range. Of course, it is possible that it may be pushed all the way into the stellar range; the likelihood of that depends on how sparsely populated the brown dwarf desert really is (see Section 6.2)

The most extensive survey for dynamical BDs has been that of Mayor and colleagues, first with the CORAVEL and then the ELODIE instruments (e.g. Mayor et al 1997, 1998). They found that several percent of solar-type systems have reflex velocities suggesting companions with lower mass limits in the substellar range. The difficulty with these candidates is exactly that they have lower mass *limits*. For a particular case, one is never quite sure whether the correction will push it into the stellar mass range. On statistical grounds one can argue that all the inclination corrections cannot be large. The extent to which this argument can be made, however, depends on the intrinsic mass function of binary companions. To see this, imagine that there are no BD companions to solar-type stars. Then one will only find BD candidates in PRV studies that are stellar systems with sufficiently low inclinations.

Indeed, about half of the Mayor BD candidates were eliminated recently by the finding of their orbital inclinations using Hipparcos data (Halbwachs et al 2000). None of the remaining candidates is incontrovertibly substellar. The PRV searches have found very few companions in the BD mass range (Marcy & Butler 1998, Mayor et al 1998) but a number in the planetary mass range (which are harder to detect). Taking all this into account, one might fairly conclude that the incidence of BD companions to stars with masses of 0.5 solar masses or more is quite low (not more than about 1%). In contrast, the incidence of stellar companions to such primaries is in the range 20–40%. This result is discussed in more detail in Section 6.2. There are no examples of unambiguous dynamical BDs at present.

5.3   Double Brown Dwarfs

The search for binary brown dwarfs (BD pairs) is barely under way. It is striking that several have already been found. Color-magnitude diagrams of Pleiades VLMS show a large spread that has been interpreted as resulting mainly from unresolved binaries (Steele & Jameson 1995, Zapatero-Osorio 1997). The presence of an unresolved substellar secondary has been inferred from infrared spectroscopy of the Pleiades VLMS HHJ54 (Steele et al 1995). A search for visible binaries among the Pleiades BDs using HST (Martín et al 1998a) identified a few such pairs (but it is turning out that they all may be non-members). If the distribution of binary frequencies among Pleiades BDs were similar to those of young stars and G dwarfs, they should have found 4.5 binaries. Dynamical stripping of wide companions of low-mass primaries should not have proceeded too far in the Pleiades, though it could explain the dearth of wide substellar companions in the Hyades (Gizis et al 1999).

There is essentially only one BD that has been searched for radial velocity variations, and that is PPl 15. Its binarity was suggested by its position in the color-magnitude diagram (Zapatero-Osorio et al 1997a). The fact that it does turn out to be a double-lined spectroscopic binary (with an eccentric orbit and a period of six days; Basri & Martín 1999) is remarkable. It seems to bode well for the discovery of a reasonable number of spectroscopic BD binaries. We do not know whether the distribution of separations for substellar binaries is different from that for stellar binaries.

Another surprisingly successful effort has been made to find field BD pairs. In only two pointings in an HST survey for binaries among the nearby field BDs, Martín et al (1999b) found that one of the three original DENIS objects is a sub-arcsec double (with a projected separation of about 5 AU). It is worth remarking that this system (DENIS-P J1228-1547) offers the first real chance for a dynamical confirmation of substellar masses. HST may be able to reveal its orbit in only a few more years. Koerner et al 1999 have discovered several similar systems among the 2MASS and DENIS objects (as yet unpublished, possibly including a second of the first three DENIS objects). Thus, searches for BD pairs with small (<50 AU) separations have been remarkably successful (though it is hardly a statistical sample). This suggests that if one looks for substellar companions in systems where the mass ratio and separation are not too high, many of them may be found. On the other hand, no wide systems have been found (even though that is easier). The scale of BD binary systems may be smaller than stellar binaries.

6.   THE SUBSTELLAR MASS FUNCTION

It should now be clear that brown dwarfs are not rare at all. A sampler of BDs in various contexts is given in Table 1. This is by no means complete, and the list is constantly growing. It is natural to ask just how many BDs are really out there, and with what masses.

6.1   The Mass Function from Clusters

The long-term goal of searching for BDs in clusters is to discover whether there is a "universal" substellar mass function among clusters (or what variations there are, and why). In doing this, one must correct for unseen binaries (and stripping of wide binaries) and for mass segregation (and eventual evaporation) of low-mass members caused by the cluster environment.

Several groups (Zapatero-Osorio 1997b, 1999b; Stauffer et al 1998; Bouvier et al 1998; Hambly et al 1999) have now found a large number of Pleiades BD candidates (selected photometrically; see Figure 7), extending down to inferred masses as low as 35 jupiters (Martín et al 1998b). Since we have successfully applied the lithium test to this cluster to determine the luminosity of the substellar boundary, it is no longer necessary to test all these candidates for lithium.

TABLE 1 A Brown Dwarf Sampler

Name	Sp. T.	Mass (jup)	Age (Gyr)	Pedigree[a]	References	Notes
Single: Cluster						
Teide 1	M8	55–60	0.12	Lithium	Reb95, Reb96	Pleiades
PIZ-1	M9	45–55	0.12	C-M[b]	Cos97	Pleiades
Roque 25	L0	35–40	0.12	C-M	Zap99b, Mar98b	Pleiades
Ap 326	M7.5	60–70	0.08	Lithium	Sta99	Alpha Per
GY 11	M7	30–50	0.001–3	C-M	Wil99	ρ Oph
ρ Oph BD 1	M8.5	20–40	0.001–3	Lithium	Luh97, Mar99a	ρ Oph
Cha Hα 1	M7.5	20–40	0.001–3	C-M	Neu98 (Xrays)	Chameleon
S Ori 47	L1	10–20	0.001–3	Lithium	Bej99, Zap99	Sigma Ori
Single: Field						
Kelu-1	L2	<60	<1	Lithium	Ruiz97	Prop. motion survey
LP944-20	M9	<60	<1	Lithium	Tin98	Prop. motion survey
2MASSW J1553214+210907	L5.5	<60	<1.5	Lithium	Kir99b	Strong lithium
2MASSW J1632291+190441	L6	>60	<1.5	v. cool	Kir99b	(ref gives L8)
DENIS-P J0255-4700	L6	>60	<1.5	v. cool	Del99, Mar99c	coolest L for now
SDSS 1624+00	T	<70	<4	Methane	Stra99	Sloan
2MASSW J1225543-273947	T	<70	<4	Methane	Bur99	10 pc
Binaries (Imaged)						
Gl 229B	T	35–50	>1.5	Methane	Nak95, Opp98	~40AU, M2
G196-3B	L2	20–25	~0.1	Lithium	Reb98	~300AU, M3e
DENIS-P J1228-1447	L4.5	<60	<1	Lithium	Delf97, Mar99b	~5AU 2BDs
GD 165B	L4	>65	>1	cool enough?	Bec88, Kir99a	comp. to WD No lithium?
LHS 102B	L4	>65	>1	cool enough?	Gol99	comp. to M3 No lithium?
GG Tau Bb	M8	50–60	0.002	Lithium	Whi99	T Tauri, quad. system
Gl 570D	T	30–70	2–10	Methane	Bur00	quad. system coolest T for now
Binaries (Radial Velocity)						
PPl 15	M6.5	60–70	0.12	Lithium	Stauf95, Bas96	Pleiades
HD114762	F9[c]	>11	>3	sin i?	Lath89	also called planet
HD29587	G2[c]	20–60	>3	dynamical	Hal00	tested by Hipparcos
HD127506	K3[c]	20–60	>3	dynamical	Hal00	tested by Hipparcos

[a]Only those with "lithium" or "methane" are certain; the others are likely.

[b]Found by its position in a color-magnitude diagram.

[c]Spectral type of primary.

Establishment of cluster membership for objects faintward of the boundary is sufficient proof that they too are BDs. One must correct for contamination by non-members (which has been estimated from the percent of spectroscopic failures among photometric candidates). These estimates are still fairly uncertain because most of the candidates have not been fully tested. Testing can be done with proper motion, radial velocity, and perhaps Hα. Lack of lithium is excellent grounds for rejecting membership below the lithium boundary. The more of these tests used, the better. The entire cluster has not been surveyed (although this is being rectified with modern wide-field cameras). We do not expect mass segregation to have gone very far in the Pleiades, although BDs should be found preferentially nearer the periphery and will be the first objects to "evaporate" away. As always, one should correct the observed MF for the effects of binaries. Unfortunately, we are still fairly ignorant of the binary fraction of these objects (see Section 5).

The substellar MF inferred from the Pleiades is gently rising. We can characterize it with the index α in the equation $dN/dM = M^{-\alpha}$. It appears that this index has a value of about $+0.5$ (with uncertainty of a few tenths) for this cluster. The stellar population is well known in this cluster, and the age of all the objects is also known (this is a major advantage over field studies). The fit of the cluster sequence to models is also good (especially after using dust in models for the lowest-mass objects). I therefore view this as the currently most reliable measurement of a substellar mass function. Work on several other clusters is rapidly approaching the point where substellar MFs can be checked in a variety of cluster environments (Section 3.4).

In order to reach all the way to the bottom of the MF one must study younger clusters, or star-forming regions. Of course, one never observes the MF directly, but rather the luminosity function. Theoretical models, tested against independently calibrated luminosity and mass observations, allow the conversion to the MF. See the article by Chabrier (this volume) for an assessment of the state-of-the-art. The recent work by Bejar et al (1999) on the σ Ori cluster suggests that the substellar MF reaches down all the way to the deuterium-burning limit (and several other groups are coming to similar conclusions for other SFRs).

6.2 The Mass Function for Binaries

The main source of BD candidates from PRV studies has been the work of Mayor et al (1997, 1998). Basri & Marcy (1997) showed that the number of BD candidates was consistent with a flat or slowly rising mass function into the substellar domain. But recently Halbwachs et al (2000) used data from the Hipparcos project to lift the ambiguity of orbital inclination in many of those cases and found that half of them are definitely stellar. They show that this result is incompatible with the MF in clusters and the field: there are too few BDs. We cannot be sure of any of the

---

Figure 7 A color-magnitude diagram for low-mass Pleiades members. Results from the central square degree of the cluster (surveyed in I-Z colors). The open symbols are stars, and the full symbols are brown dwarf candidates. Those labeled have been spectroscopically confirmed. The solid line is the main sequence, and the dashed line is a 120-Myr isochrone from the NextGen models of the Lyon group. The mass scale is shown with numbers to the right (in solar masses). The open squares are field objects with known parallax (shifted to the Pleiades distance). The downturn at the end of the sequence is better matched with dusty models.

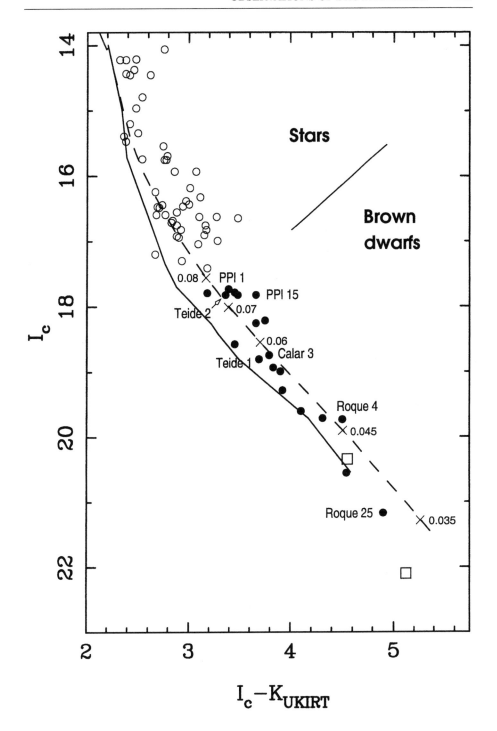

PRV BD candidates at the present time; the remaining candidates must have their orbital inclinations determined. Halbwachs et al conclude that current results are consistent with a very barren brown dwarf desert.

This means that binary companions (especially of solar-type stars) are not a good means of addressing the general substellar MF. They probably tell us more about the binary formation mechanism (itself a very interesting topic) than about the general likelihood of forming substellar objects. A review of theories of binary formation (stellar and substellar) can be found in Bodenheimer et al (2000). The metaphorical "brown dwarf desert" should now be seen as merely a "desert island" that occurs for high mass-ratio systems. The binary formation mechanism probably cares more about the mass ratio than the absolute mass of the companion. As discussed below, when one searches for BDs in other contexts, one finds verdant fields of them.

6.3   The Mass Function in the Field

Since 1997, the new NIR all-sky surveys (DENIS and 2MASS) have been un-covering nearby young BDs in the field at an increasing rate (and now the SDSS has begun to add to this tally). Close to 100 L stars are now known, though the surveys have not yet covered most of the sky. Not all of these are BDs, but some of them certainly are (those that show lithium or are cool enough). While this shows that BDs are not a rare class of object (the surveys reach out to less than 50 pc), the analysis of these results to yield a substellar MF is quite compli-cated.

The interpretation of field survey data requires two separate and difficult steps. The first is the correction of the survey for observational biases and effects. A survey with a given sensitivity will sample smaller total volumes for objects of cooler temperatures. There must also be a correction for completeness effects as a function of observed brightness in the various survey colors. One must convert observed intensities to luminosity or effective temperature. Finally, binaries must be accounted for, as they both increase the numbers of objects and increase the survey volume (because they are brighter).

The second overriding problem lies in the nature of the BDs themselves. By definition, they never come onto the main sequence and so are continually fading with time. This should give rise to a deficit of objects just below the minimum main sequence (and greater numbers where typical BDs at average Galactic ages have reached). Most BDs should have cooled into methane dwarfs. Mercifully, they all achieve similar radii as they age (slightly smaller than Jupiter), so the connection between effective temperature and luminosity is not too ambiguous. But there is a complete degeneracy in the relations between luminosity/temperature, mass, and age. Photometric observations, unfortunately, can only give us the first of these. Even that requires a spectral-type/temperature calibration, or the appropriate bolometric corrections and parallaxes. Spectroscopy cannot really resolve this problem (unless we become very precise at measuring gravity).

Most objects in the field will be older than 200 Myr (although we must account for a bias for finding younger objects). This is the maximum time required for the depletion of lithium to run its course (and most objects will finish much earlier). So it will generally be true that if we see lithium in a field object, the object must have a mass below 60 jupiters, and if we don't see lithium, the object's mass must be higher. The ambiguity between stars and BDs is removed for objects cooler than the minimum main sequence temperature—they are all BDs. Thus, if we simply want to know the ratio of VLMS to BDs (and do not demand the mass distribution), we can find it from the fraction of lithium-bearing objects cooler than spectral class M6 and the numbers of objects below the L subclass corresponding to the end of the main sequence (L3 ±1?).

An excellent preliminary attack on the mass function has been accomplished by Reid et al (1999b). They analyze the 2MASS and DENIS L star samples, carefully considering sources of observational bias. They find the mass function by modeling the luminosity function using current theory and assume a constant star formation rate over the age of the galaxy. They do not attempt to correct for binaries. The bottom line is that the observations support a mass function with α below 2 (they suggest 1.3). This implies somewhat more BDs than the cluster result. Such a mass function means that the BDs are not a dynamically important mass constituent of the disk and are unlikely to be major contributors to the baryonic dark matter (that would require α above 3).

The space density of BDs found by Delfosse et al (1998) and Reid et al (1999b) is as high as 0.1 systems per cubic parsec. The total number of BDs could then easily exceed the total number of stars. This suggests the possibility that our nearest neighbor may actually be a brown dwarf. If so, we have a pretty good chance of discovering it in the next decade (it would probably be an unusually bright methane dwarf). Such a discovery would certainly bring brown dwarfs to everyone's attention! In any case, it is clear that many astronomers will be kept happy studying these fascinating objects for some time to come.

Visit the Annual Reviews home page at www.AnnualReviews.org

LITERATURE CITED

Allard F. 1998. In *Brown Dwarfs and Extrasolar Planets*, ed. R Rebolo, EL Martín, MR Zapatero-Osorio, *A.S.P. Conf. Ser.* 134:370–82. San Francisco: Astron. Soc. Pacific

Allard F, Hauschildt P, Alexander DR, Starrfield S. 1997. *Annu. Rev. Astron. Astrophys.* 35:137–77

Allard F, Hauschildt PH, Baraffe I, Chabrier G. 1996. *Ap. J.* 465:L123–27

Bailer-Jones CAL, Mundt R. 1999. *Astron. Astrophys.* 348:800–4

Barrado y Navascués DB, Stauffer JR, Patten BM. 1999. *Ap. J.* 522:L53–56

Basri G. 1998a. In *Brown Dwarfs and Extrasolar Planets*, ed. R Rebolo, EL Martín, MR Zapatero-Osorio MR, *A.S.P. Conf. Ser.* 134:394–404. San Francisco: Astron. Soc. Pacific

Basri G. 1998b. *Cool Stars in Clusters and*

Associations: Magnetic Activity and Age Indicators, *Mem. Soc. Astron. Ital.* 68:917–24

Basri G, Marcy G, Oppenheimer B, Kulkarni SR, Nakajima T. 1996a. In *Cool Stars, Stellar Systems, and the Sun, Ninth Cambridge Workshop*, ed. R Pallavicini, AK Dupree, *A.S.P. Conf. Ser.* 109:587–88. San Francisco: Astron. Soc. Pacific

Basri G, Marcy GW. 1995. *Astron. J.* 109:762–73

Basri G, Marcy GW. 1997. In *Star Formation, Near and Far*, ed. SS Holt, LG Mundy, pp. 228–40. New York: AIP Press

Basri G, Marcy GW, Graham JR. 1996b. *Ap. J.* 458:600–9

Basri G, Martín E, Ruiz MT, Delfosse X, Forveille T, et al. 1998. In *Cool Stars, Stellar Systems, and the Sun, Tenth Cambridge Workshop*, ed. RA Donahue, JA Bookbinder, *A.S.P. Conf. Ser.* 154:1819–27. San Francisco: Astron. Soc. Pacific

Basri G, Martín EL. 1999. *Ap. J.* 510:266–73

Basri G, Mohanty S, Allard F, Hauschildt PH, Delfosse X, et al. 2000. *Ap. J.* In press

Becklin EE, Zuckerman B. 1988. *Nature* 336:656–58

Bejar VJS, Zapatero-Osorio MR, Rebolo R. 1999. *Ap. J.* 521:671–81

Bildsten L, Brown EF, Matzner CD, Ushomirsky G. 1997. *Ap. J.* 482:442–47

Black D. 1997. *Ap. J.* 490:L171–74

Bodenheimer P, Burkert A, Klein R, Boss A. 2000. *Protostars and Protoplanets IV*. Tucson: Univ. of Arizona Press. In press

Boss AP. 1997. *Science* 276:1836–39

Bouvier J, Stauffer JR, Martín EL, Barrado y Navascués D, Wallace B, Bejar VJS. 1998. *Astron. Astrophys.* 336:490–502

Briceño C, Hartmann L, Stauffer JR, Martín EL. 1998. *Astron. J.* 115:2074–91

Burgasser AJ, Kirkpatrick JD, Brown ME, Reid IN, Gizis JE, et al. 1999. *Ap. J.* 522:L65–68

Burgasser AJ, Kirkpatrick JD, Cutri RM, McCallon H, Kopan G, et al. 2000. *Ap. J.* 531:L57–60

Burrows A, Marley M, Hubbard WB, Lunine JI, Guillot T, et al. 1997. *Ap. J.* 491:856–75

Burrows A, Marley MS, Sharp CM. 1999. *Ap. J.* 531:438–46

Burrows A, Sharp C. 1999. *Ap. J.* 512:843–63

Cram LE. 1982. *Ap. J.* 253:768–72

D'Antona F, Mazzitelli I. 1985. *Ap. J.* 296:502–13

DeJager C, Nieuwenhuijzen H. 1987. *Astron. Astrophys.* 177:217–27

Delfosse X, Forveille T, Perrier C, Mayor M. 1998. *Astron. Astrophys.* 331:581–95

Delfosse X, Tinney CG, Forveille T, Epchtein N, et al. 1997. *Astron. Astrophys.* 327:L25–28

Delfosse X, Tinney CG, Forveille T, Epchtein N, Borsenberger J, et al. 1999. *Astron. Astrophys. Suppl.* 135:41–56

Fegley B, Lodders K. 1996. *Ap. J.* 472:L37–39

Forrest WJ, Barnett JD, Ninkov Z, Skrutskie M, Shure M. 1989. *P.A.S.P.* 101:877–92

Forrest WJ, Skrutskie M, Shure M. 1988. *Ap. J.* 330:L119–23

Geballe TR, Kulkarni SR, Woodward CE, Sloan GC. 1996. *Ap. J.* 467:L101–4

Giampapa MS, Rosner R, Vashyap V, Fleming TA, Schmidtt JHMM, Bookbinder JA. 1996. *Ap. J.* 463:707–25

Gizis JE, Reid IN, Monet DG. 1999. *Astron. J.* 118:997–1004

Goldman B, Delfosse X, Forveille T, Afonso C, et al. 2000. *Astron. Astrophys.* In press

Golimowski DA, Burrows CJ, Kulkarni SR, Oppenheimer BR, Brukardt RA. 1998. *Astron. J.* 115:2579–86

Griffith CA, Marley MS (organizers). 2000. *From Giant Planets to Cool Stars*. Conf. held in Flagstaff, AZ, June 8–11, 1999.

Griffith CA, Yelle RV. 1999. *Ap. J.* 519:L85–88

Griffith CA, Yelle RV, Marley MS. 1998. *Science* 282:2063–67

Halbwachs JL, Arenou F, Mayor M, Udry S, Queloz D. 2000. *Astron. Astrophys.* 355:581–94

Hambly NC, Hawkins MRS, Jameson RF. 1993. *Astron. Astrophys. Suppl.* 100:607–40

Hambly NC, Hodgkin ST, Cossburn MR, Jameson RF. 1999. *MNRAS* 303:835–44

Hawkins MRS, Ducourant C, Rapaport M, Jones HRA. 1998. *MNRAS* 294:505–12

Henry TJ, McCarthy DW. 1990. *Ap. J.* 350:334–47

Hillenbrand LA. 1997. *Astron. J.* 113:1733–68

Jameson RF, Hodgkin ST. 1997. *Contemporary Physics* 38:395–407

Jones HRA, Tsuji T. 1997. *Ap. J.* 408:39–41

Kafatos MC, Harrington RP, Maran SP. 1986. *Astrophysics of Brown Dwarfs*. New York: Cambridge Univ. Press

Kirkpatrick JD. 1998. In *Brown Dwarfs and Extrasolar Planets*, ed. R Rebolo, EL Martín, MR Zapatero-Osorio, *A.S.P. Conf. Ser.* 134: 405–15. San Francisco: Astron. Soc. Pacific

Kirkpatrick JD, Allard F, Bida T, Zuckerman B, Becklin EE, et al. 1999a. *Ap. J.* 519:834–43

Kirkpatrick JD, Reid IN, Leibert J, Cutri RM, Nelson B, et al. 1999b. *Ap. J.* 519:802–33

Koerner DW, Kirkpatrick JD, McElwain MW, Bonaventura NR. 1999. *Ap. J.* 526:L25–28

Kulkarni SR. 1997. *Science* 276:1350–54

Kumar SS. 1963. *Ap. J.* 137:1121–26

Latham DW, Mazeh T, Stefanik RP, Mayor M, Burki G. 1989. *Nature* 339:38–40

Leggett SK, Allard F, Hauschildt PH. 1998a. *Ap. J.* 509:836–47

Leggett SK, Allard FA, Dahn CC, Hauschildt PH, Rayner J. 1998b. *AAS Meeting 193*. Abstr. 98.05

Leggett SK, Toomey DW, Geballe TR, Brown RH. 1999. *Ap. J.* 517:L139–42

Leibert J, Kirkpatrick JD, Reid IN, et al. 2000. *Cool Stars, Stellar Systems, and the Sun, Eleventh Cambridge Workshop*. Submitted

Leibert J, Kirkpatrick JD, Reid IN, Fisher MD. 1999. *Ap. J.* 519:345–53

Linsky JL, Wood BE, Brown A, Giampapa MS, Ambruster C. 1995. *Ap. J.* 455:670–76

Lodders K. 1999. *Ap. J.* 519:793–801

Luhman KL, Leibert J, Rieke GH. 1997. *Ap. J.* 489:L65–68

Luhman KL, Rieke GH, Lada CJ, Lada EA. 1998. *Ap. J.* 508:347–69

Macintosh B, Zuckerman B, Becklin EE, McLean IS. 1996. *AAS Meeting 189,* Abstr. 120.05

Magazzù A, Martín EL, Rebolo R. 1993. *Ap. J.* 404:L17–20

Mannings V, Boss A, Russell SS, eds. 2000. *Protostars and Protoplanets IV*. Tucson: Univ. of Arizona Press. In press

Marcy GW, Basri G, Graham JR. 1994. *Ap. J.* 428, L57–60

Marcy GW, Butler RP. 1998. *Annu. Rev. Astron. Astrophys.* 36:57–97

Marley MS, Saumon D, Guillot T, Freedman RS. 1996. *Science* 28:1919–21

Martín EL, Basri G, Brandner W, Bouvier J, Zapatero-Osorio MR, et al. 1998a. *Ap. J.* 509:L113–16

Martín EL, Basri G, Delfosse X, Forveille T. 1997. *Astron. Astrophys.* 327:L29–32

Martín EL, Basri G, Zapatero-Osorio MR. 1999a. *Astron. J.* 118:105–14

Martín EL, Basri G, Zapatero-Osorio MR, Rebolo R, García-Lòpez RJ. 1998b. *Ap. J.* 507:L41–44

Martín EL, Brandner W, Basri G. 1999b. *Science* 238:1718–20

Martín EL, Delfosse X, Basri G, Goldman B, Forveille T, Zapatero-Osorio MR. 1999c, *Astron. J.* 118:2466–82

Martín EL, Rebolo R, Magazzù A. 1995. *Ap. J.* 436:262–69

Martín EL, Zapatero-Osorio MR, Rebolo R. 1996. In *Cool Stars, Stellar Systems, and the Sun, Ninth Cambridge Workshop*, ed. R Pallavicini, AK Dupree, *A.S.P. Conf. Ser.* 109:615–16. San Francisco: Astron. Soc. Pacific

Matthews K, Nakajima T, Kulkarni SR. 1996. *Astron. J.* 112:1678–82

Mayor M, Queloz D, Udry S. 1997. In *Brown Dwarfs and Extrasolar Planets*, ed. R Rebolo, EL Martín, MR Zapatero-Osorio, *A.S.P. Conf. Ser.* 134:140–51. San Francisco: Astron. Soc. Pacific

Mayor M, Udry S, Queloz D. 1998. In *Cool*

Stars, Stellar Systems, and the Sun, Tenth Cambridge Workshop, ed. RA Donahue, JA Bookbinder, *A.S.P. Conf. Ser.* 154:77–84. San Francisco: Astron. Soc. Pacific

McCarthy DW, Probst RC, Low FJ. 1985. *Ap. J.* 290:L9–13

Nakajima T, Oppenheimer BR, Kulkarni SR, Golimowski DA, Matthews K, Durrance ST. 1995. *Nature* 378:463–65

Nelson LA, Rappaport S, Chiang E. 1993. *Ap. J.* 413:364–67

Neuhauser R, Brice OC, Comeron F, Hearty T, Martín EL, et al. 1999. *Astron. Astrophys.* 343:883–93

Neuhauser R, Comeron F. 1999. *Astron. Astrophys.* 350:612–16

Noll K, Marley M, Freedman R, Leggett S. 1999. *AAS DPS Meeting 31*, Abstr. 02.01

Noll KS, Geballe TR, Marley MS. 1997. *Ap. J.* 489:L87–90

Oppenheimer BR. 1999. PhD thesis. Calif. Inst. of Technology. 185 pp.

Oppenheimer BR, Kulkarni SR, Matthews K, Nakajima T. 1995. *Science* 270:1478–79

Oppenheimer BR, Kulkarni SR, Matthews K, Van Kerwijk MH. 1998. *Ap. J.* 502:932–43

Oppenheimer BR, Kulkarni SR, Stauffer JR. 2000. In *Protostars and Protoplanets IV*, ed. V Mannings, A Boss, SS Russell. Tucson: Univ. of Arizona Press. In press

Patience J, Ghez AM, Reid IN, Weinberger AJ, Matthews K. 1998. *Astron. J.* 115:1972–88

Pavlenko YV, Rebolo R, Martín EL, García-Lòpez RJ. 1995. *Astron. Astrophys.* 303:807–18

Randich S. 1998. *Cool Stars in Clusters and Associations: Magnetic Activity and Age Indicators, Mem. Soc. Astr. Ital.* 68:971–84

Rebolo R, Martín EL, Basri G, Marcy GW, Zapatero-Osorio MR. 1996. *Ap. J.* 469:L53–56

Rebolo R, Martín EL, Magazzù A. 1992. *Ap. J.* 389:L83–86

Rebolo R, Martín EL, Zapatero-Osorio MR, eds. 1998a. *Brown Dwarfs and Extrasolar Planets, A.S.P. Conf. Ser.* 134. San Francisco: Astron. Soc. Pacific

Rebolo R, Zapatero-Osorio MR, Madruga S, Bejar VJS, Arribas S, et al. 1998b. *Science* 282:1309–12

Rebolo R, Zapatero-Osorio MR, Martín EL. 1995. *Nature* 377:129–31

Reid IN. 1999. *MNRAS* 302:L21–23

Reid IN, Gizis JE. 1997. *Astron. J.* 114:1992–98

Reid IN, Hawley S. 1999. *Astron. J.* 117:343–53

Reid IN, Kirkpatrick JD, Gizis JE, Leibert J. 1999a. *Ap. J.* 527:105–7

Reid IN, Kirkpatrick JD, Leibert J, Burrows A, Gizis JE, et al. 1999b. *Ap. J.* 521:613–29

Ruiz MT, Leggett SK, Allard F. 1997. *Ap. J.* 491:L107–10

Saumon D, Bergeron P, Lunine JI, Hubbard WB, et al. 1994. *Ap. J.* 424:333–44

Saumon D, Hubbard WB, Burrows A, Guillot T, Lunine JI, Chabrier G. 1996. *Ap. J.* 460:993–1018

Schulz AB, Allard F, Clampin M, McGrath M, Bruhweiler FC, et al. 1998. *Ap. J.* 492:181–84

Simons DA, Henry TJ, Kirkpatrick JD. 1996. *Astron. J.* 112:2238–48

Skrutskie MF, Forrest WJ, Shure M. 1989. *Astron. J.* 98:1409–17

Stauffer JR, Barrado y Navascués D, Bouvier J, Morrison HL, et al. 1999a. *Ap. J.* 527:219–29

Stauffer JR, Hamilton D, Probst R. 1994. *Astron. J.* 108:155–59

Stauffer JR, Herter T, Hamilton D, Rieke GH, Rieke MJ, et al. 1991. *Ap. J.* 367:L23–26

Stauffer JR, Schild R, Barrado y Navascués D, Backman DE, Angelova AM, et al. 1998. *Ap. J.* 504:805–20

Stauffer JR, Shultz G, Kirkpatrick JD. 1999b. *Ap. J.* 499:199–203

Steele IA, Jameson RF. 1995. *MNRAS* 272:630–46

Steele IA, Jameson RF, Hodgkin ST, Hambly NC. 1995. *MNRAS* 275:841–49

Strauss MA, Fan X, Gunn JE, Leggett SK, Geballe TR, et al. 1999. *Ap. J.* 522:L61–64

Stuik R, Bruls JHMJ, Rutten RJ. 1997. *Astron. Astrophys.* 322:911–23

Tamura M, Itoh Y, Oasa Y, Nakajima T. 1998. *Science* 282:1095–97

Thackrah A, Jones H, Hawkins M. 1997. *MNRAS* 284:507–12

Tinney CG. 1994. *The Bottom of the Main Sequence—and Beyond.* Heidelberg: Springer-Verlag. 309 pp.

Tinney CG. 1998. *MNRAS* 296:L42–44

Tinney CG, Delfosse X, Forveille T. 1997. *Ap. J.* 490:L95–97

Tinney CG, Delfosse X, Forveille T, Allard F. 1998. *Astron. Astrophys.* 338:1066–72

Tinney CG, Reid IN. 1998. *MNRAS* 301:1031–48

Tinney CG, Tolley AJ. 1999. *MNRAS* 304:119–26

Title AM, Schrijver CJ. 1998. In *Cool Stars, Stellar Systems, and the Sun, Tenth Cambridge Workshop*, ed. RA Donahue, JA Bookbinder, *A.S.P. Conf. Ser.* 154:345–58. San Francisco: Astron. Soc. Pacific

Tokunaga AT, Kobayashi N. 1999. *Astron. J.* 117:1010–13

Tsuji T, Ohnaka K, Aoki W. 1996a. *Astron. Astrophys.* 305:L1–4

Tsuji T, Ohnaka K, Aoki W. 1999. *Ap. J.* 520:L119–22

Tsuji T, Ohnaka K, Aoki W, Nakajima T. 1996b. *Astron. Astrophys.* 308:L29–32

Valenti JA, Johns-Krull CM, Reid IN. 2000. *Cool Stars, Stellar Systems, and the Sun, Eleventh Cambridge Workshop.* Submitted

White RJ, Ghez AM, Reid IN, Schultz G. 1999. *Ap. J.* 520:811–21

Wilking B, Greene T, Meyer MR. 1999. *Astron. J.* 117:469–82

Williams DM, Comeron F, Rieke GH, Rieke MJ. 1995. *Ap. J.* 454:144–50

Zapatero-Osorio MR, Bejar VJS, Rebolo R, Martín EL, Basri G. 1999a. *Ap. J.* 524:L115–18

Zapatero-Osorio MR, Martín EL, Rebolo R. 1997a. *Astron. Astrophys.* 323:105–12

Zapatero-Osorio MR, Martín EL, Rebolo R, Basri G, Magazzù A, et al. 1997b. *Ap. J.* 491:L81–84

Zapatero-Osorio MR, Rebolo R, Martín EL, Hodgkin ST, Cossburn MR, et al. 1999b. *Astron. Astrophys. Suppl.* 134:537–43

Annu. Rev. Astron. Astrophys. 2000. 38:521–71

PHENOMENOLOGY OF BROAD EMISSION LINES IN ACTIVE GALACTIC NUCLEI

J. W. Sulentic,[1] P. Marziani,[2] and D. Dultzin-Hacyan[3]

[1]*Department of Physics & Astronomy, University of Alabama, Tuscaloosa, AL 35487-0324*
[2]*Osservatorio Astronomico di Padova, Vicolo dell' Osservatorio 5, I-35122, Padova, Italy*
[3]*Instituto de Astronomia, Universidad Nacional Autonoma de Mexico, Apdo Postal 70-264, Mexico D. F. 04510, Mexico*

Key Words   Seyfert galaxies, quasars, accretion disks, spectroscopy, emission lines, line formation

■ **Abstract**   Broad emission lines hold fundamental clues about the kinematics and structure of the central regions in AGN. In this article we review the most robust line profile properties and correlations emerging from the best data available. We identify fundamental differences between the profiles of radio-quiet and radio-loud sources as well as differences between the high- and low-ionization lines, especially in the radio-quiet majority of AGN. An Eigenvector 1 correlation space involving FWHM Hβ, W(FeII$_{opt}$)/W(Hβ), and the soft X-ray spectral index provides optimal discrimination between all principal AGN types (from narrow-line Seyfert 1 to radio galaxies). Both optical and radio continuum luminosities appear to be uncorrelated with the E1 parameters. We identify two populations of radio-quiet AGN: Population A sources (with FWHM(Hβ) $\lesssim$ 4000 km s^{-1}, generally strong FeII emission and a soft X-ray excess) show almost no parameter space overlap with radio-loud sources. Population B shows optical properties largely indistinguishable from radio-loud sources, including usually weak FeII emission, FWHM(Hβ) $\gtrsim$ 4000 km s^{-1} and lack of a soft X-ray excess. There is growing evidence that a fundamental parameter underlying Eigenvector 1 may be the luminosity-to-mass ratio of the active nucleus (L/M), with source orientation playing a concomitant role.

1.   INTRODUCTION

Active Galactic Nuclei (AGN) are usually characterized by a strong nuclear emission-line spectrum at UV and optical wavelengths. AGN phenomenology encompasses sources that show: (1) no lines except when a highly variable continuum is in a low phase (Blazars), (2) only narrow lines (Seyfert 2, possibly, LINERS and narrow-line radio galaxies), and (3) both broad (FWHM $\approx$ 1000–15,000 km s^{-1}) and narrow (FWHM $\approx$ 200–2000 km s^{-1}) lines (Seyfert 1 galaxies, broad-line radio galaxies, (BLRGs), QSOs, and quasars). The narrow emission spectrum is dominated by forbidden lines, whereas the broad emission lines are produced mainly by permitted transitions and often shows superposed narrow components. The relative strengths of the broad and narrow components (hereafter BC and NC) formed the basis for the subclassification of Seyfert galaxies into intermediate types (Osterbrock & Koski 1976; Osterbrock 1977, 1981). See also Winkler (1992) for a revised version of the Osterbrock classification that was adopted for the AGN catalog of Véron-Cetty & Véron (1998). The AGN commonality in the context of this review involves sources that show broad emission lines most of the time. We do not consider sources with broad lines that are: (1) infrequently seen in BL Lacs (e.g. Corbett et al 1996), (2) only detected in polarized light (Seyfert 2s; Antonucci & Miller 1985, Kay & Moran 1998), and (3) only present at very low intensity levels in the nuclei of nearby galaxies (mini-Seyferts; Peimbert & Torres-Peimbert 1981, Filippenko & Sargent 1989, Ho et al 1997).

We present the AGN broad-line discussion in a slightly unconventional way by reviewing the subject in three somewhat orthogonal parts: (1) broad-line phenomenology, (2) physical interpretation of the phenomenology, and (3) confrontation with models for the structure and kinematics of the broad-line region (BLR). This division has certain advantages that will lead to a clearer picture of what we really know about the broad lines independently of the favored models. Underlying this approach is the assumption that statistical studies of lines (coupled with reverberation studies) offer the best hope for resolving BLR structure and kinematics. Since the last ARAA review (Osterbrock & Mathews 1986), large samples of high s/n optical spectra have been obtained using linear detectors. Much improved UV data have also come from the Hubble Space Telescope. AGN show an enormous range of redshift ($0.006 \lesssim z \leq 6.0$, with a peak in the redshift distribution near $z \approx 2.0$; Hewitt & Burbidge 1993, Véron-Cetty & Véron 1998). The principal broad lines can be studied in the optical domain over the following source redshift ranges: HI Hα, 0.0–0.5; Hβ, 0.0–1.0; MgIIλ2800, 0.3–2.6; CIII] λ1909, 0.8–4.2; CIVλ1549, 1.2–5.4 and HI Lyα 1.9–7.2. A more extensive list of lines is given by Burbidge & Burbidge 1967, Wilkes 1999, and Kriss et al 1999 for EUV (see also the NIST physical reference database for line identification).

Many of the largest (and highest s/n) studies have focused on two broad lines (CIVλ1549 and Hβ) that are representative of the high (HIL) and low (LIL) ionization lines (i.e. lines produced by ions with ionization potential ≥ 50 eV and ≤ 20 eV, respectively). Hβ and CIVλ1549 are also particularly useful because they are less contaminated by nearby lines and because they permit us to compare LIL and HIL properties in the same sources out to $z \approx 1.0$. Improvements in near infrared detectors offer the promise of obtaining high s/n and resolution LIL spectra at higher redshift in the near future. As will become clear later, the HIL–LIL distinction is quite important (Section 3.2). Table 1 summarizes the principal line profile data sources with resolution equal to or of at least 10 Å FWHM. The tabulation includes both original surveys (NEW) and statistical studies that used these data sources (LIT).

TABLE 1

Reference	Data[a]	NOBJ[b]	RQ[c]	RL[d]	RES (Å)[e]	Lines[f]	Zrange[g]
Gaskell 1982	NEW	33	0	33	9	HIL/LIL	0.2–2.3
Young et al. 1982	NEW	27	15	12	2.5	HIL	1.4–2.7
Wilkes 1984 et al. 1983	NEW	214	Few	Most	5–15	HIL/LIL	0–3.6
de Robertis 1985	NEW	27	26	1	≤5	LIL	0–0.3
Sargent et al. 1988	NEW	55	43	12?	2	HIL	1.8–3.6
Sulentic 1989	LIT	61	49	12	1–10	LIL	0–0.4
Baldwin et al. 1989	NEW	74	0	74	7–10	HIL	0–2.5
Sargent et al. 1989	NEW	59	45	14	4–6	HIL	2.7–4.1
Espey et al. 1989	NEW	18	9	9	3–40	HIL/LIL	1.3–2.4
Barthel et al. 1990	NEW	67	0	67	5	HIL	1.5–3.8
Steidel & Sargent 1991	NEW	92	50	42	4–6	HIL	0.7–2.7
Corbin 1991	LIT	45	0	45	2–13	HIL	1.2–3.5
Stirpe 1991	NEW	29	29	0	3	LIL	0–0.4
Kinney et al. 1991	IUE	69	34	35	2–3	HIL	0–2.2
Boroson & Green 1992	BQS	87	70	17	6–7	LIL	0–0.5
Tytler & Fan 1992	LIT	160	—	—	2–6	HIL	1.4–4
Steidel & Sargent 1992	NEW	103	50%	50%	4–6	HIL	0.2–2.2
Eracleous & Halpern 1994	NEW	94	0	94	6	LIL	0.1–0.4
Wills et al. 1993	LIT	123	94	29	5–6	HIL	0.7–2.7
Corbin 1993	LIT	55	42	19	6–7	LIL	0–0.5
Corbin & Francis 1994	LIT	79	23	56	10	HIL	1.3–2.7
Brotherton et al. 1994	LIT	85	47	38	4–6	HIL	0.9–2.2
Wills et al. 1995	FOS	31	0	31	2–3	HIL	0.3–1.3
Laor et al. 1995	FOS	13	11	2	1–2	HIL	0.2–2.1
Hewett et al. 1995+refs	LBQS	1055	10%	90%	6–10	HIL/LIL	0.2–3.4
Marziani et al. 1996	NEW	51	30	21	3–8	HIL/LIL	0–0.8
Brotherton 1996	NEW	60	0	60	3–6	LIL	0–0.95
Corbin & Boroson 1996	NEW	48	22	26	5–10	HIL/LIL	0.3–0.8
Corbin 1997	NEW	133	71	62	6–7	LIL	0.04–0.6
Grupe et al 1999	NEW	76	76	0	5	LIL	0–0.5

[a]Reference type, NEW—a sample of more than 10 new spectra. LIT—a paper that is primarily a statistical analysis of data from other references.

[b] Total number of AGN.

[c]Number of radio quiet sources.

[d]Number of radio loud sources.

[e]Approximate spectral resolution of the data.

[f] Indicates a reference presenting primarily HIL, LIL or mixed sample of source.

[g] The approximate redshift range of the spectra data presented or analyzed. A lower limit less than 0.1 is rounded to zero.

2.   LINE PROFILE PHENOMENOLOGY

In our phenomenological discussion we identify the most significant similarities and differences in the broad lines within individual sources and between different AGN populations. This is attempted without the encumbrance of uncertain emission line physics or currently popular models, both of which are considered in later sections. The principal independent and robust parameters that exist for significant samples of AGN include: (1) line width; (2) line centroid velocity shift, if possible with respect to the AGN rest frame; and (3) line shape/asymmetry (defined in slightly different ways; de Robertis 1985; Stirpe 1991; BG92[1]; Brotherton 1996; M96[2]; Corbin 1997). Our focus on two lines LIL $=$ Hβ and HIL $=$ CIVλ1549 is sufficient to compare the principal broad-line properties and differences that depend on ionization level. Accurate measures depend on our ability to extract "pure" line profiles, and this requires several data-processing steps: (1) continuum estimation and subtraction, (2) modeling and subtraction of optical (FeII$_{opt}$) or UV (FeII$_{UV}$) singly-ionized iron emission, (3) removal of superposed narrow line components, and (4) subtraction of overlapping narrow lines of other atomic species. A detailed discussion of these procedures is beyond the scope of this review. The references in Table 1 can be consulted for details. Subtraction of the blended UV and, especially, optical (broad-line) FeII emission has taken on a dual importance both to decontaminate the appropriate line and to parameterize the FeII emission strength that is emerging as an important AGN diagnostic (Section 3). It is important to point out that the FeII and other corrections require spectra with resolution of at least 10 Å (Table 1) and continuum (if we are ever really measuring it) s/n $\gtrsim$ 20.

2.1   Line Profile Widths

Full width at half maximum (FWHM) is one of the most robust line profile measures. Measures made at lower levels in the profile will be increasingly affected by uncertainties associated with the processing steps mentioned above. Figure 1 shows a histogram of the most accurate FWHM measures for Hβ_{BC} and CIVλ1549$_{BC}$ in low redshift radio-loud (RL) and radio-quiet (RQ) samples. The data for Hβ_{BC} is a combination of measures from BG92, M96, Brotherton 1996, Corbin & Boroson 1996, and Corbin 1997. This combination provides measures for 180 sources with a strong RL overrepresentation (n $=$ 96). Comparison of measures in common for individual sources indicates that, aside from a few spurious measurements, a 2σ uncertainty of $\approx$10% is a reasonable estimate of the error associated with FWHM Hβ_{BC} measures. Figure 2 illustrates the range of Hβ_{BC} width (and FeII strength) for two prototypical Seyfert (RQ) galaxies (indicated by arrows in Figure 1).

The matching CIVλ1549$_{BC}$ sample is limited to 76 objects of which n $=$ 38 are RL (M96, Laor et al 1994, Sulentic et al 2000; a 2σ uncertainty of 10–15% or more is typical; see references for specific values). Other studies do not subtract

---

[1] Abbr. for Boroson & Green 1992.
[2] Abbr. for Marziani et al 1996.

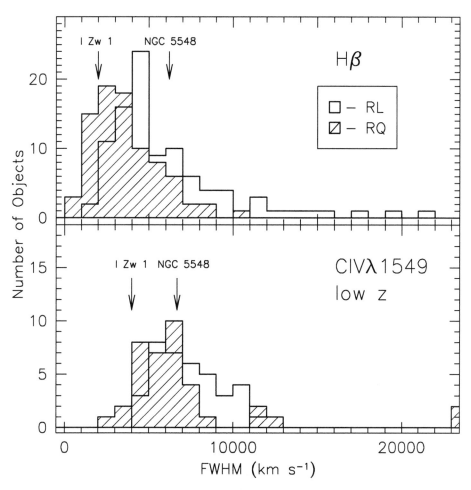

Figure 1   The distributions of rest frame FWHM for *upper*: Hβ$_{BC}$ (z ≤ 0.5; n = 180), and *lower*: CIVλ1549$_{BC}$ (z ≲ 0.8; n = 76). Radio-quiet (RQ) sources are hatched in each histogram.

a CIVλ1549$_{NC}$ component, which can yield quite different results (see Sulentic & Marziani 1999 for a recent discussion of this contentious issue). These data provide strong evidence that both HIL and LIL profiles are broader in RL sources. Sample mean and rms values are FWHM$_{RL}$ (Hβ$_{BC}$) ≈ 5900 ± 3700 km s^{-1} versus 3600 ± 1900 km s^{-1} for RQ. The corresponding CIVλ1549 values are FWHM$_{RL}$ (CIVλ1549$_{BC}$) = 7440 ± 2000 km s^{-1} versus 6200 ± 2150 km s^{-1} for RQ. A Kolmogorov-Smirnov (K–S) test indicates that the RQ–RL difference is marginally significant (probability 3 × 10^{-2} that the samples come from the same parent population). The comparison also indicates that CIVλ1549$_{BC}$ is systematically broader than Hβ$_{BC}$ in both RL and RQ sources. The HIL-LIL differences show a larger amplitude, and at least for RQ, are statistically more significant than the RL-RQ

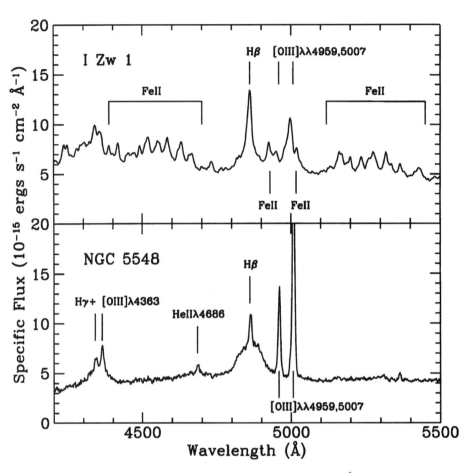

Figure 2 Spectra of typical AGN in the rest frame range λλ4200–5500 Å. *Upper:* I Zw 1, a FeII strong narrow line Seyfert 1. *Lower:* NGC5548, a FeII weak broad line Seyfert 1. Ordinate is flux in units of 10^{-15} ergs s^{-1} cm^{-2} Å^{-1} [NGC 5548 was taken with a 2 arcsec slit so absolute flux scale is below the normalized flux scale of international monitoring campaign spectra (Korista et al 1995)].

ones (M96 found K-S probabilities of 0.002 and 0.035 for the HIL-LIL differences in RQ and RL samples respectively). These differences contradict results of studies that did not subtract CIVλ1549$_{NC}$ where mean FWHM(CIVλ1549$_{BC}$) ≤ FWHM (Hβ$_{BC}$) [e.g. FWHM(Hβ) = 1.1–1.3 FWHM(CIVλ1549$_{BC}$); Corbin 1991 with IUE data, Laor et al 1995, Corbin & Boroson 1996]. There is also evidence for a difference between RL core [FWHM(Hβ$_{BC}$) ≈ 4700 ± 400 km s^{-1} (error of the mean): n = 26] and lobe [FWHM(Hβ$_{BC}$) = 6000 ± 600 km s^{-1} n = 32] dominated sources (Brotherton 1996). This result was confirmed in an analogous comparison of FWHM(Hβ$_{BC}$) in steep (SS) and flat (FS) spectrum RL sources

(9 sources in common) where $\text{FWHM}_{FS}(\text{H}\beta_{\text{BC}}) \approx 4740 \pm 2300 \text{ km s}^{-1}$ ($n = 32$) and $\text{FWHM}_{SS}(\text{H}\beta_{\text{BC}}) \approx 6130 \pm 3060 \text{ km s}^{-1}$ ($n = 30$; Corbin 1997).

Sources that have been monitored for $\text{H}\beta$ variability usually show an rms $\text{FWHM}(\text{H}\beta_{\text{BC}})$ that is within 10% of the mean $\text{FWHM}(\text{H}\beta_{\text{BC}})$ values. The 19 most reverberated sources (Wandel et al 1999) show $\langle \text{FWHM}_{mean}/\text{FWHM}_{rms} \rangle \approx 0.95 \pm 0.14$; 74% of the sources show a larger value for $\text{FWHM}_{rms}(\text{H}\beta_{\text{BC}})$, which reflects significant variation in the wings of the line. Some extreme exceptions usually involve the appearance of a new broad-line component. These include the transition from Seyfert 2 (or Liner) to Seyfert 1 (e.g. NGC 4151; Ulrich et al 1985) or from a single-peaked to a double-peaked BLR (e.g. Pictor A; Sulentic et al 1995, Halpern & Eracleous 1994). These results suggest that FWHM is a reasonable line diagnostic even in the presence of significant variations. All of these results should be viewed with caution because available data samples are often heterogeneous and subject to hidden selection effects (e.g. as we will see BG92 probably favors sources with narrow RQ profiles).

2.1.1 Line Width Dependence on Redshift

Albeit an early HIL-LIL comparison used IUE $\text{C\textsc{iv}}\lambda1549$ spectra (Corbin 1991), reliable direct HIL-LIL comparisons became possible only with the advent of the HST FOS database. High-redshift HIL-LIL FWHM comparisons are difficult for two reasons: (1) published infrared spectra for the $\text{H}\beta$ region still suffer from low resolution and s/n (e.g. Nishihara et al 1997); and (2) although large samples of high s/n $\text{C\textsc{iv}}\lambda1549$ spectra exist for high-redshift sources (Table 1), they have not been corrected for the $\text{C\textsc{iv}}\lambda1549_{\text{NC}}$ contribution. The largest available samples of RQ (BG92; M96) and RL (Brotherton 1996, Corbin 1997) data show no significant correlation between source luminosity and FWHM $\text{H}\beta_{\text{BC}}$.

We identified published spectra for a total of 53 sources ($z \gtrsim 2.0$) that showed an obvious inflection between the broad and narrow components of $\text{C\textsc{iv}}\lambda1549$ or where the line showed a rounded top consistent with the absence of any narrow component. Analog processing of this sample (obviously biased toward broader $\text{C\textsc{iv}}\lambda1549$ profiles) yielded FWHM estimates likely to be accurate within $\pm 500 \text{ km s}^{-1}$. We obtained a mean (RL+RQ) $\text{FWHM}(\text{C\textsc{iv}}\lambda1549_{\text{BC}}) \approx 7300 \pm 2100 \text{ km s}^{-1}$, which is similar to the low-redshift values presented earlier. The available evidence suggests no significant difference in profile width between high- and low-redshift samples of AGN. $\text{C\textsc{iv}}\lambda1549$ also maintains the same profile *width* as $\text{W}(\text{C\textsc{iv}}\lambda1549)$ systematically decreases with redshift/source luminosity (Section 3.1).

2.1.2 FWZI

Broad line profiles sometimes show inflections that complicate our interpretation of simple measures like FWHM. Inflections raise the possibility that broad lines are a composite of several kinematically and/or geometrically distinct emitting regions. The frequent "mismatch" in profile wings (Romano et al 1996) also point in this direction. The best evidence for a second distinct component in broad-line profiles involves a very broad component (VBC) that sometimes underlies $H\beta_{BC}$ and $CIV\lambda1549_{BC}$. It is unclear whether the VBC component sometimes seen in $HeII\lambda4686$ (Ferland et al 1990, Marziani & Sulentic 1993) shows the same properties. We are aware of only two studies that have attempted to measure the underlying VBC (Corbin 1995: $H\beta$; Laor et al 1994: $CIV\lambda1549$). The former used the mostly RQ BG92 sample and found mean FWHM($H\beta$ VBC) $\approx 9500 \pm 7500$ km s^{-1} for 67 sources, almost 3 times the mean FWHM of the classical $H\beta_{BC}$. The latter studied FOS spectra for five sources and obtained FWHM($CIV\lambda1549$ VBC) in the range 12–24,000 km s^{-1} (no $FeII_{UV}$ was subtracted).

An alternative method for studying the VBC involves available full width at zero intensity (FWZI) measures of $H\alpha$ where FeII emission is not a serious problem (Osterbrock & Shuder 1982, Shuder 1984). We analog extracted FWZI values from other samples of high s/n $H\alpha$ spectra (Stirpe 1989, Sulentic et al 1998), which yielded an additional sample of n = 47 mixed RL/RQ sources. A large complementary (n = 94) sample of FWZI ($H\alpha$) measures (Eracleous & Halpern 1994: EH94) includes all of the best known double-peaked and complex RL profiles. The mean values (±2000 km s^{-1}) for the combined RL and RQ samples are $FWZI_{RL}(H\alpha) \approx 19700$ km s^{-1} and $FWZI_{RQ}(H\alpha) \approx 17,950$ km s^{-1}. These are surprisingly similar, considering (1) a bias in the EH94 RL sample for unusually broad profiles, and (2) the significantly larger mean FWHM found for RL samples. It is unclear how we should interpret these results because some profiles with smooth, very broad wings show FWZI values just as large as obviously inflected profiles.

2.2 Line Profile Velocity Shifts

Velocity displacement of broad lines is one of the most striking and potentially model-constraining profile diagnostics. Broad-line shifts are often measured relative to one another—especially for the HIL, because strong UV forbidden lines are rare (Gaskell 1982, Wilkes 1984, Junkkarinen 1989). Because all broad lines show velocity shifts ($MgII\lambda2800$ appears to be the most stable; Junkkarinen 1989), relative measures are dangerous and difficult to interpret. Ideally, one would like to know the mean line shift and dispersion for each line relative to the rest frame. Several comparisons exist between [OIII]$\lambda\lambda4959,5007$ and radio HI and/or absorption line measures of the host galaxies (Boroson & Oke 1984, Vrtilek & Carleton 1985, Wilson & Heckman 1985, Appenzeller & Östreicher 1988). These studies suggest that the narrow lines show velocity displacements of ±100 km s^{-1} or less relative to the host rest frame. This supports the cautious use of the narrow line redshift as a measure of the local rest frame in an AGN (for exceptions see Erkens et al 1997 and Section 3.2.1).

Figure 3 shows the distribution of line shifts for $H\beta_{BC}$ and $CIV\lambda1549_{BC}$ in low redshift RQ and RL sources. The $H\beta_{BC}$ sample is a combination of data from Marziani et al 1996 (M96: n = 30 RL and n = 21 RQ), Eracleous & Halpern

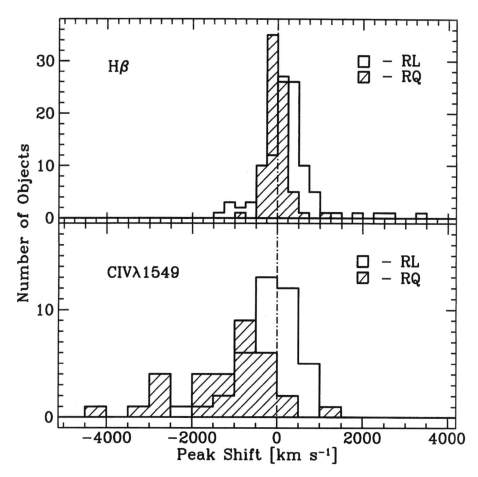

Figure 3 The distribution of rest frame peak line shifts for (*upper*) Hβ_{BC} ($z \leq 0.5$) and (*lower*) CIVλ1549$_{BC}$. Radio-quiet (RQ) sources are hatched in each histogram.

1994 (EH94: Hα measures for n = 75 RL sources), and Boroson & Green 1992 (BG92: n = 63 mostly RQ) samples. The CIVλ1549$_{BC}$ data come from M96 and Sulentic et al 2000 (plus two AGN from Laor et al 1994), who measured CIVλ1549$_{BC}$ shifts relative to [OIII]$\lambda\lambda$4959,5007 and subtracted CIVλ1549$_{NC}$. These sources are supplemented by 21 lower-resolution IUE CIVλ1549 spectra (Espey et al 1989, Rodriguez-Pascual et al 1997a). The former measures the shifts relative to MgIIλ2800 and the latter measures those relative to the CIVλ1549$_{NC}$. CIVλ1549$_{BC}$ and Hβ_{BC} show line shifts in many sources, whereas Hβ_{NC} and CIVλ1549$_{NC}$ do not. The most impressive shift behavior involves the apparent *systematic blueshift* of CIVλ1549$_{BC}$ in RQ sources. This trend was first revealed in the early 1980s (Gaskell 1982, Wilkes 1984). The mean CIVλ1549$_{BC}$ shift relative to [OIII]$\lambda\lambda$4959,5007 (n = 21 RQ sources from M96) is $\Delta v_r \approx -790 \pm$ 890 km s^{-1}. The distribution of shifts indicates that CIVλ1549$_{BC}$ is not blueshifted by a constant value relative to rest frame, but instead shows a distribution of values from zero to at least –4000 km s^{-1}. The CIVλ1549$_{BC}$ blueshift in RQ sources is usually present at all levels in the profile from peak to 1/4 maximum. In other studies, the RQ blueshift is masked/muted by the variable contribution of CIVλ1549$_{NC}$ and/or measures relative to a broad line (e.g. Junkkarinen 1989; Corbin 1990, 1991; Steidel & Sargent 1991; Corbin & Boroson 1996). Most of these samples still show evidence for a mean blueshift in the range –60 to –2000 km s^{-1}. The mean Hβ_{BC} shift values for RQ sources are –20 $\pm$ 349 and –48 $\pm$ 228 km s^{-1} (consistent with a mean of zero) for the M96 and BG92 samples, respectively. The sigma values given above are sample standard deviations that provide a measure of the range in amplitude of observed shifts. Typical (1σ) uncertainties for shift measures are $\sim$50 and 100 km s^{-1} for Hβ_{BC} and CIVλ1549$_{BC}$.

These lines, respectively, show mean (and sample standard deviation) RL shifts +500 $\pm$ 1200 and -140 ± 620 km s^{-1} (n = 40 objects from M96 and Sulentic et al 2000), the latter mean consistent with zero. Both RL values are redshifted relative to the RQ values. It may be significant that CIVλ1549$_{BC}$ consequently maintains a blueshift relative to Hβ_{BC} in almost every measured source (see M96 for a few possible exceptions). It is premature to generalize, but present data suggest that relative to the source rest frame, CIVλ1549$_{BC}$ changes from a net RQ blueshift to a mean RL shift consistent with zero. At the same time, Hβ_{BC} changes from a mean RQ shift near zero to a mean RL redshift. Although generally less than $\pm$600 km s^{-1}, the RQ Hβ_{BC} shifts are real, as profiles with a strong NC-BC inflection shown in Figure 4 make clear. Examples of a large RQ CIVλ1549$_{BC}$ blueshift and RL Hβ_{BC} redshift are shown in upper left panel Figure 5 and lower left panel Figure 6, respectively.

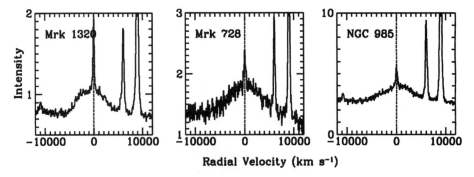

Figure 4   Examples of small Hβ_{BC} line shifts in relatively symmetric RQ sources that show a very clear inflection between the broad- and narrow-line components. From left to right, the peak line shifts are –565, +270 (marginal), and +420 km s^{-1}, respectively. Vertical dotted line marks the NLR redshift assumed to be the local rest frame.

As always happens in AGN phenomenology, there are exceptions to the rule. For example, RQ source H 1821+643 (Laor et al 1994) shows $CIV\lambda1549_{BC}$ $\Delta v_r \approx$ +1000 km s^{-1} relative to [OIII]$\lambda\lambda$4959,5007 and the RL source 3C227 (Sulentic 1989; EH94) shows a $\Delta v_r(H\beta_{BC}) \approx -1800$ km s^{-1} (RL Hβ_{BC} shows a known line shift range from -2000 to $+5000$ km/s). The more serious exception (H 1821+643) is not certain because strong FeII$_{UV}$ in this source was not subtracted. M96 concluded that the RQ–RL mean shift differences were statistically significant. The M96 RL mean $\Delta v_r(H\beta_{BC}) \approx +490$ km s^{-1} can be compared with a larger RL sample (n = 75; EH94) where $\Delta v_r \approx +186$ km s^{-1}, and a larger RQ sample (BG92; n = 63) where $\Delta v_r \approx -48$ km s^{-1}. There are a small number of sources in common between M96 and these two samples.

Other HIL are very difficult to disentangle from satellite lines (Lyα, CIII]λ1909, MgIIλ2800) and FeII$_{UV}$ emission, as a recent detailed study of IZw1 (Laor et al 1997a) makes clear. The general result is that all of the HIL are blueshifted relative to MgIIλ2800 in RQ samples. Most sources show a line shift within the envelope of the observed profile width (i.e. Δv_r/FWHM $\lesssim$ 0.5; Sulentic 1989).

2.2.1 Dependence on Redshift

Currently available measures suggest that both line width and shift show the same phenomenologies over a wide range in redshift. Espey et al (1989) was the first to compare HIL and LIL data for the same high z sources, using IR measures of Hα. Their results are fully consistent (despite a variable $CIV\lambda1549_{NC}$ contribution) with the low-redshift results of M96 (see Figure 3). Examples of unambiguous large blueshifts (e.g. Q0150-203; Sargent et al 1988) and redshifts (e.g. Q1702+289; Barthel et al 1990) can be found in the high-redshift sources listed in Table 1. The role of MgIIλ2800 may be pivotal in some sources if the results for Q1331+170 are typical (Carswell et al 1991). In that (z $\approx$ 2) source, the HIL show blueshifts ($\Delta v_r \approx -1800$ km s^{-1} for $CIV\lambda1549_{BC}$) and the LIL redshifts (+700 for Hα) all relative to [OIII]$\lambda\lambda$4959,5007. Evidence has been advanced (from studies of high-redshift samples) that line shift correlates directly with line ionization level (Tytler & Fan 1992). This requires further study because the higher ionization line shifts were measured relative to Hβ_{BC}, which is a kinematically "volatile" line.

2.2.2 Shifts at Zero Intensity

Limited studies of the underlying VBC discussed in the previous section suggest very uncertain mean shifts of $\Delta v_r \approx +800$ km s^{-1} for Hβ (not significant with an rms scatter of 3100 km s^{-1}; Corbin 1995) and +9400 $\pm$ 3100 km s^{-1} for $CIV\lambda1549$ (Laor et al 1994). Apparently contradictory results come from a large sample of RL sources (EH94) where the mean shift at ZI for Hβ is found to be only $\Delta v_r \approx$ +270 $\pm$ 1070 km s^{-1}. It is interesting that the EH94 profiles, an extreme (double-peaked or "boxy") sample, show little evidence for a redshifted VBC often seen in other samples of RL sources. M96 found Hβ_{BC} $\Delta v_r(ZI) \approx +80 \pm 2000$ km s^{-1} for RQ and $\Delta v_r(ZI) \approx +1500 \pm 1900$ km s^{-1} for RL, which is more consistent with the frequent observation of a "red shelf" underlying [OIII]$\lambda\lambda$4959,5007. The different results may indicate that a larger Balmer decrement exists in the RL VBC or that a VBC component is not present in all sources.

2.3   Line Asymmetry and Shape

Higher-order moments of the broad lines suffer from larger uncertainties while offering a better possibility to constrain models. Some of the first Hβ_{BC} measures (de Robertis 1985, Sulentic 1989) showed that large numbers of both red and blue asymmetric profiles exist. Examination of the $\sim$15 RL sources in the latter survey showed a preference for red asymmetries (the asymmetry index A.I. = $[C(\frac{3}{4})-C(\frac{1}{4})]$/FWHM, where C(i) are profile centroid measures at different levels). We made a comparison of the measured asymmetry and shape parameters for sources in common in the samples previously discussed. This comparison suggests that measures of asymmetry are good to only one significant figure (0.1).

The preference for red asymmetries in the RL AGN is the most robust result among the higher order moments. BG92 and M96 found $\langle$A.I.$\rangle \approx -0.09 \pm 0.10$ and $\approx -0.06 \pm 0.10$, respectively, for RL sources. Two large studies of RL sources confirm the RL red asymmetry excess with $\langle$A.I.$\rangle \approx -0.10$ (Brotherton 1996) and -0.125 (Corbin 1997). The lobe-dominated (steep-spectrum) RL population shows evidence for slightly more extreme red asymmetries. Both BG92 and M96 find $\langle$A.I.$\rangle \approx 0.00 \pm 0.10$ for the RQ samples, with no preference for red or blue asymmetries. RQ sources show a scatter of CIVλ1549$_{BC}$ asymmetries between ± 0.2 (one source with $+0.32$) with $\langle$A.I.$\rangle \approx +0.04$ (M96). This lack of a strong preference for red or blue asymmetries is particularly remarkable and much in need of confirmation, considering the systematic blueshift that CIVλ1549$_{BC}$ shows in RQ sources. Numerous claims for systematic CIVλ1549$_{BC}$ asymmetries, especially in RQ samples, exist in the literature (e.g. Corbin 1992; Wills et al 1993, 1995; Corbin & Boroson 1996). Measurements of the composite CIVλ1549 line with systematically blueshifted broad component and unshifted NC component would be expected to produce strong blue asymmetry measures. The RL population shows the same preference for red CIVλ1549 asymmetries (as was found for Hβ_{BC}) with $\langle$A.I.$\rangle \approx -0.06$ (M96; Sulentic et al 2000).

A qualitative description of profile shapes ranges from extremely peaked to very "boxy," which may be more common among RL AGN (e.g. the Hβ_{BC} profiles of I Zw 1, and NGC 5548, respectively; see Figure 2). This description is based on examination of reasonably symmetric profiles. The presence of a strong asymmetry makes it difficult to assign a meaningful profile shape. These differences can be quantified by taking the ratio of the line width at different fractional intensities in a profile (for example, $\frac{1}{4}$ and $\frac{3}{4}$ peak intensity). The M96 curtosis parameter (defined as FW$\frac{3}{4}$M/FW$\frac{1}{4}$M) measures between 0.2 and 0.6 in all populations. Considering the current reproducibility of such measures (± 0.1), there is little sensitivity to possible trends. Also, curtosis may be intrinsically

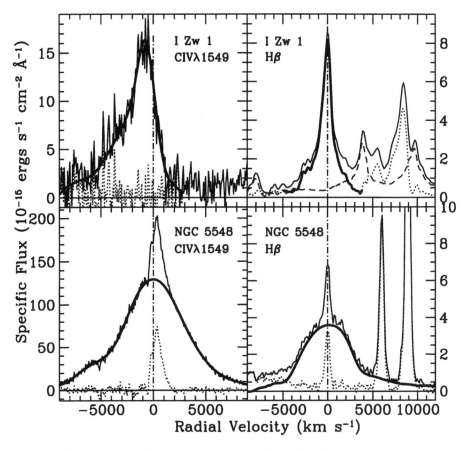

Figure 5 Emission-line profiles for (*lower*) NGC 5548, a RQ Population B Seyfert 1 source, and (*upper*) I Zw 1, a RQ Population A NLS1 source. *Left* panels show the respective CIVλ1549$_{BC}$ profiles, and *right* panels show Hβ$_{BC}$. Ordinate is flux in units of 10^{-15} ergs s^{-1} cm^{-2} AA^{-1}, and abscissa is radial velocity in km s^{-1} relative to the source rest frame (*dot-dashed line*). *Dotted lines* are the residual spectra after FeII$_{opt}$ was subtracted (*upper right*) or after the Hβ$_{BC}$ (*thick solid line*, all other panels) was subtracted. Note (*lower left*) that the apparent shift to the red of CIVλ1549$_{NC}$ in NGC 5548 is a result of self-absorption.

ambiguous if an inflection suggestive of multiple components is present in the line profile.

Studies of profile wings are another way of characterizing profile shape (van Groningen & van Weeren 1989, Stirpe 1989, Penston et al 1990, Romano et al 1996). A sample of about 100 Hα line profiles (4 to 1 RL bias; Romano et al 1996, Robinson 1995) shows not only a mix of log and power-law functions needed to fit the profile wings but also profiles that frequently show significantly different red and blue wing shapes. The latter probably reflects the presence of multiple

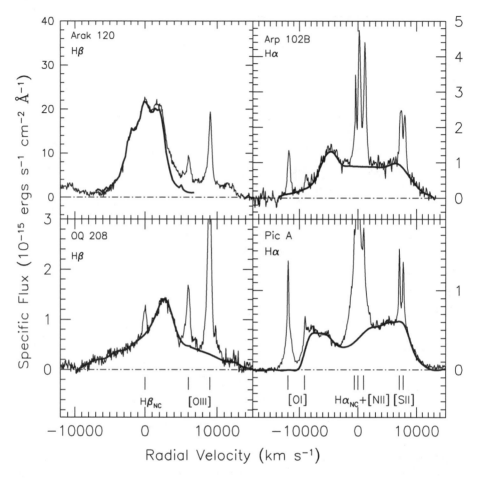

Figure 6 Examples of sources with unusual Balmer line profiles after underlying continuum subtraction. Vertical lines at bottom of the lower panels indicate positions of contaminating narrow emission lines. *Upper left*: Arakelian 120, an apparently small-separation, double-peaked RQ source. *Upper right*: Arp 102B, a prototype broad separation double-peak RL source. *Lower left*: OQ 208, a single-peaked RL source with large-peak redshift. *Lower right*: Pictor A, an RL source Balmer lines that recently changed from single- to double-peak source.

components, such as the redshifted VBC discussed earlier. Higher s/n profile samples, easily obtainable with the new generation of large telescopes (see Arav et al 1997, 1998), promise to greatly clarify the profile shapes and composite nature of broad-line profiles.

Line profiles with the most unusual shape tend to be RL sources and include the rare objects that show double-peaked structure in the Balmer lines with wide peak separation $\Delta v_r \gg 1000$ km s^{-1}, (e.g. Arp102B, see Figure 6; Chen et al 1989, Halpern et al 1996), 3C390.3 (Zheng et al 1991, O'Brien et al 1998), and

Pictor A (see Figure 6; Sulentic et al 1995). These objects show FWHM values approaching the mean FWZI values discussed earlier (EH94), indicating extreme "boxiness." Double-peaked sources provide a special opportunity to search for a corresponding HIL component in a source where the LIL have a unique shape. Above referenced IUE and HST data for Arp 102B and 3C390.3 suggest that any HIL with the same shape as the double-peaked LIL profiles is quite weak $[(I_{dp}(\text{CIV}\lambda 1549) \lesssim 0.2\ I(\text{H}\beta_{BC}); \text{Halpern et al 1996)}]$. Instead, the $\text{CIV}\lambda 1549_{BC}$ is dominated by a single-peaked component that is, again, broader than the secondary single-peaked component seen in the LIL. Very broad (FWHM $\gtrsim 10^4$ km s^{-1}) and double-peaked line profiles are primarily RL sources (EH94); however, at least two recent RQ examples of transient double-peaked Balmer lines are now known in sources that showed no previous BC (NGC1097: Storchi-Bergmann et al 1995, M81: Bower et al 1996). In RL Pictor A (Halpern & Eracleous 1994, Sulentic et al 1995), the double-peaked LIL structure appeared in a source previously showing only single-peaked BC profiles.

3. BASIC LINE-RELATED CORRELATIONS

One of the golden rules of statistical analysis involves the necessity that two variables be experimentally independent of one another so that redundant, false, or mixed correlations can be avoided. For example, a correlation between apparent magnitude and redshift for a sample of quasars does not contain useful information about BLR physics. Another golden rule is that neither of the variables should contain some obvious source of variance that could create an apparent correlation when one is correlated against the other. Anticorrelations involving FWHM versus equivalent width for HIL lines (e.g. Francis et al 1992, Wills et al 1993, Corbin & Francis 1994) are driven by the relative strength of the narrow component in that line (i.e. the same correlation would be found between FWHM and (Hβ) if the NLR component were not subtracted). Correlations between FWHM Hβ_{BC} and FWHM CIV$\lambda 1549$ (e.g. Wilkes et al 1999) are driven by the relative strengths of the broad and narrow components in CIV$\lambda 1549$. It is known that the narrow component of CIV$\lambda 1549$, like [OIII]$\lambda\lambda 4959,5007$, is stronger in RL sources (e.g. Brotherton & Francis 1999). This means that CIV$\lambda 1549$ will be measured systematically narrower in RL sources where Hβ_{BC} is broadest. If CIV$\lambda 1549_{NC}$ is not subtracted, an anticorrelation will be found between FWHM (Hβ_{BC}) and FWHM (CIV$\lambda 1549$). If there were a physically meaningful correlation between Hβ_{BC} and CIV$\lambda 1549_{BC}$ (M96 concluded there is not for RQ AGN), it could be quantified only by comparing the measured values of the broad components of the two lines without additional sources of variance.

Several correlations displayed in different contexts by different authors are driven by *biased* samples, i.e. samples for which the ranges covered by the independent variables are not sampled uniformly (e.g. Baldwin 1977a, Wang et al 1996, Corbin 1991, M96, Laor et al 1997a). Biased samples are extremely prone to spurious correlations. Consider, for example, the addition of two symmetrically displaced points, each displaced by 5σ relative to the mean, in a sample of 30 uncorrelated observations that show a Gaussian distribution of two parameters. This situation will produce a formally significant correlation 99.9% of the time. If the points are located at $\pm 4\sigma$ a significant correlation still results 91% of the time. This is not to say that correlations based on biased samples are necessarily erroneous, but assignment of a statistical significance to the correlation coefficient is invalid. Any statistical dependence based on a biased sample is of heuristic value.

3.1   Luminosity Correlations: The Baldwin Effect

Baldwin (1977a) reported an anticorrelation between rest frame W(CIVλ1549) and the 1450 Å continuum luminosity W(CIVλ1549 α $L_v^{-\frac{2}{3}}$) (the "Baldwin effect," hereafter BE). The original Baldwin (1977a) sample of $\approx$20 quasars was dominated by flat spectrum RL with a continuum luminosity range of 29.8–32.0 ergs s^{-1} Hz^{-1}. The relation was soon confirmed with a rather small rms scatter of $\approx$0.6 magnitude (Baldwin et al 1978) and later with two larger flat spectrum samples (Wampler et al 1984). Baldwin et al (1989) found W $\propto L_v^{-\frac{1}{3}}$ for a sample of BQS and PKS quasars. Combining five RQ and RL samples, Kinney et al (1990) found evidence for an even flatter correlation in the range log $L_v \sim$ 27–33 (six decades in luminosity) with a factor of 10 change in W(CIVλ1549) ($\approx$250 to 25 Å in the rest frame). Multiple observations for 5 RQ and 2 RL sources in that sample showed a source-specific anticorrelation for each source, with an average slope of –0.64, almost identical to the value originally found by Baldwin (1977a). This work led to the concept of two Baldwin effects: (1) a global BE with small slope, W $\propto L_v^{-0.18}$, and significant only over a wide range of luminosity ($L_{v,\,max}/L_{v,\,min} \gg 10^2$); and (2) an intrinsic BE reflecting the change in source equivalent width with change in continuum luminosity. The intrinsic BE contributes significantly to the scatter in the global BE (e.g. Kenney et al 1990, Pogge & Peterson 1992).

More recent studies have usually confirmed a weak effect in large samples of optically selected AGN (Cristiani & Vio 1990, Francis et al 1992, Tytler & Fan 1992, Osmer et al 1994, Francis & Koratkar 1995, Laor et al 1995, Wang et al 1998, Chavushyan 1995) with a slope of 0.1–0.2. Detection of the BE in small samples has remained controversial, with several negative results (Steidel & Sargent 1991, Wilkes et al 1999, Wills et al 1999). Attention has also been focused on other UV lines that may follow a trend similar to W(CIVλ1549). A BE has been reported in Lyα (Baldwin 1977a, Cristiani & Vio 1990, Kinney et al 1990, Green 1996), CIII]λ1909 (Tytler & Fan 1992, Green 1996, Steidel & Sargent 1991), HeIIλ1640 (Tytler & Fan 1992), HeIIλ4686 (Heckman 1980, Boroson & Green 1992-their Eigenvector 2, Zheng & Malkan 1993, Green 1996) and MgIIλ2800 (Baldwin 1977a, Tytler & Fan 1992, Thompson et al 1999, Steidel & Sargent 1991). A controversial detection for NVλ1240 (Tytler & Fan 1992) was not subsequently confirmed (Osmer et al 1994, Steidel & Sargent 1991, Laor et al 1995).

Another controversial BE involves the λ1400 blend with reports of no detection (Osmer et al 1994), and a detection that is apparently the strongest luminosity anticorrelation among the UV lines of Laor et al 1995. Zheng et al (1995) reported a BE for the HIL $OVI\lambda1034$ that was not confirmed by Green (1996), while Laor et al (1995) found a shallow trend (0.15 ± 0.08). More data, however, suggested that the strength of the BE was related to the ionization potential of the relevant ion (Espey et al 1993, Lanzetta et al 1993, Zheng & Malkan 1993): the $OVI\lambda1034$ line shows the strongest BE (steepest slope) of any previously examined line, with $MgII\lambda2800$ and the Balmer lines showing the weakest (or no) effect. These results have led to a standard scenario in which the BE occurs in all measurable HIL except $NV\lambda1240$, and slope increases with ionization potential (e.g. Osmer & Shields 1999; see Hamann & Ferland 1999 for the implications of the apparent lack of any BE for $NV\lambda1240$). No convincing BE has been reported for the Balmer lines or $FeII_{opt}$ (Espey et al 1989, Osterbrock 1978, Shuder 1981, Yee & Oke 1981, BG92, Elston et al 1994). Rather, high-redshift RQ quasars that have been observed show evidence of strong $FeII_{opt}$ and $FeII_{UV}$ (Hill et al 1993, Elston et al 1994). Observations of 12 high-redshift quasars with $z \approx 3.1$–4.7 suggest that the $FeII(2$–3000 Å$)/MgII\lambda2800$ ratio is approximately the same and possibly slightly increasing relative to the LBQS (Francis et al 1992, Thompson et al 1999; see Kawara et al 1996 for a similar result on B1422+231, a RL AGN at $z \approx 3.6$).

Kinney et al (1990) and Zheng & Malkan (1993) found no difference in the BE for RQ and RL subsamples. On the contrary, Osmer et al (1994) suggested that RL sources have a more negative power-law index than optically selected samples, although they concluded that the difference in slope was not significant. The BQS and PKS samples show an important difference: the correlation coefficient between $W(CIV\lambda1549)$ and continuum luminosity is significantly nonzero for the PKS but not for the BQS sources (Baldwin et al 1989). These last results suggest that RQ and RL AGN should be considered separately in BE studies.

3.1.1 BE Selection Effects

A BE for core-dominated quasars was criticized as a selection effect because these sources show strong intrinsic continuum variability (Murdoch 1983). Indeed, it is possible to ascribe the original BE entirely to selection effects. Core-dominated objects are a small, biased subset of a minority quasar population (RL AGN). The correlation may be stronger in small, radio-selected samples because of the increased probability of finding beamed sources. This leads to a BE-like correlation for sources with varying degrees of continuum beaming. Even the Kinney et al (1991) BE could be the result of a statistical bias. The weak, shallow anticorrelation produces a difference that is comparable in size to the observed scatter in $W(CIV\lambda1549)$ at any given luminosity. The intrinsic BE that is related to source continuum variability may enhance the scatter in UV luminosity. A spurious correlation could also arise if the dispersion in continuum luminosity is much larger than the dispersion in line luminosity (Yuan et al 1998a).

We made Monte-Carlo simulations (assuming a uniform distribution over the same equivalent width-luminosity range as for Kinney et al 1991) to test the likelihood of finding a BE in small samples. Our results suggest that randomly extracted samples of 20 objects covering two decades in luminosity will show a significant BE-like correlation $\approx 25\%$ of the time, always with a slope larger than that of the dataset from which they were extracted. Thus, the alternating positive and negative results reported above are not surprising. If we add the data obtained by Wu et al (1983) and Wilkes et al (1999) (negative results) into the Kinney et al (1990) diagram, we find that they fit the BE derived there. However, there are two important caveats; the reality of the BE depends on the real absence of (1) low EW and luminosity AGN at low z and (2) high EW and high luminosity AGN at high z. These objects could be underrepresented because of biases in selection techniques. Magnitude-limited high-redshift samples are biased toward intrinsically over-luminous quasars, which will be biased toward below-average line equivalent widths (see Section 4.7 for a possible physical interpretation). At the other extreme, low-luminosity NLS1 with low $W(\text{C\textsc{iv}}\lambda 1549)$ are found in significant numbers (e.g. Grupe et al 1999; Brotherton & Francis 1999). They have a tendency to blur the Baldwin effect at the low-luminosity end. In addition, the Kinney et al (1990) results were based on IUE spectra of AGN that often detected the UV continuum slightly above noise, a fact that tends to increase the observed $W(\text{C\textsc{iv}}\lambda 1549)$. It is important that HST archival data be used in confirmatory studies.

Another issue involves the intensity of the narrow core in the $\text{C\textsc{iv}}\lambda 1549$ line, which contributes non-negligibly to $W(\text{C\textsc{iv}}\lambda 1549)$, and which may be redshift (luminosity) dependent. It is not clear whether this is the case (Steidel & Sargent 1991, Osmer et al 1994, Brotherton & Francis 1999) but this component should not be included in $W(\text{C\textsc{iv}}\lambda 1549)$ measures evaluating the BE. This narrow component is also stronger in RL and X-ray selected AGN (see e.g. Chun et al 1999, Laor et al 1997b, Green 1998), which further heightens the danger of bias when mixing RQ–RL samples.

3.2 Eigenvector 1 Correlations

Boroson & Green (1992) identified a series of related correlations from principal component analysis (PCA) of their sample correlation matrix. The measures included $\text{Fe\textsc{ii}}_{\text{opt}}$, [O\textsc{iii}]$\lambda\lambda 4959,5007$, H$\beta_{\text{BC}}$, $\log\mathfrak{R}$ (defined as ratio of 6 cm to 4400 Å flux densities; Kellermann et al 1989), and the optical to X-ray spectral index, α_{ox}. Their sample contained 87 PG quasars with z $\lesssim$ 0.5 (17 of them radio-loud). The first PCA eigenvector (hereafter E1) strongly anticorrelates with $W(\text{Fe\textsc{ii}}\lambda 4570)/W(\text{H}\beta_{\text{BC}})$ and luminosity of [O\textsc{iii}]$\lambda\lambda 4959,5007$. In an effort to clarify the meaning of E1, we have searched for the correlation diagram that shows maximal discrimination between the various AGN subclasses. The best E1 correlation space that we can identify involves (1) Fe\textsc{ii} $\lambda 4570$ strength, defined as the ratio $R_{\text{Fe\textsc{ii}}} = W(\text{Fe\textsc{ii}}\lambda 4570)/W(\text{H}\beta_{\text{BC}})$, and (2) FWHM(H$\beta_{\text{BC}}$), supplemented by (3) the soft X-ray photon index, Γ_{soft} (measured between 0.8 and 2.4 KeV on ROSAT spectra). Appreciation of the significance of this 3D correlation space has been growing over the past few years (see e.g. Wang et al 1996, Laor et al 1997b, Lawrence et al 1997, Brotherton & Francis 1999. We focus on the BG92 sample because it is nearest to a complete sample of quasars with state-of-the-art resolution and s/n measures of $H\beta_{BC}$ and $FeII_{opt}$.

We find that variables such as W(FeII λ4570) and W($H\beta_{BC}$) show weak or ill-defined correlations with FWHM($H\beta_{BC}$). The lower right panel of Figure 7 shows a plot of W(FeII λ4570) versus FWHM($H\beta_{BC}$) for the BG92 quasars. Correlation coefficients found for W(FeII λ4570) versus FWHM($H\beta_{BC}$) include r $\approx$ −0.44 (BG92), r $\approx$ −0.54 (Wang et al 1996) and r $\approx$ −0.69 (Corbin & Boroson 1996) with considerable overlap between these samples. The diagram is obviously not a simple correlation and is best described as two clusters of sources that we will call Populations A and B. Population A [FWHM($H\beta_{BC}$) $\lesssim$ 4000 km s^{-1}] is an almost pure RQ phenomenology centered near FWHM($H\beta_{BC}$) $\approx$ 2200 km s^{-1} and W(FeII λ4570) $\approx$ 60 Å with approximate ranges 800–3600 km s^{-1} and 20–110 Å. This population includes 63% of the RQ sample. Sources within Population A show no correlation between W($FeII_{opt}$) vs. FWHM($H\beta_{BC}$). Population B [FWHM($H\beta_{BC}$) $\gtrsim$ 4000 km s^{-1}] is centered near FWHM($H\beta_{BC}$) $\approx$ 5500 km s^{-1} and W($H\beta_{BC}$) = 30 Å with approximate ranges 4000–7000 km s^{-1} and 0–60 Å. This population includes 26% of the BG92 RQ sample. There is no significant correlation within this population. The remaining BG92 sources scatter over a wide range of parameter space: W(FeIIλ4570) $\sim$ 0–110 Å and FWHM($H\beta_{BC}$) $\sim$ 3000−11,000 km s^{-1}. There may be a continuous distribution of RQ sources with $FeII_{opt}$ decreasing as FWHM($H\beta_{BC}$) increases; however, the Population A and B distinction is useful because it identifies those RQ quasars that show optical properties most similar to the bulk of RL sources (see Table 2; see also Section 4.5).

TABLE 2

AGN[a]	FWHM Hβ[b] (km^{-1})	N[c]	W Hβ[d] (Å)	W FeII[e] (Å)	RFB[f]	Γ_{soft}[g] Photon Index	W [OIII][h] (Å)
A–RQ	0–2000	20	68 ± 19	54 ± 17	0.8 ± 0.3	3.0 ± 0.4 (17)	15 ± 12
A–RQ	2–3000	18	105 ± 35	64 ± 27	0.7 ± 0.4	2.8 ± 0.2 (17)	28 ± 38
A–RQ	3–4000	11	128 ± 40	60 ± 16	0.5 ± 0.3	2.6 ± 0.1 (8)	20 ± 20
B–RQ	4–7000	17	110 ± 30	31 ± 17	0.3 ± 0.1	2.4 ± 0.2 (14)	33 ± 26
RL	2–4000	3	64 ± 36	32 ± 23	0.5 ± 0.1	2.3 ± 0.2 (3)	11 ± 6
RL	4–7000	10	83 ± 28	19 ± 19	0.2 ± 0.2	2.4 ± 0.3 (7)	27 ± 15
RL	3–20000	18	91 ± 44	20 ± 13	0.3 ± 0.2	2.1 ± 0.5 (17)	−

[a]ABN sample type. A and B refer to populations A and B. RQ and RL refer to radio-quiet and radio-loud sources.

[b] The arbitrarily chosen range of FWHM.

[c]The number of sources.

[d]Mean equivalent width and standard deviation for the broad component of $H\beta$.

[e]Mean equivalent width and standard deviation for the optical FeII$\lambda\lambda$4450-4700 Å blend.

[f] The ratio column 5/column 4.

[g] The soft X-ray spectral index from sources given in the text. Value derived with N_H as a free parameter when available.

[h]Mean equivalent width and standard deviation for [OIII]λ5007 Å.

The R_{FeII} parameter shows a significant correlation with FWHM(Hβ_{BC}) (Gaskell 1985, Zheng & O'Brien 1990, r $\approx$ −0.55 in BG92, r $\approx$ −0.74 in Wang et al 1996). The lack of any significant correlation between W(Hβ_{BC}) and FWHM (Hβ_{BC}) (r $\approx$ + 0.18) in the BG92 sample lessens concerns about correlate cross-talk in this part of E1. W(FeII λ4570) will be measured systematically low relative to W(Hβ_{BC}) because the average continuum increases toward the blue. (M96 included FeII blends on both side of Hβ, but this choice also has draw-backs).

Figure 7 shows 2D projections of the FWHM (Hβ) versus R_{FeII}, Γ_{soft} versus R_{FeII}, and FWHM(Hβ_{BC}) versus Γ_{soft}. We have supplemented the BG92 RL sample with an additional 18 sources (comparable s/n spectra) taken from M96 with W(FeII$_{opt}$) measures converted to W(FeII λ4570) as in by BG92. It has been known for some time that one can distinguish between RL and RQ quasars on the basis of both hard (Piccinotti et al 1982, Brandt et al 1997) and soft X-ray measures (Wang et al 1996, Brinkmann et al 1997a). Soft X-ray photon indices are available for a large fraction of the BG92 sample (Boller et al 1996, Brinkmann et al 1997b, Siebert et al 1998, Wang et al 1996, Yuan et al 1998b). For Figure 7, we have preferred values derived with N_H taken as a free parameter. Early claims of a correlation between Γ_{soft} and W(FeII λ4570) emission (Wilkes et al 1987) were challenged (Boroson 1989, Zheng & O' Brien 1990, Walter & Fink 1993), but a significant correlation is clearly present in the larger BG92 sample (see also Shastri et al 1993). A better-defined correlation was found between Γ_{soft} and R_{FeII} (Puchnarewicz et al 1992; Laor et al 1994, 1997b; Wang et al 1996; Boller et al 1996; Grupe et al 1999). Previous estimates of the correlation strength yielded (1) Γ_{soft} versus FWHM(Hβ_{BC}): r $\approx$ −0.73; and (2) Γ_{soft} versus R_{FeII}: r $\approx$ 0.65 (Wang et al. 1996 who did not distinguish between RL and RQ sources in their correlation analysis).

A significant difference between the R_{FeII} versus FWHM(Hβ_{BC}) and W(FeII λ4570) versus FWHM(Hβ_{BC}) correlations (Figure 7) is found because Population A quasars show a much larger scatter along the R_{FeII} axis. Population A shows a much larger range in R_{FeII} (0.1–1.5) than Population B ($R_{FeII} \approx$ 0.0–0.6). The objects in the high R_{FeII} tail are often objects with unusually low W(Hβ_{BC}) rather than unusually strong W(FeII λ4570). Low W(Hβ) in "narrow line" Seyfert 1 nuclei (see next section) was recognized a long time ago (Osterbrock & Pogge 1985, Gaskell 1985). Objects with $R_{FeII} \gtrsim$ 0.9 show mean W(Hβ) $\approx$ 67 Å while those with $R_{FeII} \lesssim$ 0.9 show W(Hβ) $\approx$ 114 Å, almost twice as large. The W (Hβ_{BC}) range in RQ Population B is also large: W(Hβ) = 23–174 Å and W(FeII λ4570) = 0–69 Å, respectively. However, the ratio R_{FeII} is surprisingly constant, with R_{FeII} >$\approx$ 0.3 ± 0.25 which is similar to the mean for RL sources. This is a justification for re-taining our Population A-B distinction. We find no significant difference in mean RQ optical luminosity between Populations A and B, which is consistent with the ortogonality between E1 and luminosity implied by the BG92 PCA analysis. Table 2 summarizes some relevant mean parameters for RQ and RL sources shown

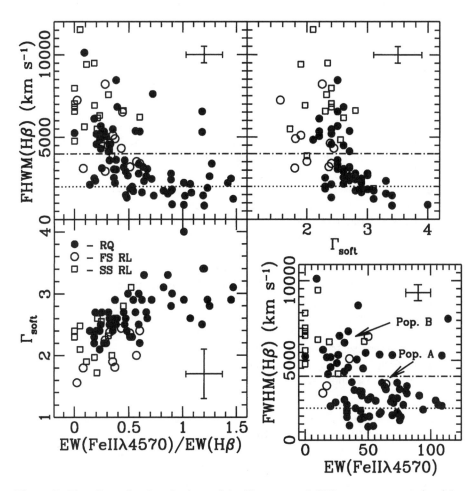

Figure 7 Two-dimensional projections of the Eigenvector 1 (E1) parameter space involving correlations between FWHM(Hβ_{BC}), R$_{FeII}$ and Γ_{soft} *Lower right*: Identification plot for RQ Populations A and B. *Lower left, upper left,* and *upper right* are the three planes of the optimal E1 parameter space. In all panels, RQ sources are *filled circles*, flat-spectrum RL are *open circles*, and step-spectrum RL are *open boxes*. Mean error bars are shown in a corner of each plot. *Dot-dash lines* in the upper panels indicate an approximate boundary between Populations A and B. The dotted line marks FWHM(Hβ_{BC}) = 2000 km s^{-1}, the maximum FWHM for NLS1. Error bars (at a 2σ confidence level) are plotted for illustrative purposes only. Errors on FWHM(Hβ_{BC}), on W(FeII λ4570), and on R$_{FeII}$ refer to an average point in the middle of the diagrams; the error bar on Γ_{soft} has been set as the median value of the errors reported for Γ_{soft} in the sample sources.

in Figure 7. The binning by profile width is somewhat arbitrary, reflecting (1) an attempt to discriminate between properties of broader and narrower line sources, as well as (2) an attempt to maintain as constant a sample size as possible. The first three lines tabulate RQ Population A, and the fourth line tabulates Population B. The next two lines represent RL sources along with the bottom line, which summarizes the values for our M96 RL sample [mean FWHM($H\beta_{BC}$) $\approx$ 6000 km s^{-1} and W(FeII$_{opt}$) converted to W(FeII λ4570) $\approx \frac{1}{3}$W(FeII$_{opt}$)].

3.2.1 Narrow-Line Seyfert 1 Galaxies

Sources in Population A with the narrowest Balmer line profiles involve the much-discussed narrow-line Seyfert 1 galaxies (NLS1). They were first identified by Osterbrock & Pogge (1985) who noted the existence of Seyfert 1 sources with unusually narrow LIL, FWHM($H\beta_{BC}$) $\lesssim$ 1000 km s^{-1}. The eighth edition of Véron-Cetty & Véron (1998) includes 119 NLS1, satisfying the defining criterion that FWHM(Balmer) $\lesssim$ 2000 km s^{-1} (Osterbrock 1987). Ninety-one of these sources fall below the canonical absolute magnitude limit dividing Seyfert galaxies and quasars (M_B = –23.0 for H_0 = 50 km s^{-1} Mpc^{-1}, q_0 = 0). NLS1 are the "extreme Population A" and constitute $\approx$15% of AGN in a sample of hard X-ray selected AGN (Stephens 1989). NLS1 are likely to be overrepresented in BG92 where they account for 27% of the RQ sample upon which Figure 7 is based. Their overrepresentation is probably related to the frequent presence of a steeply rising blue continuum (BG92, Grupe et al 1998). In soft X-ray selected samples they may represent $\approx$0.3–0.5 of all detected Seyfert 1 sources (Puchnarewicz et al 1992, Grupe et al 1999). The latter found about 33% of their AGN sample with FWHM($H\beta$) < 2000 km s$_0^{-1}$.

NLS1 Prototype I Zw 1   I Zw 1 is a prototype of the NLS1 class with FWHM ($H\beta$) $\approx$ 1240 km s^{-1} (without $H\beta_{NC}$ subtraction because $H\beta_{NC}$ and $H\beta_{BC}$ merge very smoothly), W(FeII λ4570) $\approx$ 75 Å, W([OIII]$\lambda\lambda$4959,5007) $\approx$ 22 Å, and W($H\beta$) $\approx$ 51 Å. The $H\beta$ spectral region of I Zw 1 is shown in Figure 2, and details of the $H\beta$ and CIVλ1549 profiles can be seen in Figure 5. Note that (1) [OIII]$\lambda\lambda$4959,5007 are blueshifted relative to the reference frame (determined from HI measurements) by Δv_r $\approx$ 500 km s^{-1} (see Figure 5), and (2) absorption line systems blueshifted by several hundred km s^{-1} are known, whereas $H\beta$ peaks at the rest frame radial velocity.

Figure 7, Table 2, and a recent search for new NLS1 (Moran et al 1996) suggest that FeII strength and [OIII]$\lambda\lambda$4959,5007 weakness are not the principal correlation drivers in this population, especially if the arbitrary restriction to sources with FWHM($H\beta$) $\lesssim$ 1500–2000 km s^{-1} is applied. In fact, W([OIII]$\lambda\lambda$4959,5007) and W(FeII λ4570) are almost constant throughout RQ Population A. The most sensitive NLS1 descriptors appear to be weak W($H\beta$) and a soft X-ray excess. The former correlates with FWHM($H\beta$), and the latter anticorrelates with FWHM($H\beta$). This conclusion was recently reinforced by a new study of 76 X-ray selected AGN

(Grupe et al 1999) where the mean $W(H\beta)$, $W(FeII_{opt})$, and Γ_{soft} for 26 sources with $FWHM(H\beta) \lesssim 2000$ km s^{-1} are ≈ 43 Å, 82 Å (dividing their values by 3 to approximate the BG92 measure), and ≈ 3.4, respectively.

3.2.2 Ultra Strong FeII Emitters

Relatively rare ultra-strong FeII emitters have $R_{FeII} \gtrsim 2$ and $W(FeII \, \lambda 4570) \sim$ 100 Å (Lawrence et al 1997, Grupe et al 1999; $\langle W(FeII \, \lambda 4570) \rangle \approx 67$ Å in the Fe strong optical sources of Lipari et al 1993). Most objects in the Lipari et al (1993) dataset (30 objects) are at the high-end of a continuous distribution in R_{FeII} in Figure 7. Among them there is a reasonable correlation between $W(H\beta)$ and $W(FeII)$, but with $W(H\beta)$ increasing less than $W(FeII \, \lambda 4570)$. Four outliers in the $W(H\beta)$-$W(FeII \, \lambda 4570)$ plane are extra strong FeII emitters ($R_{FeII} \gtrsim 2$), which show unusually weak $H\beta$ (all have $W(H\beta) \lesssim 45$ Å), rather than a larger $W(FeII \, \lambda 4570)$. At least two of these sources, Mark 231 and PHL 1092, are outliers in the E1 correlation space, too. Both Mark 231 and PHL 1092 are peculiar: Mark 231 is undergoing an extreme starburst (Lipari et al 1994, Taylor et al 1999, Ulvestad et al 1999). Similar mixed starburst-AGN conditions may be occurring in other ultra-luminous FIR emitters that show a UV spectrum typical of BAL QSOs, and that are also among the strongest FeII emitters [$W(FeII \, \lambda 4570) \sim 100$ Å] known (Lipari et al 1994). Mixed starburst-AGN sources may be therefore recognizable because of their outlying location in the $W(H\beta)$ – $W(FeII \, \lambda 4570)$ plane (and in the R_{FeII}- $FWHM(H\beta)$ plane as well, see next section).

3.2.3 Additional Correlates

It is interesting to note that the broad absorption line (BAL) quasars also occupy the same, or nearly the same, domain as NLS1 (see Figure 7). They are absent from correlations involving Γ_{soft} because they are soft X-ray quiet. The average parameters of the four RQ BAL quasars in the BG92 sample (PG 0043+039, PG 1001+054, PG 1700+518, and PG 2112+059) with measured CIVλ1549 BAL absorption $W \gtrsim 10$ Å (Brandt et al 1999) are $W(FeII \, \lambda 4570) = 80$ Å, $W(H\beta) \approx 76$ Å, $W([OIII]\lambda\lambda 4959,5007) = 3$ Å and $R_{FeII} \approx 1.1$). Like NLS1 sources, these objects show above-average $W(FeII_{opt})$ and below-average $W([OIII]\lambda\lambda 4959,5007)$ (see Table 2). The BG92 sample suggests that BAL quasars are above average in optical luminosity with absolute magnitude (see BG92) in the range from -24.1 to -27.3, compared with an RQ average of -23.8. This is again consistent with the concept that source luminosity is uncorrelated with E1. Source PG 0043+0354 shows an unusual $H\beta_{BC}$ profile [$FWHM(H\beta_{BC}) \approx 5300$ km s^{-1}) and may be a pathological object (Turnshek et al 1994). Most other low redshift BAL sources (e.g. Boroson & Meyers 1992, Turnshek et al 1997) show $FWHM(H\beta_{BC}) \lesssim 3000$ km s^{-1}.

$H\beta_{BC}$ asymmetry was also included by BG92 in the list of E1 correlates. In RQ Population A, 15 of the 24 largest ($\langle A.I. \rangle \gtrsim 0.07$—hence probably real) asymmetry index measures are blueward and show a mean $R_{FeII} \approx 1.0$ ($\langle W(H\beta_{BC}) \rangle \approx$

69 Å and $\langle W(\text{FeII } \lambda 4570) \rangle \approx 70$ Å), whereas the nine largest red asymmetries are concentrated to the left side of the diagram with $R_{\text{FeII}} \approx 0.4$ ($\langle W(H\beta) \rangle \approx 115$ Å and $\langle W(\text{FeII } \lambda 4570) \rangle \approx 41$ Å). Recall that RL sources show a strong preponderance of red asymmetries, whereas RQ sources are roughly evenly divided between red and blue. E1 parameterization clearly discriminates between red and blue asymmetries within the RQ Population A (see Figure 5 in BG92 for an alternative representation of this result). Three of the four BG92 BAL quasars with significant asymmetries are blue, as expected from their position in Figure 7. Boroson & Meyers (1992) found that the Balmer lines in their low redshift sample of four (low-ionization) BAL quasars (one PG source) showed large blue asymmetries. RQ Population B sources are different in the sense that more large (A.I. $\gtrsim 0.1$) asymmetries are found in this smaller sample (5 red and 4 blue). They show no clear trend within the Population B domain. Another line profile parameter that deserves further study in the context of E1 involves the shape of the lines. RQ Population B quasars show a mean shape parameter (see Section 2.3) $\langle S \rangle \approx 1.14 \pm 0.11$, whereas NLS1 in RQ Population A show $S = 1.22 \pm 0.07$. This is consistent with our visual impression that NLS1 showed more peaked profiles and Population B more boxy profiles.

3.2.4 Eigenvector 1 and Radio Loudness

RL quasars occupy an extremum of the parameter space with weak FeII_{opt}, large FWHM($H\beta_{\text{BC}}$), and an absence of a soft X-ray excess. FeII_{opt} is weaker in RL sources, and core-dominated sources have stronger emission than lobe-dominated sources (Miley & Miller 1979, Steiner 1981, Joly 1991). A comparison within the BG92 sample suggests that RQ are significantly stronger FeII emitters than RL AGN at a 98% confidence level. The steep spectrum sources show a mean FWHM($H\beta_{\text{BC}}$) that is ≈ 1300 km s^{-1} larger than flat spectrum sources (Brotherton 1996, Corbin 1997). The same studies present weaker evidence that SS sources show a smaller (~ 0.1) mean R_{FeII}. W(FeII_{opt}) measures for sources in common in the above two studies show considerable disagreement, which illustrates the difficulty in deriving accurate equivalent width measures for FeII weak sources even with above average spectroscopic data. Siebert et al (1998) found mean Γ_{soft} values of 1.95 and 1.7 for 83 flat-spectrum and 17 steep-spectrum radio galaxies, respectively. The mean values are the same if the hydrogen column density N_H is left free in the model fits. The BG92+M96 RL samples plotted in Figure 7 show the same general trends, with flat spectrum sources overlapping much more with RQ Population B sources. It is worth noting that there is no evidence for a direct continuity in measured log $\Re$ values within the RQ part of the diagram. In other words, RQ Population B quasars do not show radio power values that are intermediate between those of RL and Population A. In this sense, while concentrated in a specific region of E1, RL and lower level radio activity may both be uncorrelated with the E1 parameter space (see also Laor et al 1997b). Considering the BL Lac sources, we are only able to place BL Lac in the FWHM($H\beta_{\text{BC}}$) versus Γ_{soft} correlation plane of Figure 7 because reliable R_{FeII} measures are nonexistent. Siebert et al (1998) found a mean $\Gamma_{soft} \approx 2.4\text{--}2.5 \pm 0.3$ for a sample of 10 sources, whereas Corbett et al (1996) measured FWHM(Hα_{BC}) ≈ 4000 km s^{-1} for BL Lac (in quiescent state). If these numbers are representative, then BL Lac sources fall near the lower FWHM(Hβ_{BC})/high Γ_{soft} edge of the RL domain.

The power of the E1 parameter space can be illustrated by discussing the source PG 1211+143. It is listed as RL in BG92, but its position in the parameter space placed it in the middle of RQ Population A, where no other BG92 RL sources are found. A NASA Extragalactic Database (NED) search revealed that the RL status assigned to PG 1211+143 actually belongs to 4C14.46, located less than 1 arcmin distant. The situation quickly became confused again with the discovery of two RL NLS1 sources (Remillard et al 1986, Siebert et al 1999). When we were examining the RL behavior in the E1 correlation space, we plotted a combined sample of 88 RL sources (Brotherton 1996, Corbin 1997). All but five of these sources behaved correctly in that they occupied the region where the BG92+M96 RL sources fall in Figure 7. Five sources were displaced to much higher R_{FeII} values. Some of these outliers we find to be spurious because of s/n or other problems, but one of them is PKS 0558-504, which is one of the RL NLS1. This source and RGB J0044+193 show Γ_{soft} values of 2.7 and 3.1, respectively, which further reinforces their membership in or near the NLS1 class. At this time, all we can say is that a relatively rare population of RL NLS1 exists.

The preceding discussion of RL sources raises again the subject of a possible correlation between FWHM(Hβ_{BC}) versus log R, where R is the ratio of core-to-lobe flux density (Orr & Browne 1982, Wills & Browne 1986). Alternative definitions exist for this parameter, which discriminates between steep- and flat-spectrum radio sources (Wills & Brotherton 1995, Brotherton 1996, Corbin 1997) and has been proposed as an RL orientation indicator. Two recent correlation studies of pure RL samples (n = 60, Brotherton 1996; n = 61, Corbin 1997) find the strongest signal with FWHM or FW$\frac{1}{4}$M(Hβ_{BC}) (r ≈ 0.6). One cannot escape the impression that the correlation is caused by differences in the mean properties of steep- and flat-spectrum sources [i.e. W(FeII), FWHM(Hβ), R_{FeII} and possibly Γ_{soft}]. The individual populations show little correlation and the RL sample distribution is so broad that one cannot rule out the possibility that steep- and flat-spectrum sources are drawn from independent parent populations. If this impression is true, then the large scatter may be because of an additional physical parameter.

3.2.5 C IV λ1549 and Hβ Line Profiles

C IVλ1549 offers the promise of another correlate in E1 space involving the possibly ubiquitous blueshift associated with this line (and other HIL) in RQ sources. M96 found a wide range of C IVλ1549$_{BC}$ blueshifts (0–5000 km s^{-1}) and also noted an apparent anticorrelation between the amplitude of the shift and W(C IVλ1549). The largest C IVλ1549 blueshifts show a preference for the right side in the

FWHM(Hβ_{BC}) versus R$_{FeII}$ correlation (Figure 7). This is consistent with other results suggesting that BAL QSOs (Hartig & Baldwin 1986, Corbin 1990; but see also Weymann et al 1991) and NLS1 (Rodriguez-Pascual et al 1997a) sources show large CIVλ1549 blueshifts. Profile shapes of both Balmer and CIVλ1549 lines show evidence of correlation in E1. Balmer lines become increasingly narrow with smaller EW toward extreme Population A. W(CIVλ1549) also decreases, but the CIVλ1549 profile becomes more flat-topped and the amplitude of the blueshift increases. The spiky symmetric narrow core of the NLS1 Hβ_{BC} profile merges smoothly with the NLR component (if present) and the strong wings are in saveral cases well fit by Lorentzian models (Gonçalves et al 1999, Moran et al 1996). Such profiles are strikingly different from those observed in RQ Population B and RL AGN (Sulentic 1989, Eracleous & Halpern 1994, Romano et al 1996).

4.   INTERPRETATION: BROAD LINE REGION PHYSICS

Recent reviews have considered the broad line region (BLR) from physical (Baldwin 1997, Krolik 1999), computational (Ferland et al 1997), and structural (Osterbrock 1993, Gaskell et al 1999, Eracleous 1999, Sulentic et al 1999) perspectives. The distinction between a broad-line region (BLR) and a narrow-line region (NLR) retains its original heuristic value based on differences in density (presence or absence of forbidden emission), stability (presence or absence of variability) and kinematic (presence or absence of line shifts) properties. The alternative NLR/Intermediate Line Region/Very Broad Line Region interpretation (Brotherton et al 1994, see Section 4.1.1 and Section 4.1.2) was based primarily on inferred differences in gas density that were based on erroneous determination of line ratios (e.g. too low [CIII]λ1909/CIVλ1549 ratio for ILR) derived from average spectra. We can think of the observations connected with the resolved NLR structure as providing "boundary conditions" on the BLR. These observations provide diverse evidence for large-scale, at least partially collimated outflows: (1) superwinds (Colbert et al 1996), (2) line emission confined to "ionization cones" (Tadhunter & Tsvetanov 1989, Pogge 1988, Wilson 1996), and (3) emitting gas embedding the radio jets of nearby Seyfert galaxies (Capetti et al 1996, Falcke et al 1998). Resolution of the BLR and inner NLR will, for the foreseeable future, be possible only with spectroscopy. However, it is now clear that the emission-line spectrum is produced over a wide range of distances from the central continuum source, and under a wide range of physical conditions, as is the case for the NLR.

4.1   The Emitting Region "Stratification"

Reverberation mapping (RM) complements statistical studies by confirming that the BLR is a "stratified" and photoionized medium, as inferred earlier from profile comparisons (Osterbrock & Mathews 1986, Wandel et al 1999). A significant fraction of this medium (especially the LIL-emitting region) is optically thick to the ionizing continuum (see e.g. Baldwin 1997). Delay times for response to continuum changes appear to correlate inversely with ionization level, increasing from a minimum at $He_{II}\lambda1640$ and $He_{II}\lambda4686$ to a maximum for the Balmer lines (Peterson et al 1991, Reichert et al 1994, Peterson et al 1994, Peterson 1994, Dietrich & Kollatschny 1995, Wanders et al 1997, Malkov et al 1997, Rodriguez-Pascual et al 1997b, Kassebaum et al 1997, Peterson et al 1999). This anticorrelation, along with a correlation of time delay with line width (Antonucci & Cohen 1983, Ulrich et al 1984) are consistent with the hypothesis that the FWHM of each line is an indicator of distance from the continuum source (e.g. Peterson & Wandel 1999). RL 3C 390.3 may be an exception (or an indication that RL sources have different BLR structure) where the Balmer response appears to lead the HIL response (Wamsteker et al 1997, O'Brien et al 1998; see also Recondo-Gonzalez et al 1997 for RQ Population B source Fairall 9, and Goad et al 1999 for NGC 3516 where $Mg_{II}\lambda2800$ showed no response to continuum changes). High density ($n_e \sim 10^{11}$–10^{12} cm^{-3}) in the inner part of the BLR is required by (1) the necessity to maintain the ionization parameter U $=$ $\Phi(H^0)/(4\pi r^2 n_e c) \sim 10^{-2}$ (where $\Phi(H^0)$ is the number of HI ionizing photons, c is the speed of light, and r is the distance from the central source), and (2) the small lag times in several sources ($\lesssim$ than 30 days) that are ten times smaller than estimates based on standard photoionization calculations (Peterson & Gaskell 1991, Stirpe et al 1994, Peterson 1994; see also Table 1 of Ulrich et al 1997). If gas of any density is available at any distance from the continuum source, a distance dependence arises because different lines are radiated optimally at different values of U. We refer to this as the locally optimally emitting cloud (LOC) scenario (Baldwin et al 1995, Ferguson et al 1997). Alternatively, the stratification of line properties can be modeled by assuming that there is a fixed functional dependence between density, ionization, and distance (Rees et al 1989, Ferland et al 1992). "Stratification" may also mean a spatial separation for HIL and LIL that has been independently suggested by the phenomenological differences outlined above.

4.1.1 An Intermediate Line Region?

The distinctive $C_{IV}\lambda1549$ core profile has led to the concept of an intermediate line region (ILR), i.e. intermediate between the NLR and BLR (Brotherton et al 1994). Given stratification in radial emissivity, the FWHM($C_{IV}\lambda1549_{NC}$) is expected to decrease from a maximum for $C_{IV}\lambda1549$ through $Fe_{VII}\lambda6087$ to a minimum for [O$_{III}$]$\lambda\lambda4959,5007$ and Hβ (Sulentic & Marziani 1999). $C_{IV}\lambda1549$ is not collisionally suppressed in the inner NLR, or outer BLR, and is actually favored by large U (Ferguson et al 1997). Use of the ILR nomenclature is an overinterpretation of line taxonomy for any given AGN, which results from integration of the line profile over an unresolved source. In some cases $C_{IV}\lambda1549_{NC}$ will certainly be suppressed by reddening (De Zotti & Gaskell 1985, Wills et al 1993), although it is not clear how often.

4.1.2 Optically Thin Gas: A Very Broad Line Region

Extreme conditions are likely to be found at the innermost edge of the BLR. RM shows that the $HeII\lambda4686$ line often responds with negligible delay to continuum changes (Ulrich & Horne 1996, Dietrich & Kollatschny 1995). Peterson & Ferland (1986) observed a tenfold increase in the $HeII\lambda4686$ line in NGC5548 and no corresponding variation in $H\beta_{BC}$ which is consistent with an $HeII\lambda4686$ emitting region optically thin to the HI ionizing continuum (a similar behavior has been observed in Mark 110; Bischoff & Kollatschny 1999). Recently, Wandel et al (1999) showed that $H\beta_{BC}$ rms FWHM in some AGN are narrower than the mean FWHM, which implies that the Balmer line wings are not optically thick at 13.6 eV. The $CIV\lambda1549_{BC}$ autocorrelation function in these sources was found to be narrower than that of the continuum, which implies a negative response caused by optically thin gas (Sparke 1993). Saturation in the response of the line wings to continuum change has also been seen in Mark 590 (Ferland et al 1990). In some objects, $H\alpha$ shows evidence for a broader profile than $OI\lambda8442$ (Morris & Ward 1989). The same trend is observed for $HeII\lambda4686$ and $H\beta$ (Osterbrock & Shuder 1982, Marziani & Sulentic 1993) as well as for $Ly\alpha$ and $H\beta$ (Zheng 1992, Laor et al 1994, Netzer et al 1995). A small ratio $OVI\lambda1034/CIII\lambda977$ implies the presence of gas with $U \rightarrow 1$ (Laor et al 1994). These observations suggest that optically thin gas is associated with the highest-velocity gas at the inner edge of the BLR. Both high-density and high-ionization parameters have been suggested (Brotherton et al 1994: $n_e \sim 10^{11-13}$, Marziani & Sulentic 1993: $U \approx 0.1-1$). The different inferred physical conditions, combined with phenomenological evidence, have led some authors to point toward a distinct VBLR at the inner edge of the BLR. Evidence of an optically thin VBLR is especially for Pop. B and RL AGN showing inflected $H\beta_{BC}$ profiles (Section 2.3).

4.2 Debated Issues

While both classical and more recent studies (e.g. Shuder 1981, Peterson et al 1982, Osterbrock & Mathews 1986, M96, and RM results) point toward photoionization as the main heating source for the BLR gas, there has been a long-standing "energy budget" problem. Photoionization models are unable to account for the strength of the LIL (Netzer 1985, Collin-Souffrin et al 1986, Collin-Souffrin & Lasota 1988), and this problem has been joined by a photon deficit for the HIL production (Section 4.2.1; Korista et al 1997a). A related problem is the $Ly\alpha/H\beta$ ratio (Section 4.2.2; Baldwin 1977b, Netzer et al 1995 and references therein).

4.2.1 The Evidence of Things Not Seen: The EUV Continuum

Recent results concerning the ionizing continuum of BG92 and other quasars with $z \gtrsim 0.33$ (Laor et al 1994, Laor et al 1997b, Zheng et al 1997) suggest that it can be represented by a single power law from UV to X-ray. The interpolation of Laor et al suppresses the prominent EUV bump near 50 eV expected from Mathews & Ferland's (1987) parameterization of the typical AGN continuum [admittedly adopted ad hoc to reproduce observed W(HeIIλ4686) values]. This reduces the number of ionizing photons and makes it difficult for models to explain the observed strength of the HIL, especially HeIIλ4686 and OVIλ1034, unless some sort of anisotropy or large covering factor is invoked (Korista et al 1997a). It is unclear whether the photon deficit can be reduced in sources showing a soft X-ray excess. Markarian 335, a prototype of the NLS1 sources, shows an extremely soft EUV continuum shape with a maximum close to the Lyman edge (Zheng et al 1995). The high-ionization photon deficit itself may be ill-defined if most HeIIλ4686 is emitted in a VBLR (expected to be a strong source of HIL) rather than in the optically thick BLR.

4.2.2 The Lyα/Hβ Ratio

Recent observations confirm that the Lyα/Hβ ratio is significantly lower than expected on the basis of recombination theory (Lyα/H$\beta \sim$ 4–15; Netzer et al 1995 and references therein). In radio-loud AGN, correction for moderate internal reddening (E(B–V) $\lesssim$ 0.3) may be sufficient to bring the Lyα/Hβ ratio close to standard photoionization expectations (Netzer et al 1995). The Lyα/Hβ problem appears to be more serious for RQ AGN; however, in some AGN, especially Population A, there is evidence of physical conditions that depress Lyα. NLS1 spectra show evidence for lower ionization [e.g. low W(CIVλ 1549), strong FeII$_{opt}$] and higher density ($n_e \gtrsim 10^{11}$ cm^{-3}, Section 4.3.2) than observed in RQ Population B and RL quasars (Osterbrock & Pogge 1985, Gaskell 1985, BG92, Wills et al 1999), which agrees with Lyα/H$\beta \lesssim$ 10. Significant differences between HIL and LIL phenomenology, especially in RQ sources, may mitigate this problem if the two sets of lines arise in different clouds (Laor et al 1994, 1995; M96; Netzer et al 1995; Halpern et al 1996; Goad & Koratkar 1998; Koratkar et al 1999). On the other hand, there is no significant correlation between Lyα/Hβ and FWHM(Hβ) (Wilkes et al 1999, Wills et al 1999), as might be expected if there were a systematic trend in source BLR ionization level across E1. Interpretation may be complicated if the line ratio is different in the core and wings of a line (Netzer et al 1995). A detailed analysis of NGC 5548 (Dumont et al 1998) shows that no simple set of conditions can explain the Lyα/Hβ and Balmer decrement simultaneously. A careful decomposition of the Lyα profile must be carried out for a meaningful comparison between the two lines.

4.3 E1 Correlates

The spectra of AGN are not all the same, and matching their diversity, both in terms of line ratios and profiles, has been until now the unresolved challenge for BLR models. E1 supplements individual complexity with a parameter space that discriminates between, and suggests relationships among, diverse AGN samples. Several AGN problems and important results, especially related to FeII$_{opt}$ emission, must be reconsidered in terms of stratification and the E1 correlations.

4.3.1 FeII Emission

An explanation must be found for the wide range of $FeII_{opt}$ intensities and the $FeII_{opt}$ related E1 correlations. FeII emission has not been fully exploited as a diagnostic because the complexity of the Fe^+ atom hampers reliable modeling. It was long ago realized that FeII is emitted in the BLR; the principal evidence for this is the profile FWHM similarity between $FeII_{opt}$ and $H\beta_{BC}$ (Phillips 1978a,b). It is recognized that strong FeII emission requires high density $n_e \gg 10^9$ cm^{-3}, high column density, and low temperature ($T_e \sim 5000°$K; Collin-Souffrin et al 1980). These conditions are thought to be met in a partially ionized region created by the strong X-ray emission in AGN (Kwan & Krolik 1981, Krolik & Kallman 1988, Collin-Souffrin et al 1988).

Both collisional ionization (Joly 1987, Dultzin-Hacyan 1987) and early photoionization models failed to account for the strength of $FeII_{UV}$ and $FeII_{opt}$ in several emitters (Netzer & Wills 1983, Wills et al 1985, Joly 1987), but the disagreement between photoionization models and observation is not extremely severe and may be accounted for in the optical if we exclude the peculiar ultra-strong FeII emitters. The Wills et al (1985) computations remained state-of-the-art until very recently. Recent calculations have included (1) a large increase in the number of FeII levels considered (several hundred versus a simplified $\sim$10-level structure used in earlier work), (2) self-consistent treatment of heating and cooling, (3) fluorescence and Lyα pumping (Verner et al 1999), and (4) a largely improved atomic database (Hummer et al 1993, Verner et al 1996, Bautista et al 1998 and references therein). Although a simple photoionization scenario has been challenged on the basis of the BLR energy budget (Netzer 1985, Collin-Souffrin 1987, Dumont et al 1998), observational work has produced evidence in favor of it. The $FeII_{UV}$ response to continuum variation appears to be correlated with $H\beta_{BC}$ variations (Maoz et al 1993, Doroshenko et al 1999). Significant FeIII has also been detected in the UV spectra of some AGN (Laor et al 1997a, Vestergaard & Wilkes 1999). In following sections we argue that Population A sources are more likely to show strong FeII emission. A mixed starburst–AGN model (e.g. Williams et al 1997) may be more appropriate for Ultra Strong FeII emitters such as Mark 231. Energy sources other than photoionization may be operating in these objects as well, as was suggested by the identification of FeI emission in PHL 1092 (Kwan et al 1995).

4.3.2 Heavy Element Abundances

Early work on $FeII_{UV}$ strength suggested [Fe/H] $\sim$ 3 times [Fe/H]$_\odot$ to match the strongest emitters (Wills et al 1985). Studies of emission and narrow absorption lines suggest $Z \approx Z_\odot$ (see Hamann & Ferland 1999 and references therein). High Fe abundance has been invoked to explain the large W(Fe Kα) observed in several Seyfert galaxies (Guainazzi et al 1996, Reynolds et al 1995, George et al 1997, Lee et al 1999. [Fe/H] $\approx$ 2 [Fe/H]$_\odot$ has been inferred for several narrow line galaxies from RXTE observations as well (Netzer, Turner & George 1998. Generalized Wilcoxon tests (including upper limits) for two samples of W(Fe Kα) measures

(Sambruna et al 1999, Sulentic et al 1998) suggest that W(Fe Kα) is stronger in RQ than RL sources (at 2σ confidence level). Kaplan-Meyer estimators of the censored-sample mean equivalent widths are ≈230 and ≈380 Å, respectively. The previous results suggest that the systematic difference between RQ and RL FeII strength could be caused at least in part by a range in [Fe/H].

Soft X-ray Excess   FeII and other LIL are believed to be produced in an extended, partially ionized zone. Continuum energies that should most affect FeII strength are $h\nu \gtrsim 800$ eV (Krolik & Kallman 1988), which is consistent with observations of a genuine soft X-ray excess in FeII strong emitters such as the NLS1 and other Population A sources (Puchnarewicz et al 1992, Fiore et al 1994).

Density   Assuming solar abundance, high density and high column density can greatly enhance FeII emission (Krolik & Kallman 1988, Collin-Souffrin et al 1988) and can affect the ratio $FeII_{UV}/FeII_{opt}$. A high value of R_{FeII} can be obtained if $n_e \gtrsim 10^{10}$ cm^{-3} and $N \gtrsim 10^{24}$ cm^{-2} (Collin-Souffrin et al 1979, 1980, 1986; Ferland & Persson 1989). In the density range 10^9–10^{11} cm^{-3} some semi-forbidden lines become sensitive to density. The SiIII]λ1892/CIIIλ1909 ratio, which increases with density, becomes a useful diagnostic. Also, the strength of AlIIIλ1860 should be especially enhanced at $n_e \sim 10^{12}$ cm^{-3} (Ferland et al 1992, Korista et al 1997b). $H\beta_{BC}$ should be increasingly suppressed as the density approaches 10^{12} cm^{-3} (Gaskell 1985, Korista et al 1997b). The AlIIIλ1860 doublet is prominent in I Zw 1 (M96). Strong FeII and AlIIIλ1860 emissions are general characteristics of low-ionization BAL quasars (Weymann et al 1991). The ratio SiIII]λ1892/CIII]λ 1909 for the BAL quasars (Harting & Baldwin 1986) suggests $n_e \gg 10^9$ cm^{-3}. In I Zw 1 the ratio SiIII]λ1892/CIII]λ1909 ≈ 3.5 (Laor et al 1997a) is much larger than the typical value for quasars (≈0.3; Laor et al 1995). Because CIII]λ1909 appears to be weak, n_e must be $\sim 10^{11}$ cm^{-3}. More recent results suggest that density correlates with E1 because the diagnostic ratio SiIII]λ1892/CIII]λ1909 increases with decreasing FWHM(Hβ) (Wills et al 1999, Aoki & Yoshida 1999). Strong $FeII_{opt}$ and weak Hβ in NLS1 can be explained in terms of an average density $\gtrsim 10^{11}$ cm^{-3}, much larger than was inferred for broader (and higher-ionization) Population B quasars, which fit the "standard" AGN photoionization scenario much better.

Lyα Pumping   Lyα pumping can enhance $FeII_{UV}$ (Penston 1987, Sigut & Pradhan 1998, Verner et al 1999). The enhancement depends on the amplitude of microturbulent broadening in Lyα which governs the number of fluorescent transitions that become accessible (Verner et al 1999). The detection of appreciable emission in UV multiplet 191 supports a role for Lyα pumping in I Zw 1 (M96, Laor et al 1997a), 2226–3905 (Graham et al 1996), and some BALs (Hartig & Balwin 1986). Other emission features coming from 8 eV levels in PG 1700+518 and the detection of significant emission at 1600–1700 Å in several AGN also support fluorescence and photoionization (Wampler 1985, 1986; M96; Laor et al 1997a).

4.4   Orientation and L/M

Various authors have discussed the differences between NLS1 and "normal" AGN in terms of (1) source orientation (Osterbrock & Pogge 1985, Puchnarewicz et al 1992, Wills et al 1995, Ulvestad et al 1995, M96, Boller et al 1997), and (2) different parameters related to the physics of line and continuum formation. These parameters include (2*a*) high accretion rate (BG92; Pounds et al 1995; Boller et al 1996, 1997; Wang et al 1996; Brandt & Boller 1999; Brotherton & Francis 1999; Leighly 1999a); (2*b*) distance of BLR from the continuum (Pounds et al 1995, Giannuzzo & Stirpe 1996, Mason et al 1996, Giannuzzo et al 1998, Wandel & Boller 1998); as well as (2*c*) black hole mass (Saxton et al 1993, Boller et al 1997, Wandel & Boller 1998). In the following sections, we show that the ratio of luminosity to black hole mass L/M (convolved with orientation) is likely to be the fundamental parameter in the E1 population sequence (Section 4.5). The ratio L/M is proportional to the dimensionless accretion rate $\dot{m} = \dot{M}/\dot{M}_{Edd}$, ($\dot{M}_{Edd}$ is the accretion rate associated with the Eddington Luminosity) and to the Eddington ratio L/L_{Edd} [where $L_{Edd} = 1.5 \times 10^{38}$ (M/M$_\odot$) is the Eddington Luminosity].

4.4.1 Is FeII Emission Anisotropic?

Emission from several strong broad lines is expected to be anisotropic (Ferland et al 1992). This is even truer for FeII emission, if it is produced in a dense and large column density medium (Gaskell 1985, Ferland & Persson 1989, Collin-Souffrin et al 1988) and if this medium is plane-parallel as in an accretion disk (Dumont & Collin-Souffrin 1990a,b). In the simplest unification models, core- and lobe-dominated RL quasars should belong to the same parent population manifesting different radio morphology because of radio axis orientation (e.g. Antonucci 1993). The observation of stronger FeII in core-dominated sources (Bergeron & Kunth 1984, Joly 1991, E1) suggests that FeII emission may be anisotropic (and even more anisotropic than Hβ; Jackson & Browne 1991a,b). The anticorrelation between R_{FeII} and FWHM(Hβ) can also be interpreted in terms of orientation, with stronger FeII$_{opt}$ emitters observed when the source is oriented pole-on (Zheng & O'Brien 1990, Zheng & Keel 1991) in analogy with correlations between radio core dominance R and FWHM(Hβ) (Wills & Browne 1986). More recent studies show a decreased correlation coefficient with a scatter much beyond observational uncertainties (e.g. Brotherton 1996, Corbin 1997). A second parameter, independent from source orientation, is required to account for the large range of FWHM(Hβ) at fixed R.

4.4.2 Hβ_{BC} Line Width

Given the R-FWHM(Hβ) trend as real in a unification scenario, it is surprising that FWHM(Hβ_{BC}) does not appear to be a direct orientation indicator in RQ sources (Boroson 1992). BG92 ruled out orientation as the main driver in their correlation study because [OIII]$\lambda\lambda$4959,5007 *absolute magnitude* was inversely correlated with log R, which is incompatible with isotropic [OIII]$\lambda\lambda$4959,5007 emission.

However, both assumptions of optical continuum anisotropy and especially of [OIII]$\lambda\lambda$4959,5007 isotropy may not be appropriate. W([OIII]$\lambda\lambda$4959,5007) anti-correlates with core dominance, but the effect appears to be different for [OII]λ3727 (Baker & Hunstead 1995). Hes et al (1993) argue that [OIII]$\lambda\lambda$4959,5007 may be subject to anisotropic obscuration, whereas 3C core- and lobe-dominated RL do not show systematic differences relative to [OIII]λ3727. Intrinsic anisotropy may be related to the filamentary appearance of the NLR as revealed by HST imaging (i.e. Falcke et al 1998, Capetti et al 1996), which may suppress [OIII]$\lambda\lambda$4959, 5007 if the NLR is observed along the radio axis. The case of I Zw 1 supports this interpretation, because the [OIII]$\lambda\lambda$4959,5007 lines are systematically blueshifted with respect to the rest frame (Figure 5). We see a redistribution of [OIII]$\lambda\lambda$4959,5007 flux (associated to a decrease in peak intensity) over a broader radial velocity range in FeII strong objects, in addition to a decrease in [OIII]$\lambda\lambda$4959,5007 luminosity.

4.4.3 The Soft X-Ray Excess

The origin of the soft X-ray excess in Population A RQ sources is a major puzzle. Comptonization of thermal photons either from the base of a jet (Mannheim et al 1995) or from a disk atmosphere (Haardt & Maraschi 1991, Haardt et al 1997) is suggested by the constant temperature of the excess over a wide luminosity range (Leighly 1999b). However, a weaker warm absorber (Done et al 1995) as well as a genuine extension of the BBB (Walter & Fink 1993, Grupe et al 1998) have also been considered. In addition, Pounds et al (1995) noted the similarity between the X-ray spectra of soft excess AGN and those of galactic black hole candidates in the high phase, suggesting that $\dot{m} \to 1$. If this is the case, the traditional thin disk approximation is thought to break down (Shakura & Sunyaev 1973, Lightman & Eardley 1974) and a slim disk may replace it (Abramowicz et al 1988). The slim disk is able to produce a steep soft X-ray spectrum (Szuszkiewicz et al 1996, Wang et al 1999).

A soft X-ray excess analogous to the one observed in NLS1 sources is seen in some BL Lacs (White et al 1984; see also Padovani & Giommi 1996). Relativistic beaming is immediately suggested by the short time scale and large amplitude variability observed in these sources. Soft X-ray observations of absorption edges with energies much higher than rest provide evidence for relativistic outflows (Leighly et al 1997, Vaughan et al 1999a, Reynolds et al 1995; however, see Nicastro et al 1999 for an alternate interpretation). A moderately relativistic flow can give rise to apparent luminosity variations an order of magnitude larger than variations in the rest frame of the source (Guilbert et al 1983, Boller et al 1997). Variability in soft-excess sources seems to be enhanced at a similar level relative to other AGN (Leighly 1999a). Variability points toward a preferential orientation in at least some NLS1.

4.4.4 A Distant/Absent BLR?

It has been suggested that intense soft X-rays in NLS1 could accelerate away or highly ionize the inner accretion disk and BLR (White et al 1984, Guilbert & Rees

1988, Pounds et al 1995). However, Fe Kα emission from highly ionized iron appears to be infrequent and not confined to NLS1 (Fabian et al 1994, Comastri et al 1998, Sulentic et al 1998, Yamashita et al 1997, Nandra et al 1996, Vaughan et al 1999b). Wandel & Boller (1998) suggested that the BLR could exist at larger radii in NLS1 where a steep Γ_{soft} can enhance the ionizing continuum. Repeated observations of NLS1 suggest that $H\beta_{BC}$ may be intrinsically less variable than in other Seyferts, especially on the shortest timescales in the range 30–300^d (Giannuzzo et al 1998, Giannuzzo & Stirpe 1996), although higher frequency monitoring is needed to assess the situation. The observational evidence in favor of a larger BLR distance in NLS1 is still sparse. However, it is now obvious that NLS1 do not lack a BLR, since $CIV\lambda1549$ shows an unambiguously broad profile (M96, Rodriguez-Pascual et al 1997a).

4.5 The L/M Ratio: An E1 Main Sequence for AGN?

Wandel et al (1999) and Kaspi et al (2000) reported the most accurate $H\beta$ reverberation distances for Seyfert and low-redshift quasars, respectively, yielding data for 36 AGN, 32 of which are radio quiet. The luminosity values are derived from measured optical fluxes, and the masses are computed, under the assumption of virial motion, from the reverberation distance. The AGN luminosity to black hole mass ratio (L/M) values for NLS1 Mark 110 and 335 are among the largest. Mark 110 and 335 do not show a time delay different from the other sources, which implies that the central mass must be small. Wandel et al (1999) reported $M/M_\odot \lesssim 7.00$ for both. This is however not necessarily the general case.

Earlier we noted that source luminosity appeared to be an orthogonal variable with respect to E1. Although NLS1 are slightly under-luminous with respect to the general (BG92) quasar population, narrow lines are also found among higher-luminosity objects. Evidence includes: (1) E1346+266 (Puchnarewicz et al 1998) shows $L_X \sim 10^{47}$ ergs s^{-1} and FWHM($H\beta$) ≈ 1800 km s^{-1}); (2) PG BAL quasars, which occupy almost the same region in E1 as NLS1, yet are considerably more luminous; and (3) RL and RQ Population B sources, which inhabit the same E1 domain while the PG RL sources are considerably more luminous. It is similarly unlikely that central mass alone correlates in that space. PG 1351+640, about 10 times more massive than Mark 335 and Mark 110, has log L/M $\approx$ 3.4 in solar units, and FWHM($H\beta_{BC}$) ≈ 1200 km s^{-1}, both very similar to the values obtained for the two NLS1. The Spearman's correlation coefficient for FWHM($H\beta_{BC}$) versus log (L/M) for the objects in Kaspi et al (1999) is larger than that for FWHM versus log M (0.63 versus 0.43) in absolute value, excluding NGC 3227 and NGC 4051, two objects with weak and rather variable broad Balmer emission, suggesting (but not proving, since the FWHM was used to compute M, and the argument is therefore *in part* circular) a more fundamental relationship between FWHM($H\beta_{BC}$) and L/M. AGN with FWHM($H\beta_{BC}$) $\lesssim 2000$ km s^{-1} may be radiating at a higher L/M ratio but not necessarily with systematically smaller mass. Probably the distance in units of *gravitational radii* (which includes both the possibility of larger BLR linear distance or smaller mass) is more directly relevant.

It is clear also that reverberation techniques have limitations as long as even the first-order structure of the BLR is unknown. The transfer function depends, among other things, on the BLR geometry, which remains uncertain even for the best-studied objects such as NGC 5548 (e.g. Goad & Koratkar 1998, Dumont et al 1998, Peterson et al 1999). Therefore, results on different objects should be compared with caution (see also the discussion of Kaspi et al 2000).

4.6   A Tale of Two Emitting Regions: $CIV\lambda1549_{BC}$ and $H\beta_{BC}$

The HIL–LIL distinction has been stronly supported by theoretical arguments (Collin-Souffrin & Lasota 1988, Dumont et al 1990a,b). NLS1 are the extreme RQ Population A sources in E1 and offer the best observational evidence for a kinematic and physical decoupling of the HIL and LIL BLR (M96; Figure 5 here). $CIV\lambda1549$ and $H\beta_{BC}$ show almost no overlap in radial velocity, with almost all of the $CIV\lambda1549$ line emission blueshifted relative to the rest frame in IZw1 (see Figure 5; see also Rodriguez-Pascual et al 1997a for another example). The striking difference between $CIV\lambda1549$ and $H\beta$ in I Zw 1 suggests that the $CIV\lambda1549$ (and $[OIII]\lambda\lambda4959,5007$) emitting regions are observed along a preferential flow direction, whereas $H\beta$ is observed along a direction that minimizes the velocity dispersion. The association of the largest $CIV\lambda1549$ blueshifts with sources showing narrow $H\beta_{BC}$ can thus be explained in terms of orientation, in the context of a flattened gas distribution emitting LIL, and a wind or a system of clouds outflowing in a wide-aperture cone emitting HIL (M96, Wills et al 1995). $W(CIV\lambda1549)$ will be inclination-dependent in a wind scenario, decreasing as $i \to 0°$ (Murray et al 1995, Murray & Chiang 1998). We see two competing effects operating for $H\beta_{BC}$: an increase in peak intensity as the line profile gets narrower (because the flattened distribution of emitting gas is viewed increasingly pole-on), while $W(H\beta)$ decreases (probably because of collisional suppression or because of a stronger underlying continuum). Similar mechanisms appear to be operating for $[OIII]\lambda\lambda4959,5007$ (see Section 4.4.2). Spectropolarimetric and radio observations for two NLS1 suggest that the Balmer line polarization vector is oriented perpendicular to the radio axis. In other observed NLS1, and in all other observed Seyfert 1 and RL AGN, it is parallel (Goodrich 1989, Ulvestad et al 1995). A perpendicular orientation is consistent with line emission from a flattened geometry.

There are apparently two parallel sequences in Figure 7, which are displaced in both FWHM (larger in RL) and R_{FeII} (lower in RL). A difference in L/M can account for the FWHM displacement, whereas orientation obviously cannot. However, L/M and R_{FeII} are not obviously related in Population A, and we do not know what dynamical effect resulting from L/M is reflected in FWHM(Hβ). L/M may influence the equilibrium radius where gravitational and radiative forces approximately balance (Mathews 1993), or it may influence the distance at which radiation pressure falls below the confining pressure of the underlying disk in the case of a radiation pressure-driven wind (Murray et al 1995). Large L/M seems to imply that emission is weighted toward higher density. Higher density may in turn imply lower ionization parameter U, as was found for NLS1. Orientation is expected to have an effect concomitant with L/M, depending on assumptions about source geometry. For example, high L/M and i $\rightarrow$ 0 are both thought to favor a soft X-ray excess, small FWHM(Hβ), and stronger Fe$_{\rm II opt}$.

As we cross E1, the H$\beta_{\rm BC}$ profile broadens, and we observe systematically broader and more red asymmetric profiles in RQ Population B and RL sources. Outside Population A, evidence for decoupling is less clear. There is a difference in width between Lyα and CIVλ1549 (Laor et al 1994, 1995), and any analogous double-peaked CIVλ1549 component in sources with (H$\beta_{\rm BC}$) double-peaked profiles is weak or absent (Halpern et al 1996). These findings are apparently contradicted by the similarity of H$\beta_{\rm BC}$ and CIVλ1549$_{\rm BC}$ profiles found in many RL objects (M96). Extrapolating from 3C 390.3, we suggest that RL sources may radiate with the lowest L/M. This is consistent with a redward asymmetry that results from an increasing effect of gravitational redshift along the E1 main sequence of AGN (Anderson 1981, Corbin 1995, Popovic et al 1995, M96, Corbin & Boroson 1996, Corbin 1997). Low L/M may not permit a kinematical decoupling in RL HIL and LIL. It may instead require a single emitting region in virial equilibrium or a radial flow.

4.7 The Baldwin Effect: Evolution in L/M and Signature of a Young Universe?

Early attempts to explain the Baldwin Effect (BE) focused on the effects of embedded dust (Shuder & McAlpine 1979), a decrease in covering factor (Wu et al 1983), and/or a decrease of the ionization parameter with luminosity (Mushotzky & Ferland 1984, Zheng & Malkan 1993, Osmer et al 1994). A variation in BE slope, steepness correlating with line ionization level (with the exception of NVλ1240) favors an explanation based on a decrease in the number of CIVλ1549 ionizing photons due to a softening of the ionizing continuum with luminosity (Zheng & Malkan 1993, Osmer et al 1994, Francis & Osmer 1999, Korista 1999, Wandel 1999a,b), possibly caused by evolution of the central black hole mass (Wandel 1999a,b).

W(CIVλ1549) is possibly an E1 correlate (M96) with NLS1 galaxies showing low W(CIVλ 1549) values (Rodriquez-Pascual et al 1997) and strong FeII emission. Observations of high-redshift quasars (z $\approx$ 4.5) also show evidence for FeII emission that is surprisingly strong and comparable to low z sources (Thompson et al 1999, Hamann & Ferland 1999). One can argue that AGN become more similar in appearance to NLS1 with a typically low-ionization spectrum (see Section 3.1) at higher redshift. The similarity between NLS1 and high-luminosity quasars may reflect a similarity in L/M, with both NLS1 and high-redshift quasars radiating at a higher luminosity for a given mass (see Sections 4.5 and 5). If this interpretation is correct, the BE could be due in part to a selection effect which obviously favors the discovery of higher luminosity AGN for a given mass, due in part to an evolution of the AGN L/M ratio with redshift.

5. MODELS AND PHENOMENOLOGY

Through the late 1980s, the models were developed almost independently of the observations, to a large extent. This was in part a reflection of the degeneracy of emission-line data with respect to BLR models. Space constraints do not allow a detailed discussion of all models, so we present Table 3 as a summary of the pros and cons of contemporary BLR models. Later we discuss how E1 correlates allow us to clarify and better define the scope of competing models. Complementary material includes proceedings of several recent conferences, including IAU Colloquium 159 (Peterson et al 1997) and the Lincoln conference (Gaskell et al 1999).

5.1 Preferred Models on the Basis of Phenomenology

If the HIL blueshift is ubiquitous in RQ sources (M96), then it argues strongly for a radial outflow whose far-side component is obscured, presumably, by an accretion disk (Livio & Xu 1997). Winds are found to arise from disks in a variety of physical approaches (Perry & Dyson 1985; Mobasher & Raine 1989; Emmering et al 1992; Bottorff et al 1997; Murray & Chiang 1997, 1998; Bottorff et al 1997; Williams et al 1999). An outflowing component might be caused by a disk wind (e.g. Bottorff et al 1997, Murray & Chiang 1997, Williams et al 1999) or by some other more collimated outflow (Zheng et al 1990). The observed decrease in W(CIVλ 1549) following a continuum increase (the intrinsic BE; Pogge & Peterson 1992) can be explained in a wind framework similar to the one used for Wolf-Rayet stars (Morris et al 1993). The outflow interpretation has not been seriously challenged by reverberation mapping results. Results that suggest non-radial motion in AGN are restricted to a handful of Population B or radio-loud AGN, for which monitoring of CIVλ1549 has been carried out (O'Brien et al 1998, Korista et al 1995, Wanders et al 1997, Koratkar et al 1996; but see also Goad et al 1999, Ulrich & Horne 1996). The first comprehensive monitoring in optical and UV for an NLS1 (Arak 564) is now in progress. Korista et al (1995) and Wamsteker et al (1997) found evidence that the red wings of lines respond to continuum changes faster than the line cores or the blue wings. If radiation-transfer effects are properly taken into account, these results can be explained in a wind model (Chiang & Murray 1996, Murray & Chiang 1997).

RQ LIL show only smaller shifts that distribute stochastically relative to the rest frame defined by [OIII]$\lambda\lambda$4959,5007 (i.e. $\Delta v_r \ll$ FWHM and $\Delta v_r \ll 10^3$ km s^{-1}). FeII lines show evidence of arising from or near the same medium that produces

TABLE 3 Pros and Cons of Current BLR Models

Model	Pros	Refs.	Cons	Refs.
Binary BH	Existence of double-peaked sources and objects like NGC 5548 that show two independent RM components in wings of Hβ_{BC}	1, 2, 3 4, 5	Long-term monitoring of Arp 102B and 3C 390.3 does not show predicted v_r variations Ruled out by spectropolarimetry of several double peakers.	7 6 7
Clouds	Locally Optimized Clouds (LOC) account for BLR stratification. "Bouncing clouds" account for stochastic red/blue shifts/asymmetries. RM results for some Pop B sources support a spherically symmetric system of anisotropically illuminated clouds.	8 15 11, 12, 13	Smooth profiles require an extremely large (10^{7-8}) number of clouds in two Pop A sources Clouds are not stable if in pressure equilibrium with a confining medium Systematic blueshift of HIL profiles in Pop. A sources	9, 10 16, 17 14
Disks	Keplerian motion is compatible with RM results for Pop. A. objects. Relativistic Keplerian AD explain wide separation double peakers (mostly Pop B and RL objects). AD provide a smooth medium which meets the requirements of very high resolution Keck spectra. AD provide high n_e and N_c medium for production of FeII$_{opt}$ and other LIL such as CaII.	18 20, 21 10, 11 37, 38, 39. 40, 41, 42	General disagreement with observed profiles in line shift-asymmetry parameter space Double peakers show largest FWHM and are rare. Double-peaked lines appear suddenly in some sources: NGC 1097, Pictor A and M 81. Double peakers almost entirely Pop B and RL sources AGN, which are the ones where FeII$_{opt}$ is the weakest.	14, 19 22, 23, 24 25, 26 27, 28, 34 35, 36 25

	Evidence for	Refs	Evidence against	Refs
	Single-peaked lines possible if emission extends to large radii and/or if AD is seen almost face-on.	43, 44, 45	Line polarization is low, || to disk axis, (except in two extreme pop. A) with no dependence on λ.	8, 47, 48, 49
		46		40, 50
	-		Double-peakers show centrally peaked Hα$_{BC}$ in polarized light.	51, 8
Winds/ Bicones	Evidence for radial outflow from CIV in Pop. A sources	14, 53–57	Inconsistent with some reverberation results	3, 64, 65
	Radiation-driven/hydromagnetic winds are compatible with observation in Pop A sources		Biconical outflow models in RL require that the receding part of the flow is also seen.	58, 59, 60
	Shapes and widths of scattered lines + optical polarization vector || radio axis favors *a biconical BLR* + thick torus.	8	Double peakers are double-lobed radio sources.	30, 62, 63
	Biconical outflow can fit profile shape and variability in some double-peakers if radio axis oriented close to line of sight.	35, 61	-	

1. Gaskell 1983; 2. Gaskell 1988; 3. Gaskell 1996; 4. Marziani 1991; 5. Stockton & Farnham 1991; 6. Eracleous et al 1997; 7. Corbett et al 1998; 8. Baldwin et al 1995; 9. Arav et al 1998; 11. Wanders et al 1995; 12. Goad et al 1999; 13. Goad & Wanders 1996; 14. Marziani et al 1996 (M96); 15. Mathews 1993; 16. Nathews 1986; 17. Mathews & Doane 1990; 18. Korista et al 1995; 19. Sulentic et al 1990; 20. Chen & Halpern 1989; 21. Chen eet al 1989; 22. Eracleous & Halpern 1994 (EH94); 23. Kinney 1994; 24. Wilson 1996; 25. Gaskell & Snedden 1999; 27. Storchi-Bergmann et al 1993; 28. Storchi-Bergmann et al 1995; 29. Eracleous et al 1995; 30. Eracleous 1998; 31. Miller & Peterson 1990; 32. Zheng et al 1991; 33. Nandra et al 1999; 34. Halpern et al 1998; 35. Sulentic et al 1995; 36. Bower et al 1996; 37. Joly 1987; 38. Joly 1991; 39. Collin-Souffrin et al 1988; 40. Dumont & Joly 1992; 41. Joly 1989; 42. Dultzin-Hacyan et al 1999; 43. Dumont & Collin-Souffrin 1990a; 44. Dumont & Collin-Souffrin 1990b; 45. Dumont & Collin-Souffrin 1990c; 46. Rokaki et al 1992; 47. Stockman et al 1984; 48. Antonucci 1988; 49. Berriman et al 1990; 50. Koratkar & Blaes 1999; 51. Antonucci et al 1996; 53. Murray & Chiang 1998; 54. Chiang & Murray 1996; 55. Murray & Chiang 1997; 56. Williams et al 1999; 57. Bottorff et al 1997; 58. Livio & Pringle 1996; 59. Livio & Xu 1997; 60. Shaviv et al 1999. 61. Zheng et al 1991; 62. Valtonen 1999; 63. Appl et al 1996; 64. Goad et al. 1999; 65. O'Brien et al 1998.

the Balmer lines. The previously discussed density and column density constraints on the medium responsible for Fe II production, along with the large number of emitting clouds implied by the highest s/n observations for two RQ sources (Arav et al 1997, 1998), suggest that an accretion disk, or a flattened medium surrounding the disk, is the best candidate for LIL production (Collin-Souffrin et al 1988, M96). The approximate equality in the number of observed red and blue LIL shifts can most easily be ascribed to some structural peculiarity associated with the disk, such as a warp (Pringle 1996, Marziani et al 1997, Calvani et al 1997, Bachev 1999). Because the vast majority of RQ show single-peaked LIL, emission produced in the disk must naturally arise at large radii or with the disk oriented face-on (Collin-Souffrin & Dumont 1990; Dumont & Collin-Souffrin 1990a,b,c). In such a model, the very broad Fe $K\alpha$ line seen in many RQ could be produced simultaneously in the inner part of the disk (Sulentic et al 1998).

The situation for RL sources is unclear but is probably different from RQ AGN. If E1 is correctly interpreted as being driven primarily by L/M (convolved with a source orientation factor), then RL and possibly RQ Population B sources involve systematically lower accretion rates or larger masses. All of the well-known double-peaked sources are either RL or RQ Population B, and if there is an L/M-FWHM (Hβ) anticorrelation that can be extrapolated to the objects with broadest lines, they should be the sources with the lowest accretion rate. If the outer accretion rate is lower than the inner one, there may not be enough matter to sustain a stable emitting disk—or at least large parts of the disk might become optically thin (Collin-Souffrin & Dumont 1990, Dumont et al 1991). Note that many double-peaked and other RL/Population B sources show large amplitude variability that includes transient single- or double-peaked BLR structure. This may be evidence that these sources show only transient disk emission. The variable double-peaked line profiles in Population B source NGC 1097 have been successfully modeled as arising from an elliptical accreting ring rather than a disk (Storchi-Bergmann et al 1997, 1998; see also Eracleous et al 1995), but it is clear that a bicone model could fit this data equally well. It would be ironic if the sources in this class of AGN, which show the most "characteristic" disk signature in their line profiles (Eracleous & Halpern 1994), turn out to be the sources without a permanent emitting disk. The bicone interpretation is consistent with the observation of both large redshifts and blueshifts in the Balmer lines of RL sources, implying that both near- and far-side emission is visible (Sulentic 1989, M96, Eracleous & Halpern 1994). It might also explain the apparent lack of strong Fe $K\alpha$ emission in RL sources in contrast to the frequent detections in RQ Seyfert 1s (Nandra et al 1997a,b). Large shifts appear to be less common in Population B RQ sources. The recent appearance of broad Balmer lines with a -3500 km s^{-1} blueshift in E1 0449-1823 (Halpern et al 1998), however, reveals that such RQ sources do exist.

The simplest interpretation for such large LIL redshifts and blueshifts is some kind of frequently transient and asymmetric outflow (e.g. Marziani et al 1993). Such flows might reasonably be associated with the radio jets that are common in many of these sources. HIL emission from these sources also shows both redshifts and blueshifts but with an apparent amplitude range of $\pm 10^3$ km s^{-1} (Sulentic et al 1995, M96). Perhaps this component is associated with either (1) the remnants of the RQ HIL wind that is apparently disrupted by the RL phenomenon or (2) emission from shocked gas near the terminal velocity ends of the LIL outflow (Bottorff et al 1997). It is interesting in this context to note that double-peaked RL source 3C 390.3 (and also Population B Fairall 9) shows evidence for a faster LIL response to continuum changes (Wamsteker et al 1997, O'Brien et al 1998, Wandel et al 1999, Recondo-Gonzalez et al 1997).

6. CONCLUSIONS

A clear picture of AGN line phenomenology has begun to emerge during the past 10–15 years. Statistical studies have provided insights that are not obtainable from reverberation mapping. AGN research has been a murky terra incognita in which theoretical prejudice and observational data have often been intermingled. One clear point is that we observe unmistakable evidence for outflow at the "boundaries" of the BLR. Another clear point is that high s/n and spectral resolution are needed to obtain meaningful results. The alternative results in a mass of confusing and contradictory measures and correlations. This confusion has induced a skepticism toward phenomenology in the minds of many theorists. The most straightforward observational evidence has been emphasized in this review, including a correlation space that is likely to be the clearest representation of AGN phenomenology for the foreseeable future. We summarize some of the main phenomenological results from this review:

1. Clear phenomenological differences are found between RQ and RL AGN. These differences warrant separate consideration for these populations in all future studies. The single most striking difference involves a possibly ubiquitous RQ HIL blueshift with respect to the LIL and the AGN rest frame.

2. We find significant evidence that the high- and low-ionization lines in radio-quiet quasars arise in distinct emitting regions. The results for radio-loud sources are less clear and do not allow us to rule out a single stratified BLR.

3. We identify an E1 correlation space that provides optimal discrimination between virtually all types of broad-line AGN phenomenology. The correlates include FWHM Hβ_{BC}, R_{FeII}, and the soft X-ray spectral index Γ_{soft}. It is likely that CIVλ1549$_{BC}$ line shift represents a fourth significant correlate. Optical and radio-continuum luminosity appear to be uncorrelated with the parameters in this space.

4. The other major correlation, which does involve continuum luminosity, is the Baldwin effect. We cannot rule out the possibility that this correlation is induced by sample biases.

5. All line phenomenology including line widths, shifts, asymmetries, and heavy element abundance are strikingly similar over a wide range in redshift. The only clear luminosity dependence involves the Baldwin effect.

6. E1 phenomenology points toward the existence of two RQ AGN classes: (1) Population A, which is an almost pure RQ population with NLS1 as the most extreme subclass; and (2) Population B, which involves RQ sources whose broad-line properties are indistinguishable from RL sources. The extreme subclass of the RL/Population B class involves the steep-spectrum radio galaxies. Average or PCA derived "typical" AGN spectra must be viewed with great caution in the light of E1.

The phenomenology motivates us to advance the following ideas and interpretations:

1. If E1 is as important as we think it is, it may well represent the equivalent of an H-R diagram for AGN, with L/M rather than mass as a principal physical parameter. Orientation almost certainly plays an important role, while chemical abundance could be an important source of scatter and might account for part of the RQ–RL difference.

2. Phenomenology and our assumptions about the role of L/M suggest a trend across E1 from low mass and high accretion rate in NLS1 to high mass and/or low accretion rate in steep-spectrum RL sources.

3. Current phenomenology suggests that RQ LIL arise in an accretion disk and/or some related flattened distribution of gas. RQ HIL most likely arise in a wind associated with the disk. RL HIL and LIL may both arise in a biconical outflow that is connected with the radio jets. This scenario requires that the accretion disk be optically thin or even transient in RL sources so that the far-side outflowing gas can be observed.

4. The above results, coupled with the popular assumption that AGN grow by merger and accretion processes, suggest that we can view NLS1 as the seed population of AGN. In such a framework, RL sources would represent the most evolved population.

5. Finally, we note that there is no evidence in the current data for a difference in broad-line properties as a function of redshift. The inverse correlation between rest frame $W(CIV\lambda 1549)$ and luminosity (Baldwin effect) is the only evidence for a change correlated with distance. In this context, we point out that the *observed frame* distributions of $W(CIV\lambda 1549)$ are roughly the same for all redshifts ($z \approx 0$–4.0).

ACKNOWLEDGMENTS

JS acknowledges hospitality and support from the Padova Observatory. PM acknowledges financial support from the Italian Ministry of University and Scientific and Technological Research (MURST) through grant Cofin 98-02-32 and from the

University of Alabama where this work was completed. DD-H acknowledges financial support IN109698 from PAPIIT-UNAM as well as hospitality from the Padova Observatory and the University of Alabama.

Visit the Annual Reviews home page at www.AnnualReviews.org

LITERATURE CITED

Abramowicz MA, Czerny B, Lasota JP, Szuszkiewicz E. 1988. *Ap. J.* 332:646–58

Anderson KS. 1981. *Ap. J.* 246:13–19

Antonucci R. 1988. In *Supermassive Black Holes*, ed. M Kafatos, pp. 26–42. Cambridge: Cambridge Univ. Press

Antonucci R. 1993. *Annu. Rev. Astron. Astrophys.* 31:473–521

Antonucci RRJ, Cohen RD. 1983. *Ap. J.* 271:564–74

Antonucci RRJ, Hurt T, Agol E. 1996. *Ap. J.* 456:L25–28

Antonucci RRJ, Miller JS. 1985. *Ap. J.* 297:621–32

Aoki K, Yoshida M. 1999. See Ferland & Baldwin 1999, pp. 385–86

Appenzeller I, Östreicher R. 1988. *Astron. J.* 95:45–57

Appl S, Sol H, Vicente L. 1996. *Astron. Astrophys.* 310:419–37

Arav N, Barlow TA, Laor A, Blandford RD. 1997. *MNRAS* 288:1015–21

Arav N, Barlow TA, Laor A, Sargent WLW, Blandford RD. 1998. *MNRAS* 297:990–98

Bachev R. 1999. *Astron. Astrophys.* 348:71–76

Baker JC, Hunstead RW. 1995. *Ap. J.* 452:L95–L98

Baldwin J, Ferland G, Korista K, Verner D. 1995. *Ap. J.* 455:L119–22

Baldwin JA. 1977a. *Ap. J.* 214:679–84

Baldwin JA. 1977b. *MNRAS* 178:P67P–P74

Baldwin JA. 1997. See Peterson et al. 1997, pp. 80–93

Baldwin JA, Burke WL, Gaskell CM, Wampler EJ. 1978. *Nature* 273:431–35

Baldwin JA, Wampler EJ, Gaskell CM. 1989. *Ap. J.* 338:630–53

Barthel PD, Tytler DR, Thomson B. 1990. *Astron. Astrophys. Suppl.* 82:339–89

Bautista MA, Kallman TR, Angelini L, Liedahl DA, Smits DP. 1998. *Ap. J.* 509:848–55

Bergeron J, Kunth D. 1984. *MNRAS* 207:263–86

Berriman G, Schmidt GD, West SC, Stockman HS. 1990. *Ap. J. Suppl.* 74:869-83

Bischoff K, Kollatschny W. 1999. *Astron. Astrophys.* 345:49–58

Boller Th, Brandt WN, Fabian AC, Fink HH. 1997. *MNRAS* 289:393–405

Boller Th, Brandt WN, Fink H. 1996. *Astron. Astrophys.* 305:53–73

Boroson TA. 1989. *Ap. J.* 343:L9–L12

Boroson TA. 1992. *Ap. J.* 399:L15–L17

Boroson TA, Green RF. 1992. *Ap. J. Suppl.* 80:109–35

Boroson TA, Meyers KA. 1992. *Ap. J.* 397:442–51

Boroson TA, Oke JB. 1984. *Ap. J.* 281:535–44

Bottorff M, Korista KT, Shlosman I, Blandford RD. 1997. *Ap. J.* 479:200–21

Bower GA, Wilson AS, Heckman TM, Richstone DO. 1996. *Astron. J.* 111:1901–7

Brandt WN, Boller Th. 1999. See Gaskell et al. 1999, pp. 265–74

Brandt WN, Boller Th, Fabian AC, Ruszkowski M. 1999. *MNRAS* 303:L53–L57

Brandt WN, Mathur S, Elvis M. 1997. *MNRAS* 285:L25–L30

Brinkmann W, Siebert J, Feigelson ED, Kollgaard RI, Laurent-Muehleisen SA, et al. 1997a. *Astron. Astrophys.* 323:739–48

Brinkmann W, Yuan W, Siebert J. 1997b. *Astron. Astrophys.* 319:413–29

Brotherton M, Francis PJ. 1999. See Ferland & Baldwin 1999, pp. 395–405

Brotherton MS. 1996. *Ap. J. Suppl.* 102:1–27

Brotherton MS, Wills BJ, Steidel R, Sargent WLW. 1994. *Ap. J.* 423:131–42

Burbidge GR, Burbidge EM. 1967. *Quasi-Stellar Objects.* San Francisco: Freeman

Calvani M, Marziani P, Sulentic J. 1997. *Mem. Soc. Astron. Ital.* 68:93–94

Capetti A, Axon DJ, Macchetto F, Sparks WB, Boksenberg A. 1996. *Ap. J.* 469:554–63

Carswell RF, Mountain CM, Robertson DJ, Beard SM, Glendinning AR, et al. 1991. *Ap. J.* 381:L5–L8

Chavushyan V. 1995. *Quasi-stellar objects from the second Byurakan Survey* PhD thesis. (In Russian). Gos. Nauch. Tsentr. Spets. Astrof. Obs. RAN, Russia

Chen K, Halpern JP. 1989. *Ap. J.* 344:115–24

Chen K, Halpern JP, Filippenko AV. 1989. *Ap. J.* 339:742–51

Chiang J, Murray N. 1996. *Ap. J.* 466:704–12

Chun X, Livio M, Baum S. 1999. *Astron. J.* 118:1169–76

Colbert EJM, Baum SA, Gallimore JF, O'Dea CP, Lehnert MD, et al. 1996. *Ap. J. Suppl.* 105:75–92

Collin-Souffrin S. 1987. *Astron. Astrophys.* 179:60–70

Collin-Souffrin S, Dumont AM. 1990. *Astron. Astrophys.* 229:292–28

Collin-Souffrin S, Hameury J-M, Joly M. 1988. *Astron. Astrophys.* 205:19–25

Collin-Souffrin S, Joly M, Dumont S, Heidmann N. 1980. *Astron. Astrophys.* 83:190–98

Collin-Souffrin S, Joly M, Heidmann N, Dumont S. 1979. *Astron. Astrophys.* 72:293–308

Collin-Souffrin S, Joly M, Pequignot D, Dumont S. 1986. *Astron. Astrophys.* 166:27–35

Collin-Souffrin S, Lasota JP. 1988. *Publ. Astron. Soc. Pac.* 100:1041–50

Comastri A, Fiore F, Guainazzi M, Matt G, Stirpe GM, et al. 1998. *Astron. Astrophys.* 333:31–37

Corbett EA, Robinson A, Axon DJ, Hough JH, Jeffries RD, et al. 1996. *MNRAS* 281:737–49

Corbett EA, Robinson A, Axon DJ, Young S, Hough JH. 1998. *MNRAS* 296:721–38

Corbin MR. 1990. *Ap. J.* 357:346–52

Corbin MR. 1991. *Ap. J.* 375:503–16

Corbin MR. 1992. *Ap. J.* 391:577–88

Corbin MR. 1993. *Ap. J.* 403:L9–L11

Corbin MR. 1995. *Ap. J.* 447:496–504

Corbin MR. 1997. *Ap. J. Suppl.* 113:245–67

Corbin MR, Boroson TA. 1996. *Ap. J. Suppl.* 107:69–96

Corbin MR, Francis PJ. 1994. *Astron J.* 108:2016–24

Cristiani S, Vio R. 1990. *Astron. Astrophys.* 227:385–85

de Robertis M. 1985. *Ap. J.* 289:67–80

De Zotti G, Gaskell CM. 1985. *Astron. Astrophys.* 147:1–12

Dietrich M, Kollatschny W. 1995. *Astron. Astrophys.* 303:405–19

Done C, Pounds KA, Nandra K, Fabian AC. 1995. *MNRAS* 275:417–28

Doroshenko VT, Sergeev SG, Pronik VI, Chuvaev KK. 1999. *Astron. Lett.* 25:569–81

Dultzin-Hacyan D. 1987. *Rev. Mex. Astron. Astrofis.* 14:94–96

Dultzin-Hacyan D, Tanuguchi Y, Uranga L. 1999. See Gaskell et al. 1999, pp. 303–9

Dumont AM, Collin-Souffrin S. 1990a. *Astron. Astrophys.* 229:302–12

Dumont AM, Collin-Souffrin S. 1990b. *Astron. Astrophys.* 229:313–28

Dumont AM, Collin-Souffrin S. 1990c. *Astron. Astrophys. Suppl. Ser.* 83:71–89

Dumont AM, Collin-Souffrin S, Nazarova L. 1998. *Astron. Astrophys.* 331:11–33

Dumont AM, Joly M. 1992. *Astron. Astrophys.* 263:75–81

Dumont AM, Lasota JP, Collin-Souffrin S, King AR. 1991. *Astron. Astrophys.* 242:503–9

Elston R, Thompson KL, Hill GJ. 1994 *Nature* 367:250–51

Emmering RT, Blandford RD, Shlosman I. 1992. *Ap. J.* 385:460–77

Eracleous M. 1998. *Adv. Space Res.*, 21:33–45

Eracleous M. 1999. See Gaskell et al. 1999, pp. 175–85

Eracleous M, Halpern JP. 1994. *Ap. J. Suppl.* 90:1–30

Eracleous M, Halpern JP, Gilbert AM, Newman JA, Filippenko AV. 1997. *Ap. J.* 490:216–26

Eracleous M, Livio M, Halpern JP, Storchi-Bergmann T. 1995. *Ap. J.* 438:610–22

Erkens U, Appenzeller I, Wagner S. 1997. *Astron. Astrophys.* 323:707–16

Espey BR, Carswell RF, Bailey JA, Smith MG, Ward MJ. 1989. *Ap. J.* 342:666–76

Espey BR, Lanzetta KM, Turnshek, DA. 1993. *Bull. Am. Astron. Soc.* 25:1448

Fabian AC, Shioya Y, Iwasawa K, Nandra K, Crawford K, et al. 1994. *Ap. J.* 436:L51–L54

Falcke H, Wilson A, Simpson C. 1998. *Ap. J.* 502:199–217

Ferguson JW, Korista KT, Baldwin JA, Ferland GJ. 1997. *Ap. J.* 487:122–41

Ferland G, Baldwin J, eds. 1999. *Quasars and Cosmology, Astron. Soc. Pac. Conf. Ser.* 162. San Francisco: Astron. Soc. Pac

Ferland GJ, Korista KT, Peterson BM. 1990. *Ap. J.* 363:L21–L25

Ferland GJ, Korista KT, Verner DA. 1997. In *Astronomical Data Analysis Software and Systems* VI, ASP Conf. Ser. 125 ed. G Hunt, HE Payne, p. 213. San Francisco: *Astron. Soc. Pac.*

Ferland GJ, Persson SE. 1989. *Ap. J.* 347:656–73

Ferland GJ, Peterson BM, Horne K, Welsh WF, Nahar SN. 1992. *Ap. J.* 387:95–108

Ferland GJ, Rees MJ. 1988. *Ap. J.* 332:141–56

Filippenko AV, Sargent WLW. 1989. *Ap. J.* 342:L11–L14

Fiore F, Elvis M, McDowell JC, Siemiginowska A, Wilkes BJ. 1994. *Ap. J.* 431:515–33

Francis PJ, Hewett PC, Foltz CB, Chaffee FH. 1992. *Ap. J.* 398:476–90

Francis PJ, Hewett PC, Foltz CB, Chaffee FH, Weymann RJ, Morris SL. 1991. *Ap. J.* 373:465–70

Francis PJ, Koratkar A. 1995. *MNRAS* 274:504–18

Gaskell CM. 1982. *Ap. J.* 263:79–86

Gaskell CM. 1983. In *Liege Conference on Quasars and Gravitational Lenses*, pp. 473–77. Liege: Institute d'Astrophys.

Gaskell CM. 1985. *Ap. J.* 291:112–16

Gaskell CM. 1988. *Ap. J.* 325:114–18

Gaskell CM. 1996. *Ap. J.* 464:L107–10

Gaskell CM, Brandt WN, Dietrich M, Dultzin-Hacyan D, Eracleous M, eds. 1999. *Structure and Kinematics of Quasar Broad Line Regions. Astron. Soc. Pac. Conf. Ser.* 175. San Francisco: Astron. Soc. Pac. 163ff pp.

Gaskell CM, Snedden S. 1999. See Gaskell et al. 1999. pp. 157–62

George IM, Nandra K, Laor A, Turner TJ, Fiore F, et al. 1997. *Ap. J.* 491:508–14

Giannuzzo ME, Mignoli M, Stirpe GM, Comastri A. 1998. *Astron. Astrophys.* 330:894–900

Giannuzzo ME, Stirpe GM. 1996. *Astron. Astrophys.* 314:419–29

Goad M, Koratkar A. 1998. *Ap. J.* 495:718–39

Goad M, Wanders I. 1996. *Ap. J.* 469:113–30

Goad MR, Koratkar AP, Axon DJ, Korista KT, O'Brien PT. 1998. *Ap. J.* 512:L95–L98

Goad MR, Koratkar AP, Kim-Quijano J, Korista KT, O'Brien PT. 1999. *Ap. J.* 524:707–31

Gonçalves AC, Véron P, Véron-Cetty M-P. 1999. *Astron. Astrophys.* 341:662

Goodrich RW. 1989. *Ap. J.* 342:224–34

Graham MJ, Clowes RG, Campusano LE. 1996. *MNRAS* 279:1349–56

Green PJ. 1996. *Ap. J.* 467:61–75

Green PJ. 1998. *Ap. J.* 498:170–80

Grupe D, Beuermann K, Mannheim K, Thomas H-C. 1999. *Astron. Astrophys.* 350:805–15

Grupe D, Beuermann K, Thomas HC, Mannheim K, Fink HH. 1998. *Astron. Astrophys.* 330:25–36

Guainazzi M, Mihara T, Otani C, Matsuoka M. 1996. *Publ. Astron. Soc. Jpn.* 48:781–99

Guilbert PW, Fabian AC, Rees MJ. 1983. *MNRAS* 205:593–603

Guilbert PW, Rees MJ. 1988. *MNRAS* 233:475–84

Haardt F, Maraschi L. 1991. *Ap. J.* 380:L51–L54

Haardt F, Maraschi L, Ghisellini G. 1997. *Ap. J.* 476:620–31

Halpern JP, Eracleous M. 1994. *Ap. J.* 433:L17–L20

Halpern JP, Eracleous M, Filippenko AV, Chen K. 1996. *Ap. J.* 464:704–14

Halpern JP, Eracleous M, Forster K. 1998. *Ap. J.* 501:103

Hamann F, Ferland G. 1999. *Annu. Rev. Astron. Astrophys.* 37:487–531

Hartig GF, Baldwin JA. 1986. *Ap. J.* 302:64–80

Heckman TM. 1980. *Astron. Astrophys.* 88:311–16

Hes R, Barthel PD, Fosbury RAE. 1993. *Nature* 362:326–28

Hewett PC, Foltz CB, Chaffee FH. 1995. *Astron. J.* 109:1498–521

Hewitt A, Burbidge G. 1993. *Ap. J. Suppl.* 87:451–947

Hill GJ, Thompson KL, Elston R. 1993. *Ap. J.* 414:L1–L4

Ho LC, Filippenko AV, Sargent WLW, Peng CY. 1997. *Ap. J. Suppl.* 112:391–414

Hummer DG, Berrington KA, Eissner W, Pradhan AK, Saraph HE, et al. 1993. *Astron. Astrophys.* 279:298–309

Jackson N, Browne IWA. 1991a. *MNRAS* 250:414–21

Jackson N, Browne IWA. 1991b. *MNRAS* 250:422–31

Joly M. 1991. *Astron. Astrophys.* 242:49–57

Joly M. 1987. *Astron. Astrophys.* 184:33–42

Junkkarinen V. 1989. In *Active Galactic Nuclei*, ed. DE Osterbrock, JS Miller, pp. 122–23. Dordrecht: Kluwer Acad.

Kaspi S, Smith PS, Netzer H, Maoz D, Jannuzi BT, Giveon U. 2000. *Ap. J.* 533:631–49

Kassebaum TM, Peterson BM, Wanders I, Pogge RW, Bertram R, Wagner RM. 1997. *Ap. J.* 475:106–17

Kawara K, Murayama T, Taniguchi Y, Arimoto N. 1996. *Ap. J.* 470:L85–L88

Kay LE, Moran EC. 1998. *Publ. Astron. Soc. Pac.* 110:1003–6

Kellermann KI, Sramek R, Schmidt M, Shaffer DB, Green R. 1989. *Astron. J.* 98:1195–207

Kinney A. 1994. In The First Stromlo Symposium: *The Physics of Active Galaxies, ASP Conf. Ser.* 54, ed. GV Bicknell, MA Dopita, PJ Quinn. San Francisco: Astron. Soc. Pac.

Kinney AL, Bohlin RC, Blades JC, York DG. 1991. *Ap. J. Suppl.* 75:645–717

Kinney AL, Rivolo AR, Koratkar AP. 1990. *Ap. J.* 357:338–45

Koratkar A, Blaes O. 1999. *Publ. Astron. Soc. Pac.* 111:1–30

Koratkar A, Goad MR, O'Brien PT, Salamanca I, Wanders I, et al. 1996. *Ap. J.* 470:378–93

Korista K, Baldwin J, Ferland G, Verner D. 1997a. *Ap. J. Suppl.* 108:401–15

Korista K, Ferland G, Baldwin J. 1997b. *Ap. J.* 487:555–59

Korista KT. 1999. See Ferland & Baldwin 1999, pp. 429–45

Korista KT, Alloin D, Barr P, Clavel J, Cohen RD, et al. 1995. *Ap. J. Suppl.* 97:285–330

Kriss GA, Davidsen AF, Zheng W, Lee G. 1999. *Ap. J.* 527:683–95

Krolik JH. 1999. *Active Galactic Nuclei.* Princeton, NJ: Princeton Univ. Press

Krolik JH, Kallman TR. 1988. *Ap. J.* 324:714–20

Kwan J, Cheng F-Z, Fang L-Z, Zheng W, Ge J. 1995. *Ap. J.* 440:628–33

Kwan J, Krolik JH. 1981. *Ap. J.* 250:478–507

Lanzetta KM, Turnshek DA, Sandoval J. 1993. *Ap. J. Suppl.* 84:109–84

Laor A, Bahcall JN, Jannuzi BT, Schneider DP, Green RF, Hartig GF. 1994. *Ap. J.* 420:110–35

Laor A, Bahcall JN, Jannuzi BT, Schneider DP, Green RF. 1995. *Ap. J. Suppl.* 99:1–26

Laor A, Fiore F, Elvis M, Wilkes BJ, McDowell JC. 1997b. *Ap. J.* 477:93–113

Laor A, Jannuzi BT, Green RF, Boroson TA. 1997a. *Ap. J.* 489:656–71

Lawrence A, Elvis M, Wilkes BJ, McHardy I, Brandt N. 1997. *MNRAS* 285:879–90

Lee LC, Fabian AC, Brandt WN, Reynolds CS, Iwasawa K. 1999. *MNRAS* 310:973–81

Leighly KM. 1999a. *Ap. J. Suppl.* 125:297–316

Leighly KM. 1999b. *Ap. J. Suppl.* 125:317–48

Leighly KM, Mushotzky RF, Nandra K, Forster K. 1997. *Ap. J.* 489:L25–L28

Lightman AP, Eardley DM. 1974. *Ap. J.* 187:L1–L3

Lipari S, Colina L, Macchetto F. 1994. *Ap. J.* 427:174–83

Lipari S, Terlevich R, Macchetto F. 1993. *Ap. J.* 406:451–56

Livio M, Pringle JE. 1996. *MNRAS* 278:L35–L36

Livio M, Xu C. 1997. *Ap. J.* 478:L63–L65

Malkov YF, Pronik VI, Sergeev SG. 1997. *Astron. Astrophys.* 324:904–14

Mannheim K, Schulte M, Rachen J. 1995. *Astron. Astrophys.* 303:L41–L44

Maoz D, Netzer H, Peterson BM, Bechtold J, Bertram R, et al. 1993. *Ap. J.* 404:576–83

Marziani P. 1991. PhD thesis. Trieste: SISSA

Marziani P, Calvani M, Sulentic JW. 1997. In *Accretion Phenomena and Related Outflows*, IAU Coll. 163 *Astron. Soc. Pac. Conf. Ser.* 121. ed. DT Wickramashinge, GV Bicknell, L Ferrario pp. 761–2. San Francisco: Astron. Soc. Pac.

Marziani P, Sulentic JW. 1993. *Ap. J.* 409:612–16

Marziani P, Sulentic JW, Calvani M, Perez E, Moles M, Penston MV. 1993. *Ap. J.* 410:56–67

Marziani P, Sulentic JW, Dultzin-Hacyan D, Calvani M, Moles M. 1996. *Ap. J. Suppl.* 104:37–70

Mason KO, Puchnarewicz EM, Jones LR. 1996. *MNRAS* 283:L26–L29

Mathews WG. 1986. *Ap. J.* 305:187–203

Mathews WG. 1993. *Ap. J.* 412:L17

Mathews WG, Doane JS. 1990. *Ap. J.* 352:443–32

Mathews WG, Ferland GJ. 1987. *Ap. J.* 323:456–67

Miley GK, Miller JS. 1979. *Ap. J.* 228:L55–L58

Miller JS, Peterson BM. 1990. *Ap. J.* 361:98–100

Mobasher B, Raine DJ. 1989. *MNRAS* 237:979–94

Moran EC, Halpern JP, Helfand DJ. 1996. *Ap. J. Suppl.* 106:341–97

Morris P, Conti PS, Lamers HJGLM, Koenigsberger G. 1993. *Ap. J.* 414:L25–L28

Morris SL, Ward MJ. 1989. *Ap. J.* 340:713–28

Murdoch HS. 1983. *MNRAS* 202:987–94

Murray N, Chiang J. 1997. *Ap. J.* 474:91–103

Murray N, Chiang J. 1998. *Ap. J.* 494:125–38

Murray N, Chiang J, Grossman SA, Voit GM. 1995. *Ap. J.* 451:498–509

Mushotzky R, Ferland GJ. 1984. *Ap. J.* 278:558–63

Nandra K, George IM, Mushotzky RF, Turner TJ, Yaqoob T. 1999. *Ap. J.* 523:L17–L20

Nandra K, George IM, Mushotzky RF, Turner TJ, Yaqoob T. 1997a. *Ap. J.* 476:70–82

Nandra K, George IM, Mushotzky RF, Turner TJ, Yaqoob T. 1997b. *Ap. J.* 477:602–22

Nandra K, George IM, Turner TJ, Fukazawa Y. 1996. *Ap. J.* 464:165–69

Netzer H. 1985. *MNRAS* 216:63–78

Netzer H, Brotherton MS, Wills BJ, Han M, Wills D, et al. 1995. *Ap. J.* 448:27–40

Netzer H, Turner TJ, George IM. 1998. *Ap. J.* 504:680–92

Netzer H, Wills BJ. 1983. *Ap. J.* 275:445–60

Nicastro F, Fiore F, Matt G. 1999. *Ap. J.* 517:108–22

Nishihara E, Yamashita T, Yoshida M, Watanabe E, Okumura SI. 1997. *Ap. J.* 488:L27–L30

NIST database. http://physics.nist.gov/ PhysRefData/contents.html

O'Brien PT, Dietrich M, Leighly K, Alloin D, Clavel J, et al. 1998. *Ap. J.* 509:163–76

Orr MJL, Browne IWA. 1982. *MNRAS* 200:1067–80

Osmer PS, Porter AC, Green RF. 1994. *Ap. J.* 436:678–95

Osmer PS, Shields J. 1999. In *Quasars and Cosmology, Astron. Soc. Pac. Conf. Ser. 162.* ed. G Ferland, J Baldwin, pp. 235–64. San Francisco: Astron. Soc. pac.

Osterbrock DE. 1977. *Ap. J.* 215:733–45

Osterbrock DE. 1978. *Phys. Scr.* 17:137–43

Osterbrock DE. 1981. *Ap. J.* 249:462–70

Osterbrock DE. 1987. In *Observational Evidence of Activity in Galaxies, IAU Symp. 121*, ed. EE Khachikian, KJ Fricke, J Melnick, pp. 109–18. Dordrecht: Reidel

Osterbrock DE. 1993. *Ap. J.* 404:551–62

Osterbrock DE, Koski AT. 1976. *MNRAS* 176:P61–P66

Osterbrock DE, Mathews WG. 1986. *Annu. Rev. Astron. Astrophys.* 24:171–203

Osterbrock DE, Pogge RW. 1985. *Ap. J.* 297:166–76

Osterbrock DE, Shuder JM. 1982. *Ap. J. Suppl.* 491:49–74

Padovani P, Giommi P. 1996. *MNRAS* 279:526–34

Peimbert M, Torres-Peimbert S. 1981. *Ap. J.* 245:845–56

Penston MV. 1987. *MNRAS* 229:P1–P5

Penston MV, Robinson A, Alloin D, Appenzeller I, Aretxaga I, et al. 1990. *Astron. Astrophys.* 236:53–62

Perry JJ, Dyson JE. 1985. *MNRAS* 213:665–710

Peterson BM. 1994. In *Reverberation Mapping of the Broad-Line Region in Active Galactic Nuclei. Astron. Soc. Pac. Conf. Ser. 69*, ed. PM Gondhalekar, K Horne, BM Peterson, pp. 1–22. San Francisco: Astron. Soc. Pac.

Peterson BM, Balonek TJ, Barker ES, Bechtold J, Bertram R, et al. 1991. *Ap. J.* 368:119–37

Peterson BM, Barth AJ, Berlind P, Bertram R, Bischoff K, et al. 1999. *Ap. J.* 510:659–68

Peterson BM, Berlind P, Bertram R, Bochkarev NG, Bond D, et al. 1994. *Ap. J.* 425:622–34

Peterson BM, Cheng FZ, Wilson AS, eds. 1997. *Emission Lines in Active Galaxies: New Methods and Techniques. Astron. Soc. Pac. Conf. Ser. 113.* San Francisco: Astron. Soc. Pac.

Peterson BM, Ferland GJ. 1986. *Nature* 324:345–47

Peterson BM, Foltz CB, Byard PL, Wagner RM. 1982. *Ap. J. Suppl.* 49:469–86

Peterson BM, Gaskell CM. 1991. *Ap. J.* 368:152–57

Peterson BM, Wandel A. 1999. *Ap. J.* 521:L95–97

Phillips MM. 1978a. *Ap. J.* 226:736–52

Phillips MM. 1978b. *Ap. J. Suppl.* 38:187–203

Piccinotti G, Mushotzky RF, Boldt EA, Holt SS, Marshall FE, et al. 1982. *Ap. J.* 253:485–503

Pogge RW. 1988. *Ap. J.* 328:519–22

Pogge RW. Peterson BM. 1992. *Astron. J.* 103:1084–88

Popovic LC, Vince I, Atanackovic-Vukmanovic O, Kubicela A. 1995. *Astron. Astrophys.* 293:309–14

Pounds KA, Done C, Osborne JP. 1995. *MNRAS* 277:L5–L10

Pringle JE. 1996. *MNRAS* 281:357–61

Puchnarewicz EM, Mason KO, Córdova FA, Kartje J, Brabduardi AA, et al. 1992. *MNRAS* 256:589–623

Puchnarewicz EM, Mason KO, Siemiginowska A. 1998. *MNRAS* 293:L52–L56

Recondo-Gonzalez MC, Wamsteker W, Clavel J, Rodriguez-Pascual PM, Vio R, et al. 1997. *Astron. Astrophys. Suppl.* 121:461–87

Rees MJ, Netzer H, Ferland GJ. 1989 *Ap. J.* 347:640–55

Reichert GA, Rodriguez-Pascual PM, Alloin D, Clavel J, Crenshaw DM, et al. 1994. *Ap. J.* 425:582–608

Remillard RA, Bradt HV, Buckley DAH, Roberts W, Schwartz DA. et al. 1986. *Ap. J.* 301:742–52

Reynolds CS, Fabian AC, Inoue H. 1995. *MNRAS* 276:1311–19

Robinson A. 1995. *MNRAS* 276:933–43

Rodriguez-Pascual PM, Alloin D, Clavel J, Crenshaw DM, Horne K, et al. 1997b. *Ap. J. Suppl.* 110:9–20

Rodriguez-Pascual PM, Mas-Hesse JM, Santos-Lleo M. 1997a. *Astron. Astrophys.* 327:72–80

Rokaki E, Boisson C, Collin-Souffrin S. 1992. *Astron. Astrophys.* 253:57–73

Romano P, Zwitter T, Calvani M, Sulentic J. 1996. *MNRAS* 279:165–70

Sambruna RM, Eracleous M, Mushotzky RF. 1999. *Ap. J.* 526:60–96

Sargent WLW, Steidel CC, Boksenberg A. 1988. *Ap. J. Suppl.* 68:539–641

Sargent WLW, Steidel CC, Boksenberg A. 1989. *Ap. J. Suppl.* 69:703–61

Saxton RD, Turner MJL, Williams OR, Stewart GC, Ohashi T, Kii T. 1993. *MNRAS* 262:63–74

Seyfert CK. 1943. *Ap. J.* 97:28–40

Shakura NI, Sunyaev RA. 1973. *Astron. Astrophys.* 24:337

Shastri P, Wilkes BJ, Elvis M, McDowell J. 1993. *Ap. J.* 410:29–38

Shaviv G, Wickramasinghe D, Wehrse R. 1999. *Astron. Astrophys.* 344:639–46

Shuder JM. 1981. *Ap. J.* 244:12–18

Shuder JM. 1984. *Ap. J.* 280:491–98

Shuder JM, McAlpine GM. 1979. *Ap. J.* 230:348

Siebert J, Brinkmann W, Drinkvater MJ, Yuan W, Francis PJ, Peterson BA. 1998. *Astron. Astrophys.* 301:261–79

Siebert J, Leighly KM, Laurent-Muehleisen SA, Brinkmann W, Boller Th, et al. 1999. *Astron. Astrophys.* 348:678–84

Sigut TAA, Pradhan AK. 1998. *Ap. J.* 499:L139–42

Sparke LS. 1993 *Ap. J.* 404:570–75

Steidel CC, Sargent WLW. 1991. *Ap. J.* 382:433–65

Steidel CC, Sargent WLW. 1992. *Ap. J. Suppl.* 80:1–108

Steiner JE. 1981. *Ap. J.* 250:469–77

Stephens SA. 1989. *Astron. J.* 97:10–35

Stirpe GM. 1989. *Broad emission line profiles in active galactic nuclei.* PhD thesis. Rijksuniversiteit, Leiden. 149 pp.

Stirpe GM. 1991. *Astron. Astrophys.* 247:3–10

Stirpe GM, Winge C, Altieri B, Alloin D, Aguro EL, et al. 1994. *Ap. J.* 425:609–21

Stockman HS, Moore RL, Angel JRP. 1984. *Ap. J.* 279:485–98

Stockton A, Farnham T. 1991. *Ap. J.* 371:525–34

Storchi-Bergmann T, Baldwin JA, Wilson AS. 1993. *Ap. J.* 410:L11–L14

Storchi-Bergmann T, Eracleous M, Livio M, Wilson AS, Filippenko AV, Halpern JP. 1995. *Ap. J.* 443:617–24

Storchi-Bergmann T, Eracleous M, Ruiz MT,

Livio M, Wilson AS, Filippenko AV. 1997. *Ap. J.* 489:87–93

Storchi-Bergmann T, Ruiz MT, Eracleous M. 1998. *Adv. Space Res.* 21:47–56

Sulentic JW. 1989. *Ap. J.* 343:54–65

Sulentic JW, Calvani M, Marziani P, Zheng W. 1990. *Ap. J.* 355:L15–L18

Sulentic JW, Marziani P. 1999. *Ap. J.* 518:L9–L12

Sulentic JW, Marziani P, Dultzin-Hacyan D. 1999. See Gaskell, et al. 1999. pp. 175–85

Sulentic JW, Marziani P, Zwitter T, Calvani M. 1995. *Ap. J.* 438:L1–L4

Sulentic JW, Marziani P, Zwitter T, Calvani M, Dultzin-Hacyan D. 1998a. *Ap. J.* 501:54–68

Sulentic JW, Zwitter T, Marziani P., Dultzin-Hacyan D. 2000. *Ap. J.* 536:L5–L9

Szuszkiewicz E, Malkan MA, Abramowicz MA. 1996. *Ap. J.* 458:474–90

Tadhunter C, Tsvetanov Z. 1989. *Nature* 341:422–24

Taylor GB, Silver CS, Ulvestad JS, Carilli CL. 1999. *Ap. J.* 519:185–90

Thompson KL, Hill GJ, Elston R. 1999. *Ap. J.* 515:487–96

Turnshek DA, Espey BR, Kopko MJR, Rauch M, Weymann RJ, et al. 1994. *Ap. J.* 428:93–112

Turnshek DA, Monier EM, Sirola CJ, Espey BR. 1997. *Ap. J.* 476:40–48

Tytler D, Fan XM. 1992. *Ap. J. Suppl.* 79:1–36

Ulrich M-H, Altamore A, Perola GC, Boksemberg A, Penston MV, et al. 1985. *Nature* 313:747–51

Ulrich M-H, Boksenberg A, Bromage GE, Clavel J, Elvius A, et al. 1984. *MNRAS* 206:221–37

Ulrich M-H, Horne K. 1996. *MNRAS* 283:748–58

Ulrich M-H, Maraschi L, Urry CM. 1997. *Annu. Rev. Astron. Astrophys.* 35:445–502

Ulvestad JS, Antonucci RRJ, Goodrich RW. 1995. *Astron. J.* 109:81–86

Ulvestad JS, Wrobel JM, Roy AL, Wilson AS, Falcke H, Krichbaum TP. 1999. *Ap. J.* 1 517:L81–L84

Valtonen MJ. 1999. *Ap. J.* 520:97–104

van Groningen E, Van Weeren N. 1989. *Astron. Astrophys.* 211:318–23

van Hoof PAM. 1999. http://www.pa.uky.edu/peter/atomic

Vaughan S, Pounds KA, Reeves J, Warwick R, Edelson R. 1999b. *MNRAS* 308:L34–L38

Vaughan S, Reeves J, Warwick R, Edelson R. 1999a. *MNRAS* 309:113–23

Verner DA, Ferland GJ, Korista KT, Yakovlev DG. 1996. *Ap. J.* 465:487–98

Verner EM, Verner DA, Korista KT, Ferguson JW, Hamann F, Ferland GJ. 1999. *Ap. J. Suppl.* 120:101–12

Véron-Cetty M-P, Véron P. 1998. *ESO Sci. Rep.* 18:1–346. Munich: ESO

Vestergaard M, Wilkes BJ. 1999. See Gaskell et al. 1999, p. 297

Vrtilek JM, Carleton NP. 1985. *Ap. J.* 294:106–20

Walter R, Fink HH. 1993. *Astron. Astrophys.* 274:105–22

Wampler EJ, Gaskell CM, Burke WL, Baldwin JA, 1984. *Ap. J.* 276:403–12

Wampler EJ. 1985. *Ap. J.* 296:416–22

Wampler EJ. 1986. *Astron. Astrophys.* 161:223–31

Wamsteker W, Wang TG, Schartel N, Vio R. 1997. *MNRAS* 288:225–36

Wandel A. 1999a. *Ap. J.* 527:649–56

Wandel A. 1999b. *Ap. J.* 527:657–61

Wandel A, Boller Th. 1998. *Astron. Astrophys.* 331:884–90

Wandel A, Peterson BM, Malkan MA. 1999. *Ap. J.* 526:579–91

Wanders I, Goad MR, Korista KT, Peterson BM, Horne K, et al. 1995. *Ap. J.* 453:L87–L90

Wanders I, Peterson BM, Alloin D, Ayres TR, Clavel J, et al. 1997 *Ap. J. Suppl.* 113:69–88

Wang J-M, Szuszkiewicz E, Lu F-J, Zhou Y-Y. 1999. *Ap. J.* 522:839–85

Wang T, Brinkmann W, Bergeron J. 1996. *Astron. Astrophys.* 309:81–86

Wang TG, Lu YJ, Zhou YY. 1998 *Ap. J.* 493:1–9

Weymann RJ, Morris SL, Foltz CB, Hewett PC. 1991. *Ap. J.* 373:23–53

White NE, Fabian AC, Mushotzky RF. 1984. *Astron. Astrophys.* 133:L9–L11

Wilkes BJ. 1984. *MNRAS* 207:73–98

Wilkes BJ. 1999. Active Galactic Nuclei. In *Astrophysical Quantities.* Chapter 24, ed. A Cox. Berlin: Springer-Verlag. In press

Wilkes BJ, Elvis M, McHardy I. 1987. *Ap. J.* 321:L23–L27

Wilkes BJ, Kuraszkiewicz J, Green PJ, Mathur S, McDowell JC. 1999. *Ap. J.* 513:76–107

Wilkes BJ, Wright AE, Jauncey DL, Peterson BA. 1983. *Proc. Astron. Soc. Aus.* 5:2–83

Williams RJR, Baker AC, Perry JJ. 1997. *Rev. Mex. Astron. Astrofis. Ser. Conf.* 6:263

Williams RJR, Baker AC, Perry JJ. 1999. *MNRAS* 310:913–62

Wills BJ, Brotherton MS. 1995. *Ap. J.* 448:L81–4

Wills BJ, Brotherton MS, Fang D, Steidel CC, Sargent WLW. 1993. *Ap. J.* 415:563–79

Wills BJ, Browne IWA. 1986. *Ap. J.* 302:56–63

Wills BJ, Laor A, Brotherton MS, Wills D, et al. 1999 *Ap. J.* 515:L53–56

Wills BJ, Netzer H, Wills D. 1985. *Ap. J.* 288:94–116

Wills BJ, Thompson KL, Han M, Netzer H, Wills D, et al. 1995. *Ap. J.* 447:139–58

Wilson AS. 1996. *Vistas Astron.* 40:63–70

Wilson AS, Heckman TM. 1985. In *Astrophysics of Active Galaxies and Quasi-Stellar Objects*, ed. JS Miller, Mill Valley: University Science Books. pp. 39–109

Winkler H. 1992. *MNRAS* 257:677–88

Wu CC, Boggess A, Gull TR. 1983. *Ap. J.* 266:28–40

Yamashita A, Matsumoto C, Ishida M, Inoue H, Kii T, et al. 1997. *Ap. J.* 486:763–69

Yee HKC, Oke JB. 1981. *Ap. J.* 248:472–84

Young P, Sargent WLW, Boksenberg A. 1982. *Ap. J. Suppl.* 48:455–505

Yuan W, Brinkmann W, Siebert J, Voges W. 1998b. *Astron. Astrophys.* 330:108–22

Yuan W, Siebert J, Brinkmann W. 1998a. *Astron. Astrophys.* 334:498–504

Zheng W. 1992. *Ap. J.* 385:127–31

Zheng W, Binette L, Sulentic JW. 1990. *Ap. J.* 365:115–18

Zheng W, Keel WC. 1991. *Ap. J.* 382:121–24

Zheng W, Kriss GA, Davidsen AF. 1995. *Ap. J.* 440:606–9

Zheng W, Kriss GA, Telfer RC, Grimes JP, Davidsen AF. 1997. *Ap. J.* 475:469–78

Zheng W, Malkan MA. 1993. *Ap. J.* 415:517–23

Zheng W, O'Brien PT. 1990 *Ap. J.* 353:433–37

Zheng W, Veilleux S, Grandi SA. 1991. *Ap. J.* 381:418–25

Annu. Rev. Astron. Astrophys. 2000. 38:573–611

MASS LOSS FROM COOL STARS: Impact on the Evolution of Stars and Stellar Populations

Lee Anne Willson

Department of Physics and Astronomy, Iowa State University;
e-mail: lwillson@iastate.edu

Key Words   Stellar evolution, stellar winds, asymptotic giant stars, circumstellar dust, Mira variables

■ **Abstract**   This review emphasizes the mass loss processes that affect the fates of single stars with initial masses between one and nine solar masses. Just one epoch of mass loss has been clearly demonstrated to be important for these stars; that is the episode that ends their evolution up the asymptotic giant branch. Quite a clear picture of this evolutionary stage is emerging from current studies. Mass loss rates increase precipitously as stars evolve toward greater luminosity and radius and decreased effective temperature. As a result, empirical relationships between mass loss rates and stellar parameters are determined mostly by selection effects and tell us which stars are losing mass rather than how stars lose mass. After detailed theoretical models are found to match observational constraints, the models may be used to extrapolate to populations not available for study nearby, such as young stars with low metallicity. The fates of stars are found to depend on both their initial masses and their initial metallicities; a larger proportion of low-metallicity stars should end up with core masses reaching the Chandrasekhar limit, giving rise to Type 1.5 supernovae, and the remnant white dwarfs of low-Z populations will be both fewer and more massive than those in Population I. There are also clear indications that some stars lose one to several tenths of a solar mass during the helium core flash, but neither models nor observations reveal any details of this process yet. The observational and theoretical bases for a variety of mass loss formulae in current use are also reviewed in this article, and the relations are compared in a series of figures.

INTRODUCTION

During the past two decades, enormous strides have been taken in the detection of outflows from stars, including the massive outflows that end the nuclear burning evolution of stars like the Sun. Considerable effort has gone into understanding these observations through a variety of theoretical models. In this review, I emphasize the study of mass loss processes and events that affect the post–main-sequence evolution of stars up to about 9 M_{Sun}, and determine their final remnant masses. In Population I, these stars ascend the asymptotic giant branch (AGB) and end their

573

lives as white dwarf stars. With less efficient mass loss, as is likely for Pop II or Pop III stars, a significant fraction of the 3–10 M_{Sun} stars will end up producing super-novae of Type 1.5, with largely unexplored consequences for the early chemical evolution of our galaxy.

A number of interesting reviews relevant to this topic have appeared over the years. In the first volume of *ARAA*, Weymann reviewed mass loss from stars (Weymann 1963). More recently, Holzer & Axford (1970) covered stellar wind theory, Kwok (1993) emphasized the transition from the AGB to PNE, and Wei-demann (1990) discussed the relationship of initial mass to final mass in his review of white dwarfs. Dupree (1986) reviewed mass loss from cool stars in 1986, emphasizing spectroscopic signatures and solar-type winds, and Zucker-man (1980) reviewed envelopes around late-type giants. In another article in this volume, Kudritzki reviews mass loss from hot stars. Iben & Renzini (1983; Iben 1974) described single star evolution from the main sequence onward, includ-ing some effects of mass loss. Chiosi & Maeder (1986) reviewed massive star evolution with mass loss for this series, and Chiosi et al (1992) included dis-cussion of mass loss in their article on the HR diagram. Recent conference proceedings of note include *Physical Processes in Red Giants* (Iben & Renzini 1981), *Late Stages of Stellar Evolution* (Kwok & Pottasch 1987), and *Asymp-totic Giant Branch Stars* (Le Bertre et al 1999). *Mass Loss from Red Giants* (Morris & Zuckerman 1985) includes a number of excellent reviews that are still timely.

Interpreting observations of winds and measuring mass loss rates require an understanding of the nature of the flow and thus of the mechanisms that drive mass loss. Three types of winds that have received the most attention for cool stars are solar-type winds, dust-driven winds, and pulsation-initiated winds. These mecha-nisms are briefly reviewed in Section 1. In the development of evolutionary studies of stars and populations, if mass loss is taken into account at all, it has been done using empirical mass loss. The observational basis for these empirical formulae is discussed in Section 2, with consideration for both the mass loss rates derived from observations and the uncertainties in the other parameters of the mass-losing stars. In the final section, the mass loss epoch that ends the AGB stage of evolution is examined in detail, both because it is particularly important for the evolution of low-mass stars and because it is an example of how empirical mass loss laws can be interpreted in more than one way with very different consequences. From a careful mating of theoretical models with observational constraints it is possible to derive predictions for the mass loss history of populations not presently available for study, such as young, low-metallicity populations that were important when our galaxy was much younger.

The evolution of low- and intermediate-mass stars without mass loss is reviewed in many textbooks, and most authoritatively by Iben (1974; Iben & Renzini 1983); red giant branch development is described in detail by Sweigart & Gross (1978; also Sweigart et al 1990). Briefly, low-mass stars leave the main sequence and migrate to the red giant branch (RGB) after core H exhaustion. They evolve up the RGB with hydrogen-burning shells around growing degenerate He cores. For those stars with M $<$ 2–3 M_{Sun} (depending on composition), the RGB ends with a helium core flash, after which the star settles down to convert He to C in its core on the horizontal branch or in the red giant "clump." Slightly more massive stars loop to the left in the HRD during core He burning without such an abrupt beginning. After the He in the core has been converted to C and O, the core becomes degenerate, and shell-hydrogen burning recommences. The shell burning builds up a degenerate helium layer around the core; this layer "flashes" into renewed helium burning each time its mass reaches a critical value. Thus, H and He burning alternate as the star evolves up the asymptotic giant branch (AGB or thermally pulsing AGB). When the mass of the hydrogen-rich envelope becomes quite small—either because of conversion to He and C+O or because of mass loss—the star leaves the AGB, migrating toward hotter temperatures where it may spend some time as the central star of a planetary nebula before cooling as a white dwarf (e.g. Schönberner 1983). This general picture is well confirmed by observations, including period changes and the formation of discrete circumstellar shells consistent with variations expected during individual shell flashes (Wood & Zarro 1981; Sterken et al 1999; Olofsson et al 1990, 1996; Hashimoto et al 1998).

Mass loss determines several features of stellar populations that are important for interpreting a wide variety of observations. Mass loss determines the maximum luminosity achieved by AGB stars, the most luminous red stars in populations older than about 0.1 Gyr and younger than about 10 Gyr. Mass loss determines the mass spectrum of white dwarfs, and this in turn affects their cooling times. Mass loss determines the masses of the central stars of planetary nebulae, and this in turn determines the planetary nebula luminosity function. Mass loss determines the frequency of supernovae of type 1.5 and II, and possibly the masses of the progenitors of Type I supernovae, and this affects both the use of supernovae as probes of distant galaxy populations and distances and the chemical evolution of galaxies. Mass loss determines the maximum radius the star achieves, and thus the fates of any planets in orbit around it (LA Willson and GH Bowen 2000, in prep).

Direct observational constraints on mass loss from stars cover a limited range of populations. In our galaxy, we have young stars of high metallicity and old stars of low metallicity available to study. When our galaxy was young, however, the stars were young and of low metallicity. To understand the chemical evolution of our galaxy and the appearance of young populations of low Z outside our galaxy, we need to know the mass loss patterns of populations whose mass loss rates we cannot study directly. The solution to this apparent impasse is the same as the solution to the problem of determining the ages of stars: Rather than looking at individual objects and then trying to fit empirical relations to these for use in models, we should look at the evolution of populations of stars and fit these to theoretical models that have already been checked against cases that can be more directly observed.

1.   ELEMENTS OF THE THEORY

Theoretical studies of red giant winds are examined in this section for (a) solar-type winds in the framework of stationary outflows, (b) stationary outflows involving dust, and (c) winds from pulsating stars. The conclusion is reached that at least the massive outflows of highly evolved giants must be attributed to pulsation, acting alone in some cases and providing the opportunity for dust to form in others.

1.1   Stationary Wind Theory

The hydrodynamic theory of stationary, supersonic stellar winds was developed in a classic series of papers by Parker (1960; 1961; 1963a,b; 1964a,b,c; 1965; 1966; 1969; see also Parker 1997 for a review of the history of the subject, and Weymann 1963 for a succinct summary). Such flows are governed by an equation of the form

$$1/v \, dv/dr = (2/r)[A(r) - B(r)]/(v^2 - c_T^2), \tag{1}$$

where $c_T^2 = P/\rho$. For a spherically symmetric, isothermal, thermally driven wind, $A = c_T^2 =$ constant and $B = v_{esc}^2/4 = GM/2r$. A critical point, r_P, occurs where $A = B$. If v is less than c_T at r_P, then the flow decelerates beyond r_P. This solution requires pressure greatly exceeding typical interstellar values in most cases. A static atmosphere with $T(r)$ declining more slowly than $1/r$ also requires unreasonably high pressure at infinity to contain it. If T is proportional to $1/r$, then there is no critical point and the flow must remain subsonic; see Parker (1965) for more detailed discussion of these and related points. If v reaches c_T below r_P, then formally the acceleration becomes infinite, so such solutions are excluded. More physically, once $(1/v \, dv/dr)^{-1}$ becomes less than the gas mean free path (i.e. for a large Knudsen number), the hydrodynamic approximation breaks down, and the equations no longer apply. Oscillatory solutions around $v = c_T$ may then occur in real stars (GH Bowen, private communication), but they fall outside the theory of stationary flows.

If the outflow carries very little mass, then the mean free path is large, and the thermal velocity distribution is no longer isotropic and Maxwellian even below the critical point. In that case, the loss of mass occurs by evaporation from the "exosphere" rather than by hydrodynamic outflow. This tends to result in a slow sonic flow interior to the exosphere, along with the selective loss of light/fast particles from it, a situation frequently encountered in the context of planetary atmospheres (Chamberlain 1960, 1961, 1963).

Whether the flow is hydrodynamic or evaporative, the ratio $4c_T^2/v_{esc}^2 = 2H_s/r$ (where H_s is the static, isothermal scale height) appears as a critical parameter, and the flow may be said to originate from the region where $2H_s/r \sim 1$. For stars with atmospheres in radiative equilibrium, H_s/R^* ranges from $<10_{-4}$ for main sequence stars to $<10_{-2}$ for the most extreme giants, putting the critical point very far from the star. From this, we conclude that a stellar wind requires a driving mechanism to heat the atmosphere, accelerate the flow, and/or reduce the effective gravity substantially.

Driving mechanisms that have been invoked and studied in the formalism of the steady wind theory include thermally driven winds (with various mechanisms providing the extra heating), winds where acoustic and/or Alfvén waves are invoked to deposit momentum and energy in the material, and winds driven by the momentum of starlight ("radiation pressure" driven winds). Although thermally driven winds remain the textbook example, they are probably rarely important in practice. As was pointed out by Holzer & MacGregor (1985), the radiative losses associated with the required hot coronal material would exceed the stellar luminosity in most cool stars. This fact should perhaps not come as a surprise, now that we have clear evidence from X-ray imaging that the solar wind flows from the (cool) coronal holes and the hot coronal material is trapped in the closed magnetic loops.

Winds from hot stars are driven by the absorption and scattering of resonance line photons in the UV. Winds from cool stars could be similarly accelerated by radiation pressure acting through absorption or scattering in molecular bands, in the continuum (for a chromosphere or corona), or from dust (of which more later). The effects on cool star atmospheres and winds of radiation pressure acting in molecular lines have been investigated by Jørgensen & Johnson (1992), Maciel (1976), and Elitzur et al (1989). Unless red giants are much cooler and lower-gravity than now appears likely, this mechanism is probably important only for the effect it has on the scale height deep in the atmosphere; the wind is enhanced but not driven by radiation pressure on molecules (Höfner et al 1998b).

The power going into driving a steady wind may be written as

$$L_{wind} = \dot{M}\left(v_{esc}^2/2 + v_\infty^2/2 + \frac{3}{2}c_T^2 + X\right), \tag{2}$$

where X may include turbulent, magnetic, or wave energy per unit mass. For cool star winds, unlike hot stars, the dominant term is usually the first one that gives the gravitational potential energy that must be supplied to lift the material to infinity, and $L_{wind} < \dot{M}v_{esc}^2$. In the Sun, the mechanical energy flux into the chromosphere exceeds that involved in driving the wind by about an order of magnitude, and the power going into driving the wind directly is about 10^{-6} of L_{Sun} (Withbroe 1988, Withbroe & Noyes 1977).

Numerous studies have demonstrated that depositing momentum or energy in the atmosphere below the critical point will tend to increase the mass loss rate, whereas depositing it above the critical point tends to increase the flow velocity more than the mass loss rate. It is difficult to reproduce massive, slow outflows from red giants with stationary wind flow models; the dissipation length or other parameters have to be fine-tuned to deposit enough energy and momentum below the critical point to drive a massive flow, and not too much above it to keep the velocity low (Hartmann & Macgregor 1980, Holzer 1977, Hammer 1982).

Acoustic waves can both heat and impart momentum to the flow. It is common to treat this by including wave pressure terms in the steady wind momentum equation and/or the energy equation. Quite a few varieties of acoustic wave models have appeared in the literature in recent years (Ulmschneider et al 1978, Hammer 1982, Cuntz 1989, Cuntz et al 1998, Sutmann & Ulmschneider 1995, Rammacher & Ulmschneider 1992, Pijpers & Habing 1989). Most such models have a problem with the pattern of wave energy dissipation in the atmosphere. If the dissipation is computed from physically based theory, the result typically suggest that radiative losses dominate (because the dissipation occurs deep in the atmosphere where the density is high). If the dissipation is tuned to produce a reasonable wind model, then there is usually a very narrow range of parameters that can be made to fit.

Alfvén-wave–driven wind theories have been quite successful and are the leading candidate to explain the solar wind as well as the winds of warm giants. Alfvén waves dissipate much less readily, so they can deposit energy and momentum farther out and not lose so much to radiation. If there is a problem with Alfvén waves, it is that they may dissipate too little; at least in the linear regime, they do not provide much heating. Interesting variations on this theme may be found in Moore et al (1992) and Cranmer et al (1999). Alfvén waves may play a role in driving red giant winds, but they are not necessary and they require additional ad hoc assumptions about the nature and source of the magnetic fields involved.

1.2 Stationary Winds with Dust

Dust grains are very efficient at absorbing and scattering starlight; thus, they pick up momentum from the radiation field under conditions where the gas is effectively transparent. Collisions between dust grains and constituents of the gas transfer the momentum to the gas, thus also driving it from the star. This process may drive mass loss (under extreme conditions that rarely, if ever, occur in real stars) or it may enhance and accelerate the wind (a condition that is quite common). Important early contributions to the study of grains in winds include Gilman (1969, 1972), Salpeter (1974a,b), and Draine (1979).

Most observational studies of dusty, cool star winds have been carried out using stationary flow models. There is a minimum mass loss rate such that dust can form in the outflow, and another minimum mass loss rate that allows enough dust to form deep enough in the atmosphere for the outflow to be driven by the dust. In order for dust grains to form in an atmosphere or a stellar wind, nucleation must occur, and the grains must survive and grow. For them to drive a stellar wind, they must intercept a sufficient fraction of the starlight to accelerate outward, and they must be momentum-coupled to the gas through grain-gas collisions. These conditions have been explored by Gail & Sedlmayr (1987a,b), Netzer & Elitzur (1993), Berruyer (1991), and Hashimoto et al (1990).

In the environments where dust forms, CO is preferred over other molecules involving C or O. Thus, if there is more O than C, essentially all the C will be in CO and the excess O will be available to make oxides such as SiO in the first steps toward forming silicate grains. For detailed discussion of grain formation—equilibrium and non-equilibrium cases—see Gilman (1969), Salpeter (1974a,b; 1977), Draine & Salpeter (1979), Donn (1978), Nuth & Donn (1981), Draine (1979, 1981), Donn & Nuth (1985), and Stephens (1991). The rate of nucleation has a steep temperature dependence near the saturation temperature at any given density. The saturation temperature for SiO lies in the the range 1100–1500 K for the range of densities expected in the driving zone for massive winds. Silicate grains have lower albedos farther into the IR, so they tend to equilibrate with the radiation field at temperatures below the blackbody temperature. Carbon grains, in contrast, tend to equilibrate above T_{BB}.

S-type stars have $C \approx O$; in these stars, little C or O is left over for grain-forming molecules. This is consistent with observations. There is a higher percentage of S stars among the Miras with periods around 400 days than among Miras with shorter or longer periods (Motteran 1971). This could be because oxygen-rich stars with periods around 400 days tend to be obscured, while the S type stars have less dusty, more transparent, winds. S stars can show very large mass loss rates (Jorissen & Knapp 1998), but even then they tend to have slower and less dusty winds (Sahai & Liechti 1995).

Several conditions must be met if the grains, once formed, are to survive and grow: (*a*) The radiative equilibrium temperature (T_{RE}) of the grains must remain low enough; (*b*) the gas kinetic temperature cannot get too high for too long; and (*c*) the relative motion of the grains through the gas must not be too fast, or the grains will be abraded by gas-grain collisions. The first condition is easy to achieve, even in variable stars. The second is of interest in pulsating stars, where grains must survive the passage of shocks that raise the temperature of the gas to several thousand kelvins or more. (For a discussion of grain survival in interstellar shocks, see Tielens et al 1994, Tielens 1998, or Draine & McKee 1993.) The third condition sets the maximum size of the grains (Salpeter 1977, Draine & Salpeter 1979, Krueger & Sedlmayr 1997).

In a stationary flow, dust forms where $T_{RE} < T_{saturation}$. For silicates, $T_{sat} \sim 1450$ K if the density is a typical photospheric value of about 10^{-10} gm cm^{-3}, or ~ 1300 K if the density is below 2×10^{-12}, a more reasonable value for the dust formation region. To get $T < 1300$ K from a star with $T_{eff} > 2600$ K requires a dilution factor $W < 1/16$ or $r/R^* > 2$. Even for a star near the tip of the AGB, the density of a static atmosphere will have decreased to less than the interstellar density well below $r \sim 2R^*$. To drive a dusty wind, then, the stars must be cooler than supposed above, the temperature must be lower than T_{RE} near the star, or the true scale height must be much greater than the static value. The last two conditions are generally satisfied in such a wind once it is established, but this argument shows that it is difficult or impossible to start driving a dusty wind in a static atmosphere.

In order for dust grains to drive or accelerate a stellar wind, they must acquire momentum from the starlight and must share this momentum with the gas. Once the grains are formed, their optical properties determine how effective they will be in driving or accelerating the wind. Tabulations and discussions of the optical properties of grains may be found in Gilman (1974), Draine & Lee (1984), and Draine & Lee (1987). The dust opacity depends sensitively on the size (and structure) of the grains; for spherical grains with radius a, $\kappa_{dust} = (3Q_{pr}/4a\rho_{gr})\,X_{dust}$ where $Q_{pr} \sim 1$, $\rho_{gr} \sim 1\text{--}3$ gm/cm^3, and $X_{dust} \leq 2 \times 10^{-3}\,(Z/Z_{Sun})$ is the mass fraction of condensable material. Most calculations of grain dynamics in stellar wind conditions lead to the conclusion that the maximum size that is achieved is around 0.1 micron, and this agrees with observational estimates (Spitzer 1968, Wolff et al 1998). Thus we can't expect to get a strongly dust-driven wind in an oxygen-rich star unless $L/M > 10^4\,L_{Sun}/M_{Sun}$. This number is consistent with what has been found in detailed numerical models relevant to this case (for example, Gilman 1973; Goldreich & Scoville 1976; Kwok 1975; Tielens 1983; Gail & Sedlmayr 1987a,b; Berruyer 1991; Netzer & Elitzur 1993; Mastrodemos et al 1996). An online database for optical properties of grains is kept at http://www.astro.uni-jena.de.

The preceding condition on L/M is a strong indication that dust is unlikely to be the direct cause of stationary winds from normal stars. The coolest and most luminous stars that show signs of dusty winds are the Miras and OH-IR stars at the tip of the AGB. The theoretical limit to the AGB (where the core mass reaches the Chandrasekhar limit) is at about 50,000 L_{Sun}, and most stars only reach $L = 3000\,L_{Sun}$ (P = 200 days for the Miras at the tip of the AGB) to $L = 6300\,L_{Sun}$ (P = 400 days); see Section 3 and Figure 4.

Further complications attend the theoretical study of grains in stellar winds. In stars where dust grains form and influence the flow, the chemistry of the nucleation process is typically not that of thermodynamic equilibrium or a steady state. For example, hydrogen, generally expected to be molecular at temperatures <1500 K, may be dissociated by periodic shocks and may not have enough time to recover between the shocks (Bowen 1988). This complicates the calculations considerably. For recent work on the nucleation and growth of grains in an environment with $C/O > 1$, see Cadwell et al (1994) or Patzer et al (1998).

From the preceding discussion it appears most likely that steady winds driven by dust alone are rare or nonexistent. Perhaps that should not surprise us. Dust is observed to form primarily—perhaps only—in stars with well-established departures from a steady state, including the pulsating AGB stars, RV Tauri variables, supernovae, and novae. Even for the R CrB stars, whose erratic minima have long been understood to be associated with the formation and ejection of dust clouds, an underlying pulsation has been found to correlate in phase with the onset of a dust episode (Pugach 1977, Holm & Doherty 1988, Asplund 1995).

1.3 Winds from Pulsating Stars

Pulsation can drive or enhance a stellar wind in several ways. First, it provides a natural source of mechanical energy for heating the atmosphere. Second, the atmosphere is "levitated"—given a larger dynamical density scale height $(d \ln \rho/dr)^{-1}$—by the pulsation, so it partially overcomes the gravitational potential and makes higher density material available for an outflow. Third, departures from radiative equilibrium lead to refrigerated regions where molecules and dust can form much closer to the star than they would in a static atmosphere or stationary outflow.

Pulsation gives rise to standing or traveling waves, depending on the period of the oscillation. Longer period oscillations, trapped in a resonant cavity, are more likely to result in large-amplitude standing waves. As stars travel up the AGB, fundamental mode pulsation is increasingly favored by these and other considerations. (Wood 1974, 1990; Ostlie & Cox 1986; Bowen 1990).

Waves traveling through decreasing gas density steepen to form shocks. Even for trapped modes, nonlinear excursions of the atmosphere give rise to shocks that propagate out into the atmosphere. If the density gradient were to remain that of the static atmosphere, the shock amplitudes would increase (to conserve energy) until the shocks no longer satisfied the periodicity condition for ballistic motion (Hill & Willson 1979, Willson & Hill 1979):

$$v_{esc}\frac{P}{2r} = 38.3Q = \left[\frac{\beta}{1-\beta^2}\right] + (1-\beta^2)^{-\frac{3}{2}}\arcsin(\beta) \tag{3}$$

Here, $\beta = v_{out}/v_{esc} = (1/2)\,\Delta v_{ballistic}/v_{esc}$, since in the ballistic case $v_{out} = -v_{infall} = \Delta v_{max}/2$, and $Q = P\sqrt{\bar{\rho}/\bar{\rho}_{Sun}}$ is the "pulsation constant" expressed in days. Observed velocity curves of pulsating stars with shock discontinuities or line doubling indicating the presence of shocks, and with apparently constant acceleration between shocks, have often (but incorrectly) been interpreted as yielding a direct measure of the surface gravity, g. The maximum ballistic shock amplitude is approximately equal to g P for short periods, but for $Q \sim 0.03$–0.1 day, which is characteristic of most radially pulsating stars, $\Delta v_{max}/g\,P < 0.8$ to 0.4. The amplitude is further reduced by including the effects of pressure gradients acting between shocks. If, instead of assuming ballistic motion, we integrate the equation of motion over a cycle and again invoke periodicity, we find that the expected pattern is asymmetric, with $v_{infall} > v_{out}$, as given by Willson & Bowen (1984a):

$$v_{infall}^2/v_{esc}^2 = v_{out}^2/v_{esc}^2 + (2c_T^2/v_{esc}^2)\ln(\rho_0/\rho_f)$$

$$= (v_{out}^2/v_{esc}^2) + (H_{static}/R)\ln(\rho_0/\rho_f) \tag{4}$$

For strong adiabatic shocks, $(\rho_0/\rho_f) = 4$; for isothermal shocks it can be much larger. This term typically introduces a correction on the order of 5–10% for periodic motion in the Miras; the magnitude of this effect is less for stars of higher gravity. An indication of the importance of the pressure gradient between shocks is the inequality between the blue- and red-shifted components, which is striking for Miras but not evident for higher gravity stars (Willson et al 1982, Hinkle 1995).

Material given an impulse greater than $v_{esc}\beta$ at time t_s does not have time to return to the same position before the next shock passes at t_s+P. The result is a net outward displacement of material until the density gradient has become sufficiently shallow that the shocks can satisfy Equation 3. Thus the atmosphere is extended—levitated—by a combination of the kinematic fact that the average position of a mass element is significantly outside its minimum distance, and the need for the shallower density gradient to satisfy the energy and periodicity conditions.

Once the shock amplitude has reached the periodic limit, the material may be unable to radiate enough energy between shocks to return to the radiative equilibrium temperature. As a result, a region with persistent $T > T_{RE}$, a "calorisphere," may form, allowing all the energy dissipated in shocks to be radiated away. Farther out, the density becomes low enough that the only way the atmosphere can maintain a steady state (averaged over the pulsation cycle) is for the extra entropy introduced by each shock to be advected outward in a slow flow at roughly constant velocity (Struck et al 2000, in preparation). In Bowen's numerical models (1988; 2000 in prep) this behavior is seen in cases where dust does not form (see Figure 1, lower panel) where the outflow remains steady out to tens of stellar radii, with $v < c_T \ll v_{esc}$. Thus, circumstellar lines with low flow velocities may arise from a subsonic outflow that need not have $v > v_{esc}$ as long as there are waves propagating through it. Note that even with such slow flow there is no sign in any theoretical models of stationary layers or shells such as those supposed to exist by several groups interpreting observational data (see, for example, Hinkle et al 1982, Reid & Meuton 1997).

The temperature structure in the atmosphere is determined by a combination of the expansion (adiabatic or hydrodynamic cooling) and interaction with the radiation field. To treat the coupling of the gas with the radiation field is not a simple matter, even for local thermodynamic equilibrium (LTE). Below about 8000–10,000 K the cooling is dominated by a variety of neutral "metals" and molecules. Hydrogen becomes important around 10,000 K, and in the typical situation the Lyman lines and possibly some of the Balmer lines will be optically thick over a short distance in the atmosphere. The problem is most difficult between the simpler extremes of LTE in deep stellar atmospheres and extreme non-LTE in the interstellar medium. However, some simplification is possible. For a two-level atom with critical density ρ_x the line source function is reduced below its LTE value by $S/S_{LTE} = 1/(1 + \rho_x/\rho)$. For each transition important in the radiative transfer there is a critical density where this occurs; the gas as a whole behaves to first order as though it had a critical density (ρ_x) at a value characteristic of the most important radiative transitions. Thus, at low densities, the cooling is proportional to the collision rate in the gas. This provides a convenient parametrization of the more realistic case, where many transitions are involved, leaving only the problem of selecting an appropriate value for ρx. (For discussion of this point, see Willson & Bowen 1998; see also Woitke et al 1996, but note that their analysis implicitly assumes very narrow intrinsic line profiles and continuum opacity $\ll$ line opacity, both conditions that probably do not apply in red giant stars and that overestimate the importance of photon trapping.)

Figure 2 shows the difference in the atmospheric temperature structure for two cases that are identical except for the value chosen for ρ_x. In the bottom panel,

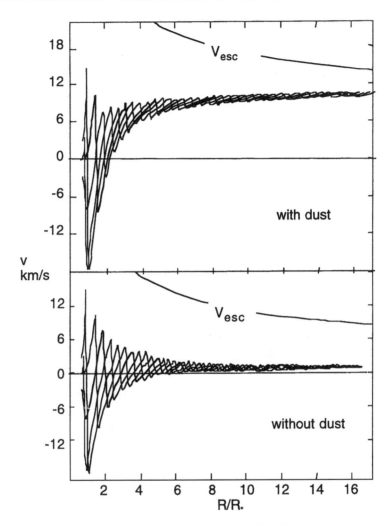

Figure 1 Velocity as a function of radius for two models differing only in that for the top panel, dust formation is allowed, whereas for the bottom panel, dust formation has been artificially suppressed. The slow, constant velocity outflow in the bottom panel is also commonly seen in models that are allowed to make dust but do not reach the right conditions—low-metallicity, low-mass, and/or low-luminosity models. (Figure courtesy of GH Bowen.)

small ρ_x, most of the atmosphere has $S \sim S_{LTE}$ and is able to return to radiative equilibrium quickly. If ρ_x is high (top panel), the partial decoupling of the gas temperature from the radiation field provides for regions with $T < T_{RE}$ as well as regions with $T > T_{RE}$. This is important for dust and molecule formation close to the star, and is discussed in more detail below.

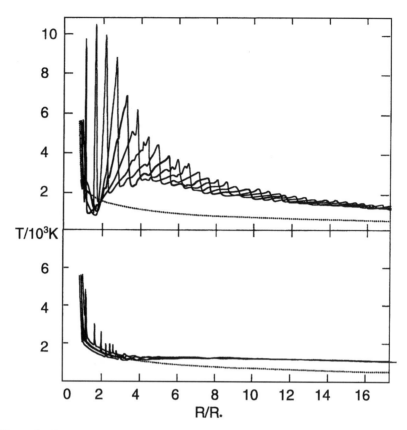

Figure 2 Temperature versus radius from two models that are identical, except that in the top panel the critical density $\rho_x = 10^{-10}$ gm cm^{-3}, while in the bottom panel $\rho_x = 10^{-18}$ gm cm^{-3}. Lower critical density results in a more extended region where T $\sim$ T$_{RE}$ between shocks. (Figure courtesy of GH Bowen.)

Given the difficulty of carrying out non-LTE radiative transfer in the context of an extended, dynamical atmosphere it is tempting to try LTE transfer using the temperature and density structure derived from a dynamic model to compute predicted spectra. However, this approach overlooks the importance of matching the assumptions in the radiative transfer code to those of the dynamical calculation. If LTE transfer is used on a hydrodynamic model that used non-LTE cooling, then the transfer code assumes that there is a lot of energy stored in excitation that in the dynamic model went into thermal kinetic energy and/or mass motions. Thus, the LTE transfer in this case will produce an overabundance of photons from some parts of the model by violating conservation of energy. For a dramatic example of the results of such a mismatch, see Bessell et al (1989, Figure 25). Similarly, because the conditions that lead to non-LTE level populations are the same as those that lead to departures from radiative equilbrium in a non-static atmosphere, it is inherently inconsistent to put non-LTE into a hydrodynamic code and then insist on radiative equilibrium.

1.4 Pulsating Stars as Dust Factories

Pulsation and dust formation together characterize the bulk of the Miras and OH-IR stars that are clearly losing mass at a rapid rate. Theoretical evidence of this is as follows: Very extreme conditions would be required to allow dust to drive the wind alone, whereas the formation of dust in a pulsating star can greatly enhance the mass loss rate over what it would be if dust did not form. Also, departures from radiative equilibrium in the atmospheres of pulsating stars can enhance the rate of dust formation. Observational evidence is as follows: All the stars in which dusty winds are evident are also pulsating. The properties of the wind correlate strongly with the pulsation characteristics, even for stars with the same period of pulsation; as Bowers & Kerr (1977) pointed out, the heaviest mass loss comes from stars with the most asymmetric light curves, all else being approximately equal. Narrow 10 μm features occur preferentially at a particular phase, indicating variation in the properties of the dust grains with phase (Monnier et al 1998).

In Figure 3, the location of the "dust factory" region is sketched for one of Bowen's (2000 in prep) model Miras. In the darkly shaded region the pressure exceeds the saturation pressure by four or more orders of magnitude, suggesting that nucleation should be rapid. The dust nucleated in one cycle acquires a small outward velocity relative to the gas; only when the dust has reached a size approaching a tenth of a micron does this relative velocity become appreciable. The coupling of the dust and the gas creates the net outflow shown in Figure 1, where it is contrasted with the velocity structure in an otherwise identical model where dust formation has been artificially suppressed. Clearly, the outflow velocity, and hence the momentum of the wind, is determined by the action of the radiation field on the dust. Thus it should be no surprise that most Miras and OH-IR stars satisfy $\dot{M} v_\infty < L/c$, and a few even exceed this condition (possible for multiple scattering in a nearly optically thick wind); see, for example, Knapp (1986). Note also that nowhere in the atmosphere is there any kind of "stationary layer"; the illusion of such a layer is presumably produced by the fact that at all phases there is material with $dr/dt \sim 0$, and that this material is in about the same phase of the expansion-cooling cycle at all phases.

The ideal model for the wind from a pulsating star would be one in which (a) the driving zone and the atmosphere are treated together in a consistent way; (b) the temperature structure is computed using full non-LTE transfer in the radiation hydrodynamics; (c) the shocks are fully resolved in space and time; (d) the non-equilibrium molecular chemistry, dust nucleation, dust-gas interactions, and resulting energy exchange are followed in full non-equilibrium; (e) the model extends to at least tens and preferably hundreds of stellar radii. Such a model is not available now nor likely to be forthcoming any time soon. The

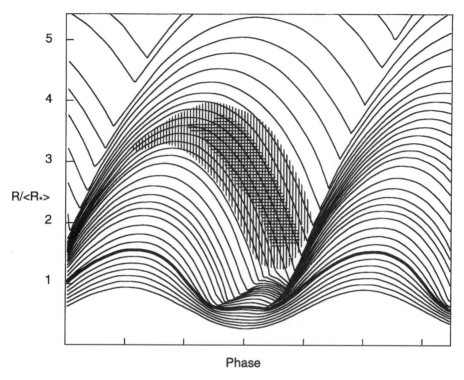

Phase

Figure 3   Radius as a function of time for a representative selection of atmospheric zones, with the "dust factory" region of a model indicated by the shaded areas. In the "dust factory," the pressure exceeds (by orders of magnitude in the darker region) the saturation pressure, and nucleation of grains is expected to be rapid. (Figure courtesy of GH Bowen.)

models that are available roughly fall into two categories: those that include some of the above processes in exhaustive detail and drastically approximate others, and those that use just-adequate approximations for as many of these processes as are needed to produce a reliable series of models.

Bowen's (1988; 2000, in preparation) models for the oxygen-rich Miras typify the second approach. His energy computation includes an Eddington-approximation grey spherical atmosphere calculation with constant opacity to determine the radiative equilibrium temperature. Relaxation toward that equilibrium temperature is determined by rates that come from a detailed computation for hydrogen cooling at high temperatures, and an approximate cooling coefficient derived from other sources that is used at lower temperatures. His dust formation is parametrized, although the region where dust forms is found after the calculation to correspond closely with the region where the SiO is supersaturated. The exchange of energy and momentum between the gas and dust is considered in some detail. The models extend to tens of stellar radii and a few have been computed to hundreds of stellar radii. All approximations used have been tested for their effects on the mass loss rates and have been found to have modest effects over a range of reasonable parameter values. Thus, Bowen's mass loss rates are probably reliable to $\sim\pm0.5$ dex for a given set of stellar parameters. As is demonstrated in Section 3, this is sufficient accuracy for deriving the pattern of late AGB mass loss and determining the properties of stars when they reach the stage of high mass loss rates.

Early studies of pulsating atmospheres in Miras by Hill & Willson (1979; also Willson & Hill 1979) and by Wood (1979; also Fox & Wood 1985) demonstrated that fast relaxation ("isothermal") shocks gave rise to little or no net outflow, whereas entirely adiabatic shocks caused the atmosphere to simply blow up. Hill & Willson recognized that slow relaxation (and thus an approach to adiabatic shock conditions with decreasing density) would give the right order of magnitude of mass loss rates. Wood was the first to explore the coexistence of shocks and dust grain formation in a Mira atmosphere; he also carried out studies of the internal pulsation and tried to establish the pulsation mode from observational constraints on the radii and masses of Miras. Hill & Willson also pointed out the difficulty of reconciling observed shock amplitudes (from infrared CO line studies; Hinkle 1978, Hinkle & Barnes 1979a,b) with overtone pulsation.

Sedlmayr and his collaborators have concentrated on the chemical reactions that lead to grain nucleation and growth, initially in the context of steady outflows and more recently for pulsating stars (for example, Winters et al 1997). They are systematically investigating a range of relevant phenomena, and their models are converging toward greater similarity with Bowen's results (making due allowance for the differences between carbon-rich and oxygen-rich chemistry) as they incorporate more of the essential physics. Most of their early work dealt with the chemistry and dynamics of atmospheres with C/O > 1 (often substantially greater than 1) because it is not possible to drive a stellar wind by the action of radiation pressure on grains unless C/O is large and the star is very luminous and very cool. The importance of carefully treating the opacity in the deeper (LTE) layers of the atmosphere is addressed in Höfner et al (1998a,b). Predicted spectral variations are compared with observations in Hron et al (1998, Loidl et al 1999).

A very strong dependence of the mass loss rate on stellar parameters emerges from theoretical studies (Bowen & Willson 1991, Höfner & Dorfi 1997). If a set of evolutionary tracks is expressed by setting the radius (or T_{eff}) equal to a function of L, M, and Z, then the parameter dependence can be reduced to mass loss rate as a function of L, M, and Z for stars evolving up the AGB. In the top panels of Figure 4, the dependence of $\dot{M}$ on L is displayed for (a) solar composition and five masses and (b) one solar mass and five metallicities, from Bowen models using evolutionary tracks from Iben (1984) with mixing length parameter $\alpha = 0.9$.[1] At a given L

---

[1]Note that the Iben mixing length formulation is slightly different from the Cox & Giuli (1968) formulation often used, so an Iben mixing length of 0.9 produces models with characteristics similar to a Cox & Giuli formulation with $\alpha = 1.5$ or more (see, for example, Becker & Iben 1979).

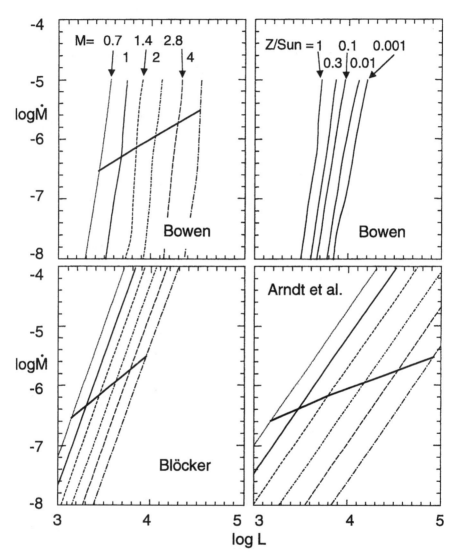

Figure 4 Mass loss rate vs. luminosity using four theoretical mass-loss relations with a single set of evolutionary tracks (Equation 7). *Top panels*: Models by Bowen (1995, unpublished; 2000, in preparation) for six masses with solar composition (*left*) and 1 M$_{Sun}$ with five metallicities (*right*). *Lower panels*: Mass loss relations by Blöcker (1995) and Arndt et al (1997) for the same six masses. In each panel the locus of critical mass loss rate, $-d \log M/dt = d \log L/dt$, is indicated.

and M, the lower Z models have much lower mass loss rates, but even very low Z stars can achieve high mass loss rates when they become sufficiently bright. Lower metallicity stars have lower mass loss rates mainly because they have smaller radii at a given M and L; the effect of dust is secondary (Bowen 2000, in preparation; Willson et al 1996). On each of these plots is a line indicating the critical mass loss rate ($-dM/dt = M/L\ dL/dt$) at which mass loss begins to dominate the evolution (see Section 3). Observational selection will favor observations of stars with long mass loss timescales and high mass loss rates—i.e. stars near the critical mass loss rate. The lower panels of Figure 4 show similar plots using the mass loss formula of Arndt et al (1997) for carbon-rich models and using the formula derived by Blöcker (1995) from the 1988 Bowen models. (The models Blöcker used were not yet constrained according to the energy condition on the "piston" amplitude that has since proved useful, therefore the results are not the same as in the current Bowen grid used in the top panels of Figure 4.) At the end of Section 2 you'll find a similar series of plots for a variety of "empirical relations" that are in common use.

2.   OBSERVATIONS OF MASS-LOSING GIANT STARS

In his review entitled "Mass loss from stars" for *ARAA* in 1963, Weymann wrote, "One gets the impression that in many aspects of the subject the number of pages expended in reviewing the subject ... probably exceeds the number of pages of material being reviewed." That is definitely not the situation now. Fortunately, there are also a number of recent reviews covering aspects of the subject very well. Knapp (1991) clearly described a variety of methods for detecting and measuring mass loss from evolved stars, and Van Der Veen & Rugers (1989) compared mass loss rates derived by molecular line and infrared techniques. Olofsson (1996a,b; 1997) reviewed the structure of the circumstellar envelope and its molecular chemistry. Dupree (1986) examined the techniques based on spectral lines in the UV and optical. Infrared methods have been reviewed by Bedijn (1987), high spatial resolution observations by Dyck (1987), and mass loss in general by Lamers & Cassinelli (1996, 1999). In this section, I stress a few points of particular relevance to the derivation of mass loss laws, the evaluation of theoretical mechanisms, and the determination of the properties of mass-losing stars.

2.1   Methods for Determining the Mass Loss Rate

The number of papers concerning mass loss from cool stars rose precipitously with the appearance of data from the IRAS satellite. The IRAS color-color plot was found to be a very useful tool for diagnosing and classifying at least the dusty winds from very evolved stars (Van Der Veen & Habing 1988). Most of the evolved sources fall along a band of increasing $\log(F_{12\mu m}/F_{25\mu m})$ that represents increasing rates of mass loss for M and C stars, with the larger amplitude pulsators also tending to cluster in the higher mass loss classes. The population of the part of the diagram with 60 μm "excess" is most readily interpreted as the result of the cessation or substantial reduction in the rate of mass loss, so that the expanding shell cools. It appears most likely that these shells represent peaks in the mass loss rate associated with shell flashes, possibly further compressed by variations in the wind speed (Chan & Kwok 1988; Olofsson et al 1990, 1998).

An unambiguous indicator of circumstellar matter, hence mass loss, is the appearance of the 10–15 μm "bump" (or dip, if in absorption) characteristic of the opacity of silicate grains (10–12 μm) and carbon grains (13–15 μm). Infrared measures of the mass loss rate generally depend on fitting models to the observations. Typically, models assume constant outflow velocity, parametrized by dust/gas ratio, optical depth, and inner dust radius or maximum dust temperature. The fits are most sensitive to the inner dust radius and temperature. Published fits have not always taken into account that the dust temperature, if silicate dust is involved, will be below the blackbody radiative equilibrium temperature (see Section 1). There is a close correlation between the appearance of the 10 μm, i.e. the optical depth of the envelope, and the K-L color; this correlation may be translated into a very convenient relation between K-L and mass loss rate (Lepine et al 1995, Jones et al 1990).

The basis for deriving mass loss rates from radio molecular lines is described and illustrated in a series of papers by Knapp (1986) and Olofsson (1996a,b; 1997), and also by Nyman et al (1992), Wannier & Sahai (1986), Wannier et al (1990), Sahai & Wannier (1992), Little-Marenin et al (1994), and Sopka et al (1989). Mass loss rates derived this way do not always agree with those derived from IR or near-IR photometry, but this may well be for good physical reasons rather than because there are flaws in any one method. For CO-line fitting, the parameter to which the result is most sensitive is the *outer* radius of the CO shell, because it is the volume of the CO shell (along with mass loss rate) that sets the amplitude of the signal. This outer radius, typically 10^{16} to 10^{17} cm, is usually set by the location where photo dissociation by interstellar UV photons becomes important—although in rare cases it could instead be determined by the time of the onset of a mass loss episode. If the flow is clumpy, the CO mass loss rate derived from the standard equations will underestimate the mass loss rate, possibly substantially (Olofsson et al 1992). Also of importance is that these lines give indications about the mass loss rate some time before the present—typically 100 to 1000 years BP—whereas mass loss rates are known to vary on timescales from decades (for reasons not well understood) to thousands of years (over the shell-flash cycle).

Many of the stars that are losing mass most rapidly are also maser sources. Maser emission has been recently reviewed by Elitzur (1992). Briefly, SiO maser lines appear to originate from about the same location in the atmosphere where grains form, around 1–3 stellar radii, where they can be pumped by collisions; this is confirmed by the correlation of the SiO maser variability with the phase of pulsation (Herpin et al 1998, Gray et al 1999, Humphreys et al 1997a,b). For Miras and semi-regular long period variables, interferometric studies have demonstrated that the H$_2$O maser emission originates from a region $\sim$5–10 $\times$ 10^{14} cm in radius centered around the star, whereas OH masers form much farther out in the wind, typically at distances of $>10^{16}$ cm, where H_2O has been photo-dissociated by interstellar UV photons. The OH maser lines have provided useful data on mass loss rates and, in particular, on outflow velocities at very great distances from the star.

Recently, images of mass-losing stars taken in the scattered light of atomic resonance lines of potassium and sodium have indicated (*a*) that these atoms are not heavily depleted onto grains in the outflow, and (*b*) that the envelope is also not heavily ionized by stellar UV from a stellar chromosphere (Gustafsson et al 1997, Guilain & Mauron 1996). The light from Miras is sometimes found to be polarized, with heavy dependence of the polarization on the wavelength (Coyne & Magalhaes 1977, 1979; McLean & Coyne 1978; Codina-Landaberry & Magalhaes 1980; Kruszewski et al 1968; Boyle et al 1986). No existing model gives a full explanation for these effects, but the scattering of photospheric photons by electrons or dust in the circumstellar envelope and wind appears most likely to be responsible.

Radio continuum measurement of mass loss rates has been a powerful tool for hot stars and has been applied with mixed success to cooler ones. Post-AGB stages were detected by Spergel et al (1983) and Knapp et al (1995). Several K and M giants were detected in a study by Drake et al (1987, 1991); 1.3 mm continuum emission has also been detected from mass-losing AGB stars by Walmsley et al (1991). Reid & Menten (1997) determined that several long-period variable stars had thermal radio emission with a spectral index ~2, characteristic of an optically thick thermal source. They interpreted this result as evidence for a "radio photosphere" and developed a detailed empirical model; however, the same data may be interpreted with quite a different model based on rising shock fronts in a dynamical atmosphere (Willson 1999).

2.2 Observations of Mass-Losing Stars with High Spatial Resolution

In theoretical models, the inner radius at which dust forms depends on the luminosity of the star, the nature of the dust, and the local gas temperature. While the nature of the dust can be constrained by general arguments, the local temperature depends strongly on the modeling methods used. Thus, observational constraints on the inner dust radius and/or the maximum temperature of the dusty shell are very useful for comparison with modeling results. Bester et al (1991) found that the inner dust radius for o Ceti was around 3 stellar radii, corresponding to a radiative equilibrium temperature around 1200 K; this is in agreement with theoretical expectations (Section 1), although this temperature is higher than has been used in many IR-fitting models (see also Danchi et al 1994, Danchi & Bester 1995).

HST observations of proto-planetary nebulae have revealed two further intriguing facets of late-AGB mass loss, also confirmed by interferometric studies (Lattanzi et al 1997). The mass loss appears to retain predominantly spherical symmetry until the very last moments of AGB outflow, when bipolar symmetry typically emerges (Skinner et al 1997). A simple examination of the ways in which an outflow may become axially symmetric reveals quickly that for a single star, until the envelope mass is very small, an impossibly large amount of angular momentum will be required to modify the flow (Willson & Bowen 1988). The more surprising fact emerging from these studies is that the outflow before this stage appears to be modulated on a timescale of decades to centuries—too short for the shell flash cycle, and too long for the pulsation cycle (Skinner et al 1997). Many of the pulsating AGB stars show modulations in their periods and/or their amplitudes on a similar timescale (Percy & Colivas 1999); these modulations are not yet well understood. This timescale is about the same as the thermal (Kelvin-Helmholtz) timescale for the envelope or the orbital periods for solar system–like planets.

2.3   Determination of the Properties of Mass-Losing Stars

Properties of mass-losing stars that we would like to know accurately include the luminosity, the radius, and the mass. Most methods for determining L and R require that we first determine the distance to the system. In addition, for studies of the evolution of stellar populations, we need to determine the dependence of the mass loss rates, or the properties of mass losing stars, on the metallicity.

2.3.1 Distances and Luminosities   The Miras in the LMC have been extensively studied (Glass et al 1987, Mould & Reid 1987, and Hughes & Wood 1990). The P-L relation for Miras in the LMC is shown in Figure 5, with the individual stars taken from Feast et al (1989); an entirely similar plot results from using data from Hughes & Wood, who observed a larger sample of stars. In this figure, the theoretically expected P-L relation (derived from the analysis in Section 3) is also shown. Feast preferred a single linear fit to the data that has been widely used as the Mira P-L relation. However, the excellent match between theoretical models and the observed P-L relation supports the contention that the "corner" in the relation is real and comes from the much weaker dependence of the effective temperature on stellar mass below about 1 M_{Sun} seen, for example, in the models of Iben (1984, and sources therein).

Obtaining the distances to Miras in our own galaxy is difficult; very few of the highly evolved AGB stars are located close enough for traditional parallax measurements. The Hipparcos list included a few Miras, as well as closely related semi-regular variables, with simultaneous ground-based observations and analyses of the light curves as a bonus. The results that have been derived from data obtained by the Hipparcos satellite have profoundly affected our confidence in the distances to a wide variety of stars. What is less widely appreciated is that the method used by the satellite to establish the positions of stars has some problems when the source becomes extended, and in particular has a very limited range of parallaxes to which it is sensitive when the object being measured is >1AU in radius. This, of course, is the case for nearly all the evolved red giants that are losing mass— radii cited in the literature range from 1 AU upward to several AU for stars near the tip of the AGB. Wallerstein & Knapp (1998) found that Hipparcos parallaxes to carbon stars were reliable only when both $\varepsilon(\pi) < 1.75$ mas and $\pi/\varepsilon(\pi) > 3$,

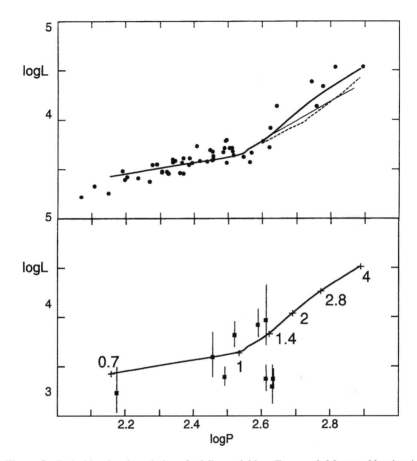

Figure 5 Period-luminosity relations for Mira variables. *Top panel*: Measured luminosities and periods for LMC stars from Feast et al (1989) with the theoretical locus derived in Section 3 superimposed. *Solid line*: Z/Zs = 1. *Dashed and dotted lines*: Z/Z$_\odot$ = 0.3, 0.1. *Lower panel*: The same theoretical locus is shown with galactic Miras, using Hipparcos parallaxes from Van Leeuwen et al (1997), selected as described in the text.

where π is the parallax and $\varepsilon(\pi)$ is the error in π. Applying the same criterion to the Miras and SR variables observed by Hipparcos and reported by Van Leeuwen et al (1997) leaves a total of seven M Miras, one S Mira, and (after adding two stars from Wallerstein & Knapp's list) three C Miras with presumably reliable Hipparcos distances. These are all shown in the lower panel of Figure 5, where the theoretical P-L line from the top panel is also reproduced. The scatter is large, as are the error bars. The only star on the line is one that is most likely to deviate: R Aql, a star experiencing rapid change in its period (and therefore its radius as well) presumably in response to a shell flash (Wood & Zarro 1981). The scatter is unlikely to result from differences in metallicity, since the P-L relations for

$Z/Z_{Sun} = 0.1–1$ nearly coincide (upper panel of Figure 5). The P-L relation from the LMC is probably still a better guide to distances for Milky Way Miras than are the Hipparcos parallaxes. The width of the relation in log L is about 0.2dex; this is not likely to be reduced much by further observations, since it is already about the width expected from the lifetime of the Miras (Section 3).

Another technique for determining the distance to a star with an OH maser is to look for time delays between maser emission coming directly to us and that bouncing off the back of the shell first. If the angular diameter is also known, this yields a direct measure of the distances (Van Langevelde 1990). The OH-IR stars have been suggested to comprise an extension of the Mira P-L relation to higher luminosities. Some of the OH-IR stars with higher-mass progenitors do appear to extend P-L relation. However, it has been clearly demonstrated for infrared LMC sources, as was expected from theoretical arguments, that most of the heavily obscured sources are at the same L but longer P, thus corresponding to evolution to lower mass at roughly constant or declining L during the heavy mass loss stage (Wood 1998). Many of the papers in the literature concerning properties of OH-IR stars have assumed that the luminosity of an OH-IR source is approximately 10^4 L_{Sun}; these analyses should be redone with better luminosities as those become available.

2.3.2 Angular and Linear Diameters

In the last decade there has been a steep increase in the number of angular diameters reported in the literature based on a range of mainly interferometric techniques, with multi-element interferometers gradually dominating over speckle and occultation methods. (See, for example, Ridgway et al 1979; Bonneau et al 1982; Haniff et al 1995; Lattanzi et al 1997; Van Belle et al 1996, 1997; Danchi et al 1995; Karovska et al 1991.) On rare occasions, these methods yield direct images or brightness distributions sampled along two perpendicular directions; more often, it is necessary to work with a one-dimensional (sampled or integrated) brightness distribution and to fit a simple model of the expected brightness distribution to the observations. Early work often used uniform disk models and/or models with a standard limb-darkening function applied. More recently, some teams have been using detailed radiative transfer computations linked to dynamical model calculations from Scholz & Takeda (1987), hoping to achieve increased accuracy; however, the fact that those models do not include the layers most likely responsible for circumstellar scattering (the ionized calorisphere and the dusty wind) is cause for concern. There is some evidence that a Gaussian profile may fit better than any of the standard models (Van Belle et al 1996), suggesting that circumstellar scattering may be an important factor. As was pointed out by Tsuji (1978), a modest amount of circumstellar scattering can have no discernable effect on the spectral energy distribution or line spectrum, while greatly affecting the determination of the angular diameter by any interferometric method. Perrin et al (1999) concluded from a detailed analysis of observations of R Leo that nearly all means of improving the fit to interferometric observations would reduce the angular diameters derived; they favored scattering over most other interpretations for their data. A great advantage of publishing uniform disk angular diameters is that it makes it much easier to reinterpret the data in the light of new models.

In theoretical models, the density declines exponentially in the vicinity of the photosphere, just as it would in a static model. The variation in the photospheric radius over the pulsation cycle is typically no more than $\pm 15\%$ (see Figure 2). The photospheric exponential decline in density goes over into a roughly $1/r^2$ decline a few tenths of a stellar radius above the photosphere; the density at which this occurs is determined by the mass loss rate. If the photosphere is truly what is being observed, then the observations should show time variability not exceeding $\pm 15\%$ from the mean, and wavelength variation significantly less than this at any given phase. The observed variations are much larger; therefore most of these observations are probing the wind, not measuring the star. The observed strong wavelength dependence of the derived stellar sizes, even for non-variable red giants (Quirrenbach et al 1993) but more extreme for Miras (Bonneau et al 1982), provides additional support for this point of view, as does the tendency for the resolved image to be non-spherical (Lattanzi et al 1997). If scattering is an important process in the atmosphere, as appears to be likely, then interferometric techniques that are not carefully controlled for polarization bias will tend to produce football-shaped images, with the scattered light being picked up where the polarization angle is right. Also, derivation of an 82% change in the radius of o Ceti over just two years (Tuthill et al 1995) should be a clear indication that the true radius is not being observed. Even for a conservative estimate of the amount of mass involved, the power needed to expand the star by that much so quickly would be greater than the stellar luminosity, and such an expansion on a timescale considerably below the Kelvin-Helmholtz time is also not plausible.

Given L and P, one should be able to obtain an estimate for the radius of the star from theoretical evolutionary models. This is also difficult for evolved, highly convective giants. The radii of AGB stars are ill-constrained by evolutionary models, since varying the mixing length or molecular opacities gives rise to models with very different radii at a given luminosity. Using period-mass-radius relations to estimate the radius depends on knowing the mass of the star and the mode of pulsation; as noted previously, uncertainties in the measured angular diameters are large enough to introduce ambiguity in the assignment of the pulsation mode.

2.3.3 Masses   Both observational studies (Clayton & Feast 1969) and theoretical models (Hughes & Wood 1990, Bowen & Willson 1991) agree that the sequence from short to long period in Miras is primarily one of increasing stellar progenitor mass, with individual stars spending $\sim 2 \times 10^5$ years evolving across the P-L band. (This is graphically displayed in Figure 8 of Wood et al 1992.) It appears most likely that the P-L relation is a kind of "main sequence" of the Miras, selected not according to the nuclear burning process or longevity of the stage but rather by the mode of pulsation and the level of mass loss. While most of these studies do not yet take into account the effects of shell flashing on the appearance, Vassiliadis & Wood (1992) have also explored the consequences for observable properties of Miras—P, $\dot{P}$, $\dot{M}$—of the shell flashes, assuming that the mass loss rate follows the period.

The period-luminosity relation may be our most reliable estimate for the present masses of Miras, and it gives values quite close to those of the progenitor stars as derived from population studies. Other direct measurements of the masses of evolved AGB stars are very few. The o Ceti visual binary system, for example, should yield a reliable mass estimate in time, but the estimated orbital period is >200 years. If the P-L relation gives correct masses, then the common assumption that the masses of Miras range from <1 to ~1.5 M_{Sun} is good for the stars with periods less than 400 days (with those <300 days most likely having masses <1 M_{Sun}), whereas the longer-period Miras probably have higher masses. Many studies of the properties of Miras and OH-IR stars as a class begin with an assumption—such as that all the stars have masses between 1 and 1.5 M_{Sun}—that is contradicted by the mass assignment shown in Figure 5. Other studies have used assumed mass loss laws and thus derived $M_{present}$ as ($M_{progenitor} - \int_{ms}^{now} \dot{M}\, dt$). Unfortunately, in some cases masses derived this way have been used to calibrate mass loss laws with quite different dependence on stellar parameters than what was used in the above integral—a reminder that it is risky to quote numbers without tracing their origin.

Masses for other cool, mass-losing stars, including the heterogeneous semi-regular variables and the supergiants, are quite uncertain at present because the classification of those stars is less definitive and their histories less clear. Most of the SRa and some of the SRb variables are probably AGB stars that have not yet reached the Mira mass loss stage. However, a given star evolves through a considerable range of L before it reaches the relatively narrow L, P of the Miras, so P will not correlate with mass for SR variables as neatly as for the Mira stage. See, for example, Kerschbaum & Hron (1996).

2.4   Modes, Periods, and Period Changes as Clues to the Nature and Evolutionary Status of Miras

An independent estimate for the mode of pulsation, and hence the likely range of radii, may be made from an estimate of the mass of the star and observation of the dynamical behavior of the atmosphere. This approach was first developed by Hill & Willson (1979; see also Willson & Hill 1979), along the lines presented in Section 1. Direct observational constraints on the shock amplitudes, the necessary ingredient for these studies, have been obtained by Hinkle (1978; Hinkle & Barnes 1979a,b; Hinkle et al 1982, 1984) for a number of Mira and SR variables using infrared CO observations. Typical shock amplitudes for Miras are 20–30 km s^{-1}, with the precise number depending on the (very uncertain) corrections made for integration over the stellar disk. The smaller visual amplitude SR variables, in contrast, show shock amplitudes approximately half as large. This strongly supports the pulsation of the Miras being in the fundamental radial mode and leads to $T_{eff} > 3000$ K, in much better agreement with the excitation temperatures derived from the infrared CO lines (Lebzelter et al 1998).

Some investigators have suggested that some Miras pulsate in the fundamental mode while others are pulsating in an overtone (e.g. Van Leeuwen et al 1997). However, the homogeneity of shock amplitudes, the characteristic presence of emission lines during part of the cycle, and the tight P-L relation all argue strongly against the Mira class including variables pulsating in more than one mode. The light curve shapes of most Miras are also asymmetric, with the rapid rise that is characteristic of fundamental mode pulsation in other classes of stars and consistent with the appearance of fundamental mode models (but not with overtone models) by Bowen (1988, 1990).

Parallel sequences in L versus P would be expected if there are stars with otherwise similar properties pulsating in different modes. Such sequences have been identified in the LMC by Wood & Sebo (1996) and confirmed by Barthes (1998), although their interpretations differed. The massive data collection from the MACHO project data base supports this result (Wood 1999), showing clearly separated sequences in L versus P. The assumption that the Mira sequence corresponds to fundamental mode pulsation assigns the smaller-amplitude SRa stars to first and second overtones, a result that is also consistent with their more nearly sinusoidal light curves.

2.5 Empirical Mass Loss Laws

In 1975, Reimers published a correlation between observed mass loss rates and the combination of parameters LR/M for an assortment of red giants and supergiants:

$$\dot{M}_R = \eta \, 4 \times 10^{-13} LR/M, \tag{5}$$

where L, R, and M are in solar units and $\dot{M}$ is in M_{Sun} yr^{-1}. This relation provided stellar evolution modelers with a tool for investigating mass loss effects on stellar evolution, which they were not slow to exploit. It soon became clear that Equation 5 overestimated the mass loss during some stages of evolution, and underestimated the rates relative to those found at the tip of the AGB; thus the variable parameter η was introduced to allow better matching between HR diagrams and models.

As noted in Section 1, $L_{wind} \sim \dot{M} \, v_{esc}^2$. If the efficiency of all winds is about the same, $L_{wind} \sim 10^{-6} \, L_*$, then we recover Reimers' relation for $\eta = 1$. However, it is certainly not obvious a priori that the efficiency should be the same, particularly when such a very small fraction of the total luminosity is involved. A very different interpretation of this observed correlation of mass loss rate with stellar parameters is described in Section 3.

With the advent of more and more sensitive methods for detecting mass loss, it has become apparent that there are many stars with $\dot{M} >$ or $\gg \dot{M}_R$, and there are certainly stars with $\dot{M} <$ or $\ll \dot{M}_R$. This has motivated a number of attempts to refine the relation, with L, R, and/or M allowed to have arbitrary exponents or more complex functional forms fitted to more extensive mass loss data. Adding more parameters reduces the scatter, of course, but does not really address the underlying issues. Invoking a "superwind" phase (Iben & Renzini 1983) near the end of the AGB improved the evolutionary models' match to observations, at the expense of adding more tunable parameters. Wood & Cahn (1977) were the first to examine the pattern of AGB evolution with mass loss. Recent examples of detailed "synthetic evolution" patterns for the AGB using a variety of mass loss laws may be found in Blöcker (1995) and Groenewegen & De Jong (1994). The result may be made to fit observable nearby populations almost arbitrarily well, but has little or no power to extend to situations not readily observable, such as young, low-metallicity stellar populations.

A different approach was taken by Vassiliadis & Wood (1993) for modeling terminal AGB evolution. They found a relationship between $\dot{M}$ and the pulsation period, P, an easily observed quantity that is also easily calculated for a given stellar model. Their fit captures the precipitous steepness of the mass loss law seen in the Bowen models, but not the spread for different masses. At a given P there is a wide range in $\dot{M}$, presumably because the stars with one P do not all have the same L, R, and M. The variation or scatter is entirely in the $\dot{M}$ axis, since P is easily observed to high accuracy. Thus, Vassiliadis & Wood's relation misses a substantial fraction of the mass dependence of $\dot{M}$, limiting its usefulness for population studies. This relation also does not include any basis for extension to metallicities other than the one used for the calibration.

Baud & Habing (1983) proposed a modification to Reimers' formula involving the envelope mass:

$$\dot{M} = (4 \times 10^{-3})(M_{env,o}/M_{env})LR/M \qquad (6)$$

where $M_{env,o}$ is the mass of the envelope at the start of the AGB. This introduces a rapid "superwind" evaporation of the last portion of the stellar envelope.

In Figure 6, four empirical mass loss relations are compared. A set of evolutionary tracks (from Iben 1984; Equation 7, Section III) and the definition of effective temperature have been used to reduce the parameters to a single one, L, for each mass and composition. L thus represents the position along the AGB or, approximately, the time elapsed on the AGB (since $\Delta t \propto \Delta \log L$). Vassiliadis & Wood's relation gives a narrower range of L, over which the envelope mass would be lost, than does the Reimers' relation, and this would still be true with larger or smaller η (since η does not change the slope of $\log \dot{M}$ versus $\log L$). Baud & Habing's relation deviates from Reimers' relation only at the highest mass loss rates. The relation by Nieuwenhuijzen & De Jager (1990) was derived from observations of luminous supergiants, and as a result predicts large mass loss rates only at large L. The panels in this figure may be directly compared with Figure 4 at the end of Section 1, where mass loss laws from theoretical computations were displayed.

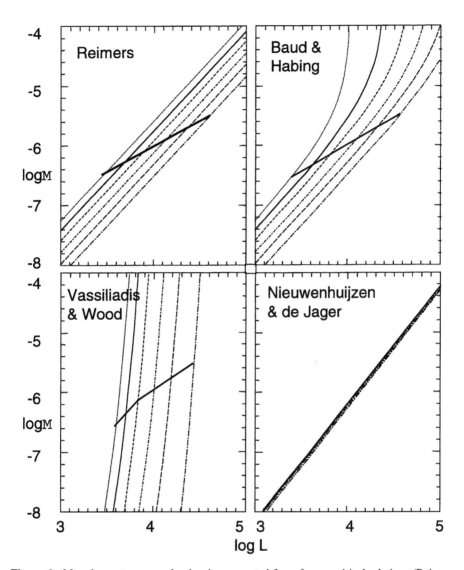

Figure 6 Mass loss rates versus luminosity computed from four empirical relations (Reimers 1975, Baud and Habing 1983, Vassiliadis & Wood 1993, Nieuwenhuijzen & De Jager 1990) in the same way as for the theoretical relations of Figure 4. In the first three panels, $M_{Sun} = 0.1$, 1, 1.4, 2, 2.8, and 4 are shown, with the heaviest line for $M_{Sun} = 1\ M_{Sun}$; in the fourth panel, $M = 5.6$, and 8 M_{Sun} have been added, since this formula was derived for more luminous and more massive stars. Across the first three panels, the location of the "cliff" or critical mass loss rate is shown. The law derived from more luminous stars predicts a higher terminal luminosity, as would be expected from the interpretation given in Section 3.

3.   EVOLUTION WITH MASS LOSS

In the introduction the current state of evolutionary modeling with and without mass loss was reviewed. In this section an alternative approach to incorporating observational constraints on mass loss into theoretical modeling is demonstrated for the case of evolution on the asymptotic giant branch, building on the material presented in the preceding sections.

3.1   The Use and Interpretation of Empirical Relationships

The approach that has been most widely used to incorporate mass loss into stellar evolution calculations has been to look for a functional dependence of the mass loss rate on stellar parameters; to use this dependence in evolutionary calculations; and, if necessary, to adjust the mass loss law and repeat the process. At the end of Sections 1 and 2, a variety of mass loss laws derived from theoretical and observational studies were compared in terms of their predictions for the relation between L and $\dot{M}$.

There is a developing consensus, among those doing detailed modeling for the mass loss at the end of the AGB phase, that this mass loss depends much more steeply on stellar parameters than was evident from the first empirical relations (Bowen & Willson 1991, Höfner & Dorfi 1997, 1998a,b, Schröder et al 1999). This leads to a picture of the final evolution where the mass remains effectively constant until the star reaches a "cliff," the locus where d log M/dt $\sim$ d log L/dt. With a small evolution in the stellar parameters—L, T_{eff}, and M—the mass loss rate rises precipitously and the star loses its envelope in a nearly exponential process. This is illustrated in Figure 7, which is based on Bowen's 1995 grid of models (also Bowen 2000, in preparation).

With a pattern such as the one in Figure 7, there will be very strong selection effects concerning stars for which mass loss will be measurable. Those not yet near the cliff will have low mass loss rates, while those beyond it will be short-lived (and probably also well obscured). Thus, we would expect a selection of stars with observed mass loss rates (such as Reimers used to establish a relation between mass loss rates and LR/M) to feature mainly stars within 1 dex in $\dot{M}$ of the cliff. Those lines (for the cliff models of Figure 7) are plotted in Figure 8, where clearly the Reimers relation is very well reproduced. Thus, the empirical mass loss laws tell us the parameters of stars that are losing mass, and not the dependence of mass loss rates on the parameters of any individual star. Two stars with the same period will have nearly the same mean density and, given that the masses are not too different, similar radii. The Miras and OH-IR stars also do not differ hugely in L and T_{eff} at a given P. However, the mass loss rates observed for Miras at a given P cover a range of several dex, illustrating again that the mass loss rates are very sensitive to stellar parameters.

In Figure 9 the "cliff" locations for populations with five different metallicities are plotted. To produce this figure, Bowen computed grids of dozens of

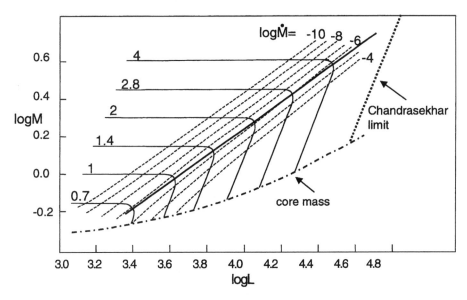

Figure 7 Evolution in mass and luminosity for solar composition stars. Contours of constant mass loss rate are indicated; the fact that these are very close together makes the "corner" on the cliff sharp. A star evolves at approximately constant mass until it is near the cliff; it leaves the cliff on an asymptote to a line of constant core mass. (Figure courtesy of GH Bowen.)

atmosphere-plus-wind models for each metallicity, using the Iben (1984) relation,

$$R = 312(L/10^4)^{0.68}(1.175/M)0^{32S}(Z/0.001)^{0.088}\alpha^{-0.52} \qquad (7)$$

(where S = 0 for M < 1.175 and S = 1 for M ≥ 1.175) to constrain the models to lie along a consistent set of evolutionary tracks. (A similar pattern would emerge for a different choice of evolutionary tracks or of α, the ratio of mixing length to scale height, as long as the following still holds true: More massive stars are hotter at a given L. Lower metallicity stars are hotter at a given L. As a star evolves to higher L, its T_{eff} decreases slowly.)

Figure 9 suggests that in young, low-metallicity populations compared with Pop. I we should expect higher-mass white dwarfs, more supernovae with the extras being of Type 1.5, less mass returned to the ISM, and most of the mass that is lost coming from a higher L stage on the AGB. This is the kind of information that is essential in order to improve our understanding of the early chemical evolution of the Milky Way. These results are also needed for modeling the appearance and chemical evolution of low-metallicity galaxies, including some starburst systems and most of the sources seen at high redshift.

The stars we classify as Miras are those that lie along the "cliff" line in Figure 7. From the models we can compute the periods (given the mode) and the luminosities of these cliff stars. In Figure 5, the resulting P-L relation (for fundamental mode)

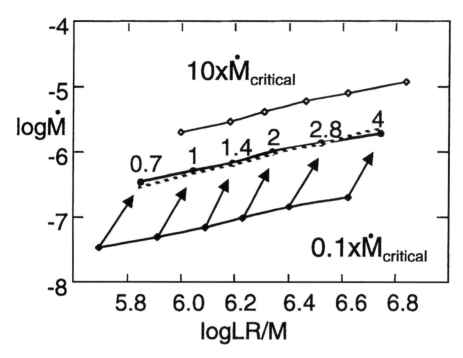

Figure 8   Mass loss rate versus LR/M for the "cliff" stars and mass loss rates 1/10, 10 times the "cliff" rates. Reimers (1975) relation is seen to coincide almost exactly with the "cliff" position. Thus Reimers' relation tells us the properties of stars when they are losing mass, not how any one star loses mass over time.

was superimposed on the observed P-L relation for LMC Miras. No parameter adjustment was made to achieve this fit.

Several conclusions that can be drawn from the above stand in contrast with current practices in putting mass loss into evolutionary and stellar population models. First, the mass loss along the AGB is abrupt, not gradual. Second, mass loss prior to the AGB is not enough to reduce all Miras to ~1 solar mass. Third, the empirical P-L relation, having been confirmed by theory, may be relied on as a distance indicator. Fourth, empirical relations give us the properties of stars that are near a critical stage of losing mass, rather than telling us how to incorporate mass loss into evolutionary calculations. Fifth, because the significant mass loss is limited to a short duration near the end of the evolution on the AGB, tabulated evolutionary models that incorporate Reimers-type mass loss laws are less useful than models without any mass loss at all.

Three very significant problems remain in the study of mass loss from evolved red giants. The first has to do with the transformation of some stars into carbon stars. Figure 9 shows where stars reach critical mass loss rates if they remain spectral type M stars, i.e. as long as the carbon abundance in their envelopes stays less than the oxygen abundance. For a star that becomes a carbon star, some of

Chamberlain JW. 1961. *Ap. J.* 133:675

Chamberlain JW. 1963. *Planet. Space Sci.* 11:901–60

Chan SJ, Kwok S. 1988. *Ap. J.* 334:362–96

Chiosi C, Bertelli G, Bressan A. 1992. *Annu. Rev. Astron. Astrophys.* 30:235–85

Chiosi C, Maeder A. 1986. *Annu. Rev. Astron. Astrophys.* 24:329–75

Clayton ML, Feast MW. 1969. *MNRAS* 146:411

Codina-Landaberry SJ, Magalhaes AM. 1980. *Astron. J.* 85:875–81

Cole PW, Deupree RG. 1981. *Ap. J.* 247:607–11.

Cox JP, Giuli RT. 1968. *Principles of Stellar Structure.* New York: Gordon & Breach. 1327 pp.

Coyne GV, Magalhaes AM. 1977. *Astron. J.* 82:908–15

Coyne GV, Magalhaes AM. 1979. *Astron. J.* 84:1200–10

Cranmer SR, Field GB, Kohl JL. 1999. *Ap. J.* 518:937–47

Cuntz M. 1989. *PASP* 101:560

Cuntz M, Ulmschneider P, Musielak ZE. 1998. *Ap. J.* 493:L117

Danchi WC, Bester M. 1995. *Astrophys. Space Sci.* 224:339–52

Danchi WC, Bester M, Degiacomi CG, Greenhill LJ, Townes CH. 1994. *Astron. J.* 107:1469–513

Danchi WC, Bester M, Greenhill LJ, Degiacomi CG, Geis N, et al. 1995. *Astrophys. Space Sci.* 224:447–48

Donn B. 1978. In *Protostars and Planets*, ed. T Gehrels, pp. 100–33. Tucson: University of Arizona Press

Donn B, Nuth JA. 1985. *Ap. J.* 288:187–90

Draine BT. 1979. *Astrophys. Space Sci.* 65:313–35

Draine BT. 1981. In *Physical Process in Red Giants*, eds. I Iben & A Renzini, pp. 317–33. Astrophysics and Space Science Library. Dordrecht/Boston/London: Reidel

Draine BT, Lee HM. 1984. *Ap. J.* 285:89–108

Draine BT, Lee HM. 1987. *Ap. J.* 318:485; erratum *Ap. J. Suppl.* 64:505

Draine BT, McKee CF. 1993. *Annu. Rev. Astron. Astrophys.* 31:373–432

Draine BT, Salpeter EE. 1979. *Ap. J.* 231:438–55

Drake SA, Linsky JL, Elitzur M. 1987. *Astron. J.* 94:1280–290

Drake SA, Linsky JL, Judge PG, Elitzur M. 1991. *Astron. J.* 101:230–36

Dupree AK. 1986. *Annu. Rev. Astron. Astrophys.* 24:377–420

Dyck M. 1987. See Kwok & Pottasch 1987, pp. 19–32

Elitzur M. 1992. *Annu. Rev. Astron. Astrophys.* 30:75–112

Elitzur M, Brown JA, Johnson HR. 1989. *Ap. J.* 341:L95–98

Feast MW, Glass IS, Whitelock PA, Catchpole RM. 1989. *MNRAS* 241:375–92

Fox MW, Wood PR. 1985. *Ap. J.* 297:455–75

Frost CA, Cannon RC, Lattanzio JC, Wood PR, Forestini M. 1998a. *Astron. Astrophys.* 332:L17–20

Frost CA, Lattanzio JC, Wood PR. 998b. *Ap. J.* 500:355

Gail HP, Sedlmayr E. 1987a. *Astron. Astrophys.* 171:197–204

Gail HP, Sedlmayr E. 1987b. *Astron. Astrophys.* 177:186–92

Gilman RC. 1969. *Ap. J.* 155:L185

Gilman RC. 1972. *Ap. J.* 178:423–26

Gilman RC. 1973. *MNRAS* 161:3P

Gilman RC. 1974. *Ap. J. Suppl.* 28:397

Glass IS, Catchpole RM, Feast MW, Whitelock PA, Reid IN. 1987. See Kwok & Pottasch 1987, pp. 51–54

Goldreich P, Scoville N. 1976. *Ap. J.* 205:144–54

Gray MD, Humphreys EML, Yates JA. 1999. *MNRAS* 304:906–24

Groenewegen MAT, De Jong T. 1994. *Astron. Astrophys.* 288:782–90

Guilain C, Mauron N. 1996. *Astron. Astrophys.* 314:585–93

Gustafsson B, Eriksson K, Kiselman D, Olander N, Olofsson H. 1997. *Astron. Astrophys.* 318:535–42

Hammer R. 1982. *Ap. J.* 259:767–91

Haniff CA, Scholz M, Tuthill PG. 1995. *MNRAS* 276:640

Härm R, Schwarzschild M. 1966. *Ap. J.* 145:496–504

Hartmann L, Macgregor KB. 1980. *Ap. J.* 242:260–82

Hashimoto O, Izumiura H, Kester DJM, Bontekoe TR. 1998. *Astron. Astrophys.* 329:213–18

Hashimoto O, Nakada Y, Onaka T, Kamijo F, Tanabe T. 1990. *Astron. Astrophys.* 227:465–72

Herpin F, Baudry A, Alcolea J, Cernicharo J. 1998. *Astron. Astrophys.* 334:1037–46

Hill SJ, Willson LA. 1979. *Ap. J.* 229:1029–45

Hinkle KH. 1978. *Ap. J.* 220:210–28

Hinkle KH. 1995. In *ASP Conf. Ser. 83, IAU Colloq. 155: Astrophysical Applications of Stellar Pulsation*, p. 399

Hinkle KH, Barnes TG. 1979a. *Ap. J.* 227:923–34

Hinkle KH, Barnes TG. 1979b. *Ap. J.* 234:548–55

Hinkle KH, Hall DNB, Ridgway ST. 1982. *Ap. J.* 252:697–714

Hinkle KH, Scharlach WWG, Hall DNB. 1984. *Ap. J. Suppl.* 56:1–17

Höfner S, Dorfi EA. 1992. *Astron. Astrophys.* 265:207–15

Höfner S, Dorfi EA. 1997. *Astron. Astrophys.* 319:648

Höfner S, Jørgensen UG, Loidl R. 1998a. *Astrophys. Space Sci.* 255:281–87

Höfner S, Jørgensen UG, Loidl R, Aringer B. 1998b. *Astron. Astrophys.* 340:497–507

Holm AV, Doherty LR. 1988. *Ap. J.* 328:726–33

Holzer TE. 1977. *J. Geophys. Res.* 82:23–35

Holzer TE, Axford WI. 1970. *Annu. Rev. Astron. Astrophys.* 8:31

Holzer TE, MacGregor KB. 1985. See Morris & Zuckerman 1985, p. 299

Hron J, Loidl R, Höfner S, Jørgensen UG, Aringer B, Kerschbaum F. 1998. *Astron. Astrophys.* 335:L69–72

Hughes SMG, Wood PR. 1990. *Astron. J.* 99:784–816

Humphreys EML, Gray MD, Field D, Yates JA, Bowen G. 1997a. *Astrophys. Space Sci.* 251:215–18

Humphreys EML, Gray MD, Yates JA, Field D. 1997b. *MNRAS* 287:663–70

Iben I, Jr. 1974. *Annu. Rev. Astron. Astrophys.* 12:215–56

Iben I, Jr. 1984. *Ap. J.* 277:333–54

Iben I, Jr, Renzini A, eds. 1981. *Physical Processes in Red Giants.* Dordrecht/Boston/Lancaster: Reidel. 488 pp.

Iben I, Jr, Renzini A. 1983. *Annu. Rev. Astron. Astrophys.* 21:271–342

Jones TJ, Bryja CO, Gehrz RD, Harrison TE, Johnson JJ, et al. 1990. *Ap. J. Suppl.* 74:785–817

Jørgensen UG, Johnson HR. 1992. *Astron. Astrophys.* 265:168–76

Jorissen A, Knapp GR. 1998. *Astron. Astrophys.* 129:363–98

Karovska M, Nisenson P, Papaliolios C, Boyle RP. 1991. *Ap. J.* 374:L51–54

Kerschbaum F, Hron J. 1996. *Astron. Astrophys.* 308:489–96

Knapp GR. 1986. *Ap. J.* 311:731–41

Knapp GR. 1991. In *ASP Conf. Ser. 20: Frontiers of Stellar Evolution*, pp. 229–63

Knapp GR, Bowers PF, Young K, Phillips TG. 1995. *Ap. J.* 455:293

Koopmann RA, Lee Y-W, Demarque P, Howard JM. 1994. *Ap. J.* 423:380–85

Krueger D, Sedlmayr E. 1997. *Astron. Astrophys.* 321:557–67

Kruszewski A, Gehrels T, Serkowski K. 1968. *Astron. J.* 73:677

Kwok S. 1975. *Ap. J.* 198:583–91

Kwok S. 1993. *Annu. Rev. Astron. Astrophys.* 31:63–92

Kwok S, Pottasch SR. 1987. *Late Stages of Stellar Evolution.* Dordrecht/Boston/Lancaster: Reidel. 426 pp.

Lamers HJGLM, Cassinelli IP. 1996. In *ASP Conf. Ser. 98, From Stars to Galaxies: the Impact of Stellar Physics on Galaxy Evolution*, p. 162

Lamers HJGLM, Cassinelli JP. 1999. *Introduction to Stellar Winds.* Cambridge, UK: Cambridge Univ. Press. 438 pp.

Lattanzi MG, Munari U, Whitelock PA, Feast MW. 1997. *Ap. J.* 485:328

Lattanzio JC. 1998. In *Stellar Evolution, Stellar Explosions and Galactic Chemical Evolution*, ed. A Mezzacappa, p. 299. Institute of Physics Publishing

Le Bertre T, Le Bertre A, Waelkens C, eds. 1999. *191 Asymptotic Giant Branch Stars.*

Lebzelter T, Hinkle KH, Hron J. 1998. *Astron. J.* 116:2520–29

Lepine JRD, Ortiz R, Epchtein N. 1995. *Astron. Astrophys.* 299:453

Lim J, White SM. 1996. *Ap. J.* 462:L91

Little-Marenin IR, Sahai R, Wannier PG, Benson PJ, Gaylard M, Omont A. 1994. *Astron. Astrophys.* 281:451–59

Loidl RH, Höfner S. Jørgensen UG, Aringer B. 1999. *Astron. Astrophys.* 342:531–41

Maciel WJ. 1976. *Astron. Astrophys.* 48:27–31

Mastrodemos N, Morris M, Castor J. 1996. *Ap. J.* 468:851

McLean IS, Coyne GV. 1978. *Ap. J.* 226:L145–48

Monnier JD, Geballe TR, Danchi WC. 1998. *Ap. J.* 502:833

Moore RL, Hammer R, Musielak ZE, Suess ST, An CH. 1992. *Ap. J.* 397:L55–58

Morris M, Zuckerman B, eds. 1985. *Mass Loss from Red Giants*. Los Angeles: Reidel. 320 pp.

Motteran M. 1971. In *Colloquium on Supergiant Stars*, ed. M Hack. Trieste:Osservatorio astronomico di Trieste, 371 pp.

Mould J, Reid N. 1987. *Ap. J.* 321:156–61

Mullan DJ. 1996. In *ASP Conf. Ser. 109, Ninth Cambridge Workshop on Cool Stars, Stellar Systems, and the Sun*, p. 461

Najita JR, Shu FH. 1994. *Ap. J.* 429:808–25

Netzer N, Elitzur M. 1993. *Ap. J.* 410:701–13

Nieuwenhuijzen H, De Jager C. 1990. *Astron. Astrophys.* 231:134–36

Nuth JA, Donn B. 1981. *Ap. J.* 247:925–35

Nyman LA, Booth RS, Carlström U, Habing HJ, Heske A, et al. 1992. *Astron. Astrophys. Suppl.* 93:121–50

Olofsson H. 1996a. *Astrophys. Space Sci.* 245:169–200

Olofsson H. 1996b. In *Molecules in Astrophysics: Probes & Processes: IAU Symposium 178*, ed. EF van Dishoeck, p. 457

Olofsson H. 1997. *Astrophys. Space Sci.* 251:31–39

Olofsson H, Bergman P, Eriksson K, Gustafsson B. 1996. *Astron. Astrophys.* 311:587–615

Olofsson H, Bergman P, Lucas R, Eriksson K, Gustafsson B, Bieging JH. 1998. *Astron. Astrophys.* 330:L1–4

Olofsson H, Carlström U, Eriksson K, Gustafsson B. 1992. *Astron. Astrophys.* 253:L17–20

Olofsson H, Carlström U, Eriksson K, Gustafsson B, Willson LA. 1990. *Astron. Astrophys.* 230:L13–16

Ostlie DA, Cox AN. 1986. *Ap. J.* 311:864–72

Ostriker EC, Shu FH. 1995. *Ap. J.* 447:813

Paczynski B. 1970. *Acta Astron.* 20:47

Parker EN. 1960. *Ap. J.* 132:821

Parker EN. 1961. *Ap. J.* 134:20

Parker EN. 1963a. *Space Sci. Rev.* 4:666–708

Parker EN. 1963b. *Interplanetary Dynamical Processes*. New York: Interscience, 272 pp.

Parker EN. 1964a. *Ap. J.* 139:72

Parker EN. 1964b. *Ap. J.* 139:93

Parker EN. 1964c. *Ap. J.* 139:690

Parker EN. 1965. *Ap. J.* 141:1463

Parker EN. 1966. *Ap. J.* 143:32

Parker EN. 1969. *Sp. Sci. Rev.* 9:325–360

Parker EN. 1997. In *Cosmic Winds and the Heliosphere*, ed. JR Jokipii, CP Sonett, MS Giampapa, pp. 3–30. Tucson: Univ. Ariz. Press

Patzer ABC, Gauger A, Sedlmayr E. 1998. *Astron. Astrophys.* 337:847–58

Percy JR, Colivas T. 1999. *PASP* 111:94–97

Perrin G, Coudé du Foresto V, Ridgway ST, Mennesson B, Ruilier C, et al. 1999. *Astron. Astrophys.* 345:221–32

Pijpers FP, Habing HJ. 1989. *Astron. Astrophys.* 215:334–46

Pugach AF. 1977. *Inf. Bull. Var. Stars* 1277:1

Quirrenbach A, Mozurkewich D, Armstrong JT, Buscher DF, Hummel CA. 1993. *Ap. J.* 406:215–19

Rammacher W, Ulmschneider P. 1992. *Astron. Astrophys.* 253:586–600

Reid MJ, Menten KM. 1997. *Ap. J.* 476:327

Reimers D. 1975. In *Problems in Stellar Atmospheres and Envelopes*, ed. B Baschek, WH Kegel, G Traving, pp. 229–56. Berlin/New York: Springer-Verlag

Ridgway ST, Wells DC, Joyce RR, Allen RG. 1979. *Astron. J.* 84:247–56

Sahai R, Liechti S. 1995. *Astron. Astrophys.* 293:198–207

Sahai R, Wannier PG. 1992. *Ap. J.* 394:320–39

Salpeter EE. 1974a. *Ap. J.* 193:579–84

Salpeter EE. 1974b. *Ap. J.* 193:585–92

Salpeter EE. 1977. *Annu. Rev. Astron. Astrophys.* 15:267–93

Schönberner D. 1983. *Ap. J.* 272:708–14

Scholz M, Takeda Y. 1987. *Astron. Astrophys.* 186:200–12

Schröder KP, Winters JM, Sedlmayr E. 1999. *Astron. Astrophys.* 349:898–906

Schwarzschild M, Härm R. 1962. *Ap. J.* 135:158–65

Shu F, Najita J, Ostriker E, Wilkin F, Ruden S, Lizano S. 1994a. *Ap. J.* 429:781–96

Shu FH, Najita J, Ruden SP, Lizano S. 1994b. *Ap. J.* 429:797–807M

Shu FH, Najita J, Ostriker EC, Shang H. 1995. *Ap. J.* 455:L155

Sopka RJ, Olofsson H, Johansson LEB, Nguyen-Rieu Q, Zuckerman B. 1989. *Astron. Astrophys.* 210:78–92

Spergel DN, Giuliani JL, Jr, Knapp GR. 1983. *Ap. J.* 275:330–41

Spitzer LJ. 1968. *Diffuse Matter in Space*. New York: Interscience Publishers. 262 pp.

Stephens JR. 1991. In *Proceedings of the International School of Physics "Enrico Fermi" Course CXI: Solid State Astrophysics*, ed. E Bussoletti and G Strazzulla, 391–402. Amsterdam/New York: Elsevier

Sterken C, Broens E, Koen C. 1999. *Astron. Astrophys.* 342:167–72

Sutmann G, Ulmschneider P. 1995. *Astron. Astrophys.* 294:241–51

Sweigart AV, Greggio L, Renzini A. 1990. *Ap. J.* 364:527–39

Sweigart AV, Gross PG. 1978. *Ap. J. Suppl.* 36:405–37

Tielens AGGM. 1983. *Ap. J.* 271:702–16

Tielens AGGM. 1998. *Ap. J.* 499:267

Tielens AGGM, McKee CF, Seab CG, Hollenbach DJ. 1994. *Ap. J.* 431:321–40

Tomasko MG. 1970. *Ap. J.* 162:125–38

Tsuji T. 1978. *Pub. Astron. Soc. J.* 30:435–54

Tuthill PG, Haniff CA, Baldwin JE. 1995. *MNRAS* 277:1541

Ulmschneider R, Schmitz F, Kalkofen W, Bohn HU. 1978. *Astron. Astrophys.* 70:487

Van Belle GT, Dyck HM, Benson JA, Lacasse MG. 1996. *Astron. J.* 112:2147

Van Belle GT, Dyck HM, Thompson RR, Benson JA, Kannappan SJ. 1997. *Astron. J.* 114:2150

Van Den Oord GHJ, Doyle JG. 1997. *Astron. Astrophys.* 319:578–88

Van Der Veen WECJ, Habing HJ. 1988. *Astron. Astrophys.* 194:125–34

Van Der Veen WECJ, Rugers M. 1989. *Astron. Astrophys.* 226:183–202

Van Langevelde HJ, Van Der Heiden R, Van Schooneveld C. 1990. *Astron. Astrophys.* 239:193–204

Van Leeuwen F, Feast MW, Whitelock PA, Yudin B. 1997. *MNRAS* 287:955–60

Van Loon JT, Groenewegen MAT, De Koter A, Trams NR, Waters LBFM, et al. 1999. *Astron. Astrophys.* 351:559–72

Vassiliadis E, Wood PR. 1992. *Proc. Astron. Soc. Austr.* 10:30–32

Vassiliadis E, Wood PR. 1993. *Ap. J.* 413:641–57

Wallerstein G, Knapp GR. 1998. *Annu. Rev. Astron. Astrophys.* 36:369–434

Walmsley CM, Chini R, Kreysa E, Steppe H, Forveille T, Omont A. 1991. *Astron. Astrophys.* 248:555–62

Wannier PG, Sahai R. 1986. *Ap. J.* 311:335–44

Wannier PG, Sahai R, Andersson BG, Johnson HR. 1990. *Ap. J.* 358:251–61

Weidemann V. 1990. *Annu. Rev. Astron. Astrophys.* 28:103–37

Weymann R. 1963. *Annu. Rev. Astron. Astrophys.* 1:97

Willson LA. 1999. In *Unsolved Problems in Stellar Evolution*, ed. M Livio, p. 227. Cambridge, UK: Cambridge Univ. Press

Willson LA, Bowen GH. 1984a. In *The Relationship Between Chromospheric/Coronal Heating and Mass Loss*, ed. R Stalio, J, Zirker, pp. 127–56. Triesta:NASA SP

Willson LA, Bowen GH. 1984b. *Nature* 312:429–31

Willson LA, Bowen GH. 1988. In *Polarized Radiation of Circumstellar Origin*, ed. G Coyne SJ, pp. 485–510. Castel Gandolfo:Univ. Ariz. Press

Willson LA, Bowen GH. 1988. In *Cyclical Variability in Stellar Winds*, Proceedings of the ESO Workshop held at Garching, Germany, 14–17 October 1997, ed. L Kaper, AW Fullerton, p. 294. Berlin/New York: Springer-Verlag

Willson LA, Bowen GH, Struck C. 1996. In *ASP Conf. Ser. 98, From Stars to Galaxies: The Impact of Stellar Physics on Galaxy Evolution*, p. 197

Willson LA, Bowen GH, Struck-Marcell C. 1987. *Comments Asrophys.* 12:17–34

Willson LA, Hill SJ. 1979. *Ap. J.* 228:854–69

Willson LA, Wallerstein G, Pilachowski CA. 1982. *MNRAS* 198:483–516

Windsteig W, Drofi EA, Hoefner S, Hron J, Kerschbaum F. 1997. *Arstron. Astrophys.* 324:617–23

Winters JM, Fleischer AJ, Le Bertre T, Sedlmayr E. 1997. *Astron. Astrophys.* 326:305–17

Withbroe GL. 1988. *Ap. J.* 325:442–67

Withbroe GL, Noyes RW. 1977. *Annu. Rev. Astron. Astrophys.* 15:363–87

Woitke P, Krueger D, Sedlmayr E. 1996. *Astron. Astrophys.* 311:927–44

Wolff MJ, Clayton GC, Gibson SJ. 1998. *Ap. J.* 503:815

Wood PR. 1974. *Ap. J.* 190:609–30

Wood PR. 1979. *Ap. J.* 227:220–31

Wood PR. 1990. In *ASP Conf. Ser. 11, Conformation Between Stellar Pulsation and Evolution*, pp. 355–63

Wood PR. 1998. *Astron. Astrophys.* 338:592–98

Wood PR. 1999. In *ASP Conf. Ser 191, Asymptotic Giant Branch Stars* ed. Bertre T, Lebre A, Waelkens C. p. 151

Wood PR, Cahn JH. 1977. *Ap. J.* 211:499–508

Wood PR, Sebo KM. 1996. *MNRAS* 282:958–64

Wood PR, Whiteoak JB, Hughes SMG, Besell MS, Gardner FF, Hyland AR. 1992. *Ap. J.* 397:552–69

Wood PR, Zarro DM. 1981. *Ap. J.* 247:247–56

Zuckerman B. 1980. *Annu. Rev. Astron. Astrophys.* 18:263–88

Annu. Rev. Astron. Astrophys. 2000. 38:613–66

WINDS FROM HOT STARS

Rolf-Peter Kudritzki and Joachim Puls

*Institut für Astronomie und Astrophysik der Universität München,
Scheinerstr. 1, D-81679 München, Germany; e-mail: kud@usm.uni-muenchen.de,
uh101aw@usm.uni-muenchen.de*

Key Words mass loss, stellar winds, massive stars, Central Stars of Planetary
Nebulae, stellar evolution

■ **Abstract** This review deals with the winds from "normal" hot stars such as
O-stars, B- and A-supergiants, and Central Stars of Planetary Nebulae with O-type
spectra. The advanced diagnostic methods of stellar winds, including an assessment of
the accuracy of the determinations of global stellar wind parameters (terminal velocities, mass-loss rates, wind momenta, and energies), are introduced and scaling relations
as a function of stellar parameters are provided. Observational results are interpreted
in the framework of the stationary, one-dimensional (1-D) theory of line-driven winds.
Systematic effects caused by nonhomogeneous structures, time dependence, and deviations from spherical symmetry are discussed. The review finishes with a brief
description of the role of stellar winds as extragalactic distance indicators and as tracers of the chemical composition of galaxies at high redshift.

1. INTRODUCTION

All hot stars have winds driven by radiation. These winds become directly observable in spectral energy distributions and spectral lines as soon as the stars are above
certain luminosity borderlines in the HRD. For massive stars of spectral type O, B,
and A, the borderline corresponds to 10^4 $L_\odot$. Above this threshold in luminosity
all massive stars show direct spectroscopic evidence of winds throughout their
lifetime (Abbott 1979). An analogous threshold exists for stars of intermediate
and low mass ($M_{ZAMS} \leq 8$ $M_\odot$), when they evolve through the post-AGB phases
toward the white dwarf final stage. Here, all objects more massive than 0.58 $M_\odot$ or
more luminous than $10^{3.6}$ $L_\odot$ exhibit direct signatures of winds in their spectra
(Pauldrach et al 1989).

Winds are able to modify the ionizing radiation of hot stars dramatically
(see Gabler et al 1989, 1991, 1992; Najarro et al 1996). Their momenta and
energies contribute substantially to the dynamics and energetics of the ambient interstellar medium in galaxies or of surrounding gaseous nebulae. Even weak winds
(i.e. those with no direct spectroscopic evidence of mass-outflow) may produce

0066-4146/00/0915-0613$14.00

613

thin gaseous envelopes around stars before they become red supergiants with dense slow winds or explode as supernovae in later stages of their evolution. The gas-dynamical interaction of stellar plasmas streaming with different velocities and densities will then eventually lead to complex surrounding structures directly observable with high spatial resolution. In binary systems, winds can be partially accreted or can produce shocks of colliding winds resulting in X-ray emission in both cases. Winds (*a*) affect the physics of stellar atmospheres by dominating the density stratification and the radiative transfer through the presence of their macroscopic transonic velocity fields, (*b*) substantially influence the evolution of stars by modifying evolutionary timescales, chemical profiles, surface abundances, and stellar luminosities, and (*c*) are vital for the evolution of galaxies by their input of energy, momentum, and nuclear-processed material to the ISM.

In addition, stellar winds are a gift of nature, allowing quantitative spectroscopic studies of the most luminous stellar objects in distant galaxies and, thus, enabling us to obtain important quantitative information about their host galaxies. Broad stellar wind lines can be identified in spectra integrating over the stellar populations of extremely distant galaxies at high redshift even when the spectral resolution and the signal-to-noise ratio are only moderate (Steidel et al 1996). Particularly in cases when the flux from these galaxies is amplified by gravitational lensing through foreground galaxy clusters, stellar wind lines can be used to estimate metallicities of starbursting galaxies in the early universe (Pettini et al 2000). In the local universe, the population synthesis of spectroscopic stellar wind features observed in the integrated light of starbursting regions of galaxies can yield important information about the population of stars formed (Leitherer 1998, Leitherer et al 1999). Finally, the observation of winds of isolated blue supergiants as individuals in spiral and irregular galaxies provides an independent new tool for the determination of extragalactic distances by means of the wind momentum–luminosity relationship (Kudritzki et al 1995, Puls et al 1996, Kudritzki 1998, Kudritzki et al 1999).

In view of the important role of winds from hot stars in so many areas of astrophysics, it is not surprising that the number of papers and conferences dealing with them is enormous. As a result, a complete review covering all aspects in theory and observation is impossible within the framework of this review series, and a restriction of the subject is needed. Therefore, we have restricted this review to the discussion of winds from "normal" hot stars in well-established evolutionary stages such as dwarfs, giants, and supergiants. We do not discuss objects with extreme winds such as Wolf-Rayet stars, luminous blue variables in outburst, Be-stars, etc. We are aware of the fact that by this restriction a significant amount of active research in interesting areas of stellar physics is omitted. On the other hand, it allows us to cover the most fundamental stages of stellar evolution in sufficient depth.

The goal of this review is to inform about the advanced diagnostic methods of stellar winds, including an assessment of the accuracy of the determinations of global stellar wind properties such as terminal velocities, mass-loss rates, wind momenta, and wind energies (Section 2). Simple empirical scaling relations of these stellar wind properties as a function of stellar parameters is provided (Section 3) and is interpreted in the framework of the stationary, 1-D theory of radiation-driven winds (Section 4). The influence of systematic effects caused by nonhomogeneous structures, time dependence, and deviations from spherical symmetry caused by rotation is discussed (Sections 4 and 5). The review finishes with a brief discussion of stellar wind lines as extragalactic distance indicators and as tracers of chemical composition of galaxies at high redshift, in particular in view of the potential of the new generation of very large ground-based telescopes (Section 6).

2.   STELLAR PARAMETERS AND GLOBAL WIND PROPERTIES

2.1   Global Stellar Wind Parameters

Winds of hot stars are characterized by two global parameters, the terminal velocity v_∞ and the rate of mass loss $\dot{M}$. Because these winds are initiated and then continuously accelerated by the absorption of photospheric photons in spectral lines, the velocity v_∞ reached at very large distances from the star, where the radiative acceleration approaches zero because of the geometrical dilution of the photospheric radiation field, corresponds to the maximum velocity of the stellar wind. If we assume that winds are stationary and spherically symmetric, then the equation of continuity yields at any radial coordinate R in the wind

$$\dot{M} = 4\pi R^2 \rho(R) V(R). \tag{1}$$

$V(R)$ and $\rho(R)$ are the velocity field and the density distribution, respectively. We regard those as local stellar wind parameters. A determination of the global parameters from the observed spectra is only possible with realistic assumptions about the stratification of the local parameters. We discuss this point carefully in the sections below.

From Equation 1 it is evident that the average mass density $\bar{\rho}$ (Equation 2 with R_* the photosperic radius) is another global parameter related to the observability of stellar winds. For instance, the optical depths of a metal resonance line [with a line absorption coefficient $\propto \rho(R)$] or a recombination line such as H_α [with a line absorption coefficient $\propto \rho^2(R)$] are proportional to Q_{res} and Q^2, respectively, defined as

$$\bar{\rho} = \frac{\dot{M}}{4\pi R_*^2 v_\infty}, \qquad Q_{res} = 4\pi \bar{\rho}\frac{R_*}{v_\infty}, \qquad Q^2 = (4\pi)^2 \bar{\rho}^2 \frac{R_*}{v_\infty}. \tag{2}$$

In the following we describe the diagnostic methods used to obtain these global parameters.

2.2 General Strategy

It is important to realize that the determination of stellar wind properties from observed spectra is never straightforward and simple. Global stellar wind parameters as defined in the previous subsection are not direct observables. Their determination relies on stellar atmosphere models, including the hydrodynamic effects of winds, which form the basis for radiative transfer calculations to be compared with the observations. In other words, whenever we discuss "observations of stellar winds," we must be aware of the fact that those observed stellar wind properties are already the result of diagnostic techniques based on a substantial amount of theoretical modeling. Depending on the degree of sophistication of the underlying model, but also depending on which part of the spectrum is used for the determination of which stellar wind quantity, the reliability of the results obtained can be vastly different.

A few examples may illustrate this point. The determination of the mass-loss rate $\dot{M}$ from a significant "thermal" radio excess in the spectral energy distribution (for details, see discussion in the corresponding subsection) caused by the Bremsstrahlung of the stellar wind envelope requires only a very simple analytical stellar wind model and radiative transfer calculation, because the thermal free-free and bound-free emission of a plasma is a simple process as long as the wind is homogeneous (not clumpy), spherically symmetric, and stationary. Mass-loss rates obtained in this way are usually regarded as sufficiently accurate. However, even such reliable "observations" may be severely affected in a systematic way as soon as winds are clumpy and deviate from spherical symmetry.

In such a situation, an alternative would be the analysis of the P-Cygni profiles of ultraviolet (UV) metal resonance lines. One can show that for these lines the influence of inhomogeneity and clumpiness is less important (see Section 5). However, to deduce $\dot{M}$, one needs an accurate estimate of the element abundance and the degree of ionization of the ion producing the analyzed P-Cygni line. In particular for the latter, a tremendous effort in terms of non-LTE multi-level and multi-line atmospheric modeling is necessary to obtain results (see Pauldrach et al 1994, 1998; Taresch et al 1997; Haser et al 1998), which are still affected by open questions such as the contribution of ionizing radiation from shocks embedded in the stellar wind flow. In consequence, mass-loss rates determined in this way are usually regarded as less reliable.

A discussion of observed global stellar wind parameters, therefore, requires a strategy concerning the underlying atmospheric models used for the diagnostics. On the one hand, for the analysis methods applied, the models have to be sophisticated enough so that the results appear to be reliable; on the other hand, the models have to be simple enough so that a large number of objects can be analyzed and discussed differentially in a parameter space of reasonable dimensions.

As a result of this consideration, the "standard model" for stellar wind diagnostics is rather simple. A stationary, spherically symmetric smooth stellar wind obeying the equation of continuity (Equation 1) is adopted with a velocity field in the supersonic region of the wind of the form

$$V(R) = v_\infty \left(1 - b\frac{R_*}{R}\right)^\beta, \qquad b = 1 - \left(\frac{V(R_*)}{v_\infty}\right)^{\frac{1}{\beta}}. \tag{3}$$

The constant b fixes the velocity at the inner boundary of the wind to the prespecified value $V(R_*)$, which is usually of the order of the isothermal sound speed. Velocity fields of this form are predicted by the theory of radiation-driven winds (see Castor et al 1975, Pauldrach et al 1986), and their use for stellar wind diagnostics is justified a posteriori by the quality of the line profile fits achieved (cf Section 2.4). v_∞ and β are treated as fit parameters to be determined from the spectrum. The selected value of $V(R_*)$ has little influence on the computed spectrum and is of minor importance.

A simple model defined by Equations 1 and 3 is usually sufficient to calculate a UV resonance line or to make a prediction about the emitted flux at radio wavelength. Of course, one additionally needs to specify the electron temperature in the wind and, for the UV diagnostics, the incident radiation field at the inner boundary. For the latter, the photospheric flux is chosen (usually from a photospheric model), and for the former a value equal to or somewhat smaller than the effective temperature of the star is adopted, assuming that the wind is in radiative equilibrium.

Although this simple approach is entirely adequate to analyze UV resonance lines with regard to the velocity field and opacity structure (Section 2.4.1) and radio fluxes with regard to mass-loss rates (Section 2.4.4), a more complex treatment is needed as soon as lines and continua in the optical and infrared are included in the analysis. In these cases, the spectral information usually originates simultaneously from the quasi-hydrostatic photospheric layers below the sonic point and from the stellar wind layers above. As has been demonstrated by Gabler et al (1989), such a situation requires the use of "unified model atmospheres," which has now become the standard treatment for model atmospheres of hot stars with winds. Unified models are stationary, in non-LTE and in radiative equilibrium, they are spherically extended, and they yield the entire sub- and supersonic atmospheric structure, either taking the density and velocity structure from the hydrodynamics of radiation-driven winds or adopting a smooth transition between an outer wind structure, as described by Equations 1 and 3 and by a hydrostatic stratification. They can be used to calculate energy distributions simultaneously with "photospheric" and "wind" lines, and most important, they can treat the multitude of "mixed cases," where a photospheric line is contaminated by wind contributions. This concept turned out to be extremely fruitful for the interpretation of hot star spectra and has led to the rapid development of significant refinements and improvements in parallel by several groups. Basic papers describing the status of the work in the different groups are those by Sellmaier et al (1993), Schaerer & Schmutz (1994), Schaerer & de Koter (1997), Hillier & Miller (1998), Santolaya-Rey et al (1997), Taresch et al (1997), and Pauldrach et al (1994, 1998). For the investigation of global stellar wind parameters from optical spectra or energy distributions, model atmospheres of this type are preferred (see discussion below).

At this stage, a short comment on the reliability of these "standard wind models" is necessary. It is certainly true that there is significant evidence for nonstationarity, clumpiness, shocks, and deviations from spherical symmetry and radiative equilibrium (see Moffat et al 1994, Kaper & Fullerton 1998, Wolf et al 1999). There are two reasons why these phenomena are ignored in the standard diagnostics. First, the amplitudes of deviations from the smooth stationary model used are normally not very large [even in the cases where spectral variability looks dramatic (see Kudritzki 1999a)], thus the standard analysis is thought to yield reliable average models of the stellar winds. Second, for obvious reasons, the appropriate inclusion of deviations from the standard model requires a significant additional effort in the diagnostics and a development of new radiative transfer methods, a task not yet completed. A discussion of the possible effects is given in Sections 4 and 5.

2.3   Stellar Parameters from Photospheric Diagnostics

To discuss the physical connection of global stellar wind parameters with the stellar properties, one needs an estimate of the stellar parameters, such as effective temperature T_{eff}, luminosity L, photospheric radius R_*, gravity $\log g$, chemical composition, etc. There are two ways to proceed. The first is to adopt general calibrations of effective temperature and bolometric correction with spectral type and to determine L, R_*, and T_{eff} from de-reddened photometry, distance, and spectral type. The stellar mass follows then from a comparison with evolutionary tracks in the HRD.

The second and more accurate way relies on model atmospheres to determine T_{eff} and $\log g$ as the most fundamental atmospheric parameters by fitting simultaneously two sets of spectral lines, one depending mostly on T_{eff} and the other on $\log g$. For the former, one uses line profiles of different ionization stages ("ionization equilibrium") and for the latter the higher (less wind-contaminated) Balmer lines. Fit curves in the ($\log g$, $\log T_{\text{eff}}$)-plane along which the calculated line profiles of the two different sets agree with the observations then yield effective temperature and gravity, together with an estimate of the uncertainties. For O-stars, the HeI/II ionization equilibrium is normally used (Kudritzki 1980; Simon et al 1983; Bohannan et al 1986; Voels et al 1989; Kudritzki et al 1983, 1992; Herrero et al 1992, 1999), which is also applied in the case of very early B-supergiants (Lennon et al 1991, McErlean et al 1998). For B-supergiants of later spectral type, T_{eff} is obtained from the SiII/III/IV ionization equilibrium (McErlean et al 1999), and for A-supergiants the MgI/II equilibrium has proven to be reliable (Venn 1995, 1999; McCarthy et al 1995, 1997). The model atmosphere approach will also yield the chemical composition of the stars from a fit of the complete line spectrum (see references above; see also Taresch et al 1997, Haser et al 1998). Using the distance modulus, the stellar radius is then obtained from the model atmosphere flux and de-reddened photometry, which then yields the luminosity. The stellar mass follows from radius and gravity.

A few problems with this approach need to be discussed. Herrero et al (1992) in their quantitative spectroscopic study of a large sample of galactic O-stars detected a systematic "mass discrepancy." They found that for evolved O-stars approaching the Eddington-limit, the stellar masses derived from evolutionary tracks and stellar luminosities are significantly larger than the masses obtained from spectroscopic gravities. Which masses are correct is still an open question. The most recent calculations by Langer & Heger (1998), Maeder (1998), and Meynet (1998) indicate that stellar evolution, including rotation and rotationally induced interior mixing, would enhance the luminosity significantly for any given mass. In addition, enhanced mass loss along the evolutionary track would further reduce the mass and bring the evolutionary masses into agreement with the spectroscopic and wind masses (Langer et al 1994). On the other hand, because Herrero et al used hydrostatic non-LTE models in their study and neglected wind contamination of the hydrogen and helium lines (Gabler et al 1989, Sellmaier et al 1993, Puls et al 1996), their spectroscopic masses may have been systematically underestimated. In addition, the neglect of metal line blanketing in their analysis could have caused a small effect (see Lanz et al 1996; but see also Hubeny et al 1998).

It is important to note here that this uncertainty of stellar mass depending systematically on the approach used for the mass determination will introduce a systematic effect in the relationship between terminal stellar wind velocity and escape velocity from the stellar photosphere discussed in Section 3.

Another systematic uncertainty affects the effective temperature of O-stars at the earliest spectral types (O3 and O4), which in the classical work (see references above) relies on the strength of one single, very weak He I line ($\lambda 4471$). Blanketing (Hubeny et al 1998), but also such exotic effects as noncoherent electron scattering of He II resonance lines in the EUV (extreme ultra-violet) (see Santolaya-Rey et al 1997), may have significant influence on this line. The introduction of new effective temperature indicators for these objects, such as metal lines in the optical and UV, leads to results that are contradictory at the moment (see de Koter et al 1998; but see also Taresch et al 1997, Pauldrach et al 1994). For a determination of the luminosity of these objects, a careful investigation of this aspect is urgently needed.

A third important uncertainty concerns the helium abundance of late O- and B-supergiants. The work by Lennon et al (1991) and Herrero et al (1992) led to the conclusion that the helium abundance in the atmospheres of such stars is significantly enhanced, indicating the presence of CNO-processed material mixed from the stellar interior. Although evolutionary tracks with rotational mixing support this result (see references above), recent analyses by Smith & Howarth (1998) and McErlean et al (1998) indicate that the helium enhancement is probably much smaller if the non-LTE radiative transfer calculations include the effect of photospheric microturbulence of the order of 10 km/s in the rate equations. The problem is not completely solved at the moment and needs further investigation. For the determination of effective temperatures and mass-loss rates, this uncertainty may turn out to be important (see Section 3).

An interesting new route to determine stellar parameters is infrared (IR) spectroscopy. Pioneering IR spectral classification work by Hanson and colleagues (1994, 1996, 1997) has demonstrated that K-band spectroscopy allows the determination of spectral type and luminosity class and, therefore, the determination of effective temperature and gravity in such completely analogous ways as optical spectroscopy. This new technique proves to be extremely valuable for massive hot stars embedded in highly dust-obscured regions of star formation, such as ultracompact or normal HII regions (Hanson 1998, Conti & Blum 1998, Drew 1998) or in the galactic center (Najarro et al 1994, 1997; Figer et al 1998). Its impact on the determination of global stellar wind parameters is discussed in Section 2.4.3.

2.4 Stellar Wind Diagnostics

In principle, two types of lines are formed in a stellar wind: P-Cygni profiles, with a blue absorption trough and a red emission peak; and pure emission profiles or absorption lines refilled by wind emission. The difference is caused by the reemission process after the photon has been absorbed within the line transition.

If the photon is immediately reemitted by spontaneous emission, then we have the case of line scattering with a source function proportional to the geometrical dilution of the radiation field (optically thin case) or an even steeper decline (roughly $\propto r^{-3}$, optically thick case), and a P-Cygni profile will result.

If the reemission occurs as a result of a different atomic process, for instance after a recombination of an electron into the upper level, after a spontaneous decay of a higher level into the upper level or after a collision, then the line source function will possibly not dilute and may stay roughly constant as a function of radius so that an emission line (or an absorption line weaker than formed by purely photospheric processes) results.

Typical examples for P-Cygni profiles are UV resonance transitions connected to the ground level, whereas excited lines of an ionization stage into which frequent recombination from higher ionization stages occurs will produce emission lines, as is the case, e.g., for H_α in O-supergiants.

Both types of lines can be used in principle to determine $\dot{M}$, v_∞, and the shape of the velocity law β. The advantages and constraints of the different methods are discussed in the following, as are the diagnostic possibilities related to continuous wind emission in the IR and radio domain.

2.4.1 UV Resonance and Subordinate Lines

The specific shape of typical UV P-Cygni profiles in hot stars (e.g. from the resonance lines of CIV, NV, SiIV, OVI, etc) (see Walborn et al 1985) can be explained as follows. The profile consists of two components: Because of its interaction with wind material in front of the stellar disk, photospheric radiation is scattered out of the observer's line of sight, and an absorption trough is formed. Because the material is moving toward the observer, the absorption is blueshifted (from zero to a frequency corresponding to some maximum velocity; see below). Superimposed onto the absorption trough is the second component, an emission profile that is formed in the wind lobes and again in front of the disk. This wind emission consists of photons that have been absorbed previously and are scattered into the observer's line of sight. Because the reemission occurs in both hemispheres of the wind (positive and negative projected velocities!), the corresponding profile is almost symmetrical with respect to line center. Note, however, that the wind emission from material behind the photospheric disk is not visible ("occulted region"). Consequently, the symmetry is slightly distorted, and the emission profile on the red side is weaker than on the blue one. By adding the blue absorption trough and the emission profile, the typical shape of a P-Cygni line is readily understood.

P-Cygni lines are usually analyzed by means of the so-called SEI method (Sobolev plus exact integration; cf Lamers et al 1987), which is based on the notion (Hamann 1981a) that the profile of such a line can be simulated with high precision if the source functions of the components (usually two due to fine-structure splitting of the lower or upper level) are calculated in the Sobolev approximation (for details, see Sobolev 1957, Castor 1970) and only the "formal integral" is done exactly, i.e. accounting for the finite profile width. For purely scattering singlets then, the source function is given by the ratio of "core penetration" to total "escape probability" (Castor 1970), with straightforward generalizations in the case of doublet formation (Hamann 1981a, Olson 1982, Puls 1987, Lamers et al 1987).

The crucial quantity that controls the line formation process (both with respect to absorption as well as for defining the escape probabilities) is the Sobolev optical depth of the specified line,

$$\tau_S(r) = \frac{\bar{\chi}(r) R_* \lambda}{v_\infty dv/dr}; \qquad \bar{\chi}(r) = \frac{\pi e^2}{m_e c} f n_1(r). \qquad (4)$$

$\bar{\chi}$ is the frequency integrated line opacity, λ the wavelength, f the oscillator-strength, and n_1 the lower occupation number of the transition, neglecting stimulated emission. Here and in the following, r is the radius in units of stellar radius R_* and $v(r)$ the velocity in units of terminal velocity v_∞.

It is important to understand that the line optical depth in expanding atmospheres or winds has an entirely different character than in hydrostatic photospheres. In winds, the interaction photon/absorbing atom is restricted to a geometrically very narrow region of thickness $\Delta r = v_{th}(v_\infty dv/dr)$ (the Sobolev length). Thus, optical depth is a local quantity in stellar winds, described to a good approximation by Equation 4. By expressing the occupation number in terms of local density $\rho = \dot{M}/(4\pi R_*^2 r^2 v_\infty v)$, this quantity is given by

$$\tau_S(r) = \frac{k(r)}{r^2 v dv/dr}; \qquad k(r) = E(r) X(r) \frac{\dot{M}}{R_* v_\infty^2} \frac{(\pi e^2)/(m_e c)}{4\pi m_H} \frac{A_k}{1 + 4Y_{He}} f\lambda, \qquad (5)$$

if E is the excitation factor of the lower level, X the ionization fraction, A_k the abundance of the element with respect to hydrogen, and Y_{He} the helium abundance.

From the above equation, it is obvious that the appropriate scaling quantity is given by $\dot{M}/(R_*v_\infty^2) = Q_{res}$ (Equation 2), as long as the ionization fraction has no external ρ dependence, i.e. as long as the ground-state occupation follows the density stratification.

Because we have assumed a velocity law of the form $V(r) = v_\infty(1 - b/r)^\beta$ (cf Section 2.2), the quantities that can be derived at maximum by a fit to observed P-Cygni lines are v_∞, β, and $k(r)$, so that, for known helium abundance, the information with respect to mass loss shows up only in the product $E(r)X(r)A_k\dot{M}/R_*$. In consequence, mass-loss rates can be determined from P-Cygni lines only if the ionization/excitation fraction (as function of radius or a mean value; see below) of the specified ion/level is known as well as the abundance!

Saturated Profiles   Saturated profiles are often found for CIV, NV in early O-stars and SiIV in late O-supergiants and are related to a large optical depth $\tau_S \geq 3$ everywhere in the wind. In contrast to photospheric profiles, this condition can be more easily achieved for wind lines due to their additional dependence on the velocity gradient dv/dr (Equation 4). For lines with a rather constant ionization/excitation fraction throughout the wind, the variation of line optical depth depends only on the steepness of the velocity law,

$$\tau_S \propto (r^2 v dv/dr)^{-1} \propto v(r)^{\frac{1}{\beta}-2}. \tag{6}$$

For typical O-star values, $\beta = 0.7 \ldots 1.5$ (see below), the variation is only mild, of the order of 10–100 in the region between the sonic point and infinity. In consequence, if a line has a large optical depth in the sonic region, the chances are good that this is also the case throughout the wind.

The absorption trough of such a saturated profile shows nearly zero intensity for all frequencies corresponding to velocities $v = 0 \ldots v_\infty$, and the profile shape depends exclusively on the emission profile, i.e. on the source function in the optically thick case. As can be easily shown, however, for large optical depths this quantity depends only on the velocity field (with an asymptotic behavior $\propto r^{-3}$), because the aforementioned ratio of escape probabilities becomes independent of line opacity, i.e independent of any information concerning occupation numbers. Consequently, saturated profiles allow in principle an easy measurement of terminal velocities from the frequency position of the blue absorption edge as well as a determination of β from the shape of the emission peak; however, only upper limits for the product of $(EXA_k\dot{M}/R_*)$ are possible.

Figure 1 gives an example for the determination of v_∞ by this method. High-precision measurements with an accuracy of 5% can be obtained in this way (for details and related methods, see Hamann 1981a,b; Lamers et al 1987; Groenewegen & Lamers 1989; Howarth & Prinja 1989; Groenewegen et al 1989; Haser et al 1995; Haser 1995), and typical values $\beta = 0.7 \ldots 1.5$ are found for OB-stars with supergiants having a clear tendency toward higher β. For A-supergiants, values as high as $\beta = 3 \ldots 4$ can be found (Stahl et al 1991) (Section 2.4.2).

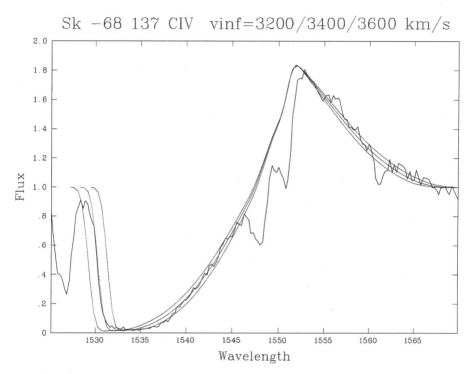

Sk −68 137 CIV vinf=3200/3400/3600 km/s

Wavelength

Figure 1 Radiative transfer calculations for the saturated ultraviolet CIV resonance doublet in the HST FOS spectrum of the large Magallanic cloud O3-star Sk-68° 137 for three different values (3200, 3400, and 3600 km/s) of v_∞. From Kudritzki (1998).

Unsaturated Profiles   Unsaturated profiles arise when τ_S becomes of the order of unity or smaller somewhere in the wind and decent diagnostics concerning $k(r)$ or $k(v)$ (Equation 5), respectively, are possible in the corresponding region: Both the depth and shape of the absorption trough as well as of the emission component [with source function(s) $\propto r^{-2}$] become dependent on line opacity. Note, however, that in this case, the terminal velocity can no longer be determined because the absorption extends only to a certain maximum velocity $<v_\infty$. Additionally, the deduction of β becomes impossible because of the intrinsic coupling of ionization fraction and velocity law in the definition of τ_S. In order to derive a unique solution for $k(v)$, the parameters v_∞ and β have to be derived independently, e.g. from saturated profiles.

Line fits to unsaturated profiles can be obtained by either parameterizing the optical depth as a function of velocity (Lamers et al 1987, Groenewegen & Lamers 1989), by parameterizing the run of $k(v)$ (Hamann 1981a), or by a parameter-free analysis based on a Newton-Raphson method (Haser et al 1994, Haser 1995). Despite the impressive quality of those fits, their diagnostical value with respect to $\dot{M}$ is marginal because the ionization fractions have to be predescribed and depend on complicated details of theoretical model computations (cf Section 2.2). In addition, a rather precise determination of the element abundances is needed.

Recently, however, Lamers et al (1999a) have suggested a significant improvement with respect to this dilemma: For unsaturated lines, it is easy to derive the column density of absorbers between velocities v_1 and v_2 from the actual fit quantity $\tau_S(v)$ itself, as well as the product of a mean ionization/excitation factor $\langle q_k \rangle$ with $\dot{M}A_k/R_*$ in dependence of this column density (cf Howarth & Prinja 1989). Extensive tables of column densities and products of $(\dot{M}\langle q_k \rangle)$ for large samples of O-stars have been determined by Howarth & Prinja (1989) (pure Sobolev analysis) and Haser (1995).

If these data are now combined with independent determinations of mass-loss rates (and abundances and stellar radii are known with reasonable accuracy), empirical mean ionization factors $\langle q_k \rangle$ can be determined. Lamers et al (1999) used the results from the UV diagnostic studies by Groenewegen & Lamers (1989) and Haser (1995), in conjunction with the latest results for mass-loss rates from H_α (Section 2.4.2) and thermal radio emission (Section 2.4.4), and derived empirical ionization/excitation fractions for all relevant UV transitions for selected O-stars covering the complete domain of spectral types and luminosity classes. Introducing calibrations of $\langle q_k \rangle$ with effective temperature, it should then be possible to derive mass-loss rates from UV profiles alone in future investigations (e.g. from FUSE spectra), provided the abundances are known. In agreement with earlier results by Haser (1995), Lamers et al (1999a) showed that SiIV is best suited for this kind of approach.

Even earlier, it was pointed out by Walborn & Panek (1984) that this line can be used as an indicator for the stellar luminosity, an effect theoretically explained by Pauldrach et al (1990) that relates to the fact that SiIV is always a trace ion in the O-star domain, with SiV (inert gas configuration) being the dominant ion. Thus, by means of a simplified non-LTE argumentation (e.g. Abbott 1982), it is easy to show that $X(r)$ has an explicit dependence proportional to density, namely $X \propto n_e/W$ with electron-density n_e and dilution factor W. For smooth winds, then, $X(v) \propto \bar{\rho}/v$ and $k(v)$ scales with Q^2 instead of Q_{res}. This increases the sensitivity of SiIV both with respect to $\dot{M}$ (see above) as well as with luminosity, anticipating the tight coupling of mass loss with luminosity in radiatively driven winds (Section 3 and 4). Note finally that the scaling of SiIV resembles that of H_α discussed in the next subsection.

Problems   By means of the discussed line fits, precise "measurements" of v_∞, β, v_{turb}, and column densities are possible, at least as long as one of the strategic lines is saturated and the contamination of unsaturated profiles with lines formed already in the photosphere is accounted for correctly (see Haser 1995).

The most severe problem of the above procedure is related to the presence of so-called black troughs in saturated profiles, which are extended regions (corresponding to several hundreds of kilometers per second) in the blue part of the profiles with almost zero residual intensity (cf Figure 1). Without further assumptions, these troughs cannot be fitted because in the usual line formation process, only a very small frequency range (of the order of v_{th}) with zero intensity is created. Furthermore, the observed blue absorption edges are much shallower than those resulting from such simple simulations. To overcome both problems, Hamann (1981a,b) introduced a highly supersonic "microturbulence" of the order of Mach 10 present in the entire wind. Because in this case the intrinsic absorption profile becomes much broader, the observed troughs as well as the blue edges can be simulated (for a comprehensive discussion, see Puls 1993). Additionally, the position of the emission peak is redshifted, in agreement with most observations (see also Section 2.4.2). Meanwhile, the assumption of a constant turbulence (used also by Groenewegen & Lamers 1989) has been relaxed (cf Haser 1995) because the justification of a highly supersonic velocity dispersion close to the photosphere is physically problematic, as well as excluded from the non-LTE analysis of photospheric lines in hot stars, resulting in values of v_{turb} below the speed of sound (e.g. Becker & Butler 1992). Thus, in order to fit the profiles, a linear increase of v_{turb} as function of the underlying velocity law is assumed, starting at roughly sonic velocities up to a maximum value v_{ta}. The actual values are found in parallel with the determination of v_∞ and β from line fits to saturated profile, and typical values for v_{ta} are of the order of $0.1 v_\infty$.

In parallel to the introduction of a velocity dispersion as described above, Lucy (1982a, 1983) suggested an alternative explanation for the black trough generation by means of enhanced backscattering in multiply nonmonotonic velocity fields. Hydrodynamical simulations of time-dependent radiatively driven winds (Owocki et al 1988, Feldmeier 1995) indicate the possibility of such nonmonotonic flows (cf Section 5), and detailed UV line formation calculations on the basis of such models have shown that black troughs are actually created in the profiles of such models (Puls et al 1993a, 1994; Owocki 1994).[1]

2.4.2 Optical Lines

Despite the continuous effort to derive mass-loss rates from UV P-Cygni lines, the basic problem remains that one has to know the mean ionization/excitation fractions as well as the element abundances accurately in order to derive reliable numbers. Fortunately, there is an alternative way to determine $\dot{M}$ using the strength of the stellar wind emission in H_α as an indicator of mass-loss rates, where the ionization correction for H_α is much simpler and the hydrogen abundance is usually much less uncertain.

First quantitative non-LTE calculations dealing with the wind emission of hydrogen (and helium lines) go back to Klein & Castor (1978). Then, Leitherer (1988), Drew (1990), Gabler et al (1990), Scuderi et al (1992), and Lamers &

---

[1] Note that the determinations of v_∞ by Howarth & Prinja (1989) rely in part on the presence of black troughs in saturated CIV lines.

Leitherer (1993) realized the potential of H_α for the determination of mass-loss rates and derived first results for a variety of hot stars. Puls et al (1996) developed a fast method to obtain mass-loss rates from H_α profiles of O-type stars, avoiding some systematic errors inherent in the approach of Leitherer (1988) and Lamers & Leitherer (1993). In the following, we consider some basic issues, including the analysis of H_α from A and late B supergiants. For a comprehensive discussion with a number of detailed profile fits, we refer the reader to Puls et al (1996), McCarthy et al (1997, 1998), Puls et al (1998), and Kudritzki et al (1999).

As pointed out in the introduction to this section, the basic difference of H_α and related lines (including the IR emission lines discussed in the next section) relative to P-Cygni lines concerns the run of the source function, which is rather constant throughout the wind. The latter fact is true as long as those lines are fed predominantly from the spontaneous decay of the upper level or recombinations into both levels, so that the involved occupation numbers are almost in LTE with respect to each other, i.e the source function is almost Planckian. This condition is met for O-stars and early B-stars. For cooler temperatures, however, the feeding of the involved levels from lines of the Lyman series becomes relevant, until in the A-supergiant domain, these lines and the Lyman continuum become optically thick and the corresponding transitions are in detailed balance. Then, the second level of hydrogen (which is the lower one of H_α) becomes the effective ground state of HI and the line behaves as a scattering line, thus displaying a typical P-Cygni signature.

Diagnostic Potential   Because neutral hydrogen is almost always a trace ion in the winds of OBA stars (excluding the outermost wind and some narrow photospheric recombination region in the A-star domain), the same argumentation as made above for SiIV applies: The opacity of H_α depends on ρ^2 (OB-stars) and on ρ^2/W (A-supergiants), so that the actual quantity that can be derived by line fits is Q^2, which manifests its sensitivity on and its diagnostic value for the determination of $\dot{M}$. Two instructive examples are given in Figures 2 and 3.

H_α Emission in O/Early B Stars   Because of the thermal character of the wind emission and the large emitting volume of the wind, H_α profiles can exhibit enormous equivalent widths and emission peaks, provided the mass-loss rate is large enough. As shown by Puls et al (1996), the equivalent width (absolute value) of the wind emission scales with $Q^{3/2}v_\infty$ for (predominantly) optically thin and with $Q^{4/3}v_\infty$ for optically thick emission. Because of this overlinear dependence on mass loss, the achieved accuracies for $\dot{M}$ (or, more correctly, for Q) from lines with large wind emission are enormous and can reach values below 10%. Additionally, because of the strong dependence of opacity on $\rho^2 \propto (r^4v^2)^{-1}$, the shape of the velocity field (i.e. β) can be determined from the shape of the emission peak. By this method, Puls et al (1996) were able to derive typical values $\beta \approx 1$ for O-type Supergiants, and the results for Q in this domain are in fair agreement with those from the simplified approach used by Leitherer (1988) and Lamers & Leitherer (1993).

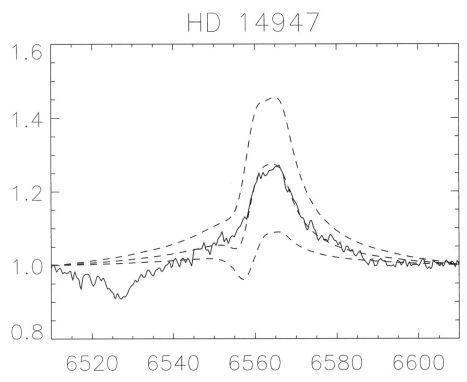

Figure 2 H_α line profile of the O 5 Iaf^+-supergiant HD 14947 compared with line fits as outlined by Puls et al (1996), adopting 10, 7.5, and 5.0 × 10^{-6} M$_\odot$/year, respectively, for the mass-loss rate. From Kudritzki (1998).

From the above scaling relations, however, it is also obvious that the emission from thin winds is small and, moreover, almost "hidden" in the normal photospheric profile with a shape dominated by rotational broadening. In these cases, then, detailed calculations on the basis of unified model atmospheres accounting for the transition to photospheric layers (cf Section 2.2) as well as an exact line formation accounting for the finite profile width are required to derive reliable numbers. Without doing this, Q^2 is easily overestimated by more than one dex.

In addition to these more theoretical considerations, even with the most elaborate methods, uncertainties up to a factor of 2 in $\dot{M}$ can arise if the absorption profile is only marginally filled and the shape of the velocity field in the lower wind part is not known (e.g. from UV resonance lines): Contrary to the case of strong emission lines, β cannot be derived in these cases because the rotationally broadened absorption corrupts all diagnostic clues concerning the wind emission. (The strong dependence of the derived values of $\dot{M}$ on β for marginal wind emission can be easily seen from the run of the radial optical depth in the line, $\tau_S \propto Q^2 v(r)^{\frac{1}{\beta}-3}/r^2$.)

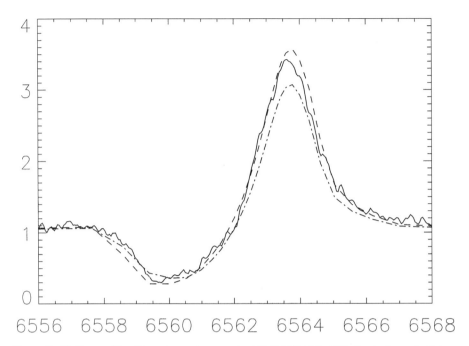

Figure 3 H_α line profile of the extreme A-supergiant 41-3654 (A3 Ia-O) in the Andromeda Galaxy M31 taken with the Keck HIRES spectrograph compared with two unified model calculations adopting $\beta = 3$, $v_\infty = 200$ km/s and $\dot{M} = 1.7$ and 2.1×10^{-6} $M_\odot$/year. Note the P-Cygni profile shape of H_α. From McCarthy et al (1997).

H_α Emission in A/Late B Supergiants   Because the H_α opacity in the winds of these stars scales with ρ^2/W and the source function is dominated by line scattering, the corresponding profiles are of P-Cygni type and usually visible until up to a frequency corresponding to v_∞, in contrast to the O/B star case, where this is true only for the most luminous objects. The latter fact also allows v_∞ to be derived in parallel with the other quantities and thus enables a complete wind analysis in the optical alone: Because of the additional factor $1/W \approx r^2$, compared with the O-star case from above, and the lower temperature, which increases the ionization fraction of HI significantly, the optical depth in H_α remains large, up to v_∞ in most cases, similar to the behavior of strong UV resonance lines. Analogously, β can be derived with significant precision from the profile shape.

A detailed parameter study (based on the unified models by Santolaya-Rey et al 1997) as well as the required fit procedure is given by McCarthy et al (1997, 1998). Kudritzki et al (1999) have used this method to obtain the wind parameters for a large sample of galactic BA-supergiants. In accordance with earlier work by Stahl et al (1991), the derived velocity fields are significantly flatter than for O-stars, with typical values $\beta = 3 \ldots 4$. It is interesting that the observed slope of the blue absorption edges as well as the position of the emission peaks made it necessary also here to assume the presence of a velocity dispersion, of the order of 10...20 km/s. Note that the terminal velocities of A-type supergiants are a factor of 10 smaller than for O-type stars (cf Section 3).

Of course, the derived values have to be consistent with the observations/simulations of other lines, the most important being the hydrogen Balmer lines. Also, these lines (especially H_β) are contaminated by wind emission, and only a simultaneous fit of all lines with identical parameters (where H_γ provides the required information concerning $\log g$, cf Section 2.3) produces results that can be trusted.

Problems   Globally, the errors in Q are of the order of 30%, with smaller values for dense and larger values (the β problem) for thin winds. The agreement with mass-loss rates from thermal radio emission (Section 2.4.4) is satisfying (cf Lamers & Leitherer 1993, Puls et al 1996). However, there are also a number of problems, which are illustrated in the following.

In O/early B-stars, the H_α line is contaminated by a blue-ward blend of HeII (separated from the line center of hydrogen by roughly 2 Å corresponding to 100 km/s), which in a number of lines mimics the presence of a P-Cygni type profile, because it is mostly in absorption. In those cases where this blend is strong, present-day simulations are partly unable to fit this component (see Herrero et al 2000). The recipe here is to concentrate on the red wing of the H/He complex, which should give more reliable results because of its weaker contamination by the helium blend, compared with the blue side.

Additionally, a certain variability, which can be severe in the blue absorption troughs of A-star profiles (cf Kaufer et al 1996 and the discussion in Section 5), influences the derived results. Test calculations (see Kudritzki 1999a) have shown that these variations indicate a significant variation for the obtained values of v_∞ and $\dot{M}$ as a function of time. The product of both, i.e. the momentum rate, remains much more constant, however, with a maximum variation of the order of 0.2 dex for the extremely variable wind of HD 92207 (A0Ia).

Another problem results from stellar rotation. The thermal Doppler width of H_α is much smaller than typical rotational velocities of early type stars (of the order of 100 km/s for OB-stars and 30...40 km/s for A-supergiants). Thus, the central emission caused by the wind will certainly be broadened by rotation. Because of conservation of specific angular momentum, the azimuthal velocity in the wind varies according to $v_{\rm rot}(R_*, \theta)/r$, where $v_{\rm rot}(R_*, \theta)$ is the photospheric rotation velocity at co-latitude θ. Thus, for the most important regime of H_α formation ($r = 1...1.5$), both the value of $v_{\rm rot}$ itself as well as its variation (i.e., the differential rotation) should have important consequences for the line shape.

Before we further comment on this problem, note that the diagnostics of UV resonance lines in OB-stars are barely affected by this effect, because the stronger lines react only mildly (if at all) to any reasonable value of $v_{\rm rot} \sin i$. In contrast to usual H_α profiles, these resonance lines are formed throughout the wind, which decreases the ratio of $v_{\rm rot} \sin i$ to total line width (v_∞). Moreover, the average rotational velocities are smaller because $v_{rot} \to 0$ in the outer wind and the intrinsic Doppler width is larger as a result of the velocity dispersion derived. Note also, however, that rotation might influence the global hydrodynamical structure, a problem we consider in Section 4.2.

With respect to the formation of H_α (and the other Balmer lines, of course), it turns out that the presence of rotation has to be accounted for in order to obtain good fits (except for very strong and very broad emission lines, which, in analogy to strong UV resonance lines, are barely influenced). The usual strategy (cf Puls et al 1996, Herrero et al 2000) is to fold the final profile with a rotational curve of width $v_{rot} \sin i$, where this value can be derived from photospheric metal lines. For a number of (fast rotating) objects, it turns out that the value of $v_{rot} \sin i$ indicated by H_α is smaller than for lines less contaminated by the wind, e.g. H_γ. This shows clearly the presence of differential rotation in some average sense, in accordance with test calculations by Petrenz & Puls (1996) based on a rigorous treatment of the problem. In conclusion, some average value $\langle v_{rot} \sin i \rangle$ to be determined in parallel with the other parameters becomes an additional part of the H_α fitting procedure, at least for lines that are neither weak ($\langle v_{rot} \sin i \rangle \to v_{rot} \sin i (R_*)$) nor strong.

Of course, the problems discussed here would be of only minor importance if our general assumption of smooth winds (at least in the lower, line-forming region) were severely violated. Because of the intrinsic dependence of the H_α line opacity on ρ^2, any significant degree of clumpiness would lead our approach to overestimate the mass-loss rates. We come back to this problem in Section 5.

2.4.3 IR Lines

The dramatic progress of IR astronomy in the past decade has opened a completely new window for the systematic investigation of winds around hot stars. In their first paper on unified model atmospheres, Gabler et al (1989) realized the importance of IR emission and absorption lines for a quantitative understanding of the physics of hot star atmospheres and winds and pointed out the potential of this new spectral range. A wealth of observational material is now available (see, for instance, Hanson et al 1996, Figer et al 1997) for quantitative work.

The theoretical framework for the analysis of IR spectra is in principle identical to that of the optical and UV lines. As for those cases, the IR spectra of hot stars also show pure emission lines, P-Cygni profiles, and wind-contaminated absorption lines. However, because the IR continuum of objects with strong winds (contrary to the optical continuum) is formed in the supersonic part of the wind, IR lines sample different depths of a stellar wind and give additional information about the shape of the velocity field. Najarro et al (1997a), Lamers et al (1996), and Hillier et al (1998) have investigated this important effect using high-quality ground-based (optical and IR) and ISO IR spectra.

The winds of hot stars of extreme luminosity and with strong IR emission characteristics in the galactic center have been investigated by Najarro et al (1994, 1997b) and Figer et al (1998). Crowther et al (1998) have provided beautiful diagnostics of late Of-supergiants and WNL stars in order to probe wind-velocity fields, effective temperatures, and blanketing effects in the photospheric EUV continua.

In the case of very thin winds with extremely low mass-loss rates, IR line profiles turn out to be outstanding diagnostic tools for the determination of mass-loss rates, clearly superior to optical lines such as H_α. Because $h\nu/kT \ll 1$ in the IR, stimulated emission becomes important, and non-LTE effects in the line (caused by the presence of the wind) are substantially amplified (Mihalas 1978, Kudritzki 1979), leading to strong emission even for very small mass-loss rates, as has been pointed out by Najarro et al (1998).

2.4.4 IR, Submillimeter, and Radio Continua

Not only the spectral lines but also the IR, (sub)millimeter, and radio continua can be used to analyze the properties of the wind plasma. Actually, they constitute one of the most reliable diagnostic tools for this purpose because they are based on simple atomic/radiation transfer processes, whereas lines usually require consideration of a number of nasty details, such as non-LTE, line broadening, etc.

The basic idea is to investigate the excess relative to the flux predicted by photospheric model atmospheres, which is attributed to free-free ("Bremsstrahlung") and bound-free emission in the winds, and which because of the λ^3 dependence of the corresponding opacities and the resulting geometrical increase of the effective photosphere becomes significant at longer wavelengths. In most cases, the emission arises from thermal electrons (deduced from the spectral index of the radiation, see below); however, there exists also a significant fraction (roughly 30%) of nonthermal emitters (Bieging et al 1989 and references therein; see also Altenhoff et al 1994). In the following, we mainly discuss thermal emission; we briefly consider nonthermal effects at the end of this subsection (see also Section 5).

The pioneering work on IR and radio free-free emission from hot star winds was presented independently by Wright & Barlow (1975) and Panagia & Felli (1975). Both publications showed that the continuum flux S_ν at long wavelengths for an optically thick, spherically symmetric and isothermal envelope expanding at constant velocity scales with $S_\nu \propto \nu^{0.6}$. Moreover, the observed radiation depends only weakly on the electron temperature in the wind and is set essentially by the distance to the star d, the mass-loss rate, and the terminal velocity via

$$S_\nu \propto \left(\frac{\dot{M}}{v_\infty}\right)^{4/3} \frac{\nu^{0.6}}{d^2}, \tag{7}$$

where we have suppressed additional dependences on the chemical composition, ionization stage, and Gaunt-factors. (For effects of different density and temperature stratifications, see Cassinelli & Hartmann 1977, Bertout et al 1985.)

The exclusive use of radio observations, however, is problematic. Because the excess flux (i.e. total flux in the radio domain) scales with $\nu^{0.6}$, OB-stars will be generally weak radio sources (which prohibits the use of this method for extragalactic work), and useful radio observations can be obtained only for nearby stars with dense winds. Moreover, because the spectral index of a nonthermal component is negative (e.g. -0.7 for synchrotron emission), the radio domain can be more easily contaminated than shorter wavelengths, e.g. the (sub)millimeter regime (see Altenhoff et al 1994, Contreras et al 1996).

By concentrating on lower frequencies, however, at least in the IR, the presence of a velocity stratification has to be accounted for, as was pointed out by Wright & Barlow (1975): In contrast to the radio/submillimeter regime, the wind becomes optically thick at velocities well below v_∞. By using a velocity law with a shape that fitted the IR continuum of P-Cygni, Barlow & Cohen (1977) took these problems into account and performed an extensive IR analysis of some 40 OBA supergiants.

The effects of a more general velocity field, together with the inclusion of bound-free opacities and an estimate concerning the influence of electron scattering (important only at very high wind densities, e.g. in the winds of Wolf-Rayets), were discussed by Lamers & Waters (1984a). They expressed the nondimensional excess in terms of a curve of growth, from which the velocity law and the mass-loss rate can be derived simultaneously. Extensive tables of curves of growth as well as the required gaunt factors as functions of chemical composition were published by Waters & Lamers (1984).

Mass-Loss Rates from Radio/Submillimeter Observations   Early radio measurements (mostly with the VLA at 2 and 6 cm) and subsequent determinations of mass-loss rates of O/early B-star winds have been published by Abbott et al (1980, 1981) and Bieging et al (1989). For a number of these stars, Howarth & Brown (1991) have added VLA fluxes at 3.6 cm. These measurements (see also the catalogue compiled by Wendker 1987) provide one of the most complete data sets and have been used, for example, in the comparison with H_α mass-loss rates by Lamers & Leitherer (1993) and Puls et al (1996), as well as for the calibration of "empirical ionization" fractions (Lamers et al 1999a) (see Section 2.4.1). Recently, Leitherer et al (1995) studied a number of (very) dense winds (including two Of-star winds) with the Australia telescope compact area at 3.5 and 6.25 cm, and Scuderi et al (1998) have detected another seven new OB radio sources with the VLA. Radio observations of later spectral types (supergiants between B2 and F8) have been performed by Drake & Linsky (1989) and were used by Kudritzki et al (1999) to compare with corresponding H_α data. Because most of these observations comprise at least two frequency points, and taking into account the 1.3-mm observations by Leitherer & Robert (1991) performed with the Swedish-ESO submillimeter telescope at La Silla, the presence of nonthermal components could be easily detected from the spectral index. Of course, only those objects with purely thermal emission have been used to derive mass-loss rates. In addition to the data by Leitherer & Robert, Altenhoff et al (1994) have observed approximately 20 OB stars at 1.2 mm with the IRAM 30-m telescope at Pico Veleta.

Because of the enormous extension of the radio photosphere for objects with large $\dot{M}$, it is, in principle, also possible to resolve the radio-emitting region.

In this way, White & Becker (1982; see also Skinner et al 1998) resolved the wind of P-Cygni at 6 cm and found an apparently spherically symmetric wind at an electron temperature of 18,000 $\pm$ 2,000 K, i.e., of the order of $T_{\text{eff}} \approx 18,500$ K.

IR Excess   Following the work by Barlow & Cohen (1977), who found significant excess for a number of objects at 10 μ (for some objects down to 2.2 μ), Tanzi et al (1981), Castor & Simon (1983), and Abbott et al (1984b) tried to derive $\dot{M}$ (and partly the shape of the velocity fields) for a number of OB stars from the IR excess in various bands. Bertout et al (1985) analyzed a large sample of early-type stars in OB associations and pointed out that reliable measurements of the excess are only possible for stars with significant wind density ($\dot{M} \geq 10^{-6} M_{\odot}/a$) and/or a flat velocity profile. IRAS observations between 10 ... 100 μ were used to analyze one of the Rosetta-stones among stellar winds, the wind of ζ Pup (O4 If). From these observations, Lamers et al (1984) could exclude older models, claiming a hot or warm base corona (thus again, wind temperature $\approx T_{\text{eff}}$), and derived a mass-loss rate and velocity profile with $\beta = 1$, in agreement with current findings from H_{α}. A recent reanalysis of most of the described data (IR excess and radio fluxes simultaneously) was performed by Runacres & Blomme (1996), who found that the majority of the observations are consistent with current theoretical work. For four of their program stars (including ζ Pup and α Cam), however, they claimed the presence of an additional emission mechanism.[2]

Problems   Generally, the agreement between the radio and IR diagnostics on the one side and the H_{α} analysis on the other is satisfying (cf Lamers & Leitherer 1993, Puls et al 1996, Kudritzki et al 1999). However, there are also problematic cases (see above) connected with too high an IR excess compared with radio and H_{α} measurements. An instructive example is discussed by Kudritzki et al (1999) for the case of HD 53138 (B3Ia), where the H_{α} measurement is consistent with the (upper limit) of the observed radio flux from Drake & Linsky (1989). In contrast, the 10-μ excess derived by Barlow & Cohen (1977), which is also peculiarly high compared with other wavelengths, would result in $\dot{M}$ a factor 20 higher. By investigating the data published in the CDS Catalog of Infrared Observations, Gezari et al (1999) discovered that many stars show a significant scatter in the IR photometry, which might be attributed to a photometric variability of B-supergiants. Thus, the "measurement" of an IR excess (requiring the "subtraction" of combined wind plus photospheric fluxes in the IR and purely photospheric fluxes in the optical/near IR) becomes difficult unless the complete energy distribution is observed simultaneously.

An alternative explanation might be given by the presence of clumping, which, in analogy to the case of H_{α}, would increase the IR excess due the ρ^2-dependence

---

[2]Investigations by F Najarro (private communications) have shown that consistent solutions can be obtained also in those cases with a β-value larger and a mass-loss rate smaller than those derived from H_{α}.

of opacity. (The inconsistency of IR and H_α fluxes would then point to depth-dependent clumping factors; cf Section 5). Simple methods to incorporate clumping in the IR diagnostics have been proposed by Abbott et al (1981) and Lamers & Waters (1984b). A first comparison with time-dependent models has been given by Puls et al (1993b), and on the basis of a simplified "partial shell" description, Blomme & Runacres (1997) have pointed out that a small amount of clumping is sufficient to reconcile the discrepancies found in their previous work.

Nonthermal Emission   As mentioned earlier, there is ample evidence of non-thermal emission in hot star winds. First observational findings were reported by White & Becker (1983) and Abbott et al (1984a), and latest measurements (for WRs) were presented by Chapman et al (1999). Although it seems clear that the nonthermal emission is due synchrotron radiation from relativistic electrons, different agents responsible for the acceleration are discussed in the literature: wind accretion onto compact objects (Abbott et al 1984a), wind-wind collisions in binary systems (Stevens 1995), magnetic reconnection in single or colliding winds (Pollock 1989), and first-order Fermi acceleration in the shocks (randomly) distributed in stellar winds (Chen & White 1994), which are thought to be responsible for the emission of X-rays as well (discussed in the following).

2.4.5 X-Ray Emission

Among the first surprising discoveries of the Einstein observatory was the detection that all O-stars are soft X-ray emitters (Harnden et al 1979, Seward et al 1979). It was soon found that the X-ray luminosity is roughly correlated with the stellar luminosity: $\log L_x/L_{\rm bol} \approx -7 \pm 1$ (Seward et al 1979, Pallavicini et al 1981, Chlebowski et al 1989). The scatter in this relation is very large, indicating a dependence on additional parameters.

The source of the O-star X-ray emission is widely believed to be shocks propagating through the stellar wind (Lucy & White 1980, Lucy 1982b, Cassinelli & Swank 1983, MacFarlane & Cassinelli 1989), where the shocks may result from a strong hydrodynamic instability of radiation-driven winds (see also Section 5). ROSAT and ASCA observations of O- and B-stars have confirmed this interpretation (Hillier et al 1993; Cassinelli et al 1994; Cohen et al 1996, 1997a,b; Berghöfer et al 1997). Assuming a simple model of randomly distributed shocks in a stellar wind, where the hot shocked gas is collisionally ionized and excited and emits spontaneously into and through an ambient "cool" stellar wind with a kinetic temperature of the order of the effective temperature, these authors were able to determine shock temperatures, filling factors, and emission measures. The diagnostic situation differs between B-stars, where the winds are usually optically thin at X-ray wavelengths, and O-stars, where the winds can become optically thick and X-ray radiative transfer is needed. For the latter case, Feldmeier et al (1997a) have developed a refined diagnostic model, including postshock cooling zones for radiative and adiabatic shocks. Kudritzki et al (1996) applied these techniques on a larger sample of O-stars observed with ROSAT and found a dependence of the filling factor of the X-ray emission on $\bar{\rho}$ (which is proportional to the inverse cooling length of the shocks), leading to an additional dependence of $\log L_x$ on the average density in the stellar wind. By a simple scaling analysis of the involved X-ray emission and absorption processes, Owocki & Cohen (1999) showed that the "natural" scaling for optically thin winds is given by $L_x \propto (\dot{M}/v_\infty)^2$ and for optically thick winds by $L_x \propto (\dot{M}/v_\infty)^{1+s}$, if one assumes a radial dependence of the filling factor as $f \propto r^s$. The loose correlation $L_x \propto L_{bol}$ for thick winds can then be reproduced, if one allows for a modest radial falloff of this quantity ($s \approx -0.25 \ldots -0.4$).

An alternative model has been suggested by Feldmeier et al (1997b) based on detailed time-dependent hydrodynamic models, where cloud collisions in an inhomogeneous wind lead to X-ray emission.

For the analysis of stellar winds, the existence of X-ray emission is important for two reasons. First, their direct spectral diagnostics—particularly in view of the X-ray telescope with spectrographs of sufficient resolving power, such as the Chandra satellite—will allow investigation of the limitations of the standard model, discussed in Section 2.1. Here a particularly promising approach is the simultaneous investigation of optical line and X-ray variability (Berghöfer et al 1996, Berghöfer & Schmitt 1994). Second, the X-rays and EUV photons emitted by the shocks severely affect the degree of ionization of highly ionized species, such as CIV, NV, and OVI and the diagnostics of their resonance lines observable in the far UV (see MacFarlane et al 1993, Pauldrach et al 1994, Taresch et al 1997, Haser et al 1998). In this sense, they introduce an additional uncertainty as long as the modeling of the shock emission is not constrained well enough.

3.   RESULTS

The diagnostic methods of stellar winds described in the previous section have been applied to a large number of early-type stars. In this section, we discuss the results of these studies, in particular the empirical correlations of the global stellar wind parameters with stellar parameters. We first discuss terminal velocities, mass-loss rates, and wind momenta of galactic stars and then we investigate the influence of stellar metallicity by a comparison with stars in the Magellanic clouds.

3.1   Terminal Velocities of Winds from Galactic Hot Stars

In two pioneering papers compiling terminal velocities of winds from massive hot stars as a function of stellar parameters, Abbott (1978, 1982) demonstrated for the first time the existence of a correlation with effective temperature and photospheric escape velocity. These papers have induced a number of systematic and comprehensive studies using refined diagnostic methods and analyzing the full UV spectroscopic material available from IUE and HST high (and medium) resolution observations. We mention here the work by Groenewegen et al (1989),

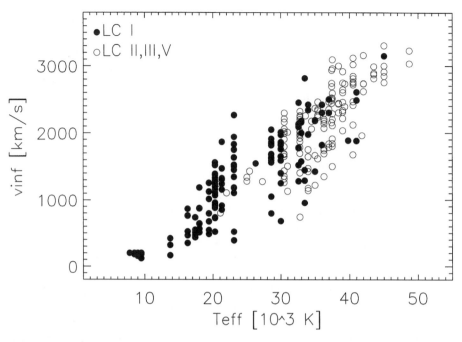

Figure 4 Terminal velocities as a function of effective temperature for massive hot stars of different luminosity classes. The data for B- and O-stars ($T_{eff} \geq 10,000$ K) are taken from Prinja et al (1990), Prinja & Massa (1998), and Howarth et al (1997), who used the effective temperature scale of Humpreys & McElroy (1984) for the conversion of spectral type to T_{eff}. The data for A-supergiants ($T_{eff} \leq 10,000$ K) are from Lamers et al (1995).

Groenewegen & Lamers (1989), Howarth & Prinja (1989), Prinja et al (1990), Lamers et al (1995), Haser (1995), and Howarth et al (1997) on massive stars in the Galaxy. These papers have convincingly confirmed the correlations proposed by Abbott and have provided quantitative coefficients to describe them on a solid statistical basis.

Figure 4 demonstrates that v_∞ and spectral type or T_{eff}, respectively, are correlated. Table 1 gives average values as a function of spectral type. Because terminal velocities can usually be measured with an accuracy of 10% (see the previous section) and the determination of spectral type is usually accurate to one subclass, the scatter must have an intrinsic physical reason. We attribute it to the fact that the dependence of v_∞ on T_{eff} is indirect through the photospheric escape velocity v_{esc}, which becomes smaller with smaller T_{eff} for main sequence stars and when massive stars evolve toward lower temperatures to become supergiants. This means that, on the average, stars with lower effective temperature are expected to have slower winds. However, at a given effective temperature, even stars of similar luminosity class will have different gravities and different photospheric escape velocities, leading to a significant spread of terminal velocities.

TABLE 1   Terminal velocities of massive hot stars: O-stars and B- and A-supergiants[a]

Sp. type	LC I	LC II	LC III	LC V	Sp. type	LC Ia	LC Ib
O3	3000			3200	B0	1450	1700
O4	2400			3000	B0.5	1450	1400
O5	2100		2800	2900	B1	1200	1000
O5.5	2000				B1.5	780	950
O6	2300		2500	2600	B2	540	840
O6.5	2200		2600	2600	B3	490	
O7	2100	2400	2600	2400	B5	270	470
O7.5	1900	2300	2300	2100	A0	160	180
O8	1500	2000	2200	1900	A1	160	
O8.5	2000		2300	1900	A2	170	
O9	1900	2100	2000	1500	A3	180	
O9.5	1700	1720	1600		A5	180	180
O9.7	1700	2000			A8		200

[a]Data for O-supergiants from Prinja et al (1900) and Haser (1995); data for B-supergiants from Prinja & Massa (1998); data for A-supergiants from Lamers et al (1995).

The direct dependence on the photospheric escape velocity is shown in Figure 5, again as a function of $T_{\rm eff}$. It is important to note that the photospheric escape velocity includes the reducing effect of Thomson scattering on the gravitational potential

$$v_{\rm esc} = (2g_* R_* (1 - \Gamma))^{0.5}, \tag{8}$$

where g_* is the photospheric gravity and Γ the ratio of radiative Thomson to gravitational acceleration.

From the definition of $v_{\rm esc}$, it is clear that the determination of stellar escape velocities requires the knowledge of stellar masses and radii. For the construction of Figure 5, this has been done by adopting effective temperature, bolometric correction, and absolute magnitude calibrations as a function of spectral type and luminosity class to determine stellar luminosities, which are then used to estimate stellar masses from evolutionary tracks. In Section 2.3 we mentioned the "mass discrepancy" between masses determined in this way and masses determined from individual NLTE spectral analyses of objects with well-known distances, which yield individual effective temperatures, gravities, and radii with presumably higher precision. Figure 6 compares results obtained by these two alternative methods in the case of supergiants of luminosity class I, where the effects of the "mass discrepancy" is expected to be largest. There are indications of small systematic effects, but the general result is similar. For effective temperatures larger than 21,000 K, the ratio of v_∞ to $v_{\rm esc}$ appears to be roughly constant. At 21,000 K, there is a sudden

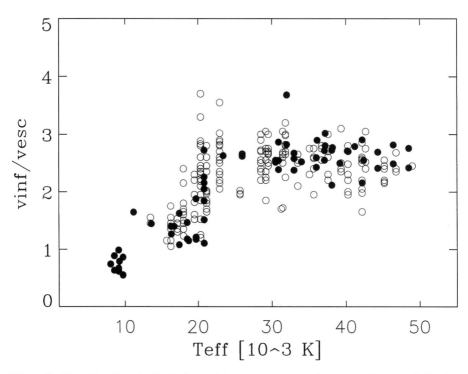

Figure 5   The ratio of terminal velocity to photospheric escape velocity as a function of effective temperature. (*Open symbols*) Prinja & Massa (1998); (*solid symbols*) Lamers et al (1995).

transition toward a significantly smaller but also constant ratio (Lamers et al 1995), which is interpreted as the result of the fact that because of changes in ionization, different ionic species start to drive the wind (for further discussion and references see Section 4). Another step in the v_∞ to v_{esc} ratio might be present at 10,000 K, although the few results from the individual spectroscopic analyses in this temperature range obtained by Kudritzki et al (1999) do not fully confirm such a conclusion. Summarizing the results from Howarth & Prinja (1989), Prinja et al (1990), Lamers et al (1995), Howarth et al (1997), Prinja & Massa (1998), Puls et al (1996), and Kudritzki et al (1999), a reasonable scaling formula for the terminal velocity is given by

$$v_\infty = C(T_{\text{eff}})\, v_{esc}, \qquad \begin{cases} C(T_{\text{eff}}) = 2.65, \ T_{\text{eff}} \geq 21{,}000\,\text{K} \\ C(T_{\text{eff}}) = 1.4, \quad 10{,}000\,\text{K} < T_{\text{eff}} < 21{,}000\,\text{K} \\ C(T_{\text{eff}}) = 1.0, \quad T_{\text{eff}} \leq 10{,}000\,\text{K}. \end{cases} \qquad (9)$$

The accuracy of $C(T_{\text{eff}})$ is roughly 20%.

A completely independent confirmation of the correlation of terminal with photospheric escape velocities comes from the investigation of Central Stars of

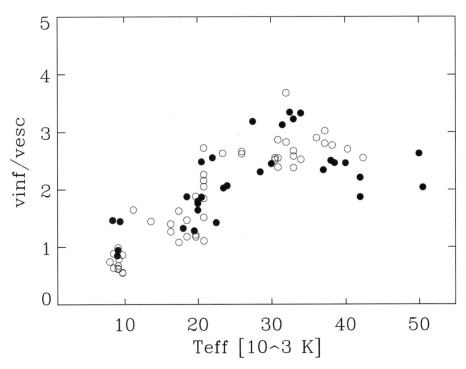

Figure 6   The ratio of terminal velocity to photospheric escape velocity as a function of effective temperature for supergiants of luminosity class I. (*Open symbols*) Lamers et al (1995); (*solid symbols*) Puls et al (1996), Kudritzki et al (1999). Note that for the *solid symbols*, the escape velocities are obtained from detailed non-LTE analyses of individual objects.

Planetary Nebulae (CSPN). Although these objects have a completely different stellar interior and are in a completely different phase of stellar evolution, as post-AGB objects of 0.5–0.7 $M_\odot$, they are certainly (very) hot stars. They evolve from the AGB toward very hot temperatures with roughly constant luminosity. Many of them have optical and UV spectra similar to O-stars. Spectroscopic non-LTE studies carried out by Méndez et al (1988), Perinotto (1993), and Kudritzki et al (1997) and using the same diagnostic techniques as outlined in Section 2 have made it possible to constrain the evolution of these objects and to estimate stellar distances and masses. They have also allowed a discussion of stellar wind properties along the evolution, with constant luminosity and constant mass from the right to the left in the HRD (see also Pauldrach et al 1989).

Figure 7 shows the striking relationship of v_∞ with T_{eff} for these objects. The interpretation of this diagram is straightforward. Because CSPN shrink during their evolution toward higher temperatures, their photospheric escape velocity increases, and the terminal velocity, which depends on v_{esc}, increases as well. Adopting an average mass of 0.58 $M_\odot$ for these post-AGB objects (comparable to

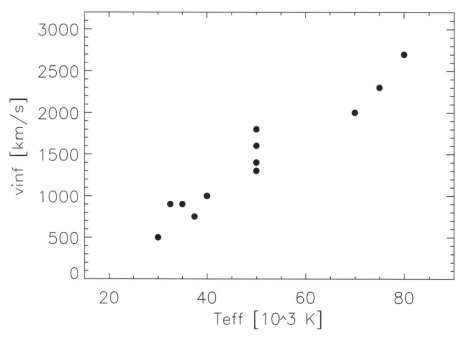

Figure 7   Terminal velocities of winds from Central Stars of Planetary Nebulae with O-type spectra as a function of effective temperature. Data are from Méndez et al (1988), Pauldrach et al (1989), Perinotto (1993), Haser (1995), and Kudritzki et al (1997).

the average mass of white dwarfs) (see Napiwotzki et al 1999) and a corresponding luminosity according to the core mass–luminosity relationship (see Schönberner 1983, Wood & Faulkner 1986), one can calculate escape velocities for each object of Figure 7 and compare the ratio of v_∞ to v_{esc} with the case of O-stars. In this way, agreement with the O-stars at least for CSPN with $T_{eff} \leq 50,000$ K is obtained. However, as stressed by Kudritzki et al (1997), much smaller escape velocities and ratios v_∞ to v_{esc} of the order of 3–5 are found if gravities and radii from their non-LTE spectroscopy are used directly. Kudritzki et al (1997) attribute this disagreement to either some unknown defect of their spectral analysis procedure or a failure of the core mass–luminosity relationship applied.

3.2   Mass-Loss Rates and Wind Momenta of Galactic Hot Stars

During the past 3 years there have been comprehensive studies to reinvestigate the mass-loss rates of galactic hot stars by analyzing H_α profiles based on unified non-LTE model atmospheres. Because the only requirement is that a good spectrum can be taken, this method can determine mass-loss rates for a much larger number of objects than the method analyzing radio emission.

Puls et al (1996) and Herrero et al (2000) have investigated the mass-loss rates of O-stars. Kudritzki et al (1997) studied CSPN, and Kudritzki et al (1999) determined mass-loss rates of galactic A- and B-supergiants. In the following, we summarize their results.

The best way to discuss the strengths of stellar winds is in terms of the wind momentum–luminosity relationship (WLR). Because the winds of hot stars are driven by radiation, it is intuitively clear that the mechanical momentum of the stellar wind flow should be mostly a function of photon momentum. Indeed, the theory of radiation-driven winds predicts (see discussion in Section 4) that the "modified stellar wind momentum" depends directly on luminosity through the WLR

$$\log D_{\mathrm{mom}} = \log D_0 + x\log(L/L_\odot), \qquad D_{\mathrm{mom}} = \dot{M}v_\infty(R_*/R_\odot)^{0.5}. \qquad (10)$$

The coefficients of the WLR, $\log D_0$ and x, are expected to vary as a function of spectral type and luminosity class.

Figure 8 shows modified wind momenta of O-stars and CSPN as a function of luminosity. It is obvious that the O-supergiants follow a tight WLR, as described by Equation 10 over at least 1 dex in luminosity. The coefficients of the corresponding

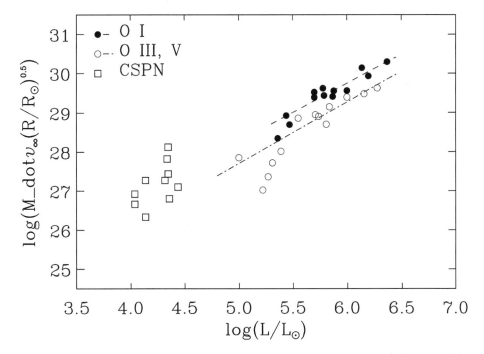

Figure 8 Modified wind momenta of galactic O-stars and Central Stars of Planetary Nebulae (CSPN) as a function of luminosity. The wind momentum–luminosity relationship for O-supergiants and O-giants/dwarfs obtained by linear regression are also shown. Data from Puls et al (1996), Herrero et al (1999b), and Kudritzki et al (1997).

TABLE 2   Coefficients of the wind momentum–luminosity
relationship for A/B-supergiants and O-stars of the solar neighborhood

Sp. type	Log D_0	x	α'
A I	14.22 ± 2.41	2.64 ± 0.47	0.38 ± 0.07
Mid B I	17.07 ± 1.05	1.95 ± 0.20	0.51 ± 0.05
Early B I	21.24 ± 1.38	1.34 ± 0.25	0.75 ± 0.15
O I	20.69 ± 1.04	1.51 ± 0.18	0.66 ± 0.06
O III, V	19.87 ± 1.21	1.57 ± 0.21	0.64 ± 0.06

linear fit are given in Table 2. The situation for O-giants and -dwarfs is similar; however, the WLR is shifted downward to lower wind momenta, and at the low luminosity end the situation is somewhat confusing. Two objects (ζ Oph and HD 13268) with $\log L/L_\odot \leq 5.3$ fall clearly below the relationship. Both are rapid rotators with very thin winds. Several additional effects may become important in such a situation, as is discussed in Section 4.

The fact that the CSPN as objects of significantly lower luminosity and of completely different evolutionary status have wind momenta corresponding to the extrapolation of the WLR of O-stars must be regarded as an encouraging success of the interpretation of winds in terms of radiative driving and of the concept of the WLR.

Figure 9 compares wind momenta of O-, B-, and A-supergiants. For all these different spectral types, a tight relationship between wind momentum and luminosity is found. However, the WLR varies as a function of spectral type. Wind momenta are strongest for O-supergiants, then decrease from early B (B0 and B1) to mid B (B1.5 to B3) spectral types and become stronger again for A-supergiants. The slope of the WLR appears to be steeper for A- and mid B-supergiants than for O-supergiants. The interpertation of this result is given in Section 4.

3.3   Observed Effects of Metallicity: The Magellanic Clouds

It is obvious that the global properties of winds must depend on metallicity. Because the winds of hot stars are driven by photon momentum transfer through metal line absorption, certainly the wind momentum rate and possibly the terminal velocities must be a function of stellar metallicity. The ideal laboratory to test the metallicity dependence are the Magellanic clouds.

Figure 10 compares terminal velocities of O-stars in the Galaxy, the large Magellanic cloud (LMC), and the small Magellanic cloud (SMC). Although there is a large scatter in this diagram, because stars of different luminosity classes and photospheric escape velocities are not disentangled, there is a clear indication that, on the average, wind velocities in the metal-poor SMC are smaller. For the LMC, the situation is not so clear.

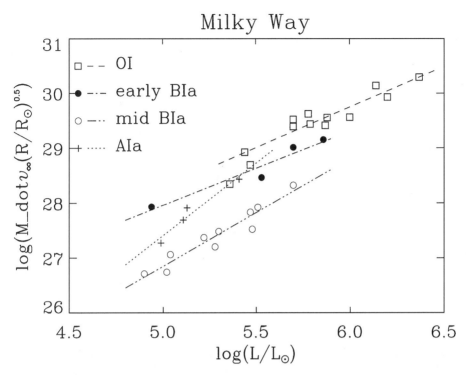

Figure 9 Wind momenta of galactic O-, early B-, mid B-, and A-supergiants as a function of luminosity together with the corresponding wind momentum–luminosity relationship obtained by linear regression. Data from Puls et al (1996), Herrero et al (1999b), and Kudritzki et al (1999).

Wind momenta of O-stars show a definite trend with metallicity. This is demonstrated by Figure 11. Although the number of objects studied in the clouds is still small, it is clear that average momenta in the LMC are smaller than in the Galaxy and average momenta in the SMC are smaller than in the LMC. Puls et al (1996) quote a difference of 0.20 and 0.65 dex between the Galaxy and the LMC or SMC, respectively.

In their study of very massive stars in the compact cluster R136a in the LMC, de Koter et al (1998) find very high wind momenta for their most luminous objects, indicating a much steeper WLR than that found by Puls et al (1996). This discrepancy is to a large extent the result of the different effective temperature determinations used. de Koter et al use the Ov λ 1371 line as a temperature indicator based on their model atmosphere calculations, whereas Puls et al used the optical HeI λ 4471 line. It is not clear which of the two methods is closer to the truth. Careful future spectroscopic work based on fully consistent hydrodynamic line-blanketed non-LTE models is needed to clarify the situation.

Large observing programs of A- and B-supergiants in the Magellanic clouds are currently under way to investigate the influence of metallicity at later spectral

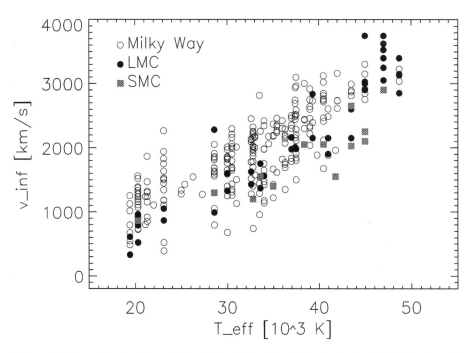

Figure 10 Terminal velocities of O-stars in the Galaxy, large Magellanic cloud (LMC), and small Magellanic cloud (SMC) as a function of effective temperature. For the conversion of spectral type into effective temperature, the scale by Humphreys & McElroy (1984) was used for simplicity. Data from Howarth et al (1997), Prinja & Crowther (1998), de Koter et al (1998), Puls et al (1996), and Haser (1995).

types. In addition, a large number of O-stars have been observed very recently in both clouds with HST. These programs will allow determination of the role of metallicity much more precisely.

4.   INTERPRETATION BASED ON THE THEORY OF LINE-DRIVEN WINDS

As mentioned in the previous sections, the basic mechanism driving the winds of hot stars is the transfer of photospheric photon momentum to the stellar atmosphere plasma through absorption by spectral lines. Obviously, the properties of stellar winds must depend on the number of lines being available to absorb significant amounts of photon momentum, in particular at wavelengths around the photospheric flux maximum. However, not only is the number of lines important; their ability to absorb, i.e. their optical thickness, is also important. In supersonically expanding winds, optical thickness is a local quantity, as described by Equations 4

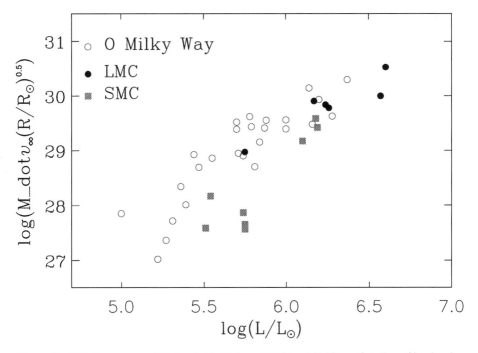

Figure 11 Wind momenta of O-stars in the Galaxy, LMC, and SMC as a function of luminosity. Data from Puls et al (1996).

and 5. Introducing the "line-strength" k_i and the Thomson optical depth parameter $t(r)$, the optical thickness τ_i of each line can be expressed as

$$k_i = \frac{n_l}{n_e \sigma_e v_{\text{therm}}} f_{lu} \lambda_i, \quad t(r) = n_e \sigma_e \frac{v_{\text{therm}}}{dv/dr} \frac{R_*}{v_\infty}, \quad \tau_i = k_i t(r). \quad (11)$$

$t(r)$ corresponds to the local optical thickness of a line for which the integrated line opacity is just equal to the one of Thomson scattering. The line-strength k_i measures line opacities in units of Thomson scattering (neglecting stimulated emission). Whether a line is optically thick or thin at a certain point in the wind depends on the line strength as well as on the local hydrodynamical situation. Optically thin and thick lines contribute to the radiative acceleration $g_{\text{rad}}^{\text{lines}}$ in a completely different way (for a detailed discussion of the following formulae, see, for instance, Kudritzki 1988, 1998; Puls et al 2000):

$$g_{\text{rad}}^{\text{thin}} \propto \frac{k_i}{r^2}, \quad g_{\text{rad}}^{\text{thick}} \propto \frac{1}{r^2 \rho} \frac{dv}{dr}. \quad (12)$$

It is, therefore, crucial to know the contribution of optically thick and thin lines throughout the wind, which can be accomplished in an elegant way by introducing a line-strength distribution function. In the theory of radiation-driven winds, such a distribution function can be calculated from the opacities of the hundreds of thousands of lines driving the wind. It turns out to be a power law in line strength (for a comprehensive discussion of the physics of the line-strength distribution function, see Puls et al 2000):

$$n(k_i)\, dk_i \propto k_i^{\alpha-2}\, dk_i, \qquad 1 \leq k_i \leq k_i^{\max}. \tag{13}$$

The exponent α of the line-strength distribution function is crucial for the calculation of the total radiative line acceleration, which (after integration over line strength) finally becomes

$$g_{\text{rad}}^{\text{lines}} \propto N_{\text{eff}} \frac{1}{r^2} t^{-\alpha}, \tag{14}$$

with N_{eff} the number of lines effectively driving the wind. This means that the line acceleration depends nonlinearly on the velocity gradient. However, for $\alpha = 0$, we would recover the optically thin case, whereas $\alpha = 1$ would correspond to the optically thick case. For O-stars, a typical value is $\alpha \sim 2/3$.

The local non-LTE ionization balance in a stellar wind depends on the ratio of electron density to the geometrically diluted radiation field (see Section 2.4.1) and may change from the photosphere to the outer layers of the stellar wind flow. In such a case we would also expect the line-strength distribution function to change so that both α and N_{eff} vary with height. Abbott (1982) introduced a parameterization of this effect by adopting $N_{\text{eff}} = N_0 (10^{-11} \text{cm}^3 n_e / W)^\delta$, with electron density n_e (in cgs units) and dilution factor W. This parameterization accounts only for ionization changes in the normalization (N_{eff}) of the line-strength distribution function, but it ignores possible simultaneous effects on its slope (α), which can become important under certain conditions (see Kudritzki et al 1998, Puls et al 2000). Typical values of δ for O-stars are $\delta \sim 0.05\text{--}0.1$.

4.1 Nonrotating Winds

Solving the equation of motion for stationary, spherically symmetric line-driven winds with line-strength distribution functions as described by Equations 13, the δ term, and corresponding radiative line accelerations (Equation 14), one obtains for terminal velocity and the mass-loss rate (see Kudritzki et al 1989)

$$v_\infty = 2.25 \frac{\alpha}{1-\alpha} v_{\text{esc}} f_1(\alpha) f_2(\delta) f_3(v_{\text{esc}}) \tag{15}$$

and

$$\dot{M} \propto N_0^{\frac{1}{\alpha'}} L^{\frac{1}{\alpha'}} (M_*(1-\Gamma))^{1-\frac{1}{\alpha'}}, \qquad \alpha' = \alpha - \delta. \tag{16}$$

The functions f_1, f_2, f_3 are of the order of unity and are obtained from numerical fits to the results found by Kudritzki et al (1989),

$$f_1(\alpha) = \frac{8}{5}\left(1 - \frac{3}{4}\alpha\right), \qquad f_2(\delta) = e^{-2\delta}, \qquad f_3(v_{\text{esc}}) = 1 - 0.3 e^{-\frac{v_{\text{esc}}}{300\,\text{km/s}}}. \tag{17}$$

Equation 15 demonstrates that the theory of line-driven winds does indeed lead to a proportionality of v_∞ to v_{esc} (the influence of f_3 is small), as indicated by the observations. The ratio v_∞/v_{esc} does obviously depend on the value of α, the power law exponent of the line-strength distribution function. This means that the physical nature of the lines driving the wind such as ionization stages, chemical composition, etc, is fundamentally important. A steep line-strength distribution function (small α, many weaker lines) will produce a slow wind, whereas a flatter function (large α, more stronger lines) will lead to a faster wind.

The mass-loss rate is influenced by both the effective number of lines N_0 driving the wind and the slope α', in particular, because of its dependence on the stellar luminosity. Unfortunately, $\dot{M}$ depends also on the "effective mass" $M_*(1 - \Gamma)$, which, for instance for O-stars and early B-supergiants, varies greatly from star to star and introduces a significant scatter, if one wants to correlate stellar mass-loss rates with luminosity.

The scatter is, however, reduced if one considers the "modified stellar wind momentum," as introduced by Equation 10 and combines Equations 15, 16, and 8 to obtain

$$D_{mom} = \dot{M}v_\infty(R_*/R_\odot)^{0.5} \propto N_0^{\frac{1}{\alpha'}} L^{\frac{1}{\alpha'}} (M_*(1 - \Gamma))^{\frac{3}{2} - \frac{1}{\alpha'}}. \tag{18}$$

The absolute value of the exponent of the effective mass is now smaller, reducing the influence of this quantity, in particular for O-stars and early B-supergiants, where the theory predicts a value of $\alpha = 2/3$ (see Puls et al 2000). This is the reason why an analysis of observed stellar wind momenta is more straightforward than a discussion of mass-loss rates. Equation 18 is the basis for the concept of the wind momentum–luminosity relationship (WLR) (see Equation 10), as introduced by Kudritzki et al (1995) and Puls et al (1996) (see Kudritzki 1998, for a simplified derivation of Equation 18).

The slope of the observed WLR can be used to determine the value of α' empirically as a function of the spectral type through $\alpha' = 1/x$. The corresponding values are given in Table 2 and indicate that α' decreases systematically with decreasing effective temperature. This conclusion is confirmed by applying Equation 15 to the observed ratios v_∞/v_{esc} (see Equation 9). Assuming $\delta = 0.05$ and $v_{esc} = 700, 350$, and 200 km/s for O-stars, B-supergiants with $T_{eff} \leq 22,000$ K, and A-supergiants, respectively, we obtain for α 0.60, 0.40, and 0.32, slightly lower than the values in Table 2 but indicating the same trend with T_{eff} (see also Lamers et al 1995, Achmad et al 1997, who were the first to point out that the terminal velocities of A- and mid-to-late B-supergiants require low values of α).

The physical reason for the change in the slope of the line-strength distribution function with effective temperature is, of course, the change in ionization of the elements contributing to the radiative line acceleration. As shown in the detailed investigation by Puls et al (2000), changes of ionization affect the line-strength distribution function mostly at small and intermediate line strengths. Here, the contribution of the iron group elements is crucial with their large number of meta-stable levels. With decreasing ionization, the number of lines from iron group elements at small and intermediate line strengths becomes larger and larger. On the other hand, the contribution to the line-strength distribution at large line strengths results mostly from light ions, whose contribution remains rather constant, as a function of temperature. As a result, the slope of the total line-strength distribution function over all line strengths becomes steeper and α becomes smaller with decreasing effective temperature.[3] The effect is, thus, understood qualitatively. A quantitative confirmation based on a detailed comparison at the low-temperature end of non-LTE wind models with observations resulting from the spectroscopic analysis has still to be carried out [see, however, first results presented by Kudritzki (1999a,b)].

The variation of the coefficient D_0 of the WLR can be used to estimate empirically the effective number of lines N_0 contributing to the stellar wind acceleration. This is, however, a more complex matter because D_0 depends on the flux-weighted total number of spectral lines as well as on α' and α itself (Puls et al 2000). Only in cases of similar slopes does a comparison give direct insight into the absolute number of effectively driving lines, whereas in all other cases their distribution with respect to line-strength (α) has a significant influence on the offset.

Because of its (partial) dependence on flux-weighted line number, D_0 varies between spectral types not only because of ionization changes, but also because of the different spectral locations of the lines with regard to the flux maximum and absorption edges, such as the hydrogen Lyman and Balmer edges. At least concerning the observed differences in D_0 between O-type and A-type supergiants (accounting for the changes in α'), D_0 can be understood in terms of the theoretically predicted effective number of lines N_0, as demonstrated by Puls et al (2000). However, to predict the full variation of D_0 over the whole range of effective temperatures quantitatively, and thus to check whether this interpretation is generally valid, will require more detailed calculations of radiation-driven wind models. As expressed by Kudritzki et al (1999), the pronounced drop in D_0 between effective temperatures of 23,500 K and 22,000 K (on the temperature scale of unblanketed models) is a particular challenge for the theory. A detailed spectroscopic study of a larger sample of objects in this transition range of temperatures to disentangle stellar wind properties in more detail will certainly be valuable. In addition, a careful reinvestigation into whether systematic effects (for instance, deviations of the helium abundances from the normal value, or metal line blanketing and blocking in this temperature range) may have influenced the results of the spectroscopic analysis will be important.

The theory of line-driven winds also makes a clear prediction about the influence of metallicity on the strengths of stellar winds. The first-order effect arises from the fact that the effective number of lines contributing to the line acceleration changes

---

[3]We note in passing that the power law fit of the line-strength distribution function with a constant exponent α is only a first-order approximation. In reality, the function shows a distinct curvature, which must be considered in realistic stellar wind calculations (see Pauldrach et al 1994, Kudritzki et al 1998).

with metallicity ϵ and, hence, causes a metallicity dependence of the mass-loss rate

$$N_0(\epsilon) = N_0^{\odot}\,(\epsilon/\epsilon_{\odot})^{1-\alpha}\,, \qquad \dot{M}(\epsilon) = \dot{M}(\epsilon_{\odot})\,(\epsilon/\epsilon_{\odot})^{\frac{1-\alpha}{\alpha'}}\,. \qquad (19)$$

Taking values of α' from Table 2 and adopting $\delta = 0.05$, we obtain exponents for the metallicity dependence of mass-loss rates of the order of 0.5, 0.8, and 1.7 for O-stars (and early B-supergiants), mid B-supergiants, and A-supergiants, respectively. Theoretical stellar wind calculations by Kudritzki et al (1987) and Leitherer et al (1992) confirm the numbers for the first two cases. The only observational test so far results from Puls et al (1996), which in the metallicity range between the Galaxy and the SMC ($\epsilon \sim \epsilon_{\odot}/5$) confirms the exponent of 0.5 for O-stars. Whether the very strong metallicity dependence predicted for A-supergiants is really observed needs to be checked by a systematic analysis of winds of those objects in the SMC. The work by McCarthy et al (1995), where the analysis of a very-metal-poor A-supergiant in the outskirts of M33 leads to a very small mass-loss rate, indicates that the large exponent might be appropriate.

A second effect might be important for the metallicity dependence of terminal velocities as well as for the overall slope ($x = 1/\alpha'$) of the WLR, as discussed by Puls et al (2000). This effect results from the curvature of the line-strength distribution function, which has a steeper slope (smaller α) for the larger line strengths, which become important for the acceleration of winds at lower metallicity, where the winds are weaker and the corresponding optical thicknesses are smaller. However, as pointed out by Puls et al, there are several competing effects of comparable magnitude in the acceleration of the outer layers of winds, such as differences in CNO abundance ratios or changes in ionization at different metallicity. It is, therefore, too early to make a clear theoretical prediction for all spectral types. The lower terminal velocities (as well as the indications of a steeper slope) observed for SMC O-stars, though, indicate that the theory is appropriate for O-stars.

4.1.1 The Problem of Thin Winds

As we have seen (cf Figure 8), the observed WLR for O-type dwarfs, if taken literally, exhibits a severe curvature toward (very) low wind momenta at luminosities lower than $\log L/L_{\odot} = 5.3$. (Note that the derived momenta are upper limits in this luminosity range.) As pointed out above, the two objects that fall below the relation are rapid rotators. Although this fact alone might initiate the deviations from the WLR (see the next subsection), there are three additional effects that are important for the physics of very thin winds from dwarfs (or in a low metallicity environment) and might induce a strong deviation from the usual WLR.

First, standard theory requires the photon momentum absorbed almost exclusively by metal ions to be transferred to the bulk of the wind plasma (hydrogen and helium) via Coulomb collision. A detailed investigation by Springmann & Pauldrach (1992) illuminated the limits of this process: For thin winds, the metal ions can decouple from the rest of the plasma, and the wind no longer reaches the same terminal velocity that would follow in the standard picture. Using a multifluid approach (including the various Coulomb collisions as function of plasma parameters), Babel (1995) presented a model for inhomogeneous radiatively driven winds of A-dwarfs, where the resulting $\dot{M}$ turned out to be very low. Second, the inclusion of shadowing by photospheric lines (not considered in the usual computation of the line force) has large consequences for the wind of B-dwarfs, resulting again in mass-loss rates well below the standard results (Babel 1996). Finally, as shown by Puls et al (1998) and Owocki & Puls (1999), curvature terms of the velocity field in the transonic region can lead to line accelerations much smaller than in the standard computations, leading to reduced mass-loss rates. This process, however, is only effective when the continuum is optically thin throughout the transonic region, i.e., it can be present only in thin winds. Together, these results imply that low-luminosity dwarfs are subject to a number of processes reducing the mass-loss rate compared with the scaling relations above. The position of the turnover point as a function of various stellar parameters (including metallicity), however, remains to be defined.

4.2 Influence of Rotation

So far, our discussion has neglected the influence of rotation on the wind structure itself. This is discussed below. Recent reviews on this topic have been given by Owocki et al (1998a,b), Bjorkman (1999), and Puls et al (1999). [For the effects of rotation on stellar evolution/structure, see Langer & Heger (1998), Maeder (1998, 1999), and Meynet (1998).]

The major aspect that has to be accounted for in addition to the "conventional" approach, at least at first glance, is the impact of centrifugal acceleration as a function of radius r and co-latitude θ. It was studied first by Friend & Abbott (1986) and Pauldrach et al (1986) in its most simple way, considering only particles in the equatorial plane. With the assumption of a purely radial line acceleration, the angular momentum remains conserved (only central forces present), and the rotational speed is given by $v_\phi(r) = v_{\rm rot}(R_*, \theta = \pi/2)/r$. Thus, the usual equation of motion is modified by the centrifugal acceleration only, which leads to an effective gravity of $GM(1 - \Gamma)/(R_* r)^2 (1 - \Omega^2/r)$, where Ω is the ratio of rotational speed at the (equatorial) surface $v_{\rm rot}(R_*)$ to the break-up velocity $v_{\rm break} = v_{\rm esc}/\sqrt{2}$, with $v_{\rm esc}$ the photospheric escape velocity defined in Equation 8. Thus, the only difference to nonrotating models is the modification of effective mass by roughly a factor of $(1 - \Omega^2)$:

$$\dot{M}(\Omega) = \dot{M}(0)(1 - \Omega^2)^{1-1/\alpha'}; \quad v_\infty(\Omega) = v_\infty(0)(1 - \Omega^2)^{\frac{1}{2}}. \quad (20)$$

$\dot{M}(0)$ is the mass-loss rate without rotation (cf Equation 16), and the second relation follows from the scaling properties of the terminal velocity, $v_\infty(\Omega = 0) \propto v_{\rm esc}$ (Section 4.1). [A more elaborate version of the latter expression has been given by Puls et al (1999).] A comparison of both scaling laws with hydrodynamical simulations shows a satisfying agreement.

Of course, the above scaling laws are valid only in the equatorial plane and can be considered as sort of a maximum effect. Accounting also for the variation of v_ϕ as a function of θ, Bjorkman & Cassinelli (1993) elaborated on the concept of so-called wind-compressed disks and zones.

The basic idea follows again from the assumption of purely radial line forces. Thus, the specific angular momentum is actually conserved for all particles, and their motion is restricted to the orbital plane to which they belong (tilted by an angle of co-latitude θ_o from which they start). Neglecting pressure forces in the supersonic region, the free flow of the particles can be simulated then by a corresponding 1-D treatment, as above, with modified rotational rate $\Omega \sin \theta_o$. Consequently, mass-loss rates and velocities in the orbital plane follow the same scaling relations as above, however with Ω^2 replaced by $\Omega^2 \sin^2 \theta_o$. In the course of the particles' motion, their azimuthal angle $\Phi'(r)$ in the orbital plane is increasing, and they are deflected toward the equator. If the ratio of $v_{\mathrm{rot}}/v_\infty$ is significant and/or the radial velocity field in the lower wind region is flat, Φ' might become $\geq \pi/2$, and the particles would cross the equator. Here, they collide supersonically with particles from the other hemisphere, and a wind-compressed disk (WCD) is formed. In those cases where the particles do not collide ($\Phi'(r \to \infty) < \pi/2$), at least a wind compressed zone (WCZ) is created, i.e. an anisotropic wind with highest densities at the equator.

The hydrodynamical stratification in stellar coordinates is obtained by following the particles' stream lines. For noncolliding winds with not too large deflection angle Φ' or generally close to the star ($\theta \approx \theta_o$), one finds

$$\dot{M}(\theta) \approx \dot{M}(0)(1 - \Omega^2 \sin^2 \theta)^{1 - 1/\alpha'}; \qquad v_\infty(\theta) = v_\infty(0)(1 - \Omega^2 \sin^2 \theta)^{\frac{1}{2}}, \quad (21)$$

i.e. $\dot{M}$ increases and v_∞ decreases toward the equator. Further consequences are discussed by Bjorkman & Cassinelli (1993) and Bjorkman (1999), and the validity of the WCD-model has been confirmed in principle by Owocki et al (1994) on the basis of time-dependent hydrodynamical simulations.

It is important to realize that the outlined results rely exclusively on the assumption of purely radially directed line forces. As determined by a detailed investigation of the other (vector) components (cf Owocki et al 1996), the polar one in particular cannot be neglected in the delicate balance of forces[4]. As long as the radial velocity law at the equator is slower than in polar regions—an almost inevitable consequence of rotation due to the reduced escape velocity at the equator—a pole-ward acceleration is created that is sufficient to stop the equatorward motion predicted by WCD/WCZ models and actually reverses its direction, an effect called the "inhibition effect": Disks can no longer be formed. Together with the distortion of the stellar surface by centrifugal forces (e.g. Cranmer & Owocki 1995) and the well-known von Zeipel theorem (surface flux scaling with effective gravity as a function of θ) (for recent improvements, cf Maeder 1998,

---

[4]Concerning the azimuthal component affecting the angular momentum and allowing for a spin-down of the wind, cf Grinin (1978) and Owocki et al (1998a).

1999) the original scaling of $\dot{M}$ (Equation 21) can be actually reversed, i.e. the inclusion of gravity darkening (increased polar radiation flux) might lead to a larger polar mass flux

$$\dot{M}(\theta) \propto (1 - \Omega^2 \sin^2 \theta)^{+1} \quad \text{with gravity darkening} \quad (22)$$

(cf. Owocki et al 1998a,b).

Recent work by Petrenz (1999) concentrated on a consistent description of hydrodynamics and statistical equilibrium (controlling the line force) in order to clarify how important the 2-D stratification of occupation numbers is and how much influence it might have on the final result. A number of models with various kinds of differing approximations have been calculated, confirming however all basic trends quoted above.

Thus, the current status of modeling rotating stellar winds can be summarized as follows: If one neglects the nonradial components of the line force, an oblate wind structure is created, with meridional velocities directed toward the equator. For large values of Ω, a WCD might show up. Accounting additionally for nonradial line forces, the resulting pole-ward acceleration reverts the direction of the meridional velocity field and "inhibits" the disk formation. Finally, the inclusion of gravity darkening allows even for a prolate wind structure.

In view of these different scenarios [for additional processes that might be relevant, cf Bjorkman (1999)] and with respect to our discussion of the WLR, it is especially interesting to investigate the dependence of wind momentum on rotation. From the simplified approach (Equation 21) and for $\alpha' = 2/3$, the wind-momentum rate becomes independent of angle, i.e., $\dot{M}v_\infty(\Omega, \theta) \approx \dot{M}v_\infty(\Omega = 0)$, because as in the corresponding 1-D models, the dependence on effective mass completely cancels out. Numerical simulations (radial forces only) verify this prediction with high accuracy (P Petrenz, private communication). The maximum influence, on the other hand, and discarding disks, is obtained for models with nonradial forces and gravity darkening. For large values of $v_{rot}(250\ldots350$ km/s), the wind-momentum rate varies typically by a factor of 2 (pole) to 0.5 (equator), compared with 1-D, nonrotating models. Because $\dot{M}$ and v_∞ depend on Ω^2, the effect of "normal" rotational rates, however, becomes minor with respect to other uncertainties. Note especially that all theoretical results derived so far unfortunately lack direct observational proof.

There exists another class of massive objects where rotation might be even more important: the B[e]-supergiants with a well-determined bimodal structure (slow and dense equatorial wind and fast, thin polar wind) (for a recent review, see Zickgraf 1999). As has been shown by Pauldrach & Puls (1990) and investigated in detail by Vink et al (1999), continuum optical depth effects in the B-supergiant domain can induce the so-called bi-stability mechanism. The decisive quantity that controls this behavior is the optical depth in the Lyman continuum, which—for rotating stars and accounting for gravity darkening—depends sensitively on rotationally induced variations of wind density and radiation temperature and thus becomes a strongly varying function of θ (Lamers & Pauldrach 1991, Lamers et al 1999b; but see also Owocki et al 1998a). Because of this behavior, the bistability effect is thought to be responsible for the B[e] phenomenon (Lamers & Pauldrach 1991) if the star rotates close to break-up and has a large average mass loss. First hydrodynamical simulations (Petrenz 1999; HJGLM Lamers, private communication) indicate that this hypothesis might actually work.

5.   TIME DEPENDENCE AND STRUCTURE

The basic philosophy underlying the previous results and conclusions has been outlined in Section 2.2 and comprises the assumption of a globally stationary wind (1-D or 2-D) with a smooth density/velocity stratification. However, as follows directly from the assumption of stationarity, these models are inherently incapable of describing a number of observational features (partly referred to in Section 2.4), which immediately show that nonstationary aspects of the wind must be important. Moreover, theoretical considerations based on a detailed investigation of the wind-driving agent, radiative-line acceleration, show clearly that this acceleration is subject to a strong instability, which, when "allowed" to operate in time-dependent models, causes a variable and strongly structured wind hardly resembling the stationary and smooth model underlying our philosophy, at least at first glance. This section outlines the status quo with respect to observations and theory and some further implications.

5.1   Observational Findings

In our review of diagnostic methods to derive global wind parameters (Section 2.4) we have met a number of issues that are difficult to reconcile with the stationary/smooth picture of winds. The following enumerations summarize the most important observational findings concerning structure and temporal variability (including those already mentioned); for further details we refer to the proceedings cited at the end of Section 2.2.

(*a*) Soft X-ray emission from hot stars is attributed to randomly distributed shocks in their winds. It is required to model the observed degree of superionization (cf Section 2.4.5). (*b*) Nonthermal radio emission is observed for roughly 30% of massive stars and may be interpreted as synchrotron radiation from first-order Fermi-accelerated electrons, requiring again the presence of shocks (cf Section 2.4.4). (*c*) Black troughs in saturated P-Cygni profiles are simulated by a velocity dispersion in the stationary description. However, a better physical basis for interpreting them is as a consequence of backscattering in multiply nonmonotonic flows (cf Section 2.4.1). (*d*) Electron-scattering wings of recombination lines in Wolf-Rayet stars are weaker than predicted by smooth models, which suggests the possibility of a clumped structure (Hillier 1991). WR models accounting for the effects of clumping succeeded in producing perfect line fits, with the implication of reduced (factor of 2) mass-loss rates (Schmutz 1997, Hamann & Koesterke 1998). (*e*) The inconsistency of IR-excess and radio flux for a small number of objects can be resolved if clumping is accounted for (Section 2.4.4). (*f*) Variability is observed in optical wind lines such as H_α (Ebbets 1982) and HeII λ 4686 (Grady et al 1983, Henrichs 1991), both for OB-stars as well as for BA-supergiants (Kaufer et al 1996). Minimum timescales are of the order of hours, i.e. wind-flow times. (*g*) Direct evidence (from HeII λ 4686) for outmoving inhomogeneities (e.g. clumps, blobs, or shocks) is reported for the wind of ζ Pup (Eversberg et al 1998). The spectropolarimetric variations detected by Lupie & Nordsieck (1987) in a sample of 10 OB-supergiants also provide direct evidence of inhomogeneous wind structure.

Predominantly photospheric lines such as HeI λ 5876 are found to be variable as well (Fullerton et al 1992), and in a spectroscopic survey of a large sample of O-stars by Fullerton et al (1996), 77% of the stars investigated (all supergiants of the sample and few dwarfs later than O7) showed photospheric line profile variability with an amplitude increasing with radius and luminosity, likely related to nonlinear stellar pulsations. (*h*) Variability is observed as well in UV P-Cygni lines, where the blue edges vary most significantly, whereas the red emission part remains relatively constant (Henrichs 1991, Prinja 1992). H_α and UV variability on the one hand (Kaper et al 1997) and H_α and X-ray variability on the other (Berghöfer et al 1996) seem to be correlated and might indicate the propagation of disturbances throughout the entire wind. (*i*) Discrete absorption components (DACs) are among the most intensively studied manifestations of wind variability and structure in OB-stars. These features are optical depth enhancements that accelerate through the absorption troughs of unsaturated P-Cygni profiles from low to high velocities much more slowly than the mean outflow (e.g. Prinja et al 1992). Lamers et al (1982) reported the presence of narrow components in 17 of 26 OB-stars at a blue velocity of $0.75v_\infty$ and discussed the presence of velocity plateaus as one of a number of possible explanations. Later studies (e.g. Henrichs 1984, Prinja & Howarth 1986) revealed that P-Cygni lines also vary at lower velocities, although over a broader velocity range. Extensive time series (Prinja et al 1987, Prinja & Howarth 1988) have shown that these broad features evolve into the high-velocity narrow components on timescales of a few days. Both features were "unified" in terms of DACs. Note that Howarth & Prinja (1989) and Prinja et al (1990) used the narrow absorption features to determine v_∞ for those objects with no black trough in CIV (cf Section 2.4.1).

Reviews of the phenomenology associated with DACs have been given by Howarth (1992), Prinja (1992), Henrichs et al (1994), and Kaper & Henrichs (1994). The latest results (also in connection with the modulation features described below) have been reported by Kaper et al (1999). Note that DACs are found not only in O- and early B-stars, but also in B-supergiants as late as B9 (Bates & Gilheany 1990) and in at least one WN7-star (Prinja & Smith 1992).

Although the nature of the DACs is still unclear, the finding that their acceleration and recurrence time is correlated with the stellar rotational period [actually, faster-developing, more frequent DACs are apparent in stars with higher $v_{rot} \sin i$ (cf Prinja 1988, 1992; Henrichs et al 1988, Kaper 1993)] turned out to be important not only by itself, but also because it inspired the "IUE MEGA Campaign" (Massa et al 1995). During 16 days of nearly continuous observation with IUE, three prototypical stars [ζ Pup (O4If), HD 64760 (B0.5Ib), and HD 50896 (WN5)] were monitored in order to assess the presumed correlation of rotation versus wind activity. The results of this campaign (together with other observational runs) have recently been summarized by Fullerton (1999). For ζ Pup, one of the derived periods [from SiIV (cf Howarth et al 1995)] is within the uncertainties associated with the rotational period; however its wind appears to be subject to a variety of perturbations. The analysis of the time series for HD 64760 (Prinja et al 1995) lead to the detection of a new type of variability, namely periodic modulations in the UV wind lines. Fullerton et al (1997) have shown that these modulations result from two quasi-sinusoidal fluctuations with periods of 1.2 and 2.4 days and can be traced throughout the emission part of the profiles, i.e. they must be caused by longitudinally extended structures. The modulations develop simultaneously toward lower and higher velocities, an effect called "phase bowing," which has been explained by co-rotating, spiral-shaped wind structures (Owocki et al 1995). These structures might be identified with so-called corotating interaction regions (CIRs) caused by the collision of fast and slow winds that emerge nearly radially from different longitudinal sectors of the stellar surface. CIRs have long been studied in the solar wind and were first proposed by Mullan (1984, 1986) as an explanation for the DACs in OB-star winds. Actually, Cranmer & Owocki (1996) have suggested that CIRs are responsible not only for the modulation features but also for the "classical" DACs (thus supporting the suggestion by Mullan) by creating velocity plateaus in the wind [see also the early suggestion by Lamers et al (1982)]. With respect to this explanation, however, a number of problems related to the treatment of the line force in the hydrodynamic simulations are still unresolved (Owocki 1999).

5.2   Theoretical Considerations

Theoretical efforts to understand the nature and origin of these observational findings have generally focused on the line-driving mechanism itself. Linear stability analyses on the basis of a non-Sobolev description (see below) have shown the line force to be highly unstable (MacGregor et al 1979; Carlberg 1980; Abbott 1980; Lucy 1984; Owocki & Rybicki 1984, 1985), with inward-propagating waves being more strongly amplified than outward-propagating ones.

Initial heuristic models by Lucy & White (1980) and Lucy (1982a) assumed that the resulting wind structure would consist of a periodic train of forward shocks. With some adjustment of parameters, this model reproduced the observed flux of soft X-rays (Lucy 1982b), although not the hard X-ray tail (Cassinelli & Swank 1983). Lucy (1982a, 1983) further showed that the multiply nonmonotonic nature of the resulting velocity field could explain the black absorption troughs observed in saturated UV resonance lines (cf Section 2.4.1).

Subsequent efforts have focused on dynamical modeling of the time-dependent wind structure from direct, numerical simulation of the nonlinear evolution of the line-driven flow instability (for reviews see Owocki 1992, Owocki 1994, Feldmeier 1999). The key aspect regards the computation of the force itself because, for unstable flows with structure at scales near or below the Sobolev length (Section 2.4.1), a local approach as given by the Sobolev treatment inevitably fails unless very refined precautions are taken (Feldmeier 1998). Instead, one must apply complex, computationally expensive integral forms that take into account the nonlocal character of radiation transfer. Owocki & Puls (1996, 1999) have discussed various levels of approximations that both account for the key effects and are computationally feasible.

The most important finding of the first simulations by Owocki et al (1988) was that the nonlinear evolution leads to strong reverse shocks, a robust result that follows directly from the much stronger amplification of inward-propagating waves (see above). So-called smooth source function models (Owocki 1991) based on a simple approximation to incorporate the potentially stabilizing effect of the "line-drag" by the diffuse, scattered radiation field (Lucy 1984, Owocki & Rybicki 1985) have proven to be significantly more stable than the first models based on pure absorption, particularly in the lower wind. The outer wind, however, still develops extensive structure that consists, as before, of strong reverse shocks separating slower, dense shells from high-speed rarefied regions in between: Only a very small fraction of material is accelerated to high speed and then shocked; for most of the matter, the major effect is a compression into narrow, dense "clumps" (shells in these 1-D models), separated by large regions of much lower density.

Feldmeier (1995) extended these models by relaxing the assumption of isothermality, accounting for the energy transport, including radiative cooling. By investigating the influence of various photospheric disturbances that induce the onset of structure formation (via exciting the line-force instability), Feldmeier et al (1997b) concluded that cloud-cloud collisions [and not the cooling of the reverse shocks themselves (e.g. Cooper & Owocki 1992)] are the actual reason for the observed X-ray emission (cf Section 2.4.5).

Current effort concentrates on 2/3-D simulations of winds in order to clarify the effects of rotation (cf Section 4.2) as well as the role of the line instability with respect to CIR models (Section 5.1). Owocki (1999) has given some impressive, however preliminary, results of first simulations, which indicate that the small-scale, intrinsic line instability can completely disrupt any kind of large-scale structure if it becomes too strong, as seems to be the case for the current model generation.

5.3   Implications with Respect to Global Parameters

Obviously, time-dependent models based on a nonlocal line force provide a satisfying explanation for a large number of observational findings. The results with respect to X-ray emission are most promising, and the presence of shocks also allows construction of a consistent description of nonthermal radio emission in winds from single stars. Because the latter scenario requires shocks that have survived until large radii, it seems at least plausible that only a fraction of massive stars are seen as nonthermal emitters (cf Chen & White 1994).

One the other hand and at first glance, these models appear to be in strong contrast with our assumptions for the standard model for wind diagnostics based on stationarity and homogeneity, especially when viewed with respect to the spatial variation of velocity and density. However, when viewed with respect to the mass distribution of these quantities, the models are not so very different (Owocki et al 1988, Owocki 1992, Puls et al 1993a). Furthermore, gross wind properties, such as terminal flow speed and time-averaged mass-loss rate, turn out to be in good agreement with those following from a stationary approach [despite the problem of (very) thin winds] (cf Section 4).

Given the intrinsic mass weighting of spectral formation (at least for resonance lines with constant ionization fraction) and the extensive temporal and spatial averaging involved, the observational properties of such structured models may be similar to what is derived from the "conventional" diagnostics, in an average sense. In detailed line-formation calculations assuming line opacities proportional to the local density, Puls et al (1993a, 1994) and Owocki (1994) showed that this conjecture is actually justified, and they were able to demonstrate that unstable winds are excellent candidates for explaining the observed black troughs.

In summary, such models seem able to offer the possibility of retaining the successes of stationary models in matching time-averaged observational properties, while also reproducing the spectral signatures (X-rays, nonthermal radio emission, black troughs, first hints on the nature of DACs, and modulation features), which suggests the existence of extensive wind structure.

However, as should be clear from Section 2.4, only part of the described diagnostic methods relies on features coupled to processes with opacities $\propto \rho$. The determination of $\dot{M}$ (and β) depends on opacities/emissivities $\propto \rho^2$, regardless of whether it is determined via H_α, IR excess, radio fluxes, or SiIV resonance lines [in so far as the calibration of mean ionization fractions of SiIV to radio and H_α mass-loss rates (Section 2.4.1) is intrinsically consistent!], so that the assumption of a stationary wind without clumping seems questionable in view of the perspectives outlined above.

Let us first concentrate on the results derived from H_α, the primary diagnostic tool also for extragalactic work. At least for the majority of OB-stars, the corresponding wind emission comes from lower wind layers, typically between 1.0 and 1.5 stellar radii. Hydrodynamical simulations of self-excited wind instabilities show that these layers seem to be unaffected by shocks and that instabilities only occur further out in the wind (Owocki 1994). This agrees with the fact that the observed X-rays most probably are emitted in the outer layers (Hillier et al 1993) and that the black troughs considered above can arise also from small-scale structures present only in the outer wind (see also Owocki 1994).

Nevertheless, instabilities producing a small-density contrast might still be present in the H_α forming region, if the wind is triggered by photospheric perturbations [sound waves, (non)radial pulsations], in accordance with observational findings that disturbances (of low amplitude) are seen throughout the wind (Section 5.1). In such cases, the location of the onset of structure formation depends crucially on the damping by the diffuse radiation force. Recent hydrodynamical simulations including photospheric perturbations (Feldmeier et al 1997b) show pronounced inhomogeneous structures only above 1.3 R_*, and even in the most elaborate models, including the effects of disturbances in the diffuse radiation field (Owocki & Puls 1999), the same result is found. Thus it seems probable that the neglect of clumping does not induce large systematic errors for OB-star mass-loss determinations. For A-supergiants, the situation remains unclear because hydrodynamical simulations in the corresponding parameter space have yet not been performed and because the lines are formed all the way out to larger radii.[5]

A second line of reasoning in favor of our assumption was given by Lamers & Leitherer (1993) and Puls et al (1993b). Briefly, they argued that because of the same ρ^2 dependence of both the radio and the H_α emission and the fact that both rates agree for those objects with H_α and radio mass-loss rates, this would imply the same clumping factor in regions close to H_α and far away from the star (radio). As this is rather unlikely because in the lower wind the formation of structure just sets in, whereas in the outer part any structure should have stabilized, it is probable that the degree of clumping in the lower wind part is small, if present at all. A scenario consistent with all observational facts has been described by Feldmeier et al (1997b), based on a simulation extended to 100 stellar radii: The lower wind remains rather smooth (damping of the instability), the intermediate part is clumpy (line-force instability "at work"), and the outermost part (above 10 R_*) is rather smooth again, related directly and indirectly to inefficient radiative cooling in this region. Note, however, that a few strong shocks actually survived up to large radii in this model, just as required to produce nonthermal radio emission.

As a consequence of this scenario, one would expect erroneous mass-loss rates only from diagnostics of the intermediate wind part, in accordance with the findings of an inconsistency of IR excess and radio mass-loss rate from a few stars described by Runacres & Blomme (1996) and the discrepancy for the mass-loss rate of HD 53138 discussed in Section 2.4.4.

If, in contrast, one were to assume the worst case of significant structure everywhere in the wind, this would imply that all current results from refined non-LTE diagnostics (Section 2.2) are simply produced by chance and are unreliable. However, because the detailed non-LTE calculations for smooth winds agree well with observed spectra for a variety of ions with different dependences on ρ, and because

---

[5]Note that at least for WR winds, clumping seems to be decisive because the analysis of the mass spectrum of the observed blobs implies a factor three overestimate in the mass loss derived from IR and radio observations (Moffat & Robert 1994), in rough agreement with the results from detailed line fits based on clumped wind models referred to in Section 5.1.

significant structure everywhere in the wind would disturb the ionization balance severely, this implication seems improbable.

Even if this were the case, however, the method to use the WLR of hot stars as an extragalactic distance indicator (see Section 6) would work because of the empirical calibration with stars at known distances as a function of spectral type and metallicity. On the other hand, all global quantities (Section 2.1) involving the mass-loss rate (linearly) would then have to be reduced by the square root of the effective clumping factor.

6.   WINDS AND EXTRAGALACTIC STELLAR ASTRONOMY—OUTLOOK

Stellar winds with their broad and easily detectable spectral features distributed over the whole wavelength range from the UV to the IR are a gift of nature. With the new ground-based telescopes of the 8-m to 10-m class, stellar wind lines can easily be identified in medium-resolution spectra of blue supergiants individually observable out to distances of 20 Mpc (Kudritzki 2000) or in the integrated spectra of starburst regions in galaxies even much more distant, out to redshifts of $z \sim 4$ (Steidel et al 1996). Using the know-how obtained from the investigation of winds in the galaxies of our local group, these stellar wind features can provide unique information about young stellar populations, chemical composition, galactic evolution, and extragalactic distances.

Figure 12 shows the observed spectrum of the gravitationally lensed, highly redshifted starburst galaxy cB58 around the wavelength of the CIV resonance line, revealing a typical stellar wind P-Cygni profile blended by interstellar lines. As demonstrated by Pettini et al (2000), this line feature can be used to constrain the star formation process and metallicity. The authors conclude from the strength of the CIV feature that the metallicity must be smaller than solar. This conclusion is confirmed by the work of Walborn et al (1995) and Haser et al (1998), which indicates that metallicity has a substantial influence on the morphology of stellar wind lines.

Figure 13 shows the "modified" stellar wind momenta of A-supergiants in the galaxy and M31 as a function of absolute magnitude. As demonstrated by Kudritzki et al (1999), such a relationship can be used to determine extragalactic distances with an accuracy of about 0.1 mag in distance modulus out to the Virgo and Fornax clusters of galaxies. A-supergiants as the optically brightest "normal stars" with absolute magnitudes between -9 and -8 are ideal for this purpose. Their wind momenta can be determined solely by optical spectroscopy at H_α. The quantitative analysis of the rest of the optical spectrum will yield effective temperature, gravity, chemical composition and—in conjunction with accurate photometry—reddening, and extinction so that an application of the wind momentum–luminosity relationship, properly calibrated as a function of metallicity in local group galaxies, will finally yield accurate distances.

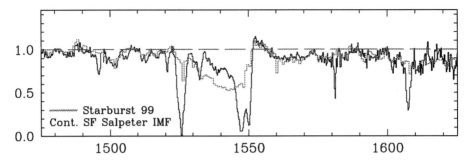

Figure 12   Comparison between the observed rest-frame spectrum of the starburst galaxy cB58 (*black histogram*) at redshift $z = 2.72$ in the region of the Civ λ 1549 stellar wind line with a spectral synthesis model, assuming continuous star formation and a Salpeter IMF. Note that the population synthesis code applied uses a library of observed spectra of galactic stars with abundances corresponding to the galactic disk at the solar neighborhood (see Leitherer et al 1999). From Pettini et al (2000).

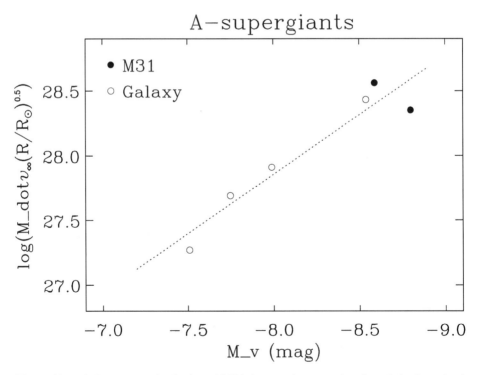

Figure 13   Wind momenta of galactic and M31 A-supergiants as a function of absolute visual magnitude. The data of the M31 objects are from McCarthy et al (1997). (*dotted*) The linear regression obtained from all objects. From Kudritzki et al (1999).

Visit the Annual Reviews home page at www.AnnualReviews.org

LITERATURE CITED

Abbott DC. 1978. *Ap. J.* 225:893

Abbott DC. 1979. *Proc. IAU Symp.* 83, p. 237

Abbott DC. 1980. *Ap. J.* 242:1183

Abbott DC. 1982. *Ap. J.* 259:282

Abbott DC, Bieging JH, Churchwell E, Cassinelli JP. 1980. *Ap. J.* 238:196

Abbott DC, Bieging JH, Churchwell E. 1981. *Ap. J.* 250:645

Abbott DC, Bieging JH, Churchwell E. 1984a. *Ap. J.* 280:671

Abbott DC, Telesco CM, Wolff SC. 1984b. *Ap. J.* 279:225

Achmad L, Lamers HJGLM, Pasquini L. 1997. *Astron. Astrophys.* 320:196

Altenhoff WJ, Thum C, Wendker H. 1994. *Astron. Astrophys.* 281:161

Babel J. 1995. *Astron. Astrophys.* 301:823

Babel J. 1996. *Astron. Astrophys.* 309:867

Barlow MJ, Cohen M. 1977. *Ap. J.* 213:737

Bates B, Gilheany S. 1990. *MNRAS* 243:320

Becker SR, Butler K. 1992. *Astron. Astrophys.* 265:647

Berghöfer TW, Schmitt JHMM. 1994. *Science* 265:1689

Berghöfer TW, Baade D, Schmitt JHMM, Kudritzki RP, Puls J, et al. 1996. *Astron. Astrophys.* 306:899

Berghöfer TW, Schmitt JHMM, Danner R, Cassinelli JP. 1997. *Astron. Astrophys.* 322:167

Bertout C, Leitherer C, Stahl O, Wolf B. 1985. *Astron. Astrophys. A* 144:87

Bieging JH, Abbott DC, Churchwell EB. 1989. *Ap. J.* 340:518

Bjorkman JE. 1999. *Proc. IAU Colloq. 169*, p. 121. Heidelberg: Springer-Verlag

Bjorkman JE, Cassinelli JP. 1993. *Ap. J.* 409:429

Blomme R, Runacres MC. 1997. *Astron. Astrophys.* 323:886

Bohannan B, Abbott DC, Voels SA, Hummer DG. 1986. *Ap. J.* 308:728

Carlberg RG. 1980. *Ap. J.* 241:1131

Cassinelli JP, Hartmann L. 1977. *Ap. J.* 212:488

Cassinelli JP, Swank JH. 1983. *Ap. J.* 271:681

Cassinelli JP, Cohen DH, MacFarlane JJ, Sanders WT, Welsh BY. 1994. *Ap. J.* 421:705

Castor JI. 1970. *MNRAS* 149:111

Castor JI, Simon T. 1983. *Ap. J.* 265:304

Castor JI, Abbott DC, Klein RI. 1975. *Ap. J.* 195:157

Chapman JM, Leitherer C, Koribalski B, Bouter R, Storey M. 1999. *Ap. J.* 518:890

Chen W, White RL. 1994. *Proc. Isle-aux-Coudre Workshop "Instability Variability Hot-Star Winds"*. *Astrophys. Space Sci.* 221:259

Chlebowski T, Harnden FR, Sciortino S. 1989. *Ap. J.* 341:427

Cohen DH, Cooper RG, McFarlane JJ, Owocki SP, Cassinelli JP. 1996. *Ap. J.* 460:506

Cohen DH, Cassinelli JP, McFarlane JJ. 1997a. *Ap. J.* 487:867

Cohen DH, Cassinelli JP, Waldron WL. 1997b. *Ap. J.* 488:397

Conti PS, Blum RD. 1998. *Proc. 2nd Boulder-Munich Workshop. PASPC* 131:24

Contreras ME, Rodriguez LF, Gomez Y, Velazquez A. 1996. *Ap. J.* 469:329

Cooper RG, Owocki SP. 1992. *Proc. "Non-isotropic Variable Outflows from Stars" PASPC* 22:281

Cranmer SR, Owocki SP. 1995. *Ap. J.* 440:308

Cranmer SR, Owocki SP. 1996. *Ap. J.* 462:469

Crowther PA, Bohannan B, Pasquali A. 1998. *Proc. 2nd Boulder-Munich Workshop. PASPC* 131:38

de Koter A, Heap SR, Hubeny I. 1998. *Ap. J.* 509:879

Drake SA, Linsky JL. 1989. *Astron. J.* 98:1831

Drew JE. 1990. *Proc. 1st Boulder-Munich Workshop. PASPC* 7:218

Drew JE. 1998. *Proc. 2nd Boulder-Munich Workshop. PASPC* 131:14

Ebbets DC. 1982. *Ap. JS* 48:399

Eversberg T, Lepine S, Moffat AFJ. 1998. *Ap. J.* 494:799

Feldmeier A. 1995. *Astron. Astrophys.* 299:523

Feldmeier A. 1998. *Astron. Astrophys.* 332:245

Feldmeier A. 1999. *Proc. IAU Colloq. 169*, p. 285. Heidelberg: Springer-Verlag

Feldmeier A, Kudritzki RP, Palsa R, Pauldrach AWA, Puls J. 1997a. *Astron. Astrophys.* 320:899

Feldmeier A, Pauldrach AWA, Puls J. 1997b. *Astron. Astrophys.* 322:878

Figer DF, McLean IS, Najarro F. 1997. *Ap. J.* 486:1117

Figer DF, Najarro F, Morris M, McLean IS, Geballe TR, et al. 1998. *Ap. J.* 506:384

Friend DB, Abbott DC. 1986. *Ap. J.* 311:701

Fullerton AW. 1999. *Proc. IAU Colloq. 169*, p. 3. Heidelberg: Springer-Verlag

Fullerton AW, Gies DR, Bolton CT. 1992. *Ap. J.* 390:650

Fullerton AW, Gies DR, Bolton CT. 1996. *Ap. J.* 103:475

Fullerton AW, Massa, DL, Prinja RK, Owocki SP, Cranmer SR. 1997. *Astron. Astrophys.* 327:699

Gabler R, Gabler A, Kudritzki RP, Pauldrach AWA, Puls J. 1989. *Astron. Astrophys.* 226:162

Gabler A, Gabler R, Kudritzki RP, Puls J, Pauldrach A. 1990. *Proc. 1st Boulder-Munich Workshop. PASPC* 7:218

Gabler R, Kudritzki RP, Méndez RH. 1991. *Astron. Astrophys.* 245:587

Gabler R, Gabler A, Kudritzki RP, Méndez RH. 1992. *Astron. Astrophys.* 265:656

Gezari DY, Pitts PS, Schmitz M. 1999. *Catalog of Infrared Observations. Ed. 5.* ftp://cdsarc.u-strasbg.fr/pub/cats/II/225/

Grady CA, Snow TP, Timothy JG. 1983. *Ap. J.* 271:691

Grinin A. 1978. *Sov. Astron.* 14:113

Groenewegen MAT, Lamers HJGLM. 1989. *Astron. Astrophys.* 79:359

Groenewegen MAT, Lamers HJGLM, Pauldrach AWA. 1989. *Astron.Astrophys.* 221:78

Hamann WR. 1981a. *Astron. Astrophys.* 93:353

Hamann WR. 1981b. *Astron. Astrophys.* 100:169

Hamann WR, Koesterke L. 1998. *Astron. Astrophys.* 335:1003

Hanson MM, Conti PS. 1994. *Ap. J. Lett.* 423:L139

Hanson MM, Conti PS, Rieke MJ. 1996. *Ap. J.* 107:281

Hanson MM, Howarth ID, Conti PS. 1997. *Ap. J.* 489:698

Hanson MM. 1998. *Proc. 2nd Boulder-Munich Workshop. PASPC* 131:1

Harnden FR, Branduardi G, Gorenstein P, Grindlay J, Rosner R, et al. 1979. *Ap. J. Lett.* 234:L51

Haser SM. 1995. *Spektroskopie heißer Sterne der Lokalen Gruppe im ultravioletten Spektralbereich.* PhD thesis. Ludwig-Maximilians Universität, München

Haser SM, Puls J, Kudritzki RP. 1994. *Space Sci. Rev.* 66:187

Haser SM, Lennon DJ, Kudritzki RP, Puls J, Pauldrach AWA, et al. 1995. *Astron. Astrophys.* 295:136

Haser SM, Pauldrach AWA, Lennon DJ, Kudritzki RP, Lennon M, et al. 1998. *Astron. Astrophys.* 330:285

Henrichs HF. 1984. *Proc. 4th Eur. IUE Conf., ESA-SP 218* , p. 43

Henrichs HF. 1991. *Proc. "Rapid Variability OB-Stars: Nature and Diagnostic Value".* *ESO Conf. Workshop Proc.* 36:199

Henrichs HF, Kaper L, Nichols JS. 1994. *Proc. IAU Symp. 162*, p. 517. Dordrecht: Kluwer

Henrichs HF, Kaper L, Zwarthoed GAA. 1988. *Proc. "A Decade UV Astron. IUE Satellite".* *ESA-SP 281*, 2:145

Herrero A, Kudritzki RP, Vilchez JM, Kunze D, Butler K, et al. 1992. *Astron. Astrophys.* 261:209

Herrero A, Corral LJ, Villamariz MR, Martin EL. 1999. *Astron. Astrophys.* 348:542

Herrero A, Puls J, Villamariz MR. 2000. *Astron. Astrophys.* 354:193

Hillier DJ. 1991. *Astron. Astrophys.* 247:455

Hillier DJ, Miller DL. 1998. *ApJ.* 496:407

Hillier DJ, Crowther PA, Najarro F, Fullerton AW. 1998. *Astron. Astrophys.* 340:438

Hillier DJ, Kudritzki RP, Pauldrach AWA,

Baade D, Cassinelli JP. 1993. *Astron. Astrophys.* 276:117

Howarth ID. 1992. *Proc. "Nonisotropic Variable Outflows Stars".* *PASPC* 22:155

Howarth ID, Brown AB. 1991. *Proc. IAU Symp. 143*, p. 289. Dordrecht: Kluwer

Howarth ID, Prinja RK. 1989. *Ap. J. Supp.* 69:527

Howarth ID, Prinja RK, Massa D. 1995. *Ap. J.* 452:L65

Howarth ID, Siebert KW, Hussain GA, Prinja RK. 1997. *MNRAS* 284:265

Hubeny I, Heap SR, Lanz T. 1998. *Proc. 2nd Boulder-Munich Workshop.* *PASPC* 131:108

Humphreys RM, McElroy DB. 1984. *Ap. J.* 284:565

Kaper L. 1993. *Wind variability in early-type stars.* PhD thesis. Univ. Amsterdam

Kaper L, Fullerton AW, eds. 1998. Cyclical variability in stellar winds. *ESO Astrophys. Symp.* Heidelberg: Springer-Verlag

Kaper L, Henrichs HF. 1994. *Proc. Isle-aux-Coudre Workshop* "Instability Variability Hot-Star Winds". *Astrophys. Space Sci.* 221:115

Kaper L, Henrichs HF, Fullerton AW, Ando H, Bjorkman KS, et al. 1997. *Astron. Astrophys.* 327:281

Kaper L, Henrichs HF, Nichols JS, Telting JH. 1999. *Astron. Astrophys.* 344:231

Kaufer A, Stahl O, Wolf B, Gäng T, Gummersbach CA, et al. 1996. *Astron. Astrophys.* 305:887

Klein RI, Castor JI. 1978. *Ap. J.* 220:902

Kudritzki RP. 1979. *Proc. 22nd Liege Int. Symp.*, p. 295

Kudritzki RP. 1980. *Astron. Astrophys.* 85:174

Kudritzki RP. 1988. *18th Adv. Course Swiss Soc. Astrophys. Astron.*, Saas-Fee Courses, Geneva Obs., 1

Kudritzki RP. 1998. Quantitative spectroscopy of the brightest blue supergiant stars in galaxies. *Proc. 8th Canary Winter Sch.*, p. 149. Cambridge Univ. Press

Kudritzki RP. 1999a. *Proc. IAU Colloq. 169*, p. 405. Heidelberg: Springer-Verlag

Kudritzki RP. 1999b. *Proc. STScI Symp.* "Unsolved Problems of Stellar Evolution", In press

Kudritzki RP. 2000. *VLT Opening Symp. ESO.* In press

Kudritzki RP, Simon KP, Hamann WR. 1983. *Astron. Astrophys.* 118:254

Kudritzki RP, Pauldrach AWA, Puls J. 1987. *Astron. Astrophys.* 173:293

Kudritzki RP, Pauldrach A, Puls J, Abbott DC. 1989. *Astron. Astrophys.* 219:205

Kudritzki RP, Hummer DG, Pauldrach AWA, Puls J, Najarro J, et al. 1992. *Astron. Astrophys.* 257:655

Kudritzki RP, Lennon DJ, Puls J. 1995. *Proc. ESO Workshop "Science VLT"*, p. 246. Heidelberg: Springer-Verlag

Kudritzki RP, Palsa R, Feldmeier A, Puls J, Pauldrach AWA. 1996. *Proc. "Roentgenstrahlung from the Universe".* *MPE Rep.* 263:9

Kudritzki RP, Méndez RH, Puls J, McCarthy JK. 1997. *Proc. IAU Symp. 180*, p. 64

Kudritzki RP, Springmann U, Puls J, Pauldrach AWA, Lennon M. 1998. *Proc. 2nd Boulder-Munich Workshop.* *PASPC* 131:299

Kudritzki RP, Puls J, Lennon DJ, Venn KA, Reetz J, et al. 1999. *Astron. Astrophys.* 350:970

Lamers HJGLM, Waters LBFM. 1984a. *Astron. Astrophys.* 136:37

Lamers HJGLM, Waters LBFM. 1984b. *Astron. Astrophys. A* 138:25

Lamers HJGLM, Pauldrach AWA. 1991. *Astron. Astrophys. A* 244:L5

Lamers HJGLM, Leitherer C. 1993. *Ap. J.* 412:771

Lamers HJGLM, Gathier R, Snow TP. 1982. *Ap. J.* 258:186

Lamers HJGLM, Waters LBFM, Wesselius PR. 1984. *Astron. Astrophys.* 134:L17

Lamers HJGLM, Cerruti-Sola M, Perinotto M. 1987. *Ap. J.* 314:726

Lamers HJGLM, Snow TP, Lindholm DM. 1995. *Ap. J.* 455:269

Lamers HJGLM, Najarro F, Kudritzki RP, Morris PW, Voors RHM, et al. 1996. *Astron. Astrophys.* 315:L225

Lamers HJGLM, Haser SM, de Koter A, Leitherer C. 1999a. *Ap. J.* 516:872

Lamers HJGLM, Vink JS, de Koter A, Cassinelli JP. 1999b. *Proc. IAU Colloq. 169*, p. 159. Heidelberg: Springer-Verlag

Langer N, Heger A. 1998. *Proc. 2nd Boulder-Munich Workshop. PASPC* 131:76

Langer N, Hamann WR, Lennon M, Najarro F, Pauldrach AWA, et al. 1994. *Astron. Astrophys.* 290:819

Lanz T, de Koter A, Hubeny I, Heap SR. 1996. *Ap. J.* 465:359

Leitherer C. 1988. *Ap. J.* 326:356

Leitherer C. 1998. *Proc. 8th Canary Winter Sch.* p. 527. Cambridge Univ. Press

Leitherer C, Robert C. 1991. *Ap. J.* 377:629

Leitherer C, Robert C, Drissen L. 1992. *Ap. J.* 401:596

Leitherer C, Chapman JM, Koribalski B. 1995. *Ap. J.* 450:289

Leitherer C, Schaerer D, Goldader JD, Gonzalez RM, Robert C, et al. 1999. *Ap. J.* 123:3

Lennon DJ, Kudritzki RP, Becker ST, Butler K, Eber F, et al. 1991. *Astron. Astrophys.* 252:498

Lucy LB, White RL. 1980. *Ap. J.* 241:300

Lucy LB. 1982a. *Ap. J.* 255:278

Lucy LB. 1982b. *Ap. J.* 255:286

Lucy LB. 1983. *Ap. J.* 274:372

Lucy LB. 1984. *Ap. J.* 284:351

Lupie OL, Nordsieck KH. 1987. *Astron. J* 93:214

MacFarlane JJ, Cassinelli JP. 1989. *Ap. J.* 347:1090

MacFarlane JJ, Waldron WL, Corcoran MF, Wolff MJ, Wang P, et al. 1993. *Ap. J.* 419:813

MacGregor KB, Hartmann L, Raymond JC. 1979. *Ap. J.* 231:514

Maeder A. 1998. *Proc. 2nd Boulder-Munich Workshop. PASPC* 131:85

Maeder A. 1999. *Astron. Astrophys.* 347:185

Massa D, Fullerton AW, Nichols JS, Owocki SP, Prinja RK, et al. 1995. *Ap. J.* 452:L53

McCarthy JK, Lennon DJ, Venn KA, Kudritzki RP, Puls J, et al. 1995. *Ap. J. Lett.* 455:L35

McCarthy JK, Kudritzki RP, Lennon DJ, Venn KA, Puls J. 1997. *Ap. J.* 482:757

McCarthy JK, Venn KA, Lennon DJ, Kudritzki RP, Puls J. 1998. *Proc. 2nd Boulder-Munich Workshop. PASPC* 131:197

McErlean ND, Lennon DJ, Dufton PL. 1998 *Astron. Astrophys.* 329:613

McErlean ND, Lennon DJ, Dufton PL. 1999 *Astron. Astrophys.*, In press

Méndez RH, Kudritzki RP, Herrero A, Husfeld D, Groth HG. 1988. *Astron. Astrophys.* 190:113

Meynet G. 1998. *Proc. 2nd Boulder-Munich Workshop. PASPC* 131:96

Mihalas D. 1978. *Stellar Atmospheres.* Freeman: San Francisco. 2nd ed.

Moffat AFJ, Robert C. 1994. *Ap. J.* 421:310

Moffat AFJ, Owocki SP, Fullerton AW, StLouis N, eds. 1994. *Proc. Isle-aux-Coudre Workshop "Instability and Variability of Hot-Star Winds". Astrophys. Space Sci.* 221:467

Mullan DJ. 1984. *Ap. J.* 283:303

Mullan DJ. 1986. *Astron. Astrophys.* 165:157

Najarro F, Hillier JD, Kudritzki RP, Krabbe A, Genzel R, et al. 1994. *Astron. Astrophys.* 285:573

Najarro F, Kudritzki RP, Cassinelli JP, Stahl O, Hillier DJ. 1996. *Astron. Astrophys.* 306:892

Najarro F, Hillier DJ, Stahl O. 1997a. *Astron. Astrophys.* 326:1117

Najarro F, Krabbe A, Genzel R, Lutz D, Kudritzki RP, et al. 1997b. *Astron. Astrophys.* 325:700

Najarro F, Kudritzki RP, Hillier DJ, Lamers HJGLM, Voors RHM, et al. 1998. *Proc. 2nd Boulder-Munich Workshop. PASPC* 131:57

Napiwotzki R, Green PJ, Saffer RA. 1999. *Ap. J.* 517:399

Olson GL. 1982. *Ap. J.* 255:267

Owocki SP. 1991. *Proc. "Stellar Atmospheres: Beyond Classical Models". NATO ASI Ser. C* 341:235. Dordrecht: Kluwer

Owocki SP. 1992. *Proc. "The Atmospheres of Early-Type Stars"*, p. 393. Berlin: Springer-Verlag

Owocki SP. 1994. *Proc. Isle-aux-Coudre Workshop "Instability and Variability of Hot-Star Winds". Astrophys. and Space Sci.* 221:3

Owocki SP. 1999. *Proc. IAU Colloq. 169*, p. 294. Heidelberg: Springer-Verlag

Owocki SP, Rybicki GB. 1984. *Ap. J.* 284:337

Owocki SP, Rybicki GB. 1985. *Ap. J.* 299:265

Owocki SP, Puls J. 1996. *Ap. J.* 462:894

Owocki SP, Puls J. 1999. *Ap. J.* 510:355

Owocki SP, Cohen DH. 1999. *Ap. J.* 520:833

Owocki SP, Castor JI, Rybicki GB. 1988. *Ap. J.* 335:914

Owocki SP, Cranmer SR, Fullerton AW. 1995. *Ap. J.* 453:L37

Owocki SP, Cranmer SR, Blondin JM. 1994. *Ap. J.* 424:887

Owocki SP, Cranmer SR, Gayley KG. 1996. *Ap. J.* 472:L1150

Owocki SP, Cranmer SR, Gayley KG. 1998a. *Proc. Workshop "B[e] Stars". Astrophys. and Space Sci.* 233:205. Dordrecht: Kluwer

Owocki SP, Gayley KG, Cranmer SR. 1998b. *Proc. 2nd Boulder-Munich Workshop. PASPC* 131:237

Pallavicini R, Golub L, Rosner R, Vaiana GS, Ayres T, et al. 1981. *Ap. J.* 248:279

Panagia N, Felli M. 1975. *Astron. Astrophys.* 39:1

Pauldrach AWA, Puls J. 1990. *Astron. Astrophys.* 237:409

Pauldrach AWA, Puls J, Kudritzki RP. 1986. *Astron. Astrophys.* 164:86

Pauldrach AWA, Puls J, Kudritzki RP, Méndez RH, Heap SR. 1989. *Astron. Astrophys.* 207:123

Pauldrach AWA, Puls J, Kudritzki RP, Butler K. 1990. *Astron. Astrophys.* 228:125

Pauldrach, AWA, Kudritzki RP, Puls J, Butler K, Hunsinger J. 1994. *Astron. Astrophys.* 283:525

Pauldrach AWA, Lennon M, Hoffmann TL, Sellmaier F, Kudritzki RP, Puls J. 1998. *Proc. 2nd Boulder-Munich Workshop. PASPC* 131:258

Perinotto M. 1993. *Proc. IAU Symp. 155*, p. 57. Dordrecht: Kluwer

Petrenz P. 1999. *Selbstkonsistente Modelle strahlungsdruckgetriebener Winde heisser Sterne unter Mitberücksichtigung der Rotation.* PhD thesis. Ludwig-Maximilians Universität, München

Petrenz P, Puls J. 1996. *Astron. Astrophys.* 312:195

Pettini M, Steidel CC, Adelberger KL, Dickinson M, Giavalisco M. 2000 *Ap. J.*, 528:96

Pollock AMT. 1989. *Proc. Int. Sch. Workshop on Reconnect. in Space Plasma*, 2:309

Prinja RK. 1988. *MNRAS* 231:21P

Prinja RK. 1992. *Proc. "Nonisotropic and Variable Outflows Stars". PASPC* 22:167

Prinja RK, Howarth ID. 1986. *Ap. JS* 61:357

Prinja RK, Howarth ID. 1988. *MNRAS* 233:123

Prinja RK, Smith LJ. 1992. *Astron. Astrophys.* 266:377

Prinja RK, Crowther PA. 1998. *MNRAS* 300:828

Prinja RK, Massa DL. 1998. *Proc. 2nd Boulder-Munich Workshop. PASPC* 131:218

Prinja RK, Howarth ID, Henrichs HF. 1987. *Ap. J.* 317:389

Prinja RK, Barlow MJ, Howarth ID. 1990. *Ap. J.* 361:607

Prinja RK, Balona LA, Bolton CT, Crowe RA, Fieldus MS, et al. 1992. *Ap. J.* 390:266

Prinja RK, Massa DL, Fullerton AW. 1995. *Ap. J.* 452:L61

Puls J. 1987. *Astron. Astrophys.* 184:227

Puls J. 1993. *Winde heißer massereicher Sterne—Theoretische Ansätze und diagnostische Methoden.* Habilitation thesis. *Ludwig-Maximilians Universität, München*

Puls J, Owocki SP, Fullerton AW. 1993a. *Astron. Astrophys.* 279:457

Puls J, Pauldrach AWA, Kudritzki RP, Owocki SP, Najarro F. 1993b. *Rev. in Mod. Astron.* 6:271.

Puls J, Feldmeier A, Springmann UWE, Owocki SP, Fullerton AW. 1994. *Proc. Isle-aux-Coudre Workshop "Instability and Variability of Hot-Star Winds". Astrophys. Space Sci.* 221:409

Puls J, Kudritzki RP, Herrero A, Pauldrach AWA, Haser SM, et al. 1996. *Astron. Astrophys.* 305:171

Puls J, Kudritzki RP, Santolaya-Rey AE,

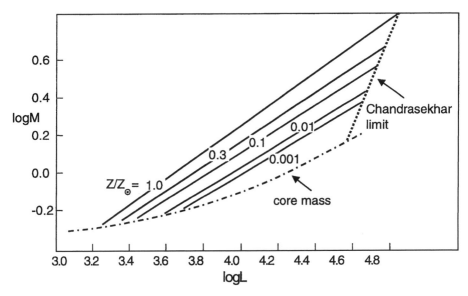

Figure 9 Cliff lines for five metallicities: 1, 0.3, 0.1, 0.01, and 0.001 times solar, with the Paczynski (1970) core mass relation also shown as a rough guide to the core mass. The shift of the cliff to higher L for lower Z comes primarily from the smaller radii at a given L of the lower-metallicity stars; the effects of dust are important only for the largest metallicities and the larger masses (Bowen 2000, in preparation). (Figure courtesy of GH Bowen.)

the relations that went into defining the cliff may need to be changed. The transition to C > O may change the relations among T_{eff} and M and L as the opacity changes in the outer envelope of the star. It greatly changes the character of the dust that plays a role in driving mass loss from some of these stars. Carbon star variables are not often classified as Mira variables, because they generally show smaller visual amplitudes, for reasons that are not yet clear. Unfortunately, in addition to the preceding uncertainties regarding the mass loss from carbon stars, we also do not yet have a reliable means of predicting which stars will become carbon stars and when, because this process involves mixing from the nuclear burning regions presumably associated with the helium shell flashes (Iben & Renzini 1983). Further, mass loss affects whether model stars become carbon stars (Frost et al 1998a,b; Lattanzio 1998). Thus we don't even know yet where to draw the gaps in the cliff lines in Figure 9, much less where to put the C* cliffs. Finally, carbon star mass loss models fully compatible with available oxygen-rich star models do not yet exist, although progress is continuing toward that goal. However, all is not lost. There are observational hints that the cliffs may not be very far apart: The P-L relation for Magellanic Cloud carbon star variables is not very different from the P-L relation for oxygen-rich Miras (Hughes & Wood 1990). Also, the observed mass loss rates of carbon stars in the LMC fall in about the same region of the Ṁ versus log L plots that we expect the oxygen-rich stars to occupy (Van Loon et al 1997), also hinting that the cliffs may not be too far separated.

The second largely unsolved problem is understanding the modulation of the mass loss that takes place over the shell flash cycle. There is interest in this problem, and attempts have been made to solve it using a variety of tools. While it is straightforward to derive a mass loss rate formula from constant L, M, R models and then apply this formula to the time-variable L and R of a shell flash, this probably does not give a very good picture of the true event. Typically, the mass loss models need to be gradually brought to full amplitude or to a new period of oscillation; if this is not done, transient mass loss events develop (Hill & Willson 1979, Bowen 1988). Such transient events are likely to occur in real stars as well, and will require linked evolutionary and pulsation models.

The third unsolved problem is that of the mass loss that occurs at or near the tip of the red giant branch. Models for the evolution of low-mass red giants prior to the He core flash do not make it into the heavy mass loss region that the stars reach on the AGB. Is there mass loss associated with the core flash event? In contrast with earlier work by Schwarzschild and Härm (1962, Härm & Schwarzschild 1966) and Tomasko (1970) arguing that the main event is slow enough to avoid becoming hydrodynamic (although they didn't rule out some surface hydrodynamic effects), Cole and Deupree (1981) found that hydrodynamic loss of some of the envelope mass in response to a core flash is possible. Models are not yet to the state of being able to predict which stars will lose how much mass during a helium core flash, however. From the observational side, the distribution of stars along the horizontal branch constrains that mass loss for low-metallicity stars below about 1 solar mass; for these, one to several tenths of a solar mass of material must come off at that time. The Mira period-luminosity relation with the interpretation given above suggests that not very much mass is removed during the helium core flash for 1–3 M_{Sun} stars. For stars above 2 or 3 M_{Sun} there is no core flash (Sweigert & Gross 1978), so we have no reason to expect higher-mass stars to have significant mass loss at the end of the RGB.

In this section, I have focused on the mass loss that occurs near the end of the AGB for intermediate-mass stars. Apart from hints of some mass loss at the core flash, this is the only stage of evolution for stars less than 9 solar masses where we have clear evidence for mass loss sufficient to alter the fate of the star. Are there other stages of evolution where mass loss is important? I have not discussed pre-main-sequence stages, where there are certainly outflows, because these may originate from circumstellar disks more than from the star (Shu et al 1994a,b, 1995; Najita & Shu 1994, Ostriker & Shu 1995). Nor have I mentioned mass loss on the main sequence, although (given the long lifetimes of main sequence stars) such mass loss could occur without being easily detected (Willson et al 1987). Willson & Bowen (1984b) suggested that mass loss associated with pulsation might play a role in reducing evolutionary masses of Cepheids, since these were in conflict with their pulsation masses. Theoretical calculations confirmed that mass loss occurring during the Cepheid stage would work to reduce the evolutionary masses (Brunish & Willson 1987, 1989); however, new opacities have improved the agreement between the evolutionary and pulsation masses, so Cepheid-stage mass loss may not be required (Andreasen 1988). Willson & Bowen (1984b) also suggested that pulsation-related mass loss from the RR Lyrae stars could send them off to the blue horizontal branch instead of up the AGB; theoretical models by Koopmann et al (1994) explored this hypothesis but found that it would not lead to the desired dispersion in color for any rates below 10^{-9} M_{Sun} yr^{-1}, and that higher mass loss rates would create other problems. Mullan (1996) derived high mass loss rates (given the evolutionary timescale) for cool dwarfs, but this result has also been questioned (Lim & White 1996, Van Den Oord & Doyle 1997).

At present the evidence is clear that mass loss at the end of the AGB is both measurable and important. There is no evidence that this is true for any other stage of evolution of stars below 9 M_{Sun}. Therefore, by studying the mass loss on the AGB we may derive most or all of what we need to know to incorporate mass loss into stellar evolution calculations for populations with ages between 10^8 and 10^{10} years.

SUMMARY/CONCLUSIONS

Stars with initial masses between about 1 and 9 solar masses that end up as white dwarf stars must lose a substantial fraction of their mass before they leave the asymptotic giant branch. Observational and theoretical studies over the past three decades have revealed that most of this mass loss occurs near the end of the AGB lifetime. Theoretical studies reveal that the dependence of the mass loss rate on stellar parameters is very steep. Observed correlations between mass loss rates and stellar parameters show much less sensitivity. The resolution of this apparent paradox lies in recognizing that where the mass loss varies over many orders of magnitude, observational selection effects will pick out stars whose mass loss rates are near the maximum value sustainable by the star.

Empirical mass loss laws cannot be derived unless the stellar parameters are well determined. For the Miras, heavily mass-losing stars at the tip of the AGB, the period-luminosity law derived from observations in the Magellanic Clouds appears to be quite good. Hipparcos data show considerably more scatter, possibly because of difficulties associated with measuring stars whose angular diameters equal or exceed the expected parallax. Angular diameters for Miras appear to be much less reliable that we would expect, likely because there is more contribution from scattering in the circumstellar envelope than has been included in any of the models. Progenitor masses deduced from kinematics of Miras are consistent with masses deduced from a theoretical fit to the P-L relation, suggesting that relatively little mass loss actually occurs before the end of the AGB.

Interpretation of observations of winds in cool stars has often been carried out using variations on the classical theory of stationary winds. However, the family of possible solutions for cool stars may be considerably more varied, since time variability (waves, noise from near-surface convection zones, pulsation), departures from spherical symmetry (bipolar flows and diverging coronal holes), and the effects of radiation pressure on (sometimes patchy) dust all play a role. To further complicate the observational side, different methods sample different distances from the star, and thus different times, with evidence for time-variable outflow on scales of decades to 10^5 years.

A framework has been developed for understanding AGB mass loss; there is now a good match between theoretical computations and observed patterns for the AGB. This will allow the extrapolation of relations for terminal AGB luminosity and remnant masses to populations that are difficult to study directly. Young, low-metallicity populations are expected to have less mass loss, and thus higher mass remnants and more supernovae of Type 1.5, as a result of the dual effects of smaller radii at a given L and less dust.

ACKNOWLEDGMENTS

The author would like to thank Jeffrey S. Willson for exorcising the bibliographic demon. Support for dynamical atmosphere and mass loss studies from NASA over the past two decades is also gratefully acknowledged, including current grant NAG58465. This paper would have been quite different without the contributions of G. H. Bowen to research in this field; his generous contribution of several key figures is also acknowledged.

Visit the Annual Reviews home page at www.AnnualReviews.org

LITERATURE CITED

Andreasen GK. 1988. *Astron. Astrophys.* 201: 72–79

Arndt TU, Fleischer AJ, Sedlmayr E. 1997. *Astron. Astrophys.* 327:614–19

Asplund M. 1995. *Astron. Astrophys.* 294: 763–72

Barthes D. 1998. *Astron. Astrophys.* 333: 647–57

Baud B, Habing HJ. 1983. *Astron. Astrophys.* 127:73–83

Becker SA, Iben I, Jr. 1979. *Ap. J.* 232:831–53

Bedijn PJ. 1987. *Astron. Astrophys.* 186: 136–52

Berruyer N. 1991. *Astron. Astrophys.* 249: 181–91

Bessell MS, Brett JM, Wood PR, Scholz M. 1989. *Astron. Astrophys.* 213:209–25

Bester M, Danchi WC, Degiacomi CG, Townes CH, Geballe TR. 1991. *Ap. J.* 367:L27

Blöcker T. 1995. *Astron. Astrophys.* 297:727

Bonneau D, Foy R, Blazit A, Labeyrie A. 1982. *Astron. Astrophys.* 106:235

Bowen GH. 1988. *Ap. J.* 329:299–317

Bowen GH. 1990. In *Numerical Modeling of Nonlinear Stellar Pulsations. Problems and Prospects*, ed. JR Buchler. Dordrecht/Boston:Kluwer. 155 pp.

Bowen GH, Willson LA. 1991. *Ap. J.* 375:L53–56

Bowers PF, Kerr FJ. 1977. *Astron. Astrophys.* 57:115–23

Boyle RP, Aspin C, McLean IS, Coyne GV. 1986. *Astron. Astrophys.* 164:310–20

Brunish WM, Willson LA. 1987. In *Stellar Pulsation*, ed. AN Cox, WM Sparks, SG Starrfield, p. 27. Berlin/New York: Springer-Verlag

Brunish WM, Willson LA. 1989. *The Use of Pulsating Stars in Fundamental Problems of Astronomy*, ed. EG Schmidt. Cambridge, UK: Cambridge Univ. Press. 252 pp.

Cadwell BJ, Wang H, Feigelson ED, Frenklach M. 1994. *Ap. J.* 429:285–99

Chamberlain JW. 1960. *Ap. J.* 131:47

Herrero A, Owocki SP, et al. 1998. *Proc. 2nd Boulder-Munich Workshop. PASPC* 131:245

Puls J, Petrenz P, Owocki SP. 1999 *Proc. IAU Colloq. 169*, p. 131. Heidelberg: Springer-Verlag

Puls J, Springmann U, Lennon M. 2000 *Astron. Astrophys.* 141:23

Runacres MC, Blomme R. 1996. *Astron. Astrophys.* 309:544

Santolaya-Rey AE, Puls J, Herrero A. 1997. *Astron. Astrophys.* 323:488

Schaerer D, Schmutz W. 1994. *Astron. Astrophys.* 288:231

Schaerer D, de Koter A. 1997. *Astron. Astrophys.* 322:598

Schmutz W. 1997. *Astron. Astrophys.* 321:268

Schönberber D. 1983. *Ap. J.* 272:708

Scuderi S, Bonanno G, Di Benedetto R, Spadaro D, Panagia N. 1992. *Ap. J.* 392:201

Scuderi S, Panagia N, Stanghellini C, Trigilio C, Umana G. 1998. *Astron. Astrophys.* 332: 251

Sellmaier F, Puls J, Kudritzki RP, Gabler A, Gabler R, et al. 1993. *Astron. Astrophys.* 273:533

Seward FD, Forman WR, Giacconi R, Griffiths RE, Harnden FR, et al. 1979. *Ap. J. Lett.* 234:L55

Simon KP, Jonas G, Kudritzki RP, Rahe J. 1983. *Astron. Astrophys.* 125:34

Skinner CJ, Becker RH, White RL, Exter KM, Barlow MJ, Davis RJ. 1998. *MNRAS* 296: 669

Smith KC, Howarth ID. 1998. *MNRAS* 299:1146

Sobolev VV. 1957. *Sov. Astron. Astrophys. J.* 1:678

Springmann UWE, Pauldrach, AWA. 1992. *Astron. Astrophys.* 262:515

Springmann U, Puls J. 1998. *Proc. 2nd Boulder-Munich Workshop. PASPC* 131:286

Stahl O, Wolf B, Aab O, Smolinski J. 1991. *Astron. Astrophys.* 252:693

Steidel CC, Giavalisco M, Pettini M, Dickinson M, Adelberger KL. 1996. *Ap. J. Lett.* 462:L17

Stevens IR. 1995. *MNRAS* 277:163

Tanzi EG, Tarenghi M, Panagia N. 1981. *Proc. 59th Colloq. Trieste*, p. 51. Dordrecht:Reidel

Taresch G, Kudritzki RP, Hurwitz M, Bowyer S, Pauldrach AWA, et al. 1997. *Astron. Astrophys.* 321:531

Venn KA. 1995. *Ap. J.* 449:839

Venn KA. 1999. *Ap. J.* 518:405

Vink JS, de Koter A, Lamers HJGLM. 1999. *Astron. Astrophys.* 350:18

Voels SA, Bohannon B, Abbott DC, Hummer DG. 1989. *Ap. J.* 340:1073

Walborn NR, Panek RJ. 1984. *Ap. J. Lett.* 280:L27

Walborn NR, Noichols-Bohlin J, Panek RJ. 1985. *International Ultraviolet Explorer Atlas of O-Type Spectra From 1200 to 1900 Å. NASA Ref. Publ.* 1155.

Walborn NR Lennon DJ, Haser SM, Kudritzki RP, Voels SA. 1995. *PASP* 107:104

Waters LBFM, Lamers HJGLM. 1984. *Astron. Astrophys.* 57:327

Waters LBFM, Wesselius PR. 1986. *Astron. Astrophys.* 155:104

Wendker HJ. 1987. *Astron. Astrophys.* 69:87

White RL, Becker RH. 1982. *Ap. J.* 262: 657

White RL, Becker RH. 1983. *Ap. J. Lett.* 272:L19

Wolf B, Stahl O, Fullerton AW. eds. 1999. *Proc. IAU Colloq. 169*, p. 169. Heidelberg: Springer-Verlag

Wood PR, Faulkner DJ. 1986. *Ap. J.* 307:659

Wright AE, Barlow MJ. 1975. *MNRAS* 170: 41

Zickgraf FJ. 1999. *Proc. IAU Colloq.* 169, p. 40 Heidelberg: Springer-Verlag

Annu. Rev. Astron. Astrophys. 2000. 38:667–715

THE HUBBLE DEEP FIELDS

Henry C. Ferguson, Mark Dickinson, and Robert Williams
Space Telescope Science Institute, Baltimore, MD 21218; e-mail: ferguson@stsci.edu

Key Words   cosmology observations, galaxy evolution, galaxy formation, galaxy photometry, surveys

■ **Abstract**   The Hubble space telescope observations of the northern Hubble deep field, and more recently its counterpart in the south, provide detections and photometry for stars and field galaxies to the faintest levels currently achievable, reaching magnitudes $V \sim 30$. Since 1995, the northern Hubble deep field has been the focus of deep surveys at nearly all wavelengths. These observations have revealed many properties of high redshift galaxies and have contributed to important data on the stellar mass function in the Galactic halo.

1.   INTRODUCTION

Deep surveys have a long history in optical astronomy. Indeed, nearly every telescope ever built has at some point in its lifetime been pushed to its practical limit for source detection. An important historical motivation for such surveys has been the desire to test cosmological models via the classical number–magnitude and angular-diameter–magnitude relations (Sandage, 1961, 1988). However, interpretation has always been hampered by the difficulty of disentangling the effects of galaxy selection and evolution from the effects of cosmic geometry. The modern era of deep surveys in optical astronomy began with the advent of CCD detectors, which allowed 4-m telescopes to reach to nearly the confusion limit allowed by the point spread function (Tyson, 1988). These images revealed a high surface density of faint, blue galaxies, and the combination of their number–magnitude relation, number–redshift relation, and angular correlation function was difficult to reconcile with the critical-density universe ($\Omega_M = \Omega_{tot} = 1$) that seemed at the time to be a robust expectation from inflationary models (Kron 1980; Fukugita et al 1990, Guiderdoni & Rocca-Volmerange, 1990; Efstathiou et al 1991).

Prior to launch, simulations based on fairly conservative assumptions suggested that the Hubble space telescope (HST) would not provide an overwhelming advantage over ground-based telescopes for studies of distant galaxies (Bahcall et al 1990):

0066-4146/00/0915-0667$14.00   **667**

Our results show that the most sensitive exposures achieved so far from the ground reveal more galaxies per unit area than will be seen by planned HST observations unless galaxy sizes decrease with the maximum rate consistent with ground-based observations. In this, the most favorable case for HST, the space exposures will show almost as many galaxy images as have been observed so far in the most sensitive ground-based data.

It was apparent from the images returned immediately after the HST refurbishment in 1993 that distant galaxies had substantially higher surface brightnesses than predicted by such simulations (Dressler et al 1994; Dickinson, 1995). Despite the pessimistic predictions, extensive field galaxy surveys were always a top priority for HST, and large amounts of observing time were awarded to deep, pointed surveys and to the medium-deep survey (Griffiths et al 1996). These observations provided valuable information on galaxy scale lengths and morphologies, on the evolution of clustering and the galaxy merger rate, and on the presence of unusual types of galaxies (Mutz et al 1994; Schade et al 1995a; Cowie et al 1995a; Griffiths et al 1996). In this context, the idea of using the HST to do an ambitious deep field survey began to look quite attractive, and was eventually adopted as one of the primary uses of the HST director's discretionary time.

1.1   This Review

The Hubble deep fields north and south (HDF-N and HDF-S, respectively) are by now undoubtedly among the most "data rich" portions of the celestial sphere. The task of reviewing progress is complicated by the fact that much of the supporting data are still being gathered, and by the wide range of uses that have been found for the HDF. It is not practical to provide a comprehensive review of all HDF-related research in a single review article. Instead we have chosen to divide the review into two parts. The first part discusses the data, both from HST and from other facilities, and summarizes measurements and phenomenology of the sources found in the field. The second part focuses specifically on distant galaxies, and attempts to provide a critical view of what the HDF has, and has not, taught us about galaxy evolution. Ferguson (1998a) has reviewed the HDF with a different focus, and the series of papers in the 1996 Herstmonceux conference and the 1997 STScI May Symposium provide a broad summary of the overall field (Tanvir et al 1996; Livio et al 1997). Throughout this review, unless explicitly stated otherwise, we adopt cosmological parameters Ω_M, Ω_Λ, $\Omega_{tot} = 0.3, 0.7, 1.0$ and $H_0 = 65\,\mathrm{km\,s^{-1}}$ $\mathrm{Mpc^{-1}}$. Where catalog numbers of galaxies are mentioned, they refer to those in Williams et al (1996).

2.   PART 1: Observations, Measurements, and Phenomenology

The HDFs were carefully selected to be free of bright stars, radio sources, nearby galaxies, etc., and to have low Galactic extinction. The HDF-S selection criteria included finding a quasi-stellar object (QSO) that would be suitable for studies of absorption lines along the line of sight. Field selection was limited to the "continuous viewing zone" around $\delta = \pm 62°$, because these declinations allow HST to observe, at suitable orbit phases, without interference by earth occultations. Apart from these criteria, the HDFs are typical high-galactic-latitude fields; the statistics of field galaxies or faint Galactic stars should be free from a priori biases. The HDF-N observations were taken in December 1995 and the HDF-S in October 1998. Both were reduced and released for study within 6 weeks of the observations. Many groups followed suit and made data from follow-up observations publicly available through the world wide web.

Details of the HST observations are set out in Williams et al (1996), for HDF-N and in a series of papers for the southern field (Williams et al 2000; Ferguson et al 2000; Gardner et al 2000; Casertano et al 2000; Fruchter et al 2000; Lucas et al 2000). The HDF-N observations primarily used the WFPC2 camera, whereas the southern observations also took parallel observations with the new instruments installed in 1997: the near-infrared camera and multi-object spectrograph (NICMOS) and the space telescope imaging spectrograph (STIS). The area of sky covered by the observations is small: 5.3 arcmin^2 in the case of WFPC2 and 0.7 arcmin^2 in the case of STIS and NICMOS for HDF-S. The WFPC2 field subtends about 4.6 Mpc at $z \sim 3$ (comoving, for Ω_M, Ω_Λ, $\Omega_{tot} = 0.3, 0.7, 1.0$). This angular size is small relative to scales relevant for large-scale structure.

The WFPC2 observing strategy was driven partly by the desire to identify high-redshift galaxies via the Lyman-break technique (Guhathakurta et al 1990; Steidel et al 1996), and partly by considerations involving scattered light within HST (Williams et al 1996). The images were taken in four very broad bandpasses (F300W, F450W, F606W, and F814W), spanning wavelengths from 2500 to 9000 Å. Although filter bandpasses and zeropoints are well calibrated,[1] no standard photometric system has emerged for the HDF. In this review, we use the notation U_{300}, B_{450}, V_{606} and I_{814} to denote magnitudes in the HST passbands on the AB system (Oke, 1974). On this system $m(AB) = -2.5 \log f_\nu(\text{nJy}) + 31.4$. Where we drop the subscript, magnitudes are typically on the Johnson-Cousins system, as defined by Landolt (1973, 1983, 1992a,b) and as calibrated for WFPC2 by Holtzman et al (1995); however we have not attempted to homogenize the different color corrections and photometric zeropoints adopted by different authors.

During the observations, the telescope pointing direction was shifted ("dithered") frequently, so that the images fell on different detector pixels. The final images were thus nearly completely free of detector blemishes, and were sampled at significantly higher resolution than the original pixel sizes of the detectors. The technique of variable pixel linear reconstruction ("drizzling") (Fruchter & Hook 1997) was developed for the HDF and is now in widespread use.

---

[1]None of the analyses to date have corrected for the WFPC2 charge-transfer inefficiency (CTE; Whitmore et al. 1999). This is likely to be a small correction $\lesssim 5\%$ for the F450W, F606W, and F814W bands. However, because of the low background in the F300W band, the correction could be significantly larger.

TABLE 1 Census of objects in the central HDF-N

Number	Type of source
~3000	Galaxies at U_{300}, B_{450}, V_{606}, I_{814}
~1700	Galaxies at J_{110}, H_{160}
~300	Galaxies at K
9	Galaxies at 3.2 μm
~50	Galaxies at 6.7 or 15 μ
~5	Sources at 850 μm
0	Sources at 450 μm or 2800 μm
6	X-ray sources
~16	Sources at 8.5 GHz
~150	Measured redshifts
~30	Galaxies with spectroscopic $z > 2$
<20	Main-sequence stars to $I = 26.3$
~2	Supernovae
0−1	Strong gravitational lenses

Both HDF campaigns included a series of shallow "flanking field" observations surrounding the central WFPC2 field. These have been used extensively to support ground-based spectroscopic follow-up surveys. Since 1995, HDF-N has also been the target of additional HST observations. Very deep NICMOS imaging and spectroscopy were carried out on a small portion of the field by Thompson et al (1999). Dickinson et al (2000b) took shallower NICMOS exposures to make a complete map of the WFPC2 field. A second epoch set of WFPC2 observations were obtained in 1997, 2 years after the initial HDF-N campaign (Gilliland et al 1999). STIS ultraviolet (UV) imaging of the field is in progress, and a third epoch with WFPC2 is scheduled.

Table 1 gives a rough indication of the different types of sources found in the HDF-N. Since 1995, the field has been imaged at wavelengths ranging from 1×10^{-3} μm (ROSAT) to 2×10^5 μm (MERLIN and VLA), with varying sensitivities, angular resolutions, and fields of view. Not all of the data have been published or made public, but a large fraction has, and it forms the basis for much of the work reviewed here. Links to the growing database of observations for HDF-S and HDF-N are maintained on the HDF world wide web sites at STScI.

3.   STARS

The nature of the faint stellar component of the Galaxy is critical to determining the stellar luminosity function and the composition of halo dark matter. In spite of the small angular sizes of the HDFs, their depth enables detection of low-luminosity objects to large distances. If the halo dark-matter is mostly composed of low-mass stars, there should be several in the HDF (Kawaler, 1996, 1998). Although HST resolution permits star-galaxy separation down to much fainter levels than ground-based observations, many distant galaxies also appear nearly point-like. Color criteria are effective at identifying likely red subdwarfs, but for $I_{814} > 25$ and $V_{606} - I_{814} < 1$ the possibility of galaxy contamination becomes significant. The counts of red main-sequence stars have been compared with galactic models by Flynn et al (1996), Elson et al (1996) and Mendez et al (1996). More detailed results have subsequently emerged from studies that have combined the HDF-N with other HST images to increase the sky coverage (Gould et al 1996, 1997; Reid et al 1996; Mendez & Guzman, 1997; Kerins, 1997; Chabrier & Mera, 1997). A general conclusion is that hydrogen-burning stars with masses less than 0.3 $M_\odot$ account for less than 1% of the total mass of the Galactic halo. The overall HST database indicates that the Galactic disk luminosity function experiences a decided downturn for magnitudes fainter than $M_v = 12(M \lesssim 0.2\ M_\odot)$. For the halo, the constraints on the luminosity function are not as good because of the limited sky coverage. However, the luminosity function clearly cannot turn up by the amount it would need for main-sequence stars to be an important constituent of halo dark matter.

In addition to the nine or so candidate red dwarfs in the HDF-N, to a limiting magnitude of $I = 28$ the field has about 50 unresolved objects with relatively blue colors. Although in principle these could be young hot white dwarfs (WDs), this possibility has appeared unlikely, as the lack of brighter point sources with similar colors would require all to be at very large distances, greater than 10 kpc away. However, recent work by Hansen (1998, 1999) has shown that at low metallicity, molecular hydrogen opacity causes the oldest, lowest luminosity WDs to become blue as they cool. Harris et al (1999) have recently discovered such an object in the Luyten proper motion survey. This result changes the way in which colors should be used to discriminate faint point-like sources in the HDFs, and may lead to the reclassification of some faint galaxies as WDs.

Proper motions are an unambiguous way to distinguish between galactic stars and galaxies. The HDFs serve as excellent first-epoch data for detection of changes in position or brightness for any objects. A second-epoch set of images of the HDF-N were obtained 2 years after the initial HDF-N campaign (Gilliland et al 1999), and were analyzed by Ibata et al (1999) to search for object motion in the images. Of 40 identified point sources with $I > 28$, five were found to have proper motions that were $>3\sigma$ above the measurement uncertainty (~ 10 mas year^{-1}), with two of the objects having proper motions exceeding 25 mas/yr and the remaining three near the detection limit. The five objects are all faint ($I \sim 28$) and of neutral color ($V - I < 0.9$), and realistic velocities require that they have distances $d < 2$ kpc. Although only a very small fraction of the total sources in the HDF-N, these objects represent a large increase over the number of stars expected from standard models of the Galaxy, which predict less than one star in the range $27 < V < 29$ with $V - I < 1.0$.

Separate evidence may be emerging that is consistent with the identification of faint blue stars in the HDFs. The HDF-S ($l = 328$, $b = -49$) points closer to the Galactic center than HDF-N ($l = 126$, $b = 55$), and hence samples a larger path length through the Galactic halo. There should be more stars in HDF-S than in HDF-N. Mendez & Minniti (1999) found roughly double the number of blue point-like sources in the HDF-S than its northern counterpart (Figure 1). This supports the hypothesis that a significant fraction of these sources are stars. A number of these stars are too faint and too blue to be on the main sequence, and could represent the old, low luminosity, blueish halo WDs proposed by Hansen (1998).

Although these developments are exciting, the evidence is far from compelling. The sources identified by Mendez & Minniti (1999) are brighter than those found to have proper motions by Ibata et al (1999). The two studies thus appear to be inconsistent, in that the closer WDs ought to have larger proper motions. Also, the enhancement in faint blue point-like objects in HDF-S relative to HDF-N is sensitive to the magnitude limit chosen. If Mendez & Minniti (1999) had included objects down to $I = 29$ in their sample, they would have found more objects in HDF-N than in HDF-S (Figure 1).

Although still very tentative, the identification of WDs in the HDFs is potentially extremely important. The results of the MACHO project (Alcock et al 1997) suggest that a substantial fraction of the halo mass is due to objects with WD masses (although this is not a unique interpretation of the microlensing statistics) (Sahu 1994). If WDs do contribute significantly to the halo mass, then the early stellar population of the halo must have formed with a peculiar initial mass function (IMF) (Reid et al 1996) deficient in both high-mass stars [to avoid over-enrichment of the halo by metals from supernovae (SN) and planetary nebulae] and low mass stars, (because stars with $M < 0.8 \, M_\odot$ would still be on the main sequence today). The presence or absence of WDs in deep fields will become more definite within the next few years via forthcoming third-epoch proper motion measurements.

3.1   Supernova Events

An estimate of the SN rate at high redshift can provide important constraints on the star-formation history of galaxies and on models for SN progenitors (Pain et al 1996; Madau et al 1998; Ruiz-Lapuente & Canal, 1998). Gilliland & Phillips (1998) took second-epoch WFPC2 observations of the HDF-N field in December 1997 and reported two SN detections, at $I = 26.0$ and 27.0. Mannucci & Ferrara (1999) report a very faint detection, at $I = 28.5$ of an object that brightened by 1 mag in 5 days in the original HDF-N images. The number of SNe expected in a single epoch of HDF-N observations is 0.5–1 (Gilliland et al 1999), consistent with the number of detections. The advanced camera for surveys (ACS) will provide substantial benefits for future SN searches, because it will have roughly twice the solid angle at five times the sensitivity of WFPC2. Based on the HDF numbers, a survey of 12 ACS fields to HDF I-band depth should identify roughly 25 SNe and would cost ∼100 orbits.

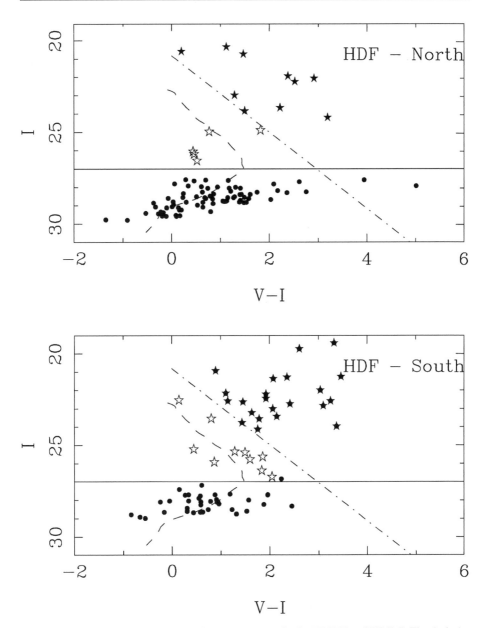

Figure 1   Color-magnitude diagrams for point sources in the HDF-N and HDF-S. The dash-dot line represents locus of disk M-dwarfs at a distance of 8 kpc, with solid star symbols representing main-sequence disk and halo stars. The horizontal line represents the (15σ) detection limit of $I = 27$ magnitude, and filled circles below that line are Galactic stars and unresolved distant galaxies. The open star symbols represent point sources which are too faint to be on the main sequence at any reasonable distance, and which may be white dwarfs. The dashed line is the 0.6 $M_\odot$ WD cooling track from Hansen (1999) for a distance of 2 kpc. (adapted from Mendez & Minniti 2000)

4.   GALAXIES

The HDFs are exquisitely deep and sharp images, detecting thousands of objects distributed throughout the observable universe, but they are also *very small* fields of view, and each is only one sightline. Figure 2 illustrates some parameters relevant for studying galaxy evolution with a single HDF. The total co-moving volume out to redshift z has been scaled (*top panel*) by the present-day normalization of the galaxy luminosity function ϕ^*. This gives a rough measure of the number of

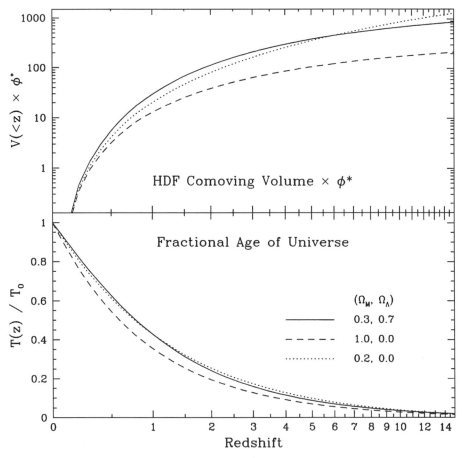

Figure 2   An illustration of volume and time in the HDF. At top, for the 5 arcmin^2 WFPC2 field of view, the co-moving volume out to redshift z is plotted for several cosmologies, scaled by the present-day normalization of the galaxy luminosity function (ϕ^*, here taken to be 0.0166) h^3 Mpc^{-3}, from Gardner et al (1997). This gives a rough measure of the number of "L^*-volumes" out to that redshift. At bottom, the fractional age of the universe versus redshift is shown. Most of cosmic time passes at low redshifts, where the HDF volume is very small.

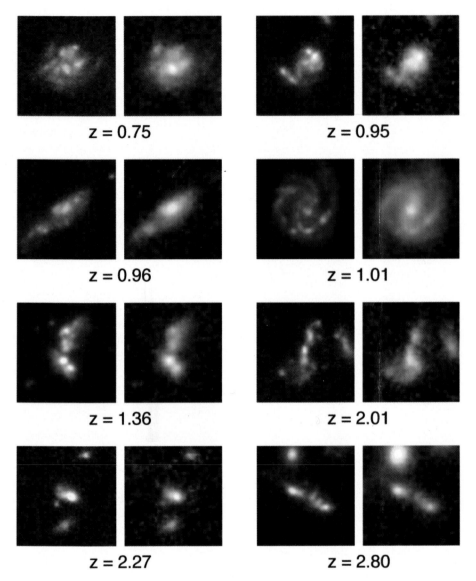

Figure 3 Selected galaxies from the HDF-N, viewed at optical and near-infrared wavelengths with WFPC2 and NICMOS. For each galaxy, the two panels show color composites made from B_{450}, V_{606}, I_{814} (*left*) and I_{814}, J_{110}, H_{160} (right). The NICMOS images have somewhat poorer angular resolution. In the giant spiral at $z = 1.01$, a prominent red bulge and bar appear in the NICMOS images, while the spiral arms and HII regions are enhanced in the WFPC2 data. For most of the other objects, including the more irregular galaxies at $z \sim 1$ and the Lyman Break galaxies at $z > 2$, the morphologies are similar in WFPC2 and NICMOS.

dominates structural measurements. Teplitz et al (1998) noted similar trends in the distributions of structural parameters from NICMOS parallel images.

Abraham et al (1999a; 1999b) have studied a variety of other issues of galaxy morphology, concluding that the fraction of barred spirals declines significantly for $z > 0.5$; that ~40% of HDF-N elliptical galaxies at $0.4 < z < 1$ have a dispersion of internal colors that implies recent, spatially localized star formation; that spiral bulges are older than disks; that the dust content of HDF spirals is similar to that locally; and that the peak of past star formation for a typical $z \sim 0.5$ spiral disk occurred at $z \sim 1$. These results are tentative because of small samples and uncertainties in stellar-population models, but they show promise for future deep surveys.

4.2 Galaxy Counts

One of the benefits derived from the ready availability of the reduced HDF data is that a variety of techniques have been used to catalog the sources (Williams et al 1996; Lanzetta er al 1996; Colley et al 1996; Sawicki et al 1997; Dell'Antonio & Tyson, 1996; Couch, 1996; Metcalfe et al 1996; Madau & Pozzetti, 1999; Chen et al 1999; Fontana et al 1999). Different algorithms have been used to construct the catalogs, but all at some level rely on smoothing the image and searching for objects above a surface brightness threshold set by the background noise. Ferguson (1998b) compared a few of the available catalogs and found reasonable agreement among them for sources brighter than $V_{606} = 28$, with systematic differences in magnitude scales of less than 0.3 (nevertheless, there *are* systematic differences at this level). The different catalogs apply different algorithms for splitting and merging objects with overlapping isophotes. These differences, together with the different schemes for assigning magnitudes to galaxies, result in overall differences in galaxy counts. At $I_{814} = 26$ the galaxy counts in the catalogs considered by Ferguson (1998b) all agree to within 25%, whereas at $I_{814} = 28$ there is a factor of 1.7 difference between them. This highlights the fact that galaxy counting is not a precise science.

Figure 4 shows the counts in four photometric bands derived from the HDF and from ground-based observations. As a fiducial comparison, a no-evolution model is shown for a spatially flat model with $\Omega_M = 0.3$ and $\Omega_\Lambda = 0.7$. In all bands the number–magnitude relation at HDF depths is significantly flatter than $N(m) \propto m^{0.4}$, but at blue wavelengths it exceeds the purely geometrical (no-evolution) predictions by roughly a factor of 3. Interpretation of the counts is addressed in more detail in Section 5.1.

4.3 Galaxy Sizes

One of the interesting and somewhat unexpected findings of HST faint-galaxy surveys has been the small angular diameters of faint galaxies. The sky is peppered with compact high-surface-brightness objects, in contrast to the expectation from $\Omega_M = \Omega_{tot} = 1$ pure-luminosity evolution (PLE) models (Section 5.1)

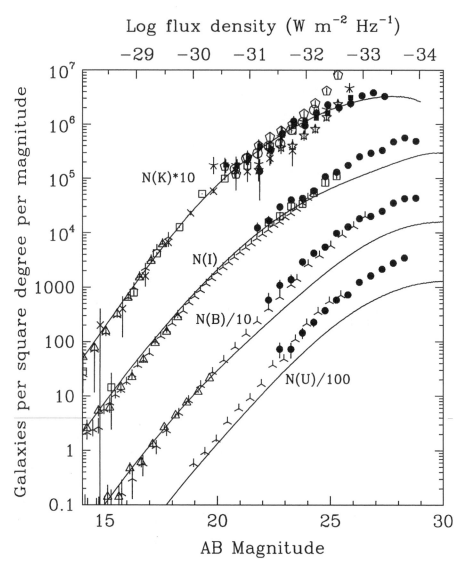

Figure 4 Galaxy counts from the HDF and other surveys. The HDF galaxy counts (solid symbols) use isophotal magnitudes and have not been corrected for incompleteness. These corrections will tend to steepen the counts at the faint end, but in a model-dependent way. For the K-band, a color correction of -0.4 mag has been applied to the NICMOS F160W band magnitudes. The HDF-N counts from Thomson et al (1999) are shown as filled squares while the HDF-S counts (Fruchter et al 2000) are filled circles. For the U, B and I bands, the HDF counts are the average of HDF-N (Williams et al 1996) and HDF-S (Casertano et al 2000) with no color corrections. The groundbased counts (open symbols) are from Mcleod & Rieke (1995), Gardner et al (1993; 1996), Postman et al (1998), Lilly Cowie & Gardner (1991), Huang et al (1997), Minezaki et al (1998), Bershady et al (1998), Moustakas et al (1997), and Djorgovski et al (1995). The smooth curves represent a no-evolution model with Ω_M, Ω_Λ, $\Omega_{tot} = 0.3, 0.7, 0.1$ based on the luminosity functions, spectral-energy distributions, and morphological type mix of the "standard NE" model of Ferguson & McGaugh (1995).

and models dominated by low-surface-brightness (LSB) galaxies (Ferguson & McGaugh, 1995).

Galaxy radii can be measured using image moments (which are subject to severe biases because of isophotal thresholds), growth curves, or profile fitting. Ferguson (1998b) considered the size distribution derived from image moments, whereas Roche et al (1998) discuss the distribution of half-light radii derived from profile fitting. Roche et al (1998) find no evidence for evolution in the rest-frame sizes or surface brightnesses of normal spirals and ellipticals out to $z = 0.35$. At higher redshift they find evidence for strong evolution; galaxies are more compact and of higher surface brightness than expected from a PLE model (see also Schade et al 1995b). The sizes of disk galaxies were shown to be reasonably well matched by a size-luminosity evolution (SLE) model wherein spirals form stars gradually from the inside out. In a study with similar goals, Simard et al (1999) analyzed a sample of 190 (non-HDF) field galaxies with spectroscopic redshifts and HST images. They find no evidence for size or luminosity evolution of disk-dominated galaxies out to $z \sim 1$. The Simard et al (1999) analysis includes a more comprehensive treatment of selection biases and an empirical comparison to local samples that makes the results quite compelling. However, Simard et al (1999) do identify nine high-surface-brightness objects in their sample at $z > 0.9$. If these were included as disk galaxies, the results would be closer to those of Roche et al (1998). A comparison of the bivariate distribution of the Simard et al (1999) luminosities and scale lengths to a $z \sim 0$ sample suggests only moderate density evolution out to $z \sim 1.2$, with a decrease in the number density of bright, large-scale-size galaxies at higher redshift (de Jong & Lacey, 1999). Galaxies in the HDF with spectroscopic or photometric redshifts greater than $z = 2$ generally appear to be more compact than present-day L^* galaxies (Lowenthal et al 1997; Roche et al 1998).

At magnitudes fainter than $I = 25$, morphological classification and profile fitting become quite difficult because most galaxies at these faint magnitudes are so small that even with HST there are few independent resolution elements within the area detected above the background. Ferguson (1998b) studied the distribution of first-moment radii and modeled the selection boundaries of the HDF in size and magnitude. Because the first-moment radii are measured above a fixed isophote, the interpretation of the observed trends is highly model dependent, and is sensitive to the assumed redshift distribution and morphological-type distribution of the galaxies. Nevertheless, down to $I \sim 27$ the HDF counts are clearly dominated by galaxies more compact than local L^* spirals. At fainter magnitudes, the locus of HDF galaxy sizes is tightly constrained by selection, and there is very little information on the *intrinsic* size distribution of the objects.

4.4 Spectroscopic and Photometric Redshifts

The HDF-N and flanking fields have been the focus of some of the most intensive and complete spectroscopic redshift survey work on faint galaxies. Essentially all spectroscopic redshifts in the HDF have come from the W.M. Keck Observatory and LRIS spectrograph (Cohen et al 1996, 2000; Lowenthal et al 1997; Phillips et al 1997; Steidel et al 1996, 1999; Dickinson, 1998; Adelberger et al 1998; Spinrad et al 1998; Weymann et al 1998; Waddington et al 1999; Zepf et al 1997; Hu et al 1998). The most recent compilation is 92% complete for $R < 24$ in the central HDF-N, and also 92% complete for $R < 23$ in the flanking fields (Cohen et al 2000). The median redshift is $z \approx 1$ at $R = 24$. Altogether, including unpublished redshifts from the Steidel group, more than 700 spectroscopic redshifts have been measured in the immediate vicinity of the HDF-N, with ~150 in the central HDF (WF and PC) alone. This latter corresponds to 30 redshifts per arcmin^2, spanning $0.089 < z < 5.60$ (and a few stars), a density unmatched by any other field galaxy survey. At the same time, it is sobering to realize that less than 5% of the galaxies from the Williams et al (1996) HDF-N catalog have spectroscopic redshifts. The vast majority are fainter than the spectroscopic limit achievable with present-day telescopes. For the HDF-S, the extant spectroscopy is more limited, consisting of about 250 redshifts in an irregular area surrounding the three HST fields (Glazebrook et al 2000; Dennefeld et al 2000; Tresse et al 1999). Three galaxies at $2.8 < z < 3.5$ have been confirmed by the VLT (Cristiani 1999). Eleven HDF-S ISO sources have redshift measurements from VLT IR spectroscopy (Rigopoulou et al 2000).

The estimation of galaxy redshifts using broad-band colors received a tremendous boost from the HDF observations and followup studies. Indeed, the advent of the HDF marked a transition for photometric redshifts, which evolved from an experimental exercise to a widely used technique. This came about for several reasons. Perhaps foremost, as noted above, the WFPC2 HDF data readily detected galaxies as much as 100 times fainter than the practical spectroscopic limits of 10-m telescopes. Therefore, in order to interpret the properties of the vast majority of HDF galaxies, non-spectroscopic techniques for estimating redshifts are necessary. Fortunately, the HDF-N WFPC2 data set offered very-high-quality, four-band photometry from 0.3 to 0.8 μm for thousands of faint galaxies, and the extensive follow-up observations that came later extended this to other wavelengths and wider fields of view. Rapid dissemination of data and source catalogs from the HDF and follow-up observations made it relatively easy for any investigator to test their photometric redshift method. Indeed, it should be noted that many photometric redshift studies to date have been tested and calibrated *solely* using the HDF-N.

Several groups (Steidel et al 1996; Lowenthal et al 1997; Clements & Couch 1996; Madau et al 1996, 1998) used two-color selection keyed to the passage of the Lyman limit and Lyman α forest breaks through the F300W and F450W passbands to identify galaxy candidates at $z > 2$. Such color-selected objects are variously referred to as Lyman-break galaxies, or UV dropouts. The Madau et al (1996, 1998) studies were seminal in combining color-selected HDF samples at $z \sim 3$ and 4 with previous (spectroscopic, non-HDF) studies at $z < 2$ to derive a measure of the global history of cosmic star formation, a goal subsequently pursued by many other groups using various kinds of HDF photometric redshifts (see Section 5.5).

More general approaches to photometric redshifts have either fit redshifted spectral templates to the photometric data (e.g. Gwyn & Hartwick 1996; Mobasher et al 1996; Lanzetta et al 1996; Sawicki et al 1997; Fernandez-Soto et al 1999; Benítez et al 1999), or have used generalized polynomial fits of redshift vs. multi-band fluxes to a spectroscopic training set (e.g. Connolly et al 1997; Wang et al 1998). Cowie & Songaila (1996) and Connolly et al (1997) were the first to include IR HDF photometry when deriving photometric redshifts. In principle, this is especially useful in the $1 < z < 2$ range where it is difficult to measure redshifts from optical spectroscopy due to the absence of strong, accessible emission or absorption features, and where the strong spectral breaks (e.g. at 4000 Å and the 912 Å Lyman limit) fall outside the optical passbands. Many other groups have since included IR photometry in their photometric redshift work, using the KPNO 4-m *JHK* images from Dickinson et al (Fernandez-Soto et al 1999), or more recently HST NICMOS data (Thompson, 1999; Budavari et al 1999; Yahata et al 2000).

Hogg et al (1998) and Cohen et al (2000) have carried out blind tests of the accuracy and reliability of HDF photometric redshifts, comparing photometric predictions by the various groups with spectroscopic redshifts unknown to those groups when the predictions were made. In general, all of the photometric techniques are quite successful at $z \lesssim 1.4$, where the spectroscopic training sets are extensive, with the best prediction schemes achieving $|z_{\mathrm{phot}} - z_{\mathrm{spec}}|/(1 + z_{\mathrm{spec}})$ $\lesssim 0.05$ for $\gtrsim 90\%$ of the galaxies. At $z > 1.9$ the results are also generally quite good, with 10 to 15% RMS in $\Delta z/(1 + z)$ after excluding the worst outliers. The intermediate redshift range, where photometric redshifts are particularly interesting, has not yet been tested due to the lack of spectroscopic calibrators.

Photometric redshifts have been used to identify galaxy candidates in both HDFs at $z > 5$. Two objects noted by Lanzetta et al (1996) and Fernández-Soto et al (1999) have been subsequently confirmed by spectroscopy, with $z = 5.34$ and 5.60 (Spinrad et al 1998; Weymann et al 1998). With the addition of near–infrared data, it becomes possible, in principle, to push to still higher redshifts. Indeed, at $z > 6.5$, galaxies should have virtually no detectable optical flux. Candidates at these very large redshifts have been identified by Lanzetta et al (1998) and Yahata et al (2000). One of these in the HDF-N was readily detected in NICMOS 1.6 μm images by Dickinson et al (2000b), but is not significantly detected at J_{110} or at optical wavelengths. Although this is a plausible candidate for an object at $z \gtrsim 12$, there are also other possible interpretations. It is curious that several of the proposed $z > 10$ candidates, including this "*J*-dropout" object in the HDF-N and several of the HDF-S/NICMOS candidates proposed by Yahata et al (2000), are surprisingly bright ($\sim 1 \mu$Jy) at 2.2 μm. These would be significantly more luminous than the brightest known Lyman-break galaxies at $2 < z < 5$. It remains to be seen whether this points to some remarkable epoch of bright objects at $z > 10$, or whether instead the photometric redshifts for these objects have been overestimated.

4.5 Galaxy Kinematics

Spectroscopy of galaxies in the HDF-N has provided kinematical information leading to constraints on mass and/or luminosity evolution. Phillips et al (1997) and Guzman et al (1997) observed 61 compact, high-surface-brightness galaxies ($r < 0\rlap{.}''5$) in the HDF-N flanking fields. The great majority show emission lines and have redshifts $0.4 < z < 1$. Masses for the systems were deduced from the line widths, which ranged from 35–150 km s^{-1}, and one-half of the sample were found to be low mass ($M < 10^{10} M_\odot$), relatively luminous systems similar to local H II galaxies. The remaining systems were a heterogeneous class similar to local starburst galaxies.

Vogt et al (1997) studied a separate magnitude-limited sample of eight high-inclination disk galaxies from the HDF-N flanking fields in the redshift range $0.15 < z < 0.75$. Reliable velocity information could be discerned to three disk scale-lengths, comparable to the extent of optical rotation curves for local galaxies. Results from this survey were combined with an earlier survey of more luminous galaxies (Vogt et al 1996) to study the Tully-Fisher (TF) relation at $z \sim 0.5$ over a span of 3 magnitudes. A luminosity–line-width relation exists for the sample, linear and of the same slope as the local TF relation, but shifted to higher luminosities by 0.4 magnitudes. This is similar to the results from some other studies (Forbes et al 1996; Bershady, 1997) but larger amounts of luminosity evolution were seen in other samples (Rix et al 1997; Simard & Pritchet, 1998). Vogt et al (1997) speculate that the differences are due to sample selection. In any case, the modest luminosity evolution is consistent with the modest surface-brightness evolution found in the analysis of structural parameters by Simard et al (1999).

4.6 Mid-IR Sources

Mid-IR observations offer the opportunity to detect emission from material heated by star formation or active galactic nuclear (AGN) activity and thus provide some indication of radiative activity that may be obscured at UV through near-IR wavelengths. The launch of the *Infrared Space Observatory* (ISO), the first space-based mid-IR telescope capable of such observations, coincided nicely with the availability of the HDF data set, and both deep fields (N and S) were the targets of deep 6.75 μm and 15 μm observations with the ISOCAM imager. The HDF-S ISO observations were made before the HST imaging was carried out, and they are currently being analyzed (e.g Oliver et al 2000). The 15 μm maps extend well beyond the central HDF to cover a portion

The HDF-N ISO observations were carried out using discretionary time allocated by the ISO director to an observing team led by Rowan-Robinson (Serjeant et al 1997; Goldschmidt et al 1997; Oliver et al 1997; Mann et al 1997; Rowan-Robinson et al 1997). Two other groups have subsequently processed and reanalyzed the ISO data using independent methods (Désert et al 1999; Aussel et al 1999a). The 15 μm maps extend well beyond the central HDF to cover a portion of the flanking fields, and have $\approx 9''$ resolution. The high-sensitivity region of the 6.7 μm maps is smaller and roughly matched to the central HDF WFPC2 area, with $\approx 4''$ resolution.

The depth of the ISO image mosaics varies over the field of view, and the data processing procedures are complex and the subject of continuing refinement. Extensive simulations have been used by the three groups to calibrate source detection, completeness and reliability, and photometry. Given these complexities, it is perhaps not unexpected that the groups have produced different source catalogs. A simple, visual inspection of the images presented by Serjeant et al (1997) or Aussel et al (1999a) demonstrates that the 15 μm data is much more straightforward to interpret—at least $\sim$20 sources are readily visible to the eye, whereas the 6.7 μm image has only a few "obvious" sources and much more low-level background structure. Indeed, the source catalogs from the three groups agree reasonably well at 15 μm, commonly detecting most of the brighter sources ($\sim$20 detections with $\gtrsim$200 μJy) with generally similar measured flux densities. Aussel et al (1999a) and Désert et al (1999), however, claim reliable source detection to fainter flux limits than do Goldschmidt et al (1997) and thus have longer source lists: The complete plus supplemental catalog of Aussel et al lists 93 sources at 15 μm. At 6.75 μm, the situation is more confused (perhaps literally). All three groups define "complete" lists of six or seven sources, but they are not all the same objects. Differences in the assumptions about ISO astrometry and geometric distortion corrections may be partially responsible. Goldschmidt et al provide a list of 20 supplemental sources at 6.7 μm, only a few of which are detected by the other groups. There also appear to be systematic differences in the derived source fluxes at this wavelength. Altogether, the 15 μm source detections and fluxes appear to be well established, but only the few brightest 6.7-μm detections are unambiguous, and the associated fluxes may be uncertain.

The optical counterparts of the HDF-N ISO sources are mostly brighter galaxies at $0 < z < 1.2$, although occasionally the identifications may be confused because of the 9$''$ PSF at 15 μm. Many of the 15 μm sources also correspond to radio sources (few or none, however, correspond to the SCUBA sources discussed below). Over this redshift range, the 15 μm measurements primarily sample emission from the "unidentified emission bands," including the strong features at 6.2–8.6 μm. These features are considered a characteristic signature of star-forming galaxies (see, e.g. Rigopoulou et al 1999). At $z > 1.4$, they shift out of the 15 μm ISO bandpass, and therefore more distant objects should become much fainter. Indeed none have been identified yet in the ISO/HDF. Warm dust ($T > 150$ K) may also contribute to the mid-IR emission, and some of the ISO sources are almost certainly AGN (e.g. HDF 2-251, a red early-type radio galaxy at $z = 0.960$, whose mid-IR emission is very likely not powered by star formation). Most of the HDF/ISO galaxies do not appear to be AGN, however, and if powered by reprocessed stellar UV radiation have obscured star formation rates that greatly exceed those derived from their UV/optical emission (Rowan-Robinson et al 1997). Assuming an M82-like spectrum to derive k-corrections, the average HDF/ISO source is roughly $10\times$ more luminous than M82 at 15 μm (Elbaz et al 1999). Rowan-Robinson (1999) estimates that roughly two thirds of the star-formation at $z \sim 0.5$ is in dust-enshrouded starbursts such as those detected by ISO.

4.7 Sub-Millimeter Sources

The commissioning of the sub-millimeter bolometer array camera SCUBA on the JCMT has opened a potentially revolutionary new window on the distant universe, permitting the detection of dust-reradiated energy from starburst galaxies and AGN out to almost arbitrarily large redshifts (thanks to the strongly negative k-correction at 850 μm, where SCUBA is most sensitive). A series of SCUBA surveys (e.g. Smail et al 1997; Barger et al 1998) have now resolved a source population that comprises a substantial fraction of the far-IR background measured by COBE and that may account for a significant fraction of the global radiative energy density from galaxies. One of the first and deepest of such surveys was carried out by Hughes et al (1998), who observed the HDF for 50 hours, detecting five sources at 850 μm to a 4.4σ limit of 2 mJy. Given the $15''$ SCUBA beam size and possible pointing uncertainties, the optical identifications for many (if not all) of these five sources remain in doubt, and some may be blends due to multiple objects. The brightest source, however, was subsequently pinpointed by a 1.3 mm interferometric observation from Institute de Radioastronomie Millimetrique (IRAM) (Downes et al 1999). Unfortunately, this accurate position (which coincides with a microjansky radio source) did not fully settle the identification of the optical counterpart, falling halfway between two HDF galaxies separated by $2''$ (and almost certainly at very different redshifts). The second brightest SCUBA source, falling just outside the primary WFPC2 field, has no obvious counterpart in the flanking field WFPC2 images, nor in the NICMOS data of Dickinson et al (2000b).

Taking advantage of the frequent (but not universal) association of submillimeter sources with centimeter radio detections, Barger et al (1999) have used SCUBA to observe 14 radio sources in the HDF flanking fields with optical and near-IR counterparts fainter than $I > 25$ and $K > 21$, respectively. They surveyed roughly half of the flanking field area to a 3σ depth of 6 mJy, detecting five of the "blank field" radio source targets in addition to two new radio-quiet sub-millimeter sources; however, none of the optical/near-IR bright sources were detected at 850 μm. For those sub-millimeter sources with radio counterparts, photometric redshifts from the 850 μm to 20 cm flux ratio (Carilli & Yun, 1999; Barger et al 1999; Cowie & Barger, 1999) suggest that most lie in the redshift range $1 < z < 3$. Sub-millimeter sources without radio counterparts may represent a higher redshift tail.

The HDF-N has also been imaged at 450 μm (Hughes et al 1998) and 2.8 mm (Wilner & Wright, 1997). No sources were detected at either wavelength, which is not surprising given the flux densities of the 850 μm detections.

4.8 Radio Sources

The HDF-N has become one of the best-studied regions of the sky at radio wavelengths, and the availability of extremely deep optical/IR imaging and extensive spectroscopy has made it arguably the most important survey for understanding the properties of "average" radio sources at the millijansky and microjansky level. The VLA and MERLIN arrays have been used to map regions including the HDF-N at 3.5 cm and 20 cm wavelengths (Fomalont et al 1997; Richards et al 1998, 1999; Muxlow et al 1999; Richards, 2000). The combined MERLIN + VLA data have been used to make high-resolution maps of 91 HDF radio sources, permitting a detailed study of radio source extent, as well as pinpointing locations for a few ambiguous identifications and confirming the blank field character of others. Altogether, 16 radio sources are located in the central HDF-N WFPC2 field, and many more in the flanking fields.

The optical counterparts for most of the radio sources (72 of 92 in the region with suitably deep imaging) are relatively bright galaxies ($\langle I \rangle \approx 22$) at $0.4 < z < 1$. Apart from a few manifestly AGN-powered objects, usually giant elliptical radio galaxies (including the $z = 1.02$ galaxy HDF 4-752.1), most are disk galaxies, sometimes disturbed or interacting. For nearly all, the radio emission is resolved with a typical size $\sim 1''$, reinforcing the evidence that it primarily arises from star formation. However, about 20% of the sources have very faint or optically invisible counterparts with $I \gtrsim 25$. Many of these have K-band identifications with very red optical-IR colors (Richards et al 1999). Waddington et al (1999) have studied one red HDF radio source which is apparently at $z = 4.4$ (from a single-line spectroscopic redshift). Several have been detected in 850 μm observations with SCUBA (Barger et al 1999). These "invisible" radio sources appear to be a population discontinuous from the majority of the optically brighter sources, and may be high-redshift, dust-obscured, vigorously star-forming systems. Finally, two VLA sources in the HDF proper have no optical counterparts to the limits of the WFPC2 data, and one (123646+621226) has no counterpart to the extremely faint limit of the NICMOS GTO HDF field.

The HDF-S was observed by the Australia Telescope National Facility (ATNF) in 1998. The first year's observations covered a 60' diameter region at 20 cm detecting 240 sources down to 100 μJy (13 in the primary WFPC2 field) and correspondingly smaller fields at 13, 6, and 3 cm. These have now been augmented by substantially deeper observations which are currently being analyzed (Norris et al 2000), but which should reach 5σ sensitivities of 40 μJy at 13 and 6 cm.

4.9 Active Galactic Nuclei

The number-density of AGN at HDF depths is of great interest because of its connection to the X-ray background and to the re-ionization of the universe at high redshift. Jarvis & MacAlpine (1998) identified 12 possible candidates with colors and morphologies consistent with AGN at $z > 3.5$. A similar study by Conti et al (1999) found no candidates above $z = 3.5$ and an upper limit of 20 candidates at lower z. These relatively low number densities support the prevailing view that the UV emission from AGN at high redshifts is insufficient to account for the ionization of the IGM, and they also suggest that black holes do not form with constant efficiency within cold-dark-matter halos (Haiman et al 1999). Fainter AGN yet to be identified may lie within the HDF. Based on the properties of many of the optical counterparts to faint ROSAT X-ray sources, Almaini & Fabian (1997) estimate that roughly 10% of the galaxies in the HDF are likely to be X-ray luminous, narrow-lined AGN. However, few of the optically identified AGN candidates in the HDF-N are detected in the X-rays by the Chandra observatory (Hornschemeier et al 2000).

4.10 Gravitational Lensing

The depths of the HDFs have made them well suited for studies of gravitational lensing. As Blandford (1998) noted in his review, based on a rule of thumb from the number of lensed quasars and radio sources in other surveys, roughly 0.2% of the galaxies, corresponding to five to ten sources, were expected to be strongly lensed (producing multiple images) in each HDF field. In addition, weak lensing by large-scale structure and individual galaxies should slightly distort most of the galaxy images. Initial visual inspection of the HDFs produced a number of possible strong lensing candidates (Hogg et al 1996; Barkana et al 1999), but none has yet been confirmed by spectroscopy. Rather, a few of the brighter examples appear not to be lensed systems, but only chance superpositions (Zepf et al 1997). A few candidates remain in each of the primary WFPC2 HDF fields (Barkana et al 1999), but it will require spectroscopy of very faint sources to confirm their nature.

An analysis of the HDF-N images by Zepf et al (1997) concluded that there were at most one to two strongly lensed sources in the entire field, although very faint objects with small angular separation could have escaped detection. On this basis they suggested that the HDF data were not compatible with a large cosmological constant. Cooray et al (1999) took this further by utilizing published photometric redshifts for the galaxies in the HDF-N to calculate the expected number of multiply-imaged galaxies in the field. They found that a limit of one detectable strongly lensed source requires $\Omega_\Lambda - \Omega_M < 0.5$. For Ω_M, Ω_λ, $\Omega_{tot} = $ 0.3, 0.7, 1, they predict 2.7 multiply-imaged galaxies. The lensing statistics are thus, in spite of earlier expectations for a larger number of lenses, not in conflict with recent determinations of Ω_Λ from high-redshift SN (Perlmutter et al 1999; Riess et al 1998).

Weak lensing is manifested by a tangential shear in the image of the more distant source, and the amount of ellipticity, or polarization p, introduced by a typical closely-spaced galaxy pair is of the order of $p \sim 10^{-2}$. It is possible to attempt a statistical detection by looking for the effect superimposed on the intrinsic morphologies of the source galaxies perpendicular to galaxy-galaxy lines of sight. As a differential effect, weak-lensing distortion is far less sensitive to the cosmological model than to the masses or surface densities of the intervening deflector galaxies. Using color and brightness as redshift indicators, Dell' Antonio & Tyson (1996) defined a sample of 650 faint background and 110 lens galaxies in the HDF-N and reported a 3σ detection equivalent to $p = 0.06$ for galaxy-lens pairs having a separation of $2''$. This corresponds to an average galaxy mass of $6 \times 10^{11} M_\odot$ inside 20 kpc, or an internal velocity dispersion of 185 km s^{-1}. A similar analysis of HDF-N images was subsequently performed by Hudson et al (1998), who used photometric redshifts to determine distances of lens and source galaxies, and who discriminated between lensing galaxy types based on colors. Limiting their analysis to separations greater than $3''$, they succeeded in measuring background shear at a 99% confidence level, finding that intermediate-redshift spiral galaxies follow the Tully-Fisher relation but are 1 mag fainter than local spirals at fixed circular velocity. Although the sense of the evolution is the opposite of that found in the kinematical studies, given the uncertainties, the lensing result is consistent with the modest luminosity-evolution observed by Vogt et al (1997) when the studies are compared using the same cosmology. The lensing results are inconsistent with the larger luminosity evolution found by Rix et al (1997) and Simard & Pritchet (1998).

4.11 Intergalactic Medium

The HDF-S STIS field was selected to contain the bright, moderate redshift quasar J2233-606. Deep HST imaging of the field around the quasar during the HDF-S campaign has provided detection and morphologies of numerous galaxies near the line of sight. Moderate- to high-resolution spectra have been obtained of the quasar both from the ground and with HST covering the wavelength regime 1140-8200 Å (Savaglio, 1998; Sealey et al 1998; Outram et al 1999; Ferguson et al 2000), whereas follow-up observations from the ground have established the redshifts of some of the brighter galaxies in the field (Tresse et al 1999). Too few redshifts have yet been measured to allow detailed study of the correspondence between galaxies and QSO absorbers. A tentative detection of Lyα emission surrounding the QSO was reported by Bergeron et al (1999).

Using the STIS UV and ground-based optical spectra, Savaglio et al (1999) found the number density of Lyα clouds with $\log N_{HI} > 14$ in the redshift interval $1.5 < z < 1.9$, to be higher than that found in most previous studies, and saw no evidence for a change in the Doppler parameters of the Lyα lines with redshift. The two-point correlation function of Lyα clouds shows clear clustering on scales less than 300 km s^{-1}, especially the higher column density systems, in agreement with results from other quasars. Metal abundances in several absorption line systems have been studied by Prochaska & Burles (1999). Petitjean & Srianand (1999) studied the metal lines in the absorption-line systems near the QSO redshift and found that relative line strengths of different species are best modeled with a multi-zone partial-covering model wherein different species (e.g. Ne VIII, O

VI) arise from gas at different distances from the AGN, and the clouds cover the central continuum emission region completely but only a fraction of the broad emission-line region.

4.12 Galaxy Clustering

The measurements of clustering using the HDF have varied in the catalog and object selection, in the angular scales considered, in the use or non-use of photometric redshifts, and in the attention to object masking and sensitivity variations. In an early paper on the HDF-N, Colly et al (1996) explored whether the galaxy counts in the HDF are "whole numbers," i.e. whether fragments of galaxies were incorrectly being counted as individual galaxies. From an analysis of the angular correlation function of objects with color-redshifts $z > 2.4$ they conclude that many of these objects are HII regions within a larger underlying galaxy. This analysis showed a strong clustering signal on scales $0''.2 < \theta < 10''$. Colley et al (1997) explored various hypotheses for the neighboring galaxies and favored a scenario in which the faint compact sources in the HDF are giant star-forming regions within small Magellanic irregulars.

It is not clear whether the Colley et al (1996, 1997) results apply to other catalogs, given the different cataloging algorithms used. In particular, programs such as SExtractor (Bertin & Arnouts, 1996) and FOCAS (Tyson & Jarvis, 1979) use sophisticated and (fortunately or unfortunately) highly tunable algorithms for merging or splitting objects within a hierarchy of isophotal thresholds. The DAOFIND algorithm used by Colley et al (1996) provides no such post-detection processing. The extent to which this affects the results can only be determined by an object-by-object comparison of the catalogs, which has not (yet) been done in any detail. Ferguson (1998b) (Figures 1–4) presents a qualitative comparison of several catalogs (not including that of Colley et al) which shows significant differences in how objects with overlapping isophotes are counted. The Colley et al (1997) analysis focused specifically on 695 galaxies with color-redshifts $z > 2.4$. This number of galaxies is considerably larger than the 69 identified by Madau et al (1996), using very conservative color selection criteria, or the 187 identified by Dickinson (1998) with somewhat less conservative criteria. For these smaller samples, we suspect that the "overcounting" problem is not as severe as Colley et al contend.

Problems separating overlapping objects, although important for clustering studies on small angular scales, do not have much effect on the overall galaxy number–magnitude relation (Ferguson, 1998b) or angular clustering measurements on scales larger than $\sim 2''$.

Other studies of clustering in the HDF have been restricted to separations larger than $2''$. The analysis has focused primarily on the angular correlation function $\omega(\theta)$, which gives the excess probability δP, with respect to a random Poisson distribution of n sources, of finding two sources in solid angles $\delta\Omega_1$, $\delta\Omega_2$ separated by angle θ:

$$\delta P = n^2 \delta\Omega_1 \delta\Omega_2 \left[1 + \omega(\theta)\right]. \tag{1}$$

The angular correlation function is normally modeled as a power law

$$\omega(\theta) = A(\theta_0)\theta^{1-\gamma}. \tag{2}$$

Because of the small angular size of the field, $\omega(\theta)$ is suppressed if the integral of the correlation function over the survey area is forced to be zero. Most authors account for this "integral constraint" by including another parameter and fitting for $\omega(\theta) = A(\theta_0)\theta^{1-\gamma} - C$, with a fixed value of $\gamma = 1.8$ and with A and C as free parameters.

Villumsen et al (1997) measured the overall angular correlation function for galaxies brighter than $R = 29$ and found an amplitude $A(\theta = 1'')$ decreasing with increasing apparent magnitude, roughly consistent with the extrapolation of previous ground-based results. The measured amplitude was roughly the same for the full sample and for a subsample of red galaxies (considered typically to be at higher redshift). The motivation for this color cut was to try to isolate the effects of magnification bias due to weak lensing by cosmological large-scale structure (Moessner et al 1998), but the predicted effect is small and could not be detected in the HDF (and will ultimately be difficult to disentangle from the evolutionary effects discussed in Section 5.6)

The remaining HDF studies have explicitly used photometric redshifts. Connolly et al (1999) considered scales $3'' < \theta < 220''$ and galaxies brighter than $I_{814} = 27$. Within intervals $\Delta z = 0.4$ the amplitude $A(\theta = 10'') \sim 0.13$ shows little sign of evolution out to $z_{\mathrm{phot}} = 1.6$ (the highest considered by Connolly et al). Roukema et al (1999) analyzed a U-band selected sample, isolated to lie within the range $1.5 < z_{\mathrm{phot}} < 2.5$ and angular separations $2'' < \theta < 40''$. The results are consistent with those of Connolly et al (1999), although Roukema et al (1999) point out that galaxy masking and treatment of the integral constraint can have a non-negligible effect on the result.

Measurements of $\omega(\theta)$ for samples extending out to $z_{\mathrm{phot}} > 4$ have been carried out by Magliocchetti & Maddox (1999), Arnouts et al (1999) and Miralles & Pello (1998). The minimum angular separations considered for the three studies were $9''$, $5''$, and $10''$, respectively: i.e. basically disjoint from the angular scales considered by Colley et al (1997). All three studies detect increased clustering at $z \gtrsim 2$, although direct comparison is difficult because of the different redshift binnings. The interpretation shared by all three studies is that the HDF shows strong clustering for galaxies with $z \gtrsim 3$, in qualitative agreement with Lyman-break galaxy studies from ground-based samples (Giavalisco et al 1998; Adelberger et al 1998). Although this may be the correct interpretation, the significant differences in the photometric redshift distributions and the derived clustering parameters from the three studies leave a considerable uncertainty about the exact value of the clustering amplitude. The two most comprehensive studies differ by more than a factor of 3 in $A(\theta = 10'')$ at $z = 3$ (Arnouts et al 1999; Magliocchetti & Maddox, 1999). Both studies use photometric redshifts to magnitudes $I_{814} = 28$, which is fainter than the detection limits in the F300W and F450W bands for even flat-spectrum galaxies. A large part of the disagreement may thus be due to scatter in z_{phot}. The analysis also hinges critically on the assumed power-law index $\gamma = 1.8$ and the necessity to fit the integral constraint. Thus, although it seems reasonably secure that a positive clustering signal has been measured in the HDF at high z_{phot}, it will require much larger data sets to constrain the exact nature of this clustering and determine its relation to clustering at lower redshift.

At brighter magnitudes, spectroscopic surveys show clear evidence for clustering in redshift space (Cohen et al 1996; Adelberger et al 1998; Cohen, 1999), with pronounced peaks even at redshifts $z > 1$. These structures, and others within the flanking fields, do not show evidence for centrally concentrated structures, and are probably analogous to walls and filaments observed locally.

5. INTERPRETATION

5.1 Galaxy Counts vs. Simple Models

The time-honored method of comparing cosmological models to field-galaxy surveys has been through the classical number–magnitude, size–magnitude, magnitude–redshift, etc. relations (Sandage, 1988). Much of this effort, including early results from the HDF, has been reviewed in Ellis (1997). There are three important changes to the scientific landscape that we must acknowledge before proceeding to discuss the more recent interpretation of the HDF number counts. The first change is the transition (motivated by high-z SNe Ia, cluster baryon fractions, etc.) from a favored cosmology with $\Omega_M = \Omega_{tot} = 1$ to one with $\Omega_M, \Omega_\Lambda, \Omega_{tot} = 0.3, 0.7, 1$. The latter cosmology comes closer to matching the galaxy counts without extreme amounts of evolution. The second change is the realization that galaxy counts at HDF depths push to depths where the *details* of galaxy formation become extremely important. Models that posit a single "epoch of galaxy formation" were never realistic, and are no longer very useful. Finally, it has become more popular to model quantities such as the metal enrichment history $\dot{\rho}_z(z)$ than it is to model galaxy counts. Consequently, to our knowledge there are no published papers that compare the overall $N(m)$ predictions of hierarchical semi-analytic models with the currently favored cosmology to the HDF galaxy counts.

5.1.1 No-Evolution (NE) Models   The assumption of no evolution is not physically reasonable, but provides a useful fiducial for identifying how much and what kinds of evolution are required to match faint-galaxy data. Traditional no-evolution models are based on estimates of the $z = 0$ luminosity functions for different types of galaxies. Bouwens et al (1997) construct a non-evolving model from the HDF itself, using 32 galaxies brighter than $I_{814} = 22.3$ to define a fiducial sample. They construct Monte-Carlo realizations of the HDF that might be seen from a universe uniformly populated with such galaxies, shifted to different redshifts and k-corrected on a pixel-by-pixel basis. This kind of simulation automatically incorporates the selection and measurement biases of the HDF at faint magnitudes. It also normalizes the model *by fiat* to match the counts at $I_{814} = 22.3$, where traditional no-evolution models, normalized to the local luminosity function, already see significant discrepancies for the Einstein–de Sitter model. The resulting models underpredict the HDF counts at $I_{814} = 27$ by factors of 4 and 7 for models with $\Omega_M = 0.1$ and 1.0 (with $\Omega_\Lambda = 0$), respectively. The angular sizes of galaxies in the models are also too big at faint magnitudes, with a median half-light radius about a factor of 1.5 larger than that observed for galaxies with $24 < I < 27.5$. Because the typical redshift of the template galaxies in this model is $z \sim 0.5$, this model comparison suggests that much of the evolution in galaxy number densities, sizes, and luminosities occurs at higher redshift. The typical small sizes of faint galaxies essentially rule out low-surface-brightness galaxies (Ferguson & McGaugh, 1995; McLeod & Rieke, 1995) as a significant contributor to the counts at magnitudes $I_{814} > 20$ (Ferguson, 1999).

5.1.2 Pure Luminosity Evolution Models

Many of the models that have been compared to the HDF number counts are variants of *pure luminosity evolution* (PLE) models (Tinsley, 1978), wherein galaxies form at some redshift z_f, perhaps varying by type, with some star-formation history $\psi(t)$. There is no merging. Metcalfe et al (1996) compare several different models to the counts and colors of galaxies in the HDF and in deep images taken at the William Herschel Telescope. For low Ω a reasonable fit to $I \approx 26$ is achieved, but the model progressively underpredicts the counts to fainter magnitudes, and the long star-formation timescales and heavily dwarf-dominated IMF adopted for ellipticals in this model seem inconsistent with the fossil evidence in local ellipticals. By including a simple prescription for dust attenuation, Campos & Shanks (1997) are able to achieve a reasonable fit to the counts for low Ω_M without resorting to a peculiar IMF. Another set of PLE models was considered by Pozzetti et al (1998), with emphasis on the near-UV counts and the effect of UV attenuation by intergalactic neutral hydrogen. From color-magnitude relations and a study of the fluctuations in the counts in the different HDF bands, Pozzetti et al (1998) conclude that at $B_{450} = 27$ roughly 30% of the sources in the HDF are at $z > 2$. The PLE model considered has $\Omega_M = 0.1$ with no cosmological constant, and with a redshift of formation $z_f = 6.3$. This model matches the counts quite well but predicts that about 80 objects brighter than $V_{606} = 28$ should disappear from the F450W band because of their high redshift, although Madau et al (1996) identify only about 15 such sources. Ferguson & Babul (1998) considered another low–Ω PLE model and encountered similar problems, predicting roughly 400 B-band Lyman-break objects, where only 15 or so were observed. The utility of PLE models clearly breaks down for $z \gtrsim 2$, where the details of galaxy formation become critical.

5.1.3 Models with Additional Galaxy Populations

The great difficulty in achieving a fit with $\Omega_M = \Omega_{tot} = 1$, even to ground-based galaxy counts, motivated investigations into different kinds of galaxies that might be missed from the census of the local universe but could contribute to the counts of galaxies at faint magnitudes (Ferguson & McGaugh, 1995; Babul & Ferguson, 1996; Koo et al 1993; McLeod & Rieke, 1995). Perhaps the most physically motivated of these more exotic possibilities is the idea that the formation of stars in low-mass galaxy halos could be inhibited until low redshifts $z \gtrsim 1$ because of photoionization by the metagalactic UV radiation field (Efstathiou, 1992; Babul & Rees, 1992). Ferguson & Babul (1998) compared the predictions of such an $\Omega_M = 1$ "disappearing dwarf" model in detail to the HDF. They found that the simplest version of the model (*a*) overpredicts the counts at faint magnitudes, and (*b*) overpredicts the sizes of very faint galaxies. These problems are caused by the fact that, for a Salpeter IMF, the dwarfs fade too slowly and would still be visible in great numbers in the HDF at redshifts $z < 0.5$. Campos (1997) considered a model with much milder evolution, with each dwarf undergoing a series of relatively long (a few $\times\ 10^8$ yr) star-formation episodes. Acceptable fits to the counts and colors of galaxies are achieved both for high and low values of Ω. Both the Ferguson & Babul (1998) and the Campos (1997) models predict that the HDF sample at $I > 25$ is dominated by galaxies with $z < 1$, a result inconsistent with existing photometric-redshift measurements. Using a volume-limited photometric redshift sample to construct the bivariate brightness distribution of galaxies with $0.3 < z < 0.5$, Driver (1999) concludes that the volume density of low-luminosity, low-surface-brightness galaxies is not sufficient to explain the faint-blue excess either by themselves or as faded remnants. Further constraints on the low-redshift, low-luminosity population can be expected from the HDF STIS UV observations.

5.2   The Morphology of High-Redshift Galaxies

As was described in Section 4.1, HST images from the HDF and other surveys have established that the fraction of irregular and peculiar galaxies increases toward faint magnitudes (down to $I \sim 25$, where galaxies become too small for reliable classification). Some of the morphological peculiarities in HDF galaxies may be the consequence of interactions, collisions, or mergers. A higher merger rate at earlier epochs is a natural consequence of models that assemble the present-day Hubble sequence galaxies by a process of hierarchical mergers (Baugh et al 1996). In the local universe, morphological asymmetry correlates reasonably well with color, with late type, blue galaxies and interacting objects showing the greatest asymmetry (Conselice et al 1999). In the HDF, however, there are also highly asymmetric galaxies with *red* rest-frame $B - V$ colors as well as blue ones, and the asymmetries persist at NICMOS wavelengths (cf. Figure 9), where longer-lived stars in a mixed-age stellar population would dominate the light and where the obscuring effects of dust would be reduced. The timescale over which an interacting galaxy will relax and regularize should be shorter ($\gtrsim 1$ Gyr) than the time over which stars formed during the interaction would burn off the main sequence, and thus if star formation occurs during collisions then blue colors should persist longer than the most extreme manifestations of morphological disturbance.

Conselice et al (2000) suggest that the large number of asymmetric HDF galaxies demonstrates the prevalence of early interactions and mergers, some of which (the bluest objects) have experienced substantial star formation during the encounter, whereas others have not (or, alternatively, have their recent star formation obscured by dust).

Although interactions seem an attractive way of explaining these morphological peculiarities, it is nevertheless worth mentioning some qualifications concerning the details of the models which have been tested to date. First, although a high merger rate at relatively low redshifts is a generic prediction of the standard cold dark matter (SCDM) model, models with lower Ω_M have a substantially lower merger rate at $z < 1$. The Baugh et al (1996) model that was explicitly compared with the HDF was based on the SCDM cosmology. Second, even with the high merger rate the Baugh et al (1996) model predicts that galaxies that have had major merger within 1 Gyr prior to the time of observation constitute less than 10% of the total population at $I = 25$ (their Figure 2). Most of the irregular galaxies in this model are simply bulgeless, late-type galaxies. Im et al (1999) have studied the redshift distribution of galaxies with $17 < I < 21.5$ and conclude that the irregular/peculiar class is a mix of low-redshift dwarf galaxies and higher-redshift ($0.4 < z < 1$) more luminous galaxies. These higher redshift galaxies are unlikely to be the progenitors of present-day dwarf irregular galaxies, but it is not clear whether they are predominantly low-mass galaxies undergoing a starburst, or more massive galaxies undergoing mergers. The scatter in colors and sizes suggests it is a mix of both, but more detailed kinematical information is needed.

5.3   Elliptical Galaxies

The simple picture of the passively evolving elliptical galaxy, formed at high redshift in a single rapid collapse and starburst, has held sway since the scenario was postulated by Eggen et al (1962) and its photometric consequences were modeled by Larson (1974) and Tinsley & Gunn (1976). This hypothesis is supported by the broad homogeneity of giant elliptical (gE) galaxy photometric and structural properties in the nearby universe, and by the relatively tight correlations between elliptical galaxy chemical abundances and mass or luminosity. Observations in the past decade have pushed toward ever higher redshifts, offering the opportunity to directly watch the evolutionary history of elliptical galaxies. Most of this work has concentrated on rich cluster environments, and is not reviewed here except to note that most observers have favored the broad interpretation of quiescent, nearly passive evolution among cluster ellipticals out to $z \approx 1$ (e.g. Aragón-Salamanca et al 1993; Stanford et al 1995, 1998; Ellis et al 1997; van Dokkum et al 1998; De Propris et al 1999).

The "monolithic" formation scenario serves as a rare example of a clearly stated hypothesis for galaxy evolution against which to compare detailed measurements and computations. The alternative "hierarchical" hypothesis is that elliptical galaxies formed mostly via mergers of comparable-mass galaxies that had at the time of merging already converted at least some of their gas into stars (e.g. Kauffmann et al 1993). The argument between the monolithic and the hierarchical camps is not about the origin of galaxies from gravitational instability within a hierarchy of structure, but rather about when gEs assemble most of their mass and whether they form their stars mostly *in situ* or in smaller galaxies that subsequently merge. At least one observational distinction is clear: In the hierarchical model the number density of elliptical galaxies should decrease with redshift. In the monolithic model the number density should remain constant and the bolometric luminosities should increase out to the epoch of formation.

Measuring the density and luminosity evolution of the *field* elliptical population, however, has proven difficult, and many different approaches, perhaps complementary but not necessarily concordant, have been used to define suitable galaxy samples. Recent debate has focused on estimates of the evolution of the co-moving density of passively-evolving elliptical galaxies in the Canada-France Redshift Survey (Lilly et al 1995b). Applying different photometric selection criteria and different statistical tests, two groups (Lilly et al 1995a; Totani & Yoshii, 1998) found no evidence for density evolution out to $z = 0.8$, whereas another group (Kauffmann et al 1996), found evidence for substantial evolution. This debate highlights the difficulties inherent to defining samples based on a color cut in the margins of a distribution, where small changes in the boundary, as well as systematic and even random photometric errors in the data, can have substantial consequences for the conclusions.

HST makes it possible to define samples of distant galaxies morphologically. The multicolor HDF images provide an attractive place to study distant ellipticals, but because of its very small volume and the inherently strong clustering of elliptical galaxies, one must be careful in drawing sweeping conclusions from HDF data alone. Thus, although we focus our attention primarily on the HDF results, they should be considered in context with results from other surveys (e.g. Driver et al 1995; Glazebrook et al 1995; Driver et al 1998; Treu et al 1999; Treu & Stiavelli 1999).

The interpretation of the high-redshift elliptical galaxy counts depends in large measure on a comparison to the *local* luminosity function (LF) of elliptical galaxies. The basic parameters of published LFs for local elliptical galaxies (or what are sometimes assumed to be elliptical galaxies)[3] span a very wide range in normalization, characteristic luminosity, and faint end behavior. Early HST studies such as that by Im et al (1996) concluded that NE or PLE models were consistent with elliptical-galaxy number counts predicted using the local LF of Marzke et al (1994).

Searches for distant elliptical galaxies in the HDFs have relied on either color or morphology. In the HDF-N, Fasano et al (1998) and Fasano & Filippi (1998), along with Schade et al (1999) for the CFRS and LDSS surveys, all examined the

---

[3]Even locally, classification uncertainties may be partly responsible for the widely diverse measurements of the elliptical galaxy LF (see Loveday et al 1992; Marzke et al 1994, 1998; Zucca et al 1994; Sandage 2000).

size-luminosity relation for morphologically-selected ellipticals, finding evolution out to $z \sim 1$ consistent with PLE models. Kodama et al (1999) found a well-defined color-magnitude sequence at $\langle z \rangle \sim 0.9$, consistent with passive evolution for approximately half the galaxies, but they also noted a substantial "tail" of bluer objects. Similarly, in the Schade et al (1999) study, about one third of the sample at $z > 0.5$ had [OII] line emission and colors significantly bluer than PLE models.

IR imaging has made it attractive to pursue gEs at redshifts $z > 1$. In the HDF-N, Zepf (1997) and Franceschini et al (1998) used ground-based K-band data, selecting samples by color and morphology, respectively, and concluded that there was an absence of the very red galaxies that would be expected if the elliptical population as a whole formed at very large redshift and evolved passively. In particular, Franceschini et al highlighted the apparently sudden disappearance of HDF ellipticals beyond $z > 1.3$, which suggests that either dust obscuration during early star formation, or morphological perturbation during early mergers, was responsible. Barger et al (1999) made an IR study of the HDF-N flanking fields, selecting objects by color without reference to morphology. To $K \approx 20$, they found few galaxies with $I - K > 4$, the expected color threshold for old ellipticals at $z \gg 1$. However, other comparably deep and wide IR surveys have reported substantially larger surface densities of $I - K > 4$ galaxies (Eisenhardt et al 1998; McCracken et al 2000), raising concerns about field-to-field variations. Menanteau et al (1999) studied the optical-to-IR color distribution for a sample of $\sim$300 morphologically early-type galaxies selected from 48 WFPC2 fields, including the HDF-N and flanking fields. They too find an absence of very red objects and a generally poor agreement with predictions from purely passive models with high formation redshifts, although the sky surface density agrees reasonably well with the sorts of models that matched the older MDS and HDF counts, i.e. those with a suitably tuned local LF.

On HST, NICMOS has provided new opportunities to identify and study ellipticals at $z \gtrsim 1$. Its small field of view, however, has limited the solid angle surveyed. Spectroscopic confirmation of the very faint, very red elliptical candidates identified so far will be exceedingly difficult, but would be well worth the effort. Treu et al (1998), Stiavelli et al (1999) and Benítez et al (1999) noted several very red, $R^{1/4}$-law galaxies in the HDF-S NICMOS field, identifying them as ellipticals at $1.4 \lesssim z_{phot} \lesssim 2$. Given the small solid angle of that image, this suggests a large space density, and Benítez et al (1999) have proposed that most early type galaxies have therefore evolved only passively since $z \sim 2$. Comparably red, spheroidal galaxies are found in the NICMOS map of the HDF-N (Dickinson et al 2000a), with photometric redshifts in the range $1.2 \lesssim z \lesssim 1.9$. Their space density appears to be well below that of HDF ellipticals with similar luminosity at $z < 1.1$, in broad agreement with Zepf (1997) and Franceschini et al (1998), although the comparison to $z = 0$ is again limited primarily by uncertainties in the local gE LF. Curiously, few fainter objects with similar colors are found in the HDF-N, although the depth of the NICMOS images is adequate to detect red ellipticals with $L \sim 0.1 \, L^*$ out to $z \approx 2$. Treu & Stiavelli (1999) find that PLE models with high

($z \sim 5$) and low ($z \sim 2$) formation redshifts over- and under-predict the observed counts, respectively, of red elliptical-like objects in 23 other NICMOS fields.

The HDF-N NICMOS images from Dickinson et al (2000a) and Thompson et al (1999) are deep enough to have detected red, evolved elliptical galaxies out to at least $z \sim 3$ if they were present, eliminating concerns about invisibility because of k-corrections (e.g. Maoz 1997). The only plausible $z > 2$ candidate is an HDF-N "J-dropout" object, whose colors resemble those of a maximally old gE at $z \sim 3$ (Lanzetta et al 1998; Dickinson et al 2000b; Lanzetta et al 1999). Other $z > 2$ HDF objects which some authors have morphologically classified as ellipticals (cf. Fasano et al 1998) are blue, mostly very small, manifestly star-forming "Lyman break" objects. Few if any appear to be passively evolving objects that have ceased forming stars, so the connection to present-day ellipticals is more speculative.

The collective evidence surveyed above suggests that mature, gE galaxies have been present in the field since $z \sim 1$ with space densities comparable to that at the present era. At $0.4 < z < 1$, however, they exhibit an increasingly broad range of colors and spectral properties, which suggests a variety of star formation histories over the proceeding few billion years (or alternatively errors in classification). The statistics seem to favor a substantial decline in their space density at $z \gg 1$, although the well-surveyed sightlines are small and few, and clustering might (and in fact, apparently does) cause large variations from field to field. Therefore, the conclusion has not been firmly established. Moreover, it is very difficult to achieve uniform selection at all redshifts, regardless of the criteria used (photometric, morphological, or both), and the existing samples of objects with spectroscopic (or at least well-calibrated photometric) redshifts are still small. Thus even the deceptively simple task of comparing elliptical galaxy evolution to the simple PLE hypothesis remains a stubborn challenge.

5.4 Obscured Populations

ISO and SCUBA observations of the HDF and other fields have revealed an energetically important population of dust-obscured objects. Interpretation of these results has been the subject of considerable debate, because of ambiguities in source identification and in distinguishing starbursts from AGN. The ISOCAM (6.7 and 15 μm) and SCUBA (850 μm) observations bracket but do not sample the wavelength regime $100 < \lambda_p < 200$ μm near the peak of the far-IR emission, making it difficult to assess reliably the source contribution to the global emissive energy budget of galaxies. Although strong mid-IR emission accompanies vigorous star formation in many nearby galaxies, the unidentified IR emission bands carry most of the energy in the wavelength range sampled by ISOCAM at $z \sim 1$. The bulk of re-emitted radiation, however, emerges near λ_p, and there is considerable diversity in $f(10 \, \mu m)/f(100 \, \mu m)$ flux ratios among nearby luminous and ultraluminous IR galaxies. Therefore, deriving star-formation rates from mid-IR measurements alone requires a substantial extrapolation and is quite uncertain. SCUBA 850 μm observations sample the re-radiated thermal emission directly, but at a wavelength well past λ_p, again requiring an extrapolation to total far-IR luminosities assuming a dust temperature and emissivity that are almost never well constrained by actual, multi-wavelength measurements. Furthermore, SCUBA and ISO do not in general detect the same sources. ISO detects objects out to $z \sim 1$, whereas most of the SCUBA sources could be at much higher redshifts. For both mid-IR and sub-millimeter sources, AGN-heated dust may also play a role. Several active galaxies (including radio ellipticals) in the HDF are detected by ISO, and some SCUBA sources (not yet in the HDF, however) have been identified with AGN. Chandra X-ray observations (Hornschemeier et al 2000) do not detect with high significance any of the sub-millimeter sources in the HDF.

Difficulties in identifying the optical counterparts to the mid-IR and sub-millimeter sources are a second source of ambiguity. The mean separation between galaxies in the HST images of the HDF is about $3''$. In comparison the ISO $15\,\mu$m PSF has full-width at half max (FWHM) $\approx 9''$, whereas the SCUBA $850\,\mu$m beam size is $\approx 15''$. Nevertheless, the HDF mid-IR sources have plausible counterparts among the brighter galaxies in the survey. SCUBA sources, on the contrary, often seem to have extremely faint, and sometimes entirely invisible, counterparts in the optical and near-infrared. Sub-mm objects sometimes correspond to microjansky radio sources at centimeter wavelengths, and sometimes have faint, very red near-IR counterparts. Many, however, do not, even with the deepest radio and near-IR data available with current instrumentation (see Section 4.7).

Other than the occasional ultra-red optical counterpart (of which there are none in the HDF-N), the non-AGN counterparts to sub-millimeter and far-IR sources rarely have particularly unusual photometric or spectral features which highlight them as the remarkable objects they must be. Counterparts to 6.7 and $15\,\mu$m ISO sources show a predominance of "post-starburst" Balmer absorption spectra (Flores et al 1999a,b; Aussel et al 1999b; Cohen et al 2000); apart from the occasional AGN, few have strong emission line spectra, which suggests the most intense star-forming regions are highly obscured. There are hints that the Lyman-break galaxies represent the faint tail of the sub-millimeter source population. Chapman et al (1999) detect 1 out of 18 galaxies in a targeted study of $z \sim 3$ Lyman-break galaxies. Peacock et al (1999) find a significant statistical correlation of Lyman-break galaxies with sky fluctuations in the HDF-N SCUBA map. A straightforward analysis suggests that the ratio of hidden star formation to star formation directly measured in the UV is about 6:1, and that Lyman-break galaxies account for at least 25% of the $850\,\mu$m background.

Given the present status of faint sub-millimeter and mid-IR surveys, the most telling information comes not from individual source identifications but from the ensemble statistics using the combined data from many surveys. The ISO $15\,\mu$m number counts show good agreement between various surveys and a strong excess over no-evolution models (Elbaz et al 1999), as do $850\,\mu$m SCUBA counts (Barger et al 1999; Blain et al 1999). As with the "faint blue galaxy" excess (Kron, 1980) that helped spark the boom in the optical study of distant field galaxies in the 1970s and 1980s, the ISO and SCUBA counts point toward strong cosmological evolution, probably manifesting the star formation history of the galaxy population. And like the faint blue galaxy problem, the robust interpretation of this number count excess will undoubtedly require extensive follow-up observations to characterize the source population and its properties. Currently, given the uncertainties in source identification, in extrapolating the current measurements to "true" far-IR luminosities, and in balancing the roles of AGN vs. star formation, it seems premature to attempt a detailed revision of our picture of galaxy evolution and cosmic star formation. But it is also abundantly clear that a true understanding will have to account for the important and perhaps dominant energetic role played by this obscured population.

5.5   Global Star-Formation History and Chemical Evolution

Interest in the integrated background light from galaxy formation is long-standing (e.g. Bondi et al 1955), and has motivated a large number of experiments aimed at measuring the diffuse extragalactic background (e.g. Spinrad & Stone 1978; Dube et al 1979; Matsumoto et al 1988; Bernstein 1997; Hauser et al 1998). Although the integrated background records the light from all galaxies, whether or not they are individually detected, the task of separating out galactic foregrounds and instrumental backgrounds is formidable, and many of the measurements are only upper limits. An alternative approach is to add up the UV emission from galaxies that are individually detected. This approach formally produces only a lower limit for the true UV luminosity density, but it provides a basis for exploring connections of the high- and low-redshift universe and for deciding which of the various selection effects are plausible and important.

The UV emission from galaxies is directly connected to their metal production (Cowie et al 1988) because the UV photons come from the same massive stars that produce most of the metals through type II SNe. The relation between UV emissivity and metal production is not strongly dependent on the initial mass function, varying by a factor of only 3.3 between the Salpeter (1955) and Scalo (1986) forms. In contrast, the relation of UV emission to the total star-formation rate is much more tenuous because the low-mass end of the IMF contains most of the mass, whereas the high-mass end produces most of the UV emission. As a result, it is easier to constrain the history of metal production in the universe than it is to constrain the overall star-formation history. Prior to the HDF, estimates of the metal-production vs. redshift, $\dot{\rho}_z(z)$ had been made from ground-based galaxy redshift surveys (Lilly et al 1996) and from QSO absorption line statistics (e.g. Lanzetta et al 1995; Pei & Fall 1995; Fall et al 1996), and the observations at the time suggested an order-of-magnitude increase in $\dot{\rho}_z(z)$ from $z = 0$ to $z = 1$.

Madau et al (1996) made the first attempt to connect the luminosity density in HDF high-redshift galaxy samples to lower-redshift surveys. The plot of the metal-formation rate vs. redshift has provided a focal point for discussion of the HDF and for comparisons to theoretical models. In the Madau et al (1996) paper, galaxies in redshift slices $2 < z < 3.5$ and $3.5 < z < 4.5$ were identified by strict color-selection criteria that were shown via simulations to provide strong rejection of objects outside the desired redshift intervals. The luminosity of the galaxies within these redshift intervals was estimated by simply summing the observed fluxes of the galaxies, and no dust or surface-brightness corrections were applied. The results were presented as lower limits because nearly all of the corrections will drive the derived metal-formation rates up. The initial Madau et al (1996) diagram showed a metal-production rate at $z \sim 4$ that was roughly a factor of 10 lower than the rate at $z \sim 1$. Madau (1997) subsequently modified the color-selection criteria and integrated down an assumed luminosity function, revising the $z > 2$ rates upward by about a factor of 3. There was remarkable agreement between $\dot{\rho}_z(z)$ derived from the galaxy luminosities, the results of Pei & Fall (1995), and the predictions of hierarchical models (e.g. White & Frenk 1991; Cole et al 1994; Baugh et al 1998), all of which show a peak in the metal production rate at $z \sim 1 - 2$. Subsequent work in this area has focused on (*a*) galaxy selection (*b*) effects of dust, and (*c*) the connection of the Madau diagram to general issues of galaxy evolution and cosmic chemical evolution, which we discuss in turn.

5.5.1 Galaxy Selection   The color-selection criteria of Madau et al (1996) are extremely conservative, and spectroscopic surveys (Steidel et al 1996; Lowenthal et al 1997) have identified at least a dozen $2 < z < 3.5$ galaxies with $U_{300} - B_{450} > 1.3$ but with $B_{450} - I_{814}$ redder than the Madau et al (1996) selection boundary. Alternative color-selection criteria (particularly in the $2 < z < 3.5$ range) have been explored in a number of studies (Clements & Couch, 1996; Lowenthal et al 1997; Madau et al 1998; Meurer et al 1999), with the result that the luminosity-density goes up as the area in color space is enlarged. The number of low-z interlopers also may go up, although spectroscopic surveys suggest that the Meurer et al (1999) selection boundary is not prone to this problem. Similar techniques have been applied to ground-based images covering much wider areas but sensitive only to brighter objects (see, e.g. Steidel et al 1999 and Giallongo et al 1998 for recent examples). Substantial progress has been made both in refining the estimates of the volume sampled by the color selection and in estimates of the high-z galaxy luminosity functions (Dickinson, 1998; Steidel et al 1999). These allow more precise (but more model-dependent) estimates of the luminosity density at $z > 3$. In Figure 5 we provide an updated plot of star-formation vs. redshift from the color-selected samples of Steidel et al (1999) and Casertano et al (2000), now including integration down the luminosity function and a correction for mean dust attenuation.

Photometric redshifts computed via template fitting can also be used to identify samples of high-z objects, and several such samples have been presented and discussed (Lanzetta et al 1996; Sawicki et al 1997; Sawicki & Yee, 1998; Mirallers & Pello, 1998; Rowan-Robinson, 1999; Fontana et al 1999). Luminosity densities at $z > 2$ from these studies are generally higher than those of Madau et al (1996), and the apparent drop in $\dot{\rho}_z(z)$ at $z \gtrsim 2$ is not universally evident. A striking example of the difference between color-selected samples and

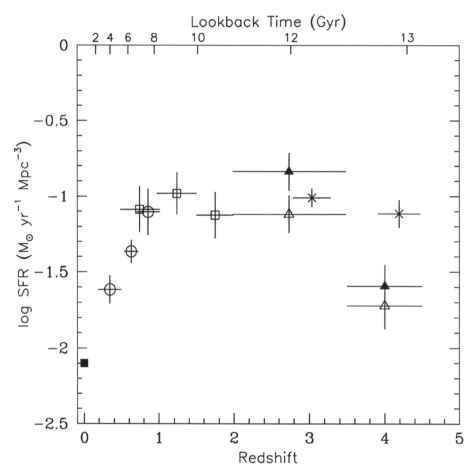

Figure 5 (left) Star-formation rate density vs. redshift derived from UV luminosity density. The $z > 2$ points are from Lyman-break objects in the HDF-N (open triangles), in the HDF-S (filled triangles) and in the Steidel et al (1999) ground-based survey (X's). The luminosity density has been determined by integrating over the luminosity function and correcting for extinction following the prescription of Steidel et al (1999). Possible contributions from far-IR and sub-millimeter sources are not included. Also not included are the upward revisions of the $z < 1$ star-formation densities suggested by Tresse & Maddox (1998) and Cowie et al (1999). For the lower-redshift points, the open squares are from HDF photometric redshifts by Connolly et al (1997), the open circles are from Lilly et al (1996), and the solid square is from the Hα survey of Gallego et al (1995).

photometric-redshift selected samples is the fact that HDF-S comes out with a higher UV luminosity-density than HDF-N in the color-selected sample shown in Figure 5, but a lower luminosity density in the photometric-redshift selected sample of Fontana et al (1999). One general concern about object selection is that galaxies at lower redshift are typically *brighter* than the high-z objects. Interlopers that make it into photometric samples can dominate the UV luminosity-density estimates.

It does appear evident that the original criteria of Madau et al (1996) were too conservative, and the *directly measured* UV luminosity density in bins at $z \sim 3$ and $z \sim 4$ is not demonstrably lower than it is at $z \sim 1$. At still higher redshifts, from the very small number of candidate objects with $z > 5$, Lanzetta et al (1998) computes an upper limit on the star-formation density at $z \sim 5 - 12$. For the cosmology adopted here, the limit on star formation in galaxies with $\dot{M} \gtrsim 100$ $h M_\odot \, \mathrm{yr}^{-2}$ is lower than the inferred rate from Lyman-break galaxies at $z = 4$. With NICMOS observations it has become possible to identify $z > 5$ candidates to much fainter limits, detecting objects with UV luminosities more like those of typical Lyman-break galaxies at $2 < z < 5$. Dickinson (2000) derives number counts for color–selected candidates at $4.5 < z < 8.5$ in the HDF–N/NICMOS survey, and finds that they fall well below nonevolving predictions based on the well-characterized $z = 3$ Lyman-break luminosity function. It appears that the space density, the luminosities, or the surface brightnesses (and hence, the detectability) of UV bright galaxies fall off at $z > 5$, at least in the HDF–N.

5.5.2 Dust and Selection Effects

5.5.2 Dust and Selection Effects   Many of the discussions of the Madau diagram have centered around selection effects and whether the data actually support a decrease in $\dot{\rho}(z)$ for $z \gtrsim 2$. Star formation in dusty or low-surface-brightness galaxies may be unaccounted for in the HDF source counts. Meurer and collaborators (Meurer et al 1997, 1999) use local starburst galaxy samples to calibrate a relation between UV spectral slope and far-IR (60–100 μm) emission, and hence compute bolometric corrections for dust attenuation. They apply these corrections to a sample of color-selected Lyman-break galaxies in the HDF and derive an absorption-corrected luminosity density at $z \sim 3$ that is a factor of 9 higher than that derived by Madau et al (1996). Not all of this comes from the dust correction; the luminosity-weighted mean dust-absorption factor for the Meurer et al (1999) sample is 5.4 at 1600 Å. Sawicki & Yee (1998) analyzed the optical/near-IR spectral energy distributions of spectroscopically confirmed Lyman-break galaxies in the HDF and found the best fit synthetic spectra involved corrections of more than a factor of 10. Smaller corrections were derived by Steidel et al (1999) for a sample of galaxies of similar luminosity. In general, the danger of interpreting UV spectral slopes as a measure of extinction is that the inferred luminosity corrections are very sensitive to the form of the reddening law at UV wavelengths. For the Calzetti et al (1994) effective attenuation law, used in most of the aforementioned studies, a small change in the UV color requires a large change in the total extinction. Photometric errors also tend to increase the dust correction, because

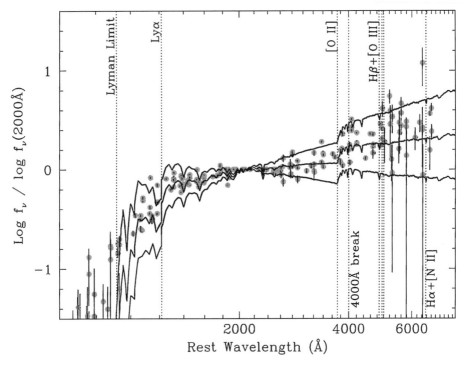

Figure 6   Spectral energy distributions of 28 spectroscopically confirmed Lyman-break galaxies with $2.5 < z < 3.5$, corresponding to a lookback time $\sim$12 billion years, superimposed on stellar population models. The models assume solar metallicity and a Salpeter IMF and include attenuation due to the IGM for $z = 3$. The models reflect constant star formation for 10^9 years, with different amounts of interstellar dust attenuation. The reddening values are $E(B - V) = 0, 0.2,$ and 0.4, with the curve that is highest at long wavelengths having the highest value. The Calzetti et al (1994) attenuation law was used.

objects that scatter to the red are assigned larger corrections that are not offset by smaller corrections for the blue objects. The corrections for dust must thus be regarded as tentative, and must be confirmed with more extensive Hα studies such as that of Pettini et al (1998) and further studies in the mid-IR, radio, and sub-millimeter.

The NICMOS HDF observations provide some additional insight into possible reddening corrections. Figure 6 shows the spectral-energy distributions of HDF galaxies with spectroscopic $2.5 < z < 3.5$, superimposed on models with a constant star-formation rate and an age of 10^9 years. The galaxies are typically best fitted with ages 10^{7-9} years and reddening $0.1 < E(B - V) < 0.4$ with the Calzetti et al (1994) attenuation law. Very young galaxies with little extinction appear to be rare, and older or more reddened galaxies would possibly escape the Lyman-break selection. However, Ferguson (1999) identified only 12 galaxies with

$V_{606} < 27$ in the HDF-N that fail the Dickinson (1998) color criteria but nevertheless have $2.5 < z_{phot} < 3.5$ from spectral-template fitting. In general these galaxies fall just outside the color selection boundaries, and the non-dropout sample contributes only 12% of the luminosity-density of the dropout sample. It thus appears unlikely that there is a population that *dominates* the metal-enrichment rate at $z \sim 3$ that is just missing from the Lyman-break samples. If the optically unidentified sub-millimeter sources (see Section 4.7) are really at $z \gtrsim 3$, they are probably a disjoint population, rather than simply the red tail of the Lyman-break population.

If sources akin to ultra-luminous IRAS galaxies were present at $z \sim 3$ in the HDF, it is likely that they would be detected, but unlikely that they would be identified as Lyman-break galaxies. Trentham et al (1999) use HST observations to extend the spectral energy distributions for three nearby ultra-luminous IR galaxies down to rest-frame 1400 Å. Although this is not far enough into the UV to predict HDF B_{450} colors at $z = 3$ with confidence, it at least gives some indication of whether such galaxies are detectable. All three galaxies would likely be detected in the $F814W$ band if shifted out to $z = 3$, although VII Zw 031 would be very near the detection limit. None of the galaxies would meet the Madau et al (1996) Lyman-break galaxy selection criteria, but some might plausibly show up as high-z objects from their photometric redshifts.

In addition to dust, cosmological surface-brightness can bias the samples of high-redshift galaxies. For fixed luminosity and physical size, i.e. no evolution, surface brightness will drop by about 1 mag between $z = 3$ and $z = 4$. This lower surface brightness will result in a decrease in the number density of objects and the inferred luminosity density, even if there is no intrinsic evolution. Using simulated images for one particular galaxy evolution model (Ferguson & Babul, 1998), Ferguson (1998b) arrived at corrections to $\dot{\rho}_z(z)$ of a factor of 1.5 at $z \sim 3$ and 4.7 at $z \sim 4$. Lanzetta et al (1999) have looked at surface-brightness selection in a less model-dependent way, computing the star-formation "intensity" (in solar masses per year per square kiloparsec) from the UV flux in each pixel of the galaxy images. The comoving volume density of the regions of highest star-formation intensity appears to increase monotonically with increasing redshift, whereas a strong selection cutoff for star-formation intensities less than 1.5 $M_\odot$ yr^{-1} kpc^{-2} affects samples beyond $z > 2$. The results suggest that surface-brightness effects produce a substantial underestimate of $\dot{\rho}_z(z)$ at high redshift.

The undetected low-surface-brightness, UV-emitting regions should contribute to the total diffuse background in the image. Bernstein (1997) has attempted to measure the mean level of the extragalactic background light (EBL) in the HDF images, while Vogeley (1997) has analyzed the autocorrelation function of the residual fluctuations after masking the galaxies. Vogeley (1997) concludes that diffuse light clustered similarly to faint galaxies can contribute no more than $\sim$20% of the mean EBL. In contrast Bernstein (1997) concludes that the total optical EBL is two to three times the integrated flux in published galaxy counts. The two results can be made marginally compatible if the fluxes of detected galaxies are corrected for light lost outside the photometric apertures and for oversubtraction of background due to the overlapping wings of galaxy profiles. If this interpretation is correct, there is not much room for the UV luminosity density to increase significantly at high redshift.

5.5.3 Connection to Galaxy Evolution

An important use of the metal-enrichment rate derived from the HDF and other surveys is the attempt to "close the loop": to show that the emission history of the universe produces the metal abundances and stellar population colors we see at $z \sim 0$. Madau et al (1996) made a first attempt at this, concluding that the metals we observe being formed [integrating $\dot{\rho}_z(z)$ over time] are a substantial fraction of the entire metal content of galaxies.

Madau et al (1998) compared the integrated color of galaxy populations in the local universe to that expected from the UV-emission history, exploring a variety of options for IMF and dust obscuration. With a Salpeter IMF and modest amounts of dust attenuation, they find that a star-formation history that rises by about an order of magnitude from $z = 0$ to a peak at $z \sim 1.5$ is compatible with the present-day colors of galaxies, with the FIR background, and with the metallicities of damped Lyα absorbers. Fall et al (1996), Calzetti & Heckman (1999) and Pei et al (1999) have attempted to incorporate dust and chemical evolution in a more self-consistent way. The models include a substantial amount of obscured star formation; more than 50% of the UV radiation is reprocessed by dust. The obscuration corrections increase the value of $\dot{\rho}_z(z)$ from samples already detected in optical surveys, but do not introduce whole classes of completely dust-obscured objects. Although there are significant differences in the inputs and assumptions of the models, in all cases a model with a peak in $\dot{\rho}_z(z)$ at $z \sim 1 - 1.5$ is found consistent with a wide variety of observations. In particular the models can simultaneously fit the COBE DMR and FIRAS measurements of the cosmic IR background (Puget et al 1996; Hauser et al 1998; Fixsen et al 1998) and the integrated light from galaxy counts. The total mass in metals at $z = 0$ is higher in these new models than in those of Madau et al (1996), but the local census now includes metals in cluster X-ray gas, which were ignored by Madau et al (1996).

Overall, the success of these consistency checks is quite remarkable. Various imagined populations of galaxies (dwarfs, low-surface-brightness galaxies, highly dust-obscured objects, etc.) now seem unlikely to be cosmologically dominant. The fact that the UV emission, gas metallicities, and IR backgrounds all appear capable of producing a universe like the one we see today leaves little room for huge repositories of gas and stars missing from either our census at $z = 0$ or our census at high redshift.

Nonetheless, there is room for caution in this conclusion. In clusters of galaxies, the mass of metals ejected from galaxies into the X-ray–emitting gas exceeds that locked inside stars by a factor of 2–5 (Mushotzky & Loewenstein, 1997). If the same factor applies to galaxies outside clusters (Renzini, 1997), then the local mass-density of metals greatly exceeds the integral of the metal-enrichment rate, implying that *most* star-formation is hidden from the UV census (although the differences in galaxy morphology in clusters and in the field suggests that clusters might not be typical regions of the universe). Various lines of evidence cited in Section 5.3 point to old ages for elliptical galaxies (both inside and outside of clusters) and the bulges of luminous early-type spirals (Renzini, 1999; Goudfrooij et al 1999). The requirements for the early formation of metals in these systems look to be at odds with the inferences from the models described above. Renzini (1999) estimates that 30% of the current density of metals must be formed by $z \sim 3$, whereas the best-fit models to the evolution of the luminosity density have only 10% formed by then. The discrepancy is interesting but is not outside of the range of error of the estimates of both $\dot{\rho}_z(z)$ and the ages of stellar populations in present-day spheroidal systems.

It also remains a challenge to ascertain how the metals got from where they are at high redshift to where they are today. The bulk of the metals locked up in stars at $z = 0$ are in luminous, normal, elliptical, and spiral galaxies. If elliptical galaxies (and spheroids in general) formed early and rapidly, they probably account for the lion's share of $\dot{\rho}_z(z)$ above $z = 2$. Thus the metals formed in the $z = 1$ peak of $\dot{\rho}_z(z)$ must end up for the most part in luminous spiral galaxies today. This is difficult to reconcile with the lack of evolution observed in number density or luminosity of luminous spirals out to $z = 1$. An order-of-magnitude decline in star-formation rate in galaxy disks since $z = 1$ also seems inconsistent with present-day colors of spiral galaxy disks, or with star-formation histories derived from chemical-evolution models (e.g. Tosi 1996; Prantzos & Silk 1998). Furthermore, at $z \sim 1$ it appears that compact-narrow-emission-line galaxies (Guzman et al 1997) and irregular galaxies account for a significant fraction of the UV luminosity density. If these galaxies fade into obscurity today, then they have not been accounted for in the census of metals in the local universe, and the $z = 0$ metallicity should be revised upwards in the global chemical-evolution models. On the other hand, if these galaxies are merging into luminous spirals and ellipticals, it is hard to understand how luminous spiral and elliptical galaxy properties can remain consistent with PLE models out to $z = 1$.

5.6 Clustering of High-Redshift Galaxies

Until relatively recently, evolution of the spatial two-point correlation function ξ has typically been modeled as a simple power-law evolution in redshift

$$\xi(r, z) = (r_0/r)^{\gamma}(1 + z)^{-(3 + \epsilon - \gamma)}, \tag{3}$$

where r and r_0 are expressed in comoving coordinates (e.g. Groth & Peebles 1977). For $\epsilon = 0$, the formula above corresponds to stable clustering (fixed in proper coordinates), while for $\epsilon = \gamma - 3$, it corresponds to a clustering pattern that simply expands with the background cosmology. Although these two cases may bound the problem at relatively low redshift, the situation becomes much more complex as galaxy surveys probe to high redshifts. In particular, the complex merging and fading histories of galaxies make it unlikely that such a simple formula could hold, and the differences in sample selection at high- and low-z make it unclear whether the analysis is comparing the same physical entities. In practice it is likely that galaxies are biased tracers of the underlying dark-matter distribution, with a bias factor b that is non-linear, scale-dependent, type-dependent, and stochastic (Magliocchetti et al 1999; Dekel & Lahav 1999).

In hierarchical models, the correlation function $\xi(r)$ of halos on linear scales is given by the statistics of peaks in a Gaussian random field (Bardeen et al 1986). Peaks at a higher density threshold $\delta\rho/\rho$ have a higher correlation amplitude, and their bias factor b, relative to the overall mass distribution, is completely specified by the number density of peaks (e.g. Mao & Mo 1998).

The comparison of the clustering of high-redshift galaxies in the HDF and in ground-based surveys provides a strong test of whether there is a one-to-one correspondence between galaxies and peaks in the underlying density field. The bias factor observed in Lyman-break samples (Steidel et al 1998; Adelberger et al 1998; Giavalisco et al 1998) is in remarkable agreement with the expectations from such a one-to-one correspondence. As the luminosity of Lyman-break galaxies decreases, the number density increases, and the clustering amplitude decreases (Giavalisco et al 2000). These trends are all in agreement with a scenario in which lower-luminosity Lyman-break galaxies inhabit lower-mass halos. Thus, at redshifts $z \sim 3$, the connection between galaxies and dark matter halos may in fact be quite simple and in good agreement with theoretical expectations.

The picture becomes less clear for the samples described in Section 4.12. In particular, the interpretation of the apparent evolution in the correlation length r_0 or the bias factor b from the Magliocchetti & Maddox (1999) and Arnouts et al (1999) studies involves a sophisticated treatment of the merging of galaxies and galaxy halos over cosmic time. Fairly detailed attempts at this have been made using either semi-analytic models (Baugh et al 1999; Kauffmann et al 1999) or more generic arguments (e.g. Matarrese et al 1997; Moscardini et al 1998; Magliocchetti et al 1999, following on earlier work by Mo & Fukugita 1996, Mo & White 1996, Jain 1997, Ogawa et al 1997 and others). These studies all assume that galaxy mass is directly related to halo mass, which may not be true in reality at lower z, but which is an assumption worth testing. A generic prediction of the models is that the effective bias factor (the value of b averaged over the mass function) should increase with increasing redshift. The correlation length r_0 is relatively independent of redshift (to within a factor of $\sim$3), in contrast to the factor of $\sim$10 decline from $z = 0$ to $z = 4$ predicted for non-evolving bias. The results shown by Arnouts et al (1999) are qualitatively consistent with this behavior. However, it is worth re-emphasizing that the HDF is a very small field. The standard definition for the bias factor b is the ratio of the root mean square density fluctuations of galaxies relative to mass on a scale of 8 h^{-1} Mpc. Fluctuations on this angular scale clearly have *not* been measured in the HDF, and the interpretation rests on (*a*) the assumption of a power-law index $\gamma = 1.8$ for the angular correlation function (which is not in fact predicted by the models) (Moscardini et al 1998), and (*b*) the fit to the integral constraint. Clearly much larger areas are needed before secure results can be obtained.

Roukema & Valls-Gabaud (1997) and Roukema et al (1999) point out that much of the measured clustering signal in the HDF comes from scales 25–250 kpc, which is within the size of a typical L^* galaxy halo at $z = 0$. The connection of the observed correlation function to hierarchical models thus depends quite strongly on what happens when multiple galaxies inhabit the same halo. Do they co-exist for a long time (e.g. as in present-day galaxy groups and clusters), or rapidly merge together to form a larger galaxy (in which case one should consider a "halo exclusion radius" in modeling the correlation function)? As the halo exclusion radius increases, the predicted correlation amplitude on scales of $5''$ decreases relative to the standard cosmological predictions. The slope of $\omega(\theta)$ also differs from $\gamma = 1.8$ for $\theta z \lesssim 20''$. Roukema et al (1999) find that the HDF data for $1.5 < z_{\mathrm{phot}} < 2.5$ are best fit with stable clustering, a halo-exclusion radius of $r_{\mathrm{halo}} = 200\,h^{-1}$ kpc, and a low-density universe.

6.   CONCLUSION

The HDFs represent an important portion of the frontier in studies of the distant universe. Although we can point to few problems that were solved by the HDF alone, the HDF has contributed in a wide variety of ways to our current understanding of distant galaxies and to shaping the debate over issues such as the origin of elliptical galaxies and the importance of obscured star formation. The scientific impact of the HDF can be attributed in part to the wide and nearly immediate access to the data. The fact that many groups and observatories followed this precedent (both for HDF data and subsequently for data from other surveys) illustrates that a deeper understanding of the universe will come not from any one set of observations but from sharing and comparing different sets of observations. Not all kinds of surveys are amenable to this kind of shared effort, but the precedent set by the HDF in the social aspects of carrying out astronomical research may ultimately rival its significance in other areas.

ACKNOWLEDGMENTS

We are indebted to our fellow HDF enthusiasts, too numerous to mention, for the many fruitful discussions that have provided input for this review. The observations themselves would not have happened without the dedicated contribution from the HST planning operations staff, to whom we owe the largest debt of gratitude. This work was based in part on observations with the NASA/ESA Hubble Space Telescope obtained at the Space Telescope Science Institute, which is operated by the Association of Universities for Research in Astronomy, Inc., under NASA contract NAS5-26555. This work was partially supported by NASA grants GO-07817.01-96A and AR-08368.01-97A.

Visit the Annual Reviews home page at www.AnnualReviews.org

LITERATURE CITED

Abraham RG. 1997. In *The Ultraviolet Universe at Low and High Redshift: Probing the Progress of Galaxy Evolution*, ed. WH Waller, MN Fanelli, JE Hollis, AC Danks, P. 195. Woodbury, New York: AIP

Abraham RG, Ellis RS, Fabian AC, Tanvir NR, Glazebrook K. 1999a. *MNRAS* 303:641

Abraham RG, Merrifield MR, Ellis RS, Tanvir NR, Brinchmann J. 1999b. *MNRAS* 308:569

Abraham RG, Tanvir NR, Santiago BX, Ellis RS, Glazebrook K, van den Bergh S. 1996. *MNRAS* 279:L47

Adelberger KL, Steidel CC, Giavalisco M, Dickinson M, Pettini M, Kellogg M. 1998. *Ap. J.* 505:18

Alcock C, Allsman RA, Alves D, Axelrod TS, Becker AC, et al. 1997. *Ap. J.* 486:697

Almaini O, Fabian A. 1997. *MNRAS* 288:L19

Aragón-Salamanca A, Ellis RS, Couch WJ, Carter D. 1993. *MNRAS* 262:764

Arnouts S, Cristiani S, Moscardini L, Matarrese S, Lucchin F, et al. 1999. *MNRAS* 310:540

Aussel H, Cesarsky CJ, Elbaz D, Starck JL. 1999a. *Astron. Astrophys.* 342:313

Aussel H, Elbaz D, Cesarsky CJ, Starck JL. 1999b. In *The Universe as Seen by ISO*, ed. P Cox MF Kessler, p. 1023. Noordwijk: ESA

Babul A, Ferguson HC. 1996. *Ap. J.* 458:100

Babul A, Rees M. 1992. *MNRAS* 255:346

Bahcall JN, Guhathakurta P, Schneider DP. 1990. *Science* 248:178

Bardeen JM, Bond JR, Kaiser N, Szalay AS. 1986. *Ap. J.* 304:15

Barger AJ, Cowie LL, Richards EA. 1999. In *Photometric Redshifts and the Detection of High-Redshift Galaxies*, ed. R Weymann, L Storrie-Lombardi, M Sawicki, & R Brunner, p. 279. San Francisco: ASP

Barger AJ, Cowie, LL, Sanders, DB. 1999. *Ap. J. Lett.* 518:L5

Barger AJ, Cowie LL, Sanders DB, Fulton E, Taniguchi Y, et al. 1998. *Nature* 394:248

Barger AJ, Cowie LL, Trentham N, Fulton E, Hu EM, et al. 1999. *Astron. J.* 117:102

Barkana R, Blandford R, Hogg DW. 1999. *Ap. J. Lett.* 513:L91

Baugh C, Benson AJ, Cole S, Frenk CS, Lacey CG. 1999. *MNRAS* 305:21

Baugh CM, Cole S, Frenk CS. 1996. *MNRAS* 282:L27

Baugh CM, Cole S, Frenk CS, Lacey CG. 1998. *Ap. J.* 498:504

Benítez N, Broadhurst T, Bouwens R, Silk J, Rosati P. 1999. *Ap. J.* 515:65

Bergeron J, Petitjean P, Cristiani S, Arnouts S, Bresolin F, Fasano G. 1999. *Astron. Astrophys.* 343:L40

Bernstein R. 1997. *"The HST/LCO Measurement of the Mean Flux of the Extragalactic Background Light (3000–8000 Å)."* PhD. Thesis. California Institute of Technology

Bershady MA. 1997. In *Dark and Visible Matter in Galaxies. ASP Conf. Ser.* 117, ed. M Persic & P Salucci, p. 537. ASP

Bershady MA, Lowenthal JD, Koo DC. 1998. *Ap. J.* 505:50

Bertin E, Arnouts S. 1996. *Astron. Astrophys. Suppl.* 117:393

Blain AW, Kneib JP, Ivison RJ, Smail I. 1999. *Ap. J. Lett.* 512:L87

Blandford R. 1998. In *The Hubble Deep Field: STScI Symposium Ser. 11*, ed. M Livio, SM Fall, & P Madau, p. 245. Cambridge: Cambridge Univ. Press

Bohlin RC, Cornett R, Hill JK, Hill RS, Landsman WB, et al. 1991. *Ap. J.* 368:12

Bondi H, Gold T, Hoyle F. 1955. *Observatory* 75:80

Bouwens R, Broadhurst T, Silk J. 1997. astro-ph/9710291

Budavari T, Szalay AS, Connolly AJ, Csabai I, Dickinson M. 1999. In *Photometric Redshifts and the Detection of High-Redshift Galaxies*, ed. R Weymann, L Storrie-Lombardi, M

Sawicki, & R Brunner, p. 19., San Francisco: ASP

Calzetti D, Heckman TM. 1999. *Ap. J.* 519:27

Calzetti D, Kinney AL, Storchi-Bergmann T. 1994. *Ap. J.* 429:582

Campos A. 1997. *Ap. J.* 488:606

Campos A, Shanks T. 1997. *MNRAS* 291:383

Carilli CL, Yun MS. 1999. *Ap. J. Lett.* 511:L13

Casertano S, de Mello D, Ferguson HC, Fruchter AS, Gonzalez RA, et al. 2000, *Astron. J.* Submitted

Casertano S, Ratnatunga KU, Griffiths RE, Im M, Neuschaefer LW, et al. 1995. *Ap. J.* 453:599

Chabrier G, Mera D. 1997. *Astron. Astrophys.* 328:83

Chapman SC, Scott D, Steidel CC, Borys C, Halpern M, et al. 1999, submitted

Chen HW, Lanzetta KM, Pascarelle S. 1999. *Nature* 398:586

Clements DL, Couch WJ. 1996. *MNRAS* 280:L43

Cohen JG. 1999. astro-ph/9910242

Cohen JG, Cowie LL, Hogg DW, Songaila A, Blandrod R, et al. 1996. *Ap. J. Lett.* 471:L5

Cohen JG, Hogg DW, Blandford R, Cowie LL, Hu EM, et al. 2000, submitted

Cole S, Aragon-Salamanca A, Frenk CS, Navarro JF, Zepf SE. 1994. *MNRAS* 271:781

Colley W, Gnedin O, Ostriker JP, Rhoads JE. 1997. *Ap. J.* 488:579

Colley WN, Rhoads JE, Ostriker JP, Spergel DN. 1996. *Ap. J. Lett.* 473:L63

Connolly AJ, Szalay AS, Brunner RJ. 1999. *Ap. J. Lett.* 499:L125

Connolly AJ, Szalay AS, Dickinson M, Sub-baRao MU, Brunner RJ. 1997. *Ap. J.* 486: 11

Conselice CJ, Bershady MA, Jangren A. 1999. ApJ, in press (astro-ph/9907399)

Conselice CJ, Bershady MD, Dickinson M, Ferguson HC, Postman M, et al. 2000, in preparation

Conti A, Kennefick JD, Martini P, Osmer PS. 1999. *Astron. J.* 117:645

Cooray A, Quashnock JM, Miller MC. 1999. *Ap. J.* 511:562

Couch WJ. 1996. http://ecf.hq.eso.org/hdf/catalogs/

Cowie LL, Barger A. 1999. astro-ph/9907043

Cowie LL, Hu EM, Songaila A. 1995a. *Nature* 377:603

Cowie LL, Hu EM, Songaila A. 1995b. *Astron. J.* 110:1576

Cowie LL, Lilly SJ, Gardner JP, McLean IS. 1988. *Ap. J. Lett.* 332:L29

Cowie LL, Songaila A. 1996. Hawaii Active Catalog http://www.ifa.hawaii.edu/~cowie/tts/tts.html

Cristiani S. 1999. *Formation and Evolution of Galaxies*, Proc. 1st Workshop of the Italian Netw., In press (astro-ph/99008165)

de Jong RS, Lacey C. 1999. astro-ph/9910066

De Propris R, Stanford SA, Eisenhardt PR, Dickinson M, Elston R. 1999. *Astron. J.* 118:719

Dekel A, Lahav O. 1999. *Ap. J.* 520:24

Dell'Antonio IP, Tyson JA. 1996. *Ap. J. Lett.* 473:L17

Dennefeld M, et al. 2000, in preparation

Désert FX, Puget JL, Clements DL, Pérault M, Abergel A, et al. 1999. *Astron. Astrophys.* 342:363

Dickinson M. 1995. In *Galaxies in the Young Universe,* ed. H Hippelein, HJ Meisen-heimer, & HJ Röser, p. 144. Berlin: Springer

Dickinson M. 1998. In *The Hubble Deep Field*, ed. M Livio, SM Fall, & P Madau, p. 219. Cambridge: Cambridge University Press

Dickinson M. 2000. Philosophical Transactions of the Royal Society, Series A, in press.

Dickinson M, et al. 2000a, in preparation

Dickinson M, Hanley C, Elston R, Eisenhardt PR, Stanford SA, et al. 2000b. *Ap. J.* 531:624

Djorgovski S, Soifer BT, Pahre MA, Larkin JE, Smith JD, et al. 1995. *Ap. J. Lett.* 438:L13

Downes D, Neri R, Greve A, Guilloteau S, Casoli F, et al. 1999. *Astron. Astrophys.* 347:809

Dressler A, Oemler AJ, Sparks WB, Lucas RA. 1994. *Ap. J. Lett.* 435:L23

Driver SP. 1999. astro-ph/9909469

Driver SP, Fernández-Soto A, Couch WJ, Odewahn SC, Windhorst RA, et al. 1998. *Ap. J. Lett.* 496:L93

Driver SP, Windhorst RA, Griffiths RE. 1995. *Ap. J.* 453:48

Driver SP, Windhorst RA, Ostrander EJ, Keel WC, Griffiths RE, Ratnatunga KU. 1995. *Ap. J. Lett.* 449:L23

Dube RR, Wickes WW, Wilkinson DT. 1979. *Ap. J.* 232:333

Efstathiou G. 1992. *MNRAS* 256:43P.

Efstathiou G, Bernstein G, Tyson JA, Katz N, Guhathakurta P. 1991. *Ap. J.* 380:L47

Eggen OJ, Lynden-Bell D, Sandage A. 1962. *Ap. J.* 136:748

Eisenhardt P, Elston R, Stanford SA, Dickinson M, Spinrad H, et al. 1998. In *Xth Rencontres de Blois: The Birth of Galaxies*, ed. B Guiderdoni, F Bouchet, TX Thuan, & JTT Van. Paris: Edition Frontieres, in press

Elbaz D, Aussel H, Cesarsky CJ, Desert FX, Fadda D, et al. 1999. In *The Universe as Seen by ISO*, ed. P Cox & MF Kessler, p. 999. Noordwijk: ESA

Ellis RS. 1997. *ARAA* 35:389

Ellis RS, Smail I, Dressler A, Couch WJ, Oemler A, et al. 1997. *Ap. J.* 483:582

Elson RAW, Santiago BX, Gilmore GF. 1996. *New Astron.* 1:1

Fall SM, Charlot S, Pei YC. 1996. *Ap. J. Lett.* 464:L43

Fasano G, Cristiani S, Arounts S, Filippi M. 1998. *Astron. J.* 115:1400

Fasano G, Filippi M. 1998. *Astron. Astrophys. Suppl.* 129:583

Ferguson HC.1998a. *Reviews in Modern Astronomy* 11:83

Ferguson HC. 1998b. In *The Hubble Deep Field*, ed. M Livio, SM Fall, & P Madau, p. 181. Cambridge: Cambridge University Press

Ferguson HC. 1999. In *Photometric Redshifts and the Detection of High-Redshift Galaxies*, ed. R Weymann, L Storrie-Lombardi, M Sawicki, & R Brunner, p. 51. San Francisco: ASP

Ferguson HC, Babul A. 1998. *MNRAS* 296: 585

Ferguson HC, Baum SA, Brown TM, Busko I, Carollo M, et al. 2000, in preparation

Ferguson HC, McGaugh SS. 1995. *Ap. J.* 440:470

Fernández-Soto A, Lanzetta KM, Yahil A. 1999. *Ap. J.* 513:34

Fixsen DJ, Dwek E, Mather JC, Bennett CL, Shafer R. 1998. *Ap. J.* 508:123

Flores H, Hammer F, Desert FX, Césarsky C, Thuan TX, et al. 1999a. *Astron. Astrophys.* 343:389

Flores H, Hammer F, Thuan TX, Césarsky C, Desert FX, et al. 1999b. *Ap. J.* 517:148

Flynn C, Gould A, Bahcall JN. 1996. *Ap. J. Lett.* 466:L55

Fomalont EB, Kellerman KI, Richards EA, Windhorst RA, Partridge RB. 1997. *Ap. J.* 475:5

Fontana A, D'Odorico S, Fosbury R, Giallongo E, Hook R, et al. 1999. *Astron. Astrophys.* 343:L19

Forbes DA, Phillips AC, Koo DC, Illingworth GD. 1996. *Ap. J.* 462:89

Franceschini A, Silva L, Fasano G, Granato GL, Bressan A, et al. 1998. *Ap. J.* 506:600

Fruchter A, Bergeron LE, Dickinson M, Ferguson HC, Hook RN, et al. 2000, in preparation

Fruchter A, Hook R. 1997. astro-ph/9708242

Fukugita M, Yamashita K, Takahara F, Yoshii Y. 1990. *Ap. J.* 361:L1

Gallego J, Zamorano J, Aragon-Salamanca A, Rego M. 1995. *Ap. J. Lett.* 455:L1

Gardner JP, Baum SA, Brown TM, Carollo CM, Christensen J, et al. 2000. *Astron. J.* 119: 486

Gardner JP, Sharples RM, Frenk CS, Carrasco BE. 1997. *Ap. J.* 480:99

Gardner JP, Cowie LL, Wainscoat RJ. 1993. *Ap. J. Lett.* 415:L9

Gardner JP, Sharples RM, Carrasco BE, Frenk CS. 1996. *MNRAS* 282:L1

Giallongo E, D'Odorico S, Fontana A, Cristiani S, Egami E, et al. 1998. *Astron. J.* 115:2169

Giavalisco M, et al. 2000, in preparation

Giavalisco M, Livio M, Bohlin RC, Macchetto FD, Stecher TP. 1996. *Astron. J.* 112:369

Giavalisco M, Steidel CC, Adelberger KL, Dickinson ME, Pettini M, Kellogg M. 1998. *Ap. J.* 492:428

Gilliland RL, Nugent PE, Phillips MM. 1999. *Ap. J.* 521:30

Gilliland RL, Phillips MM. 1998. *IAU Circular* 6810

Glazebrook K, et al. 2000, in prepraration

Glazebrook K, Ellis RS, Santiago B, Griffiths RE. 1995. *MNRAS* 275:L19

Goldschmidt P, Oliver SJ, Serjeant SB, Baker A, Eaton N, et al. 1997. *MNRAS* 289:465

Goudfrooij P, Gorgas J, Jablonka P. 1999. astro-ph/9910020

Gould A, Bahcall JN, Flynn C. 1996. *Ap. J.* 465:759

Gould A, Bahcall JN, Flynn C. 1997. *Ap. J.* 482:913

Griffiths R, Ratnatunga KU, Casertano S, Im M, Neuschaefer LW, et al. 1996. In *IAU Symposium 168: Examining the Big Bang and Diffuse Background Radiations*, ed. M Kafatos & Y Kondo. Dordrecht: Kluwer

Groth EJ, Peebles PJE. 1977. *Ap. J.* 217:385

Guhathakurta P, Tyson JA, Majewski SR. 1990. *Ap. J. Lett.* 359:L9

Guiderdoni B, Rocca-Volmerange B. 1990. *Astron. Astrophys.* 227:362

Guzman R, Gallego J, Koo DC, Phillips AC, Lowenthal JD, et al. 1997. *Ap. J.* 489:559

Gwyn SDJ, Hartwick FDA. 1996. *Ap. J. Lett.* 468:L77

Haiman Z, Madau P, Loeb A. 1999. *Ap. J.* 514:535

Hansen BMS. 1998. *Nature* 394:860

Hansen BMS. 1999. *Ap. J.* 520:680

Harris H, Dahn CC, Vrba FJ, Henden AA, Liebert J, et al. 1999. astro-ph/9909065

Hauser MG, Arendt RG, Kelsall T, Dwek E, Odegard N, et al. 1998. *Ap. J.* 508:25

Hibbard JE, Vacca WD. 1997. *Astron. J.* 114:1741

Hogg DW, Blandford R, Kundic T, Fassnacht CD, Malhotra S. 1996. *Ap. J.* 467:73

Hogg DW, Cohen JG, Blandford R, Gwyn SDJ, Hartwick FD, et al. 1998. *Astron. J.* 115:1418

Holtzman JA, Burrows CJ, Casertano S, Hester JJ, Trauger JT, et al. 1995. *Publ. Astron. Soc. Pac.* 107:1065

Hornschemeier AE, Brandt WN, Garmire GP, Schneider DP, Broos PS, et al. 2000. *Ap. J.* In press

Hu E, Cowie LL, McMahon RG. 1998. *Ap. J. Lett.* 502:L99

Huang JS, Cowie LL, Gardner JP, Hu EM, Songaila A, Wainscoat RJ. 1997. *Ap. J.* 476:12

Hudson MJ, Gwyn SDJ, Dahle H, Kaiser N. 1998. *Ap. J.* 503:531

Hughes DH, Serjeant S, Dunlop J, Rowan-Robinson M, Blain A, et al. 1998. *Nature* 394:241

Ibata R, Richer HB, Gilliland RL, Scott D. 1999. *Ap. J. Lett.* 524:L95

Im M, Griffiths RE, Naim A, Ratnatunga KU, Roche N, et al. 1999. *Ap. J.* 510:82

Im M, Griffiths RE, Ratnatunga KU, Sarajedini VL. 1996. *Ap. J.* 461:L79

Jain B. 1997. *MNRAS* 287:687

Jarvis RM, MacAlpine G. 1998. *Astron. J.* 116:2724

Karachentsev ID, Makarov DI. 1997. In *IAU Symposium 186: Galaxy Interactions at Low and High Redshift*, ed. JE Barnes & DB Sanders. Dordrecht: Kluwer

Kauffmann G, Charlot S, White SDM. 1996. *MNRAS* 283:L117

Kauffmann G, Colberg JM, Diaferio A, White SDM. 1999. *MNRAS* 307:529

Kauffmann G, White SDM, Guiderdoni B. 1993. *MNRAS* 264:201

Kawaler S. 1998. In *The Hubble Deep Field*, ed. M Livio, SM Fall, & P Madau, p. 353. Cambridge: Cambridge University Press

Kawaler SD. 1996. *Ap. J.* 467:397

Kerins EJ. 1997. *Astron. Astrophys.* 328:5

Kodama T, Bower RG, Bell EF. 1999. *MNRAS* 306:561

Koo DC, Gronwall C, Bruzual GA. 1993. *Ap. J. Lett.* 415:L21

Kron RG. 1980. *Ap. J. Suppl.* 43:305

Landolt AU. 1973. *Astron. J.* 78:959

Landolt AU. 1983. *Astron. J.* 88:439

Landolt AU. 1992a. *Astron. J.* 104:72

Landolt AU. 1992b. *Astron. J.* 104:340

Lanzetta KM, Chen HW, Fernández-Soto A,

Pascarelle S, Puetter R, et al. 1999. In *Photometric Redshifts and the Detection of High-Redshift Galaxies*, ed. R Weymann, L Storrie-Lombardi, M Sawicki, & R Brunner, p. 229., San Francisco: ASP

Lanzetta KM, Wolfe AM, Turnshek DA.1995. *Ap. J.* 440:435

Lanzetta KM, Yahil A, Fernández-Soto A. 1996. *Nature* 381:759

Lanzetta KM, Yahil A, Fernández-Soto A. 1998. *Astron. J.* 116:1066

Larson RB. 1974. *MNRAS* 166:585

Lilly S, Tresse L, Hammer F, Crampton D, Le Fevre O. 1995a. *Ap. J.* 455:108

Lilly SJ, Cowie LL, Gardner JP. 1991. *Ap. J.* 369:79

Lilly SJ, Le Fevre O, Crampton D, Hammer F, Tresse L. 1995b. *Ap. J.* 455:50

Lilly SJ, Le Févre O, Hammer F, Crampton D. 1996. *Ap. J. Lett.* 460:L1

Livio M, Fall SM, Madau P. 1997. *The Hubble Deep Field. Cambridge: Cambridge Univ. Press*

Loveday J, Peterson BA, Efstathiou G, Maddox S. 1992. *Ap. J.* 390:338

Lowenthal JD, Koo DC, Guzman R, Gallego J, Phillips AC, et al. 1997. *Ap. J.* 481:673

Lucas R, Baum SA, Brown TM, Casertano S, de Mello D, et al. 2000, in preparation

Madau P. 1997. In *Star Formation Near and Far*, ed. SS Holt & LG Mundy, p. 481. Woodbury, NY: AIP Press

Madau P, Della Valle M, Panagia N. 1998. *MNRAS* 297:L17

Madau P, Ferguson HC, Dickinson M, Giavalisco M, Steidel CC, Fruchter AS. 1996. *MNRAS* 283:1388

Madau P, Pozzetti L. 1999. astro-ph/9907315

Madau P, Pozzetti L, Dickinson M. 1998. *Ap. J.* 498:106

Magliocchetti M, Bagla JS, Maddox SJ, Lahav O. 1999. astro-ph/9902260

Magliocchetti M, Maddox SJ. 1999. *MNRAS* 306:988

Mann R, Oliver SJ, Serjeant SB, Rowan-Robinson M, Baker A, et al. 1997. *MNRAS* 289:482

Mannucci F, Ferrara A. 1999. *MNRAS* 305:L55

Mao S, Mo HJ. 1998. *MNRAS* 296:847

Maoz D. 1997. *Ap. J. Lett.* 490:L135

Marleau FR, Simard L. 1998. *Ap. J.* 507:585

Marzke RO, Da Costa LN, Pellegrini PS, Willmer CNA, Geller MJ. 1998. *Ap. J.* 503:617

Marzke RO, Geller MJ, Huchra JP, Corwin HG, Jr. 1994. *Astron. J.* 108:437

Matarrese S, Coles P, Lucchin F, Moscardini L. 1997. *MNRAS* 286:115

Matsumoto T, Akiba M, Murakami H. 1988. *Ap. J.* 332:575

McCracken HJ, Metcalfe N, Shanks T, Campos A, Gardner JP, Fong R. 2000. *MNRAS* 311:707

McLeod BA, Rieke MJ. 1995. *Ap. J.* 454:611

Menanteau F, Ellis RS, Abraham RG, Barger AJ, Cowie LL. 1999. *MNRAS* 309:208

Mendez RA, Guzman R. 1997. *Astron. Astrophys.* 333:106

Mendez RA, Minnitti D, Di Marchi G, Baker A, Couch WJ. 1996. *MNRAS* 283:666

Mendez RA, Minniti D. 1999. astro-ph/9908330

Metcalfe N, Shanks T, Campos A, Fong R, Gardner JP. 1996. *Nature* 383:236

Meurer GR, Heckman TM, Calzetti D. 1999. *Ap. J.* 521:64

Meurer GR, Heckman TM, Lehnert MD, Leitherer C, Lowenthal J. 1997. *Astron. J.* 110:54

Minezaki T, Kobayashi Y, Yoshii Y, Peterson BA. 1998. *Ap. J.* 494:111

Miralles JM, Pello R. 1998. astro-ph/9801062

Mo H, Fukugita M. 1996. *Ap. J. Lett.* 467:L9

Mo H, White SDM. 1996. *MNRAS* 282:347

Mobasher B, Rowan-Robinson M, Georgakakis A, Eaton N. 1996. *MNRAS* 282:L7

Moessner R, Jain B, Villumsen JV. 1998. *MNRAS* 294:291

Moscardini L, Coles P, Lucchin F, Matarrese S. 1998. *MNRAS* 299:95

Moustakas LA, Davis M, Graham JR, Silk J, Peterson BA, Yoshii Y. 1997. *Ap. J.* 475:455

Mushotzky R, Loewenstein M. 1997. *Ap. J. Lett.* 481:L63

Mutz SB, Windhorst RA, Schmidtke PC, Pascarelle SM, Griffiths RE, et al. 1994. *Ap. J.* 434:L55

Muxlow TWB, Wilkinson PN, Richards AMS, Kellerman KI, Richards EA, Garrett MA. 1999. *New Astronomy Reviews* 43:623

Odewahn SC, Windhorst RA, Driver SP, Keel WC. 1996. *Ap. J. Lett.* 472:L13

Ogawa T, Roukema BF, Yamashita K. 1997. *Ap. J.* 484:53

Oke JB. 1974. *Ap. J. Suppl.* 27:21

Oliver SJ, et al. 2000, in preparation

Oliver SJ, Goldschmidt P, Franceschini A, Serjeant SBG, Efstathiou A, et al. 1997. *MNRAS* 289:471

Outram PJ, Boyle BJ, Carswell RF, Hewett PC, Williams RE, Norris RP. 1999. *MNRAS* 305:685

Pain R, Hook IM, Deustua S, Gabi S, Goldhaber G, et al. 1996. *Ap. J.* 473:356

Peacock JA, Rowan-Robinson M, Blain AW, Dunlop JS, Efstathiou A, et al. 1999, submitted

Pei YC, Fall SM. 1995. *Ap. J.* 454:69

Pei YC, Fall SM, Hauser MG. 1999. *Ap. J.* 522:604

Perlmutter S, Aldering G, Goldhaber G, Knop RA, Nugent P, et al. 1999. *Ap. J.* 517:565

Petitjean P, Srianand R. 1999. *Astron. Astrophys.* 345:73

Pettini M, Kellogg M, Steidel CC, Dickinson M, Adelberger KL, Giavalisco M. 1998. *Ap. J.* 508:539

Phillips AC, Guzman R, Gallego J, Koo DC, Lowenthal JD, et al. 1997. *Ap. J.* 489:543

Pozzetti L, Madau P, Zamorani G, Ferguson H, Bruzual G. 1998. astro-ph/9803144

Prantzos N, Silk J. 1998. *Ap. J.* 507:229

Prochaska JX, Burles SM. 1999. *Astron. J.* 117:1957

Puget JL, Abergel A, Bernard JP, Boulanger F, Burton WB, et al. 1996. *Astron. Astrophys.* 308:L5

Reid IN, Yan L, Majewski S, Thompson I, Smail I. 1996. *Astron. J.* 112:1472

Renzini A. 1997. *Ap. J.* 488:35

Renzini A. 1999. In *The Formation of Bulges*, ed. CM Carollo, HC Ferguson, & RFG Wyse, p. 9. New York: Cambridge Univ. Press

Richards EA. 2000. *Ap. J.* In press

Richards EA, Fomalont EB, Kellerman KI, Windhorst RA, Partridge RB, et al. 1999. *Ap. J. Lett.* 526:L73

Richards EA, Kellerman KI, Fomalont EB, Windhorst RA, Partridge RB. 1998. *Astron. J.* 116:1039

Riess A, Filippenko AV, Challis P, Clocchiatti A, Diercks A, et al. 1998. *Astron. J.* 116:1009

Rigopoulou D, Fraceschini A, Genzel R, van der Werf P, Aussel H, et al. 2000. In *"ISO Surveys of a Dusty Universe,"* ed. D Lemke, M Stickel, K Wilke, in press. astro-ph/9912544

Rigopoulou D, Spoon HWW, Genzel R, Lutz D, Moorwood AFM, Tran QD. 1999. astro-ph/9908300

Rix HW, Guhathakurta P, Colless M, Ing K. 1997. *MNRAS* 285:779

Roche N, Ratnatunga K, Griffiths RE, Im M, Naim A. 1998. *MNRAS* 293:157

Roukema BF, Valls-Gabaud D. 1997. *Ap. J.* 488:524

Roukema BF, Valls-Gabaud D, Mobasher B, Bajtlik S. 1999. astro-ph/9901299

Rowan-Robinson M. 1999. astro-ph/9906308

Rowan-Robinson M, Mann RG, Oliver SJ, Efstathiou A, Eaton N, et al. 1997. *MNRAS* 289:490

Ruiz-Lapuente P, Canal R. 1998. *Ap. J. Lett.* 497:L57

Sahu K. 1994. *Nature* 370:275

Salpeter EE. 1955. *Ap. J.* 121:161

Sandage A. 1961. *Ap. J.* 133:355

Sandage A. 1988. *Annu. Rev. Astron. Astrophys.* 26:561

Sandage A. 2000. Preprint

Savaglio S. 1998. *Astron. J.* 116:1055

Savaglio S, Ferguson HC, Brown TM, Espey BR, Sahu KC, et al. 1999. *Ap. J. Lett.* 515:L5

Sawicki M, Yee HKC. 1998. *Ap. J.* 115:1329

Sawicki MJ, Lin H, Yee HKC. 1997. *Astron. J.* 113:1

Scalo JM. 1986. *Fundam. Cosmic Phys.* 11:1

Schade D, Lilly SJ, Crampton D, Ellis RS, Le Fevre O, et al. 1999. *Ap. J.* 525:31

Schade D, Lilly SJ, Crampton D, Hammer F, Le Fevre O, Tresse L. 1995a. *Ap. J. Lett.* 451:L1

Schade D, Lilly SJ, Crampton D, Hammer F, Le Fevre O, Tresse L. 1995b. *Ap. J. Lett.* 451:L1

Schechter P. 1976. *Ap. J.* 203:297

Sealey KM, Drinkwater MJ, Webb JK. 1998. *Ap. J. Lett.* 499:L135

Serjeant SBG, Eaton N, Oliver SJ, Efstathiou A, Goldschmidt P, et al. 1997. *MNRAS* 289: 457

Simard L, Koo DC, Faber SM, Sarajedini VL, Vogt NP, et al. 1999. *Ap. J.* 519:563

Simard L, Pritchet C. 1998. *Ap. J.* 505:96

Smail I, Ivison RJ, Blain AW. 1997. *Ap. J. Lett.* 490:L5

Spinrad H, Stern D, Bunker A, Dey A, Lanzetta K, et al. 1998. *Astron. J.* 116:2617

Spinrad H, Stone RPS. 1978. *Ap. J.* 226:609

Stanford SA, Eisenhardt PR, Dickinson M. 1995. *Ap. J.* 450:512

Stanford SA, Eisenhardt PR, Dickinson M. 1998. *Ap. J.* 492:461

Steidel CC, Adelberger KL, Dickinson M, Giavalisco M, Pettini M, Kellogg MA. 1998. *Ap. J.* 492:428

Steidel CC, Adelberger KL, Giavalisco M, Dickinson M, Pettini M. 1999. *Ap. J.* 519:1

Steidel CC, Giavalisco M, Dickinson M, Adelberger KL. 1996. *Astron. J.* 112:352

Steidel CC, Pettini M, Hamilton D. 1996. *Astron. J.* 110:2519

Stiavelli M, Treu M, Carollo CM, Rosati P, Viezzer R, et al. 1999. *Astron. Astrophys.* 343:L25

Tanvir NR, Aragón-Salamanca A, Wall JV. 1996. *The Hubble Space Telescope and the High Redshift Universe* Singapore: World Scientific

Teplitz H, Gardner J, Malmuth E, Heap S. 1998. *Ap. J. Lett.* 507:L17

Thompson RI. 1999. In *Photometric Redshifts and High Redshift Galaxies*, ed. RJ Weymann, LJ Storrie-Lombardi, M Sawicki, & RJ Brunner, p. 51. San Francisco: ASP

Thompson RI, Storrie-Lombardi LJ, Weymann RJ, Rieke MJ, Schneider, G, et al. 1999. *Astron. J.* 117:17

Tinsley BM. 1978. *Ap. J.* 220:816

Tinsley BM, Gunn J. 1976. *Ap. J.* 302:52

Tosi M. 1996. In *From Stars to Galaxies: The Impact of Stellar Physics on Galaxy Evolution, ASP conf. ser. 98*, ed. C Leitherer, U Fritze-von-Alvensleben, & J Huchra, p. 299. San Francisco: ASP

Totani T, Yoshii Y. 1998. *Ap. J. Lett.* 501:L177

Trentham N, Kormendy J, Sanders DB. 1999. *Ap. J.* 117:2152

Tresse L, Dennefeld M, Petitjean P, Cristiani S, White SDM. 1999. *Astron. Astrophys.* 346:L21

Treu M, Stiavelli M. 1999. *Ap. J. Lett.* 524:L27

Treu M, Stiavelli M, Casertano S, Møller P, Bertin G. 1999. *MNRAS* 308:1037

Treu M, Stiavelli M, Walker AR, Williams RE, Baum SA, et al. 1998. *Astron. Astrophys.* 340:L10

Tyson JA. 1988. *Astron. J.* 96:1

Tyson JA, Jarvis JF. 1979. *Ap. J.* 230:L153

van den Bergh S, Abraham RG, Ellis RS, Tanvir NR, Santiago BX, Glazebrook K. 1996. *Astron. J.* 112:359

van Dokkum P, Franx M, Kelson DD, Illingworth G, Fisher D, Fabricant D. 1998. *Ap. J.* 500:714

Villumsen J, Freudling W, da Costa LN. 1997. *Ap. J.* 481:578

Vogeley M. 1997. astro-ph/9711209

Vogt NP, Forbes DA, Phillips AC, Gronwall C, Faber SM, et al. 1996. *Ap. J.* 465:15

Vogt NP, Phillips AC, Faber SM, Gallego J, Gronwall C, et al. 1997. *Ap. J. Lett.* 479:L121

Waddington I, Windhorst RA, Cohen SH, Partridge RB, Spinrad H, Stern D. 1999. *Ap. J. Lett.* 526:L77

Wang Y, Bahcall N, Turner EL. 1998. *Astron. J.* 116:208

Weymann RJ, Stern D, Bunker A, Spinrad H, Chaffee FH, et al. 1998. *Ap. J.* 505:95

White SDM, Frenk CS. 1991. *Ap. J.* 379:52

Whitmore B, Heyer I, Casertano S. 1999. *Publ. Astron. Soc. Pac.* 111:1559

Williams RE, Baum SA, Bergeron LE, Bernstein N, Blacker BS, et al. 2000, Astron. T. Submitted

Williams RE, Blacker B, Dickinson M, Dixon WVD, Ferguson HC, et al. 1996. *Astron. J.* 112:1335

Wilner D, Wright MCH. 1997. *Ap. J. Lett.* 488:L67

Yahata N, Lanzetta KM, Chen HW, Fernandez-Soto A, Pascarelle S, et al. 2000. *Ap. J.* In press, astro-ph/0003310

Zepf S. 1997. *Nature* 390:377

Zepf SE, Moustakas LA, Davis M. 1997. *Ap. J. Lett.* 474:L1

Zucca E, Pozzetti L, Zamorani G. 1994. *MNRAS* 269:953

Annu. Rev. Astron. Astrophys. 2000. 38:717–60

MILLISECOND OSCILLATIONS IN
X-RAY BINARIES

M. van der Klis

Astronomical Institute "Anton Pannekoek," University of Amsterdam, Kruislaan 403,
1098 SJ Amsterdam, The Netherlands; e-mail: michiel@astro.uva.nl

Key Words   neutron stars, black holes, relativity, X-ray binaries, pulsars,
quasi-periodic oscillations

■ **Abstract**   The first millisecond X-ray variability phenomena from accreting compact objects have recently been discovered with the Rossi X-ray Timing Explorer. Three new phenomena are observed from low-mass X-ray binaries containing low-magnetic-field neutron stars: millisecond pulsations, burst oscillations, and kilohertz quasi-periodic oscillations. Models for these new phenomena involve the neutron star spin and orbital motion close around the neutron star, and rely explicitly on our understanding of strong gravity and dense matter. I review the observations of these new neutron-star phenomena and some possibly related phenomena in black-hole candidates, and describe the attempts to use these observations to perform measurements of fundamental physical interest in these systems.

1.   INTRODUCTION

The principal motivation for studying accreting neutron stars and black holes is that these objects provide a unique window on the physics of strong gravity and dense matter. One of the most basic expressions of the compactness of these compact objects is the short (0.1–1 msec) dynamical time scale characterizing the motion of matter under the influence of gravity near them. Millisecond variability will naturally occur in the process of accretion of matter onto a stellar-mass compact object—an insight that dates back at least to Shvartsman (1971). For example, hot clumps orbiting in an accretion disk around black holes and neutron stars will cause quasi-periodic variability on timescales of about a millisecond (Sunyaev 1973). Accreting, low-magnetic-field neutron stars will reach millisecond spin periods, which can be detected when asymmetric emission patterns form on the star's surface during X-ray bursts (Radhakrishnan & Srinivasan 1984, Alpar et al 1982, Shara 1982, Livio & Bath 1982; see also Joss 1978). These early expectations were finally verified in a series of discoveries with NASA's Rossi X-Ray Timing Explorer (RXTE; Bradt et al 1993) within 2.5 years after its launch on 30 December 1995.

In this review, I discuss these newly discovered phenomena and the attempts to use them to perform measurements of fundamental physical interest. I concentrate on millisecond oscillations—periodic and quasi-periodic variations in X-ray flux with frequencies exceeding $10^{2.5}$ Hz—but I also discuss their relations to slower variability and X-ray spectral properties. So far, millisecond oscillations have been seen nearly exclusively from low-magnetic-field neutron stars, so these will be the focus of this review; I will also compare their phenomenology to that of the black-hole candidates (Sections 6 and 7).

Accreting neutron stars and black holes occur in X-ray binaries (e.g. Lewin et al 1995a). In these systems matter is transferred from a normal (donor) star to a compact object. Thermal X-rays powered by the gravitational potential energy released are emitted by the inner regions of the accretion flow and, if present, the neutron star surface. For a compact object with a size of order 10^1 km, 90% of the energy is released in the inner $\sim 10^2$ km. It is with this inner emitting region that we are mostly concerned here. Because accreting low-magnetic-field neutron stars are mainly found in low-mass X-ray binaries (in which the donor star has a mass of $< 1\,M_\odot$), these are the systems I focus on.

The mass transfer usually occurs by way of an accretion disk around the compact object. In the disk the matter moves in near-Keplerian orbits, i.e. with an azimuthal velocity that is approximately Keplerian and a radial velocity much smaller than this. The disk has a radius of 10^{5-7} km, depending on the binary separation. The geometry of the flow in the inner emitting regions is uncertain. In most models for accretion onto low-magnetic-field neutron stars (e.g. Miller et al 1998a), at least part of the flow extends down into the emitting region in the form of a Keplerian disk. It is terminated either at the radius R of the star itself or at a radius r_{in} somewhat larger than R, by (for example) the interaction with a weak neutron-star magnetic field, radiation drag, or relativistic effects. Within r_{in} the flow is no longer Keplerian and may or may not be disk-like. Both inside and outside r_{in} matter may leave the disk and either flow in more radially or be expelled. Particularly for black holes, advective flow solutions are discussed where the disk terminates and the flow becomes more spherical at a much larger radius (e.g. Narayan 1997).

Whatever the geometry, it is clear that as the characteristic velocities near the compact object are of order $(GM/R)^{1/2} \sim 0.5c$, the dynamical timescale—the timescale for the motion of matter through the emitting region—is short: $\tau_{dyn} \equiv (r^3/GM)^{1/2} \sim 0.1$ ms for $r = 10$ km and ~ 2 ms for $r = 100$ km, near a $1.4\,M_\odot$ neutron star, and ~ 1 ms at 100 km from a $10\,M_\odot$ black hole. Therefore, the significance of millisecond X-ray variability from X-ray binaries is clear: milliseconds is the natural timescale for the accretion process in the X-ray emitting regions, and hence strong X-ray variability on such timescales is almost certainly caused by the motion of matter in these regions. Orbital motion, neutron-star spin, and disk- and neutron-star oscillations are all expected to happen on these timescales.

The inner flow is located in regions of space-time where strong-field general relativity is required to describe the motion of matter. For that reason, one expects to detect strong-field general-relativistic effects in these flows—such as, for example, the existence of a region where no stable orbits are possible. The precise interactions between the elementary particles in the interior of a neutron star that determine the equation of state (EOS) of supra-nuclear-density matter are not known. Therefore we cannot confidently predict the radius of a neutron star of given mass, or the maximum spin rate or mass of neutron stars (e.g. Cook et al 1994). So, by measuring these macroscopic quantities one constrains the EOS and tests basic ideas about the properties of elementary particles. In summary, the main motivation for studying millisecond variations in X-ray binaries is that their properties depend on untested, or even unknown, properties of space-time and matter.

Three different millisecond phenomena have now been observed in X-ray binaries. Historically, the first to be discovered were the twin kilohertz quasi-periodic oscillations (kHz QPOs), widely interpreted now as resulting from orbital motion in the inner accretion flow. Then came the burst oscillations, probably a result of the spin of a layer in the neutron star's atmosphere in near-corotation with the neutron star itself. Finally, RXTE detected the first true spin frequency of an accreting low-magnetic field neutron star, the long-anticipated accreting millisecond pulsar.

In this review, I first examine the millisecond pulsar (Section 3), then the burst oscillations (Section 4), and finally the kHz QPOs (Section 5). We will thus venture from the (relatively) well-understood accreting pulsars via the less secure regions of what happens in detail on a neutron star's surface during the thermonuclear runaway that is an X-ray burst, into the mostly uncharted territory of the innermost accretion flows around neutron stars and black holes—which obviously is "where the monsters are," but also where the greatest rewards wait. The possibly related phenomena found in black-hole candidates (Section 6.1) and at lower frequencies (Section 6.3) are then discussed, and the kHz QPO models (Section 7) are summarized.

2.   TECHNIQUES

Most of the variability measurements discussed here rely on Fourier analysis of X-ray count-rate time series with sub-millisecond time resolution (van der Klis 1989b). A quasi-periodic oscillation (QPO) in the time series stands out in the power spectrum (the square of the Fourier transform) as a broad, usually Lorentzian peak (in Figure 1 several peaks of this type can be seen), characterized by its frequency ν ('centroid frequency'), width λ (inversely proportional to the coherence time of the oscillation), and strength (the peak's area is proportional to the variance of the QPO signal). The variance is nearly always reported in terms of the root-mean-square of the signal expressed as a fraction of the count rate, the fractional rms amplitude r; the coherence is often reported in terms of a quality factor $Q = \nu/\lambda$. Conventionally, to call a local maximum in a power spectrum a QPO peak, one requires $Q > 2$. Time delays between signals simultaneously detected in different energy bands are usually measured using cross-spectra (the frequency-domain equivalent of the cross-correlation function; van der Klis et al 1987,

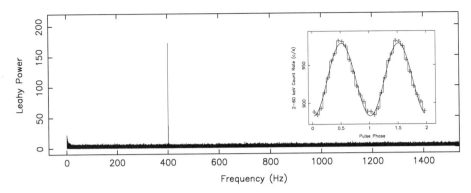

Figure 1 The discovery power spectrum and pulse profile (*inset*) of the first accreting millisecond X-ray pulsar. Note the low harmonic content evident both from the absence of harmonics in the power spectrum and the near-sinusoidal pulse profile. (After Wijnands & van der Klis 1998b)

Vaughan et al 1994a, Nowak et al 1999) and often expressed in terms of a phase lag (time lag multiplied by frequency).

The signal-to-noise of a broad power-spectral feature is $n_\sigma = \frac{1}{2}I_x r^2 (T/\lambda)^{1/2}$ (van der Klis 1989b, see van der Klis 1998 for more details), where I_x is the count rate and T the observing time (assumed $\gg 1/\lambda$). Note that n_σ is proportional to the count rate and to the signal amplitude *squared*, so that it is sufficient for the amplitude to drop by 50% for the signal-to-noise to go from, for example, a whopping 6σ to an undetectable 1.5σ—i.e. if a power-spectral feature suddenly disappears it may have only decreased in amplitude by a factor of two.

3.   MILLISECOND PULSATIONS

An accreting millisecond pulsar in a low-mass X-ray binary has sometimes been called the "Holy Grail" of X-ray astronomy. Its discovery was anticipated for nearly 20 years, because magnetospheric disk accretion theory as well as evolutionary ideas concerning the genesis of millisecond *radio* pulsars strongly suggested that such rapid spin frequencies must occur in accreting low-magnetic-field neutron stars (see Bhattacharya & van den Heuvel 1991). However, in numerous searches of X-ray binary time series (e.g. Leahy et al 1983, Mereghetti & Grindlay 1987, Wood et al 1991, Vaughan et al 1994b) such rapid pulsars did not turn up.

More than two years after RXTE's launch, the first and (as of this writing) only accreting millisecond pulsar was finally discovered on April 13, 1998 in the soft X-ray transient SAX J1808.4–3658 (Figure 2; Wijnands & van der Klis 1998a,b). The pulse frequency is 401 Hz, so this is a 2.5-millisecond pulsar. The object is almost certainly the same as the transient that burst out at the same position in

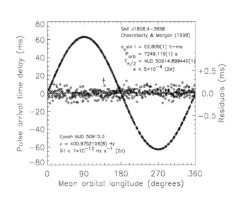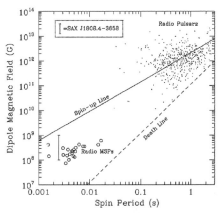

Figure 2 *Left*: the radial velocity curve and orbital elements of SAX J1808.4–3658 (Chakrabarty & Morgan 1998b). *Right*: the position of SAX J1808.4–3658 in the radio-pulsar period versus magnetic field diagram (Psaltis & Chakrabarty 1999).

September 1996 and gave the object its name (in 't Zand et al 1998). This transient showed two type 1 X-ray bursts, and SAX J1808.4–3658 is also the first genuine bursting pulsar, which breaks the long-standing rule (e.g. Lewin & Joss 1981) that pulsations and type 1 X-ray bursts are mutually exclusive.

The orbital period of this pulsar is 2 hrs (Figure 2; Chakrabarty & Morgan 1998a,b). With a projected orbital radius $a \sin i$ of only 63 light *milli*seconds and a mass function of $3.8 \ 10^{-5} \ M_\odot$, the companion star is either very low-mass, or we are seeing the orbit nearly pole-on. The amplitude of the pulsations varied between 4% and 7% and showed little dependence on photon energy (Cui et al 1998b). However, the pulsations measured at higher photon energies preceded those measured in the 2–3 keV band by a gradually increasing time interval from 20 μsec (near 3.5 keV) up to 200 μsec (between 10 and 25 keV) (Cui et al 1998b). These lags could be caused by Doppler shifting of emission from the pulsar hot spots, with higher-energy photons emitted earlier in the spin cycle as the spot approaches the observer (Ford 1999).

An accreting magnetized neutron star spinning this fast must have a weak magnetic field. If not, the radius of the magnetosphere r_M would exceed the corotation radius, and matter corotating in the magnetosphere would not be able to overcome the centrifugal barrier. A simple estimate leads to upper limits on r_M of 31 km, and on the surface field strength B of 2–$6 \ 10^8$ Gauss (Wijnands & van der Klis 1998b). A similarly simple estimate, involving the additional requirement that for pulsations to occur r_M must be larger than the radius R of the neutron star, would set a strong constraint on the star's mass-radius relation (Burderi & King 1998) and hence on the EOS (see also Li et al 1999a). However, the process of accretion onto a neutron star with such a low B is not identical to that in classical, 10^{12}-Gauss accreting pulsars. In particular, the disk model (used in calculating r_M) is different this close to the neutron star, the disk-star boundary layer may be different, and multipole components in the magnetic field become important. Conceivably, a classical magnetosphere does not even form and the 4–7% amplitude pulsation occurs as a result of milder effects of the magnetic field on either the flow or the emission. Psaltis & Chakrabarty (1999) discuss these issues and conclude that B is $(1–10)$ 10^8 Gauss, which puts the source right among the other rotation-powered, millisecond pulsars (the msec radio pulsars; Figure 2). When the accretion shuts off sufficiently for the radio pulsar mechanism to operate, the system will likely show up as a radio pulsar. This should happen at the end of the system's life as an X-ray binary—i.e. SAX J1808.4–3658 is indeed the long-sought millisecond radio-pulsar progenitor, but might also occur between the transient outbursts (Wijnands & van der Klis 1998b). So far, radio observations have not detected the source in X-ray quiescence (Gaensler et al 1999).

It is not clear what makes the neutron star spin detectable in SAX J1808.4–3658 and not (so far) in other low-mass X-ray binaries of similar and often much higher flux. Perhaps a peculiar viewing geometry (e.g. a very low inclination of the binary orbit) allows us to see the pulsations only in this system, although possible X-ray and optical modulations with binary phase make an inclination of zero unlikely (Chakrabarty & Morgan 1998b, Giles et al 1999).

With a neutron star spin frequency that is certain and good estimates of r_M and B, SAX J1808.4–3658 can serve as a touchstone in studies of low-mass X-ray binaries (LMXBs). Although no burst oscillations (Section 4) or kHz QPOs (Section 5) have been detected from the source, their absence is consistent with what would be expected from a standard LMXB in the same situation (Wijnands & van der Klis 1998c); in more intensive observations during a subsequent transient outburst, such phenomena could be detected. This would strongly test the main assumptions underlying the models for these phenomena. The X-ray spectral properties (Heindl & Smith 1998, Gilfanov et al 1998) and the slower types of variability (Wijnands & van der Klis 1998c) of the source are very similar to those of other LMXBs at low accretion rate, which suggests that either the neutron stars in those systems have similar B, or the presence of a small magnetosphere does not affect spectral and slow-variability characteristics.

4.   BURST OSCILLATIONS

Type 1 X-ray bursts are thermonuclear runaways in the accreted matter on a neutron-star surface (see Lewin et al 1995b for a review). When density and temperature in the accumulated nuclear fuel approach the ignition point, the matter ignites at one particular spot, from which a nuclear burning front then propagates around the star (see Bildsten 1998b for a review). This leads to a burst of X-ray emission with a rise time of typically <1 s, and a 10^1–10^2 s exponential decay resulting from cooling of the neutron-star atmosphere. The total amount of energy emitted is 10^{39-40} erg. In some bursts the Eddington critical luminosity is exceeded and atmospheric layers are lifted off the star's surface, leading to an increase in photospheric radius of $\sim 10^1$–10^2 km, followed by a gradual recontraction. These bursts are called radius expansion bursts.

In the initial phase, when the burning front is spreading, the energy generation is inherently very anisotropic. The occasional occurrence of multiple bursts closely spaced in time indicates that not all available fuel is burned up in each burst, which suggests that in some bursts only part of the surface participates. Magnetic fields and patchy burning (Bildsten 1995) could also lead to anisotropic emission during X-ray bursts. Anisotropic emission from a spinning neutron star leads to periodic or quasi-periodic observable phenomena, because the stellar rotation causes the viewing geometry of the brighter regions to vary periodically (unless the pattern is symmetric around the rotation axis). Searches for such periodic phenomena during X-ray bursts were performed by various groups (Mason et al 1980; Skinner et al 1982; Sadeh et al 1982; Sadeh & Livio 1982a,b; Murakami et al 1987; Schoelkopf & Kelley 1991; Jongert & van der Klis 1996), but claims of detections remained unconfirmed.

The first incontestable type 1 burst oscillation was discovered with RXTE in a burst that occurred on February 16, 1996, in the reliable burst source 4U 1728–34. An oscillation with a slightly drifting frequency near 363 Hz was evident in a power spectrum of 32 s of data starting just before the onset of the burst (Strohmayer et al 1996a,b,c; Figure 3). The oscillation frequency increased from 362.5 to 363.9 Hz in the course of ~ 10 s.

Burst oscillations have now been detected in six (perhaps seven) different sources (Table 1). They do not occur in each burst, and some burst sources have not

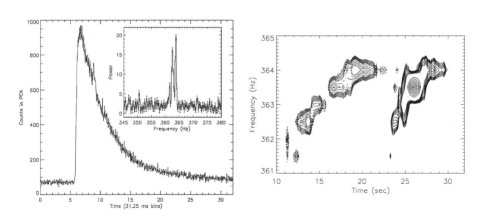

Figure 3 *Left*: A burst profile and its power spectrum (inset) showing a drifting burst oscillation in 4U 1728–34 (Strohmayer et al 1996c). *Right*: Dynamic power spectra of burst oscillations in two bursts separated by 1.6 yr in 4U 1728–34 showing near-identical asymptotic frequencies (Strohmayer et al 1998b).

TABLE 1   Burst oscillations

Source	Minimum observed frequency (Hz)	Asymptotic frequency range (Hz)		References
4U 1636–53	579.3[1]	581.47 ± 0.01	581.75 ± 0.13	1, 2, 3, 4, 5, 15
4U 1702–43	329.0	329.8 ± 0.1	330.55 ± 0.02	6, 14
4U 1728–34	362.1	363.94 ± 0.05	364.23 ± 0.05	2, 6, 7, 8
KS 1731–260	523.9	523.92 ± 0.05		9, 16
MXB 1743–29[2]	588.9	589.80 ± 0.07		10, 17
Aql X-1	547.8	548.9		11, 12, 18
Rapid Burster	154.9; 306.6[3]			13

[1]A hard to detect subharmonic exists near 290 Hz (Miller 1999a). [2]Source identification uncertain. [3]3 Marginal detections.
References: (1) Strohmayer et al 1998a; (2) Strohmayer et al 1998b; (3) Strohmayer 1999; (4) Miller 1999a; (5) Miller 1999b; (6) Strohmayer & Markwardt 1999; (7) Strohmayer et al 1996b,c; (8) Strohmayer et al 1997b; (9) Smith et al 1997; (10) Strohmayer et al 1997a; (11) Zhang et al 1998c; (12) Ford 1999; (13) Fox et al 1999; (14) Markwardt et al 1999a; (15) Zhang et al 1997a; (16) Morgan and Smith 1996; (17) Strohmayer et al 1996d; (18) Yu et al 1999.

shown oscillations at all. Sometimes the oscillations are strong for less than a second during the burst rise, then become weak or undetectable, and finally occur for ~10 s in the burst cooling tail. This can happen even in radius expansion bursts, where after the photosphere has recontracted the oscillations (re)appear (Smith et al 1997; Strohmayer et al 1997a, 1998a). Some oscillations are seen only in the burst tail and not in the rise (Smith et al 1997).

Usually the frequency increases by 1–2 Hz during the burst tail, converging to an "asymptotic frequency", which in a given source tends to be stable (Strohmayer et al 1998c; Figure 3), with differences from burst to burst of ~0.1% (Table 1). These differences are of the order of what would be expected from binary orbital Doppler shifts, strengthening an interpretation in terms of the neutron star spin. Perhaps the orbital radial velocity curve can be detected in the asymptotic frequencies, but it remains to be seen whether the asymptotic frequencies are intrinsically stable enough for this. Exceptions from the usual frequency evolution pattern do occur. Oscillations with no evidence for frequency evolution were observed in KS 1731–260 (Smith et al 1997) and 4U 1743–29 (Strohmayer et al 1997a), and in 4U 1636–53 a *decrease* in frequency was seen in a burst tail (Miller 2000, Strohmayer 1999; Figure 4).

In a widely (but not universally—see Section 7.2) accepted scenario, the burst oscillations arise as a result of a hot spot or spots in an atmospheric layer of the neutron star rotating slightly slower than the star itself because it expanded by 5–50 m in the X-ray burst but conserved its angular momentum (Strohmayer et al 1997a and references therein, Bildsten 1998b, Strohmayer 1999, Strohmayer & Markwardt 1999, Miller 2000). The frequency drifts are caused by spin-up of the atmosphere

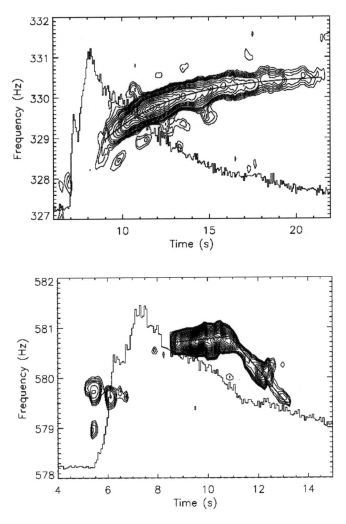

Figure 4 Dynamic power spectra of burst oscillations (contours) overlaid on burst flux profiles (traces). *Top*: Strong burst oscillation in 4U 1702–43 showing the usual asymptotic increase in frequency (Strohmayer & Markwardt 1999). *Bottom*: Oscillation in 4U 1636–53 exhibiting a drop in frequency in the burst tail (Strohmayer 1999).

as it recontracts in the burst decay. The asymptotic frequency corresponds to a fully recontracted atmosphere and is closest to the true neutron star spin frequency. From this scenario one expects a frequency drop during the burst rise, but no good evidence has been found for this yet (Strohmayer 1999). The case of a frequency drop in the burst tail (Figure 4) is explained by invoking additional thermonuclear energy input late in the burst, which also affects the burst profile (Strohmayer 1999).

If the oscillations result from a stable pattern in the spinning layer, then it should be possible to describe them as a frequency-modulated, strictly coherent signal. By applying a simple exponential model to the frequency drifts, it is possible to establish coherences of up to $Q \sim 4000$ (Strohmayer & Markwardt 1999, see also Zhang et al 1998c, Smith et al 1997, Miller 1999a, 2000). However, this is still $\sim 20\%$ less than a fully coherent signal of this frequency and duration—i.e. it is not possible to count the exact number of cycles in the way this can be done in a pulsar. It is possible that exact coherence recovery is feasible for these signals, but current signal-to-noise limits prevent us from accomplishing this because the exact frequency drift ephemeris cannot be found.

The harmonic content of the oscillations is low. In 4U 1636–53 it is just possible, by combining data from the early stages of several bursts, to detect a frequency near 290 Hz, which is half the dominant frequency (Miller 1999a, 2000); this suggests that ~ 290 Hz, not ~ 580 Hz, is the true spin frequency, and that two antipodal hot spots produce the burst oscillation in this source (its kHz peak separation is ~ 250 Hz; see Section 5). No harmonics or subharmonics have been seen in other sources, or in the burst tails of any source (Strohmayer & Markwardt 1999), with the possible exception of the marginal detections in the Rapid Burster (Fox et al 1999).

The oscillation amplitudes range from $\sim 50\%$ (rms) of the burst flux early in some bursts to between 2% and 20% (rms) in the tail (for references see Table 1). (Note that sometimes sinusoidal amplitudes—which are a factor $\sqrt{2}$ larger than rms amplitudes—are reported, and that amplitudes are expressed as a fraction of *burst flux*, defined as total flux minus the persistent flux before the burst. Early in the rise, when burst flux is low compared to total flux, the measured amplitude is multiplied by a large factor to convert it to burst flux fraction.) In KS 1731–260 (Smith et al 1997) and 4U 1743–29 (Strohmayer et al 1997a) the photon energy dependence of the oscillations was measured. Above 7 or 8 keV the amplitude was 9–18%, whereas below that energy it was undetectable at <2–4%. 4U 1636–53 shows a slight variation in spectral hardness as a function of oscillation phase (Strohmayer et al 1998c). In Aql X-1 photons below 5.7 keV lag those at higher energies by roughly 0.3 msec, which may be caused by Doppler shifts (Ford 1999).

If the burst oscillations result from hot spots on the surface, then their amplitude constrains the "compactness" of the neutron star, defined as R_G/R, where R is the star's radius and $R_G = GM/c^2$ is its gravitational radius (Strohmayer et al 1997b, 1998a; Miller & Lamb 1998). The more compact the star, the lower the oscillation amplitude, since gravitational light bending increasingly blurs the beam. In particular, when the oscillations are caused by two antipodal hot spots (cf 4U 1636–53, above, and possibly other sources; see Section 5.5), and the amplitudes are high, the constraints are strong. The exact bounds on the compactness depend on the emission characteristics of the spots; no final conclusions on the compactness of these stars have been reached yet.

Modeling of the spectral and amplitude evolution of the oscillations through the burst in terms of an expanding, cooling hot spot has been successful (Strohmayer et al 1997b). However, several issues with respect to this attractively simple interpretation have yet to be resolved. The presence of *two* burning sites as required by the description in terms of two antipodal hot spots could be related to fuel accumulation at the magnetic poles (Miller 1999a), but their simultaneous ignition seems difficult to accomplish, and obviously, in view of the frequency drifts, these sites must decouple from the magnetic field after ignition. The hot spots must survive the strong shear in layers that, from the observed phase drifts, must revolve around the star several times during the lifetime of the spots; they must even survive during radius-expansion bursts, through episodes where the photosphere becomes at least several times as large as the neutron star (which probably implies the roots of the hot spots are below the photospheric layers).

The resolution of these issues ties in with questions such as why only some burst sources show the oscillations, and why only some bursts exhibit them. This is hard to explain in a magnetic-pole accumulation scheme because the viewing geometry of the poles remains the same from burst to burst. Studies of the relation between the characteristics of the bursts and the surrounding persistent emission and presence and character of the burst oscillations could help to shed light on these various questions.

5.   KILOHERTZ QUASI-PERIODIC OSCILLATIONS

The kilohertz quasi-periodic oscillations (kHz QPOs) were discovered at NASA's Goddard Space Flight Center in February 1996, just two months after RXTE was launched (in Sco X-1: van der Klis et al 1996a,b,c; and 4U 1728–34: Strohmayer et al 1996a,b,c; see van der Klis 1998 for a historical account). Two simultaneous quasi-periodic oscillation peaks ("twin peaks") in the 300–1300 Hz range and roughly 300 Hz apart (Figure 5) occur in the power spectra of low-mass X-ray binaries containing low-magnetic-field neutron stars of widely different X-ray luminosity L_x. The frequency of both peaks usually increases with X-ray flux (Section 5.4). In 4U 1728–34 the separation frequency of the two kHz peaks is close to ν_{burst} (Section 4; Strohmayer et al 1996b,c). This commensurability of frequencies provides a powerful argument for a beat-frequency interpretation (Sections 5.1, 5.5, 7.1).

5.1 Orbital and Beat Frequencies

Orbital motion around a neutron star occurs at a frequency of

$$\nu_{orb} = \left(\frac{GM}{4\pi^2 r_{orb}^3}\right)^{1/2} \approx 1200 \, \text{Hz} \left(\frac{r_{orb}}{15 \, \text{km}}\right)^{-3/2} m_{1.4}^{1/2},$$

Figure 5   Twin kHz peaks in Sco X-1 (*top*; van der Klis et al 1997b) and 4U 1608–52 (*bottom*; Méndez et al 1998b).

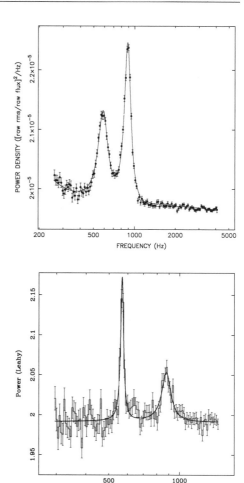

and the corresponding orbital radius is

$$r_{orb} = \left(\frac{GM}{4\pi^2 v_{orb}^2}\right)^{1/3} \approx 15 \text{ km} \left(\frac{v_{orb}}{1200 \text{ Hz}}\right)^{-2/3} m_{1.4}^{1/3},$$

where $m_{1.4}$ is the star's mass in units of $1.4 \, M_{\odot}$ (Figure 6). In general relativity, no stable orbital motion is possible within the innermost stable circular orbit (ISCO), $R_{ISCO} = 6GM/c^2 \approx 12.5 m_{1.4}$ km. The frequency of orbital motion at the ISCO, the highest possible stable orbital frequency, is $v_{ISCO} \approx (1580/m_{1.4})$ Hz.

These expressions are valid for a Schwarzschild geometry, i.e. outside a non-rotating, spherically symmetric neutron star (or black hole). Corrections to first order in $j = cJ/GM^2$, where J is the neutron-star angular momentum, have been

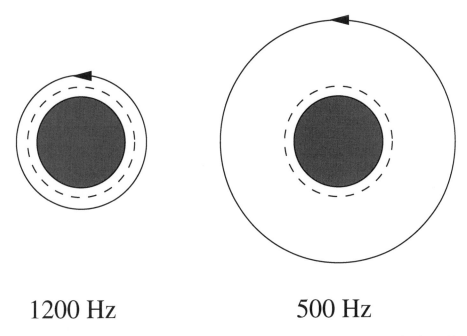

1200 Hz   500 Hz

Figure 6   A 10-km radius, 1.4 $M_\odot$ neutron star with the corresponding innermost circular stable orbit (ISCO; *dashed circles*) and orbits (*drawn circles*) corresponding to orbital frequencies of 1200 and 500 Hz, drawn to scale.

given by e.g. Miller et al (1998a) and can be several 10%. For more precise calculations, see Morsink & Stella (1999).

In spin-orbit beat-frequency models some mechanism produces an *interaction* of ν_{orb} at some preferred radius in the accretion disk with the neutron star spin frequency ν_s, so that a beat signal is seen at the frequency $\nu_{beat} = \nu_{orb} - \nu_s$. As ν_{beat} is the frequency at which a given particle orbiting in the disk overtakes a given point on the spinning star, it is the natural disk/star interaction frequency. In such a rotational interaction (with spin and orbital motion in the same sense), no signal is produced at the sum frequency $\nu_{orb} + \nu_s$ (it is a single-sideband interaction).

5.2   Early Interpretations

It was immediately realized that the observed high frequencies of kHz QPOs could arise in orbital motion of accreting matter very closely around the neutron star, or in a beat between such orbital motion and the neutron-star spin (van der Klis et al 1996a, Strohmayer et al 1996a; some of the proposals to look for such rapid QPOs with RXTE had in fact anticipated this). A magnetospheric spin-orbit beat-frequency model (Alpar & Shaham 1985, Lamb et al 1985) was already in use in LMXBs for slower QPO phenomena (Section 6.3), so when three commensurable frequencies were found in 4U 1728–34, a beat-frequency interpretation was immediately proposed (Strohmayer et al 1996c). Let us call ν_2 the frequency of the higher-frequency peak (the "upper" peak) and ν_1 that of the lower-frequency kHz peak (the "lower peak"). Then the beat-frequency interpretation asserts that ν_2 is ν_{orb} at some preferred radius in the disk, and ν_1 is the beat frequency between ν_2 and ν_s, so $\nu_1 = \nu_{beat} = \nu_{orb} - \nu_s \approx \nu_2 - \nu_{burst}$, where the approximate equality follows from $\nu_{burst} \approx \nu_s$ (Section 4). The fact that only a single sideband is observed is a strong argument for a rotational interaction (Section 5.1). Later this was worked out in detail in the form of the "sonic point beat-frequency model" by Miller et al (1996, 1998a; Section 7.1), where the preferred radius is the sonic radius (essentially, the inner edge of the Keplerian disk).

The beat-frequency interpretation implies that the observed kHz QPO peak separation $\Delta\nu = \nu_2 - \nu_1$ should be equal to the neutron star spin frequency ν_s, and should therefore be constant and nearly equal to ν_{burst}. As we will see in Section 5.5, it turned out that $\Delta\nu$ is not actually exactly constant nor precisely equal to ν_{burst} (in Section 7 we examine how a beat-frequency interpretation could deal with this), and this triggered the development of other models for the kHz QPOs. Stella & Vietri (1998) noted that the frequency ν_{LF} of low-frequency QPOs (Section 6.3) in the 15–60 Hz range that had been known in the Z sources since the 1980s (cf van der Klis 1995a), and that were being discovered with RXTE in the atoll sources as well, is approximately proportional to $\nu_2{}^2$. This triggered a series of papers (Stella & Vietri 1998, 1999; Stella et al 1999) describing what is now called the "relativistic precession model", where ν_2 is the orbital frequency at some radius in the disk, and ν_1 and ν_{LF} are frequencies of general-relativistic precession modes of a free particle orbit at that radius (Section 7.2). For further discussion of kHz QPO models see Section 7. With the exception of the photon bubble model (Klein et al 1996b; Section 7.3), all models are based on the interpretation that one of the kHz QPO frequencies is an orbital frequency in the disk.

5.3 Dependence on Source State and Type

Twenty sources have now[1] shown kHz QPOs. Sometimes only one peak is detectable, but 18 of these sources have shown two simultaneous kHz peaks; the exceptions with only a single peak are the little-studied XTE J1723–376 and EXO 0748–676. Tables 2 and 3 summarize the results. There is a remarkable similarity in QPO frequencies and peak separations across a great variety of sources.

At a more detailed level, however, some differences are evident between the different source types. The two main types are the Z sources and the atoll sources (Hasinger & van der Klis 1989; see van der Klis 1989a, 1995a,b for reviews). Z sources are named after the roughly Z-shaped tracks they trace out in X-ray color-color and hardness-intensity diagrams on a timescale of hours to days (Figure 7). They are the most luminous LMXBs, with X-ray luminosity L_x near

---

[1] 1999 October 15.

TABLE 2 Observed frequencies of kilohertz QPOs in Z sources

Source	ν_1 (Hz)	ν_2 (Hz)	$\Delta\nu$ (Hz)	ν_{burst} (Hz)	References
Sco X-1	565	870	307 ± 5		Van der Klis et al 1996a,b,c, 1997b
	845	1080	237 ± 5		
		1130			
GX 5−1	215	505			Van der Klis et al 1996e
			298 ± 11		Wijnands et al 1998c
	660	890			
	700				
GX 17+2		645			Van der Klis et al 1997a
	480	785			Wijnands et al 1997b
			294 ± 8		
	780	1080			
Cyg X-2		730			Wijnands et al 1998a
	530	855			
			346 ± 29		
	660	1005			
GX 340+0	200	535			Jonker et al 1998, 2000b
			339 ± 8		
	565	840			
	625				
GX 349+2	710	980	266 ± 13		Zhang et al 1998a
		1020[1]			Kuulkers & van der Klis 1998

Values for ν_1 and ν_2 were rounded to the nearest 5 Hz, for ν_{burst} to the nearest 1 Hz. Entries in one column for a given source indicate ranges over which the frequency was observed or inferred to vary; ranges from different observations were combined assuming the ν_1, ν_2 relation in each source is reproducible (no evidence to the contrary exists). Entries in one uninterrupted row refer to simultaneous data (except for ν_{burst} values). Values of $\Delta\nu$ straddling two rows or adjacent to a vertical line refer to measurements made over the range of frequencies indicated. Note: [1] Marginal detection.

the Eddington luminosity L_{Edd}. Atoll sources produce tracks roughly like a wide U or C (e.g. Figure 7) that are somewhat reminiscent of a geographical map of an atoll because in the left limb of the U the motion through the diagram becomes slow, so that the track is usually broken up by observational windowing into "islands." The bottom and right-hand parts of the U are traced out in the form of a curved "banana" branch on a timescale of hours to a day. In a given source, the islands correspond to lower flux levels than the banana branch. Most atoll sources are in the 0.01–0.2 L_{Edd} range; a group of four bright ones (GX 3+1, GX 9+1, GX 13+1 and GX 9+9) that is nearly always in the banana branch is more luminous than this, perhaps 0.2–0.5 L_{Edd} (the distances are uncertain). Most timing and spectral characteristics of these sources depend in a simple way on position along the Z or atoll track. Therefore, the phenomenology is essentially one-dimensional. A single quantity (usually referred to as inferred accretion rate),

TABLE 3   Observed frequencies of kilohertz QPOs in atoll sources

Source	ν_1 (Hz)	ν_2 (Hz)	$\Delta\nu$ (Hz)	ν_{burst} (Hz)[4]	References
4U 0614+09		450			Ford et al 1996, 1997a,b; Van der
	418	765	312 ± 2		Klis et al 1996d; Méndez et al
	825	1160			1997; Vaughan et al 1997, 1998;
		1215			Kaaret et al 1998; van Straaten
		1330			et al 2000
EXO 0748–676	695				Homan & van der Klis 2000
4U 1608–52	415				Van Paradijs et al 1996; Berger
	440	765	325 ± 7		et al 1996; Yu et al 1997; Kaaret
	475	800	326 ± 3		et al 1998; Vaughan et al 1997,
	865	1090^2	225 ± 12^2		1998; Méndez et al 1998a,b,
	895				1999; Méndez 1999; Markwardt
					et al 1999b
4U 1636–53	830				Zhang et al 1996a,b, 1997a; Van
	900	1150			der Klis et al 1996d; Vaughan
	950	1190	251 ± 4^2	291, 582	et al 1997, 1998; Zhang 1997;
		1230			Wijnands et al 1997a; Méndez
	1070				et al 1998c; Méndez 1999;
					Markwardt et al 1999b;
					Kaaret et al 1999a
4U 1702–43	625				Markwardt et al 1999a,b
	655	1000^2	344 ± 7^2		
	700	1040^2	337 ± 7^2	330	
	770	1085^2	315 ± 11^2		
	902				
4U 1705–44	775	1075^1	298 ± 11		Ford et al 1998a
	870				
XTE J1723–376	815				Marshall & Markwardt 1999
4U 1728–34		325			Strohmayer et al 1996a,b,c; Ford
	510	845			& van der Klis 1998; Méndez &
	875	1160	349 ± 2^3	364	van der Klis 1999; Méndez 1999;
	920		279 ± 12^3		Markwardt et al 1999b; di Salvo
					et al 1999
KS 1731–260	900	1160	260 ± 10	524	Wijnands & van der Klis 1997
		1205			
4U 1735–44	630	980	341 ± 7		Wijnands et al 1996a, 1998b;
	730	1025	296 ± 12		Ford et al 1998b
	900^1	1150	249 ± 15		
		1160			

(Continued)

TABLE 3   *(Continued)*

Source	ν_1 (Hz)	ν_2 (Hz)	$\Delta\nu$ (Hz)	ν_{burst} (Hz)[4]	References
4U 1820−30		655			Smale et al 1996, 1997; Zhang
	500	860	358 ± 42		et al 1998b; Kaaret et al 1999b;
	795	1075	278 ± 11		Bloser et al 1999
		1100			
Aql X-1	670				Zhang et al 1998c; Cui et al 1998a;
		1040[1]	241 ± 9[1]		Yu et al 1999; Reig et al 2000;
	930			549	M. Méndez et al 1999 in prep.
4U 1915−05		820			Barret et al 1997, 1998;
	515				Boirin et al 1999
	560	925			
	655	1005	348 ± 11		
	705[1]	1055			
	880				
		1265[1]			
XTE J2123−058	845	1100	255 ± 14		Homan et al 1998b, 1999a.
	855	1130	276 ± 9		Tomsick et al 1999
	870[1]	1140	270 ± 5[1]		

Caption: see Table 2. Notes: [1]Marginal detection. [2]Shift and add detection method. cf Méndez et al (1998a). [3]See Figure 10. [4]For burst oscillation references see Table 1.

varying on timescales of hours to days on the Z track and the banana branch and more slowly in the island state, must govern most of the phenomenology (but see Section 5.4).

In all Z sources and in 4U 1728–34 the kHz QPOs are seen down to the lowest inferred $\dot{M}$ levels these sources reach. The QPOs always become undetectable at the highest $\dot{M}$ levels. In the atoll sources, where the count rates are higher at higher inferred $\dot{M}$, this cannot be a sensitivity effect. In most atoll sources, the QPOs are seen in the part of the banana branch closest to the islands, i.e. near the lower left corner of the U (Figure 7); that they are often not detected in the island state may be related to low sensitivity at the low count rates there, but in one island in 4U 0614+09 the undetected lower kHz peak is really much weaker than at higher inferred $\dot{M}$ (Méndez et al 1997). No kHz QPOs have been seen in the four bright atoll sources (Wijnands et al 1998d, Strohmayer 1998, Homan et al 1998a), perhaps because they do not usually reach this low part of the banana branch, and in several faint LMXBs, probably also atoll sources in the island state (SAX J1808.4–3658: Wijnands & van der Klis 1998c and Section 3; XTE J1806–246: Wijnands & van der Klis 1999b; SLX 1735–269: Wijnands & van der Klis 1999c; 4U 1746–37: Jonker et al 2000; 4U 1323–62: Jonker et al 1999; 1E 1724–3045, SLX 1735–269, and GS 1826–238: Barret et al 2000).

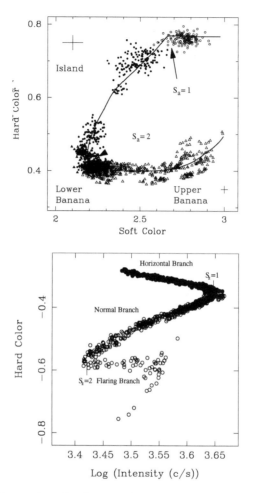

Figure 7 *Top*: X-ray color-color diagram of the atoll source 4U 1608–52 (Méndez et al 1999). *Bottom*: X-ray hardness versus intensity diagram of the Z source GX 340+0 (after Jonker et al 2000b). Values of curve-length parameters S_z and S_a and conventional branch names are indicated. Mass accretion rate is inferred to increase in the sense of increasing S_a and S_z. X-ray color is the log of a count rate ratio (3.5–6.4/2.0–3.5 and 9.7–16.0/6.4–9.7 keV for soft and hard color respectively); intensity is in the 2–16 keV band. kHz QPO detections are indicated with *filled symbols*.

The QPO frequencies increase when the sources move along the tracks in the sense of increasing inferred $\dot{M}$ (this has been seen in more than a dozen sources, and no counterexamples are known). On timescales of hours to a day, increasing inferred $\dot{M}$ usually corresponds to increasing X-ray flux, so on these timescales kHz QPO frequency usually increases with flux. When flux systematically decreases with inferred $\dot{M}$, as is the case in some parts of these tracks, the frequency is expected to maintain its positive correlation with inferred $\dot{M}$ and hence become anticorrelated to flux, and indeed this has been observed in the Z source GX 17+2 (Wijnands et al 1997b).

So, the kHz QPOs fit well within the preexisting Z/atoll description of LMXB phenomenology in terms of source types and states, including the fact that position on the tracks in X-ray color-color or hardness-intensity diagrams (inferred $\dot{M}$; see Section 5.4), and not X-ray flux, drives the phenomenology.

5.4 Dependence of QPO Frequency on Luminosity and Spectrum

Kilohertz QPOs occur at similar frequency in sources that differ in X-ray luminosity L_x by more than 2 orders of magnitude, and the kHz QPO frequency ν seems to be determined more by the difference between average and instantaneous L_x of a source than by L_x itself (van der Klis 1997, 1998, see also Zhang et al 1997b). In a plot of ν versus L_x (defined as $4\pi d^2 f_x$ with f_x the X-ray flux and d the distance; Ford et al 2000; Figure 8) a series of roughly parallel lines is seen, to first order one line per source (but see later in this section). In each source there is a definite relation between L_x and ν, but the same relation does not apply in another source with a different average L_x. Instead, that source covers the same ν-range around *its* particular average L_x. This is unexplained, and must mean that in addition to instantaneous L_x, another parameter, related to average L_x, affects the QPO frequency (van der Klis 1997, 1998). Perhaps this parameter is the neutron star magnetic field strength, which previously, on other grounds, was hypothesized to correlate to average L_x (Hasinger & van der Klis 1989, Psaltis & Lamb 1998, Konar & Bhattacharya 1999; see also Lai 1998), but other possibilities exist (van der Klis 1998, Ford et al 2000).

A similar pattern of parallel lines, but on a much smaller scale, occurs in some individual sources. When observed at different epochs, a source produces different frequency versus flux tracks that are approximately parallel (GX 5–1: Wijnands et al 1998c; GX 340+0: Jonker et al 1998; 4U 0614+09: Ford et al 1997a,b, Méndez et al 1997, van Straaten et al 2000; 4U 1608–52: Yu et al 1997, Méndez et al 1998a, 1999; 4U 1636–53: Méndez 1999; 4U 1728–34: Méndez & van der Klis 1999; 4U 1820–30: Kaaret et al 1999b; Aql X-1: Zhang et al 1998c, Reig et al 2000; Figures 8 and 9). This is most likely another aspect of the well-known fact (e.g. van der Klis et al 1990; Hasinger et al 1990; Kuulker et al 1994, 1996; van der Klis 1994a, 1995a) that while the properties of timing phenomena such as QPOs are well correlated with one another and with X-ray spectral *shape* as diagnosed by X-ray colors (and hence with position in tracks in color-color diagrams), the *flux* correlates well to these diagnostics only on short (hours to days) timescales and much less well on longer timescales. This is why color-color diagrams, independent of flux, are popular for parameterizing spectral variability in these sources. Like other timing parameters, kHz QPO

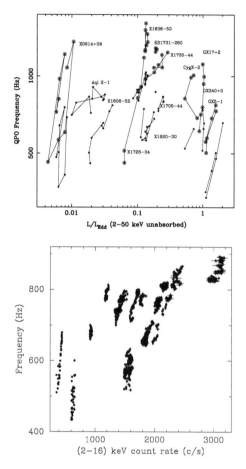

Figure 8 The parallel lines phenomenon across sources (*top*; after Ford et al 2000; upper and lower kHz peaks are indicated with different symbols) and in the source 4U 1608–52 (*bottom*; Méndez et al 1999, frequency plotted is ν_1).

frequency correlates much better with position on the track in the color-color diagram (Figure 9) than with flux (Méndez et al 1999, Méndez and van der Klis 1999, Kaaret et al 1999b, Méndez 1999, Reig et al 2000, van Straaten et al 2000, di Salvo et al 1999). Correlations of frequency with parameters describing spectral shape such as blackbody flux (Ford et al 1997b) or power-law slope (Kaaret et al 1998) in a two-component spectral model are also much better than with flux.

Usually, all this has been interpreted by saying that apparently inferred accretion rate (Section 5.3) governs both the timing and the spectral properties, but not flux (e.g. van der Klis 1995a). Of course, energy conservation suggests that total $\dot{M}$ and flux should be well correlated. Perhaps, inferred $\dot{M}$ is not the total $\dot{M}$ but only one component of it—i.e. that through the disk. While there is also a radial inflow (e.g. Fortner et al 1989, Kuulkers & van der Klis 1995, Wijnands et al 1996b,

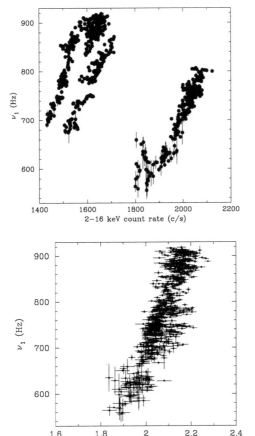

Figure 9 In 4U 1728–34 a QPO frequency (ν_1) versus count rate plot (*top*) shows no clear correlation, but instead shows a series of parallel lines. When the same data are plotted versus position in the X-ray color-color diagram (*bottom*), a single relation is observed. (After Méndez & van der Klis 1999).

Kaaret et al 1998), maybe there are large and variable anisotropies or bolometric corrections in the emission so that the flux we measure is not representative for the true luminosity (e.g. van der Klis 1995a), or possibly mass outflows destroy the expected correlation by providing sinks of both mass and (kinetic) energy (e.g. Ford et al 2000). The true explanation is unknown. We do not even know if the two different parallel-lines phenomena (across sources and within individual sources; Figure 8) have the same origin. It is possible that the quantity that everything depends on is *not* $\dot{M}$, but some other parameter such as inner disk radius r_{in}.

5.5   Peak Separation and Burst Oscillation Frequencies

In interpretations (Sections 5.1, 7.1) where the kHz peak separation $\nu_2 - \nu_1 = \Delta\nu$ is the neutron star spin frequency ν_s, one expects $\Delta\nu$ to be approximately constant.

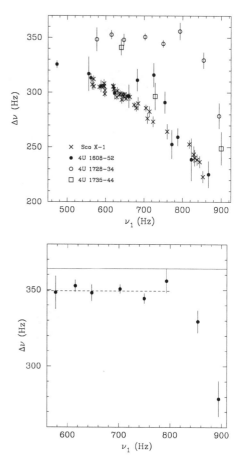

Figure 10 *Top*: The variations in kHz QPO peak separation as a function of the lower kHz frequency. *Bottom*: The data on 4U 1728–34. The burst oscillation frequency is indicated by a *horizontal drawn line* (after Méndez & van der Klis 1999).

This is not the case (see Figure 10). In Sco X-1 (van der Klis et al 1996c, 1997b), 4U 1608–52 (Méndez et al 1998a,b), 4U 1735–44 (Ford et al 1998b), 4U 1728–34 (Méndez & van der Klis 1999), and marginally also 4U 1702–43 (Markwardt et al 1999a), the separation $\Delta\nu$ decreases considerably when the kHz QPO frequencies increase. Other sources may show similar $\Delta\nu$ variations (see also Psaltis et al 1998). In a given source the ν_1, ν_2 relation seems to be reproducible.

That ν_{burst}, by interpretation close to ν_s (Section 4), is close to $\Delta\nu$ (or $2\Delta\nu$) is the most direct evidence for a beat frequency interpretation of the kHz QPOs (Section 5.1). The evidence for this is summarized in Table 4, where the highest asymptotic burst frequency (likely to be closest to the spin frequency; Section 4) is compared to the largest well-measured $\Delta\nu$ in the five sources where both have been measured. The frequency ratio may cluster at 1 and 2, but a few more examples are clearly needed; in two sources, discrepancies of $\sim$15% occur between ν_{burst}

TABLE 4 Commensurability of kHz QPO and burst oscillation frequencies

Source	Highest ν_{burst} (Hz)	Highest $\Delta\nu$ (Hz)	Ratio ($\nu_{burst}/\Delta\nu$)	Discrepancy (%)	References
4U 1702–43	330.55 ± 0.02	344 ± 7	0.96 ± 0.02	-4 ± 2	1
4U 1728–34	364.23 ± 0.05	349.3 ± 1.7	1.043 ± 0.005	$+4.3 \pm 0.5$	2
KS 1731–260	525.08 ± 0.18	260 ± 10	2.020 ± 0.078	$+1.0 \pm 3.8$	3, 6
Aql X-1	548.9	241 ± 9[1]	2.28 ± 0.09[1]	$+14 \pm 5$[1]	4
4U 1636–53	581.75 ± 0.13	254 ± 5	2.29 ± 0.04	$+15 \pm 2$	5

[1]Based on a marginal detection (see Table 3). References: (1) Markwardt et al 1999a; (2) Méndez & van der Klis 1999; (3) Wijnands & van der Klis 1997; (4) M Méndez et al 1999, in preparation; (5) Méndez et al 1998c; (6) Muno et al 2000.

and $2\Delta\nu$. Of these five sources, 4U 1728–34 also has a measured $\Delta\nu$ variation (Figure 10). When the QPO frequency drops, $\Delta\nu$ increases to saturate 4% below the burst oscillation frequency (Méndez & van der Klis 1999). In this case it certainly seems as though the kHz QPO separation "knows" the value of the burst oscillation frequency.

5.6 Strong Gravity and Dense Matter

Potentially, kHz QPOs may be used to constrain neutron-star masses and radii, and to test general relativity. Detection of the predicted innermost stable circular orbit (ISCO, Section 5.1) would constitute the first direct detection of a strong-field general-relativistic effect and prove that the neutron star is smaller than the ISCO. This possibility has fascinated researchers since observations of X-ray binaries began and was discussed well before the kHz QPOs were found (Kluźniak & Wagoner 1985, Paczyński 1987, Kluźniak et al 1990, Kluźniak & Wilson 1991, Biehle & Blandford 1993). If a kHz QPO results from orbital motion around the neutron star, its frequency cannot be larger than the frequency at the ISCO. Kaaret et al (1997) proposed that kHz QPO frequency variations that then seemed uncorrelated to X-ray flux in 4U 1608–52 and 4U 1636–53 were caused by orbital motion near the ISCO, and from this derived neutron star masses of $\sim$2 $M_\odot$. However, from more detailed studies (Section 5.4) it is now clear that on short timescales frequency does in fact correlate to flux in these sources.

The maximum kHz QPO frequencies observed in each source are constrained to a narrow range. The 12 atoll sources with twin peaks have maximum ν_2 values in the range 1074–1329 Hz; among the Z sources there are two cases of much lower maximum ν_2 values ($<$900 Hz in GX 5–1 and GX 340+0), while the other four fit in. Zhang et al (1997b) proposed that this narrow distribution is caused by the limit set by the ISCO frequency, which led them to neutron star masses near 2 $M_\odot$ as well. It is possible, in principle, that the maximum is set by some other limit on orbital radius (e.g. the neutron star surface), or that it *is* caused by the ISCO but that the frequency we observe is not orbital, in which cases no mass estimate can be made.

Miller et al (1996, 1998a) suggested that when the inner edge of the accretion disk reaches the ISCO, the QPO frequency might level off and remain constant while $\dot{M}$ continues rising. Later, an apparent leveling off at $\nu_2 = 1060 \pm 20$ Hz with X-ray count rate was found in 4U 1820–30 (Zhang et al 1998b; Figure 11, *top*). If this is the orbital frequency at the ISCO, then the neutron star has a mass of $\sim 2.2\,M_\odot$ and is smaller than its ISCO, and many equations of state are thus rejected. The leveling off is also observed as a function of X-ray flux and color (Kaaret et al 1999b) and position along the atoll track (Bloser et al 1999). However, the frequency versus flux relations are known in other sources (Section 5.4) to be

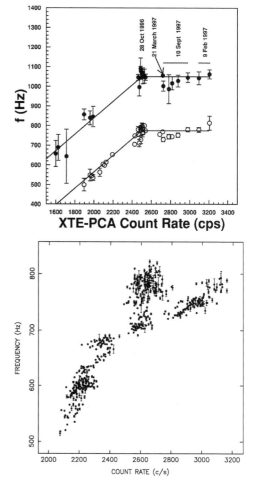

Figure 11   *Top*: Evidence for a leveling off of the kHz QPO frequency with count rate in 4U 1820–30 (Zhang et al 1998b). *Bottom*: A larger data set (see also Kaaret et al 1999b) of 4U 1820–30 at higher resolution, showing only the lower peak's frequency ν_1. (M Méndez et al, in preparation).

variable, and in 4U 1820–30 the leveling off seems not to be reproduced in the same way in all data sets (M Méndez et al, in preparation; Figure 11, *bottom*). It may also be more gradual in nature than was originally suggested. No evidence for a similar saturation in frequency was seen in other sources, and most reach higher frequencies. It is possible that the unique aspects of the 4U 1820–30 system are related to this (it is an 11-min binary in a globular cluster with probably a pure He companion star; Stella et al 1987a).

If a kHz QPO peak at frequency ν corresponds to stable Keplerian motion around a neutron star, one can immediately set limits on neutron star mass M and radius R (Miller et al 1998a). For a Schwarzschild geometry: (1) the radius R of the star must be smaller than the radius r_K of the Keplerian orbit: $R < r_K = (GM/4\pi^2\nu^2)^{1/3}$; and (2) the radius of the ISCO (Section 5.1) must *also* be smaller than r_K, as no stable orbit is possible within this radius: $r_{ISCO} = 6GM/c^2 < (GM/4\pi^2\nu^2)^{1/3}$ or $M < c^3/(2\pi 6^{3/2}G\nu)$. Condition (1) is a mass-dependent upper limit on R, and condition (2) is an upper limit on M; neither limit requires detection of orbital motion at the ISCO. Figure 12 (*top*) shows these limits in the neutron star mass-radius diagram for $\nu = 1220$ Hz, along with an indication of how the excluded area (*hatched*) shrinks for higher values of ν. The highest current value of ν_2, identified in most models with the orbital frequency, is 1329 ± 4 Hz (van Straaten et al 2000), so the hardest equations of state are beginning to be imperiled by the method. Corrections for frame dragging [shown to first order in j (see Section 5.1) in Figure 12 (*bottom*)] expand the allowed region. They depend on the neutron-star spin rate ν_s, and somewhat on the neutron star model, which sets the relation between ν_s and angular momentum. For 1329 Hz the above equations imply $M < 1.65\,M_\odot$ and $R_{NS} < 14.7$ km; with corrections for a 312 Hz spin, these numbers become $1.9\,M_\odot$ and 15.2 km (van Straaten et al 2000). Calculations exploring to what extent kHz QPOs constrain the EOS have further been performed by Miller et al (1998b), Datta et al (1998), Akmal et al (1998), Kluźniak (1998), Bulik et al (1999), Thampan et al (1999), Li et al (1999b), Schaab & Weigel (1999), and Heiselberg & Hjorth-Jensen (1999).

If $\Delta\nu$ is near ν_s then the 18 neutron stars where this quantity has been measured all spin at frequencies between ∼240 and ∼360 Hz, a surprisingly narrow range. If the stars spin at the magnetospheric equilibrium spin rates (e.g. Frank et al 1992) corresponding to their current luminosities L_x, this would imply an unlikely, tight correlation between L_x and neutron-star magnetic-field strength (White & Zhang 1997; note that a similar possibility came up in the discussion about the uniformity of the QPO frequencies themselves—see Section 5.4). White & Zhang (1997) propose that when r_M is small, as is the case here, it depends only weakly on accretion rate (as it does in some inner disk models; cf Ghosh & Lamb 1992, Psaltis & Chakrabarty 1999). Another possibility is that the spin frequency of accreting neutron stars is limited by gravitational radiation losses (Bildsten 1998a, Andersson et al 1999; see also Levin 1999). If so, then gravitational radiation is transporting angular momentum out as fast as accretion is transporting it in, making these sources the brightest gravitational radiation sources in the sky. They

Figure 12 Constraints on the mass and radius of neutron stars from the detection of orbital motion with the frequencies indicated. Graphs are for negligible neutron star angular momentum (*top*) and for the values of $j = cJ/GM^2$ (see Section 5) indicated (*bottom*). Mass-radius relations for some representative EOSs are shown (Miller et al 1998a).

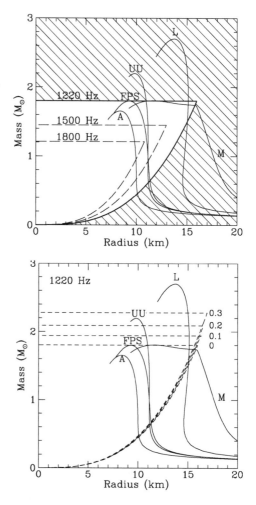

would produce a periodic signature at the neutron star spin frequency, which would facilitate their detection.

5.7 Other Kilohertz QPO Properties

The amplitudes of kHz QPOs increase strongly with photon energy (Figure 13). In similar X-ray photometric bands the QPOs tend to be weaker in the more luminous sources, with 2–60 keV amplitudes ranging from as high as 15% (rms) in 4U 0614+09 to typically a few % (rms) at their strongest in the Z sources. At high energy, amplitudes are much higher (e.g. 40% rms above 16 keV in 4U 0614+09; Méndez et al 1997). Fractional rms usually decreases with inferred $\dot{M}$ (e.g. van der Klis et al 1996b, 1997b; Berger et al 1996) but more complex behavior is

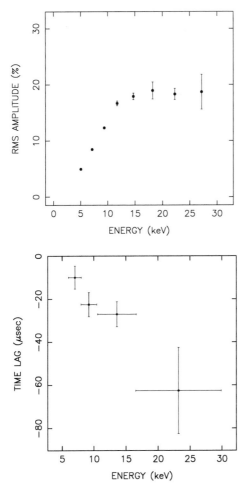

Figure 13 Energy dependence (*top*) and time lags (*bottom*) of the lower kHz peak in 4U 1608–52 (after Berger et al 1996; after Vaughan et al 1997, 1998).

sometimes seen (e.g. di Salvo et al 1999). The measured widths of QPO peaks are affected by variations in centroid frequency during the measurement, but typical values in the Z sources are 50–200 Hz. In the atoll sources, the upper peak usually has a width similar to this, although occasionally peaks as narrow as 10 or 20 Hz have been measured (e.g. Wijnands & van der Klis 1997, Wijnands et al 1998a,b) but the lower peak is clearly much narrower. It is rarely as wide as 100 Hz (Méndez et al 1997) and sometimes as narrow as 5 Hz (e.g. Berger et al 1996, Wijnands et al 1998b). Various different relations of peak width versus inferred $\dot{M}$ have been seen, but there seems to be a tendency for the upper peak to become narrower as its frequency increases.

A strong model constraint is provided by the time lags between kHz QPO signals in different energy bands (Section 2). Time-lag measurements require very high signal-to-noise ratios (Vaughan et al 1997), and have mostly been made in the very significant lower peaks observed in some atoll sources. Finite lags of 10–60 μsec occur in these peaks (Vaugan et al 1997, 1998; Kaaret et al 1999a; Markwardt et al 1999b; see Lee & Miller 1998 for a calculation of Comptonization lags relevant to kHz QPOs). Contrary to the initial report, the low-energy photons lag the high-energy ones (these are soft lags) by increasing amounts as the photon energy increases (Figure 13). The lags are of opposite sign to those expected from the inverse Compton scattering thought to produce the hard spectral tails of these sources (e.g. Barret & Vedrenne 1994), and correspond to light travel distances of only 3–20 km. From this it seems more likely that the lags originate in the QPO production mechanism than in propagation delays. Markwardt et al (1999b) reported a possible hard lag in an atoll-source upper peak.

6.   CORRELATIONS WITH LOW-FREQUENCY TIMING PHENOMENA AND WITH OTHER SOURCES

6.1 Black-Hole Candidates

The only oscillations with frequencies exceeding $10^{2.5}$ Hz (Section 1) known in black-hole candidates (BHCs) are, marginally, the 100–300 Hz oscillations in GRO J1655–40 and XTE J1550–564, and those reported very recently in Cyg X-1 and 4U 1630–47. An oscillation near 67 Hz observed in GRS 1915+105 is usually discussed together with these QPOs, although it is not clear that it is related. Usually, the phenomenology accompanying these high-frequency QPOs (spectral variations, lower-frequency variations) is complex.

The 67 Hz QPO in GRS 1915+105 (Morgan et al 1997, Remillard & Morgan 1998) varies by only a few percent in frequency when the X-ray flux varies by a factor of several. It is relatively coherent, with Q usually around 20 (but sometimes dropping to 6; Remillard et al 1999a); it has an rms amplitude of about 1%, and the signal at high photon energy lags that at lower energy by up to 2.3 radians (Cui 1999). The 300 Hz QPO in GRO J1655–40 (Remillard et al 1999b) was seen only when the X-ray spectrum was dominated by a hard power-law component. This feature was relatively broad (Q ∼ 4) and weak (0.8% rms), and did not vary in frequency by more than ∼30 Hz. The 185–285 Hz QPO in XTE J1550–564 (Remillard et al 1999c) shows considerable variations in frequency (Homan et al 1999b, Remillard et al 1999c). It is seen in X-ray spectral conditions similar to those in which the 300 Hz QPO is observed in GRO J1655–40, and has similarly low amplitudes and 3 < Q < 10. Recent reports indicate that QPOs in this frequency range may also occur in Cyg X-1 and 4U 1630–47 (Remillard 1999, Remillard & Morgan 1999).

The fact that the 300 Hz and 67 Hz oscillations were constant in frequency (which is different from anything then known to occur in neutron stars) triggered interpretations in which these frequencies depend mostly on black-hole mass and angular momentum and only weakly on luminosity—for example, orbital motion at the ISCO (Morgan et al 1997), Lense-Thirring precession there (Cui et al 1998c; see also Merloni et al 1999) or trapped-mode disk oscillations (Nowak et al 1997). However, the variations in frequency of the QPO in XTE J1550–564 have cast some doubt on the applicability of such models.

As is well known, strong similarities exist with respect to many spectral and timing phenomena between low-magnetic-field neutron stars and BHCs (e.g. van der Klis 1994a,b; Section 6.3). While the 100–300 Hz oscillations may be related to the kHz QPOs observed in neutron stars (Psaltis et al 1999a; see Section 6.3), there could also be a relation with the recently reported relatively stable QPO peaks near 100 Hz in 4U 0614+09 and 4U 1728–34 (van Straaten et al 2000, di Salvo et al 1999), which are clearly distinct from kHz QPOs. More work toward clearing up the exact phenomenology, along with more observations of black-hole transients leading to more examples of high frequency QPOs, are clearly needed.

6.2 Cen X-3

The detection of QPO features near 330 Hz and 760 Hz in the 4.8 s accreting pulsar Cen X-3 was recently reported by Jernigan et al (2000). This is the first report of millisecond oscillations from a high-magnetic-field ($\sim 10^{12}$ Gauss) neutron star. The QPO features are quite weak. Jernigan et al (2000) carefully discuss the instrumental effects, which are a concern at these low power levels, and interpret their results in terms of the photon bubble model (Section 7.3).

6.3 Low-Frequency Phenomena

Low-frequency ($<$ 100 Hz) QPOs have been studied in accreting neutron stars and black hole candidates since the 1980s, mostly with the EXOSAT and Ginga satellites (see van der Klis 1995a). Two different ones were known in the Z sources: the 6–20 Hz so-called normal and flaring-branch oscillation (NBO; Middleditch & Priedhorsky 1986) and the 15–60 Hz so-called horizontal branch oscillation (HBO; van der Klis et al 1985). The frequency of the HBO turned out to correlate well with those of the kHz QPOs (Figure 15; van der Klis et al 1997b; Wijnands et al 1997b, 1998a,c; Jonker et al 1998, 2000b), and the same is true for the NBO in Sco X-1 (van der Klis et al 1996b). Broad power-spectral bumps and, rarely, low-frequency QPOs were known in atoll sources as well (Lewin et al 1987, Stella et al 1987b, Hasinger & van der Klis 1989, Dotani et al 1989, Makishima et al 1989, Yoshida et al 1993). With RXTE, QPOs similar to HBO are often seen (Strohmayer et al 1996b, Wijnands & van der Klis 1997, Wijnands et al 1998b, Homan et al 1998a) and their frequencies also correlate well to kHz QPO frequency (Figure 15; Stella & Vietri 1998, Ford & van der Klis 1998, Markwardt et al 1999a, van Straaten et al 2000, di Salvo et al 1999, Boirin et al 1999). It is not certain yet whether these atoll QPOs are physically the same, as HBO in Z sources, but this seems likely.

It is possible that these correlations arise just because all QPO phenomena depend on a common parameter (e.g. inferred $\dot{M}$, Section 5.4), but Stella & Vietri (1998) proposed that their origin is a physical dependence of the frequencies on one another (Section 7.2). In their relativistic precession model the HBO and the similar-frequency QPOs in the atoll sources *are* the same phenomenon, and their frequency ν_{LF} is predicted to be proportional to ν_2^2. This is indeed approximately true in all Z sources except Sco X-1 (Psaltis et al 1999b), as well as in the atoll sources (references above). Psaltis et al (1999b) argue that a combination of the sonic-point and magnetospheric beat-frequency models can explain these correlations as well (Section 7.1).

Additional intriguing correlations exist between kHz QPOs and low-frequency phenomena that may link neutron stars and BHCs. It is useful to first examine a correlation between two low-frequency phenomena. At low $\dot{M}$, BHCs and atoll sources (van der Klis 1994a and references therein), the millisecond pulsar SAX J1808.4-3658 (Section 3; Wijnands & van der Klis 1998c), and perhaps even Z sources have very similar power spectra (Figure 14; Wijnands & van der Klis 1999a), with a broad-noise component that shows a break at low frequency and often a QPO-like feature above the break. Break and QPO frequency both vary in excellent correlation (Figure 15, *top*), and in a similar way in neutron stars and BHCs. This suggests that (with the possible exception of the Z sources, which are slightly off the main relation) these two phenomena are the same in neutron stars and black holes. This would exclude spin-orbit beat-frequency models and any other models requiring a material surface, an event horizon, a magnetic field, or their absence, and would essentially imply that the phenomena are generated in the accretion disk around *any* low-magnetic field compact object.

In combination with these low-frequency correlations, the good correlations between kHz QPOs and low-frequency phenomena in Z and atoll sources then suggest that kHz QPOs might also fit in with schemes linking neutron stars and BHCs (Section 6.1). However, no twin kHz QPOs have been reported from BHCs. Psaltis et al (1999a) pointed out that many Z and atoll sources, the peculiar source Cir X-1, and a few low-luminosity neutron stars and BHCs sometimes show two QPO or broad-noise phenomena whose centroid frequencies, when plotted versus each other, seem to line up (Figure 15, *bottom*). This suggests that perhaps "kHz QPOs" *do* occur in BHCs, but as features at frequencies below 50 Hz. These features have very low Q, and although the data are suggestive, they are not conclusive. The implication would be that the lower kHz QPO peak (whose frequency is the one that lines up with those seen in the BHCs) is not unique to neutron stars, but a feature of disk accretion not related to neutron star spin. The coincidence of kHz QPO frequencies with burst oscillation frequencies (Sections 4 and 5.1) would then require some other explanation. Orbital motion in the disk would remain an attractive interpretation for some of the observed frequencies. Stella et al (1999) showed that for particular choices of neutron star and black hole angular momenta, their relativistic precession model can fit these data. The phenomenology is quite

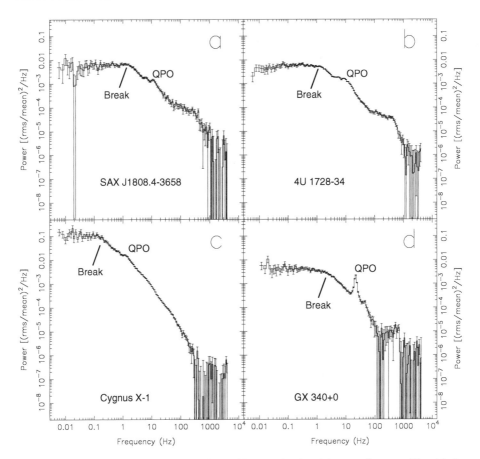

Figure 14 Broad-band power spectra of the millisecond pulsar (*a*), an atoll source (*b*), a black-hole candidate (*c*), and a Z source (*d*) (Wijnands & van der Klis 1999a).

complex; in particular, no way has been found yet to combine the Wijnands & van der Klis (1999a) work with the Psaltis et al (1999a) results in a way that works across all source types. New low-frequency phenomena are still being discovered as well (di Salvo et al 1999, Jonker et al 2000b).

7.   KILOHERTZ QPO MODELS

The possibility of deriving conclusions of a fundamental nature has led to a relatively large number of models for kHz QPOs. Most (but not all) of these involve orbital motion around the neutron star. It is beyond the scope of the present work to provide an in-depth discussion of each model. Instead, I point out some of the main issues and provide references to the literature.

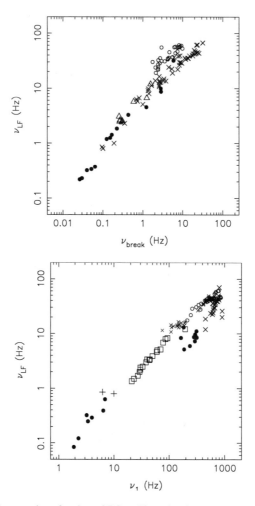

Figure 15 The frequencies of various QPO and broad-noise components seen in accreting neutron stars and black holes plotted versus each other suggest that components similar to Z-source HBO (ν_{LF}) and the lower kHz peak (ν_1) occur in all these sources and span a wide range in frequency. *Top*: ν_{LF} versus noise break frequency (after Wijnands & van der Klis 1999a); *bottom*: ν_{LF} versus ν_1 (after Psaltis et al 1999a). *Filled circles* represent black hole candidates, *open circles* represent Z sources, *crosses* represent atoll sources (the smaller crosses in the *bottom* frame are data of 4U 1728–34 where ν_1 was obtained from $\nu_1 = \nu_2 - 363$ Hz), *triangles* represent the millisecond pulsar SAX J1808.4–3658, *pluses* represent faint burst sources, and *squares* (from Shirey et al 1996) represent Cir X-1.

In early works, the magnetospheric beat-frequency model was implied when beat-frequency models were mentioned (e.g. van der Klis et al 1996b, Strohmayer et al 1996c), and this model has continued to be applied to kHz QPOs (Ford et al 1997b, Cui et al 1998a, Cui 2000, Campana 2000). Most prominent recently have been the sonic point beat-frequency model of Miller et al (1996, 1998a) and the relativistic precession model of Stella & Vietri (1998, 1999); however, the photon bubble model (Klein et al 1996b) and the disk transition layer models (Titarchuk et al 1998, 1999) have also been strongly argued for. Additional disk models have been proposed as well (Section 7.4). Neutron star oscillations have been considered (Strohmayer et al 1996c, Bildsten et al 1998, Bildsten & Cumming 1998), but probably cannot produce the required combination of high frequencies and rapid changes in frequency.

Most models have evolved in response to new observational results. The sonic point model was modified to accommodate the observed deviations from a pure beat-frequency model (Section 5.5; Lamb & Miller 1999, Miller 1999c); the relativistic precession model initially explained the lower kHz peak as a spin-orbit beat frequency (Stella & Vietri 1998) and only later by apsidal motion (Stella & Vietri 1999); the photon bubble model is based on numerical simulations that in time have become more similar to what is observed (R Klein, private communication); also, the disk transition layer models have experienced considerable evolution (e.g. Osherovich & Titarchuk 1999a).

The relativistic precession model makes the strongest predictions with respect to observable quantities, and hence allows the most direct tests, but the near-commensurability of kHz QPO and burst oscillation frequencies (Section 5.5) is unexplained in that model. The sonic-point model provides specific mechanisms to modulate the X-rays and to make the frequency vary with mass-accretion rate. Other models usually discuss at least one of these issues only generically (usually in terms of self-luminous and/or obscuring blobs, and arbitrary preferred radii in the accretion disk).

7.1   The Sonic-Point Beat-Frequency Model

Beat-frequency models involve orbital motion at some preferred radius in the disk (Section 5.1). A beat-frequency model that uses the magnetospheric radius r_M was proposed by Alpar & Shaham (1985) to explain the HBO in Z sources (Section 6.3; see also Lamb et al 1985). In the Z sources HBO and kHz QPOs have been seen simultaneously, so at least one additional model is required.

Miller et al (1996, 1998a) suggest continued use of the magnetospheric model for the HBO (see also Psaltis et al 1999b), and for the kHz QPOs they propose the sonic-point beat-frequency model. In this model the preferred radius is the sonic radius r_{sonic}, where the radial inflow velocity becomes supersonic. This radius tends to be near r_{ISCO} (Section 5.1) but radiative stresses change its location, as required by the observation that the kHz QPO frequencies vary. Comparing the HBO and kHz QPO frequencies, clearly $r_{sonic} \ll r_M$, so part of the accreting matter must remain in near-Keplerian orbits well within r_M.

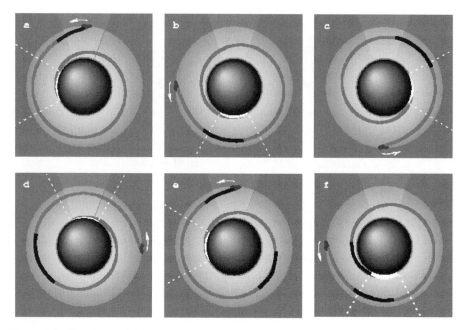

Figure 16 The clump with its spiral flow, the emission from the flow's footpoint (*dashed lines*), and the clump's interaction with the pulsar beam (*lighter shading*) in the Miller et al (1998a) model.

At r_{sonic} orbiting clumps form whose matter gradually accretes onto the neutron star following a fixed, spiral-shaped trajectory in the frame corotating with their orbital motion (Figure 16). At the "footpoint" of a clump's spiral flow, the matter hits the surface and emission is enhanced. The footpoint travels around the neutron star at the clump's orbital angular velocity, so the observer sees a hot spot moving around the surface with the Keplerian frequency at r_{sonic}. This produces the upper kHz peak at ν_2. The high Q of the QPO implies that all clumps are near one precise radius and live for several 0.01 to 0.1 s, and allows for relatively little fluctuation in the shape of the spiral flow. The beat frequency at ν_1 occurs because a beam of X-rays generated by accretion onto the magnetic poles sweeps around at the neutron star spin frequency ν_s and hence irradiates the clumps at r_{sonic} once per beat period, which modulates, at the beat frequency, the rate at which the clumps provide matter to their spiral flows, and consequently the emission from the footpoints.

So, the model predicts $\Delta\nu = \nu_2 - \nu_1$ to be constant at ν_s, contrary to observations (Section 5.5). However, if the clumps' orbits themselves gradually spiral down, then the observed beat frequency will be higher than the actual beat frequency at which beam and clumps interact, because then during the clumps' lifetime the travel time of matter from clump to surface gradually diminishes. This puts the lower kHz peak closer to the upper one, and thus decreases $\Delta\nu$.

This decrease will be larger when at higher L_x due to stronger radiation drag, the spiraling-down becomes more pronounced, as observed (Lamb & Miller 1999, Miller 1999c). Because the exact way in which this process affects the relation between the frequencies is hard to predict, this complication makes testing the model more difficult. A remaining test is that a number of specific additional frequencies is predicted to arise from the beat-frequency interaction (Miller et al 1998a); these are different from additional frequencies in, for example, the relativistic precession model.

7.2 The Relativistic Precession Model

Inclined eccentric free-particle orbits around a spinning neutron star show both nodal precession (a wobble of the orbital plane) caused by relativistic frame dragging (Thirring & Lense 1918) and relativistic periastron precession similar to Mercury's (Einstein 1915). The relativistic precession model (Stella & Vietri 1998, 1999) identifies v_2 with the frequency of an orbit in the disk, and identifies v_1 and the frequency v_{LF} of one of the observed low-frequency (10–100 Hz) QPO peaks (Section 6.3) with, respectively, periastron precession and nodal precession of this orbit.

To lowest order, the relativistic nodal precession is $v_{nod} = 8\pi^2 I v_2^2 v_s / c^2 M$, and the relativistic periastron precession causes the kHz peak separation to vary as $\Delta v = v_2 (1 - 6GM/rc^2)^{1/2}$, where I is the star's moment of inertia and r the orbital radius (Stella & Vietri 1998, 1999; see also Marković & Lamb 1998). Stellar oblateness affects both precession rates and must be corrected for (Morsink & Stella 1999, Stella et al 1999). For acceptable neutron star parameters, there is an approximate match (e.g. Figure 17) with the observed v_1, v_2 and v_{LF} relations if v_{LF} is *twice* (or perhaps occasionally four times) the nodal precession frequency, which could in principle arise from a warped disk geometry (Morsink & Stella 1999). In this model the neutron-star spin frequencies do not cluster in the 240–360 Hz range (Section 5.6), and Δv and v_{burst} are not expected to be equal as in beat-frequency interpretations. A clear prediction is that Δv should decrease not only when v_2 increases (as observed) but also when it sufficiently decreases (Figure 17).

For a precise match between model and observations, additional free parameters are required. Stella & Vietri (1999) propose that the orbital eccentricity e systematically varies with orbital frequency. A critical discussion of the degree to which the precession model and the beat-frequency model can each fit the data can be found in Psaltis et al (1999b). Vietri & Stella (1998) and Armitage & Natarajan (1999) have performed calculations relevant to the problem of sustaining the tilted orbits required for Lense-Thirring precession in a viscous disk, where the Bardeen-Petterson effect (1975) drives the matter to the orbital plane. Karas (1999a,b) calculated frequencies and light curves produced by clumps orbiting the neutron star in orbits similar to those discussed here. Miller (1999d) calculated the effects of radiation forces on Lense-Thirring precession. Kalogera & Psaltis (2000)

Figure 17 Predicted relations between ν_2 and $\Delta\nu$ (*top*) and between ν_1 and ν_{LF} as well as ν_2 (*bottom*) in the relativistic precession model compared with observed values (Stella & Vietri 1999, Stella et al 1999). See Section 6.3 for a discussion of the data in the *bottom* frame.

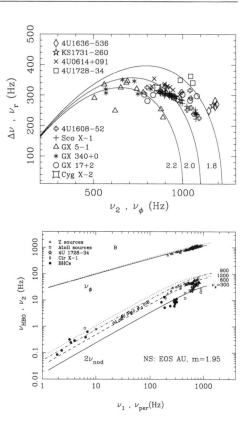

explored how the Lense-Thirring precession interpretation constrains neutron star structure.

Relatively high neutron star masses ($1.8–2\,M_\odot$), relatively stiff equations of state, and neutron star spin frequencies in the 300–900 Hz range follow from this model. Because it requires no neutron star and only a relativistic accretion disk to work, the model can also be applied to black holes. Stella et al (1999) propose that this model explains the frequency correlations discussed in Section 6.3 (Figure 17).

The idea that the three most prominent frequencies observed are in fact the three main general-relativistic frequencies characterizing a free-particle orbit is fascinating. However, questions remain. How precessing and eccentric orbits can survive in a disk is not a priori clear (see Psaltis & Norman 2000 for a possible way to obtain these frequencies from a disk). How the flux is modulated at the predicted frequencies, why the basic orbital frequency ν_2 varies with luminosity, and how the burst oscillations fit in are all open questions. With respect to the burst oscillations, the model requires explanations other than neutron star spin (Section 4), and these are being explored (Stella 1999, Kluźniak 1999, Psaltis & Norman 2000).

7.3   Photon Bubble Model

A model based on numerical radiation hydrodynamics was proposed by Klein et al (1996b) for the kHz QPOs in Sco X-1. In this model, accretion takes place by way of a magnetic funnel within which accretion is super-Eddington, so that photon bubbles form, which rise up by buoyancy, and burst at the top in quasi-periodic sequence. In some of the simulations one or two strong QPO peaks are found whose frequencies increase with accretion rate, as observed (R Klein, private communication). The model stands out by not requiring rotational phenomena to explain the QPOs, and does not naturally produce beat frequencies. In recent work, attention with respect to this model has shifted to the classical accreting pulsars for which it was originally conceived (Klein et al 1996a; Section 6.2).

7.4   Disk Mode Models

In several models the observed frequencies are identified with oscillation modes of an accretion disk. From an empirical point of view, these models fall into two classes. Some can be seen as "implementations"of a beat-frequency (e.g. Alpar & Yilmaz 1997) or precession (Psaltis & Norman 2000) model, in that they provide ways in accretion disk physics to produce signals at the frequencies occurring in those models. Others produce new frequencies. In all disk models, one of the two kHz QPOs is a Keplerian orbital frequency at some radius in the disk.

The disk transition layer models (Titarchuk & Muslimov 1997; Titarchuk et al 1998, 1999; Osherovich & Titarchuk 1999a,b; Titarchuk & Osherovich 1999) have evolved into a description (the "two-oscillator model") where ν_1 is identified with the Keplerian frequency at the outer edge of a viscous transition layer between Keplerian disk and neutron-star surface. Oscillations in this layer occur at two low frequencies, producing the noise break and a low-frequency QPO (Section 6.3). Additionally, blobs described as being thrown out of this layer into a magnetosphere oscillate both radially and perpendicular to the disk, producing two harmonics of another low-frequency QPO (in the Z sources this is the HBO) as well as the upper kHz peak. Altogether, this description provides six frequencies that can all fit observed frequencies. Further work on disk oscillations was performed by Lai (1998, 1999), Lai et al (1999), and Ghosh (1998; see also Moderski & Czerny 1999).

8.   FINAL REMARKS

RXTE has opened a window that allows us to see down to the very bottoms of the potential wells of some neutron stars, and perhaps almost to the horizons of some black holes. Three new types of millisecond phenomenon have been found, and their interpretation relies explicitly on our description of strong-field gravity and neutron-star structure. In order to take full advantage of this, we will need to observe the new phenomena with larger instruments (in the 10 m^2 class). This will allow us to follow directly the motion of clumps of matter orbiting in strong gravity and of hot spots corotating on neutron star surfaces, and thereby to map out curved space-time near accreting compact objects, and directly measure the compactness of neutron stars and the spin of black holes.

ACKNOWLEDGMENTS

It is a pleasure to acknowledge the help of many colleagues who either made data available before publication, sent originals of figures, read versions of the manuscript, or provided insightful discussion: Didier Barret, Lars Bildsten, Deepto Chakrabarty, Wei Cui, Eric Ford, Jeroen Homan, Peter Jonker, Richard Klein, Fred Lamb, Phil Kaaret, Craig Markwardt, Mariano Méndez, Cole Miller, Mike Nowak, Dimitrios Psaltis, Luigi Stella, Steve van Straaten, Tod Strohmayer, Rudy Wijnands and Will Zhang. This work was supported in part by the Netherlands Organization for Scientific Research (NWO) and the Netherlands Research School for Astronomy (NOVA).

Visit the Annual Reviews home page at www.AnnualReviews.org

LITERATURE CITED

Akmal A, Pandharipande VR, Ravenhall DG. 1998. *Phys. Rev. C* 58:1804–28

Alpar MA, Cheng AF, Ruderman MA, Shaham J. 1982. *Nature* 300:728–30

Alpar MA, Shaham J. 1985. *Nature* 316:239–41

Alpar MA, Yilmaz A. 1997. *New Astron.* 2:225–38

Andersson N, Kokkotas KD, Stergioulas N. 1999. *Ap. J.* 516:307–14

Armitage PJ, Natarajan P. 1999. *Ap. J.* 525:909–14

Bardeen JM, Petterson JA. 1975. *Ap. J.* 195:L65–67

Barret D, Olive JF, Boirin L, Done C, Skinner GK, Grindlay JE. 2000. *Ap. J.* 533:329–51

Barret D, Boirin L, Olive JF, Grindlay JE, Bloser PF, 1998. Proc. 3d Integral Workshop. *Astrophys. Lett. And Communications.* In press astro-ph/9811236

Barret D, Olive JF, Boirin L, Grindlay JE, Bloser PF, et al. 1997. *IAU Circ. No.* 6793

Barret D, Vedrenne G. 1994. *Ap. J. Suppl.* 92:505–10

Berger M, van der Klis M, van Paradijs J,

Lewin WHG, Lamb F, et al. 1996. *Ap. J. Lett.* 469:L13–16

Bhattacharya D, van den Heuvel EPJ. 1991. *Phys. Rep.* 203:1–124

Biehle GT, Blandford RD. 1993. *Ap. J.* 411:302–12

Bildsten L. 1995. *Ap. J.* 438:852–75

Bildsten L. 1998a. *Ap. J. Lett.* 501:L89–93

Bildsten L. 1998b. *The many faces of neutron stars. NATO ASI, Lipari,* C515:419–49. Dordrecht: Kluwer pp. 608

Bildsten L, Cumming A. 1998. *Ap. J.* 506:842–62

Bildsten L, Cumming A, Ushomirsky G, Cutler C. 1998. *ASP Conf. Proc.* 135:437–41

Bloser PF, Grindlay JE, Kaaret P, Smale AP, Barret D. 1999. *Ap. J.* Submitted

Boirin L, Barret D, Olive JF, Grindlay JE, Bloser PF. 1999. *Astron. Astrophys.* Submitted

Bradt HV, Rothschild RE, Swank JH. 1993. *Astron. Astrophys. Suppl.* 97:355

Bulik T, Godek-Rosińska D, Kluźniak W. 1999. *Astron. Astrophys.* 344:L71–74

Burderi L, King AR. 1998. *Ap. J.* 505:L135–37

Campana S. 2000. *Ap. J.* in press. astro-ph/0003210

Chakrabarty D, Morgan EH. 1998a. *IAU Circ. No.* 6877

Chakrabarty D, Morgan EH. 1998b. *Nature* 394:346–48

Cook GB, Shapiro SL, Teukolsky SA. 1994. *Ap. J.* 424:823–45

Cui W. 1999. *Ap. J. Lett.* 524:L59–62

Cui W. 2000. *Ap. J.* submitted. astro-ph/0003243

Cui W, Barret D, Zhang SN, Chen W, Boirin L, Swank J. 1998a. *Ap. J. Lett.* 502:L49–53

Cui W, Morgan EH, Titarchuk LG. 1998b. *Ap. J. Lett.* 504:L27–30

Cui W, Zhang SN, Chen W. 1998c. *Ap. J.* 492:L53–57

Datta B, Thampan AV, Bombaci I. 1998. *Astron. Astrophys.* 334:943–52

di Salvo T, Méndez M, van der Klis M, Ford E, Robba NR. 1999. *Ap. J.* Submitted

Dotani T, Mitsuda K, Makishima K, Jones MH. 1989. *Publ. ASJ* 41:577–89

Einstein A. 1915. *Preuss. Akad. Wiss. Berlin, Sitzber.* 47:831–39

Ford EC. 1999. *Ap. J. Lett.* 519:L73–75

Ford EC, Kaaret P, Chen K, Tavani M, Barret D, et al. 1997b. *Ap. J. Lett.* 486:L47–50

Ford E, Kaaret P, Tavani M, Barret D, Bloser P, et al. 1997a. *Ap. J. Lett.* 475:L123–26

Ford E, Kaaret P, Tavani M, Harmon BA, Zhang SN, et al. 1996. *IAU Circ. No.* 6426

Ford EC, van der Klis M. 1998. *Ap. J. Lett.* 506:L39–42

Ford EC, van der Klis M, Kaaret P. 1998a. *Ap. J. Lett.* 498:L41–44

Ford EC, van der Klis M, Méndez M, Wijnands R, Homan J, et al. 2000. *Ap. J.* in press. astro-ph/0002074

Ford EC, van der Klis M, van Paradijs J, Méndez M, Wijnands R, Kaaret P. 1998b. *Ap. J. Lett.* 508:L155–58

Fortner B, Lamb FK, Miller GS. 1989. *Nature* 342:775–77

Fox DW, Lewin WHG, et al. 1999. *IAU Circ. No.* 7081

Frank J, King A, Raine D. 1992. *Accretion Power in Astrophysics* pp. 294, Cambridge, UK: Cambridge Univ. Press.

Gaensler BM, Stappers BW, Getts TJ. 1999. *Ap. J.* 522:L117–19

Ghosh P. 1998. *Ap. J.* 506:L109–12

Ghosh P, Lamb FK. 1992. *X-ray binaries and recycled pulsars, NATO ASI, Santa Barbara.* C377:487–510. Dordrecht: Kluwer. pp. 566

Giles AB, Hill KM, Greenhill JG. 1999. *MNRAS* 304:47–51

Gilfanov M, Revnivtsev M, Sunyaev R, Churazov E. 1998. *Astron. Astrophys.* 338:L83–86

Hasinger G, van der Klis M. 1989. *Astron. Astrophys.* 225:79–96

Hasinger G, van der Klis M, Ebisawa K, Dotani T, Mitsuda K. 1990. *Astron. Astrophys.* 235:131–46

Heindl WA, Smith DM. 1998. *Ap. J.* 506:L35–38

Heiselberg H, Hjorth-Jensen M. 1999. *Ap. J. Lett.* 525:L45–48

Homan J, van der Klis M. 2000. *Ap. J.* in press. astro-ph/0003241

Homan J, Méndez M, Wijnands R, van der Klis M, van Paradijs J. 1999a. *Ap. J. Lett.* 513:L119–20

Homan J, van der Klis M, van Paradijs J, Méndez M. 1998b. *IAU Circ. No.* 6971

Homan J, van der Klis M, Wijnands R, Vaughan B, Kuulkers K. 1998a. *Ap. J. Lett.* 499:L41–44

Homan J, Wijnands R, van der Klis M. 1999b. *IAU Circ. No.* 7121

in 't Zand JJM, Heise J, Muller JM, Bazzano A, Cocchi M, et al. 1998. *Astron. Astrophys.* 331:L25–28

Jernigan JG, Klein RI, Arons J. 2000. *Ap. J.* 530:875–89

Jongert HC, van der Klis M. 1996. *Astron. Astrophys.* 310:474–76

Jonker PG, van der Klis M, Homan J, van Paradijs J, Méndez M, et al. 2000a. *Ap. J.* 531:453–57

Jonker PG, van der Klis M, Wijnands R. 1999. *Ap. J. Lett.* 511:L41–44

Jonker PG, van der Klis M, Wijnands R, Homan

J, van Paradijs J, Méndez M, et al. 2000b. *Ap. J.* in press. astro-ph/0002022

Jonker PG, Wijnands R, van der Klis M, Psaltis D, Kuulkers E, Lamb FK. 1998. *Ap. J.* 499:L191–94

Joss PC. 1978. *Ap. J. Lett.* 225:L123–27

Kaaret P, Ford E, Chen K. 1997. *Ap. J. Lett.* 480:L27–29

Kaaret P, Piraino S, Bloser PF, Ford EC, Grindlay JE, et al. 1999b. *Ap. J. Lett.* 520:L37–40

Kaaret P, Piraino S, Ford EC, Santangelo A. 1999a. *Ap. J. Lett.* 514:L31–33

Kaaret P, Yu W, Ford EC, Zhang SN. 1998. *Ap. J. Lett.* 497:L93–96

Kalogera V, Psaltis D. 2000. *Phys. Rev. D* 61:024009

Karas V. 1999a. *Publ. ASJ* 51:317–20

Karas V. 1999b. *Ap. J.* 526:953–56

Klein RI, Arons J, Jernigan JG, Hsu JJL. 1996a. *Ap. J. Lett.* 457:L85–89

Klein RI, Jernigan JG, Arons J, Morgan EH, Zhang W. 1996b. *Ap. J. Lett.* 469:L119–23

Kluźniak W. 1998. *Ap. J. Lett.* 509:L37–40

Kluźniak W. 1999. Presented at X-Ray Probes of Relativistic Effects near Neutron Stars and Black Holes, Summer Workshop, Aspen

Kluźniak W, Michelson P, Wagoner RV. 1990. *Ap. J.* 358:538–44

Kluźniak W, Wagoner RV. 1985. *Ap. J.* 297:548–54

Kluźniak W, Wilson JR. 1991. *Ap. J. Lett.* 372:L87–90

Konar S, Bhattacharya D. 1999. *MNRAS* 303:588–94

Kuulkers E, van der Klis M. 1995. *Astron. Astrophys.* 303:801–6

Kuulkers E, van der Klis M. 1996. *Astron. Astrophys.* 314:567–75

Kuulkers E, van der Klis M. 1998. *Astron. Astrophys.* 332:845–48

Kuulkers E, van der Klis M, Vaughan BA. 1996. *Astron. Astrophys.* 311:197–210

Lai D. 1998. *Ap. J.* 502:721–29

Lai D. 1999. *Ap. J.* 524:1030–47

Lai D, Lovelace R, Wasserman I. 1999. *Ap. J.* Submitted. astro-ph/9904111

Lamb FK, Miller MC. 1999. *Ap. J.* In preparation

Lamb FK, Shibazaki N, Alpar MA, Shaham J. 1985. *Nature* 317:681–87

Leahy DA, Elsner RF, Weisskopf MC. 1983. *Ap. J.* 272:256–58

Lee HC, Miller GS. 1998. *MNRAS* 299:479–87

Levin Y. 1999. *Ap. J.* 517:328–33

Lewin WHG, Joss PC. 1981. *Space Sci. Rev.* 28:3–87

Lewin WHG, van Paradijs J, Hasinger G, Penninx WH, Langmeier A, et al. 1987. *MNRAS* 226:383–94

Lewin WHG, van Paradijs J, Taam RE. 1995b. See Lewin et al 1995a, pp. 175–232

Lewin WHG, van Paradijs J, van den Heuvel EPJ. 1995a. *X-Ray Binaries.* Cambridge, UK: Cambridge Univ. Press. 662 pp.

Li XD, Bombaci I, Dey M, Dey J, van den Heuvel EPJ. 1999a. astro-ph/9905356 *Phys. Rev. Lett.* 83:3776–79

Li XD, Ray S, Dey J, Dey M, Bombaci I. 1999b. *Ap. J.* 527:L51–54. astro-ph/9908274

Livio M, Bath GT. 1982. *Astron. Astrophys.* 116:286–92

Makishima K, Ishida M, Ohashi T, Dotani T, Inoue H, et al. 1989. *Publ. ASJ* 41:531–55

Marković D, Lamb FK. 1998. *Ap. J.* 507:316–26

Markwardt CB, Lee HC, Swank JH. 1999b. *AAS HEAD* 31:15.01 (Abstr.)

Markwardt CB, Strohmayer TE, Swank JH. 1999a. *Ap. J. Lett.* 512:L125–29

Marshall FE, Markwardt CB. 1999. *IAU Circ. No.* 7103

Mason KO, Middleditch J, Nelson JE, White NE. 1980. *Nature* 287:516–18

Méndez M. 1999. Relativistic astrophysics and cosmology. *Texas Symp., 19th, Paris.* In press. astro-ph/9903469

Méndez M, van der Klis M. 1999. *Ap. J. Lett.* 517:L51–54

Méndez M, van der Klis M, Ford EC, Wijnands R, van Paradijs J. 1999. *Ap. J. Lett.* 511:L49–52

Méndez M, van der Klis M, van Paradijs J. 1998c. *Ap. J. Lett.* 506:L117–19

Méndez M, van der Klis M, van Paradijs J, Lewin WHG, Lamb FK, et al. 1997. *Ap. J. Lett.* 485:L37–40

Méndez M, van der Klis M, van Paradijs J, Lewin WHG, Vaughan BA, et al. 1998a. *Ap. J. Lett.* 494:L65–69

Méndez M, van der Klis M, Wijnands R, Ford EC, van Paradijs J, Vaughan BA. 1998b. *Ap. J. Lett.* 505:L23–26

Mereghetti S, Grindlay JE. 1987. *Ap. J.* 312:727–31

Merloni A, Vietri M, Stella L, Bini D. 1999. *MNRAS* 304:155–59

Middleditch J, Priedhorsky WC. 1996. *Ap. J.* 306:230–37

Miller MC. 1999a. *Ap. J. Lett.* 515:L77–80

Miller MC. 2000. *Ap. J.* 531:458–66

Miller MC. 1999c. Stellar endpoints. *Bologna '99 Conf.* In press

Miller MC. 1999d. *Ap. J.* 520:256–61

Miller MC, Lamb FK. 1998. *Ap. J. Lett.* 499:L37–40

Miller MC, Lamb FK, Cook GB. 1998b. *Ap. J.* 509:793–801

Miller MC, Lamb FK, Psaltis D. 1996. astro-ph/9609157v1

Miller MC, Lamb FK, Psaltis D. 1998a. *Ap. J.* 508:791–830

Moderski R, Czerny B. 1999. *Ap. J.* 513:L123–25

Morgan EH, Remillard RA, Greiner J. 1997. *Ap. J.* 482:993–1010

Morgan EH, Smith DA. 1996. *IAU Circ. No.* 6437

Morsink SM, Stella L. 1999. *Ap. J.* 513:827–44

Muno MP, Fox DW, Morgan EH, Bildsten L. 2000. *Ap. J.* in press. astro-ph/0003229

Murakami T, Inoue H, Makishima K, Hoshi R. 1987. *Publ. ASJ* 39:879–86

Narayan R. 1997. Accretion phenomena and related outflows, *IAU Coll. 163, ASP Conf. Series* 121:75–89

Nowak MA, Vaughan BA, Wilms J, Dove JB, Begelman MC. 1999. *Ap. J.* 510:874–91

Nowak MA, Wagoner RV, Begelman MC, Lehr DE. 1997. *Ap. J. Lett.* 477:L91–94

Osherovich V, Titarchuk L. 1999a. *Ap. J.* 522:L113–16

Osherovich V, Titarchuk L. 1999b. *Ap. J.* 523:L73–76

Paczyński B. 1987. *Nature* 327:303–4

Psaltis D, Belloni T, van der Klis M. 1999a. *Ap. J. Lett.* 520:262–70

Psaltis D, Chakrabarty D. 1999. *Ap. J.* 521:332–40

Psaltis D, Lamb FK. 1998. *Joint European and National Astron. Meet., 4th, Astrophys. Astron. Trans.* 18N3:447–54

Psaltis D, Méndez M, Wijnands R, Homan J, Jonker PG, et al. 1998. *Ap. J. Lett.* 501:L95–99

Psaltis D, Norman C. 2000. *Ap. J.* submitted astro-ph/0001391

Psaltis D, Wijnands R, Homan J, Jonker PG, van der Klis M, et al. 1999b. *Ap. J.* 520:763–75

Radhakrishnan V, Srinivasan G. 1984. Asia-Pacific Regional Meeting of the IAU, 2nd, Jakarta, 1981. p. 423

Reig P, Méndez M, van der Klis M, Ford EC. 2000. *Ap. J.* 530:916–22

Remillard R. 1999. See Miller 1999c, in press

Remillard RA, Morgan EH. 1999. *AAS* 195, 37.02

Remillard R, McClintock JE, Sobczak G, Bailyn C, Orosz JA, et al. 1999c. *Ap. J. Lett.* 517:L127–30

Remillard RA, Morgan EH. 1998. *Nucl. Phys. B* 69/1-3:316–23

Remillard R, Morgan E, Levine A, Muno M, McClintock J, et al. 1999a. *AAS HEAD* 31:28.08 (Abstr.)

Remillard R, Morgan EH, McClintock JE, Bailyn CD, Orosz JA. 1999b. *Ap. J.* 522:397–412

Sadeh D, Byram ET, Chubb TA, Friedman H, Hedler RL, et al. 1982. *Ap. J.* 257:214–22

Sadeh D, Livio M. 1982a. *Ap. J.* 258:770–75

Sadeh D, Livio M. 1982b. *Ap. J.* 263:823–27

Schaab C, Weigel MK. 1999. *MNRAS* 308:718–30

Schoelkopf RJ, Kelley RL. 1991. *Ap. J.* 375:696–700

Shara MM. 1982. *Ap. J.* 261:649–60

Shirey RE, Bradt HV, Levine AM, Morgan EH. 1996. *Ap. J.* 469:L21–24

Shvartsman VF. 1971. *Soviet Astron.* 15(3):377–84

Skinner GK, Bedford DK, Elsner RF, Leahy D, Weisskopf MC, Grindlay J. 1982. *Nature* 297:568–70

Smale AP, Zhang W, White NE. 1996. *IAU Circ. No.* 6507

Smale AP, Zhang W, White NE. 1997. *Ap. J. Lett.* 483:L119–22

Smith DA, Morgan EH, Bradt H. 1997. *Ap. J. Lett.* 479:L137–40

Stella L. 1999. See Kluźniak 1999

Stella L, Priedhorsky W, White NE. 1987a. *Ap. J. Lett.* 312:L17–21

Stella L, Vietri M. 1998. *Ap. J. Lett.* 492:L59–62

Stella L, Vietri M. 1999. *Phys. Rev. Lett.* 82:17–20

Stella L, Vietri M, Morskink SM. 1999. *Ap. J.* 524:L63–66

Stella L, White NE, Priedhorsky W. 1987b. *Ap. J. Lett.* 315:L49–53

Strohmayer T. 1998. *Some Like It Hot, AIP Conf. Proc.* 431:397–400

Strohmayer T, Smale A, Day C, Swank J, Titarchuk L, Lee U. 1996b. *IAU Circ. No.* 6387

Strohmayer T, Zhang W, Swank J. 1996a. *IAU Circ. No.* 6320

Strohmayer T, Zhang W, Swank JH. 1997b. *Ap. J.* 487:L77–80

Strohmayer TE. 1999. *Ap. J.* 523:L51–55

Strohmayer TE, Jahoda K, Giles AB, Lee U. 1997a. *Ap. J.* 486:355–62

Strohmayer TE, Lee U, Jahoda K. 1996d. *IAU Circ. No.* 6484

Strohmayer TE, Markwardt CB. 1999. *Ap. J. Lett.* 516:L81–85

Strohmayer TE, Swank JH, Zhang W. 1998c. *Nucl. Phys.* B 69/1-3:129–34

Strohmayer TE, Zhang W, Swank JH, Lapidus I. 1998b. *Ap. J.* 503:L147–50

Strohmayer TE, Zhang W, Swank JH, Smale A, Titarchuk L, Day C. 1996c. *Ap. J. Lett.* 469:L9–12

Strohmayer TE, Zhang W, Swank JH, White NE, Lapidus I. 1998a. *Ap. J.* 498:L135–39

Sunyaev RA. 1973. *Soviet Astron.* 16(6):941–44

Thampan AV, Bhattacharya D, Datta B. 1999. *MNRAS* 302:L69–73

Thirring H, Lense J. 1918. *Phys. Z.* 19:156–63

Titarchuk L, Lapidus I, Muslimov A. 1998. *Ap. J.* 499:315–28

Titarchuk L, Muslimov A. 1997. *Astron. Astrophys.* 323:L5–8

Titarchuk L, Osherovich V. 1999. *Ap. J. Lett.* 518:L95–98

Titarchuk L, Osherovich V, Kuznetsov S. 1999. *Ap. J.* 525:L129–32

Tomsick JA, Halpern JP, Kemp J, Kaaret P. 1999. *Ap. J.* 521:341–50

van der Klis M. 1989a. *Annu. Rev. Astron. Astrophys.* 27:517–53

van der Klis M. 1989b. Timing neutron stars, *NATO ASI, Cesme.* C262:27–69. Dordrecht: Kluwer

van der Klis M. 1994a. *Ap. J. Suppl.* 92:511–19

van der Klis M. 1994b. *Astron. Astrophys.* 283:469–74

van der Klis M. 1995a. See Lewin et al 1995a, pp. 252–307

van der Klis M. 1995b. The lives of the neutron stars, *NATO ASI, Kemer.* C450:301–30. Dordrecht: Kluwer

van der Klis M. 1997. Astronomical time series. *Wise Obs. Anniv. Symp., 25th, Tel Aviv, Kluwer Astrophys. Space Sci. Libr.* 218:121–32. Dordrecht: Kluwer

van der Klis M. 1998. See Bildsten 1998b, C515:33768. Dordrecht: Kluwer

van der Klis M, Hasinger G, Damen E, Penninx W, van Paradijs J, Lewin WHG. 1990. *Ap. J.* 360:L19–22

van der Klis M, Hasinger G, Stella L, Langmeier A, van Paradijs J, Lewin WHG. 1987. *Ap. J.* 319:L13–18

van der Klis M, Homan J, Wijnands R, Kuulkers E, Lamb FK, et al. 1997a. *IAU Circ. No.* 6565

van der Klis M, Jansen F, van Paradijs J, Lewin

WHG, van den Heuvel EPJ. 1985. *Nature* 316:225–30

van der Klis M, Swank J, Zhang W, Jahoda K, Morgan E, et al. 1996a. *IAU Circ. No.* 6319

van der Klis M, Swank JH, Zhang W, Jahoda K, Morgan EH, et al. 1996b. *Ap. J. Lett.* 469:L1–4

van der Klis M, van Paradijs J, Lewin WHG, Lamb FK, Vaughan B, et al. 1996d. *IAU Circ. No.* 6428

van der Klis M, Wijnands R, Chen W, Lamb FK, Psaltis D, et al. 1996c. *IAU Circ. No.* 6424

van der Klis M, Wijnands RAD, Horne K, Chen W. 1997b. *Ap. J. Lett.* 481:L97–100

van der Klis M, Wijnands R, Kuulkers E, Lamb FK, Psaltis D, et al. 1996e. *IAU Circ. No.* 6511

van Paradijs J, Zhang W, Marshall F, Swank JH, Augusteijn T, et al. 1996. *IAU Circ. No.* 6336

van Straaten S, Ford EC, van der Klis M, Méndez M, Kaaret P. 2000. *Ap. J.* in press. astro-ph/0001480

Vaughan B, van der Klis M, Lewin WHG, Wijers RAMJ, van Paradijs J, et al. 1994a. *Ap. J.* 421:738–52

Vaughan BA, van der Klis M, Méndez M, van Paradijs J, Wijnands RAD, et al. 1997. *Ap. J. Lett.* 483:L115–18

Vaughan BA, van der Klis M, Méndez M, van Paradijs J, Wijnands RAD, et al. 1998. *Ap. J. Lett.* 509:L145–45

Vaughan BA, van der Klis M, Wood KS, Norris JP, Hertz P, et al. 1994b. *Ap. J.* 435:362–69

Vietri M, Stella L. 1998. *Ap. J.* 503:350–60

White NE, Zhang W. 1997. *Ap. J. Lett.* 490:L87–90

Wijnands R, Homan J, van der Klis M, Kuulkers E, van Paradijs J, et al. 1998a. *Ap. J. Lett.* 493:L87–90

Wijnands R, Homan J, van der Klis M, Méndez M, Kuulkers E, et al. 1997b. *Ap. J. Lett.* 490:L157–60

Wijnands R, Méndez M, van der Klis M, Psaltis D, Kuulkers E, Lamb FK. 1998c. *Ap. J. Lett.* 504:L35–38

Wijnands R, van der Klis M. 1998a. *IAU Circ. No.* 6876

Wijnands R, van der Klis M. 1998b. *Nature* 394:344–46

Wijnands R, van der Klis M. 1998c. *Ap. J. Lett.* 507:L63–66

Wijnands R, van der Klis M. 1999a. *Ap. J. Lett.* 514:939–44

Wijnands R, van der Klis M. 1999b. *Ap. J.* 522:965–72

Wijnands R, van der Klis M. 1999c. *Astron. Astrophys.* 345:L35–38

Wijnands R, van der Klis M, van Paradijs J. 1998d. The hot universe, *IAU Symp.188,* pp. 370–71

Wijnands RAD, van der Klis M. 1997. *Ap. J. Lett.* 482:L65–68

Wijnands RAD, van der Klis M, Méndez M, van Paradijs J, Lewin WHG, et al. 1998b. *Ap. J. Lett.* 495:L39–42

Wijnands RAD, van der Klis M, Psaltis D, Lamb FK, Kuulkers E, et al. 1996b. *Ap. J. Lett.* 469:L5–8

Wijnands RAD, van der Klis M, van Paradijs J, Lewin WHG, Lamb FK, et al. 1996a. *IAU Circ. No.* 6447

Wijnands RAD, van der Klis M, van Paradijs J, Lewin WHG, Lamb FK, et al. 1997a. *Ap. J. Lett.* 479:L141–44

Wood KS, Hertz P, Norris JP, Vaughan BA, Michelson PF, et al. 1991. *Ap. J.* 379:295–309

Yoshida K, Mitsuda K, Ebisawa K, Ueda Y, Fujimoto R, et al. 1993. *Publ. ASJ* 45:605–16

Yu W, Li TP, Zhang W, Zhang SN. 1999. *Ap. J. Lett.* 512:L35–38

Yu W, Zhang SN, Harmon BA, Paciesas WS, Robinson CR, et al. 1997. *Ap. J.* 490:L153–56

Zhang, W. 1997. Presented at AAS Meeting, 190th, Winston-Salem

Zhang W, Jahoda K, Kelley RL, Strohmayer TE, Swank JH, Zhang SN. 1998c. *Ap. J.* 495:L9–12

Zhang W, Lapidus I, Swank JH, White NE, Titarchuk L. 1997a. *IAU Circ. No.* 6541

Zhang W, Lapidus I, White NE, Titarchuk L. 1996a. *Ap. J. Lett.* 469:L17–19

Zhang W, Lapidus I, White NE, Titarchuk L. 1996b. *Ap. J. Lett.* 473:L135

Zhang W, Smale AP, Strohmayer TE, Swank JH. 1998b. *Ap. J.* 500:L171–74

Zhang W, Strohmayer TE, Swank JH. 1997b. *Ap. J. Lett.* 482:L167–70

Zhang W, Strohmayer TE, Swank JH. 1998a. *Ap. J.* 500:L167–69

Annu. Rev. Astron. Astrophys. 2000. 38:761–814

EXTRAGALACTIC RESULTS FROM THE INFRARED SPACE OBSERVATORY

Reinhard Genzel[1, 2] and Catherine J. Cesarsky[3, 4]

[1]*Max-Planck Institut für extraterrestrische Physik (MPE), Garching, FRG;*
e-mail: genzel@mpe-garching.mpg.de
[2]*Department of Physics, University of California, Berkeley*
[3]*Service d'Astrophysique, DAPNIA/DSM, CEA Saclay, France*
[4]*European Southern Observatory, Garching, FRG; e-mail: ccesarsk@eso.org*

Key Words galaxies: evolution, Active Galactic Nuclei, infrared: photometry, infrared: spectroscopy, star formation

■ **Abstract** More than a decade ago the IRAS satellite opened the realm of external galaxies for studies in the 10 to 100 μm band and discovered emission from tens of thousands of normal and active galaxies. With the 1995–1998 mission of the Infrared Space Observatory[1], the next major steps in extragalactic infrared astronomy became possible: detailed imaging, spectroscopy, and spectrophotometry of many galaxies detected by IRAS, as well as deep surveys in the mid- and far-IR. The spectroscopic data reveal a wealth of detail about the nature of the energy source(s) and about the physical conditions in galaxies. ISO's surveys for the first time explore the infrared emission of distant, high-redshift galaxies. ISO's main theme in extragalactic astronomy is the role of star formation in the activity and evolution of galaxies.

1. INTRODUCTION: From IRAS to ISO

In 1983 the first cryogenic infrared astronomy satellite, IRAS[2], surveyed 96% of the sky in four broad-band filters at 12, 25, 60, and 100 μm, to limits ≤ 1 Jy (Neugebauer et al 1984, see reviews of Beichman 1987, Soifer et al 1987). IRAS detected infrared (IR) emission from about 25,000 galaxies, primarily from

---

[1]The Infrared Space Observatory (ISO) is a project of the European Space Agency (ESA). The satellite and its instruments were funded by the ESA member states (especially the PI countries: France, Germany, the Netherlands, and the United Kingdom) and with the participation of the Japanese Space Agency ISAS and NASA.
[2]The Infra-Red Astronomical Satellite was developed and operated by the US National Aeronautics and Space Administration (NASA), the Netherlands Agency for Aerospace Programs (NIVP), and the United Kingdom Science and Engineering Research Council (SERC).

0066-4146/00/0915-0761$14.00

spirals, but also from quasars (QSOs) (Neugebauer et al 1986, Sanders et al 1989), Seyfert galaxies (de Grijp et al 1985) and early type galaxies (Knapp et al 1989, 1992). IRAS discovered a new class of galaxies that radiate most of their energy in the infrared (Soifer et al 1984), many of them dusty starburst galaxies[3]. The most luminous of these infrared galaxies [(ultra-)luminous infrared galaxies: (U)LIRGs or (U)LIGs] have QSO-like bolometric luminosities (LIRGs: $L \geq 10^{11}\,L_\odot$, ULIRGs: $L \geq 10^{12}\,L_\odot$; Sanders & Mirabel 1996).

The Infrared Space Observatory (Kessler et al 1996) was the first cryogenic space infrared observatory. ISO was launched in November 1995. It was equipped with a multi-pixel near-IR/mid-IR camera (ISOCAM; Cesarsky et al 1996, Cesarsky 1999), a multiband mid- and far-IR spectro-photometer (ISOPHOT; Lemke et al 1996, Lemke & Klaas 1999), a 2.4–45 μm spectrometer (SWS; deGraauw et al 1996, de Graauw 1999) and a 43–197 μm spectrometer (LWS; Clegg et al 1996, Clegg 1999). The ISO mission lasted until April 1998, about one year longer than expected (Kessler 1999). The present review is an account of the key extragalactic results of ISO as of December 1999. We also refer the reader to the special *Astronomy and Astrophysics* issue on early ISO results (volume 315, No. 2, 1996), to the proceedings of the 1998 Paris conference (*The Universe as Seen by ISO*; Cox & Kessler 1999), and to the recent review by Cesarsky & Sauvage (2000).

2. NORMAL GALAXIES

2.1 Infrared Emission from Normal Galaxies

2.1.1 Mid-Infrared Emission: Unidentified Bands and Very Small Dust Grains
ISO studies of our Galaxy have demonstrated that the bulk of the mid-IR emission from the interstellar medium results from transient heating of small clusters of particles (Boulanger et al 1998), which confirms a long-standing hypothesis derived from ground-based and IRAS data (Sellgren 1984, Beichman 1987). The clusters are stochastically heated by single photons and as a result exhibit large temperature fluctuations. The resulting mid-IR flux is simply proportional to the underlying radiation field intensity. The spectra are surprisingly regular, exhibiting almost invariably a family of features centered at 3.3, 6.2, 7.7, 8.6, 11.3, and 12.7 μm (cf Allamandola et al 1995; D Cesarsky et al 1996a,b; Verstraete et al 1996; Mattila et al 1996, 1999; Onaka et al 1996). These unidentified infrared bands (UIBs) are

---

[3]Following the classical analysis by Rieke et al (1980) of two nearby representatives of this class, M82 and NGC 253, these galaxies are presently going through a very active but short-lived "starburst" of duration of a few tens of millions of years or less. For a Salpeter initial mass function (IMF) from 100 to 1 $M_\odot$, a luminosity of $10^{10}\,L_\odot$ corresponds to a star formation rate of roughly 1 $M_\odot$ yr^{-1}(e.g. Kennicutt 1998). Given its infrared luminosity ($\geq 4 \times 10^{10}\,L_\odot$) and central gas content ($2.5 \times 10^8\,M_\odot$), the relatively small galaxy M82 thus cannot sustain its present star formation rate for much longer than 50 million years, hence the term "starburst."

thought to result from C-C and C-H stretching/bending vibrational bands in aromatic hydrocarbons (Puget & Leger 1989, Duley & Williams 1991, Tielens et al 1999). The actual carrier of the bands is still uncertain. One possibility is that it consists of large, carbon-rich ring molecules [e.g. polycyclic aromatic hydrocarbons (PAHs), size $\leq$ a few nm]. Alternatively, the carrier may be very small, amorphous carbon dust grains that are exposed to moderately intense UV (and possibly visible) radiation (Boulade et al 1996, Uchida et al 1998, Pagani et al 1999). The most popular model is the PAH interpretation, but no rigorous identification with specific molecules has yet been established. The second component of the interstellar mid-IR emission manifests itself as a steeply rising continuum longward of 10 μm, accompanied by strong fine structure line emission, in particular [NeII] and [NeIII]. This continuum component is characteristic of active star-forming regions, such as the Galactic HII region M17 (Verstraete et al 1996, D Cesarky et al 1996b). Desert et al (1990) attributed this continuum to very small fluctuating grains (VSGs: size $\leq$10 nm); this hypothesis is consistent with the ISOCAM CVF spectral maps obtained around M17 and the photometry of NGC 7023 (Tran 1998, Laureijs et al 1996).

The transition from stellar emission to interstellar dust emission in galaxies occurs in the mid-IR (3–30 μm) band and depends on the star-forming activity. This is well demonstrated by a detailed ISOCAM study of $\sim$10^2 Virgo cluster galaxies (Boselli et al 1997, 1999). Three main components of dust emission contribute to the mid-IR spectra of galaxies. The first is a UIB-dominated, mid-IR spectral energy distribution (SED) up to 13 microns. The SED and the ratios of the different UIB features remain constant in galaxies with a wide range of radiation fields and properties (Helou 1999, Helou et al 1999). The second component, present only in some galaxies or regions of galaxies, is the steeply rising (VSG) continuum longwards of 10 microns discussed above. It is characteristic of intense star-forming regions. The third is near-radiation equilibrium emission from hot (150 to 1700 K) dust particles. This near-IR/mid-IR "bump" is characteristic of dust tori in AGNs (Section 3.3.1). A 3–5 μm continuum component may also come from a fluctuating dust component without PAH features (Helou et al 1999).

UIB emission is a good tracer of normal and moderately active star formation activity in spiral and irregular galaxies (Helou 1999, Vigroux et al 1999). In such systems the luminosity in the ISOCAM LW2 filter (containing the 6.2, 7.7, and 8.6 μm UIB features) is well correlated with the longer wavelength mid-IR (LW3 filter: 15 μm; Figure 1). Both correlate with the far-IR and the Hα line luminosities (Rouan et al 1996, Metcalfe et al 1996, Smith 1998, Vigroux et al 1999, Roussel et al 1999a). In normal galaxies with low activity, as in NGC 891, NGC 7331, M31, or in parts of the LMC, the mid-IR emission mainly traces the distribution of (molecular + atomic) gas that is bathed in the diffuse radiation field (Mattila et al 1999, Smith 1998, Pagani et al 1999, Contursi et al 1998). Under these conditions excitation of the UIB features other than by UV photons (e.g. visible photons) may play an important role (Pagani et al 1999). At higher luminosities and activity, however, the $\lambda \geq 10$ μm continuum (from warm dust in PDRs and

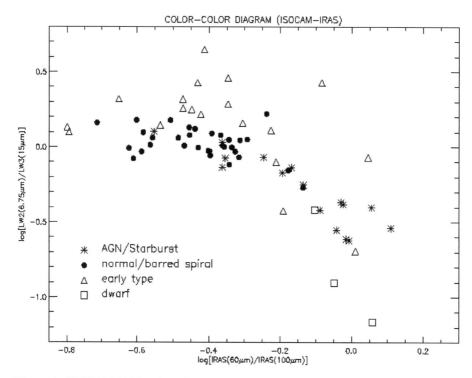

Figure 1 ISOCAM/IRAS color diagram for different types of galaxies. The decrease in LW2/LW3 ratio for 60/100 flux ratios >0.5 (AGN/starbursts and dwarf galaxies) can be explained by a combination of the destruction of the UIB features (LW2 band) in the intense radiation field and increased emission from (very small) warm grains (LW3) (from Vigroux et al 1999).

HII regions) measured in the LW3 filter increases relative to the UIB emission (Figure 1; Vigroux et al 1999). At radiation fields $\geq 10^5$ times the local interstellar radiation field, such as in the central parsec of the Galaxy (Lutz et al 1996b) or the M17 HII region (Verstraete et al 1996), in active galactic nuclei (see Section 3.1), or in metal-poor environments (such as the metal-poor blue compact dwarf galaxy SBS0335-052; Thuan et al 1999), the UIB emission strength plummets, presumably because of the destruction of its carriers. As a result, the LW3/LW2 ratio is an interesting diagnostic of the radiation environment (Figure 1). This ratio decreases outward from the nuclei in disk galaxies (Dale et al 2000).

2.1.2 Far-IR Emission: Cold Dust and Galactic Extinction   In the far-IR, the emitted spectrum results from the radiation equilibrium between absorbed short-wavelength radiation and grey-body emission with a wavelength dependent emissivity [$\varepsilon(\lambda) \propto \lambda^{-\beta}$ with $\beta \sim 1.5$ to 2 in the range $\lambda \geq 20$ μm; Draine & Lee 1984]. IRAS observations have established that the $\lambda \leq 100$ μm dust emission in normal spirals comes from $\sim$30 K dust grains with a total gas mass (HI + H2 + He) to dust mass ratio of $\geq 10^3$ (e.g. Devereux & Young 1990). This value is about an order of magnitude greater than that in the Milky Way ($M_{gas}/M_{dust} \sim$ 170; cf Haas et al 1998b). The likely source of this discrepancy is that IRAS did not pick up the coldest dust that contains most of the mass (e.g. Xu & Helou 1996).

By extending the wavelength coverage to 200 μm, ISO has resolved this puzzle. ISOPHOT observations of a number of normal, inactive spirals have uncovered a dust component with typical temperatures $\sim$20 K and a range between 10 K and 28 K (Alton et al 1998a; Davies et al 1999; Domingue et al 1999; Haas 1998; Haas et al 1998b; Hippelein et al 1996; Krügel et al 1998; Siebenmorgen et al 1999; Trewhella et al 1997, 1998; Tuffs et al 1996). With increasing activity, a second dust component (T $\sim$ 30–40 K) becomes more prominent and also dominates the 60 μm and 100 μm IRAS bands (e.g. Siebenmorgen et al 1999). The cold dust has a larger radial extent than the stars (Alton et al 1998a, Davies et al 1999) and may be partially associated with extended HI disks.

Preliminary results from the 175 μm serendipity survey indicate that this is a general result for normal spirals (Stickel et al 1999). However, Domingue et al (1999) argue against any dominant part of the dust mass being much colder than 20 K, based on a direct comparison of visual and far-IR data in three overlapping galaxies. Particularly impressive is the case of M31 where Haas et al (1998b) presented a 175 μm map of the entire galaxy (Figure 2, see color insert). Across most of the disk the far-IR SED is still rising from 100 to 200 μm with a temperature of 16 $\pm$ 2 K (see also Odenwald et al 1998). The temperature of the cold dust component is consistent with theoretical predictions from the heating/cooling balance of equilibrium dust grains in the diffuse radiation field. The gas-to-dust ratio inferred for this cold dust component is very close to the galactic value in the Milky Way.

An important consequence of the larger dust masses derived from the ISOPHOT data is that the corresponding dust extinctions in the optical become quite significant. The extinction-corrected morphologies of spirals can thus be quite different from the uncorrected ones (Trewhella et al 1997, Haas et al 1998b). For instance, by including extinction, M31 turns from an Sb into a ring galaxy (Figure 2) and NGC 6946 from an Sc into an Sb (Trewhella 1998). These findings qualitatively support earlier suggestions that spiral galaxies are significantly affected by dust (Disney et al 1989, Valentijn 1990).

2.2   Dark Matter in the Halos and Outer Parts of Disks

Evidence has been mounting for several decades that the outer parts of disk galaxies are dominated by dark matter in the form of a spheroidal halo distribution (e.g. Trimble 1987, Carr 1994). Of the various possibilities for the nature of this halo, dark baryonic matter in the form of stellar remnants, low-mass stars, and substellar objects seems the most conservative solution, and is still consistent with current Big Bang nucleosynthesis limits (Carr 1994). It is of substantial interest to complement the existing deep optical and near-infrared observations with a search for possible IR halos in edge-on galaxies. Such halos could signal the presence of cool, low-mass stars or of substellar (≤ 0.08 M$_\odot$) brown dwarfs.

With ISOCAM, Gilmore & Unavane (1998) and Beichman et al (1999) have searched for IR halos in five edge-on galaxies. No halos are detected in any of them. Gilmore & Unavane (1998) concluded that main-sequence, hydrogen-burning stars of all masses and metallicities down to the hydrogen-burning limit are excluded to dominate the halos. A mixture of low-mass stars with a mass function similar to that in the solar neighborhood can also be strongly ruled out in one galaxy (NGC 2915). Young (≤ 1 Gyr) brown dwarfs are weakly ruled out in UGC 1459. In contrast the EROS/MACHO microlensing data toward the Large Magellanic Cloud favor a mean mass of the deflectors in the halo of our own Galaxy of about 0.5 M$_\odot$ (Alcock et al 1998). A possible way out may be that the lensing objects are not in the Galactic halo but in the LMC (Sahu 1994), or in a dwarf galaxy debris between the Galaxy and the LMC (Zhao 1998). The combined ISO, HST, and MACHO/EROS results, with the additional assumption that the composition of halos is similar in the different galaxies, suggest that, with the exception of brown dwarfs between ≤ 0.01 and 0.08 M$_\odot$ (still permitted by EROS to make up to 35% of the halo mass, (Afonso et al 1999), the entire range of possible stellar and substellar entities is excluded or is highly unlikely (Gilmore 1998).

Another possible form of baryonic dark matter in the outer disks of spirals may be cold molecular gas (e.g. Pfenniger et al 1994). There is not much direct evidence for such a component from CO mm spectroscopy, but in the outer low-metallicity regions CO may have low abundance. Evidence for cold dust more extensive than the stellar distributions comes from 200 μm ISOPHOT and 450/850 S(0) and S(1) H$_2$ emission throughout the disk of the edge-on galaxy NGC 891. Remarkably, H$_2$ is detected out to ≥ 11 kpc galactocentric radius, implying the presence of warm (150–230 K) H$_2$ clouds that may represent a significant fraction of the total molecular content of the galaxy. Based on the line shapes of the two H$_2$ lines at the outermost positions, Valentijn & van der Werf argue that there must be a second cooler H$_2$ component ($\sim$80 K) that would then have to contain 4 to 10 times the mass estimated from CO and HI observations. If confirmed, such a molecular component (although much warmer than originally proposed by Pfenniger et al 1994) could make a significant contribution to the dynamical mass at that radius.

2.3 Early-Type Galaxies

IRAS observations have shown that a significant fraction of early-type galaxies contains detectable infrared emission (e.g. Knapp et al 1989, 1992). With ISO, researchers have improved their understanding of the origin of this emission. One key question is whether the mid-IR emission originates mainly in the photospheres and shells of cool asymptotic giant branch (AGB) stars or in the interstellar medium, or whether an active galactic nucleus plays a role. Several studies with ISOCAM and ISOPHOT in more than 40 early-type galaxies have addressed this issue (Madden et al 1999, Vigroux et al 1999, Knapp et al 1996, Fich et al 1999, Malhotra et al 1999b). Exploiting the imaging capability of ISOCAM at different wavelengths, these studies show that all three components play a role. Madden et al (1999) found that the 4.5/6.7 μm and 6.7/15 μm colors of about 60% of their sample galaxies are consistent with the notion of a largely stellar origin for the mid-IR emission. The other galaxies are dominated by UIB emission (17%) or an AGN (22%). Half of the sample exhibits a mid-IR excess resulting from hot interstellar dust at some level. The interstellar dust emission is associated with the optical dust lane in NGC 5266 (Madden et al 1999) and in Cen A (Mirabel et al 1999).

The detection of UIB features in early-type galaxies is of general significance because the radiation field from the old stellar population is too soft to excite the UIB emission with UV photons (Section 2.1.1). Either lower-energy photons in the visible band are sufficient to excite the mid-IR features (Boulade et al 1996, Uchida et al 1998, Vigroux et al 1999, Madden et al 1999), or there must be a young stellar component producing a stronger UV radiation field. Evidence for the latter explanation comes from the detection of 158 μm [CII] (and 63 μm [OI]) far-IR line emission in NGC 1155, NGC 1052, and NGC 6958 (Malhotra et al 1999b), and in Cen A (Madden et al 1995, Unger et al 2000). Excitation of [CII] line emission (Sections 3.1, 3.2.3) definitely requires ≥ 11.3 eV photons, and the old stellar population in these galaxies cannot account for the required far-UV radiation field (Malhotra et al 1999b). A fair amount of the [CII] emission may come from a diffuse neutral HI medium. However, an additional obscured source of UV photons is required and is probably associated with massive star formation in dense interstellar clouds. The [OI]/[CII] ratio in NGC 1155 indicates that the radiation field is about 100 times greater than that in the solar neighborhood, and the gas density is ~ 100 cm^{-3}, which is characteristic of moderately dense PDRs associated with molecular clouds (Section 3.2.3). Most of the [CII] emission in Cen A originates in the central dust disk, where ISOCAM observations provide strong evidence for ongoing star-forming activity (Mirabel et al 1999).

3.   ACTIVE GALAXIES

3.1   Analytical Spectroscopy and Spectrophotometry

The 2–200 μm band accessible to the ISO spectrometers (6.5 octaves!) contains a plethora of atomic, ionic, and molecular spectral lines spanning a wide range of excitation potential, along with various solid-state features from dust grains of different sizes (Section 2.1.1, Figure 3). These lines sample widely different

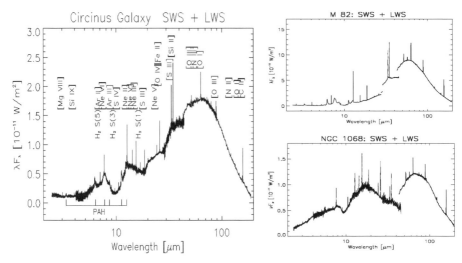

Figure 3 Combined LWS+SWS spectra of galaxies (>6 octaves). *Left*: Combined SWS/LWS spectrum of the Circinus galaxy (Moorwood 1999, Sturm et al 1999b). The H_2 lines and low-excitation atomic/ionic fine structure lines ([FeII], [SiII], [OI], [CII]) sample photodissociation regions (PDRs[4]; Sternberg & Dalgarno 1995, Hollenbach & Tielens 1997), shocks (Draine et al 1983, Hollenbach & McKee 1989), or X-ray excited gas (Maloney et al 1996). Hydrogen recombination lines and low-lying ionic fine structure lines (excitation potential <50 eV: [ArIII], [NeII], [NeIII], [SIII], [OIII], [NII]) sample mainly HII regions photoionized by OB stars (Spinoglio & Malkan 1992, Voit 1992), although ionizing shocks may contribute in some sources (e.g. Contini & Viegas 1992, Sutherland et al 1993). Ionic lines from species with excitation potentials up to ~300 eV (e.g. [OIV], [NeV], [NeVI], [SiIX]) probe highly ionized coronal gas and require very hard radiation fields (such as the accretion disks of AGNs) or fast ionizing shocks. Line ratios give information about the physical characteristics of the emitting gas. Extinction corrections are small ($A(\lambda)/A(V) \sim 0.1$ to 0.01 in the 2–40 μm region). *Right*: The starburst galaxy M82 (*top*): low excitation lines, strong UIBs/PAHs; and the AGN NGC 1068 (*bottom*): high excitation, no UIBs/PAHs (Sturm et al 1999b, Colbert et al 1999, Spinoglio et al 1999). Sudden breaks in the SEDs are the result of different aperture sizes at different wavelengths. Local bumps and unusual slopes in the Circinus spectrum (12–20 μm and 35 μm) may be caused by residual calibration uncertainties.

excitation/ionization states and are characteristic tracers of different physical regions: photodissociation regions (PDRs[4]), shocks, X-ray excited gas, HII regions

---

[4]PDRs are the origin of much of the infrared radiation from the interstellar medium (ISM). PDRs are created when far-UV radiation impinges on (dense) neutral interstellar (or circumstellar) clouds and ionizes/photodissociates atoms and molecules. The incident UV (star) light is absorbed by dust grains and large carbon molecules (such as PAHs) and is converted into infrared continuum and UIB features. As much as 0.1–1% of the absorbed starlight is converted into gas heating via photoelectric ejection of electrons from grains or UIBs (Hollenbach & Tielens 1997, Kaufman et al 1999).

photoionized by OB stars, and coronal gas photoionized by a hard central AGN source (see Genzel 1992 for a review).

As an example of the rich spectra that were obtained with ISO, we show in Figure 3 the combined SWS/LWS spectrum of the Circinus galaxy, which displays the entire set of spectral characteristics just discussed. Circinus is the closest Seyfert 2 galaxy (D = 4 Mpc) and has an active circumnuclear starburst (Moorwood et al 1996, Moorwood 1999, Sturm et al 1999b).

3.1.1 Molecular Spectroscopy
ISO has opened the realm of external galaxies for molecular infrared spectroscopy. Pure H_2 rotational emission lines have been detected in a wide range of galaxies, from normal to ultraluminous (e.g. Rigopoulou et al 1996a, 2000; Kunze et al 1996, 1999; Lutz et al 1999; Moorwood et al 1996; Sturm et al 1996; Valentijn et al 1996; Valentijn & van der Werf 1999a). The lowest rotational lines [28 μm S(0) and 17 μm S(1)] originate in warm (90–200 K; Valentijn & van der Werf 1999a,b) gas that constitutes up to 20% of the total mass of the (cold) molecular ISM in these galaxies (Kunze et al 1999, Rigopoulou et al 2000). High rotation lines [S(5), S(7)] and rovibrational lines come from a small amount of much hotter gas (≥ 1000 K). It is likely that the H_2 emission arises from a combination of shocks, X-ray illuminated clouds (in AGNs), and photodissociation regions. OH, CH, and H_2O lines are seen in emission or absorption (or a combination of the two) in a number of gas-rich starburst, Seyfert, and ultraluminous galaxies (Colbert et al 1999; Bradford et al 1999; Fischer et al 1997, 1999; Spinoglio et al 1999; Skinner et al 1997; Kegel et al 1999). The observed OH emission can generally be accounted for by infrared pumping through rotational and rovibrational transitions in sources with a warm infrared background source. The ISO SWS/LWS spectroscopy solves the long-standing puzzle of the pumping mechanism for intense radio mega-maser OH emission in luminous infrared galaxies (Baan 1993). In the mega-maser galaxies Arp 220, IRAS 20100-4156, and 3Zw35, absorption is seen in cross-ladder rotational transitions (Figure 4); this strongly supports the idea of rotational pumping models for the masers (Skinner et al 1997, Kegel et al 1999). The OH masers must therefore be located in front of the far-IR continuum source on kpc scales. The OH mega-masers originate in the extended starburst region, not in the circumnuclear environment (Skinner et al 1997). The far-IR molecular line spectra of the ultraluminous galaxies Arp 220 (Figure 4) and Mrk 231 are remarkably similar to those of the galactic center molecular cloud complex/star-forming region SgrB2 (Figure 4; Cox et al 1999; Fischer et al 1997, 1999), which indicates that the properties of the entire molecular ISM in Arp 220 (scale 1 kpc) are comparable to those of this dense galactic cloud [$N(H_2) \geq 10^{24}$ cm^{-2}, $n(H_2) \sim 10^4$ cm^{-3}].

3.1.2 Distinguishing AGNs and Starbursts: ISO Diagnostic Diagrams
One of the most powerful applications of the ISO (mid-)IR spectroscopy is as a tool for distinguishing between AGN and star formation dominated sources. The right-hand insets of Figure 3 (from Sturm et al 1999b) show two examples: the starburst

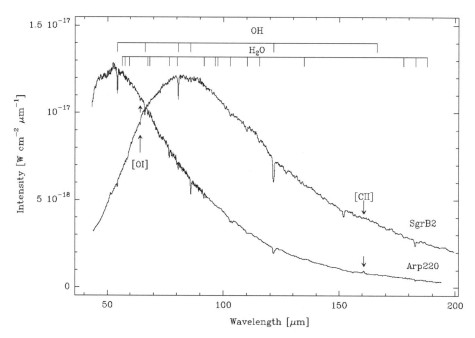

Figure 4 Comparison of the LWS spectrum of Arp 220 and the Galactic star-forming region/molecular cloud SgrB2 (Fischer et al 1999, Cox et al 1999). The spectrum of SgrB2 has been divided by a factor of 115 and shifted to the redshift of Arp 220. The main molecular absorption bands are identified, along with the [OI] and [CII] fine structure lines.

galaxy M82 and the nuclear region of the Seyfert 2 galaxy NGC 1068. The difference between the mid-IR spectra is striking, and agrees with earlier ground-based results (Roche et al 1991). M82 is characterized by strong, low-excitation, fine structure lines, prominent UIB features, and a weak $\lambda \leq 10$ μm continuum (Sturm et al 1999b). High-excitation lines are absent or weak. In contrast, NGC 1068 displays a fainter but highly excited emission line spectrum ([OIV]/ [NeII] and [NeV]/[NeII] $\geq$ 1; Lutz et al 2000 and weak or no UIB features, plus a strong mid-IR continuum (Genzel et al 1998, Sturm et al 1999b). The far-IR spectra of NGC 1068 and M82 (and Circinus) are more similar because star formation in the disks dominates the $\lambda \geq 40$ μm SED in all three galaxies (Colbert et al 1999, Spinoglio et al 1999). The deep 10 μm dip in the mid-IR SED of M82 (and other galaxies) has often been interpreted as the result of absorption by silicate dust grains [$\tau_{9.7\mu m} \sim 2$ or A(V) ~ 30]. Sturm et al (1999b) showed that the M82 spectrum can be fitted extremely well by the superposition of the (absorption-free) spectrum of the reflection nebula NGC 7023 (which exhibits strong UIB features) plus a VSG continuum at $\lambda > 10$ μm. No silicate absorption is required. Instead, the deep 10 μm dip is caused by the strong UIB emission on either side of the silicate feature.

Diagnostic diagrams can empirically characterize the excitation state of a source (Osterbrock 1989, Spinoglio & Malkan 1992, Voit 1992). Genzel et al (1998) use a combination of the 25.9 μm [OIV] to 12.8 μm [NeII] line flux ratio (or [OIV]/[SIII], or [NeV]/[NeII]) on one axis, and of the UIB strength[5] on the other axis. Figure 5a shows that this ISO diagnostic diagram clearly separates known star-forming galaxies from AGNs. The three AGN templates that are located fairly close to the starbursts in Figure 5b (CenA, NGC 7582, and Circinus; Mirabel et al 1999, Radovich et al 1999, Moorwood et al 1996) are known to contain circumnuclear starbursts; correction for the star formation activity would move these sources to the upper left (marked by arrows). The diagram is insensitive to dust extinction if the numerator and denominator in both ratios are affected by a similar amount of extinction. Lutz et al (1998b) found low-level [OIV] emission ([OIV]/[NeII] $\sim 10^{-2}$) in a number of starburst galaxies. The [OIV] emission in M82 is spatially extended and tracks galactic rotation. It cannot come from low-level AGN activity. Fast, ionizing shocks in galactic "super" winds are the most likely sources of this low-level, high-excitation gas (Lutz et al 1998b, Viegas et al 1999), but WR stars may play a role in some objects (Schaerer & Stasinska 1999).

Based on ISOCAM CVF spectra of the central part of M17 (an HII region), of the nuclear position of Cen A (an AGN; Mirabel et al 1999), and of the reflection nebula NGC 7023 (a PDR; D Cesarsky et al 1996a), Laurent et al (2000) proposed another diagnostic diagram. It is based on the 15 μm/6 μm continuum ratio on one axis and the UIB/6 μm continuum ratio on the other axis. Laurent et al (2000) demonstrated that this mid-IR diagnostic diagram also distinguishes well between AGNs and starbursts (Figure 5b).

One example of the value of the techniques proposed by Laurent et al (1999) is the study of Centaurus A, the closest radio galaxy. In the visible band, Cen A is a giant elliptical galaxy with a prominent central dust lane that was interpreted 45 years ago to be the result of a merger between the elliptical galaxy and a small gas-rich galaxy (Baade & Minkowski 1954). The central AGN is hidden in the optical band, and its presence can only be directly detected from the prominent, double-lobed, radio and X-ray jet system (Schreier et al 1998). The ISOCAM mid-IR observations penetrate the dust and reveal a strong and compact hot dust source associated with the central AGN, as well as a bisymmetric structure of circumnuclear dust extending in the plane of the dust lane (Figure 6, see color insert; Mirabel et al 1999). The mid-IR spectrum toward the nuclear position is

---

[5] The 7.7 μm (UIB) line-to-continuum ratio or "strength" is the ratio of the peak flux density in the UIB feature (continuum subtracted) to the underlying 7.7 μm continuum flux density. The latter is computed from a linear interpolation of the continuum between 5.9 μm and 11.2 μm. While the 11.2 μm continuum is often affected by silicate absorption and various emission features longward of 11 μm, 7.7 μm is close enough to the fairly clean 5.9 μm region that a simple linear extrapolation is justified as a simple and fairly robust estimate of the 7.7 μm continuum.

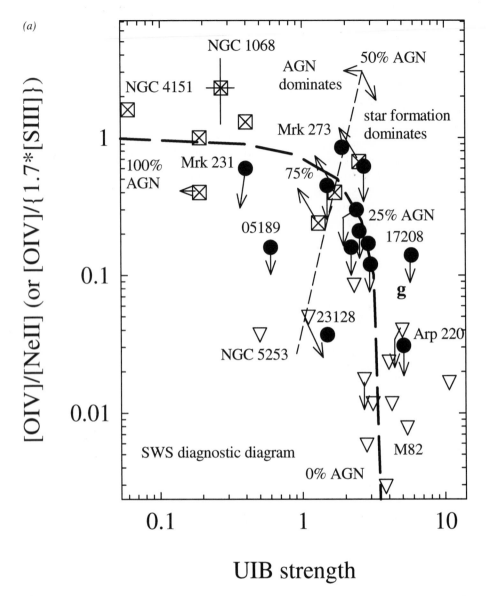

Figure 5 (*a, left*) SWS diagnostic diagram (from Genzel et al 1998). The vertical axis measures the flux ratio of high excitation to low excitation mid-IR emission lines, and the horizontal axis measures the strength (i.e. feature to continuum ratio) of the 7.7 μm UIB/PAH feature. AGN templates are marked as *rectangles with crosses*, starburst templates as *open triangles*, and UILRGs as *filled circles*. A simple mixing curve from 0% to 100% AGN is shown with long dashes. (*b, right*) ISOCAM diagnostic diagram (from Laurent et al 2000), including 35 CVF spectra from a sample of galaxies of different characteristics. The vertical axis measures the ratio of 14.5 to 5.4 μm continuum, and the horizontal axis measures the strength of the UIB feature. Active starbursts are in the upper left, AGNs in the lower left, and PDRs in the lower right. The curves from lower right to upper left denote HII/starburst fractions of 25%, 50%, and 75%. The two curves from lower left to upper right denote AGN fractions of 75% and 50%.

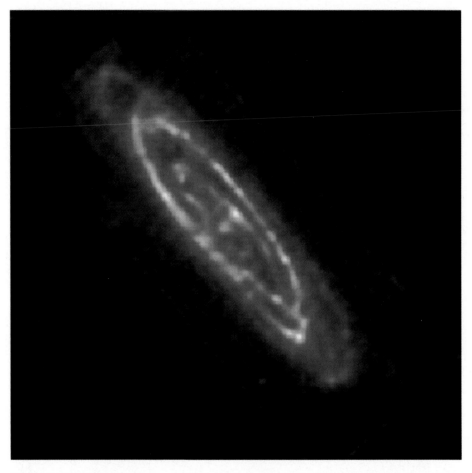

Figure 2 ISOPHOT 175 μm map of M31 (north up, east to the left). The emission is dominated by a 10 kpc ring, with numerous bright condensations and a fainter ring at 14 kpc (from Haas et al 1998b). The ratio of ring to nuclear emission is much greater than in the IRAS bands (Habing et al 1984).

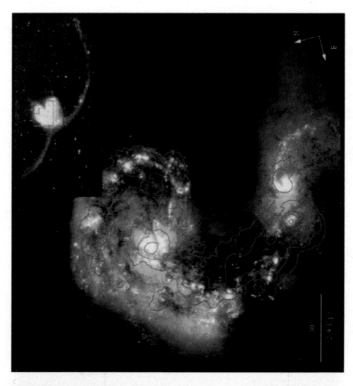

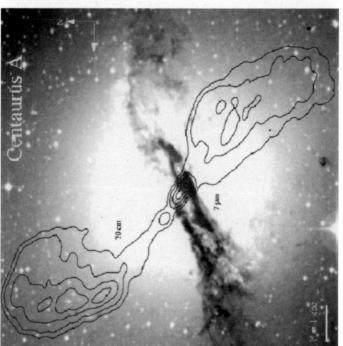

Figure 6a (left): ISOCAM LW2 map (red) of Centaurus A, superimposed on a visible image and contours (blue) of the 20 cm radio continuum. The infrared data penetrate the dust lane and reveal the presence of a barred dust spiral centered on the AGN (Mirabel et al 1999). **Figure 6b** (right): ISOCAM LW3 map (red contours) of the Antennae galaxies (upper right inset shows the larger scales and the tails) superimposed on a V/I-band HST image (Mirabel et al 1998). The nuclei of NGC 4038 and NGC 4039 are top and bottom right, respectively. The interaction region is located below the center of the image and exhibits the strongest mid-IR emission.

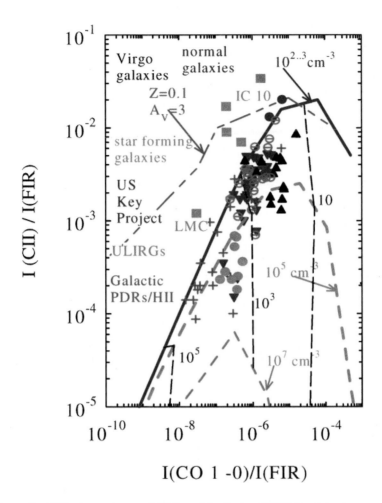

Figure 8a [CII] observations and PDR models. Ratio of [CII] line intensity to total far-infrared intensity ($Y_{[CII]}$) as a function of ratio of CO 1-0 line intensity to far-infrared intensity (Y_{CO}). KAO observations: star forming galaxies (light blue open circles, Stacey et al 1991, Lord et al 1995, 1996), normal galaxies/outer parts of spirals (filled blue circles, Stacey et al 1991, Madden et al 1993), low metallicity galaxies (pink filled rectangles, Poglitsch et al 1995, Madden et al 1997) and Galactic HII regions/PDRs (green crosses, Stacey et al 1991). LWS ISO observations: US Key Project, normal and star forming galaxies (blue, down-pointing open triangles, Malhotra et al 1997, Lord et al 2000), Virgo galaxies with normal and low activity (black open, up-pointing triangles, Smith & Madden 1997, Pierini et al 1999) and ultra-luminous galaxies (filled red circles, Luhman et al 1998, Luhman 1999). Solar metallicity PDR models for clouds of $A_V = 10$ and $n_H = 10^3$ (thick blue line), 10^5 (dashed red) and 10^7 cm^{-3} (thin dashed light blue) are from Kaufman et al (1999), with radiation field densities of 10, 10^3 and 10^5 times the solar neighborhood field (G_0) marked by black, long dashed lines. A model for a $A_V = 3$, $Z = 0.1$ cloud is given as green dashed curve.

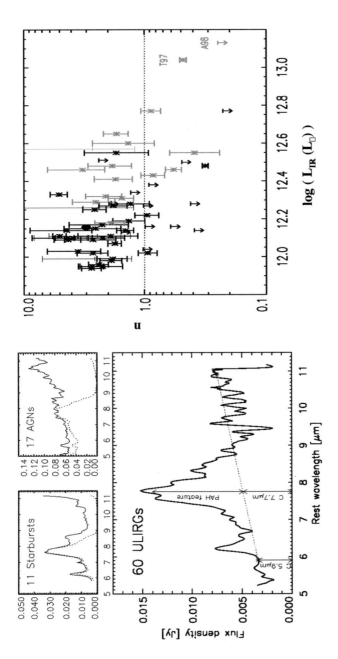

Figure 12 Mid-IR SEDs of ULIRGs. Left: Average spectrum of 60 ULIRGs with z < 0.3 (lower inset), compared to average spectra of starburst templates (upper left) and AGNs (upper right). Individual spectra were shifted to the proper rest-wavelength and normalized to give all sources equal weight. The dotted curves in the upper insets show the impact of 50 magnitudes of extinction on the SED shape (from ISOPHOT-S data of Lutz et al 1998a). Right: UIB strength as a function of IR-luminosity of ULIRGs. In addition to the Lutz et al sample (see also Rigopoulou et al 1999), the figure contains 15ULIRGs with z < 0.4 from Tran et al (2000) and two hyperluminous ULIRGs from Taniguchi et al (1997b) and Aussel et al (1998), all observed with the CVF of ISOCAM. In the text it is (arbitrarily) assumed that sources above the dotted line (UIB strength ~1) are dominated by star formation, and those below the line by AGNs.

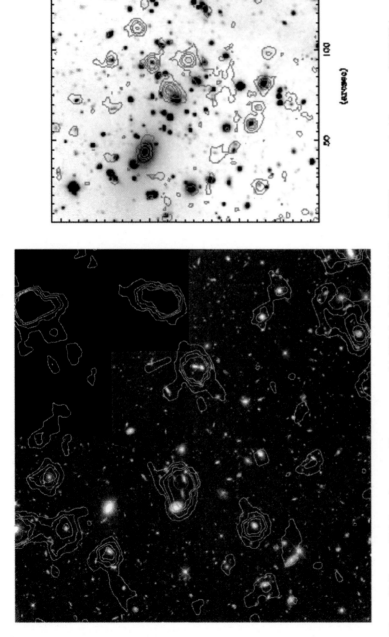

Figure 13 Deep mid-infrared surveys. Deep surveys with ISO. Left: ISOCAM 15 μm (LW3) contours (gold) on top of WFPC2 image of the HDF (N). 7 μm sources (LW2) are marked by green circles (Aussel et al 1998). Right: LW3 contours on top of R-band image of Abell 2390 ($z = 0.23$) (Altieri et al 1999). The outer regions of the image with less integration time has been down-weighted. With the aid of the foreground cluster's lensing amplification this image reaches to <30μJy for the <z> ∼ 0.7 background galaxies. It thus represents the deepest mid-infrared image obtained by ISOCAM.

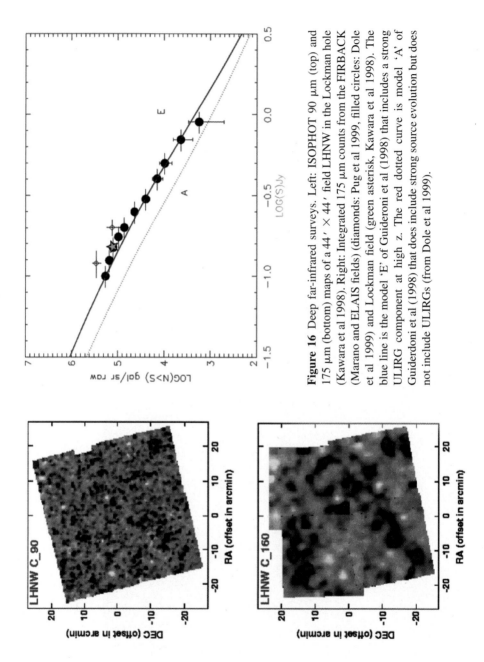

Figure 16 Deep far-infrared surveys. Left: ISOPHOT 90 μm (top) and 175 μm (bottom) maps of a 44′ × 44′ field LHNW in the Lockman hole (Kawara et al 1998). Right: Integrated 175 μm counts from the FIRBACK (Marano and ELAIS fields) (diamonds: Pug et al 1999, filled circles: Dole et al 1999) and Lockman field (green asterisk, Kawara et al 1998). The blue line is the model 'E' of Guideroni et al (1998) that includes a strong ULIRG component at high z. The red dotted curve is model 'A' of Guiderdoni et al (1998) that does include strong source evolution but does not include ULIRGs (from Dole et al 1999).

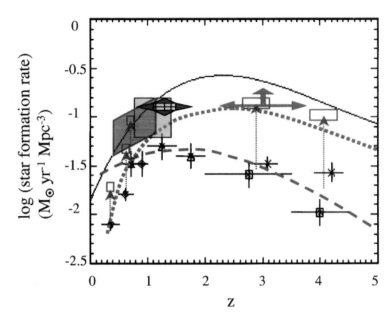

Figure 17 Cosmic star formation rate (per unit comoving volume, h = 0.6, q_0 = 0.5) as a function of redshift (the 'Madau' plot, Madau et al 1996). The black symbols (with error bars) denote the star formation history deduced from (non-extinction corrected) UV data (Steidel et al 1999 and references therein). Upward pointing dotted green arrows with open boxes mark where these points move when a reddening correction is applied. The green, four arrow symbol is derived from (non-extinction corrected) Hα NICMOS observations (Yan et al 1999). The red, three arrow symbol denotes the lower limit to dusty star formation obtained from SCUBA observations of HDF (N) (Hughes et al 1998). The continuous line marks the total star formation rate deduced from the COBE background and an 'inversion' with a starburst SED (Lagache et al 1999b). The filled hatched blue and yellow boxes denote star formation rates deduced from ISCAM (CFRS field, Flores et al 1999b) and ISOPHOT-FIRBACK (Puget et al 1999, Dole et al 1999). The light blue, dashed curve is model 'A' (no ULIRGSs) and the red dotted curve model 'E' (with ULIRGs) of Guiderdoni et al (1998).

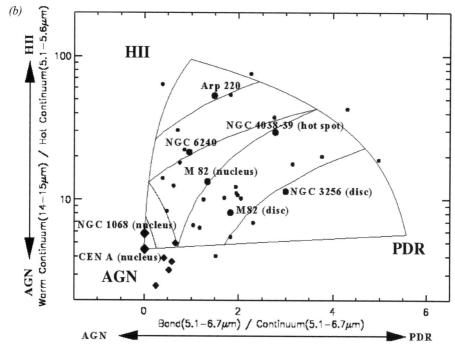

Figure 5   (*continued*)

characteristic of an AGN, as described in previous sections (Genzel et al 1998, Mirabel et al 1999, Laurent et al 2000, D Alexander et al 2000). In contrast, the off-nuclear spectra are characteristic of star-forming regions (Madden et al 1999a, Unger et al 2000). Mirabel et al interpreted the bisymmetric dust structure as a highly inclined (i = 72°) barred spiral. Its cold dust content is comparable to a small spiral galaxy. The gas bar may funnel material from kpc scales toward the circumnuclear environment (10^2 pc scale), from where it may be transported farther by a circumnuclear bar/disk inferred from near-IR polarization observations. As such, Cen A is a beautiful nearby example of how a merger can convert an (old) spheroid into a bulge+disk system.

3.2 Starburst Galaxies

3.2.1 The Dark Side of Star Formation   The beautiful HST image of the famous Antennae system of colliding galaxies (NGC 4038/39; Figure 6*b*, see color insert, from Whitmore et al 1999) summarizes one of the paradigms about active star formation on galactic scales: Spectacular fireworks of star formation with hundreds of young star clusters are triggered by strong gravitational disturbances. The interaction/collision and eventual merging of two gas-rich galaxies is the most effective perturbation of this kind. Bar-driven inflow is another such perturbation.

ISOCAM studies of several samples of interacting, barred, and collisional ring galaxies confirm this paradigm (Appleton et al 1999, Charmandaris et al 1999, Roussel et al 1999b, Wozniak et al 1999). The optical/UV data in the Antennae tell a fascinating story with evidence for a large age spread in star cluster populations. Several phases of active star formation have occurred since the first interaction ∼500 million years ago. However, the UV/optical data do not reveal the whole story. Superposed on the HST image in Figure 6*b* are the contours of 15 μm mid-IR emission (Mirabel et al 1998). The 15 μm map is an (approximate) tracer of the entire infrared emission that contains ∼80% of the bolometric luminosity of the system. About half of the entire luminosity emerges from the optically "dark" interaction region between the two galaxy nuclei. This region also contains most of the molecular gas (Stanford et al 1990, Lo et al 2000). A compact source at the southern edge of the interaction region accounts for ∼15% of the entire luminosity, yet on the HST image there is only a faint and very red compact cluster at this position. This knot is the site of the most recent star formation activity in the Antennae (see Section 3.2.3; Vigroux et al 1996, Kunze et al 1996).

Another example is the luminous merger Arp 299 (NGC 3690/IC694, Gehrz et al 1983, Satyapal et al 1999). In this source, as in the Antennae, more than 50% of the star-forming activity orginates outside the (bright) nuclei, much of it in a highly obscured, compact knot 500 pc southeast of the nucleus of NGC 3690 (Gallais et al 1999). Finally, Thuan et al (1999) have observed with ISO-CAM the blue compact dwarf galaxy SBS0335-052, which has the second lowest metallicity known in the local Universe (Z = $Z_\odot$/41). They find that this starburst system is remarkably bright in the mid-IR ($L_{12\,\mu m}/L_B$ = 2). The mid-IR spectrum can be fitted by a ∼260 K blackbody with strong 9.7/18 μm silicate absorption features and no UIB emission. Therefore, a significant fraction of the starburst activity of SBS0335-052 must be dust enshrouded—potentially a very important clue for the understanding of low metallicity starbursts in the Early Universe (Section 4).

3.2.2 Properties and Evolution of Starbursts

3.2.2 Properties and Evolution of Starbursts   Dusty starburst galaxies constitute an important class of objects (cf Moorwood 1996). About 25% of the high-mass star formation within 10 Mpc occurs in just four starburst galaxies (Heckman 1998). Key open questions of present research relate to the form of the initial stellar mass function (IMF) in starbursts and to the burst evolution. ISO has contributed to these issues mainly by studying the content of the most massive stars in dusty starbursts through nebular spectroscopy. Following pilot studies by Fischer et al (1996), Kunze et al (1996), Lutz et al (1996a), and Rigopoulou et al (1996a), Thornley et al (2000) carried out a survey with the SWS instrument (2.4 to 45 μm) of [NeIII]/[NeII] line emission in 27 starburst galaxies, with a range of luminosities from <10^8 to >10^{12} $L_\odot$. The excitation potentials of Ne$^+$ and Ne^{++} are 22 and 41 eV, and the two lines are very close in wavelength and have similar critical densities. The Neon line ratio is a sensitive tracer of the excitation

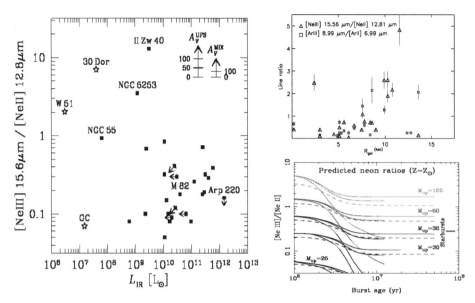

Figure 7 *Left*: SWS measurements of the [NeIII]/[NeII] ratio in starburst galaxies (*filled rectangles*) and nearby starburst templates (*asterisks*) (from Thornley et al 2000). *Top right*: [NeIII]/[NeII] line ratios (*triangles*) and [ArIII]/[ArII] ratios (*open rectangles*) as a function of galactocentric radius for HII regions in our Galaxy (Cox et al 1999). *Bottom right*: Models of [NeIII]/[NeII] ratio as a function of burst age, for different IMF upper mass cutoffs. For each M_{up} single cluster models for burst durations of 1, 5, and 20 Myr (top to bottom) are shown as continuous lines. The *dashed lines* are the corresponding 20 Myr models with a cluster mass function.

of HII regions and of the OB stars photoionizing them. The basic result for the starburst sample is that HII regions in dusty starburst galaxies on average have low excitation (left inset of Figure 7). The average [NeIII]/[NeII] ratio in starburst galaxies is a factor of 2 to 3 lower than in individual galactic compact HII regions. The SWS work establishes as a fact and puts on a safe statistical footing what was already suggested by earlier ground-based near-IR results (e.g. Rieke et al 1980, Doyon et al 1994a, Doherty et al 1995).

Thornley et al (2000) have modeled the HII regions as ionization-bounded gas clouds photoionized by central evolving star clusters. The cluster SED (as a function of IMF and evolutionary state) is used as the input for photoionization modeling of the nebular emission. The computed Ne-ratios (for a Salpeter IMF with different upper-mass cutoffs) as a function of burst age t_b and burst duration Δt_b are plotted in Figure 7 (lower right). The basic result is that on average stars more massive than about 35 to 40 $M_\odot$ are not present in starburst galaxies, either because they were never formed (upper-mass cutoff) or because they already have disappeared as a result of aging effects.

An upper-mass cutoff in the intrinsic IMF is difficult to reconcile with a growing body of direct evidence for the presence of very massive stars ($\geq 100\ M_\odot$) in nearby starburst templates (galactic center: Krabbe et al 1995, Najarro et al 1997, Serabyn et al 1998, Figer et al 1998, Ott et al 1999; R136 in 30 Doradus: Hunter et al 1995, Massey & Hunter 1998; NGC 3603: Drissen et al 1995, Eisenhauer et al 1998). In more distant HII region galaxies, the 4696 Å HeII emission line feature signals the presence of Wolf-Rayet stars of mass $\geq 60\ M_\odot$ (e.g. Conti & Vacca 1994, Gonzalez-Delgado et al 1997). The low average excitation in starbursts is thus more plausibly caused by aging (Rieke et al 1993, Genzel et al 1994). This conclusion is also supported by the large spread and substantial overlap of the Galactic and extragalactic Ne-ratios (Figure 7). The overlap region of the Antennae and the near-nuclear knot in NGC 3690 have [NeIII]/[NeII] ~ 1, comparable to nearby HII regions. Precisely in these two regions most of the flux (Section 3.2.1) comes from a single compact (and probably young) region. Everywhere else in the Antennae the Ne-ratio is much lower (Vigroux et al 1996), and at the same time the stellar cluster data (Whitmore et al 1999) indicate greater ages. Similar results emerge if, instead of the Ne-ratio, far-infrared line ratios (Fisher et al 1996, Colbert et al 1999) or the ratio L_{bol}/L_{Lyc} are used as constraints (Thornley et al 2000).

If stars of masses 50–100 $M_\odot$ are initially formed in most galaxies, the inferred burst durations must be less than 10 Myr (Figure 7). Such short-burst timescales are surprising especially for luminous, distant systems (Section 3.4.4). The dynamical time is a few million years or more. It appears that supernovae and stellar winds disrupt the ISM and prevent further star formation as soon as the first generation of O stars have formed and evolved. Starbursts must induce large negative feedback. Starburst models (Thornley et al 2000, Leitherer & Heckman 1995) as well as observations of galactic superwinds (Heckman et al 1990) indicate that 1% to 2% of the bolometric luminosity of starbursts emerges as mechanical energy in superwinds. If this mechanical energy input removes interstellar gas from the potential well with $\sim 10\%$ efficiency, the starburst is choked after the "dispersal" time t_d. For typical gas masses and starburst parameters, t_d is $\leq 10^7$ years. The burst durations thus appear to be significantly smaller than the gas consumption time scales, $t_{gc} \sim 5$ to 10×10^7 yrs. Near-IR imaging spectroscopy in M82, IC 342, and NGC 253 indicates that in the evolution of a given starburst system, there are several episodes of short-burst activity (Rieke et al 1993, Satyapal et al 1997, Böker et al 1997, Engelbracht et al 1998, Förster-Schreiber et al 2000).

Although the scenario just given is qualitatively quite plausible, the small values of the burst timescales indicated by the modeling of Thornley et al (2000) seem unphysical. In reality, burst timescales are probably somewhat larger ($\sim 10^7$ years). The galactic center starburst region has been analyzed through a direct stellar census (Krabbe et al 1995, Najarro et al 1997). The Ne-ratio computed from the starburst model ($t_b \sim 7 \times 10^6$, $\Delta t_b \sim 4 \times 10^6$ years, log U $= -1$) is more than an order of magnitude greater than the observed one (Lutz et al 1996b). The models predict that in the galactic center, the ionizing UV continuum should be dominated by O stars and hot Wolf-Rayet stars. The data show, however, that more than half of the ionizing flux comes from a population of relatively cool (20,000–30,000 K) blue supergiants (Najarro et al 1997). It would thus appear that the tracks and stellar properties adopted for the galactic center are not appropriate (Lutz 1999). Likely culprits could be the effects of metallicity and of the dense stellar winds on the UV SEDs of the massive stars. The Ne-ratio depends strongly on metallicity, as is directly demonstrated by the much higher excitation of II Zw 40 and NGC 5253 ($Z = 0.2$–$0.25 Z_{\odot}$) and of 17 HII regions in the SMC/LMC observed by Vermeij & van der Hulst (1999). The average Ne-abundance in the starburst sample is ~ 1.7 times solar. However, the models of Thornley et al (2000) are for solar metallicity. Furthermore, there appears to be a galactocentric gradient in the Ne-ratios of Cox et al (1999), again signaling dependence on metallicity. A similar discrepancy between (high-excitation) current stellar models and (low-excitation) observed nebular emission lines has also been found for Wolf-Rayet stars (Crowther et al 1999a), suggesting that their UV-spectra are softer than considered so far (Hillier & Miller 1998). Another area of concern is the impact on the interpretation of spatial variations in the various tracers that cannot be properly taken into account by single aperture spectroscopy. Crowther et al (1999b) reported SWS and ground-based infrared spectroscopy of the low-metallicity starburst galaxy NGC 5253. Although [SIV] and Brα come from a very compact (3$''$) region centered on the nucleus, the [NeII] emission is much more extended. There is a high-excitation ($T_{eff} > 38,000$ K) young (t $= 2.5$ Myr) nuclear burst surrounded by a lower-excitation ($T_{eff} \sim 35,000$ K) older (5 Myr) one. Thornley et al have raised the possibility that some of the [NeII] flux in their starburst galaxies may come from a more diffuse, low-ionization parameter zone, thus lowering the effective Ne-ratio in the SWS aperture.

3.2.3 The [CII] Line as Global Tracer of Star Formation Activity in Galaxies
Observations with the Kuiper Airborne Observatory (KAO), the COBE satellite, and balloons have established that the 158 μm [CII] emission line commonly is the strongest spectral line in the far-IR spectra of galaxies (Crawford et al 1985, Stacey et al 1991, Lord et al 1996, Wright et al 1991, Mochizuki et al 1994). The [CII] line traces photodissociation regions (PDRs) as well as diffuse HI and HII regions. It should be an excellent tracer of the global galactic star formation activity, including that of somewhat lower-mass (A+B) stars (Stacey et al 1991).

A number of studies with ISO's Long Wavelength Spectrometer (LWS) of the [CII] (and [OI]) lines have taken place. Malhotra et al (1997, 1999a), Fischer et al (1996, 1999), Lord et al (1996, 2000), Colbert et al (1999), Helou et al (1999), and Braine & Hughes (1999) have observed 70 star-forming galaxies over a wide range of activity level. Smith & Madden (1997) and Pierini et al (1999) have studied 21 Virgo cluster galaxies of normal and low star formation activity. Luhman et al (1998, and private communication) and Harvey et al (1999) have observed 14 ultraluminous galaxies. Figures 8*a* (see color insert) and 8*b* summarize these results (along with earlier work) and puts them in the context of PDR models. For most normal and moderately active star-forming galaxies, the

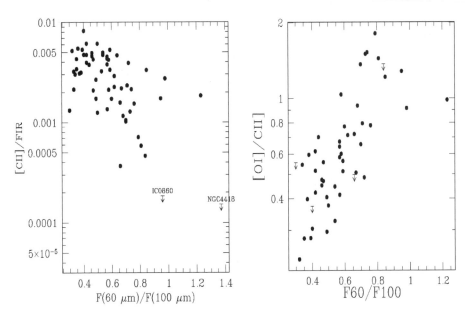

Figure 8 (b) [CII] and [OI] observations. Dependence of $Y_{[CII]}$ on IRAS 60 to 100 μm flux density ratio (~dust temperature) (from Malhotra et al 1997, 1999a). *Right*: Line flux ratio of [OI]/[CII] as a function of the IRAS 60 to 100 μm ratio (from Malhotra et al 1999a).

[CII] line is proportional to total far-IR flux and contains between 0.1 and 1% of the luminosity (Figure 8*b*; Malhotra et al 1997). In a plot[6] of the ratio of [CII] line flux to integrated far-IR continuum flux ($Y_{[CII]}$) as a function of the ratio of CO 1-0 line flux to FIR flux (Y_{CO}), most extragalactic (and galactic) data are in good overall agreeement with PDR models for hydrogen densities of $10^{3..5}$ and radiation fields of $10^{2..4}$ times the solar neighborhood field.

However, for luminous galactic HII regions and the most active star-forming galaxies, including ULIRGs, the [CII]/FIR ratio plumets to <0.1%. Malhotra et al (1997, 1999a) found that there is an inverse correlation between $Y_{[CII]}$ and dust temperature, and also between $Y_{[CII]}$ and star formation activity (as measured by the ratio of IR to B-band luminosities; Figure 8*b*). The [CII] line is faint for the most IR-luminous galaxies (Luhman et al 1998, Fischer et al 1999). In the context of standard PDR models, one major reason for the drop of $Y_{[CII]}$ is probably the lower gas heating efficiency at high radiation fields (Malhotra et al 1997). For the high UV energy densities characteristic of ULIRGs and other active galaxies, the photoelectric heating efficiency is low. This is because the dust grains are highly

---

[6]This $Y_{[CII]}$-Y_{CO} plot was first introduced by Wolfire et al (1990) to remove the dependence on the PDR filling factor in the data, and allows a direct comparison with models. These models assume that all three tracers come from the same regions, an assumption that is not always fulfilled.

positively charged, and the UIB molecules (important contributors of the photo-electrons; Bakes and Tielens 1994) are destroyed. As a consequence, the line emissivity of the PDR tracers drops sharply. Another factor may be that in such active galaxies the pressure and gas density of the ISM is significantly greater than in normal galaxies, and the $^2P_{3/2}$ level of the ground state of C^+ is collisionally de-excited. The location of the ULIRGs in Figure 8a (see color insert) can be qualitatively understood if gas densities exceed 10^5 cm^{-3}. An increase of the 63 μm [OI]/[CII] line ratio with density is, in fact, observed in the US Key Project sample (Figure 8b; Malhotra et al 1999a) but can only account for part of the decrease in $Y_{[CII]}$.

Another important effect is self-absorption in the lines. In the yet more extreme ULIRGs of the Luhman et al sample, both the 63 μm [OI] and the 145 μm [OI] lines are weak (or even in absorption) in Arp 220 (Figure 4) and Mrk 231 (Fischer et al 1997, 1999; Harvey et al 1999). The Sgr B2 star-forming region in the galactic center perhaps comes closest to the conditions in ULIRGs in terms of UV and FIR energy density, high gas density, and large H_2 column density. In Sgr B2, both the [CII] and [OI] lines show (partial) self-absorption resulting from intervening cold or subthermally excited gas (Figure 4; Cox et al 1999). The self-absorption model, does not explain the lack of 145 μm [OI] emission in Arp 220 (Luhman et al 1998), as its lower level is 228 K above the ground state and should not be populated in cold or diffuse gas. Fischer et al (1999) proposed that the radiation field in ULIRGs is soft, perhaps as a result of an aging burst. Nakagawa et al (1996) gave the same explanation for the relative weakness of the [CII] emission in the extended galactic center region. However, the moderately strong [NeII] and [SIII] lines (requiring substantial Lyman continuum luminosity) detected in the same sources make this explanation fairly implausible.

Low-metallicity galaxies often exhibit unusually intense [CII] emission, as compared with the CO mm-emission (Stacey et al 1991, Mochizuki et al 1994, Poglitsch et al 1995, Madden et al 1997; but see Israel et al 1995, Mochizuki et al 1998). They lie in the upper left of the $Y_{[CII]}$-Y_{CO} plot (Figure 8a). Low metallicity by itself cannot account for this effect, but the combination of low metallicity with small cloud column densities can (Poglitsch et al 1995, Pak et al 1998). The thin dashed curve in the upper left of Figure 8a is the model for an $A_V = 3, Z = 0.1 Z_\odot$ cloud (Kaufman et al 1999).

In the other extreme of low star-forming activity encountered in the Virgo sample, the [CII]/FIR ratio is also lower than in moderately active galaxies (Smith & Madden 1997, Pierini et al 1999). Again, this is expected in PDR models (Figure 8a) as a result of inefficient heating of the PDRs at low energy density, of a lower ratio of UV to FIR energy density, and of lower gas heating efficiency in a lower density, "HI"-medium (Madden et al 1993, 1997).

In summary, the ISO LWS results generally confirm the predictions of PDR theory (Tielens & Hollenbach 1985, Hollenbach & Tielens 1997, Kaufman et al 1999) and support the interpretation that the [CII] line is a good tool for tracing global star formation in normal and moderately active galaxies. The decrease of the [CII]/FIR ratio in very active galaxies and ULIRGs came as a surprise. In retrospect, some of that should have been expected on the basis of the theoretical predictions. Much hope has been placed on future applications of the [CII] line as a tracer of global star formation at high redshifts. The ISO data cast some doubt that these expectations will be realized for the most luminous objects unless low metallicity helps out.

3.3   Seyfert Galaxies and QSOs

3.3.1 UIB Features in Seyfert Galaxies: Confirmation of Unified Models
Clavel et al (1998) have used the UIB features as a new tool for testing unified schemes[7] in Seyfert 1 and 2 galaxies. The basic idea is that hot dust at the inner edge of the postulated circumnuclear dust/gas torus surrounding the central AGN (Antonucci 1993) re-emits the absorbed energy from the central AGN as a thermal continuum in the near- and mid-IR. In the torus models of Pier & Krolik (1992, 1993) and Granato & Danese (1994), the torus is optically thick even at mid-infrared wavelengths. The hot dust emission is predicted to be much stronger in Seyfert 1 than in Seyfert 2 galaxies, because the mid-IR continuum is angle dependent as a result of the torus blockage/shadow. The fact that the infrared energy distributions of AGNs (Seyferts and QSOs) are fairly broad and the silicate absorption/emission is relatively weak in most of the objects in the Clavel et al sample (Figure 9) argues against very compact and very thick tori. The data are in better agreement with moderately thick and extended tori, or clumpy disk configurations, as was pointed out before the ISO mission by Granato et al (1997) and Efstathiou & Rowan-Robinson (1995).

Clavel et al carried out ISOPHOT-S spectrophotometry of a sample of 20 Seyfert 1–1.5 ($\langle z \rangle = 0.036$) and 23 Seyfert 1.8–2 ($\langle z \rangle = 0.017$) galaxies drawn from the CFA sample (Huchra & Burg 1992). Their main finding is that the equivalent width of the 7.7 μm UIB feature is much stronger in Seyfert 2 galaxies than in Seyfert 1 galaxies (Figure 9). The median 7.7 μm UIB strength[5] of Seyfert 2 galaxies is ~4 times larger than for Seyfert 1s (7.7 μm strength $= 1.7 \pm 0.37$ vs. 0.4 ± 0.13). This difference is not a result of Seyfert 1s having weaker UIB emission but of their having a much stronger (red) continuum emission than Seyfert 2s (Figure 9). The distributions of UIB luminosities (and the ratios of UIB to far-IR luminosities) of the two Seyfert classes are similar (Figure 10). The UIB/far-IR ratio in Seyferts is also very similar to that in starburst galaxies (Figure 10).

The findings of Clavel et al (1998) agree with earlier ground-based, mid-IR photometry of several Seyfert samples (Maiolino et al 1995, Giuricin et al 1995,

---

[7]In unified models (e.g. Antonucci 1993), all different types of AGNs are postulated to harbor a central, accreting massive black hole surrounded by a broad line region. The AGN is proposed to be surrounded by an optically and geometrically thick gas/dust torus that anisotropically absorbs/shadows emission from the nuclear region. Broad-line Seyfert 1 galaxies (or QSOs) in this scheme are AGNs where the line of sight to the nucleus is not blocked. In narrow-line Seyfert 2 galaxies (or radio galaxies) the line of sight to the nucleus is blocked by the intervening torus.

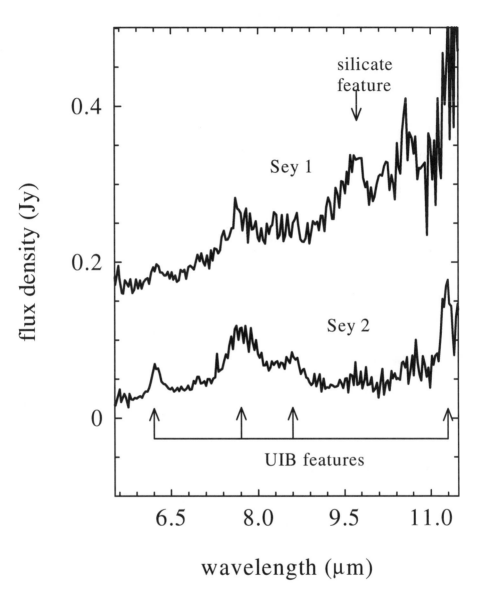

Figure 9   Average ISOPHOT-S spectra of 20 Seyfert 1 galaxies (*top*) and 23 Seyfert 2 galaxies (*bottom*) from the CFA sample (Clavel et al 1998; Schulz & Clavel, private communication). The wavelengths of the main UIB features and of the 9.7 μm silicate feature are marked by arrows. UIB features are relatively more prominent in Sey 2s than in Sey 1s, but that is mostly because of the stronger red continuum in Sey 1s. Silicate feature emission and absorption are weak.

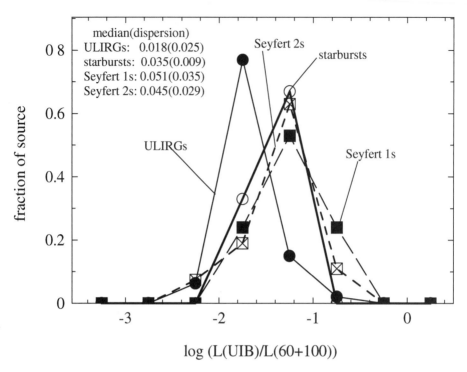

Figure 10   Fraction of sources as a function of ratio of 7.7 μm UIB luminosity to 60 + 100 μm-band IRAS luminosity for ULIRGs (*filled circles and thin continuous line*), starbursts (*open circles and thick line*), Seyfert 2s (*open rectangles and thick dashed*) and Seyfert 1s (*filled rectangles and thin long dashes*) (data from Genzel et al 1998; Rigopoulou et al 1999; Clavel et al 1998; Schulz & Clavel, private communication).

Heckman 1995). The results are in excellent agreement with the unified torus model discussed previously if the UIB emission is orientation independent. Spatially resolved ISOCAM CVF imaging in three nearby active galaxies (NGC 1068, Circinus and Cen A) shows that the UIB emission comes from kpc-scale star-forming regions in the disk, and not from the nucleus (Laurent et al 2000, Moorwood 1999). The similar UIB/far-IR ratios in Seyfert 1s, Seyfert 2s, and starburst galaxies (Figure 10) suggest that the UIB luminosity (and the far-IR continuum) in Seyfert (and other) galaxies is proportional to the sum of star formation activity and emission from the diffuse ISM (Section 3.3.2).

3.3.2 The Source of Far-Infrared Luminosity in Seyfert Galaxies and QSOs

IRAS data have shown that the infrared spectral energy distributions (SEDs) of Seyfert galaxies and (radio-quiet) QSOs are dominated by thermal (dust) emission (Sanders et al 1989, Barvainis 1990, Pier & Krolik 1993, Granato & Danese

1994, Rowan-Robinson 1995, Granato et al 1997). ISOPHOT observations of the 2–200 μm SEDs of Seyferts, radio galaxies, and QSOs have confirmed these conclusions and improved the quality and spectral detail of the data (Haas et al 1999, Wilkes et al 1999, Rodriguez Espinosa et al 1996, Perez Garcia et al 1998). As part of the European Central Quasar Program ($\sim$70 QSOs and radio galaxies), Haas et al (1998a, 1999) reported ISOPHOT/IRAS SEDs, with additional 1.3 mm points from the IRAM 30 m telescope, for two dozen (mainly radio-quiet) PG QSOs. They conclude that all QSOs in their sample, including the four radio-loud ones, have substantial amounts of cool and moderately warm (20 to 60 K) dust radiating at $\lambda \geq 60$ μm, in addition to warmer (circumnuclear) dust radiating in the mid-infrared. The derived dust masses ($10^{7\pm1}$ $M_\odot$) are typical for the total dust masses in gas-rich, normal galaxies, which suggests that the $\geq$60 μm emission is probing the large-scale disks of the host galaxies. Andreani (2000) has observed a complete subset of 34 QSOs from the Edinburgh and ESO QSO surveys that sample the bright end of the (local) QSO luminosity function. From her 11–160 μm ISOPHOT observations, she concludes that the mid-IR and far-IR fluxes are poorly correlated, as are the far-IR and blue-band fluxes. In contrast, the 60, 100 and 160 μm fluxes seem to be well correlated. This finding suggests that the far-IR emission in QSOs is a distinct physical component that may not be physically related to the shorter wavelength emission. Wilkes et al (1999) reported the first results of the US Central QSO program that contains another 70 objects and extends to z = 4.7. These data confirm as well the presence of thermal dust emission with a wide range of temperatures. The objects selected in that survey extend to the high-luminosity tail of the high-z QSO population with infrared luminosities of several 10^{15} $L_\odot$. Van Bemmel et al (1999) reported observations of matched pairs of QSOs and radio galaxies (radio power, distance, etc). They found that QSOs are actually more luminous far-IR sources than radio galaxies, contrary to simple unification schemes.

The key question that must be answered next concerns the nature of the energy source(s) powering the IR emission: direct radiation from the central AGN, or distributed star formation in the host galaxy? The near- and mid-IR emission ($\leq$30 μm) is very likely reradiated emission from the AGN accretion disk (the "Big Blue (or EUV) Bump"; Section 3.3.3). The $\lambda \leq 30$ μm SEDs can be well matched with the Pier and Krolik (1992, 1993) and Granato & Danese (1994) models of AGN heated dusty tori (scale size $\sim$100 pc), with an additional component of somewhat cooler dust.

More difficult is the answer to the question of whether the $\lambda \geq 30$ μm emission is reradiated AGN luminosity as well. A compact and thick torus, as proposed by Pier & Krolik (1992), for instance, definitely does not produce a broad enough SED to explain the far-IR emission. A clumpy, extended, and lower column density torus (Granato et al 1997), a warped disk (Sanders et al 1989), or a tapered disk (Efstathiou & Rowan-Robinson 1995) are more successful in qualitatively accounting for the observed broad SEDs. However, even these models work only marginally for a quantitative modeling of the emission at $\lambda \geq 100$ μm. For the CfA sample, the correlation between 60 + 100 μm band luminosity and (extinction-corrected) [OIII] luminosity is not impressive for Seyfert 1s, and outright poor for Seyfert 2s.

For QSOs, no clear answer has emerged yet regarding the nature of the far-IR continuum. On the one hand, Sanders et al (1989) concluded that the far-IR emission in PG quasars mainly results from AGN reradiation. On the other hand, Rowan-Robinson (1995) and Haas et al (1999) argued that the far-IR emission is caused by star-forming activity in the QSO hosts. In the Sanders et al model, a warped disk could intercept at least 10% of the luminosity of the central AGN. In their sample of $\sim$50 radio-quiet PG quasars with available IRAS luminosities, the average ratio of total infrared to total UV+visible ["Big Blue (or EUV) Bump"] luminosities is about 0.4. The $\lambda \geq 30$ μm luminosity is about half of the total infrared luminosity, thus requiring that about 15% of the nuclear luminosity be absorbed and reradiated at 10^2 to 10^3 pc from the AGN. This may be possible if substantial warps are present on that scale. In contrast, Rowan-Robinson (1995) cited the cases of three PG QSOs (0157+001, 1148+549, 1543+489) and the Seyfert 1/ULIRG Mrk 231 where the far-IR luminosity exceeds the optical+UV luminosity, a result that is not possible in the reradiation scenario (see also Section 3.4.5). An interpretation of the far-infared emission in terms of star formation is also favored by the far-IR/radio relationship in Seyferts and radio-quiet QSOs. Colina & Perez-Olea (1995, and references therein) show that the ratio of 60 + 100 μm IRAS luminosity to 5 GHz radio luminosity in these objects is in excellent agreement with the ratio found in star-forming spirals of a wide range of luminosities.

For Seyfert galaxies, a clearer picture is emerging. For the CfA Seyfert sample of Clavel et al (1998), the median ratio of 60 + 100 μm luminosity to B-band luminosity is 0.9 for Seyfert 1s and 1.6 for Seyfert 2s. Assuming that about 30% of the total UV+visible luminosity is contained in the B-band (as in the PG QSOs of Sanders et al 1989), the fraction of bolometric luminosity emerging in the far-IR in the Clavel et al Seyferts is about 40%. This is probably too large for a reradiation scenario. Perez Garcia et al (1998) have investigated the infrared spectral energy distribution of ten Seyfert galaxies with a combination of IRAS and ISOPHOT data. They decomposed the 4–200 μm spectra into a sum of blackbodies with λ^{-2} emissivities. In nine of the ten galaxies studied the infrared spectrum decomposes into three components, each with a similar, narrow range of temperatures. A 110–150 K component dominates the mid-infrared. The far-IR emission comes from a combination of a 40–50 K (30 to 100 μm) and a 10–20 K (150 to 200 μm) component. Perez Garcia et al concluded from the similarity of the temperatures of the three components that they represent well-separated spatial regions (compact torus, star-forming regions, and diffuse ISM), rather than a temperature range in a single physical component (i.e. a circumnuclear torus or warped disk). In this interpretation, the 40–50 K temperature component dominating the 60 + 100 μm IRAS band is a measure of star formation in the disk of the Seyfert galaxies. Finally, the finding (Figure 10) that the ratio of UIB luminosity to 60 + 100 μm luminosity in the Clavel et al (1998) Seyferts is essentially the same as in starburst galaxies also strongly supports an interpretation of the far-IR emission in terms of (disk) star formation.

3.3.3 Re constructing the Big Blue (EUV) Bump

In the standard paradigm, AGNs are powered by accretion onto massive black holes. From basic theoretical considerations about the size scale of the emitting region and the energy released, one expects that accretion disks emit quasi-thermal radiation with a characteristic temperature of a few 10^5 K ($\sim$40 eV). The intrinsic SEDs of AGNs cannot be directly observed between the Lyman edge and a few hundred eV, however, because of Galactic and intrinsic absorption. In Seyfert 1s and QSOs, observations shortward (X-rays) and longward (UV) of the critical region indicate that there is a break in the continuum slope. The data are broadly consistent with the existence of an emission peak [the "Big Blue Bump" (BBB)] in the extreme UV (EUV; Malkan & Sargent 1982, Sanders et al 1989, Walter et al 1994, Elvis et al 1994).

Observations of emission lines from highly excited species enable a different approach to the study of the EUV spectrum of AGNs. These "coronal" lines (Oke & Sargent 1968) originate in the narrow line region and can thus be detected even in type 2 sources, yet they may (and should, in standard unified schemes) probe the intrinsic ionizing continuum. Coronal lines are probably excited by photoionization (Oliva et al 1994). Thus, line ratios from ions in different stages of excitation, along with a photoionization model, can in principle be used to reconstruct the SED of the ionizing continuum. The tricky part is that the emission line flux depends not only on the ionizing luminosity and EUV SED, but also on the ionization parameter in the narrow line region (the ratio of ionizing flux to the local electron density). This makes the derived SED highly model-dependent on the radial and density structure of the narrow line region.

Following a first study of the ionizing continuum in the Circinus galaxy by Moorwood et al (1996a), Alexander et al (1999, 2000) have analyzed in detail three nearby Seyfert nuclei with this technique: Circinus (Seyfert 2 + starburst), NGC 4151 (Seyfert 1.5), and NGC 1068 (Seyfert 2). Starting with a (selected) compilation of ISO lines plus UV/optical/near-IR lines from the literature (typically 20 to 30 in each galaxy), Alexander et al found the best model from a scored fitting approach. In addition to the overall shape of the input SED (parameterized with 6 spectral points in the 10–500 eV range), the model depends on extinction, metallicity, and the density structure/coverage of the gas. Alexander et al explored various models of the narrow line region and concluded that there are well-defined, robust input SEDs for each of the three sources. These are shown in Figure 11. The Circinus data require a pronounced EUV bump peaking at about 70 eV and containing (in νL_ν) at least 50% of the AGN's luminosity (5×10^9 L$_\odot$; the other half is in the X-ray power-law; Moorwood et al 1996a). The black-hole mass is less than about 4×10^6 M$_\odot$ (Maiolino et al 1998), so the AGN's luminosity must be greater than 10% of the Eddington luminosity. The Circinus data are thus fully consistent with the standard AGN paradigm.

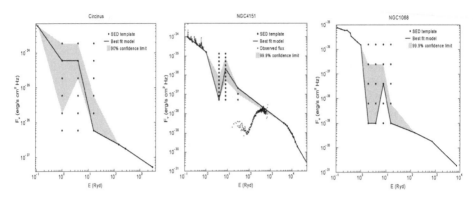

Figure 11   UV spectral energy distributions of three Seyfert galaxies reconstructed from infrared/optical narrow line region line ratios (from Alexander et al 1999, 2000; Moorwood et al 1996). The best model is indicated by a *thick continuous line*, and the shaded region marks the 99% (90% for Circinus) confidence zone. In the cases of NGC4151 and NGC1068, possible composite models with an absorbed bump are also shown.

The derived SEDs for NGC 4151 and NGC 1068 appear to be different, possibly casting doubt on unified schemes (or the technique used). In both galaxies, the SED turns down sharply just beyond the Lyman edge, comes up again equally sharply for a peak around 100 eV, and then finally drops down again and connects to the X-ray spectrum. Such a structured SED is not predicted by any accretion disk model, and it is not clear which physical mechanism might produce such a sharp emission feature at $\sim$100 eV. Alexander et al proposed as a solution that the NLR does not see the intrinsic SED of the AGN. The SEDs for both NGC 1068 and NGC 4151 can alternatively be interpreted by an EUV bump that has a deep absorption notch resulting from intervening absorption with a hydrogen column density of $\sim$4 $\times$ 10^{19} cm^{-2}. Based on ISO SWS and HST observations, Kraemer et al (1999) concluded as well that an absorber exists between the broad line region and the narrow line region of the very nearby Seyfert 1 galaxy NGC 4395. Further, there is independent evidence for such intervening absorbing gas from UV observations in NGC 4151 (Kriss et al 1992, 1995). Such absorbers may be common in Seyferts (Kraemer et al 1998). These findings are thus consistent with the standard paradigm that all Seyfert galaxies are powered by thin accretion disks with a quasi-thermal EUV bump, if in a (large) fraction of them the narrow line region sees a partially absorbed ionizing continuum.

Nevertheless, caution must be exercized before we can fully trust these very encouraging new results. Although Alexander et al explored a significant range of possible geometries of the narrow line region and density structures, it remains to be shown whether the derived SEDs are robust if more complex models are considered. For instance, Binette et al (1997) concluded that the Circinus data can be fitted with a bumpless power-law continuum if one allows for a component of optically thin (density-bounded) gas clouds.

An additional interesting aspect of the IR coronal line work is that the NGC 4151 SWS line profiles show the same blue asymmetries that are characteristic of the optical emission line profiles in this and other Seyfert galaxies (Sturm et al 1999a). This result excludes the common interpretation—at least for NGC 4151—that the profile asymmetries are caused by differential extinction in an outflow or inflow of clouds with a modest amount of mixed-in dust. Sturm et al proposed instead a geometrically thin but optically thick obscuring screen close to the nucleus of NGC 4151.

3.4 The Nature of Ultra-Luminous Infrared Galaxies

ULIRGs [$L(8-1000 \mu m) \geq 10^{12} L_{\odot}$; Soifer et al 1984, 1987] are mergers of gas-rich disk galaxies (cf Sanders & Mirabel 1996, Moorwood 1996). Luminosities and space densities of ULIRGs in the local Universe are similar to those of QSOs (Soifer et al 1987, Sanders & Mirabel 1996). In a classical paper Sanders et al (1988a) proposed that most ULIRGs are predominantly powered by dust-enshrouded QSOs in the late phases of a merger. The final state of a ULIRG merger may be a large elliptical galaxy with a massive quiescent black hole at its center (Kormendy & Sanders 1992). Despite a host of observations during the last decade, the central questions regarding what, on average, dominates the luminosity of ULIRGs, and how they evolve, are by no means answered. On the one hand, their IR, mm, and radio characteristics are similar to those of starburst galaxies (Rieke et al 1985; Rowan-Robinson & Crawford 1989; Condon et al 1991b; Sopp & Alexander 1991; Rigopoulou et al 1996b; Goldader et al 1995, 1997a,b; Acosta-Pulido et al 1996; Klaas et al 1999). Particularly compelling is the detection of a number of compact radio hypernovae in each of the two nuclei of Arp 220 (Smith et al 1998). The extended optical emission line nebulae resemble the expanding "superwind bubbles" of starburst galaxies (Armus et al 1990, Heckman et al 1990). On the other hand, a significant fraction of ULIRGs exhibits nuclear optical emission line spectra characteristic of Seyfert galaxies (Sanders et al 1988a; Armus et al 1990; Kim et al 1995, 1998; Veilleux et al 1995, 1997, 1999). Some contain compact central radio sources (Lonsdale et al 1993, 1995) and highly absorbed, hard X-ray sources (Mitsuda 1995, Brandt et al 1997, Kii et al 1997, Vignati et al 1999), all indicative of an active nucleus (AGN). With the advent of ISO, sensitive mid- and far-IR spectroscopy have become available and have allowed a fresh look at the issues of the energetics, dynamics, and evolution of ULIRGs.

3.4.1 Signature for AGN versus Starburst Activity: The ISO Diagnostic Diagram

We discussed in Section 3.1.1 that AGN and starburst template galaxies have qualitatively very different spectral characteristics (Figure 4). In the "ISO SWS diagnostic diagram," 13 bright ULIRGs studied with SWS and ISOPHOT-S (Genzel et al 1998) are located between pure AGNs and pure starbursts (Figure 5a). They appear to be composite objects, but star formation dominates in most objects (especially when taking into account that the [OIV]/[NeII] line ratios are typically upper limits). In the simple mixing model shown in Figure 5*a*, the average bright ULIRG in the study of Genzel et al (1998) has a ≤30% AGN contribution with ≥70% coming from star formation. In 4 of the 13 bright ULIRGs, an energetically significant AGN is required by the measurements, either on the basis of a detection of strong high excitation lines (Mrk 273, NGC 6240), or on the basis of low UIB strength (Mrk 231, 05189-2524). In Mrk 231, 05189-2524, and Mrk 273 the AGN contribution may range between 40 and 80%.

3.4.2 ULIRGs as a Class: The Starburst-AGN Connection   Lutz et al (1998a), Rigopoulou et al (1999a), and Tran et al (1999) extended this study to a larger sample by exploiting that the UIB (or PAH) strength criterion alone may be sufficient to distinguish between AGNs and starbursts (Figure 5*a*). The strong 6 to 11 μm UIB features can be detected in much fainter and more distant sources (up to z = 0.4 for the ISOCAM sample of Tran et al 2000). Note, however, that the UIB strength criterion rests heavily on the assumption that mid-IR features as well as continua are affected by the same extinction; otherwise, a more detailed discussion is necessary (see Sections 3.3.1, 3.4.3). The average spectrum of all 60 ULIRGs in the ISOPHOT-S sample is starburst-like (Figure 12, see color insert; Lutz et al 1998a). UIB strength and the ratio of mid-IR (5.9 μm) and far-IR (60 μm) flux densities are anticorrelated. "Warm" ULIRGs ($S_{25}/S_{60} \geq 0.2$, ∼20% of the sample) are AGN dominated, and "cold" ULIRGs are star formation dominated, in agreement with Sanders et al (1988b). The data are consistent with a model in which the UIB/FIR ratio is constant and mid-infrared extinction is one to a few magnitudes.

The UIB (PAH) strength as a qualitative tracer of AGN versus starburst nature can be exploited further to study the AGN-starburst connection as a function of luminosity and merger evolution. Figure 12 (right panel) shows the UIB strength as a function of infrared luminosity, for all ULIRGs of the ISOPHOT-S and ISOCAM samples, plus the two hyperluminous (L ∼ 10^{13} L$_\odot$) ULIRGs: FSC15307+3252 (z = 0.9; Aussel et al 1998) and P09104+4109 (z = 0.44; Taniguchi et al 1997b). The AGN fraction appears to increase with luminosity, from ∼20% for log L = 11.97–12.3, to ∼50% for log L = 12.3–13.2. Optical classifications lead to similar results. Kim et al (1998) classify as Seyfert 1/2s 26% of their ULIRGs for log L_{IR} = 12–12.3, but 45% for log L = 12.3–12.9. The ISO and optical data thus indicate that starbursts have an upper limit of ≤10^{13} L$_\odot$, consistent with the luminosities of a gas-rich galaxy whose entire molecular gas mass is converted to radiation in a dynamical time and with the nucleosynthesis efficiency $\varepsilon \sim 0.001$–0.007.

Lutz et al (1999a) found for 48 ULIRGs with ISOPHOT-S/ISOCAM data and good quality optical spectra that the optical and the ISO classifications of the individual galaxies agree very well if optical HII-galaxies and LINERs are both identified as starbursts. All but one of 23 ISO star formation dominated ULIRGs are HII/LINER galaxies, and 11 of 16 ISO AGNs are Seyferts. All but one of the Seyfert 1/2s are ISO AGNs. Identification of infrared-selected, optical LINERs as starbursts is quite plausible. LINER spectra in ULIRGs are likely caused by shock excitation in large-scale superwinds, and not by circumnuclear gas photoionized by a central AGN (Heckman et al 1987, Armus et al 1990).

The evolutionary scenario proposed by Sanders et al (1988a) postulates that interaction and merging of the ULIRG parent galaxies triggers starburst activity that later subsides, while the AGN increasingly dominates the luminosity and expels the obscuring dust. An implication of this scenario is that advanced mergers should, on average, be more AGN-like than in earlier stages when the interacting galaxies are still well separated. However, Rigopoulou et al (1999a) and Lutz et al (1998a) find that powerful starbursts are present at the smallest nuclear separations. There is no obvious trend for AGNs to dominate with decreasing separation. This suggests that local and short-term factors, in addition to the global state of the merger, play a role in determining which of the energy sources dominates the total energy output. Relevant processes include the accretion rate onto the central massive black hole(s), the radiation efficiency of the accretion flow, and the compression of the interstellar gas on scales of a few tens of parsecs or less.

3.4.3 Hidden AGNs: The Role of Dust Obscuration in ULIRGs

The active regions of ULIRGs are veiled by thick layers of dust. Fine structure and recombination line ratios imply equivalent "screen" dust extinctions in ULIRGs (and LIRGs) between AV $\sim$ 5 and 50 (Genzel et al 1998). ISOPHOT-S and ground-based data confirm this evidence for high dust extinction (Lutz et al 1998a, Dudley & Wynn-Williams 1997, Dudley 1999). The most extreme case is Arp 220, where the ISO SWS data indicate A_V(screen) $\sim$ 50 ($A_{25\mu m} \sim 1$), or an equivalent mixed model extinction of 500 to 1000 mag in the V-band[8] (Sturm et al 1996). Smith et al (1989) and Soifer et al (1999) found similar values from the depth of the silicate feature and from mid-IR imaging. These large extinctions combined with a mixed extinction model solve the puzzle that starburst models based on near-IR/optical data cannot account for the far-IR luminosities ($A_V \leq 10$; Armus et al 1995, Goldader et al 1997a,b). If the emission line data are instead corrected for the (much larger) ISO-derived extinctions, the derived L_{IR}/L_{Lyc} ratios are in reasonable agreement with starburst models (Section 3.4.4).

Submm/mm CO and dust observations imply yet larger column densities than the mid-IR data ($\geq 10^{24}$ cm^{-2}, or $A_V \geq 500$; Rigopoulou et al 1996b, Solomon et al 1997, Scoville et al 1997, Downes and Solomon 1998). Is it possible, therefore, that

---

[8]In the screen extinction model, the dust is in a homogeneous screen in front of the source, and the attenuation at λ is given by $\exp[-\tau_d(\lambda)]$, where $\tau_d(\lambda)$ is the optical depth of the dust screen at λ. An alternative and probably more plausible scenario is that obscuring dust clouds and emitting HII regions are completely spatially mixed throughout an extended region. In this case, the attenuation at λ is $\tau_d(\lambda)/\{1 - \exp[-\tau_d(\lambda)]\}$, which changes much more slowly with wavelength.

most ULIRGs contain powerful central AGNs that are missed by the mid-IR data? Hard X-rays penetrate to column densities $\geq 10^{24}$ cm^{-2}. ASCA has observed a small sample of ULIRGs in the 2–10 keV band. About a dozen sources are common between ISO and ASCA (Brandt et al 1997, Kii et al 1997, Nakagawa et al 2000, Misaki et al 1999). In Mrk 273, 05189-2524, NGC 6240, and 230605+0505, ASCA finds evidence for a hard X-ray source with $\langle L_X/L_{IR} \rangle \geq 10^{-3}$. BeppoSAX observations show that the AGN in NGC 6240 is attenuated by a Compton thick (NH $\sim 2 \times 10^{24}$ cm^{-2}) absorber (Vignati et al 1999). After correction for this absorption, and depending on its filling factor, the ratio of intrinsic AGN X-ray to IR luminosity is 2 to 6 $\times 10^{-2}$. In Mrk 231, a hard X-ray source is seen but is weak ($\langle L_X/L_{IR} \rangle \sim 10^{-3.5}$). ISO finds evidence for significant AGN activity in all of these sources as well. In sources classified by ISO as starburst dominated (Arp 220, UGC 5101, 17208-0014, 20551-4250, 23128-5919), ASCA also finds no hard X-ray source. The limit to the hard X-ray emission in Arp 220 corresponds to $\leq 10^{-4}$ of the infrared luminosity. For comparison, in Seyfert 1 and Seyfert 2 galaxies $\langle L_X/L_{IR} \rangle$ is 10^{-1} and 10^{-2}, respectively (Boller et al 1997, Awaki et al 1991). For radio-quiet QSOs, the sample averaged SEDs of Sanders et al (1989) and Elvis et al (1994), giving $L_X/L_{IR} = 0.2$ and $L_X/L_{bol} \sim 0.05$. Although the statistics are still relatively poor at this point, the hard X-ray data do not present evidence for powerful AGNs that are completely missed by the mid-IR observations. Still, there are exceptions to this reasonable agreement between IR and X-ray data. The nearby galaxy NGC 4945 fulfills all criteria of a pure starburst at optical to mid-IR wavelengths (Moorwood et al 1996b; in Figure 5a, NGC 4945 is the starburst to the bottom right of Arp 220 and top left of M82). There is no evidence for a narrow line region or any other AGN indicator at these wavelengths. The [NeIII]/[NeII] line ratio is small and indicates that the starburst is aging (Spoon, private communication). Yet ASCA and BeppoSAX data show that at its center lurks a powerful AGN, attenuated by a Compton thick foreground absorber (Iwasawa et al 1993). As in NGC 6240, both the AGN (from the X-ray data) as well as the starburst (from the optical to IR data) can account for the entire bolometric luminosity of NGC 4945. Although NGC 4945 is much less luminous than a ULIRG, the case is puzzling and requires further study.

If mid-infrared continuum and UIB features suffer different obscurations, as in Seyfert 2 galaxies (Section 3.3.1; Clavel et al 1998), the UIB strength criterion[5] loses its meaning. Instead it is necessary to directly compare the ratio of UIB luminosity to total far-IR (60 + 100 μm) luminosity, even if this ratio depends on mid-infrared extinction. The UIB/FIR ratio in ULIRGs (Figure 10) is on average half of that in starbursts (and Seyferts). Following the discussion in Section 3.3.2, this suggests that at least half of the luminosity in the average ULIRG comes from star formation if the same UIB/FIR ratio holds as in other galaxies. Correction for extinction increases this fraction. If the average mid-IR extinction is A_V(screen) ~ 15 and $A_{7.7}/A_V \sim 0.04$, the UIB/FIR ratio is fully consistent with that in starburst galaxies.

Hence, despite the large extinctions present in most ULIRGs, optical/near-IR emission line diagnostics remain useful qualitative diagnostic tools in the majority of sources. For these objects, mid-IR emission lines/UIB features can be used as quantitative tools for estimating the relative contributions of AGN and star formation. The reason is that optical and IR emission line diagnostics only rely on penetrating through dust in the disk to the narrow line region, and not through a high column density circumnuclear torus to the central AGN itself. The large-scale (≥ 100 pc) obscuring material is likely very patchy and arranged in thin (and self-gravitating) disks (as in Arp 220; Scoville et al 1998). Radiation and outflows from AGNs punch rapidly through the clumpy obscuring screen, at least in certain directions.

3.4.4 Properties and Evolution of Star Formation in ULIRGs

In the framework of starburst models with a Salpeter IMF and upper mass cutoffs of 50 to 100 M$_\odot$ (Krabbe et al 1994, Leitherer & Heckman 1995), the ULIRG data are fit with burst ages/durations of t $\sim \Delta$t $\sim$ 5 to 100 $\times$ 10^6 years (Genzel et al 1998). In the case of NGC 6240, Tecza et al (2000) have directly determined the age of the most recent star formation activity [t $= (2 \pm 0.5) \times 10^7$ years] from the fact that the K-band light in both nuclei appears to be dominated by red supergiants. For that age and the measured (low) Brγ equivalent width, the duration of the burst Δt has to be about 5 $\pm$ 2 $\times 10^6$ years (Tecza et al 2000). Rigopoulou et al (1999a) found that there is no correlation between the gas content/mass of a ULIRG and the merger phase, as measured by the separation of their nuclei, in contrast to infrared galaxies of somewhat lower luminosities (LIRGs; Gao and Solomon 1999). Even very compact ULIRGs such as Arp 220 and Mrk 231 still have a lot of molecular hydrogen.

Burst ages in ULIRGs thus are similar to those in other starbursts (Section 3.2.2), are comparable to the dynamical time scale (between peri-approaches), and are much smaller than the gas exhaustion time (Section 3.2.2) and overall merger age. There are likely several bursts during the merger evolution (Section 3.2.1). A burst may be triggered as the result of the gravitational compression of the gas during the encounters of the nuclei (Mihos & Hernquist 1996). It is terminated within an O-star lifetime by the negative feedback of supernovae and superwinds. The models of Mihos & Hernquist suggest that by far the most powerful bursts occur when the nuclei finally merge and the circumnuclear gas is compressed the most. The data indeed suggest that ULIRGs switch on only in the last $\sim$20% of the merger history (Mihos 2000). However, the data do not indicate that the most powerful starbursts in ULIRGs occur in the last phases with very small separations. Far-IR luminosities of ULIRGs do not change significantly with nuclear separation from separations of a few hundred parsecs to a few tens of kpc (Rigopoulou et al 1999a).

3.4.5 Warm ULIRGs and QSOs

"Warm" ULIRGs ($S_{25}/S_{60} \geq 0.2$) probably represent transition objects between ULIRGs and QSOs (Sanders et al 1988b). The ISO spectrophotometry discussed above strongly supports the conclusion of the earlier IRAS work that warm ULIRGs indeed contain powerful AGNs. For instance, the warm ULIRG Mrk 231 (the most luminous ULIRG within $z \leq 0.1$) is similar to the infrared-loud but optically selected QSOs I Zw1 and Mrk 1014. Most of the warm ULIRGs have Seyfert spectra (Veilleux et al 1997, 1999). High-resolution HST and ground-based observations show that warm ULIRGs typically have a compact hot dust source at the nucleus, surrounded by a luminous host galaxy with embedded compact star clusters and large-scale, tidal tails (Surace et al 1998, Surace & Sanders 1999, Lai et al 1998, Scoville et al 2000).

Although the AGN is probably dominating, circumnuclear star formation may nevertheless significantly contribute to the total luminosity of warm ULIRGs (Condon et al 1991b, Downes & Solomon 1998). In the case of Mrk 231 Taylor et al (1999) found a smooth, approximately circular radio continuum source of diameter $\sim 1''$ (1 kpc), centered on the core-jet structure of the bright AGN. The extended radio source is associated with a massive rotating molecular disk seen at low inclination (Downes & Solomon 1998) and with a luminous (2L$_*$ for no extinction correction) near-IR stellar disk (Lai et al 1998). If the far-IR–radio correlation for star-forming galaxies (Condon et al 1991a, Condon 1992) is applied to the parameters of Mrk 231's disk, the implied far-IR luminosity is 1.3×10^{12} L$_\odot$, or $\sim 70\%$ of the $60 + 100 \, \mu$m IRAS luminosity. If the red color of the disk's near-IR emission results from extinction (A$_V \sim 12$; Lai et al 1998), the corrected K-band luminosity of the disk is L$_K(0)^9 \sim 2 \times 10^{10}$ L$_\odot$. If the K-band emission comes from an old stellar population with M/L$_K \sim 100$, the mass of the disk within the central kpc has to be 2×10^{12} M$_\odot$ (an 8 M$_\odot$ disk!). This mass is two orders of magnitude greater than the dynamical mass within the central kpc derived from the CO rotation curve (Downes & Solomon 1998). It is thus more likely that the K-band light comes from a population of (young) red supergiant or AGB stars. The required mass then is 20 to 100 times smaller, and the ratio of bolometric to K-band luminosity is $\geq 10^2$ (Thatte et al 1997). The red disk may then be responsible for a significant fraction of the far-IR luminosity of Mrk 231, in accordance with the radio data. The relative weakness of the mid-infrared line emission observed by ISO-SWS would then require very high extinction or, perhaps more likely, an aged starburst. The situations for the IR-loud QSOs I Zw 1 and Mrk 1014 are similar, but even more extreme stellar masses would be required there to account for the large K-band stellar luminosities in both galaxies ($\sim 1.4 \times 10^{11}$ L$_\odot$) by an old stellar population. In both cases the dynamical masses implied by the molecular disk rotation excludes this possibility (Schinnerer et al 1998, Downes and Solomon 1998).

In summary, ISO has made substantial contributions to our understanding of ULIRGs. The resulting picture is fairly complex; there is no simple story to tell. ULIRGs do not seem to undergo an obvious metamorphosis from a starburst-powered system of colliding galaxies to a buried and then finally a naked QSO at the end of the merger phase. Average luminosity and molecular gas content do not change with merger phase. AGN activity and starburst activity often occur at the

---

[9]LK is defined here as the luminosity emitted in the K-band (in units of solar luminosities).

same time. On average, starburst activity does seem to dominate the bolometric luminosity in most objects. However, the most luminous ULIRGs appear to be AGN dominated. "Warm" ULIRGs are AGN dominated, and "cold" ULIRGs are star-formation dominated. The relative contributions of starburst and AGN activity are in part controlled by local effects. Warm ULIRGs/infrared-loud QSOs are advanced transition objects with powerful central AGNs. In a few well-studied cases, these AGNs are surrounded by luminous (aged?) starburst disks. Locally, (U)LIRGs are a spectacular curiosity. They contribute only $\sim 1\%$ to the local infrared radiation field, and $\sim 0.3\%$ to the total bolometric emissivity (Sanders & Mirabel 1996, Heckman 2000). We will show in the next chapter that this fraction increases dramatically (factor $\geq 10^2$) at higher redshift.

4. THE DISTANT UNIVERSE

In the local Universe surveyed by IRAS, only 30% of the total energy output of galaxies emerges in the mid- and far-IR (Soifer & Neugebauer 1991). If the same were true at high redshifts, optical/UV observations alone would be sufficient for determining the cosmic star formation history. Spectacular (rest frame) UV observations of star-forming galaxies at redshifts ≥ 3 have become possible in the last few years (Steidel et al 1996, Lilly et al 1996). The inferred star formation rate per (comoving) volume element increases from $z = 0$ by more than an order of magnitude to a peak at $z \sim 1$–3 (Madau et al 1996, Pettini et al 1998, Steidel et al 1999).

A significant new element in the picture of high-z star formation has emerged with the COBE detection of an extragalactic submillimeter background (Puget et al 1996). This diffuse background has an integrated intensity that is comparable to or larger than that of the integrated UV/optical light of galaxies (Dwek et al 1998, Hauser et al 1998, Lagache et al 1999). The implication is that there likely exists a very significant contribution of dust-obscured star formation at high redshifts. Clearly, the next major step is to detect these high-redshift infrared galaxies directly. With the exception of a few hyperluminous and/or lensed objects (such as F10214+4724; Rowan-Robinson et al 1991), IRAS was only able to detect infrared galaxies to moderate redshifts ($z \sim 0.3$). In the mid-infrared, ISOCAM is $\sim 10^3$ times more sensitive and has 60 times higher spatial resolution than IRAS. In the far-IR with ISOPHOT, the improvement in sensitivity is modest, but the extension to longer wavelength (175 μm) is of substantial benefit. Exploiting these improvements in deep surveys, ISO has been able to provide for the first time a glimpse of the infrared emission of galaxies at $z \geq 0.5$.

4.1 Mid-Infrared Surveys

The various ISOCAM surveys trade off depth and area (see Rowan-Robinson et al 1999 for a compilation). For high-redshift work, the most meaningful are the 15 μm (LW3 filter) surveys. At 6.75 μm (LW2), the stellar contamination is very high, and a careful screening at other wavelengths is necessary. Various strategies have been successfully developed to overcome the cosmic ray–induced glitches that mimic faint sources (Starck et al 1999, Desert et al 1999, Serjeant et al 1999). At the time of writing, the published work reports 15 μm detections of ∼1400 galaxies, with flux densities between ≥30 μJy and ≥0.3 Jy. Shallow, deep, and ultra-deep surveys were performed in the Lockman Hole and the Marano field (IGTES surveys; Cesarsky and Elbaz 1996, Elbaz et al 1999a). They are complemented on the faint end by the surveys made on deep IR fields centered on the southern and northern Hubble Deep Fields (Figure 13, see color insert) as observed by Rowan-Robinson and colleagues [HDF(N): Serjeant et al 1997, Aussel et al 1999a (Figure 13), Desert et al 1999; HDF(S): Oliver et al 2000, Elbaz et al 1999a]. Even deeper images are available from lens magnification in the direction of distant clusters [Metcalfe et al 1999, Altieri et al 1999 (Figure 13), Barvainis et al 1999]. Distant clusters also are surveyed to compare the evolution of galaxies in different environments (Lemonon et al 1998, Fadda & Elbaz 1999, Fadda et al 2000). The bright end of the luminosity function is explored by the largest area survey, the 12-square-degree ELAIS program (Rowan-Robinson et al 1999, Serjeant et al 1999). Levine et al (1998) reported the first results of a program to observe 500 candidate extragalactic sources from the IRAS faint source survey with ISOCAM and ISOPHOT (Levine et al 1998). This ISO-IRAS faint galaxy survey efficiently picks up moderate redshift (z = 0.1–0.4) (U)LIRGs.

4.1.1 Source Counts: Evidence for Strong Evolution   Figure 14 (from Serjeant, private communication; Elbaz et al 1999a) gives the integrated and differential source counts derived from these surveys. Given the difficulties of reliable faint source detection and photometry mentioned previously, it is reassuring that the counts of the different surveys agree quite well. Starting with IRAS and going to fainter flux densities, the number counts initially lie on a linear (log-log plot) extrapolation of the IRAS counts and do not require strong evolution. However, from >10 mJy down to 0.4 mJy the counts increase rapidly, with a slope in the integrated counts of $\alpha = -3$. This is significantly steeper than expected in a Euclidean model without luminosity/density evolution ($\alpha = -2.5$). Below 0.4 mJy down to the faintest flux densities sampled, the slope is flatter ($\alpha = -1.6$) and the counts appear to converge. In the differential source counts (right inset of Figure 14; Elbaz et al 1999a, 2000) this leads to a prominent hump peaking at ∼0.4 mJy. At the peak of the hump the observed source counts are an order of magnitude above non-evolution models, obtained by extrapolating the local IRAS 12 μm luminosity function (Rush et al 1993, Fang et al 1998). The mid-infrared source counts require strong cosmic evolution of the mid-infared emission of galaxies (Oliver et al 1997, Roche & Eales 1999, Clements et al 1999, Elbaz et al 1999a,b, Serjeant et al 1999).

The 15 μm integral number counts agree with model predictions over more than five orders of magnitude in flux density (Franceschini et al 1994, 2000;

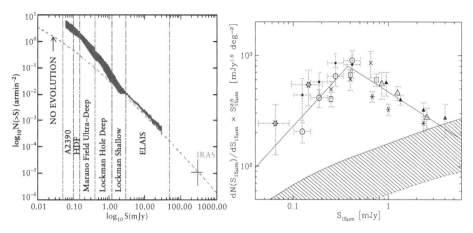

Figure 14   *Top left*: Summary of integrated 15 μm source counts from the different ISO surveys (from Serjeant, private communication; Elbaz et al 1999a; normalizing downward by a factor 1.5 the counts of Serjeant et al 1999), compared to non-evolution models (continuous) matching the IRAS counts (the Rush et al 1993 counts were renormalized downward by a factor of 2). *Top right*: Differential 15 μm counts with the shaded area marking the counts predicted with non-evolution models (from Elbaz et al 1999*a*). The counts are normalized to a Euclidean distribution of non-evolving sources, which would have a slope of index -2.5 in such a universe. *Data points*: A2390 (*open stars*; Altieri et al 1999, Metcalfe et al 1999), HDF(N) (*open circles*; Aussel et al 1999a), HDF(S) (*filled circles*; Elbaz et al 1999a), Marano (*open squares, crosses, stars*; Elbaz et al 1999a), Lockman hole (*open and filled triangles*; Elbaz et al 1999a).

Pearson & Rowan-Robinson 1996; Guiderdoni et al 1997, 1998; Xu et al 1998; Roche & Eales 1999). Common to all these models is the assumption of strong luminosity and/or density evolution of dusty star formation in bright normal spirals and starburst galaxies, with varying contributions from AGNs. We now discuss in more detail the models by Xu et al (1998), since they are based on the most recent ISO SWS-ISOPHOT-S spectra of template galaxies to determine accurate, mid-IR k-corrections. These are important because of the structure caused by UIB features in the 6–13 μm spectra of most galaxies. The models of Pearson, Rowan-Robinson, Franceschini, Roche & Eales, and Guiderdoni et al lead to similar results. Xu et al constructed a local luminosity function from IRAS observations (Shupe et al 1998) and assigned to each of these galaxies a mid-IR spectrum composed of three categories (cirrus/PDR, starburst, AGN, or a mixture). They then extrapolated the local luminosity function to higher redshift by applying the appropriate k-corrections and evolution. The predicted differential, Euclidean, normalized source counts for the LW3 ISOCAM filter show distinct differences between luminosity evolution [$L(z) \propto (1 + z)^3$] and density evolution [$\rho(z) \propto (1 + z)^4$]. For pure density evolution the predicted source counts are fairly flat, from 1 Jy to 100 μJy. For luminosity evolution, however, a characteristic bump is predicted at around 300 μJy, with a sharp fall-off of the counts at fainter flux densities. This is similar to what is actually observed by Elbaz et al (1999a; Figure 14), although the amplitude of the hump in the differential source counts requires an even steeper evolution than considered by Xu et al [e.g. in aqn Einstein-de Sitter universe combination of $(1 + z)^3$ luminosity evolution and $(1 + z)^6$ density evolution, for the starburst component only, at $z < 0.8$–0.9; Franceschini et al 2000]. The interpretation of the bump is that with luminosity evolution and flat or even "negative" k-correction (between $z = 0.5$ and 1; Elbaz et al 1999b) the number counts increase initially, with decreasing flux density much faster than Euclidean. The unfavorable k-correction and the decreasing slope of the available comoving volume cause the number counts to drop rapidly with increasing z for $z \geq 1$. For sources near the bump, the luminosity evolution model predicts a broad redshift distribution, with a median at $z \sim 0.9$ but extending with significant probability to redshifts >2. In contrast, the pure density evolution model predicts a much smaller median redshift (~ 0.5) and essentially no sources with $z > 1$. When the mid-infrared counts are compared to K-band counts at the same energy (νS_ν), the mid-infrared sources contribute only about 10% of the energy at the bright end, but more than 50% at $S_{15\mu m} \sim$ a few hundred μJy (Elbaz et al 1999a). Thus, the ISOCAM sources must represent an important subclass of the optical/near-IR galaxies that are dusty and infrared active.

The situation is less clear at 6.75 μm (Goldschmidt et al 1997, Taniguchi et al 1997a, Aussel et al 1999a,b, Sato et al 1999, Serjeant et al 1999, Flores et al 1999a). The stellar contamination is larger. The local luminosity function and template spectra appropriate for the galaxies are less well known than at 15 μm. The number of confidently detected sources is also much smaller. They are (more) compatible with a non-evolution (i.e. passive evolution) model with a significant contribution of ellipticals/S0s (E/S0s; Roche & Eales 1999). They are also consistent with the Pearson & Rowan-Robinson (1996) luminosity evolution model, but not with that of Franceschini et al (1994) (Serjeant et al 1999).

4.1.2 Nature of ISOCAM Sources: Luminous Star-Forming Galaxies at $z \sim$ 0.7
At the time of writing, the properties of the ISOCAM galaxies at other wavelength bands have been studied in detail in two fields: HDF (N) (Mann et al 1997, Rowan-Robinson et al 1997, Aussel et al 1999a) and the 1415+52 field of the Canada-France Redshift Survey (CFRS; Flores et al 1999a,b). In HDF(N), Aussel et al (1999a,b) have extracted a list of 38 galaxies (≥ 100 μJy, 99% confidence) with optical identifications in the catalog of Barger et al (1999). Of these, 26 galaxies have known redshifts. In the $(10')^2$ CFRS 1415+52 field, Flores et al (1999b) detected 78 significant ($\geq 3\sigma$) 15 μm sources (≥ 250 μJy), 22 of which have spectra and redshifts. The median redshift in both fields is near 0.7, and most redshifts range between 0.4 and 1.3. The ISOCAM galaxies have the optical colors of Sbc-Scd galaxies. The majority of the galaxies are disk or interacting systems, and the remainder are irregulars and E/S0s. The one elliptical in the HDF(N) sample is associated with an AGN ($z = 0.96$). Typical absolute K-band magnitudes HDF(N) range between -22 and -25, similar to, but on average

somewhat fainter than, an L_* galaxy [M(K) = -24]. Thus, the ISOCAM galaxies are mainly luminous disk/interacting galaxies and are definitely not part of the faint blue galaxy population responsible for the excess in faint B-band counts (Ellis 1997).

The local luminosity function at the mid-IR luminosities characteristic of the ISOCAM galaxies ($\sim$5 $\times$ 10^{10} $L_\odot$) is dominated by AGNs (Fang et al 1998). However, the rest-wavelength B-band spectroscopy in both the HDF(N) and CFRS fields (Flores et al 1999b, Aussel et al 1999b) and the rest-wavelength R-band observations of the HDF(S) field (Rigopoulou et al 1999b) suggest that most of the ISOCAM galaxies are dominated by star formation, with no more than $\sim$1/3 of them AGNs. The majority of the galaxies in the CFRS field (15% or 70%) have e(a) optical spectra[10] characteristic of post-starburst systems (age $\geq$ several hundred Myrs), or active starburst galaxies with large differential intrinsic dust extinction hiding the active burst, with some starburst activity prior to a few hundred million years ago and at the same time, the latter explanation is almost certainly the correct one (Aussel et al 1999b, Rigopoulou et al 1999b). For 19 galaxies with complete radio to UV SEDs, Flores et al (1999b) identified more than half as (highly reddened) starbursts. Dusty starburst galaxies such as M82 (Kennicutt 1992), many bright ULIRGs (Liu and Kennicutt 1995), and 50% of the (U)LIRG sample studied by Wu et al (1998) have e(a) spectra in the B-band. Local e(a) galaxies have large Hα equivalent widths, at the same time demonstrating active current star formation and differential dust extinction. Rigopoulou et al (1999b) have recently carried out near-IR, VLT-ISAAC spectroscopy of a sample of $0.6 \leq z \leq 1.3$ ISOCAM galaxies in the HDF(S) field. The ISOCAM HDF(S) galaxies have large (50–100 Å) Hα equivalent widths, which provides compelling evidence that most of them are active starbursts. The simultaneous presence of heavily dust-enshrouded present star formation and less extinct older star forming activity probably indicates several starburst episodes (Sections 3.2.1, 3.4.4).

The infrared derived star formation rates are substantially greater than those determined from the [OII] (or Hα) lines. On the basis of the SEDs, Flores et al find that the median 8–1000 μm luminosity of the CFRS sample is $\sim$3 $\times$ 10^{11} $L_\odot$ (star formation rate $\sim$30 to 50 $M_\odot$ yr^{-1}). The HDF(N) results are similar. For HDF(s) Rigopoulou et al 2000 and Franceschini 2000 (private communication) deduced star formation rates of a few tens of $M_\odot$ yr^{-1} from the Hα emission but typically three times greater values from the mid-IR data. Most of the faint ISOCAM galaxies thus appear to be LIRGs. A smaller fraction ($\sim$25%) of the CFRS sources have ULIRG-like luminosities (10^{12} $L_\odot$), in agreement with the

---

[10]e(a) galaxies (Poggianti and Wu 1999) or S+A galaxies (Hammer et al 1997) have moderate EW ([OII]) line emission and at the same time strong Balmer (Hδ, etc) absorption or (at low spectral resolution) a large 3550–3850 Å Balmer continuum break, characteristic of either A-stars or very large extinction. In the A-star model, the characteristic age of the (optically visible) star formation is $\sim$0.5–1. Gyrs. Galaxies with Balmer absorption but no emission lines are called 'E+A', or ' k+A'.

work of Rowan-Robinson et al (1997) for HDF(N). Other confirmations of the identification of the faint ISOCAM galaxies as luminous starbursts come from observations of several galaxies near z ∼ 1 that are lensed by foreground clusters (Lemonon et al 1998, Barvainis et al 1999). All e(a) galaxies in the z = 0.2 cluster A1689 are 15 μm ISOCAM sources (Duc et al 2000). Of the k+A and e(a) galaxies in Coma, only those with emission lines have excess mid-IR emission indicative of active star formation (Quillen et al 1999).

In the LW2 filter the situation is similar, but the relative number of AGNs and ellipticals is proportionally larger. Flores et al (1999a) identified 40% of the 15 CFRS/6.75 μm galaxies with spectra as AGNs, and 53% as active starbursts or S+A galaxies. Aussel et al (1999b) classified 4 of the 6 confidently detected 6.75 μm sources in HDF(N) as elliptical galaxies.

For the great majority of ISOCAM sources, the far-IR counterparts are not yet known. Still, template spectra and the overall background constraint can be used to draw first-order conclusions (Figure 15). If all the ISOCAM sources had ULIRG spectra as cool as Arp 220 [L(80 μm)/L(8 μm) ∼ 70[11]], they would significantly overproduce the far-IR/submm background and would give rise to counts at 850/175 μm, well in excess of the observations (Aussel et al 1999b, Elbaz et al 2000). Thus, on average, they must have SEDs at least as warm as LIRGs [like M82 or NGC3256: L(80 μm)/L(8 μm) ∼ 5..12], in which case they still produce 30–60% of the far-IR background at 140 μm (Figure 15). A similar contribution of the ISOCAM sources to the background would be expected if they were predominantly dust-enshrouded AGNs (Figure 15). This interpretation is not supported by the HDF(N,S) and CFRS samples, however.

4.2 Far-Infrared Surveys

The SED of actively star-forming galaxies peaks at λ ∼ 60–80 μm. For observations at $\lambda \geq 200$ μm, the "negative" k-correction is advantageous for detecting high-redshift galaxies; at the same time, such measurements give the total luminosity directly, without a model-dependent bolometric correction. For these reasons, several surveys were undertaken with the ISOPHOT at 175 μm. A 1-square-degree field in the Lockman hole was surveyed as part of the Japanese ISO cosmology project (Kawara et al 1998, 1999; see Figure 16, see color insert). Two fields in the southern Marano area and two fields in the northern ELAIS fields with a combined area of 4 square degrees constitute the FIRBACK program (Puget et al 1999, Dole et al 1999). An additional "serendipity" survey used the slews between pointed observations to survey the sky at ∼200 μm at the ≥ 2 Jy level (Stickel et al 1998, 1999). This survey extends IRAS far-IR SEDs to longer wavelengths for a few thousand nearby galaxies. The depth of the two deep surveys is ∼100 mJy (5σ). The key problems for the far-IR detection of faint

---

[11]Arp 220 has an extremely cool SED. Average ULIRGs have L(80 μm)/L(8 μm) ∼ 25, which would still exceed the far-IR/submm background (but only marginally so), yet would leave no space for a higher z population.

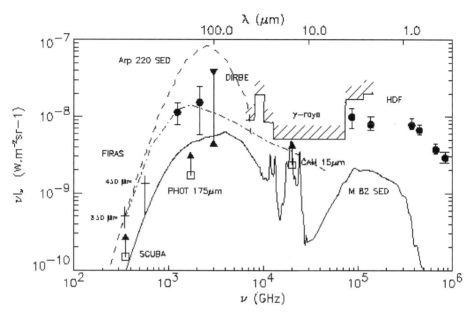

Figure 15   Cosmic UV to mm, extragalactic background. *Open squares* give the lower limits from ISOCAM 15 μm, ISOPHOT 175 μm, and SCUBA 850 μm (Blain et al 1999) sources. The optical-UV points are from Pozzetti et al (1998). The COBE FIRAS (*grey shaded*) and DIRBE 140/240 μm (*filled circles*) data are from Lagache et al (1999) (from Elbaz et al 2000). Different SEDs are shown and normalized to the 15 μm ISOCAM limit: M82 (*continuous line*), Arp 220 (*long dashes*). An Arp 220-like SED would significantly overproduce the COBE background. With an M82-like background, the ISOCAM galaxies would contribute about 30% to the COBE background. The SED of a typical Seyfert 2 galaxy (and matching the COBE background) is shown as a *dash-dotted line*. The *hatched* upper limits in the mid-IR are derived from the lack of attenuation of high-energy γ-rays (Hauser et al 1998).

individual sources are detector noise and large, spatially varying backgrounds: Galactic cirrus, extragalactic backgrounds, and (to a lesser extent) zodiacal dust emission. To minimize the impact of cirrus, the deep survey areas were chosen in regions known to have low HI-column densities and IRAS 100 μm surface brightness. In the best areas, the surveys are limited by the fluctuations in the extragalactic background (Lagache & Puget 1999). In the case of the Lockman hole and ELAIS regions, a comparison with a second map at 90 μm gives additional useful information. In the final source lists, only detections with flux densities ≥ 120 mJy were taken into account. At that level, reliability and completeness are good and far above the expected fluctuations in the cirrus and extragalactic backgrounds. The sources detected, therefore, are likely individual galaxies. At the 100 mJy cutoff level, Kawara et al reported 45 175-μm sources (and 36 95-μm sources) in the Lockman field, and Dole et al found 208 175-μm sources in the combined Marano and ELAIS fields. In a 0.25-square-degree Marano field, Puget et al detected 22 sources. Figure 16 shows the counts from all three references, in comparison with models A and E of Guiderdoni et al (1998). At the lowest flux densities, the number counts are a factor of 4 to 10 above the predictions of non-evolving models. The counts are consistent with those that include strong evolution as well as a significant population of $z \geq 1$ ULIRGs (e.g. model E of Guiderdoni et al 1998).

The integrated counts are an order of magnitude greater than the 100 μm IRAS counts at the same flux level. In the ELAIS area, 31 sources have a 90 μm counterpart. The average 175/90 flux density ratio is ≥ 2, as compared with ~ 1 for a sample of local sources (Stickel et al 1998). One part of the Marano ISOPHOT field overlaps with the deep ISOCAM survey area. At most, 50% of the ISOPHOT sources have 15 μm counterparts (Elbaz et al 2000). For those, the 175 μm/15 μm flux density ratios correspond to what one would expect for relatively nearby ($z \sim 0.5$) galaxies with Arp 220-like spectra. Scott et al (1999) observed ten FIRBACK sources with SCUBA. The average 850 (and 450) μm/175 μm flux density ratios also favor a low mean redshift (~ 0.3) although the degenerate dust temperature/ redshift relation would also permit in principle solutions with z up to 2. The far-IR sources not detected by ISOCAM could be at higher redshift. The interplay of the k-corrections is such that ISOCAM cannot see galaxies with an Arp 220 SED located at $z \sim 2$.

The integrated counts constitute less than 10% of the COBE far-IR background (Hauser et al 1998). This suggests that ISOPHOT is detecting the tip of the iceberg of a new population of very luminous, moderate to high redshift galaxies. Most of the background is in sources with 175 μm flux densities between a few and 100 mJy.

4.3 Star Formation at High Redshift: A Synthesis

From the survey of the 1415+52 CFRS field, Flores et al (1999b) concluded that $60 \pm 25\%$ of the star formation at $z \leq 1$ is associated with infrared emission. Including the dusty luminous starbursts detected by ISOCAM, the total star formation rate (per unit comoving volume) at $z \leq 1$ is 2.9 ± 1.3 times as large as that deduced from the CFRS rest-frame UV observations (without correction for extinction). Rowan-Robinson et al (1997) found an even larger correction factor from an analysis of the ISOCAM HDF(N) data. If the ISOPHOT 175 μm sources are ultraluminous starburst galaxies at $z \sim 1$, the far-IR data require a star formation rate also ~ 3 times larger than predicted by UV observations (not extinction-corrected) interpolated into this redshift regime. Finally, the 850 μm source detections with the SCUBA bolometer camera require a star formation rate at $z \sim 3$ that is ~ 3.5 times greater than that derived from Lyman-dropout galaxies (Hughes et al 1998, Barger et al 1998, Ivison et al 1998). This assumes that the 850 μm sources are powered by star formation. To explain the entire COBE far-IR/submm background requires that the star formation rate at $z \geq 1$ be greater by another factor ≥ 2 (Lagache et al 1999b). The recent

Hα-based star formation rate at z ~ 1.3 (Yan et al 1999), on the other hand, is in good agreement with the far-IR/submm estimates. The large discrepancy between the UV and IR/submm estimates disappears if the star formation rates estimated from the Lyman-dropout galaxies are corrected for extinction (Pettini et al 1998, Meurer et al 1999, Steidel et al 1999). The data are summarized in Figure 17 (the Madau plot, see color insert; adapted from Lagache et al 1999b, Steidel et al 1999).

The good agreement between the UV/optical and IR/submm star formation histories shown in Figure 17 may be a lucky coincidence, however. The IR/submm and UV/optical measurements likely trace different source populations. At the level expected from the extinction-corrected UV star formation rates, Chapman et al (1999) detected at 850 μm only one of about a dozen Lyman break galaxies. The reddening of the ISOCAM galaxies is only slighty higher than average (Elbaz et al 2000), and thus their true output cannot be correctly derived from optical and UV data alone. The extinction corrections in the UV/optical are poorly known. In the case of the far-IR/submm sources the redshifts are uncertain, and the contribution of AGNs is unknown. Based on the X-ray background, Almaini et al (1999) set a limit of ~20% to the AGN contribution to the far-IR/submm background. Only 10–40% of the far-IR/submm background are resolved into individual sources. The derivation of star formation histories from the overall far-IR/submm background depends crucially on the (uncertain) source SEDs used.

Nevertheless, a reasonably consistent picture seems to emerge. The cosmic star formation rate steeply increases between the current epoch and z ~ 1, and stays flat from z ~ 1 to at least 4. The "quiescent" mode of disk star formation (with typical gas exhaustion time scales of several Gyr) cannot explain the z ≥ 1 data. Such a scenario (model A of Guiderdoni et al 1998; dashed line in Figure 17) can definitely be excluded by the extinction-corrected UV points, the mid-infrared points, and the far-IR/submm points. A rapidly evolving "burst" mode (exhaust time scales a few hundred Myrs) is required, probably associated with the much increased merger rate at high redshift [$\propto (1 + z)^\delta$, $\delta = 2..6$; Zepf & Koo 1989, Carlberg 1990, Abraham et al 1996]. In the semi-analytic models of Guiderdoni et al (1998) the fraction of this burst mode in the numbers of stars formed increases with $(1 + z)^5$ and dominates the cosmic star formation at z ≥ 1. To fit the far-IR/submm data as well, the fraction of violent mergers within the starburst population leading to ultraluminous starbursts must rapidly increase with redshift (Guiderdoni et al 1998, Blain et al 1999). In model E of Guiderdoni et al, the fraction of star formation in ULIRGs increases from 8% at z = 1 to 27% at z = 3. Blain et al (1999) introduced an activity parameter of mergers, defined as the inverse of the product of the fraction of mergers leading to luminous starbursts/AGNs f_b and the duration of an activity event Δt_b. In the Blain et al models, the activity parameter must increase by a factor of about 3 from the present day to z ~ 0.7 [the "15 μm (ISOCAM) epoch"], by ~10 to z ~ 1 (the "ISOPHOT epoch"?) and by ~160 to z ~ 3 (the "SCUBA epoch"). Hence, the relative contribution of ULIRGs to the luminosity function increases by a factor of several hundred between the local Universe and z $\sim$ 2–3. The models of Blain et al and Guiderdoni et al are thus quite consistent with each other and with the observations. The starburst galaxies sampled by ISOCAM have characteristic ages less than a few hundred Myr (Flores et al 1999b, Aussel et al 1999b), at least an order of magnitude smaller than the gas exhaustion timescale in present-day quiescent disks. The characteristic ages of nearby ULIRGs (100 L$_*$) that may be representative of the 175 μm and 850 μm populations are $\geq$ 10 Myr (Genzel et al 1998, Tecza et al 2000), and they probably make up only a small fraction of all merger events. The ultraluminous mergers may be related to the formation of large elliptical galaxies and bulges (Kormendy & Sanders 1992). As a result of these powerful mergers, the cosmic star formation rate may stay flat or even increase to redshifts z $\geq$ 4 (Blain et al 1999, Pei & Fall 1995).

5. CONCLUSIONS AND OUTLOOK

Normal Galaxies: ISO has provided progress on many fronts. The impulsive heating mechanism of single UV (and optical?) photons impinging on very small dust grains/large molecules and emitting in the near-IR/mid-IR now appears firmly established. The mid-IR SED is a valuable diagnostic of the activity level in galaxies. The mid-IR emission from early-type galaxies is composed of stellar, interstellar, and AGN contributions which can now be separated. In nearby spirals the far-IR data require cool equilibrium dust with a dust-to-gas ratio similar to that in the Milky Way. ISO has added tantalizing results on the nature of the halos and outer disks of spiral galaxies, but has not solved the overall puzzle of the dark matter residing there.

Starburst Galaxies (and the Starburst/AGN Connection):   ISO has added infrared spectroscopy and spectrophotometry to the astronomer's arsenal of powerful analytical tools. Dust-enshrouded AGNs and starburst galaxies can be qualitatively and quantitatively distinguished. ISO has confirmed that infrared luminous starburst galaxies typically have low nebular excitation. The likely explanation is that starbursts are episodic and rapidly aging. Star formation provides strong negative feedback that tends to quench active starbursts in little more than an O-star's lifetime.

AGNs: The ISO data generally support unified accretion disk+circumnuclear torus/disk models. Emission from AGN-heated, thick tori or warped disks dominates the mid-IR emission of QSOs, radio galaxies, and Seyfert nuclei. Based on a combination of mid-IR spectrophotometry, radio/IR imaging and radiative transport models, star formation likely plays an important role in accounting for the far-IR emission of many Seyfert galaxies. However, the issue of what powers the (thermal) far-IR emission of QSOs and radio galaxies remains uncertain. Mid- and far-IR spectroscopy of powerful AGNs with SIRTF and SOFIA will provide the next major step in this field. A promising new method has emerged for reconstructing the (hidden) EUV SEDs of AGNs from optical/infrared coronal line ratios. SIRTF will tell how useful this tool is for quantitative studies of different classes of AGNs.

(U)LIRGs: With ISO, much progress has been made in understanding the properties, energy sources, and evolution of (ultra) luminous infrared galaxies, which are a spectacular curiosity locally but very common and important in the early Universe. (U)LIRGs are composite objects containing both powerful AGNs and starbursts. Star formation plausibly dominates the luminosity of most sources. At the highest luminosities, AGNs take over. The standard paradigm of ULIRGs as a (late) phase in the merger of two gas-rich galaxies is confirmed, but their evolution is more complex than was considered before. ULIRGs do not seem to undergo an obvious metamorphosis from a starburst-powered system of colliding galaxies to a buried and then finally naked QSO at the end of the merger phase. Local effects play an important role. SIRTF will increase the statistics and quality of the mid-IR spectroscopy, and will also be able to apply the new IR spectroscopic tools to higher redshift sources. XMM—Newton and Chandra will do the same at hard X-rays. SOFIA promises sensitive far-IR spectra to help tackle the problem of the weak far-IR line emission seen by ISO in nearby ULIRGs.

High-z, Dusty Star Formation: COBE, ISO, and SCUBA have clearly demonstrated that dusty star formation plays an important role in the early Universe. This is a very exciting development. However, the present data have barely scratched the surface of this important emerging field, and their interpretation is still in a stage of infancy. New models and follow-up work are in progress, and updates give an increasingly detailed picture. ISOCAM has identified an important population of active starburst galaxies (mainly LIRGs) at $z < 1.5$. This population accounts for 10–60% of the far-IR/submm background. ISOPHOT has begun to detect a ULIRG component at moderate redshift ($z \sim 0.2$ to >1) that may account for an additional 10% of the background. Submm observations (SCUBA, etc) are detecting a component of high redshift ($z \geq 2$) ULIRGs, which make up the remainder of the background. Deeper surveys with SIRTF, SOFIA, FIRST, and ALMA will be one prime objective of the next decade. Identifying counterparts at other wavelengths and studying their nature will be another. The surveys will resolve most of the far-IR/submm background into individual sources. IR spectroscopic follow-up (with ground-based telescopes, SIRTF, FIRST) and hard X-ray observations (Chandra, XMM—Newton) will establish the redshifts and will determine what fraction of the IR sources comprises active starbursts and what fraction comprises dusty Seyferts/QSOs. We can look forward to an exciting time!

ACKNOWLEDGMENTS

We would like to thank the many colleagues who have sent us reprints or preprints, or who have given us input, comments, and advice: in particular, P Andreani, H Aussel, R Barvainis, J Clavel, P Cox, H Dole, D Elbaz, J Fischer, A Franceschini, M Haas, G Helou, D Hollenbach, M Kaufman, O Laurent, D Lemke, S Lord, M Luhman, D Lutz, R Maiolino, S Malhotra, H Matsuhara, F Mirabel, AFM Moorwood, T Nakagawa, H Okuda, J-L Puget, D Rigopoulou, B Schulz, S Serjeant, B Smith, M Stickel, E Sturm, Y Taniguchi, M Thornley, and D Tran. We would like to thank Susanne Harai, Dieter Lutz, Linda Tacconi, and Lowell Tacconi-Garman for help with the manuscript and for useful comments. We are especially grateful to F Mirabel and AFM Moorwood for their comments on the manuscript.

Visit the Annual Reviews home page at www.AnnualReviews.org

LITERATURE CITED

Abraham RG, Tanvir NR, Santiago BX, Ellis RS, Glazebrook K, van den Bergh S. 1996. *MNRAS* 279:47–52

Acosta-Pulido JA, Klaas U, Laureijs RJ, Schulz B, Kinkel U, et al. 1996. *Astron. Astrophys.* 315:121–24

Afonso C, Alard C, Albert JN, Andersen J, Ansari R, et al. 1999. *Astron. Astrophys.* 344:63–66

Alcock C, et al. 1998. *Ap. J. Lett.* 499:9–12

Alexander DM, Efstathiou A, Hough JH, Aitken D, Lutz D, et al. 2000. *astro-ph/*0003142

Alexander T, Lutz D, Sturm E, Netzer H, Sternberg A, et al. 2000. *Ap. J. astro-ph/* 0002107

Alexander T, Sturm E, Lutz D, Sternberg A, Netzer H, Genzel R. 1999. *Ap. J.* 512:204–23

Allamandola LJ, Sandford SA, Hudgins DM, Witteborn FC. 1995. In *Airborne Astronomy Symposium on the Galactic Ecosystem: From Gas To Stars to Dust*, ed. MR Haas, JA Davidson, EF Erickson, pp. 23–32. San Francisco: ASP Conference series 73

Almaini O, Lawrence A, Boyle BJ. 1999. *MNRAS* 305:59–63

Altieri B, Metcalfe L, Kneib JP, McBreen B,

Aussel H, et al. 1999. *Astron. Astrophys.* 343:65–69

Alton PB, Bianchi S, Rand RJ, Xilouris EM, Davies JI, Trewhella M. 1998b. *Ap. J.* 507:L125–29

Alton PB, Trewhella M, Davies JI, Evans R, Bianchi S, et al. 1998a. *Astron. Astrophys.* 335:807–22

Andreani P. 2000. In preparation

Antonucci R. 1993. *Annu. Rev. Astron. Astrophys.* 31:473–521

Appleton PN, Charmandaris V, Horellou C, Mirabel IF, Ghigo F, Higdon JL. 1999. *Ap. J.* 527:143–53

Armus L, Heckman TM, Miley GK. 1990. *Ap. J.* 364:471–95

Armus L, Neugebauer G, Soifer BT, Matthews K. 1995. *Astron. J.* 110:2610–21

Aussel H, Cesarsky CJ, Elbaz D, Starck JL. 1999a. *Astron. Astrophys.* 342:313–36

Aussel H, Elbaz D, Cesarsky CJ, Starck JL. 1999b. See Cox & Kessler 1999, pp. 1023–26

Aussel H, Gerin M, Boulanger F, Desert FX, Casoli F, et al. 1998. *Astron. Astrophys.* 334:73–76

Awaki H, Kunieda H, Tanaka Y, Koyama K. 1991. *Publ. Astron. Soc. Jpn.* 43:37–42

Baade W, Minkowski R. 1954. *Ap. J.* 119:215–31

Baan WA. 1993. In *Astrophysical Masers*, ed. AW Clegg, GE Nedoluha, pp. 73–82. Berlin: Springer-Verlag

Bahcall JN, Kirhakos S, Saxe DH, Schneider DP. 1997. *Ap. J.* 479:642–58

Bakes EL, Tielens AGGM. 1994. *Ap. J.* 427:822–38

Barger AJ, Cowie LL, Sanders DB, Fulton E, Taniguchi Y, et al. 1998. *Nature* 394:248–51

Barger AJ, Cowie LL, Smail I, Ivison RJ, Blain AW, Kneib JP. 1999. *Astron. J.* 117:2656–65

Barnes JE, Hernquist L. 1996. *Ap. J.* 471:115–42

Barvainis R. 1990. *Ap. J.* 353:419–32

Barvainis R, Antonucci R, Helou G. 1999. *Astron. J.* 118:645–53

Beichman C, Helou G, van Buren D, Ganga K, Desert FX. 1999. *Ap. J.* 523:559–65

Beichman CA. 1987. *Annu. Rev. Astron. Astrophys.* 25:521–63

Binette L, Wilson AS, Raga A, Storchi-Bergman T. 1997. *Astron. Astrophys.* 327:909–20

Blain AW, Jameson A, Smail I, Longair MS, Kneib JP, Ivison RJ. 1999. *MNRAS* 309:715–30

Böker T, Förster-Schreiber N, Genzel R. 1997. *Astron. J.* 114:1883–98

Boller T, Bertoldi F, Dennefeld M, Voges W. 1997. *Astron. Astrophys.* 129:87–145

Boselli A, Lequeux J, Contursi A, Gavazzi G, Boulade O, et al. 1997. *Astron. Astrophys.* 324:13–16

Boselli A, Lequeux J, Sauvage M, Boulade O, Boulanger F, et al. 1999. *Astron. Astrophys.* 335:53–68

Boulade O, Sauvage M, Altieri B, Blommaert J, Gallais P, et al. 1996. *Astron. Astrophys.* 315:85–88

Boulanger F, Abergel A, Bernard JP, Cesarsky D, Puget JL, et al. 1998. In *Star Formation with ISO*, ed. JL Yun, R Liseau, pp. 15–23. San Francisco: ASP Conf. Ser. 132

Bradford CM, Stacey GJ, Fischer J, Smith HA,

Cohen RJ, et al. 1999. See Cox & Kessler 1999, pp. 861–64

Braine J, Hughes DH. 1999. *Astron. Astrophys.* 344:779–86

Brandt WN, Fabian AC, Takahashi K, Fujimoto R, Yamashita A, et al. 1997. *MNRAS* 290:617–22

Carlberg RG. 1990. *Ap. J. Lett.* 359: 1–3

Carr B. 1994. *Annu. Rev. Astron. Astrophys.* 32:531–90

Cesarsky CJ. 1999. See Cox & Kessler 1999, pp. 45–49

Cesarsky CJ, Abergel A, Agnese P, Altieri B, Augueres JJ, et al. 1996. *Astron. Astrophys.* 315:32–37

Cesarsky CJ, Elbaz D. 1996. In *Examining the Big Bang and Diffuse Background Radiations, IAU Symp. 168*, ed. Kafatos MC, Kondo Y, pp. 109–13. Kluwer: Dordrecht

Cesarsky D, Lequeux J, Abergel A, Perault M, Palazzi E, et al. 1996a. *Astron. Astrophys.* 315:305–8

Cesarsky D, Lequeux J, Abergel A, Perault M, Palazzi E, et al. 1996b. *Astron. Astrophys.* 315:309–12

Cesarsky CJ, Sauvage M. 2000. In *"Toward a New Millenium in Galaxy Morphology"*, ed. DL Block, A Stockton, D Ferreira. Dordrecht: Kluwer. In press

Chapman SC, Scott D, Steidel CC, Borys C, Halpern M, et al. 1999. astro–ph 9909092. *MNRAS*. Submitted

Charmandaris Y, Laurent O, Mirabel IF, Gallais P, Sauvage M, et al. 1999. *Astron. Astrophys.* 341:69–73

Clavel J, Schulz B, Altieri B, Barr P, Claes P, et al. 1998. *astro–ph* 9806054

Clegg PE (on behalf of the LWS Consortium). 1999. See Cox & Kessler 1999, pp. 39–43

Clegg PE, Ade PAR, Armand C, Baluteau JP, Barlow MJ. 1996. *Astron. Astrophys.* 315:38–42

Clements DL, Desert FX, Franceschini A, Reach WT, Baker AC, et al. 1999. *Astron. Astrophys.* 346:383–91

Colbert JW, Malkan MA, Clegg PE, Cox P, Fischer J, et al. 1999. *Ap. J.* 511:721–29

Colina L, Perez-Olea DE. 1995. *MNRAS* 277:845–56

Condon JJ. 1992. *Annu. Rev. Astron. Astrophys.* 30:575–611

Condon JJ, Anderson ML, Helou G. 1991a. *Ap. J.* 376:95–103

Condon JJ, Huang ZP, Yin QP, Thuan TX. 1991b. *Ap. J.* 378:65–76

Conti PS, Vacca WD. 1994. *Ap. J.* 423:97–100

Contini M, Viegas SM. 1992. *Ap. J.* 401:481–94

Contursi A, Lequeux J, Hanus M, Heydari-Malayeri M, Bonoli C, et al. 1998. *Astron. Astrophys.* 336:662–66

Cox P, Kessler MF, eds. 1999. *The Universe As Seen by ISO*, (Foreword) pp. i-ii. ESA SP–47. ESA Publ. Div., ESTEC, Noordwijk, The Netherlands

Cox P, Roelfsema PR, Baluteau JP, Peeters E, Martin-Hernandez L, et al. 1999. See Cox & Kessler 1999, pp. 631–37

Crawford MK, Genzel R, Townes CH, Watson DM. 1985. *Ap. J.* 291:755–71

Crowther PA, Beck SC, Willis AJ, Conti PS, Morris PW, Sutherland RS. 1999b. *MNRAS* 3:654–68

Crowther PA, Pasquali A, DeMarco O, Schmutz W, Hillier DJ, DeKoter A. 1999a. astro–ph 9908200

Dale D, Silbermann NA, Helou G, et al. 2000. *astro-ph*/0005083

Davies JI, Alton P, Trewhella M, Evans R, Bianchi S. 1999. *MNRAS* 304:495–500

de Graauw T. 1999. See Cox & Kessler, pp. 31–37

de Graauw T, Haser LN, Beintema DA, Roelfsema PR, van Agthoven H, et al. 1996. *Astron. Astrophys.* 315:49–54

de Grijp MKK, Miley GK, Lub J, de Jong T. 1985. *Nature* 314:240–42

Desert FX, Boulanger F, Puget JL. 1990. *Astron. Astrophys.* 237:215–36

Desert FX, Puget JL, Clements DL, Perault M, Abergel A, et al. 1999. *Astron. Astrophys.* 342:363–77

Devereux N, Young J. 1990. *Ap. J.* 359:42–56

Disney M, Davies J, Phillips S. 1989. *MNRAS* 239:939–76

Doherty RM, Puxley PJ, Lumsden SL, Doyon R. 1995. *MNRAS* 277:577–93

Dole H, Lagache G, Puget JL, Gispert R, Aussel H, et al. 1999. See Cox & Kessler, pp. 1031–35

Domingue DL, Keel WC, Ryder SD, White RE III. 1999. *Astro. J.* 119:1512

Downes D, Solomon PM. 1998. *Ap. J.* 507:615–54

Doyon R, Joseph RD, Wright GS. 1994a. *Ap. J.* 421:101–14

Doyon R, Wells M, Wright GS, Joseph RD, Nadeau D, James PA. 1994b. *Ap. J. Lett.* 437:23–26

Draine BT, Lee HM. 1984. *Ap. J.* 285:89–108

Draine BT, Roberge WG, Dalgarno A. 1983. *Ap. J.* 264:485–507

Drissen L, Moffat AFJ, Walborn NR, Shara MM. 1995. *Astron. J.* 110:2235–41

Duc PA, Poggianti BM, et al. 2000. In preparation

Dudley CC. 1999. *MNRAS* 307:553–76

Dudley CC, Wynn-Williams CG. 1997. *Ap. J.* 488:720–29

Duley WW, Williams DA. 1991. *MNRAS* 247:647–50

Dwek E, Arendt R, Hauser M, Fixsen D, Kelsall T, et al. 1998. *Ap. J.* 508:106–22

Efstathiou A, Rowan-Robinson M. 1995. *MNRAS* 273:649–61

Eisenhauer F, Quirrenbach A, Zinnecker H, Genzel R. 1998. *Ap. J.* 498:278–92

Elbaz D, Aussel H, Cesarsky CJ, Desert FX, Fadda D, et al. 1999b. See Cox & Kessler 1999, pp. 999–1006

Elbaz D, Cesarsky CJ, Aussel H, et al. 2000. In preparation

Elbaz D, Cesarsky CJ, Fadda D, Aussel H, Desert FX, et al. 1999a. *Astron. Astrophys.* 351:37–40

Ellis RE. 1997. *Annu. Rev. Astron. Astrophys.* 35:389–443

Elvis M, Wilkes BJ, McDowell JC, Green RF, Bechtold J, et al. 1994. *Ap. J. Suppl.* 95:1–68

Engelbracht CW, Rieke MJ, Rieke GH, Kelly

DM, Achtermann JM. 1998. *Ap. J.* 505:639–58

Fadda D, Elbaz D. 1999. See Cox & Kessler 1999, pp. 1037–39

Fadda D, Elbaz D, Duc PA, Flores H, Franceschini A, Cesarsky CJ. 2000. *Astron. Astrophys.* Submitted

Fang F, Shupe D, Xu C, Hacking P. 1998. *Ap. J.* 500:693–702

Fich M, Rupen MR, Knapp GR, Malyshkin L, Harper DA, Wynn-Williams CG. 1999. See Cox & Kessler 1999, pp. 877–80

Figer DF, Najarro F, Morris M, McLean IS, Geballe TR, et al. 1998. *Ap. J.* 506:384–404

Fischer J, Lord SD, Unger SJ, Bradford CM, Brauher JR, et al. 1999. See Cox & Kessler 1999, pp. 817–20

Fischer J, Satyapal S, Luhman ML, Stacey GJ, Smith HA, et al. 1997. In *Extragalactic Astronomy in the IR*, ed. GA Mamon, TX Thuan, JT Thanh Van, pp. 289–93. Paris: Ed. Front.

Fischer J, Shier LM, Luhman ML, Satyapal S, Smith HA, et al. 1996. *Astron. Astrophys.* 315:97–100

Flores H, Hammer F, Desert FX, Cesarsky CJ, Thuan TX, et al. 1999a. *Astron. Astrophys.* 343:389–98

Flores H, Hammer F, Thuan TX, Cesarsky CJ, Desert FX, et al. 1999b. *Ap. J.* 517:148–67

Förster-Schreiber N, et al. 2000. In preparation

Franceschini A, Aussel D, Elbaz D, Fadda C, Cesarsky CJ, et al. 2000. *Mem. Soc. Astron. Ital.* In press

Franceschini A, Mazzei P, deZotti G, Danese L. 1994. *Ap. J.* 427:140–54

Gallais P, Laurent O, Charmandaris V, Rouen D, Mirabel IF, et al. 1999. See Cox & Kessler 1999, pp. 881–85

Gao Y, Solomon PM. 1999. *Ap. J. Lett.* 512:99–103

Gehrz RD, Sramek RA, Weedman DW. 1983. *Ap. J.* 267:551–62

Genzel R. 1992. In *The Galactic Interstellar Medium*, ed. WB Burton, BG Elmegreen, R Genzel, D Pfenniger, P Bertholdy, pp. 275–390. Berlin: Springer-Verlag

Genzel R, Hollenbach DJ, Townes CH. 1994. *Rep. Progr. Phys.* 57:417–79

Genzel R, Lutz D, Sturm E, Egami E, Kunze D, et al. 1998. *Ap. J.* 498:579–605

Gilmore G. 1998. *Rev. Dark Matter* 98:In press

Gilmore G, Unavane M. 1998. *MNRAS* 301:813–26

Giuricin G, Mardirossian F, Mezetti M. 1995. *Ap. J.* 446:550–60

Goldader JD, Joseph RD, Doyon R, Sanders DB. 1995. *Ap. J.* 444:97–112

Goldader JD, Joseph RD, Doyon R, Sanders DB. 1997b. *Ap. J. Suppl.* 108:449–70

Goldader JD, Joseph RD, Sanders DB, Doyon R. 1997a. *Ap. J.* 474:104–20

Goldschmidt P, Oliver SJ, Serjeant SB, Baker A, Eaton N, et al. 1997. *MNRAS* 289:465–70

Gonzalez-Delgado RM, Leitherer C, Heckman TM, Cervino M. 1997. *Ap. J.* 483:705–16

Granato GL, Danese L. 1994. *MNRAS* 268:235–52

Granato GL, Danese L, Franceschini A. 1997. *Ap. J.* 486:147–59

Guiderdoni B, Bouchet F, Puget JL, Lagache G, Hivon E. 1997. *Nature* 390:257–59

Guiderdoni B, Hivon E, Bouchet F, Maffei B. 1998. *MNRAS* 295:877–98

Haas M. 1998. *Astron. Astrophys.* 337:1–4

Haas M, Chini R, Meisenheimer K, Stickel M, Lemke D, et al. 1998a. *Ap. J. Lett.* 503:109–13

Haas M, Lemke D, Stickel M, Hippelein H, Kunkel M, et al. 1998b. *Astron. Astrophys.* 338:33–36

Haas M, Mueller SAH, Chini R, Meisenheimer K, Klaas U, et al. 1999. *Astron. Astrophys.* 354:453–66

Habing H, Miley GK, Young E, Baud D, Boggess N, et al. 1984. *Ap. J. Lett.* 278:59–62

Hammer F, Flores H, Lilly SJ, Crampton D, LeFevre O, et al. 1997. *Ap. J.* 481:49–82

Harvey VI, Satyapal S, Luhman ML, Fischer J, Clegg PE, et al. 1999. See Cox & Kessler 1999, pp. 889–93

Hauser MG, Arendt RG, Kelsall T, Dwek E, Odegard N, et al. 1998. *Ap. J.* 508:25–43

Heckman TM. 1995. *Ap. J.* 446:101–7

Heckman TM. 1998. In *Origins*, ed. CE Woodward, JM Shull, HA Thronson, pp. 127–46. San Francisco: ASP Conf. Ser. 148

Heckman TM. 2000. In *ULIRGs: Monsters or Babies*, ed. D Lutz, L J Tacconi, San Francisco: ASP Conf. Ser. *Astrophys. Space Sci.* In press

Heckman TM, Armus L, Miley GK. 1987. *Astron. J.* 93:276–83

Heckman TM, Armus L, Miley GK. 1990. *Ap. J. Suppl.* 74:833–68

Helou G. 1999. See Cox & Kessler 1999, pp. 797–803

Helou G, Lu NY, Werner MW, Malhotra S, Silbermann N. 1999. *Ap. J.* 532:21–24

Hillier DJ, Miller DL. 1998. *Ap. J.* 496:407–27

Hippelein H, Lemke D, Haas M, Telesco CM, Tuffs RJ, et al. 1996. *Astron. Astrophys.* 315:82–84

Hollenbach DJ, McKee CF. 1989. *Ap. J.* 342:306–36

Hollenbach DJ, Tielens AGGM. 1997. *Annu. Rev. Astron. Astrophys.* 35:179–216

Huchra J, Burg R. 1992. *Ap. J.* 393:90–97

Hughes DH, Serjeant S, Dunlop J, Rowan-Robinson M, Balin A, et al. 1998. *Nature* 394:241–47

Hunter DA, Boyd DM, Hawley WN. 1995. *Ap. J. Suppl.* 99:551–63

Israel FP, Maloney PR, Geis N, Herrmann F, Madden SC, et al. 1995. *Ap. J.* 465:738–47

Ivison RJ, Smail I, Le Borgne JF, Blain AW, Kneib JP, et al. 1998. *MNRAS* 298:583–93

Iwasawa K, Koyama K, Awaki H, Kunieda H, Makishima K, et al. 1993. *Ap.J.* 409:155–61

Kaufman MJ, Wolfire MG, Hollenbach DJ, Luhman ML. 1999. *Ap. J.* 527:795–813

Kawara K, Sato Y, Matsuhara H, Taniguchi Y, Okuda H, et al. 1998. *Astron. Astrophys.* 336:9–12

Kawara K, Sato Y, Matsuhara H, Taniguchi Y, Okuda H, et al. 1999. See Cox & Kessler 1999, pp. 1017–18

Kegel WH, Hertenstein T, Quirrenbach A. 1999. *Astron. Astrophys.* 351:472–76

Kennicutt RC. 1992. *Ap. J.* 388:310–27

Kennicutt RC. 1998. *Annu. Rev. Astron. Astrophys.* 36:189–232

Kessler M. 1999. See Cox & Kessler 1999, pp. 23–29

Kessler MF, Steinz JA, Anderegg ME, Clavel J, Drechsel G, et al. 1996. *Astron. Astrophys.* 315:27–31

Kii T, Nakagawa T, Fujimoto R, Ogasaka T, Miyazaki T, et al. 1997. In *X-ray Imaging and Spectroscopy of Cosmic Hot Plasmas*, ed. F Makino, K Mitsuda, pp. 161–64. Tokyo: Universal Acad. Press

Kim DC, Sanders DB, Veilleux S, Mazarella JM, Soifer BT. 1995. *Ap. J. Suppl.* 98:129–70

Kim DC, Veilleux S, Sanders DB. 1998. *Ap. J.* 508:627–47

Klaas U, Müller TG, Laureijs RJ, Clavel J, Lagerros JSV, et al. 1999. See Cox & Kessler 1999, pp. 77–80

Knapp GR, Guhathakarta P, Kim D-W, Jura M. 1989. *Ap. J. Suppl.* 70:329–87

Knapp GR, Gunn JE, Wynn-Williams CG. 1992. *Ap. J.* 399:76–93

Knapp GR, Rupen MP, Fich M, Harper DA, Wynn-Williams CG. 1996. *Astron. Astrophys.* 315:75–78

Kormendy J, Sanders DB. 1992. *Ap. J. Lett.* 390:53–56

Krabbe A, Genzel R, Eckart A, Najarro F, Lutz D, Cameron M, et al. 1995. *Ap. J. Lett.* 447:95–98

Krabbe A, Sternberg A, Genzel R. 1994. *Ap. J.* 425:72–90

Kraemer SB, Ho LC, Crenshaw DM, Shields JC, Filippenko AU. 1999. *Ap. J.* 520:564–73

Kraemer SB, Ruiz JR, Crenshaw DM. 1998. *Ap. J.* 508:232–42

Kriss GA, Davidsen AF, Blair WP, Bowers CW, Dixon WV, et al. 1992. *Ap. J.* 392:485–91

Kriss GA, Davidsen AF, Zheng W, Kruk JW, Espey BR. 1995. *Ap. J. Lett.* 454:7–10

Krügel E, Siebenmorgen R, Zota V, Chini R. 1998. *Astron. Astrophys.* 331:9–12

Kunze D, Rigopoulou D, Genzel R, Lutz D. 1999. See Cox & Kessler 1999, pp. 909–12

Kunze D, Rigopoulou D, Lutz D, Egami E,

Feuchtgruber H, et al. 1996. *Astron. Astrophys.* 315:101–4

Lagache G, Abergel A, Boulanger F, Desert FX, Puget JL. 1999a. *Astron. Astrophys.* 344:322–32

Lagache G, Puget JL. 1999. *Astron. Astrophys.* 355:17–22

Lagache G, Puget JL, Gispert R. 1999b. *astro–ph* 9910258

Lai O, Rouan D, Rigaut F, Arsenault E, Gendron E. 1998. *Astron. Astrophys.* 334:783–88

Laureijs RJ, Acosta-Pulido J, Abraham P, Kinkel U, Klaas U, et al. 1996. *Astron. Astrophys.* 315:313–16

Laurent O, Mirabel IF, Charmandaris V, Gallais P, Madden SC, et al. 2000. *Astron. Astrophys.* In press

Leitherer C, Heckman TM. 1995. *Ap. J. Suppl.* 96:9–38

Lemke D, Klaas U. 1999. See Cox & Kessler 1999, pp. 51–60

Lemke D, Klaas U, Abolins J, Abraham P, Acosta-Pulido J, et al. 1996. *Astron. Astrophys.* 315:64–70

Lemonon L, Pierre M, Cesarsky CJ, Elbaz D, Pello R, et al. 1998. *Astron. Astrophys.* 334:21–25

Levine DA, Lonsdale CJ, Hurt RL, Smith HE, Helou G, et al. 1998. *Astron. J.* 504:64–76

Lilly SJ, LeFevre O, Hammer F, Crampton D. 1996. *Ap. J. Lett.* 460:1–4

Liu CT, Kennicutt RC. 1995. *Ap. J.* 450:547–58

Lo KY, et al. 2000. In preparation

Lonsdale CJ, Smith HE, Lonsdale CJ. 1993. *Ap. J. Lett.* 405:9–12

Lonsdale CJ, Smith HE, Lonsdale CJ. 1995. *Ap. J.* 438:632–42

Lord SD, et al. 2000. In preparation

Lord SD, Hollenbach DJ, Colgan SWJ, Haas MR, Rubin RH, et al. 1995. In *Airborne Astronomy Symposium on the Galactic Ecosystem:From Gas to Stars to Dust*, ed. MR Haas, JA Davidson, EF Erickson, pp. 151–58. San Francisco: ASP Conf. Ser. 73

Lord SD, Malhotra S, Lim T, Helou G, Rubin RH, et al. 1996. *Astron. Astrophys.* 315:117–20

Luhman ML, Satyapal S, Fischer J, Wolfire MG, Cox P, et al. 1998. *Ap. J.* 504:11–15

Lutz D. 1999. See Cox & Kessler 1999, pp. 623–26

Lutz D, et al. 2000. *Ap. J.* 536:In press

Lutz D, Feuchtgruber H, Genzel R, Kunze D, Rigopoulou D, et al. 1996b. *Astron. Astrophys.* 315:269–72

Lutz D, Genzel R, Sternberg A, Netzer H, Kunze D, et al. 1996a. *Astron. Astrophys.* 315:137–40

Lutz D, Kunze D, Spoon HWW, Thornley MD. 1998b. *Astron. Astrophys.* 333:75–78

Lutz D, Spoon HWW, Rigopoulou D, Moorwood AFM, Genzel R. 1998a. *Ap. J. Lett.* 505:103–7

Lutz D, Veilleux S, Genzel R. 1999a. *Ap. J.* 517:13–17

Lutz D, Sturm E, Genzel R, Moorwood AFM, Alexander T, Netzer H, Sternberg A. 2000. *Ap. J.* In press (*astro-ph*/0002008)

Madau P, Ferguson HC, Dickinson ME, Giavalisco M, Steidel CC, Fruchter A. 1996. *MNRAS* 283:1388–404

Madden SC, Geis N, Genzel R, Herrmann F, Jackson J, et al. 1993. *Ap. J.* 407:579–87

Madden SC, Geis N, Townes CH, Genzel R, Herrmann F, et al. 1995. In *Airborne Symposium on the Galactic Ecosystem:From Gas to Stars to Dust*, ed. MR Haas, JA Davidson, EF Erickson, pp. 181–84. San Francisco: ASP Conf. Ser. 73

Madden SC, Poglitsch A, Geis N, Stacey GJ, Townes CH. 1997. *Ap. J.* 483:200–9

Madden SC, Vigroux L, Sauvage M. 1999. See Cox & Kessler 1999, pp. 933–36

Maiolino R, Krabbe A, Thatte N, Genzel R. 1998a. *Ap. J.* 493:650–65

Maiolino R, Ruiz M, Rieke GH, Keller LD. 1995. *Ap. J.* 446:561–73

Malhotra S, Helou G, Hollenbach DJ, Kaufman M, Lord SD, et al. 1999a. See Cox & Kessler 1999, pp. 813–16

Malhotra S, Helou G, Stacey GJ, Hollenbach DJ, Lord SD, et al. 1997. *Ap. J. Lett.* 491:27–30

Malhotra S, Hollenbach DJ, Helou G, Valjavec E, Hunter D, et al. 1999b. See Cox & Kessler 1999, pp. 937–40

Malkan MA, Sargent WLW. 1982. *Ap. J.* 254:22–37

Maloney PR, Hollenbach DJ, Tielens AGGM, 1996. *Ap. J.* 466:561–84

Mann RG, Oliver S, Serjeant SB, Rowan-Robinson M, Baker A, et al. 1997. *MNRAS* 289:482–89

Massey P, Hunter DA. 1998. *Ap. J.* 493:180–94

Mattila K, Lehtinen K, Lemke D. 1999. *Astron. Astrophys.* 342:643–54

Mattila K, Lemke D, Haikala LK, Laureijs RJ, Leger A, et al. 1996. *Astron. Astrophys.* 315:353–56

Metcalfe L, Altieri B, McBreen B, Kneib JP, Delaney M, et al. 1999. See Cox & Kessler 1999, pp. 1019–22

Metcalfe L, Stell SJ, Barr P, Clavel J, Delaney M, et al. 1996. *Astron. Astrophys.* 315:105–8

Meurer GR, Heckman TM, Calzetti D. 1999. *Ap. J.* 521:64–80

Mihos JC. 2000. In *ULIRGs: Monsters or Babies,* ed. D Lutz, LJ Tacconi. San Francisco: ASP Conf. Ser. *Astrophys. Space Sci.* In press

Mihos JC, Hernquist L. 1996. *Ap. J.* 464:641–63

Mirabel IF, Laurent O, Sanders DB, Sauvage M, Tagger M, et al. 1999. *Astron. Astrophys.* 341:667–74

Mirabel IF, Vigroux L, Charmandaris V, Sauvage M, Gallais P, et al. 1998. *Astron. Astrophys.* 333:1–4

Misaki K, Iwasawa K, Taniguchi Y, Terashima Y, Kunieda H, Watarai H. 1999. *Adv. Space Res.* 23(5):1051–55

Mitsuda K. 1995. *Proc. 17th Texas Symp.*, ed. H Böhringer, GE Morfill, J Trümper, pp. 213–16. New York: NY Acad. Sci.

Mochizuki K, Nakagawa T, Doi Y, Yui YY, Okuda H, et al. 1994. *Ap. J. Lett.* 430:37–40

Mochizuki K, Onaka T, Nakagawa T. 1998. In *Star Formation with ISO,* ed. J Yun, R Liseau, pp. 386–89. San Francisco: ASP Conf. Ser. 132

Moorwood AFM. 1996. *Space Sci. Rev.* 77:303–66

Moorwood AFM. 1999. See Cox & Kessler 1999, pp. 825–31

Moorwood AFM, Lutz D, Oliva E, Marconi A, Netzer H, et al. 1996a. *Astron. Astrophys.* 315:109–12

Moorwood AFM, van der Werf PP, Kotilainen JK, Marconi A, Oliva E. 1996b. *Astron. Astrophys.* 308:L1–L5

Najarro F, Krabbe A, Genzel R, Lutz D, Kudritzki RP, Hillier DJ. 1997. *Ap. J.* 325:700–8

Nakagawa T, Doi Y, Yui Y, Okuda H, Mochizuki K, et al. 1996. *Ap. J. Lett.* 455:35–38

Nakagawa T, Kii T, Fujimoto R, Miyazaki T, Inoue I, et al. 2000. In *ULIRGs: Monsters or Babies,* ed. D Lutz, LJ Tacconi. San Francisco: ASP Conf. Ser. *Astrophys. Space Sci.* In press

Neugebauer G, Habing HJ, van Duinen R, Aumann HH, Baud B, et al. 1984. *Ap. J. Lett.* 278:1–6

Neugebauer G, Miley GK, Soifer BT, Clegg PE. 1986. *Ap. J.* 308:815–28

Odenwald S, Newmark J, Smoot G. 1998. *Astron. J.* 500:554–68

Oke JB, Sargent WLW. 1968. *Ap. J.* 151:807–23

Oliva E, Salvati M, Moorwood AFM, Marconi A. 1994. *Astron. Astrophys.* 288:457–65

Oliver SJ, et al. 2000. Preprint

Oliver SJ, Goldschmidt P, Franceschini A, Serjeant SB, Efstathiou A, et al. 1997. *MNRAS* 289:471–81

Onaka T, Yamamura I, Tanabe T, Roellig T, Yuen L. 1996. *Publ. Astron. Soc. Jpn.* 48:59–63

Osterbrock D. 1989. *Astrophysics of Gaseous Nebulae and Active Galactic Nuclei*, pp. 308–77. Mill Valley: Univ. Sci. Books

Ott T, Eckart A, Genzel R. 1999. *Ap. J.* 523:248–64

Pagani L, Lequeux J, Cesarsky D, Donas J, Milliard B, et al. 1999. *Astron. Astrophys.* 351:447–58

Pak S, Jaffe DT, van Dishoeck E, Johansson LEB, Booth RS. 1998. *Ap. J.* 498:735–56

Pearson C, Rowan-Robinson M. 1996. *MNRAS* 283:174–92

Pei YC, Fall SM. 1995. *Ap. J.* 454:69–76

Perez Garcia AM, Rodriguez Espinosa JM, Santolaya Rey AE. 1998. *Ap. J.* 500:685–92

Pettini M, Kellogg M, Steidel CC, Dickinson M, Adelberger KL, Giavalisco M. 1998. *Ap. J.* 508:539–50

Pfenniger D, Combes F, Martinet L. 1994. *Astron. Astrophys.* 285:79–93

Pier EA, Krolik JH. 1992. *Ap. J.* 401:99–109

Pier EA, Krolik JH. 1993. *Ap. J.* 418:673–86

Pierini D, Leech KJ, Tuffs RJ, Völk HJ. 1999. *MNRAS* 303:29–33

Poggianti BM, Wu H. 1999. *Ap. J.* 529:157–69

Poglitsch A, Krabbe A, Madden SC, Nikola T, Geis N, et al. 1995. *Ap. J.* 454:293–306

Pozzetti L, Madau P, Zamorani G, Ferguson HC, Bruzual GA. 1998. *MNRAS* 298:1133–44

Puget JL, Abergel A, Bernard JP, Boulanger F, Burton WB, et al. 1996. *Astron. Astrophys.* 308:5–8

Puget JL, Lagache G, Clements DL, Reach WJ, Aussel H, et al. 1999. *Astron. Astrophys.* 345:29–35

Puget JL, Leger A. 1989. *Annu. Rev. Astron. Astrophys.* 27:161–98

Quillen AC, Rieke GH, Rieke MJ, Caldwell N, Engelbracht CW. 1999. *Ap. J.* 518:632–40

Radovich M, Klaas U, Acosta-Pulido J, Lemke D. 1999. *Astron. Astrophys.* 348:705–10

Rice W, Lonsdale CJ, Soifer BT, Neugebauer G, Koplan EL, et al. 1988. *Astron. Astrophys. Suppl.* 68:91–127

Richards EA, Kiellermann KI, Fomalont EB, Windhorst RA, Partridge RB. 1998. *Astron. J.* 116:1039–54

Rieke GH, Cutri R, Black JH, Kailey WF, McAlary CW, et al. 1985. *Ap. J.* 290:116–24

Rieke GH, Lebofsky MJ, Thompson RI, Low FJ, Tokunaga AT. 1980. *Ap. J.* 238:24–40

Rieke GH, Loken K, Rieke MJ, Tamblyn P. 1993. *Ap. J.* 412:99–110

Rigopoulou D, Francheschini A, Aussel H, Genzel R, van der Werf PP, et al. 2000. *Ap. J. Lett.* Submitted

Rigopoulou D, Franceschini A, Genzel R, van der Werf PP, Aussel H, et al. 1999b. *astro–ph* 9912544

Rigopoulou D, Lawrence A, Rowan-Robinson M. 1996b. *MNRAS* 278:1049–68

Rigopoulou D, Lutz D, Genzel R, Egami E, Kunze D, et al. 1996a. *Astron. Astrophys.* 315:125–28

Rigopoulou D, Spoon HWW, Genzel R, Lutz D, Moorwood AFM, Tran QD. 1999a. *Astron. J.* 118:2625–45

Roche N, Eales S. 1999. *MNRAS* 307:111–21

Roche PF, Aitken DK, Smith C, Ward M. 1991. *MNRAS* 248:606–29

Rodriguez Espinosa JM, Perez Garcia AM, Lemke D, Meisenheimer K. 1996. *Astron. Astrophys.* 315:129–32

Rouan D, Tiphene D, Lacombe F, Boulade O, Clavel J, et al. 1996. *Astron. Astrophys.* 315:141–44

Roussel H, Sauvage M, Vigroux L. 1999a. *Astron. Astrophys.* Submitted

Roussel H, Vigroux L, Sauvage M, Bonoli C, Bosma A, et al. 1999b. See Cox & Kessler 1999, pp. 957–60

Rowan-Robinson M. 1995. *MNRAS* 272:737–48

Rowan-Robinson M, Broadhurst T, Oliver SJ, Taylor AN, Lawrence A, et al. 1991. *Nature* 351:719–21

Rowan-Robinson M, Crawford J. 1989. *MNRAS* 238:523–58

Rowan-Robinson M, Mann RG, Oliver SJ, Efstathiou A, Eaton N, et al. 1997. *MNRAS* 289:482–90

Rowan-Robinson M, Oliver S, Efstathiou A, Gruppioni C, Serjeant S, et al. 1999. See Cox & Kessler 1999, pp. 1011–15

Rush B, Malkan MA, Spinoglio L. 1993. *Ap. J. Suppl.* 89:1–33

Sahu KC. 1994. *PASP* 106:942–48

Sanders DB, Mirabel IF. 1996. *Annu. Rev. Astron. Astrophys.* 34:749–92

Sanders DB, Phinney ES, Neugebauer G, Soifer BT, Matthews K. 1989. *Ap. J.* 347:29–51

Sanders DB, Soifer BT, Elias JH, Madore BF, Matthews K, et al. 1988a. *Ap. J.* 325:74–91

Sanders DB, Soifer BT, Elias JH, Neugebauer G, Matthews K. 1988b. *Ap. J. Lett.* 328:35–39

Sato Y, Cowie LL, Taniguchi Y, Sanders DB, Kawara K, Okuda H. 1999. See Cox & Kessler 1999, pp. 1055–58

Satyapal S, Watson DM, Pipher JL, Forrest WJ, Fischer J, et al. 1999. *Ap. J.* 516:704–15

Satyapal S, Watson DM, Pipher JL, Forrest WJ, Greenhouse MA, et al. 1997. *Ap. J.* 483:148–60

Schaerer D, Stasinska G. 1999. *Astron. Astrophys.* 345:L17–L21

Schinnerer E, Eckart A, Tacconi LJ. 1998. *Ap. J.* 500:147–61

Schreier EJ, Marconi A, Axon Dj, Caon N, Machetto D, et al. 1998. *Ap. J. Lett.* 499:143–47

Scott D, Lagache G, Boys C, Chapman SC, Halpern M, et al. 1999. *astro–ph* 9910428

Scoville NZ, Evans AS, Dinshaw N, Thompson R, Rieke M, et al. 1998. *Ap. J. Lett.* 492:107–10

Scoville NZ, Evans AS, Thompson R, Rieke M, Hines DC, et al. 2000. *Ap. J.* In press

Scoville NZ, Yun MS, Bryant PM. 1997. *Ap. J.* 484:702–19

Sellgren K. 1984. *Ap. J.* 277:623–33

Serabyn E, Shupe D, Figer DF. 1998. *Nature* 394:448–51

Serjeant S, Oliver S, Rowan-Robinson M, Crocke HH, Missoulis V, et al. 1999. *MNRAS.* In press

Serjeant SB, Eaton N, Oliver S, Efstathiou A, Goldschmidt P, et al. 1997. *MNRAS* 289:457–64

Shupe DL, Fang F, Hacking PB, Huchra JP. 1998. *Ap. J.* 501:597–607

Siebenmorgen R, Krügel E, Chini R. 1999. *Astron. Astrophys.* 351:495–505

Skinner CJ, Smith HA, Sturm E, Barlow MJ, Cohen RJ, Stacey GJ. 1997. *Nature* 386:472–74

Smail I, Ivison TJ, Blain AW. 1997. *Ap. J. Lett.* 490:5–8

Smith BJ. 1998. *Ap. J.* 500:181–87

Smith BJ, Madden SC. 1997. *Astron. J.* 114:138–46

Smith CH, Aitken DK, Roche PF. 1989. *MNRAS* 241:425–31

Smith HE, Lonsdale CJ, Lonsdale CJ, Diamond PJ. 1998. *Ap. J. Lett.* 493:13–16

Soifer BT, Houck JR, Neugebauer G. 1987. *Annu. Rev. Astron. Astrophys.* 25:187–230

Soifer BT, Neugebauer G. 1991. *Astron. J.* 101:354–61

Soifer BT, Neugebauer G, Matthews K, Becklin EE, Ressler M, et al. 1999. *Ap. J.* 513:207–14

Soifer BT, Rowan-Robinson M, Houck JR, de Jong T, Neugebauer G, et al. 1984. *Ap. J. Lett.* 278:71–74

Solomon PM, Downes D, Radford SJE, Barrett JW. 1997. *Ap. J.* 478:144–61

Sopp HM, Alexander P. 1991. *MNRAS* 251:112–18

Spinoglio L, Malkan MA. 1992. *Ap. J.* 399:504–20

Spinoglio L, Suter J-P, Malkan MA, Clegg PE, Fischer J, et al. 1999. See Cox & Kessler 1999, pp. 969–72

Stacey GJ, Geis N, Genzel R, Lugten JB, Poglitsch A, et al. 1991. *Ap. J.* 373:423–44

Stanford SA, Sargent AI, Sanders DB, Scoville NZ. 1990. *Ap. J.* 349:492–96

Starck JL, Abergel A, Aussel H, Sauvage M, Gastaud R, et al. 1999. *Astron. Astrophys. Suppl.* 134:135–48

Steidel CC, Adelberger KL, Giavalisco M, Dickinson M, Pettini M. 1999. *Ap. J.* 519:1–17

Steidel CC, Giavalisco M, Pettini M, Dickinson M, Adelberger KL. 1996. *Ap. J. Lett.* 462:17–21

Sternberg A, Dalgarno A. 1995. *Ap. J. Suppl.* 99:565–667

Stickel M, Bogun S, Lemke D, Klaas U, Toth LV, et al. 1998. *Astron. Astrophys.* 336:116–22

Stickel M, Lemke D, Klaas U, Beichman CA, Kessler MF, et al. 1999. Preprint

Sturm E, Alexander T, Lutz D, Sternberg A, Netzer H, Genzel R. 1999a. *Ap. J.* 512:197–203

Sturm E, Lutz D, Genzel R, Sternberg A, Egami E, et al. 1996. *Astron. Astrophys.* 315:133–36

Sturm E, Lutz D, Tran D, Feuchtgruber H, Genzel R, et al. 1999b. *Astron. Astrophys.* In press. (*astro-ph/*0002195)

Surace JA, Sanders DB. 1999. *Ap. J.* 512:162–77

Surace JA, Sanders DB, Vacca WD, Veilleux S, Mazarella JM. 1998. *Ap. J.* 492:116–36

Sutherland RS, Bicknell GV, Dopita MA. 1993. *Ap. J.* 414:510–26

Taniguchi Y, Cowie LL, Sato Y, Sanders DB, Kawara K, et al. 1997a. *Astron. Astrophys.* 328:9–12

Taniguchi Y, Sato Y, Kawara K, Murayama T, Mouri H. 1997b. *Astron. Astrophys.* 318:1–4

Taylor GB, Silver CS, Ulvestad JS, Carilli CL. 1999. *Ap. J.* 519:185–90

Tecza M, Genzel R, Tacconi LJ, Tacconi-Garman LE 2000. *Ap. J.* In press. (*astro-ph/*0001424)

Thatte N, Quirrenbach A, Genzel R, Maiolino R, Tecza M. 1997. *Ap. J.* 490:238–46

Thornley MD, Förster-Schreiber N, Lutz D, Genzel R, Spoon HWW, et al. 2000. *Ap. J.* In press. (*astro-ph/*0003334)

Thuan TX, Sauvage M, Madden SC. 1999. *Ap. J.* 516:783–87

Tielens AGGM, Hollenbach DJ. 1985. *Ap. J.* 291:722–54

Tielens AGGM, Hony S, van Kerckhoven C, Peeters E. 1999. See Cox & Kessler 1999, pp. 579–87

Tran DQ. 1998. PhD thesis. Paris Univ.

Tran DQ, Lutz D, Genzel R, Rigopoulou D, Spoon HWW, et al. 2000. *Ap. J.* Submitted

Trewhella M. 1998. *MNRAS* 297:807–16

Trewhella M, Davies JB, Disney MJ, Jones HGW. 1997. *MNRAS* 288:397–403

Trimble V. 1987. *Annu. Rev. Astron. Astrophys.* 25:425–72

Tuffs RJ, Lemke D, Xu C, Davies JI, Gabriel C, et al. 1996. *Astron. Astrophys.* 315:149–52

Uchida K, Sellgren K, Werner MW. 1998. *Ap. J.* 493:L109–12

Unger SJ, Clegg PE, Stacey GJ, Cox P, Fischer J, et al. 2000. *Astron. Ap.* 355:885–90

Valentijn EA. 1990. *Nature* 346:153–55

Valentijn EA, van der Werf PP. 1999a. See Cox & Kessler 1999, pp. 821–24

Valentijn EA, van der Werf PP. 1999b. *Ap. J.* 522:L29–L33

Valentijn EA, van der Werf PP, de Graauw T, de Jong T. 1996. *Astron. Astrophys.* 315:145–48

van Bemmel IM, Barthel PD, de Graauw Th. 1999. See Cox & Kessler 1999, pp. 981–84

Veilleux S, Kim DC, Sanders DB, Mazarella JM, Soifer BT. 1995. *Ap. J. Suppl.* 98:171–217

Veilleux S, Sanders DB, Kim DC. 1997. *Ap. J.* 484:92–107

Veilleux S, Sanders DB, Kim DC. 1999. *Ap. J.* 522:139–56

Vermeij R, van der Hulst JM. 1999. See Cox & Kessler 1999, pp. 985–88

Verstraete L, Puget J-L, Falgarone E, Drapatz S, Wright CM, Timmermann R. 1996. *Astron. Astrophys.* 315:337–40

Viegas SM, Contini M, Contini T. 1999. *Astron. Astrophys.* 347:112–18

Vignati P, Molendi S, Matt G, Guainazzi M, Antonelli LA, et al. 1999. *Astron. Astrophys.* 349:57–60

Vigroux L, Charmandaris V, Gallais P, Laurent O, Madden SC, et al. 1999. See Cox & Kessler 1999, pp. 805–11

Vigroux L, Mirabel IF, Altieri B, Boulanger F, Cesarsky D, et al. 1996. *Astron. Astrophys.* 315:93–96

Voit GM. 1992. *Ap. J.* 399:495–503

Walter R, Orr A, Courvoisier RJ-L, Fink HH, Makino F, et al. 1994. *Astron. Astrophys.* 285:119–31

Whitmore BC, Zhang Q, Leitherer C, Fall SM, Schweizer F, Miller BW. 1999. *Astron. J.* 118:1551–76

Wilkes BJ, Hooper EJ, McLeod KK, Elvis MS, Impey CD, et al. 1999. *(astro–ph 9902084)*

Wolfire MG, Tielens AGGM, Hollenbach DJ. 1990. *Ap. J.* 358:116–31

Wozniak H, Friedli D, Martinet L, Pfenniger D. 1999. See Cox & Kessler 1999, pp. 989–91

Wright EL, Mather JC, Bennett CL, Cheng ES, Shafer RA, et al. 1991. *Ap. J.* 381:200–9

Wu H, Zou ZL, Xiu XY, Deng ZG. 1998. *Astron. Astrophys. Suppl.* 127:521–26

Xu C, Hacking PB, Fang F, Shupe DJ, Lonsdale CJ, et al. 1998. *Ap. J.* 508:576–89

Xu C, Helou G. 1996. *Ap. J.* 456:163–73

Yan L, McCarthy PJ, Freudling W, Teplitz HI, Malumuth EM, et al. 1999. *Ap. J.* 519:L47–L50

Zepf SE, Koo DC. 1989. *Ap. J.* 337:34–44

Zhao H-S. 1998. *MNRAS* 294:139–46

SUBJECT INDEX

A

Abell 1795 cluster, 278, 280
Abell 2190 cluster, 278, 280
Abell 4038 cluster, 278, 280
Abundance(s)
  cometary, 462
  globular cluster, 65
  heavy element
    and active galactic
    nuclei (AGN) emission
    lines, 550–51
  in interstellar gas and ice
    and cometary volatiles,
    434–35
  in low-mass stars (LMS)
    and substellar objects
    (SSOs), 346–48
  and rotating stars, 161–66
  in stars, 91–98
  of stellar coronae, 242–44
  in supergiants, 162–63
  and W-R star formation,
    174, 175
  See also Chemical
  abundances
Accretion
  disk(s)
    and active galactic
    nuclei (AGN), 26–27
    around compact objects,
    718
    onto a neutron star,
    721–22
Active galactic nuclei (AGN),
  26–27, 31
  broad emission lines in,
    521–62
  hidden, 791–93
  and Hubble deep fields
    (HDFs), 685–86
  and Infrared Space

Observatory (ISO),
  771–75, 782–89, 799,
  800, 804–5
  Palomar research on,
    104–5
  and starburst galaxies,
    789–91, 804
  stars
    chemical abundances,
      166
  and ultra-luminous infrared
    galaxies (ULIRGs),
    793–95
Additive-volume law, 341
Advanced Satellite for
  Cosmology and
  Astrophysics (ASCA),
  243, 289, 292–95, 303,
  305–9
  data, 792
Alfvén waves
  and stellar mass loss, 577,
    578
Allende meteorite,
  464–65
ALMA, 441, 805
Amino acids
  in carbonaceous
    chondrites, 466–67
Angular diameters
  stellar, 594, 595, 605
Angular momentum, 169
  and binary systems,
    114–19, 121–23, 127,
    130, 132–33, 135, 137
  and mass loss, 152
  of neutron stars, 741
  and Rayleigh's criterion for
    stability, 148
  and rotational velocities,
    159

and stellar evolution, 178
and stellar winds, 157, 629,
  650, 651
transport of
  by gravity waves, 150
  and rotation stars,
    144–45, 150–53, 161,
    180
and white dwarfs, 366,
  367, 369
Angular velocity
  and binary systems, 119,
    123
  evolution, 154
  and stellar evolution, 142,
    143
Antennae system
  of colliding galaxies,
    775–76
Apollo-Soyuz mission, 232,
  250, 251
ASCA
  See Advanced Satellite for
    Cosmology and
    Astrophysics (ASCA),
    792
Astronomy
  ground-based, 36
  optical
    deep surveys in, 667
  space, 36
    with Hipparcos, 37–39
Asymptotic giant branch
  (AGB) stars, 142
  3.4 μm feature in, 448
  and binary systems, 130
  and infrared emission, 769
  and mass loss, 573–606
ATLAS, 43, 44
Atmosphere(s)
  of G1 229B, 503–5

815

Cumulative Indexes

CONTRIBUTING AUTHORS, VOLUMES 27–38

CHAPTER TITLES, VOLUMES 27–38

Prefatory Chapters

Solar System Astrophysics

Solar Physics

Dynamical Astronomy

Interstellar Medium

The Galaxy

Extragalactic Astronomy and Cosmology

High Energy Astrophysics

Instrumentation and Techniques

New Areas of Research, History